FOUNDATIONS OF
Electricity
AND
Magnetism

Basudev Ghosh

CELEBRATING 66 YEARS 1960-2026

Since BOOKS AND ALLIED 1960

Dedicated to

my Father

who inspired me to take up

Teaching Profession

Preface to the Fifth Edition

Keeping in mind the latest CBCS syllabus of UGC the book has been updated with the addition of new text matter and some additional important questions and problems set in the current university examinations.

I take this opportunity to thank the readers for their appreciation of the book. Hope this edition of the book will also be equally appreciated.

January 1, 2020

Department of Physics
Jadavpur University
Kolkata 700032
India

Basudev Ghosh

Preface to the Fourth Edition

I am happy to present the Fourth Edition of the book with the addition of some questions and problems which will definitely enrich the book. Typographical errors that have came to my notice have been corrected.

Once again, I take this opportunity to thank the readers for their appreciation of the book.

July 11, 2015

Department of Physics
Jadavpur University
Kolkata 700032
India

Basudev Ghosh

Preface to the Third Edition

Being encouraged by the overwhelming response of the readers I am presenting this edition of the book with the addition of some important questions and problems set in the current university examinations. Also, the typographical errors that have come to my notice have been corrected.

I take this opportunity to thank the readers for their appreciation of the book.

August 9, 2012

Department of Physics
Jadavpur University
Kolkata 700032
India

Basudev Ghosh

Preface to the Second Edition

Being encouraged by the overwhelming response of the readers I am presenting this edition of the book with the addition of some important questions and problems set in the current university examinations. Also, the typographical errors that have come to my notice have been corrected.

I take this opportunity to thank the readers for their appreciation of the book.

September 9, 2009

Department of Physics
Jadavpur University
Kolkata 700032
India

Basudev Ghosh

Preface to the First Edition

Electricity, magnetism and *electromagnetic theory* form an important part of the syllabus for undergraduate physics, engineering and technical courses. While teaching this subject, particularly in physics honours classes, we very often refer to different standard books for different topics. Many Indian students do not get the opportunity to consult so many books. Some students, though lucky enough to get this opportunity, become confused with different styles of communication used by different authors and become unable to derive interest out of them. The purpose of the present book is to provide a coherent piece of exhaustive text covering the current syllabi of physics honours courses of most Indian Universities.

The book is primarily based on the lectures given to the physics honours students during the last two decades. Here I have used my own method of communication that I have developed in me during my long career as a physics teacher. Greater emphasis has been given on the conceptual understanding, mathematical formulation and solution of problems. I have tried to develope the topics from first principles so that the students can learn through self-study. They are expected to gain some basic knowledge on vector analysis, analytical geometry and some relevant mathematics before going through this text. As a ready reference some of the mathematical relations, formulas and theorems used in this text have been appended at the end of the book. Keeping in mind the present international practice, SI units have been consistently used throughout the book. However, considering that Gaussian units are still in use I have included in the appendix SI-Gaussian conversion formulas.

A number of thought provoking multiple-choice-questions (MCQ) have been included at the end of each chapter. This will help the students to check immediately their level of understanding.

Each chapter contains a good number of solved problems some of which contain subtle points, not explicit in the text, and hence will help the reader to attain a deeper understanding of the ideas presented in the book. A large variety of problems given with answers at the end of each chapter for practice of the students, will surely help to grow self-confidence.

It is hoped that the book will be able to meet up all the requirements of undergraduate students. However, any constructive criticism and suggestion for the improvement of the book will always be highly appreciated.

I take this opportunity to express my gratitude to the renowned authors like Feynmann, Jackson, Jeans, Corson, Lorrain, Coulson, Stratton, Yarwood, Starling, Smith, Griffith, Jordan, Panofsky, Phillips, Reitz, Milford, Tamm, Irodov, Nayfeh, Brussel and many others whose writings have contributed to my understanding of the subject of this book. I am indebted to my colleagues, Prof. M.K. Mukherjee and Prof. C.M. Pal for providing me with the solutions of a few problems.

I thank my students who inspired me in writing the book and provided me with stimulating questions. Thanks are due to Books & Allied (P) Ltd. for making sincere efforts towards timely publication of the book.

Finally it is a pleasure for me to thank my family members—my wife, Mrs. Shyamali Ghosh and son, Soumyadip Ghosh — for sparing me from the regular household duties during the preparation of the manuscript.

October 17, 2006

Department of Physics
R.K. Mission Vidyamandira
Belur Math, Howrah 711202
India

Basudev Ghosh

Contents

Chapter **1**

Electrostatics in Vaccum

1.1 Electric Charge

Electrostatics is a branch of physics, which deals with the behaviour of *electric charges* at rest. So it is important to seek the answer to the question: What is electric charge? Many substances such as ebonite, glass, rubber, etc. when rubbed with silk, flannel or other suitable substances acquire the property of attracting light bodies like bits of paper. The substances in such state are said to be charged and they are said to possess what is called *electric charge.* Simple experiments show that there are two types of charges. One variety is called *positive charge* and the other negative charge. Like charges are found to repel each other and unlike charges attract each other. According to our present knowledge of the atomic structure of matter, all elements consist of positively charged protons, negatively charged electrons and neutral neutrons. The charge of a proton is numerically equal to that of the electron. In an atom or an ordinary piece of matter there are equal number of protons and electrons. As a result, matter as a whole is electrically neutral. Whenever two different bodies are rubbed with each other some of the electrons from one body may be transferred to the other body. The body which gets excess electrons becomes negatively charged and the body which loses electrons becomes positively charged. Matter as a whole or elementary particles of matter like protons, electrons can carry charge. But it is found that electric charge cannot exist without matter. Charge is thus a fundamental and characteristic property of elementary particles. It is a scalar quantity. It adds up like real numbers. While adding charges one must take care of their signs also.

• Quantization of charge

In nature all electric charges are found to be integral multiples of one basic unit, the magnitude of electronic charge (e). This occurrence of electric charge in discrete units is called the *quantization of charge.* Since the electrification of a body is generally caused by taking up a number of electrons or by giving up a number of electrons the charge on the body will be equal to an integral multiple of electronic charge. Experiments show that the magnitude of charge of all elementary charged particles is equal to the magnitude of electronic charge. Very recently the existence of elementary particles (called *quarks*) with fractional electronic charge has been suggested theoretically. However, its existence in free state has not yet been proved experimentally.

In the macroscopic level the charge we encounter in *charged bodies* is so large compared with electronic charge that to a good approximation we can regard charge as continuous.

• Conservation of charge

Total charge i.e., the algebraic sum of positive and negative charges in an isolated system remains constant for all times. This is known as the law of conservation of charge. Here by *isolated system* we mean that no charge is allowed to cross the boundary of the system. The law of conservation of charge allows simultaneous *creation* or *annihilation* of equal and opposite amount of charges within the isolated system. For example, a high energy neutral photon may produce an electron of charge $-e$ and a positron of charge $+e$ or an electron-positron pair may combine to produce a photon. This does not violate the law of conservation of charge. But a single positive charge or a single negative charge can neither be created nor destroyed.

1.2 Coulomb's Law

From an experimental study Coulomb in 1785 discovered a law of force between charge particles. This is known as *Coulomb's law.* According to this law *the force of interaction between two stationary point charges is proportional to the product of the charges, inversely proportional to the square of the distance between them and it acts along the line joining the charges.*

Suppose $\vec{r}_1$ and $\vec{r}_2$ be the position vectors of two point charges q_1 and q_2 with respect to some coordinate system as shown in Fig. 1.2-1.

Let $\vec{r}_{21} = \vec{r}_2 - \vec{r}_1$ be the vector from q_1 to q_2, $r_{21} = |\vec{r}_{21}|$ be the distance between q_1 and q_2 and $\hat{r}_{21} = \vec{r}_{21}/|\vec{r}_{21}|$ be the unit vector in the direction from q_1 to q_2. Then from Coulomb's law the force on q_2 due to q_1 is

$$\vec{F}_{21} = k \cdot \frac{q_1 q_2}{r_{21}^2} \hat{r}_{21},$$

where k is a proportionality constant, its value depends on the nature of surrounding medium and the system of units used. In vacuum in SI units F_{21} is in newtons (N),

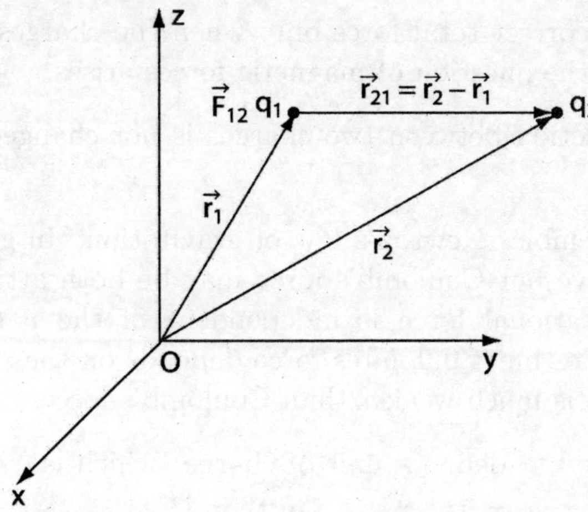

Fig. 1.2-1

q_1, q_2 in coulombs (C), r_{21} in metres and $k = \frac{1}{4\pi\varepsilon_0}$. Here ε_0 is called the *permittivity of free space.* The measured value of ε_0 is 8.85×10^{-12} $C^2/N\text{-}m^2$. For simplified numerical calculations one can approximately use

$$\frac{1}{4\pi\varepsilon_0} = 9 \times 10^9 \text{ N-m}^2/\text{C}^2.$$

Thus the force on q_2 due to q_1 is

$$\vec{F}_{21} = \frac{1}{4\pi\varepsilon_0} \cdot \frac{q_1 q_2}{r_{21}^2} \hat{r}_{21} \tag{1.2-1}$$

The force on q_1 due to q_2 is

$$\vec{F}_{12} = \frac{1}{4\pi\varepsilon_0} \cdot \frac{q_2 q_1}{r_{12}^2} \hat{r}_{12},$$

where $\hat{r}_{12}$ is the unit vector in the direction from q_2 to q_1.

So $\hat{r}_{12} = -\hat{r}_{21}$ and hence,

$$\vec{F}_{12} = -\vec{F}_{21}.$$

Thus Coulomb force is consistent with Newton's third law of motion.

• Discussion

(i) Coulomb's law is strictly applicable to *point charges.* Practically the separation between the charges should be large compared with their spatial dimensions. Coulomb's law is found to be valid for any arbitrarily large distance between the charges. We have no experimental evidence or any theoretical reasoning to expect a breakdown of this law at large distances. However, in the domain of very small distances there is experimental evidence to believe that the law breaks down for distances of the order of a nuclear diameter, i.e., 10^{-15} m.

(ii) The law gives the correct total force only when the charges are at rest. In case of moving charges the question of magnetic forces arises.

(iii) The force of interaction between two charges is not changed by the presence of other charges.

(iv) Coulomb's law resembles Newton's law of gravitation. In gravitation the forces are always attractive but Coulomb's force may be both attractive and repulsive in nature. Gravitational force is independent of the nature of the medium between the particles but Coulomb's force depends on the surrounding medium. Gravitational force is much weaker than Coulomb's force.

(v) The law can be used to define a unit of charge (which is coulomb in SI). In Eq. (1.2-1) if we put $q_1 = q_2 = 1C$, $r_{21} = 1\,\text{m}$ then $\left|\vec{F}_{21}\right| = \frac{1}{4\pi\varepsilon_0} = 9 \times 10^9 N$. Thus $1C$ is the charge which when placed at a distance of 1 m from another similar charge in vacuum experiences an electrostatic force of 9×10^9 N.

1.3 Principle of Superposition

The total force on a charge due to a number of other charges is given by the vector sum of the Coulomb forces exerted on the charge due to each of the other charges acting separately. This fact is called the principle of superposition. Thus if we have N point charges $q_1, q_2, q_3, \ldots, q_N$ the total force acting on a charge q_0 due to all these N charges can be found out by calculating separately the forces $\vec{F}_{01}, \vec{F}_{02}, \ldots, \vec{F}_{0N}$ on q_0 due to $q_1, q_2, \ldots, q_N$, respectively and then taking their vector sum. So total force on q_0 due to the N-charges would be

$$\vec{F}_0 = \vec{F}_{01} + \vec{F}_{02} + \cdots + \vec{F}_{0N} = \sum_{j=1}^{N} \vec{F}_{0j}.$$

1.4 Electric Field

Suppose we have a point charge Q at the position $\vec{r}\,'$ and another small point charge q placed at the position $\vec{r}$ with respect to some origin O (Fig. 1.4-1). We are to find the force on q due to Q. Here Q is called the *source charge* and q is called *test charge.* From Coulomb's law this force is

$$\vec{F} = \frac{1}{4\pi\varepsilon_0} \cdot \frac{Q \cdot q}{R^2} \hat{R} = q \cdot \vec{E}, \tag{1.4-1}$$

where

$$\vec{E} = \frac{1}{4\pi\varepsilon_0} \cdot \frac{Q}{R^2} \hat{R}; \tag{1.4-2}$$

R is the distance between Q and q, $\hat{R}$ = unit vector in the direction from Q to q.

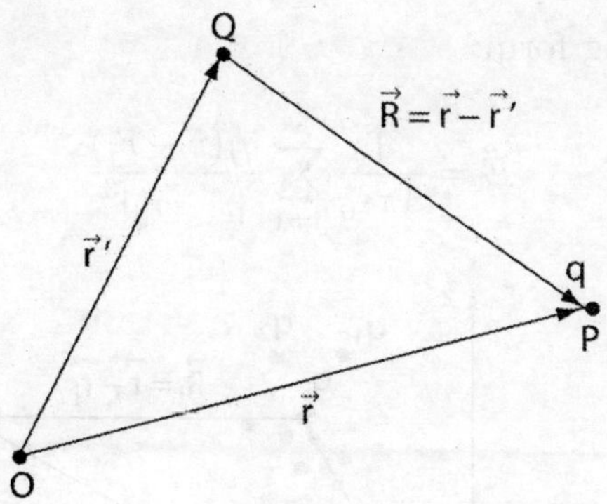

Fig. 1.4-1

Obviously $\vec{E} = \vec{F}/q$, i.e., the force per unit charge placed at the point P. $\vec{E}$ is a function of the position vector $\vec{R}$ of P with respect to the source charge Q. It depends on Q but is independent of the test charge q. Thus in the region of space surrounding the source charge we can uniquely specify a vector physical quantity $\vec{E}$ at every point. This means that $\vec{E}$ defines a vector field.

Thus the force $\vec{F}$ on q due to Q may be considered as arising in two steps. (i) The charge Q sets up an electrical environment, called the *electric field* $\vec{E}$, in the surrounding space. (ii) When a charge q is placed at any point without disturbing the position of Q it experiences a force, which equals the charge q multiplied by $\vec{E}$ at that point.

From Eq. (1.4-2) *electric field at a point in space due to a charge Q may be defined as the force on a unit positive charge placed at that point.* The unit of $\vec{E}$ is N.C^{-1}. A unit charge may be high enough to disturb the configuration of the charge producing the field. A way out of this difficulty is to choose the test charge q negligibly small. Then we may define the electric field at a point as the limiting force per unit charge placed at that point, i.e.,

$$\vec{E} = \lim_{q\to 0} \frac{\vec{F}}{q}.$$

Note that $q \to 0$ is an idealisation because $q \to 0$ contradicts the quantization of charge. In case of immobile source charges we need not require this idealisation.

Our definition of electric field in Eq. (1.4-2) assumes that the source of the field is a point charge Q. If the source is a set of discrete point charges $q_1, q_2, \ldots, q_N$ whose position vectors with respect to the origin O of some coordinate system are respectively $\vec{r}_1, \vec{r}_2, \ldots, \vec{r}_N$ then the electric field at any point $P(\vec{r})$ can be calculated by using the principle of superposition as

$$\vec{E} = \vec{E}_1 + \vec{E}_2 + \cdots + \vec{E}_N = \sum_{j=1}^{N} \vec{E}_j = \frac{1}{4\pi\varepsilon_0} \sum_{j=1}^{N} \frac{q_i}{R_j^2} \hat{R}_j, \qquad (1.4\text{-}3)$$

where R_j is the distance of the point of observation P from the location of the point charge q_j and $\hat{R}_j$ is the unit vector in the direction from q_j to P. Equation (1.4-3) can also be put in the following form:

$$\vec{E} = \frac{1}{4\pi\varepsilon_0} \sum_{j=1}^{N} \frac{q_j(\vec{r} - \vec{r}_j)}{|\vec{r} - \vec{r}_j|^3} \tag{1.4-4}$$

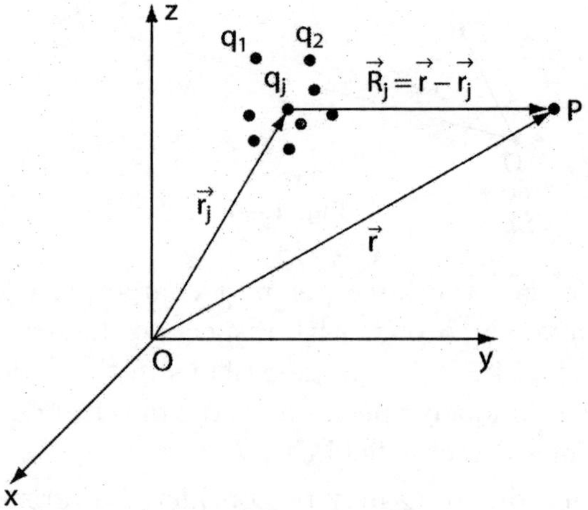

Fig. 1.4-2

• Continuous charge distribution

In macroscopic physics the charge we encounter is so large compared with the basic unit of charge that to a good approximation we can neglect the quantized-nature of electric charge and consider it as a continuous distribution. In a macroscopic charge distribution even a small volume element contains large number of elementary charges. In such situations it is impractical to account for these charges individually and we introduce, for convenience, the idea of charge densities. If charge is distributed over a volume V then we talk about volume density of charge. It is defined as the limiting charge per unit volume. If Δq be the amount of charge in a small volume element ΔV about the point $\vec{r}$ then volume density of charge at this point is defined as

$$\rho = \lim_{\Delta V \to 0} \frac{\Delta q}{\Delta V} = \frac{dq}{dV}.$$

Sometimes charges spread over a surface and it is described by surface charge density. It is defined as the limiting charge per unit area. If Δq be the amount of charge in a small surface element ΔS about a point then surface charge density at this point is defined as

$$\sigma = \lim_{\Delta S \to 0} \frac{\Delta q}{\Delta S} = \frac{dq}{dS}.$$

Similarly if charges spread out along a line we describe it by line charge density (λ). This is defined as the limiting charge per unit length. Thus if Δq be the charge on a line element Δl then line charge density

$$\lambda = \lim_{\Delta l \to 0} \frac{\Delta q}{\Delta l} = \frac{dq}{dl}.$$

Let us see that the electric field $\vec{E}$ defined by the Eq. (1.4-3) for a set of discrete point charges can be easily generalised to various charge distributions. Suppose charge is distributed over a volume V and $\rho(\vec{r}\,')$ is the charge density at the point $\vec{r}\,'(x', y', z')$. Then electric field $\vec{E}$ at the point $P(\vec{r})$ due to this volume distribution of charge can be written as

$$\vec{E}(\vec{r}) = \frac{1}{4\pi\varepsilon_0} \int_V \hat{R} \frac{\rho(\vec{r}\,')\, dV}{R^2}, \tag{1.4-5}$$

where R is the distance of the point of observation P from the volume element dV $(= dx'dy'dz')$ at $\vec{r}\,'$ and $\hat{R} = \vec{R}/R$ is the unit vector in the direction from the volume element to the point P (Fig. 1.4-3);

$$\vec{R} = \vec{r} - \vec{r}\,' = \hat{i}(x - x') + \hat{j}(y - y') + \hat{k}(z - z').$$

In a similar way electric field due to a surface distribution of charge (Fig. 1.4-4) can be written as

$$\vec{E}(\vec{r}) = \frac{1}{4\pi\varepsilon_0} \int_S \hat{R} \cdot \frac{\sigma\, dS}{R^2} \tag{1.4-6}$$

and that due to a line charge distribution (Fig. 1.4-5) can be written as

$$\vec{E}(\vec{r}) = \frac{1}{4\pi\varepsilon_0} \int_l \hat{R} \cdot \frac{\lambda dl}{R^2}. \tag{1.4-7}$$

At this point it is important to mention that a point charge can be described with a charge density by means of Dirac delta function δ, which has the following properties (see Appendix C):

(i) $\delta(\vec{r} - \vec{r}\,') = 0$ for $\vec{r}\,' \neq \vec{r}$. (1.4-8)

(ii) $\displaystyle\int_V \delta(\vec{r} - \vec{r}\,')\, dV = \begin{cases} 1 & \text{if } \vec{r}\,' = \vec{r} \text{ is included in } V \\ 0 & \text{otherwise.} \end{cases}$ (1.4-9)

(iii) $\displaystyle\int_V f(\vec{r}\,')\delta(\vec{r} - \vec{r}\,')\, dV = f(\vec{r})$. (1.4-10)

These properties show that the delta function is a very high singular function; it is zero everywhere except at a particular point; yet it has a nonzero integral. It is not a continuous function but is found to be very useful.

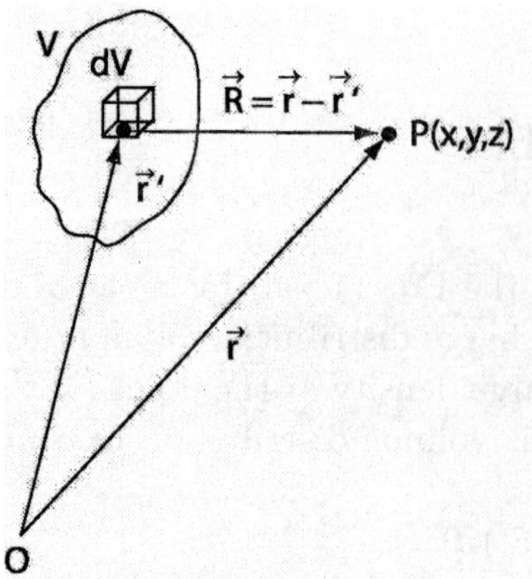

Fig. 1.4-3

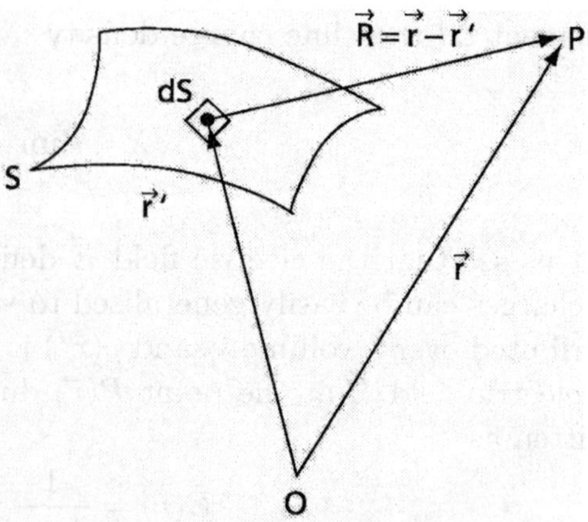

Fig. 1.4-4

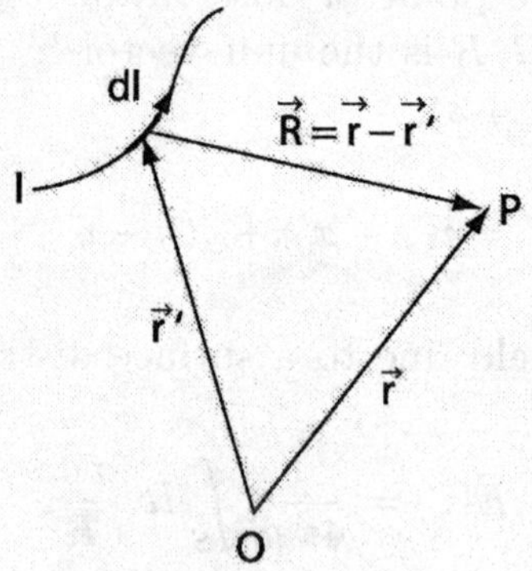

Fig. 1.4-5

A point charge q_1 situated at the position $\vec{r}_1$ can be described by a volume charge density $\rho(\vec{r}\,') = q_1\delta(\vec{r}\,' - \vec{r}_1)$. This is justified because:

(i) Charge density, $\rho(\vec{r}\,') = 0$ for $\vec{r}\,' \neq \vec{r}_1$.

(ii) Total charge, $\displaystyle\int_V \rho(\vec{r}\,')\,dV = \int q_1\delta(\vec{r}\,' - \vec{r}_1)\,dV = q_1$.

(iii) Following Eq. (1.4-5) the electric field at the point $P(\vec{r})$ due to a point charge q_1 at $\vec{r}_1$ described by the above charge density can be calculated and shown to be identical with that calculated by considering it a point charge.

$$\begin{aligned}\vec{E}(\vec{r}) &= \frac{1}{4\pi\varepsilon_0}\int_V \frac{(\vec{r} - \vec{r}\,')}{|\vec{r} - \vec{r}\,'|}\frac{\rho(\vec{r}\,')\,dV}{|\vec{r} - \vec{r}\,'|^2} \\ &= \frac{1}{4\pi\varepsilon_0}\int_V \frac{(\vec{r} - \vec{r}\,')}{|\vec{r} - \vec{r}\,'|^3} q_1\delta(\vec{r}\,' - \vec{r}_1)\,dV \\ &= \frac{1}{4\pi\varepsilon_0}\cdot\frac{q_1(\vec{r} - \vec{r}_1)}{|\vec{r} - \vec{r}_1|^3} \quad \text{[using delta function property]}\end{aligned}$$

• Electric lines of force or field lines

An electric field associated with a particular distribution of charges can be represented geometrically by using the concept of electric lines of force. This concept was introduced by Faraday. *An electric line of force is an imaginary curve drawn in such a way that the tangent to this curve at any point gives the direction of electric field at the point.* For a single positive point charge the field lines are radially outward and for a single negative point charge they are radially inward (Fig. 1.4-6). Such diagrams are also useful in representing more complicated fields. The field lines around two positive charges give a vivid pictorial description of their mutual repulsion.

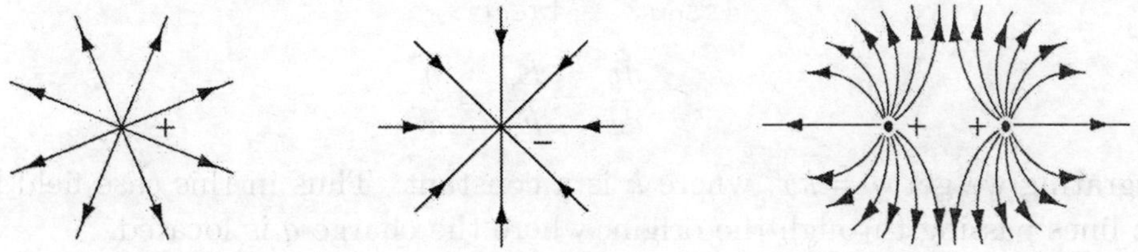

Fig. 1.4-6

The magnitude of the electric field can be represented by the density of field lines, i.e., the number of lines passing through unit area perpendicular to the lines. Suppose we assume that q/ε_0 number of lines emanate from a charge q symmetrically in all directions. Imagine a sphere of radius r with the charge at its centre. All the lines q/ε_0 pass isotropically through the area $4\pi r^2$. So the number of field lines passing through unit area of this imaginary sphere would be

$$\frac{q/\varepsilon_0}{4\pi r^2} = \frac{1}{4\pi\varepsilon_0} \cdot \frac{q}{r^2}.$$

This is exactly the magnitude of electric field at a distance r from a point charge.

While sketching the field lines the following properties of the field lines should be kept in mind:

(i) Field lines are continuous curves.

(ii) Two field lines cannot intersect each other.

(iii) Field lines start from a positive charge and end in a negative charge. For a single charge they can extend up to infinity.

• Differential equation for lines of force

Let $\vec{r} = \vec{r}(s)$ represents a space curve. $d\vec{r}/ds$ is a vector tangent to this curve at every point. If this curve is to represent a line of force, $d\vec{r}/ds$ must be parallel to the local electric field, i.e., $d\vec{r}/ds = \alpha\vec{E}(\vec{r})$, where α is a constant.

In component form:

$$\frac{dx}{ds} = \alpha E_x, \quad \frac{dy}{ds} = \alpha E_y, \quad \frac{dz}{ds} = \alpha E_z.$$

$$\therefore \quad \frac{dx}{E_x} = \frac{dy}{E_y} = \frac{dz}{E_z}. \tag{1.4-11}$$

For example, let us consider the field lines of a point charge q in xy-plane.

Here

$$\vec{E} = \frac{q}{4\pi\varepsilon_0}\frac{\vec{r}}{r^3} = \frac{q}{4\pi\varepsilon_0 r^3}(x\hat{\imath} + y\hat{\jmath}).$$

$$\therefore \quad \frac{dy}{dx} = \frac{E_y}{E_x} = \frac{y}{x}.$$

Integrating we get $y = kx$, where k is a constant. Thus in this case field lines are straight lines passing through the origin, where the charge q is located.

1.5 Electric Potential

From vector analysis it is well-known that if the curl of a vector vanishes, then the vector may be expressed as the gradient of a scalar. It is shown below that $\vec{\nabla} \times \vec{E} = 0$. This means that the electrostatic field $\vec{E}$ is conservative and it can be written as the gradient of some scaler ϕ as $\vec{E} = -\vec{\nabla}\phi$, where the negative sign is chosen for convenience. The scalar function ϕ thus introduced is known as *electrostatic potential.*

To show that $\vec{\nabla} \times \vec{E} = 0$ we first consider the electric field at the position $\vec{r}$ due a point charge q located at $\vec{r}\,'$. This is

$$\vec{E} = \frac{q}{4\pi\varepsilon_0} \cdot \frac{\vec{r} - \vec{r}\,'}{|\vec{r} - \vec{r}\,'|^3}. \tag{1.5-1}$$

Now

$$\begin{aligned} \vec{\nabla} \times \vec{E} &= \frac{q}{4\pi\varepsilon_0}\vec{\nabla} \times \frac{(\vec{r} - \vec{r}\,')}{|\vec{r} - \vec{r}\,'|^3}, \quad \text{where } \vec{\nabla} \text{ operates on } \vec{r} \\ &= \frac{q}{4\pi\varepsilon_0}\left\{\vec{\nabla}\frac{1}{|\vec{r} - \vec{r}\,'|^3} \times (\vec{r} - \vec{r}\,') + \frac{1}{|\vec{r} - \vec{r}\,'|^3}\vec{\nabla} \times (\vec{r} - \vec{r}\,')\right\}, \end{aligned} \tag{1.5-2}$$

where we have used the vector identity $\vec{\nabla} \times \phi\vec{A} = \vec{\nabla}\phi \times \vec{A} + \phi\vec{\nabla} \times \vec{A}$.

Now using Cartesian coordinates it can be shown by direct calculation that

$$\vec{\nabla}\frac{1}{|\vec{r} - \vec{r}\,'|^3} = -\frac{3(\vec{r} - \vec{r}\,')}{|\vec{r} - \vec{r}\,'|^5} \quad \text{and} \quad \vec{\nabla} \times (\vec{r} - \vec{r}\,') = 0$$

Substituting these results in Eq. (1.5-2) we get

$$\vec{\nabla} \times \vec{E} = 0. \tag{1.5-3}$$

From the principle of superposition we known that the total field due to any charge distribution is given by the vector sum of the fields due to individual charges. So we can say that the result (1.5-3) holds for any charge distribution. Therefore, we can write

$$\vec{E} = -\vec{\nabla}\phi. \tag{1.5-4}$$

Since $\vec{\nabla}(\phi + \text{constant}) = \vec{\nabla}\phi$, the potential ϕ defined by (1.5-4) is arbitrary by some additive constant. Addition of a constant to ϕ yields the same electric field. Moreover, it does not affect the potential difference between two points because the constants cancel out. The absolute value of potential is of no importance, only potential differences have physical significance. To get a physical significance of the electrostatic potential let us calculate the work done against the field in moving a unit positive charge from a reference point a to the point b. This is

$$\begin{aligned} W_{ab} &= -\text{work done by the field} \\ &= -\int_a^b \vec{E} \cdot d\vec{r} = \int_a^b \vec{\nabla}\phi \cdot d\vec{r} = \int_a^b d\phi = \phi(b) - \phi(a). \end{aligned} \tag{1.5-5}$$

Obviously the work done depends only on the positions a and b and not on the path connecting them. This indicates that the work done over any close path is zero. This result can be easily obtained from Eq. (1.5-3) by using Stokes' theorem,

$$\oint_C \vec{E} \cdot d\vec{r} = \int_S \left(\vec{\nabla} \times \vec{E}\right) \cdot d\vec{S} = 0, \tag{1.5-6}$$

where C is the contour bounding an open surface S.

This must be so because electrostatic force is conservative. Equation (1.5-5) suggests that the electrostatic potential may be considered as potential energy per unit charge. If the charge distribution that creates the electric field is localised in a finite region of space the electric field vanishes at infinite distances from the charge distribution. Then one usually takes the reference point a at infinity, where the potential is taken to be zero; $\phi(\infty) = 0$. In this case Eq. (1.5-5) gives

$$\phi(b) = -\int_\infty^b \vec{E} \cdot d\vec{r}. \tag{1.5-7}$$

Now, as the electric field is the force per unit positive charge, *the electrostatic potential at any point* (b) *may be defined as the work done by an external agency in bringing a unit positive charge from infinity to that point.*

If the same reference point is chosen for the electrostatic potential and potential energy then the potential energy of a charge is given by the product of the charge and the electric potential at the location of the charge.

• Potential due to a point charge

Electric field at the position $\vec{r}$ due to a charge q located at $\vec{r}'$ is given by the Eq. (1.5-1). Now from vector analysis it can be shown that

$$\vec{\nabla}\frac{1}{|\vec{r}-\vec{r}'|} = \frac{\vec{r}-\vec{r}'}{|\vec{r}-\vec{r}'|^3}, \quad \text{where } \vec{\nabla} \text{ operates on } \vec{r}.$$

Therefore, Eq. (1.5-1) can be rewritten as,

$$\vec{E} = -\vec{\nabla}\left[\frac{1}{4\pi\varepsilon_0}\cdot\frac{q}{|\vec{r}-\vec{r}'|}\right]$$

Comparing it with the relation $\vec{E} = -\vec{\nabla}\phi$ we get the electrostatic potential due to the point charge as

$$\phi = \frac{1}{4\pi\varepsilon_0}\cdot\frac{q}{|\vec{r}-\vec{r}'|}. \tag{1.5-8}$$

• Potential due to charge distributions

Just like the electrostatic force ($\vec{F}$) on a test charge and the electrostatic field ($\vec{E}$), the electrostatic potential (ϕ) also obeys the principle of superposition. According to the superposition principle total force on a test charge q due to a charge distribution is given by the sum of the forces due to individual charges. Symbolically,

$$\vec{F} = \vec{F}_1 + \vec{F}_2 + \cdots$$

Dividing throughout by q we get

$$\vec{E} = \vec{E}_1 + \vec{E}_2 + \cdots$$

$$\text{or } -\vec{\nabla}\phi = -\vec{\nabla}\phi_1 - \vec{\nabla}\phi_2 - \cdots$$

$$\text{or } \vec{\nabla}\phi\cdot d\vec{r} = \vec{\nabla}\phi_1\cdot d\vec{r} + \vec{\nabla}\phi_2\cdot d\vec{r} + \cdots$$

Integrating this equation from ∞ to the position $\vec{r}$ we get

$$\phi = \phi_1 + \phi_2 + \cdots$$

Thus the potential at any point is the sum of the potentials due to individual charges. So using Eq. (1.5-8) and this principle it is possible to calculate potential due to arbitrary charge distributions. Considering Figs. 1.4-3–5 we can write for potentials due volume, surface and line charge distributions as

$$\phi(\vec{r}) = \frac{1}{4\pi\varepsilon_0}\int_V \frac{\rho(\vec{r}')\,dV}{|\vec{r}-\vec{r}'|} \quad \text{(Volume charge distribution)}$$

$$\phi(\vec{r}) = \frac{1}{4\pi\varepsilon_0}\int_S \frac{\sigma(\vec{r}')\,dS}{|\vec{r}-\vec{r}'|} \quad \text{(Surface charge distribution)}$$

$$\phi(\vec{r}) = \frac{1}{4\pi\varepsilon_0}\int_l \frac{\lambda(\vec{r}')\,dl}{|\vec{r}-\vec{r}'|} \quad \text{(Line charge distribution)}$$

• Equipotential surfaces

An equipotential surface is the locus of points having the same potential. In case of a point charge q located at the origin, the potential at a distance r from the origin is given by

$$\phi(\bar{r}) = \frac{1}{4\pi\varepsilon_0} \cdot \frac{q}{r}.$$

Obviously a sphere of radius r with the centre at q will be an equipotential surface. In this case, all concentric spheres with the centre at q will be equipotential surfaces (Fig. 1.5-1). In general the shape of the equipotential surface depends on the charge configuration. But whatever may be its shape the relation $\vec{E} = -\vec{\nabla}\phi$ indicates that the *electric field* $\vec{E}$ *is always perpendicular to the equipotential surface,* $\phi =$ constant, at every point.

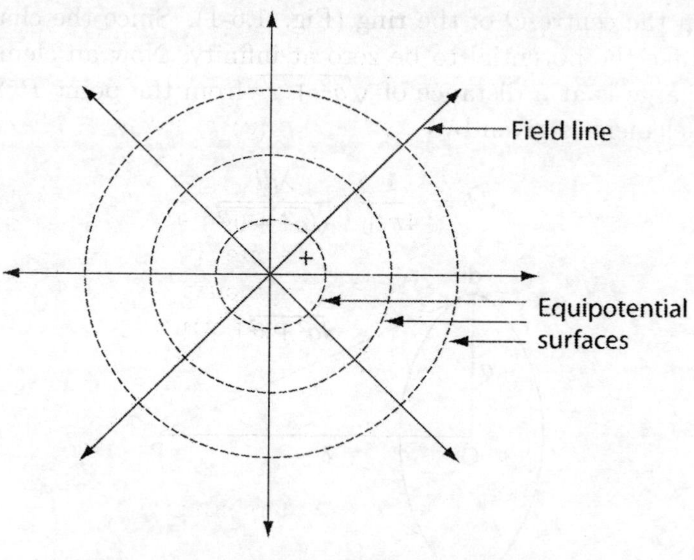

Fig. 1.5-1 Field lines and equipotential surfaces for a point charge

• Advantage of the potential concept

Electric field $\vec{E}$ is a vector quantity. Its direct calculation very often becomes tedious and cumbersome. On the other hand, potential ϕ is a scalar quantity. In many cases its calculation is found to be easier. The potential concept reduces a vector problem down to a scalar one. So in practice it is often preferable to determine $\vec{E}$ by first calculating ϕ and then using the relation $\vec{E} = -\vec{\nabla}\phi$, instead of determining $\vec{E}$ directly.

• Units of electric field and potential

The SI unit of potential is volt (V). The potential at a point is said to be 1 V if one joule (J) of work is to be done in taking one coulomb (C) of charge from infinity up to that point. Thus one volt equals one joule per coulomb.

By definition $\vec{E}$ is the force per unit charge. Therefore, unit of $\vec{E}$ may be newton per coulomb (N/C). Its most common SI unit is volt per metre ($V.m^{-1}$).

1.6 Calculations of Potential and Field —Some Common Examples

1. A uniformly charged ring

Suppose we have a thin circular ring of radius a having a uniform line charge density of λ per unit length. We are to calculate the field and potential at an axial point P at a distance x from the centre O of the ring (Fig. 1.6-1). Since the charge distribution is finite we can take the potential to be zero at infinity. Now an element of length dl containing λdl charge is at a distance of $\sqrt{a^2+x^2}$ from the point P. So the potential at P due to this element is given by

$$d\phi = \frac{1}{4\pi\varepsilon_0} \cdot \frac{\lambda dl}{\sqrt{a^2+x^2}}.$$

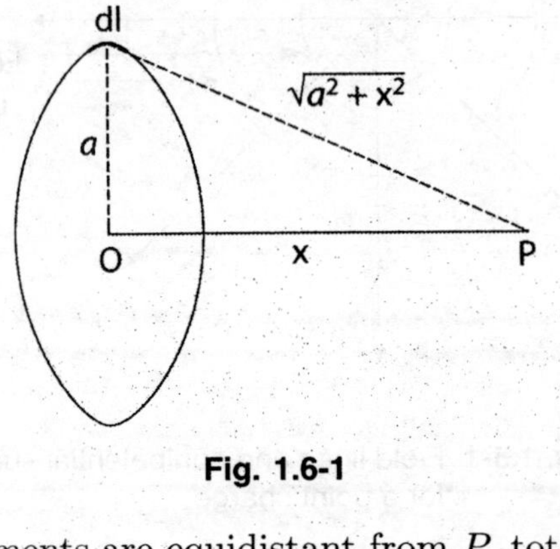

Fig. 1.6-1

Since all the length elements are equidistant from P, total potential at P would be given by

$$\phi = \frac{1}{4\pi\varepsilon_0} \int \frac{\lambda dl}{\sqrt{a^2+x^2}}$$

$$= \frac{1}{4\pi\varepsilon_0} \cdot \frac{Q}{\sqrt{a^2+x^2}}, \tag{1.6-1}$$

where $Q = \int \lambda dl$ is the total charge on the ring.

Because of perfect symmetry about the axis the electric field at P is directed along the axis and is given by

$$\vec{E} = -\vec{\nabla}\phi = -\hat{x}\frac{d\phi}{dx} = \frac{1}{4\pi\varepsilon_0} \cdot \frac{Qx}{(a^2+x^2)^{3/2}}\hat{x}, \tag{1.6-2}$$

where $\hat{x}$ is a unit vector along OP.

Fig. 1.6-2 shows the variation of electric potential $\phi(x)$ and field $E(x)$ with distance x along the axis of the ring. The field is maximum at a distance x for which $\frac{dE}{dx} = 0$. This gives $x = \pm a/\sqrt{2}$.

Note that if the point P is far away from the ring then $x \gg a$ and we can write approximately,

$$\vec{E} \approx \frac{1}{4\pi\varepsilon_0} \cdot \frac{Q}{x^2}\hat{x}.$$

In this case the charged ring behaves like a point charge Q placed at the centre of the ring.

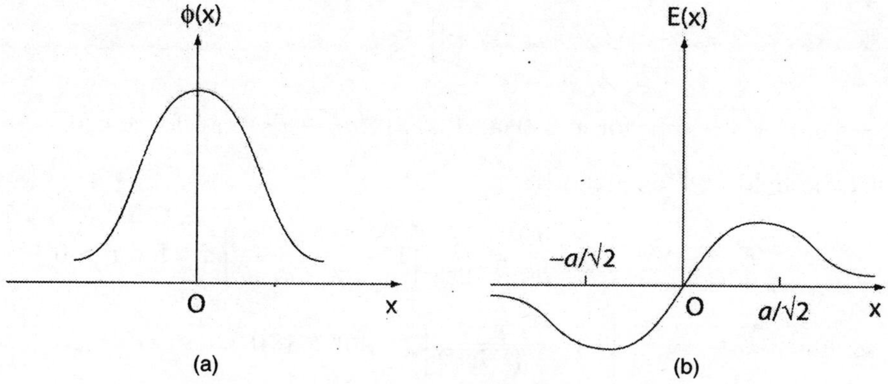

Fig. 1.6-2 Variation of (a) potential and (b) field with distance on the axis of a charged ring

2. A uniformly charged disc

Suppose we have a thin circular disc of radius a having a uniform surface charge density of σ on it. We are to calculate the field and potential at an axial point P at a distance x from the centre O of the disc (Fig. 1.6-3). Let us consider a thin ring of radius r and thickness dr. The charge on this ring is $\sigma.2\pi r\, dr$. Each element of this ring is almost equidistant from P and hence the potential at P due to this ring is given by

$$d\phi = \frac{1}{4\pi\varepsilon_0} \cdot \frac{\sigma \cdot 2\pi r\, dr}{\sqrt{r^2+x^2}}.$$

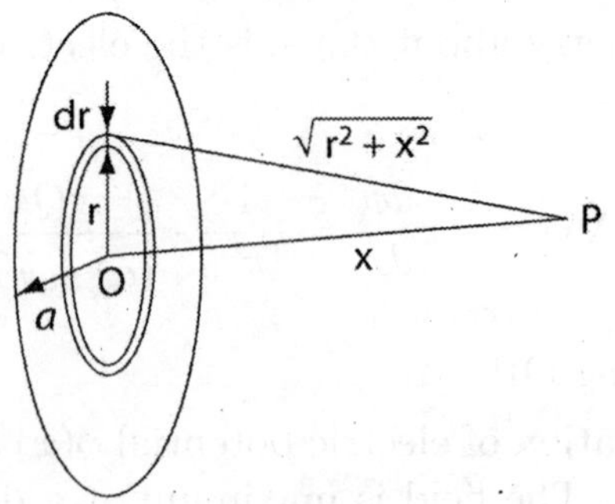

Fig. 1.6-3

Therefore, total potential at P due to the whole disc is obtained by summing over all rings from $r = 0$ to $r = a$,

$$\phi = \frac{\sigma}{2\varepsilon_0}\int_0^a \frac{r\,dr}{\sqrt{r^2+x^2}}$$

$$= \frac{\sigma}{2\varepsilon_0}\int_x^{\sqrt{a^2+x^2}} \frac{Z\,dZ}{Z} \quad [\text{Putting } r^2 + x^2 = Z^2]$$

$$= \frac{\sigma}{2\varepsilon_0}\left[\sqrt{a^2+x^2} - |x|\right] \tag{1.6-3}$$

$\therefore\ \phi = \dfrac{\sigma}{2\varepsilon_0}\left[\sqrt{a^2+x^2} - x\right]$ for $x > 0$ and $\frac{\sigma}{2\varepsilon_0}\left[\sqrt{a^2+x^2} + x\right]$ for $x < 0$.

The electric field at P is given by

$$\vec{E} = -\vec{\nabla}\phi = -\hat{x}\frac{d\phi}{dx} = \frac{\sigma}{2\varepsilon_0}\left[1 - \frac{x}{\sqrt{a^2+x^2}}\right]\hat{x} \quad \text{for } x > 0 \tag{1.6-4}$$

$$\text{and} \quad \vec{E} = -\frac{\sigma}{2\varepsilon_0}\left[1 + \frac{x}{\sqrt{a^2+x^2}}\right]\hat{x} \quad \text{for } x < 0 \tag{1.6-5}$$

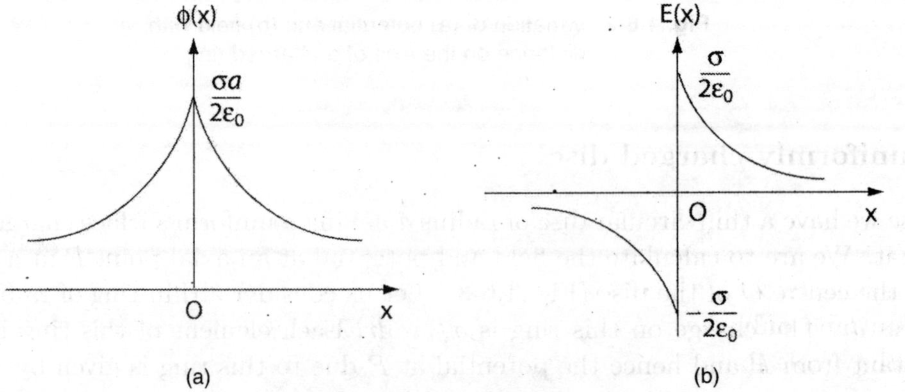

Fig. 1.6-4 Variation of (a) potential and (b) on the axis of a uniformly charged disc

Fig. 1.6-4 shows the variation of electric potential $\phi(x)$ and field $E(x)$ with the distance x. Obviously at the surface $x = 0$, $\phi(x)$ is continuous whereas $E(x)$ is discontinuous. In fact this is true for any surface containing charges.

If the point P is far away from the disc then $|x| \gg a$ and using binomial theorem we can write approximately,

$$\sqrt{a^2 + x^2} = x\left(1 + \frac{a^2}{x^2}\right)^{1/2} \approx x\left(1 + \frac{a^2}{2x^2}\right) = x + \frac{a^2}{2x}.$$

Substituting this in Eq. (1.6-3) we get

$$\phi(x) = \frac{\sigma}{2\varepsilon_0} \cdot \frac{a^2}{2x} = \frac{1}{4\pi\varepsilon_0} \cdot \frac{Q}{x},$$

where $Q = \pi a^2.\sigma$ is the total charge on the disc. This shows that for points far away from the disc it behaves like a point charge Q placed at the centre of the disc.

If the disc is of infinite extent $(a \to \infty)$ then

$$\vec{E} = \begin{cases} \dfrac{\sigma}{2\varepsilon_0}\hat{x} & \text{for } x > 0 \\[2ex] -\dfrac{\sigma}{2\varepsilon_0}\hat{x} & \text{for } x < 0 \end{cases}$$

In this case the field is uniform and everywhere normal to the plane of the disc.

3. A line of charge

Suppose a uniform line charge of density λ per unit length exists on the y-axis from $y = -L$ to $y = +L$. We are to calculate the field and potential at any point P at a distance x from the line of charge and lying on the perpendicular bisector of the line (Fig. 1.6-5). The electric field $d\vec{E}$ at P due to an element dy is of magnitude

$$\left|d\vec{E}\right| = \frac{1}{4\pi\varepsilon_0} \cdot \frac{\lambda dy}{R^2}$$

and directed along $\overrightarrow{PA}$. It can be resolved into two components $\left|d\vec{E}\right| \cos\theta$ along $\overrightarrow{OP}$ and $\left|d\vec{E}\right| \sin\theta$ perpendicular to $\overrightarrow{OP}$. Now, if we consider another element of line charge placed symmetrically about O it becomes clear that the perpendicular components cancel out and only the horizontal components contribute. Therefore, total field at P due to the whole line of charge would be

$$\vec{E} = \frac{1}{4\pi\varepsilon_0} \cdot \int \frac{\lambda\, dy}{R^2} \cos\theta \cdot \hat{x}.$$

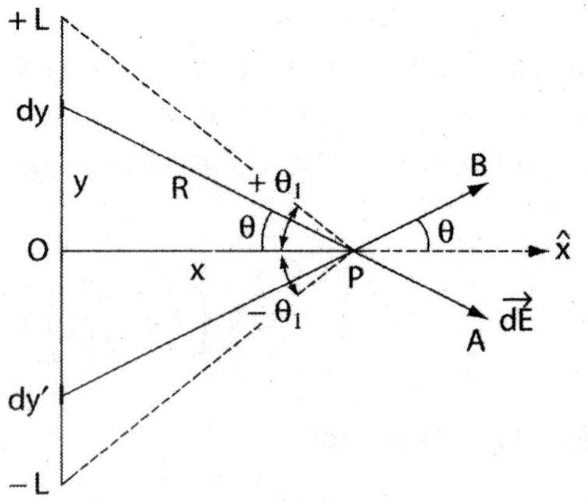

Fig. 1.6-5

Now from Fig. 1.6-5,

$$y = x\tan\theta, \quad dy = x\sec^2\theta\, d\theta \quad \text{and} \quad x/R = \cos\theta \quad \text{or} \quad R = x\sec\theta.$$

Substituting all these we get,

$$\vec{E} = \frac{\hat{x}\lambda}{4\pi\varepsilon_0}\int_{-\theta_1}^{+\theta_1}\frac{\cos\theta\, d\theta}{x} = \frac{\hat{x}\lambda}{4\pi\varepsilon_0 x}\cdot 2\sin\theta_1 = \frac{\hat{x}}{4\pi\varepsilon_0}\cdot\frac{2\lambda L}{x\sqrt{x^2+L^2}}. \qquad (1.6\text{-}6)$$

If the point P is far away from the line charge then $x \gg L$ and we get approximately,

$$\vec{E} = \frac{\hat{x}}{4\pi\varepsilon_0}\cdot\frac{2\lambda L}{x^2} = \frac{\hat{x}}{4\pi\varepsilon_0}\cdot\frac{Q}{x^2},$$

where $Q = 2\lambda L$ is the total line charge. In this case the line charge behaves like a point charge Q placed at the mid-point of the line.

For a line charge of infinite extent or for points very close to the line charge $L \gg x$ or $\theta_1 = \pi/2$ and we can write

$$\vec{E} \approx \frac{\hat{x}}{4\pi\varepsilon_0}\cdot\frac{2\lambda}{x} = \frac{\hat{x}\lambda}{2\pi\varepsilon_0 x}. \qquad (1.6\text{-}7)$$

• Potential of an infinite line charge

So far in calculating the potential due to a charge distribution we have assumed the potential to be zero at infinity. This is allowed for charge distributions located within finite region of space. In this example charge distribution extends to infinity and we cannot take the potential to be zero at infinity. If we try to do so the integral determining the potential diverges and we get infinite result. The work done in taking a unit charge from some reference point x_0 to the point x is given by

$$\phi(x) - \phi(x_0) = -\int_{x_0}^{x} E(x)\, dx = -\int_{x_0}^{x}\frac{\lambda}{2\pi\varepsilon_0 x}dx = -\frac{\lambda}{2\pi\varepsilon_0}\ln x + \frac{\lambda}{2\pi\varepsilon_0}\ln x_0. \qquad (1.6\text{-}8)$$

This shows that the electric potential due to an infinite line charge can be taken as

$$\phi(x) = -\frac{\lambda}{2\pi\varepsilon_0} \ln x + \text{constant}. \tag{1.6-9}$$

Note that the constant $\frac{\lambda}{2\pi\varepsilon_0} \ln x_0$ in (1.6-8) has no effect when we take $-\vec{\nabla}\phi$ to calculate $\vec{E}$:

$$\vec{E} = -\vec{\nabla}\phi = -\vec{x}\frac{d\phi}{dx} = \frac{\vec{x}\lambda}{2\pi\varepsilon_0 x}.$$

1.7 Flux of Electric Field and Gauss's Law

The electric field due to any given charge distribution can, in principle, be determined by using Coulomb's law and the superposition principle. The process is very often found to be tedious. The Gauss's law, which will be discussed now provides a very powerful method of solution of some symmetric electrostatic problems. It is not an independent law but can be derived from Coulomb's law. However, in many cases Gauss's law is found to be more powerful than Coulomb's law. Before introducing this law let us explain two useful ideas *flux of electric field* and *solid angle.*

• Flux of electric field

Consider a surface element $d\vec{S} = \hat{n}\, dS$ in an electric field $\vec{E}$ (Fig. 1.7-1), where $\hat{n}$ is the outward unit vector normal to the surface element. The quantity $d\Phi = \vec{E} \cdot d\vec{S} = E \cos\theta\, dS$ is called the *flux* of $\vec{E}$ through $d\vec{S}$. If we consider that q/ε_0 number of lines of force emanate isotropically from a point charge q then the idea of electric flux becomes meaningful. If we use this picture of electric field then $d\Phi = \vec{E} \cdot d\vec{S}$ becomes equal to the number of lines of force passing through the area $d\vec{S}$. The flux of $\vec{E}$ over any arbitrary surface S is given by the integral

$$\Phi = \int d\Phi = \int_S \vec{E} \cdot d\vec{S}$$

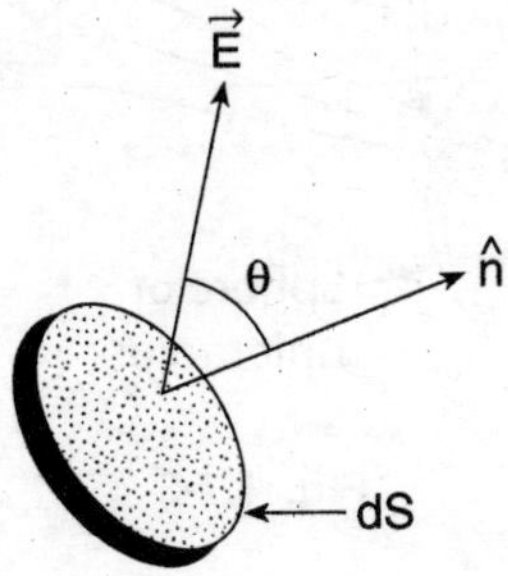

Fig 1.7-1

• Solid angle

The concept of solid angle arises as the three dimensional extension of the idea of two dimensional planar angle. The plane angle $d\theta$ subtended by an arc dl of a circle of radius r, at the centre of the circle is defined by

$$d\theta = \frac{dl}{r}.$$

Similarly in three dimension the solid angle $d\Omega$ subtended by a surface element dS of a sphere of radius r, at the centre of the sphere is defined by

$$d\Omega = \frac{dS}{r^2}.$$

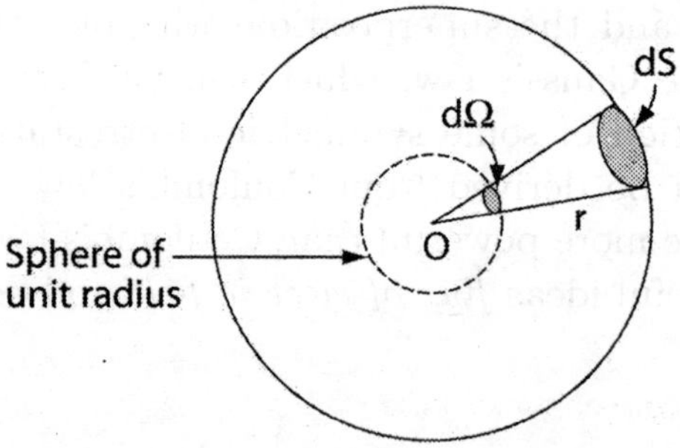

Fig. 1.7-2

The solid angle is a dimensionless quantity. Its unit is steradian (sr). If the peripheral points of the surface element dS are connected to O by straight lines, a cone is generated whose base is dS and apex O (Fig. 1.7-2). The area that this constructed cone intercepts from the surface of a sphere of unit radius drawn about O is numerically equal to the solid angle $d\Omega$.

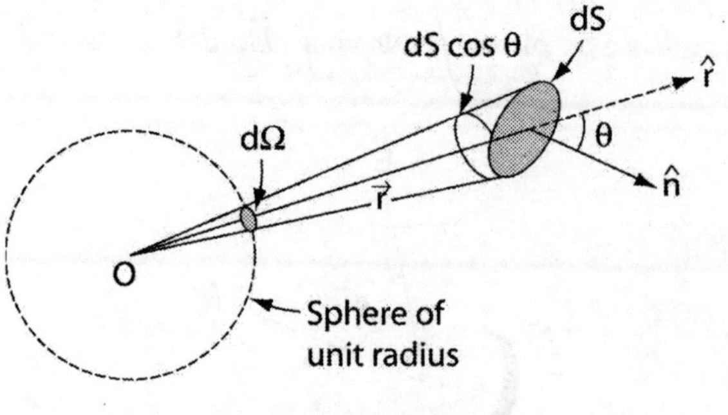

Fig. 1.7-3

Now we consider an arbitrary surface element $d\vec{S} = \hat{n}\, dS$ whose position vector with respect a point O is $\vec{r}$ (Fig. 1.7-3). If the vector $\vec{r}$ makes an angle θ with $\hat{n}$, the unit vector normal to dS then the projection of the surface element on a plane normal to $\vec{r}$ would be $dS\cos\theta = \hat{r}\cdot\hat{n}\,dS$. In this case the solid angle subtended by $d\vec{S}$ at O would be

$$d\Omega = \frac{dS\cos\theta}{r^2} = \frac{\hat{r}\cdot\hat{n}\,dS}{r^2}. \tag{1.7-1}$$

Remember that total solid angle subtended by a closed surface at any internal point is 4π but at any external point is zero.

• Gauss's law

Gauss showed that an important relationship exists between the total electric flux over a closed surface and the total charge enclosed by the surface. This relationship is known as *Gauss's law.* Using SI units the law may be stated as follows:

In an arbitrary electrostatic field (in vacuum) the total electric flux over any closed surface is equal to $1/\varepsilon_0$ times the total charge enclosed by the surface, where ε_0 is the free space permittivity.

For a closed surface S enclosing N number of point charges $q_1, q_2, \ldots, q_N$ the law may be expressed mathematically as

$$\oint_S \vec{E}\cdot d\vec{S} = \frac{1}{\varepsilon_0}\times\sum_{i=1}^{N} q_i. \tag{1.7-2}$$

For a continuous distribution of charge within a volume V enclosed by S and characterised by a volume charge density ρ the law may be written as

$$\oint_S \vec{E}\cdot d\vec{S} = \frac{1}{\varepsilon_0}\int_V \rho\,dV. \tag{1.7-3}$$

This is known as the *integral form of Gauss's law.* The closed surface on which the Gauss's law is applied is called a *Gaussian surface.*

• Proof of Gauss's law

Suppose we consider a point charge q located at the point O and draw a close surface S enclosing q as shown in Fig. 1.7-4. Let $d\vec{S} = \hat{n}\,dS$ be a surface element on this surface and the position vector of any point P on this element with respect to O be $\vec{r}$. The electric field $\vec{E}$ at P due to q, i.e., the force on unit positive charge placed at P is given by Coulomb's law as,

$$\vec{E} = \frac{q}{4\pi\varepsilon_0}\frac{\hat{r}}{r^2}.$$

The flux of $\vec{E}$ through $d\vec{S}$ is $\vec{E}\cdot d\vec{S} = \vec{E}\cdot\hat{n}\,dS.$

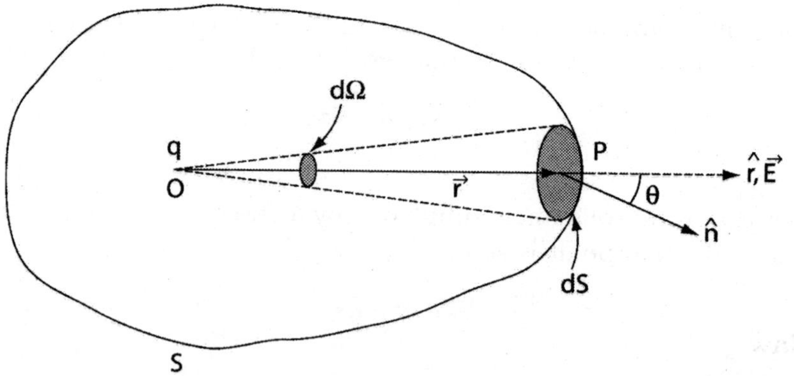

Fig. 1.7-4

Therefore, total electric flux over S is

$$\oint_S \vec{E}\cdot\hat{n}\,dS = \frac{q}{4\pi\varepsilon_0}\oint_S \frac{\hat{r}\cdot\hat{n}\,dS}{r^2}.$$

Now the quantity $\dfrac{\hat{r}\cdot\hat{n}dS}{r^2} = \dfrac{dS\cos\theta}{r^2} = d\Omega$ is the solid angle subtended by dS at q. Therefore,

$$\oint \vec{E}\cdot\hat{n}\,dS = \frac{q}{4\pi\varepsilon_0}\cdot\oint d\Omega = \frac{q}{4\pi\varepsilon_0}\times 4\pi = \frac{q}{\varepsilon_0} \qquad (1.7\text{-}4)$$

[∵ Total solid angle subtended by S at O is 4π]

Charges may lie outside the surface S. Consider a charge q at a point O outside a closed surface S as shown in Fig. 1.7-5. If an elementary cone of solid angle $d\Omega$ be drawn from q, it would cut the surface twice in areas dS_1 and dS_2 at P_1 and P_2.

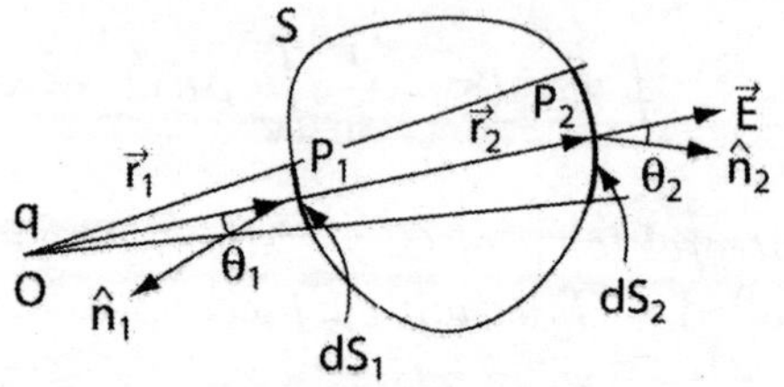

Fig. 1.7-5

The electric field lines enter the close surface through dS_1 and leave it through dS_2. As the solid angles subtended by dS_1 and dS_2 at O are the same we can say that the electric flux through dS_1 and dS_2 will be equal and opposite and hence, they contribute nothing to the total electric flux over S. This is true for all cones drawn from O to cover the whole surface S. Thus the contribution to the total electric flux over S will be zero when charges lie outside S. If there are a number of charges q_1, q_2, ... inside S then by the principle of superposition total field

$$\vec{E} = \sum_i \vec{E}_i,$$

where $\vec{E}_i$ is the field due to i-th charge q_i. Then Eq. (1.7-4) takes the form

$$\oint_S \vec{E} \cdot \hat{n}\, dS = \frac{1}{\varepsilon_0} \sum_i q_i. \tag{1.7-5}$$

As the charges lying outside S do not contribute to the total electric flux over S the electric field $\vec{E}$ in Eq. (1.7-5) may be considered as the total field due to all charges lying within as well as outside S. The result (1.7-5) can also be generalised to the case of a continuous distribution of charges within S. If S encloses a volume V and ρ is the charge density within V then Eq. (1.7-5) changes to

$$\oint_S \vec{E} \cdot \hat{n}\, dS = \frac{1}{\varepsilon_0} \int_V \rho\, dV. \tag{1.7-6}$$

• Differential form of Gauss's law

Using divergence theorem left-hand side of the Eq. (1.7-6) can be changed into a volume integral,

$$\int_V \vec{\nabla} \cdot \vec{E}\, dV = \frac{1}{\varepsilon_0} \int_V \rho\, dV.$$

Since this is true for any volume V, the integrands must be equal, i.e.,

$$\vec{\nabla} \cdot \vec{E} = \frac{1}{\varepsilon_0} \rho. \tag{1.7-7}$$

This is the *differential form of Gauss's law.* It is a basic equation in electrostatics. It relates electric field at a point with the charge density at that point.

Since $\vec{E} = -\vec{\nabla}\phi$, Eq. (1.7-7) may be rewritten as

$$\vec{\nabla} \cdot \left(-\vec{\nabla}\phi\right) = \frac{1}{\varepsilon_0} \rho \text{ or } \nabla^2 \phi = -\frac{\rho}{\varepsilon_0}. \tag{1.7-8}$$

This is known as *Poisson's equation.* When $\rho = 0$, it reduces to

$$\nabla^2 \phi = 0, \tag{1.7-9}$$

which is known as *Laplace's equation.*

1.8 Application of Gauss's Law

Gauss's law is always true in electrostatics but useful when there is sufficient symmetry.

Electrostatic problems having sufficient symmetry can be solved in a very easy way by using Gauss's law. Calculations become very simple if it is possible to choose a Gaussian surface such that the normal component of electric field on it is either zero or has a fixed value at every point on the surface. A few symmetric problems are discussed here.

1. A uniformly charged sphere

(a) **Field outside the sphere:** Consider a uniformly charged sphere of radius a carrying a total charge Q. We are to find the electric field at an external point P at a distance r from the centre of the sphere (Fig. 1.8-1). We construct the Gaussian surface, which in this case is a concentric sphere of radius r passing through the point P. For this surface (S) Gauss's law states that

$$\int_S \vec{E} \cdot d\vec{S} = \frac{1}{\varepsilon_0} \cdot Q. \tag{1.8-1}$$

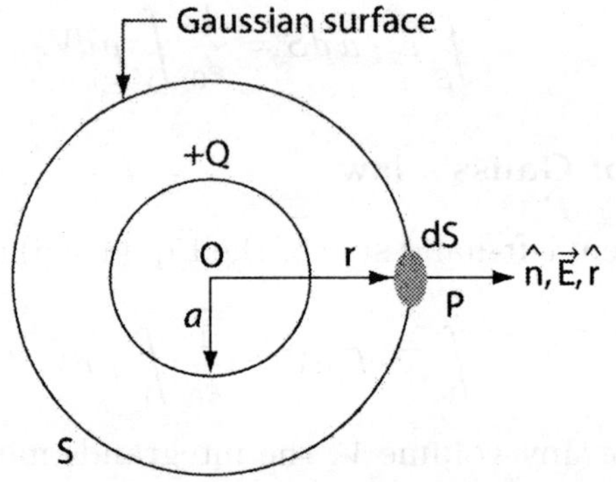

Fig. 1.8-1

Because of spherical symmetry $\vec{E}$-field lines will be normal to the Gaussian surface at every point and $\vec{E}$ will also be of constant magnitude all over the surface. Therefore,

$$\int_S \vec{E} \cdot d\vec{S} = E \int_S dS = E \cdot 4\pi r^2$$

and hence, from (1.8-1)

$$E \cdot 4\pi r^2 = \frac{1}{\varepsilon_0} Q \quad \text{or} \quad E = \frac{1}{4\pi\varepsilon_0} \frac{Q}{r^2}. \tag{1.8-2}$$

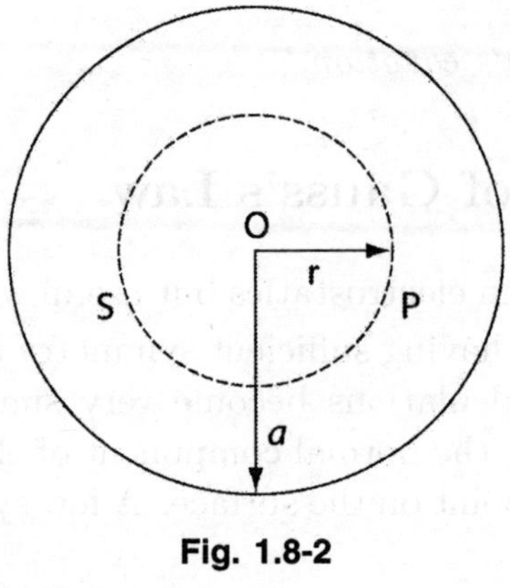

Fig. 1.8-2

In vector form, $$\vec{E} = \frac{1}{4\pi\varepsilon_0}\frac{Q}{r^2}\hat{r}. \tag{1.8-3}$$

This is the same field as produced by a point charge Q placed at the centre O. Thus for an outside point the whole charge on the sphere may be assumed to be concentrated at the centre of the sphere.

(b) **Field inside the sphere :** To find the field at any internal point P (Fig. 1.8-2), we again construct the Gaussian surface, which is a concentric sphere of radius r passing through P. The charge enclosed by this surface is

$$Q\,\text{enc} = \tfrac{4}{3}\pi r^3 \cdot \rho,$$

where ρ is the charge density given by

$$\rho = \frac{Q}{\frac{4}{3}\pi a^3}. \tag{1.8-4}$$

Now from Gauss's law,

$$\int_S \vec{E}\cdot d\vec{S} = \frac{1}{\varepsilon_0}\cdot\frac{4}{3}\pi r^3\rho. \tag{1.8-5}$$

As before, the spherical symmetry of the problem indicates that $\vec{E}$-field lines are normal to the constructed Gaussian surface and $\vec{E}$ is of constant magnitude all over the surface. Hence, we can write from (1.8-5)

$$E\int dS = \frac{1}{\varepsilon_0}\frac{4}{3}\pi r^3\rho$$

$$\text{or } E\cdot 4\pi r^2 = \frac{1}{\varepsilon_0}\frac{4}{3}\pi r^3\rho$$

$$\text{or } E = \frac{\rho r}{3\varepsilon_0} = \frac{1}{4\pi\varepsilon_0}\cdot\frac{Qr}{a^3} \quad \text{[using Eq. (1.8-4)]} \tag{1.8-6}$$

In vector form

$$\vec{E} = \frac{1}{4\pi\varepsilon_0}\cdot\frac{Qr}{a^3}\hat{r}. \tag{1.8-7}$$

Thus the field inside the sphere increases linearly with the distance from the centre and outside sphere it decrease inversely as the square of the distance from the centre (Fig. 1.8-3).

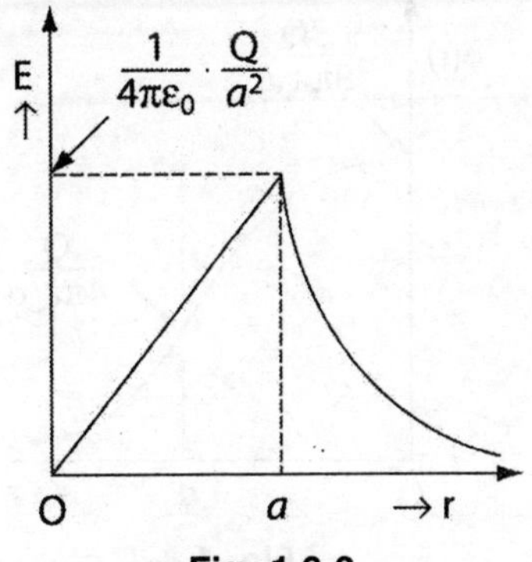

Fig. 1.8-3

• Calculation of potential

As the electric field $\vec{E}$ is now known the corresponding potential can be found by using the relation $\vec{E} = -\vec{\nabla}\phi$.

For an external point $(r \geq a)$

$$\frac{1}{4\pi\varepsilon_0} \cdot \frac{Q}{r^2}\hat{r} = -\hat{r}\frac{d\phi}{dr}$$

$$\text{or} \quad \phi(r) = -\frac{Q}{4\pi\varepsilon_0}\int \frac{dr}{r^2} + C_1 = \frac{Q}{4\pi\varepsilon_0} \cdot \frac{1}{r} + C_1, \tag{1.8-8}$$

where C_1 is a constant of integration. As the charge is located within a finite region of space we can take $\phi = 0$ at $r = \infty$. Therefore, $C_1 = 0$ in Eq. (1.8-8) and we get

$$\phi(r) = \frac{1}{4\pi\varepsilon_0} \cdot \frac{Q}{r} \quad \text{for } r \geq a. \tag{1.8-9}$$

For an internal point $(r \leq a)$

$$\frac{1}{4\pi\varepsilon_0} \cdot \frac{Qr}{a^3}\hat{r} = -\hat{r}\frac{d\phi}{dr}$$

$$\text{or} \quad \phi(r) = -\frac{Q}{4\pi\varepsilon_0 a^3}\int r\,dr + C_2 = -\frac{Q}{4\pi\varepsilon_0 a^3} \cdot \frac{r^2}{2} + C_2, \tag{1.8-10}$$

where C_2 is another constant of integration. To find C_2 we apply the condition that $\phi(r)$ is continuous across the boundary $r = a$. Therefore, from Eqs. (1.8-9) and (1.8-10) we get

$$\frac{Q}{4\pi\varepsilon_0 a} = -\frac{Q}{4\pi\varepsilon_0 a} \cdot \frac{1}{2} + C_2 \text{ or } C_2 = \frac{Q}{4\pi\varepsilon_0 a} \cdot \frac{3}{2}.$$

So from Eq. (1.8-10),

$$\phi(r) = \frac{Q}{4\pi\varepsilon_0 a}\left[\frac{3}{2} - \frac{r^2}{2a^2}\right] \quad \text{for } r \leq a \tag{1.8-11}$$

Fig. 1.8-4 shows the variation of potential $\phi(r)$ with distance from O.

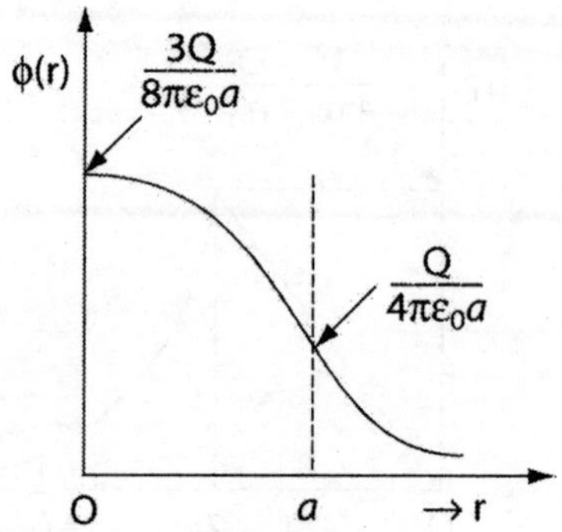

Fig. 1.8-4

2. A uniformly charged long cylinder

(a) **Field outside the cylinder :** Consider a long cylinder of radius a carrying a uniform charge density of λ per unit length. We are to find the electric field at an external point P at a distance r from the axis of the cylinder (Fig. 1.8-5). We construct the Gaussian surface, which in this case is a coaxial cylinder of radius r and length l.

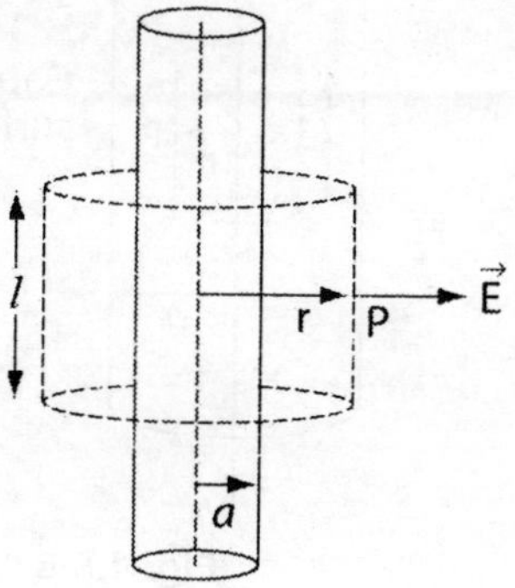

Fig. 1.8-5

For this surface (S) Gauss's law states that

$$\int_S \vec{E}\cdot d\vec{S} = \frac{1}{\varepsilon_0}\cdot \lambda l,$$

where λl is the charge enclosed by the surface.

Because of cylindrical symmetry of the problem $\vec{E}$-field lines will be normal to the curved portion of the constructed Gaussian surface and $\vec{E}$ will also be of constant magnitude all over it. As the $\vec{E}$-field lines are tangential to the two end faces $\vec{E}\cdot d\vec{S}$ will be zero on these faces. Thus,

$$\int_S \vec{E}\cdot d\vec{S} = E\int dS = E\cdot 2\pi r l = \frac{1}{\varepsilon_0}\lambda l$$

$$\text{or}\quad E = \frac{\lambda}{2\pi\varepsilon_0}\cdot\frac{1}{r} = \frac{1}{4\pi\varepsilon_0}\cdot\frac{2\lambda}{r}.$$

In vector form

$$\vec{E} = \frac{1}{4\pi\varepsilon_0}\cdot\frac{2\lambda}{r}\hat{r}. \tag{1.8-12}$$

(b) **Field inside the cylinder :** To find the field at any internal point P (Fig, 1.8-6) we again construct the Gaussian surface, which is a coaxial cylinder of radius r, length l and passes through P. The charge enclosed by it is

$$Q\,\text{enc} = \pi r^2 l\cdot\rho, \tag{1.8-13}$$

where ρ is the volume density of charge. It is related to λ by

$$\pi a^2 \cdot 1 \cdot \rho = \lambda \quad \text{or} \quad \rho = \lambda/\pi a^2. \tag{1.8-14}$$

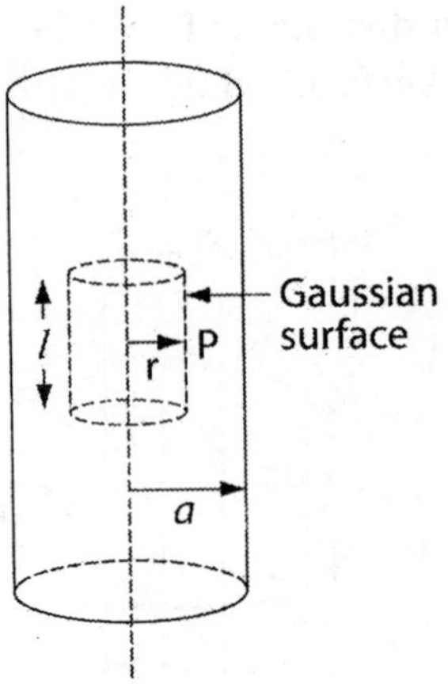

Fig. 1.8-6

Now from Gauss's law,

$$\int_S \vec{E} \cdot d\vec{S} = \frac{1}{\varepsilon_0} \cdot Q_{\text{enc}}.$$

Applying the same arguments as in (a) we find that

$$\int_S \vec{E} \cdot d\vec{S} = E \cdot 2\pi r l.$$

$$\therefore\ E \cdot 2\pi r l = \frac{1}{\varepsilon_0} \cdot Q\ \text{enc} = \frac{1}{\varepsilon_0} \cdot \pi r^2 l \rho = \frac{1}{\varepsilon_0} \pi r^2 l \cdot \frac{\lambda}{\pi a^2}$$

$$\text{or}\ \ E = \frac{1}{4\pi\varepsilon_0} \cdot \frac{2\lambda r}{a^2}.$$

In vector form

$$\vec{E} = \frac{1}{4\pi\varepsilon_0} \cdot \frac{2\lambda r}{a^2} \cdot \hat{r}. \tag{1.8-15}$$

3. Uniformly charged infinite plane

Let us consider a uniformly charged infinite plane having a surface density σ. Suppose we are to find field at any point P (Fig. 1.8-7). We construct the Gaussian surface (S), which in this case is a cylinder extending equal distances on the two sides of the plane. From Gauss's law,

$$\int_S \vec{E} \cdot d\vec{S} = \frac{1}{\varepsilon_0} \cdot Q_{\text{enc}}. \tag{1.8-16}$$

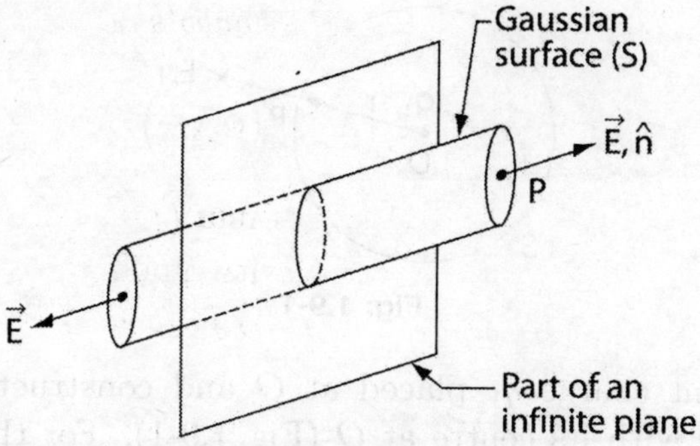

Fig. 1.8-7

If ΔS be the cross-sectional area of the Gaussian cylinder then $Q_{\text{enc}} = \sigma \Delta S$. As the plane is of infinite extent the $\vec{E}$-field lines will be perpendicular to the plane and they point away from the plane on its both sides; also $\vec{E}$ will have constant magnitude all over the end faces.

So flux through the curved surface of the Gaussian cylinder will be zero and considering two end faces we get

$$\int_S \vec{E} \cdot d\vec{S} = E \cdot \Delta S + E \cdot \Delta S = 2E\Delta S.$$

Therefore, from (1.8-16) we get

$$2E\Delta S = \frac{1}{\varepsilon_0}\sigma\Delta S \quad \text{or} \quad E = \frac{\sigma}{2\varepsilon_0}.$$

In vector form $\vec{E} = \sigma/(2\varepsilon_0)\hat{n}$ on the right-hand side of the plane and $\vec{E} = -\sigma/2\varepsilon_0 \cdot \hat{n}$ on the left-hand side of the plane. Note that the field E is independent of the distance from the plane. Therefore, it is constant on both sides of the plane, but oppositely directed. There is a discontinuity equal to

$$\frac{\sigma}{2\varepsilon_0} - \left(-\frac{\sigma}{2\varepsilon_0}\right) = \frac{\sigma}{\varepsilon_0} \quad \text{at the plane.}$$

1.9 Coulomb's Law from Gauss's Law

We have seen that Gauss's Law is not an independent law but can be established from Coulomb's law. Conversely starting from Gauss's law we can arrive at Coulomb's law. In fact these two laws express the same fact of nature in two different ways. Coulomb's law helps us to determine the electric field from a knowledge of electric charge distribution, whereas Gauss's law enables us to determine the charge distribution from a knowledge of electric field.

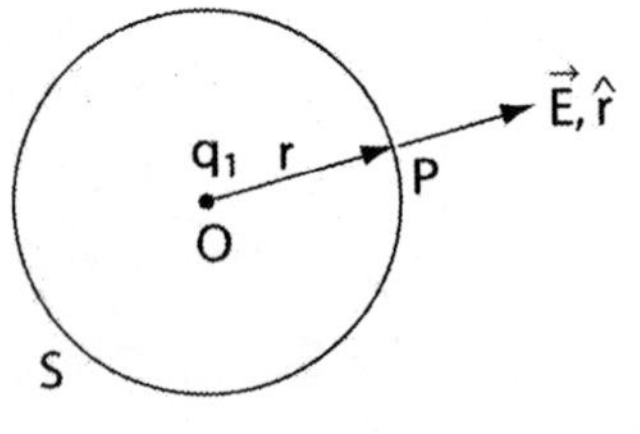

Fig. 1.9-1

We consider a point charge q_1 placed at O and construct a spherical Gaussian surface (S) of radius r with its centre at O (Fig. 1.9-1). For this surface Gauss's law states that

$$\int_S \vec{E} \cdot d\vec{S} = \frac{1}{\varepsilon_0} \cdot q_1. \tag{1.9-1}$$

Because of spherical symmetry electric field lines will be normal to the Gaussian surface and $\vec{E}$ will have constant magnitude all over the Gaussian surface.

$$\int_S \vec{E} \cdot d\vec{S} = E \int dS = E \cdot 4\pi r^2.$$

Now from (1.9-1) we get

$$E \cdot 4\pi r^2 = \frac{1}{\varepsilon_0} q_1 \quad \text{or} \quad E = \frac{1}{4\pi\varepsilon_0} \frac{q_1}{r^2}.$$

In vector form the electric field at any point P with position vector $\vec{r}$ with respect to the position of q_1 is given by

$$\vec{E} = \frac{1}{4\pi\varepsilon_0} \frac{q_1}{r^2} \hat{r}.$$

If a second point charge q_2 is placed at P then the force on it due to q_1 will be

$$\vec{F} = q_2 \vec{E} = \frac{1}{4\pi\varepsilon_0} \frac{q_1 q_2}{r^2} \hat{r}.$$

This is the Coulomb's law of force between two point charges. At this point we may point out that the two laws may not be equivalent always. For example, for a moving charge electric field is not isotropic about the charge. Then, it can be shown that Gauss's law is still valid for this field but Coulomb's law is not.

1.10 Earnshaw's Theorem

An important consequence of Gauss's law concerns the equilibrium of a charged particle in an electrostatic field. It is shown that a freely movable charge cannot exist in stable equilibrium in free space under the influence of electrostatic fields alone. This fact is often given the name *Earnshaw's theorem.*

The potential energy of a charge q at a point $P(x, y, z)$ having potential ϕ is $W = q\phi$. Now stable equilibrium of q requires that W be minimum at P. In free space the charge density $\rho = 0$ and ϕ satisfies Laplace's equation,

$$\nabla^2\phi = 0 \quad \text{or} \quad \frac{\partial^2\phi}{\partial x^2} + \frac{\partial^2\phi}{\partial y^2} + \frac{\partial^2\phi}{\partial z^2} = 0. \tag{1.10-1}$$

Multiplying by q we get

$$\frac{\partial^2 W}{\partial x^2} + \frac{\partial^2 W}{\partial y^2} + \frac{\partial^2 W}{\partial z^2} = 0. \tag{1.10-2}$$

For W to be minimum at the point (x, y, z) the second order derivatives

$$\frac{\partial^2 W}{\partial x^2}, \quad \frac{\partial^2 W}{\partial y^2} \quad \text{and} \quad \frac{\partial^2 W}{\partial z^2}$$

must each be positive at that point. But this condition is inconsistent with the Eq. (1.10-2). Poison's equation $\nabla^2\phi = -\rho/\varepsilon_0$ or $\nabla^2 W = -q\rho/\varepsilon_0$ suggests that this condition can only be satisfied if the point is occupied by a charge opposite in sign to that of q. This contradicts the assumption of free space. Thus we prove Earnshaw's theorem.

The validity of the theorem can also be understood physically. If a test charge is in stable equilibrium at any point P in an electrostatic field then for any slight displacement of the charge from P some restoring electrostatic forces must bring it back to P. This requires the presence of a charge of opposite sign at P which contradicts the assumption of free space.

• Importance of the theorem

According to the theorem no purely stationary system of charges can be in stable equilibrium under their mutual interactions only. This suggests that in atoms or molecules electrons cannot be at rest. The theorem also indicates that confinement of plasma (fully ionised gas) particles needed for controlled fusion is not possible by electrostatic fields alone.

SOLVED PROBLEMS

▶ **Problem 1.** *Suppose the electron cloud around the nucleus of charge $+e$ in an atom can be represented by the following spherical charge distribution*

$$\rho(r) = \frac{-e}{\pi r} \cdot e^{-2r/r_0},$$

where r_0 is called the Bohr radius. Calculate the total charge of the cloud.

Solution. We consider a thin spherical shell of radius r and thickness dr. Its volume is $4\pi r^2 dr$ and charge within it is $\rho(r) \cdot 4\pi r^2 dr$. Therefore, total charge of the cloud is

$$Q = \int_0^\infty \rho(r) 4\pi r^2 dr = \frac{-e}{\pi r_0} \cdot 4\pi \int_0^\infty r^2 e^{-2r/r_0} dr.$$

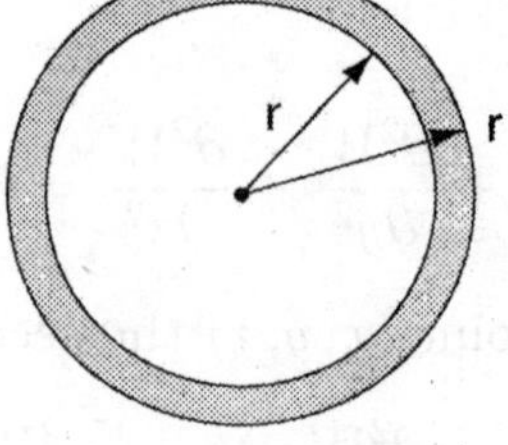

Fig. 1.P-1

Putting $2r/r_0 = x$ we get

$$Q = \frac{-e}{2} \int_0^\infty x^2 \cdot e^{-x} dx = \frac{-e}{2} \Gamma(3) = \frac{-e}{2} \times 2 = -e,$$

where we have used *Gamma-function* $\Gamma(n)$ defined by

$$\Gamma(n) = \int_0^\infty e^{-x} x^{n-1} dx$$

and having properties

$$\Gamma(1) = 1, \quad \Gamma(n+1) = n\Gamma(n), \quad \Gamma\left(\frac{1}{2}\right) = \sqrt{\pi}, \text{ etc.}$$

► **Problem 2.** *A fixed ring of radius a has a positive charge Q uniformly distributed over it. A particle of negative charge $-e$ and mass m is placed on the axis of the ring at a small distance from the centre of the ring. Show that the particle, if free to move, will under a simple harmonic motion. Find the time period of oscillation.*

Solution. Electric field on the axis of the ring at a distance x from its centre is

$$E = \frac{1}{4\pi\varepsilon_0} \cdot \frac{Qx}{(a^2 + x^2)^{3/2}}.$$

Force on the particle of charge $-e$ is $F = -eE$ and its acceleration is

$$\frac{d^2x}{dt^2} = \frac{F}{m} = -\frac{eE}{m} = -\frac{eQ}{4\pi\varepsilon_0 m} \cdot \frac{x}{(a^2 + x^2)^{3/2}}$$

$$\approx -\frac{eQ}{4\pi\varepsilon_0 m a^3} x \quad [\text{for } x \ll a]$$

Thus acceleration is proportional to the displacement and hence the motion will be simple harmonic. If ω be the angular frequency then

$$\omega^2 = \frac{eQ}{4\pi\varepsilon_0 ma^3}.$$

$\therefore$ Time period of oscillation

$$T = \frac{2\pi}{\omega} = 2\pi\sqrt{\frac{4\pi\varepsilon_0 ma^3}{eQ}}.$$

▶ **Problem 3.** *Suppose a charge Q is distributed within a sphere of radius R in such a way that the charge density $\rho(r)$ at a distance r from the centre of the sphere is*

$$\rho(r) = \begin{cases} K(R-r) & \textit{for } 0 < r < R \\ 0 & \textit{for } r > R \end{cases}$$

Determine constant K in terms of Q and R. Calculate the electric field at any point inside the sphere. Find the value of r for which field is maximum. What is the value of this maximum field? **[J.U. 2010]**

Solution. Consider a thin spherical shell of radius r and thickness dr. Charge within it is $\rho(r)\cdot 4\pi r^2 dr$. Therefore, total charge

$$Q = \int_0^R \rho(x) 4\pi r^2 dr = 4\pi K \int_0^R (R-r) r^2 dr = \frac{\pi K R^4}{3}.$$

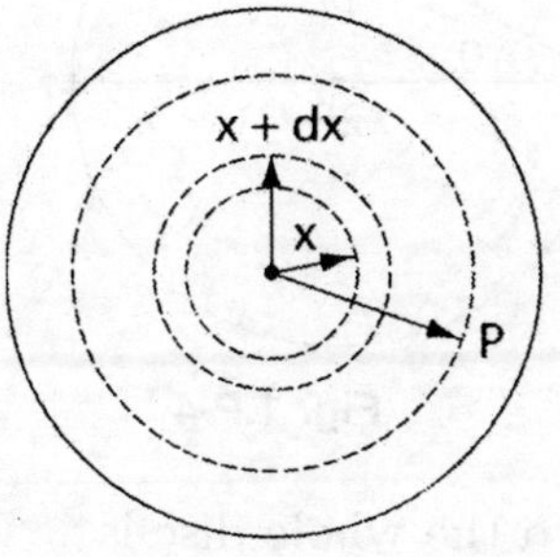

Fig. 1.P-3

$$\therefore \quad K = \frac{3Q}{\pi R^4} \quad \text{or} \quad \rho(r) = \frac{3Q}{\pi R^4}(R-r). \tag{1}$$

Suppose we are to find the electric field at any internal point P at a distance r from the centre. We construct the Gaussian surface which in this case is a concentric sphere passing through P. From Gauss's law,

$$\int_S \vec{E} \cdot d\vec{S} = \frac{1}{\varepsilon_0} \cdot Q\,\text{enc} = \frac{1}{\varepsilon_0} \int_0^r \rho(x) 4\pi x^2 dx$$

$$\text{or } E \cdot 4\pi r^2 = \frac{1}{\varepsilon_0} \cdot \int_0^r \frac{3Q}{\pi R^4}(R - x) \cdot 4\pi x^2 dx = \frac{12Q}{\varepsilon_0 R^4}\left(\frac{Rr^3}{3} - \frac{r^4}{4}\right)$$

$$\text{or } E = \frac{1}{4\pi\varepsilon_0} \cdot \frac{12Q}{R^4}\left(\frac{Rr}{3} - \frac{r^2}{4}\right). \tag{2}$$

For maximum E we must have $dE/dr = 0$. This gives $r = (2/3)R$. This is the position of maximum field. The maximum field at that point is obtained by putting $r = (2/3)R$ in Eq. (2). Thus

$$E_{\text{max}} = \frac{Q}{3\pi\varepsilon_0 R^2}.$$

▶ **Problem 4.** *Consider a thin circular disc of radius R having a uniform surface charge density σ. Calculate the potential at any point on its circumference. Comment on the case of a conducting disc.* **[cf. C.U. 2009]**

Solution. Consider a surface element $rd\theta \cdot dr$ in plane polar coordinates as shown in Fig. 1.P-4. The potential at O due to charges on this surface element is

$$d\phi = \frac{1}{4\pi\varepsilon_0} \cdot \frac{\sigma r d\theta dr}{r}.$$

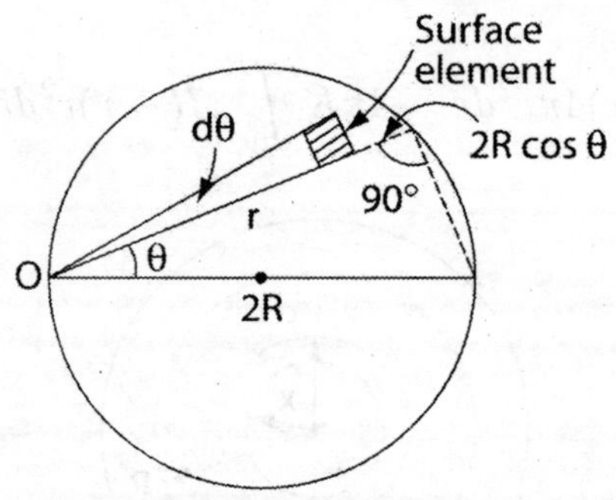

Fig. 1.P-4

∴ Total potential at O due to the whole disc is

$$\phi = \frac{\sigma}{4\pi\varepsilon_0} \int_{\theta=-\pi/2}^{+\pi/2} \int_{r=0}^{2R\cos\theta} dr\, d\theta = \frac{\sigma}{4\pi\varepsilon_0} \int_{-\pi/2}^{+\pi/2} 2R\cos\theta d\theta$$

$$= \frac{\sigma}{4\pi\varepsilon_0} \cdot 2R \times 2 = \frac{\sigma R}{\pi\varepsilon_0}.$$

If the disc is made of conducting material then it will be an equipotential surface and in that case the distribution of charge will not be uniform as we are considering.

▶ **Problem 5.** *The electric potential in a certain region of space is given by*

$$\phi(\vec{r}) = \frac{q}{4\pi\varepsilon_0} \cdot \frac{e^{-r/\lambda}}{r},$$

where λ is a constant and $\vec{r}$ is the position vector. Find the corresponding electric field $\vec{E}(\vec{r})$ and charge density $\rho(\vec{r})$. **[cf. C.U. 2014, 2017; B.U. 2017]**

Solution. The electric field is given by the relation $\vec{E} = -\vec{\nabla}\phi$.

Here

$$\vec{E} = -\hat{r}\frac{d\phi}{dr} = -\frac{q}{4\pi\varepsilon_0}\hat{r}\left[-\frac{1}{\lambda}e^{-r/\lambda} \cdot \frac{1}{r} - e^{-r/\lambda} \cdot \frac{1}{r^2}\right]$$

$$= \frac{q}{4\pi\varepsilon_0}\left(\frac{1}{r} + \frac{1}{\lambda}\right)e^{-r/\lambda}\frac{\vec{r}}{r^2}. \quad \left[\because \hat{r} = \frac{\vec{r}}{r}\right].$$

From the differential form of Gauss's law

$$\vec{\nabla} \cdot \vec{E} = \frac{\rho}{\varepsilon_0}.$$

$\therefore$ Charge density

$$\rho = \varepsilon_0 \vec{\nabla} \cdot \vec{E} = \frac{q}{4\pi}\vec{\nabla} \cdot \left[e^{-r/\lambda}\frac{\vec{r}}{r^3} + \frac{1}{\lambda}e^{-r/\lambda}\frac{\vec{r}}{r^2}\right].$$

Using the vector identity

$$\vec{\nabla} \cdot \phi\vec{A} = \vec{\nabla}\phi \cdot \vec{A} + \phi\vec{\nabla} \cdot \vec{A}.$$

We can write

$$\rho = \frac{q}{4\pi}\left[\vec{\nabla}(e^{-r/\lambda}) \cdot \frac{\vec{r}}{r^3} + e^{-r/\lambda}\vec{\nabla} \cdot \frac{\vec{r}}{r^3} + \frac{1}{\lambda}\vec{\nabla}(e^{-r/\lambda}) \cdot \frac{\vec{r}}{r^2} + \frac{1}{\lambda}e^{-r/\lambda}\vec{\nabla} \cdot \frac{\vec{r}}{r^2}\right].$$

Now

$$\vec{\nabla}(e^{-r/\lambda}) = -\frac{1}{\lambda}e^{-r/\lambda}\hat{r},$$

$$\vec{\nabla} \cdot \frac{\vec{r}}{r^3} = -\vec{\nabla} \cdot \vec{\nabla}\frac{1}{r} \quad \left[\because \vec{\nabla}\frac{1}{r} = -\frac{\vec{r}}{r^3}\right]$$

$$= -\nabla^2\frac{1}{r} = 4\pi\delta(\vec{r}),$$

where $\delta(\vec{r})$ is the Dirac delta function (See, Appendix C) and $\vec{\nabla} \cdot \frac{\vec{r}}{r^2} = \frac{1}{r^2}$. Therefore,

$$\rho = \frac{q}{4\pi}\left[-\frac{1}{\lambda}e^{-r/\lambda}\frac{1}{r^2} + e^{-r/\lambda}4\pi\delta(\vec{r}) - \frac{1}{\lambda^2}e^{-r/\lambda} \cdot \frac{1}{r} + \frac{1}{\lambda}e^{-r/\lambda}\frac{1}{r^2}\right]$$

$$= q\left[\delta(\vec{r}) - \frac{1}{4\pi\lambda^2 r}\right]e^{-r/\lambda}.$$

► **Problem 6.** *Consider a cylindrical electron beam having a circular cross section of radius a and charge density* $\rho(r) = \rho_0(1 + r^2/a^2)$. *Find the electric field at any internal point* $(r < a)$. **[J.U. 2012]**

Solution. Suppose we are to find the field at the point P at a distance r from the axis of the beam. We construct the Gaussian surface which in this case is a coaxial cylinder of length l and radius r. For this surface (S) from Gauss's law, we can write

$$\int_S \vec{E} \cdot d\vec{S} = \frac{1}{\varepsilon_0} \cdot Q\,\text{enc}.$$

Now enclosed charge

$$Q\,\text{enc} = \int \rho\, dV.$$

In cylindrical coordinates (r, θ, z) the volume element $dV = r dr d\theta dz$

$$\therefore \quad Q_{\text{enc}} = \int_0^r \rho_0 \left(1 + \frac{r^2}{a^2}\right) r dr \cdot \int_0^{2\pi} d\theta \int_0^l dz = 2\pi l \rho_0 \left(\frac{r^2}{2} + \frac{r^4}{4a^2}\right).$$

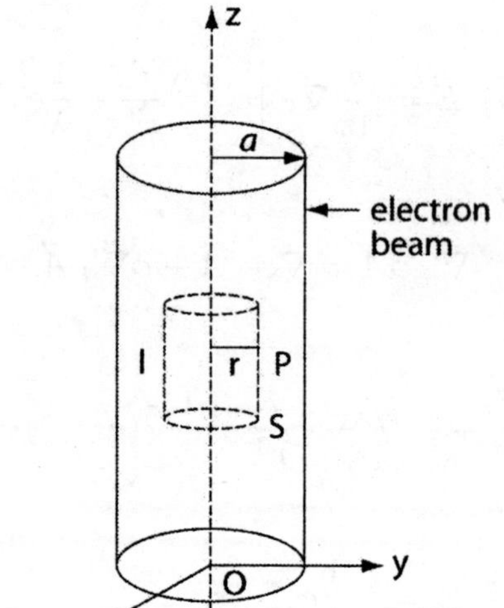

Fig. 1.P-6

Because of cylindrical symmetry $\vec{E}$ is in radially outward directions. So

$$\int_{\text{flat surface}} \vec{E} \cdot d\vec{S} = 0 \quad \text{and} \quad \int_{\text{curved surface}} \vec{E} \cdot d\vec{S} = E \int_{\text{curved surface}} dS = E \cdot 2\pi r l.$$

Thus

$$E \cdot 2\pi r l = \frac{1}{\varepsilon_0} \cdot 2\pi l \rho_0 \left(\frac{r^2}{2} + \frac{r^4}{4a^2}\right) \text{ or } E = \frac{\rho_0}{2\varepsilon_0}\left(r + \frac{r^3}{2a^2}\right).$$

In vector form

$$\vec{E} = \frac{\rho_0}{2\varepsilon_0}\left(r + \frac{r^3}{2a^2}\right)\hat{r}.$$

▶ **Problem 7.** *Suppose there is a small spherical cavity of radius r_1 within a large sphere of radius r_2 carrying uniform charge density ρ. Show that the field within the cavity is uniform.*

Solution. Let P be any point within the cavity (Fig. 1.P-7). If there were no cavity, the electric field at P would be

$$\vec{E}_2 = \frac{\rho r}{3\varepsilon_0}\hat{r}, \quad \text{[See Eq. (1.8-6)]}$$

where $\vec{r} = r\hat{r} = \overrightarrow{O_2P}$. Thus $\vec{E}_2 = \frac{\rho}{3\varepsilon_0}\overrightarrow{O_2P}$.

If the cavity region has charge density ρ then field at P due this would be

$$\vec{E}_1 = \frac{\rho}{3\varepsilon_0}\overrightarrow{O_1P}.$$

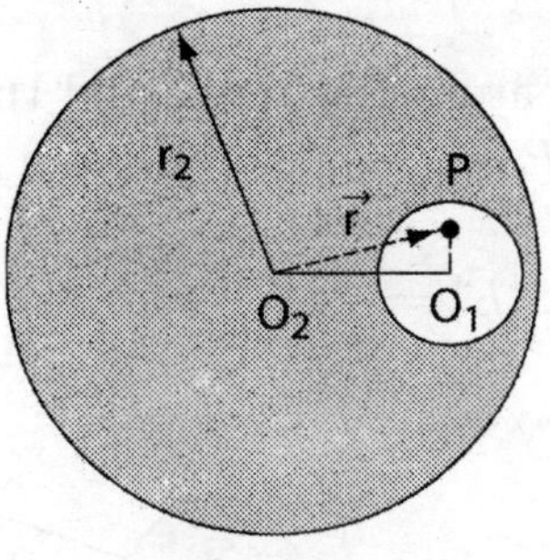

Fig. 1.P-7

Using the principle of superposition we get the required field at P as

$$\vec{E} = \vec{E}_2 - \vec{E}_1 = \frac{\rho}{3\varepsilon_0}\left(\overrightarrow{O_2P} - \overrightarrow{O_1P}\right) = \frac{\rho}{3\varepsilon_0}\overrightarrow{O_2O_1},$$

which is independent of the position of P. Thus the field inside the cavity is uniform.

▶ **Problem 8.** *A point charge of 1C is located at $(3\hat{i} + 4\hat{j} + 5\hat{k})m$. Find the electric field due to it at the point $(5\hat{i} + 4\hat{j} + 3\hat{k})\,m$.*

Solution. Let $\vec{r}' = 3\hat{i} + 4\hat{j} + 5\hat{k}$ and $\vec{r} = 5\hat{i} + 4\hat{j} + 3\hat{k}$

∴ The required field

$$\vec{E} = \frac{1}{4\pi\varepsilon_0}\cdot\frac{q(\vec{r} - \vec{r}')}{|\vec{r} - \vec{r}'|^3}.$$

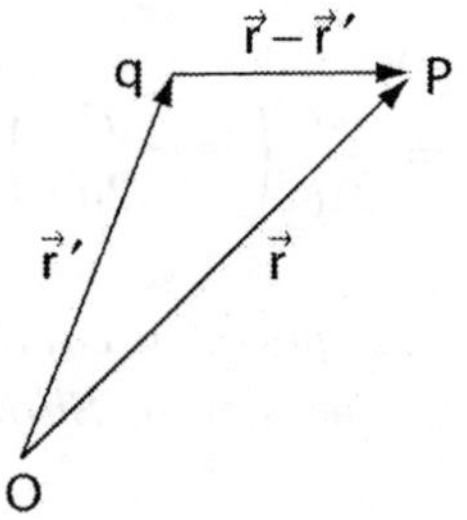

Fig. 1.P-8

Here $q = 1C$, $\vec{r} - \vec{r}' = 2\hat{i} - 2\hat{k}$, $|\vec{r} - \vec{r}'| = \sqrt{2^2 + 2^2} = 2\sqrt{2}$

$$\therefore \quad \vec{E} = \frac{q}{4\pi\varepsilon_0} \cdot \frac{2(\hat{i} - \hat{k})}{(2\sqrt{2})^3} = \frac{9 \times 10^9 \times 1 \times 2}{16\sqrt{2}}(\hat{i} - \hat{k})\,\text{N.C}^{-1}$$

$$= 7.9 \times 10^8 (\hat{i} - \hat{k})\,\text{N.C}^{-1}.$$

▶ **Problem 9.** *Suppose a point charge q is placed on the axis of a circular area of radius a at a distance x from its centre. Calculate the flux of the electric field due to q through the circular area.* **[C.U. 2003]**

Solution. Consider an annular area of radius r and thickness dr on the circular area. The electric field at any point P on this ring due to q is

$$|\vec{E}| = \frac{1}{4\pi\varepsilon_0} \frac{q}{(r^2 + x^2)}.$$

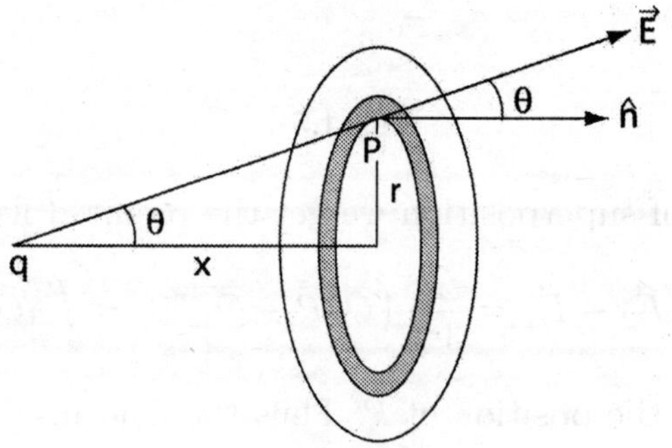

Fig. 1.P-9

Flux of $\vec{E}$ through the annular area is

$$d\Phi = \vec{E} \cdot d\vec{S} = \frac{q}{4\pi\varepsilon_0} \cdot \frac{1 \times \cos\theta \times 2\pi r dr}{(r^2 + x^2)}$$

$$= \frac{q}{2\varepsilon_0} \cdot \frac{x r dr}{(r^2 + x^2)^{3/2}}. \quad \left[\because \ \cos\theta = \frac{x}{\sqrt{r^2 + x^2}}\right]$$

$\therefore$ Total flux through the circular area is

$$\Phi = \frac{qx}{2\varepsilon_0}\int_{r=0}^{a}\frac{rdr}{(r^2+x^2)^{3/2}}.$$

Putting $r^2 + x^2 = z^2$, we get

$$\Phi = \frac{qx}{2\varepsilon_0}\int_{x}^{\sqrt{a^2+x^2}}\frac{dz}{z^2} = \frac{qx}{2\varepsilon_0}\left(\frac{1}{x} - \frac{1}{\sqrt{a^2+x^2}}\right) = \frac{q}{2\varepsilon_0}\left(1 - \frac{x}{\sqrt{a^2+x^2}}\right).$$

▶ **Problem 10.** *Consider a circular disc of radius a whose surface density of charge at any point* (r, θ) *is* $\sigma(r,\theta) = \sigma_0 r^2 \sin^2\theta$. *Find* (i) *total charge on the disc and* (ii) *potential at an axial point at a distance x from the centre of the disc.*

Solution. Consider a surface element $rd\theta dr$ on the disc about the point (r, θ). The charge on it is $\sigma(r,\theta)rd\theta dr$. So the total charge on the disc is

$$Q = \int \sigma(r,\theta)rd\theta dr = \sigma_0\int_0^a r^3 dr\int_0^{2\pi}\sin^2\theta d\theta = \sigma_0\cdot\frac{a^4}{4}\cdot\pi = \pi\sigma_0\frac{a^4}{4}.$$

The potential at P due to charges on the chosen surface element is

$$d\phi(x) = \frac{1}{4\pi\varepsilon_0}\frac{\sigma(r,\theta)rd\theta dr}{\sqrt{r^2+x^2}}.$$

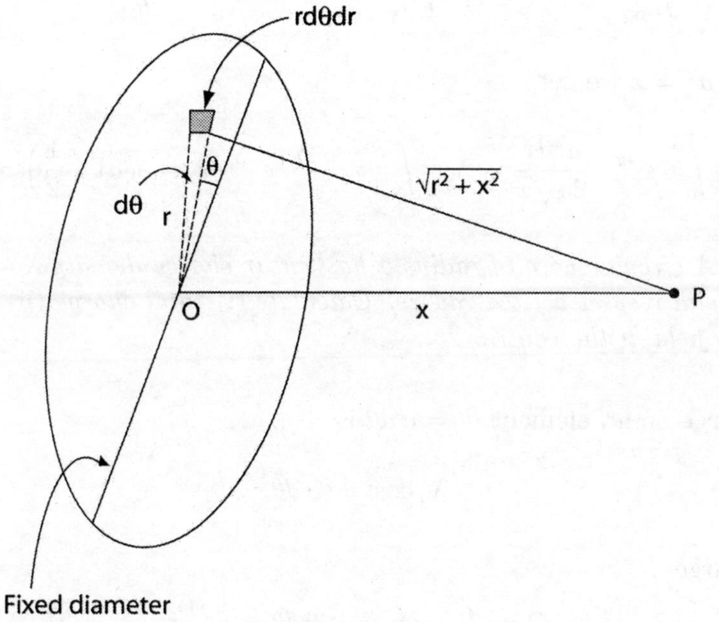

Fig. 1.P-10

$\therefore$ Total potential at P due to the whole disc is

$$\phi(x) = \frac{\sigma_0}{4\pi\varepsilon_0}\int_0^a \frac{r^3 dr}{\sqrt{r^2+x^2}}\int_0^{2\pi}\sin^2\theta d\theta.$$

Putting $r^2 + x^2 = z^2$ we get

$$\begin{aligned}\phi(x) &= \frac{\sigma_0}{4\pi\varepsilon_0}\int_x^{\sqrt{a^2+x^2}}(z^2 - x^2)\,dz\cdot\pi \\ &= \frac{\sigma_0}{4\varepsilon_0}\left[\frac{z^3}{3} - x^2 z\right]_x^{\sqrt{a^2+x^2}} \\ &= \frac{\sigma_0}{12\varepsilon_0}\left[(a^2+x^2)^{3/2} + 2x^3 - 3x^2\sqrt{a^2+x^2}\right].\end{aligned}$$

▶ **Problem 11.** *Consider a line of infinite extent along z-axis with a line charge density*

$$\lambda(z) = \lambda_0\cdot e^{-z^2/a^2},$$

where λ_0 are a are constants. Find the total charge of the distribution.

Solution. Total charge

$$Q = \int_{-\infty}^{+\infty}\lambda(z)\,dz = \lambda_0\int_{-\infty}^{+\infty}e^{-z^2/a^2}dz = 2\lambda_0\int_0^{\infty}e^{-z^2/a^2}dz.$$

Putting $z^2/a^2 = x$ we get

$$Q = 2\lambda_0\int_0^\infty e^{-x}\cdot\frac{a^2dx}{2a\sqrt{x}} = \lambda_0 a\int_0^\infty x^{-1/2}\cdot e^{-x}\,dx = \lambda_0 a\Gamma\left(\frac{1}{2}\right) = \lambda_0 a\sqrt{\pi}.$$

▶ **Problem 12.** *A circular wire of radius a has linear charge density $\lambda = \lambda_0\cos\theta$, where θ is the angle with respect a fixed radius. Calculate* (i) *total charge* (ii) *potential at the centre and* (iii) *field at the centre.*

Solution. Charge on an element $dl = a\,d\theta$ is

$$\lambda_0\cos^2\theta\cdot a\,d\theta.$$

$\therefore$ Total charge

$$Q = \int_0^{2\pi}a\lambda_0\cos^2\theta\,d\theta = a\lambda_0\pi.$$

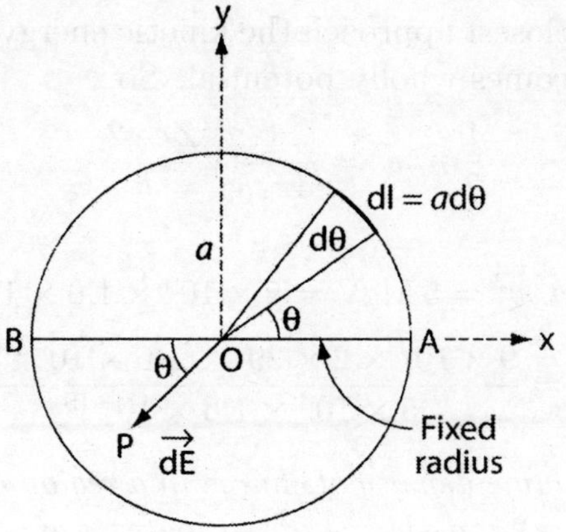

Fig. 1.P-12

Potential at the centre O due to the element dl is

$$d\phi = \frac{1}{4\pi\varepsilon_0} \cdot \frac{\lambda\, dl}{a} = \frac{1}{4\pi\varepsilon_0} \cdot \frac{\lambda_0 \cos^2\theta \cdot a\, d\theta}{a}.$$

$\therefore$ Total potential at O is

$$\phi = \frac{\lambda_0}{4\pi\varepsilon_0} \int_0^{2\pi} \cos^2\theta\, d\theta = \frac{\lambda_0}{4\varepsilon_0}.$$

Field at O due to the element dl is

$$d\vec{E} = \frac{1}{4\pi\varepsilon_0} \cdot \frac{\lambda dl}{a^2} \text{ along } \overrightarrow{OP} \text{ (Fig. 1.P-12)}$$

$$= \frac{\lambda_0}{4\pi\varepsilon_0 a} \cos^2\theta\, d\theta \text{ along } \overrightarrow{OP}.$$

Let us resolve $d\vec{E}$ into two components $dE\cos\theta$ along negative x-axis and $dE\sin\theta$ along negative y-axis. Therefore, total x-component of the field is

$$E_x = -\int dE\cos\theta = -\frac{\lambda_0}{4\pi\varepsilon_0 a} \int_0^{2\pi} \cos^3\theta\, d\theta = 0.$$

Total y-component of the field is

$$E_y = -\int dE\sin\theta = -\frac{\lambda_0}{4\pi\varepsilon_0 a} \int_0^{2\pi} \cos^2\theta \sin\theta\, d\theta = 0.$$

$\therefore$ Total field at O is zero.

► **Problem 13.** *An α-particle of energy* 5 MeV *is directed straight towards a* Cu *nucleus* ($Z = 29$). *Because of Coulomb's repulsion the α-particle will be momentarily stopped at a certain distance b from the nucleus and then it retraces its path. Find this distance b of closest approach.*

Solution. At the point of closest approach the kinetic energy of the α-particle decreases to zero and the energy becomes wholly potential. So

$$\frac{1}{2}m_\alpha v^2 = \frac{1}{4\pi\varepsilon_0}\cdot\frac{Ze\cdot 2e}{b}.$$

Now

$$(1/2)m_\alpha v^2 = 5\,\text{MeV} = 5\times 10^6\times 1.6\times 10^{-19}\,\text{J}.$$

$$\therefore\quad b = \frac{1}{4\pi\varepsilon_0}\cdot\frac{2Ze^2}{\frac{1}{2}m_\alpha v^2} = \frac{9\times 10^9\times 2\times 29\times(1.6\times 10^{-19})^2}{5\times 10^6\times 1.6\times 10^{-19}}\,\text{m} = 1.67\times 10^{-14}\,\text{m}.$$

► **Problem 14.** *Find the volume density of charge in a region of space, where electrostatic potential is given by $\phi(x,y,z) = a - b(x^2+y^2) - c\log(x^2+y^2)$, where a, b, c are constants.*

[C.U. 2000]

Solution. From Poisson's equation $\nabla^2\phi = -\rho/\varepsilon_0$

$$\therefore\quad \rho = -\varepsilon_0\left[\frac{\partial^2\phi}{\partial x^2}+\frac{\partial^2\phi}{\partial y^2}+\frac{\partial^2\phi}{\partial z^2}\right].$$

Direct differentiations show that $\rho = 4\varepsilon_0 b$.

► **Problem 15.** *The electric field in a certain region is given as $\vec{E} = Ar^3\hat{r}$. Prove that charge contained within a spherical surface of radius a centred at the origin is $4\pi\varepsilon_0 Aa^5$.*

[J.U. 2012, cf. C.U. 2013]

Solution. From the differential form of Gauss's law $\vec{\nabla}\cdot\vec{E} = \rho/\varepsilon_0$.

$\therefore$ Charge density

$$\rho = \varepsilon_0\vec{\nabla}\cdot\left(Ar^3\hat{r}\right) = \varepsilon_0 A\vec{\nabla}\cdot\left(r^2\vec{r}\right)$$
$$= \varepsilon_0 A\left[\frac{\partial}{\partial x}\left(xr^2\right)+\frac{\partial}{\partial y}\left(yr^2\right)+\frac{\partial}{\partial z}\left(zr^2\right)\right].$$

Simple calculations show that $\rho = 5\varepsilon_0 Ar^2$.

$\therefore$ The required charge

$$Q = \int_0^a \rho(r)\cdot 4\pi r^2\,dr = 20\pi\varepsilon_0 A\int_0^a r^4\,dr = 4\pi\varepsilon_0 Aa^5.$$

► **Problem 16.** *Examine, which of the following functions may be a possible electrostatic field:*

(i) $\vec{E} = A\left(4y\hat{\imath} - 2x\hat{\jmath} - \hat{k}\right)$, **[J.U. 2014, C.U. 2019]**

(ii) $\vec{E} = A\left[2x\hat{\imath} - yz^2\hat{\jmath} - (1+y^2z\hat{k})\right]$;

where A is a constant with suitable dimensions. For the field, which you find permissible, find how charge density changes with position.

Solution. For $\vec{E}$ to be a possible electrostatic field we must have $\vec{\nabla} \times \vec{E} = 0$.

For the case (i) $\vec{\nabla} \times \vec{E} \neq 0$ (show it), so it cannot represent an electrostatic field.

For the case (ii) $\vec{\nabla} \times \vec{E} = 0$ (show it), so it represents an electrostatic field.

Now $\vec{\nabla} \cdot \vec{E} = \rho/\varepsilon_0$

$$\therefore \quad \rho = \varepsilon_0 \left\{ \frac{\partial}{\partial x}(2Ax) + \frac{\partial}{\partial y}(-Ayz^2) + \frac{\partial}{\partial z}\left(-A - Ay^2 z\right) \right\} = \varepsilon_0 A\left(2 - y^2 - z^2\right).$$

▶ **Problem 17.** *Two point charges $-q$ and $+nq$ are situated at the origin and at the point $(a, 0, 0)$ respectively. Show that the equipotential surface of zero potential is a sphere. Find the centre and radius of the sphere. What happens to these parameters in the limit $n \to 1$.* **[C.U. 2007]**

Solution. Potential at any point (x, y, z) is

$$\phi(x, y, z) = \frac{1}{4\pi\varepsilon_0}\left[\frac{-q}{\sqrt{x^2 + y^2 + z^2}} + \frac{nq}{\sqrt{(x-a)^2 + y^2 + z^2}}\right].$$

Putting $\phi = 0$ we get

$$(x-a)^2 + y^2 + z^2 = n^2(x^2 + y^2 + z^2)$$

$$\text{or} \quad x^2 + y^2 + z^2 + \frac{2xa}{n^2 - 1} = \frac{a^2}{n^2 - 1}$$

$$\text{or} \quad \left(x + \frac{a}{n^2 - 1}\right)^2 + y^2 + z^2 = \left(\frac{an}{n^2 - 1}\right)^2.$$

It represents a sphere radius $\dfrac{an}{n^2 - 1}$ with its centre at $\left(-\dfrac{a}{n^2 - 1}, 0, 0\right)$.

In the limit $n \to 1$, the radius becomes infinite and the centre moves to $(-\infty, 0, 0)$. Actually in this limit Eq. (1) reduces to $x = a/2$, which represents a plane, parallel to the yz-plane.

▶ **Problem 18.** *Find the electric field at a distance z above the centre of a square loop of side a carrying a uniform line charge λ per unit length.* **[C.U. 2016]**

Solution. To solve the problem let us first calculate the electric field at a distance z from the mid-point of a line charge of length a having charge λ per unit length. Considering two symmetrically placed pairs, we note from Fig. 1.P-18(a) that the horizontal components of the two fields cancel out and hence, we need to consider the vertical components only. The vertical field at P due to an element dx is

$$dE = \frac{1}{4\pi\varepsilon_0} \cdot \frac{\lambda dx}{(x^2 + z^2)} \cos\theta = \frac{1}{4\pi\varepsilon_0} \cdot \frac{\lambda z}{(x^2 + z^2)^{3/2}} dx \quad \left[\because \cos\theta = \frac{z}{\sqrt{x^2 + z^2}}\right].$$

$\therefore$ Integrating from $x = -\frac{a}{2}$ to $\frac{a}{2}$ we get the total field,

$$E = \frac{1}{4\pi\varepsilon_0}\frac{\lambda a}{z\sqrt{z^2 + a^2/4}}. \tag{1}$$

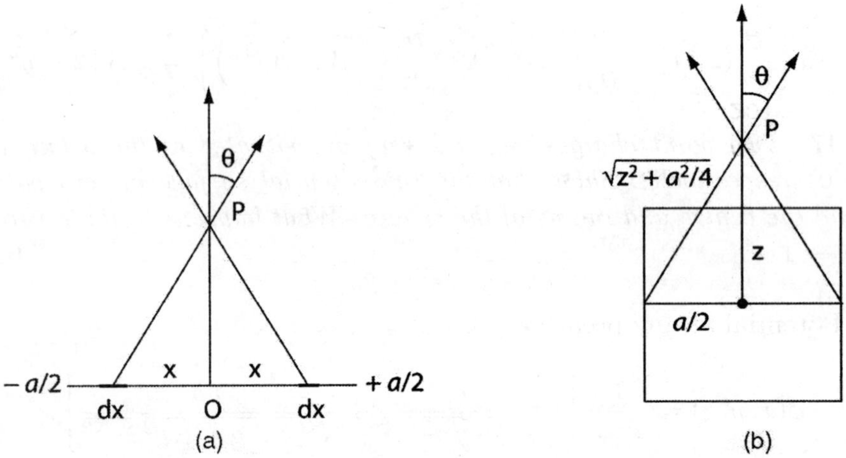

Fig. 1.P-18

Now for a square loop [Fig. 1.P-17(b)] the distance of the point P from the mid-point of any side is $\sqrt{z^2 + a^2/4}$.

$\therefore$ Electric field due to one side can be obtained by replacing z by $\sqrt{z^2 + a^2/4}$ in Eq. (1). Thus,

$$E_1 = \frac{1}{4\pi\varepsilon_0} \cdot \frac{\lambda a}{\sqrt{z^2 + a^2/4} \cdot \sqrt{z^2 + \frac{a^2}{4} + \frac{a^2}{4}}}.$$

Fig. 1.P-18(b) indicates that we need to consider the vertical components only. So considering four sides we get total field at P as

$$\vec{E} = \hat{z}4 \times E_1 \cos\theta = \hat{z}\frac{1}{4\pi\varepsilon_0} \cdot \frac{4\lambda a \cdot z}{\left(z^2 + \frac{a^2}{4}\right)\sqrt{z^2 + \frac{a^2}{2}}}.$$

▶ **Problem 19.** *A conical surface carries a uniform surface charge of density σ. The height of the cone and the radius of the top (open face) both are a. Find the potential difference between the vertex and the the centre of the top open face.* **[C.U. 2017]**

Solution. Referring to Fig. 1.P-19 we can write for the potential at A as

$$\phi(A) = \frac{1}{4\pi\varepsilon_0}\int \frac{\sigma \cdot 2\pi z\, dl}{l},$$

where $l = \sqrt{z^2 + z^2} = \sqrt{2}z$.

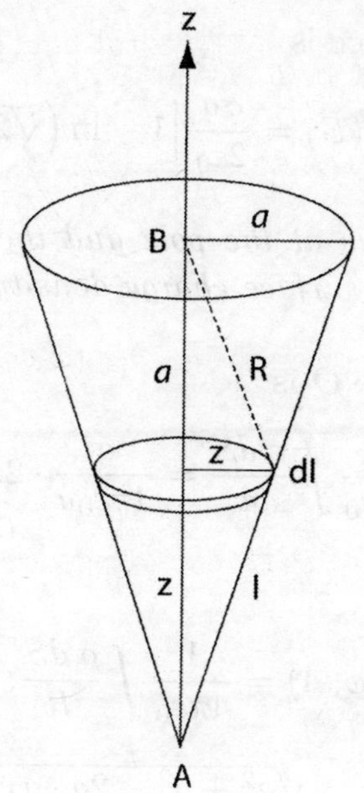

Fig. 1.P-19

$$\therefore\ \phi(A) = \frac{2\pi\sigma}{4\pi\varepsilon_0}\int_0^a \frac{\sqrt{2}z\,dz}{\sqrt{2}z} = \frac{\sigma a}{2\varepsilon_0}.$$

$$\phi(B) = \frac{1}{4\pi\varepsilon_0}\int \frac{\sigma\cdot 2\pi z\,dl}{R} = \frac{\sigma}{2\varepsilon_0}\int_0^a \frac{z\cdot\sqrt{2}dz}{\sqrt{(a-z)^2+z^2}}$$

$$= \frac{\sigma\sqrt{2}}{2\varepsilon_0}\int_0^a \frac{z\,dz}{\sqrt{a^2+2z^2-2az}}.$$

Put $a^2+2z^2-2az = p$ and $z\,dz = \frac{dp}{4} + \frac{a\,dz}{2}$.
Then

$$\phi(B) = \frac{\sigma\sqrt{2}}{2\varepsilon_0}\left[\int \frac{dp}{4\sqrt{p}} + \int_0^a \frac{a\,dz}{2\sqrt{a^2+2z^2-2az}}\right]$$

$$= \frac{\sigma\sqrt{2}}{2\varepsilon_0}\left[\frac{1}{2}\sqrt{a^2+2z^2-2az}\Big|_0^a + \frac{a}{2\sqrt{2}}\int_0^a \frac{dz}{\sqrt{z-\frac{a}{2}^2+\frac{a^2}{4}}}\right]$$

$$= \frac{\sigma\sqrt{2}}{2\varepsilon_0}\cdot\frac{a}{2\sqrt{2}}\cdot\ln\left|\left(z-\frac{a}{2}\right)+\sqrt{\left(z-\frac{a}{2}\right)^2+\frac{a^2}{4}}\right|_0^a$$

$$= \frac{\sigma a}{4\varepsilon_0}\ln\frac{\frac{a}{2}+\frac{a}{\sqrt{2}}}{-\frac{a}{2}+\frac{a}{\sqrt{2}}} = \frac{\sigma a}{4\varepsilon_0}\ln\frac{2+\sqrt{2}}{2-\sqrt{2}}$$

$$= \frac{\sigma a}{4\varepsilon_0}\ln\frac{\sqrt{2}+1}{\sqrt{2}-1} = \frac{\sigma a}{4\varepsilon_0}\ln\frac{\left(\sqrt{2}+1\right)^2}{\left(\sqrt{2}\right)^2-1^2} = \frac{\sigma a}{2\varepsilon_0}\ln\left(\sqrt{2}+1\right).$$

∴ The required potential difference is

$$\phi(A) - \phi(B) = \frac{\sigma a}{2\varepsilon_0}\left[1 - \ln\left(\sqrt{2}+1\right)\right].$$

► **Problem 20.** *Find the potentials at the pole and the centre of a thin hemispherical shell of radius a having uniform surface charge density σ.*

Solution. Potential at the centre O is

$$\phi(O) = \frac{1}{4\pi\varepsilon_0}\int \frac{\sigma\, dS}{a} = \frac{\sigma}{4\pi\varepsilon_0 a}\cdot 2\pi a^2 = \frac{\sigma a}{2\varepsilon_0}.$$

Potential at the pole A is

$$\phi(A) = \frac{1}{4\pi\varepsilon_0}\int \frac{\sigma\, dS}{R},$$

where $dS = 2\pi a \sin\theta \cdot a\, d\theta$ and $r = \sqrt{a^2 + a^2 - 2a\cdot a\cos\theta} = \sqrt{2}a\sqrt{1-\cos\theta}$.

$$\therefore \quad \phi(A) = \frac{\sigma a}{2\sqrt{2}\varepsilon_0}\int_0^{\pi/2} \frac{\sin\theta\, d\theta}{\sqrt{1-\cos\theta}} = \frac{\sigma a}{2\sqrt{2}\varepsilon_0} 2\sqrt{1-\cos\theta}\Big|_0^{\pi/2} = \frac{\sigma a}{\sqrt{2}\,\varepsilon_0}.$$

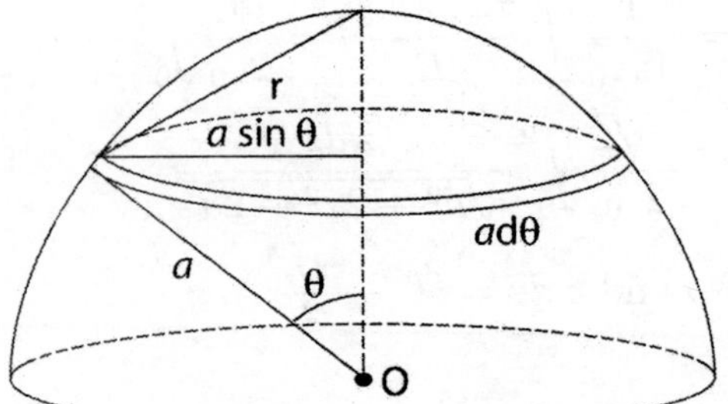

Fig. 1.P-20

► **Problem 21.** *Show that the average field produced inside a sphere by charges lying entirely outside is equal to that they produce at the centre of the sphere.*

Solution. Field produced by the point charge q (at P) at the volume element dV is

$$\vec{E} = \frac{1}{4\pi\varepsilon_0}\cdot\frac{q}{R^3}\vec{R},$$

where $\vec{R}$ is the vector from P to dV. The average field within the sphere is

$$\langle\vec{E}\rangle = \frac{1}{\frac{4}{3}\pi a^3}\int \vec{E}\, dV = \frac{1}{4\pi\varepsilon_0}\int \frac{(-q/\frac{4}{3}\pi a^3)(-\vec{R})}{R^3}\, dV.$$

As the vector connecting dV to P is $-\vec{R}$ right-hand side may be interpreted as the field at P due to a uniformly charged sphere with charge density $\rho = -q/\frac{4}{3}\pi a^3$. Now, it is easy to prove by using Gauss's law that the field at an external point at a distance r from the centre of a uniformly charged sphere may be calculated by assuming that the whole charge is concentrated at the centre of the sphere. Thus

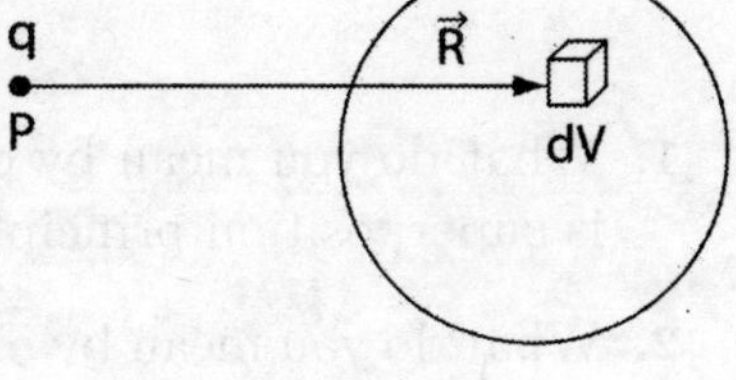

Fig. 1.P-21

$$\langle \vec{E} \rangle = \frac{1}{4\pi\varepsilon_0} \frac{\left(\frac{4}{3}\pi a^3 \rho\right)}{r^3} \vec{r} = \frac{1}{4\pi\varepsilon_0} \cdot \frac{q(-\vec{r})}{r^3},$$

which is exactly the field produced by q at the centre of the sphere. Thus the average field produced inside the sphere by a point charge q lying entirely outside is equal to field that the point charge produces at the centre of the sphere. By the principle of superposition this result is true for any arbitrary charge distribution lying outside the sphere.

▶ **Problem 22.** *Consider a uniformly charged sphere of radius a carrying a total charge Q. Calculate the force which one half of the sphere exerts on the other half.*

Solution. Using Gauss's theorem we can find the electric field at any inside point P as

$$\oint \vec{E} \cdot d\vec{S} = \frac{1}{\varepsilon_0} \cdot \frac{4}{3}\pi r^3 \rho \quad \text{or} \quad E \cdot 4\pi r^2 = \frac{4\pi r^3}{3\varepsilon_0} \cdot \frac{Q}{\frac{4}{3}\pi a^3}.$$

$$\therefore \quad \vec{E} = \frac{1}{4\pi\varepsilon_0} \cdot \frac{Q}{a^3}\vec{r}.$$

Force on an element of charge $\rho\, dV$ at P is $\vec{E}\rho\, dV$.

By symmetry total force on the upper hemisphere will be along z-axis. So considering z-component of this force, we can calculate the total force on the hemisphere as

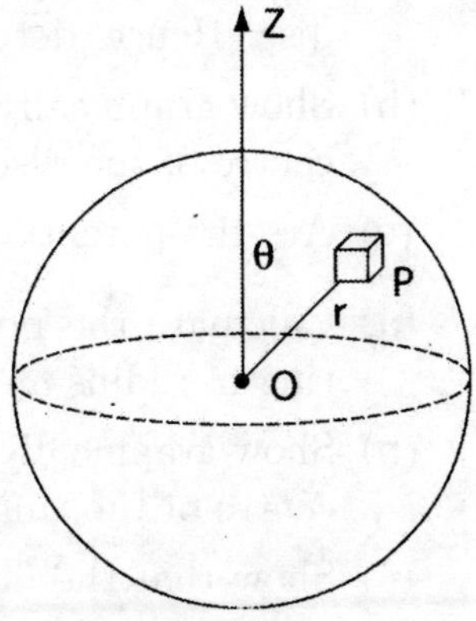

Fig. 1.P-22

$$F_z = \int E_z \rho\, dV = \frac{1}{4\pi\varepsilon_0} \frac{Q}{a^3} \int r\cos\theta \cdot \rho \cdot r^2 \sin\theta\, d\theta\, d\varphi dr$$

$$= \frac{3}{\varepsilon_0}\left(\frac{Q}{4\pi a^3}\right)^2 \int_0^a r^3 dr \int_0^{\pi/2} \sin\theta\cos\theta\, d\theta \int_0^{2\pi} d\varphi$$

$$= \frac{3Q^2}{64\pi\varepsilon_0 a^2}.$$

QUESTIONS

1. What do you mean by quantization of charge and conservation of charge? What is superposition principle as applied in electrostatics?

2. What do you mean by electric field and potential? Calculate the electric potential at an axial point due to a uniformly charged disc and hence, determine electric field.

3. State and prove Gauss's law in electrostatics. Apply this law to calculate the electric field due to

   (a) a uniformly charged sphere,
   (b) a uniformly charged infinite plane sheet of charge,
   (c) a uniformly charged infinite cylinder. **[C.U. 2016]**

   Obtain Coulomb's law from Gauss's law. **[C.U. 2010]**

4. Starting from Coulomb's law establish Poisson's equation in electrostatics. What is Laplace's modification of the same? What connection between an electrostatic field and its sources is revealed by Coulomb's law and Gauss's law?

5. (a) Calculate the potential at an axial point due to a thin uniformly charged disc. Hence, determine the electric field due to it.

   (b) Show graphically the variations of potential and field with distance from the centre of the disc.

   (c) Are the potential and field continuous across the surface of the disc?

6. (a) Calculate the potential at an axial point due to a thin uniformly charged ring of radius a. Hence, determine the electric field due to it.

   (b) Show graphically the variations of potential and field with distance from the centre of the ring.

   (c) Show that the field is maximum at a distance $\pm a/\sqrt{2}$ from the centre.

7. (a) An amount of charge q is distributed uniformly inside a sphere of radius a. Apply Gauss's law to find electric field at any internal and external points. Hence, calculate the potentials. Show graphically the variation of field and potential with distance measured from the centre of the sphere.

   (b) Calculate the potential and field inside a uniformly charged spherical shell. **[C.U. 2017]**

8. (a) Show that the electrostatic field of a point charge is irrotational.

   (b) Show that the electrostatic field is conservative. **[C.U. 2009; J.U. 2010]**

(c) Starting from Coulomb's law prove that $\vec{\nabla} \times \vec{E} = 0$. **[C.U. 2015, 2018]**

(d) Show that electric field is always perpendicular to a equipotential surface. **[C.U. 2019]**

9. State and explain Earnshaw's theorem. What is its importance? **[J.U. 2013, B.U. 2017]**

10. Starting from Gauss's law obtain Poisson's equation in terms of electrostatic potential. **[C.U. 2013]**

EXERCISES

1. Compare the electrostatic and gravitational forces acting between the nucleus and the electron rotating in the first Bohr orbit of a hydrogen atom. Given radius of the orbit = 0.53Å, $G = 6.7 \times 10^{-11}$N.m^2.kg^{-2}, mass of a proton = 1.7×10^{-27} kg. Comment on the result. [*Ans.* $F_e/F_g \approx 10^{39}$.]

2. A movable particle of mass m and charge $+q$ is placed midway between two fixed positive charges $+Q$ each kept separated by a distance $2d$. If the particle of charge $+q$ is slightly moved towards one of the fixed charges and then released, show that it will undergo a simple harmonic motion with a time period $T = 2\pi\sqrt{\pi\varepsilon_0 m d^3/Qq}$.

3. (a) Show that the electric field at the centre of curvature of a uniformly charged semicircular arc is $\lambda/2\pi\varepsilon_0 R$, where R is the radius and λ charge per unit length of the arc.

(b) A circular wire with its centre at the origin is lying in the xy plane. It carries a uniform line density of charge λ. Find the electric field at the origin due to the portion of the wire lying in the first quadrant.

[**Hints.** See, Solved Problem 12] [*Ans.* $\vec{E} = -\frac{\lambda}{4\pi\varepsilon_0 a}(\hat{i} + \hat{j})$]

4. A static charge distribution produces an electric field

$$\vec{E} = \frac{1}{4\pi\varepsilon_0}\frac{e^{-br}}{r^2}\hat{r},$$

where b is a constant and $\hat{r}$ is the unit vector along position vector $\vec{r}$. Using Gauss's law find the total charge of the distribution. [*Ans.* 0.]

[**Hints.** $Q_{\text{enc}} = \varepsilon_0 \oint \vec{E} \cdot \hat{n}\, dS = \frac{1}{4\pi} \cdot \frac{e^{-bR}}{R^2} \cdot 4\pi R^2 = e^{-bR}$. Assuming a Gaussian surface as a sphere of radius $R = \infty$, $Q_{\text{enc}} = 0$]

5. (a) Consider a uniformly charged sphere of radius a and charge density ρ. Show that the electric potential ϕ at an internal point is given by

$$\phi(r) = \frac{\rho a^2}{6\varepsilon_0}\left(3 - \frac{r^2}{a^2}\right).$$ **[J.U. 2013]**

(b) A spherical charge distribution consists of a uniform charge density ρ_1 from $r = 0$ to $r = a/2$ and ρ_2 from $r = a/2$ to $r = a$. Find electrostatic potentials at $r = 0, a/2, a$. [*Ans.* $\frac{a^2}{24\varepsilon_0}(3\rho_1 + 9\rho_2)$, $\frac{a^2}{24\varepsilon_0}(2\rho_1 + 9\rho_2)$, $\frac{a^2}{24\varepsilon_0}(\rho_1 + 7\rho_2)$]

6. A charge Q is distributed uniformly over a spherical volume. Show that the electric field at any inside point is directly proportional to its distance from the centre of the sphere.

7. The electric field along the length of a discharge tube is given by $E(x) = kx$, where k is a constant and x is the distance from one end of the tube. Using Poisson's equation find the charge density. [*Ans.* $k\varepsilon_0$]

8. Consider a line charge of infinite extent along the z-axis with a line charge density $\lambda(z) = \lambda_0/(1 + z^2)$. Find the total charge. [*Ans.* $\lambda_0\pi$.]

9. Consider a rod of length L with a charge Q distributed uniformly over it. A test charge q_0 is placed on the axis of the rod at a distance d from its nearest end. Find the force on q_0 due to the rod. [*Ans.* $\frac{Qq_0}{4\pi\varepsilon_0} \cdot \frac{1}{d(L+d)}$.]

10. Consider a cylindrical surface defined by $x^2 + y^2 = a^2$, $z = 0$, $z = 1$. It is placed in an electric field $\vec{E} = E_0\left[\hat{i}x + \hat{j}y + \hat{k}(z^2 - 1)\right]$. Calculate the total flux of $\vec{E}$ across the cylindrical surface. [*Ans.* $3E_0\pi a^2$]

[**Hints.** Calculate $\int_S \vec{E}\cdot d\vec{S}$ directly or by using $\int_V \left(\vec{\nabla}\cdot\vec{E}\right) dV$.]

11. Consider a long cylinder with a charge density proportional to the distance from its axis, i.e., $\rho(r) = kr$, where k is a constant. Using Gauss's law find the electric field at any point inside the cylinder. [*Ans.* $\vec{E} = \frac{k}{3\varepsilon_0} r^2\hat{r}$.]

12. A spherical shell of inner radius a and outer radius b has a charge distribution of density $\rho(r) = a + br$ (for $a < r < b$), where r is the distance from the centre. Find expressions for electric field for (a) $r < a$, (b) $a < r < b$ and (c) $r > b$.

[*Ans.* (a) 0, (b) $\frac{1}{\varepsilon_0 r^2}\left\{\frac{a}{3}(r^3 - a^3) + \frac{b}{4}(r^4 - a^4)\right\}\hat{r}$,

(c) $\frac{1}{\varepsilon_0 r^2}\left\{\frac{a}{3}(b^3 - a^3) + \frac{b}{4}(b^4 - a^4)\right\}\hat{r}$]

13. Consider two spheres, each of radius a and carrying uniform charge densities $+\rho$ and $-\rho$ respectively. The spheres are so placed that they partially overlap. Show that the field inside the region of overlap is uniform.

14. An electric field is represented by $\vec{E} = \hat{i}y + \hat{j}x$. Find the corresponding potential function ϕ. Assume $\phi = 0$ at $x = y = 0$. **[J.U. 2014]** [*Ans.* $\phi = -xy$]

15. Calculate the electric field on the surface of $^{238}_{92}$U nucleus. Assume the nucleus to be a sphere of radius 7×10^{-15} m. [*Ans.* 27×10^{20} N.C^{-1}]

16. Two uniformly charged infinite planes of charge densities $+\sigma$ and $-\sigma$ intersect each other at right angles. Find the electric field at any point and hence, sketch the field lines. [*Ans.* $E = \frac{\sqrt{2}\sigma}{2\varepsilon_0}$.]

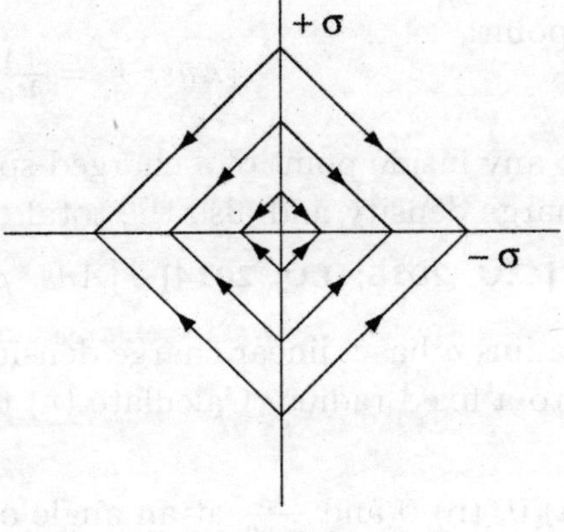

Fig. 1.E-16

17. The electric field at any point (x, y, z) in free space is $\vec{E} = x\hat{i} + y\hat{j} + z\hat{k}\,\text{Nm}^{-1}$. Find the charge density $\rho(x, y, z)$. [*Ans.* 26.5×10^{-12} C.m^{-3}]

18. (a) A point charge Q is kept at the centre of a cube. Calculate the electric flux through one face of the cube.

[**Hints.** Total Flux $\int_S \vec{E} \cdot d\vec{S} = \frac{1}{\varepsilon_0} \cdot Q$. $\therefore$ Required flux $= \frac{1}{6} \times \frac{1}{\varepsilon_0} \times Q$.]

(b) A point charge q is located at the corner of a cube. Show that the electric flux due to q through the three faces of the cube, which do not meet at q is $q/8\varepsilon_0$.

[**Hints.** Imagine 8 cubes meeting at the charge point. Now the charge point is at the centre of a bigger cube.]

19. The electric potential at a point (x, y, z) is

$$\phi(x, y, z) = \frac{A}{\sqrt{x^2 + y^2 + z^2}},$$

where A is a constant. Find the electric field at (x, y, z).

[*Ans.* $A(\hat{i}x + \hat{j}y + \hat{k}z(\big/(x^2 + y^2 + z^2)^{3/2}$]

20. Consider a point charge q placed on the axis of a circular area of radius a at a distance $\frac{3}{4}a$ from the centre. Show that the total electric flux through the area is $q/5\varepsilon_0$.

21. The electric potential at a point (x, y, z) due to a certain charge distribution is given by $\phi(x, y, z) = -A(x^2yz + xy^2z + xyz^2)$, where A is a constant. Find the electric field and charge density at the point $(1, 1, 1)$.

[*Ans.* $\phi(1, 1, 1) = 4A(\hat{i} + \hat{j} + \hat{k})$, $\rho(1, 1, 1) = 6A\varepsilon_0$]

22. Consider a very long cylinder of volume charge density $\rho(r) = Are^{-2r}$, where A is a constant and r is the distance from axis. Using Gauss's law determine the field at any inside point.

[*Ans.* $\vec{E} = \frac{A}{4\varepsilon_0 r}\left\{1 - 2e^{-2r}\left(r^2 + r + \frac{1}{2}\right)\right\}\hat{r}$]

23. The electric field at any inside point of a charged sphere of radius a is $\vec{E} = Ar^3\hat{r}$. Find the volume charge density and also the total charge within the sphere.

[C.U. 2015; J.U. 2014] [*Ans.* $\rho(r) = 5A\varepsilon_0 r^2$, $Q = 4\pi\varepsilon_0 Aa^5$]

24. A circular wire of radius a has a linear charge density $\lambda = \lambda_0 \cos\theta$, where θ is the angle with respect to a fixed radius. Calculate (a) total charge (b) potential and field at the centre. **[cf. C.U. 2004]**

[*Ans.* a) 0, (b) 0 and $\frac{\lambda_0}{4\varepsilon_0 a}$ at an angle of 180° with the fixed radius.]

25. In a certain region the electric field in spherical polar coordinates is given by

$$\vec{E} = a\sin\theta\hat{r} + b\cos\theta\hat{\theta},$$

where a and b are constants. Find the charge density that produces this field.

[**Hints.** Use $\dfrac{\rho}{\varepsilon_0} = \vec{\nabla}\cdot\vec{E} = \dfrac{1}{r^2}\dfrac{\partial}{\partial r}(r^2E_r) + \dfrac{1}{r\sin\theta}\dfrac{\partial}{\partial\theta}(\sin\theta E_\theta)$.

Thus $\rho = \varepsilon_0(2a\sin^2\theta + b\cos 2\theta)/r\sin\theta$.]

26. Consider a sphere of radius a having a charge density $\rho(r) = A/r$. Find the electric potentials at any internal and external points.

[*Ans.* $\phi(r) = \frac{A}{\varepsilon_0}\left(a - \frac{r}{2}\right)$ for $r \le a$ and $\phi(r) = \frac{Aa^2}{2\varepsilon_0 r}$ for $r \ge a$]

[**Hints.** Find $\vec{E}$ by using Gauss's law. Then find ϕ by integrating the relation $\vec{E} = -\vec{\nabla}\phi$.]

27. The radial electric field in a certain region is given by

$$\vec{E} = \frac{q}{4\pi\varepsilon_0}\frac{e^{-br}}{r^2}\hat{r},$$

where $\hat{r} = \vec{r}/r$ is a unit vector along $\vec{r}$ and b is a constant. Find the charge density producing this field. Also calculate the total charge. **[cf. B.U. 2016]**

[*Ans.* $\rho(\vec{r}) = q\left[\delta(\vec{r}) - \frac{b}{4\pi r^2}\right]e^{-br}$, $Q_{\text{tot}} = 0$]

28. A charge Q is split into two parts q and $Q - q$. What should be the relationship between Q and q such that a maximum repulsive force is obtained between them, when placed at a given distance apart? [*Ans.* $q = \frac{1}{2}Q$]

29. Can the vector $\vec{E}(x, y, z) = (x^2 + y^2)\hat{\imath} + (y^2 + x^2)\hat{\jmath} + (x^2 + y^2)\hat{k}$ represent an electrostatic field? **[J.U. 2010]** [*Ans.* No, as $\vec{\nabla} \times \vec{E} \neq 0$]

30. Consider a spherical shell of internal and external radii a and b, respectively. It contains a volume charge density $\rho(r) = A/r$, where A is constant and r is the radial distance. A point charge q is placed at the centre $r = 0$. Find the value of A for which electric field in the region $a < r < b$ becomes constant.

[*Ans.* $A = \frac{q}{2\pi a^2}$]

31. A hollow cylindrical conductor of radius R and length $2L$, open at both ends, carries a charge Q, which is uniformly distributed over its curved surface. Calculate the potential at the mid-point of the cylinder axis. Use the result for a circular charged loop.

[*Ans.* $\frac{1}{4\pi\varepsilon_0} \cdot \frac{Q}{L} \ln \frac{\sqrt{a^2+L^2}+L}{\sqrt{a^2+L^2}-L}$]

32. If the charge density within a spherical charge distribution varies as the distance from its centre, show that the electric field inside has a constant magnitude everywhere.

33. Find the electric field $\vec{E}$ at any point (r, θ), where potential is $\phi(r, \theta) = r^2 \cos\theta$.

[J.U. 2011] [*Ans.* $\vec{E} = -2r\cos\theta\,\hat{r} + r\sin\theta\,\hat{t}\theta$]

34. The surface density of charge on a thin circular disc of radius R is $\sigma(r) = Ar$, where A is a constant and r is the radial distance from the centre of the disc. Calculate the electric field at any point on the axis of the disc. **[C.U. 1990]**

[*Ans.* $\vec{E} = \hat{x}\frac{Ax}{2\varepsilon_0}\left(\ln \frac{R+\sqrt{R^2+x^2}}{x} - \frac{R}{\sqrt{R^2+x^2}}\right)$]

35. An amount of charge q is distributed within a sphere with volume density $\rho(r) = \rho_0 r$, where ρ_0 is a constant and r is the distance from the centre. Applying Gauss's law find the electric fields and then potentials. Show that

$$\vec{E} = \frac{1}{4\pi\varepsilon_0}\frac{q}{r^2}\hat{r}, \qquad \phi = \frac{1}{4\pi\varepsilon_0}\frac{q}{r} \quad \text{for } r > a$$

$$\vec{E} = \frac{1}{4\pi\varepsilon_0} \cdot \frac{qr^2}{a^4}\hat{r}, \quad \phi = \frac{1}{4\pi\varepsilon_0} \cdot \frac{q}{3a}\left(4 - \frac{r^3}{a^3}\right) \quad \text{for } r < a$$

36. (a) The electric field in a certain region of space is $\vec{E} = 2x\hat{\imath} + 2y\hat{\jmath} + z\hat{k}$. Find the corresponding charge density.

[**Hints.** Use $\nabla \cdot \vec{E} = \rho/\varepsilon_0$. This gives $\rho = 5\varepsilon_0$.]

(b) The electric field in xy-plane is given by $\vec{E} = 2ax\hat{\imath} + by\hat{\jmath}$ where a and b are constants. What is the charge density responsible for this field?

[C.U. 2014, 2017] [*Ans.* $\varepsilon_0(2a+b)$]

37. The total charge in a sphere of radius r about the centre of a spherical charge cloud of radius a is

$$\frac{qr^2}{a^2} \cdot \left(e^{-r/a} - e^{-2r/a}\right).$$

Find out the corresponding field and potential. Show that in the limit $r \to 0$, charge density $\rho \to 3q/4\pi a^3$.

[**Hints.** Applying Gauss's law $\int \vec{E}\cdot d\vec{S} = \dfrac{Q\,\text{tot.}}{\varepsilon_0}$, show that $E = \dfrac{q}{4\pi\varepsilon_0 a^2}\left(e^{-r/a} - e^{-2r/a}\right)$

Using $E = -\dfrac{d\phi}{dr}$ show that $\phi(r) = \dfrac{q}{4\pi\varepsilon_0 a}\left\{e^{-r/a} - \dfrac{1}{2}e^{-2r/a}\right\}$

Using $\vec{\nabla}\cdot\vec{E} = \dfrac{\rho}{\varepsilon_0}$ or $\rho = \dfrac{\varepsilon_0}{r^2}\dfrac{d}{dr}(r^2E)$

show that $\rho = \dfrac{q}{4\pi a^2 r}\left[\left(2 - \dfrac{r}{a}\right)e^{-r/a} - \left(2 - \dfrac{2r}{a}\right)e^{-2r/a}\right]$

Using the approximation $e^x \approx 1 + x$ show that $\rho \to \dfrac{3q}{4\pi a^3}$ for $r \to 0$.]

38. (a) The electric field in a certain region is given by $\vec{E} = A\dfrac{x\hat{\imath} + y\hat{\jmath}}{x^2+y^2}$, where A is a constant. Find the total charge within a sphere of radius R with the centre at the origin. [*Ans.* $4\pi\varepsilon_0 AR$]

(b) An eletric field $\vec{E}(\vec{r}) = \frac{1}{r}(\alpha\hat{r} + \beta\sin\theta\cos\varphi\hat{\varphi})$ exists in space. Find the total charge enclosed in a sphere of unit radius about the origin. [*Ans.* $4\pi\varepsilon_0\alpha$]
[**Hints.** Use $\oint \vec{E}\cdot d\vec{S} = q/\varepsilon_0$]

39. The potential in a certain region is given by $\phi(r) = ar^2 + b$, where a and b are constants. Find the corresponding volume distribution of charge $\rho(r)$.

[*Ans.* $\rho(r) = -6\varepsilon_0 a$]

40. A charge q is distributed about the origin in such a way that the charge density at a distance r is given $\rho(r) = \rho_0 e^{-\alpha r}$, where ρ_0 and α are constants. Show that the potential due to this charge distribution is given by

$$\phi(r) = \frac{q}{4\pi\varepsilon_0 r}(1 - e^{-\alpha r}) - \frac{q\alpha}{8\pi\varepsilon_0}e^{-\alpha r}.$$

41. A charge Q is uniformly distributed over a thin rod lying between $x = -a$ to $x = +a$. Find the potential due to it at any point on the x-axis.

[*Ans.* $\frac{Q}{8\pi\varepsilon_0 a}\ln\frac{x+a}{x-a}$]

42. Calculate the electrostatic potential on the axis of a hemispherical cap carrying a uniform surface charge density σ.

[**Hints.** With reference to Fig. 1.E-42.

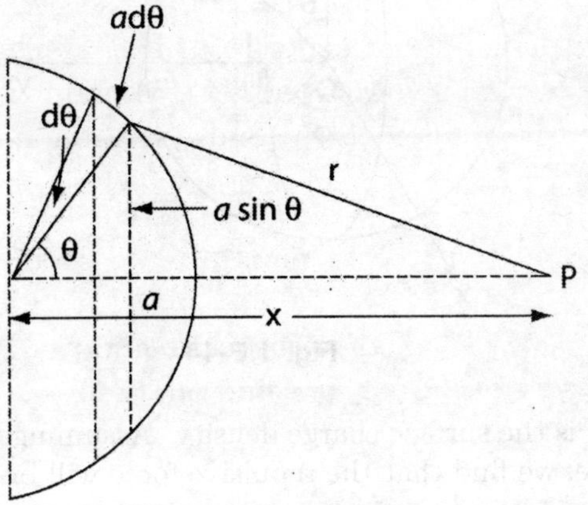

Fig. 1.E-42

$$\phi(P) = \frac{1}{4\pi\varepsilon_0} \cdot \int \frac{2\pi a \sin\theta \cdot a\,d\theta \cdot \sigma}{r}$$

$r^2 = a^2 + x^2 - 2ax\cos\theta$ or $r\,dr = ax\sin\theta\,d\theta$

$$\therefore \phi(P) = \frac{2\pi a\sigma}{4\pi\varepsilon_0 x} \int dr$$

$$\text{For } x > a, \quad \phi(P) = \frac{2\pi a\sigma}{4\pi\varepsilon_0 x} \int_{x-a}^{\sqrt{x^2+a^2}} dr = \frac{2\pi a\sigma}{4\pi\varepsilon_0 x}\left[\sqrt{x^2+a^2} - (x-a)\right]$$

$$\text{For } x < a, \quad \phi(P) = \frac{2\pi a\sigma}{4\pi\varepsilon_0 x} \int_{a-x}^{\sqrt{a^2+x^2}} dr = \frac{2\pi a\sigma}{4\pi\varepsilon_0 x}\left[\sqrt{a^2+x^2} - (a-x)\right].]$$

43. In a region of space the electrostatic field is everywhere directed parallel to x-axis. Show that the filed in this region is independent of y- and z-coordinates. Further, if there is no charge in this region, prove that the field is also independent of x.

[**Hints.** Use $\vec{E} = E(x, y, z)\hat{i}$; $\vec{\nabla} \times \vec{E} = 0$ and $\vec{\nabla} \cdot \vec{E} = 0$.]

44. A conducting spherical shell carrying a total charge Q is cut into two halves. What is the force of repulsion between the two halves.

[**Hints.** (Refer Fig. 1.E-44, next page) Force on a surface element $d\vec{S}$ at (r, θ, φ) is $\frac{\sigma^2}{2\varepsilon_0} d\vec{S}$,

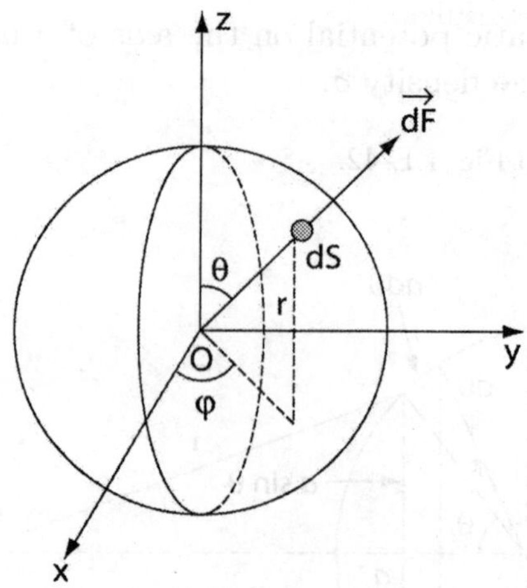

Fig. 1.E-44

where $\sigma = Q/4\pi r^2$ is the surface charge density. Assuming the plane, where the sphere is cut to be xoz-plane, we find that the repulsive force will be along y-axis. The magnitude of the force is

$$F = \int dF \sin\theta \sin\varphi = \frac{\sigma^2 r^2}{2\varepsilon_0} \int_0^\pi \sin^2\theta \, d\theta \cdot \int_0^\pi \sin\varphi \, d\varphi = \frac{\pi\sigma^2 r^2}{2\varepsilon_0} = \frac{Q^2}{32\pi\varepsilon_0 r^2},$$

where we use $dS = r^2 \sin\theta \, d\theta \, d\varphi$ in spherical polar coordinates.]

45. Suppose a spherical shell of inner radius a and outer radius b carries a charge density $\rho = A/r^2$, where A is some constant. Find the electric field in three regions (a) $r < a$, (b) $a < r < b$ and (c) $r > b$.

[*Ans.* (a) 0, (b) $\frac{A}{\varepsilon_0} \cdot \frac{r-a}{r^2}\hat{r}$, (c) $\frac{A}{\varepsilon_0}\frac{b-a}{r^2}\hat{r}$.]

46. (a) Which one of the following can represent an electrostatic field?

(i) $\vec{E}_1 = A\left[xy\hat{t}i + 2yz\hat{t}j + 3xz\hat{t}k\right]$; **[C.U. 2007, 2010; J.U. 2011]**

(ii) $\vec{E}_2 = A\left[y^2\hat{t}i + (2xy + z^2)\hat{t}j + 2yz\hat{t}k\right]$. [*Ans.* $\vec{E}_2$]

(b) Can an electrostatic field vector have the following form

$$\vec{E} = x\hat{\imath} - y\hat{\jmath}$$

Give reasons for your answers. **[C.U. 2012]**

(c) Find a and b such that $\vec{E} = -by^2\hat{\imath} + (2xy + az^2)\hat{\jmath} + 2yz\hat{k}$ becomes of electrostatic type. **[C.U. 2016]** [*Ans.* $a = 1, b = -1$]

47. Two particles each of charge $-Q$ are fixed at a distance $2d$ apart. A particle of mass m and charge $+q$ is placed midway between them. Suppose the particle m is displaced slightly along the perpendicular to the line joining the $-Q$ charges. Find the period of small oscillation of the particle. [*Ans.* $T = 2\pi\sqrt{\frac{2\pi\varepsilon_0 m d^3}{qQ}}$]

48. Consider a thin annular disc of inner radius b and outer radius a, having a uniform surface charge density σ. Find the potential and field at an axial point at a distance x from the centre of the disc. Discuss the limits $x \to 0$ and $b \to 0$.

[*Ans.* $\phi(x) = \frac{\sigma}{2\varepsilon_0}(\sqrt{a^2+x^2} - \sqrt{b^2+x^2})$, $\vec{E}(x) = \hat{x}\frac{\sigma x}{2\varepsilon_0}(\frac{1}{\sqrt{b^2+x^2}} - \frac{1}{\sqrt{a^2+x^2}})$]

49. Eight equal positive charges are placed at the corners of an imaginary cube. A ninth positive charge placed at the centre of the cube will be repelled away from each corner and net force on it will be zero. Do you think that this ninth charge will be in stable equilibrium?

[**Hints.** No, because of Earnshaw's theorem.]

50. Show that $\phi(x,y) = e^{-y} \cdot \sin x$ represents an electrostatic field in free space.
[**Hints.** Show that $\nabla^2\phi = 0$] **[C.U. 2008]**

51. The potential in a certain region is given by $\phi(x,y,z) = x^2 y(z+3)V$. Calculate the total charge enclosed within the cube $0 \le x, y, z \le 1$.

[C.U. 2008] [*Ans.* $-7\varepsilon_0/2$]

52. (a) A cube of side a has one corner placed at the origin and sides parallel to the cartesian axes. There are point charges $+q$ located at each corner except that at the origin. Find the electric field at the centre of the cube.

[**Hints.** Place two charges $+q$ and $-q$ at the origin. The new system of charges is electrostatically equivalent to the original one. Field due to eight $+q$ charges is zero. So the net field will be due to $-q$ placed at the origin. $\therefore$ $\vec{E} = -\frac{q}{4\pi\varepsilon_0}\frac{\vec{r}}{r^3}$ where $\vec{r} = \frac{a}{2}\hat{i} + \frac{a}{2}\hat{j} + \frac{a}{2}\hat{k}$. Thus $\vec{E} = -\frac{q}{3\sqrt{3}\pi\varepsilon_0 a^2}(\hat{i}+\hat{j}+\hat{k})$.]

(b) Suppose there is a circular hole of radius 'a' at the centre of an infinite plane having a uniform charge σ per unit area. Find the potential at an axial distance x from the centre of the hole.

[**Hints.** By superposition principle the required potential is

$$\frac{\sigma}{2\varepsilon_0} - \frac{\sigma}{2\varepsilon_0}(\sqrt{a^2+x^2} - |x|)\ ]$$

53. n equal charges are placed at the corners of a regular polygon of n-sides. Show by using Coulomb's law that the electric field at the centre of the polygon is zero.

[**Hints.** Taking the sum of components of field (E_1) due each charge along a particular line joining the centre and one corner we get

$$E_x = E_1 + E_1\cos\theta + E_1\cos 2\theta + \cdots + E_1\cos(n-1)\theta$$
$$= E_1\frac{\cos\frac{n-1}{2}\theta}{\sin\theta/2}\cdot\sin\frac{n\theta}{2} \quad \text{where } \theta = \frac{2\pi}{n}. \quad \text{Thus } E_x = 0 \text{ for all } n.$$

The perpendicular component is

$$E_y = 0 + E_1\sin\theta + E_1\sin 2\theta + \cdots + E_1\sin(n-1)\theta$$
$$= \frac{\sin\frac{n-1}{2}\theta}{\sin\theta/2}\cdot\sin\frac{n\theta}{2} = 0 \text{ for all } n]$$

54. Two particles each of mass m carrying charge q are suspended by massless strings of length L from a common point. If θ be the angle that each string makes with the vertical line and g is the acceleration due to gravity, show that $16\pi\varepsilon_0 mgL^2 \sin^3\theta = q^2 \cos\theta$. **[C.U. 2012]**

55. The electric field in a region of space is given by $\vec{E} = 8x\hat{\imath} - 4y\hat{\jmath} - 4z\hat{k}$. Find the equation of the lines of force in the $z = 0$ plane. **[C.U. 2014, 2017]**

[**Hints.** Use equation $\frac{dx}{E_x} = \frac{dy}{E_y}$. Integrate and find that $xy^2 = $ constant.]

OBJECTIVE TEST

1. If there exists a body with total charge 1.3e, it will violate the principle of:

(a) conservation of charge,
(b) quantization of charge,
(c) superposition,
(d) conservation of energy.

2. SI units of the absolute permittivity of a medium is:

(a) $N.m^2C^{-2}$, (b) $N.C^{-1}$, (c) $N^{-1}.m^{-2}C^2$, (d) F.m.

3. Dimensions of electric field $\vec{E}$ are:

(a) $[MLT^{-3}I]$, (b) $[MLT^{-3}I^{-1}]$, (c) $M^{-1}L^{-1}T^3I$, (d) $[MLT^3I^{-1}]$.

4. An electric charge can be held in a position of stable equilibrium by

(a) a purely electrostatic field,
(b) a purely mechanical force,
(c) both (a) and (b),
(d) neither (a) nor (b).

5. Total electric flux emanating from a charge q coulomb placed in air is:

(a) q/ε_0, (b) $\varepsilon_0 q$, (c) q, (d) $4\pi q$.

6. A point charge q situated at the position $\vec{r}$ can be described by a volume charge density $\rho(\vec{r}\,')$ given by:

(a) $q\delta(\vec{r}\,' - \vec{r})$, (b) $q/\delta(\vec{r}\,' - \vec{r})$, (c) $\frac{1}{q}\delta(\vec{r}\,' - \vec{r})$, (d) $q/|\vec{r}\,' - \vec{r}|^3$.

7. Electric field lines are always:

(a) tangential to an equipotential surface,
(b) normal to an equipotential surface,
(c) closed lines,
(d) none of the above.

8. Electrostatic field is always:

(a) solenoidal, (c) harmonic in character,

(b) irrotational, (d) none of the above.

9. The unit of electric flux is:

(a) gauss, (b) weber, (c) $N.m^2.C^{-1}$, (d) $N.C^{-1}$.

10. Work done in displacing a charge $2e$ through a distance of 0.1 m on an equipotential surface is:

(a) 0.2 J, (b) 20 J, (c) zero, (d) none of these.

11. If a proton, a deuton and an α-particle are all accelerated through the same potential difference, their kinetic energies would be in the ratio of:

(a) 1 : 2 : 4, (b) 1 : 1 : 1, (c) 2 : 2 : 1, (d) 1 : 1 : 2.

12. If the electric field and potential at a point are $\vec{E}$ and ϕ, respectively then:

(a) for $\phi = 0$, $\vec{E}$ must be zero, (c) for $\vec{E} \neq 0$, $\phi \neq 0$,

(b) for $\vec{E} = 0$, ϕ must be zero, (d) none of the above.

13. An electric dipole is placed at the centre of a sphere. Then:

(a) electric field is zero at every point on the sphere,

(b) electric field is zero everywhere within the sphere,

(c) electric field is zero outside the sphere,

(d) total electric flux through the surface is zero.

14. An electron travelling in a region of electrostatic potential V_1 passes into a region of higher potential V_2. The change in kinetic energy of the electron is proportional to:

(a) $\sqrt{V_2 - V_1}$, (b) $V_2 - V_1$, (c) $(V_2 - V_1)^2$, (d) none of these.

15. The integral $\oint \vec{E} \cdot d\vec{l} = 0$ if the electric field $\vec{E}$ is caused by:

(a) a static charge, (c) both (a) and (b),

(b) a time varying magnetic field, (d) none of these.

16. An electric field in a certain region has the components $E_x = ax - bz$, $E_y = -ay + bz$ and $E_z = b(y - x)$. Then which of the following statements is incorrect?

(a) $\vec{E}$ is solenoidal, (c) $\vec{E}$ is an electrostatic field,

(b) $\vec{E}$ is irrotational, (d) there is free charge in space.

17. In reference to a cylindrical polar coordinates (r, θ, z), a particle of a charge q and momentum $\vec{p}$ is moving in an electric potential $\phi(r)$. Then, which of the following is conserved?

(a) p_r, (b) p_θ, (c) p_z, (d) rp_θ.

18. Equipotential surfaces corresponding to a particular charge distribution are given by $4x^2 + (y-2)^2 + z^2 =$ constant. The electric field at the origin is

(a) $\vec{E} = 0$, (b) $\vec{E} = 2\hat{\imath}$, (c) $\vec{E} = 4\hat{\jmath}$, (d) $\vec{E} = -4\hat{\jmath}$.

19. A charge q is placed at the corner of a cube. Then electric flux through a face that does not touch the charge is

(a) $q/6\varepsilon_0$, (b) $q/8\varepsilon_0$, (c) $q/12\varepsilon_0$, (d) $q/24\varepsilon_0$.

20. Number of stable equilibrium points for a charged particle in an electrostatic field is

(a) one, (b) two, (c) zero, (d) infinite.

ANSWERS

1. (b)	**2.** (c)	**3.** (b)	**4.** (d)	**5.** (a)	**6.** (a)	**7.** (b)
8. (b)	**9.** (c)	**10.** (c)	**11.** (d)	**12.** (d)	**13.** (d)	**14.** (b)
15. (a)	**16.** (d)	**17.** (d)	**18.** (d)	**19.** (d)	**20.** (c)	

Chapter 2

Electrostatics in Conductors and Dielectrics

2.1 Conductors and Dielectrics

On the basis of electrical conductivity we generally classify substances into two main categories (*i*) *conductors* and (*ii*) *insulators* or *dielectrics*. A conductor is a substance in which there are plenty of free carriers of charge.

For example metals are usually available in the form of crystalline solids in which atoms are held by interatomic forces in a regular pattern or space lattice. Here atoms are so closely spaced that due to interaction of nearest neighbouring atoms the valence electrons become free and they can wander freely from one atom to another through interatomic spaces. But normally they cannot go outside the metal because of the existence of a surface potential energy barrier. In liquid conductors such as electrolytes it is the ions that move. An ideal conductor is one which has an unlimited supply of completely free charges. The presence of even a small electric field inside a conductor causes movement of charge carriers and this changes the field inside.

Insulators or dielectrics are substances which do not contain mobile charges as the conductors have. Here all charges are attached to specific atoms or molecules. In presence of electric field these charges can only move over microscopic distances within the molecules and cannot go away from specific molecules over macroscopic distances. However, the effects of small displacements of charges within each molecule add up and it can significantly change the original field.

Thus, it becomes important to study the effects of electric fields on conductors and dielectrics.

2.2 Electrostatics of Conductors

1. Electric field inside a conductor is zero

In a perfect conductor there is no resistance to the movement of charges within it. So if there were any field inside a conductor the free charges would move and no static equilibrium would be possible. So we may conclude that the electrostatic field inside a conductor is always zero. Let us examine what happens when a conductor is placed in an external electrostatic field. It causes the electrons to move opposite to the field direction (Fig. 2.2-1) and thus, produces *induced* positive and negative charges. This movement continues until the field $\vec{E}_i$ produced by the induced charges cancels the external field and the total field inside becomes zero.

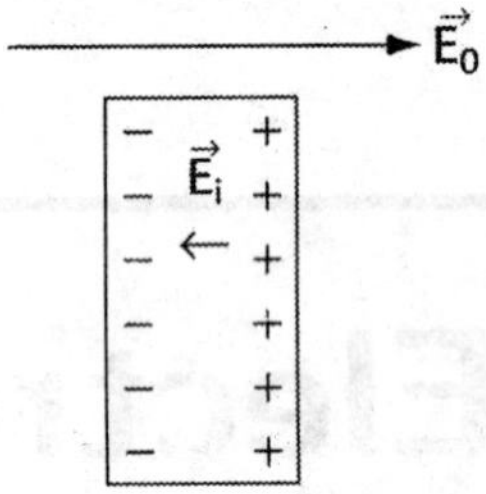

Fig. 2.2-1

2. Charges given to a conductor reside on its outer surface

If charge density ρ is nonzero at any point inside a conductor then Gauss's law $\vec{\nabla} \cdot \vec{E} = \rho/\varepsilon_0$ indicates that an electric field $\vec{E}$ exists at that point. As the charges are free to move they move and ultimately take up positions such that electric field inside becomes zero. Now the surface is the only place where the charges can go. They cannot leave the surface because of the existence of surface potential energy barrier.

3. Conductor forms an equipotential region

Any net charge on a conductor lies on it surface and they are distributed in such a way that the electric field $\vec{E}$ inside becomes zero. Then integrating the relation $\vec{E} = -\vec{\nabla}\phi = 0$ we find that $\phi =$ constant, i.e., potential ϕ is constant throughout. The conductor surface forms an equipotential surface otherwise charges would flow from regions of higher potential to those of lower potential.

4. The electric field outside of a conductor is normal to the surface

If the electric field is not normal to the surface there will be tangential component and electrons will move around the surface to destroy this tangential component. So in the static equilibrium condition the electric field must be normal to conductor surface. In other words, conductor surface is an equipotential surface, $\phi =$ constant. As $\vec{\nabla}\phi$ is a vector normal to the surface $\phi =$ constant. So we can say that $\vec{E} = -\vec{\nabla}\phi$ is a vector normal to the conductor surface.

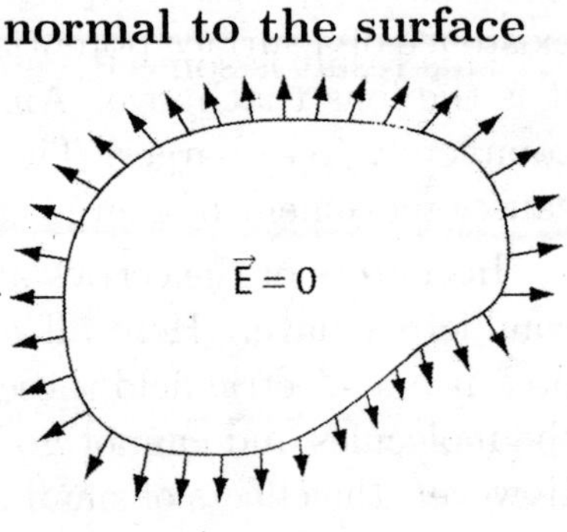

Fig. 2.2-2

5. Charge distribution on the surface of a conductor may not be uniform

Conductor surface forms an equipotential surface $\phi = \text{constant}$. Poisson's equation $\nabla^2\phi = -\rho/\varepsilon_0$ indicates that the charge density would depend on the curvature of the surface. The surface density of charge on a conductor surface is greater at points having more curvature (Fig. 2.2-3).

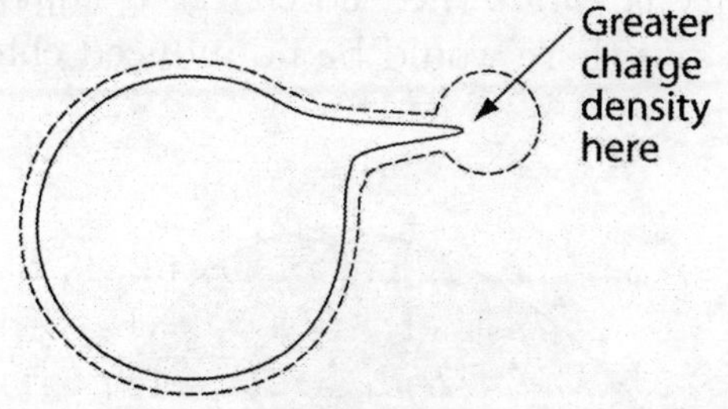

Fig. 2.2-3

6. Strength of electric field just outside a conductor

It has a magnitude σ/ε_0, where σ is the *local* surface density of charge. This can be easily proved by using Gauss's law.

Let us construct a Gaussian surface in the form of a small cylindrical box half inside and half outside, with its flat surfaces parallel and curved surface perpendicular to the conductor surface (Fig. 2.2-4). As $\vec{E} = 0$ inside the conductor and $\vec{E}$ is perpendicular to the conductor surface the electric flux through curved surface and lower flat surface would be zero. So there is contribution to the total flux only from the flat surface outside the conductor. Now from Gauss's law

$$\vec{E}\cdot\hat{n}\Delta S = \frac{1}{\varepsilon_0}\sigma\Delta S \quad \text{or} \quad E\Delta S = \frac{1}{\varepsilon_0}\sigma\Delta S \quad \text{or} \quad E = \frac{\sigma}{\varepsilon_0}.$$

In vector form

$$\vec{E} = \frac{\sigma}{\varepsilon_0}\hat{n}. \tag{2.2-1}$$

This result is sometimes called *Coulomb's theorem.*

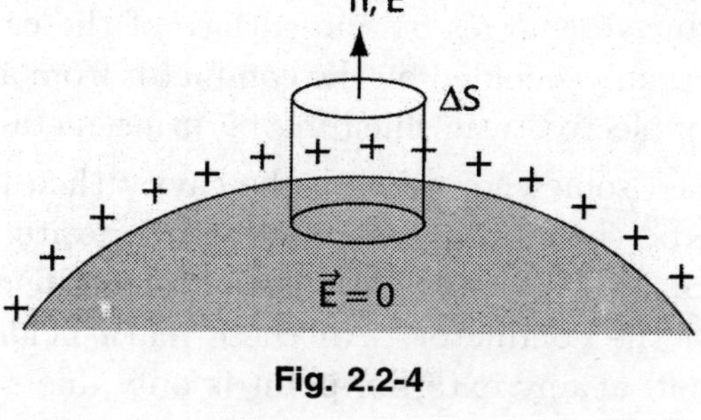

Fig. 2.2-4

7. The field in the cavity inside a conductor

We have seen that $\vec{E} = 0$ inside the material of a conductor. Let us now examine $\vec{E}$ inside the cavity of a conductor (Fig. 2.2-5). If any charge is given to the conductor it will completely lie on the outer surface of the conductor and it cannot produce any field inside the cavity. If the conductor is placed in an external electrostatic field, the field cannot penetrate the conductor; it is cancelled at the outer surface by the induced charges there. If the cavity is *empty* (i.e., no charge is kept within it) the field within the cavity would be zero and there would be no induced charges on the surface (S) of the cavity. Let us prove it.

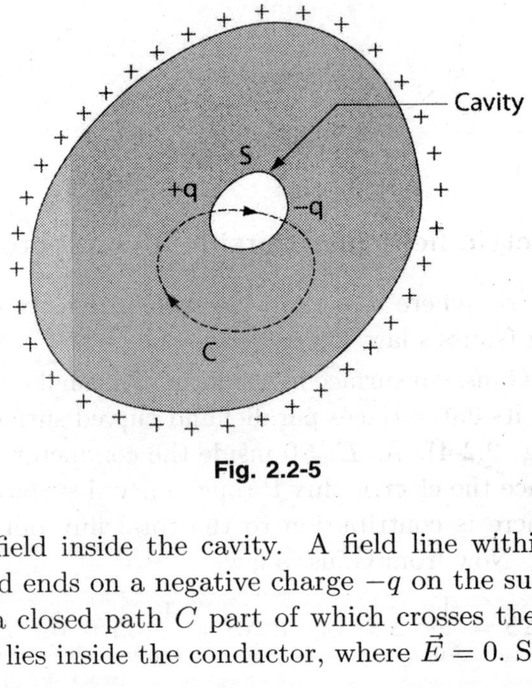

Fig. 2.2-5

Suppose there is field inside the cavity. A field line within the cavity starts at positive charge $+q$ and ends on a negative charge $-q$ on the surface (S) of the cavity. Let us now consider a closed path C part of which crosses the cavity along the field line and rest of which lies inside the conductor, where $\vec{E} = 0$. So the line integral

$$\oint_C \vec{E} \cdot \vec{dl}$$

would obviously be nonzero. But we know that for an electrostatic field this line integral must be zero. So we may conclude that there can be no fields inside the empty cavity or there can be no 'induced' charges on the surface of the cavity. The volume of the cavity is thus electrostatically shielded by the conductor from any charges outside. This explains the principle of electrostatic shielding by using metal case.

If a charge q is placed somewhere within the cavity then there can be fields inside the cavity. It can then be shown that an equal and opposite charge $-q$ is induced on the wall of the cavity and as the conductor is electrically neutral, $+q$ charge appears on the outer surface of the conductor. All these make field at any point inside the conductor zero. The field at any external point is only due to the $+q$ charge induced on the outer surface. So it will be

$$\frac{1}{4\pi\varepsilon_0} \cdot \frac{q}{r^2}\hat{r}.$$

8. Mechanical pressure on the surface of a charged conductor

Electric field at any external point P just outside a charged conductor is given by

$$\vec{E} = \frac{\sigma}{\varepsilon_0}\hat{n}, \tag{2.2-2}$$

where σ is the local surface density of charge and $\hat{n}$ is the outward unit vector normal to the surface. This field may be considered as the sum of two fields:

$$\vec{E} = \vec{E}_1 + \vec{E}_2, \tag{2.2-3}$$

where $\vec{E}_1$ is due to charges on a very small area ΔS around P and $\vec{E}_2$ is due to rest of the charges. For an internal point P' just on opposite of ΔS the contribution $\vec{E}_2$ remains same but the contribution $\vec{E}_1$ changes sign such that total field becomes zero, i.e.,

$$0 = -\vec{E}_1 + \vec{E}_2. \tag{2.2-4}$$

Solving Eqs. (2.2-3) and (2.2-4) we get

$$\vec{E}_1 = \vec{E}_2 = \frac{\vec{E}}{2} = \frac{\sigma}{2\varepsilon_0}\hat{n}. \tag{2.2-5}$$

The charge $\sigma\Delta S$ on ΔS cannot exert a force on itself but this charge $\sigma\Delta S$ experiences a force due to field $\vec{E}_2$ produced by other charges. Therefore, force on the charges contained in ΔS is

$$\vec{F} = \sigma\Delta S\vec{E}_2 = \frac{\sigma^2\Delta S}{2\varepsilon_0}\hat{n}.$$

So the force on unit area or electrostatic pressure is

$$\vec{P} = \frac{\sigma^2}{2\varepsilon_0}\hat{n}.$$

Obviously the pressure acts in the outward direction irrespective of the sign of σ. The pressure can also be expressed in terms of the field given by Eq. (2.2-2) at the conductor surface,

$$\vec{P} = \tfrac{1}{2}\varepsilon_0 E^2\hat{n}.$$

2.3 Electric Dipole

Two equal and opposite charges separated by a very small distance are said to constitute an *electric dipole.* The molecules of a dielectric medium placed in an electrostatic field behave like an electric dipole. So it is important to study the electrostatics of electric dipole before considering the electrostatics in dielectric.

For example, in a H_2O molecule the positive and negative charge centres do not coincide but are separated from each other by a very small distance. Hence, a H_2O molecule can be considered as an electric dipole. We define *dipole moment* $\vec{p}$ of a dipole constituted by two charges $+q$ and $-q$ separated by a distance l as a vector quantity of magnitude ql and directed from $-q$ to $+q$. It has the unit of coulomb-metre. The value of dipole moment for a molecule is usually very small. For this a smaller unit called the *debye unit* is used. 1 debye unit $= 1D = 3.336 \times 10^{-30}$ C.m.

Potential and field due to an electric dipole

Suppose that we have a dipole consisting of a charge $-q$ at the point $\vec{r}'$ and a charge $+q$ at $\vec{r}' + \vec{l}$ as shown in Fig. 2.3-1. The electric potential at any arbitrary point $\vec{r}$ is given by

$$\phi(\vec{r}) = \frac{1}{4\pi\varepsilon_0} \cdot \frac{q}{\left|\vec{r} - \vec{r}' - \vec{l}\right|} - \frac{1}{4\pi\varepsilon_0} \cdot \frac{q}{|\vec{r} - \vec{r}'|}$$

$$= \frac{1}{4\pi\varepsilon_0} \cdot \frac{q}{\left|\vec{R} - \vec{l}\right|} - \frac{1}{4\pi\varepsilon_0} \cdot \frac{q}{|\vec{R}|}. \tag{2.3-1}$$

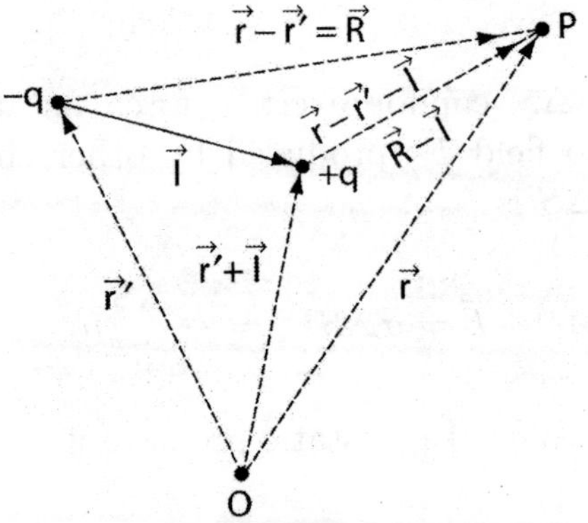

Fig. 2.3-1

If the distance $|\vec{R}| = R$ is large compared with the dipole length l then we can make the following expansion:

$$\frac{1}{\left|\vec{R} - \vec{l}\right|} = \frac{1}{\left|\left(\vec{R} - \vec{l}\right) \cdot \cdot \left(\vec{R} - \vec{l}\right)\right|^{1/2}}$$

$$= \frac{1}{\left|R^2 - 2\vec{R}\cdot\vec{l} + l^2\right|^{1/2}}$$

$$= \frac{1}{R}\left[1 + \frac{\vec{R}\cdot\vec{l}}{R^2} + \cdots\right]. \tag{2.3-2}$$

$\therefore$ From Eq. (2.3-1) we get

$$\phi(\vec{r}) = \frac{q}{4\pi\varepsilon_0}\cdot\frac{\vec{R}\cdot\vec{l}}{R^3}, \tag{2.3-3}$$

where we have kept only terms linear in l. This expression is valid only for $|\vec{R}| \gg l$. In many cases the concept of *point* or *pure dipole* becomes useful. Here $l \to 0$ and $q \to \infty$ in such a away that the product ql remains constant. In this case, Eq. (2.3-3) becomes exact. Now as $q\vec{l} = \vec{p}$ is the dipole moment Eq. (2.3-3) can be written as

$$\phi(\vec{r}) = \frac{1}{4\pi\varepsilon_0}\frac{\vec{p}\cdot\vec{R}}{R^3}. \tag{2.3-4}$$

As
$$\vec{\nabla}\frac{1}{R} = -\frac{\vec{R}}{R^3} \quad \text{and} \quad \vec{\nabla}'\frac{1}{R} = \frac{\vec{R}}{R^3}.$$

Equation (2.3-4) can also be written as

$$\phi(\vec{r}) = \frac{1}{4\pi\varepsilon_0}\vec{p}\cdot\vec{\nabla}'\frac{1}{R} \quad \text{or} \quad \phi(\vec{r}) = -\frac{1}{4\pi\varepsilon_0}\vec{p}\cdot\vec{\nabla}\frac{1}{R} = -\vec{p}\cdot\vec{\nabla}\phi_0, \tag{2.3-5}$$

where $\phi_0 = 1/4\pi\varepsilon_0 R$ is the potential of a unit point charge.

The electric field at $P(\vec{r})$ can be obtained as

$$\vec{E}(\vec{r}) = -\vec{\nabla}\phi(\vec{r}).$$

Using Eq. 2.3-4 we get

$$\vec{E}(\vec{r}) = -\frac{1}{4\pi\varepsilon_0}\left[\frac{1}{R^3}\vec{\nabla}(\vec{p}\cdot\vec{R}) + \vec{p}\cdot\vec{R}\vec{\nabla}\frac{1}{R^3}\right].$$

Now from vector analysis

$$\vec{\nabla}\left(\vec{p}\cdot\vec{R}\right) = \vec{p} \quad \text{and} \quad \vec{\nabla}\frac{1}{R^3} = -\frac{3\vec{R}}{R^5}$$

$$\therefore \quad \vec{E}(\vec{r}) = \frac{1}{4\pi\varepsilon_0}\left[\frac{3(\vec{p}\cdot\vec{R})\vec{R}}{R^5} - \frac{\vec{p}}{R^3}\right]. \tag{2.3-6}$$

The electric field can also be calculated directly by using Coulomb's law.

For convenience let us now choose the charge $-q$ at the origin (i.e., $\vec{r}' = 0$, and $\vec{R} = \vec{r}$ and the charges lie along z-axis (Fig. 2.3-2). Let θ be the angle between the axis of the dipole and the radius vector to the point of observation P. In this case, Eq. (2.3-4) takes the form

$$\phi(\vec{r}) = \frac{1}{4\pi\varepsilon_0}\frac{\vec{p}\cdot\vec{r}}{r^3} = \frac{1}{4\pi\varepsilon_0}\frac{p\cos\theta}{r^2}. \tag{2.3-7}$$

Note that the *dipolar potential varies as* $1/r^2$, *whereas the potential due an isolated charge (monopole) varies as* $1/r$.

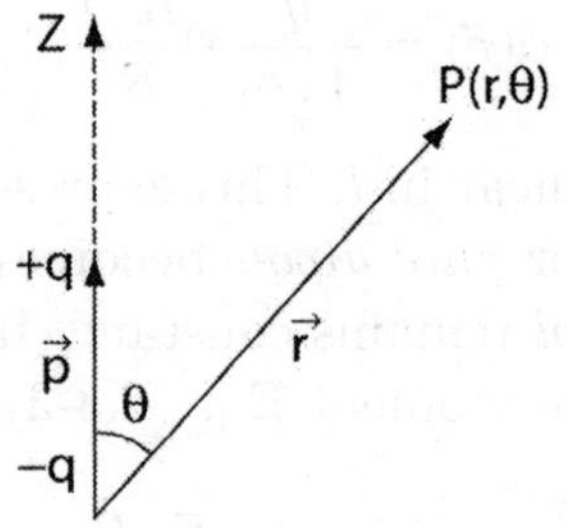

Fig. 2.3-2

Very often it becomes convenient to express electric field at P in terms of spherical polar coordinates (r, θ, φ). Using Eq. (2.3-7) we get

$$E_r = -\frac{\partial\phi}{\partial r} = \frac{1}{4\pi\varepsilon_0}\frac{2p\cos\theta}{r^3} \qquad \left[\because\ \vec{\nabla} = \hat{r}\frac{\partial}{\partial r} + \hat{\theta}\frac{1}{r}\frac{\partial}{\partial\theta} + \hat{\varphi}\frac{1}{r\sin\theta}\frac{\partial}{\partial\varphi}\right]$$

$$E_\theta = -\frac{1}{r}\frac{\partial\phi}{\partial\theta} = \frac{1}{4\pi\varepsilon_0}\cdot\frac{p\sin\theta}{r^3}$$

$$E_\varphi = -\frac{1}{r\sin\theta}\frac{\partial\phi}{\partial\varphi} = 0. \tag{2.3-8}$$

Thus,

$$\vec{E} = \frac{p}{4\pi\varepsilon_0 r^3}\left(2\cos\theta\hat{r} + \sin\theta\hat{\theta}\right). \tag{2.3-9}$$

The magnitude of $\vec{E}$ is

$$|\vec{E}| = \frac{p}{4\pi\varepsilon_0 r^3}\sqrt{4\cos^2\theta + \sin^2\theta} = \frac{p}{4\pi\varepsilon_0 r^3}\sqrt{1 + 3\cos^2\theta}.$$

If β is the angle made by $\vec{E}$ with the radius vector then

$$\tan\beta = \frac{E_\theta}{E_r} = \frac{1}{2}\tan\theta \quad \text{or} \quad \beta = \tan^{-1}\left(\frac{1}{2}\tan\theta\right).$$

Note that the *dipole field falls off as* $1/r^3$ *whereas the field of an isolated charge (monopole) falls off as* $1/r^2$.

Equipotential surfaces and field lines of a dipole

The equation of lines of force may be obtained from Fig. 2.3-3(a) which shows a line element $d\vec{l} = \hat{r}dr + \hat{\theta}rd\theta$ and electric field $\vec{E} = \hat{r}E_r + \hat{\theta}E_\theta$ parallel to each other. Therefore,

$$\frac{rd\theta}{dr} = \frac{E_\theta}{E_r} = \frac{\sin\theta}{2\cos\theta} \quad \text{or} \quad \frac{dr}{r} = \frac{2d(\sin\theta)}{\sin\theta}.$$

Integrating we get.

$$r = C\sin^2\theta.$$

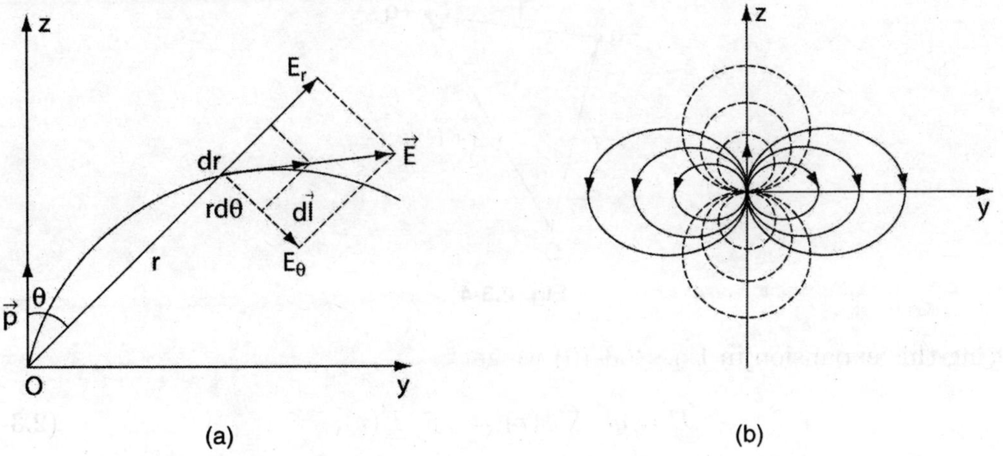

Fig 2.3-3 Equipotential lines (dotted lines) and lines of force (lines with arrows) of the dipole shown in (b)

This equation gives a family of lines of force, where the constant C is different for different lines.

Equipotential surfaces are everywhere perpendicular to the lines of force. Traces of equipotential surfaces in the yz-plane can be obtained as follows. The potential at any point (r, θ) is

$$\phi(r,\theta) = \frac{1}{4\pi\varepsilon_0}\frac{p\cos\theta}{r^2}.$$

Therefore, for an equipotential line,

$$r(\theta) = A\sqrt{\cos\theta},$$

where $A = \sqrt{p/4\pi\varepsilon_0\phi}$ = constant. This equation gives a family of equipotential lines, where A is different for different lines.

Potential energy of a dipole placed in an external electric field

Suppose we have an electric dipole consisting of a charge $-q$ at the point $\vec{r}$ and a charge $+q$ at $\vec{r}+\vec{l}$ as shown in Fig. 2.3-4. It is placed in an external electric field $\vec{E}$ described by the potential function $\phi(\vec{r})$. So the potential energy of the dipole is

$$U = -q\phi(\vec{r}) + q\phi(\vec{r}+\vec{l}). \tag{2.3-10}$$

If $|\vec{l}| \ll |\vec{r}|$ then $\phi(\vec{r}+\vec{l})$ can be expanded in a Taylor series as

$$\phi(\vec{r}+\vec{l}) = \phi(\vec{r}) + \vec{l}\cdot\vec{\nabla}\phi(\vec{r}), \tag{2.3-11}$$

where we neglect higher order smaller terms.

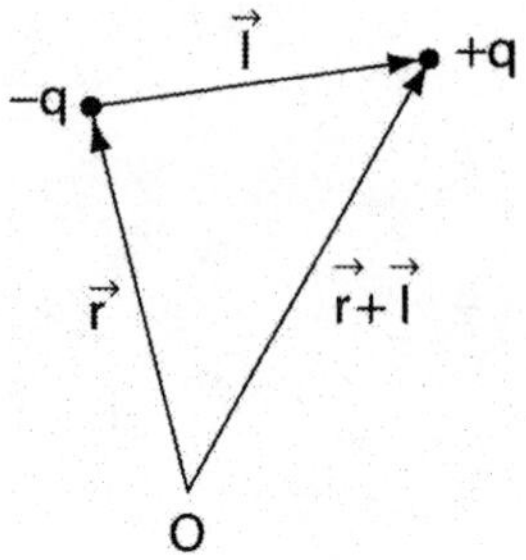

Fig. 2.3-4

Using this expansion in Eq. (2.3-10) we get

$$U = q\vec{l}\cdot\vec{\nabla}\phi(\vec{r}) = -\vec{p}\cdot\vec{E}(\vec{r}), \tag{2.3-12}$$

where $\vec{p} = \overrightarrow{ql}$ is the dipole moment and $\vec{E}(\vec{r}) = -\vec{\nabla}\phi(\vec{r})$ is the external electric field.

If the angle between $\vec{p}$ and $\vec{E}$ be θ then potential energy

$$U = -pE\cos\theta. \tag{2.3-13}$$

Note that $\theta = 90°$ is chosen as the position of zero potential energy.

Force and couple on a dipole placed in an electric field

The potential energy W of a dipole of dipole moment $\vec{p}$ placed in an external electric field $\vec{E}$ is

$$U = -\vec{p}\cdot\vec{E}.$$

The force on the dipole due to the electric field is given by[1]

$$\vec{F} = -\vec{\nabla}U = \vec{\nabla}\left(\vec{p}\cdot\vec{E}\right).$$

[1]Work done in an elementary displacement $d\vec{r}$ is

$$\begin{aligned}\vec{F}\cdot d\vec{r} &= -dU \quad \text{(decrease in P.E.)}\\ &= -\vec{\nabla}U\cdot d\vec{r}.\end{aligned}$$

Since $d\vec{r}$ is arbitrary we have $\vec{F} = -\vec{\nabla}U$.

Now $\vec{p}$ is a constant vector and $\vec{\nabla}\times\vec{E}=0$. So we can write from the vector identity

$$\vec{\nabla}\left(\vec{A}\cdot\vec{B}\right)=\left(\vec{A}\cdot\vec{\nabla}\right)\vec{B}+\left(\vec{B}\cdot\vec{\nabla}\right)\vec{A}+\vec{A}\times\left(\vec{\nabla}\times\vec{B}\right)+\vec{B}\times\left(\vec{\nabla}\times\vec{A}\right)$$

that

$$\vec{\nabla}\left(\vec{p}\cdot\vec{E}\right)=\left(\vec{p}\cdot\vec{\nabla}\right)\vec{E}.$$

Therefore,

$$\vec{F}=\left(\vec{p}\cdot\vec{\nabla}\right)\vec{E}. \qquad (2.3\text{-}14)$$

It is a general expression valid for both uniform and nonuniform field.

If the field $\vec{E}$ is uniform then $\vec{F}=0$ and there will be no translational motion of the dipole. For a uniform field the force on charge $+q$ is $q\vec{E}$ and that on $-q$ is $-q\vec{E}$. These two equal and opposite forces make the resultant force zero.

The torque acting on the dipole is given by

$$\Gamma=-\frac{\partial U}{\partial\theta}=-\frac{\partial}{\partial\theta}(pE\cos\theta) \quad [\text{where } \theta \text{ is the angle between } \vec{p} \text{ and } \vec{E}]$$

$$=pE\sin\theta.$$

In vector form

$$\vec{\tau}=\vec{p}\times\vec{E}. \qquad (2.3\text{-}15)$$

Work done in rotating a dipole in a uniform field

Torque $\vec{\tau}$ acting on a dipole of moment $\vec{p}$ making an angle θ with the direction of a uniform electric field $\vec{E}$ is

$$\vec{\tau}=\vec{p}\times\vec{E} \quad \text{or} \quad |\vec{\tau}|=pE\sin\theta.$$

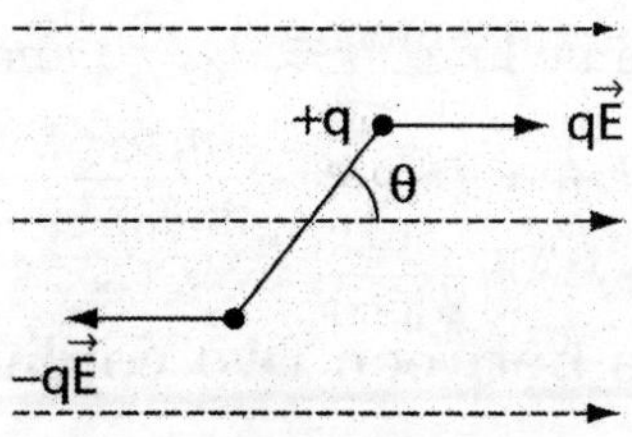

Fig. 2.3-5

Work done in deflecting the dipole further through an angle $d\theta$ is

$$dW=pE\sin\theta\,d\theta.$$

∴ Work done in deflecting the dipole from a position making an angle θ_1 to the position making an angle θ_2 with $\vec{E}$ is

$$\int_{\theta_1}^{\theta_2} pE\sin\theta\,d\theta=pE\left(\cos\theta_1-\cos\theta_2\right). \qquad (2.3\text{-}16)$$

Mutual potential energy between two coplanar dipoles

Let us consider two dipoles 1, 2 of dipole moments $\vec{p}_1$ and $\vec{p}_2$ lying on a single plane, $\vec{r}_{21}$ is the position vector of dipole 2 with respect to the dipole 1 (Fig. 2.3-6). The electric field at the centre of the dipole 2 due to dipole 1 is

$$\vec{E}_1 = \frac{1}{4\pi\varepsilon_0}\left[\frac{3(\vec{p}_1 \cdot \vec{r}_{21})\vec{r}_{21}}{r_{21}^5} - \frac{\vec{p}_1}{r_{21}^3}\right]$$

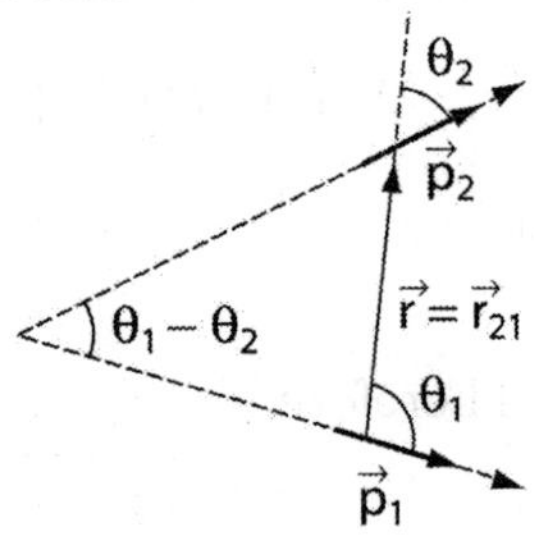

Fig. 2.3-6

The potential energy of the dipole 2 in this field $\vec{E}_1$ is

$$U_{21} = -\vec{p}_2 \cdot \vec{E}_1 = \frac{1}{4\pi\varepsilon_0}\left[\frac{\vec{p}_1 \cdot \vec{p}_2}{r_{21}^3} - \frac{3(\vec{p}_1 \cdot \vec{r}_{21})(\vec{p}_2 \cdot \vec{r}_{21})}{r_{21}^5}\right], \tag{2.3-17}$$

where we have assumed that the dipoles are very small in size compared to their separation so that $\vec{E}_1$ can be taken to be uniform over the length of the dipole 2.

Note that the expression for U_{21} remains unchanged when the subscripts 1 and 2 are interchanged. This means that the potential energy of dipole 1 in the field of dipole 2 is also given by the Eq. (2.3-17). The Eq. (2.3-17) is a general one and valid for both coplanar and non-coplanar dipoles. For two coplanar dipoles as shown in Fig. 2.3-6 we can write

$$\begin{aligned} U_{21} = U_{12} &= \frac{p_1 p_2}{4\pi\varepsilon_0 r^3}[\cos(\theta_2 - \theta_1) - 3\cos\theta_1\cos\theta_2] \\ &= \frac{p_1 p_2}{4\pi\varepsilon_0 r^3}[\sin\theta_1\sin\theta_2 - 2\cos\theta_1\cos\theta_2], \end{aligned} \tag{2.3-18}$$

where we have put $r_{21} = r$. This is the interaction energy of two coplanar dipoles separated by a distance r.

If the dipoles lie along the same line, $\theta_1 = \theta_2 = 0$ and then

$$U_{21} = U_{12} = -\frac{p_1 p_2}{2\pi\varepsilon_0 r^3}. \tag{2.3-19}$$

Mutual forces and torques between two coplanar dipoles

Suppose dipole 1 is kept fixed and dipole 2 undergoes a general type of small displacement which alters r, θ_1 and θ_2. If the radial and transverse components of force on dipole 2 due to dipole 1 are R_{21} and T_{21} and torque acting on dipole 2 is of magnitude τ_{21} then work done in an elementary displacement is given by

$$R_{21}dr + T_{21}rd\theta_1 + \tau_{21}d((\theta_1 - \theta_2)). \tag{2.3-20}$$

This must be equal to decrease in P.E. $-dU_{21}$. Since $U_{21} = U_{21}(r, \theta_1, \theta_2)$

$$-dU_{21} = -\frac{\partial U_{21}}{\partial r}dr - \frac{\partial U_{21}}{\partial \theta_1}d\theta_1 - \frac{\partial U_{21}}{\partial \theta_2}d\theta_2. \tag{2.3-21}$$

Since the displacement is arbitrary we can equate coefficients of dr, $d\theta_1$ and $d\theta_2$ in Eqs. (2.3-20) and (2.3-21). Thus,

$$R_{21} = -\frac{\partial U_{21}}{\partial r} = \frac{3p_1p_2}{4\pi\varepsilon_0 r^4}(\sin\theta_1\sin\theta_2 - 2\cos\theta_1\cos\theta_2) \tag{2.3-22}$$

$$T_{21}r + \tau_{21} = -\frac{\partial U_{21}}{\partial \theta_1} = -\frac{p_1p_2}{4\pi\varepsilon_0 r^3}(\cos\theta_1\sin\theta_2 + 2\sin\theta_1\cos\theta_2) \tag{2.3-23}$$

$$\tau_{21} = \frac{\partial U_{21}}{\partial \theta_2} = \frac{p_1p_2}{4\pi\varepsilon_0 r^3}(\sin\theta_1\cos\theta_2 + 2\cos\theta_1\sin\theta_2). \tag{2.3-24}$$

Solving Eqs. (2.3-23) and (2.3-24) we get

$$T_{21} = -\frac{3p_1p_2}{4\pi\varepsilon_0 r^4}\sin(\theta_1+\theta_2). \tag{2.3-25}$$

In a similar way we can calculate the forces and torques on dipole 1 due to dipole 2. It can be shown that $R_{12} = -R_{21}$, $T_{12} = -T_{21}$ and

$$\tau_{12} = \frac{\partial U_{12}}{\partial \theta_1} = \frac{p_1p_2}{4\pi\varepsilon_0 r^3}(\cos\theta_1\sin\theta_2 + 2\sin\theta_1\cos\theta_2). \tag{2.3-26}$$

Thus, the forces are equal and opposite but torques are not so. The resultant torque is

$$\tau = \tau_{21} + \tau_{12} = \frac{3p_1p_2}{4\pi\varepsilon_0 r^3}\sin(\theta_1+\theta_2).$$

So it appears that if the two dipoles are fixed on a board, which is free to rotate then the board would rotate with increasing velocity. But this is not the actual picture because the transverse components of forces T_{21} and T_{12} constitute a couple.

$$T_{21}\cdot r = \frac{3p_1p_2}{4\pi\varepsilon_0 r^3}\sin(\theta_1+\theta_2),$$

which is equal and opposite to τ. Thus, there is no net force or torque acting on the system.

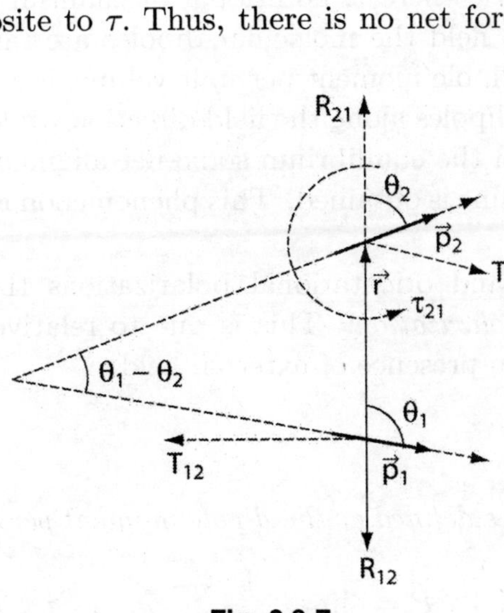

Fig. 2.3-7

2.4 Polar and Nonpolar Molecules

The molecules of a dielectric can be classified as polar and nonpolar molecules. In a molecule the protons or positive charges may be imagined to be concentrated in effect at a specific point within the molecule. This may be called the *centre of gravity* of positive charges.

Likewise there is a 'centre of gravity' of negative charges or the distributed electrons. If these two charge centres coincide then the molecule does not possess any intrinsic dipole moment. Such molecules are called *nonpolar molecules.* Symmetric molecules like H_2, N_2, O_2, CCl_4 and monoatomic molecules like He, Ne, Ar fall into this class.

On the other hand, if the two charge centres of a molecule do not coincide but are separated from each other by a small distance, the molecule possesses an intrinsic dipole moment and it is called a *polar molecule.* Asymmetric molecules like H_2O, H_2S, etc., fall into this class.

2.5 Dielectric Polarization

When any dielectric is placed in an electrostatic field the charge distribution within the molecules may become modified. The electrons are pulled in one direction and the nuclei in the opposite way. This causes a slight displacement (on the molecular scale) of the centres of positive and negative charges. The electrostatic force between positive and negative charges opposes it and ultimately an equilibrium is set in with a small dipole moment induced in nonpolar molecules. For polar molecules this induces an extra dipole moment. This phenomenon is known as *electronic polarization* or *deformation polarization.* The medium in which such charge separation has taken place is said to be polarized.

In case of polar dielectric there is additional mechanism of polarization. In the absence of external electric field the molecular dipoles are randomly oriented due to thermal agitation and net dipole moment per unit volume is zero. If an electric field is applied it tries to align the dipoles along the field direction while thermal agitation tries to oppose the alignment. In the equilibrium some net alignment and hence, some net dipole moment per unit volume is obtained. This phenomenon is known as *orientational polarization.*

Apart from electronic and orientational polarizations there is another kind of polarization, called *ionic polarization.* This is due to relative displacement between positive and negative ions in presence of external field.

Polarization $\vec{P}$

The electric polarization $\vec{P}$ is defined as the dipole moment per unit volume. Therefore,

$$\vec{P} = \lim_{\Delta V \to 0} \frac{1}{\Delta V} \sum_{i=1}^{N} \vec{p_i}, \tag{2.5-1}$$

where $\vec{p}_i$ is the dipole moment of the i-th molecule and N is the number of molecules in volume ΔV. The dimensions of $\vec{P}$ are charge per unit area so that $\vec{P}$ can be measured in units of coulomb/metre^2.

Bound or polarization charges

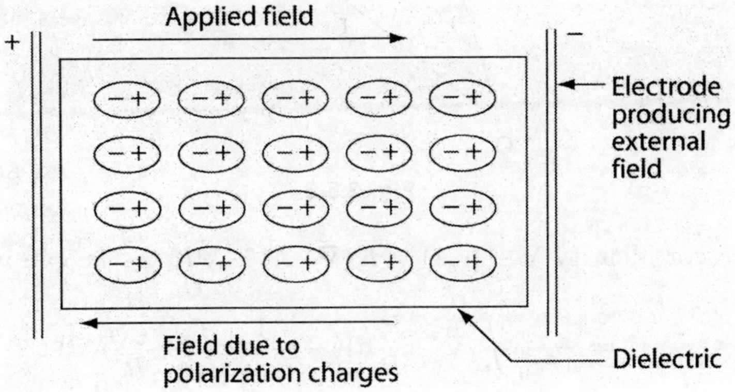

Fig. 2.5-1

Due to polarization of any dielectric medium some kind of charges appear in the volume and on the surface of the dielectric medium. These charges are called *bound* or *polarization charges.* Unlike the free charges of a conductor these charges cannot move over macroscopic distances. These bound charges give rise to a field in a direction opposite to the applied field. This makes the electric field within the dielectric smaller than the applied field. In case of uniform polarization the positive and negative charges of neighbouring molecular dipoles tend to cancel their effects. But on the dielectric boundary there is no such cancellation and surface charges appear (Fig. 2.5-1). For nonuniform polarization we get accumulation of bound charges on the surface as well as within the volume of the material.

To find expressions for these bound charges let us consider the potential at any external point $\vec{r}$ due to a polarized medium (Fig. 2.5-2). The potential $d\phi$ at P due to an element of volume dV of the polarized medium is given by Eq. (2.3-5) as

$$d\phi(\vec{r}) = \frac{1}{4\pi\varepsilon_0}\left(\vec{P}\,dV\right)\cdot\vec{\nabla}'\frac{1}{R},$$

where $\vec{\nabla}'$ is with respect to source point coordinates (x', y', z').

∴ Total potential at P due to the whole polarized medium is

$$\phi(\vec{r}) = \frac{1}{4\pi\varepsilon_0}\int_V \vec{P}\cdot\vec{\nabla}'\frac{1}{R}\,dV, \tag{2.5-2}$$

where the integration is over the whole volume V of the dielectric.

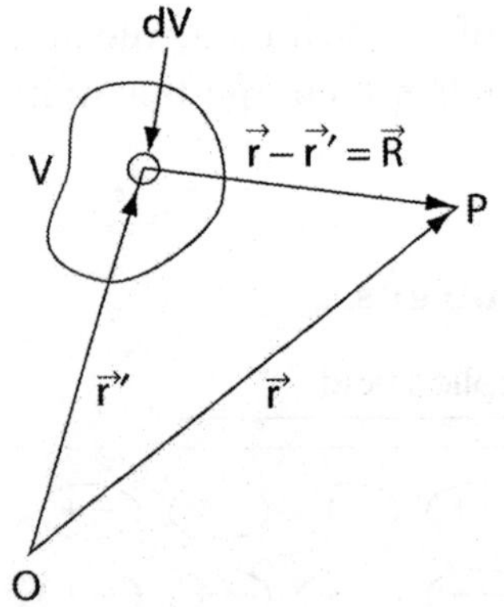

Fig. 2.5-2

Using the vector identity $\vec{\nabla} \cdot (K\vec{A}) = K(\vec{\nabla} \cdot \vec{A}) + \vec{\nabla}K \cdot \vec{A}$ we can write

$$\phi(\vec{r}) = \frac{1}{4\pi\varepsilon_0} \int_V \vec{\nabla}' \cdot \frac{\vec{P}}{R} dV - \frac{1}{4\pi\varepsilon_0} \int_V \frac{\vec{\nabla}' \cdot \vec{P}}{R} dV. \tag{2.5-3}$$

Using Gauss's divergence theorem we get

$$\phi(\vec{r}) = \frac{1}{4\pi\varepsilon_0} \int_S \frac{\vec{P} \cdot \hat{n}\, dS}{R} + \frac{1}{4\pi\varepsilon_0} \int_V \frac{(-\vec{\nabla}' \cdot \vec{P})}{R} dV, \tag{2.5-4}$$

where S is the surface enclosing volume V, $\hat{n}$ is the outward unit normal to the surface element dS. Equation (2.5-4) shows that the potential at any external point due to a polarized dielectric may be considered as due to a surface distribution of charge of density $\sigma = \vec{P} \cdot \hat{n}$ over the surface of the dielectric and a volume distribution of charge of density $\rho = -\vec{\nabla}' \cdot \vec{P}$ throughout the dielectric. These charges are called *bound charges* as they are not mobile like free charges. They are also called *polarization charges* as they appear due to polarization of the dielectric. These charges really exist in the medium. They are produced due to separation of charges under the action of electric field. In case of uniform polarization $\vec{\nabla}' \cdot \vec{P} = 0$ and there is no volume distribution of charge.

2.6 Gauss's Law in Dielectric Medium

In free space according to Gauss's law total electric flux over any closed surface equals $1/\varepsilon_0$ times the total charge Q_t enclosed, i.e.,

$$\oint_S \vec{E} \cdot d\vec{S} = \frac{1}{\varepsilon_0} \times Q_t. \tag{2.6-1}$$

For dielectric medium we must consider both free and bound charges. Suppose an imaginary surface S in a dielectric medium encloses bound charges Q_b as well as free charges Q_f distributed on the surfaces S_1, S_2, ... of a number of conductors (Fig. 2.6-1). Then

$$Q_t = Q_b + Q_f = \int_{S_1+S_2+\cdots} \vec{P}\cdot\hat{n}\,dS + \int_V (-\vec{\nabla}\cdot\vec{P})\,dV + Q_f, \qquad (2.6\text{-}2)$$

where V is the volume of the dielectric enclosed by the surface S.

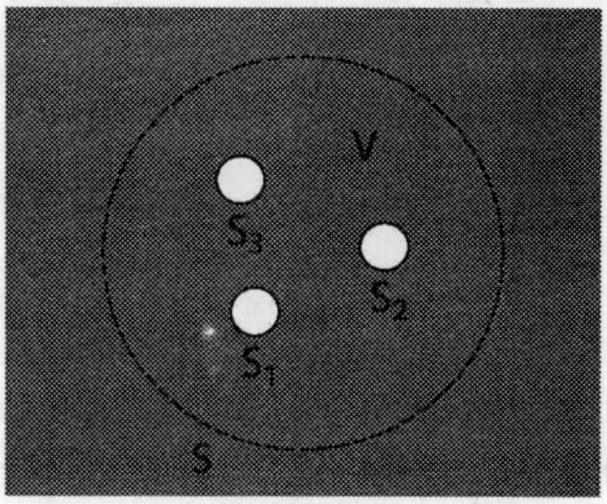

Fig. 2.6-1

Note that S is not a boundary of the dielectric and, therefore, no surface charge appears on it. Using Gauss's divergence theorem we can write

$$\begin{aligned} Q_t &= \int_{S_1+S_2+\cdots} \vec{P}\cdot\hat{n}\,dS - \int_{S+S_1+S_2+\cdots} \vec{P}\cdot\hat{n}\,dS + Q_f \\ &= -\int_S \vec{P}\cdot\hat{n}\,dS + Q_f. \end{aligned}$$

[∵ Volume V is bounded by the surfaces S, S_1, S_2, etc.]

Therefore, from Eq. (2.6-1)

$$\oint_S \vec{E}\cdot d\vec{S} = \frac{1}{\varepsilon_0}\times Q_t = -\frac{1}{\varepsilon_0}\int_S \vec{P}\cdot\hat{n}dS + \frac{1}{\varepsilon_0}Q_f$$

$$\text{or} \quad \oint_S (E_0\vec{E}+\vec{P})\cdot d\vec{S} = Q_f. \qquad (2.6\text{-}3)$$

This equation states that the flux of the vector $\varepsilon_0\vec{E}+\vec{P}$ through any closed surface is equal to only the free charge enclosed by the surface. Because of historical reason and some practical advantages in the problems involving dielectrics the vector $\varepsilon_0\vec{E}+\vec{P}$ is given a new name, the *electric displacement vector* $\vec{D}$. Thus, we define $\vec{D}$ as

$$\vec{D} = \varepsilon_0\vec{E} + \vec{P}. \qquad (2.6\text{-}4)$$

Obviously $\vec{D}$ has the same dimensions as that of $\vec{P}$, i.e., charge per unit area (C.m^{-2}).

In terms of $\vec{D}$ Eq. (2.6-3) becomes

$$\oint_S \vec{D}\cdot d\vec{S} = Q_f. \qquad (2.6\text{-}5)$$

This is the *integral form of Gauss's law in dielectric.* If the free charge has a volume distribution ρ_f inside S then

$$\oint_S \vec{D} \cdot d\vec{S} = \int_V \rho_f \, dV.$$

Using divergence theorem we get

$$\oint_V \vec{\nabla} \cdot \vec{D} \, dV = \int_V \rho_f \, dV.$$

Since this is true for any volume V we must have

$$\vec{\nabla} \cdot \vec{D} = \rho_f. \tag{2.6-6}$$

This result is known as the *differential form of Gauss's law.*

It is to be noted that $\vec{D}$ is the field of free charges only whereas $\vec{E}$ is the field of both free and bound charges. *$\vec{E}$ is the fundamental electric vector,* which determines the force character of electric field whereas $\vec{D}$ is an auxiliary vector introduced for convenience. $\vec{D}$ has no direct connection with the forces exerted on charges.

The equation $\vec{\nabla} \cdot \vec{E} = \rho_t/\varepsilon_0$, where $\rho_t = \rho_f + \rho_b$ is the total charge density, is sufficient to find $\vec{E}$. If we try to find $\vec{E}$ by integrating the equation it will involve a constant of integration. This constant can be evaluated by using the condition $\vec{\nabla} \times \vec{E} = 0$. But the equation $\vec{\nabla} \cdot \vec{D} = \rho_f$ is not always sufficient to find $\vec{D}$ because it is not always true that

$$\vec{\nabla} \times \vec{D} = \varepsilon_0 \vec{\nabla} \times \vec{E} + \vec{\nabla} \times \vec{P} = \vec{\nabla} \times \vec{P} = 0.$$

2.7 Susceptibility, Permittivity and Dielectric Constant

There exists a class of dielectric materials for which polarization $\vec{P}$ is found to be proportional to the macroscopic electric field $\vec{E}$, provided the field is not very strong. The proportionality is usually written as

$$\vec{P} = \varepsilon_0 \mathcal{X} \vec{E}, \tag{2.7-1}$$

where $\mathcal{X}$ is called the *electric susceptibility of the dielectric.* The factor ε_0 is chosen to make $\mathcal{X}$ dimensionless. The value of $\mathcal{X}$ depends on the density of the medium and the polarizability of the molecules. If $\mathcal{X}$ does not depend on E then the material is called *linear dielectric.* If $\mathcal{X}$ depends on $\vec{E}$ then the material is called *nonlinear dielectric.* If χ does not depend on the direction of $\vec{E}$ then the material is called *isotropic.* If $\mathcal{X}$ does not change from point to point in a piece of material, the material is called *homogeneous.* Most liquids and gases are homogeneous, isotropic and linear materials at least at low electric fields. They are called *simple* dielectrics. For linear medium

$$\begin{aligned}\vec{D} = \varepsilon_0 \vec{E} + \vec{P} &= \varepsilon_0 \vec{E} + \varepsilon_0 \mathcal{X} \vec{E} \\ &= \varepsilon_0 (1 + \mathcal{X}) \vec{E} = \varepsilon \vec{E},\end{aligned} \tag{2.7-2}$$

where

$$\varepsilon = \varepsilon_0(1 + \mathcal{X}). \tag{2.7-3}$$

Thus, $\vec{D}$ is also proportional to $\vec{E}$. ε is called the *permittivity of the medium.* In vacuum there is no material to polarize. Hence, for vacuum $\mathcal{X} = 0$ and $\varepsilon = \varepsilon_0$. For this ε_0 is called *permittivity of free space.* The dimensionless quantity $K = \varepsilon/\varepsilon_0 = 1 + \mathcal{X}$ is called *relative permittivity* or *dielectric constant* of the medium. For free space $K = 1$. The electrical behaviour of dielectrics can be described in terms $\mathcal{X}$ or ε. However, it is more convenient to work with K.

For many crystalline substances $\mathcal{X}$ depends on the direction of $\vec{E}$ in the material. In this case, the directions of $\vec{P}$ and $\vec{E}$ or of $\vec{D}$ and $\vec{E}$ do not coincide except for certain specific orientations. The materials get easily polarized in some directions than others. Such materials are called *anisotropic.* In this case, in order to preserve the linear relationship between $\vec{P}$ and $\vec{E}$ each component of $\vec{P}$ is assumed to be linearly related to the components of $\vec{E}$, i.e.,

$$\begin{aligned}
P_x &= \varepsilon_0 \left[\mathcal{X}_{xx}E_x + \mathcal{X}_{xy}E_y + \mathcal{X}_{xz}E_z\right] \\
P_y &= \varepsilon_0 \left[\mathcal{X}_{yx}E_x + \mathcal{X}_{yy}E_y + \mathcal{X}_{yz}E_z\right] \\
P_z &= \varepsilon_0 \left[\mathcal{X}_{zx}E_x + \mathcal{X}_{zy}E_y + \mathcal{X}_{zz}E_z\right].
\end{aligned}$$

The set of nine constants $\mathcal{X}_{ij}$ is known as the *susceptibility tensor.*

Similarly for anisotropic material the relation between $\vec{D}$ and $\vec{E}$ can be written as

$$D_i = \sum_{k=x,y,z} \varepsilon_{ik}E_k \quad (i = x, y, z),$$

where the set of nine constants ε_{ik} constitute the *permittivity tensor.* Here $\mathcal{X}$ and ε are not scalar quantities. When operated on $\vec{E}$ they change not only the magnitude but also the direction of $\vec{E}$. In this case, $\mathcal{X}$ and ε are tensors of rank two.

2.8 Poisson's and Laplace's Equations in Dielectrics

In a dielectric medium the differential form of Gauss's law is

$$\vec{\nabla} \cdot \vec{D} = \rho_f, \tag{2.8-1}$$

where ρ_f is the free charge density. For a linear homogeneous and isotropic dielectric $\vec{D} = \varepsilon\vec{E}$, where ε is a constant, called the *permittivity of the medium.*

Therefore, from Eq. (2.8-1) we can write

$$\vec{\nabla} \cdot \vec{E} = \frac{\rho_f}{\varepsilon}. \tag{2.8-2}$$

Using the relation $\vec{E} = -\vec{\nabla}\phi$ we get

$$\nabla^2\phi = -\frac{\rho_f}{\varepsilon}. \tag{2.8-3}$$

This is known as *Poisson's equation* for a dielectric. If $\rho_f = 0$ then Eq. (2.8-3) reduces to *Laplace's equation*

$$\nabla^2\phi = 0. \tag{2.8-4}$$

2.9 Point Charge in a Dielectric Fluid

Suppose a point $+q$ is placed in a homogeneous isotropic dielectric medium of infinite extent. Let us assume that the medium is linear and characterised by a dielectric constant K. A point charge in vacuum produces pure radial field. As $\vec{E}$, $\vec{D}$ and $\vec{P}$ are all parallel to one another at each point, the radial nature of the field will not be changed due to the presence of the dielectric. We construct a spherical Gaussian surface of radius r with its centre at the charge $+q$. Now, applying Gauss's law to this surface we get

$$\oint_S \vec{D}\cdot\hat{n}\,dS = q \quad \text{or} \quad D\cdot 4\pi r^2 = q \quad \text{or} \quad D = \frac{q}{4\pi r^2}.$$

Using $D = \varepsilon E$ we can write,

$$\begin{aligned} E = \frac{D}{\varepsilon} &= \frac{1}{4\pi\varepsilon}\cdot\frac{q}{r^2} \\ &= \frac{1}{4\pi\varepsilon_0}\frac{q}{Kr^2} \qquad \left[\because K = \frac{\varepsilon}{\varepsilon_0}\right] \end{aligned}$$

In vector form

$$\vec{E} = \frac{1}{4\pi\varepsilon_0}\cdot\frac{q}{Kr^2}\hat{r} = \frac{1}{4\pi\varepsilon}\cdot\frac{q}{r^2}\hat{r}. \tag{2.9-1}$$

Obviously the electric field in a dielectric medium is reduced by a factor K from its value in free space. Physical reason for this reduction of electric field is the polarization

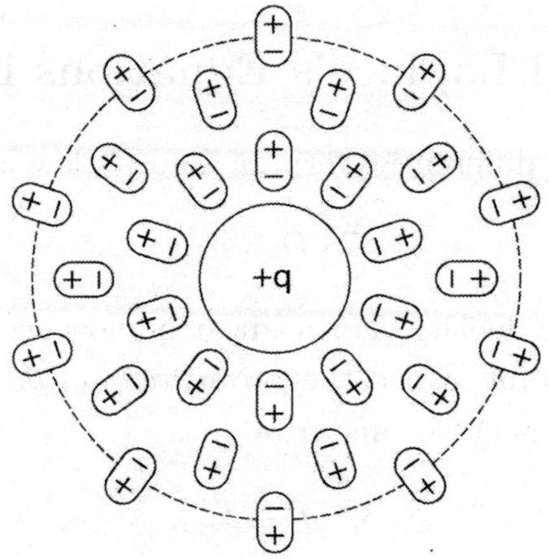

Fig. 2.9-1

of the dielectric. Under the influence of the charge $+q$ the negative charge centres of the constituent molecules of the medium are pulled towards $+q$ and the positive charge centres are pushed away (Fig. 2.9-1). As a result bound charges appear on the surface of the dielectric in contact with $+q$. This makes the effective positive charge producing the field smaller.

2.10 Nonpolar Liquid Dielectrics and the Clausius-Mossotti Relation

In a rarefied dielectric medium where molecular separations are large there is little difference between the macroscopic electric field $\vec{E}$ and that responsible for polarizing a molecule. But in dense media the polarization of neighbouring molecules makes the actual field acting on a molecule different from the average macroscopic field.

The electric field that is responsible for polarizing a molecule of the dielectric is called the *molecular field* or *local field*. For simple nonpolar liquid dielectrics we may expect that a molecule finds itself, on the average, within a spherical hole. To calculate the molecular field at a point, let us imagine to remove a small sphere of polarized material from the region surrounding the point under consideration. Let $\vec{E}$ be the uniform macroscopic field within the dielectric before the sphere was removed. By the superposition principle $\vec{E}$ may be considered as the sum the field $\vec{E}_1$ within the cavity due to all charges outside the cavity and the field $\vec{E}_2$ due to all charges within the removed polarized sphere, i.e.,

$$\vec{E} = \vec{E}_1 + \vec{E}_2. \tag{2.10-1}$$

To find $\vec{E}_2$ we consider a uniformly polarized sphere as shown in Fig. 2.10-1. Consider an elementary area $d\vec{S} = \hat{n}\,dS$ at any point $A(r,\theta,\varphi)$ on the surface of the sphere and take the z-axis along the direction of polarization $\vec{P}$. The surface density of polarization charge at A is $\vec{P}\cdot\hat{n} = P\cos\theta$. The field at O due to charges on dS is

$$d\vec{E}_2 = -\frac{1}{4\pi\varepsilon_0}\cdot\frac{P\cos\theta\,dS}{r^2}\hat{r}. \tag{2.10-2}$$

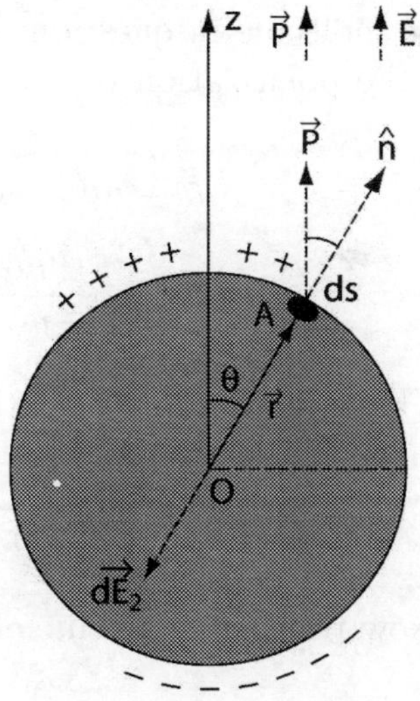

Fig 2.10-1 Uniformly polarized sphere

It is along the direction AO.

Because of symmetry about z-axis only the component of $d\vec{E}_2$ along the negative z-axis will contribute to the integral of Eq. (2.10-2) over the complete surface; the other component perpendicular to z-axis will give zero contribution when integrated over the whole surface. Since $dS = r^2 \sin\theta \, d\theta d\varphi$,

$$\vec{E}_2 = -\frac{1}{4\pi\varepsilon_0}\int \frac{P\cos\theta \, dS}{r^2}\cos\theta\,\hat{z}$$

$$= -\frac{\vec{P}}{4\pi\varepsilon_0}\int_0^{2\pi} d\varphi \int_0^{\pi} \cos^2\theta \sin\theta \, d\theta \quad [\because \vec{P} = P\hat{z}]$$

$$= -\frac{\vec{P}}{4\pi\varepsilon_0}\cdot 2\pi \cdot \frac{2}{3} = -\frac{\vec{P}}{3\varepsilon_0}. \tag{2.10-3}$$

Thus, the field inside is uniform. It is a depolarizing field.

Now from Eq. (2.10-1) we get the field $\vec{E}_1$ within the cavity as

$$\vec{E}_1 = \vec{E} - \vec{E}_2 = \vec{E} + \frac{\vec{P}}{3\varepsilon_0}. \tag{2.10-4}$$

For nonpolar dielectrics with simple structures such as for most dielectric liquids, the molecular field that polarizes a molecule may be taken as $\vec{E}_1$. Under the action of this field the molecule will be polarized. The dipole moment $\vec{p}$ induced in the molecule will be proportional to $\vec{E}_1$, i.e.,

$$\vec{p} = \alpha\vec{E}_1, \tag{2.10-5}$$

where the constant of proportionality α is called the *molecular polarizability.* It is the dipole moment per unit polarizing field. Its value depends on the detailed structure of the molecule in question. If there are n molecules per unit volume of the dielectric then the polarization vector

$$\vec{P} = n\vec{p} = n\alpha\vec{E}_1 = n\alpha\left(\vec{E} + \frac{\vec{P}}{3\varepsilon_0}\right) \quad \text{[Using Eq. (2.10-4)]}$$

$$\text{or} \quad \vec{P} = \frac{n\alpha}{1 - n\alpha/3\varepsilon_0}\vec{E}. \tag{2.10-6}$$

This equation may be written in terms of the dielectric constant K of the medium. We have the relation

$$\vec{D} = \varepsilon_0\vec{E} + \vec{P} \quad \text{or} \quad \vec{P} = \varepsilon\vec{E} - \varepsilon_0\vec{E} = \varepsilon_0(K-1)\vec{E} \quad [\because K = \varepsilon/\varepsilon_0]$$

Now putting this value of $\vec{P}$ in Eq. (2.10-6) we get

$$\varepsilon_0(K-1) = \frac{n\alpha}{1 - \frac{n\alpha}{3\varepsilon_0}} \quad \text{or} \quad \alpha = \frac{3\varepsilon_0}{n}\cdot\frac{K-1}{K+2}. \tag{2.10-7}$$

This is known as *Clausius-Mossotti* equation. This equation relates molecular polarizability (i.e., a molecular property) with a measurable macroscopic quantity K. Thus, from a measurement of K, it is possible to get important quantitative informations about molecular structures.

2.11 Polar Dielectrics and the Langevin-Debye Formula

Let us consider a polar gaseous or liquid dielectric in which each molecule has a permanent dipole moment $\vec{p}$. In the absence of any electric field the dipoles because of random thermal agitation, point in random directions and net dipole moment per unit volume is zero. When an electric field is applied two things happen: (a) Individual dipoles experience torques, which tend to align them along the field direction. Thermal agitation opposes the alignment and at ordinary temperatures and electric fields the alignment is far from complete. But there is some net alignment and so some net dipole moment per unit volume results. (b) There is some extra dipole moment induced due to increase in separation between positive and negative charge centres as in the case of nonpolar molecules. We shall neglect this second effect for the moment but its effect will be added in later.

The polarization that occurs due to orientation effect can be calculated as follows. According to a principle of statistical mechanics, in thermal equilibrium, the probability of finding a molecule with potential energy U is proportional to $e^{-U/kT}$, where k is the Boltzmann's constant and T is the absolute temperature. The potential energy of a dipole of moment $\vec{p}$ in an electric field $\vec{E}$ is

$$U = -\vec{p} \cdot \vec{E} = -pE\cos\theta, \qquad (2.11\text{-}1)$$

where θ is the angle between $\vec{p}$ and $\vec{E}$.

Note that the local field acting on a molecule in a gas is almost the same as the external field $\vec{E}$.

So the probability that a dipole will have orientation θ with respect to the field is proportional to $e^{pE\cos\theta/kT}$. The probability that the axis of a dipole lies between the angles θ and $\theta + d\theta$, i.e., within the conical solid angle (Fig. 2.11-1)

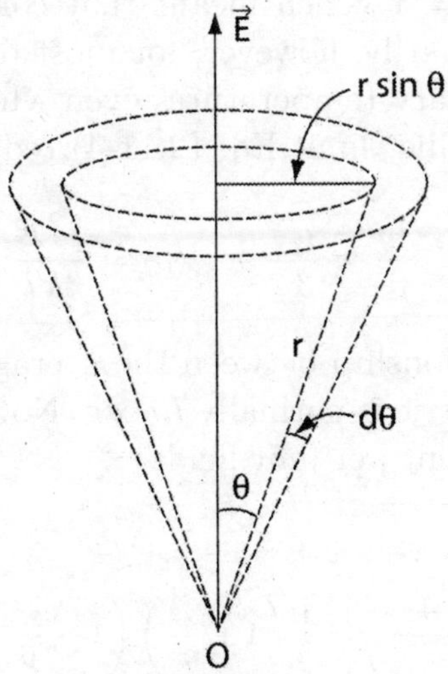

Fig. 2.11-1

$$d\Omega = \frac{2\pi r \sin\theta \cdot r d\theta}{r^2} = 2\pi \sin\theta d\theta \tag{2.11-2}$$

is proportional to $e^{pE\cos\theta/kT} d\Omega$.

The effective value of dipole moment is its component along the field direction, i.e., $p\cos\theta$. Now its average value is

$$\langle p \rangle = \frac{\int p\cos\theta e^{pE\cos\theta/kT} d\Omega}{\int e^{pE\cos\theta/kT} d\Omega} = p\frac{\int_0^\pi \cos\theta e^{pE\cos\theta/kT} \sin\theta d\theta}{\int_0^\pi e^{pE\cos\theta/kT} \cdot \sin\theta d\theta}. \tag{2.11-3}$$

Putting

$$\frac{pE}{kT} = a \text{ and } \quad \cos\theta = x,$$

we get

$$\langle p \rangle = p\frac{\int_1^{-1} xe^{ax} dx}{\int_1^{-1} e^{ax} dx} = p\frac{\left[\frac{xe^{ax}}{a} - \frac{e^{ax}}{a^2}\right]_1^{-1}}{\left[\frac{e^{ax}}{a}\right]_1^{-1}} = p\left[\frac{e^a + e^{-a}}{e^a - e^{-a}} - \frac{1}{a}\right] = p\left[\coth a - \frac{1}{a}\right]$$

$$\therefore \quad \frac{\langle p \rangle}{p} = \coth a - \frac{1}{a}, \tag{2.11-4}$$

where $a = pE/kT$.

Equation (2.11-4) is known as *Langevin formula* because it was first derived by Langevin for magnetic dipoles. Fig. 2.11-2 shows the variation of $\langle p \rangle / p$ as a function of a. At large field strengths and/or low temperatures, i.e., when $a = \frac{pE}{kT} \gg 1$ Langevin formula predicts that $\langle p \rangle / p = 1$ which means that complete alignment of dipoles or saturation takes place. Ordinarily, however, for most dielectric materials the value of p is such that $a \ll 1$ at ordinary temperatures even when E approaches the dielectric strength of the material. In this[2] limit Eq. (2.11-4) reduces to

$$\frac{\langle p \rangle}{p} \approx \frac{a}{3} \quad \text{or} \quad \langle p \rangle = \frac{p^2 E}{3kT}, \tag{2.11-5}$$

which indicates a linear relationship between the average effective dipole moment and the field. Thus, a polar dielectric is normally *linear.* Now the polarizability α is defined as the molecular dipole moment per unit field.

[2]N.B. $e^a = 1 + a + \frac{a^2}{2!} + \frac{a^3}{3!} + \cdots, \quad e^{-a} = 1 - a + \frac{a^2}{2!} - \frac{a^3}{3!} + \cdots$

$\therefore \quad \coth a = \frac{e^a + e^{-a}}{e^a - e^{-a}} = \frac{1 + \frac{a^2}{2!} + \cdots}{a + \frac{a^3}{3!} + \cdots} = \frac{1}{a}\left(1 + \frac{a^2}{2}\right)\left(1 - \frac{a^2}{6}\right) \approx \frac{1}{a} + \frac{a}{3}$

$\therefore \quad \coth a - \frac{1}{a} \approx \frac{a}{3}$

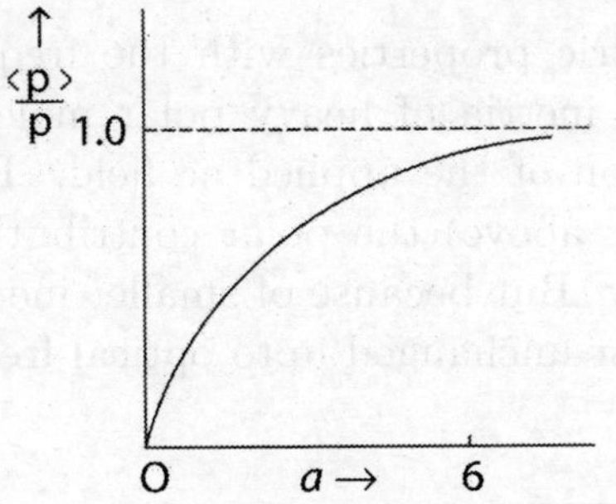

Fig. 2.11-2

Therefore,

$$\alpha = \frac{\langle p \rangle}{E} = \frac{p^2}{3kT}. \tag{2.11-6}$$

In deriving this formula we have considered only the orientational polarization. For this α is called *orientational polarizability.* If we consider deformational or electronic polarizability in addition to the orientational effect then the total molecular polarizability becomes

$$\alpha = \alpha_0 + \frac{p^2}{3kT}. \tag{2.11-7}$$

This is known as *Langevin-Debye equation.* It indicates that α *depends inversely on temperatures.* This is expected because at higher temperatures there will be more disalignment of dipoles due to collisions.

For polar molecules if α is plotted as a function of $1/T$ it will be a straight line. But Eq. (2.10-7) shows that for nonpolar molecules α versus $1/T$ graph will be a straight line parallel to $1/T$-axis (Fig. 2.11-3).

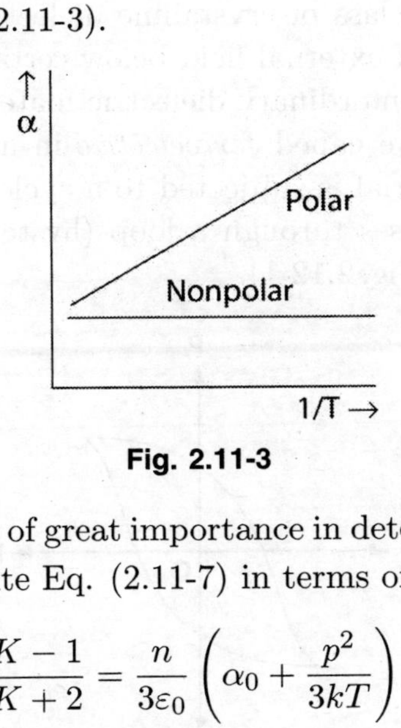

Fig. 2.11-3

Equation (2.11-7) has been of great importance in determining molecular structures. Using Eq. (2.10-7) we can write Eq. (2.11-7) in terms of dielectric constant as

$$\frac{K-1}{K+2} = \frac{n}{3\varepsilon_0}\left(\alpha_0 + \frac{p^2}{3kT}\right). \tag{2.11-8}$$

This equation is known as *Debye equation.*

The variation of dielectric properties with the frequency of an applied ac field is also interesting. Due to inertia of heavy polar molecules they cannot follow the rapid change in the direction of the applied ac field. For this at higher frequencies (in the microwave region or above) the polar contribution to the dielectric constant begins to fall with frequency. But because of smaller inertia of electrons the electronic polarizability remains almost unchanged upto optical frequencies.

2.12 Solid Dielectrics

There are some solid dielectrics which have permanent electric moments even in the absence of external electric fields. There is an interesting class of polar dielectrics which in the liquid state can have their molecules aligned with an external electric field. Now, if they are allowed to solidify with the field on, their molecules get locked in their aligned position. As a result the solid material will have a permanent polarization even in the absence of external electric field. Such a solid is called an *electret.* The polarization charges appearing on their surface are neutralised by free charges from air and there appears no visible external electric field. In some crystalline substances the internal polarization can be changed by thermal expansion or by mechanical stresses. Then external electric field appears because in these cases no sufficient time is available for neutralisation of polarization charges. The phenomenon of the appearance of electric effect due to thermal expansion of dielectric is known as *pyroelectricity.* The phenomenon of the appearance of electric effect due to mechanical stress on dielectric is known as *piezoelectric effect.*

There is another special class of crystalline dielectric which exhibits spontaneous polarization in the absence of external field below certain temperature and above this temperature it behaves like an ordinary dielectric material with a very large dielectric constant. Such dielectrics are called *ferroelectric* in analogy with the ferromagnetic effect. If a ferroelectric material is subjected to a cycle of polarizing and depolarizing fields the polarization P passes through a loop (hysteresis curve) and it always lags behind the polarizing field (Fig. 2.12-1).

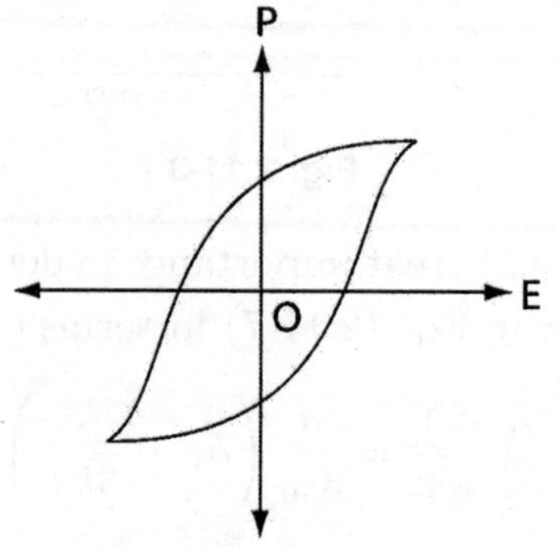

Fig. 2.12-1

2.13 The Boundary Conditions on $\vec{E}$ and $\vec{D}$

The boundary conditions on $\vec{E}$ and $\vec{D}$ tell us how these fields change across the boundary surface separating two media.

Let us consider two media, 1 and 2, with S_1S_2 as the surface of separation (Fig. 2.13-1). Let there is a free surface density of charges σ on the surface. We construct a small Gaussian pillbox on the surface at O. Its flat surface area ΔS is assumed to be small and its height approaches zero such that the flat faces are arbitrarily close to the boundary. Therefore, from Gauss's law,

$$\int_S \vec{D}\cdot\hat{n}\,dS = \sigma\cdot\Delta S \quad \text{or} \quad \vec{D}_2\cdot\hat{n}_2\Delta S + \vec{D}_1\cdot\hat{n}_1\Delta S = \sigma\Delta S \quad \text{or} \quad \left(\vec{D}_2 - \vec{D}_1\right)\cdot\hat{n} = \sigma,$$

where we take $\hat{n}_2 = -\hat{n}_1 = \hat{n}$, unit vector pointing from medium 1 to medium 2 normal to the interface.

$$\therefore \quad \vec{D}_2\cdot\hat{n} - \vec{D}_1\cdot\hat{n} = \sigma \tag{2.13-1}$$

$$\text{or} \quad D_2\cos\theta_2 - D_1\cos\theta_1 = \sigma \tag{2.13-2}$$

Thus, the normal components of $\vec{D}$ are discontinuous across the boundary by the surface charge density. If the two media have permittivities ε_1 and ε_2 then condition (2.13-1) can be expressed as

$$\varepsilon_2\vec{E}_2\cdot\hat{n} - \varepsilon_1\vec{E}_1\cdot\hat{n} = \sigma \quad \text{or} \quad -\varepsilon_2\vec{\nabla}\phi_2\cdot\hat{n} + \varepsilon_1\vec{\nabla}\phi_1\cdot\hat{n} = \sigma. \tag{2.13-3}$$

If there is no free surface charge on the boundary then $\vec{D}_2\cdot\hat{n} = \vec{D}_1\cdot\hat{n}$, i.e., the normal components of $\vec{D}$ are continuous across the boundary.

If the medium 1 is a conducting medium then $\vec{D}_1 = 0$ and the relevant boundary condition is

$$\vec{D}_2\cdot\hat{n} = \sigma. \tag{2.13-4}$$

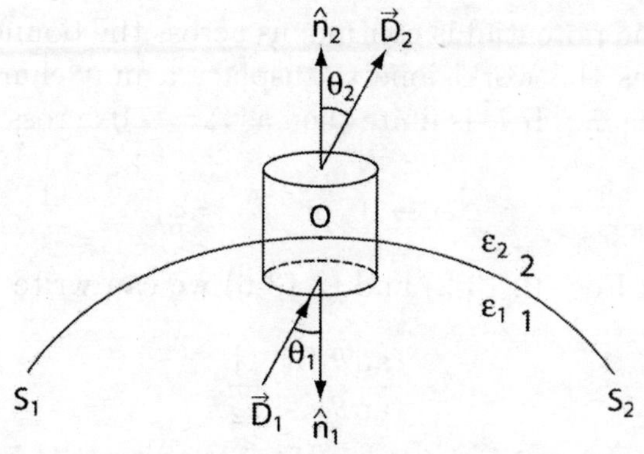

Fig. 2.13-1

To find conditions on $\vec{E}$ let us consider a small rectangular path $ABCD$ with sides AB and CD being perpendicular to the surface of separation S_1S_2 and other sides AD and BC being parallel to the surface at O (Fig. 2.13-2). Let the sides $AD = BC = \Delta l$ and AB and $CD(= \Delta h)$ negligibly small. Now for any electrostatic field $\vec{E}$

$$\oint \vec{E} \cdot d\vec{l} = 0.$$

Applying this result to the rectangular path $ABCD$ of Fig. 2.13-2 we get

$$\vec{E}_2 \cdot \overrightarrow{\Delta l} - \vec{E}_1 \cdot \overrightarrow{\Delta l} = 0$$

$$\text{or } \vec{E}_2 \cdot \hat{\tau} = \vec{E}_1 \cdot \hat{\tau} \tag{2.13-5}$$

$$\text{or } E_2 \sin\theta_2 = E_1 \sin\theta_1, \tag{2.13-6}$$

where $\hat{\tau}$ is a unit vector along the tangent to the interface. Thus, *the tangential component of the electric is continuous across the boundary.*

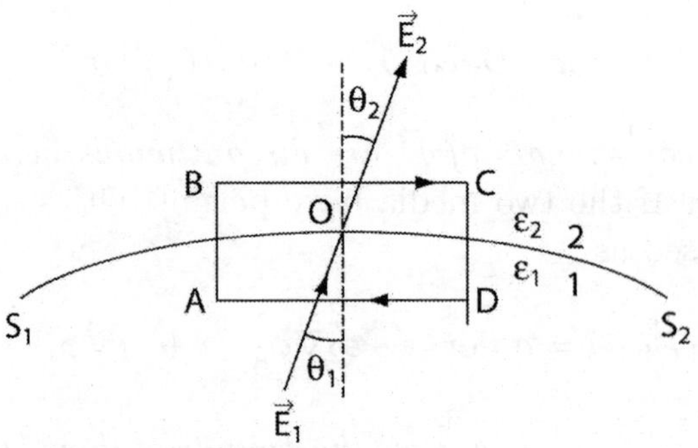

Fig. 2.13-2

In terms of potentials condition (2.13-6) takes the form

$$\phi_2 = \phi_1.$$

This means that the potential is continuous across the boundary. This follows from the definition of $\Delta\phi$ as the work done to displace a unit charge between two points separated by a distance $\Delta\vec{r}$. If $\vec{E}$ is finite then as $\Delta\vec{r} \to 0$ across the surface $\vec{E} \cdot \Delta\vec{r} \to 0$, implying that

$$\Delta\phi \to 0 \quad \text{or} \quad \phi_2 = \phi_1.$$

If $\sigma = 0$ then from Eqs. (2.13-2) and (2.13-6) we can write

$$\frac{\tan\theta_1}{\tan\theta_2} = \frac{\varepsilon_1}{\varepsilon_2}.$$

This is the law of refraction of the electric field at a boundary free of charge.

SOLVED PROBLEMS

▶ **Problem 1.** *An electric charge Q is given to a conducting sphere of radius R. Show that the force on a small charge element ΔQ on its surface is $\frac{Q \cdot \Delta Q}{8\pi\varepsilon_0 R^2}$.*

Solution. The charge Q gets distributed over the surface of the sphere and surface charge density becomes $\sigma = \frac{Q}{4\pi R^2}$. If ΔF be the force on an element of area ΔS containing $\Delta Q = \sigma \Delta S$ charge then outward mechanical pressure becomes

$$\frac{\Delta F}{\Delta S} = \frac{\sigma^2}{2\varepsilon_0}.$$

$$\therefore \quad \Delta F = \frac{\sigma}{2\varepsilon_0} \cdot \sigma \Delta S = \frac{\sigma}{2\varepsilon_0} \Delta Q = \frac{Q \cdot \Delta Q}{8\pi\varepsilon_0 R^2}.$$

▶ **Problem 2.** *If the radius of a spherical soap bubble be a show that the charge required to expand the bubble to twice its radius would be $8\pi \left[\varepsilon_0 a^3 (12T + 7aP)\right]^{1/2}$, where T is the surface tension of soap solution and P is the atmospheric pressure.*

Solution. In an uncharged soap bubble total internal pressure

$$P_i = \text{atmospheric pressure} + \text{excess pressure due to surface tension}$$
$$= P + \frac{4T}{a}$$

As P_i is inversely proportional to volume we can write

$$P_i = \frac{k}{a^3} \quad \text{where } k \text{ is a constant.}$$

$$\therefore \quad P + \frac{4T}{a} = \frac{k}{a^3} \tag{1}$$

Soap solution is a conductor of electricity. So when a charge Q is put on the bubble the charge is uniformly distributed on the bubble's surface and an *outward* mechanical pressure of value $\sigma^2/2\varepsilon_0$ acts on the bubble's surface, where σ is the surface charge density. So for equilibrium of the charged bubble of radius $2a$ is

$$P + \frac{4T}{2a} - \frac{\sigma^2}{2\varepsilon_0} = \frac{k}{(2a)^3} \tag{2}$$

Eliminating k from Eqs. (1) and (2) we get

$$7P + \frac{12T}{a} = \frac{4\sigma^2}{\varepsilon_0}.$$

Now using $\sigma = \dfrac{Q}{4\pi(2a)^2} = \dfrac{Q}{16\pi a^2}$ we get

$$Q = 8\pi \left[\varepsilon_0 a^3 (12T + 7aP)\right]^{1/2}$$

► **Problem 3.** *A soap bubble of radius r is electrified with a small charge Q. Assuming internal pressure within the bubble before charging to be P show that the increase in radius due to electrification is* $Q^2/96\varepsilon_0 P\pi^2 r^3$.

Solution. As the total internal pressure is inversely proportional to volume, we can write $P = k/r^3$, where k is a constant

$$\therefore \quad \frac{dP}{dr} = -\frac{3k}{r^4} = -\frac{3P}{r}$$

$$\text{or} \quad dr = -\frac{r}{3P}dP. \tag{1}$$

Charging generates an outward mechanical pressure and hence, a drop in internal pressure. Thus,

$$-dP = \frac{\sigma^2}{2\varepsilon_0} = \frac{(Q/4\pi r^2)^2}{2\varepsilon_0} = \frac{Q^2}{32\pi^2\varepsilon_0 r^4}. \tag{2}$$

From (1) and (2) we get

$$dr = \frac{Q^2}{96\varepsilon_0 P\pi^2 r^3}.$$

► **Problem 4.** *Suppose the potential at any point* (r, θ) *outside a conducting sphere of radius a is given by* $\phi(r,\theta) = -E_0 \cos\theta \cdot (r - a^3/r^2)$, *where* E_0 *is a constant and* (r, θ) *are the polar coordinates of the point with respect to the centre as origin and a fixed diameter as polar axis. Find the surface density of charge at any point on the sphere. Also, find the total charge on the sphere.* **[J.U. 2013]**

Solution. Normal component of electric field just outside the surface is

$$E_n = -\left.\frac{\partial\phi}{\partial r}\right|_{r=a} = E_0 \cos\theta \left.\left(1 + \frac{2a^3}{r^3}\right)\right|_{r=a} = 3E_0 \cos\theta.$$

From Coulomb's theorem this equals $\sigma(\theta)/\varepsilon_0$, where σ is the local surface density of charge. Therefore

$$\sigma(\theta) = 3\varepsilon_0 E_0 \cos\theta.$$

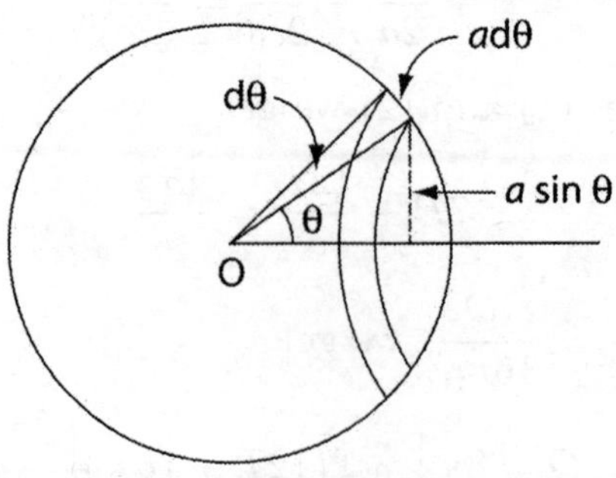

Fig. 2.P-4

Considering an annular ring between θ and $\theta + d\theta$ as shown in Fig. 2.P-4 we find that the total charge on the sphere is

$$\int_0^{\pi} 2\pi(a \sin\theta)a\, d\theta \cdot \sigma(\theta) = 6\pi a^2 \varepsilon_0 E_0 \int_0^{\pi} \sin\theta \cos\theta\, d\theta = 0$$

▶ **Problem 5.** *A point charge q is placed in a medium whose permittivity ε changes with the distance r from q as $\varepsilon = 1 + A/r$, where A is a constant. Show that the potential at any point is given by $\phi(r) = \frac{q}{4\pi A}\ln\left(1 + \frac{A}{r}\right)$.*

Solution. Considering a spherical Gaussian surface with the charge at its centre we can write from Gauss's law

$$\int_S \vec{D} \cdot d\vec{S} = q.$$

Because of spherical symmetry $\vec{D}$ is constant all over the Gaussian surface and is normal to it at every point. So,

$$D \cdot 4\pi r^2 = q \quad \text{or} \quad D = \frac{q}{4\pi r^2}$$

$$\therefore \quad E = \frac{D}{\varepsilon} = \frac{q}{4\pi\varepsilon r^2} = \frac{q}{4\pi\left(1 + \frac{A}{r}\right)r^2} = \frac{q}{4\pi r(r + A)} = \frac{q}{4\pi A}\left[\frac{1}{r} - \frac{1}{r + A}\right]$$

As $\vec{E}$ is in radial direction, $E = -\frac{d\phi}{dr}$.

$$\therefore \quad \phi(r) = -\int E dr = -\frac{q}{4\pi A}[\ln r - \ln(r + A)] + C = \frac{q}{4\pi A}\ln\left(1 + \frac{A}{r}\right) + C,$$

where C is a constant of integration.

Now applying the condition $\phi \to 0$ as $r \to \infty$ we get $C = 0$

$$\therefore \quad \phi(r) = \frac{q}{4\pi A}\ln\left(1 + \frac{A}{r}\right).$$

▶ **Problem 6.** *Consider two coplanar dipoles of moments $\vec{p}_1$ and $\vec{p}_2$. The first dipole is fixed at the origin and the second dipole, placed at the position $\vec{r}$, is free to rotate. If $\vec{r}$ makes angles θ_1 and θ_2 with $\vec{p}_1$ and $\vec{p}_2$, respectively then show that for equilibrium* $\tan\theta_1 = -2\tan\theta_2$. **[C.U. 2012, B.U. 2014]**

Solution. Referring to Fig. 2.3-6, the potential energy of the system is

$$U = \frac{p_1 p_2}{4\pi\varepsilon_0 r^3}\left[\sin\theta_1 \sin\theta_2 - 2\cos\theta_1 \cos\theta_2\right]$$

For equilibrium we must have $\frac{dU}{d\theta_2} = 0$. This gives $\tan\theta_1 = -2\tan\theta_2$.

► **Problem 7.** *Find the mutual potential energy of two dipoles of equal moment* 6×10^{-30} *C.m lying along the same line at a separation of* 3×10^{-10} *m.*

Solution. General expression for P.E. of two coplanar dipoles is

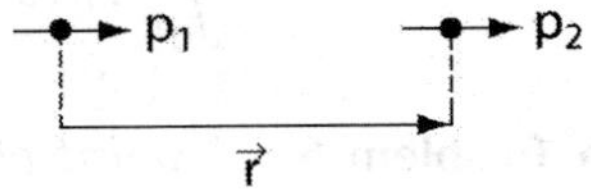

Fig. 2.P-7

$$U=\frac{1}{4\pi\varepsilon_0 r^3}\left[\frac{\vec{p}_1\cdot\vec{p}_2}{r^3}-\frac{3(\vec{p}_1\cdot\vec{r})(\vec{p}_2\cdot\vec{r})}{r^5}\right].$$

Here $\vec{p}_1=\vec{p}_2$, $\vec{p}_1\parallel^l\vec{r}$

$$\therefore\quad U=-\frac{1}{4\pi\varepsilon_0}\cdot\frac{2p^2}{r^3}=\frac{-9\times10^9\times2\times(6\times10^{-30})^2}{(3\times10^{-10})^3}\,\mathrm{J}$$

$$=-24\times10^{-21}\,\mathrm{J}$$

$$=-\frac{24\times10^{-21}}{1.6\times10^{-19}}\,\mathrm{eV}=-0.15\,\mathrm{eV}.$$

► **Problem 8.** *Consider a point charge q situated at the origin and a dipole of moment* $\vec{p}$ *is situated at* $\vec{r}$ *with* $\vec{p}$ *oriented along* $\vec{r}$*. Find the force and torque on the dipole due to q.*

Solution. Field at the position of the dipole is

$$\vec{E}=\frac{1}{4\pi\varepsilon_0}\frac{q}{r^3}\vec{r}$$

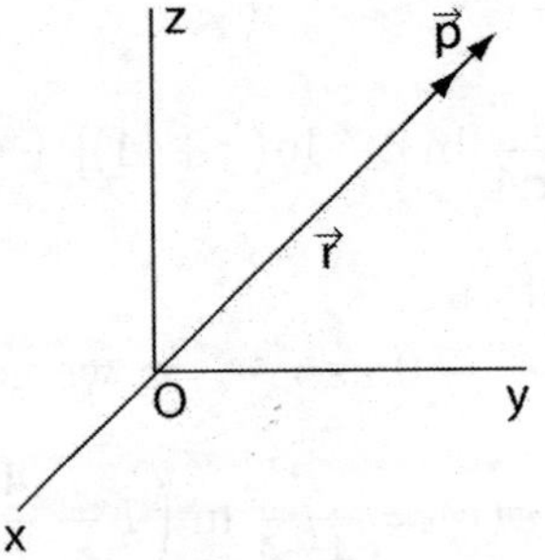

Fig. 2.P-8

Now force on the dipole is

$$\vec{F}=\left(\vec{p}\cdot\vec{\nabla}\right)\vec{E}=\frac{q}{4\pi\varepsilon_0}\left(p_x\frac{\partial}{\partial x}+p_y\frac{\partial}{\partial y}+p_z\frac{\partial}{\partial z}\right)\left(\frac{x\hat{i}+y\hat{j}+z\hat{k}}{r^3}\right)$$

$$=\frac{q}{4\pi\varepsilon_0}\left[p_x\left(\frac{\hat{i}}{r^3}-\frac{3\vec{r}}{r^4}\cdot\frac{x}{r}\right)+p_y\left(\frac{\hat{j}}{r^3}-\frac{3\vec{r}}{r^4}\cdot\frac{y}{r}\right)+p_z\left(\frac{\hat{k}}{r^3}-\frac{3\vec{r}}{r^4}\cdot\frac{z}{r}\right)\right]$$

$$=\frac{q}{4\pi\varepsilon_0}\left[\frac{\vec{p}}{r^3}-\frac{3\vec{r}}{r^5}(\vec{p}\cdot\vec{r})\right].$$

Here $\vec{p}=p\hat{r}$.

$\therefore\ \vec{F} = -\dfrac{q}{4\pi\varepsilon_0} \cdot \dfrac{2p}{r^3}\hat{r}$, which is attractive in nature.

Torque $\vec{c\tau} = \vec{p} \times \vec{E} = p\hat{r} \times \dfrac{q}{4\pi\varepsilon_0} \dfrac{1}{r^2}\hat{r} = 0$

► **Problem 9.** *The dielectric constant of a monoatomic gas at N.T.P. is* 1.000538. *Calculate the dipole moment induced in each atom when the gas is placed in an external electric field of* 30 kV/m.

Solution. Dielectric constant $K = 1 + \mathcal{X}$, where $\mathcal{X}$ is the susceptibility

$$\therefore\ \mathcal{X} = K - 1 = 0.000538.$$

Now polarization $P = \varepsilon_0 \mathcal{X} E$ is the dipole moment per unit volume. At N.T.P. 22.4 litres of the gas contains 6.023×10^{23} atoms. Therefore, number of atoms per unit volume is

$$n = \frac{6.023 \times 10^{23}}{22.4 \times 10^{-3}} \text{ per m}^3 = 2.7 \times 10^{25} \text{ per m}^3.$$

$\therefore$ Induced dipole moment per atom

$$= \frac{P}{n} = \frac{\varepsilon_0 \mathcal{X} E}{n} = \frac{8.85 \times 10^{-12} \times 0.000538 \times 30 \times 10^3}{2.7 \times 10^{25}} = 5.3 \times 10^{-38} \text{ C.m.}$$

► **Problem 10.** *Consider a sphere of radius a made of inhomogeneous dielectric whose dielectric constant is given by*

$$K = \frac{1}{1 + Ar},$$

where A is a constant and r is the radial distance. A charge q is placed at the centre of the sphere. Find $\vec{D}$, $\vec{E}$ *and* $\vec{P}$ *at any internal point. Also, find the polarization charge densities.*

Solution. Considering a concentric spherical Gaussian surface and spherical symmetry of the problem we may write

$$\int \vec{D} \cdot d\vec{S} = q \quad \text{or} \quad D = \frac{q}{4\pi r^2}.$$

In vector form $\vec{D} = \dfrac{q}{4\pi r^2}\hat{r}$.

Electric field, $\vec{E} = \dfrac{\vec{D}}{\varepsilon} = \dfrac{\vec{D}}{\varepsilon_0 K} = \dfrac{q}{4\pi\varepsilon_0 r^2}(1 + Ar)\hat{r}$.

Polarization,

$$\vec{P} = \vec{D} - \varepsilon_0\vec{E} = (\varepsilon - \varepsilon_0)\vec{E} = \varepsilon_0(K - 1)\vec{E}$$

$$= \varepsilon_0 \frac{(-Ar)}{1 + Ar} \cdot \frac{q}{4\pi\varepsilon_0 r^2}(1 + Ar)\hat{r} = -\frac{qA}{4\pi r}\hat{r}.$$

Volume density of polarization charge is

$$\rho_p = -\vec{\nabla} \cdot \vec{P} = -\frac{1}{r^2}\frac{d}{dr}(r^2 P) = \frac{1}{r^2}\frac{d}{dr}\left(\frac{qAr}{4\pi}\right) = \frac{qA}{4\pi r^2}.$$

Surface density of polarization charge at $r = a$ (surface of the sphere) is

$$\sigma_p = \vec{P} \cdot \hat{n} = \vec{P} \cdot \hat{r} = -\frac{qA}{4\pi a}.$$

▶ **Problem 11.** *A small spherical cavity is cut in a dielectric where there is a uniform macroscopic field $\vec{E}$. If $\vec{P}$ be the uniform polarization in the dielectric show that the electric field at the centre of the cavity is $\vec{E} + \vec{P}/3\varepsilon_0$.*

Solution. The field $\vec{E}_0$ within the cavity will be given by the sum of the field $\vec{E}$ and the field $\vec{E}_p$ due to polarization charges appearing on the surface of the cavity. The surface density of polarization charge appearing on the surface of the cavity is $\sigma_p = \vec{P} \cdot \hat{n}$, where $\hat{n}$ is the outward normal to the dielectric. Obviously here $\hat{n}$ is radially inward to the centre of the cavity. Therefore, the charge density on an elementary area $d\vec{S} = \hat{n}\, dS$ chosen at any point $A(r, \theta, \varphi)$ on the surface of the cavity will be

$$\sigma_p = -P\cos\theta. \tag{1}$$

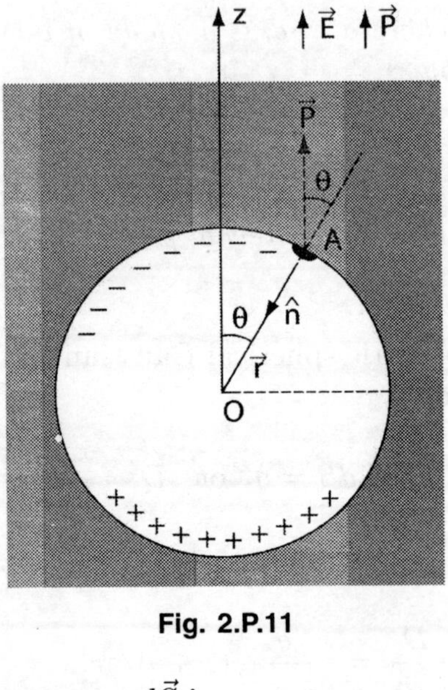

Fig. 2.P.11

The field at O due to charges on $d\vec{S}$ is

$$d\vec{E}_p = \frac{1}{4\pi\varepsilon_0}\frac{P\cos\theta\, dS}{r^2}\hat{r}, \tag{2}$$

which is directed along $\overrightarrow{OA}$.

Because of symmetry about z-axis only the component of $d\vec{E}_p$ along z-axis will contribute to the integral of (2) over the complete surface.

Using $dS = r^2 \sin\theta\, d\theta\, d\varphi$ we get

$$\vec{E}_p = \frac{1}{4\pi\varepsilon_0}\int \frac{P\cos\theta\, dS}{r^2}\cos\theta\,\hat{z} = \frac{\vec{P}}{4\pi\varepsilon_0}\int_0^{2\pi} d\varphi \int_0^{\pi} \cos^2\theta \sin\theta\, d\theta$$

$$= \frac{\vec{P}}{4\pi\varepsilon_0}\cdot 2\pi \cdot \frac{2}{3} = \frac{\vec{P}}{3\varepsilon_0}.$$

Therefore,

$$\vec{E}_0 = \vec{E} + \frac{\vec{P}}{3\varepsilon_0}.$$

▶ **Problem 12.** *Consider a simple atom model consisting of a point nucleus of charge $+Ze$ surrounded by the electronic charge $-Ze$ distributed uniformly over a sphere of radius R. When placed in an external polarizing field $\vec{E}$ the nucleus moves relative to the centre of the electronic charge. Assuming that the electron cloud moves rigidly without deformation by the polarizing field show that the atomic polarizability is $\alpha = 4\pi\varepsilon_0 R^3$.*

Solution. Under the influence of the field $\vec{E}$ the nucleus moves relative to the centre of the negative charge cloud until the attractive Coulomb force between the cloud and the nucleus balances the force $Ze \cdot E$. Let x be the separation between the charge centres in equilibrium. By Gauss's law the negative charge attracting the nucleus is that part of the cloud within the sphere of radius x. This charge is equal to

$$\frac{-Ze}{\frac{4}{3}\pi R^3} \times \frac{4}{3}\pi x^3 = \frac{-Ze \cdot x^3}{R^3}.$$

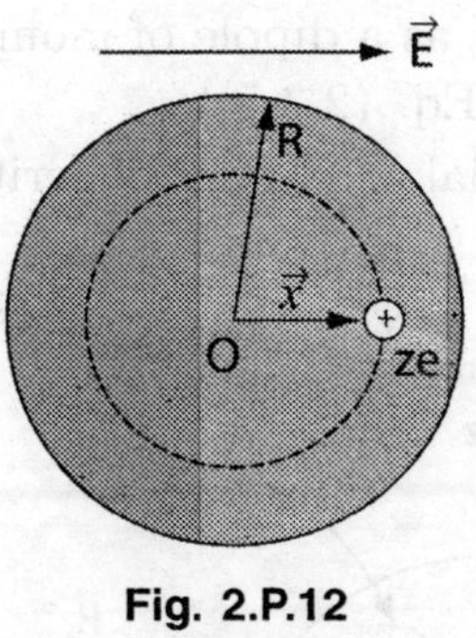

Fig. 2.P.12

Therefore, for equilibrium

$$Ze \cdot E = \frac{1}{4\pi\varepsilon_0}\cdot\frac{Ze \cdot \frac{Zex^3}{R^3}}{x^2} \quad \text{or} \quad x = \frac{4\pi\varepsilon_0 R^3 E}{Ze}.$$

So the atomic dipole moment induced in the process is

$$\vec{p} = Ze\vec{x} = 4\pi\varepsilon_0 R^3 \vec{E} = \alpha\vec{E},$$

where $\alpha = 4\pi\varepsilon_0 R^3$ is the required atomic polarizability.

► **Problem 13.** *The interior of a cylinder defined by* $x^2 + y^2 = R^2$, $z = 0$, $z = h$ *is occupied by a dielectric material with polarization* $\vec{P} = x^2\hat{\imath} + y^2\hat{\jmath}$. *Find the volume and surface densities of polarization charges.*

Solution. Volume density of polarization charge is

$$\rho_p = -\vec{\nabla} \cdot \vec{P} = -2(x + y).$$

For the plane surfaces, $\sigma_p = \vec{P} \cdot \hat{n} = 0$ because $\hat{n} = \hat{k}$ or $-\hat{k}$

For the curved surface,

$$\hat{n} = \frac{\vec{\nabla}\psi}{|\vec{\nabla}\psi|} \quad \text{where } \psi = x^2 + y^2 - R^2$$

$$= \frac{2x\hat{\imath} + 2y\hat{\jmath}}{\sqrt{4x^2 + 4y^2}} = \frac{1}{R}(x\hat{\imath} + y\hat{\jmath}).$$

$\therefore$ On the curved surface,

$$\sigma_p = \vec{P} \cdot \hat{n} = \frac{1}{R}(x^3 + y^3).$$

► **Problem 14.** *Consider a dielectric sphere of radius a with uniform polarization* $\vec{P} = P\hat{k}$. *Find the electric potential at any external point.*

Solution. Referring to Fig. 2.P-14 we can write for the potential at any point $A(\vec{r})$ due to an elementary volume element dV at the point $\vec{r}\,'$ as

$$d\phi = \frac{1}{4\pi\varepsilon_0}(\vec{P}dV) \cdot \vec{\nabla}\,'\frac{1}{R},$$

where we consider the element dV as a dipole of moment $\vec{P}\,dV$ and use the expression for potential due to a dipole [See Eq. (2.3-5)].

As $\vec{\nabla}\,'\frac{1}{R} = -\vec{\nabla}\frac{1}{R}$, total potential at A may be written as

$$\phi(\vec{r}) = -\frac{\vec{P}}{4\pi\varepsilon_0} \cdot \int_{\text{Source}} \vec{\nabla}\frac{1}{R}\,dV \quad [\because\ \vec{P} \text{ is a constant vector}]$$

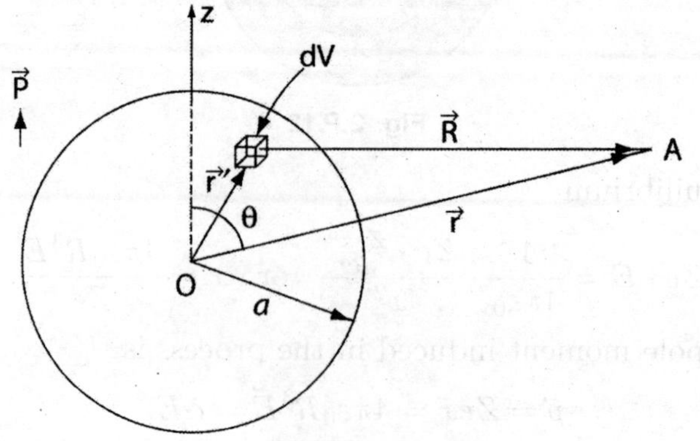

Fig. 2.P.14

Since $\vec{\nabla}$ refers to differentiation with respect to field point coordinates (x, y, z) and the integration is with respect to source point coordinates (x', y', z') we can write

$$\phi(\vec{r}) = -\frac{\vec{P}}{4\pi\varepsilon_0} \cdot \vec{\nabla} \int_{\text{Source}} \frac{dV}{R}.$$

The integration involved is not a trivial one. However, we observe that the quantity

$$\psi = \frac{1}{4\pi\varepsilon_0} \int_{\text{Source}} \frac{dV}{R}$$

may be identified as the potential at A if the sphere were uniformly charged with unity volume density and $-\vec{\nabla}\psi$ is the corresponding field. Also, we know that for calculation of field at an external point the whole charge on the sphere may be assumed to be concentrated at the centre. So we may write

$$-\frac{1}{4\pi\varepsilon_0}\vec{\nabla}\int \frac{dV}{R} = \frac{1}{4\pi\varepsilon_0} \cdot \frac{\frac{4}{3}\pi a^3 \cdot 1}{r^3}\vec{r}.$$

Thus,

$$\phi(\vec{r}) = \frac{1}{4\pi\varepsilon_0}\frac{\vec{P}_t \cdot \vec{r}}{r^3},$$

where $\vec{P}_t = \frac{4}{3}\pi a^3 \vec{P}$ is the total dipole moment of the sphere.

► **Problem 15.** *Show that at the interface between two ideal dielectrics the electric field lines are refracted in such away that $\varepsilon \cot\theta$ remains invariant, where θ is the angle which the direction of the field makes with the normal to the interface.*

Solution. For an uncharged interface between two dielectrics with permittivities ε_1 and ε_2 we have from boundary conditions on $\vec{E}$ and $\vec{D}$,

$$E_1 \sin\theta_1 = E_2 \sin\theta_2 \tag{1}$$

$$\text{and} \quad \varepsilon_1 E_1 \cos\theta_1 = \varepsilon_2 E_2 \cos\theta_2 \tag{2}$$

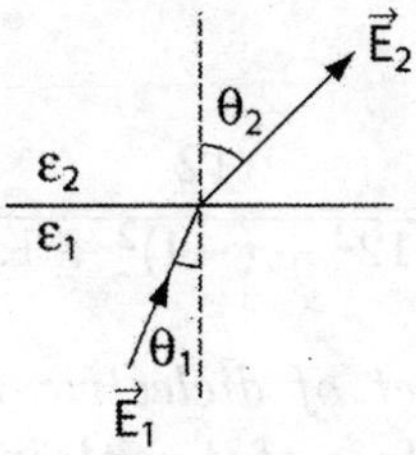

Fig. 2.P.15

Dividing (2) by (1) we get $\varepsilon_1 \cot\theta_1 = \varepsilon_2 \cot\theta_2 = \text{constant}.$

► **Problem 16.** *A dielectric sphere of radius a and permittivity ε_1 is surrounded by a dielectric shell of inner and outer radii a and b, respectively. The potential inside the sphere and the shell are given, respectively, by $\phi_1(r,\theta) = Ar\theta$ and $\phi_2(r,\theta) = Aa^2\theta/r$. Find the free charge density on the interface.*

Solution. From boundary conditions on $\vec{D}$ we have

$$\vec{D}_2 \cdot \hat{n} - \vec{D}_1 \cdot \hat{n} = \sigma_f$$

$$\text{or} \quad -\varepsilon_2 \frac{\partial \phi_2}{\partial r}\bigg|_{r=a} + \varepsilon_1 \frac{\partial \phi_1}{\partial r}\bigg|_{r=a} = \sigma_f$$

$$\text{or} \quad -\varepsilon_2 \left(-\frac{Aa^2\theta}{r^2}\right)_{r=a} + \varepsilon_1 (A\theta)_{r=a} = \sigma_f$$

$$\text{or} \quad \sigma_f = (\varepsilon_2 + \varepsilon_1) A\theta$$

► **Problem 17.** *Suppose $x = 0$ represents the interface of a dielectric medium and vacuum. The region 1, $x \le 0$, is occupied by a medium of dielectric constant 2.5 while region 2, $x \ge 0$, represents the vacuum. If $\vec{D}_1 = 12\hat{\imath} - 10\hat{\jmath} + 4\hat{k}\,n\ \mathrm{C/m^2}$, find $\vec{D}_2$ and θ_2 (angle made by $\vec{D}_2$ with the normal to the interface)* **[cf. C.U. 2016]**

Solution. Continuity of the normal component of $\vec{D}$ requires that

$$\vec{D}_{2n} = \vec{D}_{1n} = 12\hat{\imath} \quad [\because \hat{\imath} \text{ is normal to the interface}]$$

Continuity of the tangential component of $\vec{E}$ requires that

$$\vec{E}_{2t} = \vec{E}_{1t} = \frac{\vec{D}_{1t}}{\varepsilon_1} = \frac{1}{2.5\varepsilon_0}\left[-10\hat{\jmath} + 4\hat{k}\right]$$

$$\therefore \quad \vec{D}_{2t} = \varepsilon_0 \vec{E}_{2t} = -4\hat{\jmath} + 1.6\hat{k}$$

$$\therefore \quad \vec{D}_2 = \vec{D}_{2n} + \vec{D}_{2t} = 12\hat{\jmath} - 4\hat{\jmath} + 1.6\hat{k}$$

Angle made by $\vec{D}_2$ with $\hat{\imath}$ is

$$\theta_2 = \cos^{-1} \frac{12}{\sqrt{12^2 + (-4)^2 + 1.6^2}} = 19.75^\circ$$

► **Problem 18.** *Consider a thin sheet of dielectric material which has been polarized normally everywhere such that one face of it contains all positive charges and the other all negative charges. If P_s be the dipole moment per unit area show that the potential due to this 'electric double layer' at an external point is given by $\phi = \frac{P_s \Omega}{4\pi\varepsilon_0}$, where Ω is the solid angle subtended by the whole sheet at the point.*

Solution. The dipole moment associated with an elementary area $d\vec{S} = \hat{n}\,dS$ of the sheet is $d\vec{P} = P_s dS\hat{n}$. The potential at P due to this elementary dipole is

$$d\phi = \frac{1}{4\pi\varepsilon_0}\frac{d\vec{P}\cdot\vec{r}}{r^3} = \frac{P_s}{4\pi\varepsilon_0}\frac{d\vec{S}\cdot\vec{r}}{r^3} = \frac{P_s}{4\pi\varepsilon_0}d\Omega,$$

where $d\Omega$ is the solid angle subtended by $d\vec{S}$ at P. Note that $d\Omega$ is taken to be positive if $d\vec{P}\cdot\vec{r} \geq 0$, otherwise it is to be taken as negative. Therefore, the total potential at P due to the whole sheet is

$$\phi = \frac{1}{4\pi\varepsilon_0}\int P_s d\Omega.$$

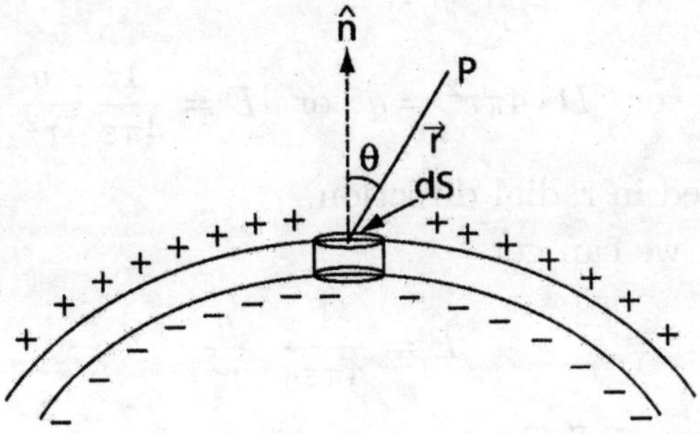

Fig. 2.P-18

If P_s is constant all over the sheet then $\phi = \frac{P_s}{4\pi\varepsilon_0}\Omega$, where Ω is the total solid angle subtended by the whole sheet at the point of observation.

▶ **Problem 19.** *For a dielectric specimen of volume V and surface S, prove the following relationship between polarization $\vec{P}$ and the polarization charge densities ρ_p and σ_p :*

$$\int_V \vec{P}\,dV = \int_V \rho_p\vec{r}\,dV + \int_S \sigma_p\vec{r}\,dS,$$

where $\vec{r}$ is the position vector from any fixed origin.

Solution. From vector analysis

$$\vec{\nabla}\cdot x\vec{P} = \vec{\nabla}x\cdot\vec{P} + x\vec{\nabla}\cdot\vec{P} = \hat{i}\cdot\vec{P} + x\vec{\nabla}\cdot\vec{P} = P_x + x\vec{\nabla}\cdot\vec{P}$$

$$\therefore \quad \int_V \vec{\nabla}\cdot x\vec{P}\,dV = \int P_x dV + \int x\vec{\nabla}\cdot\vec{P}\,dV$$

$$\text{or} \quad \int_S x\vec{P}\cdot\hat{n}\,dS = \int P_x dV + \int x\vec{\nabla}\cdot\vec{P}\,dV \quad \text{[Using divergence theorem]}$$

Since $\rho_p = -\vec{\nabla}\cdot\vec{P}$ and $\sigma_p = \hat{n}\cdot\vec{P}$,

$$\int P_x\,dV = \int_V x\rho_p\,dV + \int_S x\sigma_p\,dS.$$

Similarly, we can find

$$\int P_y\,dV \quad \text{and} \quad \int P_z\,dV.$$

Then combining all these we get

$$\int_V \vec{P}\,dV = \int_V \rho_p \vec{r}\,dV + \int_S \sigma_p \vec{r}\,dS.$$

Note that it gives the dipole moment of the polarized body.

► **Problem 20.** *A conducting sphere of radius a carries a charge q. It is coated with a linear dielectric material of permittivity ε out to a radius b. Find the potential at the centre of the conductor. What is the capacitance of the system?* **[cf. C.U. 2015]**

Solution. Applying Gauss's law to a spherical surface of radius r, where $a < r < b$ we get

$$\oint \vec{D}\cdot d\vec{S} = q \quad \text{or} \quad D\cdot 4\pi r^2 = q \quad \text{or} \quad E = \frac{1}{4\pi\varepsilon}\cdot\frac{q}{r^2} \quad \text{for } a < r < b,$$

where $\vec{D}$ or $\vec{E}$ is directed in radial direction.

Similarly, for $r > b$, we can get

$$E = \frac{1}{4\pi\varepsilon_0}\cdot\frac{q}{r^2}$$

Inside the metal sphere, $\vec{E} = 0$.

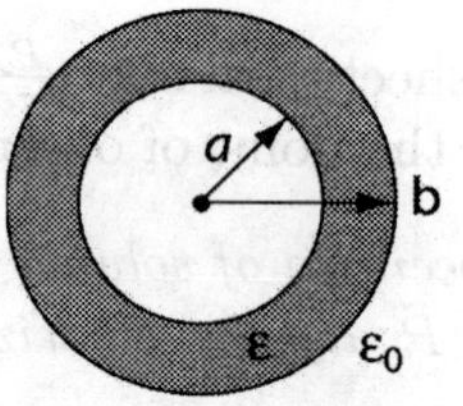

Fig. 2.P-20

Now the potential at the centre of the sphere is

$$\phi = -\int_\infty^a \vec{E}\cdot d\vec{r} = -\int_\infty^b \frac{1}{4\pi\varepsilon_0}\frac{q}{r^2}dr - \int_b^a \frac{1}{4\pi\varepsilon}\cdot\frac{q}{r^2}dr - \int_a^o 0\cdot dr$$

$$= \frac{q}{4\pi\varepsilon_0 b} + \frac{q}{4\pi\varepsilon}\left(\frac{1}{a} - \frac{1}{b}\right)$$

This is the potential at any point inside the conducting sphere.

The capacitance of the system $C = q/\phi = 4\pi\varepsilon\varepsilon_0 ab/(\varepsilon a + \varepsilon_0 b - \varepsilon_0 a)$.

► **Problem 21.** *An electret sphere of radius a centred at the origin carries a frozen-in polarization $\vec{P} = k\vec{r}$, where k is constant and r is the distance from the centre*

(a) *Calculate the polarization charge densities and electric field both inside and outside the sphere due to polarization charges.* **[C.U. 2014, 2016; J.U. 2013]**

(b) *Find $\vec{D}$ by using Gauss's law and then find $\vec{E}$.*

Solution. (a) Using spherical coordinates we find that polarization volume charge density is

$$\rho_p = -\vec{\nabla}\cdot\vec{P} = -\frac{1}{r^2}\frac{d}{dr}(r^2 P_r) = -\frac{1}{r^2}\frac{d}{dr}(r^2\cdot kr) = -3k.$$

Polarization surface charge density is $\sigma_p = \vec{P}\cdot\hat{n} = ka\hat{r}\cdot\hat{r} = ka$. The problem has a spherical symmetry. So considering a spherical Gaussian surface of radius $r(r < a)$ we can write

$$\oint \vec{E}\cdot d\vec{S} = \frac{1}{\varepsilon_0}\left(q_{\text{free}} + q_{\text{polarization}}\right).$$

In this problem $q_{\text{free}} = 0$,

$$q_{\text{polarization}} = \int \rho_p dV = -3k\cdot\frac{4}{3}\pi r^3 = -4\pi k r^3.$$

Therefore,

$$E\cdot 4\pi r^2 = \frac{1}{\varepsilon_0}(0 - 4\pi k r^3) \quad \text{or} \quad E = -\frac{kr}{\varepsilon_0}.$$

Vectorially,

$$\vec{E} = -\frac{kr}{\varepsilon_0}\hat{r} \text{ for } r < a.$$

Total polarization charge

$$\int \rho_p\, dV + \int \sigma_p\, dS = -3k\cdot\frac{4}{3}\pi a^3 + ka\cdot 4\pi a^2 = 0.$$

So for an external point

$$\oint \vec{E}\cdot d\vec{S} = 0.$$

This indicates that $\vec{E} = 0$ for $r > a$.

(b) $\oint \vec{D}\cdot d\vec{S} = Q_{\text{free}} = 0$

$\therefore$ $\vec{D} = 0$ everywhere.

Now

$$\vec{D} = \varepsilon_0\vec{E} + \vec{P} = 0.$$

$\therefore$ Inside the sphere

$$\vec{E} = -\frac{\vec{P}}{\varepsilon_0} = \frac{kr}{\varepsilon_0}\hat{r}.$$

Outside the sphere $\vec{P} = 0$ and hence $\vec{E} = 0$.

▶ **Problem 22.** *When a neutral dielectric is polarized, the polarization volume and surface charges appear. Show that total charge, however, remains zero.* **[B.U. 2017]**

Solution. Total polarization charge

$$= \int_V \rho_p\, dV + \int_S \sigma_p\, dS = \int_V (-\vec{\nabla}\cdot\vec{P})\, dV + \int_S \vec{P}\cdot\hat{n}\, dS$$

$$= -\int_S \vec{P}\cdot\hat{n}\, dS + \int_S \vec{P}\cdot\hat{n}\, dS \quad \text{[Using divergence theorem]}$$

$$= 0$$

QUESTIONS

1. Explain the following facts:
   (a) Electric field inside a conductor is zero. **[C.U. 2019]**
   (b) Charges given to a conductor reside on its outer surface.

2. Show that the electric field just outside a conductor has a magnitude σ/ε_0 and is directed normal to the conductor surface, where σ is the local surface charge density. **[J.U. 2013]**

3. Show that the mechanical pressure on the surface of a charged conductor is given by $\sigma^2/2\varepsilon_0$, where σ is the local surface charge density.

4. (a) What is an electric dipole? Define its dipole moment. Find an expression for the field and the potential due to a short electric dipole. Also find the equations to the lines of force and equipotential lines due to a dipole.
   (b) Calculate the force exerted on a dipole of moment $\vec{p}$ placed in a nonuniform electric field $\vec{E}$. **[C.U. 2015, 2017; J.U. 2011]**
   (c) An electric dipole of moment $\vec{p}$ is placed in a uniform electric field $\vec{E}$. Show that its potential energy is $-\vec{p}\cdot\vec{E}$. **[C.U. 2007, 2011]**

5. (a) Show that the mutual interaction energy of two dipoles of moments $\vec{p}_1$ and $\vec{p}_2$ is given by

$$U = \frac{1}{4\pi\varepsilon_0}\left[\frac{\vec{p}_1\cdot\vec{p}_2}{r^3} - \frac{3(\vec{p}_1\cdot\vec{r})(\vec{p}_2\cdot\vec{r})}{r^5}\right],$$

   where $\vec{r}$ is the radius vector joining the centres of the two dipoles. **[C.U. 2007; J.U. 2010]**
   (b) An electric dipole with dipole moment $\vec{p}_1 = p_1\hat{t}k$ is located at the origin of a coordinate system. A second dipole of dipole moment $\vec{p}_2 = p_2\hat{k}$ is located at (i) on the $+z$-axis at a distance r from the origin, (ii) on the $+y$ axis at a distance r from the origin. Show that the force between the dipoles is attractive in case (i) and repulsive in case (ii). Calculate the magnitude of the force in the two cases. **[B.U. 2010]**

6. (a) What do you mean by polar and non-polar dielectrics? Explain physically what is meant by the polarization of a dielectric.
   (b) Show that the potential at any external point due to a polarized dielectric may be considered as due to a certain surface and volume distribution of charges.

OR Show that the electric field produced by a polarized dielectric can be given by the contributions from a bound surface charge density, $\sigma_b' = \vec{P}\cdot\hat{n}$ and a volume charge density $\rho_b = -\vec{\nabla}\cdot\vec{P}$ where $\vec{P}$ is the polarization and $\hat{n}$ is the unit vector normal to the surface of the dielectric. **[C.U. 2013,2015]**

7. What is a dielectric? State and establish Gauss's law in presence of dielectric.

8. Explain the necessity of introducing the vector $\vec{D}$ in presence of dielectric medium. How is $\vec{D}$ related to $\vec{P}$ and $\vec{E}$? Is the equation $\vec{\nabla}\cdot\vec{D} = \rho_{\text{free}}$ sufficient to determine $\vec{D}$?

9. What do you mean by the polarizability of a molecule? Establish Clausius-Mossotti relation for a nonpolar dielectric? What is the importance of this equation? OR, Derive the Clausius-Mossotti relation expressing the relationship between dielectric constant and atomic polarizability. **[B.U. 2014]**

10. What is a polar dielectric? Deduce Langevin-Debye relation for the polarizability for a gaseous dielectric subjected to an electrostatic field. How does it change with temperature?

11. State and prove boundary conditions on $\vec{E}$ and $\vec{D}$ prevailing at the interface of two dielectric media.

12. Define electric susceptibility and dielectric constant. How are they related? **[C.U. 2009]**

13. Define dielectric susceptibility of a medium. What is nonlinear dielectric? **[C.U. 2014, 2018]**

14. A sphere of linear dielectric material is placed in an originally uniform electric field $\vec{E}_0$. Using the concept of polarization and depolarizing field calculate the new field inside. **[J.U. 2012]**

EXERCISES

1. A soap bubble of radius 2 cm is charged to a potential ϕ such that the pressure inside the bubble becomes equal to that of the surrounding atmosphere. If the surface tension of the soap solution is 24×10^{-3}N/m, find the value of ϕ. [*Ans.* 21 kV]

2. To what voltage must a spherical conductor of radius 1 cm be charged in order that the mechanical pressure on its surface becomes equal to the normal atmospheric pressure (10^5 N/m^2)? [*Ans.* 15×10^5 V]

3. (a) A point charge Q is placed on the axis of a short dipole of moment p at a large distance r from the centre of the dipole. Find the forces acting on the charge and on the dipole. Comment on the result.

[*Ans.* $\frac{1}{4\pi\epsilon_0}\cdot\frac{2p}{r^3}$. Forces are equal and opposite in accordance with Newton's third law.]

(b) A point dipole with moment $\vec{p} = p\hat{k}$ is situated at the origin. Find the force on a point charge q placed at $(a, 0, 0)$. What would be the force if the point charge is placed at $(0, 0, a)$? Calculate the work done in moving q from $(a, 0, 0)$ to $(0, 0, a)$. [*Ans.* $\vec{E}(a,0,0) = -\frac{1}{4\pi\epsilon_0}\cdot\frac{pq}{a^3}\hat{k}, \vec{E}(0,0,a) = \frac{1}{4\pi\epsilon_0}\cdot\frac{2pq}{a^3}\hat{k}$,

$W = q\{\phi(0,0,a) - \phi(a,0,0)\} = \frac{pq}{4\pi\epsilon_0 a^2}\{\cos 0° - \cos 90°\} = \frac{pq}{4\pi\epsilon_0 a^2}$.]

(c) Two dipoles of moments $p\hat{k}$ and $\frac{1}{4}p\hat{k}$ are located at $(0, 0, 0)$ and $(0.1, 0, 0.2)$ m respectively. Show that the resultant field is zero at the point $(0, 0, 0.2)$ m.

[B.U. 2013]

4. An electric dipole formed by two equal and opposite charges each of magnitude 1μC separated by a distance of 1 cm is placed in a uniform electric field of 10^5 V/m. Calculate the maximum torque acting on the dipole. Also, calculate the work done in rotating the dipole end to end starting from a position parallel to the field.

[*Ans.* $\tau_{\max} = 10^{-3}$N-m, Work $= 2 \times 10^{-3}$J.]

5. Consider a point charge q located at the origin and a dipole of moment $\vec{p}$ placed at $\vec{r}$. Find the force and torque acting on the dipole due q.

[cf. C.U. 2012, 2018] [*Ans.* $\vec{F} = \frac{q}{4\pi\epsilon_0 r^3}[\vec{p} - 3(\vec{p}\cdot\hat{r})\hat{r}], \vec{\tau} = \frac{q}{4\pi\epsilon_0 r^2}(\vec{p}\times\hat{r})$.]

6. (a) Two coplanar dipoles of moments $\vec{p}_1$ and $\vec{p}_2$ are at a distance r apart. Find the force and torque that the second dipole experiences due to the first dipole. Assume that $\vec{p}_1$ and $\vec{p}_2$ are parallel and lie along the same line.

[*Ans.* Force $= -\frac{6p_1p_2}{4\pi\epsilon_0 r^4}$, Torque $= 0$.]

(b) Two water molecules each having a dipole moment 6.2×10^{-30} C.m point in the same direction and are inclined at an angle 60° to the line joining their centres. Calculate the potential energy due to their dipole-dipole interactions when their centres are 3.1×10^{-10} m apart. [*Ans.* 18.14×10^{-3} eV.]

(c) Two electric dipoles are placed at $(0, 0, 0)$ and $(1, 0, 0)$ respectively with both of them pointing in the $+z$ direction. The second dipole is moved to $(0, 2, 0)$ without changing its orientation. Find the ratio of final and initial potential energy. [*Ans.* 1/8]

7. Suppose a thin metal foil having mass per unit area 0.5 mg/mm^2 is placed on a horizontal charged plate. Find the surface charge density on the plate for which the metal foil will just float in air. [*Ans.* $\sigma = 9.3 \times 10^{-6}$ C/m^2.]

8. Suppose the electronic charge in a hydrogen atom is distributed in the following way:

$$\rho(r) = -\frac{e}{\pi a_0^3} \cdot e^{-2r/a_0},$$

where a_0 is the first Bohr radius and e is the magnitude of electronic charge. The proton is located at the centre. When placed in a small external field the proton moves relative to the centre of electronic charge cloud and the charge cloud moves rigidly without distortion. Using the model find the polarizability of a hydrogen atom.

[**Hints.** See, Solved Problem 12.

$$e.E = \frac{1}{4\pi\varepsilon_0} \cdot \frac{e \cdot \int_0^x \rho(r) 4\pi r^2\, dr}{x^2} = \frac{e^2}{\pi\varepsilon_0 a_0^3 x^2} \int_0^x r^2 e^{-2r/a_0}\, dx.$$

If E is small then displacement $x \ll a_0$. So using this approximation we get $x = \frac{3\pi\varepsilon_0 a_0^3}{e} E$.

$\therefore\ p = ex = 3\pi\varepsilon_0 a_0^3 E = \alpha E$ or $\alpha = 3\pi\varepsilon_0 a_0^3$.]

9. Consider a nonpolar dielectric material of dielectric constant $K = 4$, density 1000 kg/m^3 and molecular weight 59. Find the molecular polarizability.

[*Ans.* $\alpha = 1.3 \times 10^{-39}$ C.m^2.V^{-1}]

10. Consider a dielectric hemisphere with its plane surface in the xy-plane. It has a polarization $\vec{P} = A\hat{k}$ in the z-direction. Find the volume and surface charge densities of polarization charges.

[*Ans.* $\rho_p = 0$, $\sigma_p = -A$ on the plane surface and $A\cos\theta$ on the curved surface, where θ is the angle between z-axis and radius vector.]

11. Consider a dielectric sphere of radius a with a uniform polarization $\vec{P} = P\hat{k}$. Electrically it is equivalent to a system of two uniformly charged spheres of radius a with equal and opposite charge densities and slightly displaced from each other. Using this idea calculate the field and potential due to the polarized sphere.

[**Hints.** Using Eq. (1.8-7) we can write for field at any inside point A as

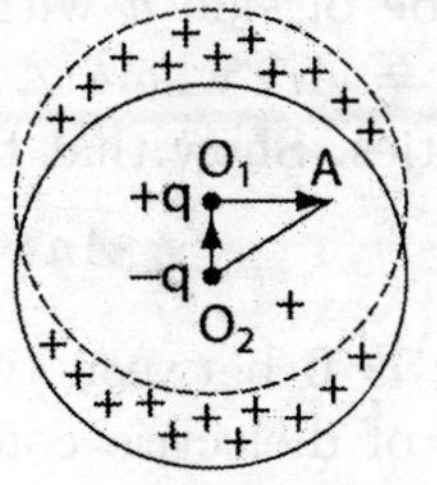

Fig. 2.E-11

$$\vec{E} = \frac{q}{4\pi\varepsilon_0 a^3}\overrightarrow{O_1A} - \frac{q}{4\pi\varepsilon_0 a^3}\overrightarrow{O_2A} = -\frac{q}{4\pi\varepsilon_0 a^3}\vec{c\delta},$$

where $\vec{c\delta} = \overrightarrow{O_2O_1}$, $q =$ total charge of positive sphere. Polarization $\vec{P} = \frac{q\vec{c\delta}}{\frac{4}{3}\pi a^3}$.

$$\therefore \ \vec{E} = -\frac{\vec{P}}{3\varepsilon_0}.$$

For outside points the system may be considered as a dipole of moment

$$\vec{p} = q\vec{\delta} = \frac{4}{3}\pi a^3 \vec{P}.$$

So for an outside point the dipole potential is

$$\phi(\vec{r}) = \frac{1}{4\pi\varepsilon_0}\frac{\vec{p}\cdot\vec{r}}{r^3}.]$$

12. The space between two concentric spherical conductor is filled with an inhomogeneous dielectric with dielectric constant $K = \frac{1}{a-br}$, where a and b are constants. A charge q is placed on the inner sphere and the outer sphere is grounded. Determine the volume density of polarization charge. [*Ans.* $-\frac{qb}{4\pi r^2}$.]

13. (a) A conducting sphere of radius a is placed in a homogeneous isotropic dielectric medium of dielectric constant K. Determine $\vec{D}$, $\vec{E}$ and $\vec{P}$ inside the dielectric. Hence determine the volume and surface densities of polarization charges.

[*Ans.* $\vec{D} = \frac{q}{4\pi r^2}\hat{r}$, $\vec{E} = \frac{q}{4\pi\varepsilon_0 K r^2}\hat{r}$, $\vec{P} = \frac{q(K-1)}{4\pi K r^2}\hat{r}$, $\rho_p = 0$, $\sigma_p = -\frac{q(K-1)}{4\pi K a^2}$.]

(b) A sphere of radius 'a' is made of linear dielectric material of dielectric constant K. A point charge q is placed at the centre of the sphere. Find $\vec{D}$, $\vec{E}$ and $\vec{p}$ at any internal point. Also find polarisation charge densities.

[*Ans.* $\vec{D} = \frac{q}{4\pi r^2}\hat{r}$, $\vec{E} = \frac{q}{4\pi\varepsilon_0 K r^2}\hat{r}$, $\vec{p} = \frac{q}{4\pi}\cdot\frac{K-1}{Kr^2}\hat{r}$,

$\rho_b = -q\frac{K-1}{K}\delta(\vec{r})$, $\sigma_b = \frac{q}{4\pi}\cdot\frac{K-1}{Ka^2}$]

14. The potential inside a dielectric sphere of radius a and permittivity ε is $\phi(r,\theta) = Ar\theta$. Find the volume density of polarization charge.[*Ans.* $-A(\varepsilon-\varepsilon_0)(2\theta+\cot\theta)/r$.]

15. Consider a dielectric cube of side a with its centre at the origin. It carries a frozen-in polarization $\vec{P} = A\vec{r}$, where A is constant. Find polarization volume and surface charge densities. Show that total polarization charge is zero.

[*Ans.* $\rho_p = -3A$, $\sigma_p = Aa/2$ on all six surfaces.]

16. Consider the interface $z = 0$ between two dielectrics. The region $1(z \geq 0)$ is occupied by a dielectric of dielectric constant $K_1 = 4$ and region $2(z \leq 0)$ is occupied by another dielectric of constant $K_2 = 3$. If $\vec{E}_1 = 5\hat{\imath} - 2\hat{\jmath} + 3\hat{k}$ kV/m find $\vec{E}_2$ for $z \leq 0$. [*Ans.* $\vec{E}_2 = 5\hat{\imath} - 2\hat{\jmath} + 4\hat{k}$ kV/m.]

17. A dielectric sphere of radius a carries a surface density of polarization charge $\sigma_p = \sigma_0\cos\theta$. Show that for all external points this is equivalent to a point dipole of moment $\frac{4}{3}\pi a^3\sigma_0\hat{k}$ situated at the centre of the sphere.

18. A block of carbon is uniformly polarized with $P = 4.8 \times 10^{-4}$ C.m^2. If the density of carbon is 3500 kg.m^{-3}, what would be the separation between the positive and negative charge centres in a carbon atom? [*Ans.* 2.85×10^{-13} m.]

19. A right circular cone of semivertical angle θ has uniform polarization $\vec{P}$ parallel to its axis and directed from the vertex to its base. Find the polarization charges on the plane and curved surfaces. Show that the total charge on its surface is zero. [*Ans.* $\sigma_p = P$ and $-P\sin\theta$.]

20. A cavity of radius a is made in a large block of dielectric of dielectric constant K. If a point charge q is placed at the centre of the cavity show that the electric field at a distance $r(r > a)$ from the centre of the cavity is given by $\vec{E} = \dfrac{1}{4\pi\varepsilon_0}\dfrac{q}{Kr^2}\hat{r}$.

21. Consider a large block of dielectric with uniform polarization $\vec{P}$. A thin disc-shaped cavity is cut within this block the axis of the cavity being parallel to $\vec{P}$. Show that the electric field within the cavity due to *surface charges on the cavity* is given by $\vec{P}/\varepsilon_0$. What would be the corresponding value of electric field when a long needle-shaped cavity is cut with its axis parallel to $\vec{P}$? [*Ans.* Zero.]

22. A thin dielectric rod is placed along x-axis with its ends at $x = 0$ and $x = l$. The rod has uniform polarization $\vec{P} = (ax^2 + b)\hat{\imath}$. Find volume and surface density of polarization charges. Show that total polarization charge is zero.

[*Ans.* $\sigma_p = -b$ at $x = 0$, $(al^2 + b)$ at $x = l$; $\rho_p = -2ax$.]

23. A plane slab of dielectric with dielectric constant K is placed in free space. Suppose there exists an electric field $\vec{E}_0$ in the free space, where $\vec{E}_0$ is uniform and perpendicular to the boundary of the slab. Find field $\vec{E}$ and polarization $\vec{P}$ within the slab. [*Ans.* $\vec{E} = \vec{E}_0/K, \vec{P} = \varepsilon_0(1 - 1/K)\vec{E}_0$.]

24. Consider a dielectric rod of length $2l$ and radius a, which has a uniform polarization $\vec{P}$ parallel to its axis. Find the electric field due to this rod at an axial point at a distance x from the centre of the rod.

[*Ans.* $\vec{E} = \frac{\vec{P}}{2\varepsilon_0}\left[\frac{l-x}{\sqrt{(l-x)^2+a^2}} + \frac{l+x}{\sqrt{(l+x)^2+a^2}}\right]$.]

[**Hints.** Consider field due to two parallel discs of area πa^2 having surface charge densities $-P$ and $+P$, respectively.]

25. Show that the electric field $\vec{E}_0$ within a long narrow needle-shaped cavity cut in a dielectric is equal to the field $\vec{E}$ in the dielectric when the length of the cavity is along $\vec{E}$. But if the length of the cavity is perpendicular to $\vec{E}$ then $\vec{E}_0$ becomes equal to $\vec{D}$ in the dielectric.

26. Suppose an electric field line moves from a medium having dielectric constant 6 into a second medium having dielectric constant 2. If the electric field line in the first medium makes an angle 60° with the normal to the interface what would be the angle made by the field line with the normal in the second medium?

[*Ans.* 30°.]

27. The angle between two OH bonds in a H_2O molecule is 108°. Assuming that each OH bond is associated with a dipole moment is 1.51D (1D= 3.336×10^{-30} C.m), calculate the dipole moment of a water molecule. [*Ans.* 1.78D.]

28. Show that the field lines due to a dipole in the plane of the dipole can be represented by the polar equation $\sin^2\theta/r$ = constant, where θ is the angle between the dipole axis and the radius vector to the point under consideration. Also, show that the equipotentials are given by $\cos\theta/r^2$ = constant.

[**Hints.** $d\vec{r} = \hat{r}dr + \vec{\varepsilon}\theta r d\theta$ and $\vec{E} = E_r\hat{r} + E_\theta\hat{\theta}$ are parallel. Therefore,

$$\frac{dr}{rd\theta} = \frac{E_r}{E_\theta} = 2\cot\theta.$$

Integrating, $\sin^2\theta/r$ = constant.]

29. The potential due to a point dipole is $\phi = \dfrac{1}{4\pi\varepsilon_0}\dfrac{\vec{p}\cdot\vec{r}}{r^3}$. Using the relation $\vec{E} = -\vec{\nabla}\phi$, find $\vec{E}$.

30. Suppose that in the vicinity of a point on an uncharged vacuum-dielectric interface the electric field in vacuum is $\vec{E}_0$. If θ_0 be the angle between $\vec{E}_0$ and the unit normal drawn from dielectric to vacuum show that the magnitude of electric field $\vec{E}$ in the dielectric, near the point, is given by $E = E_0\left(\sin^2\theta + \cos^2\theta/K^2\right)^{1/2}$, where K is the dielectric constant of the dielectric.

31. Suppose the surface separating two dielectrics 1 and 2 (having permitivities ε_1 and ε_2) has surface charge density σ. If electric fields, $\vec{E}_1$ and $\vec{E}_2$, on the two sides of the surface make angles θ_1 and θ_2 with the unit normal to the surface drawn from medium 1 to medium 2 show that

$$\varepsilon_1\cot\theta_1 = \varepsilon_1\cot\theta_2(1 - \sigma/\varepsilon_2 E_2\cos\theta_2).$$

32. Suppose a point charge q is placed inside a cavity in the interior of a conducting body. Using Gauss law prove that a charge $-q$ is induced on the surface of the cavity.

[**Hints.** As the field in the interior of the conductor is zero, we can write for a closed surface S' lying wholly within the material of the conductor and enclosing the cavity as

$$\oint \vec{E}\cdot d\vec{S} = q_{\text{total}}/\varepsilon_0 = 0 \quad \text{or} \quad q + q' = 0 \quad \text{or} \quad q' = -q,$$

which is the charge on the surface of the cavity.]

33. Two semi-infinite blocks of dielectric are placed almost in contact such that there exists a thin narrow gap of constant separation between them. The polarization $\vec{P}$ is constant throughout all the dielectric material and it makes an angle θ with the normal to the planes bounding the gap. Find the electric field in the gap.

[**Hints.** The polarization charge appearing per unit area of the bounding surfaces is given by $\sigma_p = \hat{n} \cdot \vec{P}$. So the bounding surfaces will have charge densities $\pm P \cos\theta$. It is a case very similar to a parallel plate capacitor and hence, $E = P\cos\theta/\varepsilon_0$.]

34. A dielectric cylinder of radius a carries a polarization of $\vec{P} = Ar\hat{r}$, where r is the distance from the axis of the cylinder and A is a constant. Find the polarization charge densities. [*Ans.* $\rho_p = -2A,\ \rho_s = Aa$.]

35. A point charge q is situated at a large distance r from a neutral atom of polarizability α. Find the force of attraction between them. [*Ans.* $\left(\frac{q}{4\pi\varepsilon_0}\right)^2 \cdot \frac{2\alpha}{r^5}$.]

[**Hints.** Induced dipole moment $\vec{p} = \alpha\vec{E} = \alpha \cdot \frac{1}{4\pi\varepsilon_0} \cdot \frac{q}{r^2}\hat{r}$. Field due to this dipole at q is $\vec{E}_{\text{dip}} = \frac{1}{4\pi\varepsilon_0}\frac{2p}{r^3}\hat{r}$. $\therefore$ Force $= q\vec{E}_{\text{dip}}$]

36. A spherical shell of dielectric having inner radius a and outer radius b carries a 'frozen-in' polarization $\vec{P} = \frac{A}{r}\hat{r}$, where A is a constant and $r(a < r < b)$ is the distance from the centre. Calculate the polarization charge densities and electric field everywhere.

[*Ans.* $\rho_p = -\frac{A}{r^2}, \sigma_p = \frac{A}{b}$ at $r = b$ and $-\frac{A}{a}$ at $r = a$; $\vec{E} = 0$ for $r < a$, $\vec{E} = -\frac{A}{\varepsilon_0 r}\hat{r}$ for $a < r < b$, $\vec{E} = 0$ for $r > b$.]

OBJECTIVE TEST

1. The electric field inside a uniformly charged conducting sphere is:

(a) uniform but nonzero,

(b) zero,

(c) dependent on the distance from the centre,

(d) none of the above.

2. The electric field just outside a charged conductor having a surface density σ is:

(a) σ/ε_0, (b) $\varepsilon_0\sigma$, (c) $2\sigma/\varepsilon_0$, (d) zero.

3. Mechanical pressure on the surface of a charged conductor having surface charge density σ is:

(a) $\varepsilon_0\sigma^2$, (b) σ^2/ε_0, (c) $\sigma^2/2\varepsilon_0$, (d) $\sigma/2\varepsilon_0$.

4. Dipolar potential varies with distance r as:

(a) $\frac{1}{r}$, (b) $\frac{1}{r^2}$, (c) $\frac{1}{r^3}$, (d) none of these.

5. Dipolar field varies with distance as

(a) $\frac{1}{r}$, (b) $\frac{1}{r^2}$, (c) $\frac{1}{r^3}$, (d) $\frac{1}{r^4}$.

6. In a dielectric medium if $\vec{P}$ is the polarization vector and $\vec{E}$ is the electric field then in the relation $\vec{P} = \alpha\vec{E}$, α is in general a:

(a) scalar,
(b) vector,
(c) tensor of rank 2,
(d) none of these.

7. Gauss's law in a dielectric medium takes the form $\oint_S \vec{D} \cdot d\vec{S} = q$ where q is:

(a) total free charge enclosed,
(b) polarization charges,
(c) total of both free and polarization charges,
(d) zero.

8. Atomic polarizability α has the dimensions of:

(a) $C.m^{-1}.V^{-1}$,
(b) $C.m^2.V^{-1}$,
(c) $C.m^{-2}$,
(d) none of these.

9. The interaction energy of an electric dipole of moment $\vec{p}$ in an external field $\vec{E}$ is:

(a) $|\vec{p} \times \vec{E}|$,
(b) $-|\vec{p} \times \vec{E}|$,
(c) $\vec{p} \cdot \vec{E}$,
(d) $-\vec{p} \cdot \vec{E}$.

10. The net force on an electric dipole placed in a uniform electric field:

(a) depends on the dipole moment,
(b) depends on the strength of electric field,
(c) depends on the orientation of the dipole,
(d) is always zero.

11. The bound volume charge density in a sphere of radius a carrying polarization $\vec{P} = k\vec{r}$ is: **[NET 2009]**

(a) kr,
(b) k,
(c) zero,
(d) $-3k$.

12. A dielectric slab is placed partly into a charged insulated parallel plate air capacitor. The slab, in this case, feels:

(a) no force,
(b) a force pulling it in,

(c) a force pushing it out,

(d) none of these.

13. A dielectric material has nonuniform polarization $\vec{P}$. The polarization volume charge density is given by:

(a) $-\vec{\nabla}\cdot\vec{P}$, (c) $|\vec{P}|/\varepsilon_0$,

(b) $\vec{\nabla}\cdot\vec{P}$, (d) $|\vec{P}|^2/\varepsilon_0$.

14. A point charge q is held at a distance $a/2$ from the centre of a thin conducting shell of radius a. The potential of the shell is:

(a) $\dfrac{1}{2\pi\varepsilon_0}\cdot\dfrac{q}{a}$, (c) $\dfrac{1}{8\pi\varepsilon_0}\cdot\dfrac{q}{a}$,

(b) $\dfrac{1}{4\pi\varepsilon_0}\cdot\dfrac{q}{a}$, (d) zero.

15. A medium in which the relations $\vec{\nabla}\cdot\vec{D}=\varepsilon\vec{\nabla}\cdot\vec{E}$ and $\vec{\nabla}\cdot\vec{J}=\sigma\vec{\nabla}\cdot\vec{E}$ are true, is said to be:

(a) linear and inhomogeneous,

(b) linear and isotropic,

(c) linear and homogeneous,

(d) homogeneous and isotropic.

16. An electric dipole of moment $\vec{p}=p_0\hat{k}$ is situated at the origin. Then work done in moving a point charge q from the point $(d,0,0)$ to $(0,0,d)$ would be

(a) $\dfrac{p_0 q}{4\pi\varepsilon_0 d^2}$, (b) $\dfrac{p_0 q}{4\pi\varepsilon_0 d^3}$, (c) $\dfrac{p_0 q}{2\pi\varepsilon_0 d^2}$, (d) zero.

17. Suppose yz plane is the interface between two dielectric media 1 and 2 having dielectric constants $K_1 = 3$ and $K_2 = 4$ respectively. If the electric field at the interface in medium 1 is $\vec{E}_1 = 4\hat{\imath}+3\hat{\jmath}+4\hat{k}$ then assuming no free charge at the interface the electric field at the interface in medium 2 would be

(a) $\vec{E}_2 = 3\hat{\imath}+3\hat{\jmath}+4\hat{k}$, (c) $4\hat{\imath}+3\hat{\jmath}+3\hat{k}$

(b) $4\hat{\imath}+\frac{9}{4}\hat{\jmath}+4\hat{k}$, (d) $3\hat{\imath}+\frac{9}{4}\hat{\jmath}+3\hat{k}$.

18. A dielectric cylinder of length L and radius R is placed with its axis along z-direction and one end at the origin. If the cylinder carries a polarization $\vec{p}=(5z^2+7)\hat{k}$, the volume bound charge inside the dielectric is

(a) zero, (b) $10\pi R^2 L$, (c) $-5\pi R^2 L$, (d) $-5\pi R^2 L^2$.

[GATE 2009]

ANSWERS

1. (b)	**2.** (a)	**3.** (c)	**4.** (b)	**5.** (c)	**6.** (c)	**7.** (a)
8. (b)	**9.** (d)	**10.** (d)	**11.** (d)	**12.** (b)	**13.** (a)	**14.** (a)
15. (c)	**16.** (a)	**17.** (a)	**18.** (d)			

Chapter 3

Multipole Expansion

3.1 Multipole

A single point charge is called *a monopole.* If we start from a monopole $+q$ and then take another monopole $-q$ and fix it at a small distance from the first monopole, we get a *dipole.* Similarly, if we start from a dipole of moment $\vec{p}$ and then take another dipole of moment $-\vec{p}$ and fix it at a small distance from the first dipole a *quadrupole* is formed (Fig. 3.1-1). It is possible to extend this dipole and quadrupole concepts to a system of large number of positive and negative charges located at very small distances from each other. For example, a system of two quadrupoles separated from each other by a small distance and all the charges of one quadrupole having same magnitude but of opposite sign to that of the corresponding charges in the other quadrupole, is said to form an *octupole.*

In this way the above concept can be continued indefinitely to get multipoles of higher order. The importance of this multipole idea lies in the fact that the potential and field due to an arbitrary charge distribution localised in a small region of space can be expressed as a sum of different multipole terms. It is found that the potential due to a monopole varies with distance (r) as $1/r$, that due to a dipole as $1/r^2$, that due to a quadrupole as $1/r^3$ and so on. So for many practical purposes first two or three terms alone give sufficient accuracy.

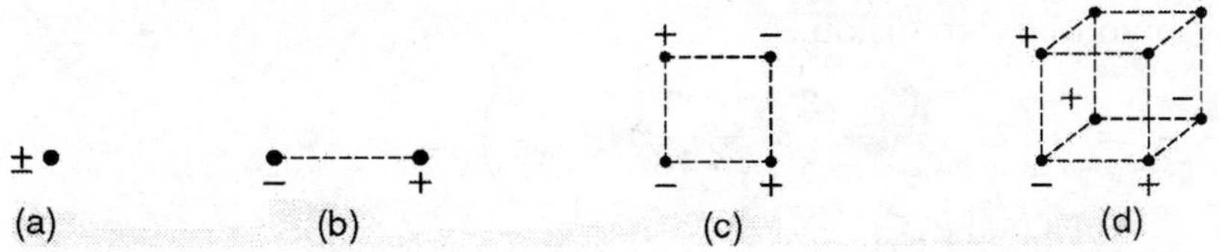

Fig. 3.1-1 : (a) Monopole, (b) Dipole, (c) Quadrupole, (d) Octopole

3.2 The Multipole Expansion

Let us consider an arbitrary charge distribution of density $\rho(\vec{r}\,')$ localised in a rather small region of volume V. We wish to find potential $\phi(\vec{r})$ at an outside point $P(\vec{r})$. We choose the origin O in or near the charge distribution and assume that the distance r is large compared with the maximum value of r', i.e., the maximum distance the charge extends away from O. From Fig. 3.2-1 potential at $\vec{r}$ is given by

$$\phi(\vec{r}) = \frac{1}{4\pi\varepsilon_0}\int_V \frac{\rho(\vec{r}\,')}{|\vec{r}-\vec{r}\,'|}\,dV. \tag{3.2-1}$$

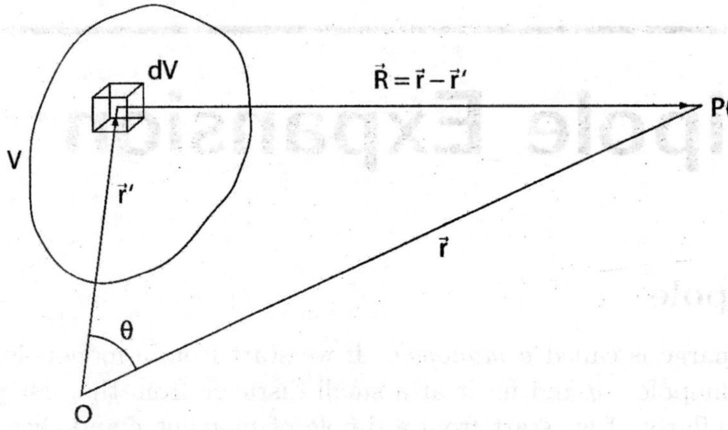

Fig. 3.2-1

Now

$$\begin{aligned}
\frac{1}{|\vec{r}-\vec{r}\,'|} &= \frac{1}{\sqrt{(\vec{r}-\vec{r}\,')\cdot(\vec{r}-\vec{r}\,')}} \\
&= \frac{1}{(r^2 - 2\vec{r}\cdot\vec{r}\,' + r'^2)^{1/2}} \\
&= \frac{1}{r}\left(1 + \frac{r'^2}{r^2} - \frac{2r'}{r}\cos\theta\right)^{-1/2} \\
&= \frac{1}{r}(1+\delta)^{-1/2} \quad \text{where } \delta = \frac{r'}{r}\left(\frac{r'}{r} - 2\cos\theta\right).
\end{aligned} \tag{3.2-2}$$

In view of our assumption made above $r'/r \ll 1$ and $\delta \ll 1$ and hence we can make the following binomial expansion.

$$\begin{aligned}
\frac{1}{|\vec{r}-\vec{r}\,'|} &= \frac{1}{r}\left(1 - \frac{1}{2}\delta + \frac{3}{8}\delta^2 - \frac{5}{16}\delta^3 + \cdots\right) \\
&= \frac{1}{r}\left[1 - \frac{1}{2}\frac{r'}{r}\left(\frac{r'}{r} - 2\cos\theta\right) + \frac{3}{8}\left(\frac{r'}{r}\right)^2\left(\frac{r'}{r} - 2\cos\theta\right)^2 + \cdots\right]
\end{aligned}$$

$$= \frac{1}{r}\left[1 + \frac{r'}{r}\cos\theta + \left(\frac{r'}{r}\right)^2 \frac{1}{2}(3\cos^2\theta - 1) + \cdots\right], \tag{3.2-3}$$

where we have grouped like powers of r'/r.

Therefore, from Eq. (3.2-1) we can write

$$\begin{aligned}\phi(\vec{r}) &= \frac{1}{4\pi\varepsilon_0 r}\int_V \rho\, dV + \frac{1}{4\pi\varepsilon_0 r^2}\int_V r'\cos\theta\rho\, dV \\ &\quad + \frac{1}{4\pi\varepsilon_0 r^3}\int_V r'^2 \frac{1}{2}(3\cos^2\theta - 1)\rho\, dV + \cdots \\ &= \phi_0(\vec{r}) + \phi_1(\vec{r}) + \phi_2(\vec{r}) + \cdots,\end{aligned} \tag{3.2-4}$$

where

$$\phi_0(\vec{r}) = \frac{1}{4\pi\varepsilon_0 r}\int_V \rho\, dV \tag{3.2-5}$$

$$\phi_1(\vec{r}) = \frac{1}{4\pi\varepsilon_0 r^2}\int_V r'\cos\theta\rho\, dV \tag{3.2-6}$$

$$\phi_2(\vec{r}) = \frac{1}{4\pi\varepsilon_0 r^3}\int_V r'^2 \cdot \frac{1}{2}(3\cos^2\theta - 1)\rho\, dV. \tag{3.2-7}$$

Note that for $r \gg r'$ the successive terms in the series (3.2-4) become smaller and smaller. In fact the series (3.2-3) and hence the *multipole expansion* (3.2-4) *is convergent* for any $r > r'$. For many practical purposes $r \gg r'$ and hence the first two or three terms alone give sufficiently accurate value of $\phi(\vec{r})$. Equation (3.2-4) is known as *multipole expansion* for the potential in power of $1/r$.

Monopole term $\phi_0(\vec{r})$

The first term $\phi_0(\vec{r})$ is known as *monopole potential.* It is the potential which we would have at P if the whole charge were concentrated at the origin. The quantity $\int_V \rho\, dV$ is called *the monopole moment* which is simply the total charge in the distribution.

Dipole term $\phi_1(\vec{r})$

The term $\phi_1(\vec{r})$ is called *the dipole term.* As $\vec{r}' \cdot \hat{r} = r'\cos\theta$ it can be rewritten as follows:

$$\phi_1(\vec{r}) = \frac{1}{4\pi\varepsilon_0}\frac{\hat{r}\cdot\vec{p}}{r^2}, \tag{3.2-8}$$

where

$$\vec{p} = \int \vec{r}'\rho\, dV$$

is called the *dipole moment of the charge distribution.*

For a collection of point charges the dipole moment is given by

$$\vec{p} = \sum_i q_i \vec{r}_i'$$

The importance of the dipole term is that when the total charge is zero then $\phi_0(\vec{r}) = 0$ and the dipole term $\phi_1(\vec{r})$ dominates in the expansion (3.2-4). $\phi_1(\vec{r})$ is the potential that would result if a 'point or pure' dipole of moment $\vec{p}$ were located at the origin.

Quadrupole term $\phi_2(\vec{r})$

The term $\phi_2(\vec{r})$ is the potential due to quadrupole moment of the charge distribution. It can be rewritten in the following form:

$$\phi_2(\vec{r}) = \frac{1}{4\pi\varepsilon_0} \cdot \frac{1}{2r^5} \int \left[3(\vec{r}\cdot\vec{r}')^2 - r^2 r'^2\right] \rho \, dV \tag{3.2-9}$$

Now using Cartesian coordinates as $x = x_1$, $y = x_2$ and $z = x_3$, for convenience, we can write

$$(\vec{r}\cdot\vec{r}')^2 = \left(\sum_{i=1}^{3} x_i x_i'\right)^2 = \sum_i \sum_j x_i x_j x_i' x_j'$$

$$\text{and } r^2 r'^2 = r'^2 \sum_i x_i^2 = r'^2 \sum_i \sum_j x_i x_j \delta_{ij},$$

where δ_{ij} is the Kronecker delta,

$$\delta_{ij} = \begin{cases} 1, & \text{if } i = j \\ 0, & \text{if } i \neq j \end{cases}$$

Therefore,

$$\phi_2(\vec{r}) = \frac{1}{4\pi\varepsilon_0} \sum_i \sum_j \frac{x_i x_j}{2r^5} \int_V \left(3x_i' x_j' - \delta_{ij} r'^2\right) \rho \, dV \tag{3.2-10}$$

$$= \frac{1}{4\pi\varepsilon_0} \sum_i \sum_j \frac{x_i x_j}{2r^5} Q_{ij}, \tag{3.2-11}$$

where

$$Q_{ij} = \int_V \left(3x_i' x_j' - \delta_{ij} r'^2\right) \rho \, dV. \tag{3.2-12}$$

The nine components Q_{ij} $(i, j = 1, 2, 3)$ are said to form the quadrupole moment tensor for the charge distribution. It forms a real symmetric tensor, i.e., $Q_{ij} = Q_{ji}$. Thus, out of its nine components six are equal in pairs and hence, six components are distinct. Also, there is another property that the sum of the diagonal elements is zero, i.e., $Q_{11} + Q_{22} + Q_{33} = 0$. By proper choice of the coordinate systems it is possible to make the 'off diagonal' terms zero (i.e., $Q_{ij} = 0$ for $i \neq j$). If the charge distribution is symmetric about an axis, say z-axis (for example, a cylindrically symmetric charge distribution) then all the quadrupole elements are expressible in terms of a single element $Q_{33}(= Q_{zz})$. For such a distribution $Q_{11} = Q_{22} = -\frac{1}{2}Q_{33}$. Generally Q_{33} is quoted for such a distribution and it is called the *quadrupole moment of the distribution:*

$$Q_{33} = \int_V \left(3z'^2 - r'^2\right) \rho \, dV. \tag{3.2-13}$$

For a charge distribution having symmetry about z-axis and elongated along this axis $Q_{33} > 0$. Such a distribution is called *prolate distribution.* For a distribution, which spreads more in the xy-plane and has symmetry about z-axis $Q_{33} < 0$. It is called *oblate distribution.*

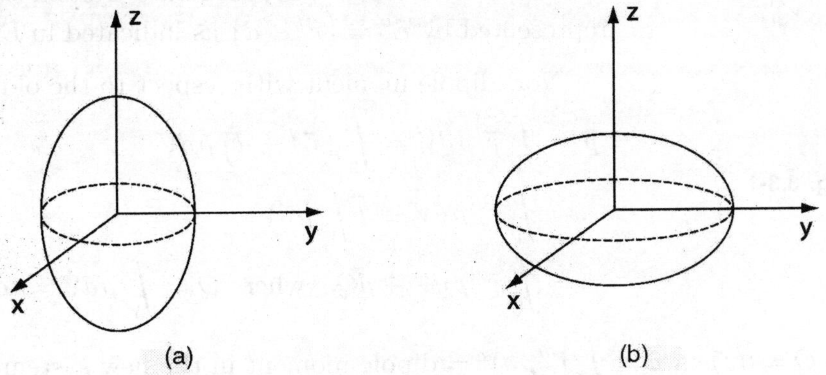

Fig. 3.2-2 : (a) Prolate, (b) Oblate distribution

A spherically symmetric charge distribution $(Q_{11} = Q_{22} = Q_{33})$ has zero quadrupole moment. In fact *for a spherically symmetric charge distribution only the monopole moment has nonzero value.* For this the quadrupole moment of a nucleus may be considered as a measure of its deviation from spherical shape.

There are also higher order moments, which are generated if we keep higher order terms in the expansion Eq. (3.2-3). Such higher order moments are called *multipoles,* which are found to be useful in nuclear physics.

In summary, we note that the potential or field due to an arbitrary charge distribution may be considered as the sum of the potentials or fields due to a point charge (monopole), a point dipole, a point quadrupole and higher order multipole moments of the charge distribution, all placed at the origin. The contribution of higher order multipole moments decreases rapidly with increase in order.

So for many practical purposes first few multipole terms alone give sufficient accuracy.

3.3 Dependence of Multipole Moments on the Choice of Origin

In general various multipole terms depend on the choice of the origin, except the first one. The monopole moment, i.e., $\int_V \rho\, dV$ is simply the total charge in the distribution. For this it does not depend on the choice of the coordinate system. The dipole moment $\vec{p}$ generally depends on the position of the origin if the total charge of the distribution is not zero. So, in general, one must mention the dipole moment of a charge distribution about some point. However, *if the total charge is zero* (i.e., monopole moment is zero) then $\vec{p}$ is independent of the choice of origin of the coordinate system. This can be proved as follows:

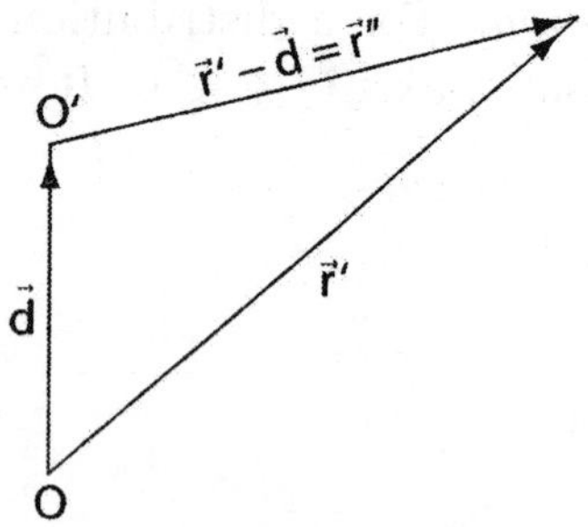

Fig. 3.3-1

Suppose we consider a new coordinate system whose origin O' is at $\vec{d}$ with respect to the origin O of the old coordinate system. So a point which is denoted by $\vec{r}\,'$ with respect to the old coordinate system will now be represented by $\vec{r}\,'' = \left(\vec{r}\,' - \vec{d}\right)$ as indicated in Fig. 3.3-1.

Now dipole moment with respect to the old system is

$$\begin{aligned}\vec{p} &= \int_V \vec{r}\,'\rho\, dV = \int_V \left(\vec{r}\,'' + \vec{d}\right)\rho\, dV \\ &= \int_V \vec{r}\,''\rho\, dV + \vec{d}\int_V \rho\, dV \\ &= \int_V \vec{r}\,''\rho\, dV + \vec{d}Q \quad \text{where } Q = \int_V \rho\, dV = \text{total charge}\end{aligned}$$

Thus if $Q = 0$ then $\vec{p} = \int_V \vec{r}\,''\rho\, dV$ = dipole moment in the new system.

It can be proved that if monopole and dipole moments vanish the quadrupole moment is independent of origin. In general the lowest nonzezro multipole is always independent of origin.

At this point it may be mentioned that even if any multipole moment does not depend on the choice of origin its multipole expansion in general will depend on the choice of origin. For example, the multipole expansion for a point charge at the origin contains only a monopole term and therefore a point charge at the origin may be termed as 'pure' monopole. But if the point charge is not at the origin the multipole expansion will contain a series of terms in inverse powers of the distance to the origin, not just the first. Thus a *point charge not at the origin cannot be considered as a 'pure' monopole.*

3.4 Some Dipole Moments

1. A point charge

The dipole moment $\vec{p}$ of a point charge q located the position $\vec{r}\,'$, about the origin is $\vec{p} = q\vec{r}\,'$.

2. Two equal and opposite point charge

Consider a dipole consisting of two equal and opposite charges separated by a small displacement $\vec{l}$ as shown in Fig. 3.4-1. Its dipole moment with respect to the origin O is

$$\vec{p} = \int \vec{r}'dq = \vec{r}'_-(-q) + \vec{r}'_+(+q) = q\left(\vec{r}'_+ - \vec{r}'_-\right) = q\vec{l}$$

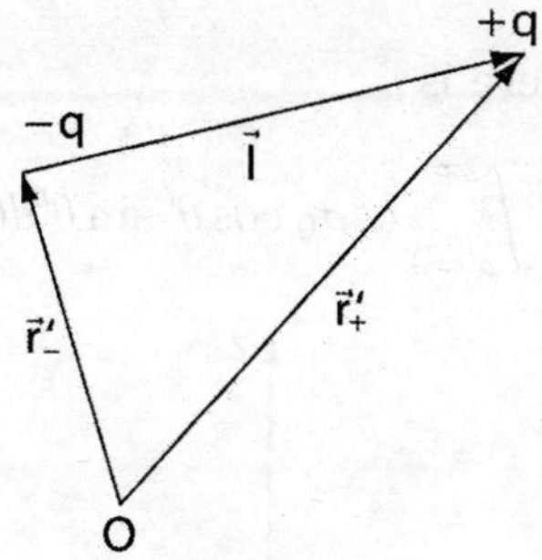

Fig. 3.4-1

As the total charge is zero, the dipole moment $\vec{p}$ is independent of the choice of origin.

3. N-point charges

Suppose N point charges q_1, q_2, ..., q_N are located at $\vec{r}'_1$, $\vec{r}'_2$, ..., $\vec{r}'_N$, respectively. The dipole moment of this charge distribution about the origin is

$$\vec{p} = \sum_{i=1}^{N} q_i \vec{r}'_i.$$

4. Three charges $+q$, $+q$ and $-2q$ placed at the vertices of an equilateral triangle

As the total charge is zero the dipole moment $\vec{p}$ will be independent of the choice of origin. Let the charges be placed at the corners of an equilateral triangle ABC lying in the xy-plane as shown in Fig. 3.4-2. If each side is a then $OB = OC = \frac{a}{2}$ and $OA = \frac{\sqrt{3}}{2}a$. Therefore, $\overrightarrow{OB} = \hat{i}\frac{a}{2}$, $\overrightarrow{OC} = -\hat{i}\frac{a}{2}$ and $\overrightarrow{OA} = \hat{j}\frac{\sqrt{3}}{2}a$. So the dipole moment with respect O is

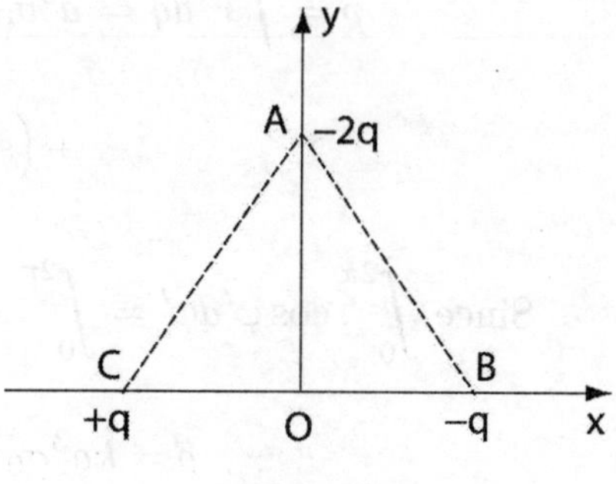

Fig. 3.4-2

$$\vec{p} = q\left(\hat{i}\frac{a}{2}\right) + q\left(-\hat{i}\frac{a}{2}\right) - 2q\left(\hat{j}\frac{\sqrt{3}}{2}a\right) = -\hat{j}\sqrt{3}aq.$$

5. A sphere with a given surface charge distribution

Suppose we have sphere of radius a with a surface charge density $\sigma(\theta) = \sigma_0 \cos\theta$, where θ is the angle which the position vector of the point under consideration with respect to the centre of the sphere makes with z-axis. We are to calculate its dipole moment. Charge on a surface element $a^2 \sin\theta' d\theta' d\varphi'$ is

$$dq = a^2 \sin\theta' d\theta' d\varphi' \sigma(\theta') = a^2 \sigma_0 \cos\theta' \sin\theta' d\theta' d\varphi'$$

and the total charge on the sphere is

$$Q = \int_{\theta'=0}^{\pi} \int_{\varphi'=0}^{2\pi} a^2 \sigma_0 \cos\theta' \sin\theta' d\theta' \, d\varphi' = 0.$$

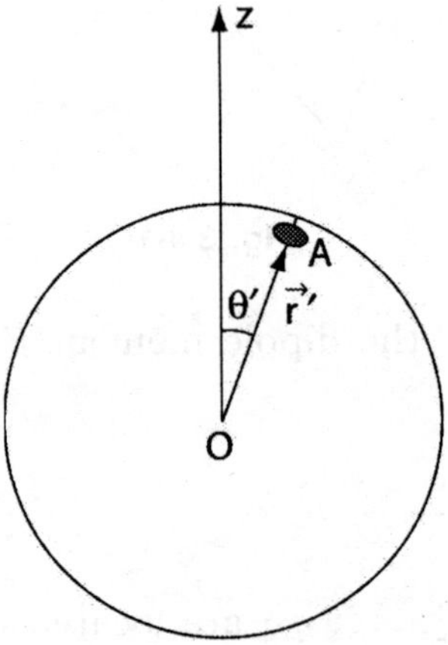

Fig. 3.4-3

So the dipole moment will not depend on the choice of origin. Now a point on the surface element may be located by

$$\vec{r}' = a \sin\theta' \cos\varphi' \hat{\imath} + a \sin\theta' \sin\varphi' \hat{\jmath} + a \cos\theta' \hat{k}.$$

Therefore, dipole moment

$$\vec{p} = \int \vec{r}' dq = a^3 \sigma_0 \int_{\theta'=0}^{\pi} \int_{\varphi'=0}^{2\pi} \Big[\left(\sin^2\theta' \cos\theta' \cos\varphi'\right) \hat{\imath} + \left(\sin^2\theta' \cos\theta' \sin\varphi'\right) \hat{\jmath} + \cos^2\theta' \sin\theta' \hat{k} \Big] d\theta' d\varphi'.$$

Since $\int_0^{2\pi} \cos\varphi' d\varphi' = \int_0^{2\pi} \sin\varphi' d\varphi = 0$ the first two terms vanish.

$$\therefore \quad \vec{p} = \hat{k} a^3 \sigma_0 \int_0^{\pi} \cos^2\theta' \sin\theta' d\theta' \int_0^{2\pi} d\varphi' = \hat{k} \frac{4\pi a^3 \sigma_0}{3}.$$

3.5 Some Quadrupole Moments

1. Linear quadrupole

A linear quadrupole consists of four charges each of same magnitude $|q|$ two of which are positive and two are negative, arranged in the manner shown in Fig. 3.5-1. We can calculate the potential at any point $P(\vec{r})$ due to this quadrupole by using Eqs. (3.2-11) and (3.2-12). Here we have only point charges so we can write for the quadrupole tensor components as

$$Q_{ij} = \sum_k \left(3x'_{ik}x'_{jk} - \delta_{ij}r'^2_k\right) q_k, \tag{3.5-1}$$

where the sum is over the number of charges; q_k is the magnitude and x_{ik} is the ith coordinate of the kth charge located at r'_k.

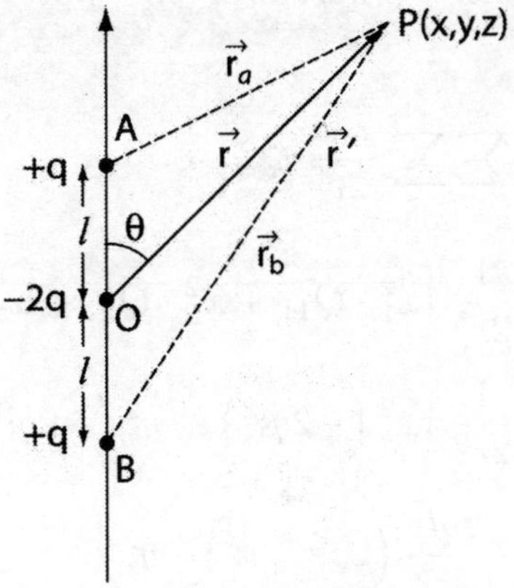

Fig. 3.5-1

In Fig. 3.5-1 for the charge $q_1 = -2q$ at O, $\vec{r}'_1 = 0$, for the charge $q_2 = +q$ at A, $\vec{r}'_2 = \overrightarrow{OA} = l\hat{k}$ and for the change $q_3 = +q$ at B, $\vec{r}'_3 = \overrightarrow{OB} = -l\hat{k}$. Therefore, $q_1 = -2q$, $q_2 = +q$, $q_3 = +q$ and their coordinates are

$$(x'_{11}, x'_{21}, x'_{31}) \equiv (x'_1, y'_1, z'_1) = (0, 0, 0)$$
$$(x'_{12}, x'_{22}, x'_{32}) \equiv (x'_2, y'_2, z'_2) = (0, 0, l)$$
$$(x'_{13}, x'_{23}, x'_{33}) \equiv (x'_3, y'_3, z'_3) = (0, 0, -l)$$

For the point of observation, $(x_1, x_2, x_3) \equiv (x, y, z)$.
Thus we get

$$Q_{11} = \sum_k \left(3x'_{1k}x'_{1k} - r'^2_k\right) q_k$$
$$= \sum_k \left(3x'^2_k - r'^2_k\right) q_k$$

$$= \sum_k \left(2x_k'^2 - y_k'^2 - z_k'^2\right) q_k$$
$$= -2q \cdot 0 + q\left(0^2 - 0^2 - l^2\right) + q\left(0^2 - 0^2 - l^2\right)$$
$$= -2ql^2.$$

Similarly,

$$Q_{22} = -2ql^2,$$
$$Q_{12} = Q_{13} = Q_{23} = Q_{21} = Q_{31} = Q_{32} = 0$$
$$\text{and} \quad Q_{33} = 4ql^2 \tag{3.5-2}$$

Note that the charge distribution has symmetry about z-axis and Q_{33} is sometimes called its quadrupole moment. Therefore, the potential $\phi_2(\vec{r})$ at P due to the quadrupole for $r \gg l$ is

$$\phi_2(\vec{r}) = \frac{1}{4\pi\varepsilon_0} \sum_i \sum_j \frac{x_i x_j}{2r^5} Q_{ij} \tag{3.5-3}$$
$$= \frac{1}{4\pi\varepsilon_0} \frac{1}{2r^5}\left[x_1^2 \cdot Q_{11} + x_2^2 \cdot Q_{22} + x_3^2 \cdot Q_{33}\right]$$
$$= \frac{1}{4\pi\varepsilon_0} \frac{1}{2r^5}\left[x^2\left(-2ql^2\right) + y^2\left(-2ql^2\right) + z^2\left(4ql^2\right)\right]$$
$$= \frac{1}{4\pi\varepsilon_0} \cdot \frac{2ql^2}{2r^5}\left(2z^2 - x^2 - y^2\right)$$
$$= \frac{ql^2}{4\pi\varepsilon_0 r^5}\left(3z^2 - r^2\right)$$
$$= \frac{ql^2}{4\pi\varepsilon_0 r^3}\left(3\cos^2\theta - 1\right). \tag{3.5-4}$$

Thus at large distances the electric potential due to a linear quadrupole varies inversely as the cube of the distance. For a dipole this potential varies inversely as the square of the distance.

Note that the charge distribution shown in Fig. 3.5-1 has zero monopole and dipole moments.

Direct calculation

The electric potential at P due to the quadrupole can also be calculated directly as follows:

$$\phi(\vec{r}) = \frac{1}{4\pi\varepsilon_0}\left[\frac{q}{r_a} + \frac{q}{r_b} - \frac{2q}{r}\right] = \frac{q}{4\pi\varepsilon_0 r}\left[\frac{r}{r_a} + \frac{r}{r_b} - 2\right] \tag{3.5-5}$$

Now assuming $r \gg l$ we can make the following expansions:

$$r_a = \left|\vec{r} - \vec{l}\right| = \left[\left(\vec{r} - \vec{l}\right) \cdot \left(\vec{r} - \vec{l}\right)\right]^{1/2}$$

$$= r\left[1 + \frac{l^2}{r^2} - \frac{2l\cos\theta}{r}\right]^{1/2} \qquad \left[\because \vec{r} \cdot \vec{l} = rl\cos\theta\right]$$

$$\therefore \quad \frac{r}{r_a} = \left[1 + \frac{l^2}{r^2} - \frac{2l\cos\theta}{r}\right]^{-1/2}$$

$$= 1 - \frac{1}{2}\left[\frac{l^2}{r^2} - \frac{2l\cos\theta}{r}\right] + \frac{3}{8}\left[\frac{l^2}{r^2} - \frac{2l\cos\theta}{r}\right]^2 + \cdots$$

$$= 1 + \frac{l}{r}\cos\theta + \left(\frac{l}{r}\right)^2 \frac{1}{2}\left(3\cos^2\theta - 1\right) + \cdots \tag{3.5-6}$$

Similarly,

$$\frac{r}{r_b} = 1 - \frac{l}{r}\cos\theta + \left(\frac{l}{r}\right)^2 \frac{1}{2}\left(3\cos^2\theta - 1\right) + \cdots \tag{3.5-7}$$

Now using Eqs. (3.5-6) and (3.5-7) and neglecting terms higher order then l^2/r^2 we get from Eq. (3.5-5),

$$\phi\left(\vec{r}\right) = \frac{ql^2}{4\pi\varepsilon_0 r^3}\left(3\cos^2\theta - 1\right) \tag{3.5-8}$$

2. Two dimensional quadrupole

Consider a two dimensional quadrupole consisting of four charges as shown in Fig. 3.5-2. Here the monopole and dipole moments are zero. The quadrupole tensor components can be calculated by using Eq. (3.5-1). Here $q_1 = q$, $q_2 = -q$, $q_3 = q$, $q_4 = -q$ and their coordinates are

$$\left(x'_{11}, x'_{21}, x'_{31}\right) \equiv \left(x'_1, y'_1, z'_1\right) = \left(\frac{a}{2}, \frac{a}{2}, 0\right)$$

$$\left(x'_{12}, x'_{22}, x'_{32}\right) \equiv \left(x'_2, y'_2, z'_2\right) = \left(\frac{a}{2}, -\frac{a}{2}, 0\right)$$

$$\left(x'_{13}, x'_{23}, x'_{33}\right) \equiv \left(x'_3, y'_3, z'_3\right) = \left(-\frac{a}{2}, -\frac{a}{2}, 0\right)$$

$$\left(x'_{14}, x'_{24}, x'_{34}\right) \equiv \left(x'_4, y'_4, z'_4\right) = \left(-\frac{a}{2}, \frac{a}{2}, 0\right)$$

For the point of observation, $(x_1, x_2, x_3) = (x, y, z)$.

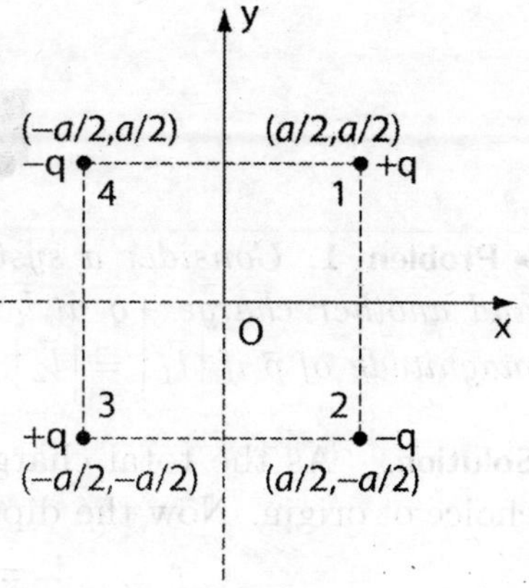

Fig. 3.5-2

Using these we get from Eq. (3.5-1),

$$\begin{aligned}
Q_{11} &= \sum_k \left(3x'_{1k}x'_{1k} - r'^2_k\right) q_k \\
&= \sum_k \left(3x'^2_k - r'^2_k\right) q_k \\
&= \sum_k \left(2x'^2_k - y'^2_k - z'^2_k\right) q_k \\
&= q\left(2\cdot\frac{a^2}{4} - \frac{a^2}{4}\right) - q\left(2\frac{a^2}{4} - \frac{a^2}{4}\right) + q\left(2\frac{a^2}{4} - \frac{a^2}{4}\right) - q\left(2\frac{a^2}{4} - \frac{a^2}{4}\right) \\
&= 0.
\end{aligned}$$

Similarly, $Q_{22} = Q_{33} = Q_{13} = Q_{31} = Q_{23} = Q_{32} = 0$

$$\begin{aligned}
Q_{12} = Q_{21} &= \sum_k 3x'_{1k}x'_{2k}q_k \\
&= 3\left(x'_1y'_1q_1 + x'_2y'_2q_2 + x'_3y'_3q_3 + x'_4y'_4q_4\right) \\
&= 3\left(\frac{a^2}{4}q + \frac{a^2}{4}q + \frac{a^2}{4}q + \frac{a^2}{4}q\right) = 3a^2q.
\end{aligned}$$

Therefore, the quadrupole potential $\phi_2(\vec{r})$ for $r \gg a$ is

$$\begin{aligned}
\phi_2(\vec{r}) &= \frac{1}{4\pi\varepsilon_0}\sum_i\sum_j \frac{x_ix_j}{2r^5}Q_{ij} \\
&= \frac{1}{4\pi\varepsilon_0}\cdot\frac{1}{2r^5}\left[x_1x_2Q_{12} + x_2x_1Q_{21}\right] \\
&= \frac{1}{4\pi\varepsilon_0}\cdot\frac{1}{2r^5}\cdot 2xy\cdot 3a^2q \\
&= \frac{qa^2}{4\pi\varepsilon_0}\cdot\frac{3xy}{r^5}.
\end{aligned}$$

SOLVED PROBLEMS

► **Problem 1.** *Consider a system of three charges: $-2q$ charge at the origin, $+q$ at $\vec{l}_1$ and another charge $+q$ at $\vec{l}_2$. Find the dipole moment $\vec{p}$ of the system. What is the magnitude of $\vec{p}$ if $\left|\vec{l}_1\right| = \left|\vec{l}_2\right| = l$ and angle between $\vec{l}_1$ and $\vec{l}_2$ is θ.*

Solution. As the total charge is zero the dipole moment will be independent of the choice of origin. Now the dipole moment with respect to the origin is

$$\vec{p} = \int \vec{r}'dq = \vec{l}_1\cdot(+q) + \vec{l}_2\cdot(+q) + 0\cdot(-2q) = q\left(\vec{l}_1 + \vec{l}_2\right).$$

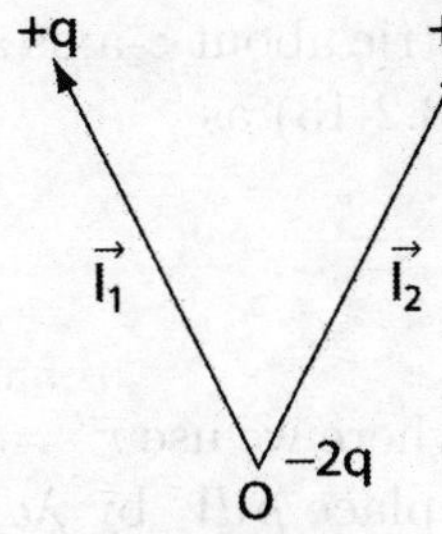

Fig. 3.P.-1

Its magnitude $p = |\vec{p}| = q\sqrt{l_1^2 + l_2^2 + 2l_1 l_2 \cos\theta} = ql\sqrt{2(1+\cos\theta)}$.

▶ **Problem 2.** *Consider a thin rod with its one end at $z = 0$ and the other end at $z = l$. Its charge density is given by $\rho(z) = \alpha(z - l/2)$ for $0 \le z \le l$. Area of cross-section of the rod is A. Find the total charge on the rod and dipole moment of the system.*

Solution. Total charge on the rod is

$$\int_0^l \rho(z)\cdot A dz = \alpha A \int_0^l \left(z - \frac{l}{2}\right) dz = 0.$$

So the dipole moment is independent on the choice of the origin. Now the dipole moment with respect the origin O (Fig. 3.P-2) is

$$\vec{p} = \int \vec{r}' dq = \hat{k}\int_0^l z\cdot(\rho A dz) = \hat{k}\alpha A \int_0^l z\left(z - \frac{l}{2}\right) dz = \hat{k}\alpha A \cdot \frac{l^3}{12}.$$

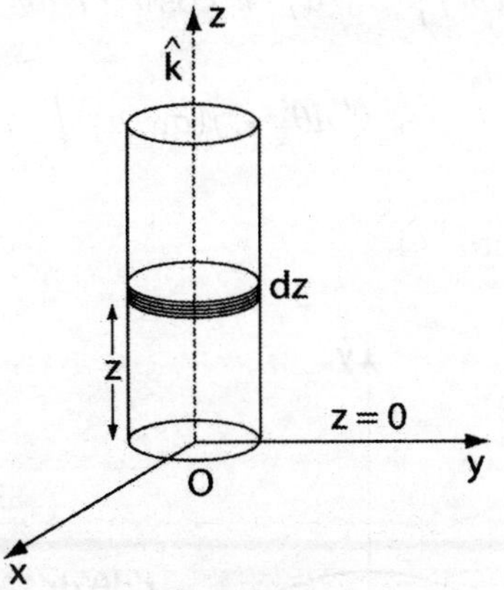

Fig. 3.P-2

▶ **Problem 3.** *Consider a uniform line distribution of charge extending from $z = -z_0$ to $z = z_0$ and having a charge λ per unit length. Find the dipole and quadrupole moments of the charge distribution.*

Solution. Dipole moment

$$\vec{p} = \int \vec{r}' dq = \int_{-z_0}^{+z_0} \hat{k}z' \cdot (\lambda dz') = \hat{k}\lambda \left[\frac{z'^2}{2}\right]_{-z_0}^{+z_0} = 0.$$

The charge distribution is symmetric about z-axis and we define the quadrupole moment, from Eq. (3.2-13) as

$$Q_{33} = \int_V \left(3z'^2 - r'^2\right) \rho\, dV$$

$$= \int_{-z_0}^{+z_0} \left(3z'^2 - z'^2\right) \lambda dz' \quad \text{[where we use } \vec{r}\,' = z'\hat{k} \text{ and replace } \rho\, dV \text{ by } \lambda dz'\text{]}$$

$$= 2\lambda \cdot \left[\frac{z'^3}{3}\right]_{-z_0}^{+z_0} = \frac{4}{3}\lambda z_0^3.$$

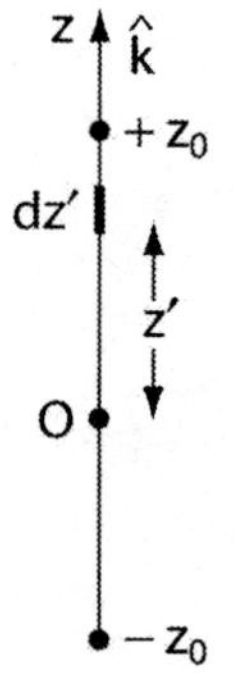

Fig. 3.P-3

▶ **Problem 4.** *Consider a circular disc of radius a in the xy-plane centred on the origin and containing a surface charge density* $\sigma(r,\theta) = \sigma_0(r/a)\cos\theta'$, *where* θ *is the angle made by the position vector with a fixed diameter chosen as x-axis. Find the dipole moment of the charge distribution.* **[C.U. 2015, 2017]**

Solution. Considering a surface element $r'd\theta' \cdot dr'$ about the point $\vec{r}\,'$ we can write for the dipole moment about the origin as

$$\vec{p} = \int \vec{r}\,' dq = \int \left(x'\hat{\imath} + y'\hat{\jmath}\right) \sigma \cdot r'd\theta'\, dr'$$

$$= \int \left(r'\cos\theta'\hat{\imath} + r'\sin\theta'\hat{\jmath}\right) (\sigma_0/a) \cdot r'\cos\theta' \cdot r'd\theta'\, dr'$$

$$= \hat{\imath}(\sigma_0/a) \int_0^a r'^3 dr' \int_0^{2\pi} \cos^2\theta'\, d\theta' + \hat{\jmath}(\sigma_0/a) \int_0^a r'^3 dr' \int_0^{2\pi} \sin\theta' \cos\theta'\, d\theta'$$

$$= \hat{\imath}\sigma_0 \frac{a^3}{4} \cdot \pi + 0 = \hat{\imath}\sigma_0 \pi a^3/4.$$

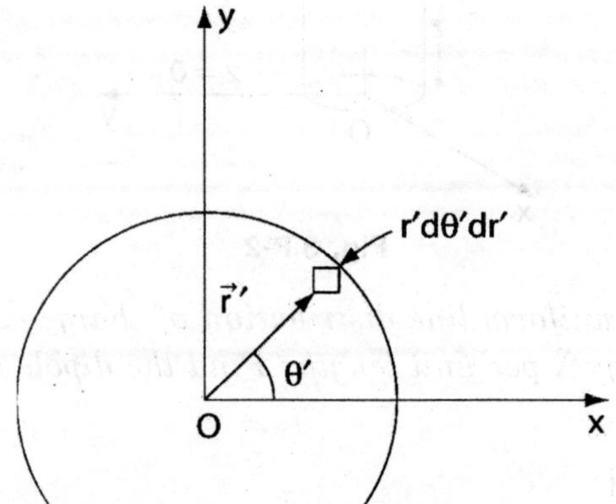

Fig. 3.P-4

▶ **Problem 5.** *Consider a spherical shell of radius a carrying a surface charge density $\sigma(\theta) = \sigma_0 \cos\theta$ where θ is the polar angle, which the position vector of the point under consideration with respect to the centre of the sphere makes with z-axis. Calculate the dipole moment of this charge distribution. Find the approximate potential considering two lowest orders in multipole expansion. The exact answer, obtained by some other method, is $\phi(r,\theta) = \frac{\sigma_0 a^3}{3\varepsilon_0} \cdot \frac{\cos\theta}{r^2}$. What is your conclusion regarding the higher multipoles?*

Solution. See, Section 3.4. Total charge is zero. So the monopole potential $\phi_0(\vec{r})$ is zero. The dipole moment is

$$\vec{p} = \frac{4\pi a^3 \sigma_0}{3}\hat{k}.$$

So the dipole potential is

$$\phi_1(\vec{r}) = \frac{1}{4\pi\varepsilon_0} \cdot \frac{\vec{p}\cdot\vec{r}}{r^3} = \frac{\sigma_0 a^3}{3\varepsilon_0} \cdot \frac{\cos\theta}{r^2}.$$

This is also the exact potential. So we can say that all multipole moments of this charge distribution except the dipole are zero.

▶ **Problem 6.** *Four charges q, −2q, 3q and −2q are placed respectively on the corners of a square of side a. Find an approximate expression for potential, correct to the order of r^{-2}, for points far away from the centre of the square.*

Solution. Total charge is zero, so the monopole potential term $\phi_0(\vec{r})$ is zero. The dipole moment of the system with respect to the origin O (Fig. 3.P-6) is

$$\vec{p} = \frac{a}{\sqrt{2}} \cdot 3q\hat{k} - \frac{a}{\sqrt{2}} \cdot q\hat{k} + \frac{a}{\sqrt{2}}(-2q)\hat{j} + \frac{a}{\sqrt{2}}(-2q)\left(-\hat{j}\right) = \frac{2aq}{\sqrt{2}}\hat{k}.$$

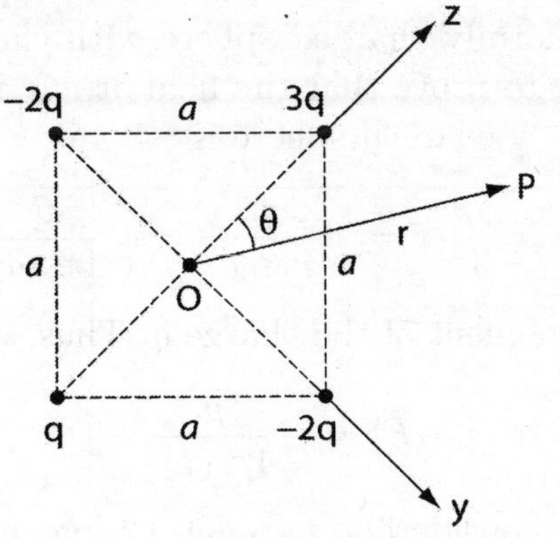

Fig. 3.P-6

So the dipole potential term is

$$\phi_1(\vec{r}) = \frac{1}{4\pi\varepsilon_0}\cdot\frac{\vec{p}\cdot\vec{r}}{r^3} = \frac{1}{4\pi\varepsilon_0}\cdot\frac{p\cos\theta}{r^2} = \frac{1}{4\pi\varepsilon_0}\cdot\frac{2aq}{\sqrt{2}}\cdot\frac{\cos\theta}{r^2}.$$

This represent the potential correct to the order of r^{-2}.

► **Problem 7.** *Show that the average field inside a sphere of radius a containing an arbitrary charge distribution is given by* $\langle\vec{E}\rangle = -\dfrac{\vec{p}}{4\pi\varepsilon_0 a^3}$*, where* $\vec{p}$ *is the total dipole moment of the charge distribution.*

Solution. Field produced by a point charge q at P at the volume element dV is

$$\vec{E} = \frac{1}{4\pi\varepsilon_0}\cdot\frac{q}{R^3}\vec{R},$$

where $\vec{R}$ is the vector from P to the volume element. So the average field within the sphere due to the charge q at P is

$$\langle\vec{E}\rangle = \frac{1}{\frac{4}{3}\pi a^3}\int \vec{E}\,dV = \frac{1}{4\pi\varepsilon_0}\int\frac{\left(\frac{q}{\frac{4}{3}\pi a^3}\right)\vec{R}}{R^3}\,dV = \frac{1}{4\pi\varepsilon_0}\int\frac{\left(\frac{-q}{\frac{4}{3}\pi a^3}\right)\left(-\vec{R}\right)}{R^3}\,dV.$$

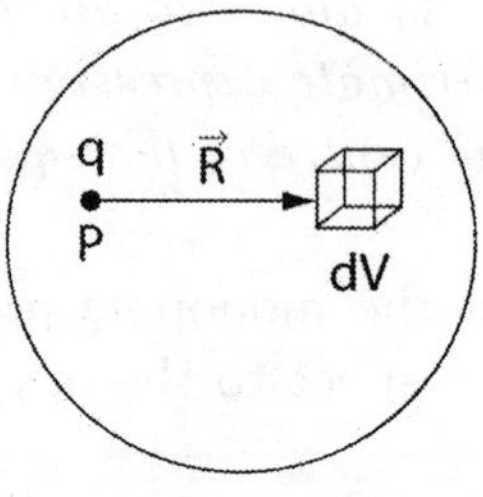

Fig. 3.P-7

As the vector connecting dV to P is $-\vec{R}$, the above expression may interpreted as the field at P due to a uniformly charged sphere with charge density $\rho = -q/\frac{4}{3}\pi a^3$. Using Gauss's law it is easy to prove that the field at any inside point at a distance r from the centre of a uniformly charged sphere is

$$\vec{E}_\rho = \frac{\rho}{3\varepsilon_0}\vec{r} = -\frac{q\vec{r}}{4\pi\varepsilon_0 a^3} = -\frac{\vec{p}}{4\pi\varepsilon_0 a^3},$$

where $\vec{p} = q\vec{r}$ is the dipole moment of the charge q. Thus, we prove that

$$\langle\vec{E}\rangle = -\frac{\vec{p}}{4\pi\varepsilon_0 a^3}.$$

We have proved it by considering a point charge only. By the principle of superposition above result is true for any arbitrary charge distribution.

▶ **Problem 8.** *Consider a physical electric dipole consisting of two equal and opposite charges* $+q$ *located at* $(0,0,a/2)$ *and* $-q$ *located at* $(0,0,-a/2)$. *Find potential at a point* (r,θ,φ) *to the order* r^{-4} *due to this dipole. What is the potential in the point dipole limit?*

Solution.

$$\phi(\vec{r}) = \frac{1}{4\pi\epsilon_0}\left(\frac{q}{AP} - \frac{q}{BP}\right).$$

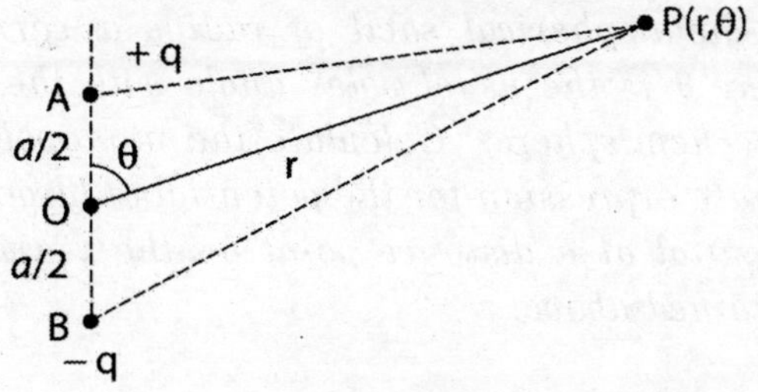

Fig. 3.P-8

From Fig. 3.P-8,

$$AP^2 = r^2 + \left(\frac{a}{2}\right)^2 - 2\cdot r\cdot\frac{a}{2}\cos\theta$$

$$= r^2\left[1 - 2\cdot\frac{a}{2r}\cdot\cos\theta + \left(\frac{a}{2r}\right)^2\right]$$

$$\therefore \quad \frac{1}{AP} = \frac{1}{r}\cdot\frac{1}{\sqrt{1 - 2\cdot\frac{a}{2r}\cdot\cos\theta + \left(\frac{a}{2r}\right)^2}}$$

$$= \frac{1}{r}\sum_{n=0}^{\infty}\left(\frac{a}{2r}\right)^n P_n(\cos\theta).$$

Similarly,

$$\frac{1}{BP} = \frac{1}{r}\cdot\frac{1}{\sqrt{1 + 2\cdot\frac{a}{2r}\cdot\cos\theta + \left(\frac{a}{2r}\right)^2}} = \frac{1}{r}\sum_{n=0}^{\infty}\left(\frac{a}{2r}\right)^n P_n(-\cos\theta).$$

$$\therefore \quad \phi(\vec{r}) = \frac{q}{4\pi\epsilon_0}\cdot\frac{1}{r}\cdot\sum_{n=0}^{\infty}\left(\frac{a}{2r}\right)^n\left[P_n(\cos\theta) - P_n(-\cos\theta)\right]$$

$$= \frac{2q}{4\pi\epsilon_0 r}\sum_{n=1,3,5}\left(\frac{a}{2r}\right)^n P_n(\cos\theta) \quad [\text{Using } P_n(-x) = (-1)^n P_n(x)]$$

$$= \frac{qa}{4\pi\epsilon_0}\cdot\frac{\cos\theta}{r^2} + \frac{qa^3}{4\pi\epsilon_0}\cdot\frac{P_3(\cos\theta)}{8r^4} + \cdots$$

(dipole term) (octopole term)

Obviously the quadrupole term is zero.

In the limit of a point dipole $a \to 0$, $q \to \infty$ such that the product qa remains constant. In this limit the contributions of higher terms become zero and the potential is exactly given by

$$\phi(\vec{r}) = \frac{1}{4\pi\varepsilon_0}\frac{\vec{p}\cdot\hat{r}}{r^2} = \frac{1}{4\pi\varepsilon_0}\cdot\frac{p\cos\theta}{r^2},$$

where $\vec{p}$ is the dipole moment.

▶ **Problem 9.** *Consider a hemispherical shell of radius a carrying a surface charge density $\sigma = \sigma_0\cos\theta$, where θ is the usual polar angle with the z-axis passing through the centre and pole of the hemisphere. Calculate the monopole and dipole moments. Hence, write an approximate expression for the potential at large distance from it. Also, calculate directly the potential at a distance point on the z-axis and comment on the validity of your result obtained above.*

Solution. The monopole moment

$$\begin{aligned} &= \int dq = \int_0^{\pi/2} 2\pi a\sin\theta'\cdot a d\theta'\cdot\sigma_0\cos\theta' \\ &= \pi a^2\sigma_0\int_0^{\pi/2}\sin 2\theta' d\theta' \\ &= \pi a^2\sigma_0. \end{aligned}$$

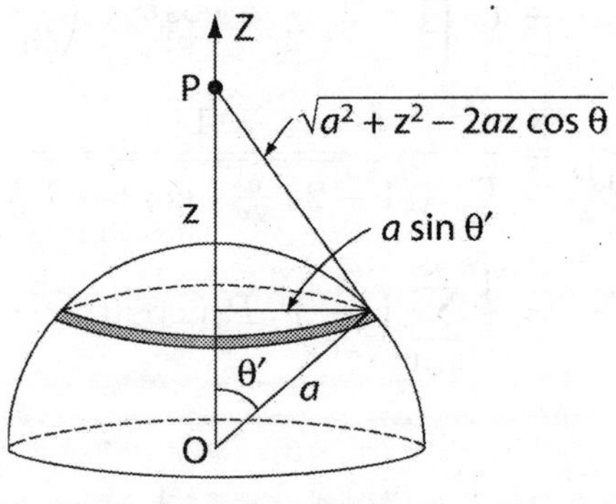

Fig. 3.P-9

Symmetry of the problem indicates that the dipole moment will be in the z-direction and will be of magnitude

$$\begin{aligned} p &= \int r'\cos\theta'\cdot\sigma dS \\ &= \int a\cos\theta'\cdot\sigma_0\cos\theta'\cdot a^2\sin\theta' d\theta' d\varphi' \\ &= \sigma_0 a^3\int_0^{\pi/2}\cos^2\theta'\sin\theta' d\theta'\int_0^{2\pi} d\varphi' \\ &= \frac{2\pi}{3}\sigma_0 a^3. \end{aligned}$$

∴ The potential at any external point (r, θ) is given by

$$\phi(\vec{r}) = \phi_{\text{monopole}} + \phi_{\text{dipole}} + \cdots$$

$$= \frac{1}{4\pi\varepsilon_0}\frac{\pi a^2\sigma_0}{r} + \frac{1}{4\pi\varepsilon_0}\cdot\frac{\frac{2\pi}{3}\sigma_0 a^3\cos\theta}{r^2} + \cdots$$

For any point P on the z-axis $(r = z, \theta = 0)$

$$\phi(P) = \frac{1}{4\pi\varepsilon_0}\cdot\frac{\pi a^2\sigma_0}{z} + \frac{1}{4\pi\varepsilon_0}\cdot\frac{\frac{2\pi}{3}\sigma_0 a^3}{z^2} + \cdots$$

Let us now calculate the potential at P directly.

$$\phi(P) = \frac{1}{4\pi\varepsilon_0}\cdot\int_0^{\pi/2}\frac{\sigma_0\cos\theta' 2\pi a^2\sin\theta' d\theta'}{\sqrt{a^2+z^2-2az\cos\theta'}}.$$

Putting $a^2 + z^2 - 2az\cos\theta' = s^2$ and $az\sin\theta' d\theta' = sds$ we get

$$\phi(P) = \frac{1}{4\pi\varepsilon_0}\cdot\frac{2\pi a\sigma_0}{z}\int_{z-a}^{\sqrt{z^2+a^2}}\frac{s\cos\theta' ds}{s}$$

$$= \frac{1}{4\pi\varepsilon_0}\cdot\frac{2\pi a\sigma_0}{z}\int_{z-a}^{\sqrt{z^2+a^2}}\frac{a^2+z^2-s^2}{2az}ds$$

$$= \frac{1}{4\pi\varepsilon_0}\cdot\frac{\pi\sigma_0}{z^2}\left[\left(a^2+z^2\right)s - \frac{s^3}{3}\right]_{z-a}^{\sqrt{z^2+a^2}}.$$

$$\text{or } \phi(P) = \frac{1}{4\pi\varepsilon_0}\cdot\frac{\pi\sigma_0}{z^2}\left[\left(a^2+z^2\right)\left(\sqrt{z^2+a^2}-z+a\right) - \frac{1}{3}\left(z^2+a^2\right)^{3/2} + \frac{1}{3}(z-a)^3\right].$$

For $z \gg a$ we can write

$$\phi(P) = \frac{1}{4\pi\varepsilon_0}\cdot\frac{\pi r_0}{z^2}\left[z^3\left(1+\frac{3}{2}\frac{a^2}{z^2}+\cdots\right) - a^2z + a^3 - z^3 + az^2\right.$$

$$\left. -\frac{1}{3}z^3\left(1+\frac{3}{2}\frac{a^2}{z^2}+\cdots\right) + \frac{1}{3}\left(z^3 - 3z^2a + 3za^2 - a^3\right)\right]$$

$$= \frac{1}{4\pi\varepsilon_0}\cdot\frac{\pi a^2\sigma_0}{z} + \frac{1}{4\pi\varepsilon_0}\frac{\frac{2}{3}\pi\sigma_0 a^3}{z^2} + \cdots$$

$$= \phi_{\text{monopole}} + \phi_{\text{dipole}} + \cdots$$

This result agrees with that obtained by multipole expansion. Thus, the example justifies the use of multipole expansion in calculating the potential due to arbitrary charge distribution.

► **Problem 10.** *If a charge distribution is spherically symmetric show that all but the monopole term of the multipole expansion corresponding to this distribution vanish.*

[C.U. 2017]

Solution. We know that the potential $\phi(\vec{r})$ at any external point $P(\vec{r})$ due to an arbitrary charge distribution can be expressed as a sum of multiple terms.

$$\phi(\vec{r}) = \frac{1}{4\pi\varepsilon_0}\left[\frac{1}{r}\int \rho(\vec{r}')\,dV + \frac{1}{r^2}\int r'\cos\theta\rho(\vec{r}')\,dV + \frac{1}{r^3}\int \frac{1}{2}\left(3r'^2\cos^2\theta - r'\right)^2 dV + \cdots\right]$$
$$= \phi_{\text{monopole}} + \phi_{\text{dipole}} + \phi_{\text{quadrupole}} + \cdots \qquad (1)$$

Because of spherical symmetry we can directly determine field and potential by using Gauss's law. Thus,

$$\oint_S \vec{E}\cdot d\vec{S} = \frac{1}{\varepsilon_0}q_{\text{enc}} \quad \text{or} \quad E\cdot 4\pi r^2 = \frac{1}{\varepsilon_0}\cdot q_{\text{enc}}$$
$$\vec{E} = \frac{1}{4\pi\varepsilon_0}\cdot\frac{q_{\text{enc}}}{r^2}\hat{r},$$

where q_{enc} is the charge enclosed by the spherical surface S.

Potential is given by

$$\phi(\vec{r}) = -\int_\infty^r \vec{E}\cdot d\vec{r} = \frac{1}{4\pi\varepsilon_0}\cdot\frac{q_{\text{enc}}}{r}.$$

This is an exact relation. Since $\int \rho(\vec{r}')\,dV = q_{\text{enc}}$, Eq. (1) indicates that only the monopole term gives the correct potential at P. Hence, dipole, quadrupole and other higher multipole moments must be zero for spherically symmetric charge distribution.

A nucleus cannot have any electric dipole moment and for this any deviation of it from spherical symmetry is detected by the measurement of its quadrupole moment.

QUESTIONS

1. (a) Show that the potential due to an arbitrary charge distribution of finite extent at a large distance can be expanded as a sum of multipole potentials. Comment on the convergence of this expansion. **[C.U. 2018]**

   (b) On the basis of it, explain the terms monopole moment, dipole moment and quadrupole moment.

2. (a) Show that the dipole moment of an arbitrary charge distribution is independent of the choice of origin if the total charge of the distribution is zero. **[C.U. 2015, 2017]**

   (b) A point charge at the origin constitutes a 'pure' monopole. But a point charge not at the origin cannot be considered as a 'pure' monopole. Explain. **[J.U. 2012]**

3. Consider a linear quadrupole formed by a charge $-2q$ at the origin, $+q$ at $(0, 0, +l)$ and $+q$ at $(0, 0, -l)$. Calculate the first three multipole moments of the charge distribution. Hence show that the potential $\phi(\vec{r})$ at a distance large compared with l is approximately given by

$$\phi(\vec{r}) = \frac{ql^2}{4\pi\varepsilon_0 r^3}\left(3\cos^2\theta - 1\right),$$

where θ is the angle between $\vec{r}$ and the axis of the quadrupole. **[C.U. 2016]**

4. A spherical shell of radius a carries a surface charge of density $\sigma = \sigma_0 \cos\theta$, where θ is the angle which the position vector of the point under consideration makes with z-axis. (a) Show that the dipole moment of the system is $\frac{4}{3}\pi\sigma_0 a^3\hat{k}$. (b) Show that the monopole potential does not exist for this charge distribution. **[C.U. 2016]**

5. Show that the quadrupole moment of a physical dipole is zero. **[C.U. 2015]**

6. For a physical dispole i.e., two point charges $+q$ and $-q$ separated by a finite distance 'a' determine the quadrupole moment and octopole terms in the potential. **[J.U. 2012]**

7. Find an expression for the quadrupole moment of an orbitrary charge distribution. **[B.U. 2015]**

EXERCISES

1. Suppose there is a uniform surface density of charge σ over the curved surface of a hemisphere of radius a. Calculate the dipole moment of the distribution about the centre of curvature. [*Ans.* $\pi\sigma a^3\hat{k}$, $\hat{k}$ being unit normal to the plane face.]
2. Show that the dipole moment of a uniformly charged circular ring with respect to a origin at its centre is zero.
3. (a) Three point charges $+2q$, $+q$ and $-q$ are located in xy-plane at points $(1, 1)$, $(-1, 1)$ and $(0, -1)$, respectively. Find the dipole moment of the system with respect to the origin. [*Ans.* $\vec{p} = q\left(\vec{i} + 4\vec{j}\right)$.]

   (b) Three point charges are locatd at the following points: $+2q$ at $(0, a, a)$; $+q$ at $(0, -a, a)$ and $-q$ at $(0, 0, -a)$. Calculate the dipole moment about the origin. [*Ans.* $qa(\hat{j} + 4\hat{k})$] **[C.U. 2012]**

   (c) Three charges $+q$, $+q$ and $-2q$ are placed at the vertices of an equilateral triangle of side 'a' as shown in Fig. 3.E-3. Calculate the total dipole moment of this configuration. [*Ans.* $-\sqrt{3}qa\hat{j}$] **[C.U. 2014, 2018]**

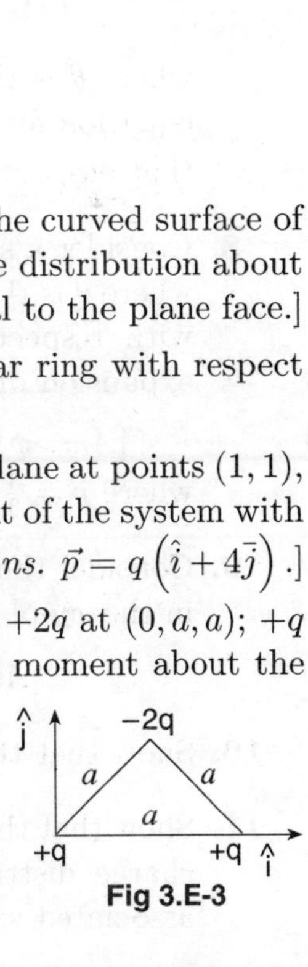

Fig 3.E-3

4. Find the quadrupole tensor components of the charge distribution consisting of three point charges $+2q$, $+q$ and $-q$ located in the xy-plane at points (a, a), $(-a, a)$ and $(0, -a)$, respectively.

[*Ans.* $Q_{11}=4qa^2$, $Q_{22}=qa^2$, $Q_{33}=-5qa^2$, $Q_{12}=Q_{21}=3qa^2$, $Q_{23}=Q_{32}=0$, $Q_{13}=Q_{31}=0$.]

5. Suppose a charge q is uniformly distributed over a cylindrical volume of radius R and height H centred on the origin and with its axis along z-axis. Find the quadrupole moment Q_{33} of the system. [*Ans.* $\frac{q}{6}\left(H^2 - 3R^2\right)$.]

6. Consider a two dimensional quadrupole consisting of four charges $+q$, $-q$, $+q$ and $-q$ located at the points $(a\sqrt{2}, 0)$, $(0, a\sqrt{2})$, $(-a\sqrt{2}, 0)$ and $(0, -a\sqrt{2})$, respectively.

Show that the quadrupole moment tensor takes the form

$$\begin{pmatrix} 12qa^2 & 0 & 0 \\ 0 & -12qa^2 & 0 \\ 0 & 0 & 0 \end{pmatrix}$$

7. Consider a linear quadrupole consisting of a charge $-2q$ at the origin, $+q$ at $(a, 0)$ and $+q$ at $(-a, 0)$. The potential at any point (r, θ) is approximately given by

$$\phi(r, \theta) = \frac{qa^2}{4\pi\epsilon_0 r^3}(3\cos^2\theta - 1),$$

where θ is the angle between $\vec{r}$ and the axis of the quadrupole. Show that the equation for the lines of force is given by $\sin^2\theta \cdot \cos\theta / r^2 = $ constant. Also show that electric field $E(r, \theta) = \frac{3qa^2}{4\pi\epsilon_0 r^4}\sqrt{5\cos^4\theta - 2\cos^2\theta + 1}$.

8. Consider a spherical shell of radius a with a surface charge density $\sigma = \sigma_0 \cos\theta$, where θ is the angle which the position vector, $\vec{r}$, of the point under consideration with respect to the centre of the shell makes with z-axis. Using multipole expansion find the potential at any outside point $(r > a)$.

[*Ans.* $\phi(\vec{r}) = \phi_0(\vec{r}) + \phi_1(\vec{r}) + \phi_2(\vec{r}) + \cdots = 0 + \phi_1(\vec{r}) + 0 + \cdots = \frac{1}{4\pi\epsilon_0}\frac{p\cos\theta}{r^2}$ where $p = \frac{4}{3}\pi a^3 \sigma_0$.]

9. Consider the charge distribution in problem 6 and find the potential at any point in the xy-plane.

[*Ans.* $\phi(x, y) = \frac{1}{4\pi\epsilon_0} \cdot \frac{12qa^2}{2r^5}\left(x^2 - y^2\right)$ or $\phi(r, \theta) = \frac{1}{4\pi\epsilon_0} \cdot \frac{12qa^2}{2r^3}\cos 2\theta$.]

10. Show that the quadrupole moment vanishes for a spherical charge distribution.

11. Show that the average electric field inside a sphere of radius a containing arbitrary charge distribution is given by $\vec{E} = -\frac{\vec{p}}{4\pi\epsilon_0 a^3}$, where $\vec{p}$ is the dipole moment associated with the charge distribution.

12. Suppose three point charges $-q$, q and $-q$ are respectively placed at the points $(0, a, 0)$, $(0, 0, a)$ and $(0, -a, 0)$. Find the approximate potential, correct up to order r^{-2}, and electric field correct up to order r^{-3} at points far away from the origin. [*Ans.* $\phi(r, \theta) = \frac{q}{4\pi\epsilon_0}\left\{-\frac{1}{r} + \frac{a\cos\theta}{r^2}\right\}$, $\vec{E}(r, \theta) = \frac{q}{4\pi\epsilon_0}\left\{-\frac{1}{r^2}\hat{r} + \frac{a}{r^3}\left(2\cos\theta\hat{r} + \sin\theta\hat{\theta}\right)\right\}$, where θ is the angle between $\hat{z}$-axis and $\vec{r}$.]

[**Hints.** Here total charge is not zero. So monopole term exists]

13. Find the dipole moment about the centre of a uniformly charged disc of radius a having a surface charge density σ. [*Ans.* 0]

14. Suppose a sphere of radius a centred at the origin, carries a charge density $\rho(r, \theta) = \rho_0 \cdot \frac{a(a-2r)}{r^2}\sin\theta$, where ρ_0 is a constant and (r, θ) are the usual spherical coordinates. Show that both the monopole and dipole moments of the charge distribution are zero.

15. A sphere of radius R centred at the origin carries a charge density $\rho(r, \theta) = K \cdot \frac{R}{r^2}(R - 2r)\sin\theta$ where K is a constant and r, θ are usual spherical coordinates. Find the approximate potential for points on z-axis far from the sphere. **[C.U. 2012]**

OR, show that quadrupole term is the most dominant term in the expression of potential far from the sphere. **[J.U. 2011]**

[**Hints.** Monopole term $\phi_0 \propto \int \rho dV = \int_{r=0}^{R}\int_{\theta=0}^{\pi}\int_{\phi=0}^{2\pi} \rho(r', \theta)\cdot r'^2 \sin\theta d\theta\, d\phi = 0$

Dipole term $\phi_1 \propto \int r' \cos\theta \rho dV = 0$

$$\text{Quadrupole term} = \frac{1}{4\pi\varepsilon_0 r^3}\int r'^2 \cdot \frac{1}{2}(3\cos^2\theta - 1)\rho dV \approx \frac{1}{4\pi\varepsilon_0}\cdot\frac{K\pi^2 R^5}{48z^3} \quad (r \to z)\ ]$$

16. A solid sphere of radius a centred at the origin has uniform charge density $+\rho_0$ for one hemisphere and $-\rho_0$ for the other hemisphere. Find the dipole moment of the system. [*Ans.* $\frac{1}{2}\pi\rho_0 a$]

OBJECTIVE TEST

1. The deviation of charge distribution of a nucleus from spherical symmetry can be estimated by measuring its:

(a) electric dipole moment, (c) electric quadrupole moment,
(b) magnetic dipole moment, (d) electric charge.

2. The electric field due to an electric quadrupole varies with distance r as:

(a) $\frac{1}{r^2}$, (b) $\frac{1}{r^3}$, (c) $\frac{1}{r^4}$, (d) none of these.

3. If the total charge of a system is zero then its dipole moment is:

(a) zero, (c) independent on the choice of origin,
(b) dependent on the choice of origin, (d) none of these.

4. Four charges q, q, $-2q$ and q are placed at the corners of a square of side a having coordinates $(0,0)$, $(a,0)$, (a,a) and $(0,a)$, respectively. The electric dipole moment of the configuration about the origin is:

(a) $qa\left(\hat{\imath}+\hat{\jmath}\right)$, (b) $-qa\left(\hat{\imath}+\hat{\jmath}\right)$, (c) $qa\left(\hat{\imath}-\hat{\jmath}\right)$, (d) $-qa\left(\hat{\imath}-\hat{\jmath}\right)$.

[C.U. 2016]

5. The electric field due to an electric dipole varies with distance r as:

(a) $\frac{1}{r^2}$, (b) $\frac{1}{r^3}$, (c) $\frac{1}{r^4}$, (d) none of these.

6. For a spherically symmetric charge distribution quadrupole moment is:

(a) always positive, (b) always negative, (c) always zero, (d) undefined.

7. Number of component/components sufficient to describe the electric quadrupole moment tensor for a cylindrically symmetric charge distribution is/are:

(a) 1, (b) 3, (c) 6, (d) 9.

8. Suppose two charges $+q$ and $-q$ are placed at the points $(a,0)$ and $(-a,0)$ in xy plane. If the origin is moved to the point (x,y) the magnitude of dipole moment with respect to new origin would be

(a) $q(a-x)-q(-a+x)$, (b) $q\sqrt{(a-x)^2+y^2}-q\sqrt{(a+x)^2+y^2}$,
(c) $2qa$, (d) zero.

9. Four point charges $+q,+q,-q,-q$ are placed successively at the four corners of a square of side a. The dipole moment of the system is of magnitude

(a) $2qa$ (b) $2\sqrt{2}qa$, (c) qa, (d) 0

10. Four point charges $+q,-q,+q$ and $-q$ are successively at the four corners of a square of side a. At a large distance $r \gg a$ from the square the dominant multipole potential term will be propertional to

(a) $1/r$, (b) $1/r^2$, (c) $1/r^3$, (d) $1/r^4$

ANSWERS

1. (c) **2.** (c) **3.** (c) **4.** (b) **5.** (b) **6.** (c) **7.** (a)
8. (c) **9.** (a) **10.** (c)

Chapter **4**

Electrostatic Energy and Capacitors

4.1 Introduction

The solution of many mechanical problems gets very much simplified by means of energy consideration. The study of the mechanical behaviour of an electrical system is also found to be simplified by the use of energy consideration. In this chapter we shall calculate the energy associated with various electrostatic charge distributions. For an electrostatic system no kinetic energy is imparted to the charges and the energy is wholly potential in nature. The work necessary to assemble a system of charges against coulomb forces is stored in the system as a potential energy. This is known as *electrostatic energy.* Here it is assumed that the charges are brought to their positions at rest from their initial positions at rest at infinite distance apart. From a knowledge of this electrostatic energy it is possible to calculate the forces and torques acting on the system.

4.2 Electrostatic Energy of an Assembly of Point Charges

Let us calculate the electrostatic energy of an assembly of point charges. We can do this by calculating the work done in assembling the system by bringing the charges in one by one from positions at infinite distance apart. To place the first charge q_1 at the position $\vec{r}_1$ we require no work ($u_1 = 0$) because there is no interacting coulomb field.

Next to bring the charge q_2 to the position $\vec{r}_2$ we require work to be done against the Coulomb repulsion due to q_1. This work equals

$$u_{12} = -\frac{1}{4\pi\varepsilon_0}\int_{\infty}^{r_{12}} \frac{q_1 q_2}{r^2}\,dr = \frac{1}{4\pi\varepsilon_0}\frac{q_1 q_2}{r_{12}}, \tag{4.2-1}$$

where $r_{12} = |\,\vec{r}_1 - \vec{r}_2\,|$ is the distance of q_2 from q_1.

Next to bring the charge q_3 to the position $\vec{r}_3$ work is to be done against the field of both q_1 and q_2. So the work required is

$$\frac{1}{4\pi\varepsilon_0}\frac{q_1 q_3}{r_{13}} + \frac{1}{4\pi\varepsilon_0}\frac{q_2 q_3}{r_{23}} = u_{13} + u_{23}. \tag{4.2-2}$$

Thus, total electrostatic energy of a system of three charges is $u_{12}+u_{13}+u_{23}$, which is equal to the sum of potential energies of various pairs.

Proceeding in this way to bring the charges one after another we find that the total work involved in assembling N point charges is

$$U = u_{12} + u_{13} + u_{23} + \cdots = \sum_{\text{all pairs}} \frac{1}{4\pi\varepsilon_0}\frac{q_i q_j}{r_{ij}}, \tag{4.2-3}$$

where q_i and q_j are a pair of charges separated by a distance r_{ij}.

For N point charges the expression for U can also be written as

$$U = \frac{1}{2}\sum_{i=1}^{N}\sum_{\substack{j=1\\ j\neq i}}^{N} \frac{1}{4\pi\varepsilon_0}\frac{q_i q_j}{r_{ij}}, \tag{4.2-4}$$

where the factor $\frac{1}{2}$ is included to avoid double counting of each pair. Note that the terms with $j = i$ are excluded because it represents self-terms.

The electrostatic energy U can also be written in terms of the electrostatic potential. Thus,

$$U = \frac{1}{2}\sum_{i=1}^{N} q_i \phi_i, \tag{4.2-5}$$

where

$$\phi_i = \sum_{\substack{j=1\\ j\neq i}}^{N} \frac{1}{4\pi\varepsilon_0}\frac{q_j}{r_{ij}}, \tag{4.2-6}$$

is the potential at the location of the ith charge due to all other charges excepting q_i.

4.3 Electrostatic Energy of a Continuous Charge Distribution

For a region of space having a continuous charge distribution characterised by a volume density $\rho(\vec{r})$ over a volume V the expression (4.2-5) for U can be written in the form

$$U = \frac{1}{2}\int_V \rho(\vec{r})\,\phi(\vec{r})\,dV. \tag{4.3-1}$$

For a general type of charge distribution characterised by a volume density $\rho(\vec{r})$ over a volume V, surface density $\sigma(\vec{r})$ over a surface S, line density $\lambda(\vec{r})$ over a length l and points charges $q_i (i = 1, 2, \ldots, N)$ total electrostatic energy of the system will be given by

$$U = \frac{1}{2}\int_V \rho(\vec{r})\,\phi(\vec{r})\,dV + \frac{1}{2}\int_S \sigma(\vec{r})\,\phi(\vec{r})\,dS + \frac{1}{2}\int_l \lambda(\vec{r})\,\phi(\vec{r})\,dl + \frac{1}{2}\sum_{i=1}^{N} q_i\phi_i. \tag{4.3-2}$$

4.4 Electrostatic Energy in Terms of Field Distribution

In some cases it becomes important to express electrostatic energy in terms of field vectors $\vec{E}$ and $\vec{D}$ of the system. Suppose we have a finite region of space V in a dielectric medium of permittivity ε, characterised by a volume density ρ of free charges. The electrostatic energy for this system is given by

$$U = \frac{1}{2}\int_V \rho(\vec{r})\,\phi(\vec{r})\,dV. \tag{4.4-1}$$

Now using the differential form of Gauss's law, $\vec{\nabla}\cdot\vec{D} = \rho$, we can write

$$U = \frac{1}{2}\int_V \left(\vec{\nabla}\cdot\vec{D}\right)\phi\,dV. \tag{4.4-2}$$

Using the vector identity

$$\vec{\nabla}\cdot\left(\phi\vec{D}\right) = \vec{\nabla}\phi\cdot\vec{D} + \phi\left(\vec{\nabla}\cdot\vec{D}\right)$$

we get

$$U = \frac{1}{2}\int_V \vec{\nabla}\cdot\left(\phi\vec{D}\right)dV - \frac{1}{2}\int_V \vec{\nabla}\phi\cdot\vec{D}\,dV. \tag{4.4-3}$$

By using divergence theorem and the relation $\vec{E} = -\vec{\nabla}\phi$ we get

$$U = \frac{1}{2}\oint_S \left(\phi\vec{D}\right)\cdot d\vec{S} + \frac{1}{2}\int_V \vec{E}\cdot\vec{D}\,dV, \tag{4.4-4}$$

where S is the surface enclosing the volume V. Let us now allow the region of integration V to expand to infinity. This is permissible since from Eq. (4.4-1) we find that the contribution to the integral will be zero wherever $\rho = 0$. Now at large distances from the charge distribution ϕ falls off as $1/r$, $\vec{D}$ as $1/r^2$ while the surface area S grows like r^2. So, roughly speaking, the surface integral in Eq. (4.4-4) goes down like $1/r$. Thus, as the surface S is expanded to include all of space the surface integral vanishes and we are left with

$$U = \frac{1}{2}\int_{\text{all space}} \vec{E}\cdot\vec{D}\,dV. \tag{4.4-5}$$

This equation suggests that we may consider the electrostatic energy as stored in the electric field with an energy density (i.e., energy per unit volume)

$$u = \tfrac{1}{2}\vec{E}\cdot\vec{D}. \tag{4.4-6}$$

In case of linear isotropic dielectrics $\vec{D} = \varepsilon\vec{E}$ and we can write $u = \frac{1}{2}\varepsilon E^2$. For charges in vacuum, $u = \frac{1}{2}\varepsilon_0 E^2$.

4.5 Where is the Electrostatic Energy Stored?

Equations (4.4-1) and (4.4-5) show that the electrostatic energy can be expressed either as an integral over charge distribution or as an integral over the electrostatic field. Thus, one can assume that the electrostatic energy is stored either in the charge or in the field. Both the ideas lead to the same total electrostatic energy. From the stand point of electrostatics alone it is not possible to tell where the energy is actually stored. However, in non-static case (e.g., in radiation theory) it is found to be more useful to consider that the energy is stored in the field. When a charged particle oscillates it emits electromagnetic wave. The wave travels from one point to another and carries energy with it but there is no charge in the wave. Thus, it seems reasonable to locate energy within the field and not at the charges, which produce the field.

4.6 Self-Energy of a Point Charge

Let us examine whether the energy expressions (4.2-5) and (4.4-5) are equivalent or not. According to Eq. (4.4-5) U is always positive (since $\vec{E}\cdot\vec{D} = \varepsilon E^2$ is always positive) whereas according to Eq. (4.2-5) U can take up positive as well as negative values. This means that there is some intriguing difference between the two equations. In Eq. (4.2-5) we do not include the work required to fabricate the point charges. Here we start with ready made point charges and take into account only the work done in bringing them to their desired positions. Equation (4.4-5), on the other hand, gives the total energy stored in a charge distribution including the energy required to fabricate finite point charges from infinitesimal parts. Energy given by the Eq. (4.4-5) is sometimes called the *self-energy* of the charge distribution. Let us calculate the self-energy U of a point charge q placed in vacuum by applying Eq. (4.4-5). Thus,

$$U = \tfrac{1}{2}\varepsilon_0 \int\limits_{\text{all space}} E^2\,dV = \tfrac{1}{2}\varepsilon_0 \int_{r=0}^{\infty} \left(\frac{1}{4\pi\varepsilon_0}\cdot\frac{q}{r^2}\right)^2 \cdot 4\pi r^2 dr = -\frac{q^2}{8\pi\varepsilon_0}\left[\frac{1}{r}\right]_0^{\infty}.$$

Obviously the result comes out to be infinite, which is physically absurd. One way out of the difficulty is to consider finite radius of point charges. This leads to the idea of the *classical* radius of an electron.

Discussion

We have derived Eq. (4.4-5) by using Eq. (4.2-5). Then naturally the question arises *why do these equations describe two different situations?.* In Eq. (4.2-5) ϕ_i is the potential due to all other charges excepting q_i. But when we go over to continuous distribution of charge in Eq. (4.3-1) $\phi(\vec{r})$ represents the total potential at the point $\vec{r}$ because for a continuous distribution the charge right at the point can be taken to be vanishingly small.

4.7 Electrostatic Self-Energy of a Uniformly Charged Sphere

Method I (Direct-method)

Suppose the sphere is built up by assembling a succession of thin spherical shells of infinitesimal thickness. Let at any stage the radius is r. Now to increase the radius from r to $r + dr$ we require to bring an amount of charge $dq = \rho \cdot 4\pi r^2 dr$ from infinity, where ρ is the density of charge. When the radius of the sphere is r, its charge will be $q = \frac{4}{3}\pi r^3 \rho$ and the potential at its surface will be $q/4\pi\varepsilon_0 r$. To add the dq amount of charge the work necessary will be

$$dU = \frac{q}{4\pi\varepsilon_0 r} \times dq = \frac{\frac{4}{3}\pi r^3 \rho}{4\pi\varepsilon_0 r} \times \rho \cdot 4\pi r^2 dr = \frac{4\pi\rho^2}{3\varepsilon_0} \cdot r^4 dr.$$

Therefore, total work required to build up the sphere up to a radius a is

$$U = \frac{4\pi\rho^2}{3\varepsilon_0} \int_0^a r^4 dr = \frac{4\pi\rho^2}{3\varepsilon_0} \cdot \frac{a^5}{5} = \frac{1}{4\pi\varepsilon_0} \cdot \frac{3Q^2}{5a}, \qquad (4.7\text{-}1)$$

where $Q = \frac{4}{3}\pi a^3 \rho$ is the total charge on the sphere.

Method II [using Eq. (4.4-5)]

$$U = \frac{1}{2}\varepsilon_0 \int_{\text{inside}} E^2\, dV + \frac{1}{2}\varepsilon_0 \int_{\text{outside}} E^2\, dV$$

Now

$$E = \frac{1}{4\pi\varepsilon_0} \cdot \frac{Q_r}{r^2} = \frac{1}{4\pi\varepsilon_0} \frac{\frac{4}{3}\pi r^3 \rho}{r^2} = \frac{1}{4\pi\varepsilon_0} \cdot \frac{Q \cdot r}{a^3} \text{ for } r < a.$$

$$E = \frac{1}{4\pi\varepsilon_0} \cdot \frac{Q}{r^2} \text{ for } r > a$$

$$\therefore \quad U = \frac{\varepsilon_0}{2}\int_0^a \left(\frac{Qr}{4\pi\varepsilon_0 a^3}\right)^2 \cdot 4\pi r^2 dr + \frac{\varepsilon_0}{2}\int_a^\infty \left(\frac{Q}{4\pi\varepsilon_0 r^2}\right)^2 \cdot 4\pi r^2 dr$$

$$= \frac{Q^2}{2}\cdot\frac{1}{4\pi\varepsilon_0}\left(\frac{1}{5a}+\frac{1}{a}\right) = \frac{1}{4\pi\varepsilon_0}\cdot\frac{3Q^2}{5a}$$

Method III [using Eq. (4.3-1)]

From Eq. (1.8-11) the potential inside a uniformly charged sphere of radius a is given by

$$\phi(r) = \frac{Q}{4\pi\varepsilon_0 a}\left(\frac{3}{2}-\frac{r^2}{2a^2}\right),$$

where Q is the total charge.

Now

$$U = \frac{1}{2}\int \rho\phi(r)\, dV = \frac{Q\rho}{8\pi\varepsilon_0 a}\int_0^a \left(\frac{3}{2}-\frac{r^2}{2a^2}\right)\cdot 4\pi r^2 dr$$

$$= \frac{Q\rho}{2\varepsilon_0 a}\left[\frac{3}{2}\cdot\frac{r^3}{3}-\frac{r^5}{10a^2}\right]_0^a = \frac{Q}{2\varepsilon_0 a}\cdot\frac{Q}{\frac{4}{3}\pi a^3}\cdot\frac{2a^3}{5}$$

$$= \frac{1}{4\pi\varepsilon_0}\cdot\frac{3Q^2}{5a}.$$

Classical radius of an electron

Suppose we consider the electron as a uniformly charged sphere of radius r_0 containing a total charge $-e$. The energy required to assemble this sphere of charge is

$$U = \frac{1}{4\pi\varepsilon_0}\cdot\frac{3e^2}{5r_0}.$$

Experiments on the phenomenon of pair production show that it takes about 0.51 MeV energy to create an electron. This is equal to the rest mass energy, mc^2, of the electron. Let us assume that the electron radius can be estimated by equating its electrostatic self-energy to the rest mass energy, i.e.,

$$\frac{1}{4\pi\varepsilon_0}\cdot\frac{3e^2}{5r_0} = mc^2 \quad \text{or} \quad r_0 = \frac{3e^2}{20\pi\varepsilon_0 mc^2}. \tag{4.7-2}$$

Substituting the values of different quantities we get $r_0 \approx 1.7\times 10^{-15}$ m. The actual radius of the electron is found to be much smaller then r_0. This indicates that the actual structure of the electron is not so simple as described above. It is rather complicated.

4.8 Electrostatic Energy of a System of Charged Conductors; Coefficient of Potential and Capacitance

Let us consider a system of N conductors where all charges reside on the surface of the conductors. In terms of surface density of charge σ and the potential ϕ total electrostatic energy of the system can be calculated as

$$U = \frac{1}{2}\int_S \sigma(\vec{r})\,\phi(\vec{r})\,dS. \tag{4.8-1}$$

Since each conductor surface is an equipotential surface the above integration can be written as

$$U = \frac{1}{2}\sum_{i=1}^{N}\phi_i \oint_{S_i}\sigma\,dS = \frac{1}{2}\sum_{i=1}^{N}\phi_i Q_i, \tag{4.8-2}$$

where Q_i is the charge on the ith conductor and ϕ_i is its potential. It is found that a linear relationship exists between the potentials and charges on various conductors in the system. In fact for a system of N conductors we can write

$$\phi_i = \sum_{j=1}^{N} p_{ij}Q_j \tag{4.8-3}$$

$$\text{and}\quad Q_i = \sum_{j=1}^{N} c_{ij}\phi_j, \tag{4.8-4}$$

where p_{ij} are called the *coefficients of potential,* c_{ii} are the coefficients of capacitance and $c_{ij}(i \neq j)$ are the coefficients of inductance. p_{ij} and c_{ij} depend only on the geometry.

The electrostatic energy of N conductors can now be written in terms of p_{ij} or c_{ij} :

$$U = \frac{1}{2}\sum_{i=1}^{N}\sum_{j=1}^{N} p_{ij}Q_iQ_j \tag{4.8-5}$$

$$U = \frac{1}{2}\sum_{i=1}^{N}\sum_{j=1}^{N} c_{ij}\phi_i\phi_j. \tag{4.8-6}$$

Some of the important properties of p_{ij} and c_{ij} are the following:

$$p_{ij} = p_{ji},\ p_{ii} > 0 \quad\text{and}\quad p_{ii} \geq p_{ij}$$

$$c_{ij} = c_{ji},\ c_{ii} > 0 \quad\text{and}\quad c_{ij} \leq 0 \quad \text{for } i \neq j.$$

4.9 Forces and Torques from Electrostatic Energy

From a knowledge of the electrostatic energy it is possible to calculate the forces and torques acting on the system. Suppose we have an isolated system of charges whose electrostatic energy depends on the particular charge configuration. Now we allow one part of the system to undergo a small displacement $d\vec{r}$ under the influence of the electrical forces $\vec{F}$ acting upon it. The work done by $\vec{F}$ on the system under these circumstances is

$$dW = \vec{F} \cdot d\vec{r}. \tag{4.9-1}$$

As the system is isolated the work is done at the expense of the electrostatic energy U. Thus,

$$dW = -dU = -\vec{\nabla}U \cdot d\vec{r}. \tag{4.9-2}$$

Therefore,

$$\vec{F} = -\vec{\nabla}U \tag{4.9-3}$$

or in components form,

$$F_x = -\left.\frac{\partial U}{\partial x}\right|_Q, \quad F_y = -\left.\frac{\partial U}{\partial y}\right|_Q \quad \text{and} \quad F_z = -\left.\frac{\partial U}{\partial z}\right|_Q, \tag{4.9-4}$$

where a subscript Q has been added to emphasize that the system is isolated and its total charge remains constant during the displacement.

If the part of the system under consideration is constrained to rotate about an axis and it undergoes a small rotation $d\vec{\theta}$ under the action of a torque $\vec{\tau}$ then,

$$dW = \vec{\tau} \cdot d\vec{\theta} = -dU = -\vec{\nabla}_\theta U \cdot d\vec{\theta},$$

where $\vec{\nabla}_\theta$ represents derivatives with respect to θ_x, θ_y and θ_z. Therefore,

$$\vec{\tau} = -\vec{\nabla}_\theta U \tag{4.9-5}$$

and in components form,

$$\tau_x = -\left.\frac{\partial U}{\partial \theta_x}\right|_Q, \quad \tau_y = -\left.\frac{\partial U}{\partial \theta_y}\right|_Q \quad \text{and} \quad \tau_z = -\left.\frac{\partial U}{\partial \theta_z}\right|_Q. \tag{4.9-6}$$

There is an important class of problems in which Q is not maintained constant. All charges reside on the surfaces of conductors and the potentials of all conductors are kept constant during displacement by some external sources of energy such as batteries. If we allow a part of such a system to undergo a small displacement $d\vec{r}$ the work done by the electrical forces $\vec{F}$ acting on it will be $dW = \vec{F} \cdot d\vec{r}$. Additional work dW_B must also be done by the battery to maintain all the conductors at a constant potential. So for the conservation of energy,

Work done on the system = (Increase in potential energy of the system)
+(work done by the system)

$$\therefore \quad dW_B = dU + dW.$$

If the change in charge of the jth conductor is dQ_j then

$$dW_B = \sum_j \phi_j dQ_j. \tag{4.9-7}$$

The electrostatic energy U of the system of charged conductors is given by

$$U = \tfrac{1}{2}\sum_j \phi_j Q_j. \tag{4.9-8}$$

So at constant potential

$$dU = \tfrac{1}{2}\sum_j \phi_j dQ_j.$$

Therefore, $dW_B = 2dU$, i.e., energy supplied by the battery to maintain the potential constant is equal to double the increase in the electrical energy of the system. Using this result we get from Eq. (4.9-7),

$$dW = dU|_\phi \quad \text{or} \quad \vec{F}\cdot d\vec{r} = \vec{\nabla}U\cdot d\vec{r}|_\phi.$$

Thus, $\vec{F} = \vec{\nabla}U|_\phi$. In component forms,

$$F_x = \left.\frac{\partial U}{\partial x}\right|_\phi, \quad F_y = \left.\frac{\partial U}{\partial y}\right|_\phi \quad \text{and} \quad F_z = \left.\frac{\partial U}{\partial z}\right|_\phi. \tag{4.9-9}$$

In a similar way we can show that

$$\tau_x = \left.\frac{\partial U}{\partial \theta_x}\right|_\phi, \quad \tau_y = \left.\frac{\partial U}{\partial \theta_y}\right|_\phi \quad \text{and} \quad \tau_z = \left.\frac{\partial U}{\partial \theta_z}\right|_\phi. \tag{4.9-10}$$

An example of the above energy method

Suppose we are to find the force on the plates of a parallel plate or capacitor. If Q be the charge on the positive plate, V be the potential difference between the two plates and C be the capacitance then electrostatic energy stored in the capacitor is given by (See, Section 4.13)

$$U = \frac{Q^2}{2C} \tag{4.9-11}$$

$$\text{or} \quad U = \tfrac{1}{2}CV^2. \tag{4.9-12}$$

where the capacitance C is given by

$$C = \frac{\varepsilon_0 A}{x}$$

in which A is the area of each plate, x is the plate separation and ε_0 is the permittivity of the space (assumed to be air) between the plates.

Now if the charge Q is maintained constant then force on the plate will be

$$F_x = -\left.\frac{\partial U}{\partial x}\right|_Q = -\frac{\partial}{\partial x}\left(\frac{Q^2}{2C}\right)_Q = -\frac{Q^2}{2}\frac{\partial}{\partial x}\left(\frac{x}{\varepsilon_0 A}\right) = -\frac{Q^2}{2\varepsilon_0 A}.$$

If the potential V is kept constant then the force on the plate will be

$$F_x = \left.\frac{\partial U}{\partial x}\right|_V = \frac{\partial}{\partial x}\left(\frac{1}{2}CV^2\right)_V = \frac{1}{2}V^2\frac{\partial C}{\partial x} = \frac{1}{2}V^2\frac{\partial}{\partial x}\left(\frac{\varepsilon_0 A}{x}\right) = -\frac{\varepsilon_0 A V^2}{x^2} = -\frac{C^2V^2}{2\varepsilon_0 A}.$$

4.10 Capacitors

A capacitor is a system of conductors and dielectric, which can store electric charge. Usually it consists of a pair of conductors containing equal and opposite charges ($\pm Q$) with a potential difference (V) between them, which is independent of the presence of other conductors.

Let us consider a capacitor formed by two such conductors 1 and 2 carrying charges $Q_1 = +Q$ and $Q_2 = -Q$, respectively. Total charge of the system is zero but by convention the charge on the positive conductor is called the *charge on the capacitor.* Now the potentials of the conductors in terms of the coefficients p_{ij} of potential can be written as

$$\phi_1 = p_{11}Q_1 + p_{12}Q_2 = Q\left(p_{11} - p_{12}\right)$$
$$\phi_2 = p_{21}Q_1 + p_{22}Q_2 = Q\left(p_{21} - p_{22}\right).$$

Therefore, potential difference

$$V = \phi_1 - \phi_2 = Q\left(p_{11} - p_{12} - p_{21} + p_{22}\right).$$

Thus, the potential difference between the conductors is proportional to the charge on the capacitor. The above relation is conventionally written as

$$Q = CV,$$

where $C = \left(p_{11} - p_{12} - p_{21} + p_{22}\right)^{-1}$ is called the *capacitance of the capacitor.* Occasionally the word *condenser* is also used for capacitor. It is a purely geometrical quantity determined by the shape, size and separation of the two conductors. The capacitance C of a capacitor is numerically equal to the charge required to be placed on the capacitor to raise its potential by unity. The SI unit of capacitance is the coulomb per volt, called the *farad* (F). *One farad is the capacitance of a capacitor, which requires 1C charge to establish a pd of one volt across it.* For practical purposes it is a too large unit. For this units like micro farad (μF), nano farad (nF) or pico farad (pF) are used: 1 μF $= 10^{-6}$F, 1nF$= 10^{-9}$F, 1pF$= 10^{-12}$F.

Sometimes we speak of the capacitance of an isolated single conductor. Such a conductor is assumed to be the part of a capacitor whose other conductor is at infinity.

4.11 Calculation of Capacitance for Different Geometries

1. Parallel plate capacitor

A parallel plate capacitor consists of two parallel metal plates separated by a dielectric medium. Suppose the plates are of area A each and separated by a distance d. We assume that d is very small compared with the linear dimensions of the plates so that the end effects may be neglected and the electric field between the plates may be taken to be uniform. Suppose that equal and opposite charges $\pm Q$ are put on the plates. The charges get distributed on the plates in such a way that the field inside the thickness of the plates becomes zero. Thus, charges spread only over the inner surfaces of the plates and give uniform charge densities $+\sigma = Q/A$ and $-\sigma = -Q/A$ as shown in Fig. 4.11-1. The electric field will be uniform, normal to the plates and of magnitude σ/ε between the plates and zero elsewhere, where ε is the permittivity of the medium.

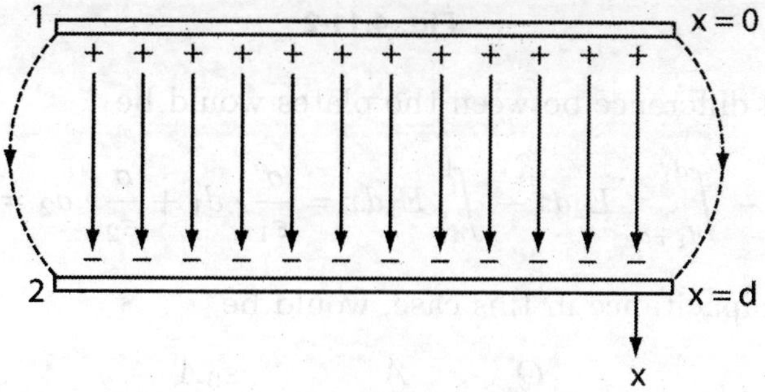

Fig. 4.11-1

The potential difference between the plates is given by

$$V = \phi_1 - \phi_2 = -\int_2^1 \vec{E} \cdot d\vec{r} = -\int_2^1 \frac{\sigma}{\varepsilon}\hat{i} \cdot d\vec{r} = \int_2^1 \frac{\sigma}{\varepsilon} dx$$

$$= -\left[\frac{\sigma}{\varepsilon}x\right]_d^0 = \frac{\sigma}{\varepsilon} \cdot d = \frac{Qd}{\varepsilon A}.$$

Now by definition the capacitance of this capacitor is

$$C = \frac{Q}{V} = \frac{\varepsilon A}{d} \qquad (4.11\text{-}1)$$

Special cases

(i) Air capacitor

If the space between the plates is air then its capacitance would be

$$C = \frac{\varepsilon_0 A}{d}. \qquad (4.11\text{-}2)$$

(ii) Capacitor with a composite dielectric medium

Let the medium between the plates consists of two linear dielectrics of thickness d_1 and d_2 and permittivities ε_1 and ε_2, respectively (Fig. 4.11-2). The field intensities within the first and second media are given by

$$E_1 = \frac{\sigma}{\varepsilon_1} \text{ and } E_2 = \frac{\sigma}{\varepsilon_2}.$$

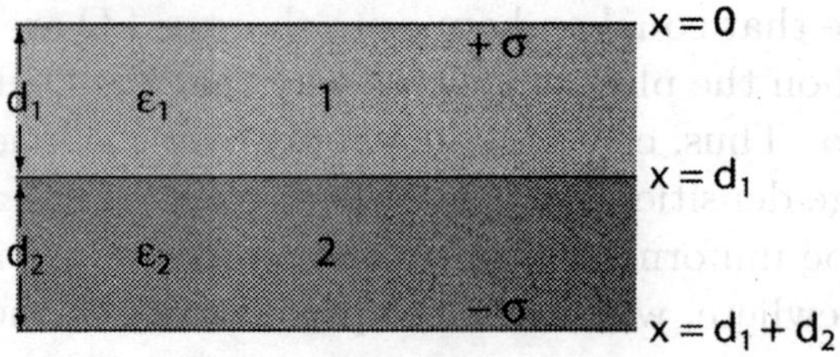

Fig. 4.11-2

So the potential difference between the plates would be

$$V = \phi_1 - \phi_2 = -\int_{d_1+d_2}^{d_1} E_2 dx - \int_{d_1}^{0} E_1 dx = \frac{\sigma}{\varepsilon_1} \cdot d_1 + \frac{\sigma}{\varepsilon_2} \cdot d_2 = \frac{Q}{A}\left(\frac{d_1}{\varepsilon_1} + \frac{d_2}{\varepsilon_2}\right).$$

Therefore, the capacitance in this case, would be

$$C = \frac{Q}{V} = \frac{A}{\frac{d_1}{\varepsilon_1} + \frac{d_2}{\varepsilon_2}} = \frac{\varepsilon_0 A}{\frac{d_1}{K_1} + \frac{d_2}{K_2}}, \tag{4.11-3}$$

where K_1 and K_2 are the dielectric constants of the two media. For a capacitor with several dielectric media,

$$C = \frac{\varepsilon_0 A}{\sum_i \frac{d_i}{K_i}}. \tag{4.11-4}$$

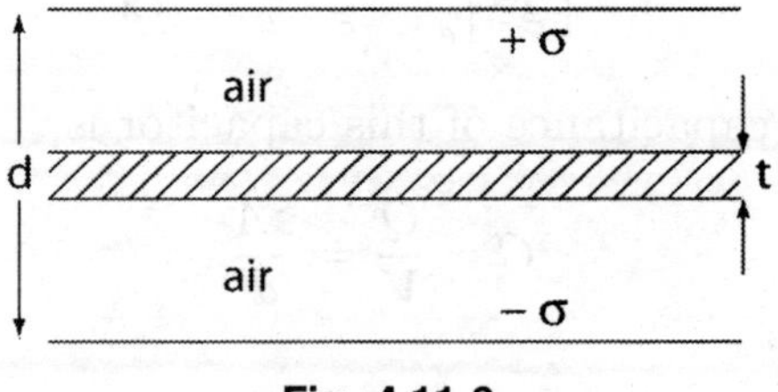

Fig. 4.11-3

As a special case let us consider a parallel plate air capacitor with a dielectric slab of thickness $t (t < d)$ and permittivity ε introduced between the plates (Fig. 4.11-3). In this case, pd between the plates would be

$$V = \frac{\sigma}{\varepsilon} \cdot t + \frac{\sigma}{\varepsilon_0}(d - t) = \frac{Q}{A}\left[\frac{t}{\varepsilon} + \frac{d-t}{\varepsilon_0}\right]$$

and capacitance

$$C = \frac{Q}{V} = \frac{\varepsilon_0 A}{d - t + \frac{t}{K}} = \frac{\varepsilon_0 A}{d - t\left(1 - \frac{1}{K}\right)}. \tag{4.11-5}$$

As the dielectric constant $K > 1$, comparison with Eq. (4.11-2) shows that the capacitance increases with the introduction of the dielectric slab.

(*iii*) Capacitor with a nonuniform dielectric

Suppose the space between the plates is filled with a nonuniform dielectric whose dielectric constant varies linearly from one plate to the other. Let the dielectric constant at P at a distance x from the plate 1 be K and $\frac{dK}{dx} = \alpha$ (constant). Therefore, $K = \alpha x + \beta$ (constant).

Now at $x = 0$, let $K = K_1$ and at $x = d$, $K = K_2$. So $\beta = K_1$ and $\alpha = \frac{K_2 - K_1}{d}$. Thus,

$$K = \frac{K_2 - K_1}{d} x + K_1.$$

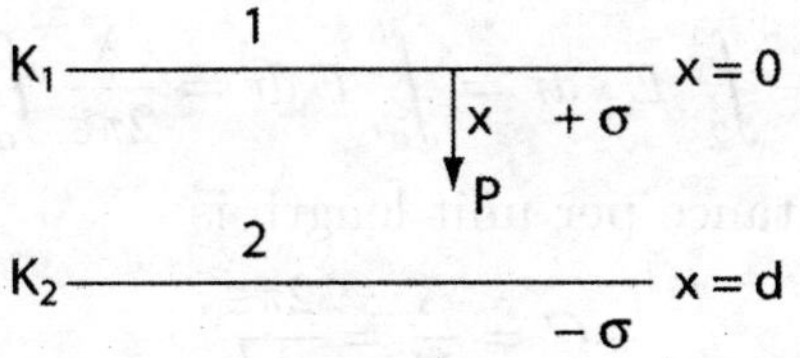

Fig. 4.11-4

Now field at P is

$$E = \frac{\sigma}{\varepsilon} = \frac{\sigma}{\varepsilon_0 K} = \frac{\sigma}{\varepsilon_0 \left[\frac{K_2 - K_1}{d} x + K_1\right]}$$

Therefore, pd between the plates is

$$\begin{aligned} V &= -\int_2^1 E\,dx = -\frac{\sigma}{\varepsilon_0}\int_d^0 \frac{dx}{\frac{K_2 - K_1}{d} x + K_1} \\ &= -\frac{\sigma}{\varepsilon_0}\cdot\frac{d}{K_2 - K_1}\ln\left(\frac{K_2 - K_1}{d} x + K_1\right)\Bigg|_d^0 \\ &= \frac{Qd}{\varepsilon_0 A(K_2 - K_1)}\ln\frac{K_2}{K_1}. \end{aligned}$$

Now by definition the capacitance is

$$C = \frac{Q}{V} = \frac{\varepsilon_0 A(K_2 - K_1)}{d\ln(K_2/K_1)}. \tag{4.11-6}$$

2. Cylindrical capacitor

A cylindrical capacitor consists of a pair of long concentric metal cylinders, the space between them being filled with a dielectric. The coaxial cable used in communication system is a common example of such a capacitor. Let a and b be the radii of the inner and outer cylinders respectively and ε be the permittivity of the dielectric medium (Fig. 4.11-5). Suppose equal and opposite charges $+Q$ and $-Q$ are put on the inner and outer cylinders respectively. By symmetry the electric field will be radial from the axis if we are far away from the ends. If λ be the charge per unit length on the inner cylinder then electric field is given by

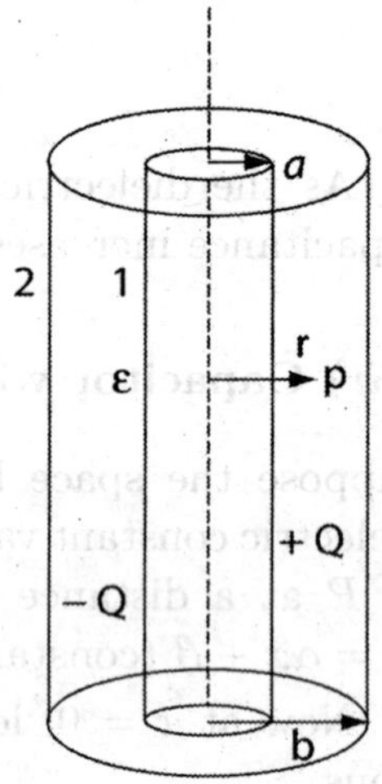

Fig. 4.11-5

$$\vec{E} = \frac{\lambda}{2\pi\varepsilon r}\hat{r} \quad \text{for } a < r < b$$

and $\vec{E} = 0$ for $r < a$ and $r > b$.

Now the potential difference between the two cylinders is

$$V = \phi_1 - \phi_2 = -\int_2^1 \vec{E}\cdot d\vec{r} = \int_a^b E\,dr = \frac{\lambda}{2\pi\varepsilon}\int_a^b \frac{dr}{r} = \frac{\lambda}{2\pi\varepsilon}\ln\frac{b}{a}.$$

By definition the capacitance per unit length is

$$C = \frac{\lambda}{V} = \frac{2\pi\varepsilon}{\ln\frac{b}{a}}. \tag{4.11-7}$$

It is interesting to note that if the gap between the cylinders is very small compared with a, i.e., if $b - a \ll a$ then we can write

$$C = \frac{2\pi\varepsilon}{\ln\left(1 + \frac{b-a}{a}\right)} \approx \frac{2\pi\varepsilon}{\frac{b-a}{a}} = \frac{\varepsilon\cdot 2\pi a}{b-a}.$$

Since $2\pi a$ is the area per unit length, the formula can be identified to be similar to that of a parallel plate capacitor.

3. Spherical capacitor

A spherical capacitor consists of a pair concentric metal spheres, the space between them being filled with a dielectric. Let a and b be the radii of the inner and outer spheres respectively and ε be the permittivity of the dielectric medium (Fig. 4.11-6). Suppose equal and opposite charges $+Q$ and $-Q$ are put on the inner and outer spheres respectively. By symmetry the field lines will be radial. The electric field at a distance r from the common centre O is

$$\vec{E} = \begin{cases} \dfrac{Q}{4\pi\varepsilon r^2}\hat{r}, & \text{for } a < r < b \\ 0, & \text{for } r < a \text{ and } r > b. \end{cases}$$

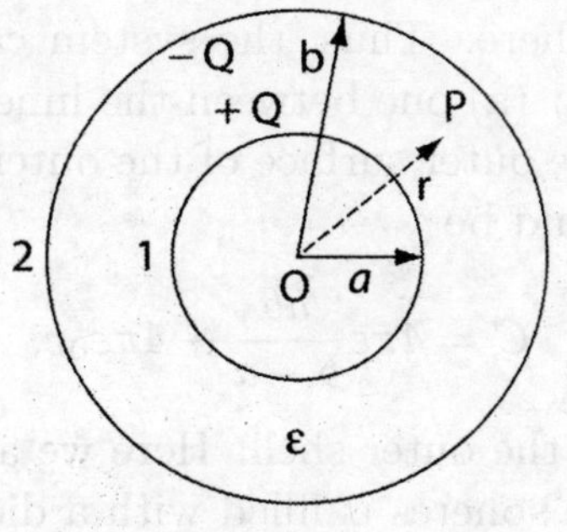

Fig. 4.11-6

Now the pd between the spheres is

$$V = \phi_1 - \phi_2 = -\int_2^1 \vec{E}\cdot d\vec{r} = \int_a^b E\,dr = \frac{Q}{4\pi\varepsilon}\int_a^b \frac{dr}{r^2} = \frac{Q}{4\pi\varepsilon}\left(\frac{1}{a} - \frac{1}{b}\right)$$

By definition the capacitance of the capacitor is

$$C = \frac{Q}{V} = \frac{4\pi\varepsilon}{\frac{1}{a} - \frac{1}{b}} = \frac{4\pi\varepsilon ab}{b - a} \tag{4.11-8}$$

Special cases

(i) Isolated sphere

An isolated sphere may be assumed to form a spherical capacitor whose outer spherical shell is at infinity. So putting $b = \infty$ in Eq. (4.11-8) we may get the capacitance of an isolated sphere as

$$C = 4\pi\varepsilon a \tag{4.11-9}$$

(ii) Outer sphere is earthed

If the outer sphere is grounded its charge does not flow to the ground because it is held by the opposite charge of the inner sphere. If the system is initially uncharged and we put a charge $+Q$ on the inner sphere, this will induce a charge $-Q$ on the outer sphere. So in this case, capacitance will be the same as given by the Eq. (4.11-8).

(iii) Inner sphere is earthed and outer sphere is charged

Suppose that the total charge on the outer sphere be $+Q$. Due to the presence of inner grounded sphere a part of $+Q$, say, $+q_1$ will lie on the inner surface of the outer sphere and the rest $Q - q_1$ will lie on its outer surface (Fig. 4.11-7). The charge $+q_1$ will induce a charge $-q_1$ on the inner sphere. Thus, the system can be considered as a parallel combination of two capacitors: (a) one between the inner sphere and the inner surface of the outer sphere and (b) the outer surface of the outer sphere and the earth. Hence, capacitance of the system would be

$$C = 4\pi\varepsilon\frac{ab}{b-a} + 4\pi\varepsilon_0 c, \tag{4.11-10}$$

where c is the outer radius of the outer shell. Here we assume that outside medium is air and the space between the spheres is filled with a dielectric of permittivity ε.

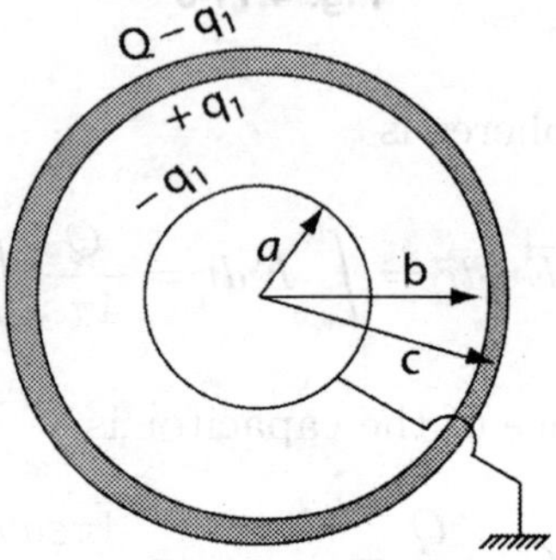

Fig. 4.11-7

4. Capacitance between two long thin parallel wires in air

The overhead transmission line is a common example of such a capacitor system. Let a be the radius of each wire, d be the distance between them ($d \gg a$) and $+\lambda$ be the charge per unit length of wire 1 and $-\lambda$ be the charge per unit length of wire 2. The total electric field at a point P at a distance x from the wire 1 is

$$\vec{E} = \left[\frac{\lambda}{2\pi\varepsilon_0 x} + \frac{\lambda}{2\pi\varepsilon_0(d-x)}\right]\hat{x}.$$

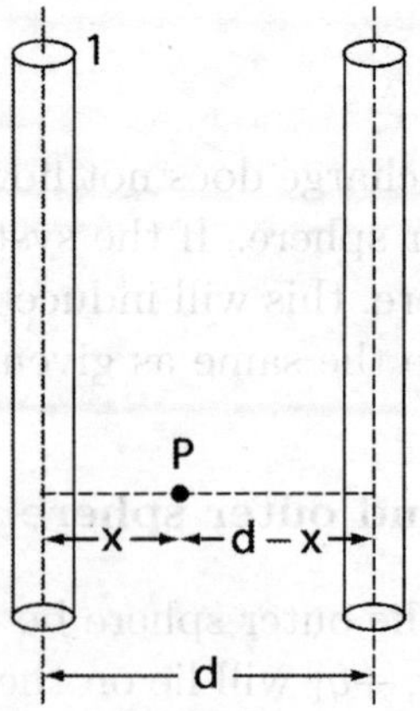

Fig. 4.11-8

Hence, the pd between the wires is given by

$$V = \phi_1 - \phi_2 = -\int_{d-a}^{a} E\,dx = \frac{\lambda}{2\pi\varepsilon_0}\int_a^{d-a}\left(\frac{1}{x}+\frac{1}{d-x}\right)dx$$
$$= \frac{\lambda}{2\pi\varepsilon_0}\left[\ln x - \ln(d-x)\right]_a^{d-a}$$
$$= \frac{\lambda}{\pi\varepsilon_0}\ln\frac{d-a}{a}.$$

∴ Capacitance per unit length

$$C = \frac{\lambda}{V} = \frac{\pi\varepsilon_0}{\ln\frac{d-a}{a}} \tag{4.11-11}$$
$$\approx \frac{\pi\varepsilon_0}{\ln\frac{d}{a}}. \tag{4.11-12}$$

4.12 Combination of Capacitors

Capacitors are often used in combinations. There are two simple combinations—*series* and *parallel.* Fig. 4.12-1 shows the parallel combination of two capacitors. Here pd between the plates of each capacitor is the same and charge stored by a capacitor depends on its capacitance. Let $V = V_A - V_B$ be the pd across each capacitor. Charge on C_1 is $q_1 = C_1V$ and that on C_2 is $q_2 = C_2V$. The total charge on the two capacitors is $Q = q_1 + q_2 = C_1V + C_2V = (C_1 + C_2)V$. In terms of a single equivalent capacitor C, which stores a charge Q for a pd V we can write

$$C = \frac{Q}{V} = C_1 + C_2. \tag{4.12-1}$$

In the general case of N capacitors in parallel we have the equivalent capacitor,

$$C = \sum_{i=1}^{N} C_i. \tag{4.12-2}$$

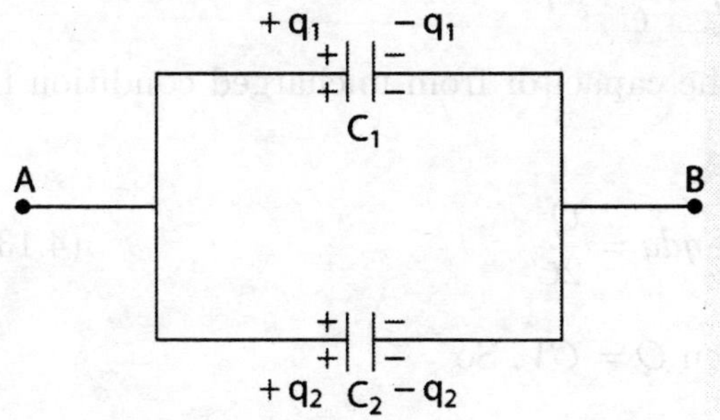

Fig. 4.12-1

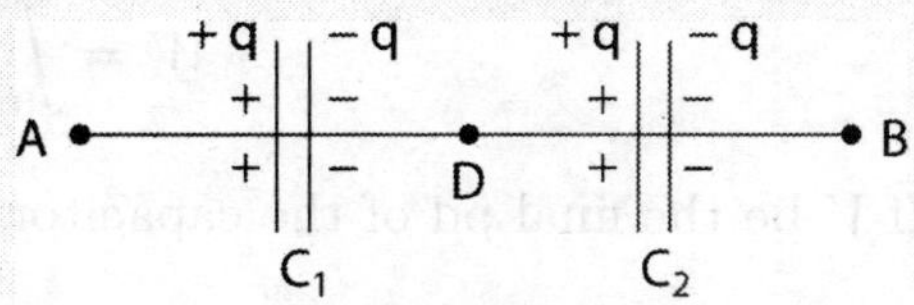

Fig. 4.12-2

Fig. 4.12-2 shows the series combination of two capacitors. Here, each capacitor acquires the same charge q but pd across each is different. From Fig. 4.12-2,

$$\phi_A - \phi_D = \frac{q}{C_1} \quad \text{and} \quad \phi_D - \phi_B = \frac{q}{C_2}.$$

$\therefore$ pd across the combination is

$$V = \phi_A - \phi_B = (\phi_A - \phi_D) + (\phi_D - \phi_B) = \frac{q}{C_1} + \frac{q}{C_2} = q\left(\frac{1}{C_1} + \frac{1}{C_2}\right).$$

In terms of an equivalent single capacitor C, which stores a charge q for a pd V we can write

$$\frac{1}{C} = \frac{V}{q} = \frac{1}{C_1} + \frac{1}{C_2}. \tag{4.12-3}$$

In the general case of N capacitors in series we can write

$$\frac{1}{C} = \sum_{i=1}^{N} \frac{1}{C_i}. \tag{4.12-4}$$

Note that the equivalent capacitance in parallel combination is greater than any single one of them, whereas in series combination the equivalent capacitance is smaller than any single one.

4.13 Energy Stored in a Charged Capacitor

In order to charge up a capacitor work is to be done against the electrostatic forces present. This work is stored in the capacitor as electrostatic potential energy. Let us start with an uncharged capacitor. To charge it up we must remove electrons from one conductor and carry them to the other conductor. Suppose at any intermediate stage during charging the charge on the positive conductor is q and the pd across the capacitor is $\phi = q/C$, where C is the capacitance. Work done to increase the charge by an amount dq is

$$dW = \phi dq = \frac{q}{C} \cdot dq.$$

Therefore, total work done in charing the capacitor from uncharged condition to a state with charge Q would be

$$W = \int_0^Q \frac{1}{C} q dq = \frac{Q^2}{2C}. \tag{4.13-1}$$

If V be the final pd of the capacitor then $Q = CV$. So

$$W = \tfrac{1}{2} C V^2. \tag{4.13-2}$$

4.14 Dielectric Strength and Maximum Operating Voltage of a Capacitor

When a dielectric is placed in a sufficiently large electric field it begins to pull electrons completely from the atoms and molecules and the dielectric starts conducting. Then dielectric breakdown is said to have occurred. Dielectric breakdown depends on the nature of material, temperature, humidity etc. *The maximum electric field that a dielectric can withstand or tolerate without breaking down is called the dielectric strength.*

This problem of dielectric breakdown is of considerable importance in electrical technology. Capacitor specifications include the maximum voltage that can be safely applied without having dielectric breakdown. This voltage depends on the nature and thickness of the dielectric. If the capacitor is used beyond the specified voltage the dielectric is likely to breakdown. For air at normal pressure the dielectric strength is about 3×10^6 V/m. By connecting capacitors in series they can be used over higher voltages. By the specification "10μF, 50 V" we mean a capacitor of 10μF and 50 V is the maximum voltage that can be safely applied across it.

SOLVED PROBLEMS

▶ **Problem 1.** *Find the electrostatic energy stored in the space surrounding a uniformly charged, spherical shell (or a charged conducting sphere) of radius R carrying a total charge Q.* **[J.U. 2010, cf. C.U. 2019]**

Solution. Suppose the shell is placed in air. Now the electric field at a point at a distance r from the centre of the shell is

$$\vec{E} = \begin{cases} \dfrac{1}{4\pi\varepsilon_0} \cdot \dfrac{Q}{r^2}\hat{r}, & \text{for } r > R \\ 0, & \text{for } r < R \end{cases}$$

The electrostatic energy density is $\frac{1}{2}\varepsilon_0 E^2$. Now considering a spherical shell of radii r and $r + dr$ we can calculate the total energy as

$$U = \int_R^\infty \frac{1}{2}\varepsilon_0 E^2 \cdot 4\pi r^2 dr = \frac{Q^2 \cdot 4\pi}{32\pi^2\varepsilon_0}\int_R^\infty \frac{dr}{r^2} = \frac{1}{4\pi\varepsilon_0} \cdot \frac{Q^2}{2R}.$$

▶ **Problem 2.** *Assuming the nucleus to be a uniformly charged sphere calculate its Coulomb energy.* **[J.U. 2014]**

Solution. Coulomb energy of a nucleus is defined as the amount of work that must be done against the repulsive Coulomb's force to assemble the nucleus. Assuming the nucleus to be a uniformly charged sphere of radius a it is easy to show that (See, Section 4.7) its Coulomb energy is

$$U = \frac{1}{4\pi\varepsilon_0} \cdot \frac{3Q^2}{5a} = \frac{1}{4\pi\varepsilon_0} \frac{3Z^2e^2}{5a} \quad \text{where } Z = \text{atomic number.}$$

In this calculation even if there were only one proton ($Z = 1$) we would obtain

$$U = \frac{1}{4\pi\varepsilon_0} \frac{3e^2}{5a}.$$

But it is obvious that no work is necessary in assembling a single proton. Applying this correction for Z protons we get the Coulomb energy of the nucleus as

$$U = \frac{1}{4\pi\varepsilon_0} \frac{3Z^2e^2}{5a} - Z \times \frac{1}{4\pi\varepsilon_0} \frac{3e^2}{5a} = \frac{1}{4\pi\varepsilon_0} \frac{3}{5} \frac{Z(Z-1)e^2}{a}.$$

▶ **Problem 3.** *Consider two identical spheres of radius a carrying charges q_1 and q_2 respectively and separated by a distance $d(\gg a)$. Find the electrostatic energy of the system. Is it equal to the sum of self-energies of two sphere? Comment on the result.*

Solution. Potentials of the two spheres are given by

$$\phi_1 = \frac{1}{4\pi\varepsilon_0} \cdot \frac{q_1}{a} + \frac{1}{4\pi\varepsilon_0} \cdot \frac{q_2}{d} = p_{11}q_1 + p_{12}q_2$$

$$\phi_2 = \frac{1}{4\pi\varepsilon_0} \cdot \frac{q_1}{d} + \frac{1}{4\pi\varepsilon_0} \cdot \frac{q_2}{a} = p_{21}q_1 + p_{22}q_2,$$

where p_{ij} are the coefficients of potential: $p_{11} = p_{22} = 1/(4\pi\varepsilon_0 a)$, $p_{12} = p_{21} = 1/(4\pi\varepsilon_0 d)$.

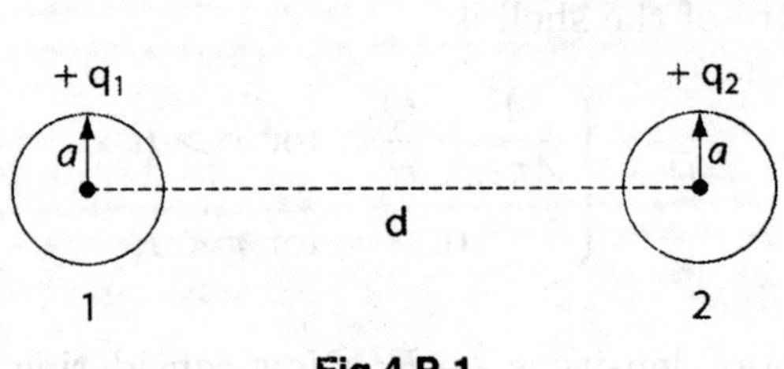

Fig 4.P-1

Now total energy of the system is

$$U = \frac{1}{2}\sum_{i=1}^{2}\sum_{j=1}^{2} p_{ij}q_iq_j = \frac{1}{2}\sum_{i}(p_{i1}q_iq_1 + p_{i2}q_iq_2)$$

$$= \frac{1}{2}\left[p_{11}q_1^2 + p_{22}q_2^2 + 2p_{12}q_1q_2\right] = \frac{1}{8\pi\varepsilon_0 a}\left(q_1^2 + q_2^2\right) + \frac{1}{4\pi\varepsilon_0 d} \cdot q_1q_2.$$

Self-energies of the two spheres are

$$U_1 = \frac{1}{4\pi\varepsilon_0} \cdot \frac{q_1^2}{2a} \quad \text{and} \quad U_2 = \frac{1}{4\pi\varepsilon_0} \cdot \frac{q_2^2}{2a} \text{ (See, \textit{Solved Problem} 1).}$$

Obviously, $U \neq U_1 + U_2$. Actually, the energy is a quadratic function of $\vec{E}$ and for this superposition principle is not obeyed by it.

▶ **Problem 4.** *Consider a parallel plate capacitor with plate separation d and each plate having a length l and width a. A dielectric slab of permittivity ε, thickness d, length l and width a is partially inserted between the plates. If a constant pd V is maintained between the plates, find the force exerted on the dielectric slab when a length x of the slab remains between the plates.*

Solution. The system may be considered as the parallel combination of two capacitors C_1 and C_2, where,

$$C_1 = \frac{\varepsilon a x}{d} \quad \text{and} \quad C_2 = \frac{\varepsilon_0 a(l-x)}{d}.$$

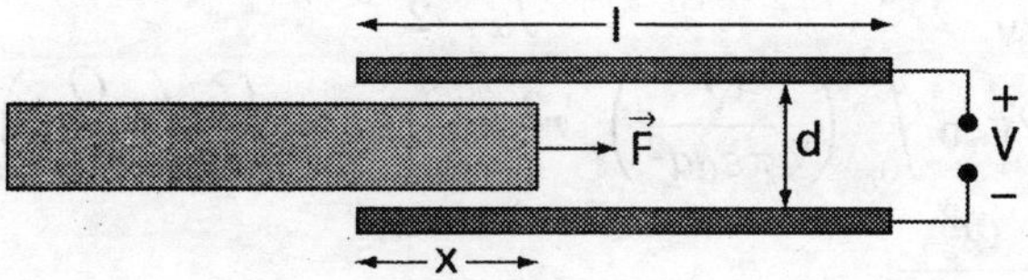

Fig 4.P-4

The total capacitance is then

$$C = C_1 + C_2 = \frac{a}{d}\left[(\varepsilon - \varepsilon_0)x + \varepsilon_0 l\right].$$

The electrostatic energy of the system is

$$U = \tfrac{1}{2}CV^2 = \frac{a}{2d}\left[(\varepsilon - \varepsilon_0)x + \varepsilon_0 l\right] V^2.$$

As V is constant the force experienced by the slab will be

$$\vec{F} = \left.\frac{\partial U}{\partial x}\right|_V \hat{x} = \frac{a}{2d}(\varepsilon - \varepsilon_0)V^2\hat{x},$$

where $\hat{x}$ is a unit vector in the direction of increasing x. This means that the force is attractive.

▶ **Problem 5.** *Consider a spherical charge distribution with a volume charge density*

$$\rho(r) = \begin{cases} \rho_0 \cdot \frac{a}{r} & \text{for } 0 < r < a \\ 0 & \text{for } r > a \end{cases}$$

Show that the total electrostatic energy of the system is $Q^2/6\pi\varepsilon_0 a$, where Q is the total charge in the system.

Solution. Total charge in the system is

$$Q = \int_0^a \rho(r)\cdot 4\pi r^2 dr = 4\pi\rho_0 a \int_0^a r\, dr = 2\pi\rho_0 a^3.$$

The electric field at any outside point can be obtained from Gauss's law as

$$\int \vec{E}_0 \cdot d\vec{S} = \frac{Q}{\varepsilon_0} \quad \text{or} \quad E_0 \cdot 4\pi r^2 = \frac{Q}{\varepsilon_0} \quad \text{or} \quad E_0 = \frac{Q}{4\pi\varepsilon_0 r^2}.$$

For an internal point,

$$\int \vec{E}_i \cdot d\vec{S} = \frac{Q_r}{\varepsilon_0} = \frac{1}{\varepsilon_0}\int_0^r \rho(r)\cdot 4\pi r^2 dr = \frac{2\pi\rho_0 a r^2}{\varepsilon_0}$$

$$\therefore\ E_i \cdot 4\pi r^2 = \frac{Qr^2}{\varepsilon_0 a^2} \quad \text{or} \quad E_i = \frac{Q}{4\pi\varepsilon_0 a^2}.$$

Now the total energy in the system is

$$\begin{aligned} U &= \int_0^a \frac{1}{2}\varepsilon_0 E_i^2 \cdot 4\pi r^2 dr + \int_a^\infty \frac{1}{2}\varepsilon_0 E_0^2 \cdot 4\pi r^2 dr \\ &= 2\pi\varepsilon_0 \int_0^a \left(\frac{Q}{4\pi\varepsilon_0 a^2}\right)^2 r^2 dr + 2\pi\varepsilon_0 \int_a^\infty \left(\frac{Q}{4\pi\varepsilon_0}\right)^2 \cdot \frac{dr}{r^2} \\ &= \frac{Q^2}{6\pi\varepsilon_0 a}. \end{aligned}$$

► **Problem 6.** *A spherical capacitor has inner radius a and outer radius b. It is filled with an inhomogeneous dielectric with permittivity* $\varepsilon = \varepsilon_0 K/r^2$ *for* $a < r < b$. *The outer sphere is grounded and a charge is placed on the inner sphere. Find the capacitance of the system.*

Solution. The electric field at any inside point is

$$\vec{E} = \frac{Q}{4\pi\varepsilon r^2}\hat{r} = \frac{Q}{4\pi\varepsilon_0 K}\hat{r},$$

where Q is the charge on the inner sphere.

Now the pd between the spheres is

$$V = -\int_b^a \vec{E}\cdot d\vec{r} = \int_a^b \frac{Q}{4\pi\varepsilon_0 K} dr = \frac{Q}{4\pi\varepsilon_0 K}(b-a).$$

$$\therefore\ \text{Capacitance } C = \frac{Q}{V} = \frac{4\pi\varepsilon_0 K}{b-a}.$$

► **Problem 7.** *Suppose you are to construct a parallel plate capacitor of* 1μF *by using paper sheets of thickness 0.05 mm as dielectric and a number of circular parallel metal foils connected alternately. If the dielectric constant of paper is 4 and radius of each foil is 10 cm then find the number of metal foils needed for this purpose.*

Solution. If there are n foils connected alternately as shown in Fig. 4.P-7 then it can be considered as the parallel combination of $(n-1)$ capacitors. Thus, capacitance,

$$C = (n-1)\frac{\varepsilon A}{d}.$$

Putting $C = 1\mu\text{F} = 10^{-6}\text{F}$, $\varepsilon = K\varepsilon_0 = 4\times 8.854\times 10^{-12}$ F/m, $A = \pi r^2 = \pi(0.1)^2$ m^2 and $d = 0.05 \times 10^{-3}$m we get $n = 46$.

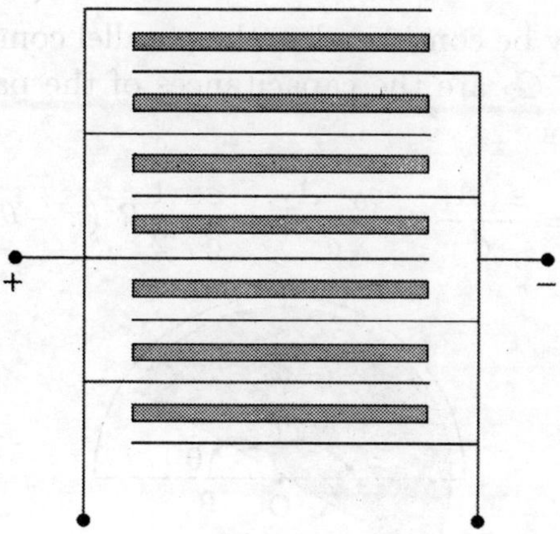

Fig 4.P-7

▶ **Problem 8.** *A parallel plate air capacitor has the plates of area A each.*

(a) *Assuming that the charge Q on it is kept constant find the work done against the electrical forces to increase the plate separation from d_1 to d_2.*

(b) *Assuming that the pd V across it is maintained constant find the change in energy of the system.*

Solution. The energy of the system is

$$U = \frac{Q^2}{2C} = \frac{Q^2 d}{2\varepsilon_0 A} \quad \text{or} \quad U = \tfrac{1}{2}CV^2 = \frac{\varepsilon_0 A V^2}{2d}$$

(*a*) When the charge is kept constant the work done against the electrical forces is completely spent for increasing the energy of the system. Thus, work done in this case, would be

$$W = \Delta U = \frac{Q^2}{2\varepsilon_0 A}(d_2 - d_1).$$

(*b*) When the pd is maintained constant the change in energy of the system would be

$$\Delta U = \frac{\varepsilon_0 A V^2}{2}\left(\frac{1}{d_2} - \frac{1}{d_1}\right).$$

This is negative. To understand it we must consider the work done against the electrical forces as well as the energy supplied by the source, which maintains the pd constant.

▶ **Problem 9.** *Consider a capacitor consisting of two fixed semicircular plates of radius R separated by a distance d. A semicircular dielectric slab of dielectric constant K almost fills up the space between the plates. The slab is free to rotate about an axis normal to the plates and passing through the centre. A constant pd V is maintained between the plates. If the slab is now rotated through an angle θ show that the torque tending to pull the slab back inside the capacitor would be* $(K-1)\varepsilon_0 R^2 V^2/4d$.

Solution. The system may be considered as the parallel combination of two capacitors C_1 and C_2, where C_1 and C_2 are the capacitances of the parts with and without the dielectric. Thus, capacitance

$$C = C_1 + C_2 = \frac{\varepsilon \cdot A_1}{d} + \frac{\varepsilon_0 \cdot A_2}{d} = \frac{\varepsilon}{d}\cdot\frac{1}{2}R^2(\pi-\theta) + \frac{\varepsilon_0}{d}\cdot\frac{1}{2}R^2\theta.$$

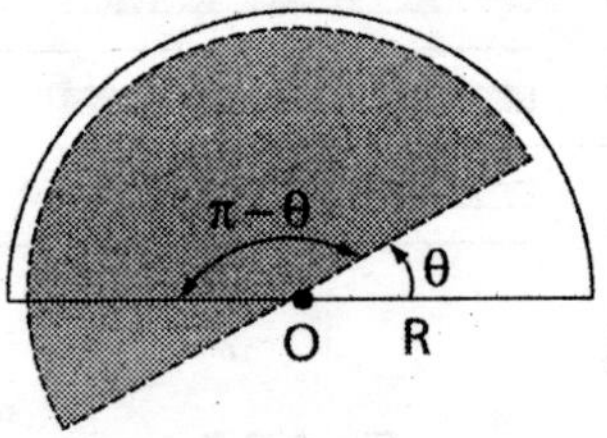

Fig 4.P-9

The electrostatic energy of the system is

$$U = \tfrac{1}{2}CV^2.$$

As the pd V is constant the torque acting on the system is

$$\tau = \frac{\partial U}{\partial\theta} = \frac{V^2}{2}\cdot\frac{\partial C}{\partial\theta} = \frac{V^2}{2}\cdot\frac{R^2}{2d}(\varepsilon_0-\varepsilon) = -(K-1)\varepsilon_0 R^2V^2/4d.$$

The negative sign indicates that the torque tries to decrease the angle θ.

▶ **Problem 10.** *In a capacitor with square plates each of side 'a, one plate is making a small angle θ with the other. Plate separation at one end is d and at other end d + Δd. Find the capacitance of the system.*

Solution. The system may be considered as the parallel combination of infinitesimal parallel plate capacitors of increasing separations. One such infinitesimal capacitor at a distance x from left end (Fig. 4.P-10) has plate separation

$$d(x) = d + x\tan\theta \approx d + x\theta.$$

The plate of this capacitor is a differential strip of area adx. So its capacitance is

$$dC = \frac{\varepsilon_0 adx}{d + x\theta}.$$

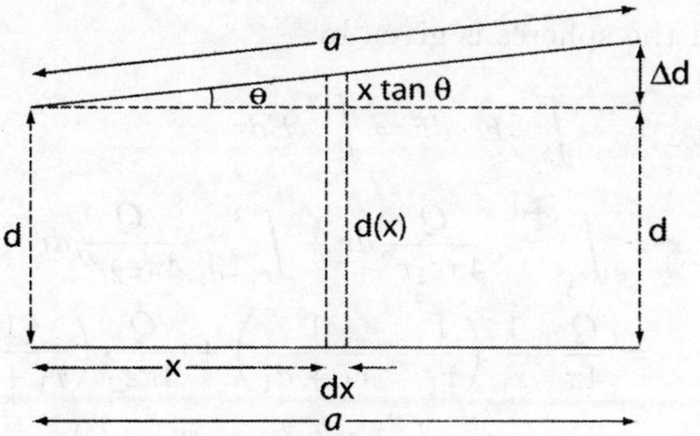

Fig 4.P-10

Therefore, total capacitance is

$$C = \int_0^a \frac{\varepsilon_0 a dx}{d + x\theta} = \frac{\varepsilon_0 a}{\theta} \cdot \ln \frac{d + a\theta}{d}.$$

From figure $\theta \approx \dfrac{\Delta d}{a}$

So

$$C = \frac{\varepsilon_0 a^2}{\Delta d} \ln\left(1 + \frac{\Delta d}{d}\right).$$

As θ is small, $\dfrac{\Delta d}{d} \ll 1$, therefore,

$$\ln\left(1 + \frac{\Delta d}{d}\right) \approx \frac{\Delta d}{d} - \frac{1}{2}\left(\frac{\Delta d}{d}\right)^2.$$

Thus,

$$C = \frac{\varepsilon_0 A}{d}\left(1 - \frac{\Delta d}{2d}\right),$$

where $A = a^2$ is the area of each plate.

▶ **Problem 11.** *Estimate the maximum amount of electrostatic energy that can be stored in 1 m^3 volume of air. Given the dielectric strength of air 3×10^6 V/m.*

Solution. Maximum energy per/m^3 volume is

$$\tfrac{1}{2}\varepsilon_0 E_{\max}^2 = \tfrac{1}{2} \times 8.85 \times 10^{-12} \times (3 \times 10^6)^2 \text{ J/m}^3 = 39.8 \text{ J/m}^3.$$

▶ **Problem 12.** *Two concentric spheres of radii r_1 and r_2 ($r_2 > r_1$) carry electric charges $+Q$ and $-Q$, respectively. The region between the spheres is filled with two insulating layers of dielectric constants K_1 and K_2. The first layer is of thickness d_1 and the second layer fills up the remaining space between the spheres. Compute the capacitance of the system. Also, calculate the electrostatic energy stored in the region $r_1 < r < r_1 + d_1$.*

[C.U. 2004]

Solution. pd between the spheres is given by

$$V = \phi_1 - \phi_2 = -\int_{r_2}^{r_1} \vec{E} \cdot d\vec{r} = \int_{r_1}^{r_2} E dr$$

$$= \int_{r_1}^{r_1+d_1} \frac{Q}{4\pi\varepsilon_1 r^2} dr + \int_{r_1+d_1}^{r_2} \frac{Q}{4\pi\varepsilon_2 r^2} dr$$

$$= \frac{Q}{4\pi} \cdot \frac{1}{\varepsilon_1} \left(\frac{1}{r_1} - \frac{1}{r_1 + d_1} \right) + \frac{Q}{4\pi\varepsilon_2} \left(\frac{1}{r_1 + d_1} - \frac{1}{r_2} \right)$$

$$= \frac{Q}{4\pi} \cdot \frac{\varepsilon_2 r_2 \cdot d_1 + \varepsilon_1 r_1 (r_2 - r_1 - d_1)}{\varepsilon_1 \varepsilon_2 r_1 r_2 (r_1 + d_1)}.$$

$$\text{Capacitance } C = \frac{Q}{V} = \frac{4\pi\varepsilon_1\varepsilon_2 r_1 r_2 (r_1 + d_1)}{\varepsilon_2 r_2 d_1 + \varepsilon_1 r_1 (r_2 - r_1 - d_1)}$$

$$= \frac{4\pi\varepsilon_0 K_1 K_2 r_1 r_2 (r_1 + d_1)}{K_2 r_2 d_1 + K_1 r_1 (r_2 - r_1 - d_1)}.$$

The required energy stored is

$$\int \tfrac{1}{2}\varepsilon_1 E^2 \, dV = \int_{r_1}^{r_1+d_1} \tfrac{1}{2}\varepsilon_1 \left(\frac{Q}{4\pi\varepsilon_1 r^2} \right)^2 \cdot 4\pi r^2 dr = \frac{Q^2}{8\pi\varepsilon_0 K_1} \cdot \frac{d_1}{r_1(r_1 + d_1)}.$$

QUESTIONS

1. What do you mean by energy density in electrostatic field? Show that it is given by $\frac{1}{2}\vec{E} \cdot \vec{D}$. Apply this formula to find the total energy in the field of a point charge. Discuss the result.

2. Assuming that the electrostatic energy density is given by $u = \frac{1}{2}\varepsilon_0 E^2$ find the electrostatic energy of a uniformly charged sphere. Apply the result to find the Coulomb energy of a nucleus.

3. Consider the electron as a uniformly charged sphere of radius r_0 containing a total charge $-e$. Equating the electrostatic self-energy of this electron model with the rest mass energy mc^2 of the electron estimate the radius r_0.

4. Establish a relation between the force acting on a static charge configuration and the electrostatic energy of the system. Use the result to find the force between the plates of a parallel plate capacitor when the charge (or potential) is maintained constant.

5. When a system of conductors is maintained at a constant potential by means a battery show that the energy supplied by the battery during a virtual displacement of the system equals twice the increase in electrostatic energy of the system.

6. (a) In a parallel plate capacitor the space between the plates is filled with a medium whose dielectric constant increases uniformly with distance. If d be the distance between the plates and K_1 and K_2 be the dielectric constants of the medium at the two plates respectively, determine the capacitance per unit area.

(b) In a parallel plate capacitor the space between the plates is filled with a medium composed of two linear dielectrics of thickness d_1 and d_2 and dielectric constants K_1 and K_2, respectively. Determine the capacitance per unit area.

7. Show that the electrostatic energy stored in capacitance C charged to a voltage V is $\frac{1}{2}CV^2$.

8. (a) Calculate the capacitance of a coaxial cylindrical capacitor. Cite a practical example of such a configuration.

(b) A capacitor is made of two coaxial metallic cylinders of radii r_1 and $r_2 (r_1 < r_2)$ and length $L (L \gg r_2)$. The region between $r = r_1$ and $r = \sqrt{r_1 r_2}$ is filled with a medium of dielectric constant K_1 and the remaining space is filled with a medium of dielectric constant K_2. Find the capacitance of the system.

9. Explain why the introduction of a dielectric medium between the plates of a capacitor changes its capacitance.

10. Calculate the capacitance of a spherical capacitor consisting of two concentric spheres of radii a and b, respectively, the space between them being filled with a dielectric of relative permittivity ε_r.

11. (a) What do you mean by dielectric strength?

(b) Discuss its importance in determining the maximum operating voltage of a capacitor?

(c) What do you mean by a "10μF, 50V" capacitor? **[J.U. 2010]**

EXERCISES

1. Suppose eight charges $+q$ are placed at the corners a cube of side a. If a charge $-2q$ is placed at the centre of the cube what would be the electrostatic energy of the system. $\left[Ans. \ \frac{4.32q^2}{4\pi\varepsilon_0 a}\right]$

2. Consider two concentric spherical metal shells of radii a and $b(a < b)$. Charge $+Q$ is placed on the inner shell and a charge $-Q$ is placed on the outer shell. Find the electrostatic energy stored in the system. Hence, determine the capacitance of the system. [*Ans.* $\frac{1}{4\pi\epsilon_0} \cdot \frac{Q^2}{2}\left(\frac{1}{a} - \frac{1}{b}\right)$; $\frac{4\pi\epsilon_0 \cdot ab}{b-a}$]

3. Consider a parallel plate capacitor with plate separation d, each plate of length l and width a and the space between the plates is filled up with a dielectric slab of permittivity ϵ, thickness d, length l and width a. The capacitor is charged with a charge Q by connecting to a battery. The battery is then disconnected and the dielectric slab is now withdrawn along the l dimension until only a length x remains between the plates. Find the force on the slab tending to bring it back to the original position. [*Ans.* $\frac{Q^2 d(\epsilon - \epsilon_0)}{2a[(\epsilon-\epsilon_0)x + \epsilon_0 l]^2}$]

4. Assuming the earth to be a conducting sphere of radius 6400 km, find its capacitance in units of farad. [*Ans.* 7.1×10^{-4}F]

5. A conducting sphere of radius a is surrounded by a thick spherical shell having inner and outer radii b and c, respectively. The outer shell is isolated and initially unchanged. Charge $+Q$ is placed on the inner sphere. Now find the capacitance of the system. [*Ans.* $4\pi\epsilon_0 \left(\frac{1}{a} - \frac{1}{b} + \frac{1}{c}\right)^{-1}$]

6. A parallel plate air capacitor is partly immersed in a liquid dielectric of dielectric constant K such that one-half of each plate lies vertically inside the liquid. Show that the change in capacitance due to this is $\frac{\varepsilon_0 A}{2d}(K-1)$, where A is the area of each plate and d is the separation between the plates.

7. Consider a cylindrical capacitor with the space between the two cylinders filled up with an inhomogeneous dielectric of dielectric constant $K(r) = 1 + \frac{A}{r}$ for $a < r < b$, where a and b are the radii of the two cylinders. Find the capacitance per unit length of the capacitor. [*Ans.* $\frac{2\pi\varepsilon_0}{\ln\frac{A+b}{A+a}}$]

8. The capacitance of a parallel plate air capacitor increases due to insertion of a dielectric slab between the plates. If t be the thickness of the slab and K be the dielectric constant show that in order to restore the original capacitance the distance between the plates should be increased by $t(K-1)/K$.

9. A pd of 400 V is applied across the series combination of two capacitors 10μF and 30μF. Find the pd across the 10μF capacitor and also the energy stored in it. [*Ans.* 300 V, 0.45 J]

10. (a) Four identical point charges q are placed at the four corners of a square of side a. Find the electrostatic energy of the system. **[C.U. 2018]**

[**Hints.** Use Eq. (4.2-4) $U = \dfrac{q^2}{4\pi\varepsilon_0 a}(4+\sqrt{2})$]

(b) Four identical charges q are placed at the vertices of a tetrahedron with the length of an edge a. Find the electrostatic energy of the system. Also, calculate the work required to move these charges to the corners of a tetrahedron of side $a/2$. [*Ans.* Both $\frac{6q^2}{4\pi\varepsilon_0 a}$]

11. Consider a parallel plate capacitor with a dielectric slab of dielectric constant K filling the entire space between the plates. It is given a charge Q and is kept isolated. Show that the work done against the electric forces to remove the slab is $\frac{Q^2(K-1)}{2CK}$, where C is the capacitance in the absence of the slab.

12. Consider a parallel plate capacitor C_1 whose one plate is grounded and the other plate is charged by a potential V_1 and then insulated. The separation between the plates is now increased and the capacitance decreases to C_2. What would the final potential of the insulated plate? [*Ans.* C_1V_1/C_2]

13. A parallel plate capacitor C is charged to a pd V and is then disconnected from the battery. How much work would be required now to double the plate separation? [*Ans.* $\frac{1}{2}CV^2$]

14. A spherical capacitor consists of two concentric conducting spheres of radii a and b $(b > a)$. The outer sphere is grounded and a charge Q is placed on the inner sphere. The outer sphere is now made to contract to a radius b' . Find the work required. [*Ans.* $\frac{Q^2}{8\pi\varepsilon_0}\left(\frac{1}{b'} - \frac{1}{b}\right)$]

15. A cylindrical capacitor of length L consists of an inner cylinder of radius a and an outer hollow cylinder of radius b, the space between them is filled up with a solid dielectric of dielectric constant K. Suppose the dielectric is pull out by a distance $x(x < L)$. Find the force acting on the dielectric tending to pull it back into the capacitor. Assume that the capacitor is connected to a source of constant voltage V. [*Ans.* $(K-1)\pi\varepsilon_0 V^2/\ln(b/a)$]

16. The space between two concentric spherical conductors of radii a and b $(b > a)$ is filled with an in homogeneous dielectric with permittivity $\varepsilon = \frac{\varepsilon_0}{1+Ar}$, where A is a constant and r is the radial distance. The outer sphere is grounded and a charge Q is placed on the inner sphere. Find the capacitance of the system.

[*Ans.* $4\pi\varepsilon_0/\left(\frac{1}{a} - \frac{1}{b} + A\ln\frac{b}{a}\right)$]

17. Show that the capacitance of a parallel plate capacitor does not change when a thin conducting plate is inserted parallelly in halfway between the plates.

18. Consider a parallel plate capacitor with a inhomogeneous dielectric whose permittivity varies with the distance x from one plate as $\varepsilon(x) = \varepsilon_0/(1 - x^2/3d^2)$, where d is the plate separation. Show that the capacitance of the system is $9/8$ times the capacitance without the dielectric medium.

19. Consider a cylindrical capacitor of inner and outer radii a and b, respectively and a composite dielectric with dielectric constant K_1 for $a < r < c$ and K_2 for $c < r < b$. Find the capacitance per unit length of the system.

[*Ans.* $2\pi\varepsilon_0 \left(\frac{1}{K_1}\ln\frac{c}{a} + \frac{1}{K_2}\ln\frac{b}{c}\right)^{-1}$]

20. A spherical conductor of radius a is coated with a uniform thickness d of a dielectric material of dielectric constant K. Show that the capacitance of the conductor is increased in the ratio $Ka(a+d)/(Ka+d)$ due to the coating.

[cf. C.U. 2015]

21. A coaxial cable consists of a copper core of 1 mm radius and an outer metal sheath of radius 1 cm, separated by a dielectric material of dielectric constant 5. Calculate the energy stored in 1 km length of this cable when 10 kV pd is applied between core and the sheath. [*Ans.* 6.04 J]

22. (a) Consider a point charge q located at the centre of a spherical conducting shell of inner and outer radii a and b, respectively. Find the electrostatic energy of the system. [*Ans.* $\frac{1}{4\pi\varepsilon_0}\cdot\frac{q^2}{2}\left(\frac{1}{b}-\frac{1}{a}\right)$]

[**Hints.** Note that $\vec{E} = \dfrac{1}{4\pi\varepsilon_0}\cdot\dfrac{q}{r^2}\hat{r}$ for $r < a$ & $r > b$ and $\vec{E} = 0$ for $a < r < b$.
$U = \int \frac{1}{2}\varepsilon_0 E^2\, dV -$ self-energy of $q = \int_0^a \frac{1}{2}\varepsilon_0 E^2\, dV + \int_b^\infty \frac{1}{2}\varepsilon_0 E^2\, dV - \int_0^\infty \frac{1}{2}\varepsilon_0 E^2\, dV$.]

(b) Electric field in cylindrical coordinates is given by $\vec{E} = \frac{K}{r}\hat{r}$. Find work done to move a point charge Q to a point at twice the radial distance.

[*Ans.* $KQ\ln 2$]

(c) A sphere of radius a carries a charge density, $\rho(r) = Kr$ where K is a constant. Find the total electrostatic energy of the system.

[*Ans.* $\pi^2K^2a^7/7\varepsilon_0$]

23. Three equal point charges q are located at the vertices of an equilateral triangle. Find the electrostatic energy of the system. Show that the insertion of an additional charge $-q/\sqrt{3}$ at the centroid makes the total electrostatic energy of the system zero.

24. Assuming the electron to be a tiny ball of radius r_0 with its total charge $-e$ distributed uniformly over the surface of the ball calculate its total electrostatic self-energy. Estimate the radius r_0 by equating this energy to the rest mass energy mc^2 of the electron. [*Ans.* 2.8×10^{-15} m]

25. The distance between the plates of a parallel plate air capacitor is d. A dielectric slab of thickness x is introduced in the air gap. Show that the capacitance of the capacitor will be doubled if the dielectric constant of the material of the slab is $K = 2x/(2x-d)$. **[C.U. 2019]**

26. Consider the model of a hydrogen atom with a proton surrounded by an electron cloud with charge density $\rho(r) = \frac{-e}{\pi a_0^3}\cdot e^{-2r/a_0}$, where a_0 is the first Bohr radius.

Show that the interaction energy between the electron cloud and the proton is given by $U = -\frac{e^2}{4\pi\varepsilon_0 a_0}$.

27. (a) Two protons inside a nucleus are 3×10^{-15} m apart. Find their mutual electrostatic interaction energy. [*Ans.* 7.68×10^{-14} J]

(b) Consider an infinite straight chain of alternating charges $+q$ and $-q$. Distance between alternating charges is d. Find the interaction energy of each charge with all others. **[C.U. 2017]** [*Ans.* $-\frac{q^2}{4\pi\varepsilon_0}\ln 2$]

(c) Electrostatic potential at any point is $\phi = 2x + 4y$ V. Find the electrostatic energy density at that point. [*Ans.* $10\varepsilon_0\,\mathrm{J/m^3}$]

28. A parallel plate air capacitor is charged by connecting it to a 90 V battery. The battery is then disconnected and a dielectric liquid is poured between the plates to fill the air gap. The voltage across the capacitor is now found to be reduced to 25 V. Find the dielectric constant of the liquid. [*Ans.* 3.6]

29. A coaxial cable consists of a metal wire of radius a surrounded by a concentric hollow metal cylinder of radius b. The space between them is filled with a dielectric of strength E_m. Show that in order to avoid dielectric breakdown the voltage V applied between the wire and the hollow cylinder should satisfy the relation $V \le E_m a \ln b/a$.

30. Consider a spherical capacitor consisting of two concentric spherical shells of radii a and b ($a < b$). The outer one is grounded and the inner one is given a charge Q. Show that half of the total energy is stored within the spherical shell of radii between a and r, where r is given by $\frac{2}{r} = \frac{1}{a} + \frac{1}{b}$.

31. A spherical conductor of radius a carries a charge q. The conductor is coated with a uniform thickness b of a dielectric material of permittivity ε. Show that the total energy stored in electrostatic field is $U = \frac{q^2}{8\pi\varepsilon_0 b} + \frac{q^2(b-a)}{8\pi\varepsilon ab}$.

[**Hints.** $U = \int_a^b \frac{1}{2}\varepsilon\left(\frac{1}{4\pi\varepsilon}\cdot\frac{q}{r^2}\right)^2 \cdot 4\pi r^2 dr + \int_b^\infty \frac{1}{2}\varepsilon_0\left(\frac{1}{4\pi\varepsilon_0}\cdot\frac{q}{r^2}\right)^2 \cdot 4\pi r^2 dr$]

32. The space between the metal plates of a parallel plate capacitor is filled with two sheets of dielectric material having dielectric constants K_1 and K_2, respectively. Thickness of each sheet is d. The plates carry free charge densities $+\sigma$ and $-\sigma$, respectively. Find $\vec{D}$ and $\vec{P}$ in each dielectric.

[*Ans.* $\vec{D}_1 = \vec{D}_2 = \sigma\hat{n}, \vec{P}_1 = \left(1 - \frac{1}{K_1}\right)\sigma\hat{n}, \vec{P}_2 = \left(1 - \frac{1}{K_2}\right)\sigma\hat{n}$, $\hat{n}$ being unit vector from positive plate towards negative plate]

33. A capacitor is formed by two square metal plates each of side a separated by a distance d. The gap is filled by dielectrics of dielectric constants K_1 and K_2 as shown in Fig. 4.E-33. Show that the capacitance of the capacitor is

$$C = \frac{\epsilon_0 a^2 K_1 K_2}{(K_1 - K_2)d}\ln(K_1/K_2)$$

[C.U. 2012]

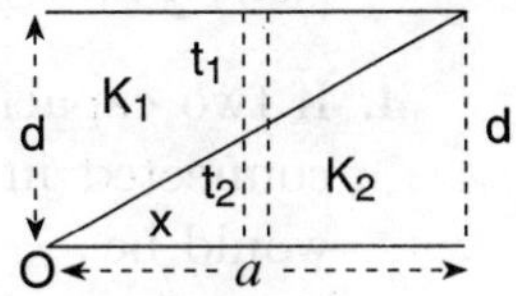

Fig. 4.E-33

[**Hints.** Capacitance of a strip of width dx at a distance x from edge O is

$$dc = \frac{\varepsilon_0 a dx}{\frac{d \cdot x}{K_2 a} + \frac{1}{K_1}\left(d - \frac{d \cdot x}{a}\right)}$$

Note that dielectric thickness at the stips: $\frac{t_2}{x} = \frac{d}{a}$ and $t_1 = d - t_2$

Integrating over $x = 0$ to $x = a$ get the result.]

34. Four 1μF capacitors are connected in parallel charged to 200 V and then discharged through a 5 mm length of find copper wire at 25°C. This wire has resistance $4\,\Omega$/m and mass 0.045 g/m. Would you expect the wire to melt? Given melting point of copper = 1356°C, specific heat of copper 382 J.kg^{-1}.K^{-1}.

[*Ans.* Yes]

35. (a) Find the maximum charge density that a conducting surface can carry in air. Given dielectric strength of air = 3×10^6 V/m. [*Ans.* 2.65×10^{-5} C/m^2]

(b) What maximum charge can be stored on a metallic sphere of radius 10 cm in air. [*Ans.* 3.32 μC]

OBJECTIVE TEST

1. If the separation between two charges is increased the electric potential energy:

(a) always decreases,
(b) always increases,
(c) may increase or decrease,
(d) remains the same.

2. Energy density of the electric field produced by a point charge varies with the distance r from the point charge as:

(a) $\frac{1}{r^2}$, (b) $\frac{1}{r^3}$, (c) $\frac{1}{r^4}$, (d) $\frac{1}{r^5}$.

3. If two capacitors having the same capacitance C and breakdown voltage V are connected in series, their equivalent capacitance and the breakdown voltage would be:

(a) $\frac{C}{2}$ and $\frac{V}{2}$, (b) $\frac{C}{2}$ and 2V, (c) 2C and 2V (d) 2C and $\frac{V}{2}$.

4. If two capacitors having the same capacitance C and breakdown voltage V are connected in parallel, their equivalent capacitance and the breakdown voltage would be

(a) $\frac{C}{2}$ and $\frac{V}{2}$, (b) $\frac{C}{2}$ and $2V$, (c) $2C$ and $2V$ (d) $2C$ and V.

5. A dielectric slab is introduced between the plates of an isolated charged parallel plate air capacitor. Which of the following quantities will remain unchanged?

(a) charge on the capacitor, (c) energy of the capacitor;
(b) pd across the capacitor, (d) electric field inside the capacitor.

6. Capacitance has the dimensions

(a) $[M^{-1}L^{-2}T^4I^2]$, (c) $[MLT^{-3}I^{-1}]$,
(b) $[ML^2T^{-4}I^{-2}]$, (d) $[M^{-1}L^{-1}T^3I]$.

7. A fixed pd V is applied between the plates of a parallel plate capacitor. Now a conducting plate is introduced between the plates such that it divides the space between positive and negative plates in the ratio $d_1 : d_2$. Then the voltage of the conducting plate with respect to the negative plate would be:

(a) zero, (b) V, (c) $\frac{d_2}{d_1 + d_2}V$, (d) $\frac{d_1}{d_1 + d_2}V$.

8. The pd across a capacitor is kept constant. If a dielectric slab of dielectric constant K is introduced between the plates, the stored energy will:

(a) decrease by a factor K,
(b) increase by a factor K,
(c) remain constant,
(d) increase or decrease depending on the nature of the dielectric.

9. A parallel plate capacitor of plate area A and plate separation d is charged to a pd of V_0. It is now disconnected from source of pd and then its plate separation is increased to $3d$. Work necessary for this is:

(a) $3\varepsilon_0 AV_0^2/d$, (b) $\varepsilon_0 AV_0^2/2d$, (c) $\varepsilon_0 AV_0^2/3d$, (d) $\varepsilon_0 AV_0^2/d$.

10. A metal sphere of radius r_1 carrying a charge q_1 is connected by a thin wire to another metal sphere of radius r_2 carrying a charge q_2. The common potential of the system is:

(a) $\frac{1}{4\pi\varepsilon_0} \cdot \frac{q_1 + q_2}{r_1 + r_2}$, (c) $\frac{1}{4\pi\varepsilon_0} \cdot \left(\frac{q_1}{r_1} - \frac{q_2}{r_2}\right)$,
(b) $\frac{1}{4\pi\varepsilon_0} \cdot \left(\frac{q_1}{r_1} + \frac{q_2}{r_2}\right)$, (d) $\frac{1}{4\pi\varepsilon_0} \cdot \frac{q_1 q_2}{\sqrt{r_1 r_2}}$.

11. N identical mercury drop each of capacitance C coalesce into a larger drop. The capacitance of the large drop is

(a) nC, (b) C/n, (c) $n^{1/3}C$, (d) $n^{5/3}C$.

ANSWERS

1. (c) **2.** (c) **3.** (b) **4.** (d) **5.** (a) **6.** (a) **7.** (c)
8. (b) **9.** (d) **10.** (a) **11.** (c)

Chapter **5**

Electrostatic Boundary Value Problems

5.1 Introduction

The primary problem in electrostatic is to find the electric field due to a given stationary charge distribution. If the charge distribution is specified everywhere then, in principle, the electric field can be found by using Coulomb's law as the integral over the charge distributions. In some cases it is easier to find the potential ϕ first and then to calculate $\vec{E}$ by using the relation $\vec{E} = -\vec{\nabla}\phi$. In symmetric cases Gauss's law simplifies the solution. However, there is a large class of problems in which we do not know the charge distribution in advance and consequently we cannot directly determine $\vec{E}$ and ϕ. For example, in the problem of a system of conductors the potential on the surface or total charge on each conductor may be given but the distribution of surface charges may not be known in advance. In such cases the solution of Poisson's equation $\nabla^2\phi = -\rho/\epsilon_0$ or Laplace's equation $\nabla^2\phi = 0$ with appropriate boundary conditions is found to be useful. Very often we need to find potential in a region where charge density $\rho = 0$. In such cases we need to solve Laplace's equation with appropriate boundary conditions. In this chapter we shall consider the solution of these so-called boundary value problems.

5.2 Uniqueness Theorem

The solution of an electrostatic problem is obtained if the potential is determined at all points. The problems in electrostatics generally involve finite bounded regions. If the solution to a potential problem is found which obeys Poisson's equation and the given boundary conditions then the solution is unique. This is known as *uniqueness theorem.* The theorem is of great utility in solving electrostatic problems because one can use the power of guess work and intuition. If the solution is obtained by whatever means and it satisfies Poisson's equation and given boundary conditions then it can be accepted as the true solution without further arguments.

Suppose there exist two solutions ϕ_1 and ϕ_2 of the given electrostatic problem satisfying Poisson's equation $\nabla^2\phi = -\rho/\epsilon_0$ and the given boundary conditions within a volume V bounded by the surface S. Physical experience leads us to two types of boundary conditions yielding unique solutions. Either the potential ϕ is specified on the boundary (*Dirichlet condition*) or the normal derivative of ϕ, i.e., $\partial\phi/\partial n$ is specified on the boundary (*Neumann condition*). Let $\phi = \phi_1 - \phi_2$. As $\nabla^2\phi_1 = -\rho/\epsilon_0$ and $\nabla^2\phi_2 = -\rho/\epsilon_0$, therefore, $\nabla^2\phi = \nabla^2\phi_1 - \nabla^2\phi_2 = 0$ inside V and $\phi = 0$ or $\partial\phi/\partial n = 0$ on the surface S for Direchlet and Neumann boundary conditions, respectively. Now applying Gauss's divergence theorem to the vector $\phi\vec{\nabla}\phi$ we find

$$\begin{aligned}\int_S \phi\vec{\nabla}\phi\cdot\hat{n}\,dS &= \int_V \vec{\nabla}\cdot\left(\phi\vec{\nabla}\phi\right)dV \\ &= \int_V \left[\left|\vec{\nabla}\phi\right|^2 + \phi\nabla^2\phi\right]dV \\ &= \int_V \left|\vec{\nabla}\phi\right|^2 dV \quad [\because\ \nabla^2\phi = 0]\end{aligned}$$

The integration on the left-hand side vanishes for both types of boundary conditions. Therefore,

$$\int_V \left|\vec{\nabla}\phi\right|^2 dV = 0.$$

Since the integrand is positive definite quantity it must be zero at every point in V for the integral to vanish. Thus,

$$\vec{\nabla}\phi = 0 \quad \text{or} \quad \phi_1 - \phi_2 = \text{constant inside } V.$$

For Dirichlet boundary conditions $\phi_1 - \phi_2 = 0$ on the surface S, so we have $\phi_1 = \phi_2$ throughout, i.e., the solution is unique.

For Neumann's boundary conditions

$$\begin{aligned}\vec{\nabla}(\phi_1 - \phi_2)\cdot\hat{n} &= 0 \text{ on } S \\ \text{or } \phi_1 - \phi_2 &= \text{constant on } S.\end{aligned}$$

Since the constant is arbitrary we can take it to be zero and the solution is again unique: $\phi_1 = \phi_2$.

5.3 Laplace's Equation in One Dimension

For highly symmetric problem the potential ϕ may be a function of a single variable. In these cases we need to solve Laplace's equation in one dimension. If the potential ϕ depends only on one coordinate x of an (x, y, z) system of Cartesian coordinates then

Laplace's equation becomes

$$\frac{d^2\phi}{dx^2} = 0.$$

The general solution to this equation is

$$\phi(x) = c_1 x + c_2, \tag{5.3-1}$$

where c_1, c_2 are constants to be evaluated from the given boundary conditions.

Spherical symmetry

In case of spherical symmetry the potential ϕ is a function of r only and independent of θ and φ coordinates of an (r, θ, φ) spherical coordinate system. Then Laplace's equation $\nabla^2\phi = 0$ reduces to

$$\frac{1}{r^2}\frac{d}{dr}\left(r^2\frac{d\phi}{dr}\right) = 0. \tag{5.3-2}$$

Integrating we get for $r \neq 0$,

$$r^2\frac{d\phi}{dr} = c_1 \quad \text{where } c_1 \text{ is a constant.}$$

Integrating again we get

$$\phi = -\frac{c_1}{r} + c_2, \tag{5.3-3}$$

where c_2 is another constant. The constant c_1 and c_2 are to be evaluated from the given boundary conditions.

For example, let us consider two concentric spherical conductors of radii a and b, respectively ($a < b$); the inner one is kept at a potential ϕ_0 and the outer one is grounded. The potential at any inside point is given by (5.3-3), where $a < r < b$. Now

$$\phi = \phi_0 \quad \text{at} \quad r = a$$

$$\text{and } \phi = 0 \quad \text{at} \quad r = b.$$

Applying these boundary conditions and solving for c_1 and c_2 we get

$$c_1 = -\frac{\phi_0 ab}{b-a} \quad \text{and} \quad c_2 = -\frac{\phi_0 a}{b-a}.$$

Thus,

$$\phi(r) = \frac{\phi_0 ab}{b-a}\cdot\frac{1}{r} - \frac{\phi_0 a}{b-a} \tag{5.3-4}$$

Cylindrical symmetry

In case of cylindrical symmetry the potential ϕ is a function of r only and independent of θ and z coordinates of an (r, θ, z) cylindrical coordinate system. Then Laplace's equation $\nabla^2\phi = 0$ reduces to

$$\frac{1}{r}\frac{d}{dr}\left(r\frac{d\phi}{dr}\right) = 0.$$

Integrating we get for $r \neq 0$,

$$r\frac{d\phi}{dr} = c_1, \text{ where } c_1 \text{ is a constant.}$$

Integrating again we get

$$\phi = c_1 \ln r + c_2, \tag{5.3-5}$$

where c_2 is another constant.

The constants c_1 and c_2 are to be evaluated from the given boundary conditions.

For example, let us consider two long coaxial cylinders of radii a and b, respectively ($a < b$), inner one of which is earthed and the outer one is at a potential ϕ_0. The potential at any inside point is given by Eq. (5.3-5).

Now

$$\phi = 0 \quad \text{at } r = a$$
$$\text{and } \phi = \phi_0 \text{ at } r = b.$$

Applying these boundary conditions to Eq. (5.3-5) and solving we get

$$c_1 = \phi_0 / \ln(b/a) \quad \text{and} \quad c_2 = -\phi_0 \ln a / \ln(b/a).$$

Thus,

$$\phi(r) = \frac{\phi_0}{\ln(b/a)} \ln r - \frac{\phi_0}{\ln(b/a)} \ln a = \phi_0 \frac{\ln(r/a)}{\ln(b/a)}. \tag{5.3-6}$$

5.4 Laplace's equation in Two Dimensions: Spherical Coordinates

In some problems the potential may a function of two variables. For example in case of azimuthal symmetry the potential ϕ will be a function of r, θ and independent of azimuthal angle φ of an (r, θ, φ) spherical coordinate system. In this case Laplace's equation reduces to

$$\frac{1}{r^2}\frac{\partial}{\partial r}\left(r^2 \frac{\partial \phi}{\partial r}\right) + \frac{1}{r^2 \sin\theta}\frac{\partial}{\partial \theta}\left(\sin\theta \frac{\partial \phi}{\partial \theta}\right) = 0. \tag{5.4-1}$$

To solve the equation by the *method of separation of variables,* let us assume that

$$\phi(r, \theta) = F_1(r) \cdot F_2(\theta), \tag{5.4-2}$$

where $F_1(r)$ is a function of r only and $F_2(\theta)$ is a function of θ only. Substituting (5.4-2) in (5.4-1), dividing throughout by $F_1 F_2$ and then multiplying it by r^2 we get

$$\frac{1}{F_1} \cdot \frac{d}{dr}\left(r^2 \frac{dF_1}{dr}\right) = -\frac{1}{F_2 \sin\theta}\frac{d}{d\theta}\left(\sin\theta \frac{dF_2}{d\theta}\right). \tag{5.4-3}$$

This left-hand side of this equation is a function of r only and the right-hand side is a function of θ only. So for the validity of the identity (5.4-3) for all values of r and θ each side must be equal to a constant c(say), which is called the *separation constant.* Thus, we get

$$\frac{d}{dr}\left(r^2\frac{dF_1}{dr}\right) = cF_1. \tag{5.4-4}$$

$$\frac{1}{\sin\theta}\cdot\frac{d}{d\theta}\left(\sin\theta\frac{dF_2}{d\theta}\right) + cF_2 = 0. \tag{5.4-5}$$

Putting $\cos\theta = x$ Eq. (5.4-5) can be written as

$$(1-x^2)\frac{dF_2}{dx^2} - 2x\frac{dF_2}{dx} + cF_2 = 0. \tag{5.4-6}$$

It is the well-known Legendre's differential equation and it has physically acceptable well behaved solutions for all values of θ from 0 to π only if we take $c = n(n+1)$, where n is a positive integer. Corresponding solutions are written as $F_2(\theta) \sim P_n(\cos\theta)$, called *Legendre's polynomials* or *zonal harmonics.* A few of these polynomials are $P_0(x) = 1$, $P_1(x) = x$, $P_2(x) = \frac{1}{2}(3x^2-1)$, $P_3(x) = \frac{1}{2}(5x^3-3x)$.

Now with $c = n(n+1)$ Eq. (5.4-4) becomes

$$\frac{d}{dr}\left(r^2\frac{dF_1}{dr}\right) = n(n+1)F_1. \tag{5.4-7}$$

To solve this equation let us put $r = e^s$. Thus, we get

$$\frac{d^2F_1}{ds^2} + \frac{dF_1}{ds} - n(n+1)F_1 = 0. \tag{5.4-8}$$

Let $F_1 = e^{\lambda s}$.

Then for nonzero solution,

$$\lambda^2 + \lambda - n(n+1) = 0 \quad \text{or} \quad (\lambda - n)(\lambda + n + 1) = 0 \quad \text{or} \quad \lambda = n, -(n+1).$$

Therefore, for each n Eq. (5.4-7) will have two independent solutions

$$F_1 = e^{ns} = r^n \quad \text{and} \quad F_1 = e^{-(n+1)s} = \frac{1}{r^{n+1}}. \tag{5.4-9}$$

Combining solutions for F_1 and F_2 we can say that Eq. (5.4-1) will have two independent solutions

$$\phi_n(r,\theta) = r^n P_n(\cos\theta) \quad \text{and} \quad \frac{P_n(\cos\theta)}{r^{n+1}} \quad \text{for each } n.$$

Therefore, the most general solution of Eq. (5.4-1) will be obtained by superposing the linear combinations of all possible solutions as

$$\phi(r,\theta) = \sum_{n=0}^{\infty}\left(a_n r^n + \frac{b_n}{r^{n+1}}\right)P_n(\cos\theta), \tag{5.4-10}$$

where a_n and b_n are constants to be evaluated from the boundary conditions.

5.5 Laplace's Equation in Two Dimensions: Cylindrical Coordinates

In problems with cylindrical configuration having symmetry along z-axis the potential ϕ will be a function of r and θ and independent of z of an (r, θ, z) cylindrical coordinate system. In this case Laplace's equation takes the form,

$$\frac{1}{r}\frac{\partial}{\partial r}\left(r\frac{\partial \phi}{\partial r}\right) + \frac{1}{r^2}\frac{\partial^2 \phi}{\partial \theta^2} = 0. \tag{5.5-1}$$

To solve the equation by the method of separation of variables we assume that

$$\phi(r, \theta) = F_1(r) \cdot F_2(\theta), \tag{5.5-2}$$

where $F_1(r)$ is function of r only and $F_2(\theta)$ is a function of θ only. Substituting (5.5-2) in Eq. (5.5-1), dividing throughout by F_1F_2 and then multiplying it by r^2 we get

$$\frac{r}{F_1}\frac{d}{dr}\left(r\frac{dF_1}{dr}\right) = -\frac{1}{F_2}\frac{d^2F_2}{d\theta^2} \tag{5.5-3}$$

The left-hand side is a function of r only and the right-hand is a function of θ only. So for validity of the identity (5.5-3) each side must be equal to some constant n^2 (say), which is called the *separation constant.* Thus, we get

$$\frac{d}{dr}\left(r\frac{dF_1}{dr}\right) - \frac{n^2F_1}{r} = 0 \tag{5.5-4}$$

$$\text{and}\quad \frac{d^2F_2}{d\theta^2} + n^2F_2 = 0. \tag{5.5-5}$$

For $n = 0$, we have the solutions

$$\begin{aligned} F_1 &= A_0 \ln r + B_0 \\ \text{and}\quad F_2 &= C_0\theta + D_0. \end{aligned} \tag{5.5-6}$$

For $n \neq 0$, putting $x = \ln r$ in Eq. (5.5-4) we get

$$\frac{d^2F_1}{dx^2} - n^2F_1 = 0.$$

Therefore,

$$F_1 = A_ne^{nx} + B_ne^{-nx} = A_n \cdot r^n + \frac{B_n}{r^n}. \tag{5.5-7}$$

Now

$$F_2 = C_n \cos n\theta + D_n \sin n\theta. \tag{5.5-8}$$

Combining these solutions we can write for the total potential ϕ as

$$\phi_0 = (A_0 \ln r + B_0)(C_0\theta + D_0). \tag{5.5-9}$$

$$\phi_n = \left(A_n r^n + \frac{B_n}{r^n}\right)(C_n \cos n\theta + D_n \sin n\theta). \tag{5.5-10}$$

For physically meaningful solutions ϕ_n must be single valued, i.e., we must have

$$\phi_n(r, \theta) = \phi_n(r, \theta + 2\pi).$$

This requires that n can take up only integer values and $C_0 = 0$. The most general solution is obtained by summing over solutions for all possible values of n in the form

$$\phi(r, \theta) = a_0 \ln r + \sum_{n=1}^{\infty} (a_n \cos n\theta + b_n \sin n\theta)\, r^n$$

$$+ \sum_{n=1}^{\infty} (c_n \cos n\theta + d_n \sin n\theta) \cdot \frac{1}{r^n} + e_0, \tag{5.5-11}$$

where a_n, b_n, c_n, d_n and e_0 are constants to be evaluated from the boundary conditions. These solutions of Laplace's equation are called *circular harmonics* of degree n.

5.6 Some Typical Boundary Value Problems

1. An uncharged grounded sphere in a uniform electric field

Suppose a conducting sphere of radius a is placed in an initially uniform electric field $\vec{E}_0 = E_0 \hat{z}$ directed along z-axis (polar direction). We choose the centre of the sphere as origin of a spherical polar coordinate system (r, θ, φ). The electric field induces charges on the conductor and these charges distorts the field near the conductor. We are to find the potential ϕ at any external point $P(r, \theta, \varphi)$. Let

$$\phi = \phi_0 + \phi_1, \tag{5.6-1}$$

where ϕ_0 is the potential associated with $\vec{E}_0$ and ϕ_1 is due to induced charges. ϕ_0 is given by

$$E_0 \hat{z} = -\frac{d\phi_0}{dz}\hat{z} \quad \text{or} \quad \phi_0 = -E_0 z + c = -E_0 r \cos\theta + c, \tag{5.6-2}$$

where c is a constant.

Symmetry about z-axis suggests that ϕ_1 at any outside point will be a function of r and θ only. It satisfies Laplace's equation and is given by

$$\phi_1(r, \theta) = \sum_{n=0}^{\infty} \left(a_n r^n + \frac{b_n}{r^{n+1}}\right) P_n(\cos\theta). \tag{5.6-3}$$

At large distances the induced charges will have no effect. So as $r \to \infty$, $\phi_1 \to 0$. This requires that $a_n = 0$.

Therefore, total potential at any external point is

$$\phi(r,\theta) = \phi_0 + \phi_1 = c - E_0 r \cos\theta + \sum_{n=0}^{\infty} \frac{b_n}{r^{n+1}} P_n(\cos\theta). \tag{5.6-4}$$

As the sphere is grounded potential on its surface is zero, i.e., $\phi = 0$ at $r = a$. Thus,

$$0 = c - E_0 a \cos\theta + \sum_{n=0}^{\infty} \frac{b_n}{a^{n+1}} P_n(\cos\theta). \tag{5.6-5}$$

Equating coefficients of $P_n(\cot\theta)$ on both sides we get

$$c + \frac{b_0}{a} = 0, \quad b_1 = E_0 a^3 \quad \text{and} \quad b_n = 0 \quad \text{for } n \geq 2. \tag{5.6-6}$$

Hence Eq. (5.6-4) can be written as

$$\phi(r,\theta) = b_0\left(\frac{1}{r} - \frac{1}{a}\right) - E_0 r\cos\theta + \frac{E_0 a^3}{r^2}\cos\theta. \tag{5.6-7}$$

To find the constant b_0 we shall use the fact that total charge on the sphere is zero. The charge density on the sphere is given by

$$-\left.\frac{\partial\phi}{\partial r}\right|_a = \frac{\sigma}{\varepsilon_0} \quad \text{or} \quad \sigma = \frac{\varepsilon_0 b_0}{a^2} + 3\varepsilon_0 E_0 \cos\theta. \tag{5.6-8}$$

Now total charge

$$\begin{aligned} Q = \int \sigma\, dS &= \int_0^\pi \sigma \cdot 2\pi a \sin\theta \cdot d\theta \\ &= \frac{\varepsilon_0 b_0}{a^2} \cdot 2\pi a \int_0^\pi \sin\theta\, d\theta + 3\varepsilon_0 E_0 \int_0^\pi 2\pi a \sin\theta\cos\theta\, d\theta \\ &= \frac{4\pi\varepsilon_0 b_0}{a} + 0. \end{aligned}$$

As the sphere is uncharged, $b_0 = 0$.

Therefore, from Eq. (5.6-7), we get

$$\phi(r,\theta) = -E_0 r\cos\theta + \frac{E_0 a^3}{r^2}\cos\theta. \tag{5.6-9}$$

The electric field is given by

$$\begin{aligned} \vec{E} = -\vec{\nabla}\phi &= E_0\hat{z} - E_0 a^3 \left(\hat{r}\frac{\partial}{\partial r} + \hat{\theta}\frac{1}{r}\frac{\partial}{\partial\theta}\right)\left(\frac{\cos\theta}{r^2}\right) \\ &= E_0\hat{z} + \frac{E_0 a^3}{r^3}\left(\hat{r}2\cos\theta + \hat{\theta}\sin\theta\right). \end{aligned} \tag{5.6-10}$$

The first term on the right-hand side is associated with the external field and the second term is due to a dipole of moment $\vec{p} = 4\pi a^3 \varepsilon_0 E_0 \hat{z}$ formed by the induced charges on the sphere.

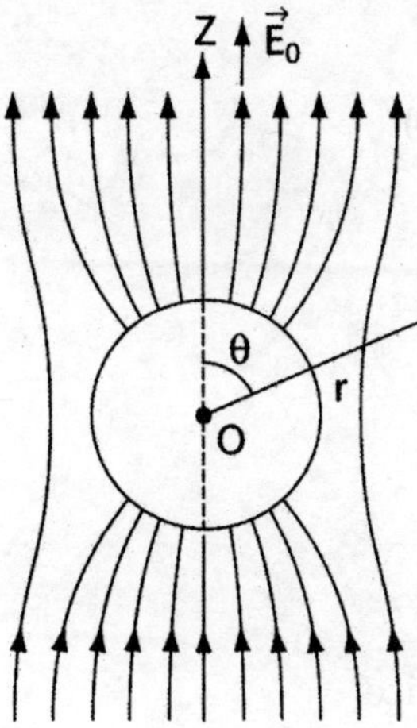

Fig. 5.6-1 : Lines of force in presence of the conducting sphere

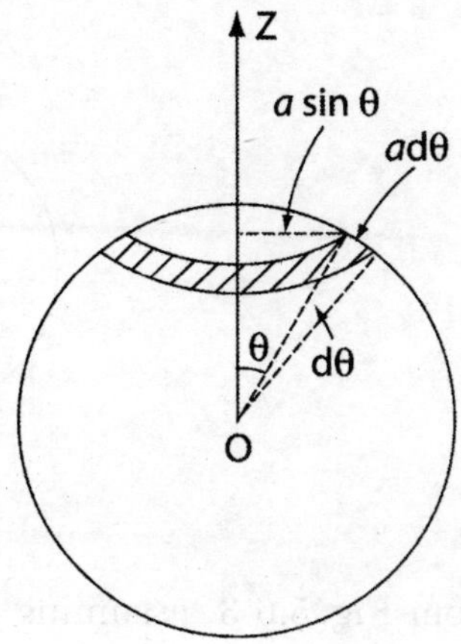

Fig. 5.6-2 : Calculation of total charge

2. A point charge in front of a grounded conducting sphere

Suppose we have a grounded conducting sphere of radius a and a point charge $+q$ placed at the point A at a distance d from the centre of the sphere. We choose z-axis along the line joining the centre of the sphere O and the charge $+q$. There will be some charges induced on the sphere and total potential $\phi(r, \theta)$ at any point will be

$$\phi(r,\theta) = \frac{1}{4\pi\varepsilon_0} \cdot \frac{q}{r_1} + \phi_1, \tag{5.6-11}$$

where $r_1 = AP$ and ϕ_1 is due to induced charges. At any outside point ϕ_1 satisfies Laplace's equation. Because of symmetry about z-axis ϕ_1 will be independent of the azimuthal angle. So ϕ_1 will be given by the solution of the Laplace's equation as

$$\phi_1 = \sum_{n=0}^{\infty} \left(a_n r^n + \frac{b_n}{r^{n+1}} \right) P_n(\cos\theta). \tag{5.6-12}$$

Since $\phi_1 \to 0$ as $r \to \infty$, we must have $a_n = 0$.
Therefore,

$$\phi_1 = \sum_{n=0}^{\infty} \frac{b_n}{r^{n+1}} P_n(\cos\theta) \tag{5.6-13}$$

and

$$\phi(r,\theta) = \frac{1}{4\pi\varepsilon_0} \cdot \frac{q}{r_1} + \sum_{n=0}^{\infty} \frac{b_n}{r^{n+1}} P_n(\cos\theta). \tag{5.6-14}$$

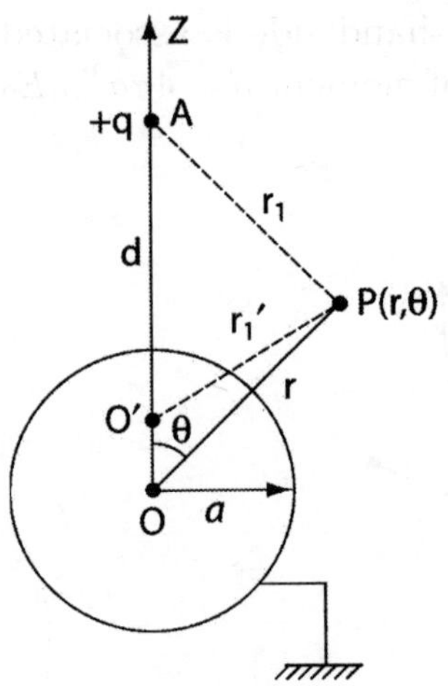

Fig. 5.6-3

Now from Fig. 5.6-3, assuming $r/d < 1$, we can write

$$\frac{1}{r_1} = \frac{1}{\sqrt{r^2 + d^2 - 2rd\cos\theta}} = \frac{1}{d} \cdot \frac{1}{\sqrt{1 + \left(\frac{r}{d}\right)^2 - 2\frac{r}{d}\cos\theta}}$$

$$= \frac{1}{d}\sum_{n=0}^{\infty}\left(\frac{r}{d}\right)^n P_n(\cos\theta), \tag{5.6-15}$$

where we use the generating function of Legendre polynomials.

Therefore, from Eqs. (5.6-14) and (5.6-15),

$$\phi(r,\theta) = \sum_{n=0}^{\infty}\left[\frac{q}{4\pi\varepsilon_0} \cdot \frac{r^n}{d^{n+1}} + \frac{b_n}{r^{n+1}}\right] P_n(\cos\theta). \tag{5.6-16}$$

Since the sphere is grounded $\phi = 0$ at $r = a$. Therefore, from (5.6-16)

$$b_n = -\frac{q}{4\pi\varepsilon_0} \cdot \frac{a^{2n+1}}{d^{n+1}}.$$

Thus, from Eq. (5.6-13) the potential due to induced charges becomes

$$\phi_1 = -\frac{qa}{4\pi\varepsilon_0 d}\sum_{n=0}^{\infty}\frac{a^{2n}}{d^n} \cdot \frac{P_n(\cos\theta)}{r^{n+1}}. \tag{5.6-17}$$

Now we consider a point O', where $OO' = d'$ and $dd' = a^2$. It is called the *inverse point* of $z = d$ with respect to the sphere. If $O'P = r_1'$ then

$$\frac{1}{r_1'} = \frac{1}{\sqrt{r^2 + d'^2 - 2rd'\cos\theta}}$$

$$\frac{1}{r} = \frac{1}{r\sqrt{1+\left(\frac{d'}{r}\right)^2 - 2\left(\frac{d'}{r}\right)\cos\theta}}$$

$$= \frac{1}{r}\sum_{n=0}^{\infty}\left(\frac{d'}{r}\right)^n P_n(\cos\theta) \quad [\text{since } \frac{d'}{r} < 1]$$

$$= \sum_{n=0}^{\infty}\frac{a^{2n}}{d^n}\frac{P_n(\cos\theta)}{r^{n+1}} \quad [\text{using } dd' = a^2] \tag{5.6-18}$$

So Eq. (5.6-17) may be rewritten as

$$\phi_1 = \frac{(-qa/d)}{4\pi\varepsilon_0 r_1'}. \tag{5.6-19}$$

This indicates that the potential at any external point due to the induced charges may be considered as due to an imaginary point charge $-qa/d$ located at the inverse point $z = d' = a^2/d$. It is known as *image charge.*

3. A dielectric sphere in a uniform electric field

Let us consider a dielectric sphere of radius a and permittivity ε_1 placed in a homogeneous isotropic dielectric of permittivity ε_2. Let there is an initially uniform electric field $\vec{E}_0 = \vec{E}_0\hat{z}$ along z-axis (polar direction). We choose the centre of the sphere as the origin of a spherical coordinate system (r, θ, φ). The electric field polarizes the dielectric sphere and this distorts the field near the sphere. Let the potential at any external point be

$$\phi_e = \phi_0 + \phi_2, \tag{5.6-20}$$

where ϕ_0 is the potential associated with $\vec{E}_0$ and ϕ_2 is due to polarization charges. ϕ_0 is given by

$$E_0\hat{z} = -\frac{d\phi_0}{dz}\hat{z} \quad \text{or} \quad \phi_0 = -E_0 z + c = -E_0 r\cos\theta + c, \tag{5.6-21}$$

where c is a constant.

Symmetry of the problem about z-axis suggests that ϕ_2 at any outside point will be a function of r and θ only. It is given by the solution of Laplace's equation as

$$\phi_2(r,\theta) = \sum_{n=0}^{\infty}\left(a_n r^n + \frac{b_n}{r^{n+1}}\right)P_n(\cos\theta). \tag{5.6-22}$$

Since the polarization charges will have no effect at infinity, $\phi_2 \to 0$ as $r \to \infty$. Therefore, we must have $a_n = 0$. Thus, total potential at any external point is

$$\phi_e(r,\theta) = c - E_0 r\cos\theta + \sum_{n=0}^{\infty}\frac{b_n}{r^{n+1}}P_n(\cos\theta). \tag{5.6-23}$$

For any internal point the potential ϕ_i satisfies Laplace's equation. Since ϕ_i is finite at $r = 0$, solution (5.6-22) indicates that the coefficients like b_n must be zero. So we can write for ϕ_i,

$$\phi_i(r,\theta) = \sum_{n=0}^{\infty} c_n r^n P_n(\cos\theta). \tag{5.6-24}$$

To find the constants b_n and c_n we use the following boundary conditions:

(i) $\phi_i = \phi_e$ at $r = a$

(ii) $\varepsilon_1 \dfrac{\partial \phi_i}{\partial r} = \varepsilon_2 \dfrac{\partial \phi_e}{\partial r}$ at $r = a$ (since there is no free charge on the surface).

Applying these conditions to the Eqs. (5.6-23) and (5.6-24) and then equating coefficients of $P_n(\cos\theta)$ from both sides and solving the resulting equations it can be shown that $c_0 = c$, $b_0 = 0$, $c_n = b_n = 0$ for $n \geq 2$ and

$$c_1 = -\frac{3\varepsilon_2}{\varepsilon_1 + 2\varepsilon_2} E_0, \quad b_1 = \frac{\varepsilon_1 - \varepsilon_2}{\varepsilon_1 + 2\varepsilon_2} E_0 a^3. \tag{5.6-25}$$

Therefore, from Eqs. (5.6-23) and (5.6-24) we get

$$\phi_i(r,\theta) = c - \frac{3\varepsilon_2}{\varepsilon_1 + 2\varepsilon_2} E_0 r\cos\theta \quad \text{for } r < a \tag{5.6-26}$$

$$\phi_e(r,\theta) = c - E_0 r\cos\theta + \frac{\varepsilon_1 - \varepsilon_2}{\varepsilon_1 + 2\varepsilon_2} E_0 a^3 \frac{\cos\theta}{r^2} \quad \text{for } r > a. \tag{5.6-27}$$

Using $r\cos\theta = z$ we find from (5.6-26) that the electric field inside the sphere is

$$\vec{E}_i = -\hat{z}\frac{d\phi_i}{dz} = \frac{3\varepsilon_2}{\varepsilon_1 + 2\varepsilon_2} E_0 \hat{z}, \tag{5.6-28}$$

which is uniform and parallel to the initial field $\vec{E}_0$. $\vec{E}_i$ can also be written as

$$\vec{E}_i = E_0\hat{z} - \frac{\varepsilon_1 - \varepsilon_2}{\varepsilon_1 + 2\varepsilon_2} E_0 \hat{z}. \tag{5.6-29}$$

If the sphere has a higher permittivity than the surrounding, i.e., $\varepsilon_1 > \varepsilon_2$ then $E_i < E_0$. The reduction of $\vec{E}_0$ is due to polarization charges, which produces a uniform field

$$\vec{E}_d = -\frac{\varepsilon_1 - \varepsilon_2}{\varepsilon_1 + 2\varepsilon_2} E_0 \hat{z}$$

in opposite direction to the external field $\vec{E}_0$. $\vec{E}_d$ is called *depolarization field* because it tends to opposite the polarization.

On the other hand, if $\varepsilon_1 < \varepsilon_2$ (e.g., a cavity inside a dielectric medium) $E_i > E_0$. Here polarization charges produce field in the direction of $\vec{E}_0$.

From the third term on the right-hand side of Eq. (5.6-27) we note that the potential at any external point due to induced charges is that of a dipole situated at O and of moment.

$$\vec{p} = 4\pi\varepsilon_2 \cdot \frac{\varepsilon_1 - \varepsilon_2}{\varepsilon_1 + 2\varepsilon_2} a^3 E_0 \hat{z}. \tag{5.6-30}$$

The polarization vector $\vec{P}$, defined as dipole moment per unit volume, is given by

$$\vec{P} = \frac{\vec{p}}{\frac{4}{3}\pi a^3} = 3\varepsilon_2 \cdot \frac{\varepsilon_1 - \varepsilon_2}{\varepsilon_1 + 2\varepsilon_2} E_0 \hat{z}. \tag{5.6-31}$$

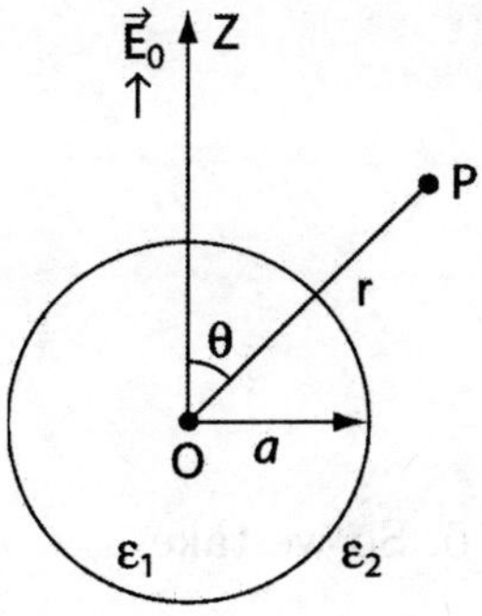

Fig. 5.6-4

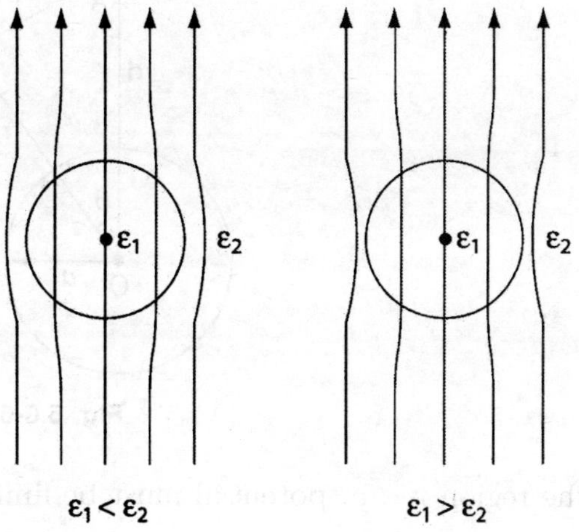

Fig. 5.6-5

The electric field at any outside point is given by

$$\begin{aligned}\vec{E}_e = -\vec{\nabla}\phi_e &= E_0\hat{z} - \frac{\varepsilon_1 - \varepsilon_2}{\varepsilon_1 + 2\varepsilon_2} E_0 a^3 \left(\hat{r}\frac{\partial}{\partial r} + \hat{\theta}\frac{1}{r}\frac{\partial}{\partial \theta}\right)\left(\frac{\cos\theta}{r^2}\right) \\ &= E_0\hat{z} + \frac{\varepsilon_1 - \varepsilon_2}{\varepsilon_1 + 2\varepsilon_2} \cdot \frac{E_0 a^3}{r^3}\left(\hat{r} 2\cos\theta + \hat{\theta}\sin\theta\right).\end{aligned} \tag{5.6-32}$$

The nature of the distribution of electric lines of force will be as shown in Fig. 5.6-5. The dielectric polarization modifies the field within the sphere. It is convenient to express this change directly in terms of polarization $\vec{P}$. A *depolarizing factor* L is defined as the ratio of the change in field inside from the original field to the polarization of the sphere, i.e.,

$$L = \frac{\left|\vec{E}_0\right| - \left|\vec{E}_i\right|}{\left|\vec{P}\right|} = \frac{1}{3\varepsilon_2}.$$

4. A point charge in front of a dielectric sphere

The problem is explained in Fig. 5.6-6. Because of axial symmetry about z-axis potential will be a function r and θ only. Now in the region $r > a$, potential must vanish at infinity. So we can write,

$$\phi_2(r,\theta) = \sum_{n=0}^{\infty} \frac{b_n}{r^{n+1}} P_n(\cos\theta) + \frac{1}{4\pi\varepsilon_0} \cdot \frac{q}{r_1}. \tag{5.6-33}$$

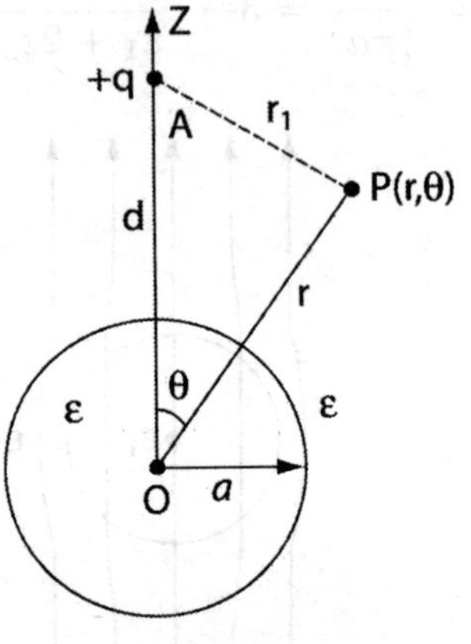

Fig. 5.6-6

For the region $r < a$, potential must be finite at $r = 0$. So we take

$$\phi_1(r,\theta) = \sum_{n=0}^{\infty} a_n r^n P_n(\cos\theta). \tag{5.6-34}$$

As the sphere has no net charge there should not be any term $\sim 1/r$ in the potential given by (5.6-33). So we choose $b_0 = 0$. For $r < d$ we can write

$$\frac{1}{r_1} = \frac{1}{\sqrt{r^2 + d^2 - 2rd\cos\theta}} = \frac{1}{d} \cdot \frac{1}{\sqrt{1 + \left(\frac{r}{d}\right)^2 - 2\frac{r}{d}\cos\theta}}$$

$$= \frac{1}{d} \sum_{n=0}^{\infty} \left(\frac{r}{d}\right)^n P_n(\cos\theta) \tag{5.6-35}$$

$$\therefore \quad \phi_2(r,\theta) = \sum_{n=1}^{\infty} \frac{b_n}{r^{n+1}} P_n(\cos\theta) + \frac{q}{4\pi\varepsilon_0 d} \sum_{n=0}^{\infty} \left(\frac{r}{d}\right)^n P_n(\cos\theta). \tag{5.6-36}$$

Now we apply the boundary conditions

$$\phi_1 = \phi_2 \quad \text{and} \quad \varepsilon \frac{\partial \phi_1}{\partial r} = \varepsilon_0 \frac{\partial \phi_2}{\partial r} \quad \text{at} \quad r = a.$$

Thus, we get from Eqs. (5.6-34) and (5.6-36),

$$\sum_{n=0}^{\infty} a_n a^n P_n(\cos\theta) = \sum_{n=1}^{\infty} \frac{b_n}{a^{n+1}} P_n(\cos\theta) + \frac{q}{4\pi\varepsilon_0} \cdot \frac{1}{d} \sum_{n=0}^{\infty} \left(\frac{a}{d}\right)^n P_n(\cos\theta)$$

$$\varepsilon \sum_{n=0}^{\infty} n a_n \cdot a^{n-1} P_n(\cos\theta) = -\varepsilon_0 \sum_{n=1}^{\infty} \frac{(n+1)b_n}{a^{n+2}} P_n(\cos\theta) + \frac{q}{4\pi d} \sum_{n=0}^{\infty} \frac{n a^{n-1}}{d^n} P_n(\cos\theta).$$

Equations coefficients of $P_n(\cos\theta)$ we get

$$a_0 = \frac{q}{4\pi\varepsilon_0 d}, \quad a_n a^n = \frac{b_n}{a^{n+1}} + \frac{q a^n}{4\pi\varepsilon_0 d^{n+1}} \quad \text{for } n \geq 1$$

$$\text{and} \quad \varepsilon \cdot n a_n \cdot a^{n-1} = -\frac{\varepsilon_0 (n+1) b_n}{a^{n+2}} + \frac{q n a^{n-1}}{4\pi d^{n+1}} \quad \text{for } n \geq 1.$$

Solving these equations we can find the constants a_n and b_n. Hence, the potentials ϕ_1 and ϕ_2 are now determined.

SOLVED PROBLEMS

▶ **Problem 1.** *The potential on the surface of a hollow sphere of radius a is given by $\phi_0(\theta) = A\sin^2\theta/2$, where θ is the polar angle. Using Laplace's equation, find the potential at any inside point.*

Solution. The potential at any inside point satisfies Laplace's equation and it is given by the solution

$$\phi(r,\theta) = \sum_{n=0}^{\infty} \left(a_n r^n + \frac{b_n}{r^{n+1}}\right) P_n(\cos\theta) \quad \text{for } r < a \tag{1}$$

For ϕ to be finite at $r = 0$, we must have $b_n = 0$. Thus,

$$\phi(r,\theta) = \sum_{n=0}^{\infty} a_n r^n P_n(\cos\theta).$$

On the surface of the sphere $r = a$, $\phi = \phi_0$. Therefore,

$$\phi_0 = \sum_{n=0}^{\infty} a_n a^n P_n(\cos\theta) \tag{2}$$

$$\text{or} \sum_{n=0}^{\infty} a_n a^n P_n(\cos\theta) = A\sin^2\theta/2 = \frac{A}{2}(1-\cos\theta) = \frac{A}{2}P_0(\cos\theta) - \frac{A}{2}P_1(\cos\theta).$$

Equating coefficients of $P_n(\cos\theta)$ from both sides we get

$$a_0 = \frac{A}{2}, \quad a_1 = -\frac{A}{2a} \quad \text{and} \quad a_n = 0 \quad \text{for } n > 1. \tag{3}$$

Thus, from Eq. (2),

$$\phi(r,\theta) = \frac{A}{2} - \frac{A}{2a} r\cos\theta = \frac{A}{2}\left(1 - \frac{r}{a}\cos\theta\right) \quad \text{for } r < a.$$

▶ **Problem 2.** *Consider a spherical shell of radius a with a surface charge distribution $\sigma(\theta) = \sigma_0 \cos\theta$, where σ_0 is a constant and θ is the polar angle. Using the solutions of Laplace's equation and appropriate boundary conditions find the potentials at any inside and outside points.* **[cf. C.U. 2008, 2013]**

Solution. Let

$$\phi_1(r,\theta) = \sum_{n=0}^{\infty}\left(a_n r^n + \frac{b_n}{r^{n+1}}\right) P_n(\cos\theta), \quad r < a \tag{1}$$

$$\phi_2(r,\theta) = \sum_{n=0}^{\infty}\left(c_n r^n + \frac{d_n}{r^{n+1}}\right) P_n(\cos\theta), \quad r > a. \tag{2}$$

The potential ϕ_1 must be finite at $r = 0$, which requires that $b_n = 0$. The potential ϕ_2 must be zero at $r = \infty$, which requires that $c_n = 0$. Therefore,

$$\phi_1 = \sum_{n=0}^{\infty} a_n r^n P_n(\cos\theta) \tag{3}$$

$$\phi_2 = \sum_{n=0}^{\infty} \frac{d_n}{r^{n+1}} P_n(\cos\theta). \tag{4}$$

Now we apply the boundary conditions:

(i) $\phi_1 = \phi_2$ at $r = a$

(ii) $\vec{D}_2 \cdot \hat{n} - \vec{D}_1 \cdot \hat{n} = \sigma$ or $-\varepsilon_0 \dfrac{\partial \phi_2}{\partial r} + \varepsilon_0 \dfrac{\partial \phi_1}{\partial r} = \sigma_0 \cos\theta$ at $r = a$.

Thus, we get

$$\sum_{n=0}^{\infty} a_n a^n P_n(\cos\theta) = \sum_{n=0}^{\infty} \frac{d_n}{a^{n+1}} P_n(\cos\theta)$$

and $$\sum_{n=0}^{\infty} \frac{\varepsilon_0 d_n \cdot (n+1)}{a^{n+2}} P_n(\cos\theta) + \sum_{n=0}^{\infty} \varepsilon_0 a_n \cdot n a^{n-1} P_n(\cos\theta) = \sigma_0 P_1(\cos\theta).$$

Equating coefficients of $P_n(\cos\theta)$ from both sides we get

$$\left.\begin{aligned} a_1 \cdot a - \frac{d_1}{a^2} &= 0 \\ a_1 + \frac{2}{a^3} d_1 &= \frac{\sigma_0}{\varepsilon_0} \end{aligned}\right\} \quad \text{for } n = 1$$

$$\left.\begin{aligned} a_n \cdot a^n - \frac{d_n}{a^{n+1}} &= 0 \\ a_n \cdot n \cdot a^{n-1} + \frac{d_n(n+1)}{a^{n+2}} &= 0 \end{aligned}\right\} \quad \text{for } n \neq 1$$

Solving we get $a_n = d_n = 0$ for $n \neq 1$ and

$$a_1 = \frac{\sigma_0}{3\varepsilon_0} \quad \text{and} \quad d_1 = \frac{\sigma_0 a^3}{3\varepsilon_0}.$$

Thus, from (3) and (4) we get

$$\phi_1(r,\theta) = \frac{\sigma_0}{3\varepsilon_0} r\cos\theta, \quad r < a \tag{5}$$

$$\phi_2(r,\theta) = \frac{\sigma_0 a^3}{3\varepsilon_0} \cdot \frac{\cos\theta}{r^2}, \quad r > a. \tag{6}$$

Note that the potential inside depends only on $z(= r\cos\theta)$ and is independent of the size of the shell. The potential outside is a dipole potential with a dipole moment $\vec{p} = \frac{4}{3}\pi a^3 \sigma_0 \hat{z}$.

▶ **Problem 3.** *Consider a system of two coaxial conducting cones with their vertices at $r = 0$ and axes along z-axis. The semi-vertical angles are θ_1 and θ_2 ($\theta_2 > \theta_1$). The cones are insulated from each other, the inner one is kept at a potential ϕ_0 and the outer one grounded. Find the potential in the region between the cones by solving the Laplace's equation. Assume that the cones are sufficiently large so that the end effects can be neglected.*

Solution. Here the potential ϕ may be taken to be independent of r and φ of an (r, θ, φ) spherical coordinate system. So Laplace's equation takes the form

$$\frac{1}{r^2 \sin\theta}\frac{d}{d\theta}\left(\sin\theta \frac{d\phi}{d\theta}\right) = 0.$$

Integrating we get

$$\sin\theta \frac{d\phi}{d\theta} = c_1,$$

where c_1 is a constant. Integrating again we get

$$\phi(\theta) = c_1 \ln\left(\tan\frac{\theta}{2}\right) + c_2, \tag{1}$$

where c_2 is another constant. To find the constants c_1 and c_2 we apply the boundary conditions $\phi = \phi_0$ at $\theta = \theta_1$ and $\phi = 0$ at $\theta = \theta_2$. Thus,

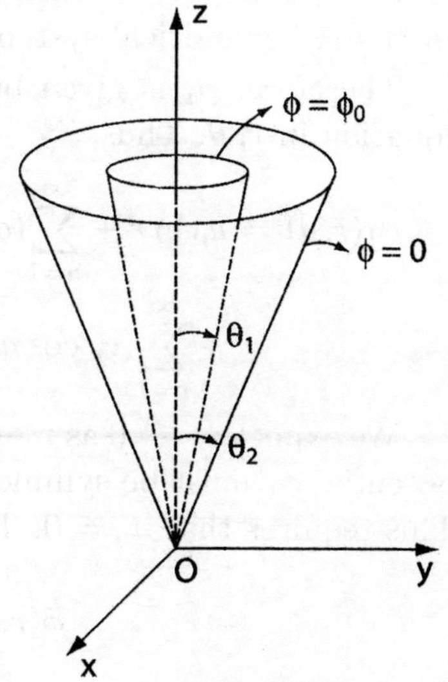

Fig 5.P-3

$$\phi_0 = c_1 \ln\left(\tan\frac{\theta_1}{2}\right) + c_2 \quad \text{and} \quad 0 = c_1 \ln\left(\tan\frac{\theta_2}{2}\right) + c_2.$$

Solving we get

$$c_1 = \frac{\phi_0}{\ln\dfrac{\tan\theta_1/2}{\tan\theta_2/2}} \quad \text{and} \quad c_2 = -\frac{\phi_0 \ln(\tan\theta_2/2)}{\ln\dfrac{\tan\theta_1/2}{\tan\theta_2/2}}.$$

Therefore, from (1),

$$\phi(\theta) = \phi_0 \cdot \frac{\ln\dfrac{\tan\theta/2}{\tan\theta_2/2}}{\ln\dfrac{\tan\theta_1/2}{\tan\theta_2/2}}.$$

▶ **Problem 4.** *A long uncharged grounded conducting cylinder of radius a is placed with its axis perpendicular to an initially uniform electric field $\vec{E}_0$. Find the potential at any outside point.*

Solution. Suppose the electric field is along x-axis, i.e., $\vec{E}_0 = E_0\hat{x}$ and the axis of the cylinder is chosen as z-axis. Let the potential at any outside point is given by $\phi = \phi_0 + \phi_1$, where ϕ_0 is associated with $\vec{E}_0$ and ϕ_1 is due to induced charges. ϕ_0 is given by

$$E_0\hat{x} = -\hat{x}\frac{d\phi_0}{dx} \quad \text{or} \quad \phi_0 = -E_0 x + c = -E_0 r\cos\theta + c,$$

where c is a constant.

ϕ_1 satisfies Laplace's equation. Here, ϕ_1 is independent of z and can be taken as a function of r and θ of an (r, θ, z) cylindrical system.

Therefore, ϕ_1 is given by the solution of Laplace's equation in r, θ. Thus,

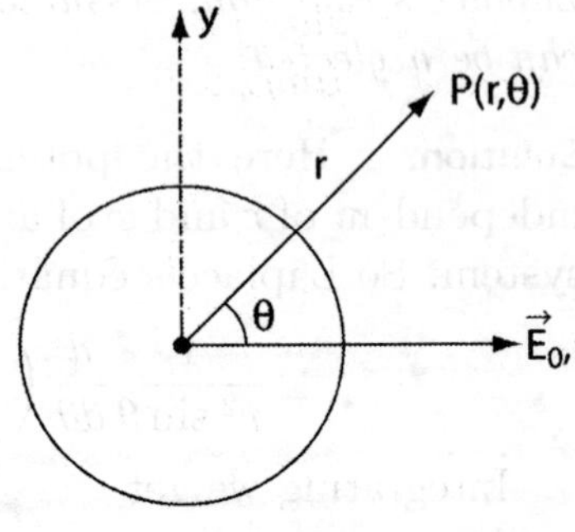

Fig 5.P-4

$$\begin{aligned}\phi_1(r,\theta) &= a_0 \ln r + \sum_{n=1}^{\infty}(a_n\cos n\theta + b_n \sin n\theta)r^n \\ &\quad + \sum_{n=1}^{\infty}(c_n\cos n\theta + d_n\sin n\theta)\cdot\frac{1}{r^n} + c_0.\end{aligned}$$

We expect $\phi_1 \to 0$ as $r \to \infty$. So we must have $a_n = b_n = a_0 = c_0 = 0$. By symmetry potential ϕ_1 must be symmetric about x-axis and hence, it must be even function of θ. This requires that $d_n = 0$. Thus,

$$\phi(r,\theta) = c - E_0 r\cos\theta + \sum_{n=1}^{\infty}\frac{c_n}{r^n}\cos n\theta.$$

Now at $r = a$, $\phi = 0$ because the cylinder is grounded.

$$\therefore \quad 0 = c - E_0 a\cos\theta + \sum_{n=1}^{\infty}\frac{c_n}{a^n}\cos n\theta.$$

Since cosine functions are linearly independent we can equate on both sides of the equation the coefficients of $\cos n\theta$. This yields,

$$c = 0, \quad c_1 = E_0 a^2 \quad \text{and} \quad c_n = 0 \quad \text{for } n > 1.$$

Thus, finally we can write

$$\phi(r,\theta) = -E_0 r\cos\theta + \frac{E_0 a^2}{r}\cos\theta = -E_0\left(1 - \frac{a^2}{r^2}\right) r\cos\theta.$$

▶ Problem 5. *Consider a rectangular box of dimensions a, b and c along x, y and z-directions, with one of its corners at the origin. The face $z = c$ is kept at a potential ϕ_0 and all the other surfaces are at zero potential. Find the potential at any point inside the box.*

Solution. Laplace's equation in Cartesian coordinate

$$\frac{\partial^2\phi}{\partial x^2} + \frac{\partial^2\phi}{\partial y^2} + \frac{\partial^2\phi}{\partial z^2} = 0. \qquad (1)$$

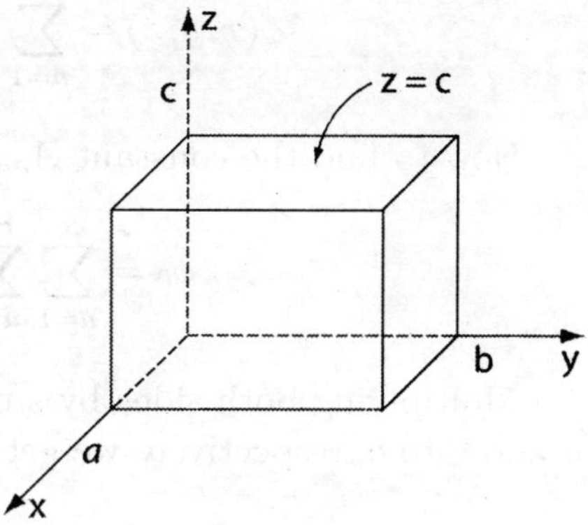

Fig 5.P-5

To solve it by the method of separation of variables we put

$$\phi(x,y,z) = F_1(x)\cdot F_2(y)\cdot F_3(z). \qquad (2)$$

Substituting (2) in (1) and dividing throughout by $F_1F_2F_3$ we get

$$\frac{1}{F_1}\frac{d^2F_1}{dx^2} + \frac{1}{F_2}\frac{d^2F_2}{dy^2} + \frac{1}{F_3}\frac{d^2F_3}{dz^2} = 0.$$

For validity of this equation for all values of x, y, z each term must be equal to some constant. In the given problem the potential is periodic along x- and y-directions. So we may choose

$$\frac{1}{F_1}\frac{d^2F_1}{dx^2} = -\alpha^2, \quad \frac{1}{F_2}\frac{d^2F_2}{dy^2} = -\beta^2 \quad \text{and} \quad \frac{1}{F_3}\frac{d^2F_3}{dz^2} = \gamma^2, \qquad (3)$$

where $\gamma^2 = \alpha^2 + \beta^2$ and each of α^2, β^2 and γ^2 are positive constants.

Solution of the Eqs. (3) are

$$\begin{aligned} F_1 &= A_1\sin\alpha x + A_2\cos\alpha x \\ F_2 &= B_1\sin\beta y + B_2\cos\beta y \\ F_3 &= C_1e^{\gamma z} + C_2e^{-\gamma z}. \end{aligned}$$

Now the boundary conditions that $\phi = 0$ at $x = 0$, $x = a$, $y = 0$ and $y = b$ indicate that $A_2 = B_2 = 0$ and

$$\alpha = \frac{m\pi}{a} \quad \text{and} \quad \beta = \frac{n\pi}{b},$$

where m, n are positive integers.

The condition that $\phi = 0$ at $z = 0$ gives $C_2 = -C_1$ and we can write

$$F_3 = 2C_1 \sinh \gamma z. \tag{4}$$

Thus,

$$\phi(x, y, z) = A_{mn} \cdot \sin \frac{m\pi x}{a} \sin \frac{n\pi y}{b} \sinh \gamma_{mn} z, \tag{5}$$

where

$$\gamma_{mn} = \sqrt{\alpha^2 + \beta^2} = \sqrt{\frac{m^2\pi^2}{a^2} + \frac{n^2\pi^2}{b^2}}. \tag{6}$$

The general solution is obtained by superposing all possible solutions with different values of m an n. Therefore,

$$\phi(x, y, z) = \sum_{n=1}^{\infty} \sum_{m=1}^{\infty} A_{mn} \sin \frac{m\pi x}{a} \sin \frac{n\pi y}{b} \sinh \gamma_{mn} z. \tag{7}$$

Now to find the constant A_{mn} we apply the condition $\phi = \phi_0$ at $z = c$.

$$\therefore \quad \phi_0 = \sum_{n=1}^{\infty} \sum_{m=1}^{\infty} A_{mn} \sin \frac{m\pi x}{a} \sin \frac{n\pi y}{b} \cdot \sinh \gamma_{mn} c. \tag{8}$$

Multiplying both sides by $\sin \frac{m\pi x}{a} \cdot \sin \frac{n\pi y}{b}$ and integrating over x and y from 0 to a and 0 to b, respectively we get

$$\phi_0 \cdot \int_0^a \sin \frac{m\pi x}{a} dx \int_0^b \sin \frac{n\pi y}{b} dy = A_{mn} \sinh \gamma_{mn} c \cdot \frac{a}{2} \cdot \frac{b}{2}, \tag{9}$$

where we use the following orthogonal properties of sine and cosine functions:

$$\int_0^a \sin \frac{m\pi x}{a} \cdot \sin \frac{n\pi x}{a} dx = \frac{a}{2}\delta_{mn} = \begin{cases} 0, & \text{if } m \neq n \\ a/2, & \text{if } m = n \end{cases}$$

(Here δ_{mn} is the Kronecker delta function, $\delta_{mn} = 0$ for $m \neq n$ and 1 for $m = n$)

$$\begin{aligned} \therefore \quad A_{mn} &= \frac{4\phi_0}{ab \sinh \gamma_{mn} c} \int_0^a \sin \frac{m\pi x}{a} dx \int_0^b \sin \frac{n\pi y}{b} dy \\ &= \frac{16\phi_0}{mn\pi^2 \sinh \gamma_{mn} c} \qquad \text{for odd } m, n \qquad (10) \\ &= 0 \qquad \text{for even } m, n \end{aligned}$$

Therefore, from (7) we get

$$\phi(x, y, z) = \frac{16\phi_0}{\pi^2} \sum_{\text{odd } n} \sum_{\text{odd } m} \frac{1}{mn \sinh \gamma_{mn} c} \cdot \sin \frac{m\pi x}{a} \sin \frac{m\pi y}{b} \cdot \sinh \gamma_{mn} z. \tag{11}$$

▶ **Problem 6.** *Using Laplace's equation find the capacitance of a parallel plate capacitor.*

Solution. Consider a parallel plate capacitor formed by two parallel plates kept at $x = 0$ and $x = d$ in air. The plate at $x = 0$ is at potential V and the plate at $x = d$ is grounded. If the plates are large enough the potential $\phi(x, y, z)$ will be independent of y and z coordinates. Thus, Laplace's equation reduces to

$$\frac{d^2\phi}{dx^2} = 0.$$

Integrating twice we get $\phi(x) = Ax + B$, where A and B are constants.
Now at $x = 0$, $\phi = V$ and at $x = d$, $\phi = 0$
Therefore, $A = -\frac{V}{d}$, $B = V$ and $\phi(x) = V\left(1 - \frac{x}{d}\right)$
Now charge density σ near the plate at $x = 0$ is given by

$$\frac{\sigma}{\varepsilon_0} = \vec{E}\cdot\hat{n} = -\left.\frac{d\phi}{dx}\right|_{x=0} = \frac{V}{d}$$

$$\therefore \quad \sigma = \frac{\varepsilon_0 V}{d}$$

Total charge on the plate of area A is $Q = \frac{\varepsilon_0 AV}{d}$.
Therefore, capacitance is $C = \frac{Q}{V} = \frac{\varepsilon_0 A}{d}$.

▶ **Problem 7.** *The potential on the surface of a hollow sphere of radius a is given by $V(\theta) = V_0\cos^2\theta$, where V_0 is a constant and θ is the polar angle. Using Laplace's equation find the potential at any inside and outside points. Also find the surface charge density on the sphere.*

Solution. The potentials are given by the solutions of the Laplace's equation as

$$\phi_1(r,\theta) = \sum_{n=0}^{\infty}\left(a_n r^n + \frac{b_n}{r^{n+1}}\right)P_n(\cos\theta) \text{ for } r < a$$

$$\phi_2(r,\theta) = \sum_{n=0}^{\infty}\left(c_n r^n + \frac{d_n}{r^{n+1}}\right)P_n(\cos\theta) \text{ for } r > a$$

Now ϕ_1 must be finite at $r = 0$ and ϕ_2 must vanish at $r = \infty$. These requires that $b_n = c_n = 0$. Thus,

$$\phi_1(r,\theta) = \sum_n a_n r^n P_n(\cos\theta) \quad \text{and} \quad \phi_2(r,\theta) = \sum_n \frac{d_n}{r^{n+1}}P_n(\cos\theta).$$

On the surface of the sphere

$$\phi_1(a,\theta) = \phi_2(a,\theta) = V_0\cos^2\theta = \frac{2V_0}{3}P_2(\cos\theta) + \frac{V_0}{3}P_0(\cos\theta),$$

where we use the definitions $P_2(\cos\theta) = \frac{1}{2}(3\cos^2\theta - 1)$ and $P_0(\cos\theta) = 1$

$$\therefore \quad \sum_{n=0}^{\infty} a_n\cdot a^n\cdot P_n(\cos\theta) = \sum_{n=0}^{\infty}\frac{d_n}{a^{n+1}}P_n(\cos\theta) = \frac{2V_0}{3}P_2(\cos\theta) + \frac{V_0}{3}P_0(\cos\theta).$$

Equating coefficients of $P_n(\cos\theta)$ we get

$$a_0 = \frac{V_0}{3}, \quad a_2 = \frac{2V_0}{3a^2}, \quad d_0 = \frac{V_0 a}{3}, \quad d_2 = \frac{2V_0 a^3}{3}, \quad a_n = d_n = 0 \quad \text{for } n \neq 0, 2.$$

Thus,

$$\phi_1(r,\theta) = \frac{V_0}{3} + \frac{2V_0}{3a^2} r^2 P_2(\cos\theta) \quad \text{for } r < a$$

$$\phi_2(r,\theta) = \frac{V_0 a}{3r} + \frac{2V_0 a^3}{3r^3} P_2(\cos\theta) \text{ for } r > a.$$

From boundary condition we get the charge density as

$$\sigma(\theta) = \varepsilon_0 \frac{\partial\phi_1}{\partial r}\bigg|_a - \varepsilon_0 \frac{\partial\phi_2}{\partial r}\bigg|_a = \frac{\varepsilon_0 V_0}{3a}\left[1 + 5P_2(\cos\theta)\right].$$

▶ **Problem 8.** *The potential on the surface of a spherical shell of radius R is given by $V_0(\theta)$, where θ is the polar angle. Show that the potential at any inside and outside points may be represented by*

$$\phi_1(r,\theta) = \sum_{n=0}^{\infty} a_n r^n P_n(\cos\theta) \;\; \textit{for } r < R$$

$$\phi_2(r,\theta) = \sum_{n=0}^{\infty} \frac{b_n}{r^{n+1}} P_n(\cos\theta) \;\; \textit{for } r > R,$$

where

$$a_n = \frac{2n+1}{2R^n} \int_0^{\pi} V_0(\theta) \cdot P_n(\cos\theta) \sin\theta \, d\theta$$

$$\textit{and } b_n = \frac{(2n+1)R^{n+1}}{2} \int_0^{\pi} V_0(\theta) \cdot P_n(\cos\theta) \sin\theta \, d\theta$$

Solution. See problem 7. As ϕ_1 is finite at $r = 0$ and $\phi_2 \to 0$ as $r \to \infty$. So ϕ_1 and ϕ_2 may be represented as

$$\phi_1(r,\theta) = \sum_{n=0}^{\infty} a_n r^n \cdot P_n(\cos\theta) \tag{1}$$

$$\phi_2(r,\theta) = \sum_{n=0}^{\infty} \frac{b_n}{r^{n+1}} P_n(\cos\theta). \tag{2}$$

To find the constants a_n and b_n we are to use the orthogonal properties of Legendre polynomials:

$$\int_{-1}^{+1} P_n(x) P_m(x) dx = \int_0^{\pi} P_n(\cos\theta) P_m(\cos\theta) \sin\theta \, d\theta$$

$$= \frac{2}{2n+1}\delta_{mn} = 0 \text{ for } m \neq n$$

$$= \frac{2}{2n+1} \text{ for } m = n.$$

On the surface of the sphere ($r = R$),

$$V_0(\theta) = \sum_{n=0}^{\infty} a_n R^n P_n(\cos\theta).$$

Multiplying both sides by $P_n(\cos\theta)\sin\theta$ and integrating with respect to θ from 0 to π we get

$$\int_0^{\pi} V_0(\theta) P_n(\cos\theta)\sin\theta\, d\theta = a_n R^n \cdot \frac{2}{2n+1}$$

$$\therefore \quad a_n = \frac{2n+1}{2R^n}\int_0^{\pi} V_0(\theta) P_n(\cos\theta)\sin\theta\, d\theta.$$

Similarly we can show that

$$b_n = \frac{(2n+1)R^{n+1}}{2}\int_0^{\pi} V_0(\theta) P_n(\cos\theta)\sin\theta\, d\theta.$$

▶ **Problem 9.** *Using Laplace's equation find the potential at any point due to a uniformly charged circular ring of radius a carrying a total charge q.*

Solution. Let us choose the centre of the ring as origin and its axis as z-axis. Because of symmetry about z-axis potential ϕ at any outside point $P(r, \theta)$ will be a function of r and θ only. So from the solution of Laplace's equation in spherical polar coordinates we can write

$$\phi(r,\theta) = \sum_{n=0}^{\infty}\left(a_n r^n + \frac{b_n}{r^{n+1}}\right) P_n(\cos\theta). \qquad (1)$$

For $r > a$ we have the condition that $\phi \to 0$ as $r \to \infty$, so in this case $a_n = 0$ and we get

$$\phi(r,\theta) = \sum_{n=0}^{\infty}\frac{b_n}{r^{n+1}} P_n(\cos\theta). \qquad (2)$$

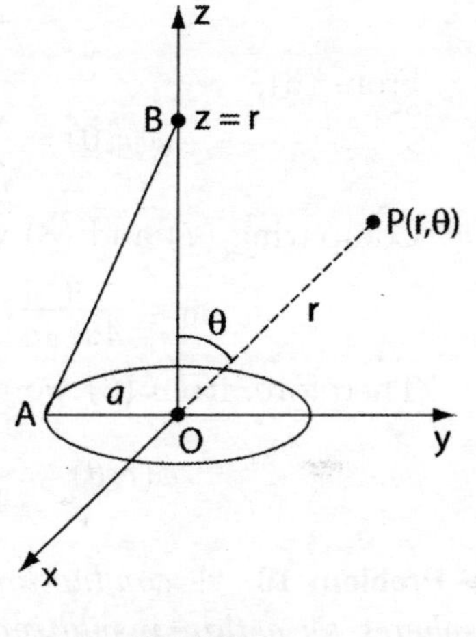

Fig 5.P-9

Now for any point $B(z = r)$ on the z-axis ($\theta = 0$) we find that all the charges are equidistant from B and hence, potential at B is given by

$$\phi(r, 0) = \frac{1}{4\pi\varepsilon_0}\cdot\frac{q}{AB} = \frac{q}{4\pi\varepsilon_0}\cdot\frac{1}{\sqrt{a^2+r^2}}$$

$$= \frac{q}{4\pi\varepsilon_0 r}\left(1 + \frac{a^2}{r^2}\right)^{-1/2}$$

$$= \frac{q}{4\pi\varepsilon_0 r}\left(1 - \frac{a^2}{2r^2} + \frac{3}{8}\frac{a^4}{r^4} - \cdots\right). \qquad (3)$$

Again from (2)

$$\phi(r,0) = \sum_{n=0}^{\infty} \frac{b_n}{r^{n+1}} P_n(1) = \frac{b_0}{r} + \frac{b_1}{r^2} + \frac{b_2}{r^3} + \cdots \tag{4}$$

Comparing (3) and (4) we get $b_0 = \frac{q}{4\pi\varepsilon_0}$,

$$b_1 = 0, \quad b_2 = -\frac{q}{4\pi\varepsilon_0} \cdot \frac{a^2}{2}, \quad b_3 = 0, \ldots$$

Therefore, from (2), we have for $r > a$,

$$\phi(r,\theta) = \frac{q}{4\pi\varepsilon_0}\left[\frac{1}{r}P_0(\cos\theta) - \frac{a^2}{2r^3}P_2(\cos\theta) + \cdots\right] \tag{5}$$

For $r < a$, we require that ϕ is finite at $r = 0$, so b_n must be zero. In this case

$$\phi(r,\theta) = \sum_{n=0}^{\infty} a_n r^n P_n(\cos\theta). \tag{6}$$

For $r < a$,

$$\phi(r,0) = \frac{q}{4\pi\varepsilon_0}\frac{1}{\sqrt{a^2+r^2}} = \frac{q}{4\pi\varepsilon_0 a}\left(1 + \frac{r^2}{a^2}\right)^{-1/2}$$

$$= \frac{q}{4\pi\varepsilon_0 a}\left(1 - \frac{r^2}{2a^2} + \frac{3}{8}\frac{r^4}{a^4} - \cdots\right). \tag{7}$$

From (6),

$$\phi(r,0) = \sum_{n=0}^{\infty} a_n r^n = a_0 + a_1 \cdot r + a_2 r^2 + a_3 r^3 + \cdots \tag{8}$$

Comparing (7) and (8) we find that

$$a_0 = \frac{q}{4\pi\varepsilon_0 a}, \quad a_1 = 0, \quad a_2 = -\frac{q}{4\pi\varepsilon_0} \cdot \frac{1}{2a^3}, \quad a_3 = 0, \cdots$$

Therefore, from (6) we have for $r < a$,

$$\phi(r,\theta) = \frac{q}{4\pi\varepsilon_0 a}\left[P_0(\cos\theta) - \frac{1}{2}\cdot\frac{r^2}{a^2}P_2(\cos\theta) + \cdots\right] \tag{9}$$

▶ **Problem 10.** *A conducting spherical shell of radius R is separated into two hemispheres by a thin insulating ring. The hemispheres are maintained at potentials +V and −V. Find the potential at any inside and outside points. You may assume the solutions of Laplace's equation.*

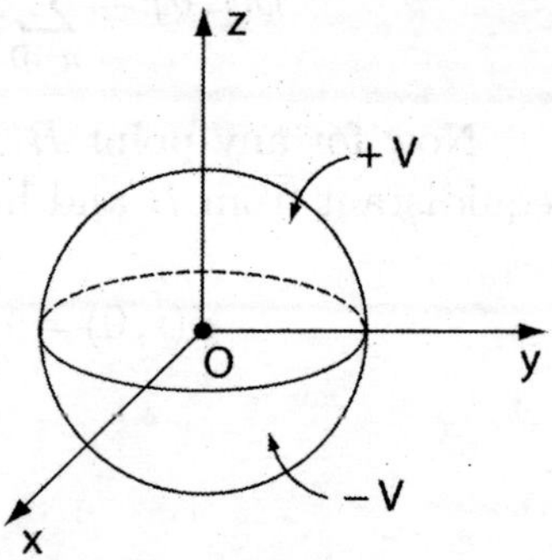

Fig 5.P-10

Solution. Let the insulating ring lies in the $z = 0$ plane as shown in Fig. 5.P-10. Following *Solved Problem* 8, let the potentials be

$$\phi_1(r,\theta) = \sum_{n=0}^{\infty} a_n r^n P_n(\cos\theta) \quad \text{for } r < a \tag{1}$$

$$\phi_2(r,\theta) = \sum_{n=0}^{\infty} \frac{b_n}{r^{n+1}} P_n(\cos\theta) \quad \text{for } r > a. \tag{2}$$

On the surface of the sphere $r = R$. Therefore, from (1)

$$\sum_n a_n \cdot R^n P_n(\cos\theta) = \phi_1(R,\theta) = \begin{cases} +V \text{ for } 0 \le \theta < \pi/2 \\ -V \text{ for } \pi/2 < \theta \le \pi \end{cases}$$

The constants a_n are given by (See, *Solved Problem* 8)

$$a_n = \frac{2n+1}{2 \cdot R^n} \int_0^\pi \phi_1(R,\theta) P_n(\cos\theta) \sin\theta \, d\theta$$

$$= \frac{2n+1}{2R^n} \left[\int_0^{\pi/2} V \cdot P_n(\cos\theta) \sin\theta \, d\theta - \int_{\pi/2}^{\pi} V \, P_n(\cos\theta) \sin\theta \, d\theta \right]$$

Writing x for $\cos\theta$ we get

$$a_n = \frac{2n+1}{2R^n} \left[-\int_1^0 V \, P_n(x) dx + \int_0^{-1} V \, P_n(x) dx \right]$$

$$= \frac{2n+1}{2R^n} V \left[\int_0^1 P_n(x) dx - (-1)^n \int_0^1 P_n(x) dx \right]$$

$$[\because \; P_n(-x) = (-1)^n P_n(x)]$$

Therefore, $a_n = 0$ for n even, i.e., $a_0 = a_2 = a_4 = \ldots = 0$. For n odd we note that $P_1(x) = x$, $P_3(x) = \frac{5}{2}x^3 - \frac{3}{2}x$, etc.

$$\therefore \quad a_1 = \frac{3V}{2R} \times 2 \int_0^1 x \, dx = \frac{3V}{2R}$$

$$a_3 = \frac{7V}{2R^3} \times 2 \int_0^1 \left(\frac{5}{2}x^3 - \frac{3}{2}x \right) dx = -\frac{7V}{8R^3},$$

etc.

Thus, potential at any inside point is

$$\phi_1(r,\theta) = a_1 \cdot r \cdot P_1(\cos\theta) + a_3 \cdot r^3 P_3(\cos\theta) + \cdots$$

$$= \frac{3V}{2} \left(\frac{r}{R} \right) P_1(\cos\theta) - \frac{7V}{8} \left(\frac{r}{R} \right)^3 P_3(\cos\theta) + \cdots$$

To find potential at any outside point, we note from (2) that on the surface

$$\sum_{n=0}^{\infty} \frac{b_n}{R^{n+1}} P_n(\cos\theta) = \phi_2(R,\theta) = \begin{cases} +V \text{ for } 0 \le \theta < \pi/2 \\ -V \text{ for } \pi/2 < \theta \le \pi/2 \end{cases}$$

The coefficients b_n are given by (See, *Solved Problem* 8)

$$b_n = \frac{(2n+1)R^{n+1}}{2} \int_0^\pi \phi_2(R,\theta) P_n(\cos\theta) \sin\theta \, d\theta.$$

Proceeding as before we find that $b_0 = b_2 = b_4 = \; \ldots \; = 0$.

$$b_1 = \frac{3V}{2} R^2, \quad b_3 = -\frac{7V}{8} R^4, \text{etc.}$$

Thus, potential at any outside point is

$$\begin{aligned}\phi_2(r,\theta) &= \frac{b_1}{r^2}P_1(\cos\theta) + \frac{b_3}{r^4}P_3(\cos\theta)\\ &= \frac{3V}{2}\cdot\left(\frac{R}{r}\right)^2 P_1(\cos\theta) - \frac{7V}{8}\left(\frac{R}{r}\right)^4 P_3(\cos\theta) + \cdots\end{aligned}$$

► **Problem 11.** *Show that the field is uniquely determined when the charge density ρ is specified within a bounded region and either the potential ϕ or its normal derivative $\frac{\partial\phi}{\partial n}$ is specified on each boundary.*

Solution. Suppose, if possible, there are two fields $\vec{E}_1$ and $\vec{E}_2$ satisfying the conditions of the problem. They satisfy the Gauss's law in differential form, i.e.,

$$\vec{\nabla}\cdot\vec{E}_1 = \frac{\rho}{\varepsilon_0} \quad \text{and} \quad \vec{\nabla}\cdot\vec{E}_2 = \frac{\rho}{\varepsilon_0}.$$

Therefore, $\vec{E}_3 = \vec{E}_1 - \vec{E}_2$ satisfies the equation

$$\vec{\nabla}\cdot\vec{E}_3 = \vec{\nabla}\cdot E_1 - \vec{\nabla}\cdot E_2 = \frac{\rho}{\varepsilon_0} - \frac{\rho}{\varepsilon_0} = 0.$$

Now if ϕ_3 be the potential corresponding to $\vec{E}_3$ then $\vec{E}_3 = -\vec{\nabla}\phi_3$ and we can write

$$\vec{\nabla}\cdot\left(\phi_3\vec{E}_3\right) = \phi_3\vec{\nabla}\cdot\vec{E}_3 + \vec{\nabla}\phi_3\cdot E_3 = \vec{\nabla}\phi_3\cdot\vec{E}_3 = -E_3^2.$$

Integrating over the entire region in question

$$\int_V \vec{\nabla}\cdot\left(\phi_3\vec{E}_3\right)\,dV = -\int_V E_3^2\,dV$$

$$\text{or} \int_S \phi_3\vec{E}_3\cdot d\vec{S} = -\int_V E_3^2\,dV.$$

Now on the surface $\phi_1 = \phi_2$ or $\phi_3 = 0$ or $(E_1)_n = (E_2)_n$ or $(E_3)_n = 0$.

So $\int_V E_3^2 dV = 0$.

As E_3^2 is always positive the integral can vanish only if $E_3 = 0$ everywhere. Hence $\vec{E}_1 = \vec{E}_2$.

► **Problem 12.** *A charge density $\sigma(\theta) = a\sin 5\theta$ is glued over the surface of an infinite cylinder of radius R, where a is a constant. Find the potential at any inside and outside points in cylindrical coordinates (r,θ,z).* **[J.U. 2011]**

Solution. . Here the potential ϕ is independent of z and can be taken as function of r and θ. From the solution of Laplace's equation the potential at any internal and external points are given by

$$\phi_1(r,\theta) = a_0 + \sum_n r^n(a_n\cos n\theta + b_n\sin n\theta) \;\text{ for }\; r < R \tag{1}$$

$$\phi_2(r,\theta) = b_0 + \sum_n \frac{1}{r^n}(c_n\cos n\theta + d_n\sin n\theta) \;\text{ for }\; r > R \tag{2}$$

Boundary conditions at $r = R$,

$$\sigma = -\varepsilon_0 \left(\frac{\partial \phi_2}{\partial r} - \frac{\partial \phi_1}{\partial r} \right)_{r=R}$$

$$\text{or,} \quad a \sin 5\theta = -\varepsilon_0 \sum_n \left[-\frac{n}{R^{n+1}} (c_n \cos n\theta + d_n \sin \theta) \right.$$

$$\left. -nR^{n-1} (a_n \cos n\theta + b_n \sin n\theta) \right] \quad (3)$$

Evidently,
$$a_n = c_n = 0 \text{ for all } n$$
$$b_n = d_n = 0 \text{ except for } n = 5$$

For $n = 5$,

$$a = 5\varepsilon_0 \left[\frac{d_5}{R^6} + R^4 b_5 \right] \quad (4)$$

Also, we have

$$\phi_2 = \phi_1 \text{ at } r = R \quad (5)$$

$$\therefore \quad b_0 + \frac{d_5}{R^5} \sin 5\theta = a_0 + b_5 R^5 \sin 5\theta$$

$$\therefore \quad a_0 = b_0$$

which may be chosen as zero and

$$\frac{d_5}{R^5} = R^5 b_5 \quad (6)$$

Solving (4) and (6) we get

$$b_5 = \frac{a}{10\varepsilon_0 R^4} \quad \text{and} \quad d_5 = \frac{aR^6}{10\varepsilon_0}$$

Thus

$$\phi_1(r, \theta) = \frac{ar^5 \sin 5\theta}{10\varepsilon_0 R^4} \quad \text{for } r < R$$

$$\phi_2(r, \theta) = \frac{aR^6 \sin 5\theta}{10\varepsilon_0 r^5} \quad \text{for } r > R$$

QUESTIONS

1. (a) State and establish uniqueness theorem in electrostatics.

[cf. C.U. 2013; J.U. 2011, 2014]

(b) Show that a harmonic function is uniquely determined at all points within a given region by its value at the boundary.

2. Use Laplace's equation to find the capacitance of a

(a) parallel-plate, (b) spherical, (c) cylindrical capacitor.

3. An uncharged grounded conducting sphere is placed in an initially uniform electric field $\vec{E}_0$. Using Laplace's equation find the potential at any external point (r, θ). Also find the induced surface charge density.

4. A point charge q is placed at a distance d from the centre of a conducting grounded sphere of radius a $(d > a)$. Use Laplace's equation to find the potential at any external point due to induced charges on the sphere. Show that this potential may be considered as due to an imaginary charge $-qa/d$ located at the inverse point at a distance a^2/d from the centre of the sphere.

5. A dielectric sphere of radius a and permittivity ε_1 is placed in a uniform electric field $\vec{E}_0$ in a medium of permittivity ε_2. Using Laplace's equation show that the polarization induced in the sphere is given by

$$\vec{P} = 3\varepsilon_2 \cdot \frac{\varepsilon_1 - \varepsilon_2}{\varepsilon_1 + 2\varepsilon_2}\vec{E}_0.$$

6. Show that the electric field inside a spherical cavity in a dielectric medium of dielectric constant K placed in an initially uniform electric field $\vec{E}_0$ is given by

$$\vec{E} = \frac{3K}{2K+1}\vec{E}_0.$$

7. A spherical grounded conductor of radius a is placed in a uniform electric field $\vec{E}_0 = E_0\hat{z}$. The primary field induces charges on the sphere and when seen from external points these charges appear to be equivalent to a dipole located at the centre. Assuming expression for dipolar potential and applying boundary conditions find the resultant potential at any external point and surface charge density on the sphere.

[**Hints.** Assume $\phi(r,\theta) = -E_0 r\cos\theta + \dfrac{A\cos\theta}{r^2}$ and apply the condition $\phi(a,\theta) = 0$.]

EXERCISES

1. Consider two concentric spheres of radii a and b $(b > a)$. The potential all over the inner sphere is zero and that on surface of the outer sphere is given by $\phi(\theta) = \phi_0 \cos\theta$, where ϕ_0 is a constant. Show that the potential at any inside point $(a < r < b)$ is given by

$$\phi(r,\theta) = \frac{\phi_0 b^2}{b^3 - a^3}\left(1 - \frac{a^3}{r^3}\right) r\cos\theta.$$

2. Consider a spherical conductor of radius a whose surface is at a constant potential ϕ_0. Using Laplace's equation with proper boundary conditions, find the potential at any inside and outside points. [*Ans.* $\phi_i = \phi_0, \phi_e = \dfrac{\phi_0 a}{r}$]

3. Show that a sphere of infinite permittivity placed in a uniform electric field behaves like a conducting sphere.

[**Hints.** Show that in the limit $\varepsilon_1 \to \infty$ the field in presence of a dielectric sphere as given by the Eq. (5.6-32) reduces to the field in presence of a conducting sphere as given by the Eq. (5.6-10), also from Eq. (5.6-28) as $\varepsilon_1 \to \infty$, $\vec{E}_i \to 0$]

4. A dielectric sphere of permittivity ε is placed in air, where there is an initially uniform field $\vec{E}_0$ along polar direction (z-axis). Show that the induced surface charge density is

$$\sigma(\theta) = \frac{\varepsilon - \varepsilon_0}{\varepsilon + 2\varepsilon_0} \cdot 3\varepsilon_0 E_0 \cos\theta.$$

[**Hints.** *Using Eq.* (5.6-31) *find* $\sigma = \hat{n} \cdot \vec{P}$; *replace* ε_2 *by* ε_0 *and* ε_1 *by* ε.]

5. Suppose a point charge q is placed at the centre of a sphere of radius a. There is no other charges inside and outside the sphere. The potential of the surface of the sphere is $\phi = \phi_0 \cos\theta$, where θ is the polar angle. Find the potential at any inside and outside point.

[*Ans.* $\frac{q}{4\pi\varepsilon_0}\left(\frac{1}{r} - \frac{1}{a}\right) + \frac{\phi_0 \cos\theta}{a} \cdot r$ for $r < a$ and $\frac{\phi_0 a^2}{r^2} \cos\theta$ for $r > a$]

6. A conducting sphere of radius a carrying a total charge q is placed in an initially uniform electric field $E_0 \hat{z}$. Find the electric potential at any outside point.

[*Ans.* $\frac{q}{4\pi\varepsilon_0} \cdot \frac{1}{r} - E_0 r \cos\theta + \frac{E_0 a^3}{r^2} \cos\theta$]

7. A uniform electric field $\vec{E}_0$ is set up in a medium of dielectric constant K. Show that the electric field inside a spherical cavity in the medium is

$$\frac{3K}{2K+1} \vec{E}_0.$$

8. If ϕ_1 and ϕ_2 are two solutions of Laplace's equation, show that $C_1\phi_1 + C_2\phi_2$, where C_1 and C_2 are arbitrary constants, is also a solution.

9. Consider two large plates forming a wedge of angle α. The plates are insulated, one is kept at zero potential and the other at ϕ_0. Using Laplace's equation in cylindrical coordinates (r, θ, φ) show that the potential and field at any point between the plates are given by $\phi = \phi_0 \cdot \frac{\varphi}{\alpha}$ and $\vec{E} = -\frac{\phi_0}{\alpha r}\hat{\varphi}$, respectively, where $\hat{\varphi}$ is the unit vector in the φ-direction.

10. Using Laplace's equation find the capacitance of a spherical capacitor consisting of two concentric spheres of radii a and b $(b > a)$ outer one grounded and inner one kept at a potential ϕ_0. **[C.U. 2015]**

[**Hints.** Use Eq. (5.3-4). Surface charge density on the sphere is

$$\sigma = \varepsilon_0 E_r \Big|_{r=a} = -\varepsilon_0 \frac{d\phi}{dr}\Big|_{r=a} = \frac{\varepsilon_0 b\phi_0}{a(b-a)}.$$

Total charge on inner sphere $\quad Q = 4\pi a^2 \sigma = \dfrac{4\pi\varepsilon_0 ab}{b-a}\phi_0.$

Therefore, $C = \dfrac{Q}{\phi_0} = \dfrac{4\pi\varepsilon_0 ab}{b-a}.$]

11. Using Laplace's equation find the capacitance per unit length of a cylindrical capacitor consisting of two concentric cylinders of radii a and b outer one grounded and the inner one kept at a potential ϕ_0. [*Ans.* $2\pi\varepsilon_0/\ln(b/a)$]

[**Hints.** Use Eq. (5.3-6)]

12. Consider a point charge q placed on the plane interface between two homogeneous infinite dielectrics with permittivities ε_1 and ε_2. Show that the potential at a distance r from the charge is given by $\phi(r) = \dfrac{1}{2\pi(\varepsilon_1+\varepsilon_2)}\cdot\dfrac{q}{r}$ in both the dielectrics.

13. Consider a long uncharged cylinder of radius a made of dielectric material of permittivity ε, placed in a uniform electric field that is normal to its axis. Using Laplace's equation find the potentials at any inside and outside points.

[*Ans.* $\phi_i(r,\theta) = c\,(\text{constant})$; $\phi_e(r,\theta) = c - E_0 r\cos\theta + \frac{\varepsilon-\varepsilon_0}{\varepsilon+\varepsilon_0}\frac{E_0 a^2}{r}\cos\theta$]

14. Suppose a charge q is uniformly distributed over a sphere of radius a with a constant volume charge density ρ. Solve the corresponding Poisson's equation and find the potential at any inside point. Assume that the potential at any external point is $\phi(r) = q/4\pi\varepsilon_0 r$. [*Ans.* $\phi(r) = \frac{\rho a^2}{2\varepsilon_0}\left(1 - \frac{r^2}{3a^2}\right)$ for $r < a$]

15. An insulating thin spherical shell of radius a is kept at a potential $V(\theta) = V_0\cos\theta$, where V_0 is a constant and θ is the angle relative to a fixed diameter. Find the potentials inside and outside the sphere. [*Ans.* $\phi_i = \frac{V_0\cos\theta}{a}\cdot r$, $\phi_e = \frac{V_0 a^2}{r^2}\cos\theta$]

16. Consider a sphere of radius a having a uniform surface density of charge σ_0. Assuming the solution of Laplace's equation in spherical coordinates show that the potential at any inside point is $\sigma_0 a/\varepsilon_0$ and that at any outside point is $\sigma_0 a^2/\varepsilon_0 r$.

17. A surface charge density $\sigma(\theta) = \sigma_0\cos 2\theta$ is glued to the surface of a spherical shell of radius a, where σ_0 is a constant and θ is the polar angle. Find the potential at any outside point (r,θ).

[**Hints.** See, *Solved Problem* 2. Use $\cos 2\theta = 2\cos^2\theta - 1 = \frac{4}{3}P_2(\cos\theta) - \frac{1}{3}P_0(\cos\theta)$.]

18. A hemisphere of radius a is charged uniformly with a total charge q. Show that the potential at any external point is given by

$$\phi(r,\theta) \approx \frac{q}{4\pi\varepsilon_0 r}\left[1+\frac{3a}{2r}\cos\theta\right],$$

where θ is the polar angle and r is the distance from the centre of the base.

19. The potential on the axis of a circular disc of radius a carrying a uniform surface charge density σ is given by $\phi(r,0) = \frac{\sigma}{2\varepsilon_0}\left(\sqrt{r^2+a^2}-r\right)$. Using this result and the solution of Laplace's equation find the potential at any point (r,θ) off the axis (for $r > a$).

[**Hints.** $\phi(r,\theta) = \sum_{n=0}^{\infty} \frac{b_n}{r^{n+1}} P_n(\cos\theta)$

$$\therefore \quad \phi(r,0) = \sum_{n=0}^{\infty} \frac{b_n}{r^{n+1}} P_n(1) = \frac{\sigma}{2\varepsilon_0}\left(\sqrt{r^2+a^2}-r\right) = \frac{\sigma}{2\varepsilon_0}\left[r\left(1+\frac{a^2}{r^2}\right)^{1/2} - r\right]$$

$$\approx \frac{\sigma}{2\varepsilon_0}\left(\frac{a^2}{2r} - \frac{a^4}{8r^3} + \cdots\right)$$

Comparing equal powers of r, we get

$$b_0 = \frac{\sigma a^2}{4\varepsilon_0}, \quad b_1 = 0, \quad b_2 = -\frac{\sigma a^4}{16\varepsilon_0}$$

$$\therefore \ \phi(r,\theta) = \frac{\sigma a^2}{4\varepsilon_0}\left\{\frac{1}{r}P_0(\cos\theta) - \frac{a^2}{4r^3}P_2(\cos\theta) + \cdots\right\} = \frac{\sigma a^2}{4\varepsilon_0 r}\left\{1 - \frac{a^2}{8r^2}\left(3\cos^2\theta - 1\right) + \cdots\right\}.]$$

20. A spherical shell of dielectric material of dielectric constant 3 is placed in an initially uniform electric field $\vec{E}_0$. The inner and outer radii of the shell are a and $2a$, respectively. Show that the field inside the empty space of the shell is given by $\vec{E}_i = \frac{27}{34}\vec{E}_0$.

21. A long cylinder of radius a, made of a dielectric material of dielectric constant K, is placed with its axis perpendicular to an initially uniform electric field $\vec{E}_0$. Show that the field inside the cylinder is given by $\vec{E}_i = \frac{2}{K+1}\vec{E}_0$. Assume that the cylinder contains no external charge. Given the solution to Laplace's equation for cylindrically symmetric system as

$$\phi(r,\theta) = a_0 \ln r + \sum_{n=1}^{\infty}\left(a_n\cos n\theta + b_n \sin n\theta\right) r^n + \sum_{n=1}^{\infty}\left(c_n \cos n\theta + d_n \sin n\theta\right)\frac{1}{r^n} + c_0.$$

22. A dielectric sphere of radius a carries a uniform polarization $\vec{P} = P\hat{z}$. Calculate the potential at any internal and external points. Also, find the corresponding electric fields.

[**Hints.** $\rho_p = -\vec{\nabla} \cdot \vec{P} = 0$, $\sigma_p = \vec{P} \cdot \hat{n} = P\cos\theta$. Following the *Solved Problem* 2, find that

$$\begin{aligned}
\phi(r,\theta) &= \frac{P}{3\varepsilon_0} r\cos\theta \quad \text{for } r \le a, \\
&= \frac{P}{3\varepsilon_0} \cdot \frac{a^3}{r^2}\cos\theta \quad \text{for } r \ge a. \\
\vec{E} &= -\frac{\vec{P}}{3\varepsilon_0} \quad \text{for } r < a, \\
&= \frac{p}{4\pi\varepsilon_0 r^3}\left(2\cos\theta\hat{r} + \sin\theta\hat{\theta}\right) \quad \text{for} \quad r > a, \quad \text{where } \vec{p} = \frac{4}{3}\pi a^3 \vec{P}.]
\end{aligned}$$

23. The potential on the surface of a hollow sphere of radius a is given by $\phi_0(\theta) = A\cos 3\theta$, where θ is the polar angle. Find the potential inside and outside the sphere. Also, find the surface charge density on the sphere.

[**Hints.**

$$\begin{aligned}
\cos 3\theta = 4\cos^3\theta - 3\cos\theta &= \frac{8}{5} \cdot \left(\frac{5}{2}\cos^3\theta - \frac{3}{2}\cos\theta\right) - \frac{3}{5} \cdot \cos\theta \\
&= \frac{8}{5}P_3(\cos\theta) - \frac{3}{5}P_1(\cos\theta)
\end{aligned}$$

See, *Solved Problem* 7.

$$\begin{aligned}
\phi_1(r,\theta) &= -\frac{3A}{5a} \cdot r \cdot P_1(\cos\theta) + \frac{8A}{5a^3} \cdot r^3 P_3(\cos\theta) \\
\phi_2(r,\theta) &= -\frac{3Aa^2}{5r^2} P_1(\cos\theta) + \frac{8Aa^4}{5r^4} P_3(\cos\theta) \\
\sigma(\theta) &= \frac{\varepsilon_0 A}{5a}\{-9P_1(\cos\theta) + 56P_3(\cos\theta)\}]
\end{aligned}$$

24. A conducting sphere of radius a carrying a charge Q is placed in a uniform electric field $\vec{E}_0$. Find the potential at any external point. Take the potential to be zero on equatorial plane, far from the sphere. [*Ans.* $\phi(r,\theta) = -E_0\left(r - \frac{a^3}{r^2}\right)\cos\theta + \frac{1}{4\pi\varepsilon_0} \cdot \frac{Q}{r}$]

25. Consider a charge $+Q$ spread uniformly along a line from $z = -a$ to $z = +a$. Find the potential at any point $(z > a)$ on the z-axis.

Assuming the solution of Laplace's equation for an axially symmetric system as well as the result

$$\log_e \frac{1+x}{1-x} = 2\left[x + \frac{x^3}{3} + \frac{x^5}{5} + \cdots\right] \quad \text{for} \quad |x| < 1,$$

calculate the potential at any point (r,θ) for $r > a$ at least up to r^{-4}.

[**Hints.** From *Exercise Problem* 41 of chapter 1,

$$\begin{aligned}
\phi(r,0) &= \frac{1}{4\pi\varepsilon_0} \cdot \frac{Q}{2a} \ln\frac{r+a}{r-a} \\
&= \frac{1}{4\pi\varepsilon_0} \cdot \frac{Q}{2a} \ln\frac{1+a/r}{1-a/r} \\
&= \frac{1}{4\pi\varepsilon_0} \cdot \frac{Q}{2a} \cdot 2\left\{\frac{a}{r} + \frac{a^3}{3r^3} + \frac{a^5}{5r^5} + \cdots\right\} \qquad (1)
\end{aligned}$$

From the solution of Laplace's equation

$$\phi(r,\theta) = \sum_{n=0}^{\infty}\left(a_n r^n + \frac{b_n}{r^{n+1}}\right) P_n(\cos\theta)$$

For the region $r > a$, we take $a_n = 0$ as $\phi \to 0$ for $r \to \infty$.

$$\therefore \quad \phi(r,0) = \sum_{n=0}^{\infty} \frac{b_n}{r^{n+1}} P_n(1) \qquad (2)$$

Comparing (1) and (2) we get

$$b_0 = \frac{Q}{4\pi\varepsilon_0}, \quad b_1 = 0, \quad b_2 = \frac{Q}{4\pi\varepsilon_0}\cdot\frac{a^2}{3}, \quad b_3 = 0, \quad b_4 = \frac{Q}{4\pi\varepsilon_0}\cdot\frac{a^4}{5}$$

$$\therefore \quad \phi(r,\theta) = \frac{b_0}{r}P_0(\cos\theta) + \frac{b_2}{r^3}P_3(\cos\theta) + \cdots$$

$$= \frac{Q}{4\pi\varepsilon_0}\left\{\frac{1}{r}P_0(\cos\theta) + \frac{a^2}{3r^3}P_2(\cos\theta) + \frac{a^4}{5r^5}P_4(\cos\theta) + \cdots\right\}].$$

26. Consider a long rectangular metal pipe running parallel to z-axis. Its three sides at $x = 0$, $y = 0$ and $y = b$ are grounded and the fourth side at $x = a$ is insulated and maintained at constant potential ϕ_0. Determine the potential at any point inside the pipe. **[J.U. 2012]**

[**Hints.** See *Solved Problem* 5, Here potential is independent of z. Solving two dimensional Laplace's equation by the method of separation of variables get

$$\phi(x,y) = (A_1 e^{\alpha x} + A_2 e^{-\alpha x})(B_1 \sin\alpha y + B_2 \cos\alpha y)$$

Boundary conditions, $\phi(0,y) = \phi(x,0) = \phi(x,b) = 0$ require that $B_2 = 0$, $A_2 = -A_1$ and $\alpha = \frac{n\pi}{b}$ (n is an integer)

$\therefore$ General solution is $\qquad \phi(x,y) = \sum_n A_n \sinh\frac{n\pi x}{b} \cdot \sin\frac{n\pi y}{b}$

To find the constant A_n apply the boundary condition $\phi(a,y) = \phi_0$. Thus

$$\phi_0 = \sum_n A_n \sinh\frac{n\pi a}{b} \sin\frac{n\pi y}{b}$$

Multiply both sides by $\sin\frac{m\pi y}{b}$ and integrate over y from 0 to b and get

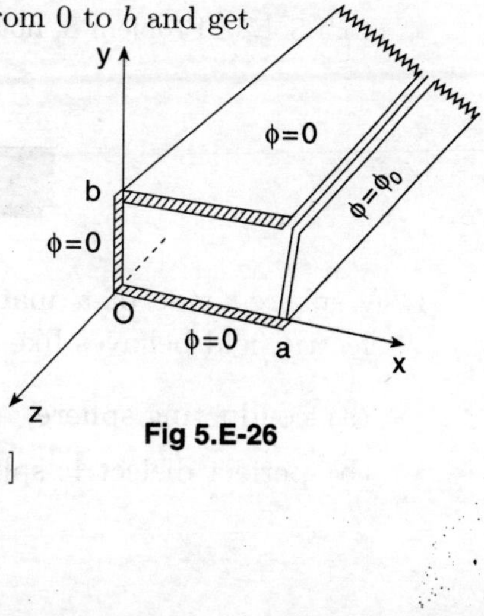

Fig 5.E-26

$$\phi_0 \int_0^b \sin\frac{m\pi y}{b}dy = A_m \sinh\frac{m\pi a}{b}\cdot\frac{b}{2}$$

$$\text{or,} \quad \phi_0 \frac{b}{m\pi}(1 - \cos m\pi) = A_m \sinh\frac{m\pi a}{b}\cdot\frac{b}{2}$$

$$\therefore \quad A_m = 0 \text{ if } m \text{ is even}$$

$$= \frac{4\pi_0}{\pi}\frac{1}{m\sin\frac{m\pi a}{b}}, \quad \text{if } m \text{ is odd}$$

$$\text{Finally,} \quad \phi(x,y) = \frac{4\phi_0}{\pi}\sum_{m=1,3,5,\cdots} \frac{1}{m}\frac{\sinh\frac{m\pi x}{b}}{\sinh\frac{m\pi a}{b}} \sin\frac{m\pi y}{b}\,]$$

27. Consider two infinite grounded metal plates lying parallel to the xz-plane one at $y = 0$ and the other at $y = b$ (Fig. 5.E-27). The boundary at $x = 0$ consists of an infinite strip. This strip is insulated from the two plates and maintained at a constant potential ϕ_0. Find the potential at any point inside this 'slot'.

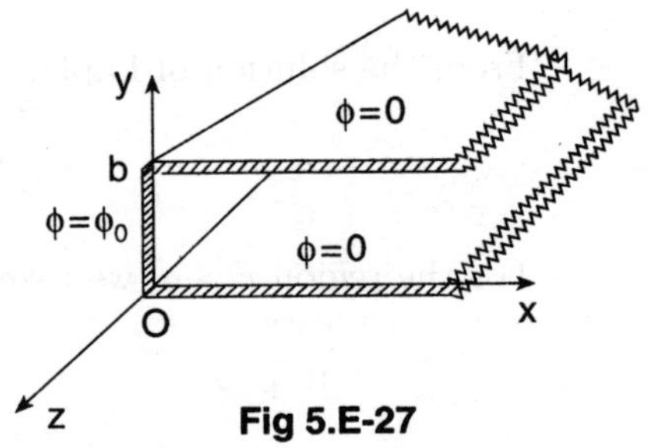

Fig 5.E-27

[*Hints.* See, Hints of Problem 26. Here apply the boundary conditions:

$$\phi(x, 0) = \phi(x, b) = 0, \quad \phi(0, y) = \phi_0$$

and $\phi \to 0$ as $x \to \infty$.

Finally, $\phi(x, y) = \dfrac{4\phi_0}{\pi} \displaystyle\sum_{m=1,3,5} \frac{1}{m} e^{-m\pi x/b} \sin(m\pi y/b)$]

28. Suppose the potential on the surface of a sphere of radius R is specified as $V_0(\theta)$ and there is no charge inside or outside the sphere. Show that the charge density on the sphere is given by

$$\sigma(\theta) = \frac{\varepsilon_0}{2R} \sum_{n=0}^{\infty} (2n+1)^2 c_n P_n(\cos\theta)$$

where $$c_n = \int_0^{\pi} V_0(\theta) P_n(\cos\theta) \sin\theta \, d\theta$$ **[J.U. 2012]**

[**Hints.** See Solved Problem 8.

$$\sigma(\theta) = -\varepsilon_0 \left[\frac{\partial \phi_2}{\partial r} - \frac{\partial \phi_1}{\partial r}\right]_{r=R}$$

$$= \varepsilon_0 \left[\frac{n+1}{R^{n+2}} a_n P_n(\cos\theta) + n\, a_n R^{n-1} P_n(\cos\theta)\right]$$

From Solved Problem 8, note that $a_n = \dfrac{b_n}{R^{2n+1}}$. Using this get the result.]

OBJECTIVE TEST

1. A sphere made of a material of infinite permittivity and placed in a uniform electric field behaves like a:

(a) conducting sphere,
(b) perfect dielectric sphere,
(c) free space
(d) none of these.

2. Electric fields $\vec{E}_1$ and $\vec{E}_2$ on the two sides of an interface of two dielectric media having dielectric constants K_1 and K_2, make angles θ_1 and θ_2 with the normal to the interface. Which one of the following relations is true?

(a) $K_1 \sin\theta_1 = K_2 \sin\theta_2$, (c) $K_1 \tan\theta_1 = K_2 \tan\theta_2$,

(b) $K_1 \cos\theta_1 = K_2 \cos\theta_2$, (d) $K_1 \cot\theta_1 = K_2 \cot\theta_2$.

3. The function $P_n(x)$ is even if n is:

(a) even, (c) zero only,

(b) odd, (d) none of these.

4. A function is said to be harmonic if it satisfies:

(a) Hermite differential equation, (c) Laplace's equation,

(b) Legendre's differential equation, (d) Wave equation.

5. Circular harmonics are the solution of Laplace's equation in:

(a) plane polar coordinates, (c) Cartesian coordinates,

(b) spherical polar coordinates, (d) none of these.

6. Two potential functions ϕ_1 and ϕ_2 satisfy Laplace's equation within a closed volume and assume the same values on its surface. Then

(a) $\phi_1 = \phi_2$, (c) $\phi_1 - \phi_2 \neq 0$,

(b) ϕ_1 may or may not be equal to ϕ_2, (d) none of these.

ANSWERS

1. (a) **2.** (d) **3.** (a) **4.** (c) **5.** (a) **6.** (a)

Chapter 6

Method of Electrical Image

6.1 Introduction

Various methods are available for the solution of electrostatic problems. Most straight forward method is to solve Laplace's or Poisson's equations with proper boundary conditions. However, the method is not found to be always simple. There are many special problems in electrostatics, which can be solved with greater ease by the method of images, invented by Lord Kelvin. The essence of the method consists in determining the potential due to a given charge distribution by considering a different charge configuration, which satisfies the given boundary conditions and then exploiting the uniqueness theorem. The method is particularly useful for problems consisting of point charges near conductors. To find the potential outside the conductor the induced charges on the conductor are replaced by one or more fictitious point charges of suitable magnitudes at suitable locations such that the equipotential conditions at the conductor surfaces and Laplace's or Poisson's equation is satisfied everywhere in the region outside the conductors. According to uniqueness theorem the potential thus found will be the correct one for the entire region outside the conductors. These fictitious, unreal, imaginary charges are called *image charges* (from the analogy with optical images in mirror reflection). The method is illustrated in this chapter with a few examples.

6.2 Point Charge in Front of an Earthed Conducting Plane

Suppose we have a large grounded conducting plane forming xy-plane and a point charge $+q$ placed on the z-axis at $z = d$ (Fig. 6.2-1). We wish to find the potential and field in the region $z \geq 0$. The potential will be partly due to q and partly due to induced charges on the plane. To find the latter contribution we must have a knowledge about the induced charge distribution. In the method of electrical images we try to replace this induced charges by image charges. In this attempt the following conditions must be obeyed:

(i) Potential $\phi = 0$ all over the conducting plane ($z = 0$).

(ii) ϕ must be zero at infinity.

(iii) ϕ satisfies Laplace's equation in the region $z > 0$ except at the point A where the charge q is located.

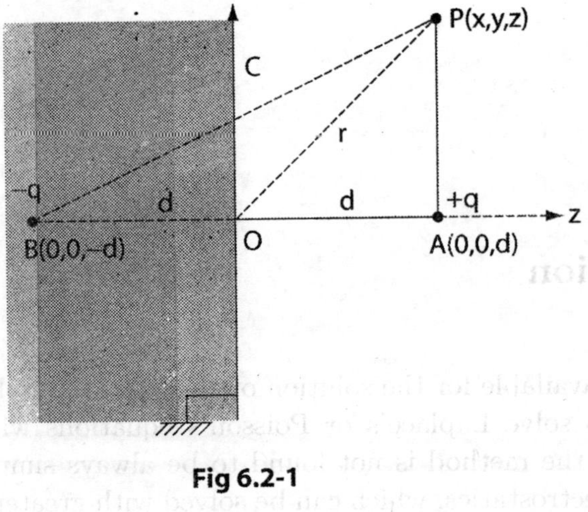

Fig 6.2-1

The symmetry of the problem suggests an image charge $-q$ located at $z = -d$. Let us now see that this choice satisfies all the above conditions (i)–(iii). The potential at any point C on the plane due to q at A and the image charge $-q$ at B is

$$\phi_c = \frac{1}{4\pi\varepsilon_0}\left(\frac{q}{AC} - \frac{q}{BC}\right) = 0.$$

The placing of $-q$ at a finite distance does not affect the condition (ii) at infinity. As the image charge $-q$ is located in the region $z < 0$, it will not affect the condition (iii) in the region $z > 0$. Thus, our choice satisfies all the conditions. According to the uniqueness theorem the solution thus found is unique. So instead of the actual induced charges the image charge can be used to find the potential and field in the region $z > 0$. The actual induced charge distribution can also be found.

Potential and field at any point $P(x, y, z)$

$$\phi(x,y,z) = \frac{1}{4\pi\varepsilon_0}\left(\frac{q}{AP} - \frac{q}{BP}\right)$$
$$= \frac{q}{4\pi\varepsilon_0}\left(\frac{1}{\sqrt{x^2+y^2+(z-d)^2}} - \frac{1}{\sqrt{x^2+y^2+(z+d)^2}}\right). \quad (6.2\text{-}1)$$

The electric field can be determined by using the relation

$$\vec{E} = -\vec{\nabla}\phi = -\left[\hat{i}\frac{\partial\phi}{\partial x} + \hat{j}\frac{\partial\phi}{\partial y} + \hat{k}\frac{\partial\phi}{\partial z}\right]. \quad (6.2\text{-}2)$$

Lines of force

The electric lines of force will be the same as that between the charge $+q$ at A and $-q$ at B (Fig. 6.2-2). The lines of force on the side $z < 0$ is shown by dotted lines because they do not exist really, the field there is zero. The field lines meet the plane normally.

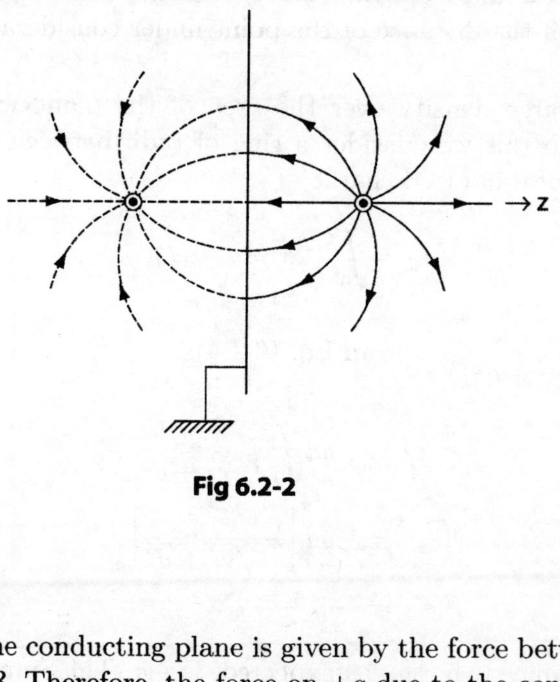

Fig 6.2-2

Force on $+q$

Force on $+q$ due to the conducting plane is given by the force between $+q$ at A and its image charge $-q$ at B. Therefore, the force on $+q$ due to the conducting plane is

$$\vec{F} = -\frac{1}{4\pi\varepsilon_0}\cdot\frac{q^2}{(2d)^2}\hat{k}, \quad (6.2\text{-}3)$$

where $\hat{k}$ is the unit vector along OA.

The force is attractive in nature. This force is largely responsible for preventing the free electrons from leaving the surface of conductor and is associated with the work function of the conducting material.

Induced charges

The actual induced surface charge density σ can be obtained as follows

$$\sigma = -\varepsilon_0 \frac{\partial \phi}{\partial z}\bigg|_{z=0}.$$

From Eq. (6.2-1) we have

$$\frac{\partial \phi}{\partial z} = \frac{q}{4\pi\varepsilon_0}\left[\frac{-(z-d)}{\{x^2+y^2+(z-d)^2\}^{3/2}} + \frac{z+d}{\{x^2+y^2+(z+d)^2\}^{3/2}}\right]$$

$$\therefore \quad \sigma(x,y) = -\frac{q}{2\pi}\cdot\frac{d}{(x^2+y^2+d^2)^{3/2}}. \tag{6.2-4}$$

The induced charge, as expected, is negative. It has a maximum value at $x = y = 0$ on the plane. The distance of any point $(x, y, 0)$ on the plane from $A(0, 0, d)$ is $\sqrt{x^2+y^2+d^2}$. Thus, we find from Eq. (6.2-4) that the induced charge density varies inversely as the cube of the distance of the point under consideration from the charge $+q$ at A.

Integrating the charge density over the area of the plane we can find the total induced charge. To do this we consider a ring of radii between r and $r + dr$, where $r = \sqrt{x^2+y^2}$. Then total induced charge

$$Q = \int_0^\infty 2\pi r\, dr \cdot \sigma(r),$$

where $\sigma(r) = -\dfrac{q}{2\pi}\cdot\dfrac{d}{(r^2+d^2)^{3/2}}$ [from Eq. (6.2-4)]

$$\begin{aligned}\therefore \quad Q &= -qd\int_0^\infty \frac{r\,dr}{(r^2+d^2)^{3/2}}\\ &= -qd\left[-\frac{1}{\sqrt{r^2+d^2}}\right]_0^\infty\\ &= -q.\end{aligned} \tag{6.2-5}$$

Thus, the total induced charge, as expected, is $-q$. This implies that all the lines of force ultimately terminate on the plane.

6.3 A Point Charge and a Grounded Conducting Sphere

Consider a point charge q placed at a distance d from the centre of a grounded conducting sphere of radius a (Fig. 6.3-1). The potential at any external point P will be partly due to q and partly due to induced charges on the sphere. To solve the

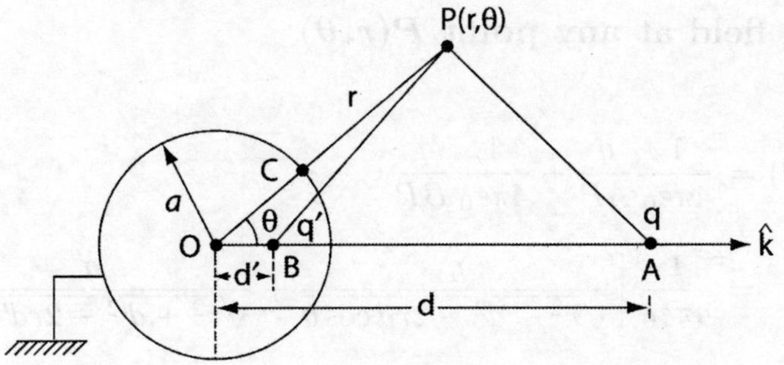

Fig 6.3-1

problem by the method of image we are to replace the induced charges by suitable image charges. The symmetry of the problem suggests that the image charge must be placed symmetrically about the line of symmetry OA. Let us try a single image charge q' located within the sphere at some point B (with $OB = d'$) on the line OA. In this attempt the following conditions must be obeyed:

(i) Potential ϕ is zero all over the surface of the sphere.

(ii) Potential ϕ must be zero at infinity.

(iii) Laplace's equation must be satisfied at all outside points excepting at A.

The choice of image charge q' inside the sphere does not affect the conditions (ii) and (iii). Now the potential at any point $C(a, \theta)$ on the surface of the sphere must be zero, i.e.,

$$\frac{1}{4\pi\varepsilon_0}\left[\frac{q}{AC} + \frac{q'}{BC}\right] = 0$$

$$\text{or} \quad \frac{q}{\sqrt{a^2 + d^2 - 2ad\cos\theta}} + \frac{q'}{\sqrt{a^2 + d'^2 - 2ad'\cos\theta}} = 0$$

$$\text{or} \quad \frac{q/a}{\sqrt{1 + \frac{d^2}{a^2} - 2\frac{d}{a}\cos\theta}} = \frac{-q'/d'}{\sqrt{1 + \frac{a^2}{d'^2} - 2\frac{a}{d'}\cos\theta}}.$$

To make this equality valid for all values of θ we may choose

$$\frac{q}{a} = -\frac{q'}{d'} \quad \text{and} \quad \frac{d}{a} = \frac{a}{d'}.$$

Therefore,

$$q' = -\frac{qa}{d} \quad \text{and} \quad d' = \frac{a^2}{d}. \tag{6.3-1}$$

This choice of image charge satisfies all the boundary conditions for our original problem in the region outside the sphere. According to the uniqueness theorem the solution thus obtained is the only correct solution to the problem. So the image charge can now be used instead of the induced charges to find potential, field etc.

Potential and field at any point $P(r, \theta)$

$$\phi(r,\theta) = \frac{1}{4\pi\varepsilon_0}\frac{q}{AP} + \frac{1}{4\pi\varepsilon_0}\frac{q'}{BP}$$

$$= \frac{1}{4\pi\varepsilon_0}\left[\frac{q}{\sqrt{r^2+d^2-2rd\cos\theta}} + \frac{q'}{\sqrt{r^2+d'^2-2rd'\cos\theta}}\right], \tag{6.3-2}$$

where q' and d' are given by the Eq. (6.3-1).

Corresponding field is given by

$$\vec{E} = -\vec{\nabla}\phi = -\left[\hat{r}\frac{\partial\phi}{\partial r} + \hat{\theta}\frac{1}{r}\frac{\partial\phi}{\partial\theta}\right].$$

Therefore, the components of $\vec{E}$ are:

$$E_r = -\frac{\partial\phi}{\partial r} = \frac{1}{4\pi\varepsilon_0}\left[\frac{q(r-d\cos\theta)}{(r^2+d^2-2rd\cos\theta)^{3/2}} + \frac{q'(r-d'\cos\theta)}{(r^2+d'^2-2rd'\cos\theta)^{3/2}}\right] \tag{6.3-3}$$

$$E_\theta = -\frac{1}{r}\frac{\partial\phi}{\partial\theta} = \frac{1}{4\pi\varepsilon_0}\left[\frac{qd\sin\theta}{(r^2+d^2-2rd\cos\theta)^{3/2}} + \frac{q'd'\sin\theta}{(r^2+d'^2-2rd'\cos\theta)^{3/2}}\right] \tag{6.3-4}$$

Lines of force

Lines of force will be the same as that between q at A and $q' = -qa/d$ at B (Fig. 6.3-2). The lines meet the conducting sphere normally. The lines of force within the sphere are shown by dotted lines because field inside the sphere is zero and they do not exist really.

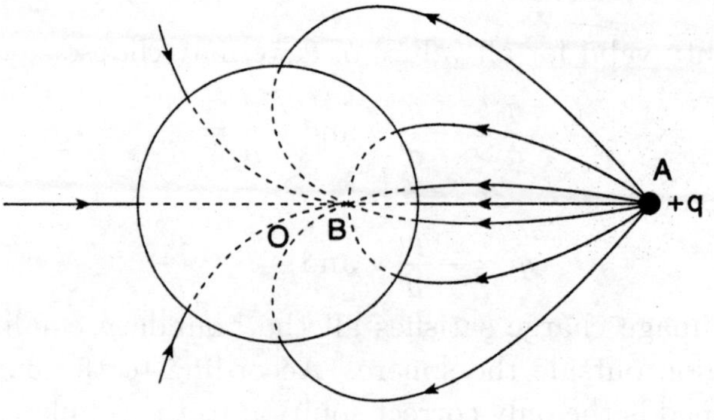

Fig 6.3-2

Force on $+q$

Force between $+q$ at A and the sphere will be given by the force between $+q$ at A and the image charge q' placed at B. Therefore, force on $+q$ due to the sphere would be

$$\vec{F} = \frac{1}{4\pi\varepsilon_0} \cdot \frac{qq'}{(BA)^2}\hat{k} \quad \text{where } \hat{k} \text{ is a unit vector along } OA$$

$$= -\frac{1}{4\pi\varepsilon_0} \cdot \frac{q^2 a/d}{(d - a^2/d)^2}\hat{k} \quad \left[\because\ BA = OA - OB = d - a^2/d\right]$$

$$= -\frac{q^2 ad}{4\pi\varepsilon_0 (d^2 - a^2)^2}\hat{k}. \tag{6.3-5}$$

The force is attractive in nature.

If the charge q is very close to the sphere, i.e., $d = a + \delta(\delta \ll a)$ then

$$\vec{F} \to -\frac{1}{4\pi\varepsilon_0} \cdot \frac{q^2}{(2\delta)^2}\hat{k} \tag{6.3-6}$$

This result is similar to the case of a point charge placed in front of a conducting plane.

Induced charges

The surface density of induced charges at any point $C(a, \theta)$ on the sphere is given by

$$\sigma(\theta) = -\varepsilon_0 \frac{\partial \phi}{\partial r}\bigg|_{r=a}$$

$$= -\frac{q}{4\pi a} \cdot \frac{d^2 - a^2}{(a^2 + d^2 - 2ad\cos\theta)^{3/2}} \quad \text{[Using Eq. (6.3-3)]} \tag{6.3-7}$$

As expected the induced charge density is negative. Note that $(a^2 + d^2 - 2ad\cos\theta)^{3/2} = AC^3$. Thus, we can say that charge density at a point is inversely proportional to the cube of the distance of the point from q. The charge density is thus maximum at $\theta = 0°$ and minimum at $\theta = 180°$.

$$\sigma_{\max} = -\frac{q}{4\pi a} \cdot \frac{d + a}{(d - a)^2} \quad \text{and} \quad \sigma_{\min} = -\frac{q}{4\pi a}\frac{d - a}{(d + a)^2}. \tag{6.3-8}$$

Note that σ cannot be zero at any point on the sphere.

Total induced charge on the sphere can be obtained by integrating the charge density over the surface of the sphere. Thus, total induced charge

$$Q = \int_0^\pi \sigma(\theta) \cdot 2\pi a \sin\theta \cdot a d\theta = -\frac{q\,(d^2 - a^2)\,a}{2}\int_0^\pi \frac{\sin\theta\, d\theta}{(a^2 + d^2 - 2ad\cos\theta)^{3/2}}.$$

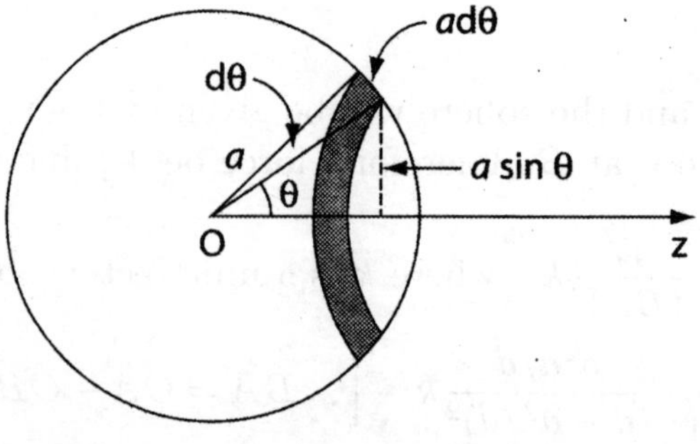

Fig. 6.3-3: Calculation of induced charge

Putting $a^2 + d^2 - 2ad\cos\theta = p^2$ we get

$$Q = -\frac{q\,(d^2 - a^2)}{2d}\int_{d-a}^{d+a}\frac{pdp}{p^3} = \frac{q\,(d^2 - a^2)}{2d}\left[\frac{1}{p}\right]_{d-a}^{d+a} = -\frac{qa}{d}, \qquad (6.3\text{-}9)$$

which is, as expected, equal to the image charge.

6.4 A Point Charge and an Uncharged Insulated Conducting Sphere

Suppose a point charge q is placed at a distance d from the centre an insulated conducting sphere of radius a. The geometry of the problem is sketched in Fig. 6.4-1. The potential ϕ at any external point $P(r, \theta)$ will be partly due to q and partly due to induced charges on the sphere. To solve the problem by the method of images we are to replace the induced charges by suitable image charges. In this attempt the following conditions must be satisfied:

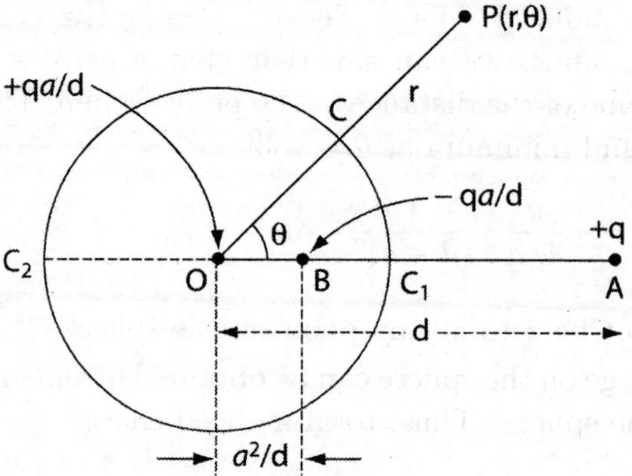

Fig. 6.4-1

(i) ϕ is constant all over the surface of the sphere.

(ii) Net charge on the conductor is zero.

(iii) $\phi = 0$ at infinity.

(iv) ϕ satisfies Laplace's equation at all outside points except A.

If the sphere is grounded the induced charges can be replaced by an image charge $-\frac{qa}{d}$ at B at a distance a^2/d from the centre O. This image charge together with $+q$ at A makes potential on the surface of the sphere zero. Here, to satisfy the conditions (i) and (ii) an additional image charge $+qa/d$ may be placed at O, the centre of the sphere. These image charges satisfy all the specified boundary conditions and hence they can be used instead of the actual induced charges to find field, potential etc.

Potential and field

The potential at any point $P(r, \theta)$ is given by

$$\phi(r,\theta) = \frac{1}{4\pi\varepsilon_0}\left[\frac{q}{\sqrt{r^2+d^2-2rd\cos\theta}} - \frac{qa/d}{\sqrt{r^2+\left(\frac{a^2}{d}\right)^2 - 2r\cdot\frac{a^2}{d}\cos\theta}} + \frac{qa/d}{r}\right] \tag{6.4-1}$$

The electric field will be given by $\vec{E} = -\vec{\nabla}\phi = -\hat{r}\dfrac{\partial\phi}{\partial r} - \hat{\theta}\dfrac{1}{r}\dfrac{\partial\phi}{\partial\theta}$

Force on $+q$ due to the sphere

The force $\vec{F}$ on $+q$ due to the sphere is given by the total force on $+q$ due to the image charges $-qa/d$ placed at B and $+qa/d$ at O. Thus,

$$\vec{F} = \frac{1}{4\pi\varepsilon_0}\frac{q(-qa/d)}{(d-a^2/d)^2}\hat{k} + \frac{1}{4\pi\varepsilon_0}\cdot\frac{q\cdot qa/d}{d^2}\hat{k} = -\frac{q^2}{4\pi\varepsilon_0}\left(\frac{a}{d}\right)^3\cdot\frac{2d^2-a^2}{(d^2-a^2)^2}\hat{k}. \tag{6.4-2}$$

Obviously, the force is attractive in nature.

Induced charges

The surface density of induced charges at any point $C(a, \theta)$ on the sphere is given by

$$\sigma(\theta) = -\varepsilon_0\left.\frac{\partial\phi}{\partial r}\right|_{r=a} = -\frac{q}{4\pi a}\cdot\frac{d^2-a^2}{(a^2+d^2-2ad\cos\theta)} + \frac{q}{4\pi ad} \tag{6.4-3}$$

[Using Eq. (6.4-1)]

$$= -\frac{q}{4\pi a}\cdot\left[\frac{d^2-a^2}{(AC)^3} - \frac{1}{d}\right]. \tag{6.4-4}$$

At the point C_1 on the sphere nearest to $+q$ the surface charge density is given by

$$\sigma_{C_1} = -\frac{q}{4\pi a}\left[\frac{d^2 - a^2}{(AC_1)^3} - \frac{1}{d}\right] = -\frac{q}{4\pi a}\left[\frac{d^2 - a^2}{(d-a)^3} - \frac{1}{d}\right] = -\frac{q}{4\pi}\frac{3d - a}{d(d-a)^2}, \quad (6.4\text{-}5)$$

which is negative. Now the surface charge density at C_2 on the sphere farthest from $+q$ is

$$\sigma_{C_2} = -\frac{q}{4\pi a}\left[\frac{d^2 - a^2}{(AC_2)^3} - \frac{1}{d}\right] = -\frac{q}{4\pi a}\left[\frac{d^2 - a^2}{(d+a)^3} - \frac{1}{d}\right] = +\frac{q}{4\pi}\cdot\frac{3d + a}{d(d+a)^2}, \quad (6.4\text{-}6)$$

which is positive.

Circle of no electrification

Since the surface charge density at C_1 is negative and at C_2 positive there must be a *line of no electrification* on the sphere, where surface charge density is zero. Putting $\sigma(\theta) = 0$ in Eq. (6.4-4) we get

$$(AC)^3 = d(d^2 - a^2) = d^2\left(d - \frac{a^2}{d}\right) = OA^2 \cdot OB. \quad (6.4\text{-}7)$$

This represents a circle on the sphere separating positive and negative charges.

6.5 A Point Charge and a Charged Insulated Conducting Sphere

Suppose a point charge q is placed at a distance d from the centre of an insulated spherical conductor of radius a and carrying a charge Q. The geometry of the problem is sketched in Fig. 6.4-1. The potential ϕ at any outside point $P(r, \theta)$ will be due to q and the charges on the sphere including induced charges. Let us replace the charges on the sphere by suitable image charges. In this attempt the following conditions must be satisfied:

(i) ϕ is constant all over the surface of the sphere.

(ii) Net charge on the sphere is Q.

(iii) $\phi = 0$ at $r = \infty$.

(iv) ϕ satisfies Laplace's equation at all outside points except A.

In case of an uncharged insulated sphere the induced charges are replaced by two image charges $-qa/d$ at B and $+qa/d$ at O. In the present case of charged sphere we are to introduce an additional image charge Q at O. This choice satisfies all the boundary conditions as stated above and hence according to the uniqueness theorem will give the only correct solution to the problem.

Potential and field

The potential at any external point $P(r, \theta)$ is given by

$$\phi(r,\theta) = \frac{1}{4\pi\varepsilon_0}\left[\frac{q}{\sqrt{r^2+d^2-2rd\cos\theta}} - \frac{qa/d}{\sqrt{r^2+\left(\frac{a^2}{d}\right)^2 - 2r\cdot\frac{a^2}{d}\cdot\cos\theta}} + \frac{Q+qa/d}{r}\right] \tag{6.5-1}$$

The electric field can be obtained by using the relation $\vec{E} = -\vec{\nabla}\phi$.

Force on q due to the sphere

The force $\vec{F}$ is given by the total force on $+q$ due to image charges $-qa/d$ placed at B and $Q + qa/d$ at O. Thus,

$$\vec{F} = \frac{1}{4\pi\varepsilon_0}\cdot\frac{q(-qa/d)}{(d-a^2/d)^2}\hat{k} + \frac{1}{4\pi\varepsilon_0}\cdot\frac{q(Q+qa/d)}{d^2}\hat{k}. \tag{6.5-2}$$

The first term on the right-hand side is negative and the second term is positive. So the force may be attractive or repulsive in nature depending on their relative values. In the limit $d \gg a$ the force reduces to the usual Coulomb's law of force between two small charged bodies. But close to the sphere the force is modified due to induced charges on the sphere.

Suppose the point A is very close to the sphere. Let $d = a + \delta$, where $\delta \ll a$. Then $d^2 - a^2 \approx 2a\delta$ and we can write from Eq. (6.5-2),

$$\begin{aligned}\vec{F} &\simeq \frac{q}{4\pi\varepsilon_0}\left[\frac{-qad}{(2a\delta)^2} + \frac{Q+qa/d}{d^2}\right]\hat{k} \\ &\approx \frac{q}{4\pi\varepsilon_0}\left[-\frac{q}{4\delta^2} + \frac{Q+q}{d^2}\right]\hat{k} \quad \text{where we use } d/a \approx 1\end{aligned} \tag{6.5-3}$$

Since $\delta \ll d$ and Q may be of the order of q, $\dfrac{q}{4\delta^2} > \dfrac{Q+q}{d^2}$. Thus, there is a force of attraction if the point charge q is located very near to the sphere (even if q and Q are of same sign). In fact the force is attractive if $d < d_0$ and repulsive if $d > d_0$, where d_0 is given by the condition that $\vec{F} = 0$. Therefore,

$$\frac{qa/d_0}{\left(d_0 - \frac{a^2}{d_0}\right)^2} = \frac{Q + \frac{qa}{d_0}}{d_0^2}. \tag{6.5-4}$$

In case $Q \gg q$, the point of zero force is very close to the sphere and is at

$$d_0 \approx a\left(1 + \frac{1}{2}\sqrt{q/Q}\right).$$

Induced charges

The surface density of induced charges at any point $C(a, \theta)$ on the sphere is given by

$$\sigma(\theta) = -\varepsilon_0 \frac{\partial \phi}{\partial r}\bigg|_{r=a} \quad \text{where } \phi \text{ is given by the Eq. (6.5-1)}$$

$$\therefore \quad \sigma(\theta) = -\frac{q}{4\pi a} \cdot \frac{d^2 - a^2}{(AC)^3} + \frac{(Q + qa/d)}{4\pi a^2}. \qquad (6.5\text{-}5)$$

At the point C_1, $AC = AC_1 = d - a$ and we have

$$\sigma_{C_1} = -\frac{q(3d - a)}{4\pi d(d - a)^2} + \frac{Q}{4\pi a^2} \qquad (6.5\text{-}6)$$

and at the point C_2, $AC = AC_2 = d + a$ and we have

$$\sigma_{C_2} = \frac{q(3d + a)}{4\pi d(d + a)^2} + \frac{Q}{4\pi a^2}. \qquad (6.5\text{-}7)$$

Note that σ_{C_2} is always positive but the sign of σ_{C_1} depends on the magnitude of Q.

6.6 A Point Charge in Front of a Semi-infinite Dielectric

Consider a point charge $+q$ placed in air at A at a distance d from the plane surface of a semi-infinite dielectric medium of permittivity ε (Fig. 6.6-1).

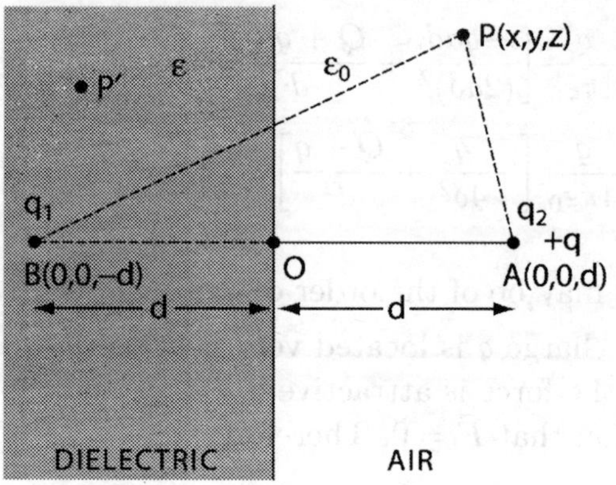

Fig. 6.6-1

The field of q polarizes the dielectric and negative bound charges appear on the surface. The resultant field at any point can be calculated by considering these charges together with $+q$. To solve the problem by the method of images we are to choose suitable image charges at suitable locations. These choices must be such that the produced fields satisfy the boundary conditions appropriate to the problem. Let the potential in air be represented by the charge q and an image charge q_1 at B located within the dielectric at a distance $OB = d$. For any point within the dielectric let us represent the potential only by an image charge q_2 placed at A and the dielectric occupying the whole space. Thus, for any point $P(x, y, z)$ in air

$$\phi_0(r) = \frac{1}{4\pi\varepsilon_0}\left(\frac{q}{AP} + \frac{q_1}{BP}\right) \quad \text{for } z > 0$$

$$= \frac{1}{4\pi\varepsilon_0}\left[\frac{q}{\sqrt{x^2+y^2+(z-d)^2}} + \frac{q_1}{\sqrt{x^2+y^2+(z+d)^2}}\right]. \tag{6.6-1}$$

For any point $P'(x, y, z)$ within the dielectric

$$\phi(r) = \frac{1}{4\pi\varepsilon} \cdot \frac{q_2}{\sqrt{x^2+y^2+(z-d)^2}}. \tag{6.6-2}$$

Now, our choice must satisfy the boundary conditions that (i) $\phi_0 = \phi$ at $z = 0$ and (ii) $\varepsilon_0 \dfrac{\partial \phi_0}{\partial z} = \varepsilon \dfrac{\partial \phi}{\partial z}$ at $z = 0$. Thus,

$$\frac{1}{\varepsilon_0}(q + q_1) = \frac{q_2}{\varepsilon} \tag{6.6-3}$$

$$\text{and } q - q_1 = q_2. \tag{6.6-4}$$

Solving Eqs. (6.6-3) and (6.6-4) we get

$$q_1 = -\frac{\varepsilon - \varepsilon_0}{\varepsilon + \varepsilon_0} q \quad \text{and} \quad q_2 = \frac{2\varepsilon}{\varepsilon + \varepsilon_0} q. \tag{6.6-5}$$

Note that we have chosen the image charge always outside the region in which the field is to be determined. So our choice does not violate the condition that the potential must satisfy Laplace's equation (except at A) in the region of interest. Moreover, as the image charges are located at finite distances they will not violate the conditions at infinity. As all the necessary conditions are satisfied with our choice, according to uniqueness theorem it will give the correct solution. Thus, potentials are given by

$$\phi_0(r) = \frac{q}{4\pi\varepsilon_0}\left[\frac{1}{\sqrt{x^2+y^2+(z-d)^2}} - \frac{(\varepsilon-\varepsilon_0)/(\varepsilon+\varepsilon_0)}{\sqrt{x^2+y^2+(z+d)^2}}\right] \quad \text{for } z > 0 \tag{6.6-6}$$

$$\phi(r) = \frac{q}{2\pi(\varepsilon + \varepsilon_0)} \cdot \frac{1}{\sqrt{x^2+y^2+(z-d)^2}} \quad \text{for } z < 0 \tag{6.6-7}$$

$$\text{or} \quad \phi_0(r) = \frac{q}{4\pi\varepsilon_0}\left[\frac{1}{R} - \frac{\varepsilon - \varepsilon_0}{\varepsilon + \varepsilon_0} \cdot \frac{1}{R_1}\right] \tag{6.6-8}$$

$$\phi(r) = \frac{q}{2\pi(\varepsilon + \varepsilon_0)} \cdot \frac{1}{R}, \tag{6.6-9}$$

where R and R_1 are the distances of the point of observation from the charge q (or q_2) and q_1, respectively.

Force on q due to the dielectric

It is a matter of interest in the region $z > 0$ and is given by the force between q and q_1 placed in air. Thus,

$$\vec{F} = \frac{1}{4\pi\varepsilon_0}\frac{qq_1}{(2d)^2}\hat{k} = -\frac{1}{4\pi\varepsilon_0}\cdot\frac{\varepsilon-\varepsilon_0}{\varepsilon+\varepsilon_0}\cdot\frac{q^2}{4d^2}\hat{k}. \tag{6.6-10}$$

Obviously it an attractive force. In the limit $\varepsilon \to \infty$,

$$\vec{F} \to -\frac{1}{4\pi\varepsilon_0}\cdot\frac{q^2}{4d^2}\hat{k}.$$

This result is the same as in the case of a point charge in front of a conducting plane. So from electrostatic point of view we can say that a conductor behaves like a dielectric with infinite permittivity.

Field lines

In the air ($z > 0$) the lines of force will be the same as that between $+q$ and $q_1\left(= -\dfrac{\varepsilon-\varepsilon_0}{\varepsilon+\varepsilon_0}q\right)$. Within the dielectric ($z < 0$) the line of force will be as due $q_2\left(= \dfrac{2\varepsilon}{\varepsilon+\varepsilon_0}q\right)$ only, i.e., radial in direction emanating from the point A (Fig. 6.6-2).

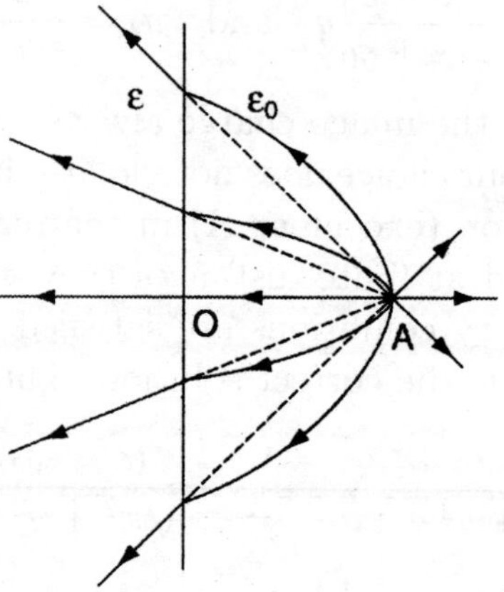

Fig. 6.6-2: Lines of force for a point change near a dielectric slab

6.7 Conducting Sphere in a Uniform Electric Field by the Method of Images

Let us consider a grounded conducting sphere of radius a placed in a uniform electric field $\vec{E}_0 = E_0\hat{k}$. We shall solve this problem by the method of electrical images. The geometry of the problem is sketched in Fig. 6.7-1. A uniform field can be thought of as being produced by appropriate positive and negative charges at infinity. For example, a charge $+q$ located at $z = -d$ and a charge $-q$ at $z = +d$ may produce in a small region near O an approximately uniform field

$$\vec{E}_0 = \frac{1}{4\pi\varepsilon_0} \cdot \frac{2q}{d^2}\hat{k}$$

along z-axis. This approximation becomes exact in the limit $q \to \infty$, $d \to \infty$ with q/d^2 constant.

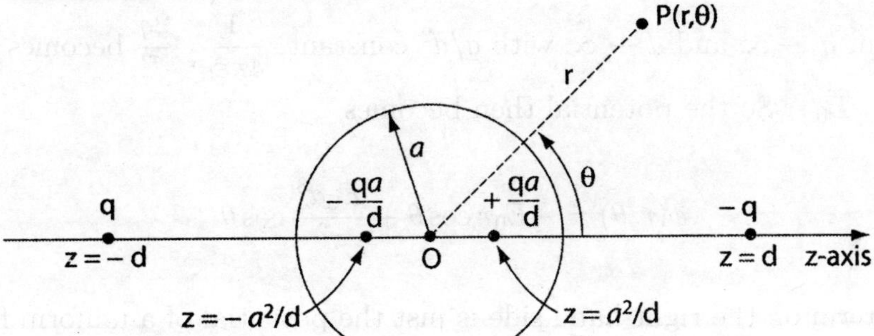

Fig. 6.7-1 : Conducting sphere in a uniform field by the method of images

If now the conducting sphere is placed at the origin O then the potential at any outside point will be due to the charges $+q$ at $z = -d$, $-q$ at $z = +d$ and induced charges on the sphere. These induced charges can be replaced by two image charges $-qa/d$ at $z = -\frac{a^2}{d}$ and $+qa/d$ at $z = +\frac{a^2}{d}$.

The total potential at any external point $P(r, \theta)$ will be due to all these four point charge. Therefore,

$$\phi(r,\theta) = \frac{1}{4\pi\varepsilon_0}\left[\frac{q}{\sqrt{r^2+d^2+2rd\cos\theta}} - \frac{q}{\sqrt{r^2+d^2-2rd\cos\theta}} - \frac{qa/d}{\sqrt{r^2+\left(\frac{a^2}{d}\right)^2+2\cdot r\cdot\frac{a^2}{d}\cos\theta}} + \frac{qa/d}{\sqrt{r^2+\left(\frac{a^2}{d}\right)^2-2\cdot r\cdot\frac{a^2}{d}\cos\theta}}\right]. \qquad (6.7\text{-}1)$$

Now as $d \gg r$ by our assumption we can write

$$\frac{1}{\sqrt{r^2+d^2\pm 2rd\cos\theta}} = \frac{1}{d}\left(1+\frac{r^2}{d^2}\pm 2\frac{r}{d}\cos\theta\right)^{-1/2} \approx \frac{1}{d}\left(1\mp\frac{r}{d}\cos\theta\right)$$

$$\frac{1}{\sqrt{r^2 + \left(\frac{a^2}{d}\right)^2 \pm 2r \cdot \frac{a^2}{d} \cdot \cos\theta}} = \frac{1}{r}\left(1 + \frac{a^4}{r^2 d^2} \pm 2\frac{a^2}{rd}\cos\theta\right)^{-1/2} \approx \frac{1}{r}\left(1 \pm \frac{a^2}{rd}\cos\theta\right)$$

Using these approximations Eq. (6.7-1) becomes

$$\phi(r,\theta) = \frac{1}{4\pi\varepsilon_0}\left[-\frac{2q}{d^2} r\cos\theta + \frac{2q}{d^2} \cdot \frac{a^3\cos\theta}{r}\right]. \tag{6.7-2}$$

In the limit $q \to \infty$ and $d \to \infty$ with q/d^2 constant, $\frac{1}{4\pi\varepsilon_0} \cdot \frac{2q}{d^2}$ becomes the applied uniform field $\left|\vec{E}_0\right|$. So the potential then becomes

$$\phi(r,\theta) = -E_0 r\cos\theta + \frac{E_0 a^3}{r^2}\cos\theta. \tag{6.7-3}$$

The first term on the right-hand side is just the potential of a uniform field $\vec{E}_0$ and the second term is a dipolar potential formed by the induced charges or equivalently the image charges. Note that the image charges form a dipole of moment

$$\vec{P} = \frac{qa}{d} \cdot \frac{2a^2}{d}\hat{k} = 4\pi\varepsilon_0 a^3 E_0 \hat{k}. \tag{6.7-4}$$

The surface density of induced charges is given by

$$\sigma = -\varepsilon_0 \frac{\partial\phi}{\partial r}\bigg|_{r=a} = 3\varepsilon_0 E_0 \cos\theta. \tag{6.7-5}$$

SOLVED PROBLEMS

▶ **Problem 1.** *A point charge q of mass m is released from rest at a distanced d from an infinite grounded conducting plate. Find the time taken by the charge to hit the plate.*

(J.U. 2010, 2014, C.U. 2013)

Solution. When the charge q is at a distance x from the plate the force on it due to the conducting plate is given by the force between q and its image charge $-q$ located at $-x$. So the equation of motion is given by

$$m \cdot \frac{d^2x}{dt^2} = -\frac{1}{4\pi\varepsilon_0} \cdot \frac{q^2}{(2x)^2}.$$

Multiplying both sides by $2\frac{dx}{dt}$ and integrating we get

$$m\left(\frac{dx}{dt}\right)^2 = -\frac{q^2}{8\pi\varepsilon_0}\int\frac{dx}{x^2}$$
$$= \frac{q^2}{8\pi\varepsilon_0}\cdot\frac{1}{x} + c,$$

where c is a constant.

Now at $x = d$, $\frac{dx}{dt} = 0$. Therefore, $c = -\frac{q^2}{8\pi\varepsilon_0}\cdot\frac{1}{d}$ and

$$m\left(\frac{dx}{dt}\right)^2 = \frac{q^2}{8\pi\varepsilon_0}\left(\frac{1}{x} - \frac{1}{d}\right) \quad \text{or} \quad dt = -\sqrt{\frac{8\pi\varepsilon_0 m}{q^2}}\frac{dx}{\sqrt{\frac{1}{x}-\frac{1}{d}}}.$$

Integrating we get the required time as

$$T = \int dt = -\sqrt{\frac{8\pi\varepsilon_0 m}{q^2}}\int_d^0 \frac{dx}{\sqrt{\frac{1}{x}-\frac{1}{d}}} = \sqrt{\frac{8\pi\varepsilon_0 md}{q^2}}\int_0^d \frac{\sqrt{x}}{\sqrt{d-x}}dx.$$

Substituting $x = d\sin^2\theta$ we find that $\int_0^d \frac{\sqrt{x}}{\sqrt{d-x}}dx = d\frac{\pi}{2}$.

$$\therefore \quad T = \sqrt{\frac{8\pi\varepsilon_0 md}{q^2}}\cdot\frac{d\pi}{2} = \frac{\pi}{q}\sqrt{2\pi\varepsilon_0 md^3}.$$

▶ **Problem 2.** *A point charge $+q$ is placed at a distance d from the centre of an insulated conducting sphere of radius a. Find the minimum charge Q that should be put on the sphere to make the force on the point charge zero.*

Solution. From Eq. (6.5-2) we find that $\vec{F} = 0$ if

$$\frac{Q + \frac{qa}{d}}{d^2} = \frac{\frac{qa}{d}}{\left(d - \frac{a^2}{d}\right)^2} \quad \text{or} \quad Q = \frac{qa^3}{d}\cdot\frac{2d^2 - a^2}{(d^2 - a^2)^2}.$$

▶ **Problem 3.** *Two similar point charges q, q are kept separated by a distance 2d in air. Now an insulated uncharged conducting sphere of radius a is placed midway between them. If d $\gg$ a show that the introduction of the sphere reduces the force experienced by either point charge to $(1 - 24a^5/d^5)$ of its initial value.* **[J.U. 2013; C.U. 2014]**

Solution. The initial force between the point charges is

$$F_1 = \frac{1}{4\pi\varepsilon_0}\cdot\frac{q^2}{(2d)^2}.$$

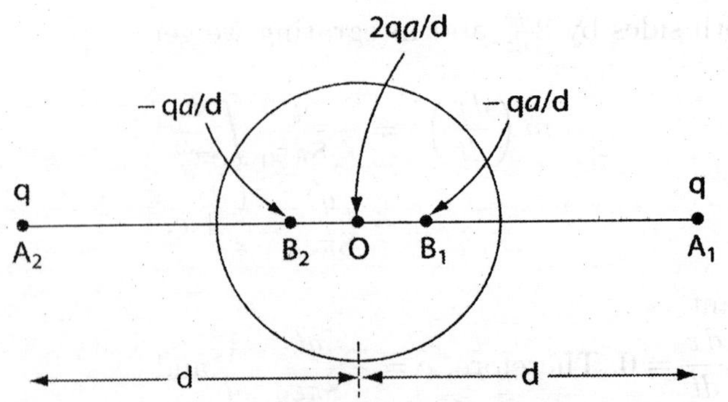

Fig. 6.P-3

In presence of the conducting sphere we are to consider the effects of induced charges. The induced charges can be replaced by image charges $-(qa/d)$ at B_1, $-(qa/d)$ at B_2 and $2qa/d$ at the centre O, where $OB_1 = OB_2 = a^2/d$. Now the total force experienced by q at A_1 is

$$F_2 = \frac{1}{4\pi\varepsilon_0}\left[\frac{q^2}{(2d)^2} - \frac{q^2a/d}{\left(d+\frac{a^2}{d}\right)^2} + \frac{2q^2a/d}{d^2} - \frac{q^2a/d}{\left(d-\frac{a^2}{d}\right)^2}\right]$$

$$= \frac{d^2}{4\pi\varepsilon_0 d^2}\left[\frac{1}{4} - \frac{a}{d}\left(1+\frac{a^2}{d^2}\right)^{-2} + 2\frac{a}{d} - \frac{a}{d}\left(1-\frac{a^2}{d^2}\right)^{-2}\right]$$

$$\approx \frac{q^2}{4\pi\varepsilon_0 d^2}\left[\frac{1}{4} - \frac{a}{d}\left(1 - \frac{2a^2}{d^2} + \frac{3a^4}{d^4}\right) + \frac{2a}{d} - \frac{a}{d}\left(1+\frac{2a^2}{d^2}+\frac{3a^4}{d^4}\right)\right]$$

$$\approx \frac{q^2}{4\pi\varepsilon_0 d^2}\left[\frac{1}{4} - \frac{6a^5}{d^5}\right] = \frac{q^2}{4\pi\varepsilon_0(2d)^2}\left(1 - \frac{24a^5}{d^5}\right).$$

$$\therefore \quad F_2 = \left(1 - 24a^5/d^5\right) F_1.$$

► **Problem 4.** *A point charge q is placed at a distance d from the centre of an earthed conducting sphere of radius a ($d > a$). Find the density of induced surface charge on the sphere. Hence show that the ratio of charge induced on the part visible from q to that of the remaining part of the sphere is $\sqrt{(d+a)/(d-a)}$.*

Solution. From Eq. (6.3-7) surface density at any point $C(a, \theta)$ on the sphere is given by

$$\sigma(\theta) = -\frac{q}{4\pi a}\cdot\frac{d^2 - a^2}{(a^2 + d^2 - 2ad\cos\theta)^{3/2}}$$

Let

$$a^2 + d^2 - 2ad\cos\theta = CA^2 = x^2 \quad \text{and} \quad 2ad\sin\theta\, d\theta = 2x\, dx$$

$$\therefore \quad \sigma = -\frac{q}{4\pi a}\cdot\frac{d^2 - a^2}{x^3}.$$

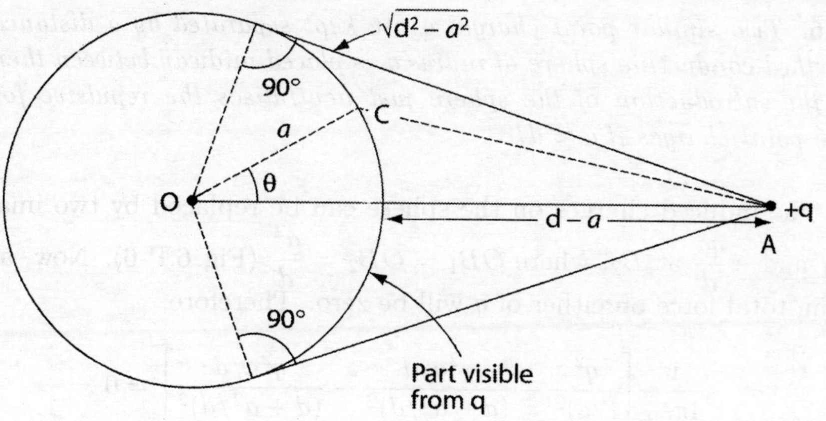

Fig 6.P-4

Now considering an annular strip between θ and $\theta + d\theta$ on the surface we find that the charge on the part visible from q is given by

$$\begin{aligned} Q_1 &= \int \sigma(\theta) 2\pi a^2 \sin\theta d\theta \\ &= -\frac{q\left(d^2 - a^2\right)}{2d} \cdot \int_{d-a}^{\sqrt{d^2-a^2}} \frac{dx}{x^2} \\ &= -\frac{q\left(d^2 - a^2\right)}{2d} \left[\frac{1}{d-a} - \frac{1}{\sqrt{d^2 - a^2}}\right]. \end{aligned}$$

The charge on the rest part is

$$Q_2 = -\frac{q\left(d^2 - a^2\right)}{2d} \int_{\sqrt{d^2-a^2}}^{d+a} \frac{dx}{x^2} = -\frac{q\left(d^2 - a^2\right)}{2d} \left[\frac{1}{\sqrt{d^2 - a^2}} - \frac{1}{d+a}\right].$$

On simplification we find that

$$\frac{Q_1}{Q_2} = \sqrt{\frac{d+a}{d-a}}$$

▶ **Problem 5.** *A point charge q is at a distance d from an infinite conducting plane. Find the work necessary to remove the charge to infinite distance from the plane.*

[B.U. 2010; J.U. 2012]

Solution. Force on the charge q due to the plane is given by the force between q and its image charge $-q$. So when the charge q is at a distance z from the plane, the force acting on it is

$$\vec{F} = -\frac{1}{4\pi\varepsilon_0} \cdot \frac{q^2}{(2z)^2} \hat{z}.$$

$$\text{Therefore, the required work} = \int_{z=d}^{\infty} \frac{1}{4\pi\varepsilon_0} \cdot \frac{q^2}{(2z)^2} \cdot dz = \frac{q^2}{16\pi\varepsilon_0 d}.$$

► **Problem 6.** *Two similar point charges q are kept separated by a distance 2d in air. Now an earthed conducting sphere of radius a is placed midway between them. If $d \gg a$ show that the introduction of the sphere just neutralises the repulsive force existing between the point charges if $a \approx d/8$.*

Solution. The induced charges on the sphere can be replaced by two image charges. $-\dfrac{qa}{d}$ at B_1 and $-\dfrac{qa}{d}$ at B_2, where $OB_1 = OB_2 = \dfrac{a^2}{d}$ (Fig. 6.P-6). Now, according to the problem, total force on either of q will be zero. Therefore,

$$\frac{1}{4\pi\varepsilon_0}\left[\frac{q^2}{(2d)^2} - \frac{q^2a/d}{(d-a^2/d)^2} - \frac{q^2a/d}{(d+a^2/d)^2}\right] = 0$$

$$\text{or} \quad \frac{1}{4d^2} = \frac{a}{d^3}\left(1-\frac{a^2}{d^2}\right)^{-2} + \frac{a}{d^3}\left(1+\frac{a^2}{d^2}\right)^{-2}$$

$$= \frac{a}{d^3}\left(1+\frac{2a^2}{d^2}+\frac{3a^4}{d^4}\right) + \frac{a}{d^3}\left(1-\frac{2a^2}{d^2}+\frac{3a^4}{d^4}\right) \approx \frac{2a}{d^3}.$$

$$\therefore \quad a \approx \frac{d}{8}.$$

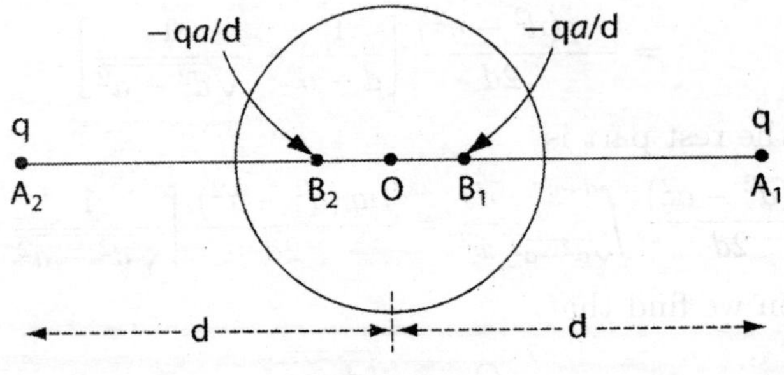

Fig. 6.P-6

► **Problem 7.** *A point charge q is placed at a distance d from the centre of a conducting sphere of radius a maintained at a constant potential V_0. Using the method of image find the force on the charge q due to the sphere.*

Solution. We know that the point charge q and an image charge $-qa/d$ placed at B at a distance a^2/d from the centre O make the potential on the surface zero (Fig. 6.P-7). Now the constant potential V_0 on the surface can be produced by considering another image charge q' at the centre, which is given by

$$V_0 = \frac{1}{4\pi\varepsilon_0}\frac{q'}{a} \quad \text{or} \quad q' = 4\pi\varepsilon_0 a V_0.$$

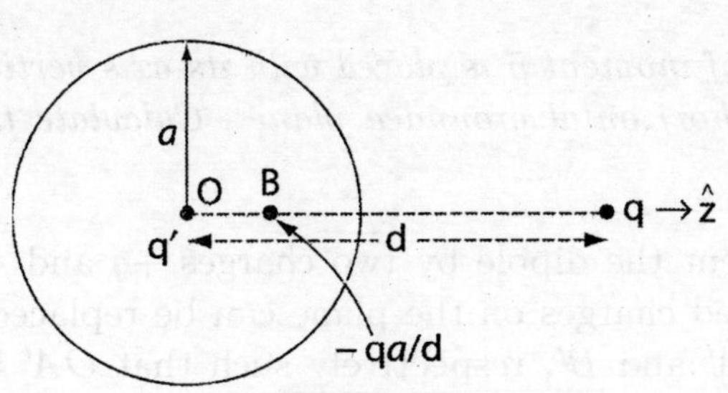

Fig. 6.P-7

Thus, force on q due to the sphere will be given by

$$\vec{F} = \frac{1}{4\pi\varepsilon_0} \frac{q(-qa/d)}{(d - a^2/d)^2}\hat{z} + \frac{q \cdot (4\pi\varepsilon_0 a V_0)}{4\pi\varepsilon_0 d^2}\hat{z}$$

$$= \left[\frac{qaV_0}{d^2} - \frac{q^2 ad}{4\pi\varepsilon_0 (d^2 - a^2)^2}\right]\hat{z}.$$

▶ **Problem 8.** *A long straight wire of radius a having a charge λ per unit length runs parallel to an infinite earthed conducting plane. The radius a is small compared with the height d of the wire above the plane. Assume that the potential at a distance r from the axis of the wire in the absence of the plane is given by $\phi(r) = \frac{\lambda}{2\pi\varepsilon_0} \ln \frac{r_0}{r}$, where r_0 is a reference point, where potential is zero. Find the potential difference between the surface of the wire and the conducting plane. Hence find the capacitance per unit length of the wire-ground system.*

Solution. The problem can be solved by considering an electric image in the form of a long straight wire carrying a charge $-\lambda$ per unit length symmetrically placed on the other side of the plane. Then the capacitance of the wire-ground system will be equal to that between two straight wires separated by a distance $2d$. Now the potential at a distance r from the axis of the original wire and at a distance $2d - r$ from the image wire is

$$\phi(r) = \frac{\lambda}{2\pi\varepsilon_0} \ln \frac{r_0}{r} - \frac{\lambda}{2\pi\varepsilon_0} \ln \frac{r_0}{2d - r} = \frac{\lambda}{2\pi\varepsilon_0} \ln \frac{2d - r}{r}.$$

So the potential difference between the surface of the original wire and the plane is

$$V = \phi(a) - \phi(d) = \frac{\lambda}{2\pi\varepsilon_0} \ln \frac{2d - a}{a} \approx \frac{\lambda}{2\pi\varepsilon_0} \ln \frac{2d}{a}.$$

Therefore, the required capacitance per unit length of the wire is

$$C = \frac{\lambda}{V} = \frac{2\pi\varepsilon_0}{\ln \frac{2d}{a}}.$$

▶ **Problem 9.** *A dipole of moment $\vec{p}$ is placed with its axis vertical at a distance d from an infinite conducting horizontal grounded plane. Calculate the force exerted on the plane by the dipole.*

Solution. If we represent the dipole by two charges $+q$ and $-q$ placed at A and B, respectively then induced charges on the plane can be replaced by two image charges $-q$ and $+q$ placed at A' and B', respectively such that $OA' = OA$ and $OB' = OB$ ((Fig. 6.P-9). These image charges also constitute a dipole of same moment $\vec{P}$ placed at a distance d just beneath the plane. The force on the plane exerted by the dipole will be equal to the force exerted by the dipole on the image dipole. Now the mutual interaction energy of two coaxial dipoles each of moment $\vec{p}$ is given by the Eq. (2.3-19)

$$U = -\frac{p^2}{2\pi\varepsilon_0 r^3},$$

where r is the distance between the centres of the two dipoles.

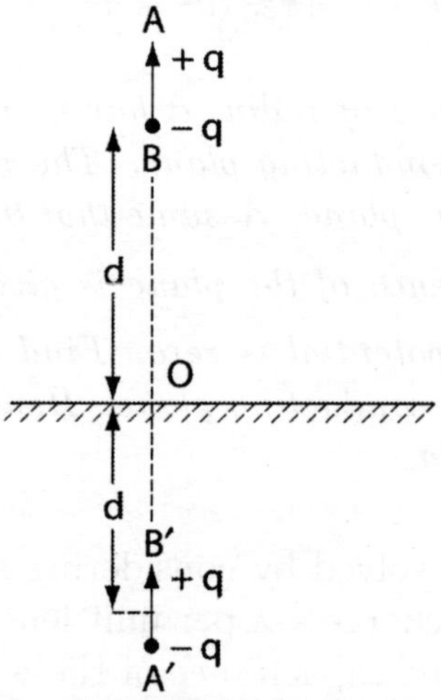

Fig. 6.P-9

Therefore, the required force

$$F = -\frac{\partial U}{\partial r} = -\frac{3p^2}{2\pi\varepsilon_0 r^4}.$$

Here, $r = 2d$ and hence $F = -\dfrac{3p^2}{32\pi\varepsilon_0 d^2}$. The negative sign indicates that the force is attractive.

▶ **Problem 10.** *A point dipole of moment $\vec{p}$ is placed at a distance d above an infinite grounded conducting plane. The dipole makes an angle θ with the perpendicular to the conducting plane. Find the torque acting on the dipole. If the dipole is free to rotate, find its equilibrium position.*

Solution. Let us represent the dipole by two charges $+q$ and $-q$. The image of $+q$ is $-q$ and placed beneath the plane at an equal distance. Similarly, the image of $-q$ is $+q$ and placed beneath the plane at an equal distance. Clearly, the image charges constitute a dipole of dipole moment $\vec{p}'$ such that $p'_z = p\cos\theta$ and $p'_x = -p\sin\theta$. The field produced by the image dipole at the position of the original dipole is

$$\vec{E}' = \frac{1}{4\pi\varepsilon_0}\left[\frac{3\,(\vec{p}'\cdot\vec{r})\,\vec{r}}{r^5} - \frac{\vec{p}'}{r^3}\right],$$

where $\vec{r} = 2d\hat{z}$ is the position vector of the dipole with respect to its image. The torque on $\vec{p}$ is, therefore,

$$\vec{\tau} = \vec{p}\times\vec{E}' = \frac{1}{4\pi\varepsilon_0}\left[\frac{3\,(\vec{p}'\cdot\vec{r})\,(\vec{p}\times\vec{r})}{r^5} - \frac{\vec{p}\times\vec{p}'}{r^3}\right].$$

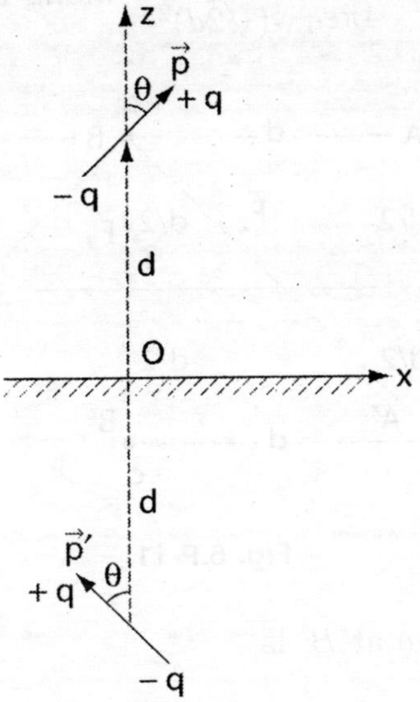

Fig. 6.P-10

The magnitude of $\vec{\tau}$ is

$$|\,\vec{\tau}\,| = \frac{1}{4\pi\varepsilon_0}\left[\frac{3pr\cos\theta\cdot pr\sin\theta}{r^5} - \frac{p^2\sin 2\theta}{r^3}\right]$$

$$= \frac{1}{4\pi\varepsilon_0}\cdot\frac{p^2}{r^3}\,[3\sin\theta\cos\theta - \sin 2\theta]$$

$$= \frac{p^2}{8\pi\varepsilon_0 r^3}\sin 2\theta.$$

In the equilibrium the torque must be zero. Therefore, we must have $\theta = 0$, i.e., the dipole will set itself perpendicular to the grounded plane.

► **Problem 11.** *Two equal charges $+q$ are placed a distance d apart in front of an infinite earthed conducting plane such that each charge is at a distance $d/2$ from the plane. Show that the force on each charge is of magnitude $3q^2/8\pi\varepsilon_0 d^2$.* **[C.U. 2011]**

Solution. The induced charges on the plane can be replaced by two image charges at A' and B' as shown in (Fig. 6.P-11. The total force on B is given by the vector sum of the forces due $+q$ at A, $-q$ at A' and $-q$ at B'. Now the force on $+q$ at B due to $+q$ at A is

$$\vec{F_1} = \frac{1}{4\pi\varepsilon_0} \cdot \frac{q^2}{d^2} \quad \text{along } \overrightarrow{AB}$$

Force on $+q$ at B due to $-q$ at A' is

$$\vec{F_2} = \frac{1}{4\pi\varepsilon_0} \cdot \frac{q^2}{(\sqrt{2}d)^2} \quad \text{along } BA'$$

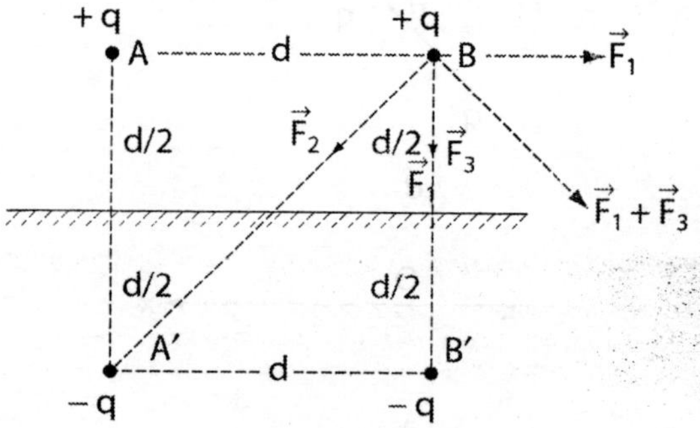

Fig. 6.P-11

Force on $+q$ at B due to $-q$ at B' is

$$\vec{F_3} = \frac{1}{4\pi\varepsilon_0} \cdot \frac{q^2}{d^2} \quad \text{along } BB'$$

$\vec{F_1} \perp^r F_3$ and their sum is $\dfrac{q^2}{4\pi\varepsilon_0 d^2} \cdot \sqrt{2}$ in a direction at right-angle to $\vec{F_2}$. Therefore, the magnitude of total force is

$$\frac{q^2}{4\pi\varepsilon_0 d^2}\sqrt{\left(\sqrt{2}\right)^2 + \left(\frac{1}{2}\right)^2} = \frac{3q^2}{8\pi\varepsilon_0 d^2}.$$

► **Problem 12.** *Two infinite parallel grounded conducting plates are held a distance d apart. A point charge q is placed between the plates at a distance x from one plate. Find the location of image charges. Express the force on q by an infinite series. What is the magnitude of the force for $x = d/2$? Discuss the case $d \to \infty$.*

Solution. Positions of the image are shown in Fig. 6.P-12. The charge q is assumed to be at the origin O. The charge $-q$ at A_1 is image of q due to plate 1 and $-q$ at B_1 is that due to plate 2. So $OA_1 = 2x$ and $OB_1 = -2(d-x)$.

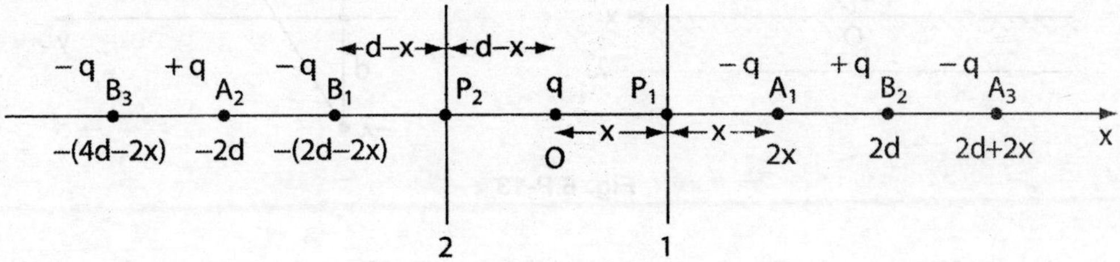

Fig. 6.P-12

Now as A_1 is in front of plate 2 we are to consider its image $+q$ at A_2, where $A_2P_2 = A_1P_2 = 2x+d-x = d+x$. Therefore, $OA_2 = A_2P_2 + OP_2 = d+x+d-x = 2d$. Similarly, as B_1 is in front of plate 1 we are to consider its image at B_2 such that $B_1P_1 = P_1B_2 = 2(d-x)+x$. Therefore, $OB_2 = x + 2(d-x) + x = 2d$. Proceeding in this way we can find other image charges. Now force on q at O is given by the resultant force due to all the image charges. Thus,

$$\vec{F} = \hat{x}\frac{q^2}{4\pi\varepsilon_0}\left[\frac{1}{(2x)^2} - \frac{1}{(2d)^2} + \frac{1}{(2d+2x)^2} - \cdots \right.$$
$$\left. - \frac{1}{(2d-2x)^2} + \frac{1}{(2d)^2} - \frac{1}{(4d-2x)^2} + \cdots \right].$$

If $x = d/2$ then $\vec{F} = 0$.

If $d \to \infty$ then $\vec{F} = \hat{x}\dfrac{1}{4\pi\varepsilon_0} \cdot \dfrac{q^2}{(2x)^2}$, same as for only one plane.

▶ **Problem 13.** *A long straight thin wire having a charge λ per unit length runs parallel to an infinite grounded conducting plane. Find the potential in the region above the plane.*

Solution. The problem can be solved by considering an image charge in the form of a long wire carrying $-\lambda$ charge per unit length and placed symmetrically on the other side of the plane. Suppose the conducting plane is the xy-plane and the wire runs parallel to x-axis (Fig. 6.P-13). The potential at P due to the original wire is

$$\phi_1(y,z) = \frac{\lambda}{2\pi\varepsilon_0}\ln\frac{r_0}{r_1}$$

and that due to the image wire is

$$\phi_2(y,z) = \frac{-\lambda}{2\pi\varepsilon_0}\ln\frac{r_0}{r_2},$$

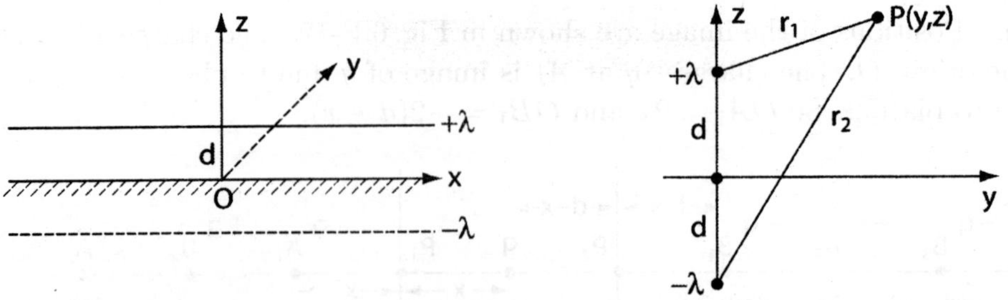

Fig. 6.P-13

where r_0 is a reference point, where potential is zero. Therefore, total potential at P is

$$\phi(y, z) = \phi_1 + \phi_2 = \frac{\lambda}{2\pi\varepsilon_0} \ln \frac{r_2}{r_1} = \frac{\lambda}{2\pi\varepsilon_0} \ln \left\{ \frac{y^2 + (z+d)^2}{y^2 + (z-d)^2} \right\}.$$

► **Problem 14.** *A straight wire of length 2l having a 'rigid' charge distribution of λ per unit length is placed at a distance d from the centre of a grounded conducting sphere of radius 'a' such that the ends of the wire are equidistant from the centre of the sphere. Show that the total charge induced on the sphere is $-2\lambda a \sinh^{-1} l/d$.*

Solution. Let λdx be the charge on an element dx of the wire at a distance x from the mid-point of the wire ((Fig. 6.P-14). This element induces a charge

$$-\frac{(\lambda dx)a}{OA} = -\frac{\lambda a dx}{\sqrt{d^2 + x^2}}$$

on the sphere.

Therefore, total induced charge is

$$-\int_{-l}^{+l} \frac{\lambda a dx}{\sqrt{d^2 + x^2}} = -2\lambda a \sinh^{-1} \frac{l}{d}.$$

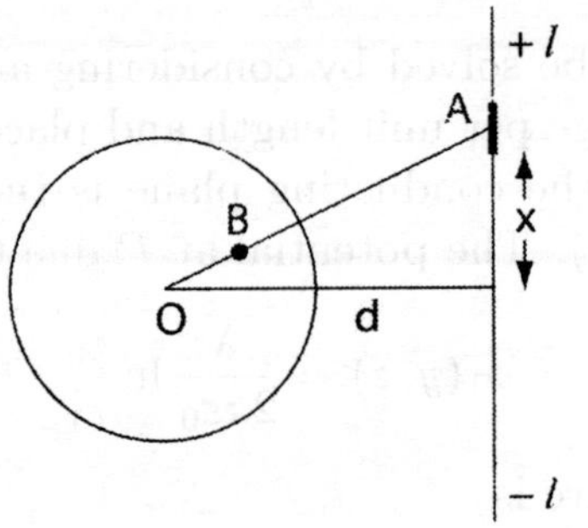

Fig. 6.P-14

QUESTIONS

1. What is an electrical image? A point charge is placed in front of a grounded conducting infinite plane. Using the method of images find

   (a) the surface density of induced charge on the plane,

   (b) the force between the plane and the charges,

   (c) also calculate by direct integration the force between the point charge and the induced charge on the plane obtained by the method of image.

2. (a) Explain the method of electrical image for the solution of electrostatic problems. **[C.U. 2009]**

   (b) Obtain the electrical image for the system consisting of a point charge q placed at a distance d from the centre of a grounded conducting sphere of radius a.

3. A point charge $+q$ is placed in front of a grounded conducting sphere of radius 'a' at a distance d from the centre. Find (i) the location and magnitude of image charge (ii) the potential and radial component of electric field at an internal point due to the induced charge on the sphere (iii) the induced charge density on the sphere. **[C.U. 2014]**

4. A point charge q is placed at distance d from the centre of a sphere of radius a maintained at zero potential. Find the potential at an external point due to the charges induced on the sphere by the method of image and also by using the solution of Laplace's equation. Show that the results are consistent with each other.

5. A point charge q is placed at a distance d from the centre of a grounded conducting sphere of radius a $(d > a)$. Using the method of images find the density of induced surface charge on the sphere. **[J.U. 2011]**

6. A point charge q is placed at a distance d from the centre of an insulated conducting sphere of radius a $(d > a)$, carrying a charge Q. Obtain the electrical images of the system. Show that there is always a force of attraction between q and the charged sphere, irrespective of the signs of q and Q, when q is very close to the sphere.

7. A point charge q is placed in air at a distance d from the surface of a 'semi-infinite' dielectric medium of dielectric constant K. Obtain the electrical images of the problem. Hence show that the charge q is attracted to the dielectric with a force

$$\frac{1}{16\pi\epsilon_0} \cdot \frac{K-1}{K+1} \cdot \frac{q^2}{d^2}.$$

8. A grounded conducting sphere of radius a is placed in a uniform electric field $\vec{E}_0 = E_0\hat{k}$. Using the method of image show that the potential at any external point (r, θ) is given by

$$\phi(r, \theta) = -E_0 r \cos\theta + \frac{E_0 a^3}{r^2} \cos\theta.$$

EXERCISES

1. (a) A point charge $+q$ is placed in front of an infinite grounded conducting plane. Show that the surface density of induced charge at any point on the plane varies inversely as the cube of the distance of the point from the point charge q. Also show that the total charge induced on the plane is $-q$.

   (b) A charged particle is released from rest at a distance d from an infinite grounded conducting plane. The particle attains a speed u at a distance $d/2$ from the plane. At what distance from the plane the particle will attain the speed $2u$? **[C.U. 2017]** [*Ans.* $d/5$]

2. A point charge $+q$ is placed at a distance d from the centre of an earthed conducting sphere of radius a. Show that the total induced charge on the sphere is $-qa/d$.

3. An insulated conducting sphere of radius a carries a charge $+Q$. A point charge $+q$ is placed at a distance d from the centre of the sphere. Find the minimum value of Q, which makes surface density of charge on the sphere positive everywhere.
[*Ans.* $Q_{\min} = \frac{q(3d-a)a^2}{d(d-a)^2}$; use Eq. (6.5-6)]

4. A point charge q is placed near a spherical conductor of radius a carrying charge Q. Show that the charge q will experience a repelling force unless its distance from the centre of the sphere is less than $a\left(1 + \frac{1}{2}\sqrt{q/Q}\right)$. Assume that $Q \gg q$.

5. A point charge q is placed at a distance d from the centre of an earthed conducting sphere of radius a. Show that the ratio of maximum and minimum surface densities of induced charge is $[(d+a)/(d-a)]^3$.

6. A thin conducting ring of radius a having a charge q is placed with its plane parallel to an infinite conducting plane at a distance d from it. Show that the potential at the centre of the ring is

$$\phi = \frac{q}{4\pi\varepsilon_0}\left(\frac{1}{a} - \frac{1}{\sqrt{a^2 + 4d^2}}\right)$$

and the surface density of induced charges on the plane at a point directly below the centre of the ring is

$$\frac{qd}{2\pi\,(a^2 + d^2)^{3/2}}.$$

7. (a) Two semiinfinite earthed conducting planes meet at right-angles. In the region between them there is a point charge q, which is at a distance d from each plane. Set up the image charges. Hence show that the force on q due to the planes is attractive in nature and of magnitude $\frac{q^2(2\sqrt{2}-1)}{32\pi\varepsilon_0 d^2}$. Sketch some lines of force between q and the planes. **[cf. C.U. 2015, C.U. 2019]**

[**Hints.** To satisfy the condition of zero potential at the planes we require three image charges as shown in Fig. 6.E-7(a).]

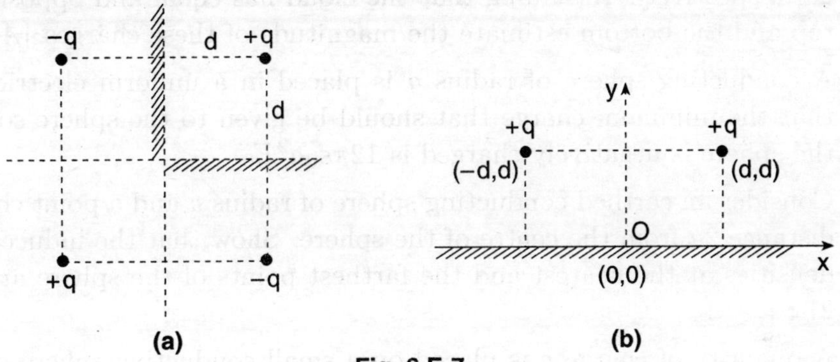

Fig. 6.E-7

(b) Suppose a large conducting grounded plane form xz-plane and two point changes q are placed at the points (d, d) and $(-d, d)$ as shown in Fig. 6.E-7(b). Find the net force on the charge at (d, d).

$\left[Ans.\ \vec{F} = \frac{q^2}{4\pi\varepsilon_0(4d)^2}\left[\frac{2\sqrt{2}-1}{2\sqrt{2}}\hat{i} - \frac{2\sqrt{2}+1}{2\sqrt{2}}\hat{j}\right]\right]$

8. A point charge q is placed at a distance d from the centre of an insulated conducting sphere of radius a. Show that the induced positive and negative charge-zones are separated by a circle and the distance r of any point of this circle from q is given by $r^3 = d(d^2 - a^2)$.

9. An earthed conducting sphere of radius a is placed with its centre on the straight line joining two charges q and q'. The distances of q and q' from the centre of the sphere are $2a$ and $4a$, respectively. Show that the charge q' experiences a repelling force, if $q' < \frac{25q}{144}$.

10. A point charge q is placed inside a hollow grounded metal sphere of radius a at a distance d from the centre $(d < a)$. Show that the force on the charge is

$$-\frac{q^2 ad}{4\pi\varepsilon_0\,(a^2 - d^2)^2}.$$

[**Hints.** Here $d < a$ and the image charge $-qa/d$ is placed at an outside point at a distance a^2/d from the centre of the sphere.]

11. Suppose a dipole of moment $\vec{p}$ is placed at a distance d from the centre of an earthed conducting sphere of radius a, with the vector $\vec{p}$ pointing in the radially outward direction. Show that the image configuration consists of a dipole of moment $\left(\frac{a}{d}\right)^3 \vec{p}$ and a charge ap/d^2 located at a distance a^2/d from the centre.

12. A point charge of 10 μC is placed at a distance 10 cm from an earthed infinite conducting plane. Find the work required to move the charge to infinity.
[*Ans.* 2.25J]

13. When a cloud passes over a place on the surface of the earth a vertical electric field E_0 is recorded there. The bottom and top of the cloud are at heights d and $2d$, respectively. Assuming that the cloud has equal and opposite charges at its top and the bottom estimate the magnitude of these charges. [*Ans.* $8\pi\varepsilon_0 d^2 E_0/3$]

14. A conducting sphere of radius a is placed in a uniform electric field $\vec{E}_0$. Show that the minimum charge that should be given to the sphere so that no part of the sphere is negatively charged is $12\pi\varepsilon_0 a^2 E_0$.

15. Consider an earthed conducting sphere of radius a and a point charge placed at a distance $2a$ from the centre of the sphere. Show that the induced surface charge densities at the nearest and the farthest points of the sphere are in the ratio of 27 : 1.

16. A quantity of charge q is placed on a small conducting sphere of radius a. Now it is taken towards a large earthed conducting plane and placed at a distance d from the plane ($d \gg a$). Show that as a result of this the potential of the sphere reduces to $(1 - a/2d)$ of its original value.

17. A point charge q is placed at a distance d from an infinite earthed conducting plane. Let O is the foot of the perpendicular from q on the plane. Find the radius of the circle on the plane centred on O, which contains half of the total induced charges.
[*Ans.* $\sqrt{3}d$]

18. A point charge $q = 10\mu$C is placed at a distance 25 cm from the centre of an earthed conducting sphere of radius 10 cm. Find the magnitude and position of the possible image charge. Also, calculate the maximum surface density of charge on the sphere. [*Ans.* -4μC at a distance 4 cm from the centre, $-123.8\mu\text{C/m}^2$]

19. A point charge q is located between two earthed intersecting conducting planes that are inclined at 60° to each other. Determine the number of image charges and their locations.
[*Ans.* 5]

OBJECTIVE TEST

1. A charge placed in front of a grounded conducting plane:

   (a) is repelled by the plane,

   (b) is attracted by the plane,

(c) is not affected by the presence of the plane,

(d) none of these.

2. A point charge q is held at a distance $2a$ from the centre of an isolated, uncharged conducting sphere of radius a. The potential of the sphere is:

(a) zero, (b) $\frac{q}{2\pi\varepsilon_0 a}$, (c) $\frac{q}{4\pi\varepsilon_0 a}$, (d) $\frac{q}{8\pi\varepsilon_0 a}$.

3. A charge $+q$ is placed at a distance d from an earthed conducting plane. The force on $+q$ due to the plane is:

(a) $\frac{1}{4\pi\varepsilon_0}\frac{q^2}{(2d)^2}$, (c) $\frac{1}{4\pi\varepsilon_0}\cdot\frac{q^2}{d^2}$,

(b) $-\frac{1}{4\pi\varepsilon_0}\cdot\frac{q^2}{(2d)^2}$, (d) $-\frac{1}{4\pi\varepsilon_0}\cdot\frac{q^2}{d^2}$.

4. Two semi-infinite earthed conducting planes meet at right-angles. In the region between them there is a point charge. To solve the problem by the method of image the number of images required is:

(a) 1, (b) 2, (c) 3, (d) 4.

5. Two conducting plates are inclined at an angle 30° to each other. A point charge is placed between them. The number of image charges required is:

(a) 12, (b) 11, (c) 2, (d) 1. **[GATE 2015]**

6. Which of the following statements is true for electrical image?

(a) It is always equal and opposite to the real charge,

(b) It may be real or virtual,

(c) It is always virtual,

(d) It is always equal in magnitude to the real charge.

7. A charge $+q$ is kept at a distance $2a$ from the centre of a grounded conducting sphere of radius 'a'. The magnitude and position of image charge with respect to the centre of the sphere are

(a) $-\frac{q}{2}$ and $\frac{a}{2}$, (b) $-\frac{q}{2}$ and $\frac{a}{4}$, (c) $-q$ and $\frac{a}{2}$, (d) $+\frac{q}{2}$ and $\frac{a}{2}$.

8. The work done in brining a charge $+q$ from infinity to a position at a distance d in front of a semi-infinite grounded metal surface is

(a) $-\frac{q^2}{4\pi\varepsilon_0 d}$, (b) $-\frac{q^2}{8\pi\varepsilon_0 d}$, (c) $-\frac{q^2}{16\pi\varepsilon_0 d}$, (d) $-\frac{q^2}{24\pi\varepsilon_0 d}$.

9. Two charges q and $2q$ are placed in front of a grounded infinite conducting plane at distances 0.5 m and 1.5 m respectively. Then total force on q is

(a) $\frac{1}{4\pi\varepsilon_0}\cdot q^2$, (b) $\frac{1}{4\pi\varepsilon_0}\cdot\frac{7q^2}{2}$, (c) $\frac{1}{4\pi\varepsilon_0}\cdot\frac{q^2}{2}$, (d) $\frac{1}{4\pi\varepsilon_0}\cdot 2q^2$

ANSWERS

1. (b) **2.** (d) **3.** (b) **4.** (c) **5.** (b) **6.** (c) **7.** (a)
8. (c) **9.** (b)

Chapter 7

Steady Current

7.1 Electric Current

So far we have considered only static conditions in which electric charges are at rest. Now we shall consider charges in uniform motion. Moving charges constitute an electric current. A current is said to be *steady current* if it is independent of time, though it may vary spatially. Quantitatively electric current (I) is defined as the rate of flow of charge through a given surface in a conducting medium. Thus if dq amount of charge is transported in time dt through the given cross-section then electric current,

$$I = \frac{dq}{dt}. \tag{7.1-1}$$

The unit of current in SI system is the ampere(A). According to Eq. (7.1-1)

$$1 \text{ ampere} = 1 \text{ coulomb/second.}$$

The current may be due to flow of both positive and negative charges. For example in metals charge carriers are negatively charged free electrons, in electrolytes charge carriers are positive and negative ions. In semiconductors charge carriers are electrons and holes.

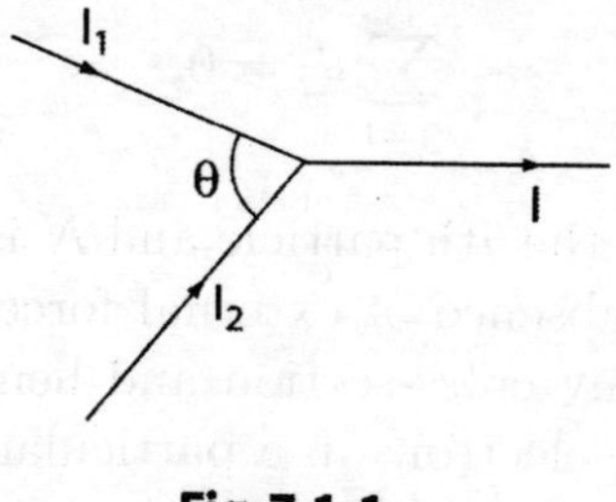

Fig.7.1-1

Conventionally the direction of motion of positive charges is taken as the direction of electric current. So in metals the direction of current is opposite to the actual direction of motion of the electrons. Here it is important to mention that though, conventionally, a direction is associated with *electric current, it is not a vector quantity.* It does not obey the rules of vector addition. For example in Fig. 7.1-1 the current $I = I_1 + I_2$ irrespective of the angle θ. This means that the current behaves like a scalar.

Current density ($\vec{J}$)

Electric current I is a macroscopic quantity. It is related to the average rate of flow charges through any cross-section. For the microscopic description of the flow of charges in a medium we introduce the idea of a current density ($\vec{J}$). It is defined as the current per unit area, assumed to be perpendicular to the flow. Its unit is A/m^2. It is a vector point function. Its direction at a point is the same as the direction of current at that point.

If dI be the current through any surface element $d\vec{S}$ then

$$dI = \vec{J} \cdot d\vec{S} \tag{7.1-2}$$

and total current through the surface S is

$$I = \int_S \vec{J} \cdot d\vec{S}. \tag{7.1-3}$$

Here it is important to point out that in the general case the current density $\vec{J}$ may not be the same over the entire cross-section of a conductor though current I is the same through every cross-section of a single conductor.

Drift velocity and electric current

Let us consider electrical conduction in metals. In metals there are plenty of free electrons, which move randomly and haphazardly through the interatomic spaces inside the metals just like the gas molecules. They collide with each other and with atoms in the lattice. They are said to form classical electron gas. The velocity of a particular electron varies randomly and the average velocity is expected to be zero. Thus

$$\sum_{j=1}^{N} \vec{u}_j = 0, \tag{7.1-4}$$

where $\vec{u}_j$ is thermal velocity of the jth particle and N is the number of electrons. This is the state of motion in the absence of external forces. In this case there is no net transport of charge through any cross-section and hence current is zero. In order to have current we must push the electrons in a particular direction by applying external forces. In presence of an electric field $\vec{E}$ each electron experience a force $q\vec{E}$, where

$q = -e$ is the electronic charge, and gets accelerated along the field direction. Thus in addition to the random thermal motion electrons acquire additional velocity $\vec{u}'_j$ (say) in the field direction. Hence, total velocity of the jth electron will be $\vec{v}_j = \vec{u}_j + \vec{u}'_j$ and the average velocity of the electrons will be

$$\vec{v} = \frac{1}{N}\sum_{j=1}^{N}\vec{v}_j = \frac{1}{N}\sum_{j=1}^{N}\left(\vec{u}_j + \vec{u}'_j\right) = \frac{1}{N}\sum_{j=1}^{N}\vec{u}'_j \quad \text{[Using Eq. (7.1-4)]} \qquad (7.1\text{-}5)$$

This velocity $\vec{v}$ is called *drift velocity.* Thus the drift velocity *is the average velocity with which charged particles move or drift across the system in presence of electric field.* The random thermal velocity ($\vec{u}_j$) is of the order of 10^8m/s whereas the drift velocity is of the order of a few mm/s only. The current through any conductor flows almost with the speed of light. Actually the impulse that is created with the application of electric field propagates with such high speed.

Suppose there are n number of charged particles, each of charge q, moving with drift velocity $\vec{v}$ under the influence of an electric field $\vec{E}$. If we consider an elementary area $d\vec{S}$ around a point P then the number of charged particles crossing this area in time dt will be equal to that contained in a cylinder of volume $dV = d\vec{S} \cdot \vec{v}dt$ (Fig. 7.1-2). Therefore, the number of charged particles crossing this elementary area in time dt will be

$$n\,dV = n\,d\vec{S} \cdot \vec{v}\,dt.$$

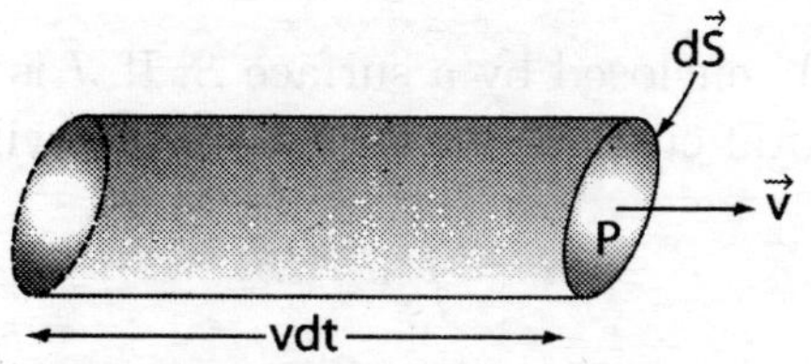

Fig. 7.1-2

So the amount of charge crossing the elementary area $d\vec{S}$ in time dt will be

$$dq = qnd\vec{S} \cdot \vec{v}\,dt.$$

Therefore, current through $d\vec{S}$ is

$$I = \frac{dq}{dt} = nq\vec{v} \cdot d\vec{S} \qquad (7.1\text{-}6)$$

and current density

$$\vec{J} = nq\vec{v}. \qquad (7.1\text{-}7)$$

For electrons $q = -e$ and Eq. (7.1-7) takes the form

$$\vec{J} = -ne\vec{v}. \qquad (7.1\text{-}8)$$

The quantity nq represents volume charge density ρ and hence, Eq. (7.1-7) can also be expressed as

$$\vec{J} = \rho \vec{v}. \tag{7.1-9}$$

If there are several types of current carriers then

$$\vec{J} = \sum_j n_j q_j \vec{v}_j = \sum_j \rho_j \vec{v}_j, \tag{7.1-10}$$

where the subscript j refers to the jth type of carrier.

Surface current

When the charges flow over a surface the idea of surface current is introduced. If we consider a line of unit length on the surface of a current carrying body normal to the direction of flow then charge crossing this line per unit time is called surface current density ($\vec{K}$). If σ_f be the free surface charge density and $\vec{v}$ is the velocity of flow then

$$\vec{K} = \sigma_f \vec{v}. \tag{7.1-11}$$

7.2 Equation of Continuity and Local Conservation of Charge

Let us consider a volume V enclosed by a surface S. If $\vec{J}$ is the current density at any surface element $d\vec{S}$ then total current flowing through S will be

$$I = \oint_S \vec{J} \cdot d\vec{S}. \tag{7.2-1}$$

It gives the rate at which charge is flowing out of the volume enclosed by S. According to the local conservation of charge, the charge within the volume must decrease at the same rate. If ρ is the charge density at any volume element dV then total charge within V is

$$q = \int_V \rho \, dV.$$

Therefore,

$$\oint_S \vec{J} \cdot d\vec{S} = -\frac{dq}{dt} = -\frac{d}{dt}\int_V \rho \, dV = -\int_V \frac{\partial \rho}{\partial t} \, dV.$$

From Gauss's divergence theorem

$$\oint_S \vec{J} \cdot d\vec{S} = \int_V \left(\vec{\nabla} \cdot \vec{J}\right) dV.$$

Thus

$$\int_V \left(\vec{\nabla} \cdot \vec{J} + \frac{\partial \rho}{\partial t}\right) dV = 0.$$

This is true for any volume V and hence we can write

$$\vec{\nabla} \cdot \vec{J} + \frac{\partial \rho}{\partial t} = 0. \tag{7.2-2}$$

This is known as the *equation of continuity.* It is the mathematical statement of local conservation of charge which assumes that charge can be neither created nor destroyed. It is a basic equation of electromagnetism relating charge density at a point to the current density at that point.

If the charge density ρ is constant in time at every point then

$$\vec{\nabla} \cdot \vec{J} = 0. \tag{7.2-3}$$

This is the case for a *steady current* flow. So for steady current

$$\oint_S \vec{J} \cdot d\vec{S} = 0,$$

which means that net current entering or leaving any closed surface is zero. Thus for steady current there cannot be any accumulation of charge anywhere.

In case of surface current equation of continuity (7.2-2) takes the form,

$$\frac{\partial \sigma_f}{\partial t} + \vec{\nabla} \cdot \vec{K} = 0, \tag{7.2-4}$$

where σ_f is the free surface charge density and $\vec{K}$ is the surface current density.

7.3 Lorentz-Drude Theory of Electric Conduction and Ohm's Law

To understand the mechanism of electric conduction in metals we start with the following assumptions:

(i) Free electrons inside metals move randomly and haphazardly with all sorts of speed. In the absence of any external field average velocity is zero.

(ii) They collide elastically with each other as well as with the crystal lattice. Between two successive collisions electrons move in straight line.

(iii) In the presence of an external applied field electrons acquire an additional velocity in the direction opposite to the applied field. If an electric field $\vec{E}$ is applied, it acts on every electron and produces an acceleration

$$\vec{a} = \frac{q\vec{E}}{m},$$

where $q = -e$ is the electronic charge and m is its mass.

Let τ be the average time between two successive collisions. τ is called mean *free time*. We assume that after each collision additional velocity acquired is reduced to zero. So the additional velocity acquired just before the next collision is $\vec{a}\tau = \frac{q\vec{E}}{m}\tau$. Therefore, the average or drift velocity is

$$\vec{v} = \frac{0 + \frac{q\tau}{m}\vec{E}}{2} = \frac{q\tau}{2m}\vec{E}. \tag{7.3-1}$$

Drift velocity per unit field is called *mobility*. Thus mobility μ of an electron is given by $\mu = \frac{q\tau}{2m}$.

Now current density

$$\vec{J} = nq\vec{v} = \frac{ne^2\tau}{2m}\vec{E} = \sigma\vec{E}, \tag{7.3-2}$$

where $\sigma = \frac{ne^2\tau}{2m}$ is called the *electrical conductivity*. Its practical unit is siemens (S) per metre. A siemens was formerly called a mho.

Equation (7.3-2) is known as the *differential form of Ohm's law*. It does not include any derivatives yet the term 'differential' is used to mean that Eq. (7.3-2) represents relationship between quantities ($\vec{J}$ and $\vec{E}$) characterising a single point of a conductor.

Consider a conducting specimen in the form of a rod of length L and of uniform cross-section S whose ends are maintained at constant potential difference V. Here the electric field $\vec{E}$ is uniform along the length of the rod and is given by $\left|\vec{E}\right| = \frac{V}{L}$.

Now current through any cross-section of the rod is

$$I = \int \vec{J}\cdot d\vec{S} = \int \sigma\vec{E}\cdot d\vec{S} = \sigma ES,$$

where S is the cross-sectional area of the rod; σ is constant for a homogeneous material. Now using the relation $E = -\frac{d\phi}{dx}$ we get

$$-\frac{d\phi}{dx} = \frac{I}{\sigma S}$$

$$\text{or} \quad -\int_1^2 d\phi = \int_1^2 \frac{I}{\sigma S}dx$$

$$\text{or} \quad V = \phi_1 - \phi_2 = \frac{I}{\sigma S}(x_2 - x_1) = \frac{IL}{\sigma S},$$

where $x_2 - x_1 = L$ =length of the rod.

Thus

$$I = \frac{\phi_1 - \phi_2}{L/\sigma S} = \frac{V}{R}, \tag{7.3-3}$$

where $R = \frac{L}{\sigma S} = \frac{\rho L}{S}$ is called the *resistance of the conductor*; $\rho = \frac{1}{\sigma}$ is the resistivity, which is measured in units of ohm-metre (Ω m). Equation (7.3-3) is the familiar version of *Ohm's law*.

Note that Ohm's law is valid if $\sigma(= ne^2\tau/2\text{m})$ is constant independent of $\vec{E}$. The mean free time

$$\tau = \frac{\text{Mean free path } (\lambda)}{\text{Total velocity (thermal + drift)}}$$

For low $\vec{E}$, drift velocity $(v) \ll$ thermal velocity (u) and τ is almost constant at a particular temperature and hence Ohm's law is valid for low $\vec{E}$. If $\vec{E}$ becomes very large of the order of 10^8 V/m then τ becomes a function of $\vec{E}$ and hence in that case Ohm's law becomes inapplicable. For very high frequency $\vec{E}$-field when the time period becomes comparable with $\tau(\sim 10^{-10}\text{s})$ Ohm's law fails. Thus for ac voltages of frequencies greater than 10^{10} Hz Ohm's law becomes inapplicable. Ohm's law is also found to be not applicable to vacuum tube devices. In these devices electron density is very low; collisions are very few and far between. Here some of the electrons may go from one electrode to another with accelerated motion without suffering collision. So the concept of drift velocity becomes inapplicable and Ohm's law fails here.

Power consideration

The passage of current through a conductor having a resistance is accompanied by the liberation of heat. Suppose there exists a potential difference V between the two ends of a homogeneous conductor of resistance R. According to Ohm's law the current is $I = V/R$. The work done in taking dq amount of charge from one end to the other is $dw = V dq$. Therefore, the rate at which energy is dissipated in the resistance is

$$P = \frac{dw}{dt} = V\frac{dq}{dt} = VI = I^2R \tag{7.3-4}$$

This is the expression for the well-known *Joule's law.*

The law can also be expressed in the differential form. Let us consider an elementary cylinder of length dL, area of cross section dS and axis parallel to $\vec{J}$ at a point. The rate of energy dissipation in this elementary volume is given by

$$dP = I^2 dR = (JdS)^2 \frac{dL}{\sigma\, dS} = \frac{J^2}{\sigma} \cdot dV,$$

where $dV = dS\, dL$ is the volume of the elementary cylinder. Therefore, the amount of power dissipation per unit volume or the thermal power density of current is

$$\frac{dP}{dV} = \frac{J^2}{\sigma} = \vec{J} \cdot \vec{E}. \tag{7.3-5}$$

This is the *Joule's law in differential form.* It characterises the liberation of heat at different points of a conducting medium. It is applicable to any conductor regardless of the shape, homogeneity etc. Also, it is applicable to both steady and variable currents.

7.4 Steady Current Problems and its Analogy with Electrostatics

It is found that a close analogy exists between problems involving steady currents and electrostatic problems involving conductors and dielectrics.

For steady currents, though the charges are in motion, local macroscopic charge density $\rho(x, y, z)$ does not change with time, i.e., $\partial\rho/\partial t = 0$. Hence, from the equation of continuity (7.2-2) we have

$$\vec{\nabla} \cdot \vec{J} = 0. \tag{7.4-1}$$

Using Ohm's law (7.3-2) we get

$$\vec{\nabla} \cdot (\sigma \vec{E}) = 0.$$

For a homogeneous medium it reduced to

$$\vec{\nabla} \cdot \vec{E} = 0. \tag{7.4-2}$$

The electric field of steady currents satisfies the condition $\vec{\nabla} \times \vec{E} = 0$ and hence is derivable from a scalar potential ϕ :

$$\vec{E} = -\vec{\nabla}\phi. \tag{7.4-3}$$

Combining Eqs. (7.4-2) and (7.4-3) we get

$$\nabla^2\phi = 0, \tag{7.4-4}$$

which is Laplace's equation. Therefore, steady current problems can be solved in the same way as electrostatic problems. In principle we can solve any steady current problem by solving the Laplace's equation with appropriate boundary conditions which specifies ϕ or equivalently $\vec{E}$ or $\vec{J}$ at each point on the surface of the conducting medium. Once ϕ is known then $\vec{E}$ and hence $\vec{J}$ can be determined at each point in the conducting medium by using Eq. (7.4-3) and Ohm's law (7.3-2).

Boundary conditions

For steady currents $\vec{\nabla} \cdot \vec{J} = 0$ and hence,

$$\oint_S \vec{J} \cdot d\vec{S} = 0.$$

This is similar to

$$\oint_S \vec{D} \cdot d\vec{S} = 0$$

in electrostatics. So by analogy we can say that the normal component of $\vec{J}$ must be continuous across the interface of two media. Thus

$$J_{1n} = J_{2n}, \tag{7.4-5}$$

where $\vec{J_1}$ and $\vec{J_2}$ are the current densities in the first and second media and n refers to their normal components. In terms of electric field condition (7.4-5) becomes

$$\sigma_1 E_{1n} = \sigma_2 E_{2n}, \tag{7.4-6}$$

where σ_1 and σ_2 represent conductivity of the respective medium.

The boundary condition on the tangential components of $\vec{J}$ follows directly from the continuity condition on the tangential components of $\vec{E}$ (i.e., $E_{1t} = E_{2t}$). Thus,

$$\frac{J_{1t}}{\sigma_1} = \frac{J_{2t}}{\sigma_2}. \tag{7.4-7}$$

7.5 Kirchhoff's Laws and Network Analysis

In many practical problems we come across *electrical circuits* which are nothing but a complete path for electric current to flow. Each individual circuit component such as resistor, inductor, capacitor etc. is called a *circuit element.* A group of such elements, usually in series, and having two terminals is called a *branch* of the circuit. A combination of circuit elements or branches interconnected in some ways is called a *network.* Any closed path formed by a number of branches in a network is called a *loop.* A *mesh* is the simplest closed path. A common point where two or more circuit components meet is called a *node* or *junction.* In order to have steady current in a circuit it is usually provided with a battery or an electronic power supply which acts as a source of currents and voltages. In network analysis usually a network is given and we are to find the current in each element or potential drop in each element.

Simple networks can be analysed by using Ohm's law. But in practice most circuits are usually complicated. Analysis of such circuits can be considerably simplified by using the following two Kirchhoff's laws:

1. Kirchhoff's current law(KCL)

It states that *the algebraic sum of the currents meeting at a junction point in a network is zero.* The law is a consequence of the conservation of charge. In the steady state the rate at which charge enters a junction must be equal to the rate at which it leaves the junction. Mathematically the law may be stated as

$$\sum_k I_k = 0. \tag{7.5-1}$$

If the current entering a junction is taken as positive then the current leaving the junction should be taken as negative. Thus, when applied to Fig. 7.5-1, Eq. (7.5-1) assumes the form: $I_1 + I_2 - I_3 - I_4 = 0$.

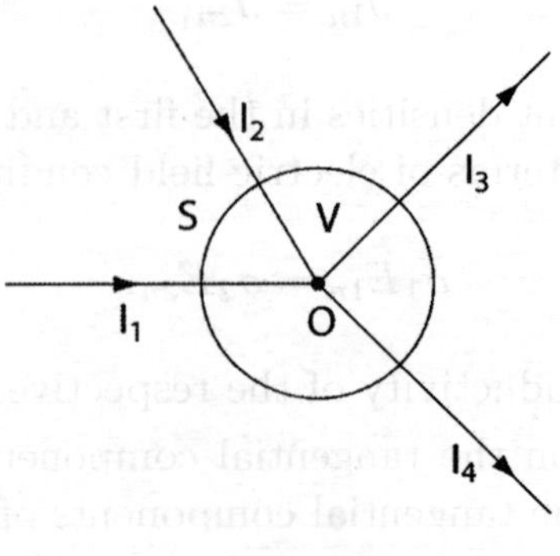

Fig. 7.5-1

2. Kirchhoff's voltage law (KVL)

It states that *the algebraic sum of the products of current and resistance in each part of any loop in a network is equal to the algebraic sum of the electromotive forces (emfs) in the same loop.* The law is a consequence of the law of conservation of energy. Mathematically the law may be sated as

$$\sum_k I_k R_k = \sum_k \mathcal{E}_k, \tag{7.5-2}$$

where R_k is the resistance in the kth branch of the loop I_k is the current through it and $\mathcal{E}_k$ the emf.

KCL and conservation of charge

In Fig. 7.5-1, let us imagine a closed surface S about the node O enclosing a volume V. From the equation of continuity, which follows from the principle of conservation of charge we have for steady current,

$$\vec{\nabla} \cdot \vec{J} = 0,$$

where $\vec{J}$ is the current density.

Therefore,

$$\int_V \vec{\nabla} \cdot \vec{J}\, dV = 0$$

or $\oint_S \vec{J} \cdot d\vec{S} = 0$ [using divergence theorem]

or $\sum_{k=1}^{N} I_k = 0$ where N is the number of conductors meeting at O.

This is the basis of Kirchhoff's current law. Conversely, we can establish the principle of conservation of charge from KCL. Writing $I_k = \frac{dq_k}{dt}$ we get from KCL

$$\sum_k I_k = \sum_k \frac{dq_k}{dt} = 0.$$

Integrating we get $\sum_k q_k$ = constant, which is nothing but the mathematical statement of the principle of conservation of charge.

KVL and conservation of energy

The net emf acting in a closed path in a network is given by the work done in taking a unit charge once around the closed path. Thus, net emf

$$\mathcal{E} = \sum_j \mathcal{E}_j = \oint \vec{E} \cdot d\vec{l} = \oint \frac{1}{\sigma} \vec{J} \cdot d\vec{l} \quad [\because \ \vec{J} = \sigma \vec{E} \text{ from Ohm's law}]$$

$$= \oint \frac{1}{\sigma} J \, dl$$

$$= \oint JS \cdot \frac{dl}{\sigma S} \quad [S = \text{area of cross-section for current flow}]$$

$$= \oint I \, dR$$

$$= \sum_{\hat{k}} I_k R_k.$$

This is the basis of Kirchhoff's voltage law.

Conversely, we can show that the KVL is consistent with the principle of conservation of energy.

Multiplying the above equation by q we get

$$\sum q\mathcal{E}_j = \sum qI_k R_k = \sum qV_k,$$

where $V_k = I_k R_k$ is the pd in the kth branch. Let us imagine a charge q is taken once around the loop. Then $\sum q\mathcal{E}_j$ is the work done by the source of emf and $\sum qV_k$ is the total energy dissipated in the loop. This is in accordance with the conservation of energy.

To apply Kirchhoff's laws to any network the following procedure may be adopted:

(i) Select an arbitrary closed loop and assume a direction and a magnitude to the current following through each branch.

(ii) Choose a direction, say clockwise to go around the loop.

(iii) If the assumed direction of any current coincides with the direction we go around take the corresponding IR term as positive; in the opposite case take the IR term as negative. For any emf, take it as positive if it tries to send a current in the direction we go around, i.e., in the clockwise direction; use negative sign in the opposite case.

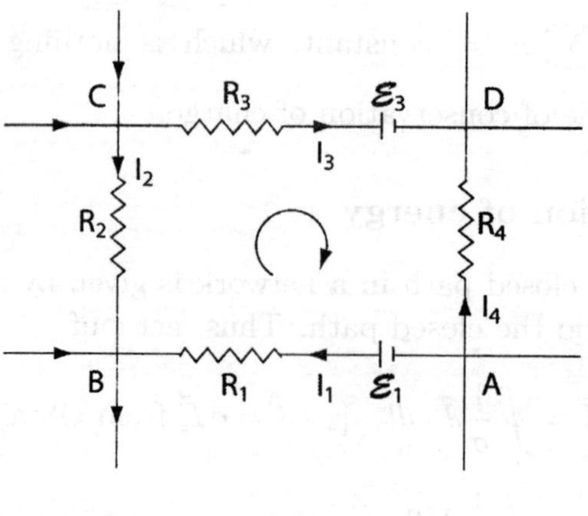

Fig. 7.5-2

For example we consider Fig. 7.5-2, which shows one of the loops of a complex network. Let us assume arbitrary direction and magnitude of current through each branch of the loop as shown in Fig. 7.5-2. Let us start from A and go around the loop clockwise. So the IR products in R_1 and R_3 should be taken positive as the direction of current and the direction we go around are the same. But the corresponding products in R_2 and R_4 should be taken as negative because here current direction is opposite. Thus,

$$\sum_k I_k R_k = I_1R_1 - I_2R_2 + I_3R_3 - I_4R_4.$$

The source of emf $\mathcal{E}_1$ tries to send current in the direction we go around and hence should be taken as positive whereas $\mathcal{E}_2$ tries to send current in the opposite direction and should be taken as negative. Thus,

$$\sum_k \mathcal{E}_k = \mathcal{E}_1 - \mathcal{E}_2.$$

Therefore, for the chosen loop we get the equation,

$$I_1R_1 - I_2R_2 + I_3R_3 - I_4R_4 = \mathcal{E}_1 - \mathcal{E}_2.$$

Thus we may write a equation for each loop of the given network. We need to consider only the independent loops. The resulting equations are algebraic and hence can be easily solved for unknown currents. If any current comes out to be negative, its correct direction would be opposite to that we have assumed.

7.6 Kirchhoff's Laws Applied to Wheatstone Bridge

Wheatstone bridge consists of four resistances R_1, R_2, R_3 and R_4 as shown in Fig. 7.6-1. E is a battery of emf $\mathcal{E}$ and internal resistance r_i. G is a galvanometer of resistance r_g. The bridge is widely used for the measurement of unknown resistances or for comparison of resistances. Let us analyse the circuit by applying Kirchhoff's laws. The current I entering at A is divided into I_1 and I_2. So we have from KCL,

$$I = I_1 + I_2.$$

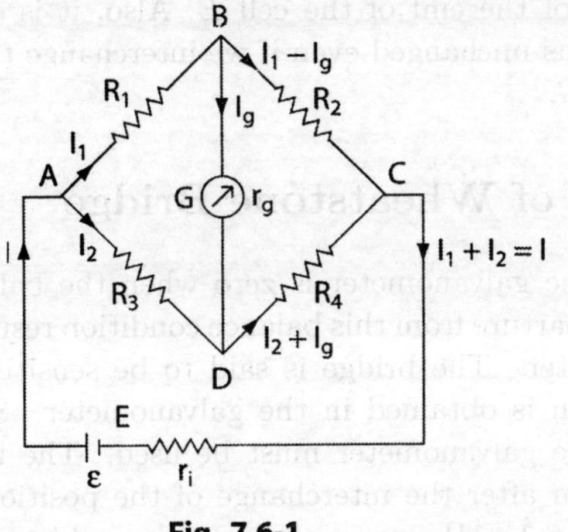

Fig. 7.6-1

Similarly, we assume other current directions as shown in Fig. 7.6-1.

Now applying KVL to the loops $ABDA$, $BCDB$ and $ABCEA$ we get

$$I_1R_1 + I_gr_g - I_2R_3 = 0$$
$$(I_1 - I_g)\,R_2 - (I_2 + I_g)\,R_4 - I_gr_g = 0$$
$$I_1R_1 + (I_1 - I_g)\,R_2 + (I_1 + I_2)\,r_i = \mathcal{E}.$$

Rearranging these equations we get

$$I_1R_1 - I_2R_3 + I_gr_g = 0 \tag{7.6-1}$$
$$I_1R_2 - I_2R_4 - I_g\,(R_2 + R_4 + r_g) = 0 \tag{7.6-2}$$
$$I_1\,(R_1 + R_2 + r_i) + I_2r_i - I_gR_2 = \mathcal{E}. \tag{7.6-3}$$

These linear equations can be solved conveniently by using Cramer's rule. Thus, we get the unbalanced current through the galvanometer as

$$I_g = \frac{\begin{vmatrix} R_1 & -R_3 & 0 \\ R_2 & -R_4 & 0 \\ R_1 + R_2 + r_i & r_i & \mathcal{E} \end{vmatrix}}{\begin{vmatrix} R_1 & -R_3 & r_g \\ R_2 & -R_4 & -(R_2 + R_4 + r_g) \\ R_1 + R_2 + r_i & r_i & -R_2 \end{vmatrix}}. \tag{7.6-4}$$

To work with the bridge the ratio R_1/R_2 is kept constant and R_3 is varied until the galvanometer shows no deflection. The bridge is then said to be balanced. So for balance $I_g = 0$ or,

$$\mathcal{E}(-R_1R_4 + R_2R_3) = 0$$

$$\text{or} \quad \frac{R_1}{R_2} = \frac{R_3}{R_4}. \tag{7.6-5}$$

This equation is used to find the unknown resistance R_4. The balance condition (7.6-5) is independent of the emf of the cell E. Also, it is obvious that the balance condition (7.6-5) remains unchanged even if we interchange the positions of the cell E and the galvanometer G.

7.7 Sensitivity of Wheatstone Bridge

The current through the galvanometer is zero when the balance condition (7.6-5) is satisfied. Any slight departure from this balance condition results in unbalanced current through the galvanometer. The bridge is said to be sensitive if for a given want of balance larger deflection is obtained in the galvanometer. So for good sensitiveness of the bridge a sensitive galvanometer must be used. The balance condition (7.6-5) remains unchanged even after the interchange of the positions of the cell E and the galvanometer G (Fig. 7.6-1). However, these two possible arrangements may not be equally sensitive. To understand it let us follow the treatment given by Prof. H. L. Callendar: In the Wheatstone bridge shown in Fig. 7.7-1 let I be the current through the unknown resistance $R_4 = r$, I_g the current through the galvanometer and currents in other branches are as shown. Applying KVL to the loops $ABDA$ and $BCDB$ we get

$$I_1mnr_1 - I_g \cdot R_g - (I + I_g)nr_1 = 0 \tag{7.7-1}$$

$$(I_1 + I_g)mr_1 - Ir + I_gR_g = 0 \tag{7.7-2}$$

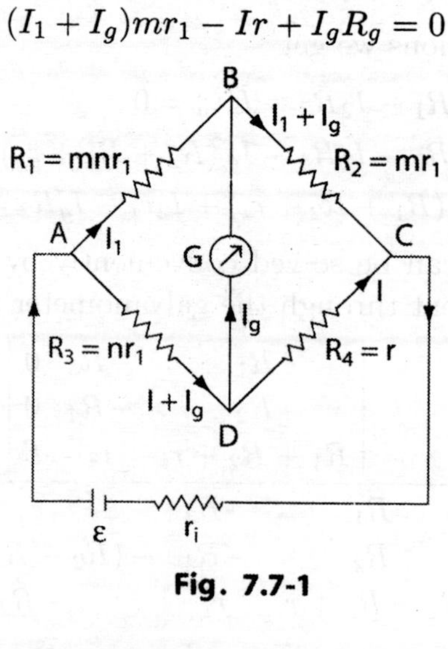

Fig. 7.7-1

Eliminating I_1 from these equations we can write

$$\frac{I_g}{I} = \frac{r - r_1}{R_g\left(1 + \frac{1}{n}\right) + r_1(1 + m)}. \tag{7.7-3}$$

The bridge is balanced, i.e., $I_g = 0$ for $r = r_1$. Equation (7.7-3) gives the ratio of galvanometer current and the current through the unknown resistance in terms of want of balance $r - r_1$. Hence, it is useful in discussing the sensitivity of the bridge. Note that I_g/I increases as the ratio $R_1/R_2 = n$ increases and the ratio $m = R_1/R_3$ decreases. The limiting case $n \to \infty$ and $m \to 0$ is impractical and not of great advantage because with $m = n = 1$ we can get the sensitiveness as great as half the limiting value:

$$\left(\frac{I_g}{I}\right)_{\substack{m=1\\n=1}} = \frac{1}{2}\left(\frac{I_g}{I}\right)_{\substack{m=0\\n=\infty}}.$$

It is a common procedure to keep n large and m small as far as practicable.

Again we note from Eq. (7.7-3) that for a given want of balance $(r - r_1)$, the galvanometer current I_g increases linearly with the current I through the unknown resistance. However, we cannot make I large because of its heating effect and consequent change in resistance due to it. This effect is particularly important when the material of the resistance to be measured differs from that of the resistances forming other arms of the bridge. According to Prof. Calendar it is advisable to connect the battery in such a way that the resistance in series with the resistance to be measured is greater than the resistance in parallel with it. This makes the current through the unknown resistance small and at the same time makes $n > m$.

For accurate detection of the balanced condition the galvanometer should be highly sensitive. The deflection of the galvanometer involves power. So for best sensitivity the galvanometer resistance R_g should be so chosen that the power delivered to the galvanometer becomes maximum. From maximum power transfer theorem (Refer Section 7.9) R_g should be equal to the Thevenin resistance (Refer Section 7.8) measured between the terminals B and D. Assuming the battery resistance $r_i = 0$ we can write for maximum sensitivity,

$$\begin{aligned} R_g &= R_1 \parallel^l R_2 + R_3 \parallel^l R_4 \\ &= \frac{R_1R_2}{R_1 + R_2} + \frac{R_3R_4}{R_3 + R_4} \\ &= \frac{m+1}{1+\frac{1}{n}} r_1. \end{aligned}$$

If $n = m = 1$ then $R_g = r_1$, if $n = 1$, $m = 0$ then $R_g = r_1/2$. Unless n is made less than unity or m greater than unity the value of R_g lies between $r_1/2$ and $2r_1$. So the galvanometer should have the same order of resistance as r.

In practical Wheatstone bridge it is a common practice to try to make $R_1 = R_2 = R_3 = R_4 = R_g$.

7.8 Network Theorems

To simplify the analysis of complex networks several network theorems have been developed. These theorems are applicable to both dc and ac. These theorems are no new physical laws but can be derived from Kirchhoff's laws. Here we consider a few of them.

1. Superposition theorem

It states that *in any linear bilateral network containing several energy sources the current or voltage in any element is the algebraic sum of the effects produced by each source acting separately.* While considering the effect of a particular source other sources must be replaced by their internal resistances.

As an example, let us consider the circuit of Fig. 7.8-1(a). Suppose we are to find the current through the resistance R_3. It can be done by considering the effects of the two sources simultaneously or separately. Applying KVL to the circuit of Fig. 7.8-1(a)

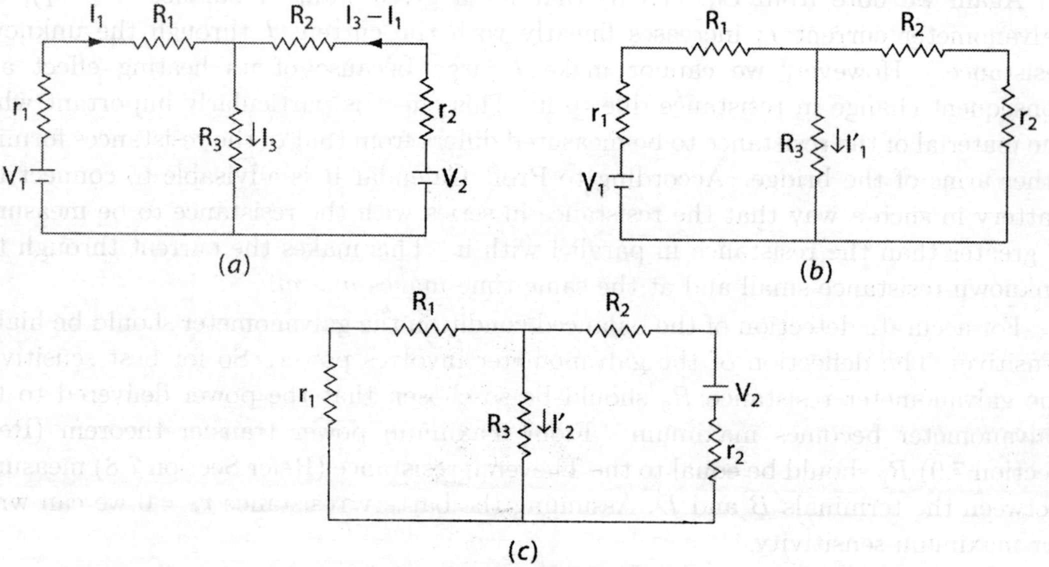

Fig. 7.8-1

we get the following loop equations:

$$I_1(r_1 + R_1) + I_3R_3 = V_1 \tag{7.8-1}$$

$$(I_3 - I_1)(r_2 + R_2) + I_3R_3 = V_2$$

$$\text{or} \quad -I_1(r_2 + R_2) + I_3(r_2 + R_2 + R_3) = V_2 \tag{7.8-2}$$

Solving Eqs. (7.8-1) and (7.8-2) we get by Cramer's rule

$$I_3 = \frac{1}{\Delta}\begin{vmatrix} r_1 + R_1 & V_1 \\ -(r_2 + R_2) & V_2 \end{vmatrix}, \tag{7.8-3}$$

where

$$\Delta = \begin{vmatrix} r_1 + R_1 & R_3 \\ -(r_2 + R_2) & r_2 + R_2 + R_3 \end{vmatrix} \tag{7.8-4}$$

In Fig. 7.8-1(b) the source V_2 has been replaced by its internal resistance r_2. So putting $V_2 = 0$ in Eq. (7.8-3) we can get the current I_1' as

$$I_1' = \frac{1}{\Delta}\begin{vmatrix} r_1 + R_1 & V_1 \\ -(r_2 + R_2) & 0 \end{vmatrix}. \tag{7.8-5}$$

In Fig. 7.8-1(c) the source V_1 has been replaced by its internal resistance r_1. As before we can get

$$I_2' = \frac{1}{\Delta}\begin{vmatrix} r_1 + R_1 & 0 \\ -(r_2 + R_2) & V_2 \end{vmatrix}. \tag{7.8-6}$$

It is now easy to show that $I_3 = I_1' + I_2'$. This is as expected from the superposition theorem.

2. Thevenin's theorem

This theorem is very useful in the analysis of complex network. According to this theorem *any two-terminal linear network containing energy sources and resistances can be replaced by a single equivalent circuit containing a voltage source V_T together with a series resistance R_T, where V_T is the open circuit voltage between the terminals of the network and R_T is the resistance measured between the terminals with all the energy sources replaced by their internal resistances.* The term *voltage source* means a

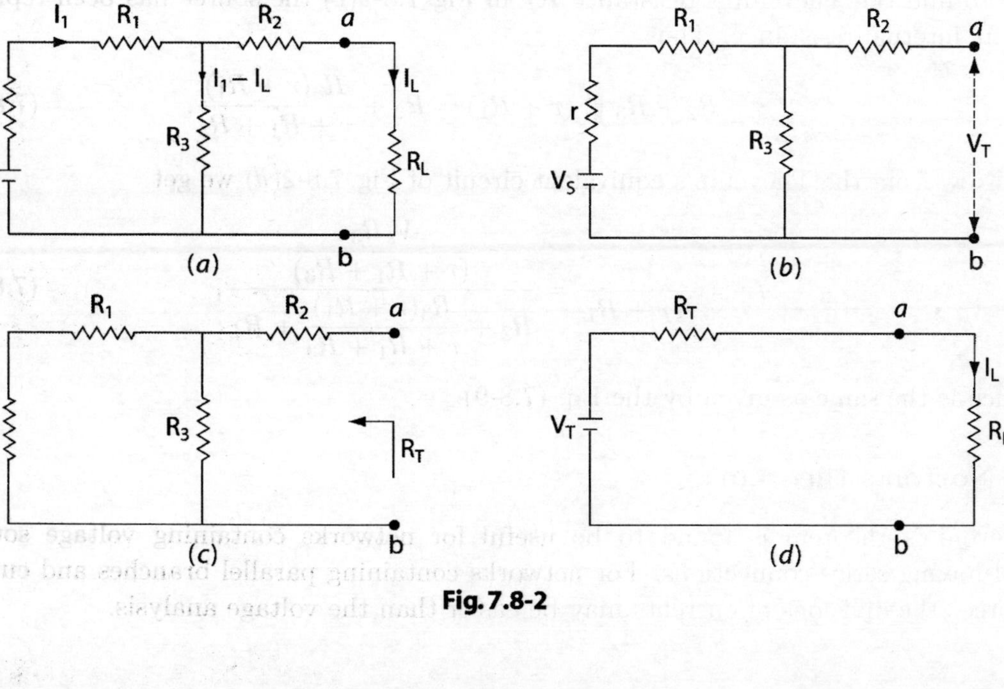

Fig. 7.8-2

source of supply whose output voltage is constant independent of the load resistance connected. Such a source must have zero internal resistance.

As an example, let us consider the circuit of Fig. 7.8-2. Suppose we are to find the current through the external resistance R_L, called *load resistance*, driven by the circuit to the left of the terminal points a, b. It can be done directly by writing the loop equations:

$$I_1(r+R_1)+(I_1-I_L)R_3=V_s$$

$$\text{and}\quad I_L(R_2+R_L)-(I_1-I_L)R_3=0$$

$$\text{or}\quad I_1(r+R_1+R_3)-I_LR_3=V_s \tag{7.8-7}$$

$$\text{and}\quad -I_1R_3+I_L(R_2+R_3+R_L)=0 \tag{7.8-8}$$

Solving Eqs. (7.8-7) and (7.8-8) we get

$$I_L=\frac{\dfrac{V_sR_3}{(r+R_1+R_3)}}{R_2+\dfrac{R_3(r+R_1)}{r+R_1+R_3}+R_L}. \tag{7.8-9}$$

Now let us solve the problem by applying Thevenin's theorem. In Fig. 7.8-2(b) the load has been disconnected (i.e., opened) to find the open circuit voltage between a and b. As there is no current through R_2 in this open circuit condition, the Thevenin's voltage is

$$V_T=\frac{V_s\cdot R_3}{r+R_1+R_3}. \tag{7.8-10}$$

To find the Thevenin's resistance R_T in Fig. 7.8-2(c) the source has been replaced by its internal resistance. Thus

$$R_T=R_2+R_3\,\|^l\,(r+R_1)=R_2+\frac{R_3(r+R_1)}{r+R_1+R_3}. \tag{7.8-11}$$

Now from the Thevenin's equivalent circuit of Fig. 7.8-2(d) we get

$$I_L=\frac{V_T}{R_T+R_L}=\frac{\dfrac{V_sR_3}{(r+R_1+R_3)}}{R_2+\dfrac{R_3(r+R_1)}{r+R_1+R_3}+R_L}, \tag{7.8-12}$$

which is the same as given by the Eq. (7.8-9).

3. Norton's theorem

Thevenin's theorem is found to be useful for networks containing voltage sources and having series connections. For networks containing parallel branches and current sources the divisions of currents may be easier than the voltage analysis.

In this case another theorem, called *Norton's Theorem,* is found to be more useful. According to this theorem *any two-terminal linear network containing energy sources and resistances can be replaced by a simple equivalent circuit containing a current source I_N in parallel with a resistance R_N, where I_N is the short-circuit current between the terminals of the network and R_N is the resistance measured between the terminals with all the energy sources replaced by their internal resistances.* The term *current source* means a source which supplies a constant current independent of the voltage across its terminals. An ideal current source must have an infinite internal resistance. If two terminals are connected by a wire of almost zero resistance they are said to be short circuited.

As an example, let us again consider the circuit of Fig. 7.8-2(a).

We have seen by direct application of KVL that the current I_L through R_L is given by the Eq. (7.8-9). Now let us solve the problem by applying Norton's theorem. In Fig. 7.8-3(a) the load R_L has been replaced by a short and hence Norton's current is obtained by using the usual rule of current division as,

$$I_N = I_1 \cdot \frac{R_3}{R_3 + R_2} = \frac{V_s}{r + R_1 + (R_3 \parallel^l R_2)} \cdot \frac{R_3}{R_3 + R_2}$$

$$= \frac{V_s R_3}{(r + R_1)(R_2 + R_3) + R_2 R_3}$$

$$= \frac{V_s R_3}{(r + R_1 + R_3)R_2 + R_3(r + R_1)}.$$

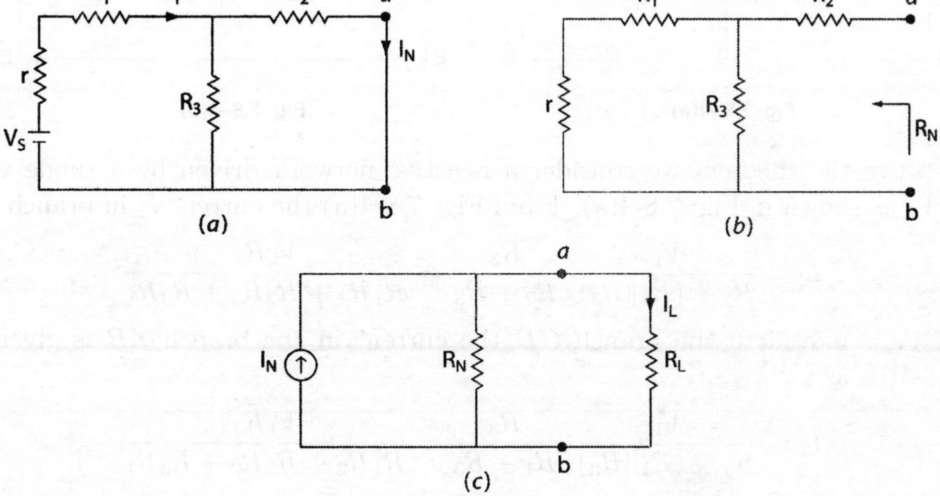

Fig. 7.8-3

To find Norton's resistance R_N we consider Fig. 7.8-3(b) where the source has been replaced by its internal resistance. Thus

$$R_N = R_2 + R_3 \parallel^l (r + R_1) = R_2 + \frac{R_3(r + R_1)}{r + R_1 + R_3}$$

Fig. 7.8-3(c) shows the Norton's equivalent circuit of the original circuit of Fig. 7.8-2(a). Therefore, load current

$$I_L = I_N \frac{R_N}{R_N + R_L} = I_N \frac{1}{1 + \frac{R_L}{R_N}}$$

$$= \frac{V_s R_3}{(r + R_1 + R_3)R_2 + R_3(r + R_1)} \cdot \frac{1}{1 + \frac{R_L(r+R_1+R_3)}{(r+R_1+R_3)R_2+R_3(r+R_1)}}$$

$$= \frac{V_s R_3}{(r + R_1 + R_3)R_2 + R_3(r + R_1) + R_L(r + R_1 + R_3)},$$

which is the same as given by the Eq. (7.8-9).

4. Reciprocity theorem

The reciprocity theorem in its simplest form can be stated as follows: *In any linear bilateral network if a voltage source V in any branch produces a current I in another branch then if the voltage source V is moved from the first branch to the second it will cause the same current in the first branch in which the source was originally located.*

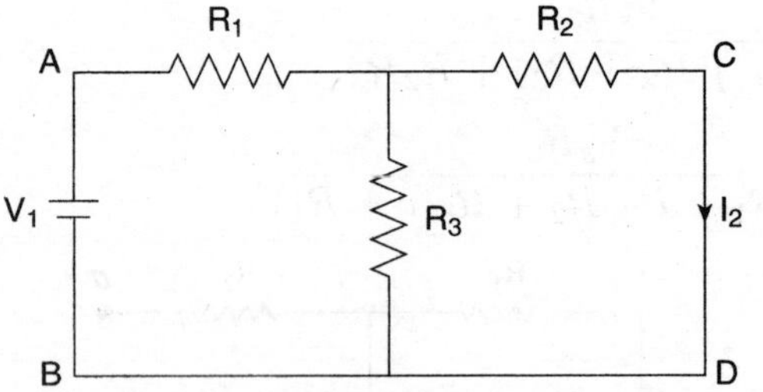

Fig. 7.8–4(a)

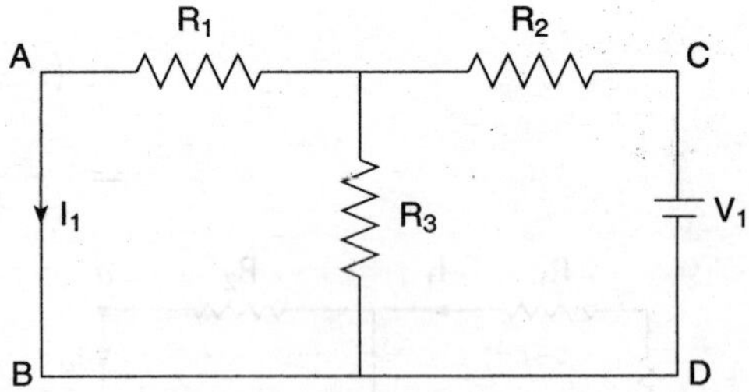

Fig. 7.8–4(b)

To prove the theorem we consider a resistive network driven by a single voltage source V_1 as shown in Fig. 7.8-4(a). From Fig. 7.8-4(a) the current I_2 in branch CD is

$$I_2 = \frac{V_1}{R_1 + (R_2 \| R_3)} \cdot \frac{R_3}{R_2 + R_3} = \frac{V_1 R_3}{R_1 R_2 + R_2 R_3 + R_3 R_1}$$

Now if V_1 is moved to the branch CD the current in the branch AB is given from Fig. 7.8-4(b) as

$$I_1 = \frac{V_1}{R_2 + (R_1 \| R_3)} \cdot \frac{R_3}{R_1 + R_3} = \frac{V_1 R_3}{R_1 R_2 + R_2 R_3 + R_3 R_1}$$

From the above equations it is obvious that $I_1 = I_2$. This proves the theorem.

Networks for which the reciprocity theorem holds are said to be reciprocal networks. Reciprocity phenomenon is also found in other areas of science including mechanics and acoustics. Most generally in a physical system where reciprocity theorem is applicable we can change the excitation and the response points without changing the response of the system to a give excitation.

7.9 Maximum Power Transfer Theorem

Maximum power is delivered from a resistive network to a load R_L when R_L is equal to the internal resistance measured looking back into the terminals of the network. This is known as the maximum power transfer theorem for a resistive network.

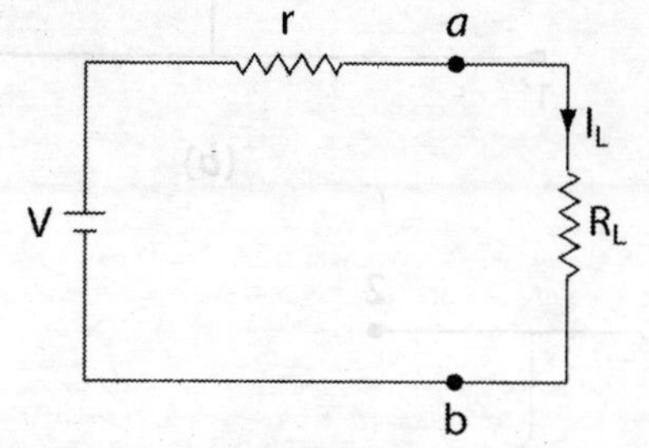

Fig. 7.9-1

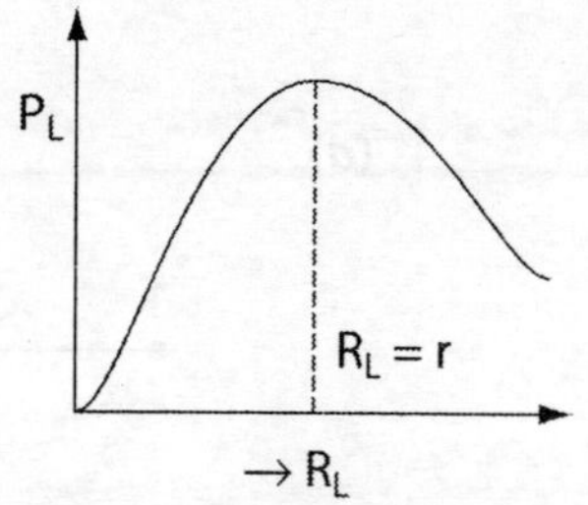

Fig. 7.9-2

To prove the theorem let us consider the circuit of Fig. 7.9-1, which is assumed to be the Thevenin's equivalent of an arbitrary two-terminal linear network. The power delivered to the load is

$$P_L = I_L^2 R_L = \left(\frac{V}{r + R_L}\right)^2 R_L.$$

If R_L is varied, P_L varies and for certain value of R_L, P_L becomes maximum (Fig. 7.9-2). For maximum power,

$$\frac{dP_L}{dR_L} = 0,$$

which requires that $R_L = r$.

The ratio of the power P_L delivered to the load and the total power, $P_T = I_L^2(R_L + r)$, supplied by the source is called the *power transfer efficiency*. Under maximum power transfer condition this efficiency is only 50%.

This theorem is of great importance in many electronic circuits. For example the load resistance connected to an amplifier output must match the output resistance of the amplifier, otherwise maximum power will not be delivered from the amplifier to the load.

7.10 T- and π-Networks

Two four-terminal or two-port networks are said to be equivalent to each other if they have identical terminal properties, i.e., for the same input they yield identical outputs. Sometimes the analysis of a complex circuit can be simplified by considering a simpler equivalent circuit. Here we show that any two-port linear resistive network can be replaced by an equivalent T-network or a π-network.

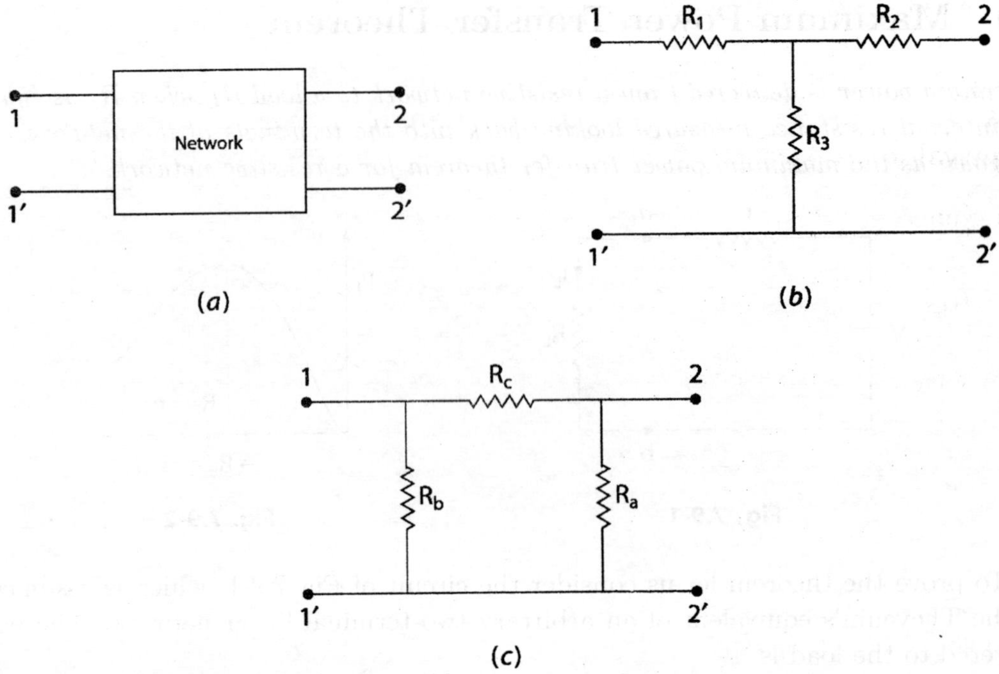

Fig. 7.10.1 : (a) Arbitrary network, (b) T-equivalent, (c) p-equivalent

Fig. 7.10-1 shows an arbitrary complicated network and its T- and π-equivalent circuits. Now for the circuits to be equivalent the values of terminal voltages and currents must be identical. So the open and short-circuit resistances measured at the terminals must also be identical.

T-equivalent network

Resistance at $1, 1'$ with $2, 2'$ short circuited is

$$R_{1S} = R_1 + \frac{R_2 R_3}{R_2 + R_3}. \tag{7.10-1}$$

Resistance at $1, 1'$ with $2, 2'$ open circuited is

$$R_{1O} = R_1 + R_3. \tag{7.10-2}$$

Resistance at $2, 2'$ with $1, 1'$ open circuited is

$$R_{2O} = R_2 + R_3. \tag{7.10-3}$$

Solving these equations we can find the required values of R_1, R_2 and R_3 in terms of the measurements R_{1S}, R_{1O}, R_{2O} of the original network. Subtracting Eq. (7.10-1) from Eq. (7.10-2) and then using Eq. (7.10-3) we get

$$R_{1O} - R_{1S} = \frac{R_3^2}{R_{2O}} \quad \text{or} \quad R_3 = \sqrt{R_{2O}(R_{1O} - R_{1S}}. \tag{7.10-4}$$

Now using Eqs. (7.10-2) and (7.10-3) we can find R_1 and R_2 as

$$R_1 = R_{1O} - R_3 \quad \text{and} \quad R_2 = R_{2O} - R_3. \tag{7.10-5}$$

π-equivalent network

For π-equivalent network of Fig. 7.10-1(c) we have similarly,

$$R_{1S} = \frac{R_b R_c}{R_b + R_c}, \quad R_{1O} = \frac{R_b(R_c + R_a)}{R_a + R_b + R_c}, \quad R_{2O} = \frac{R_a(R_b + R_c)}{R_a + R_b + R_c} \tag{7.10-6}$$

To solve these equations we note that

$$R_{1S} R_{2O} = \frac{R_a R_b R_c}{R_a + R_b + R_c} \tag{7.10-7}$$

$$\text{and } \sqrt{R_{2O}(R_{1O} - R_{1S})} = \frac{R_a R_b}{R_a + R_b + R_c}. \tag{7.10-8}$$

Thus

$$R_c = \frac{R_{1S} R_{2O}}{\sqrt{R_{2O}(R_{1O} - R_{1S})}},$$

$$R_b = \frac{R_{1S} R_{2O}}{R_{2O} - \sqrt{R_{2O}(R_{1O} - R_{1S})}},$$

$$R_a = \frac{R_{1S} R_{2O}}{R_{1O} - \sqrt{R_{2O}(R_{1O} - R_{1S})}}. \tag{7.10-9}$$

T-π transformations

Sometimes it becomes helpful to convert T-network to π-network and vice versa. For the T- and π-networks of Fig. 7.10-1 to be equivalent the short and open circuited measurements must be identical. Equating the two values each of R_{1S}, R_{1O} and R_{2O} as given by the Eqs. (7.10-1)–(7.10-3) and (7.10-6) and then solving we may get the transformation formulas:

$$R_1 = \frac{R_b R_c}{R_a + R_b + R_c}, \quad R_2 = \frac{R_c R_a}{R_a + R_b + R_c}, \quad R_3 = \frac{R_a R_b}{R_a + R_b + R_c} \tag{7.10-10}$$

and

$$R_a = (R_1 R_2 + R_2 R_3 + R_3 R_1)/R_1,$$
$$R_b = (R_1 R_2 + R_2 R_3 + R_3 R_1)/R_2,$$
$$R_c = (R_1 R_2 + R_2 R_3 + R_3 R_1)/R_3. \tag{7.10-11}$$

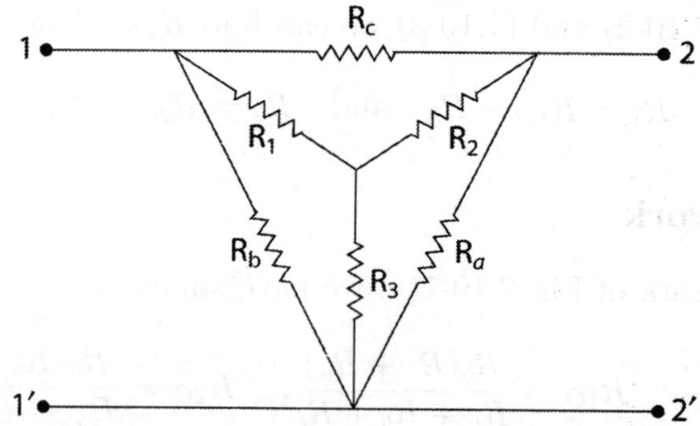

Fig. 7.10.2: Helps to memorize T-π transformation

These formulas can be easily memorized by considering Fig. 7.10-2, where we draw the T-network in the form of Y and π-network in the form of Δ (delta). Note that

$$R_T = \frac{\text{Product of two adjacent } R \text{ in } \pi}{\text{Sum of all three } R \text{ in } \pi}$$

$$R_\pi = \frac{\text{Sum of all double products of } R \text{ in } T}{R \text{ in } T \text{ arm opposite to the desired } \pi\text{-arm}}.$$

SOLVED PROBLEMS

► Problem 1.

(a) *A thin disk of radius a carrying uniform surface charge density σ is rotating with a constant angular velocity ω about its axis (z-axis). Find the surface current density $\vec{K}$ at a distance r from the centre.*

(b) *A uniformly charged sphere of radius a and total charge Q is centred at the origin and spinning at a constant angular velocity ω about z-axis. Find the current density at any point (r, θ, φ) inside the sphere.*

Solution. (a) Consider a ring of radius r and radial thickness dr. The charge on the ring is $\sigma \cdot 2\pi r dr$. It constitutes a current loop of current

$$dI = \frac{\sigma \cdot 2\pi r dr}{T} = \sigma \omega r dr.$$

As the length perpendicular to the flow is dr, the current density is $K = \frac{dI}{dr} = \sigma\omega r$ in circumferential direction, i.e., $\vec{K} = \sigma\omega r\hat{\theta}$.

(b) Linear velocity $\vec{v}$ at any point (r, θ, φ) with position vector

$$\vec{r} = x\hat{r} + y\hat{j} + z\hat{k} = r\sin\theta\cos\varphi\hat{i} + r\sin\theta\sin\varphi\hat{j} + r\cos\theta\hat{k}$$

is given by

$$\vec{v} = \vec{\omega} \times \vec{r} = \begin{vmatrix} \hat{i} & \hat{j} & \hat{k} \\ 0 & 0 & \omega \\ r\sin\theta\cos\varphi & r\sin\theta\sin\varphi & r\cos\theta \end{vmatrix}$$

$$= \omega r \sin\theta \left(-\hat{i}\sin\varphi + \hat{j}\cos\varphi\right)$$

$$= \omega r \sin\theta \hat{\varphi} \quad \left[\because \hat{\varphi} = -\hat{i}\sin\varphi + \hat{j}\cos\varphi\right]$$

$\therefore$ Current density $\vec{J} = \rho\vec{v} = \dfrac{Q}{\frac{4}{3}\pi a^3} \cdot \omega r \sin\theta \hat{\varphi}$.

▶ **Problem 2.**

(a) *An electric field of 0.01* $V.m^{-1}$ *exists between two points in a conducting wire of cross-sectional area 1* mm^2*. The resistivity* (ρ) *of the material of the wire is* $1.70 \times 10^{-8}\ \Omega.m$*. Find the current density and current through the wire.*

(b) *In a cloud of electrons having* 5×10^{28} *electrons per* m^3*, 80% of the electrons drift along x-axis with a velocity 0.02* $m.s^{-1}$ *and 20% of the electrons drift along y-axis with the same speed. Find the current density.*

Solution. (a) Current density $\left|\vec{J}\right| = \sigma E = \dfrac{E}{\rho} = \dfrac{0.01}{1.70 \times 10^{-8}} = 0.59 \times 10^6\ \text{A/m}^2$

$\therefore$ Current $I = JA = 0.59 \times 10^6 \times 10^{-6}\ \text{A} = 0.59\ \text{A}$

(b) Current density

$$\vec{J} = -n_x \cdot e \cdot \vec{v}_{dx} - n_y \cdot e \cdot \vec{v}_{dy}$$

$$= -1.6 \times 10^{-19} \left[\frac{80}{100} \times 5 \times 10^{28} \times 0.02\hat{i} + \frac{20}{100} \times 5 \times 10^{28} \times 0.02\hat{j}\right]$$

$$= -3.2 \times 10^7 \left(4\hat{i} + \hat{j}\right) A.m^{-2}.$$

▶ **Problem 3.** *Show that any initial charge distribution within a medium characterised by the conductivity* σ *and permittivity* ε *decays exponentially with time at every point in accordance with the law* $\rho(t) = \rho(o)e^{-\sigma t/\varepsilon}$*, where* $\rho(t)$ *is the charge density at any time* t*, and tries to attain equilibrium.*

Solution. From the continuity equation,

$$\vec{\nabla} \cdot \vec{J} + \frac{\partial \rho}{\partial t} = 0.$$

Now $\vec{J} = \sigma\vec{E}$ and $\vec{\nabla} \cdot \vec{E} = \rho/\varepsilon$. Therefore,

$$\frac{\sigma\rho}{\varepsilon} + \frac{\partial \rho}{\partial t} = 0.$$

Integrating this equation we get $\rho(t) = \rho(o)e^{-\frac{\sigma}{\varepsilon}t}$.

The quantity ε/σ represents the time required for the charge density to decay to $1/e$ of its initial value. It is called the *charge relaxation time.* For good conducting media this time is very small, which means that in conductors the equilibrium is attained very quickly.

▶ **Problem 4.** *The surface current density in the xy-plane is given by*

$$\vec{K}(x,y) = K_0\left(\frac{x^2}{a^2}\hat{i} + \frac{y^2}{b^2}\hat{j}\right).$$

Using corresponding equation of continuity calculate the time rate of change of charge contained in the first quadrant of a circle of unit radius centred on the origin.

Solution. We have the equation of continuity for surface charges as

$$\vec{\nabla}\cdot\vec{K} + \frac{\partial\sigma}{\partial t} = 0 \quad (\text{Here } \sigma = \text{ surface charge density})$$

$$\therefore \quad \frac{\partial\sigma}{\partial t} = -\vec{\nabla}\cdot\vec{K} = -2K_0\left(\frac{x}{a^2} + \frac{y}{b^2}\right) = -2K_0\left(\frac{r\cos\theta}{a^2} + \frac{r\sin\theta}{b^2}\right),$$

where r, θ are the polar coordinates. Therefore, the required time rate of change of charge would be

$$\frac{\partial Q}{\partial t} = \int\limits_{r=0}^{1}\int\limits_{\theta=0}^{\pi/2}\frac{\partial\sigma}{\partial t}\cdot r\,d\theta\,dr = -\frac{2K_0}{3}\left(\frac{1}{a^2} + \frac{1}{b^2}\right).$$

▶ **Problem 5.**

(a) *The space between two long coaxial conducting cylinders of radii a and b $(a < b)$ and length L is filled with a slightly conducting material of conductivity σ. A pd is maintained between the cylinders such that a cylindrically symmetric current flows between them. Show that the resistance against such a current flow is*

$$R = \frac{\ln(b/a)}{2\pi\sigma L}.$$

(b) *If the conductivity of the material is not uniform but varies with the distance r from the common axis by the relation $\sigma(r) = \frac{k}{r}$, where k is a constant, what would be the value of the resistance?*

Solution. (a) **Method 1:** Because of cylindrical symmetry $\vec{J}$ is radial and depends only on r, i.e.,

$$\vec{J}(\vec{r}) = J(r)\hat{r}.$$

Now total current out of a coaxial cylindrical surface of radius $r(a < r < b)$ is

$$\int \vec{J} \cdot d\vec{S} = J(r) \cdot 2\pi r L = I$$

$$\therefore \quad \vec{J}(r) = \frac{I}{2\pi L r}\hat{r}.$$

The electric field in the medium can be calculated by using $\vec{J} = \sigma \vec{E}$. Thus,

$$\vec{E} = \frac{I}{2\pi L \sigma r}\hat{r}.$$

The potential difference between the cylinders is

$$V = -\int_b^a \vec{E} \cdot d\vec{r} = \frac{I}{2\pi L \sigma} \ln b/a.$$

Therefore, the resistance between the cylinders is

$$R = \frac{V}{I} = \frac{\ln(b/a)}{2\pi L \sigma}.$$

Method 2 : Consider a cylindrical shell of radii r and $r + dr$. The area of the shell through which current flows is $A(r) = 2\pi r L$ and length of current flow is dr. Therefore, resistance of this shell is

$$dR = \frac{1}{\sigma} \cdot \frac{dr}{2\pi r L}.$$

Total resistance

$$R = \int dR = \int_a^b \frac{1}{\sigma} \cdot \frac{dr}{2\pi r L} = \frac{\ln(b/a)}{2\pi L \sigma}.$$

(b) In this case by the method 2 we get

$$R = \int_a^b \frac{1}{\sigma(r)} \cdot \frac{dr}{2\pi r L} = \frac{1}{2\pi L k}\int_a^b dr = \frac{b-a}{2\pi L k}.$$

► **Problem 6.** *Current density at any time instant t at a distance r from the centre of a spherical region is given by $\vec{J}(r) = Ar^2 e^{-kt}\hat{r}$, where A and k are constants. Using the equation of continuity find the charge density $\rho(r)$ at time t.* **[J.U. 2011]**

Solution. $\vec{\nabla} \cdot \vec{J} = Ae^{-kt}\vec{\nabla} \cdot (r^2\hat{r}) = Ae^{-kt}\vec{\nabla} \cdot (r\vec{r}) = Ae^{-kt} \cdot 4r$

Now

$$\vec{\nabla} \cdot \vec{J} + \frac{\partial \rho}{\partial t} = 0 \quad \text{or} \quad \frac{\partial \rho}{\partial t} = -\vec{\nabla} \cdot \vec{J} = -4Are^{-kt}.$$

Integrating we get

$$\rho = \frac{4Ar}{k}e^{-kt}.$$

▶ **Problem 7.** *Suppose a spherical conductor is placed in an infinite slightly conducting medium with conductivity σ and dielectric permittivity ε. Show that the product $RC = \varepsilon/\sigma$, where R is the resistance and C is the capacitance of the sphere-medium system.*

Solution. Suppose we place a charge q on the sphere. Let V be its potential as result of q. Because of spherically symmetry we expect a spherically symmetric electric field and current density distribution. Considering a closed surface S adjoining the sphere we can calculate R and C as follows:

$$R = \frac{V}{I} = \frac{V}{\oint_S \vec{J} \cdot d\vec{S}} = \frac{V}{\sigma \oint_S \vec{E} \cdot d\vec{S}} \quad [\because \ \vec{J} = \sigma \vec{E}]$$

$$C = \frac{q}{V} = \frac{1}{V} \oint_S \vec{D} \cdot d\vec{S} = \frac{\varepsilon}{V} \oint_S \vec{E} \cdot d\vec{S} \quad \text{[Using Gauss's law]}$$

Therefore, $RC = \varepsilon/\sigma$.

▶ **Problem 8.** *A conducting sphere of radius a is placed in an infinite slightly conducting medium of conductivity σ and permittivity ε. An electric charge q_0 is placed on the sphere at time $t = 0$. Show that the charge leaks away from the sphere according to the law $q(t) = q_0 e^{-\sigma t/\varepsilon}$.*

Solution. The leakage current at any time instant t is

$$-\frac{dq(t)}{dt} = I = \int_{\text{sphere}} \vec{J} \cdot d\vec{S} = J 4\pi a^2 = \sigma E \cdot 4\pi a^2,$$

where E is the normal component of electric field at the spherical surface. Now from Coulomb's theorem,

$$E = \frac{\text{Charge density}}{\varepsilon} = \frac{q(t)}{\varepsilon \cdot 4\pi a^2}.$$

Therefore,

$$-\frac{dq(t)}{dt} = \frac{\sigma}{\varepsilon} q(t) \quad \text{or} \quad \frac{dq(t)}{q(t)} = -\frac{\sigma}{\varepsilon} dt.$$

Integrating we get $q(t) = q_0 e^{-\frac{\sigma}{\varepsilon}t}$, where we use the condition that $q(o) = q_0$.

▶ **Problem 9.** *In a conducting wire of radius a current density varies with radial distance according to the law $\vec{J} = (r/a)\vec{J}_0$, where $\vec{J}_0$ is parallel to the axis. Calculate the total current through the wire.* **[cf. C.U. 2014]**

Solution. Consider a ring of radii r and $r + dr$ on a given cross-section. Area of this ring is $dS = 2\pi r dr$ and current density on it is $\vec{J} = (r/a)\vec{J}_0$. So current through this ring is $dI = \vec{J} \cdot d\vec{S} = \frac{r}{a} \cdot J_0 \cdot 2\pi r dr$.

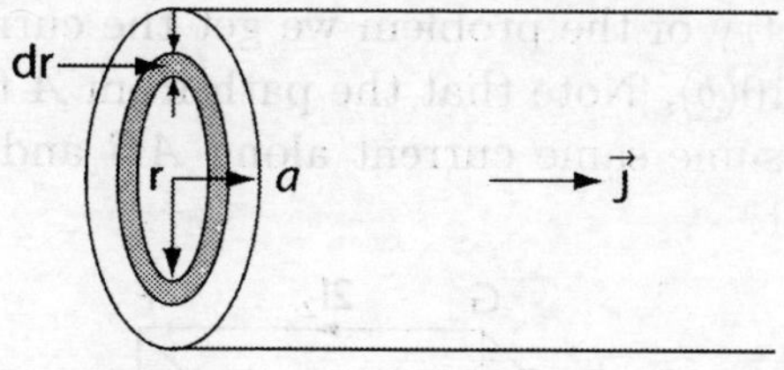

Fig 7.P-9

$$\therefore \text{ Total current } I = \frac{2\pi J_0}{a}\int_0^a r^2 dr = \frac{2\pi}{3}\cdot a^2 J_\circ.$$

▶ **Problem 10.** *Twelve identical wires, each of resistance R, are connected to form a cube. Calculate the equivalent resistance between* (a) *the diagonally opposite corners* (b) *the two corners of the same edge of the cube.*

Solution. (a) As the wires are identical the current is divided symmetrically in various branches. Using KCL and considering symmetry the current distribution in various

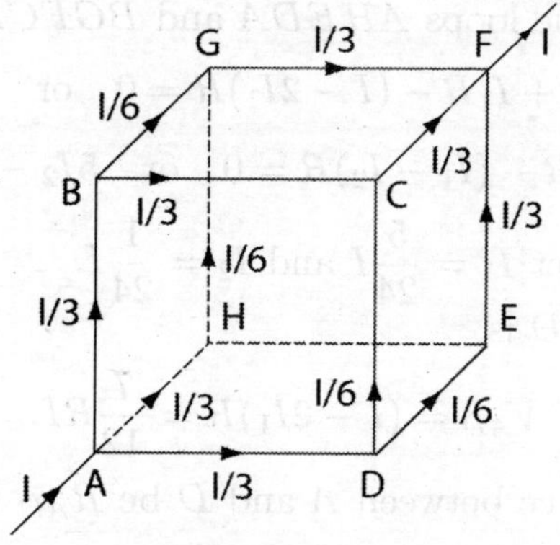

Fig 7.P-10(*a*)

branches are determined as shown in Fig. 7.P-10(a). Now the pd between the points A and F is

$$\begin{aligned} V_{AF} &= V_{AD} + V_{DE} + V_{EF} \\ &= \frac{I}{3}R + \frac{I}{6}R + \frac{I}{3}R \\ &= \frac{5}{6}RI. \end{aligned}$$

If the equivalent resistance between A and F be R_{AF} then

$$V_{AF} = IR_{AF}.$$

Thus,

$$R_{AF} = \frac{5}{6}R.$$

(b) Considering symmetry of the problem we get the current distribution by using KCL as shown in Fig. 7.P-10(*b*). Note that the path from A to D via B and via H are equivalent. For this we assume same current along AB and AH. Similarly, branches BC and HE are equivalent.

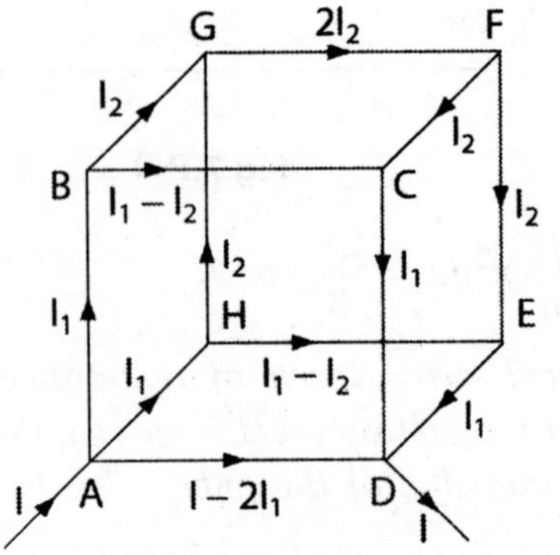

Fig 7.P-10(*b*)

Now applying KVL to the loops $AHEDA$ and $BGFCB$ we get

$$I_1R + (I_1 - I_2)R + I_1R - (I - 2I_1)R = 0 \quad \text{or} \quad 5I_1 - I_2 = I \tag{1}$$

$$\text{and} \quad I_2R + 2I_2R + I_2R - (I_1 - I_2)R = 0 \quad \text{or} \quad 5I_2 - I_1 = 0 \tag{2}$$

Solving (1) and (2) we get $I_1 = \frac{5}{24}I$ and $I_2 = \frac{1}{24}I$.

Now pd between A and D is

$$V_{AD} = (I - 2I_1)R = \frac{7}{12}RI.$$

If the equivalent resistance between A and D be R_{AD} then

$$V_{AD} = R_{AD} \cdot I.$$

Thus,

$$R_{AD} = \frac{7}{12}R.$$

► **Problem 11.** *Consider the circuit shown in* Fig. 7.P-11. *Use the principle of superposition to find the current through the* $5\,\Omega$ *resistor.*

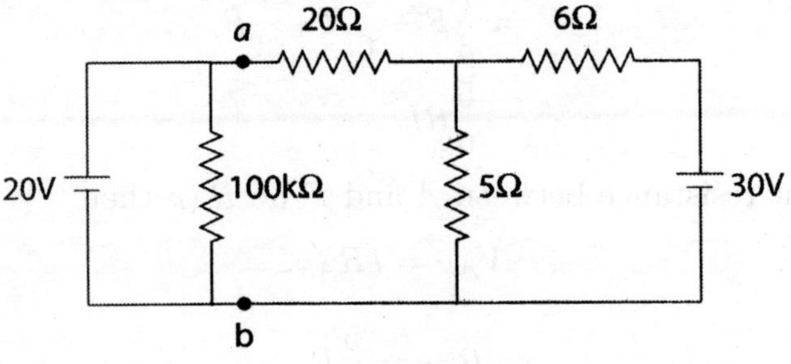

Fig. 7.P-11

Solution. At first short the voltage source 30 V and find the current through 5 Ω resistor due to 20 V source only. Note that the circuit to the left of the points *a*, *b* is simply equivalent to a voltage source 20 V. Thus we get the circuit of Fig. 7.P-11(*a*). Therefore,

$$I_1 = \frac{20}{20 + (5 \parallel^l 6)} \times \frac{6}{5+6} A = \frac{20}{20 + \frac{5\times 6}{5+6}} \times \frac{6}{5+6} A = \frac{12}{25} A.$$

Now to find the effect of 30 V, short circuit 20 V and get the circuit of Fig. 7.P-11(*b*)

$$\therefore \quad I_2 = \frac{30}{6 + (5 \parallel^l 20)} \times \frac{20}{20+5} A = \frac{30}{6 + \frac{5\times 20}{5+20}} \times \frac{20}{25} A = \frac{60}{25} A.$$

So, from superposition theorem total current through 5 Ω in Fig. 7.P-11 is

$$I_1 + I_2 = \frac{72}{25} A.$$

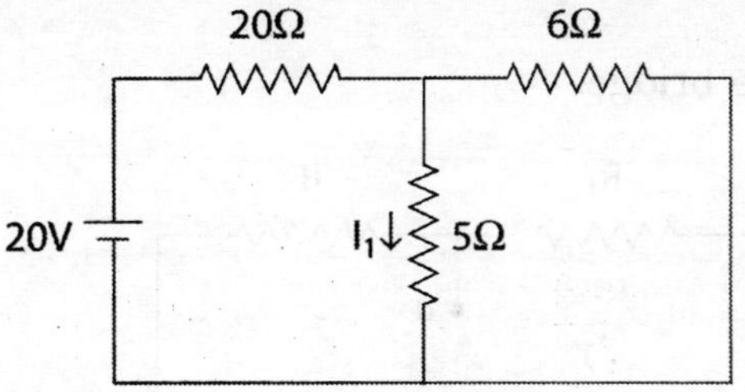

Fig. 7.P-11(*a*)

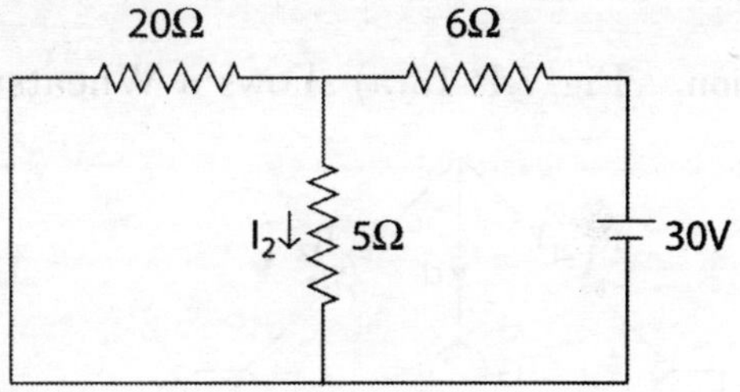

Fig. 7.P-11(*b*)

► **Problem 12.** *Consider the circuit of* Fig. 7.P-12. *Using the principle of superposition find the potential drop across the* 5 kΩ *resistor.* **[C.U. 2019]**

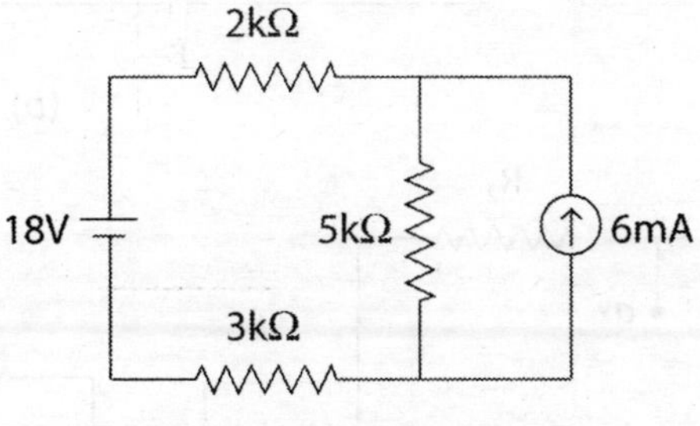

Fig. 7.P-12

Solution. At first replace the current source by its infinite internal resistance (i.e., we make it open) and get the circuit of Fig. 7.P-12(a). So pd across 5 kΩ due to 18 V only is

$$V_1 = \frac{18 \times 5}{2+3+5} V = 9V.$$

Fig. 7.P-12(*b*) shows the effect of 6 mA source only. So pd across 5 kΩ resistor due to 6 mA source only is $V_2 = 5\,\text{k}\Omega \times 3\,\text{mA} = 15$ V.

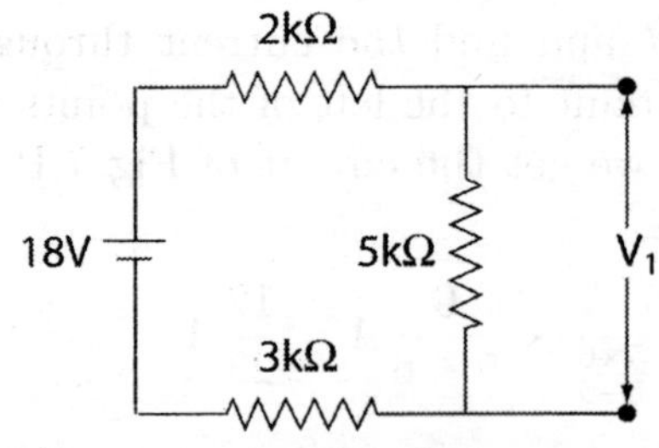

Fig. 7 P-12(*a*)

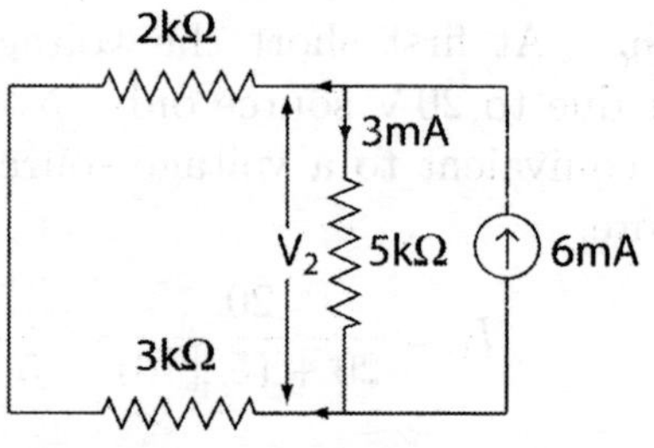

Fig. 7.P-12(*b*)

So, from superposition theorem total pd across 5 kΩ resistor would be

$$V = V_1 + V_2 = 9V + 15V = 24V.$$

► **Problem 13.** *Using Thevenin's theorem find the unbalanced current through the galvanometer of a Wheatstone bridge supplied with an ideal voltage source. Hence, find the balance condition of the bridge.*

Solution. Fig. 7.P-13(a) shows a Wheatstone bridge.

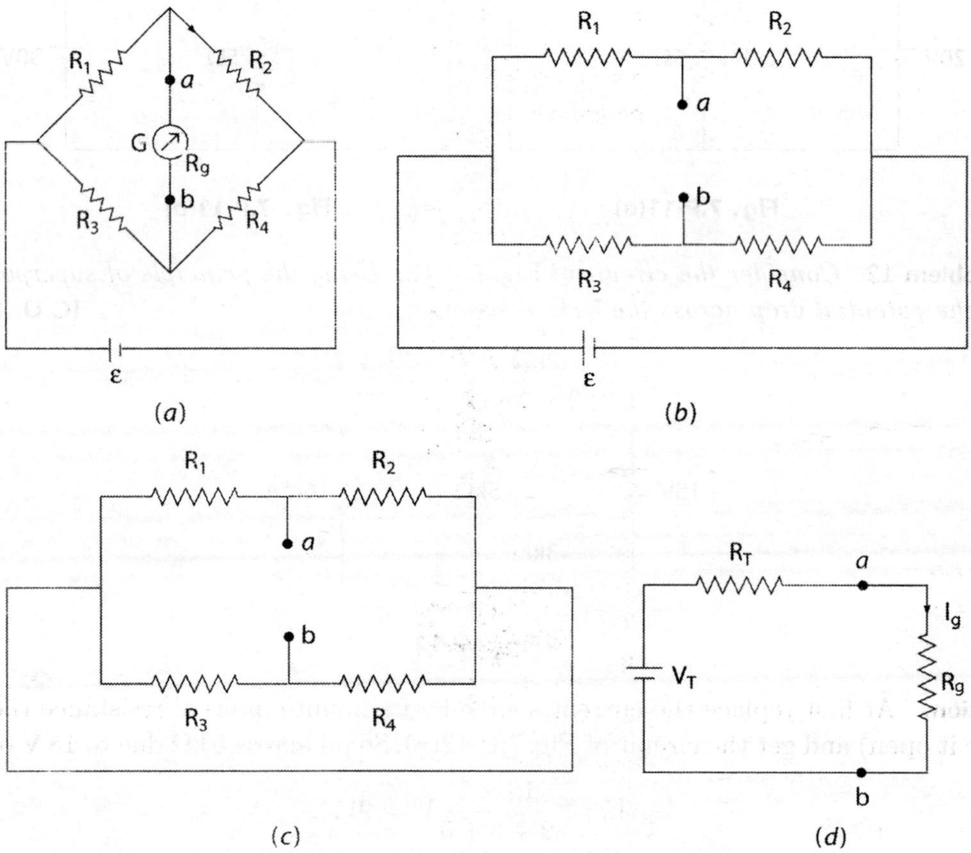

Fig. 7.P-13 : (a) Wheatstone bridge, (b) G is disconnected to find V_T, (c) Source is short circuited to find R_T, (d) Thevenin's equivalent

In Fig. 7.P-13(b) the galvanometer G is disconnected to find the Thevenin's voltage V_T. Thus

$$V_T = V_a - V_b = \frac{\mathcal{E}\cdot R_2}{R_1+R_2} - \frac{\mathcal{E}\cdot R_4}{R_3+R_4}.$$

To find Thevenin's resistance the source is short circuited in Fig. 7.P-13(c). Thus,

$$R_T = R_1 \parallel^l R_2 + R_3 \parallel^l R_4 = \frac{R_1R_2}{R_1+R_2} + \frac{R_3R_4}{R_3+R_4}.$$

From the Thevenin's equivalent circuit of Fig. 7.P-13(d) we find that the current through the galvanometer would be

$$I_g = \frac{V_T}{R_T+R_g} = \mathcal{E}\frac{\dfrac{R_2}{R_1+R_2} - \dfrac{R_4}{R_3+R_4}}{\dfrac{R_1R_2}{R_1+R_2} + \dfrac{R_3R_4}{R_3+R_4} + R_g}.$$

Note that I_g will be zero for

$$\frac{R_2}{R_1+R_2} = \frac{R_4}{R_3+R_4} \quad \text{or} \quad \frac{R_1}{R_2} = \frac{R_3}{R_4}.$$

► **Problem 14.** *Consider the circuit of* Fig. 7.P-11. (a) *Use Norton's theorem to find the current through the* 5 Ω *resistor.* (b) *Calculate the power dissipated in the* 5 Ω *resistor. Suppose this* 5 Ω *resistor is replaced by a variable resistor* R. *What should be the value of* R *such that maximum power is dissipated in it.*

Solution. (a) Applying Norton's theorem to the circuits lying on the left and right-hand sides of 5 Ω resistor we get the circuit of Fig. 7.P-14(a). Now combining the two current sources we get the circuit of Fig. 7.P-14(b).

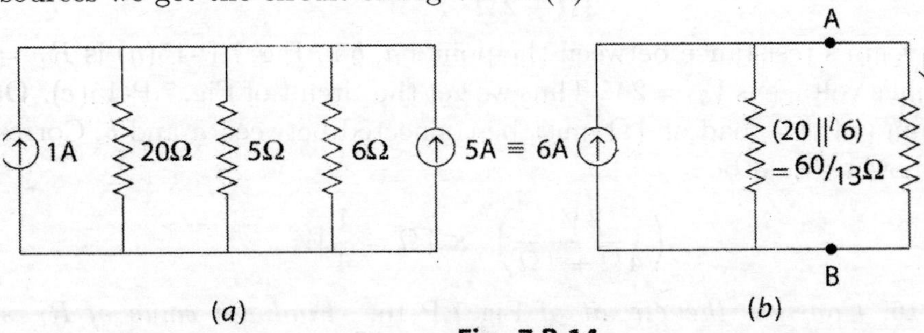

Fig. 7.P-14

Now current through 5 Ω resistor is

$$I = 6 \times \frac{\frac{60}{13}}{5+\frac{60}{13}} A = \frac{72}{25} A$$

(Same as obtained by using superposition principle in *Solved Problem* 11)

(b) The power dissipated in the 5 Ω resistor is

$$P = I^2 \cdot R = \left(\frac{72}{25}\right)^2 \cdot 5\,\text{W} = 41.5\,\text{W}.$$

From maximum power transfer theorem we can say from Fig. 7.P-14(b) that for maximum power the load resistor R, if connected in place of $5\,\Omega$ must be equal to $\frac{60}{13}\,\Omega$.

▶ **Problem 15.** *Consider the circuit of* Fig. 7.P-15. (a) *Find the pd* V_{ab} *between the points a and b* (b) *Use Thevenin's theorem to find the current that would flow from a to b if the points are joined by a short.* (c) *What resistance must be connected between a and b such that maximum power is dissipated in that resistor? Calculate this maximum power.*

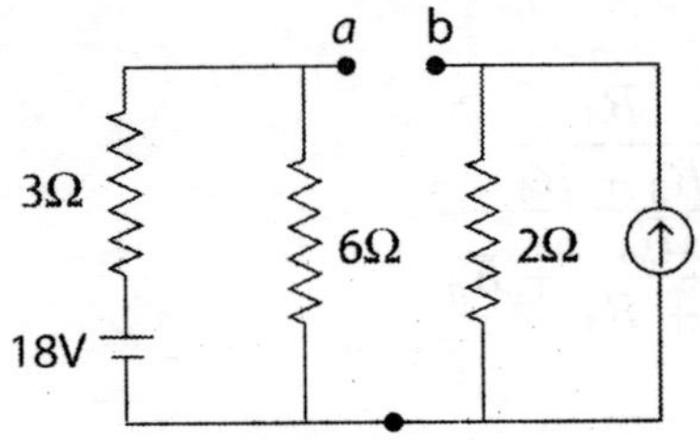

Fig. 7.P-15

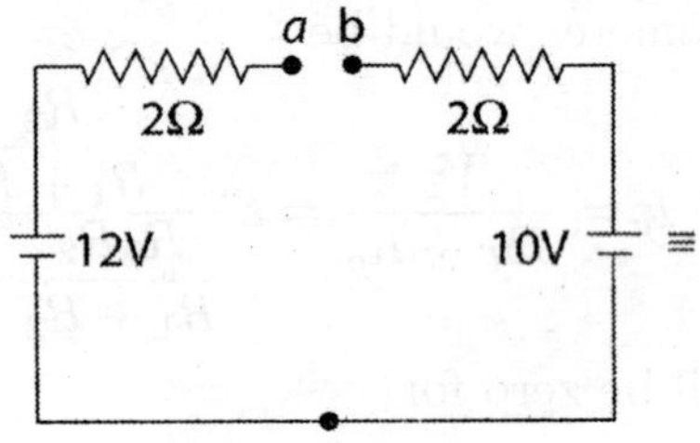

Fig. 7.P-15(a)

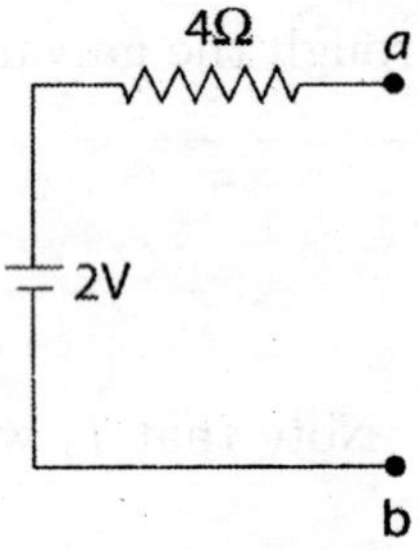

Fig. 7.P-15(b)

Solution. (a) $V_{ac} = \frac{18}{3+6} \times 6\,V = 12\,V$, $V_{bc} = 5 \times 2\,V = 10\,V$

$$\therefore \quad V_{ab} = V_{ac} - V_{bc} = 2V.$$

(b) Applying Thevenin's theorem to the circuits lying left of the point a, c and right of the points b, c in Fig. 7.P-15 we get the circuit of Fig. 7.P-15(a). So, if a, b are joined by a short the current would be

$$\frac{12\,V - 10\,V}{2\,\Omega + 2\,\Omega} = 0.5\,\text{A}.$$

(c) Thevenin's resistance between the points a, b in Fig. 7.P-15(a) is $R_T = 4\,\Omega$ and the Thevenin's voltage is $V_T = 2\,V$. Thus we get the circuit of Fig. 7.P-15(c). Obviously, for maximum power a load of $4\,\Omega$ must be connected between a and b. Corresponding maximum power would be

$$\left(\frac{2\,V}{4\,\Omega + 4\,\Omega}\right)^2 \times 4\,\Omega = \frac{1}{4}\,W.$$

▶ **Problem 16.** *Consider the circuit of* Fig. 7.P-16. *Find the value of* R_L *such that maximum power is delivered to it. Also, find the value of maximum power.*

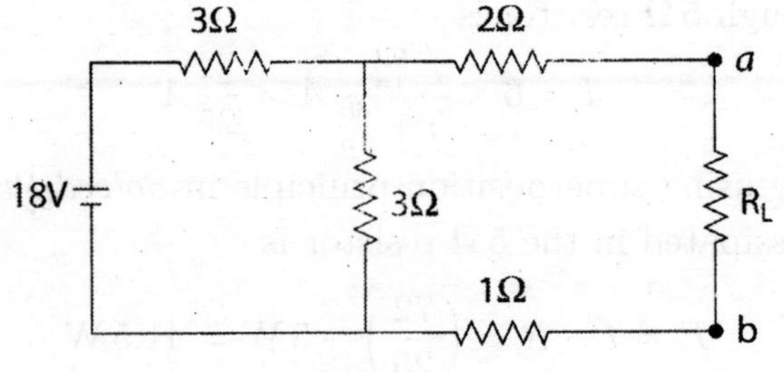

Fig. 7.P-16

Solution. Let us first simplify the circuit by using Thevenin's theorem to the circuit left of the points a, b. If R_L is disconnected no current passes through the $2\,\Omega$ and $1\,\Omega$ resistors and hence Thevenin voltage

$$V_T = \frac{18 \times 3}{3+3} V = 9V$$

Thevenin resistance $R_T = 2 + (3 \parallel^l 3) + 1\,\Omega = 4.5\,\Omega$. So the given circuit may be replaced by the Thevenin's equivalent circuit as shown in Fig. 7.P-16(a)

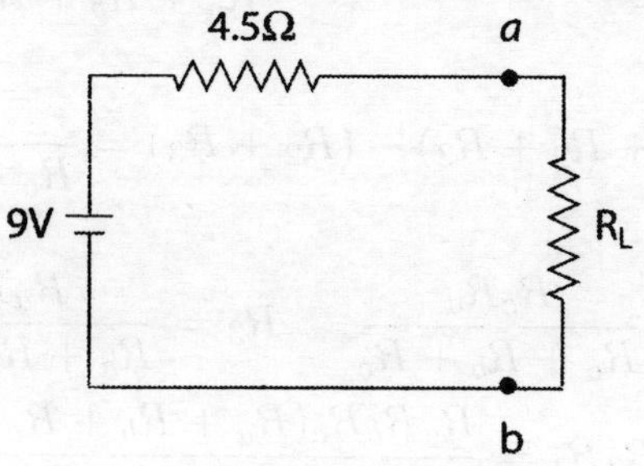

Fig. 7.P-16(a)

Obviously, from the maximum power transfer theorem R_L should be equal to $4.5\,\Omega$ for maximum power. Under this condition power supplied to the load is

$$\left(\frac{9}{4.5+4.5}\right)^2 \times 4.5\,W = 4.5\,W.$$

▶ **Problem 17.** *Consider the T- and π-networks of resistances of* Fig. 7.10-1(b) *and* Fig. 7.10-1(c). *Show that these networks will be equivalent in the sense that the resistances between corresponding pair of terminals will be identical provided*

$$R_1 = \frac{R_b R_c}{R_a + R_b + R_c}, \quad R_2 = \frac{R_c R_a}{R_a + R_b + R_c} \quad \textit{and} \quad R_3 = \frac{R_a R_b}{R_a + R_b + R_c} \quad \textit{or equivalently}$$

$$R_a = \frac{R_1 R_2 + R_2 R_3 + R_3 R_1}{R_1},$$

$$R_b = \frac{R_1 R_2 + R_2 R_3 + R_3 R_1}{R_2}$$

$$\textit{and} \quad R_c = \frac{R_1 R_2 + R_2 R_3 + R_3 R_1}{R_3}$$

Solution. Resistance between the terminals 1 and 1′ is

$$R_{11'} = R_1 + R_3 = \frac{R_b(R_a + R_c)}{R_a + R_b + R_c}.$$

Similarly,

$$R_{12} = R_1 + R_2 = \frac{R_c(R_a + R_b)}{R_a + R_b + R_c}$$

$$R_{22'} = R_2 + R_3 = \frac{R_a(R_b + R_c)}{R_a + R_b + R_c}$$

Adding we get

$$2(R_1 + R_2 + R_3) = \frac{2(R_aR_b + R_bR_c + R_cR_a)}{R_a + R_b + R_c}.$$

Now

$$R_1 = (R_1 + R_2 + R_3) - (R_2 + R_3) = \frac{R_bR_c}{R_a + R_b + R_c}.$$

Similarly,

$$R_2 = \frac{R_cR_a}{R_a + R_b + R_c}, \quad R_3 = \frac{R_aR_b}{R_a + R_b + R_c}$$

$$R_1R_2 + R_2R_3 + R_3R_1 = \frac{R_aR_bR_c(R_a + R_b + R_c)}{(R_a + R_b + R_c)^2} = \frac{R_aR_bR_c}{R_a + R_b + R_c}$$

Now

$$R_a = \frac{R_1R_2 + R_2R_3 + R_3R_1}{R_1},$$

$$R_b = \frac{R_1R_2 + R_2R_3 + R_3R_1}{R_2}$$

$$\text{and} \quad R_c = \frac{R_1R_2 + R_2R_3 + R_3R_1}{R_3}.$$

The given networks can also be redrawn as follows:

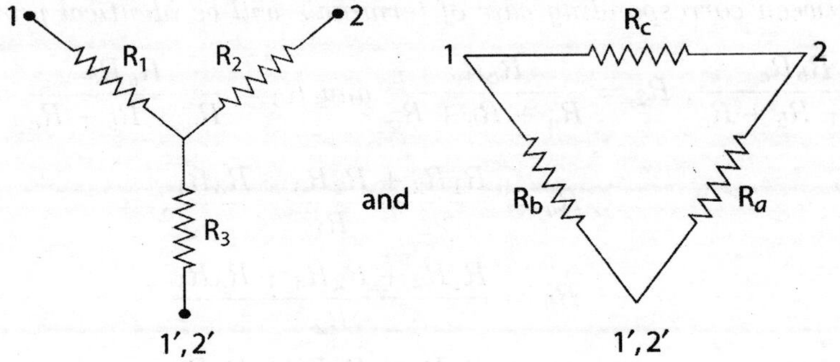

Fig. 7 P-17

For this the above connections are called *star* and *delta* connections. In the analysis of complicated networks sometimes it becomes convenient to replace a delta mesh (π-mesh) by a star mesh (T-mesh) and vice versa (See, *Solved Problem* 18). This is known as *star-delta* or *T-π transformation.*

▶ **Problem 18.** *Find the equivalent resistance between the points A and C of the network in* Fig. 7.P-18 *by using star-delta transformation.*

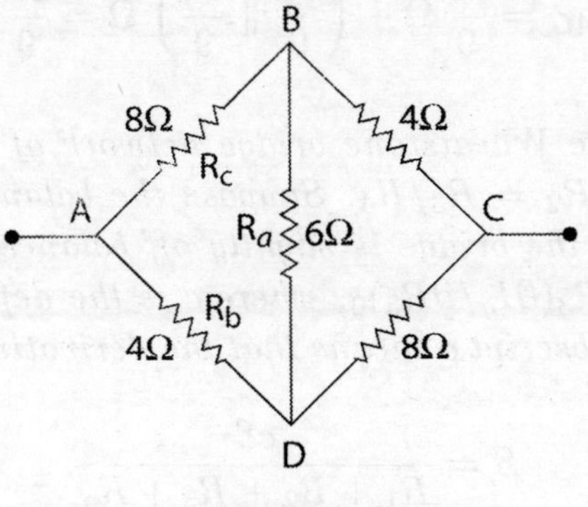

Fig. 7.P-18

Solution. The delta (π)-mesh ABD of the given network can be converted into a star (T) mesh ABD as shown in Fig. 7.P-18(a)

Here

$$R_1 = \frac{R_b R_c}{R_a + R_b + R_c} = \frac{8 \times 4}{8 + 4 + 6} = \frac{16}{9}\ \Omega,$$

$$R_2 = \frac{R_c R_a}{R_a + R_b + R_c} = \frac{48}{18}\ \Omega = \frac{24}{9}\ \Omega,$$

$$R_3 = \frac{R_a R_b}{R_a + R_b + R_c} = \frac{24}{18}\ \Omega = \frac{12}{9}\ \Omega.$$

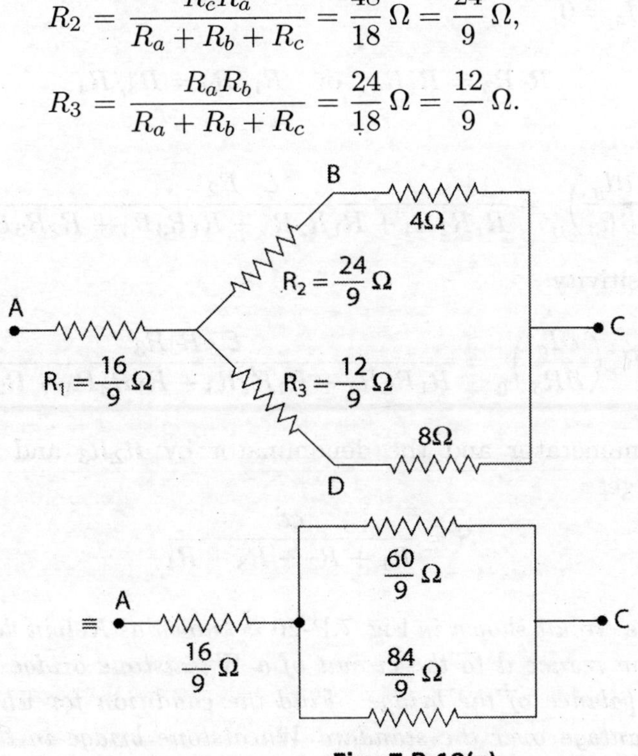

Fig. 7.P-18(*a*)

Therefore, the required equivalent resistance as seen from Fig. 7.P-18(a) would be

$$R_{AC} = \frac{16}{9}\,\Omega + \left(\frac{60}{9} \parallel \frac{84}{9}\right)\,\Omega = \frac{51}{9}\,\Omega.$$

► **Problem 19.** *Consider the Wheatstone bridge network of* Fig. 7.6-1 *with* $r_i = r_g = 0$. *Show that* $I_g = 0$ *for* $R_1/R_2 = R_3/R_4$. *Suppose the balance condition is obtained by varying* R_3. *Find* I_g *when the bridge is slightly off balanced. The sensitivity* S *of the bridge is defined by* $S = cR_3(\partial I_g/\partial R_3)_0$, *where* c *is the deflection of the galvanometer per unit current and the subscript o means that the derivative is taken at balance. Show that*

$$S = \frac{c\mathcal{E}}{R_1 + R_2 + R_3 + R_4}.$$

Solution. From Eq. (7.6-4) with $r_i = r_g = 0$, we get

$$\begin{aligned} I_g &= \frac{\mathcal{E}(R_2R_3 - R_1R_4)}{R_1R_2R_4 + R_3(R_1 + R_2)(R_2 + R_4) - R_3R_2^2} \\ &= \frac{\mathcal{E}(R_2R_3 - R_1R_4)}{R_1R_2R_4 + R_1R_2R_3 + R_1R_3R_4 + R_2R_3R_4}. \end{aligned}$$

Obviously for $I_g = 0$,

$$R_2R_3 = R_1R_4 \quad \text{or} \quad R_1/R_2 = R_3/R_4$$

Now

$$\left(\frac{\partial I_g}{\partial R_3}\right)_0 = \frac{\mathcal{E} \cdot R_2}{R_1R_2R_4 + R_1R_2R_3 + R_1R_3R_4 + R_2R_3R_4}.$$

Therefore, sensitivity

$$S = cR_3\left(\frac{\partial I_g}{\partial R_3}\right)_0 = \frac{\mathcal{E} \cdot R_2R_3}{R_1R_2R_4 + R_1R_2R_3 + R_1R_3R_4 + R_2R_3R_4}.$$

Dividing the numerator and the denominator by R_2R_3 and using the relation $R_2R_3 = R_1R_4$ we get

$$S = \frac{c\mathcal{E}}{R_1 + R_2 + R_3 + R_4}.$$

► **Problem 20.** *The circuit shown in* Fig. 7.P-20 *is known as Kelvin double bridge. Using T-π transformation reduce it to the circuit of a Wheatstone bridge and hence find out the condition for balance of the bridge. Find the condition for which the circuit will provide extra advantage over the standard Wheatstone bridge in the measurement of low resistance.*

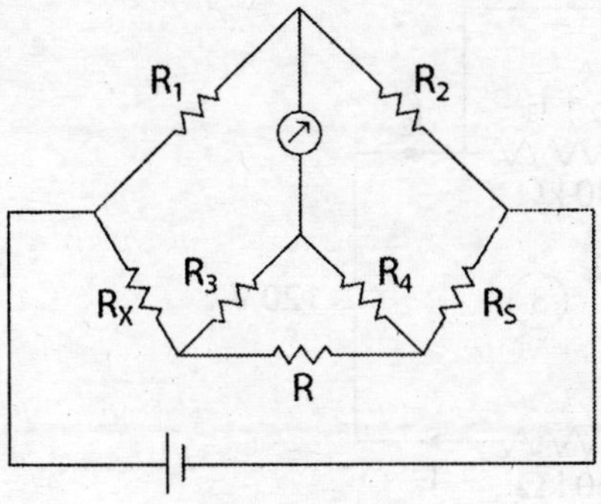

Fig. 7.P-20

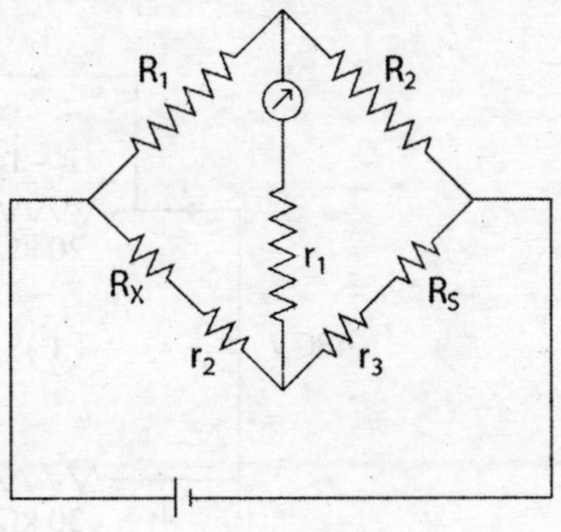

Fig. 7.P-20(a)

Fig. 7.P-18(a) is obtained by using π-T transformation in the circuit of Fig. 7.P-20. Thus

$$r_1 = \frac{R_3R_4}{R_3+R_4+R}, \quad r_2 = \frac{RR_3}{R_3+R_4+R} \quad \text{and} \quad r_3 = \frac{RR_4}{R_3+R_4+R}.$$

Now applying the condition of balance of a Wheatstone bridge we get

$$\frac{R_1}{R_2} = \frac{R_x+r_2}{R_s+r_3}.$$

$$\begin{aligned} \therefore \quad R_x &= \frac{R_1}{R_2}\cdot R_s + \frac{R_1}{R_2}r_3 - r_2 \\ &= \frac{R_1}{R_2}\cdot R_s + \frac{RR_4}{R_3+R_4+R}\left(\frac{R_1}{R_2} - \frac{R_3}{R_4}\right). \end{aligned}$$

If we choose

$$\frac{R_1}{R_2} = \frac{R_3}{R_4}$$

then

$$R_x = \frac{R_1}{R_2}R_s.$$

In this case the effect of R is entirely eliminated. R is usually in the form of a thick wire of low resistance. Thus the bridge can eliminate the effects of the resistances of contact and connecting leads. This makes the bridge more suitable for the measurement of low resistances below 0.1 Ω. Note that Wheatstone bridge is not suitable for the measurement of low resistance because here junction and contact resistances need be considered.

▶ **Problem 21.** *Consider the circuit of* Fig. 7.P-21. *Use Kirchhoff's laws to find current through R when* $R = 12$ kΩ. **[C.U. 2011]**

What is the power dissipated in this resistor? For which other value of R will the same power be dissipated in it?

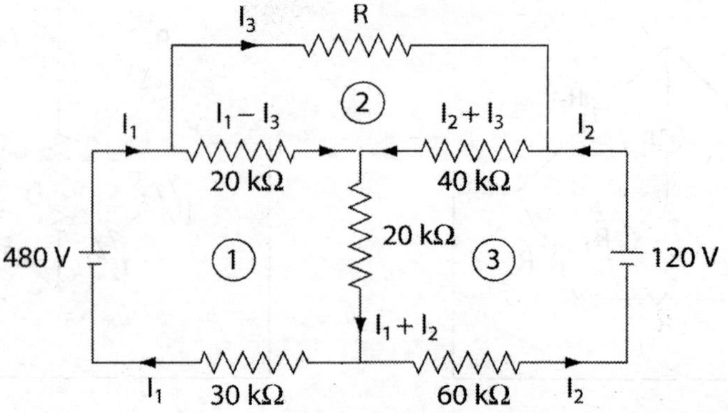

Fig. 7.P-21

Solution. Using KCL we assume the current distribution to be as shown in Fig. 7.P-21. Apply KVL to the meshes ①, ② and ③ we get

$$20(I_1 - I_3) + 20(I_1 + I_2) + 30I_1 = 480$$

$$\text{or} \quad 7I_1 + 2I_2 - 2I_3 = 48 \tag{1}$$

$$RI_3 + 40(I_2 + I_3) - 20(I_1 - I_3) = 0$$

$$\text{or} \quad -5I_1 + 10I_2 + \left(\frac{R}{4} + 15\right) I_3 = 0 \tag{2}$$

$$40(I_2 + I_3) + 20(I_1 + I_2) + 60I_2 = 120$$

$$\text{or } I_1 + 6I_2 + 2I_3 = 6. \tag{3}$$

Solving (1), (2) and (3) by Cramer's rule we get

$$I_3 = \frac{\begin{vmatrix} 7 & 2 & 48 \\ -5 & 10 & 0 \\ 1 & 6 & 6 \end{vmatrix}}{\begin{vmatrix} 7 & 2 & -2 \\ -5 & 10 & \frac{R}{4} + 15 \\ 1 & 6 & 2 \end{vmatrix}} = \frac{1440}{360 + 10R} = \frac{144}{36 + R} \text{ mA.}$$

For $R = 12\,\text{k}\Omega$, $I_3 = 3\,\text{mA}$ and hence power dissipation

$$P = \left(3 \times 10^{-3}\right)^2 \times 12\,W = 108 \times 10^{-6}\,\text{W} = 108\,\mu\text{W}.$$

Now for the other resistance R with same power dissipation we must have

$$I_3^2 R = 108$$

$$\left(\frac{144}{36 + R}\right)^2 R = 108$$

$$\text{or} \quad (36 + R)^2 = 192R$$

$$\text{or} \quad R^2 - 120R + 1296 = 0.$$

This gives $R = 12\,\text{k}\Omega$ and $108\,\text{k}\Omega$.

QUESTIONS

1. (a) What do you mean by electric current and current density? Are they vector quantities?

   (b) What are the differences between drift current and diffusion current? **[B.U. 2010]**

2. (a) What do you mean by drift velocity and random thermal velocity of electrons in a conductor? What is the speed of current flow in a conductor? What are the limitations of Ohm's law?

   (b) Establish the differential form $\left(\vec{J} = \sigma\vec{E}\right)$ of Ohm's law by using Lorentz-Drude theory of electrical conduction. From it derive the common form of Ohm's law. Also, show that the resistance of a wire of length l, cross sectional area S, made of a material of conductivity σ is $R = l/\sigma S$.

3. What is meant by conservation of charge? Express it in the form of a differential equation. What is the name of this equation? What is the form of this equation for steady current? OR, Establish the continuity equation relating the charge density and the current density at a point in a medium. **[C.U. 2011]**

4. State and explain Kirchhoff's laws for the distribution of currents and voltages in a network. Show that the first law is consistent with the principle of conservation of charge and the second law is consistent with the principle of conservation of energy. **[cf, C.U. 2019]**

5. (a) Applying Kirchhoff's laws find the current through the galvanometer in an unbalanced Wheatstone bridge. Hence find the condition of balance. **[J.U. 2010]**

   (b) Discuss the sensitivity of a Wheatstone bridge in terms of current through the unknown resistance, positions of galvanometer and battery and the internal resistance of the galvanometer. **[J.U. 2013]**

   (c) Four resistances P, Q, R, S are joined sequentially to form a quadrilateral. P and Q have higher values than R and S. How should the battery and the galvanometer be connected to construct a Wheatstone bridge of high sensitivity—Explain with a diagram. **[C.U. 2011]**

6. State superposition theorem as applied to network analysis. Considering a simple resistive T-network and two dc sources verify the validity of the theorem.

7. State Thevenin's and Norton theorems. Are these theorems applicable to both dc and ac? Are they independent laws or can be derived from Kirchhoff's laws? Distinguish between a voltage source and a current source.

8. A battery of emf $\mathcal{E}$ and internal resistance r is connected to an external variable load R. Find the condition for maximum power transfer from the battery to the load.

9. What is maximum power transfer theorem? Using it show that power transfer efficiency cannot exceed 50%. **[C.U. 2019]**

10. (a) What do you mean by an equivalent circuit? Show that any complicated two-port network containing linear elements can be replaced by a simple T-or π-network.
    (b) Derive the T-π transformation relationships. **[cf. C.U. 2017]**

EXERCISES

1. A current of $10A$ flows through a wire of cross-section 1mm^2. The density of free electrons within the wire is 10^{27}per m^3. Calculate the drift velocity of the electrons. Given $e = 1.6 \times 10^{-19}\text{C}$. [*Ans.* 0.0625 m/s]
2. In a certain region of space there are equal number of electrons and singly charged positive ions. There exists an electric field along x-axis. As a result ions are moving along positive x-axis with a drift velocity of 0.1 m/s and electrons are moving along negative x-axis with a drift velocity 0.4 m/s. The number density of electrons or ions is 5×10^{13}per m^3. Calculate the current density. [*Ans.* 4.0 $\mu\text{A/m}^2$]
3. The passage of electric current through a conductor is accompanied by the liberation of heat. Show that the power dissipation per unit volume can be expressed as $\vec{J} \cdot \vec{E}$ or J^2/σ, where the symbols have their usual meaning.
4. Show that the equivalent resistance of the parallel combination of n-resistors is smaller than the smallest of them.
5. The space between two concentric spherical conductors of radii a and b is filled with a slightly conducting material of conductivity σ and dielectric permittivity ε. A constant pd is maintained between the spheres such that a spherical symmetric current flows between them. Show that the resistance between the spheres is given by

$$R = \frac{1}{4\pi\sigma}\left(\frac{1}{a} - \frac{1}{b}\right).$$

   Also, show that the product $RC = \varepsilon/\sigma$, where C is the capacitance of the system.

   [**Hints.** $R = \displaystyle\int_a^b \frac{1}{\sigma} \cdot \frac{dr}{4\pi r^2}$, where dr is the thickness of a spherical shell of radii r and $r+dr$]
6. A source of emf is connected to an external resistance which is K times higher than the internal resistance of the source. Show that the pd across the source is $K/(1+K)$ times the emf of the source.

7. A source of emf e_1 and internal resistance r_1 is connected in parallel with another source of emf e_2 and internal resistance r_2 with similar poles connected together. The combination is now used to drive current through a wire of resistance R. Show that the current through the wire is $(e_1 r_2 + e_2 r_1)/(Rr_1 + Rr_2 + r_1 r_2)$.

8. In a Wheatstone bridge the resistances of the four arms are $R_1 = 2\,\Omega$, $R_2 = 1\,\Omega$, $R_3 = 4\,\Omega$ and $R_4 = 3\,\Omega$. A cell of emf 2 V and negligible internal resistance is connected between the junctions of (R_1, R_3) and (R_2, R_4). A galvanometer of internal resistance $10\,\Omega$ is connected between the two other junctions. Find the current through the galvanometer. [*Ans.* $1/65$ A]

9. The space between two long coaxial conducting cylinders of radii a and b $(a < b)$ and length L is filled with a slightly conducting medium whose resistivity ρ varies with radial distance r as $\rho(r) = Kr$. Calculate the resistance between the cylinders in radial direction. [*Ans.* $\frac{K(b-a)}{2\pi L}$]

10. Consider a cylinder of length L, area of cross-section A made of a material whose resistivity varies as $\rho(x) = \rho(o)\sqrt{x}/L$, where x is the distance along the axis from one end. A pd of V is applied between its two ends. Find the resistance of the cylinder along the axis and the current density. [*Ans.* $\frac{2\rho_o\sqrt{L}}{3A}, \frac{3V}{2\rho_o\sqrt{L}}$]

11. Twelve identical wires, each of resistance R, are connected to form a cube. Calculate the equivalent resistance between the two opposite corners of a face of the cube. [*Ans.* $\frac{3}{4}R$]

12. Consider a Wheatstone bridge formed by four resistance $r_1 = 5\,\Omega$, $r_2 = 2\,\Omega$, $r_3 = 2\,\Omega$ and $r_4 = 4\,\Omega$. The junction points (r_1, r_2) and (r_3, r_4) is connected by another resistance $3\,\Omega$. Find the equivalent resistance between the junction points (r_1, r_3) and (r_2, r_4). [*Ans.* $3\,\Omega$]

13. Two cells each of emf $\mathcal{E}$ and internal resistance r are connected in parallel across a variable load R. Find the value of the load for which power delivered becomes maximum. Also, find the maximum power. [*Ans.* $r/2$, $\mathcal{E}^2/2r$]

14. Six identical wires each of resistance R are connected in the form of a regular tetrahedron. Find the equivalent resistance between any two corners of the tetrahedron. [*Ans.* $R/2$]

15. Consider two resistances r_1 and r_2 connected in parallel and a current I is divided between them as I_1 and I_2, respectively. Using Kirchhoff's laws find the currents I_1 and I_2. Show that this current division satisfies the requirement of minimum power dissipation.

16. Consider the circuit of Fig. 7.E-16 and apply Thevenin's theorem to find the current through the $4\,\Omega$ resistor. [*Ans.* $\frac{3}{22}A$]

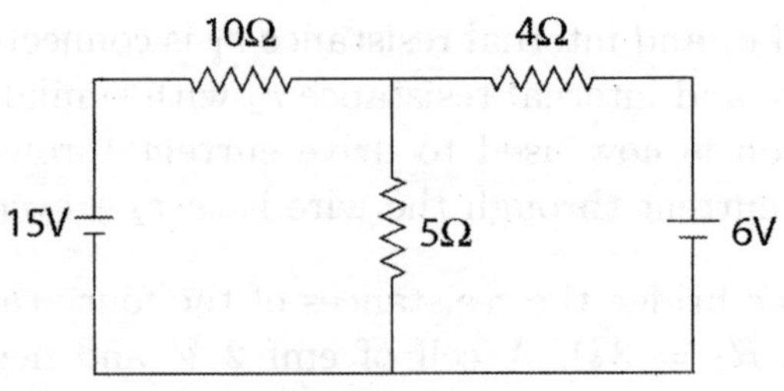

Fig 7.E-16

17. Consider the circuit of Fig. 7.E-16. Using superposition theorem find the current through $5\,\Omega$ resistor. [*Ans.* $1.09\,A$]

18. The four arms of a Wheatstone bridge have resistances $r_1 = 100\,\Omega$, $r_2 = 10\,\Omega$, $r_3 = 50\,\Omega$ and $r_4 = 4\,\Omega$. A galvanometer of resistance $20\,\Omega$ is connected between the junctions (r_1, r_2) and (r_3, r_4). A cell of emf $10\ V$ is connected between the junctions (r_1, r_3) and (r_2, r_4). Apply Thevenin's theorem to find the current through the galvanometer. **[cf. C.U. 2011, 2014]** [*Ans.* $5.1mA$]

19. Consider the circuit of Fig. 7.8-1(a) with $V_1 = 120\,\text{V}$, $r_1 = 0$, $R_1 = 40\,\Omega$, $R_2 = 60\,\Omega$, $R_3 = 24\,\Omega$, $V_2 = 180\,\text{V}$, $r_2 = 0$. Using Norton's theorem, find the current through the resistor R_3. [*Ans.* $3A$]

20. Find the Thevenin's and Norton's equivalent circuits for the circuit of Fig. 7.E-20.

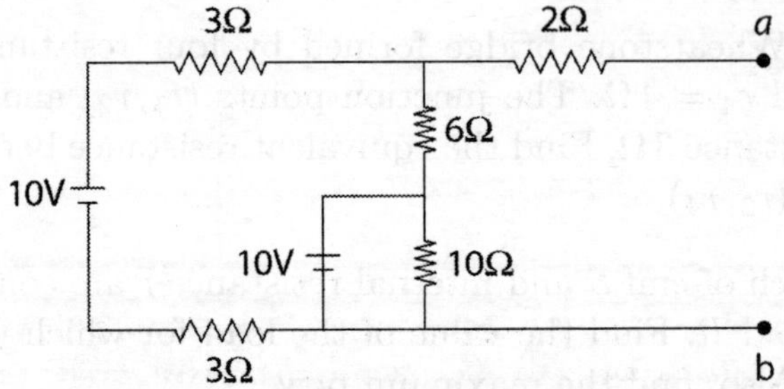

Fig 7.E-20

[*Ans.*

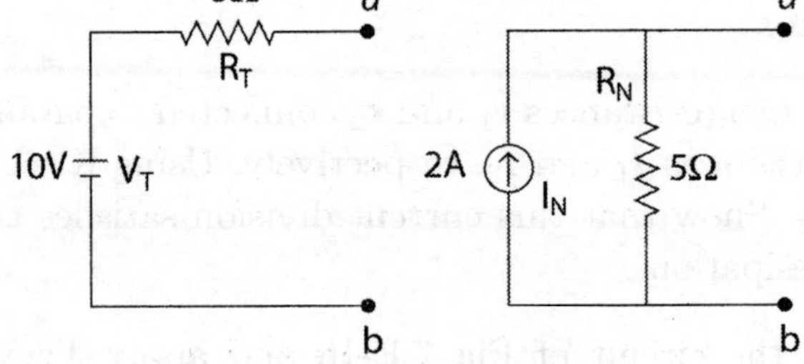

]

Fig 7.E-20(*a*)

21. A voltage source of internal resistance r delivers same power when connected separately to the loads R_1 and R_2. Show that $r = \sqrt{R_1 R_2}$.

22. Consider a Wheatstone bridge formed by four wires of resistances $R_1 = 1\,\Omega$, $R_2 = 1\,\Omega$, $R_3 = 2\,\Omega$ and $R_4 = 1\,\Omega$. A galvanometer of internal resistance $1\,\Omega$ is connected between the junctions (R_1, R_2) and (R_3, R_4). Using star/delta (or T-π) transformation find the equivalent resistance between the other two junctions.

[*Ans.* $\frac{13}{11}\,\Omega$]

23. Six wires are connected in the form of a tetrahedron. Five wires are identical and each has a resistance R. The sixth wire has a resistance R_1. A pd is applied across one of the resistors adjoining R_1. Show that for maximum power dissipation in the sixth wire we must have $R_1 = 3R/5$.

24. The current-voltage relationship in a non-Ohmic device is given by

$$I = I_0\left(e^{eV/kT} - 1\right),$$

where I_0 is a constant and T is the temperature. Show that the incremental resistance of the device is inversely proportional to the current.

25. Find the equivalent resistance between the terminals a, b of the infinite ladder network shown in Fig. 7.E-25, where all the resistances are equal to R.

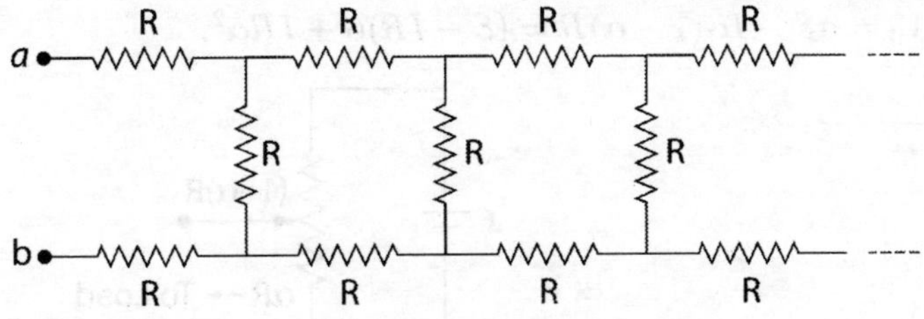

Fig. 7.E-25

[*Ans.* $\left(\sqrt{3}+1\right) R$]

26. Consider a Wheatstone bridge with four arms having resistances R_1, R_2, R_3 and R_4. An ideal source of emf $\mathcal{E}$ is connected between the junctions (R_1, R_3) and (R_2, R_4). A galvanometer of internal resistance r_g is connected between the two other junctions. Calculate the current through the galvanometer using Norton's theorem. Hence, deduce the balance condition for the bridge.

27. Consider the bridge-network of Fig. 7.E-27. Use T-π (star-delta) transformation to calculate the effective resistance R_{ab} between the points a and b as a function of the variable resistance R. Verify that your result coincides with that obtained by direct inspection in the two limiting cases $R \to 0$ and $R \to \infty$.

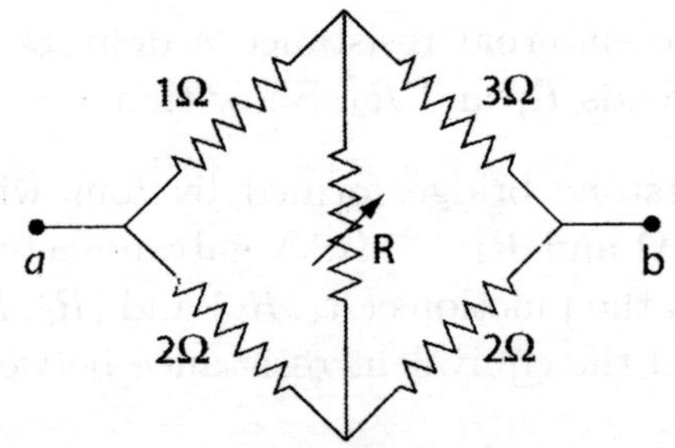

Fig.7.E-27

[**Hints.** Proceed as in the *Solved Problem* 18 and show that

$$R_{ab} = \frac{16R^2 + 76R + 84}{8R^2 + 39R + 45}.$$

Now $R_{ab} \to \frac{28}{15}\,\Omega$ as $R \to 0$ and $R_{ab} \to 2\,\Omega$ as $R \to \infty$.]

28. Consider the standard potential divider arrangement shown in Fig. 7.E-28, using a single wire-wound potentiometer of resistance R. Find the Thevenin's equivalent of this arrangement in terms of the fraction α. Hence find the voltage supplied to the load which draws a current I from it. Sketch the variation of this voltage with the fraction α, for a fixed I. Using your results comment on the choice of the size of the potentiometer, which helps to achieve a smooth variation of the voltage supplied to a load of (roughly) fixed size.

[**Hints.** $V_0 = \alpha\mathcal{E} - I\alpha(1-\alpha)R = (\mathcal{E} - IR)\alpha + IR\alpha^2$.

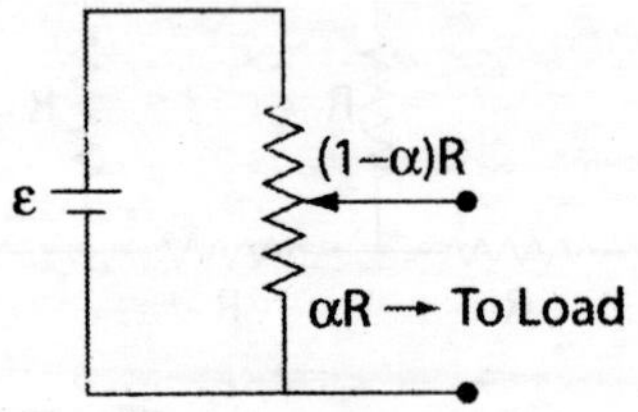

Fig.7-E-28

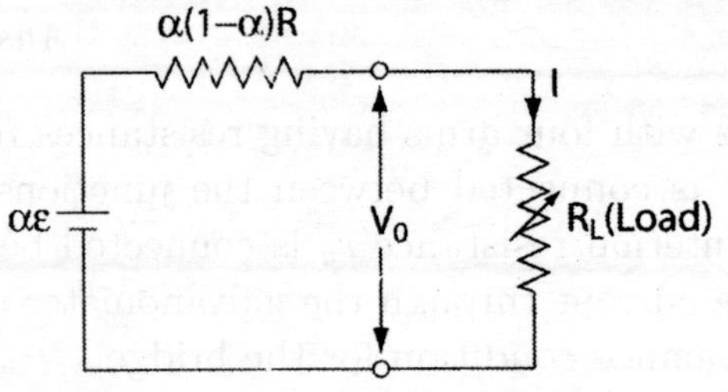

Fig.7-E-28(*a*): Thevenin equivalent

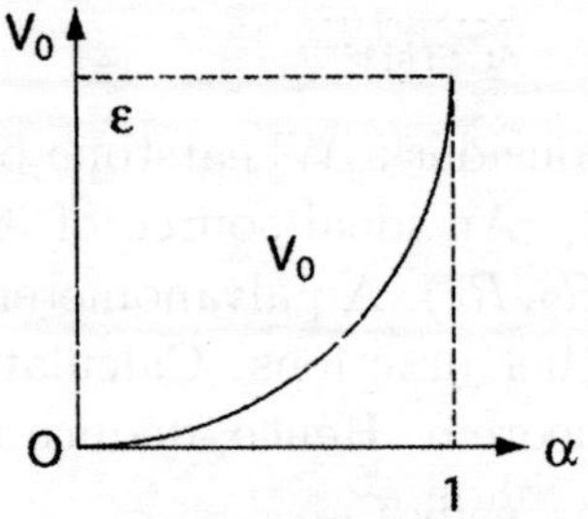

Fig.7-E-28(*b*)

The second terms causes the variation of V_0 with α non-linear. So for linear smooth variation, we must choose, $\mathcal{E} - IR \gg IR$ or $R \ll \mathcal{E}/2I$]

29. Consider the circuit of Fig. 7.P-11 and use Kirchhoff's laws to calculate the current flowing through each of the resistors. **[C.U. 2007]**

[*Ans.* $I_{100\,\mathrm{k}\Omega} = 0.2\,\mathrm{mA}$, $I_{20\,\mathrm{k}\Omega} = 0.,28\,\mathrm{mA}$, $I_{5\,\mathrm{k}\Omega} = 2.88\,\mathrm{mA}$, $I_{6\,\mathrm{k}\Omega} = 2.6\,\mathrm{mA}$]

30. Find out the voltage V_0 in the circuit shown in Fig. 7.E-30. Take $R = 1\Omega$.

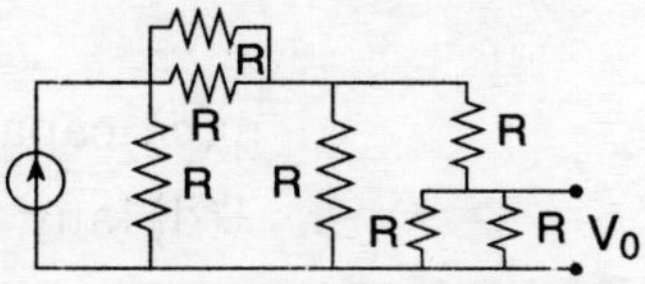

Fig 7.E-30

[*Ans.* 10/11 V] **[B.U. 2011]**

31. Find out the value of R in the circuit shown in Fig. 7.E-31.

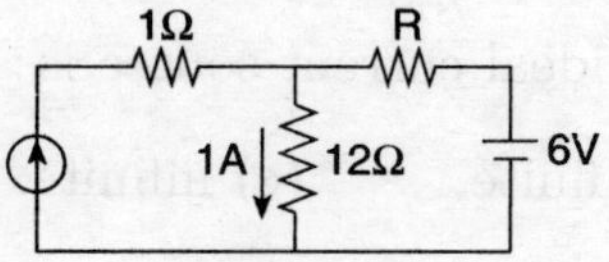

Fig 7.E-31

[*Ans.* 6Ω] **[GATE 2010]**

OBJECTIVE TEST

1. Steady electric current I may be considered as a:

(a) scalar, (b) vector, (c) tensor, (d) none of these.

2. Differential equation for the conservation of charge is:

(a) $\vec{\nabla} \cdot \vec{J} = \dfrac{\partial \rho}{\partial t}$, (c) $\vec{J} = \sigma \vec{E}$,

(b) $\vec{\nabla} \cdot \vec{J} = -\dfrac{\partial \rho}{\partial t}$, (d) none of these.

3. Differential form of Ohm's law is:

(a) $\dfrac{dV}{dI} = R$, (b) $\vec{J} = \sigma \vec{E}$, (c) $\vec{\nabla} \cdot \vec{J} = 0$, (d) $R = \dfrac{l}{\sigma A}$.

4. For steady current:

(a) $\vec{\nabla} \cdot \vec{J} = 0$, (b) $\vec{\nabla} \times \vec{J} = 0$, (c) $\nabla^2 \vec{J} = 0$, (d) none of these.

5. Kirchhoff's current law is a consequence of the principle of conservation of:

(a) energy, (c) momentum,

(b) charge, (d) none of these.

6. Kirchhoff's voltage law is a consequence of the principle of conservation of:

(a) energy, (b) momentum, (c) charge, (d) none of these.

7. Ohm's law is not valid for:

(a) alternating currents, (b) inductor, (c) capacitor, (d) any non-linear circuit element.

8. The internal resistance of an ideal voltage source is:

(a) infinite, (b) zero, (c) large but finite, (d) small but finite.

9. The internal resistance of an ideal current source is:

(a) zero, (b) small but finite, (c) infinite, (d) large but finite.

10. A load resistor R is connected to an ideal voltage source. If R is increased the power dissipated in it will:

(a) increase, (b) decrease, (c) remain the same, (d) none of these.

11. SI units of electrical conductivity (σ) is:

(a) siemens per metre, (b) siemens, (c) mho, (d) ohm-metre.

12. Two resistances r and $2r$ are connected in parallel in an electrical circuit. Power dissipations in r and $2r$ will be in the ratio:

(a) $2:1$, (b) $1:2$, (c) $4:1$, (d) $1:4$.

13. Dimensions of resistance R are:

(a) $[ML^2T^{-3}I^2]$, (b) $[ML^2T^{-3}I^{-2}]$, (c) $[ML^2T^3I^{-2}]$, (d) $[ML^2T^3I^2]$.

14. A wire of length l_1 and conductivity σ_1 is connected in series with another wire of length l_2 and conductivity σ_2. Free end of the first wire is maintained at a constant potential V_1 and that of the second wire at V_2. Then the potential at the junction of the wires is:

(a) $\left(\frac{\sigma_1 V_1}{l_1}+\frac{\sigma_2 V_2}{l_2}\right)\Big/\left(\frac{\sigma_1}{l_1}+\frac{\sigma_2}{l_2}\right)$,

(b) $\left(\frac{\sigma_1 V_1}{l_1}-\frac{\sigma_2 V_2}{l_2}\right)\Big/\left(\frac{\sigma_1}{l_1}-\frac{\sigma_2}{l_2}\right)$,

(c) $(\sigma_1 V_1 l_1+\sigma_2 V_2 l_2)/(\sigma_1 l_1+\sigma_2 l_2)$,

(d) $(\sigma_1 V_1 l_1-\sigma_2 V_2 l_2)/(\sigma_1 l_1-\sigma_2 l_2)$.

15. Three identical resistors each of 10 Ω, 20 W are connected in such a way that two are in parallel and third one is in series with it. Then, the maximum power that the combination can dissipate is:

(a) 60 W, (b) 15 W, (c) 30 W, (d) 40 W.

16. The mobility of a charge carrier can be expressed in units of:

(a) volt-sec/m^2, (c) volt-sec/m^2,
(b) m^2/volt-sec, (d) m^2-sec/volt.

17. If a steady pd is applied across the ends of a wire:

(a) all electrons move with a constant acceleration,

(b) all electrons move with a constant velocity,

(c) electrons move randomly but on the average, each electron move with a constant velocity,

(d) none of these.

18. Which one of the following situation does not have a constant charge density $\rho(\vec{r}, t)$ in space?

(a) $\vec{J} = \text{constant}$, (c) $\vec{J} = \vec{\nabla} \times \frac{\vec{r}}{r}$,

(b) $\vec{J} = 0$, (d) $\vec{J} = c\frac{\vec{r}}{r}$; $c = \text{constant}, r \neq 0$.

19. In the following circuit the value of R for maximum power transfer to the load $R_L = 3\Omega$ is **[GATE 2011]**

(a) zero, (c) 6Ω
(b) 3Ω (d) infinity.

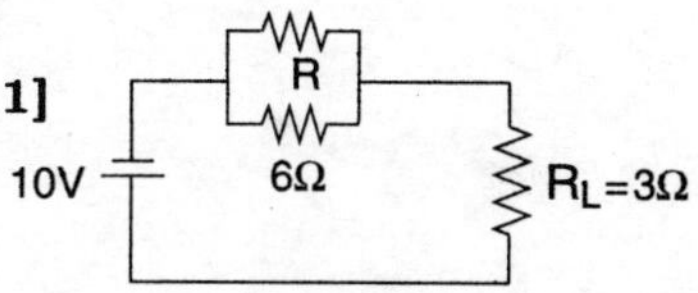

Fig 7.O-19

20. A long hollow cylinder of radius R carrying a surface charge density σ is rotating about its own axis with a constant speed ω. The magnitude of surface current density is

(a) $\sigma R\omega$, (b) $2\sigma R\omega$, (c) $\pi\sigma R\omega$, (d) $2\pi\sigma R\omega$.

21. In the circuit of Fig. 7.O.21 voltage and current sources are ideal. The voltage V_{out} in volt is

(a) 0, (b) 5, (c) 10, (d) 20.

[GATE 2016]

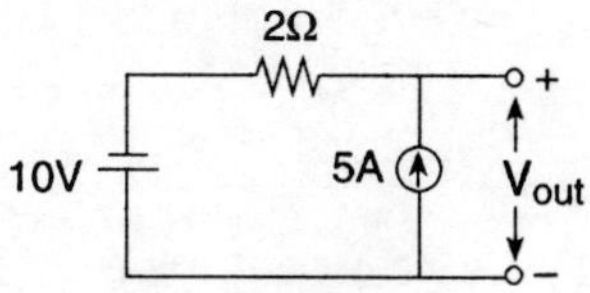

Fig 7.O-21

ANSWERS

1. (a)	**2.** (b)	**3.** (b)	**4.** (a)	**5.** (b)	**6.** (a)	**7.** (d)
8. (b)	**9.** (c)	**10.** (b)	**11.** (a)	**12.** (a)	**13.** (b)	**14.** (a)
15. (c)	**16.** (b)	**17.** (c)	**18.** (d)	**19.** (c)	**20.** (a)	**21.** (d)

Chapter 8

Magnetostatics

8.1 Introduction

Magnetic phenomenon has been known since ancient times. The naturally occurring magnets can attract pieces of iron and certain other materials. When suspended freely these magnets align themselves along the north-south direction. These properties created great curiosity. But magnetism was a less understood phenomenon until in 1820 Oersted discovered that an electric current produces a magnetic field. Further works of Biot, Savart, Ampere, Faraday and others helped to establish the laws of magnetism as we know them today. In the earlier studies of magnetism it was supposed that just as an electric charge produces an electric field, in the same way there are magnetic monopoles which produce magnetic field. However, all the theoretical and practical attempts to confirm the existence of such magnetic monopoles have failed. According to modern view all magnetic effects are associated with electric current. The magnetic field produced by a bar magnet can be explained as the resultant field produced by the tiny current loops formed by the electrons revolving around the nucleus of each atom.

Experiments show that the force $\vec{F}_m$ on a point charge q moving with a velocity $\vec{v}$ due to a magnetic field depends on the velocity $\vec{v}$. This force is always perpendicular to velocity vector $\vec{v}$. The magnitude of $\vec{F}_m$ at a point is proportional to the component of velocity perpendicular to a fixed direction in space. All these properties can be described by introducing the field vector $\vec{B}$. $\vec{B}$ is called *magnetic induction, magnetic flux density* or simply *magnetic field.* Steady currents produce magnetic fields that are constant in time. For this the study of magnetic effects of steady current is known as *magnetostatics.* The purpose of the present chapter is to consider the basic definition of $\vec{B}$ and to study the magnetic effects of steady current.

8.2 Magnetic Forces

Magnetic field can exert force on a moving charge or a current carrying conductor. A magnetic field can be measured by measuring these magnetic forces.

The force on a charge q moving with velocity $\vec{v}$ in a magnetic field $\vec{B}$ is

$$\vec{F}_m = q\left(\vec{v} \times \vec{B}\right). \tag{8.2-1}$$

If, in addition, there is an electric field $\vec{E}$ then total electromagnetic force on the charge will be

$$\vec{F} = q\left[\vec{E} + \vec{v} \times \vec{B}\right]. \tag{8.2-2}$$

This is called *Lorentz force.* Obviously, $\vec{B}$ has the unit of N.s/C-m or V.s/m^2. In SI system it is called weber/m^2 or tesla (T). Another unit in common use is gauss. 1 T=10^4 gauss. Dimensions of $\vec{B}$ are $\left[MT^{-2}I^{-1}\right]$.

From Eq. (8.2-1) it is clear that magnetic field does not exert any force on an electric charge at rest. The magnetic force $\vec{F}_m$ is always perpendicular to the velocity vector of the charge. Therefore, *it cannot do any work on the charge* $\left(\because \frac{dW}{dt} = \vec{F}_m \cdot \vec{v} = 0\right)$.

Equation (8.2-1) can also be used to define $\vec{B}$ *to be numerically equal to the force per unit charge moving with unit velocity perpendicular to the magnetic field.*

Magnetic force on a current carrying conductor

Suppose a conducting wire, carrying a current I, is placed in a magnetic field $\vec{B}$. Consider a small current element $Id\vec{l}$ of the wire. Let the current consists of charge carriers moving with a velocity $\vec{v}$ along the wire. Force on each charge is $q\left(\vec{v} \times \vec{B}\right)$. If there are n number of charge carriers per unit volume and A is the area of cross-section of the wire then number of carriers in the element $d\vec{l}$ would be $nAdl$. Thus total magnetic force on the element $d\vec{l}$ is

$$d\vec{F} = nAdl \cdot q\left(\vec{v} \times \vec{B}\right) = I\left(d\vec{l} \times \vec{B}\right), \tag{8.2-3}$$

where $I = nqvA$ is the current through the wire. For a conductor of finite length total force is given by

$$\vec{F} = I\int d\vec{l} \times \vec{B}.$$

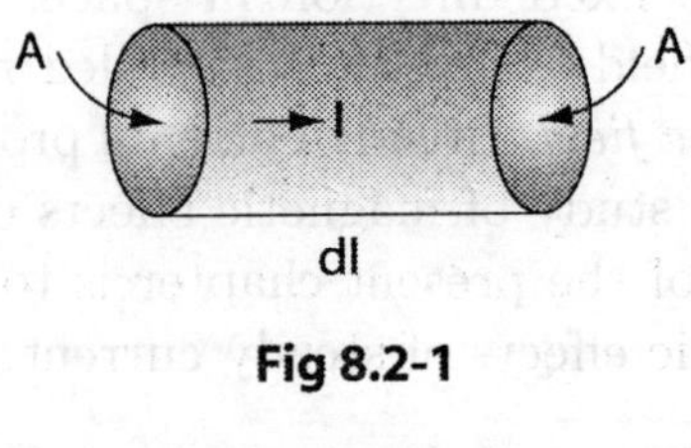

Fig 8.2-1

For a wire of length l placed in a uniform field $\vec{B}$, the force is

$$\vec{F} = I\vec{l} \times \vec{B}.$$

For a closed loop placed in a *uniform field* $\vec{B}$, the force is

$$\vec{F} = I\left(\oint d\vec{l}\right) \times \vec{B} = 0 \qquad \left[\because \oint d\vec{l} = 0\right] \tag{8.2-4}$$

From Eq. (8.2-3), we may define $\vec{B}$ *to be numerically equal to the force per unit current element placed perpendicular to the magnetic field.*

Motion of a charged particle in a uniform magnetic field

Suppose a particle of charge q and mass m enters with a velocity $\vec{v}$ into a region of uniform magnetic field $\vec{B}$. Its equation of motion is

$$m\frac{d\vec{v}}{dt} = q\left(\vec{v} \times \vec{B}\right). \tag{8.2-5}$$

Assuming that $\vec{B} = B\hat{k}$ the above equation can be written in the following component forms:

$$m\frac{dv_x}{dt} = qBv_y \tag{8.2-6}$$

$$m\frac{dv_y}{dt} = -qBv_x \tag{8.2-7}$$

$$m\frac{dv_z}{dt} = 0. \tag{8.2-8}$$

From Eq. (8.2-8) it follows that v_z = constant = $v_{\parallel}$(say). Thus the velocity component parallel to $\vec{B}$ remains constant. Now $v_z = dz/dt$ and hence $dz/dt = v_{\parallel}$. Integrating we get

$$z = v_{\parallel}t + z_0. \tag{8.2-9}$$

The coupled Eqs. (8.2-6) and (8.2-7) can be combined as

$$\frac{d}{dt}(v_x + jv_y) = -j\omega(v_x + jv_y), \tag{8.2-10}$$

where $\omega = qB/m$ is called *cyclotron frequency.*

Solution of Eq. (8.2-10) is

$$v_x + jv_y = Ce^{-j\omega t}.$$

Integrating we get

$$x + jy = C_1e^{-j\omega t} + C_2,$$

where C_1 and C_2 are constants to be determined from initial conditions. Let us choose the constants in the form $C_1 = Ae^{-j\alpha}$ and $C_2 = x_0 + jy_0$.

Then

$$x + jy = Ae^{-j(\omega t+\alpha)} + x_0 + jy_0.$$

Now equating real and imaginary parts separately we get

$$\left.\begin{aligned} x &= A\cos(\omega t + \alpha) + x_0 \\ y &= -A\sin(\omega t + \alpha) + y_0 \end{aligned}\right\}. \tag{8.2-11}$$

Equations (8.2-11) and (8.2-9) give the trajectory of the particle in a uniform magnetic field. From Eq. (8.2-11),

$$(x - x_0)^2 + (y - y_0)^2 = A^2,$$

which represents a circle in the xy-plane with the centre at (x_0, y_0). Often one has the condition $v_{\parallel} = 0$ and the particle describes a circular path in the plane perpendicular to $\vec{B}$.

The radius of the circle is given by

$$v_{\perp}^2 = \dot{x}^2 + \dot{y}^2 = A^2\omega^2 \quad \text{or} \quad A = \frac{v_1}{\omega} = \frac{v_{\perp}m}{qB}. \tag{8.2-12}$$

It is known as the *radius of gyration.*

Thus, if a charged particle enters a uniform magnetic field normally it describes a circular path. The radius r of the circular path can be easily determined by equating magnetic force to the necessary centripetal force, i.e.,

$$\frac{mv^2}{r} = qvB \quad \text{or} \quad r = \frac{mv}{qB}. \tag{8.2-13}$$

In this case, the momentum of the particle is given by

$$p = mv = qBr. \tag{8.2-14}$$

This relation is often used to find the momentum of a particle. Just by measuring the radius of the circular trajectory of the particle in a magnetic field it is possible to determine its momentum.

It is interesting to note that the time period of revolution in the circular orbit, i.e.,

$$T = \frac{2\pi r}{v} = \frac{2\pi m}{qB} \tag{8.2-15}$$

is independent of the speed of the particle. This fact is used in a particle accelerator, called *cyclotron.* This property is also used in devices called *magnetic lenses* for focusing beams of charged particles.

If $v_{\parallel} \neq 0$ the particle moves parallel $\vec{B}$ with constant speed $v_{\parallel}$ and at the same time gyrates about $\vec{B}$. Therefore, the particle moves in a helix with $\vec{B}$ as axis (Fig. 8.2-2).

The pitch of the helix is the distance between two successive turns. It is given by the distance travelled by the particle along the direction of $\vec{B}$ in one time period

$$\therefore \quad \text{Pitch} = v_{\parallel}T = \frac{2\pi m}{qB} \cdot v_{\parallel}$$

Note that the direction of gyration is such that the magnetic field generated by the moving charged particle is opposite to the externally applied magnetic field.

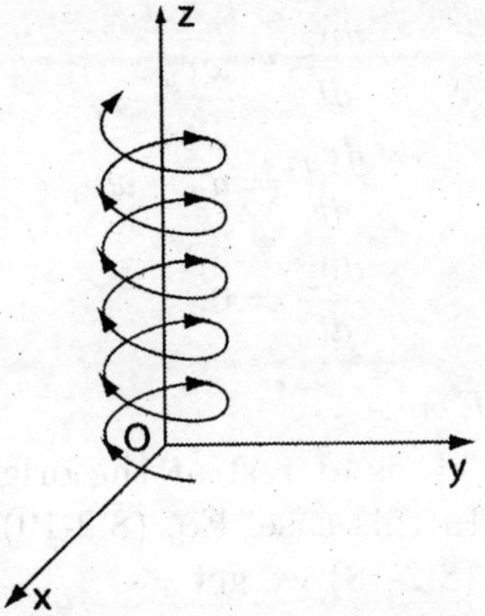

Fig. 8.2-2: Helical path of charged particle

Motion of a charged particle in crossed electric and magnetic fields

Suppose a particle of charge q and mass m enters with a velocity $\vec{v}$ into a region of crossed $\vec{E}$ and $\vec{B}$ fields. Let $\vec{B}$ is along z-axis and $\vec{E}$ along y-axis (Fig. 8.2-3). The equation of motion of the particle in such a field is

$$m\frac{d\vec{v}}{dt} = q\left(\vec{E} + \vec{v} \times \vec{B}\right). \tag{8.2-16}$$

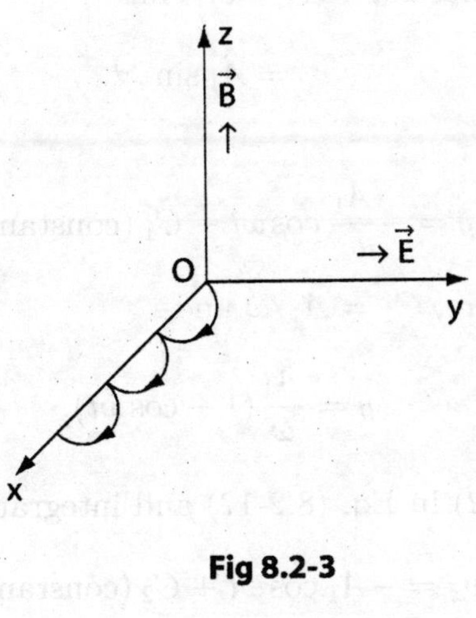

Fig 8.2-3

Let $\vec{B} = B\hat{k}$ and $\vec{E} = E\hat{\jmath}$. Then

$$\vec{v} \times \vec{B} = \begin{vmatrix} \hat{\imath} & \hat{\jmath} & \hat{k} \\ v_x & v_y & v_z \\ 0 & 0 & B \end{vmatrix}.$$

So Eq. (8.2-16) can be written in the following component forms:

$$\frac{dv_x}{dt} = \omega v_y \tag{8.2-17}$$

$$\frac{dv_y}{dt} = a_y - \omega v_x \tag{8.2-18}$$

$$\frac{dv_z}{dt} = 0, \tag{8.2-19}$$

where $\omega = qB/m$ and $a_y = qE/m$.

Suppose the charged particle is at rest at the origin at time $t = 0$ and we are to find the subsequent motion. In this case, Eq. (8.2-19) yields that $v_z = z = 0$ always. Combining Eqs. (8.2-17) and (8.2-18) we get

$$\frac{d^2v_x}{dt^2} = \omega a_y - \omega^2 v_x \tag{8.2-20}$$

$$\frac{d^2v_y}{dt^2} = -\omega^2 v_y. \tag{8.2-21}$$

Equation (8.2-21) has a solution

$$v_y = A_1 \sin \omega t + A_2 \cos \omega t.$$

Now at $t = 0$, $v_y = 0$, therefore, $A_2 = 0$. Thus

$$v_y = A_1 \sin \omega t. \tag{8.2-22}$$

Integrating we get

$$y = -\frac{A_1}{\omega} \cos \omega t + C_1 \text{ (constant)}.$$

At $t = 0$, $y = 0$, therefore, $C_1 = A_1/\omega$ and

$$y = \frac{A_1}{\omega}(1 - \cos \omega t). \tag{8.2-23}$$

Substituting Eq. (8.2-22) in Eq. (8.2-17) and integrating we get

$$v_x = -A_1 \cos \omega t + C_2 \text{ (constant)}.$$

At $t = 0$, $v_x = 0$, therefore, $C_2 = A_1$ and thus

$$v_x = A_1(1 - \cos\omega t). \tag{8.2-24}$$

Integrating we get,

$$x = \frac{A_1}{\omega}(\omega t - \sin\omega t), \tag{8.2-25}$$

where the constant of integration is chosen to be zero as $x = 0$ at $t = 0$.

Equations (8.2-23) and (8.2-25) are the parametric equations of a *cycloid.* This is the path generated by a point on the rim of a wheel rolling along a line.

The constant A_1 can be determined from Eqs. (8.2-22) and (8.2-18):

$$\frac{dv_y}{dt} = A_1\omega\cos\omega t = a_y - \omega v_x \tag{8.2-26}$$

At $t = 0$, $v_x = 0$, therefore,

$$A_1 = \frac{a_y}{\omega} = \frac{E}{B}. \tag{8.2-27}$$

If initially the particle is at the origin and moving along x-axis with a velocity $v_{xo}\hat{i}$ then from (8.2-26)

$$A_1\omega = a_y - \omega v_{xo} \quad \text{or} \quad A_1 = \frac{E}{B} - v_{xo}.$$

In this case,

$$v_z = 0 \quad \text{and} \quad v_y = \left(\frac{E}{B} - v_{xo}\right)\sin\omega t.$$

Thus, if $v_{xo} = E/B$ then the particle goes undeflected. So crossed fields can act as a *velocity filter* for charge particles.

8.3 Biot-Savart Law

Soon after the discovery by Oersted that electric current produces magnetic effects, Ampere performed a series of experiments. From the results of these experiments came out a definite law determining the magnetic field due to a current element. This law is known as *Biot-Savart law.* According to this law the magnetic field at any point due to a current element $I d\vec{l}$ is given by

$$d\vec{B} = \frac{\mu_0}{4\pi}\frac{I d\vec{l} \times \vec{R}}{R^3}, \tag{8.3-1}$$

where $\vec{R}$ is the vector from the current element to the point of observation and the constant $\mu_0 = 4\pi \times 10^{-7}\mathrm{N/A^2}$ is called the *permeability of free space.*

In terms of current density $\vec{J}$ we can write

$$I d\vec{l} = \left(\vec{J} \cdot \hat{n}\, dS\right) d\vec{l} = \left(d\vec{l} \cdot \hat{n} dS\right) \vec{J} = \vec{J}\, dV, \tag{8.3-2}$$

where $\hat{n}\, dS$ is the cross-sectional area and dV is the volume of the current element; we use the fact that $\vec{J}$ is parallel to $d\vec{l}$ because by definition $d\vec{l}$ is parallel to the direction of current in it.

So replacing $Id\vec{l}$ in Eq. (8.3-1) by $\vec{J}\, dV$ we get for a volume distribution of current,

$$d\vec{B} = \frac{\mu_0}{4\pi} \frac{\vec{J} \times \vec{R}}{R^3}\, dV. \tag{8.3-3}$$

Formulas (8.3-1) and (8.3-3) express the Biot-Savart law.

According to the principle of superposition total field $\vec{B}$ due to the entire line current I or the volume distribution of current is given by

$$\vec{B} = \frac{\mu_0}{4\pi} \int \frac{I d\vec{l} \times \vec{R}}{R^3} \quad \text{or} \quad \vec{B} = \frac{\mu_0}{4\pi} \int_V \frac{\vec{J} \times \vec{R}}{R^3}\, dV, \tag{8.3-4}$$

where the integration is over the whole current path or the whole volume of the current distribution $\vec{J}(x', y', z')$.

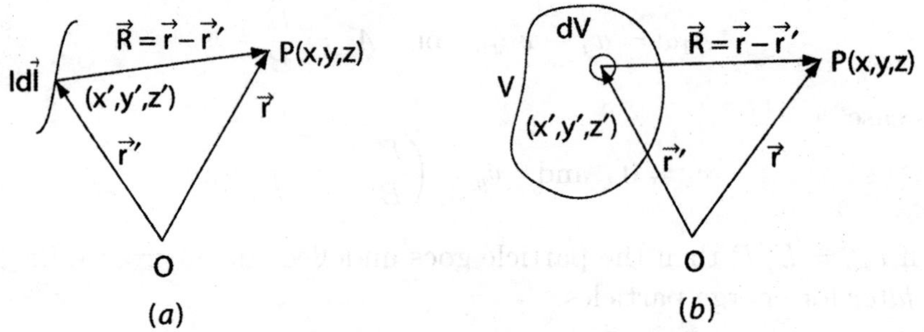

Fig. 8.3-1 : Biot-Savart law

For surface current Biot-Savart law takes the form:

$$\vec{B} = \frac{\mu_0}{4\pi} \int \frac{\vec{K} \times \vec{R}}{R^3} dS. \tag{8.3-5}$$

Note that here $\vec{B} = \vec{B}(x, y, z)$, $dV = dx'dy'dz'$, $\vec{R} = \hat{i}(x - x') + \hat{j}(y - y') + \hat{k}(z - z')$.

8.4 Applications of Biot-Savart Law

Calculation of magnetic field due to arbitrary current distribution by using Biot-Savart law may become complicated. However, for current distributions having certain symmetry the calculation may be simple. Here we consider a few such problems.

1. Long straight current carrying wire

Consider a long straight wire AB carrying current I as shown in Fig. 8.4-1. Suppose we are to find the magnetic field at any point P at a distance a from the wire. Let us consider a small element $d\vec{l}$ at a distance l from O, the foot of the perpendicular from the point P on the wire. According to Biot-Savart law the magnetic field at P due to this element is

$$d\vec{B} = \frac{\mu_0}{4\pi}\frac{Id\vec{l} \times \vec{R}}{R^3}.$$

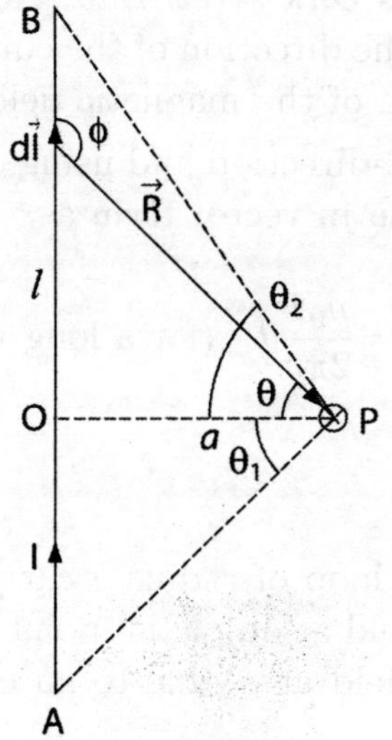

Fig 8.4-1

Obviously $d\vec{B}$ points towards the plane of the paper. This is true for the field produced by all current elements. Hence, total magnetic field at P due to the whole wire would be

$$\vec{B} = \frac{\mu_0}{4\pi}\int \frac{Id\vec{l} \times \vec{R}}{R^3}.$$

Now from Fig. 8.4-1

$$\left| d\vec{l} \times \vec{R} \right| = dlR\sin\phi = dlR\sin(90^\circ + \theta) = dlR\cos\theta \quad \text{and} \quad l = a\tan\theta,$$

so $dl = a\sec^2\theta d\theta$. Also $a = R\cos\theta$ or $R = a\sec\theta$. Thus

$$\left| \vec{B} \right| = \frac{\mu_0 I}{4\pi a}\int_{-\theta_1}^{\theta_2} \cos\theta d\theta = \frac{\mu_0 I}{4\pi a}(\sin\theta_2 + \sin\theta_1), \tag{8.4-1}$$

where θ_1 and θ_2 are the values of the angle θ corresponding to the end points of the wire.

For an infinitely long wire $\theta_1 = \theta_2 = \pi/2$ and we obtain

$$\left|\vec{B}\right| = \frac{\mu_0}{4\pi} \cdot \frac{2I}{a} = \frac{\mu_0 I}{2\pi a}. \tag{8.4-2}$$

Note that the field is inversely proportional to the distance from the wire. The result is similar to the electric field produced by a long line charge. The direction of $\vec{B}$ at any point P is tangential to the circle of radius a drawn through P with centre on the wire and lying on a plane perpendicular to the wire. The direction $\vec{B}$ can be easily found by using *Maxwell's cork screw rule.* According to this rule if we drive a right-handed cork screw along the direction of the current then the direction of rotation of the thumb gives the direction of the magnetic field.

Taking the current along $+z$-direction and using cylindrical coordinates the field at any point (r, θ, z) can be written in vector form as

$$\vec{B} = \frac{\mu_0 I}{2\pi r}\hat{\theta} \quad \text{(for a long wire)}.$$

2. Circular loop

Suppose we consider a circular loop of radius a carrying a current I (Fig. 8.4-2). We are to calculate the magnetic field at any axial point P at a distance x from the centre O of the loop. The magnetic field at P due to an element $d\vec{l}$ is given by Biot-Savart law as,

$$d\vec{B} = \frac{\mu_0}{4\pi}\frac{I d\vec{l} \times \vec{R}}{R^3}. \tag{8.4-3}$$

The field $d\vec{B}$ is perpendicular to the plane containing $d\vec{l}$ and $\vec{R}$. When Eq. (8.4-3) is integrated over the whole coil the vector $d\vec{B}$ describes a cone with OP as axis. As a result, the components of $d\vec{B}$ normal to OP, the axis of the loop, cancel out and the components parallel to OP combine to give a resultant magnetic field along the OP (x-axis).

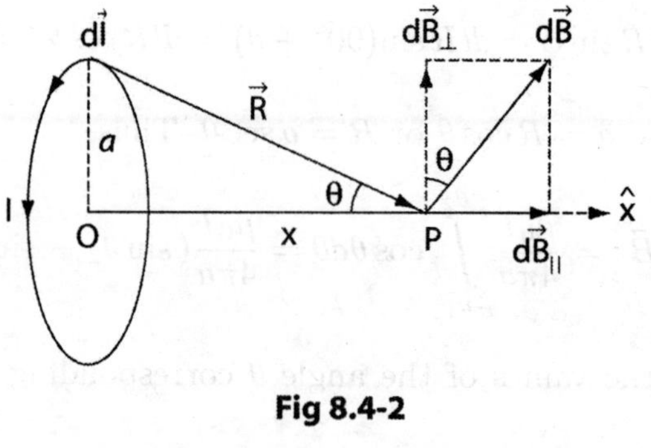

Fig 8.4-2

Thus we can write for the total field at P as

$$\vec{B} = \hat{x}\frac{\mu_0 I}{4\pi}\int \frac{dl \cdot R \cdot \sin 90^\circ}{R^3} \cdot \sin\theta \quad \left[\because\ d\vec{l} \perp^r \vec{R}\right]$$

$$= \hat{x}\frac{\mu_0 I}{4\pi}\int \frac{dl}{R^2}\sin\theta.$$

Now $\sin\theta$ and R^2 are constants and $\int dl = 2\pi a$, so

$$\vec{B} = \hat{x}\frac{\mu_0 I}{4\pi} \cdot \frac{\sin\theta}{R^2} \cdot 2\pi a = \hat{x} \cdot \frac{\mu_0}{4\pi} \cdot \frac{2\pi I a^2}{(a^2+x^2)^{3/2}}$$

$$= \hat{x}\frac{\mu_0 I}{2}\frac{a^2}{(a^2+x^2)^{3/2}}, \qquad (8.4\text{-}4)$$

where we use

$$\sin\theta = \frac{a}{R} = \frac{a}{\sqrt{a^2+x^2}}.$$

If there are N turns in the loop then the expression for $\vec{B}$ is to be multiplied by N. Field at the centre of the loop ($x = 0$) is

$$\vec{B} = \hat{x} \cdot \frac{\mu_0 I}{2a}.$$

3. Helmholtz coil

The magnetic field due to a single current loop is found to be maximum at the centre and decreases as we go away from the centre on both sides of the loop. In an arrangement due to Helmholtz a uniform magnetic field can be obtained midway between two identical coaxial circular coils placed exactly parallel to each other with their centres at a distance apart equal to the radius of either coil; the coils are wound in the same sense and same current is passed through the coils in series such that the axial magnetic fields at the mid-point are additive.

Let us consider two such coils 1 and 2 each of radius a, same number of turns N and carrying same current I. If $2d$ be the separation between the coils the resultant axial magnetic field at P, at a small distance $x(x \ll a)$ from the mid-point O is obtained by using Eq. (8.4-4) as,

$$B = \frac{\mu_0 NI}{2}\left[\frac{a^2}{\{a^2+(d+x)^2\}^{3/2}} + \frac{a^2}{\{a^2+(d-x)^2\}^{3/2}}\right]. \qquad (8.4\text{-}5)$$

For B to be uniform near about the mid-point we must have $dB/dx = 0$. This gives,

$$\frac{d+x}{\{a^2+(d+x)^2\}^{5/2}} = \frac{d-x}{\{a^2+(d-x)^2\}^{5/2}}$$

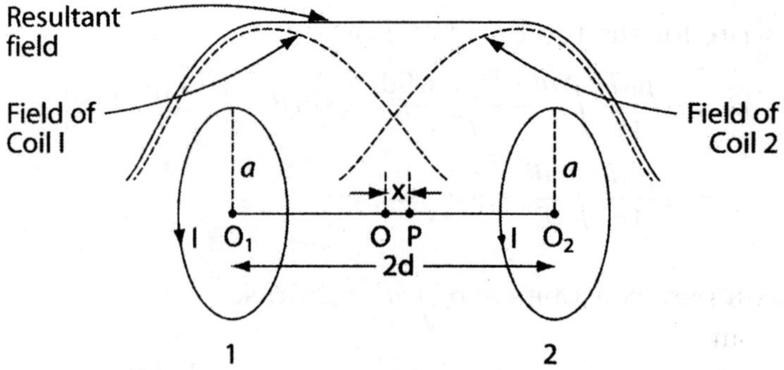

Fig 8.4-3

$$\text{or} \quad \frac{d+x}{d-x} = \left\{\frac{d^2+a^2+2dx}{d^2+a^2-2dx}\right\}^{5/2} \quad [\because x^2 \ll d^2+a^2]$$

$$= \left\{\frac{1+\dfrac{2dx}{d^2+a^2}}{1-\dfrac{2dx}{d^2+a^2}}\right\}^{5/2} \approx \frac{1+\dfrac{5dx}{d^2+a^2}}{1-\dfrac{5dx}{d^2+a^2}}$$

$$\text{or} \quad \frac{x}{d} = \frac{5dx}{d^2+a^2}$$

$$\text{or} \quad a^2 = 4d^2, \quad \text{or} \quad 2d = a. \tag{8.4-6}$$

So for uniform magnetic field in the neighbourhood of the mid-point O the separation between the coils must be equal to the radius of either coil. The field at the centre is obtained by putting $x = 0$ in Eq. (8.4-5) as

$$B = \frac{\mu_0 NI}{2}\left[\frac{a^2}{\left\{a^2+\left(\frac{a}{2}\right)^2\right\}^{3/2}} + \frac{a^2}{\left\{a^2+\left(\frac{a}{2}\right)^2\right\}^{3/2}}\right] = \frac{8\mu_0 NI}{5\sqrt{5}a} \tag{8.4-7}$$

4. Solenoid

A coil of wire wound uniformly over a hollow nonconducting cylinder having a length large compared with its radius, is said to constitute a solenoid. When a current is passed through the coil a uniform magnetic field is produced inside the coil parallel to its axis.

Let us consider a solenoid of length l, radius a having n number of turns per unit length and carrying a current I. We consider an element of length dx containing ndx number of turns at a distance x along the axis from the point of observation P. The magnetic field at P due to this elementary part is

$$d\vec{B} = \hat{x}\frac{\mu_0(ndx)Ia^2}{2(a^2+x^2)^{3/2}}.$$

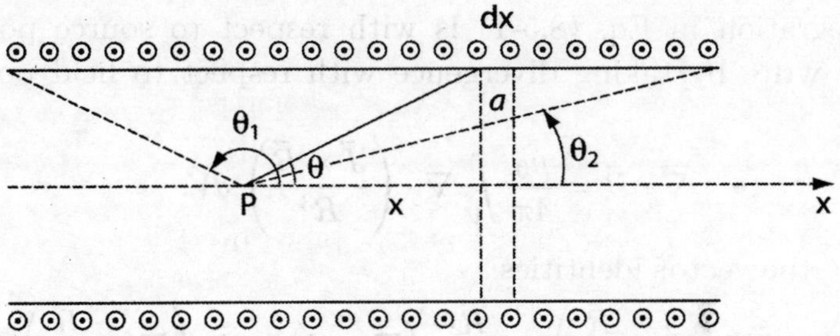

Fig 8.4-4

From Fig. 8.4-4, $x = a\cot\theta$ and so, $dx = -a\,\text{cosec}^2\theta d\theta$.

$$\therefore \quad d\vec{B} = -\hat{x}\frac{\mu_0 nI}{2}\sin\theta d\theta.$$

If the end points of the solenoid correspond to the angles θ_1 and θ_2 then total magnetic field at P is given by

$$\vec{B} = -\hat{x}\frac{\mu_0 nI}{2}\int_{\theta_1}^{\theta_2}\sin\theta d\theta = \hat{x}\frac{\mu_0 nI}{2}(\cos\theta_2 - \cos\theta_1). \tag{8.4-8}$$

For a long solenoid ($l \gg a$) we have for any point well inside; $\theta_1 = \pi$ and $\theta_2 = 0$, then

$$\vec{B} = \hat{x}\mu_0 nI. \tag{8.4-9}$$

If P is at the extreme end of the solenoid $\theta_1 = \pi$ and $\theta_2 = \pi/2$ then

$$\vec{B} = \hat{x}\frac{\mu_0 nI}{2},$$

which is half the field at the central part of the solenoid.

8.5 The Divergence of the Magnetic Field $\vec{B}$

In order to characterise the magnetic field $\vec{B}$ we must know its divergence and curl. Here we shall show that the divergence of $\vec{B}$ is always zero. According to Biot-Savart law the magnetic field $\vec{B}$ at any point $P(x, y, z)$ due to a volume distribution of current $\vec{J}(x', y', z')$ [Fig. 8.3-1(b)] is given by

$$\vec{B} = \frac{\mu_0}{4\pi}\int_V \frac{\vec{J}\times\vec{R}}{R^3}\,dV, \tag{8.5-1}$$

where the integration is over the whole volume V of the source; here

$$dV = dx'dy'dz' \quad \text{and} \quad \vec{R} = \vec{r} - \vec{r}\,' = \hat{i}(x - x') + \hat{j}(y - y') + \hat{k}(z - z').$$

Since the integration in Eq. (8.5-1) is with respect to source point coordinates (x', y', z') we can write by taking divergence with respect to field point coordinates (x, y, z) as,

$$\vec{\nabla} \cdot \vec{B} = \frac{\mu_0}{4\pi} \int_V \vec{\nabla} \cdot \left(\frac{\vec{J} \times \vec{R}}{R^3} \right) dV. \tag{8.5-2}$$

Now let us use the vector identities

$$\vec{\nabla} \cdot \left(\vec{J} \times \frac{\vec{R}}{R^3} \right) = \frac{\vec{R}}{R^3} \cdot \left(\vec{\nabla} \times \vec{J} \right) - \vec{J} \cdot \left(\vec{\nabla} \times \frac{\vec{R}}{R^3} \right) \tag{8.5-3}$$

$$\text{and} \quad \vec{\nabla} \times \frac{\vec{R}}{R^3} = \frac{1}{R^3} \left(\vec{\nabla} \times \vec{R} \right) + \vec{\nabla} \frac{1}{R^3} \times \vec{R} \tag{8.5-4}$$

Since $\vec{J} = \vec{J}(x', y', z')$, $\vec{\nabla} \times \vec{J} = 0$. From vector calculus,

$$\vec{\nabla} \times \vec{R} = 0 \quad \text{and} \quad \vec{\nabla} \frac{1}{R^3} = -\frac{3\vec{R}}{R^5}.$$

Thus we find that

$$\vec{\nabla} \cdot \left(\vec{J} \times \frac{\vec{R}}{R^3} \right) = 0.$$

Hence, from Eq. (8.5-2),

$$\vec{\nabla} \cdot \vec{B} = 0. \tag{8.5-5}$$

This equation is valid for all magnetic fields whether produced by steady currents or varying currents. This result is known as *Gauss's law in magnetostatics.* Integrating this equation over any volume V and applying divergence theorem we get

$$\int_V \vec{\nabla} \cdot \vec{B} \, dV = \oint_S \vec{B} \cdot d\vec{S} = 0, \tag{8.5-6}$$

where S is a surface enclosing volume V.

This result shows that the flux of $\vec{B}$ through any closed surface S is always zero. That is, magnetic lines of force are always closed and there is no *isolated magnetic monopole* or *free magnetic charge* similar to the free electric charges in an electrostatic field.

8.6 The Curl of the Magnetic Field $\vec{B}$

According to Biot-Savart law the magnetic field $\vec{B}$ at any point $P(x, y, z)$ due to a volume distribution of current $\vec{J}(x, y, z)$ [Fig. 8.3-1(b)] is given by

$$\vec{B} = \frac{\mu_0}{4\pi} \int_V \frac{\vec{J} \times \vec{R}}{R^3} dV, \tag{8.6-1}$$

where the integration is over the whole volume V of the source,

$$dV = dx'dy'dz' \quad \text{and} \quad \vec{R} = \vec{r} - \vec{r}' = \hat{\imath}(x - x') + \hat{\jmath}(y - y') + \hat{k}(z - z').$$

Since

$$\vec{\nabla}\frac{1}{R}=-\frac{\vec{R}}{R^3},$$

Eq. (8.6-1) can be written as

$$\vec{B}=-\frac{\mu_0}{4\pi}\int\vec{J}\times\vec{\nabla}\frac{1}{R}\,dV, \tag{8.6-2}$$

where $\vec{\nabla}$ is with respect to field point coordinates (x, y, z).

Now

$$\begin{aligned}\vec{\nabla}\times\frac{1}{R}\vec{J}&=\frac{1}{R}\vec{\nabla}\times\vec{J}+\vec{\nabla}\frac{1}{R}\times\vec{J}\\&=\vec{\nabla}\frac{1}{R}\times\vec{J}\quad\left[\because\ \vec{J}=\vec{J}(x',y',z')\text{ and so }\vec{\nabla}\times\vec{J}=0\right]\\&=-\vec{J}\times\vec{\nabla}\frac{1}{R}.\end{aligned}$$

Therefore, from (8.6-2),

$$\vec{B}=\frac{\mu_0}{4\pi}\vec{\nabla}\times\int\frac{1}{R}\vec{J}\,dV. \tag{8.6-3}$$

Taking curl and using the vector identity $\vec{\nabla}\times\vec{\nabla}\times\vec{A}=\vec{\nabla}\left(\vec{\nabla}\cdot\vec{A}\right)-\nabla^2\vec{A}$ we can write

$$\begin{aligned}\vec{\nabla}\times\vec{B}&=\frac{\mu_0}{4\pi}\vec{\nabla}\left(\vec{\nabla}\cdot\int\frac{1}{R}\vec{J}dV\right)-\frac{\mu_0}{4\pi}\nabla^2\int\frac{1}{R}\vec{J}\,dV\\&=\frac{\mu_0}{4\pi}\vec{\nabla}\int\vec{J}\cdot\vec{\nabla}\frac{1}{R}\,dV-\frac{\mu_0}{4\pi}\int\vec{J}\nabla^2\frac{1}{R}\,dV\end{aligned}$$

Now

$$\nabla^2\frac{1}{R}=-4\pi\delta\left(\vec{r}-\vec{r}'\right)\quad\text{and}\quad\vec{\nabla}\frac{1}{R}=-\vec{\nabla}'\frac{1}{R}. \tag{8.6-4}$$

Therefore,

$$\begin{aligned}\vec{\nabla}\times\vec{B}&=-\frac{\mu_0}{4\pi}\vec{\nabla}\int\vec{J}\cdot\vec{\nabla}'\frac{1}{R}\,dV+\frac{\mu_0}{4\pi}\int\vec{J}4\pi\delta\left(\vec{r}-\vec{r}'\right)dV\\&=-\frac{\mu_0}{4\pi}\vec{\nabla}\int\left[\vec{\nabla}'\cdot\frac{1}{R}\vec{J}-\frac{1}{R}\vec{\nabla}'\cdot\vec{J}\right]dV+\mu_0\vec{J}(\vec{r}).\end{aligned}$$

For steady currents $\vec{\nabla}'\cdot\vec{J}=0$. So

$$\vec{\nabla}\times\vec{B}=\mu_0\vec{J}\left(\vec{r}\right)-\frac{\mu_0}{4\pi}\vec{\nabla}\int\vec{\nabla}'\cdot\frac{1}{R}\vec{J}\,dV. \tag{8.6-5}$$

The integral on the right-hand side can be converted using the divergence theorem to a surface integral. If the surface is chosen to lie outside the bounded region, where $\vec{J}$ is nonvanishing, then obviously the integral vanishes. Thus,

$$\vec{\nabla} \times \vec{B} = \mu_0 \vec{J}(\vec{r}) . \tag{8.6-6}$$

This equation is called the differential form of *Ampere's law.* It is valid for magnetic field of steady currents and in general requires modification if currents change in time. Equation (8.6-6) can be converted into integral form by applying Stokes' theorem:

$$\int_S \left(\vec{\nabla} \times \vec{B}\right) \cdot d\vec{S} = \mu_0 \int_S \vec{J} \cdot d\vec{S}$$

$$\text{or} \quad \oint_C \vec{B} \cdot d\vec{l} = \mu_0 I, \tag{8.6-7}$$

where I is the total current passing through the surface S enclosed by the contour C. Equation (8.6-7) is known as the *Ampere's Circuital law.* In words, the line integral of the magnetic field around any closed path is equal to μ_0 times the total current enclosed by the path.

In Eq. (8.6-7) the direction of integration around the contour C is associated with the direction of a positive current through right-hand screw rule. Thus if a right-hand screw is rotated in the direction of integration then its direction of movement defines a positive current. The current having opposite direction is considered to be negative.

8.7 Application of Ampere's Law

Ampere's circuital law allows us to determine magnetic field in a very simple way for currents having special symmetry, where it is possible to find a closed path over which $\vec{B}$ is constant in magnitude. Here we consider a few practically important examples.

1. Long straight current carrying wire

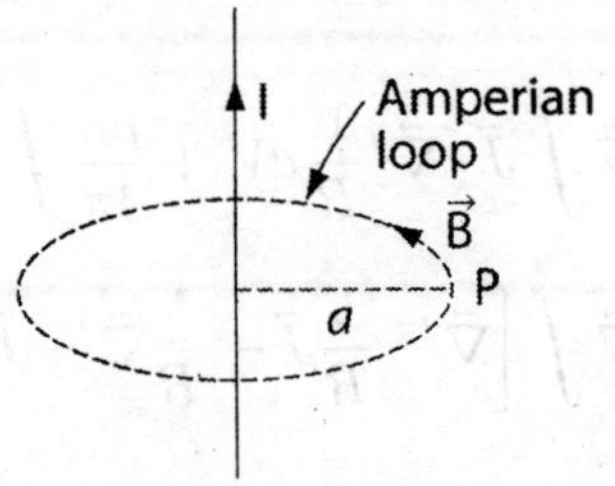

Fig 8.7-1

Suppose we have a long straight wire carrying current I. We are to find the magnetic field at any outside point P at a distance a from the axis of the wire. We draw a circle of radius a through the point P and with centre on the axis of the wire. It is called an *Amperian loop.* It follows from the symmetry of the problem that in this case, the magnitude of the vector $\vec{B}$ will be the same all over the circular path and the direction of $\vec{B}$ will be tangential to the every path element $d\vec{l}$. Therefore, from Ampere's circuital law,

$$\oint \vec{B}\cdot d\vec{l} = B\oint dl = B\cdot 2\pi a = \mu_0 I \quad \text{or} \quad B = \frac{\mu_0}{4\pi}\cdot\frac{2I}{a} = \frac{\mu_0 I}{2\pi a}. \tag{8.7-1}$$

The direction of $\vec{B}$ circles around the wire in accordance with the right-hand cork screw rule. That is, if a right-handed screw is driven along the direction of current the direction of rotation of the thumb points in the direction of the magnetic field.

2. Long solenoid

Let us consider a long solenoid consisting of n turns per unit length closely wound on a circular cylinder of radius a and carrying a steady current I.

Symmetry consideration of the problem indicates that the magnetic field inside the solenoid is directed along its axis; it cannot have any radial or cross radial component. Vector $\vec{B}$ forms a right-handed system with the direction of current in the solenoid; the magnetic field outside the solenoid is practically zero. Let us consider an Amperian loop in the form of a rectangle *abcd* which lies entirely outside the solenoid with its sides at distances r_1 and r_2 from the axis of the solenoid (Fig. 8.7-2). From Ampere's circuital law

$$\oint \vec{B}\cdot d\vec{l} = [B(r_2) - B(r_1)]\cdot l = \mu_0 I_{\text{enclosed}} = 0,$$

where we assume that $\vec{B}$ is perpendicular to *ad* and *bc*.

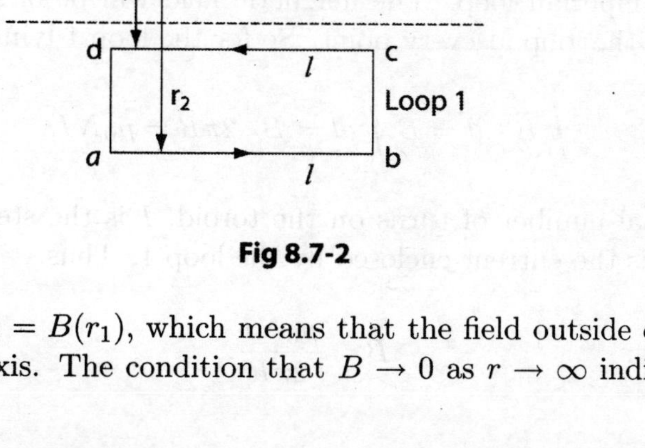

Fig 8.7-2

Therefore, $B(r_2) = B(r_1)$, which means that the field outside does not depend on distance from the axis. The condition that $B \to 0$ as $r \to \infty$ indicates that the field outside must be zero everywhere. Now we consider another rectangular Amperian loop $PQRS$ which is half inside and half outside as shown in Fig. 8.7-2. Now from Ampere's circuital law

$$\oint \vec{B} \cdot d\vec{l} = \mu_0 n l I. \tag{8.7-2}$$

Now

$$\begin{aligned}\oint \vec{B} \cdot d\vec{l} &= \int_{PQ} \vec{B} \cdot d\vec{l} + \int_{QR} \vec{B} \cdot d\vec{l} + \int_{RS} \vec{B} \cdot d\vec{l} + \int_{SP} \vec{B} \cdot d\vec{l} \\ &= Bl + 0 + 0 + 0 \\ &= Bl,\end{aligned}$$

where we use the fact that $\vec{B}$ is constant over PQ, zero over RS, zero or perpendicular over QR and SP.

Therefore, the field inside the solenoid is given by

$$Bl = \mu_0 n l I \quad \text{or,} \quad B = \mu_0 n I, \tag{8.7-3}$$

which is directed along the axis of the solenoid.

Vectorially,

$$\vec{B} = \begin{cases} \mu_0 n I \hat{z}, & \text{inside the solenoid} \\ 0, & \text{outside the solenoid} \end{cases}$$

3. A toroid (endless solenoid)

A long wire wound closely around a circular ring forms a toroid. Considering symmetry of the problem it can be argued that the lines of force would be circles whose centres lie on the axis of the toroid passing through O. To apply Ampere's law we consider one such circle as an Amperian loop. The magnetic field will be of same magnitude and directed tangent to the loop at every point. So for the loop 1 lying inside the solenoid,

$$\oint \vec{B} \cdot d\vec{l} = B \oint dl = B \cdot 2\pi R = \mu_0 N I,$$

where N is the total number of turns on the toroid, I is the steady current through each turn and NI is the current enclosed by the loop 1. Thus,

$$B = \frac{\mu_0 N I}{2\pi R}.$$

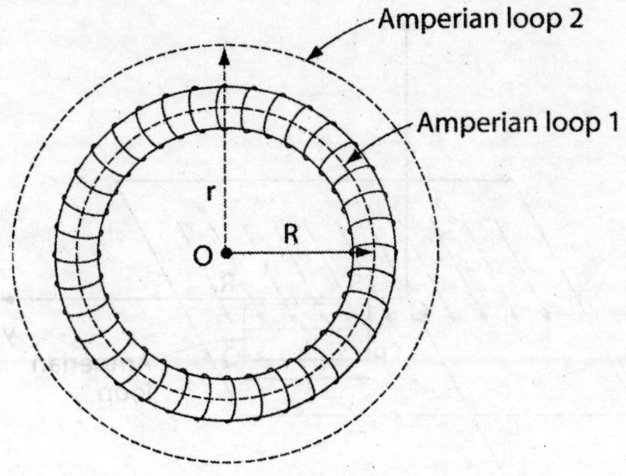

Fig 8.7-3

If $N/2\pi R = n$ be the number of turns per unit length then

$$B = \mu_0 n I. \tag{8.7-4}$$

For the Amperian loop 2 chosen outside the toroid total enclosed current is $NI - NI = 0$ and hence,

$$\oint \vec{B} \cdot d\vec{l} = 0 \quad \text{or} \quad B \cdot 2\pi r = 0 \quad \text{or} \quad B = 0.$$

This means that outside the toroid magnetic field is zero.

Note that the formula (8.7-3) is the same as that for a long solenoid. Thus a straight solenoid may be regarded as a special toroid with radius $R \to \infty$.

4. Infinite sheet of current

Consider an infinite plane $z = 0$ on which a uniform surface current density $\vec{K} = K\hat{i}$ flows as shown in Fig. 8.7-4. Dividing the current sheet into pairs of current filaments at $+y$ and $-y$ it is easy to show that the resultant magnetic field $\vec{B}$ will be directed along $-y$-axis for $z > 0$ and along $+y$-axis for $z < 0$. With this in mind we choose a rectangle $PQRS$ parallel to yz-plane as an Amperian loop Fig. 8.7-4. Now, applying Ampere's law we get,

$$\oint \vec{B} \cdot d\vec{l} = \mu_0 K l$$

$$\text{or} \quad \int_{PQ} \vec{B} \cdot d\vec{l} + \int_{QR} \vec{B} \cdot d\vec{l} + \int_{RS} \vec{B} \cdot d\vec{l} + \int_{SP} \vec{B} \cdot d\vec{l} = \mu_0 K l$$

$$\text{or} \quad Bl + 0 + Bl + 0 = \mu_0 K l$$

$$\text{or} \quad B = \frac{\mu_0}{2} K. \tag{8.7-5}$$

Note that B is independent of the distance from the current sheet.

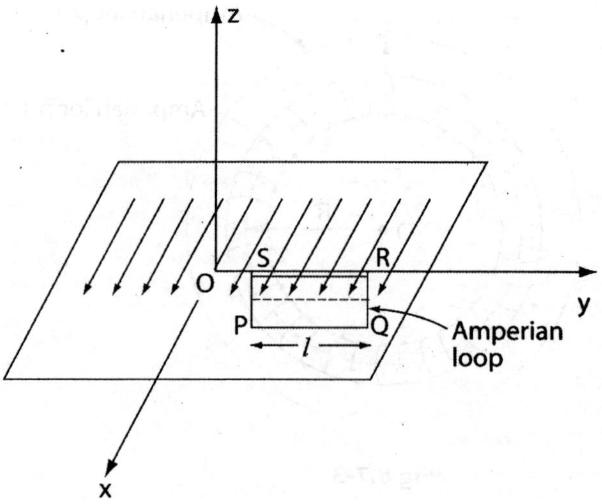

Fig 8.7-4

Vectorially

$$\vec{B} = \begin{cases} -\frac{\mu_0 K}{2}\hat{j} & \text{for } z > 0 \\ \\ +\frac{\mu_0 K}{2}\hat{j} & \text{for } z < 0 \end{cases} \tag{8.7-6}$$

In general

$$\vec{B} = \frac{\mu_0}{2}\vec{K} \times \hat{n}, \tag{8.7-7}$$

where $\hat{n}$ is the unit normal from the current sheet to the point of interest.

8.8 Magnetic Vector Potential

Calculation of electric fields gets very much simplified with the introduction of the electric potential. Similarly, in magnetostatics calculation $\vec{B}$ may get simplified with the introduction of magnetic potential. For a magnetic field we have $\vec{\nabla} \cdot \vec{B} = 0$ always. Since the divergence of curl of any vector is always zero, it is possible to write

$$\vec{B} = \vec{\nabla} \times \vec{A}, \tag{8.8-1}$$

where $\vec{A}$ is called *magnetic vector potential.* From Ampere's law

$$\vec{\nabla} \times \vec{B} = \mu_0 \vec{J}. \tag{8.8-2}$$

Therefore,

$$\vec{\nabla} \times \left(\vec{\nabla} \times \vec{A}\right) = \mu_0 \vec{J} \quad \text{or} \quad \vec{\nabla}\left(\vec{\nabla} \cdot \vec{A}\right) - \nabla^2 \vec{A} = \mu_0 \vec{J}. \tag{8.8-3}$$

The potential defined by (8.8-1) is not unique because we can add to $\vec{A}$ any function whose curl is zero with no effect on $\vec{B}$. We may exploit this freedom to choose

$$\vec{\nabla} \cdot \vec{A} = 0. \tag{8.8-4}$$

This choice is referred to as the *Coulomb gauge* on the vector potential $\vec{A}$. Then from Eq. (8.8-3),

$$\nabla^2 \vec{A} = -\mu_0 \vec{J}, \tag{8.8-5}$$

which is called *vector Poisson's* equation.

$\vec{A}$ can be obtained by solving this equation. Each component of $\vec{A}$ satisfies the equation $\nabla^2 A_i = -\mu_0 J_i$ $(i = x, y, z)$, which is of the form of Poisson's equation $\nabla^2 \phi = -\rho/\varepsilon_0$ in electrostatics, which has a solution for volume charge distribution as

$$\phi(\vec{r}) = \frac{1}{4\pi\varepsilon_0} \int \frac{\rho\, dV}{R},$$

where ϕ goes to zero at infinity.

Using this as a guide the solution of Eq. (8.8-5) can be written, when $\vec{J}$ does not extend to infinity, as

$$\vec{A}(\vec{r}) = \frac{\mu_0}{4\pi} \int \frac{\vec{J}\, dV}{R}, \tag{8.8-6}$$

where $R = |\vec{r} - \vec{r}\,'|$.

The solution (8.8-6) can also be obtained easily from Biot-Savart law, that is

$$\vec{B} = \frac{\mu_0}{4\pi} \int \frac{\vec{J} \times \vec{R}}{R^3}\, dV. \tag{8.8-7}$$

Since

$$\vec{\nabla} \frac{1}{R} = -\frac{\vec{R}}{R^3} \quad \text{and} \quad \vec{\nabla} \times \frac{1}{R}\vec{J} = \vec{\nabla}\frac{1}{R} \times \vec{J} + \frac{1}{R}\vec{\nabla} \times \vec{J}$$

we can write,

$$\begin{aligned} \frac{\vec{J} \times \vec{R}}{R^3} &= \vec{\nabla}\frac{1}{R} \times \vec{J} = \vec{\nabla} \times \frac{1}{R}\vec{J} - \frac{1}{R}\vec{\nabla} \times \vec{J} \\ &= \vec{\nabla} \times \frac{1}{R}\vec{J} \quad \left[\because\ \vec{J} = \vec{J}(x', y', z') \text{ and so } \vec{\nabla} \times \vec{J} = 0\right] \end{aligned} \tag{8.8-8}$$

Thus from (8.8-7) and (8.8-8),

$$\vec{B} = \vec{\nabla} \times \frac{\mu_0}{4\pi} \int \frac{\vec{J}\, dV}{R},$$

where $\vec{\nabla}$ operator has been taken outside the integral sign. This is allowed because $\vec{\nabla}$ is with respect to field point coordinates (x, y, z) and the integration is over source coordinates x', y', z'.

Since $\vec{B} = \vec{\nabla} \times \vec{A}$, therefore,

$$\vec{A} = \frac{\mu_0}{4\pi} \int \frac{\vec{J}\, dV}{R}.$$

For line current,

$$\vec{A} = \frac{\mu_0}{4\pi}\int \frac{I d\vec{l}}{R} \tag{8.8-9}$$

and for surface current

$$\vec{A} = \frac{\mu_0}{4\pi}\int \frac{\vec{K}\, dS}{R}. \tag{8.8-10}$$

Here it is important to point out that the vector potential $\vec{A}$ is not as useful as electrostatic potential in calculating field. Use of $\vec{A}$ in the calculation of $\vec{B}$ is limited. Its main use is in problems involving electromagnetic radiations.

8.9 Derivation of Biot-Savart Law and Ampere's Law from the Concept of Magnetic Vector Potential

Biot-Savart law

The magnetic vector potential $\vec{A}(\vec{r})$ at a point $\vec{r}(x, y, z)$ due to a volume distribution of current $\vec{J}(\vec{r}')$ or $\vec{J}(x', y', z'))$ is given by

$$\vec{A} = \frac{\mu_0}{4\pi}\int \frac{\vec{J}\, dV}{R}, \tag{8.9-1}$$

where $R = |\vec{r} - \vec{r}'|$ and $dV = dx'dy'dz'$.

Therefore,

$$\vec{B} = \vec{\nabla} \times \vec{A} = \frac{\mu_0}{4\pi}\int \vec{\nabla} \times \frac{1}{R}\vec{J}\, dV,$$

where $\vec{\nabla}$ is with respect to field point coordinates.

Now

$$\vec{\nabla} \times \frac{1}{R}\vec{J} = \vec{\nabla}\frac{1}{R} \times \vec{J} + \frac{1}{R}\vec{\nabla} \times \vec{J} = -\frac{\vec{R}}{R^3} \times \vec{J} \quad \left[\because\ \vec{\nabla} \times \vec{J} = 0\right].$$

Therefore,

$$\vec{B} = \frac{\mu_0}{4\pi}\int \frac{\vec{J} \times \vec{R}}{R^3}\, dV. \tag{8.9-2}$$

Each volume current element $\vec{J}\, dV$ contributes a quantity

$$d\vec{B} = \frac{\mu_0}{4\pi}\frac{\vec{J} \times \vec{R}}{R^3}\, dV \tag{8.9-3}$$

to the magnetic field $\vec{B}$. For a line current, $\vec{J}\, dV$ is replaced by $I d\vec{l}$ and then

$$d\vec{B} = \frac{\mu_0}{4\pi}\frac{I d\vec{l} \times \vec{R}}{R^3}. \tag{8.9-4}$$

Equations (8.9-3) and (8.9-4) are known as *Biot-Savart law.*

Ampere's law

From Stokes' theorem,

$$\oint_C \vec{B}\cdot d\vec{l} = \oint_C \left(\vec{\nabla}\times\vec{A}\right)\cdot d\vec{l} = \oint_S \vec{\nabla}\times\left(\vec{\nabla}\times\vec{A}\right)\cdot d\vec{S}. \tag{8.9-5}$$

Now

$$\vec{\nabla}\times\left(\vec{\nabla}\times\vec{A}\right) = \vec{\nabla}\left(\vec{\nabla}\cdot\vec{A}\right) - \nabla^2\vec{A} = -\nabla^2\vec{A} = \mu_0\vec{J}, \tag{8.9-6}$$

where we choose $\vec{\nabla}\cdot\vec{A} = 0$ and use Eq. (8.8-5) satisfied by $\vec{A}$. Therefore, from Eq. (8.9-5) we get

$$\oint_C \vec{B}\cdot d\vec{l} = \int_S \mu_0\vec{J}\cdot d\vec{S} = \mu_0 I,$$

which is *Ampere's circuital law.*

8.10 Calculation of $\vec{A}$ in a few Simple Cases

1. A straight current carrying conductor

Let us consider a long straight wire AB carrying a steady current I. We are to calculate the magnetic vector potential $\vec{A}$ at any point P. We choose cylindrical coordinates (r,θ,z) with z-axis along the wire and the foot O of the perpendicular from P on the wire as the origin. Let us consider an element $d\vec{l} = \hat{z}dz$ at a distance z from O. Now by definition

$$\begin{aligned}\vec{A} = \frac{\mu_0}{4\pi}\int\frac{Id\vec{l}}{R} &= \hat{z}\frac{\mu_0 I}{4\pi}\int_{-l_1}^{+l_2}\frac{dz}{\sqrt{r^2+z^2}}\\ &= \hat{z}\frac{\mu_0 I}{4\pi}\left[\ln\left(z+\sqrt{r^2+z^2}\right)\right]_{-l_1}^{+l_2}\\ &= \hat{z}\frac{\mu_0 I}{4\pi}\ln\left[\frac{l_2+\sqrt{r^2+l_2^2}}{-l_1+\sqrt{r^2+l_1^2}}\right].\end{aligned} \tag{8.10-1}$$

If the wire is of length $2L$ and the point P is just above the centre of the wire then $l_1 = l_2 = L$ and we can write

$$\vec{A} = \hat{z}\frac{\mu_0 I}{4\pi}\ln\left[\frac{L+\sqrt{r^2+L^2}}{-L+\sqrt{r^2+L^2}}\right].$$

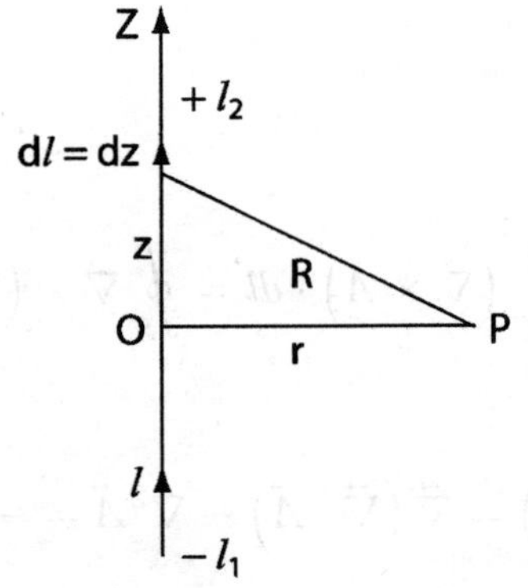

Fig 8.10-1

Now if $r/L \ll 1$ then we can approximate $\vec{A}$ as follows:

$$\vec{A} = \hat{z}\frac{\mu_0 I}{4\pi}\ln\left[\frac{1+\left(1+\frac{r^2}{L^2}\right)^{1/2}}{-1+\left(1+\frac{r^2}{L^2}\right)^{1/2}}\right] \approx \hat{z}\frac{\mu_0 I}{4\pi}\ln\left[\frac{2+\frac{r^2}{2L^2}}{\frac{r^2}{2L^2}}\right]$$

$$= \hat{z}\frac{\mu_0 I}{4\pi}\ln\left(\frac{4L^2}{r^2}+1\right) \approx \hat{z}\frac{\mu_0 I}{2\pi}\ln\left(\frac{2L}{r}\right). \qquad (8.10\text{-}2)$$

Using Eq. (8.10-2) we can calculate the magnetic field due to a long straight wire as follows:

$$\vec{B} = \vec{\nabla}\times\vec{A} = \frac{1}{r}\begin{vmatrix} \hat{r} & r\hat{\theta} & \hat{z} \\ \frac{\partial}{\partial r} & \frac{\partial}{\partial \theta} & \frac{\partial}{\partial z} \\ 0 & 0 & A_z \end{vmatrix} = -\hat{\theta}\frac{\partial A_z}{\partial r}$$

$$= -\hat{\theta}\frac{\partial}{\partial r}\left[\frac{\mu_0 I}{2\pi}\ln\left(\frac{2L}{r}\right)\right] = \hat{\theta}\frac{\mu_0 I}{2\pi r}. \qquad (8.10\text{-}3)$$

2. Two long straight parallel wires carrying same current

Suppose we have two long parallel wires each of length $2L$ carrying same current I in opposite directions. The magnetic vector potential $\vec{A}$ at any point P due to this system can be calculated by using the result (8.10-2). Thus,

$$\vec{A} = \vec{A}_1 + \vec{A}_2 = \hat{z}\frac{\mu_0 I}{2\pi}\ln\left(\frac{2L}{r_1}\right) - \hat{z}\frac{\mu_0 I}{2\pi}\ln\left(\frac{2L}{r_2}\right) = \hat{z}\frac{\mu_0 I}{2\pi}\ln\left(\frac{r_2}{r_1}\right), \qquad (8.10\text{-}4)$$

where r_1 and r_2 are the distances of the point of observation P from the wires and $\hat{z}$ is a unit vector parallel to the wires as shown in Fig. 8.10-2.

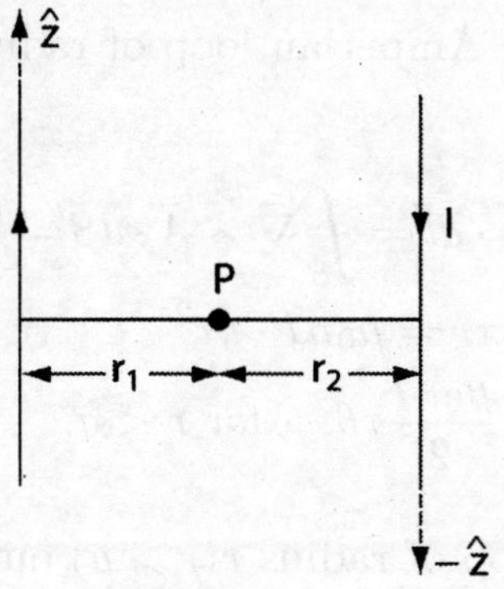

Fig 8.10-2

Now we shall consider two more cases, where current distribution itself extends to infinity and Eqs. (8.8-6), (8.8-9) or (8.8-10) cannot be used directly to calculate $\vec{A}$. However, if we have a knowledge of $\vec{B}$ then $\vec{A}$ can be calculated as shown in the following examples:

3. Infinite sheet of current

Consider Fig. 8.7-4, which shows an infinite current sheet. From Eq. (8.7-6) we can write

$$\vec{B} = \vec{\nabla} \times \vec{A} = \pm \frac{\mu_0 K}{2} \hat{\jmath},$$

where + and − signs apply to $z < 0$ and $z > 0$ regions, respectively. As the current is along x-axis we expect $\vec{A}$ to be also x-axis. Thus

$$\begin{vmatrix} \hat{\imath} & \hat{\jmath} & \hat{k} \\ \frac{\partial}{\partial x} & \frac{\partial}{\partial y} & \frac{\partial}{\partial z} \\ A_x & 0 & 0 \end{vmatrix} = \pm \frac{\mu_0 K}{2} \hat{\jmath} \quad \text{or} \quad \frac{\partial A_x}{\partial z} = \pm \frac{\mu_0 K}{2}.$$

Integrating we get,

$$A_x = \pm \frac{\mu_0 K z}{2} + \text{constant}, \tag{8.10-5}$$

where the constant may be used to define the position at which $A = 0$.

4. Long solenoid

Consider an infinite solenoid with n turns per unit length, radius a and carrying current I. Symmetry of the problem suggests that $\vec{A}$ should have the direction of the current, i.e., the circumferential direction $(\hat{\theta})$. The magnetic field inside is along the axis (z-axis) of the solenoid and is given by

$$\vec{B} = \mu_0 n I \hat{z}.$$

Now, considering a circular Amperian loop of radius r $(r < a)$ inside the solenoid in xy-plane we can write

$$\oint_C \vec{A} \cdot d\vec{l} = \int_S \vec{\nabla} \times \vec{A} \cdot d\vec{S} = \int_S \vec{B} \cdot d\vec{S}$$

$$\text{or } A \cdot 2\pi r = \mu_0 n I \cdot \pi r^2$$

$$\text{or } \vec{A} = \frac{\mu_0 n I}{2} r\hat{\theta}, \quad \text{for } r < a. \tag{8.10-6}$$

For a circular Amperian loop of radius $r (r > a)$ outside the solenoid,

$$\int_S \vec{B} \cdot d\vec{S} = \mu_0 n I \cdot \pi a^2,$$

since the field extends only up to $r = a$. Thus,

$$A \cdot 2\pi r = \mu_0 n I \cdot \pi a^2 \quad \text{or} \quad \vec{A} = \frac{\mu_0 n I a^2}{2r}\hat{\theta} \quad \text{for } r > a. \tag{8.10-7}$$

It is easy to show that $\vec{A}$ as given by the Eq. (8.10-6) or (8.10-7) satisfies the relations $\vec{\nabla} \times \vec{A} = \vec{B}$ and $\vec{\nabla} \cdot \vec{A} = 0$. For example Eq. (8.10-6) can be written in Cartesian coordinates as

$$\vec{A} = \frac{\mu_0 n I}{2} r(-\sin\theta\hat{i} + \cos\theta\hat{j}) = \frac{\mu_0 n I}{2}(-y\hat{i} + x\hat{j}). \tag{8.10-8}$$

Now

$$\vec{\nabla} \times \vec{A} = \frac{\mu_0 n I}{2} \begin{vmatrix} \hat{i} & \hat{j} & \hat{k} \\ \frac{\partial}{\partial x} & \frac{\partial}{\partial y} & \frac{\partial}{\partial y} \\ -y & x & 0 \end{vmatrix} = \hat{k}\mu_0 n I = \vec{B} \quad \left(\because \hat{z} = \hat{k}\right)$$

and

$$\vec{\nabla} \cdot \vec{A} = \frac{\mu_0 n I}{2}\left[\frac{\partial}{\partial x}(-y) + \frac{\partial}{\partial y}(x)\right] = 0.$$

8.11 Magnetic Field at a Large Distance due to a Small Current Loop

In this case, it is easier first to calculate the magnetic vector potential $\vec{A}$ and then to find $\vec{B}$ by using curl operation.

Referring to Fig. 8.11-1 we can write for $\vec{A}$ at P due to the current loop as

$$\vec{A}(\vec{r}) = \frac{\mu_0}{4\pi} \oint_C \frac{I d\vec{l}}{R}. \tag{8.11-1}$$

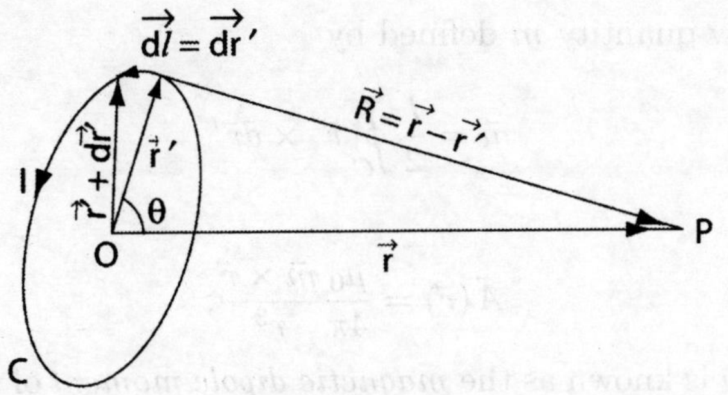

Fig 8.11-1

From figure $d\vec{l} = d\vec{r}'$ and

$$R = |\vec{r} - \vec{r}'| = \sqrt{(\vec{r} - \vec{r}') \cdot (\vec{r} - \vec{r}')} = \left(r^2 - 2\vec{r} \cdot \vec{r}' + r'^2\right)^{1/2}.$$

Considering the dimensions of the loop to be small compared with r and taking the origin O in or near the current loop we can make the following binomial expansion:

$$\frac{1}{R} = \frac{1}{r}\left(1 + \frac{\vec{r} \cdot \vec{r}'}{r^2} + \cdots\right) \quad \text{to first order in } \frac{r'}{r}.$$

Using this we get from Eq. (8.11-1),

$$\vec{A}(\vec{r}) = \frac{\mu_0 I}{4\pi}\left[\frac{1}{r}\oint_C d\vec{r}' + \frac{1}{r^3}\oint_C d\vec{r}'\,(\vec{r} \cdot \vec{r}') + \cdots\right]. \tag{8.11-2}$$

The first integral which represents total vector displacement around a closed loop is simply zero. To put the second integral in a useful form we note that

$$\vec{r} \times (d\vec{r}' \times \vec{r}') = d\vec{r}'\,(\vec{r} \cdot \vec{r}') - \vec{r}'\,(\vec{r} \cdot d\vec{r}') \tag{8.11-3}$$

and for a small change in $\vec{r}'$

$$d\left[\vec{r}'\,(\vec{r} \cdot \vec{r}')\right] = d\vec{r}'\,(\vec{r} \cdot \vec{r}') + \vec{r}'\,(\vec{r} \cdot d\vec{r}'). \tag{8.11-4}$$

Adding Eqs. (8.11-3) and (8.11-4) and dividing by 2 we get

$$d\vec{r}'\,(\vec{r} \cdot \vec{r}') = \frac{1}{2}\vec{r} \times (d\vec{r}' \times \vec{r}') + \frac{1}{2}d\left[\vec{r}'\,(\vec{r} \cdot \vec{r}')\right]. \tag{8.11-5}$$

The last term is an exact differential and hence, it contributes nothing to the second integral in Eq. (8.11-2). Thus, substituting (8.11-5) in Eq. (8.11-2) we may get

$$\vec{A}(\vec{r}) \simeq \frac{\mu_0 I}{4\pi r^3}\oint_C \frac{1}{2}\vec{r} \times (d\vec{r}' \times \vec{r}') = \frac{\mu_0}{4\pi}\left[\frac{I}{2}\oint_C (\vec{r}' \times d\vec{r}')\right] \times \frac{\vec{r}}{r^3}. \tag{8.11-6}$$

Introducing a new quantity $\vec{m}$ defined by

$$\vec{m} = \frac{I}{2}\oint_C \vec{r}\,' \times d\vec{r}\,'. \tag{8.11-7}$$

We can write

$$\vec{A}\,(\vec{r}) = \frac{\mu_0}{4\pi}\frac{\vec{m}\times\vec{r}}{r^3}, \tag{8.11-8}$$

where the quantity $\vec{m}$ is known as the *magnetic dipole moment* of the loop. Its direction is given by the direction of movement of a right-handed screw when rotated in the direction of current. In the special case of a plane loop the definition (8.11-7) of the magnetic moment could be given a simple geometrical meaning. The quantity

$$\frac{1}{2}\oint (\vec{r}\,' \times d\vec{r}\,') = \vec{S}$$

is the area of the whole loop. $\vec{S}$ points perpendicular to the plane of the loop in a direction given by the usual right-hand rule. Therefore, for a plane loop the magnetic moment equals current multiplied by area, i.e., $\vec{m} = I\vec{S}$. Note that a small enough current loop may be assumed to be planar. Now the magnetic field can be determined by taking the curl of Eq. (8.11-8). Thus,

$$\begin{aligned}\vec{B} = \vec{\nabla}\times\vec{A} &= \frac{\mu_0}{4\pi}\vec{\nabla}\times\left(\frac{\vec{m}\times\vec{r}}{r^3}\right)\\ &= \frac{\mu_0}{4\pi}\left[\frac{1}{r^3}\vec{\nabla}\times(\vec{m}\times\vec{r}) + \vec{\nabla}\frac{1}{r^3}\times(\vec{m}\times\vec{r})\right].\end{aligned} \tag{8.11-9}$$

To evaluate the right-hand side we note that

$$\left[\vec{\nabla}\times(\vec{m}\times\vec{r})\right]_{x-\text{comp}} = \frac{\partial}{\partial y}(m_x y - m_y x) - \frac{\partial}{\partial z}(m_z x - m_x z) = 2m_x.$$

Considering other components similarly we get

$$\vec{\nabla}\times(\vec{m}\times\vec{r}) = 2\vec{m}.$$

Since

$$\vec{\nabla}\frac{1}{r^3} = -\frac{3\vec{r}}{r^5},$$

the second term can be written as

$$\begin{aligned}\vec{\nabla}\frac{1}{r^3}\times(\vec{m}\times\vec{r}) &= -\frac{3}{r^5}\vec{r}\times(\vec{m}\times\vec{r}) = -\frac{3}{r^5}\left[\vec{m}r^2 - \vec{r}(\vec{r}\cdot\vec{m})\right]\\ &= -\frac{3\vec{m}}{r^3} + \frac{3\,(\vec{m}\cdot\vec{r})\,\vec{r}}{r^5}.\end{aligned}$$

Thus, using all these results in Eq. (8.11-9) we get finally

$$\vec{B}(\vec{r}) = \frac{\mu_0}{4\pi}\left[-\frac{m}{r^3} + \frac{3(\vec{m}\cdot\vec{r})\,\vec{r}}{r^5}\right]. \tag{8.11-10}$$

This equation shows that the magnetic field of a small current loop at a distant point does not depend on its detailed geometry but only on its magnetic moment $\vec{m}$. The expression for $\vec{B}$ is of the same form as the electric field due to an electric dipole. Thus we may conclude that a small current loop behaves as a magnetic dipole. It can be shown that *the magnetic dipole moment is always independent of the origin.* This indicates that total *magnetic charge* or pole strength is always zero. Note that in the corresponding electrical case the electric dipole moment is independent of origin only when the total charge is zero.

8.12 Magnetic Scalar Potential

Ampere's law for steady currents, i.e., $\vec{\nabla}\times\vec{B} = \mu_0\vec{J}$ indicates that $\vec{B}$ is not a conservative field and thus it cannot be described by a scalar potential. However, *if the current density is zero in some region of space* then $\vec{\nabla}\times\vec{B} = 0$ and $\vec{B}$ can be described as a conservative field. In such regions $\vec{B}$ can be expressed as the gradient of a scalar potential:

$$\vec{B} = -\mu_0\vec{\nabla}\phi, \tag{8.12-1}$$

where ϕ is called the *magnetic scalar potential.*

Now

$$\vec{\nabla}\cdot\vec{B} = 0 \quad \text{or} \quad \vec{\nabla}\cdot\left(\vec{\nabla}\phi\right) = 0 \quad \text{or} \quad \nabla^2\phi = 0. \tag{8.12-2}$$

Thus magnetic scalar potential satisfies Laplace's equation.

We can find unique solutions for it for various situations by applying suitable boundary conditions.

Let us find an explicit expression for the scalar potential of a small current loop. The magnetic field due to a small current loop at a distance point P is given by the Eq. (8.11-10), and it may be written as

$$\vec{B}(\vec{r}) = -\mu_0\vec{\nabla}\left(\frac{\vec{m}\cdot\vec{r}}{4\pi r^3}\right). \tag{8.12-3}$$

Comparing Eqs. (8.12-1) and (8.12-3) we find that the magnetic scalar potential due to a small current loop is given by

$$\phi = \frac{1}{4\pi}\cdot\frac{\vec{m}\cdot\vec{r}}{r^3}, \tag{8.12-4}$$

where $\vec{m}$ is the magnetic moment associated with the current loop. $\vec{m}$ is given by the product of current and area. So *magnetic scalar potential has the dimensions of current.*

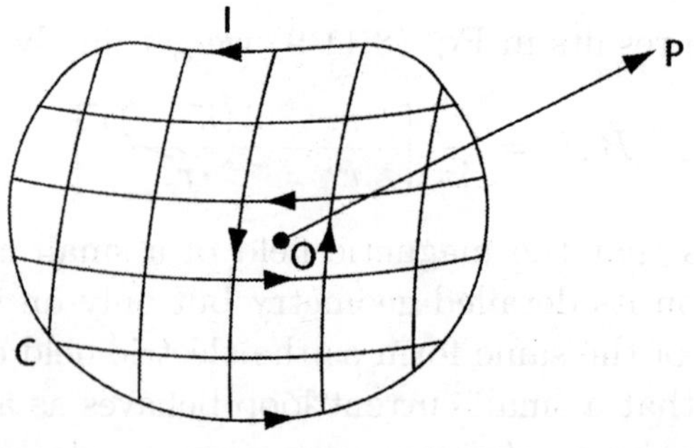

Fig. 8.12-1 : A large circuit is equivalent to a large number of elementary loops

Let us now find the magnetic scalar potential due to a large current loop of arbitrary shape. The large circuit loop C may be imagined to be divided into many small loops around each of which there flows the same current I. Due to cancellation of currents in the common branch of adjacent loops the net effect would be the same as that due to current flow in original circuit (Fig. 8.12-1). If the loops are chosen infinitesimally small, it can be regarded as a plane element with associated magnetic dipole moment

$$d\vec{m} = Id\vec{S},$$

where $d\vec{S}$ is the area of the elementary loop. Contribution to the scalar potential at P due to an elementary loop is

$$d\phi = \frac{d\vec{m}\cdot\vec{r}}{4\pi r^3} \tag{8.12-5}$$

and consequently total potential at P is obtained by integrating over the surface S bounded by C,

$$\phi = \frac{1}{4\pi}\int \frac{d\vec{m}\cdot\vec{r}}{r^3} = \frac{I}{4\pi}\int_S \frac{d\vec{S}\cdot\vec{r}}{r^3} = \frac{I}{4\pi}\int d\Omega,$$

where $d\Omega = \dfrac{d\vec{S}\cdot\vec{r}}{r^3}$ is the solid angle subtended by the elementary loop at P. Thus,

$$\phi = \frac{I}{4\pi}\Omega, \tag{8.12-6}$$

where $\Omega = \int d\Omega$ is the solid angle subtended by the boundary of the circuit at P.

In the above, we note that a large circuit is equivalent to a large number of elementary loops each of which behaves like a magnetic dipole with its axis normal to the surface. Thus, the whole circuit its equivalent to a layer of large number of dipoles or a sheet magnetized normally and bounded by the circuit. Such a layer is known as *magnetic double layer* or a *magnetic shell.*

From Eq. (8.12-6) it is obvious that the unit of ϕ is ampere. Also, the value of ϕ depends only the boundary of the loop and current through it and not on the shape of the loop.

8.13 Magnetic Shell and Ampere's Equivalence Theorem

A thin sheet of magnetic material, which is magnetized everywhere at right angles to its surface is called a *magnetic shell.* The strength (τ) of the shell is defined as its magnetic moment per unit facial area. If $\vec{M}$ be the magnetization (i.e., magnetic dipole moment per unit volume) of a shell of thickness t then strength of the shell is $\tau = Mt$.

Now we know that a current loop, as regards its magnetic effect, behaves like a magnetic double layer or a magnetic shell whose boundary coincides with the boundary of the current loop and has a magnetic moment equal to current × area. Thus, the strength of the equivalent magnetic shell would be

$$\tau = \frac{\text{magnetic moment}}{\text{area}} = \frac{\text{current} \times \text{area}}{\text{area}} = \text{current}.$$

Therefore, the strength of the equivalent magnetic shell equals the current in the loop. Thus we may conclude that *a current carrying loop produces the same magnetic field as is produced by a magnetic shell whose boundary coincides with the boundary of the loop and whose strength is equal to the current flowing in the loop.* This fact was stated by Ampere in 1823 and is known as *Ampere's equivalence theorem.* The magnetic scalar potential ϕ at any point P due to a magnetic shell of strength τ can thus be written by using Eq. (8.12-6) as

$$\phi = \frac{\tau \cdot \Omega}{4\pi}, \tag{8.13-1}$$

where Ω is the solid angle subtended by the shell at P.

8.14 Magnetic Scalar Potential and Field due to a Circular Current Loop

Suppose we consider a circular loop AB of radius a carrying current I. We are to find the magnetic field at an axial point P at a distance x from its centre O (Fig. 8.14-1). The magnetic scalar potential at P, on the positive side, is given by

$$\phi = \frac{I \cdot \Omega}{4\pi},$$

where Ω is the solid angle subtended by the boundary of the loop at P. To find Ω we note that the loop AB may be considered as the boundary of a thin disc cut out of a sphere as shown in Fig. 8.14-1(b). Now the solid angle Ω subtended by the loop AB at P, centre of the sphere of radius $r(= AP)$, is given by

$$\Omega = \frac{\text{area of the cap } ACB}{r^2} = \frac{\int_0^\pi 2\pi r \sin\theta \cdot r d\theta}{r^2} = 2\pi(1 - \cos\theta).$$

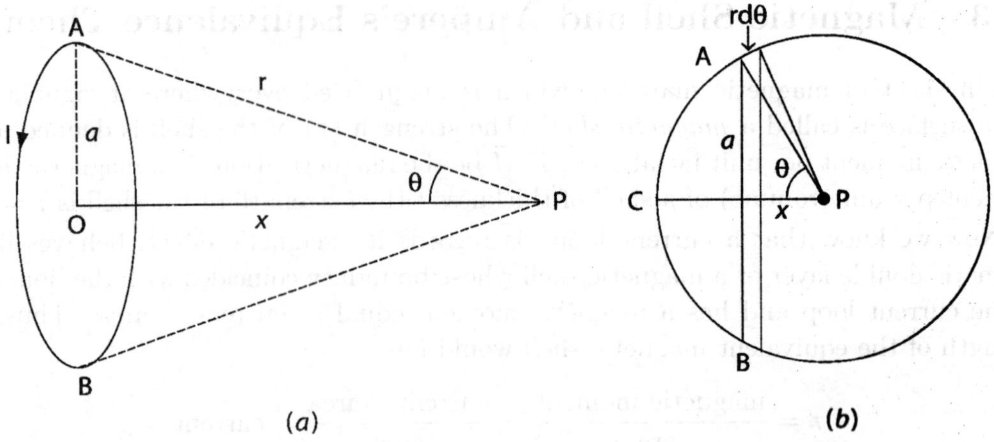

Fig 8.14-1

Therefore,

$$\phi = \frac{I}{4\pi} \cdot 2\pi(1 - \cos\theta) = \frac{I}{2}\left(1 - \frac{x}{\sqrt{a^2 + x^2}}\right).$$

Now the magnetic field at P is given by

$$\vec{B} = -\mu_0 \vec{\nabla}\phi = -\hat{x}\mu_0 \frac{d\phi}{dx} = \hat{x}\frac{\mu_0 I}{2}\left[\frac{1}{\sqrt{a^2 + x^2}} - \frac{x^2}{(a^2 + x^2)^{3/2}}\right]$$

$$= \hat{x}\frac{\mu_0 I a^2}{2\,(a^2 + x^2)^{3/2}},$$

which is exactly the same result as obtained earlier [See, Eq. (8.4-4)].

8.15 Force between Two Current Carrying Conductors

In Section 8.2 we have studied the force a magnetic field exerts on charged particles, current-elements and loops. We have already considered the basic laws and techniques in calculating magnetic field $\vec{B}$ due to current elements. Let us now apply these ideas to study the force between current carrying conductors.

1. Force between two long straight parallel conductors carrying current

Let us consider two long parallel wires (1 and 2) carrying currents I_1 and I_2, separated by a distance a as shown in Fig. 8.15-1. The conductor 1 produces a magnetic field

$$B_1 = \frac{\mu_0}{4\pi} \cdot \frac{2I_1}{a}$$

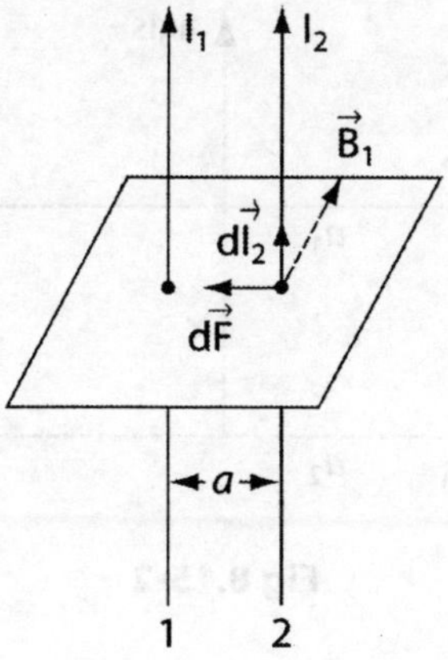

Fig 8.15-1

at the position of conductor 2. It is directed perpendicular to the conductor 2 as shown in Fig. 8.15-1. The force $d\vec{F}$ acting on the element $d\vec{l}_2$ of the conductor 2 due to $\vec{B}_1$ is

$$d\vec{F} = I_2\left(d\vec{l}_2 \times \vec{B}_1\right) \quad \text{or} \quad \left|d\vec{F}\right| = dF = \frac{\mu_0}{4\pi} \cdot \frac{2I_1 I_2}{a} dl_2.$$

This force is directed towards the conductor 1, i.e., attractive if I_1 and I_2 flow in the same direction. On the other hand, the force is directed away from the conductor 1 if I_1 and I_2 flow in opposite directions.

The force per unit length of each wire is of magnitude

$$\left|\vec{F}\right| = \frac{\mu_0}{4\pi} \cdot \frac{2I_1 I_2}{a} \tag{8.15-1}$$

and directed perpendicularly to their lengths.

2. Mutual force between two current carrying parallel circular coaxial coils of nearly the same radius

Let us consider two circular coaxial coils 1 and 2 of radii a_1 and a_2 and carrying currents I_1 and I_2, respectively. We assume a_1 to be nearly equal to a_2 and the distance of separation x between the coils to be small. Now the force per unit length of coil 1 due to coil 2 is

$$\frac{\mu_0}{4\pi} \cdot \frac{2I_1 I_2}{PQ}$$

and it acts along the directions such as QP. By symmetry the components of this force normal to the axis will cancel out when we integrate the force over the entire coil. But the components parallel to the axis, i.e.,

$$\frac{\mu_0}{4\pi}\frac{2I_1 I_2}{PQ}\sin\theta = \frac{\mu_0}{4\pi} \cdot \frac{2I_1 I_2 x}{(PQ)^2}$$

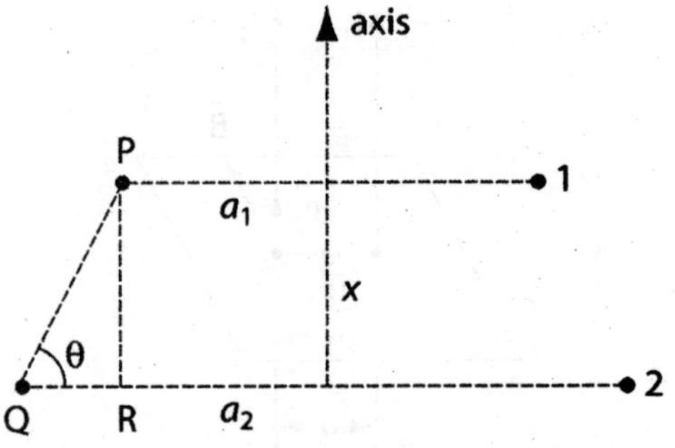

Fig 8.15-2

will add up. So the total force on coil 1 due to coil 2 will be

$$F = \frac{\mu_0}{4\pi} \cdot \frac{2I_1 I_2 x}{(PQ)^2} \cdot 2\pi a_1 = \frac{\mu_0}{4\pi} \cdot \frac{4\pi I_1 I_2 a_1 x}{(a_2 - a_1)^2 + x^2}. \qquad (8.15\text{-}2)$$

This is also the force on coil 2 due to coil 1 but in opposite direction. The force is attractive if I_1 and I_2 flow in the same sense, otherwise it will be repulsive.

The force F is zero for $x = 0$. For F to be maximum we must have $dF/dx = 0$. This gives $x = a_2 - a_1$. This means that the force will be maximum when the distance between the centres of the two coils equals the difference between their radii. The maximum value of F is then

$$F_{\text{max}} = \frac{\mu_0}{4\pi} \cdot \frac{2\pi I_1 I_2 a_1}{a_2 - a_1}. \qquad (8.15\text{-}3)$$

3. Force between two arbitrary current loops

Let us consider two current elements $I_1 d\vec{l}_1$ and $I_2 d\vec{l}_2$ of two arbitrary current loops 1 and 2(Fig. 8.15-3). According to Biot-Savart law the magnetic field $d\vec{B}_1$ produced by the element $I_1 d\vec{l}_1$ on the element $I_2 d\vec{l}_2$ is

$$d\vec{B}_1 = \frac{\mu_0}{4\pi} \frac{I_1 d\vec{l}_1 \times \vec{R}_{12}}{R_{12}^3}.$$

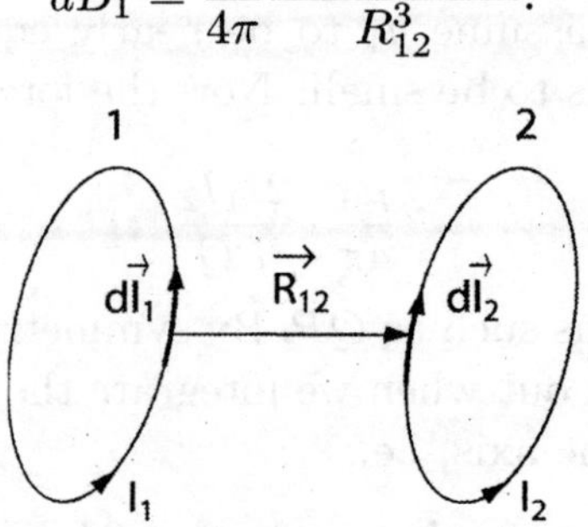

Fig 8.15-3

The force experienced by the element $I_2 d\vec{l_2}$ in this field $d\vec{B_1}$ is

$$d\vec{F}_{21} = I_2 d\vec{l_2} \times d\vec{B_1} = \frac{\mu_0}{4\pi} \cdot \frac{2I_1I_2}{R_{12}^3} d\vec{l_2} \times \left(d\vec{l_1} \times \vec{R}_{12}\right). \tag{8.15-4}$$

This is the law of force between two current elements and is analogous to Coulomb's law in electrostatics.

Total force on loop 2 due to loop 1 is

$$\vec{F}_{21} = \frac{\mu_0 I_1 I_2}{4\pi} \oint_1 \oint_2 \frac{d\vec{l_2} \times \left(d\vec{l_1} \times \vec{R}_{12}\right)}{R_{12}^3}. \tag{8.15-5}$$

Similarly, the force on loop 1 due to loop 2 can be obtained as

$$\vec{F}_{12} = \frac{\mu_0 I_1 I_2}{4\pi} \oint_1 \oint_2 \frac{d\vec{l_1} \times \left(d\vec{l_2} \times \vec{R}_{21}\right)}{R_{12}^3}. \tag{8.15-6}$$

It is important to point out here that the Eqs. (8.15-4)–(8.15-6) were experimentally established by Oersted and Ampere. Biot-Savart law is actually based on it.

Interactions between current loops and Newton's third law

As $d\vec{l_2} \times \left(d\vec{l_1} \times \vec{R}_{12}\right) \neq d\vec{l_1} \times \left(d\vec{l_2} \times \vec{R}_{21}\right)$, it appears that $d\vec{F}_{12} \neq -d\vec{F}_{21}$. This means that Newtons third law cannot be applied to force between two current elements. This is because isolated steady current elements are non-existent. However, for force between two real current loops we expect that $\vec{F}_{12} = -\vec{F}_{21}$. Using the rules of vector triple product we can write

$$d\vec{l_2} \times \left(d\vec{l_1} \times \vec{R}_{12}\right) = d\vec{l_1} \left(d\vec{l_2} \cdot \vec{R}_{12}\right) - \vec{R}_{12} \left(d\vec{l_2} \cdot d\vec{l_1}\right).$$

Therefore,

$$\vec{F}_{21} = \frac{\mu_0 I_1 I_2}{4\pi} \oint_1 d\vec{l_1} \oint_2 \frac{d\vec{l_2} \cdot \vec{R}_{12}}{R_{12}^3} - \frac{\mu_0 I_1 I_2}{4\pi} \oint_1 \oint_2 \frac{\vec{R}_{12} \left(d\vec{l_2} \cdot d\vec{l_1}\right)}{R_{12}^3}.$$

As

$$\vec{\nabla} \frac{1}{R_{12}} = -\frac{\vec{R}_{12}}{R_{12}^3}$$

the first integral can be written in the form

$$-\oint_1 d\vec{l_1} \oint_2 \vec{\nabla} \frac{1}{R_{12}} \cdot d\vec{l_2} = -\oint_1 d\vec{l_1} \int_{S_2} \vec{\nabla} \times \vec{\nabla} \frac{1}{R_{12}} \cdot d\vec{S_2} \quad \text{[Using Stokes' law]}$$

$$= 0 \quad [\because \text{ Curl of gradient of any scalar is always zero}]$$

where S_2 is any surface bounded by the loop 2 and $\vec{\nabla}$ involves derivatives with respect to any point on the loop 2. Thus,

$$\vec{F}_{21} = -\frac{\mu_0 I_1 I_2}{4\pi} \oint_1 \oint_2 \frac{\vec{R}_{12}\left(d\vec{l}_2 \cdot d\vec{l}_1\right)}{R_{12}^3}.$$

Since $d\vec{l}_2 \cdot d\vec{l}_1 = d\vec{l}_1 \cdot d\vec{l}_2$ and $\vec{R}_{12} = -\vec{R}_{21}$. Thus, we find that Newton's third law is consistent with physically realizable complete current loops.

8.16 Current Loop in an External Magnetic Field

If a current loop is placed in an external magnetic field $\vec{B}$, it experiences forces and torques. The force on a current element $Id\vec{l}$ is given by $d\vec{F} = Id\vec{l} \times \vec{B}$. The torque on the element chosen at $\vec{r}'$ is given by $d\vec{\tau} = \vec{r}' \times d\vec{F}$ and the total torque on the loop is

$$\vec{\tau} = I \oint_C \vec{r}' \times \left(d\vec{l} \times \vec{B}\right). \tag{8.16-1}$$

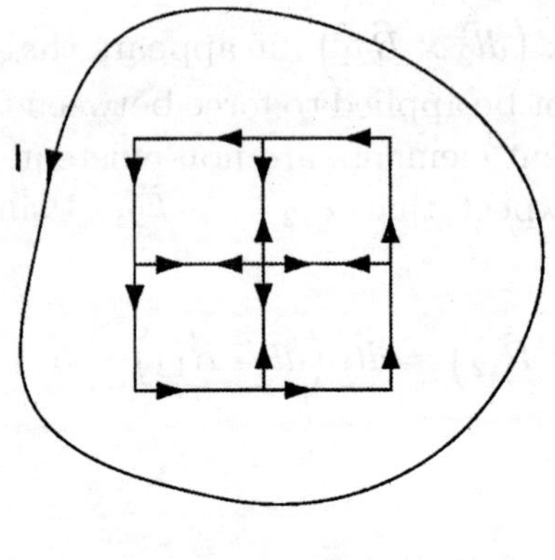

Fig 8.16-1

It can be shown that the torque on any arbitrary planar current loop placed in a uniform field $\vec{B}$ is given by $\vec{\tau} = \vec{m} \times \vec{B}$, where $\vec{m}$ is the magnetic dipole moment of the current loop. However, for simplicity we consider a square current loop of side a placed in a uniform $\vec{B}$. There is no loss of generality in assuming a square loop, because any arbitrary current loop can be imagined to be divided into large number of small square loops around each of which flows the same current, where due to cancellation of currents in the common internal sides the net effect remains the same as that due to current flow in the original loop.

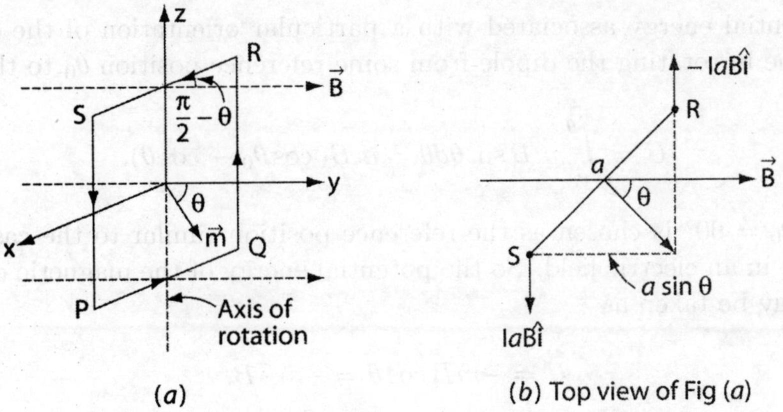

Fig 8.16-2

Let us consider a small square loop $PQRS$ carrying current I and placed in a uniform magnetic field $\vec{B} = B\hat{\jmath}$. Let the normal to the plane of the loop makes an angle θ with $\vec{B}$. The sides PQ and RS make an angle $\frac{\pi}{2} - \theta$ with the field lines. The forces acting on PQ and RS are equal, opposite and act along the line joining the centres of PQ and RS. Thus they cancel each other. The arms SP and QR are perpendicular to the field lines. The force on side SP is

$$I\overrightarrow{SP} \times \vec{B} = I\left(-a\hat{k}\right) \times B\hat{\jmath} = IaB\hat{\imath}$$

and that on QR is

$$I\overrightarrow{QR} \times \vec{B} = I\left(a\hat{k}\right) \times B\hat{\jmath} = -IaB\hat{\imath}.$$

These two forces are equal and opposite but their line of action is different. They generate a torque

$$\vec{\tau} = IaB \cdot a\sin\theta \cdot \hat{k} = Ia^2B\sin\theta\hat{k} = mB\sin\theta\hat{k} = \vec{m} \times \vec{B}, \qquad (8.16\text{-}2)$$

where

$$m = |\,\vec{m}\,| = Ia^2 = \text{current} \times \text{area}$$

is the magnetic dipole moment of the loop, whose direction is given by the usual right-hand rule, i.e., $\vec{m}$ is in the direction in which a right-handed screw advances when rotated along I.

As argued earlier the Eq. (8.16-2) though derived for a square loop is valid for a loop of any arbitrary shape placed in a uniform $\vec{B}$. For a nonuniform $\vec{B}$, Eq. (8.16-2) can be used only for loops of infinitesimal dimensions.

Note that the Eq. (8.16-2) is similar to the torque experienced by an electric dipole $\vec{\tau} = \vec{p} \times \vec{E}$. The torque is in such a direction as to line the dipole parallel to the field.

Since a torque acts on a current loop or equivalently a magnetic dipole when placed in a magnetic field, work is needed to change the orientation of such a dipole. Thus there is potential energy associated with a particular orientation of the dipole. Now the work done in rotating the dipole from some reference position θ_0 to the position θ is given by

$$U = \int_{\theta_0}^{\theta} mB \sin\theta d\theta = mB(\cos\theta_0 - \cos\theta).$$

Usually $\theta_0 = 90°$ is chosen as the reference position similar to the case of electric dipole placed in an electric field. So the potential energy of the magnetic dipole in any position θ may be taken as

$$U = -mB\cos\theta = -\vec{m}\cdot\vec{B}. \tag{8.16-3}$$

The result is similar to the potential energy of an electric dipole placed in electrostatic field; $U = -\vec{p}\cdot\vec{E}$.

From Eq. (8.2-4) we find that force on a closed loop placed in a uniform magnetic field $\vec{B}$ is zero. But this is not the case when $\vec{B}$ is nonuniform. The force on a current loop of dipole moment $\vec{m}$ placed in a nonuniform field $\vec{B}$ can be written as the negative gradient of the potential energy,

$$\begin{aligned}\vec{F} &= -\vec{\nabla}U = \vec{\nabla}\left(\vec{m}\cdot\vec{B}\right) \\ &= \left(\vec{m}\cdot\vec{\nabla}\right)\vec{B} + \left(\vec{B}\cdot\vec{\nabla}\right)\vec{m} + \vec{m}\times\left(\vec{\nabla}\times\vec{B}\right) + \vec{B}\times\left(\vec{\nabla}\times\vec{m}\right).\end{aligned} \tag{8.16-4}$$

Since $\vec{m}$ is a constant vector, $\left(\vec{B}\cdot\vec{\nabla}\right)\vec{m} = 0$ and $\vec{\nabla}\times\vec{m} = 0$. Therefore,

$$\vec{F} = \left(\vec{m}\cdot\vec{\nabla}\right)\vec{B} + \vec{m}\times\left(\vec{\nabla}\times\vec{B}\right).$$

For cases with $\vec{\nabla}\times\vec{B} = 0$ (i.e., the magnetic field has no sources at the position of the dipole), we have

$$\vec{F} = \left(m\cdot\vec{\nabla}\right)\vec{B}. \tag{8.16-5}$$

This result is also similar to the case of an electric dipole placed in an electric field; $\vec{F} = \left(\vec{p}\cdot\vec{\nabla}\right)\vec{E}$.

We should note that the torque as given by the Eq. (8.16-2) tries to rotate the loop so that the loop embraces maximum flux of $\vec{B}$. The magnetic force as given by Eq. (8.16-5) tries to push the loop into a region of stronger field.

More general proof of the relation $\vec{\tau} = \vec{m}\times\vec{B}$

Here we try to prove that the torque on any planar current carrying loop of any arbitrary shape (not just a square loop as done earlier) placed in a uniform magnetic field is given by $\vec{\tau} = \vec{m}\times\vec{B}$, where $\vec{m}$ is the dipolement of the current loop. Force on a current element $Id\vec{l}$ of the loop is $d\vec{F} = Id\vec{l} \times \vec{B}$ and the torque is

$$d\vec{\tau} = \vec{r} \times d\vec{F} = I\vec{r} \times \left(d\vec{l} \times \vec{B}\right). \tag{8.16-6}$$

From vector analysis,

$$\vec{r} \times \left(d\vec{l} \times \vec{B}\right) + d\vec{l} \times \left(\vec{B} \times \vec{r}\right) + \vec{B} \times \left(\vec{r} \times d\vec{l}\right) = 0$$

$$\text{or} \quad \vec{r} \times \left(d\vec{l} \times \vec{B}\right) = -d\vec{l} \times \left(\vec{B} \times \vec{r}\right) - \vec{B} \times \left(\vec{r} \times d\vec{l}\right) \tag{8.16-7}$$

Also, we have

$$d\left[\vec{r} \times \left(\vec{r} \times \vec{B}\right)\right] = d\vec{r} \times \left(\vec{r} \times \vec{B}\right) + \vec{r} \times \left(d\vec{r} \times \vec{B}\right) \quad [\because \ \vec{B} \text{ is uniform}]$$

Since $d\vec{r} = d\vec{l}$ we can write

$$\vec{r} \times \left(d\vec{l} \times \vec{B}\right) = d\left[\vec{r} \times \left(\vec{r} \times \vec{B}\right)\right] - d\vec{l} \times \left(\vec{r} \times \vec{B}\right). \tag{8.16-8}$$

Adding (8.16.7) and (8.16.8) we get

$$2\vec{r} \times \left(d\vec{l} \times \vec{B}\right) = d\left[\vec{r} \times \left(\vec{r} \times \vec{B}\right)\right] - \vec{B} \times \left(\vec{r} \times d\vec{l}\right). \tag{8.16-9}$$

Therefore, from (8.16.6) and (8.16.9),

$$d\vec{\tau} = \frac{1}{2}I\left\{d\left[\vec{r} \times \left(\vec{r} \times \vec{B}\right)\right] - \vec{B} \times \left(\vec{r} \times d\vec{l}\right)\right\}.$$

So total torque on the loop is

$$\vec{\tau} = \frac{1}{2}I\left\{\oint_C d\left[\vec{r} \times \left(\vec{r} \times \vec{B}\right)\right] - \vec{B} \times \oint_C \left(\vec{r} \times d\vec{r}\right)\right\}.$$

The first integral on the right-hand is zero because the integrand is an exact differential. Therefore,

$$\vec{\tau} = \left[\frac{1}{2}I\oint_C \left(\vec{r} \times d\vec{r}\right)\right] \times \vec{B} = \vec{m} \times \vec{B},$$

where

$$\vec{m} = \tfrac{1}{2}I\oint_C \left(\vec{r} \times d\vec{r}\right)$$

is the dipole moment of the current loop.

SOLVED PROBLEMS

► **Problem 1.** *A proton at a distance* 10 cm *from a long straight wire carrying current* 1*A, is moving parallel to the direction of current flow with a speed* 0.6*c. Calculate the force on the proton.*

Solution. The magnetic field at the position of the proton due to the wire is of magnitude

$$\left|\vec{B}\right| = \frac{\mu_0}{4\pi}\cdot\frac{2I}{a} = 10^{-7}\cdot\frac{2\times 1}{0.1}T = 2\times 10^{-6}T.$$

Its direction is determined by the usual right-hand rule.

Now the force on the proton is of magnitude

$$\left|\vec{F}\right| = \left|e\vec{v}\times\vec{B}\right| = evB \quad \left(\because\ \vec{v}\perp^{r}\vec{B}\right)$$

$$= 1.6\times 10^{-19}\times 0.6\times 3\times 10^{8}\times 2\times 10^{-6}\,\mathrm{N}$$

$$= 5.76\times 10^{-17}\,\mathrm{N}.$$

It is directed towards the wire.

► **Problem 2.** *An electron of charge e is rotating n times per see around the nucleus in a circular orbit of radius a. Find the magnetic field B at the position of the nucleus. If* $a = 5.1\times 10^{-11}$ *m and* $n = 6.8\times 10^{15}$*times/sec, calculate the value of B.*

[C.U. 2014, 2018]

Solution. If T is the time period of revolution then the equivalent current is

$$I = \frac{e}{T} = e.n.$$

So the magnetic field at the centre

$$B = \frac{\mu_0}{4\pi}\cdot\frac{2\pi I}{a} = \frac{\mu_0 en}{2a}.$$

Now putting the values of a and n we get

$$B = \frac{4\pi\times 10^{-7}\times 1.6\times 10^{-19}\times 6.8\times 10^{15}}{2\times 5.1\times 10^{-11}}\,\mathrm{T} = 13.4\,\mathrm{T}.$$

► **Problem 3.** *A wire carrying current I is bent into the form of a regular n-sided polygon. The distance of any vertex from the centre of the polygon is r. Find the magnetic field at its centre. Discuss the case* $n\to\infty$. **[J.U. 2017, C.U. 2018]**

Solution. An arm of a polygon of n sides makes an angle of $2\pi/n$ at the centre (O). Now the field at O due the entire wire will be equal to n times the field due to a single side. Thus,

$$B = n \cdot \frac{\mu_0 I}{4\pi a} \cdot \left(\sin\frac{\pi}{n} + \sin\frac{\pi}{n}\right) = \frac{\mu_0 n I}{4\pi} \cdot \frac{2\sin\frac{\pi}{n}}{r\cos\frac{\pi}{n}} \quad \text{or} \quad B = \frac{\mu_0 n I}{2\pi r}\tan\frac{\pi}{n}.$$

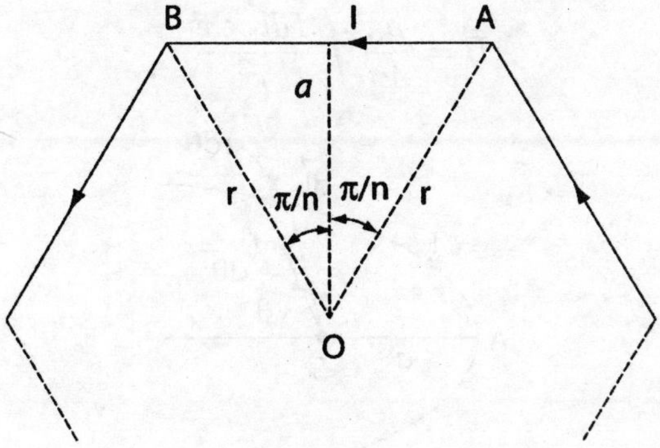

Fig 8.P-3

In the limit $n \to \infty$, $\tan\frac{\pi}{n} \to \frac{\pi}{n}$ and then

$$B = \frac{\mu_0 I}{2r},$$

which is same as the magnetic field at the centre of a circular loop of radius r carrying current I.

▶ **Problem 4.** *An infinitely long wire carries a current I. It is bent so as to have a semicircular detour around the origin, with radius r. Calculate the magnetic field at the origin.*

Solution. The straight parts of the wire do not contribute to the field at O since for them $Id\vec{l} \times \vec{r} = 0$. Considering the semicircular part we get for the field at O as

$$\vec{B} = \frac{\mu_0}{4\pi}\int \frac{Id\vec{l} \times \vec{r}}{r^3}.$$

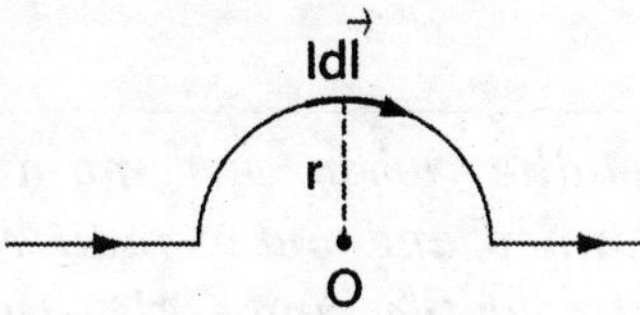

Fig 8.P-4

Obviously, $d\vec{l} \perp^r \vec{r}$ and $\vec{B}$ points into the page. The magnitude of $\vec{B}$ is

$$B = \frac{\mu_0 I}{4\pi r^2}\int dl = \frac{\mu_0 I}{4\pi r^2} \cdot \pi r = \frac{\mu_0 I}{4r}.$$

► **Problem 5.** *An infinite wire carrying current I is bent in the form of a parabola. Find the magnetic field at the focus of the parabola. Take the distance of the focus from the apex as a.*

Solution. From Biot-Savart law the magnetic field at S (Fig. 8.P-5) is given by

$$\vec{B} = \frac{\mu_0}{4\pi} \int \frac{I d\vec{l} \times \vec{r}}{r^3}.$$

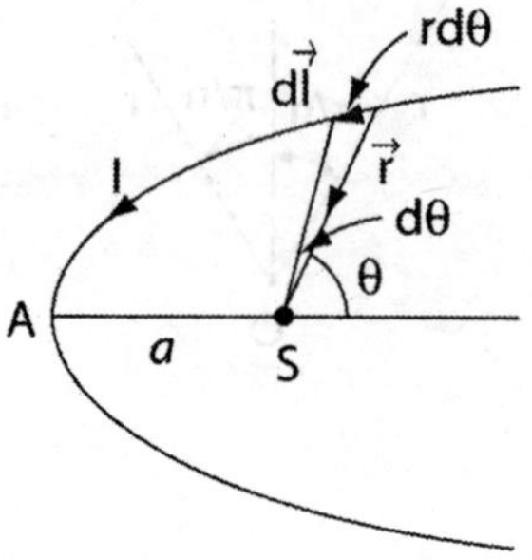

Fig 8.P-5

From figure we note that

$$\left| d\vec{l} \times \vec{r} \right| = \text{area of the parallelogram generated by } d\vec{l} \text{ and } \vec{r}$$
$$= 2 \times \tfrac{1}{2} r \cdot r d\theta = r^2 d\theta.$$

$$\therefore \quad \left| \vec{B} \right| = \frac{\mu_0 I}{4\pi} \int_0^{2\pi} \frac{d\theta}{r}.$$

Using $r(1 - \cos\theta) = 2a$ as the equation to the parabola we get

$$\left| \vec{B} \right| = \frac{\mu_0 I}{4a}.$$

$\vec{B}$ is directed perpendicular to the plane of the parabola in the usual right-hand sense.

► **Problem 6.** *Suppose a thin metallic ribbon carrying a steady current I is bent into the form of a circular ring of inner and outer radii r_1 and r_2, respectively. Find the magnetic field $\vec{B}$ at the centre of the ring. Also calculate the magnetic moment associated with it.*

Solution. Current through an elementary ring of radius r and thickness dr is

$$dI = \frac{I}{r_2 - r_1} dr.$$

Magnetic field at the centre due to it is

$$dB = \frac{\mu_0 dI}{2r} = \frac{\mu_0 I}{2(r_2 - r_1)} \frac{dr}{r}.$$

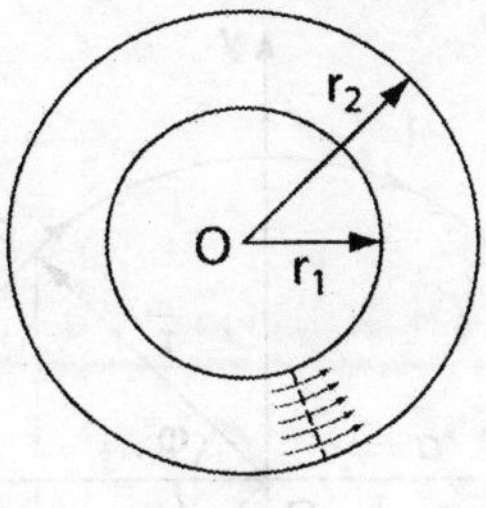

Fig 8.P-6

∴ Total field at the centre would be

$$B = \int dB = \frac{\mu_0 I}{2(r_2 - r_1)} \int_{r_1}^{r_2} \frac{dr}{r} = \frac{\mu_0 I}{2(r_2 - r_1)} \ln \frac{r_2}{r_1}.$$

The magnetic moment associated with the elementary ring is

$$dm = dI \times \pi r^2 = \frac{\pi I}{r_2 - r_1} r^2 dr.$$

∴ Total magnetic moment

$$m = \int dm = \frac{\pi I}{r_2 - r_1} \int_{r_1}^{r_2} r^2 dr$$

$$= \frac{\pi I}{r_2 - r_1} \frac{(r_2^3 - r_1^3)}{3} = \frac{\pi I}{3} \left(r_1^2 + r_1 r_2 + r_2^2 \right).$$

Both $\vec{B}$ and $\vec{m}$ are directed perpendicular to the plane of the ribbon in the usual right-hand sense.

▶ **Problem 7.** *A semicircular wire carries a steady current I. Show that the magnetic field at any point P on the other half of the circle is given by*

$$\vec{B} = \hat{k} \frac{\mu_0 I}{8\pi a} \ln \left[\frac{\tan \frac{\theta + \pi}{4}}{\tan \frac{\theta}{4}} \right],$$

where a is the radius of the circle, $\hat{k}$ unit outward normal to the plane of the circle and θ is the angle which the line joining the centre of the circle and P makes with the diameter of the semicircle.

Solution. Referring to Fig. 8.P-7, we can write for the position vectors of the source point P' as $\vec{r}\,' = a\cos\varphi\hat{i} + a\sin\varphi\hat{j}$ and that of the field point P as $\vec{r} = a\cos\theta\hat{i} - a\sin\theta\hat{j}$.

Therefore,

$$\vec{R} = \vec{r} - \vec{r}' = a(\cos\theta - \cos\varphi)\hat{i} - a(\sin\theta + \sin\varphi)\hat{j}$$

and $d\vec{l} = -a\sin\varphi d\varphi\hat{i} + a\cos\varphi d\varphi\hat{j}$.

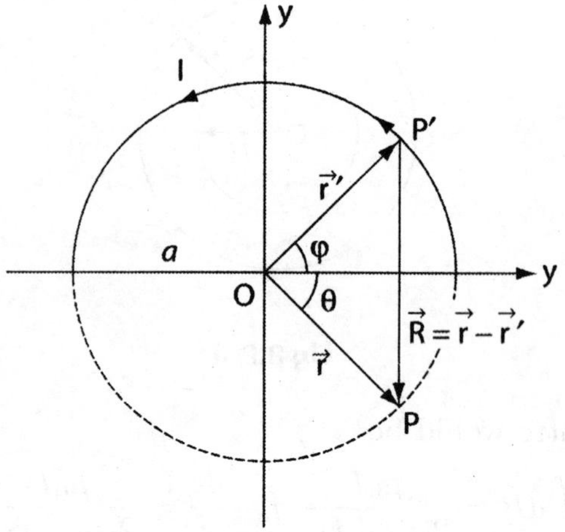

Fig 8.P-7

$$\therefore \quad d\vec{l} \times \vec{R} = a^2 d\varphi \begin{vmatrix} \hat{i} & \hat{j} & \hat{k} \\ -\sin\varphi & \cos\varphi & 0 \\ \cos\theta - \cos\varphi & -(\sin\theta + \sin\varphi) & 0 \end{vmatrix}$$

$$= \hat{k}a^2 d\varphi \{1 - \cos(\theta + \varphi)\}.$$

Now

$$\vec{B} = \frac{\mu_0 I}{4\pi}\int \frac{d\vec{l} \times \vec{R}}{R^3} = \hat{k}\frac{\mu_0 I a^2}{4\pi}\int_0^{\pi} \frac{\{1 - \cos(\theta + \varphi)\}\, d\varphi}{\{a^2(\cos\theta - \cos\varphi)^2 + a^2(\sin\theta + \sin\varphi)\}^{3/2}}$$

or,

$$\vec{B} = \hat{k}\frac{\mu_0 I}{4\pi a}\int_0^{\pi} \frac{\{1 - \cos(\theta + \varphi)\}\, d\varphi}{\{2 - 2\cos(\theta + \varphi)\}^{3/2}} = \hat{k}\frac{\mu_0 I}{4\pi a \cdot 2^{3/2}}\int_0^{\pi} \frac{d\varphi}{\sqrt{1 - \cos(\theta + \varphi)}}$$

$$= \hat{k}\frac{\mu_0 I}{4\pi a \cdot 2^{3/2}}\int_0^{\pi} \frac{d\varphi}{\sqrt{2} \cdot \sin\frac{\theta + \varphi}{2}} = \hat{k}\frac{\mu_0 I}{16\pi a} \cdot \left[2\ln\left|\tan\frac{\theta + \varphi}{4}\right|\right]_0^{\pi}$$

$$= \hat{k}\frac{\mu_0 I}{8\pi a}\ln\left[\frac{\tan\frac{\theta + \pi}{4}}{\tan\frac{\theta}{4}}\right].$$

► **Problem 8.** *Find the magnetic field at a distance z above the centre of a square loop of side a carrying a current I. Show that for $z \gg a$, it reduces to the field of a dipole with the appropriate dipole moment.*

Solution. Referring to Fig. 8.P-8, we can write for the magnetic field at P due to one side as

$$B_1 = \frac{\mu_0 I}{4\pi\sqrt{z^2+\frac{a^2}{4}}}(\sin\theta_1 + \sin\theta_2),$$

where

$$\sin\theta_1 = \sin\theta_2 = \frac{A_1O_1}{A_1P} = \frac{A_1O_1}{\sqrt{A_1O_1^2 + O_1P^2}} = \frac{A_1O_1}{\sqrt{A_1O_1^2 + OP^2 + O_1O^2}}$$

$$= \frac{a/2}{\sqrt{\frac{a^2}{4} + z^2 + \frac{a^2}{4}}} = \frac{a/2}{\sqrt{z^2+\frac{a^2}{2}}}$$

$$\therefore \quad B_1 = \frac{\mu_0 I}{4\pi}\frac{a}{\sqrt{z^2+\frac{a^2}{4}}\cdot\sqrt{z^2+\frac{a^2}{2}}}.$$

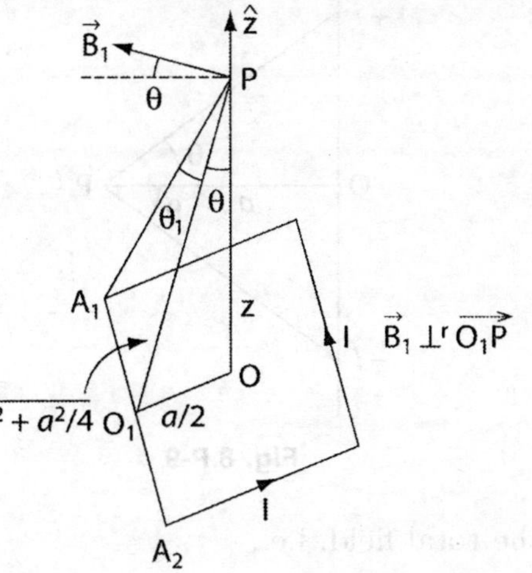

Fig 8.P-8

This field is normal to O_1P and hence its component along $\hat{z}$ is

$$B_1 \sin\theta = B_1 \cdot \frac{a/2}{\sqrt{z^2+\frac{a^2}{4}}}.$$

So considering the contributions of all four sides we find that the components normal to $\hat{z}$ cancel out and components parallel to $\hat{z}$ are added up to give the resultant field as

$$\vec{B} = \hat{z}4B_1\sin\theta = \hat{z}\frac{\mu_0 I}{2\pi}\cdot\frac{a^2}{\left(z^2+\frac{a^2}{4}\right)\sqrt{z^2+\frac{a^2}{2}}}$$

For $z \gg a$,

$$\vec{B} \approx \hat{z}\frac{\mu_0 I}{2\pi} \cdot \frac{a^2}{z^3} = \hat{z}\frac{\mu_0}{4\pi}\frac{2m}{z^3} \quad \text{where } m = Ia^2.$$

This is the magnetic field at an axial point at a distance z from a magnetic dipole of moment $\vec{m} = Ia^2\hat{z}$, placed at the centre of the square.

► **Problem 9.** *Consider the magnetic field produced by a long straight wire carrying current I, at a point P at a distance a from it. Calculate the length of current that contributes 90% of the total field at P.*

Solution. Referring to Fig. 8.P-9, the magnetic field at P due to a current length $2l$ is given by

$$B = \frac{\mu_0 I}{4\pi a}(\sin\theta + \sin\theta) = \frac{\mu_0 I}{2\pi a} \cdot \sin\theta = \frac{\mu_0 I}{2\pi a} \cdot \frac{l}{\sqrt{a^2 + l^2}}.$$

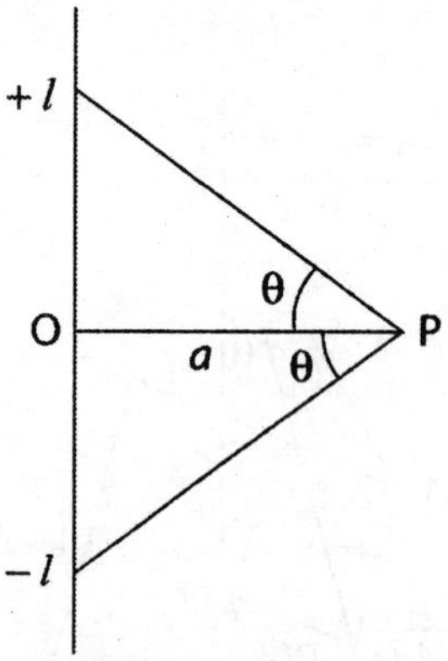

Fig. 8.P-9

This equals 90% of the total field, i.e.,

$$\frac{\mu_0 I}{2\pi a} \cdot \frac{l}{\sqrt{a^2 + l^2}} = \frac{90}{100} \times \frac{\mu_0 I}{2\pi a}.$$

Solving this we get the required length as $2l = 4.13a$.

► **Problem 10.** *A straight wire of length $2l$ carries a charge λ per unit length. It rotates uniformly with an angular velocity ω about an axis passing through its mid-point and perpendicular to its length. Show that the equivalent magnetic dipole moment is of magnitude $(1/3)\lambda\omega l^3$.*

Solution. Consider an element dx at a distance x from the centre O (Fig. 8.P-10). Charge on it is λdx and it is rotating $\omega/2\pi$ times per sec in a circle of radius x. So the current produced is

$$I = \lambda dx \times \frac{\omega}{2\pi}$$

and the associated magnetic moment is

$$I \times \pi x^2 = \tfrac{1}{2}\lambda\omega x^2 dx.$$

The total dipole moment is, therefore,

$$\int_{-l}^{+l} \tfrac{1}{2}\lambda\omega x^2 dx = \tfrac{1}{3}\lambda\omega l^3.$$

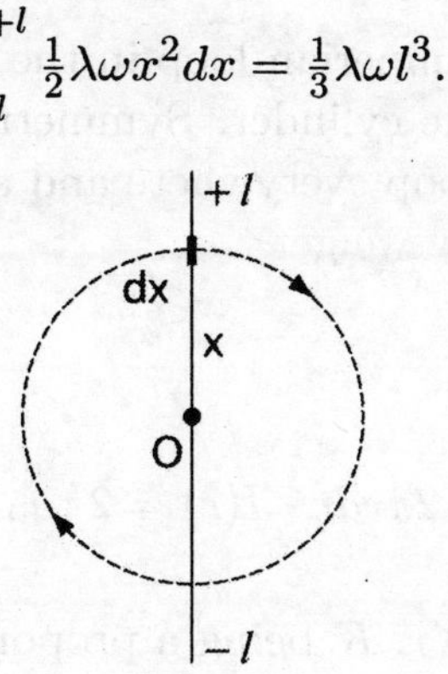

Fig 8.P-10

▶ **Problem 11.** *A solenoid 1 m long and radius 4 cm has 1000 turns and is carrying a current of 1 A. Find the magnetic field at the centre.*

Solution. The magnetic field at P is given by

$$B = \frac{\mu_0 n I}{2}(\cos\theta_2 - \cos\theta_1).$$

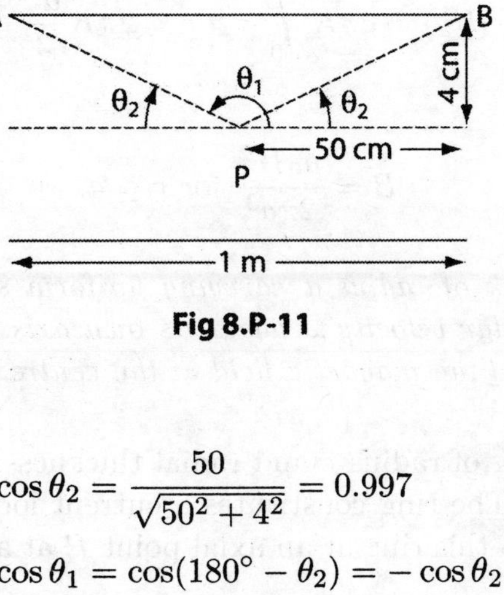

Fig 8.P-11

From Fig. 8.P-11,

$$\cos\theta_2 = \frac{50}{\sqrt{50^2 + 4^2}} = 0.997$$

$$\cos\theta_1 = \cos(180° - \theta_2) = -\cos\theta_2.$$

$$\therefore \quad B = \mu_0 n I \cdot \cos\theta_2 = 4\pi \times 10^{-7} \times \frac{1000}{1} \times 1 \times 0.997\,\text{T} = 12.5 \times 10^{-4}\,\text{T}.$$

► **Problem 12.** *A steady current I flows down a long cylindrical conductor of radius a. The current density at a distance r from the axis of the conductor is proportional to r. Calculate the magnetic field both inside and outside the wire as a function of r.*

[C.U. 2011]

Solution. Let us consider an Amperian loop in the form of a circle of radius $r(r < a)$ with its centre on the axis of the cylinder. Symmetry of the problem indicates that $\vec{B}$ is tangential to the Amperian loop everywhere and also of constant magnitude all over it. So by applying Ampere's law we get

$$\oint \vec{B} \cdot d\vec{l} = \mu_0 I_{\text{enc}}$$

$$\text{or} \quad B \cdot 2\pi r = \mu_0 \int_0^r 2\pi r dr \cdot J(r) = 2\pi\mu_0 K \int_0^r r^2 dr = 2\pi\mu_0 K \frac{r^3}{3},$$

where we assume that $J(r) = Kr$, K being a proportionality constant.

$$\therefore \quad B = \mu_0 K \frac{r^3}{3} \text{ for } r \leq a.$$

For any external point $r > a$, $I_{\text{enc}} = I$ and then

$$B \cdot 2\pi r = \mu_0 I \quad \text{or} \quad B = \frac{\mu_0 I}{2\pi r} \text{ for } r \geq a.$$

Total current

$$I = \int_0^a 2\pi r dr \cdot J(r) = 2\pi K \int_0^a r^2 dr = 2\pi K \frac{a^3}{3} \quad \text{or} \quad K = \frac{3I}{2\pi a^3}.$$

Thus,

$$B = \frac{\mu_0 I r^2}{2\pi a^3} \text{ for } r \leq a.$$

► **Problem 13.** *A thin disk of radius a carrying uniform surface charge density σ, is rotating at constant angular velocity ω about its own axis. Find the magnetic field at any axial point. Also, find the magnetic field at the centre.*

Solution. Consider a ring of radius r and radial thickness dr. The speed of any point on this ring is $v = \omega \cdot r$. The ring constitutes a current loop of current $dI = \sigma \cdot \omega r dr$. The magnetic field due to this ring at an axial point P at a distance x from the centre is

$$dB = \frac{\mu_0}{4\pi} \cdot \frac{2\pi \cdot dI \cdot r^2}{(r^2 + x^2)^{3/2}} = \frac{\mu_0 \sigma \omega}{2} \cdot \frac{r^3 dr}{(r^2 + x^2)^{3/2}}.$$

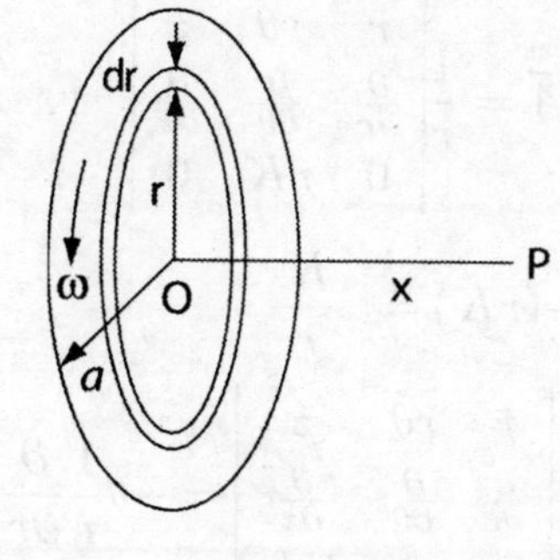

Fig 8.P-13

Therefore, total field at P will be

$$B = \frac{\mu_0 \sigma \omega}{2} \int_0^a \frac{r^3 dr}{(r^2 + x^2)^{3/2}}.$$

Substituting $r^2 + x^2 = z^2$ we get

$$B = \frac{\mu_0 \sigma \omega}{2} \int_{|x|}^{\sqrt{a^2+x^2}} \left(1 - \frac{x^2}{z^2}\right) dz = \frac{\mu_0 \sigma \omega}{2} \left[\frac{a^2 + 2x^2}{\sqrt{a^2 + x^2}} - 2|x|\right].$$

If σ is positive then $\vec{B}$ will be directed along $\overrightarrow{OP}$. At the centre, $x = 0$ and

$$B = \frac{\mu_0 \sigma \omega a}{2}.$$

▶ **Problem 14.** *A current distribution gives rise to the magnetic vector potential*

$$\vec{A}(x, y, z) = x^2 y \hat{\imath} + y^2 x \hat{\jmath} - xyz \hat{k} \;\; T.m.$$

Find the corresponding magnetic field $\vec{B}$ at $(-1, 2, 5)$. **[C.U. 2017]**

Solution. By definition

$$\vec{B}(x, y, z) = \vec{\nabla} \times \vec{A} = \begin{vmatrix} \hat{\imath} & \hat{\jmath} & \hat{k} \\ \frac{\partial}{\partial x} & \frac{\partial}{\partial y} & \frac{\partial}{\partial z} \\ x^2 y & y^2 x & -xyz \end{vmatrix} = -xz\hat{\imath} + yz\hat{\jmath} + (y^2 - x^2)\hat{k}$$

$$\therefore \quad \vec{B}(-1, 2, 5) = 5\hat{\imath} + 10\hat{\jmath} + 3\hat{k}.$$

▶ **Problem 15.** *Certain volume current distribution $\vec{J}$ gives rise to the magnetic vector potential (a)$\vec{A}(r, \theta, z) = K\hat{\theta}$ (b) $\vec{A}(r, \theta, z) = \frac{K}{r^2}\hat{z}$, where K is a constant. Find the corresponding current distribution $\vec{J}$.*

Solution. (a)

$$\vec{B} = \vec{\nabla} \times \vec{A} = \frac{1}{r}\begin{vmatrix} \hat{r} & r\hat{\theta} & \hat{z} \\ \frac{\partial}{\partial r} & \frac{\partial}{\partial \theta} & \frac{\partial}{\partial z} \\ 0 & rK & 0 \end{vmatrix} \quad [\because A_\theta = K]$$

$$= \hat{z}\frac{1}{r}\frac{\partial}{\partial r}(rK) = \hat{z}\frac{K}{r}.$$

$$\therefore \quad \vec{\nabla} \times \vec{B} = \frac{1}{r}\begin{vmatrix} \hat{r} & r\hat{\theta} & \hat{z} \\ \frac{\partial}{\partial r} & \frac{\partial}{\partial \theta} & \frac{\partial}{\partial z} \\ 0 & 0 & K/r \end{vmatrix} = -\hat{\theta}\frac{1}{r}\frac{\partial}{\partial r}(K/r) = \hat{\theta}\frac{K}{r^2}.$$

So from the equation $\vec{\nabla} \times \vec{B} = \mu_0 \vec{J}$ we get

$$\vec{J} = \hat{\theta}\frac{K}{\mu_0 r^2}.$$

(b) The equation satisfied by $\vec{A}$ is

$$\nabla^2 \vec{A} = -\mu_0 \vec{J}.$$

As $\vec{A}$ has the only z-component, $\vec{J}$ will also have the z-component only.

$$\therefore \quad -\mu_0 J_Z = \nabla^2 A_z = \frac{1}{r}\cdot\frac{\partial}{\partial r}\left(r\frac{\partial A_z}{\partial r}\right) + \frac{1}{r^2}\frac{\partial^2 A_z}{\partial \theta^2} + \frac{\partial^2 A_z}{\partial z^2}$$

$$= \frac{1}{r}\frac{\partial}{\partial r}\left(r\frac{\partial}{\partial r}\frac{K}{r^2}\right) = \frac{4K}{r^4} \quad \text{or} \quad J_z = -\frac{4K}{\mu_0 r^4}.$$

In vector form,

$$\vec{J} = J_z\hat{z} = -\frac{4K}{\mu_0 r^4}\hat{z}.$$

▶ **Problem 16.** *Using the relation* $\vec{B} = -\mu_0 \vec{\nabla}\phi$, *determine the magnetic scalar potential* ϕ *due to a long straight wire carrying current* I.

Solution. Suppose the current is flowing along $+z$-direction. The magnetic field at any point $P(r, \theta, z)$ with cylindrical coordinates (r, θ, z) is given by

$$\vec{B} = \frac{\mu_0}{4\pi}\cdot\frac{2I}{r}\hat{\theta}.$$

Now

$$d\phi = \vec{\nabla}\phi \cdot d\vec{r} = -\frac{1}{\mu_0}\vec{B}\cdot d\vec{r} = -\frac{I}{2\pi}\cdot\frac{\hat{\theta}\cdot d\vec{r}}{r}$$

$$= -\frac{I}{2\pi}\cdot\frac{\hat{\theta}\cdot(dr\hat{r} + rd\theta\hat{\theta} + dz\hat{z})}{r} = -\frac{I}{2\pi}d\theta.$$

Therefore,

$$\phi = \int d\phi = -\frac{I}{2\pi}\int_0^{2\pi} d\theta = -\frac{I}{2\pi}\cdot\theta.$$

▶ **Problem 17.** *Consider a long straight wire carrying current I current along $+z$-direction. Assuming the expression for $\vec{B}$ due to it and the relation $\vec{B} = \vec{\nabla} \times \vec{A}$, determine the vector potential $\vec{A}$ at any outside point. Check that $\vec{\nabla} \cdot \vec{A} = 0$ and $\vec{\nabla} \times \vec{A} = \vec{B}$. Also, find $\vec{A}$ at any inside point. Assume that the current is uniformly distributed over the cross-section of the wire of radius a.*

Solution. Using cylindrical coordinates the magnetic field at any outside point (r, θ, z) can be written in the form

$$\vec{B} = \frac{\mu_0}{4\pi} \cdot \frac{2I}{r}\hat{\theta}.$$

As the current is along $+z$-direction $\vec{A}$ can be written in the form $\vec{A} = A\hat{z}$. Now

$$\vec{B} = \vec{\nabla} \times \vec{A} \text{ or } \frac{\mu_0 I}{2\pi r}\hat{\theta} = \frac{1}{r}\begin{vmatrix} \hat{r} & r\hat{\theta} & \hat{z} \\ \frac{\partial}{\partial r} & \frac{\partial}{\partial \theta} & \frac{\partial}{\partial z} \\ 0 & 0 & A \end{vmatrix}$$

$$\text{or} \quad \frac{\mu_0 I}{2\pi r} = -\frac{\partial A}{\partial r}$$

$$\text{or} \quad A = -\frac{\mu_0 I}{2\pi}\int \frac{dr}{r} = -\frac{\mu_0}{2\pi} I \ln r + \text{constant}.$$

$\vec{A}$ points in the direction of the current and hence we can write

$$\vec{A} = -\hat{z}\frac{\mu_0 I}{2\pi} \ln \frac{r}{b},$$

where b is an arbitrary constant.

$$\vec{\nabla} \cdot \vec{A} = \frac{\partial A_z}{\partial z} = 0, \quad \vec{\nabla} \times \vec{A} = -\hat{\theta}\frac{\partial A_z}{\partial r} = \hat{\theta}\frac{\mu_0 I}{2\pi r} = \hat{\theta}\frac{\mu_0}{4\pi} \cdot \frac{2I}{r}.$$

The magnetic field at any inside point is given by Ampere's law as

$$\oint \vec{B} \cdot d\vec{l} = \mu_0 I \text{ enc} = \mu_0 \cdot J \cdot \pi r^2 \text{ or } B \cdot 2\pi r = \mu_0 J \pi r^2$$

$$\text{or } B = \frac{\mu_0}{2} Jr = \frac{\mu_0 I}{2\pi a^2} \cdot r.$$

Vectorially,

$$\vec{B} = \frac{\mu_0 I}{2\pi a^2} r\hat{\theta}.$$

As before

$$\frac{\mu_0 I}{2\pi a^2} r = -\frac{\partial A}{\partial r} \quad \text{or} \quad A = -\frac{\mu_0 I}{4\pi a^2} r^2 + \text{constant}$$

$$\therefore \quad \vec{A} = -\hat{z}\frac{\mu_0 I}{4\pi a^2} r^2 + \text{constant}.$$

▶ **Problem 18.** *Verify Ampere's circuital law for a long straight current carrying conductor.*

Solution. Suppose current through the wire is flowing along $+z$-direction. Then the magnetic field at any point with cylindrical coordinates (r, θ, z) is given by

$$\vec{B} = \frac{\mu_0 I}{2\pi r}\hat{\theta}.$$

Now considering an arbitrary closed path around the wire we find that

$$\oint \vec{B}\cdot d\vec{l} = \frac{\mu_0 I}{2\pi}\int \frac{1}{r}\hat{\theta}\cdot\left(\hat{r}dr + \hat{\theta}rd\theta + \hat{z}dz\right) = \frac{\mu_0 I}{2\pi}\int_0^{2\pi} d\theta = \mu_0 I,$$

which is Ampere's circuital law.

▶ **Problem 19.** *If $\vec{B}$ is uniform show that $\vec{A} = -\frac{1}{2}\left(\vec{r}\times\vec{B}\right)$, where $\vec{r}$ is the position vector of the point in question. Check that $\vec{\nabla}\cdot\vec{A} = 0$ and $\vec{\nabla}\times\vec{A} = \vec{B}$.*

[C.U. 2014, 2018; B.U. 2017]

Solution. Let $\vec{B} = B\hat{k}$ and $\vec{C} = \vec{r}\times\vec{B}$.

$$\therefore\ \vec{\nabla}\times\vec{C} = \vec{\nabla}\times\left(\vec{r}\times\vec{B}\right) = \left(\vec{B}\cdot\vec{\nabla}\right)\vec{r} - \left(\vec{r}\cdot\vec{\nabla}\right)\vec{B} + \vec{r}\left(\vec{\nabla}\cdot\vec{B}\right) - \vec{B}\left(\vec{\nabla}\cdot\vec{r}\right).$$

Now

$$\left(\vec{B}\cdot\vec{\nabla}\right)\vec{r} = B\frac{\partial}{\partial z}\left(x\hat{i} + y\hat{j} + z\hat{k}\right) = B\hat{k} = \vec{B}$$

$$\left(\vec{r}\cdot\vec{\nabla}\right)\vec{B} = 0 \quad \text{as } \vec{B} \text{ is uniform}$$

$$\vec{\nabla}\cdot\vec{B} = 0$$

$$\text{and}\quad \vec{\nabla}\cdot\vec{r} = 3$$

Therefore,

$$\vec{\nabla}\times\vec{C} = \vec{B} - 0 + 0 - 3\vec{B} = -2\vec{B}$$

$$\text{or,}\quad \vec{B} = -\tfrac{1}{2}\vec{\nabla}\times\vec{C} = \vec{\nabla}\times\left(-\tfrac{1}{2}\vec{C}\right) = \vec{\nabla}\times\vec{A}.$$

$$\therefore\ \vec{A} = -\tfrac{1}{2}\vec{C} = -\tfrac{1}{2}\left(\vec{r}\times\vec{B}\right).$$

$$\vec{\nabla}\cdot\vec{A} = -\tfrac{1}{2}\vec{\nabla}\cdot\left(\vec{r}\times\vec{B}\right) = -\tfrac{1}{2}\left[\vec{B}\cdot\left(\vec{\nabla}\times\vec{r}\right) - \vec{r}\cdot\left(\vec{\nabla}\times\vec{B}\right)\right]$$

Now $\vec{\nabla}\times\vec{r} = 0$; $\vec{\nabla}\times\vec{B} = 0$ as $\vec{B}$ is uniform.

Therefore, $\vec{\nabla}\cdot\vec{A} = 0$

$$\vec{\nabla}\times\vec{A} = -\tfrac{1}{2}\vec{\nabla}\times\left(\vec{r}\times\vec{B}\right)$$

$$= -\tfrac{1}{2}\left[\vec{r}\left(\vec{\nabla}\cdot\vec{B}\right) - \vec{B}\left(\vec{\nabla}\cdot\vec{r}\right) + \left(\vec{B}\cdot\vec{\nabla}\right)\vec{r} - \left(\vec{r}\cdot\vec{\nabla}\right)\vec{B}\right].$$

Now $\vec{\nabla} \cdot \vec{B} = 0$ and $\left(\vec{r} \cdot \vec{\nabla}\right) \vec{B} = 0$ because $\vec{B}$ is uniform, $\vec{\nabla} \cdot \vec{r} = 3$ and

$$\left(\vec{B} \cdot \vec{\nabla}\right) \vec{r} = \left(B_x \frac{\partial}{\partial x} + B_y \frac{\partial}{\partial y} + B_z \frac{\partial}{\partial z}\right)\left(x\hat{i} + y\hat{j} + z\hat{k}\right)$$

$$= B_x\hat{i} + B_y\hat{j} + B_z\hat{k} = \vec{B}.$$

Thus,

$$\vec{\nabla} \times \vec{A} = -\tfrac{1}{2}\left[-3\vec{B} + \vec{B}\right] = \vec{B}.$$

▶ **Problem 20.** *A thin disk of radius a carrying uniform surface charge density σ is rotating with constant angular velocity ω about its axis (z-axis). Suppose there is a uniform magnetic field $\vec{B} = \vec{B}\hat{j}$. Show that the torque acting on the disk is of magnitude $\frac{1}{4}\pi\sigma\omega B a^4$.*

Solution. Consider a ring of radius r and radial thickness dr. The charge on the ring is $\sigma \cdot 2\pi r dr$. It constitutes a current loop of current

$$dI = \frac{\sigma \cdot 2\pi r dr}{T} = \sigma 2\pi r dr \cdot \left(\frac{\omega}{2\pi}\right) = \sigma\omega r dr.$$

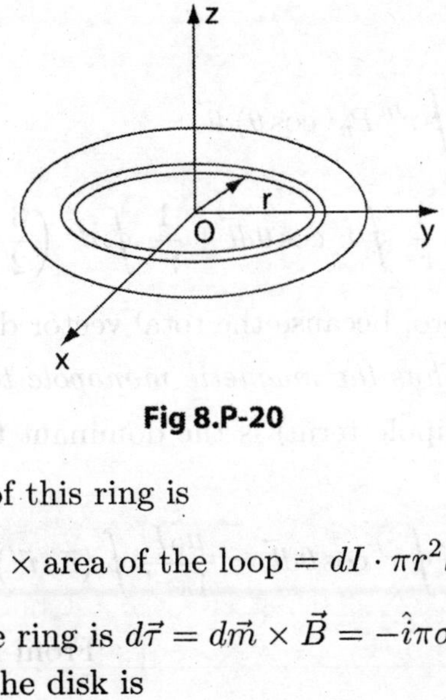

Fig 8.P-20

The magnetic moment of this ring is

$$d\vec{m} = \text{current} \times \text{area of the loop} = dI \cdot \pi r^2 \hat{k} = \pi\sigma\omega r^3 dr\hat{k}.$$

Therefore, torque on the ring is $d\vec{\tau} = d\vec{m} \times \vec{B} = -\hat{i}\pi\sigma\omega B r^3 dr$.
Hence, total torque on the disk is

$$\vec{\tau} = -\hat{i}\pi\sigma\omega B \int_0^a r^3 dr = -\hat{i}\tfrac{1}{4}\pi\sigma\omega B a^4.$$

▶ **Problem 21.** *Show that in Coulomb gauge $\left(\vec{\nabla} \cdot \vec{A} = 0\right)$ the magnetic vector potential at a large distance due to a small loop carrying current I may be expressed in the form*

$$\vec{A}(\vec{r}) = \frac{\mu_0 I}{4\pi}\left[\frac{1}{r^2}\oint r' \cos\theta d\vec{l} + \frac{1}{r^3}\oint r'^2 \left(\frac{3}{2}\cos^2\theta - \frac{1}{2}\right) d\vec{l} + \cdots\right],$$

where the integrals are over the loop; r' and θ are the coordinates of the source point, the angle θ being measured with respect to the position vector $\vec{r}$ of the field point. You can use the generating function for the Legendre polynomials. Give a physical interpretation of the first term in the expression for $\vec{A}(\vec{r})$.

Solution. Referring to Fig. 8.11-1 we can write

$$\vec{A}(\vec{r}) = \frac{\mu_0}{4\pi}\oint \frac{I d\vec{l}}{R},$$

where

$$R^2 = r^2 + r'^2 - 2rr'\cos\theta = r^2\left(1 + \frac{r'^2}{r^2} - 2\frac{r'}{r}\cos\theta\right)$$

$$\text{or} \quad \frac{1}{R} = \frac{1}{r\sqrt{1 - 2\frac{r'}{r}\cos\theta + \frac{r'^2}{r^2}}} = \frac{1}{r}\sum_{n=0}^{\infty}\left(\frac{r'}{r}\right)^n P_n(\cos\theta),$$

where we assume $|r'/r| < 1$ and use the generating functions for the Legendre polynomials.

Therefore,

$$\begin{aligned}\vec{A}(\vec{r}) &= \frac{\mu_0 I}{4\pi}\sum_{n=0}^{\infty}\frac{1}{r^{n+1}}\oint r'^n P_n(\cos\theta)d\vec{l} \\ &= \frac{\mu_0 I}{4\pi}\left[\frac{1}{r}\oint d\vec{l} + \frac{1}{r^2}\oint r'\cos\theta d\vec{l} + \frac{1}{r^3}\oint r'^2\left(\frac{3}{2}\cos^2\theta - \frac{1}{2}\right)d\vec{l} + \cdots\right].\end{aligned}$$

The first integral is simply zero, because the total vector displacement around a closed loop is zero, i.e., $\oint d\vec{l} = 0$. *Thus the magnetic monopole term is always zero.* In most cases the second term (the dipole term) is the dominant term in the above multipole expansion of $\vec{A}$. Thus

$$\vec{A}_{\text{dipole}}(\vec{r}) = \frac{\mu_0 I}{4\pi r^2}\oint r'\cos\theta d\vec{l} = \frac{\mu_0 I}{4\pi r^3}\oint (\vec{r}\cdot\vec{r}')\, d\vec{r}'$$

[$\because$ From Fig 8.11-1 $d\vec{l} = d\vec{r}'$]

Now proceeding as in Section 8.11 we can show that

$$\vec{A}_{\text{dipole}}(\vec{r}) = \frac{\mu_0}{4\pi}\frac{\vec{m}\times\vec{r}}{r^3},$$

where

$$\vec{m} = \frac{I}{2}\oint \vec{r}'\times d\vec{r}'$$

is the dipole moment of the loop.

▶ **Problem 22.** *A circular coil of radius a carrying current I is lying in the xy-plane with its centre at the origin. There is an external uniform field* $\vec{B} = \left(\frac{1}{\sqrt{2}}\hat{\imath} + \frac{1}{\sqrt{2}}\hat{\jmath}\right) B_0$. *Find the force and the torque acting on the coil.*

Solution. Let the current is flowing in such a direction that the magnetic dipole moment of the coil becomes $\vec{m} = I\pi a^2 \hat{k}$.

Therefore, force on the coil

$$\vec{F} = \left(\vec{m} \cdot \vec{\nabla}\right) \vec{B} = I\pi a^2 \frac{\partial \vec{B}}{\partial z} = 0.$$

Torque on the coil is

$$\vec{\tau} = \vec{m} \times \vec{B} = I\pi a^2 \cdot \frac{B_0}{\sqrt{2}}\left(\hat{k} \times \hat{\imath} + \hat{k} \times \hat{\jmath}\right) = \frac{\pi a^2 I B_0}{\sqrt{2}}\left(\hat{\jmath} - \hat{\imath}\right).$$

▶ **Problem 23.** *Consider two coaxial circular loops of wire of radii a and b separated by a distance x and carrying currents* I_1 *and* I_2, *respectively. Assuming* $b \ll a$ *find the force between the two loops.*

Solution. The small loop may be considered as a magnetic dipole of moment $m = I_2 \cdot \pi b^2$. The magnetic field produced by the larger loop at the position of the small loop is

$$B = \frac{\mu_0}{4\pi} \cdot \frac{2\pi I_1 a^2}{(a^2 + x^2)^{3/2}},$$

which is directed parallel to x-axis. B can be assumed to be uniform all over the small loop. Now the force on the small loop is given by

$$\vec{F} = \left(\vec{m} \cdot \vec{\nabla}\right) \vec{B} \quad \text{or} \quad F = m\left|\frac{\partial B}{\partial x}\right| = \frac{3\pi \mu_0 a^2 b^2 I_1 I_2 x}{2\,(a^2 + x^2)^{5/2}}.$$

If the currents in the two coils are in the same direction the force will be attractive. On the other hand, if the currents are in opposite directions the force will be repulsive.

▶ **Problem 24.** *Find the force between two ideal magnetic dipoles of moments* $\vec{m}_1$ *and* $\vec{m}_2$ *separated by a distance r. Assume that* $\vec{m}_1$ *and* $\vec{m}_2$ *point in the direction of the vector joining them.* **[C.U. 2013]**

Solution. The force on the dipole $\vec{m}_2$ due to $\vec{m}_1$ is given by

$$\vec{F} = \left(\vec{m}_2 \cdot \vec{\nabla}\right) \vec{B}_1,$$

where

$$\vec{B}_1 = \frac{\mu_0}{4\pi}\left[-\frac{\vec{m}_1}{r^3} + \frac{3\left(\vec{m}_1 \cdot \vec{r}\right)\vec{r}}{r^5}\right]$$

is the magnetic field at the position of $\vec{m}_2$ due to $\vec{m}_1$. Now from Fig. 8.P-24

$$\vec{r} = x\hat{\imath} + y\hat{\jmath} + z\hat{k} = r\hat{\imath}, \quad \vec{m}_1 = m_1\hat{\imath}, \quad \vec{m}_2 = m_2\hat{\imath}$$

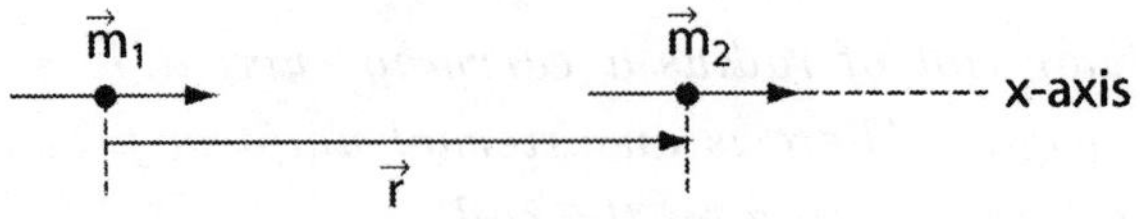

Fig 8.P-24

$$\therefore \quad \vec{F} = \left(m_2 \frac{\partial}{\partial x}\right)\left[\frac{\mu_0}{4\pi r^3}\left(-m_1 + 3m_1\right)\right]\hat{i} = \hat{i}\frac{\mu_0 m_1 m_2}{2\pi}\frac{\partial}{\partial x}\left(\frac{1}{x^3}\right)$$

$$= -\hat{i}\frac{3\mu_0}{2\pi}\frac{m_1 m_2}{x^4} = -\hat{i}\frac{\mu_0}{4\pi}\cdot\frac{6m_1 m_2}{x^4},$$

a force of attraction along the line joining the dipoles.

▶ **Problem 25.** *Find the force between two parallel ideal coplanar magnetic dipoles of moments $\vec{m}_1$ and $\vec{m}_2$ separated by a distance r. Assume that the vector joining them is perpendicular to the directions of the dipole, which point in the same direction.*

Solution. As in the *Solved Problem* 24, we can write

$$\vec{F} = \left(\vec{m}_2 \cdot \vec{\nabla}\right)\vec{B}_1 = \frac{\mu_0}{4\pi}\left(\vec{m}_2 \cdot \vec{\nabla}\right)\left[-\frac{m_1}{r^3} + \frac{3\left(\vec{m}_1 \cdot \vec{r}\right)\vec{r}}{r^5}\right].$$

Here $\vec{m}_1 = m_1\hat{j}$, $\vec{m}_2 = m_2\hat{j}$, $\vec{r} = x\hat{i} + y\hat{j} + z\hat{k} = r\hat{i}$

$$\therefore \ \vec{F} = \frac{\mu_0}{4\pi}m_2\frac{\partial}{\partial y}\left[-\frac{m_1}{x^3}\hat{j} + \frac{3(m_1 y)x\hat{i}}{x^5}\right]$$

[Note that $\vec{m}_1 \cdot \vec{r} = 0$ but $\vec{\nabla}\left(\vec{m}_1 \cdot \vec{r}\right) \neq 0$]

$$= \frac{\mu_0 m_2}{4\pi}\cdot\frac{3m_1}{x^4}\hat{i} = \frac{\mu_0}{4\pi}\cdot\frac{3m_1 m_2}{r^4}\hat{i},$$

a force of repulsion along the line joining the dipoles.

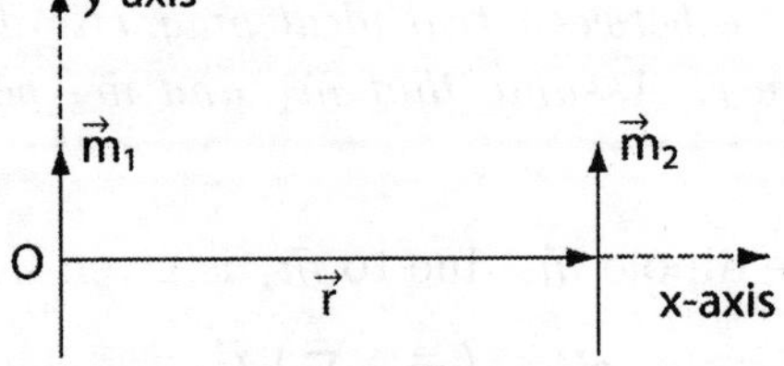

Fig 8.P-25

▶ **Problem 26.** *A thin nonconducting rod of length l and radius a, carrying a uniform surface charge of density σ is set spinning about its axis at an angular speed ω. Find the magnetic field at a point on the perpendicular bisector plane at a large distance from the centre of the rod.*

Solution. Current per dz length is

$$dI = \sigma \cdot v dz = \sigma\omega a dz,$$

where v is the linear speed of any point on the surface.

Magnetic moment per dz length $dm = \sigma\omega a \cdot \pi a^2 \cdot dz = \pi\sigma\omega a^3 dz$.

The whole rod may be considered as a stack of such magnetic dipoles. For a dipole $d\vec{m} = dm\hat{k}$ at the origin the magnetic field at any point (r, θ) in the xz-plane is given by

$$\begin{aligned} d\vec{B} &= \frac{\mu_0}{4\pi}\frac{dm}{r^3}\left(2\cos\theta\hat{r} + \sin\theta\hat{\theta}\right) \\ &= \frac{\mu_0 dm}{4\pi r^3}\left[2\cos\theta\left(\sin\theta\hat{i} + \cos\theta\hat{k}\right) + \sin\theta\left(\cos\theta\hat{i} - \sin\theta\hat{k}\right)\right] \\ &= \frac{\mu_0 dm}{4\pi r^3}\left[3\sin\theta\cos\theta\hat{i} + \left(2\cos^2\theta - \sin^2\theta\right)\hat{k}\right] \end{aligned}$$

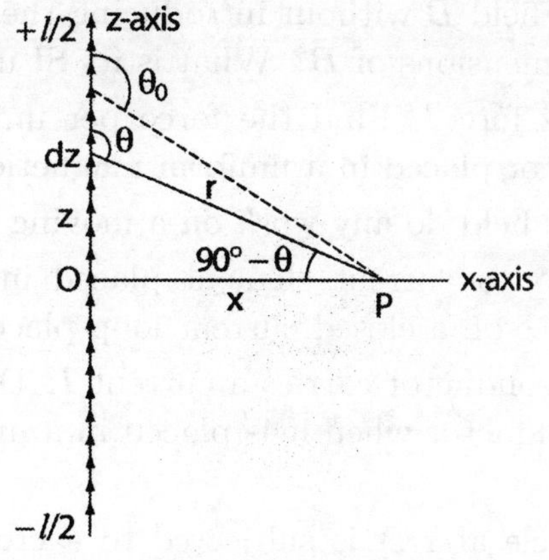

Fig 8.P-26

Symmetry of the problem indicates that when we integrate $d\vec{B}$ for $z = -l/2$ to $z = +l/2$, $\hat{i}$-components cancel out. Hence, total field at P is directed along z-axis and is of magnitude

$$\begin{aligned} \left|\vec{B}\right| &= 2 \times \int_0^{l/2} \frac{\mu_0}{4\pi r^3}(\pi\sigma\omega a^3 dz)\left(2\cos^2\theta - \sin^2\theta\right) \\ &= \int_0^{l/2} \frac{\mu_0\sigma\omega a^3}{2r^3}\left(3\cos^2\theta - 1\right)dz. \end{aligned}$$

Now from Fig. 8.P-26 $\sin\theta = x/r$, $z = -x\cot\theta$ or $dz = x\,\text{cosec}^2\theta d\theta$

$$\therefore \quad \left|\vec{B}\right| = \frac{\mu_0\sigma\omega a^3}{2x^2}\int_{\pi/2}^{\theta_0}(3\cos^2\theta - 1)\sin\theta d\theta,$$

where,

$$\cos\theta_0 = \frac{-l/2}{\sqrt{x^2 + \frac{l^2}{4}}} \quad \text{and} \quad \sin\theta_0 = \frac{x}{\sqrt{x^2 + \frac{l^2}{4}}}.$$

So

$$\left|\vec{B}\right| = \frac{\mu_0\sigma\omega a^3}{2x^2}\left[-\cos^3\theta + \cos\theta\right]_{\pi/2}^{\theta_0} = \frac{\mu_0\sigma\omega a^3}{2x^2}\cos\theta_0(1-\cos^2\theta_0)$$

$$= \frac{\mu_0\sigma\omega a^3}{2x^2}\cdot\cos\theta_0\cdot\sin^2\theta_0 = \frac{\mu_0\sigma\omega a^3}{2}\cdot\frac{l}{(x^2+l^2/4)^{3/2}}.$$

In vector form,

$$\vec{B} = -\frac{\mu_0\sigma\omega a^3 l}{4(x^2+l^2/4)^{3/2}}\,\hat{k}$$

QUESTIONS

1. (a) Define magnetic field $\vec{B}$ without introducing the concept of magnetic poles. What are the dimensions of $\vec{B}$? What is its SI unit?

(b) What is Lorentz force? Find the force per unit length of a long current carrying conductor placed in a uniform magnetic field. **[C.U. 2019]**

(c) Does a magnetic field do any work on a moving charge? **[C.U. 2013]**

(d) Find the force on a current element placed in a magnetic field. Hence, calculate the force on a closed current loop placed in a uniform field $\vec{B}$.

(e) A long straight conductor carries a current I. Determine the force per unit length of the conductor when it is placed in a uniform magnetic field. **[C.U. 2011]**

(f) A charged particle at rest is subjected to a crossed electric and magnetic field. Show that the path of the particle is a cycloid. **[N.B.U. 2017,, C.U. 2018, cf. B.U. 2017]**

2. State and explain Biot-Savart law in vector form for line, surface and volume currents. Apply the law to find the magnetic field due to:

(a) a long straight current carrying conductor,

(b) a circular current carrying loop at an axial point,

(c) a current carrying solenoid of finite length at an internal point.

3. Starting from Biot-Savart law for the general case of a volume current show that $\vec{\nabla}\cdot\vec{B} = 0$ always. What is the physical significance of the equation $\vec{\nabla}\cdot\vec{B} = 0$? **[C.U. 2012, cf. C.U., 2017; J.U. 2012]**

4. Starting from Biot-Savart law establish:

(a) Ampere's circuital law; **[C.U. 2013]**

(b) the relation $\vec{\nabla}\times\vec{B} = \mu_0\vec{J}$. **[N.B.U. 2017; C.U. 2011]**

5. State Ampere' circuital law. Apply the law to find the magnetic field due to:

(a) a long straight current carrying conductor,

(b) a long current carrying solenoid at an internal and external points, **[B.U. 2011]**

(c) a current carrying toroid at an internal and external points,

(d) an infinite sheet of current.

6. Calculate the magnetic field at an axial point of a circular current carrying coil. Show that a uniform magnetic field can be produced by the use of a pair of such coils. Cite an example where the above principle is used for the measurement of current.

7. (a) What is magnetic vector potential $\vec{A}$? Is it unique? Justify your answer. **[B.U. 2011]**

(b) Using the concept of $\vec{A}$ establish Biot-Savart law and Ampere's circuital law. **[B.U. 2016]**

(c) From the multipole expansion of magnetic vector potential $\vec{A}$, show that the monopole term vanishes. **[J.U. 2012]**

8. (a) Find the magnetic vector potential $\vec{A}$ at a distance r from an infinite straight wire carrying current I. Check that $\vec{\nabla}\cdot\vec{A} = 0$. **[C.U. 2012]**

(b) Hence calculate the magnetic field. **[J.U. 2012]**

(c) Find the magnetic vector potential inside and outside of an infinite solenoid with n turns per unit length, radius R and current I. **[C.U. 2010]**

9. (a) Determine the magnetic vector potential and hence the magnetic field at a large distance due to a small current loop placed in air. What important conclusion can be drawn from this result? OR, Establish the equivalence between a current loop and a magnetic dipole. **[B.U. 2011]**

OR, Show that the magnetic field at a large distance $(\vec{r})$ due to a small current loop having magnetic moment $\vec{m}$ is given by $\vec{B}(\vec{r}) = \frac{\mu_0}{4i}\left[-\frac{\vec{m}}{r^3} + \frac{3(\vec{m}\cdot\vec{r})\vec{r}}{r^5}\right]$. **[C.U. 2013]**

(b) Consider a small loop of radius a and carrying current I. Write down the integral that represents, in Coulomb gauge, the magnetic vector potential due to this loop at an arbitrary point. Expanding the integral in powers of $1/r$ show that for points far away from the loop the integral reduces to the vector potential of a magnetic dipole. Find this dipole moment.

10. (a) What do you mean by magnetic scalar potential? Find an expression for the magnetic scalar potential due to a circular current loop at an axial point.

(b) Hence calculate the corresponding magnetic field.

(c) A uniformly magnetized shell is equivalent to a current carrying loop-explain. **[C.U. 2010]**

11. (a) Find an expression for force per unit length between two parallel current carrying conductors.

(b) Find the mutual force between two current carrying parallel circular coaxial coils of nearly the same radius. For what distance between the coils the force would be maximum?

(c) Find a general expression for the mutual force between two arbitrary current loops. Show that the force is in accordance with Newton's third law of motion.

12. Verify that the magnetic vector potential $\vec{A}$ due to uniform magnetic field $\vec{B}$ is given by $\vec{A} = -\frac{1}{2}\left(\vec{r} \times \vec{B}\right)$.

13. (a) Show that the torque $\vec{\tau}$ acting on a rectangular current carrying coil placed in a uniform magnetic field $\vec{B}$ is given by $\vec{\tau} = \vec{m} \times \vec{B}$, where $\vec{m}$ is the magnetic dipole moment of the coil. Is the expression for $\vec{\tau}$ true for current loop of any arbitrary shape?

(b) Show that the potential energy of a current loop of magnetic dipole moment $\vec{m}$ placed in a uniform magnetic field $\vec{B}$ is given by $U = -\vec{m} \cdot \vec{B}$.

14. What is a magnetic dipole? Derive an expression for the force on a dipole placed in a nonuniform magnetic field. **[B.U. 2010]**

EXERCISES

1. (a) Two particles having equal charge and kinetic energy enter normally into a uniform magnetic field. Compare the radii of their paths. Assume that the mass of one particle is twice that of the other. [*Ans.* $1 : \sqrt{2}$]

(b) If a particle of charge q enters a magnetic field $\vec{B}$ along x-axis with speed v the force on it is $\frac{qvB}{2}(-\hat{j} + \sqrt{3}\hat{k})$; but if it enters along y-axis with same speed the force is $\frac{qvB}{2}\hat{j}$. Find $\vec{B}$. [*Ans.* $\vec{B} = \frac{B}{2}(\sqrt{3}\hat{j} + \hat{k})$]

(c) A particle of charge q enters with a velocity $\vec{v}$ into a region of uniform magnetic field $\vec{B}$ (pointing into the page) as shown in Fig. 8.E-1. The field which extends over a distance a, deflects the particle a distance d from the original line of flight. In terms of q a, d, and B find the momentum of the particle at the point of exit from the field.

[**Hints.** Within the field the particle describes a circular arc of radius r given by

$$\frac{mv^2}{r} = qvB \quad \text{or} \quad mv = qBr.$$

From Fig. 8.E-1(a),

$$r^2 = (r-d)^2 + a^2 \quad \text{or} \quad r = \frac{d^2+a^2}{2d}.$$

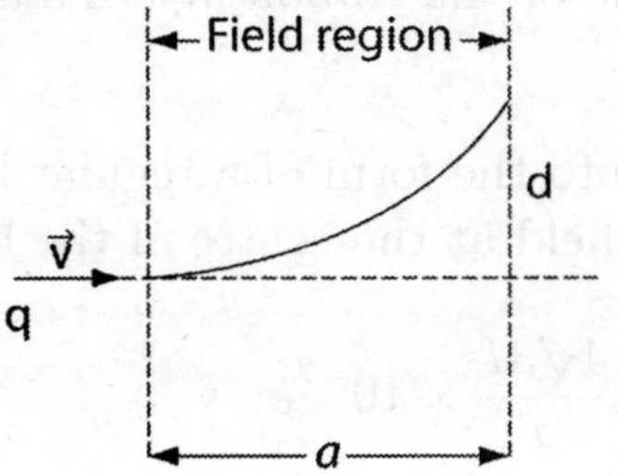

Fig 8.E-1

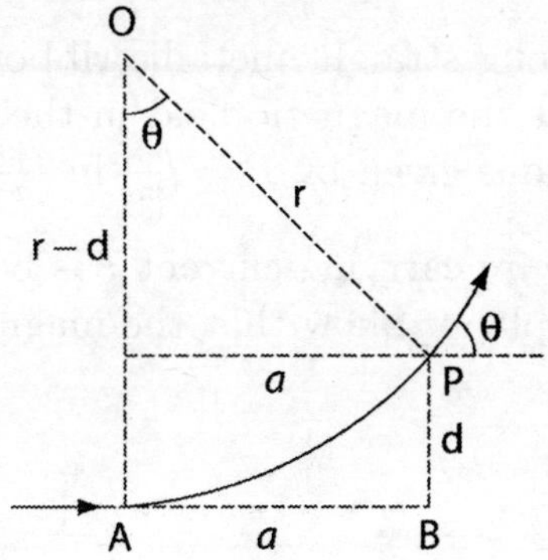

Fig 8.E-1(*a*)

Therefore, the momentum of the particle at P is of magnitude

$$mv = \frac{qB}{2d}(d^2+a^2)$$

and it makes an angle θ with the original line of flight given by $\sin\theta = \frac{a}{r} = \frac{2ad}{d^2+a^2}$.]

(c) In a cartesian frame there is a uniform electromagnetic field: $\vec{E} = 3A\hat{i}$, $\vec{B} = 5A\hat{k}$. A particle of rest mass m and charge q is released from rest at $(0, y_0, 0)$. What time elapses before it returns to the y-axis?

[**Hints.** Equation of motion: $m\dfrac{d\vec{v}}{dt} = q(\vec{E} + \vec{v} \times \vec{B})$

Write it in component forms, apply initial conditions and obtain

$x = \dfrac{3m}{25Aq}(1 - \cos\omega t)$ where $\omega = \dfrac{5Aq}{m_0}$. Required time $T = \dfrac{2\pi}{\omega} = \dfrac{2\pi m}{5Aq}$]

2. (a) Find the magnetic field at a distance of 1 cm from a long straight wire carrying current of 3A. [*Ans.* 60 μT]

(b) A current element $Id\vec{l}$ is directed along z-axis and is located at the origin. What is the x-component of the magnetic field at the point (x, y, z)?

[*Ans.* $-i\frac{\mu_0}{4\pi} \cdot ydl/(x^2+y^2+z^2)^{3/2}$]

3. Two long straight parallel wires placed d distance apart are carrying equal currents I in opposite directions. Find the magnetic field $\vec{B}$ at a distance x from one wire.

[*Ans.* $\frac{\mu_0}{4\pi} \cdot \frac{2Id}{x(d-x)}$]

4. A square loop of wire of edge a lying in the $z = 0$ plane carries a current I in anticlockwise direction. Find the magnetic field at the centre of the loop.

[*Ans.* $\frac{\mu_0 I 2\sqrt{2}}{\pi a}\hat{k}$]

5. Show that the magnetic moment associated with the orbital motion of the electron in the first Bohr orbit of a hydrogen atom is of magnitude $\frac{eh}{4\pi m}$, where m =mass of electron and h =Planck's constant. Is the magnetic moment quantized? (Note that the quantity $\frac{eh}{4\pi m}$ is called the *Bohr magneton.*)

6. A long straight metallic ribbon of width b is carrying a steady current I. Show that the magnetic field in the plane of the ribbon at a distance x from its near edge is given by $B = \frac{\mu_0 I}{2\pi b} \ln \frac{a+b}{a}$.

7. A wire carrying current I is bent into the form of a regular hexagon of side a in xy-plane. Show that the magnetic field at the centre of the hexagon is given by

$$\vec{B} = \frac{4\sqrt{3}I}{a} \times 10^{-7}\hat{k}.$$

8. (a) A particle of charge q is projected with a speed v along x-axis in a region of space having a magnetic field $\vec{B} = c_1\hat{j} + c_2\hat{k}$, where c_1 and c_2 are constants. Find the force on the particle. [*Ans.* $qv\left(-c_2\hat{j} + c_1\hat{k}\right)$]

   (b) A proton placed at a distance 10 cm from a long wire carrying a current of 1 A starts moving with a velocity $0.6C$ parallel to the direction of current flow. Find the force on the proton. **[C.U. 2015]** [*Ans.* 5.76×10^{-17} N]

9. Consider a circular coil of radius a carrying current I. Find the axial distance from the centre of the coil at which the magnetic field falls to half its maximum value. [*Ans.* $\pm 0.77a$]

10. Suppose two semicircular wires of radii r_1 and r_2 are joined to form a complete loop and a current I is passed through it (Fig. 8.E-1). Find the magnetic field at the centre.

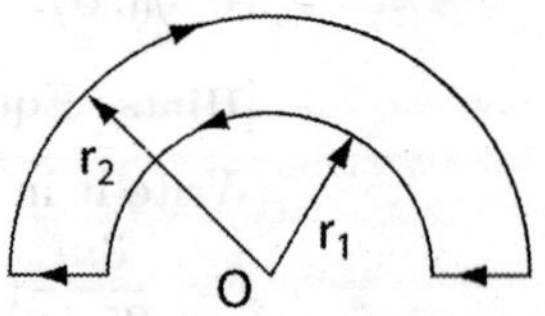

Fig 8.E-10

[*Ans.* $\frac{\mu_0 I}{4}\left(\frac{1}{r_1} - \frac{1}{r_2}\right)$]

11. (a) A particle of mass m and charge q is rotating in a circle of radius a with an angular velocity ω. Show that the ratio of its magnetic moment to mechanical angular momentum is $q/2m$. (Note that this ratio is called *gyromagnetic ratio*).

    (b) If the magnitude of the angular momentum of an electron rotating in a circular orbit is L, find its magnetic moment. [*Ans.* $eL/2m$] **[C.U. 2012]**

12. (a) A spherical shell of radius a, carrying a uniform surface charge density σ, is set spinning with an angular velocity ω about a diameter (considered to

be x-axis). Find the magnetic field at the centre of the shell. Also, find the equivalent magnetic dipole moment. **[cf. J.U. 2012]**

$\left[Ans.\ \frac{2\mu_0\omega a\sigma}{3}\hat{x};\ \frac{4}{3}\pi\sigma\omega a^4\hat{x}\right]$

(b) A uniformly charged sphere is rotating with a constant angular velocity ω about z-axis. Show that the magnetic field at the centre is $\vec{B} = \frac{1}{3}\mu_0\rho\omega a^2\hat{z}$ and the magnetic dipole moment of the system is $\vec{m} = \frac{4}{15}\pi\omega\rho a^5\hat{z}$, where ρ is the charge density. **[cf. C.U. 2007]**

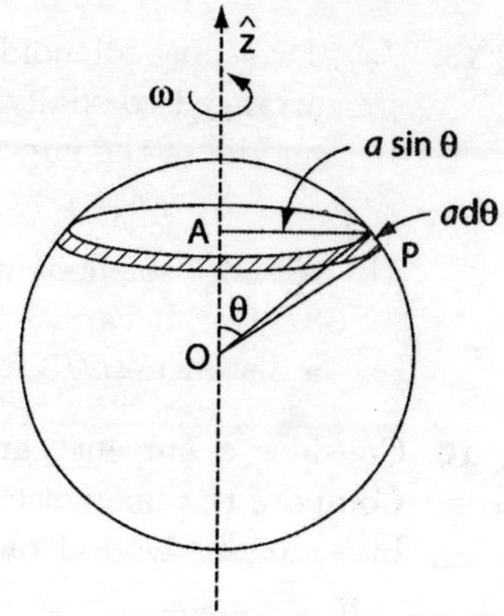

Fig 8-E-12

[**Hints.** (a) Charge on the elementary ring shown in Fig. 8.E-12 is $2\pi a\sin\theta\cdot ad\theta\cdot\sigma$. Current associated with this ring is

$$dI = 2\pi a^2\sigma\sin\theta d\theta(\omega/2\pi) = \omega\sigma a^2\sin\theta d\theta.$$

Dipole moment of the current ring is

$d\vec{m}$ = current×area $= \omega\sigma a^2\sin\theta\cdot d\theta\times\pi(a\sin\theta)^2\hat{z}.$

$\therefore$ Total magnetic moment

$$\vec{m} = \hat{z}\pi\omega\sigma a^4\int_0^\pi \sin^3\theta d\theta = \hat{z}\frac{4}{3}\pi\sigma\omega a^4$$

The magnetic field at O due to the current ring is

$$d\vec{B} = \hat{z}\cdot\frac{\mu_0}{4\pi}\cdot\frac{2\pi(dI)\cdot(AP)^2}{(AP^2+AO^2)^{3/2}} = \hat{z}\cdot\frac{\mu_0}{2}\cdot\frac{\omega\sigma a^2\sin\theta d\theta\cdot a^2\sin^2\theta}{a^3} = \hat{z}\cdot\frac{\mu_0}{2}\cdot\omega\sigma a\sin^3\theta d\theta.$$

$\therefore$ Total magnetic field at O is

$$\vec{B} = \hat{z}\frac{\mu_0}{2}\cdot\omega\sigma a\int_0^\pi \sin^3\theta d\theta = \hat{z}\frac{2\mu_0\omega a\sigma}{3}$$

(b) This problem can be solved by using the above result. Consider a thin spherical shell of radius r and thickness dr. Now replacing σ by ρdr we can write for the magnetic moment and field at the centre due to this shell as

$$d\vec{m} = \hat{z}\frac{4}{3}\pi\omega r^4\cdot\rho dr \quad\text{and}\quad d\vec{B} = \hat{z}\frac{2}{3}\mu_0\omega r\cdot\rho dr.$$

$\therefore$ Total quantities are obtained by integrating over the range $r = 0$ to $r = a$ as

$$\vec{m} = \hat{z}\frac{4}{15}\pi\omega\rho a^5 \quad\text{and}\quad \vec{B} = \hat{z}\frac{1}{3}\mu_0\omega\rho a^2.$$

(c) A thin disc of radius a has a uniform surface charge of density $-\sigma$ from its centre upto a radius, $r = a/\sqrt{2}$ and of density $+\sigma$ from $r = a/\sqrt{2}$ to $r = a$. The disc is rotating about its own axis (z-axis) with angular velocity ω. Find the equivalent magnetic dipole moment. Note that here though total charge is zero but the system can have magnetic moment.

[**Hints.** See, Solved Problem 20.] $\left[Ans.\ \frac{\pi\sigma\omega a^4}{8}\right]$

13. A standard solenoid of length 2 m and radius 4 cm has 5000 turns. A current of 1 A is passed through it. Calculate the magnetic field at an interior point of the solenoid. [*Ans.* 3.14×10^{-3}T]

14. The magnetic vector potential at any point $\vec{r}$ in a magnetic field is given by $\vec{A} = r^2\hat{\imath}$. Calculate the magnetic field at that position. [*Ans.* $2\left(z\hat{\jmath} - y\hat{k}\right)$]

15. (a) Two short solenoids each of radius 10 cm, length 20 cm and 30 turns/cm are arranged coaxially with a gap of 6 cm. Calculate the magnetic field at the middle of the gap when 1 A current passes through the solenoids in series. [*Ans.* 23.7×10^{-4}T]

(b) A short solenoid with n turns per unit length has diameter D and length $8D/15$. It carries a current I. Find the magnetic field at an axial point at a distance $2D/3$ from the near end of the solonoid [*Ans.* $\frac{4}{65}\mu_0 nI$]

16. Consider a current carrying wire bent into the form of a square of side $2a$. Compare the magnetic field at the centre of the square with the field produced by a circular loop of radius a carrying the same current at its centre. [*Ans.* $2\sqrt{2} : \pi$]

17. A vertical current carrying wire produces a neutral point with the earth's magnetic field at a distance of 5 cm from the wire. Find the current through the wire. Assume the horizontal component of earth's magnetic field $B_h = 36\mu$T. [*Ans.* 9A]

18. A circular loop of radius a carrying current I is placed with its plane normal to a uniform magnetic field $\vec{B}$. Find the work done in rotating the loop through 180∘. [*Ans.* $2\pi a^2 IB$]

19. Suppose a wire carrying current I is bent into the form an equilateral triangle of side a. Find the magnetic field at the centroid of the triangle. [*Ans.* $\frac{9\mu_0 I}{2\pi a}$]

20. Consider a long straight wire carrying current I. A rectangular loop is placed near the wire such that its two sides of length l become parallel to the wire with one placed at a distance a and the other at a distance b from the wire. Find the magnetic flux through the rectangular loop. [*Ans.* $\frac{\mu_0 Il}{2\pi}\ln\left(\frac{b}{a}\right)$]

21. A wire carrying current I is bent into the form of an equilateral triangle of side a. It is placed in a uniform magnetic field B with its plane parallel to the direction of the magnetic field. Calculate the torque acting on the loop. [*Ans.* $BIa^2\sqrt{3}/4$]

22. A singly ionised atom enters the region between two parallel metal plates with a velocity v parallel to the plates. The separation between the plates is d and a pd V is applied between the plates. Find the magnetic field required to keep the ion moving undeflected between the plates. [*Ans.* V/vd]

23. An infinitely long wire carrying a current IA is bent into an L-shape in xy-plane. Find the magnetic field at the point $(2, 2)$ with respect to the bending point as origin. Assume that the current flows in the directions of negative x-axis and positive y-axis. [*Ans.* $-\frac{\mu_0 I}{4\pi} \cdot \frac{\sqrt{2}+1}{\sqrt{2}} \hat{k}$]

24. A long straight wire is carrying a current of 2.5 A in the $+z$-direction. Find the magnetic field $\vec{B}$ at the point $(3, 4, 5)$. [*Ans.* 10^{-7}T]

25. A current distribution gives rise to the magnetic vector potential $\vec{A}(r, \theta, z) = Ke^{-r} \sin\theta \hat{z}$, where K is a constant. Find the corresponding magnetic field $\vec{B}(r, \theta, z)$ at the point with cylindrical coordinates (r, θ, z).

[*Ans.* $\vec{B} = \hat{r}\frac{K}{r}e^{-r} \cos\theta + \hat{t}\theta K e^{-r} \sin\theta$]

26. Show that the magnetic scalar potential ϕ due to a circular loop of radius a carrying current I, at an axial point at a distance x from the centre of the loop is given by

$$\phi = \frac{I}{2}\left(1 - \frac{x}{\sqrt{a^2 + x^2}}\right).$$

27. Find the magnetic field $\vec{B}$ near a long straight conductor carrying current I. Hence, show by direct calculation that $\vec{\nabla} \cdot \vec{B} = 0$.

28. Consider a long nonmagnetic hollow cylinder carrying a current of I. The inner and outer radii of the cylinder are a and b, respectively. Assuming the current density to be uniform find the magnetic field as a function of radial distance:

(a) inside the hollow space $(r < a)$;
(b) within the material of the conductor $(a < r < b)$;
(c) outside the conductor $(r > b)$.

[*Ans.* (a) $\vec{B} = 0, r < a$, (b) $\vec{B} = \frac{\mu_0 I}{2\pi r} \cdot \frac{r^2 - a^2}{b^2 - a^2}\hat{\theta}, a < r < b$ (c) $\vec{B} = \frac{\mu_0 I}{2\pi r}\hat{\theta}, r > b$]

29. Suppose that the earth's magnetic field is caused by a small current loop located the centre of the earth. The axis of the loop coincides with the axis of rotation of the earth. If the magnetic filed at the pole is B_0 and the radius of the earth is R then show that the magnetic moment of the current loop is of magnitude $2\pi R^3 B_0/\mu_0$.

30. (a) A magnetic dipole of moment $\vec{m} = m\vec{j}$ is placed in a magnetic field

$$\vec{B} = (x^2 - y^2)\hat{\imath} - 2xy\hat{\jmath}.$$

Find the force on the dipole. [*Ans.* $-2m\left(y\hat{\imath} + x\hat{\jmath}\right)$]

(b) A circular coil of a radius a carrying current I is fixed in the xy plane with its centre at the origin. Find the force on a dipole $\vec{m} = m\hat{k}$ placed at the point $(0, 0, z)$. [*Ans.* $\hat{k}3\mu_0 I a^2 m z/R(a^2 + z^2)^{5/2}$]

31. An electron circulates around the nucleus in a circular orbit of radius 0.51Å at a frequency 6.8×10^{15} rev/sec. Find the magnetic field B produced at the nucleus. Also, calculate the equivalent magnetic dipole moment.

[*Ans.* 13.4T, $8.9 \times 10^{24}\text{Am}^2$]

32. Show that $\vec{B} = K\left(\hat{i}x + \hat{j}y\right)$, where K is a constant, cannot represent a magnetic field.

[**Hints.** Show that $\vec{\nabla} \cdot \vec{B} \neq 0$]

33. Find the steady current density that can give rise to a magnetic field

$$\vec{B} = K\left(\hat{i}y - \hat{j}x\right),$$

where K is a constant. [*Ans.* $\vec{J} = -\frac{2K}{\mu_0}\hat{k}$]

[**Hints.** Use $\vec{\nabla} \times \vec{B} = \mu_0\vec{J}$.]

34. A long straight vertical wire is carrying a current of 15 A in the upward direction. The horizontal component of earth's magnetic field is 30μT. Find the position of the neutral point. [*Ans.* At a point 10 cm from the wire]

35. A spherical shell of radius a carrying a uniform surface charge density σ is spinning with an angular velocity ω about a diameter (z-axis). If a uniform magnetic field, $\vec{B} = B\hat{j}$, exists along y-axis show that the torque acting on it is of magnitude $\frac{4}{3}\pi\sigma\omega Ba^4$ and is along negative x-axis.

36. The magnetic scalar potential in a certain region is given by

$$\phi(x, y, z) = B_0 z(1 + x/a)/\mu_0,$$

where B_0 and a are constants. Find the static magnetic field which can be associated with this potential. [*Ans.* $-B_0\left\{\hat{i}z/a + \hat{k}(1 + x/a)\right\}$]

37. (a) A long cylindrical conductor of radius a is carrying a uniform steady current density $\vec{J} = J\hat{z}$. Applying Ampere's law find the magnetic field at any inside point at a distance r from its axis ($r < a$). [*Ans.* $\frac{\mu_0 Jr}{2}\hat{\theta}$ or $\frac{\mu_0}{2}\vec{J} \times \vec{r}$]

(b) If the cylinder in the above problem has an eccentric hole of radius $b(b < a)$ show that the magnetic field in the hole is uniform.

[**Hints.** Use superposition principle: Field due to $\vec{J}$ in the entire conductor of radius a+ Field due to $-\vec{J}$ in the volume of the cavity]

(c) A steady current is flowing through a long wire of radius 'a' in such a way that the current density at a distance r from the axis is $J = kr$ where k is a constant. Find the magnetic field at any inside and outside point.

[*Ans.* $\frac{1}{3}\mu_0 kr^2\hat{\theta}$, $\frac{\mu_0 ka^3}{3r}\hat{\theta}$]

38. Suppose the magnetic field $\vec{B}$ is given by $\vec{B} = 3xy^2\hat{k}$ Wb/m^2. Find the magnetic flux passing through the portion of xy-plane lying between $x = 0$, $x = 1$, $y = 0$ and $y = 1$. [*Ans.* 0.5Wb]

39. A beam of monoenergetic electrons is passed between two parallel plates 1 cm apart with a pd of 730 V across them. The electrons go undeflected when a magnetic field of 28×10^{-4}T is applied parallel to the plane of the plates. In a second experiment the same electrons are found to move in a circle of radius 12 cm about a magnetic field of 12.5×10^{-4}T. Find e/m of the electron.

[*Ans.* 1.76×10^{-11}C/kg]

40. A large number N of closely spaced turns of fine wire are wound in a single layer upon the surface of a wooden sphere of radius a, such that they completely cover its surface. The planes of the turns are perpendicular to a fixed diameter of the sphere. If the current in the turns is I, determine the magnetic field at the centre of the sphere. [*Ans.* $\frac{\mu_0 NI}{4a}$ along the fixed diameter]

41. Show that the magnetic flux through a surface bounded by the contour C is given by $\oint_C \vec{A} \cdot d\vec{l}$, where $\vec{A}$ is the magnetic vector potential at the element $d\vec{l}$ of the contour C.

42. Write down an expression for the magnetic vector potential $\vec{A}$ due to a small finite, current element $I\delta\vec{l}$ of a straight wire carrying current I. Find $\vec{\nabla} \times \delta\vec{A}$ and hence establish Biot-Savart law.

43. Write down an expression for the axial magnetic field $B(x)$ due to a circular current carrying loop. Find the integral $\int_{-\infty}^{+\infty} B(x)dx$. Does the result has any relationship with Ampere's circuital law.

[**Hints.** $B(x) = \frac{\mu_0}{4\pi} \cdot \frac{2\pi I a^2}{(a^2+x^2)^{3/2}}$: Substituting $x = a\tan\theta$ show that

$$\int_{-\infty}^{+\infty} B(x)dx = 2\int_0^{\infty} B(x)dx = \mu_0 \cdot I.$$

If we assume that the two ends of x-axis meet at infinity, then the integral may be considered as the line integral of $\vec{B}$ over a closed path linked with the current loop. So the result is just a confirmation of Ampere's law.]

44. Suppose a square loop of side a is lying in the yz-plane with its centre at the origin. A current I flows through the loop. It appears to be counter-clockwise when looked down the x-axis. There exists a magnetic field $\vec{B} = Kz\hat{i}$, where K is a constant. Find the force on the loop.

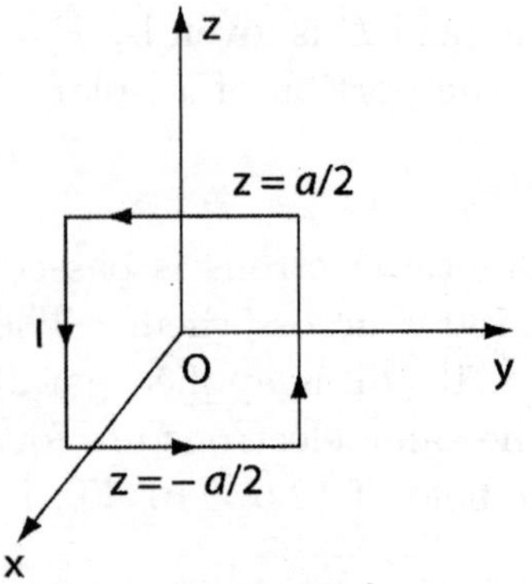

Fig 8.E-44

[**Hints.** Refer to Fig. 8.E-44. Forces on the left and right sides are equal and opposite. Force on the top side $= I\left(-a\hat{j}\right) \times \vec{B} = -Ia\hat{j} \times K\frac{a}{2}\hat{i} = \frac{1}{2}KIa^2\hat{k}$.

Force on the bottom side $= Ia\hat{j}\times\vec{B} = Ia\hat{j}\times(-K\frac{a}{2})\,\hat{i} = \frac{1}{2}KIa^2\hat{k}$. $\therefore$ Total force $\vec{F} = KIa^2\hat{k}$.]

45. (a) Show that the interaction energy of two magnetic dipoles of moments $\vec{m}_1$ and $\vec{m}_2$ is given by

$$U = \frac{\mu_0}{4\pi} \cdot \frac{1}{r^3} \left[\vec{m}_1 \cdot \vec{m}_2 - 3\left(\vec{m}_1 \cdot \hat{r}\right)\left(\vec{m}_2 \cdot \hat{r}\right)\right],$$

where $\vec{r} = r\hat{r}$ is the vector joining the centres of the dipoles.

(b) Two magnetic dipoles of magnitude 'm' each are placed in a plane as shown in Fig. 8.E-45. Calculate the interaction energy.

[C.U. 2018] [*Ans.* $-\frac{3\mu_0 m^2}{8\pi d^3}$]

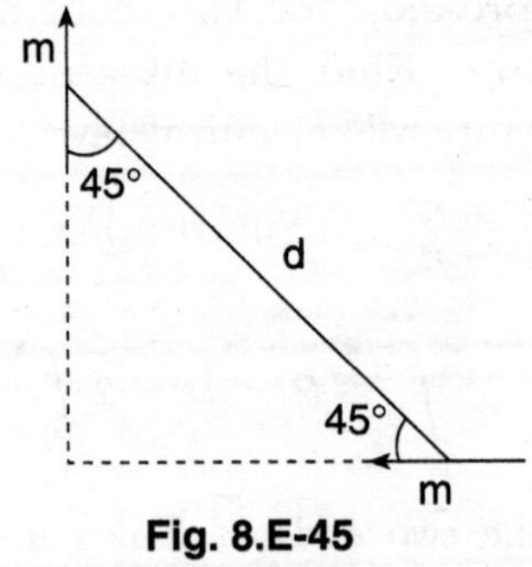

Fig. 8.E-45

46. A pure point dipole of moment $\vec{m}$ is placed at the origin pointing along z-axis. Write down the magnetic vector potential due to it in spherical polar coordinates. Hence show that the magnetic field due to it is given by

$$\vec{B} = \frac{\mu_0 m}{4\pi r^3}\left[2\cos\theta\hat{r} + \sin\theta \cdot \hat{\theta}\right]$$

[**Hints.** $\vec{A} = \frac{\mu_0}{4\pi}\frac{\vec{m}\times\hat{r}}{r^2} = \frac{\mu_0 m \sin\theta}{4\pi r^2}\hat{\varphi}$. Now find $\vec{B} = \vec{\nabla} \times \vec{A}$.]

47. Suppose a point magnetic dipole of moment $\vec{m}_1 = m_1\hat{k}$ is fixed at the origin and another dipole of moment $\vec{m}_2 = m_2\hat{\jmath}$ is placed at the point $(0, d, 0)$. Find the torque on $\vec{m}_2$ due to $\vec{m}_1$. If the second dipole is free to rotate what will its equilibrium orientation be?

[*Ans.* $-\frac{\mu_0}{4a}\frac{m_1 m_2}{d^2}\hat{\imath}$; In equilibrium the second dipole will point towards negative z-axis]

48. Show that the magnetic dipole moment of a planar current carrying loop is independent of the choice of origin. **[N.B.U. 2017]**

[**Hints.** $\vec{m} = \frac{I}{2}\oint(\vec{r}\times d\vec{r})$. With respect to a new origin at $\vec{r}_0$, $\vec{r} = \vec{r}_0 + \vec{r}\,'$ and

$$\begin{aligned}\vec{m}\,' = \frac{I}{2}\oint(\vec{r}\,'\times d\vec{r}\,') &= \frac{I}{2}\oint(\vec{r}-\vec{r}_0)\times d\vec{r} \quad [\because \vec{r}_0 \text{ is a fixed vector}]\\ &= \frac{I}{2}\oint\vec{r}\times d\vec{r} - \frac{I}{2}\vec{r}_0\times\oint d\vec{r}\\ &= \frac{I}{2}\oint\vec{r}\times d\vec{r} = \vec{m}.]\end{aligned}$$

49. Using Ampere's circuital law calculate the magnetic field at any internal point and external point due to a long straight wire of radius a carrying a current I with uniform density $\vec{J}$. Show that your results are consistent with the differential form of Ampere's law.

[**Hints.** Outside field $\vec{B} = \frac{\mu_0 I}{2\pi r}\hat{\theta}$. Show that $\vec{\nabla}\times\vec{B} = 0$. Inside field $\vec{B} = \frac{\mu_0 J}{2}r\hat{\theta}$. Show that $\vec{\nabla}\times\vec{B} = \mu_0 J\hat{z} = \mu_0\vec{J}$.]

50. Using magnetic vector potential $\vec{A} = e^{-x}\sin y\,\hat{\imath} + (1+\cos y)\hat{\jmath}$, calculate the magnetic induction. [*Ans.* $\vec{B} = -e^{-x}\cos y\,\hat{k}$] **[C.U. 2015; J.U. 2013]**

51. (a) A charged particle moves with uniform velocity $\vec{v} = 4\hat{\imath}$ m/s in a region where $\vec{E} = 20\hat{\jmath}$ V/m and $\vec{B} = B_0\hat{k}$ Wb/m^2. Determine B_0 such that the velocity of the particle remains constant. [*Ans.* 5 Wb/m^2] **[C.U. 2008]**

(b) An electron enters into a region where exist a magnetic field $\vec{B} = \hat{\imath}+2\hat{\jmath}-4\hat{k}$T and an electric field $\vec{E}$. If force on the electron is zero, find $\vec{E}$.

[C.U. 2017] [*Ans.* $\vec{E} \equiv 20(\hat{\imath}+\hat{k})$]

52. Find the force per unit length between two long parallel wires each carrying 1A current and placed 10 cm apart. [*Ans.* 2×10^{-6} N/m] **[cf. C.U. 2007]**

53. How much force would act on a 10 cm long wire carrying a current 10 A placed at an angle 30° with the direction of a uniform magnetic field of strength 1 Wb/m^2.

[*Ans.* 0.5 N] **[B.U. 2011]**

54. Consider three coplanar straight, equally spaced, long thin parallel wires each carrying a current I in the same direction. Outer two wires are held fixed and the middle wire is given a small displacement in the same plane towards one of the outer wires. Show that the motion of the middle wire will be simple harmonic with a time period $\frac{2\pi d}{I}\sqrt{\frac{\pi m}{\mu_0}}$ where m is the mass per unit length of middle wire and d is the separation between two wires.

55. Suppose that the earth's magnetic field is caused by a small current loop located at the centre of the earth. If the magnetic field at the pole is 80 μT and the radius of the earth is 6000 km find the magnetic moment of this small current loop.

[*Ans.* 8.64×10^{-26} Am^2]

56. Lorentz force on a particle of charge q and mass m is $\vec{F} = q(\vec{E} + \vec{v} \times \vec{B})$.

If $\vec{E} = -\vec{\nabla}\phi(x, y, z)$ is a time-independent electric field then show that the energy $\frac{1}{2}mv^2 + q\phi$ is a constant.

[**Hints.** $\vec{F} = m\frac{d\vec{v}}{dt}$. Thus $\vec{v}\cdot\left(m\frac{d\vec{v}}{dt} - q\vec{E}\right) = \vec{v}\cdot(\vec{v} \times \vec{B})q = 0$

$$\frac{d}{dt}\left[\frac{1}{2}mv^2 + q\phi\right] = m\vec{v}\cdot\frac{d\vec{v}}{dt} + q\frac{d\phi}{dt}$$

Now, $$\frac{d\phi}{dt} = \frac{\partial\phi}{\partial x}\frac{dx}{dt} + \frac{\partial\phi}{\partial y}\cdot\frac{dy}{dt} + \frac{\partial\phi}{\partial z}\cdot\frac{dz}{dt} = \vec{v}\cdot\vec{\nabla}\phi = -\vec{v}\cdot\vec{E}$$

$$\therefore \quad \frac{d}{dt}\left[\frac{1}{2}mv^2 + q\phi\right] = \vec{v}\cdot\left(m\frac{d\vec{v}}{dt} - q\vec{E}\right) = 0]$$

57. A point particle of mass m and magnetic dipole moment $\vec{m}_1$ moves in a circular orbit of radius R about a fixed magnetic dipole of moment $\vec{m}_2$ located at the centre of the circle. The vectors $\vec{m}_1$ and $\vec{m}_2$ are antiparallel to each other and perpendicular to the plane of the orbit. Show that the speed v of the orbiting dipole is $\sqrt{3\mu_0 m_1 m_2/4\pi m R^3}$.

OBJECTIVE TEST

1. A charged particle is projected with certain velocity parallel to a magnetic field. It will:

(a) trace a circular path, (b) trace a helical path, (c) come to rest instantly,

(d) continue its motion along a straight line without any change.

2. Two parallel wires are carrying currents I_1 and I_2 ($I_1 > I_2$) when the currents flow in the same direction the magnetic field at the mid-point between them is $10\,\mu$T. But if the current flow in the opposite directions then the corresponding field is $40\,\mu$T. Then I_1/I_2 is

(a) 4/1, (b) 5/3, (c) 2/1, (d) 3/2.

3. Magnetic flux has the dimensions:

(a) $[ML^2T^{-2}I^{-1}]$, (c) $[ML^2T^{-1}I^{-1}]$,
(b) $[MLT^2I]$, (d) $[ML^2T^2I]$.

4. Time taken by a charged particle to move once around a circular orbit in a magnetic field is independent of:

(a) mass, (c) strength of magnetic field,
(b) charge, (d) speed.

5. That the magnetic lines of force are always closed may be expressed by

(a) $\vec{\nabla} \times \vec{B} = 0$, (b) $\vec{\nabla} \cdot \vec{B} = 0$, (c) $\vec{\nabla} \times \vec{B} = \mu_0 \vec{J}$, (d) $\vec{\nabla} \times \vec{A} = 0$.

6. A magnetic dipole of moment $\vec{m}$ placed in a nonuniform magnetic field $\vec{B}$ experiences a force,

(a) $\vec{\nabla}\left(\vec{m} \cdot \vec{B}\right)$, (b) $\vec{m} \times \vec{B}$, (c) $-\vec{m} \cdot \vec{B}$, (d) none of these.

7. Four particles—*electron*, *proton*, He^+ and Li^+ are projected with the same velocity perpendicular to a given magnetic field. Then the circle described will be smallest for:

(a) electron, (b) proton, (c)] He^+, (d) Li^+.

8. A uniformly charged spherical metal ball of mass M, radius R and charge q is rotating with an angular velocity ω about one of its diameter. Then the ratio of its magnetic dipole moment to the mechanical moment (angular momentum) is

(a) $q/2MR$, (b) $q/2M$, (c) q/M, (d) $q/M\omega$.

9. A moving charge produces

(a) $\vec{E}$-field only, (c) both $\vec{E}$ and $\vec{B}$-fields,
(b) $\vec{B}$-field only, (d) none of these.

10. A long straight wire is carrying a current distributed uniformly over its cross-section. Then the magnetic field is

(a) minimum at the axis,
(b) maximum at the axis,
(c) minimum at the surface,
(d) maximum at the half-way from the axis to the surface.

11. Differential form of Ampere's law for a steady current is

(a) $\vec{\nabla} \times \vec{H} = \vec{J} + \dfrac{\partial \vec{D}}{\partial t}$, (c) $\oint \vec{B} \cdot d\vec{l} = \mu_0 I_{\text{enc}}$,

(b) $\vec{\nabla} \times \vec{B} = \mu_0 \vec{J}$, (d) $\vec{\nabla} \cdot \vec{B} = 0$.

12. The magnetic field of a small current loop at a large distance r varies as:

(a) $1/r$, (b) $1/r^2$, (c) $1/r^3$, (d) none of these.

13. Magnetic force on a closed current loop placed in a uniform magnetic field

(a) depends on the orientation of the loop,
(b) depends on the strength of magnetic field,
(c) depends both on orientation and shape of the loop,
(d) is always zero.

14. Which one of the following is not a source of magnetostatic field?

(a) An accelerated charge, (b) A steady current carrying wire,
(c) A bar magnet, (d) An electric field varying linearly with time.

15. Which one of the following is not a characteristic of magnetostatic field?

(a) It is solenoidal,
(b) It is conservative,
(c) Flux lines are always closed,
(d) Net magnetic flux through a closed surface is always zero.

16. Two identical coaxial coils carry the same current (I) but in opposite directions. The magnitude of the magnetic field $\vec{B}$ at a point on the axis midway between the coils is:

(a) the same as that produced by a single coil,
(b) twice that produced by a single coil,
(c) zero, (d) none of these.

17. The vector potential $\vec{A}$ corresponding to a constant magnetic field B along z-direction can be represented by: **[cf. NET 2015]**

(a) $-Bz\hat{k}$, (c) $\frac{B}{2}\left(\hat{i}x - \hat{j}y\right)$,

(b) $B\left(\hat{j}x - \hat{i}y\right)$, (d) $\frac{B}{2}\left(\hat{j}x - \hat{i}y\right)$.

18. The field represented by $\vec{F} = yz\hat{i} + zx\hat{j} + xy\hat{k}$ may be

(a) only an electrostatic field
(b) only a magnetostatic field

(c) both electrostatic and magnetostatic fields

(d) Neither electrostatic not magnetic field.

19. Two long parallel wires are carrying same current I but in opposite directions. The separation between the wires is d. The magnetic field at the mid point between the wires is

(a) zero, (b) $\dfrac{\mu_0 I}{2\pi d}$, (c) $\dfrac{2\mu_0 I}{\pi d}$, (d) $\dfrac{\mu_0 I}{\pi d}$. **[JAM 2011]**

20. In a region of space there are uniform electric field and magnetic field mutually parallel to each other. A charged particle released from rest in this region will move in a

(a) circle, (b) helix, (c) cycloid, (d) straight line

21. A particle of charge $+q$ enters into a region where there is a uniform $\vec{E}$-field and a uniform $\vec{B}$-field both directed along $+y$ axis At time $t = 0$ the particle is at origin and has a speed v_0 along $+x$ axis. y-coordinate of the particle after one time period T is

(a) $\pi^2 mE/qB^2$, (b) $2\pi^2 mE/qB^2$, (c) $2\pi^2 mv_0/qB$, (d) $\pi^2 mv_0/2qB$. **[JAM 2015]**

22. A conducting wire in the shape of a regular hexagon is inscribed inside an imaginary circle of radius r. The magnetic field of the centre of the circle is

(a) $\dfrac{\sqrt{3}\mu_0 I}{\pi r}$, (b) $\dfrac{3\mu_0 I}{2\pi r}$, (c) $\dfrac{\mu_0 I}{2\sqrt{3}\pi r}$, (d) $\dfrac{\sqrt{3}\mu_0 I}{2\pi r}$. **[JAM 2015]**

23. A charged particle is moving in a helical path under the influence of a uniform magnetic field. If the initial velocity component parallel to magnetic field is twice the velocity component normal to magnetic field, the ratio of pitch and radius of helical path would be

(a) $\pi/2$, (b) π, (c) 2π, (d) 4π

ANSWERS

1. (d)	**2.** (b)	**3.** (a)	**4.** (d)	**5.** (b)	**6.** (a)	**7.** (a)
8. (b)	**9.** (c)	**10.** (a)	**11.** (b)	**12.** (c)	**13.** (d)	**14.** (a)
15. (b)	**16.** (c)	**17.** (d)	**18.** (c)	**19.** (c)	**20.** (c)	**21.** (b)
22. (a)	**23.** (d)					

Chapter 9

Magnetic Field in Material Media

9.1 Introduction

So far we have considered the magnetic field $\vec{B}$ produced by currents in vacuum. In this chapter we shall consider the effect of the presence of a material medium in a magnetic field of currents. All matter consists of atoms which may possess magnetic dipole moments and interact with the applied magnetic field. Thus, a material medium is likely to affect and produce magnetic field. The purpose of this chapter is to consider the microscopic changes in matter, when placed in a magnetic field and their effects on the macroscopic properties of matter.

9.2 Magnetization

All matter consists ultimately of atoms and in each atom electrons revolve round the atomic nucleus and also spin about their own axis. These motions are associated with magnetic dipole moments. An atom may have many electrons whose magnetic moments tend to cancel each other. Atoms with *unpaired* electrons possess an intrinsic magnetic dipole moment whose magnitude is of the order of $\mu_B = eh/4\pi m = 9.2 \times 10^{-24}\,\text{A.m}^2$. μ_B is called the *Bohr magneton.* In the absence of any external magnetic field the intrinsic atomic dipoles remain randomly oriented due to thermal agitation and the net dipole moment of a macroscopic element of matter becomes zero. In the presence of a magnetic field the atomic dipoles experience a torque and to tend to orient themselves along the field direction. Thus, a net alignment of the dipoles is obtained along the field direction. The medium is then said to be *magnetized.* In cases where the atoms do not possess any intrinsic magnetic dipole moment the orbital speed of the electrons is altered under the influence of external magnetic field and a dipole moment is induced in a direction opposite to the field. Whatever may be the cause the state of a magnetized matter can be characterised by a vector $\vec{M}$, called the *magnetization.*

The magnetization $\vec{M}$ at a point is defined as the magnetic dipole moment per unit volume at that point and is given by

$$\vec{M} = \lim_{\Delta V \to 0} \frac{1}{\Delta V} \sum_{i=1}^{N} \vec{m}_i, \tag{9.2-1}$$

where $\vec{m}_i$ is the magnetic dipole moment of the ith atom and N is the number of atoms in volume ΔV. From purely macroscopic point of view all magnetic effects of a magnetized matter can be described adequately in terms of $\vec{M}$ and its derivatives. $\vec{M}$ provides us with a macroscopic description of the atomic currents in the matter. In an unmagnetized state $\vec{M} = 0$. In presence of external magnetic field $\vec{M}$ usually depends on this field. If the vector $\vec{M}$ is the same at each point of a matter it is said that the matter is uniformly magnetized. For uniform magnetization the atomic currents in various loops tend to cancel each other and there is no effective current in the interior of the matter. For nonuniform magnetization the cancellation is not complete and we get some equivalent transport current density which produces the same magnetic field as $\vec{M}$ itself. This is known as *magnetization* or *bound current density* $\left(\vec{J}_m\right)$. Unlike conduction current it neither transports charge nor produces Joule heating. It is current in the sense that it can produce magnetic field in the same way any other current does.

In order to understand physically how magnetization currents appear, let us consider a simplified picture of magnetic material consisting of atomic current loops with currents circulating in the same direction (Fig. 9.2-1).

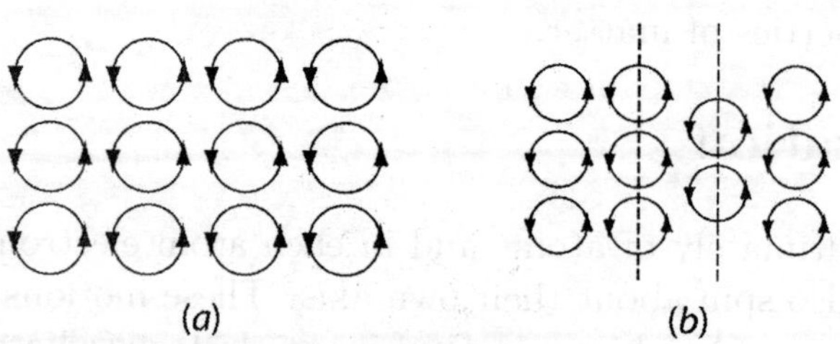

(a) (b)

Fig 9.2-1

In case of uniform magnetization [Fig. 9.2-1(a)] the currents in the adjacent current loops at the points of their contact have opposite directions and cancel each other. In this case uncompensated current appears on the surface of the body. Here every charge is attached or bound to a particular atom and cannot move over macroscopic distance.

Nevertheless the net effect is a macroscopic surface current. In case of nonuniform magnetization the cancellation of internal currents is only partial and as a result a macroscopic volume density of current appears. Such a situation is shown in Fig. 9.2-1(b). It is evident that in the region between the dashed lines there is more charge moving up than that moving down. This gives rise to the so-called magnetization current.

To find a quantitative relation between $\vec{M}$ and $\vec{J}_m$ let us consider the magnetic vector potential due a magnetized body (Fig. 9.2-2). The vector potential due to a single current loop of magnetic moment $\vec{m}$ is given by

$$\vec{A} = \frac{\mu_0}{4\pi}\frac{\vec{m}\times\vec{R}}{R^3},$$

where $\vec{R}$ is a radius vector from the loop to the point of observation. Now if $\vec{M}(\vec{r}\,')$ be the magnetization at the point $\vec{r}\,'$ in a magnetized body then the element dV carries a dipole moment $\vec{M}\,dV$ and so the total vector potential at $P(\vec{r})$ due to the magnetized body will be given by

$$\vec{A}(\vec{r}) = \frac{\mu_0}{4\pi}\int_V \frac{\vec{M}\,dV\times\vec{R}}{R^3} = \frac{\mu_0}{4\pi}\int_V \vec{M}\times\vec{\nabla}'\frac{1}{R}\,dV, \tag{9.2-2}$$

where we use the result that $\vec{\nabla}'\cdot\frac{1}{R} = \frac{\vec{R}}{R^3}$.

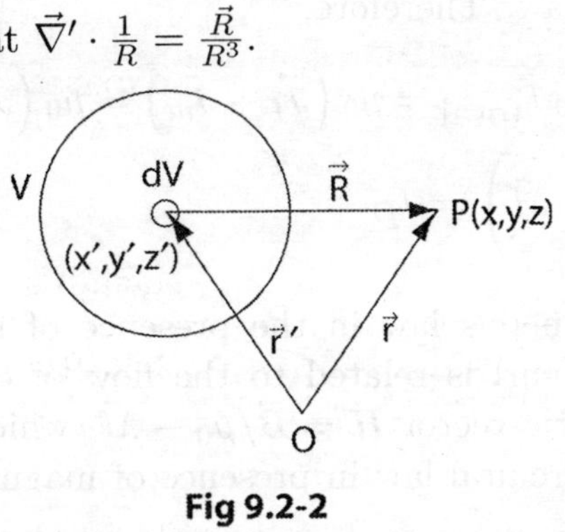

Fig 9.2-2

Using the vector identity

$$\vec{\nabla}'\times\frac{1}{R}\vec{M} = \vec{\nabla}'\frac{1}{R}\times\vec{M} + \frac{1}{R}\vec{\nabla}'\times\vec{M},$$

we get

$$\vec{A} = \frac{\mu_0}{4\pi}\int_V \frac{\vec{\nabla}'\times\vec{M}}{R}\,dV - \frac{\mu_0}{4\pi}\int_V \vec{\nabla}'\times\frac{1}{R}\vec{M}\,dV. \tag{9.2-3}$$

From vector calculus,

$$\int_V \vec{\nabla}'\times\frac{\vec{M}}{R}\,dV = -\int_S \frac{\vec{M}\times d\vec{S}}{R}. \tag{9.2-4}$$

Therefore,

$$\vec{A} = \frac{\mu_0}{4\pi}\int_V \frac{\vec{\nabla}'\times\vec{M}}{R}\,dV + \frac{\mu_0}{4\pi}\int_S \frac{\vec{M}\times d\vec{S}}{R}. \tag{9.2-5}$$

This expression for $\vec{A}$ suggests that a distribution of atomic currents inside magnetized matter produces the same magnetic vector potential as produced by a volume density of current $\vec{J}_m = \vec{\nabla}' \times \vec{M}$ and a surface density of current $\vec{K}_m = \vec{M} \times \hat{n}$, where $\hat{n}$ is the unit normal to the surface element dS. Thus instead of integrating the effects of individual atomic dipoles, we can use $\vec{J}_m$ and $\vec{K}_m$ to calculate the field they produce.

It is important to note that

$$\vec{\nabla}' \cdot \vec{J}_m = \vec{\nabla}' \cdot \left(\vec{\nabla}' \times \vec{M}\right) = 0.$$

Hence this magnetization current like any other steady current will not affect the continuity equation or the charge conservation law.

9.3 Auxiliary Field $\vec{H}$

In presence of magnetizable media both the conventional free (or conduction) current density $\vec{J}_f$ and the magnetization current density $\vec{J}_m$ must be counted as a source of magnetic field in Ampere's law. Therefore,

$$\vec{\nabla} \times \vec{B} = \mu_0 \vec{J}_{\text{total}} = \mu_0 \left(\vec{J}_f + \vec{J}_m\right) = \mu_0 \left(\vec{J}_f + \vec{\nabla} \times \vec{M}\right) \tag{9.3-1}$$

$$\text{or } \vec{\nabla} \times \left(\frac{\vec{B}}{\mu_0} - \vec{M}\right) = \vec{J}_f.$$

This is the form of Ampere's law in the presence of magnetizable media. The quantity $\vec{B}/\mu_0 - \vec{M}$ whose curl is related to the flow of conduction current only is used to define a new magnetic vector $\vec{H} = \vec{B}/\mu_0 - \vec{M}$, which is often called *magnetic intensity.* Then Ampere's circuital law in presence of magnetizable medium takes the simple form

$$\vec{\nabla} \times \vec{H} = \vec{J}_f. \tag{9.3-2}$$

In the integral form

$$\oint_C \vec{H} \cdot d\vec{l} = I_{\text{enc}},$$

where I_{enc} is the total free current passing through the loop C. $\vec{H}$ is related to the free current in the same way as $\vec{B}$ is related to the total current, i.e., free as well as magnetization current. Calculation of $\vec{B}$ requires a knowledge of magnetization current over which we have much less control. On the other hand, we can directly control free current $\vec{J}_f$ and hence the introduction of $\vec{H}$ becomes advantageous. It is to be noted that $\vec{B}$ is the fundamental magnetic vector whereas $\vec{H}$ is an auxiliary vector introduced for convenience. It is $\vec{B}$, not $\vec{H}$, which determines the force character of magnetic field. The dimensions of $\vec{H}$ are IL^{-1}, where I stands for current and L for length. Its SI unit is A.m^{-1}.

9.4 Magnetic Pole Density

Considering Fig. 9.2-2 we can write for the magnetic scalar potential at P due to a magnetized body as

$$\phi(\vec{r}) = \frac{1}{4\pi}\int_V \frac{\left(\vec{M}\,dV\right)\cdot\vec{R}}{R^3}.$$

Since $\frac{\vec{R}}{R^3} = \vec{\nabla}'\frac{1}{R}$ we can write

$$\phi(\vec{r}) = \frac{1}{4\pi}\int_V \vec{M}\cdot\vec{\nabla}'\frac{1}{R}\,dV.$$

Using the vector identity $\vec{\nabla}\cdot K\vec{A} = K\left(\vec{\nabla}\cdot\vec{A}\right) + \vec{\nabla}K\cdot\vec{A}$ we get

$$\phi = \frac{1}{4\pi}\int_V \vec{\nabla}'\cdot\frac{1}{R}\vec{M}\,dV - \frac{1}{4\pi}\int_V \frac{1}{R}\left(\vec{\nabla}'\cdot\vec{M}\right)dV.$$

Using divergence theorem we get

$$\phi = \frac{1}{4\pi}\int_S \frac{\vec{M}\cdot\hat{n}\,dS}{R} + \frac{1}{4\pi}\int_V \frac{\left(-\vec{\nabla}'\cdot\vec{M}\right)}{R}\,dV,$$

where S is the surface enclosing volume V of the magnetized body.

By analogy with the potential due to a polarized dielectric we may define $\rho_m = -\vec{\nabla}'\cdot\vec{M}$ as *magnetic pole density* and $\sigma_m = \vec{M}\cdot\hat{n}$ as the *surface density of magnetic pole strength.* The units of ρ_m and σ_m are A/m^2 and A/m, respectively. The quantities ρ_m and σ_m though rather artificial are useful in the theory of magnetism. In case of uniform magnetization $\rho_m = 0$. The total pole strength of any magnetized body is always zero. This can be seen by applying divergence theorem as

$$\int_V\left(-\vec{\nabla}'\cdot\vec{M}\right)dV + \int_S \vec{M}\cdot\hat{n}\,dS = -\int_S \vec{M}\cdot\hat{n}\,dS + \int_S \vec{M}\cdot\hat{n}\,dS = 0.$$

In case of a uniformly magnetized bar magnet $\rho_m = 0$ but σ_m does not vanish on the end faces. These are called the *poles of the bar magnet.*

We have the relation, $\vec{B} = \mu_0\vec{H} + \mu_0\vec{M}$.

Since $\vec{\nabla}\cdot\vec{B} = 0$, $\vec{\nabla}\cdot\vec{H} = -\vec{\nabla}\cdot\vec{M} = \rho_m$.

Therefore,

$$\int_V \vec{\nabla}\cdot\vec{H}\,dV = \int_V \rho_m\,dV \quad \text{or} \quad \int_S \vec{H}\cdot d\vec{S} = \int_V \rho_m\,dV.$$

This is Gauss's law for $\vec{H}$, which is analogous to the corresponding law for $\vec{E}$ in electrostatics.

9.5 Magnetic Susceptibility and Permeability

To solve the problems in magnetic theory we require relationship between $\vec{B}$ and $\vec{H}$ or between $\vec{M}$ and $\vec{H}$. These relationships depend on the nature of the material and are usually determined by experiment. For most magnetic materials the magnetization $\vec{M}$ is found to be proportional to the magnetizing field $\vec{H}$, provided the field is not too high. Such a material is called a *linear material.* For an isotropic linear magnetic material,

$$\vec{M} = \chi_m \vec{H}, \tag{9.5-1}$$

where χ_m is a dimensionless scalar called the *magnetic susceptibility* of the material. It is a characteristic of the material. Linear magnetic materials are of two types; materials with $\chi_m > 0$ are called *paramagnetic* and that with $\chi_m < 0$ are *diamagnetic.* Thus magnetic field $\vec{B}$ is strengthened by the presence of paramagnetic materials and it is weakened by the presence of the diamagnetic materials. It is found that paramagnetic materials are feebly attracted to magnetic fields, whereas diamagnetic materials are feebly repelled. The values of χ_m for some common paramagnetic and diamagnetic materials are given in **Table 9.1.**

Table 9.1

Material	χ_m	Material	χ_m
Paramagnetic:		Diamagnetic:	
Aluminium	2.1×10^{-5}	Bismuth	-16.4×10^{-5}
Sodium	0.85×10^{-5}	Gold	-3.5×10^{-5}
Oxygen(1 atm)	193×10^{-8}	Silver	-2.4×10^{-5}
Tungsten	7.6×10^{-5}	Copper	-0.98×10^{-5}
Magnesium	1.2×10^{-5}	Hydrogen(1 atm)	-0.22×10^{-8}

A linear relationship between $\vec{M}$ and $\vec{H}$ implies also a linear relationship between $\vec{B}$ and $\vec{H}$:

$$\vec{B} = \mu_0 \left(\vec{H} + \vec{M}\right) = \mu_0(1 + \chi_m)\vec{H} \tag{9.5-2}$$

for linear media. Thus, $\vec{B}$ is also proportional to $\vec{H}$:

$$\vec{B} = \mu \vec{H}, \tag{9.5-3}$$

where $\mu = \mu_0(1+\chi_m)$ is called *permeability of the material.* For free space $\chi_m = 0$ and $\mu = \mu_0$. The quantity $\mu_r = \mu/\mu_0 = 1 + \chi_m$ is called *relative permeability.* For dia- and paramagnetic materials $|\chi_m| \ll 1$. Hence μ_r is very close to unity for such materials.

There is another class of magnetic materials which responds to external magnetic field in a non-linear fashion. Here μ becomes a function of $\vec{H}$ and the relations (9.5-1) and (9.5-3) with constant χ_m and μ do not apply. These materials are called *ferromagnetic.* Examples of such materials are iron, cobalt, nickel and their alloys. Such materials exhibit much higher degree of magnetization compared to paramagnetic materials. These materials can also show spontaneous magnetization and an irreversible phenomenon known as *hysteresis.* For a ferromagnetic material there exists a particular temperature, called *Curie point,* above which the ferromagnetic material becomes paramagnetic. The non-linear permeability μ of these materials may be tabulated or plotted as a function of $\vec{H}$. In the literature, however, this information is supplied indirectly in the form of the dependence of $\vec{B}$ on $\vec{H}$.

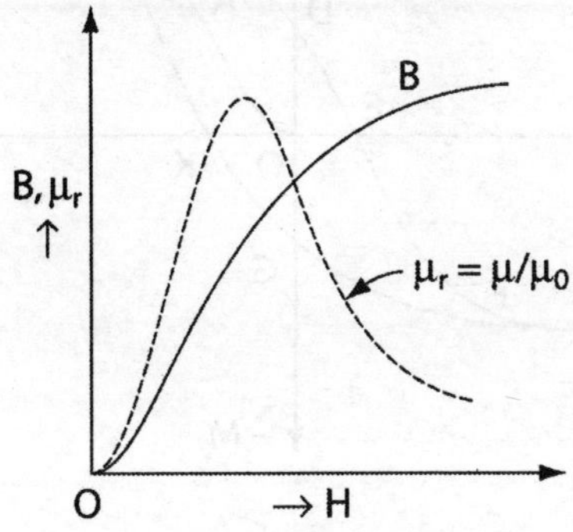

Fig 9.5-1

Fig. 9.5-1 shows the magnetization curve and relative permeability of a typical ferromagnetic material, where μ is calculated through the relation $\vec{B} = \mu\vec{H}$. In some cases a small alternating H-field superposed on a large constant H-field is applied to a magnetic material. In this case one defines incremental permeability μ_{in} by $\mu_{\text{in}} = \Delta B/\Delta H$, where ΔB is the small change in B produced by a small change ΔH in the H-field. It is approximately given by the slope of the B-H loop at the point in question.

9.6 Ferromagnetic Materials and Hysteresis

Let us consider an unmagnetized sample of ferromagnetic material. It is placed in an external magnetizing field H, which is increased monotonically from zero. The magnetization M developed in the sample is found to follow a curve like $OACC'$ as shown in Fig. 9.6-1. It is known as the *magnetization curve* of the material. Eventually there is magnetic saturation at C with the all the dipoles aligned. Further increase in H has no effect on M. If H is now gradually reduced to zero M does not retrace the curve instead it follows a path CD. At D, though $H = 0$, some amount of magnetization OD is retained by the sample. This is known as the *residual* or *remanent magnetization* or the *retentivity* or *remanence of the material.* On reversing the magnetizing field H the value of M falls and becomes zero at E. OE represents the reverse field necessary to destroy the residual magnetization. It is known as *coercive force* or *coercivity* of the material. If the reverse field is increased further saturation occurs at F in the reverse direction. If H is now again reversed and increased in magnitude M follows the path $F'FGKCC'$. Thus, when the sample is subjected to a cycle of magnetization, M follows a closed loop. This closed loop is called *hysteresis loop.* At each step M lags behind the corresponding magnetizing field H. This phenomenon is called *hysteresis.*

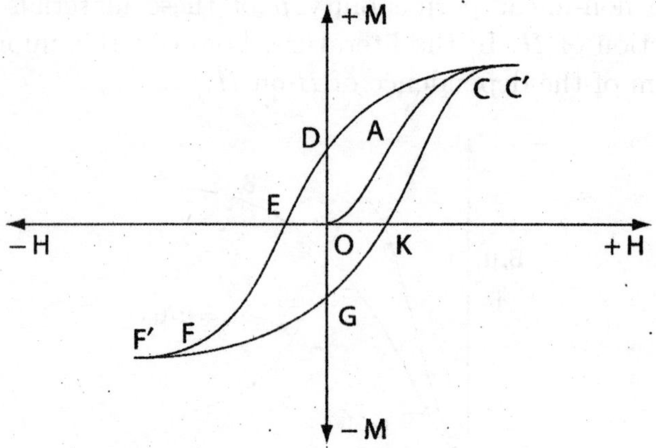

Fig 9.6-1

If instead of M the magnetic induction B is plotted against H a similar loop is obtained which is called *B-H loop.* It will be shown latter that the area of the *B-H* loop equals energy loss per unit volume per cycle of magnetization. It is called *hysteresis loss.*

Demagnetization process

A ferromagnetic sample once magnetized could not be brought back to the unmagnetized state by just reducing the external magnetizing field to zero. A possible way of demagnetizing a ferromagnetic sample is to take the sample through a numerous cycles

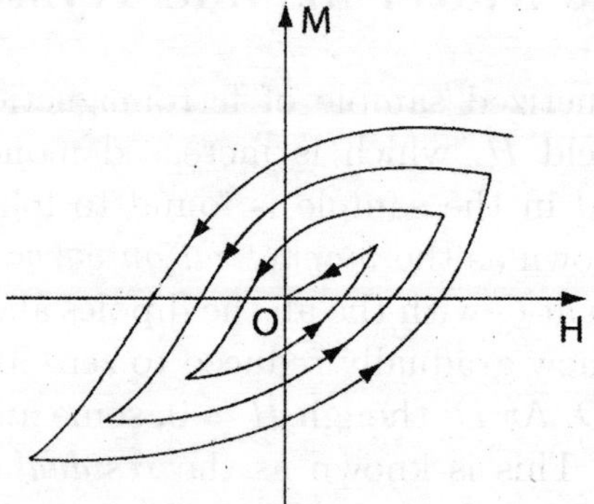

Fig. 9.6-2 : Demagnetization

of magnetization in which the maximum value of magnetizing field is progressively decreased and finally the field is reduced to zero. In this way the sample is taken round a spiral curve of diminishing amplitude as shown in Fig. 9.6-2 and finally returns to O the unmagnetized state.

Hard and soft magnetic materials

The shape and size of the hysteresis loop of a ferromagnetic material determine the suitability of the material in a technical application. Materials having small coercive forces are called *soft magnetic materials.* They can be easily magnetized and demagnetized. On the other hand, materials having large coercive forces are called *hard magnetic materials.* They cannot be easily magnetized and demagnetized. Magnetically soft materials also have greater permeability, retentivity and smaller hysteresis loop area. Consequently soft materials have smaller energy losses during cyclic magnetization. On the other hand, magnetically hard materials have smaller permeability, retentivity and larger hysteresis loop area and consequently hard materials have greater energy losses during cyclic magnetization. Most common soft magnetic material is the soft iron and hard magnetic material is the steel. Fig. 9.6-3 shows the nature of B-H loops for soft iron and steel. A close scrutiny of these loops reveals the following information: (*i*) The retentivity of soft iron is greater than that of steel. (*ii*) The coercivity of soft iron is smaller than that of steel. (*iii*) Area of B-H loop and hence hysteresis loss for soft iron is smaller than that of steel.

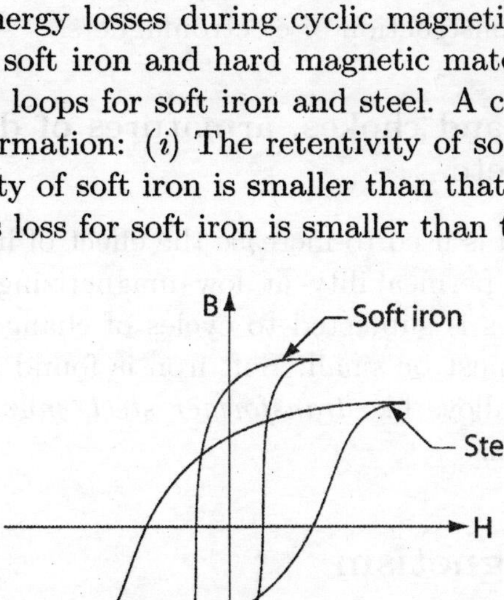

Fig 9.6-3

Choice of magnetic material for different applications

The suitability of a ferromagnetic material for a particular application can be decided from the nature of its hysteresis loop. Here we give some examples.

(i) Permanent magnets

The materials used for the construction of a permanent magnet should have high retentivity and high coercivity so that the magnet becomes strong and it becomes able to resist loss of magnetism due to rough handling or temperature changes. Hysteresis loss is unimportant here because a permanent magnet is never subjected to cycles of magnetization. All these considerations indicate that steel is more suitable than soft iron for the construction of permanent magnet. There are some alloys like *Alni* (Fe, Al, Ni), *Alnico* (Al, Ni, Co, Fe) which are more suitable for the construction of permanent magnets.

(ii) Electromagnet

The materials used for the construction of an electromagnet should have high magnetization at low magnetizing field, high initial permeability and low hysteresis loss. Soft iron possesses all these properties and hence is best suited for the construction of electromagnet. Also, some alloys like permalloy (Fe, Ni, Mn), *mu* metal (Fe, Ni, Cr, Cu) etc. are used for the construction of electromagnets.

Cores of transformer and chokes, armatures of dynamos and motors, telephone diaphragm etc.

Here the magnetic material is used to increase the effect of induction. So the material selected should have high permeability at low magnetizing field. As the magnetic material used in these cases is subjected to cycles of changes the hysteresis loss and hence the *B-H* loop area must be small. Soft iron is found to be better than steel in these applications. Some alloys like *transformer steel, mumetal, permalloy,* etc. are also used for these purposes.

9.7 Origin of Magnetism

So far we have discussed the properties of magnetic materials in terms of the macroscopic variable magnetization $\vec{M}$. Let us now look at the microscopic aspect of magnetization by analysing the response of individual atoms or molecules to external magnetic fields. A complete discussion of this subject is beyond the scope of this book. Here we shall discuss simple ideas that describe the qualitative nature of different classes of magnetic materials and predict an order of magnitude of their susceptibilities.

All matter consists ultimately of atoms and in each atom electrons revolve round the atomic nucleus and also spin about their own axes. These motions are associated with magnetic dipole moments and the atoms interact with external fields via this moments. Total magnetic moment of an atom is given by the vector sum of the orbital and spin magnetic moments of all the electrons within the atom. Thus the atoms may or may not have an intrinsic dipole moment. There may or may not have strong interaction among these dipoles. Thus different materials will behave differently in external fields and we can classify magnetic materials as diamagnetic, paramagnetic and ferromagnetic.

A. Langevin's theory of diamagnetism

The origin of diamagnetism can be explained by using the simple model of the electronic motion around the nucleus. According to the classical theory of Langevin in diamagnetic materials atoms do not possess any intrinsic magnetic dipole moment. When placed in a magnetic field a current is induced in these atoms, which produces a magnetic effect in opposition to the applied field. Suppose an electron of charge $-e$, mass m_e is rotating around the nucleus in a circular orbit of radius r with angular frequency ω in the absence of any magnetic field. The equivalent current is $-\frac{e}{T} = -\frac{e\omega}{2\pi}$ and magnetic moment

$$\vec{m} = \text{current} \times \text{area} = -\frac{e\omega}{2\pi} \cdot \pi r^2 \hat{z} = -\frac{e\omega}{2} r^2 \hat{z}, \tag{9.7-1}$$

where $\hat{z}$ is a unit vector normal to the plane of motion. Now suppose a magnetic field $\vec{B} = B\hat{z}$ is applied perpendicular to the plane of the orbit. Assuming that the areas of the orbits remain unaltered, the magnetic field will cause a change in velocity of the electron in the orbit and a modified value of equivalent current is produced. In the absence of magnetic field the centripetal force on the electron is supplied by the electrical force F_e between the electron and the nucleus:

$$m_e \omega^2 r = F_e. \tag{9.7-2}$$

Application of magnetic field exerts an additional force $-e\vec{v} \times \vec{B}$ on the electron. With reference to Fig. 9.7-1 the electron angular speed will now be greater than ω,

$$m_e(\omega + \Delta\omega)^2 r = F_e + evB = m_e\omega^2 r + e\omega r B.$$

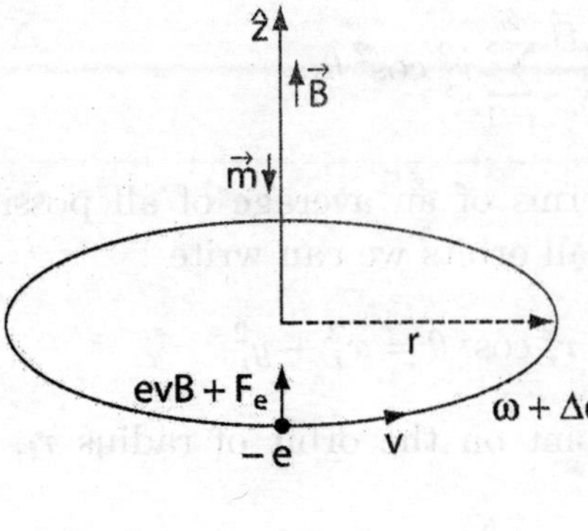

Fig 9.7-1

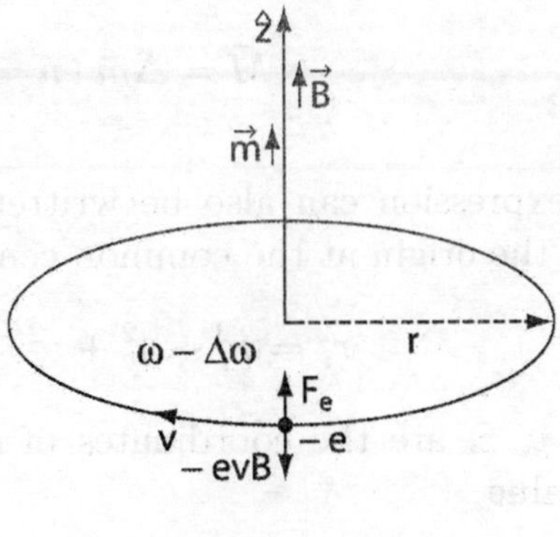

Fig 9.7-2

Assuming the change $\Delta\omega$ to be small we can get

$$m_e\left(\omega^2 + 2\omega\Delta\omega\right)r = m_e\omega^2 r + e\omega r B \quad \text{or} \quad \Delta\omega = \frac{eB}{2m_e}. \tag{9.7-3}$$

This change in frequency is known as *Larmor frequency.*.

A change in angular frequency means a change in the dipole moment (9.7-1),

$$\Delta\vec{m} = -\frac{er^2}{2}\Delta\omega\hat{z} = -\frac{e^2r^2}{4m_e}\vec{B}. \tag{9.7-4}$$

Note that the change in dipole moment is opposite to the direction of $\vec{B}$. If the electron rotates in a direction opposite to that shown in Fig. 9.7-1 its dipole moment would be $\vec{m} = \frac{e\omega r^2}{2}\hat{z}$ and in the presence of magnetic field its motion would be slowed down (Fig. 9.7-2):

$$\Delta\omega = -\frac{eB}{2m_e}.$$

As a result the change in magnetic dipole moment will again be in a direction opposite to $\vec{B}$.

If the electronic orbit is not perpendicular to the applied field then r is to be replaced by r_1, the projection of the radius r on the plane perpendicular to the magnetic field, i.e., in the xy-plane. Then

$$\Delta\vec{m} = -\frac{e^2r_1^2}{4m_e}\vec{B}. \tag{9.7-5}$$

If there are Z electronic orbits in an atom and the normal to the ith orbit of radius r_i makes an angle θ_i with the field then the change in the magnetic moment of an atom would be

$$\Delta\vec{m} = -\frac{e^2\vec{B}}{4m_e}\sum_{i=1}^{Z} r_i^2\cos\theta_i^2.$$

If there are n atoms per unit volume then the induced magnetization will be given as

$$\vec{M} = \Delta\vec{m}\cdot n = -\frac{e^2n\vec{B}}{4m_e}\sum_{i=1}^{Z} r_i^2\cos^2\theta_i. \tag{9.7-6}$$

This expression can also be written in terms of an average of all possible radii. Choosing the origin at the common centre of all orbits we can write

$$r_i^2 = x_i^2 + y_i^2 + z_i^2 \quad \text{and} \quad r_i^2\cos^2\theta = x_i^2 + y_i^2,$$

where x_i, y_i, z_i are the coordinates of any point on the orbit of radius r_i. Now the average value

$$\overline{x_i^2} = \overline{y_i^2} = \overline{z_i^2} = \frac{\overline{r_i^2}}{3}.$$

Therefore,

$$\overline{r_i^2 \cos^2\theta} = \overline{x_i^2} + \overline{y_i^2} = \frac{2}{3}\overline{r_i^2} = \frac{2}{3}\overline{r^2}, \tag{9.7-7}$$

where $\overline{r^2}$ is the mean square distance between the electrons and the nucleus of an atom. Thus, from (9.7-6)

$$\vec{M} = -\frac{Ze^2 n\overline{r^2}}{6m_e}\vec{B} = -\frac{\mu_0 n Ze^2\overline{r^2}}{6m_e}\vec{H}, \tag{9.7-8}$$

where we assume the permeability of a diamagnetic material to be μ_0. So the diamagnetic susceptibility per unit volume is

$$\chi = \frac{M}{H} = -\frac{\mu_0 n Ze^2\overline{r^2}}{6m_e}. \tag{9.7-9}$$

Obviously *the diamagnetic susceptibility is negative and independent of the field strength and temperature,* which is in accordance with experimental results.

It is to be noted that the mechanism responsible for diamagnetism is a universal phenomenon affecting all atoms whether it possesses atomic dipole moment or not. However, diamagnetism is much weaker than paramagnetism or ferromagnetism. For this diamagnetism is observed mainly in atoms with no intrinsic magnetic dipole moment.

B. Langevin's theory of paramagnetism

The classical theory of paramagnetism as developed by Langevin is based on the idea that the atoms or molecules of paramagnetic materials possess intrinsic magnetic dipole moment $\vec{m}$ and that the interaction among these dipoles is negligible. In the absence of any magnetic field the dipoles point in random directions due to thermal agitation and the vector sum of the dipole moments of all the atoms vanishes. When placed in a magnetic field the field exerts torques on the individual dipoles and tend to align them. Thermal agitation tries to oppose it. At any given temperature (T) a state of equilibrium is set up with some net alignment along the field direction. This gives rise to the magnetization of the medium.

Let us consider a molecular gas containing n molecules per unit volume and each molecule having an intrinsic magnetic dipole moment. The potential energy of a dipole of moment $\vec{m}$ making an angle θ with the applied field $\vec{B}$ is

$$U = -\vec{m}\cdot\vec{B} = -mB\cos\theta \tag{9.7-10}$$

According to Boltzmann distribution law the number of molecules per unit volume having energy in the range U to $U + dU$ is

$$dn = C\cdot e^{-U/kT}\cdot dU, \tag{9.7-11}$$

where k is the Boltzmann constant and C is a constant to be determined. Using (9.7-10) we get

$$dn = Ce^{mB\cos\theta/kT}\cdot mB\sin\theta d\theta. \tag{9.7-12}$$

To find C we integrate this expression over all values of θ. Thus,

$$n = CmB \int_0^\pi e^{mB\cos\theta/kT} \cdot \sin\theta d\theta.$$

Thus,

$$dn = \frac{ne^{mB\cos\theta/kT} \cdot \sin\theta d\theta}{\int_0^\pi e^{mB\cos\theta/kT} \sin\theta d\theta} \tag{9.7-13}$$

These dn molecules each have a component of magnetic moment $m\cos\theta$ in the field direction. The components perpendicular to $\vec{B}$ will obviously cancel out. Hence the magnetic dipole moment per unit volume or intensity of magnetization will be given by

$$\begin{aligned} M &= \int_0^\pi m\cos\theta \cdot dn \\ &= nm \frac{\int_0^\pi \cos\theta e^{mB\cos\theta/kT} \cdot \sin\theta d\theta}{\int_0^\pi e^{mB\cos\theta/kT} \cdot \sin\theta d\theta} \quad \text{[Using Eq. (9.7-13)].} \end{aligned}$$

Putting $\dfrac{mB}{kT} = \alpha$ and $\cos\theta = x$ we get

$$\begin{aligned} M &= nm \frac{\int_{-1}^{+1} xe^{\alpha x} dx}{\int_{-1}^{+1} e^{\alpha x} dx} = nm \frac{\left[\frac{xe^{e^{\alpha x}}}{\alpha} - \frac{e^{\alpha x}}{\alpha^2}\right]_{-1}^{+1}}{\left[\frac{e^{\alpha x}}{\alpha}\right]_{-1}^{+1}} \\ &= nm \left[\frac{e^\alpha + e^{-\alpha}}{e^\alpha - e^{-\alpha}} - \frac{1}{\alpha}\right] = nm\left(\coth\alpha - \frac{1}{\alpha}\right) \end{aligned} \tag{9.7-14}$$

The quantity nm represents saturation magnetization (M_s) when all the dipoles are fully aligned in the field direction.

Therefore, we can write

$$M = M_s\left(\coth\alpha - \frac{1}{\alpha}\right) \tag{9.7-15}$$

$\coth\alpha - 1/\alpha$ is known as *Langevin's function* and is denoted by $L(\alpha)$. Its variation with α is shown in Fig. 9.7-3. If $\alpha = mB/kT$ is very large, i.e., if T is very low, or the applied field B is large and/or the intrinsic dipole moment m is very large then in the limit $\alpha \to \infty$, $L(\alpha) \to 1$. In this case $M = M_s$. Saturation effects would, therefore, be expected at low temperatures.

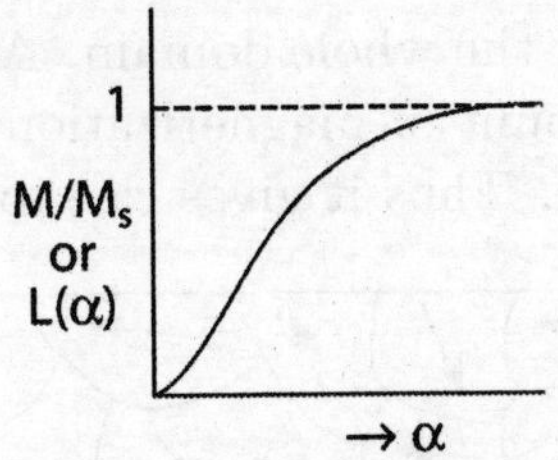

Fig. 9.7-3: Variation of Langevin function $L(\alpha)$ with α

If $\alpha = mB/kT$ is small, i.e., if T is high or the applied field B is small and/or the basic dipole moment m is small then $L(\alpha)$ can be approximated by $\alpha/3$. In this case

$$M = M_s \cdot \frac{\alpha}{3} = nm \cdot \frac{mB}{3kT} = \frac{m^2 nB}{3kT}. \tag{9.7-16}$$

For paramagnetic material the permeability may be taken as μ_0. Then $\vec{B} = \mu_0 \vec{H}$ and from (9.7-16) we get the susceptibility for a paramagnetic material as

$$\chi = \frac{M}{H} = \frac{m^2 n \mu_0}{3kT}. \tag{9.7-17}$$

This shows that *the susceptibility of a paramagnetic gas varies inversely as the absolute temperature. This result is known as Curie law.*

For paramagnetic solids and liquids mutual interactions among the dipoles is not negligible. Many of these materials are found to obey the *Curie-Weiss law:*

$$\chi \propto \frac{1}{T - \theta},$$

where θ is a characteristic constant of the substance.

C. Weiss theory of ferromagnetism

In ferromagnetic materials each atom or molecule possesses intrinsic magnetic dipole moment. But unlike paramagnetic materials in this case there is a strong interaction among the neighbouring dipoles. This interaction is of quantum mechanical nature and is known as *exchange interaction.* Due to this interaction a large group of neighbouring dipoles get aligned parallel to each other even in the absence of external magnetizing field. This group of parallelly oriented atomic dipoles is called a *domain.* Thus, a ferromagnetic sample may be assumed to consist of large number of domains. In an unmagnetized sample domains are randomly oriented such that the total magnetic moment of the sample becomes zero (Fig. 9.7-4). Now if an external magnetizing field is applied the sample may be magnetized in two ways: (*i*) *Domain wall displacement–* in weak field the domains parallel to the field grow at the expense of unfavourably oriented domains. (*ii*) *Domain rotation–* in strong field there may be rotation of the direction of the magnetization of the whole domain. As the process of magnetization is not entirely reversible some amount of magnetization is retained by the sample when the magnetizing field is removed. Thus it gives rise to the phenomenon of hysteresis.

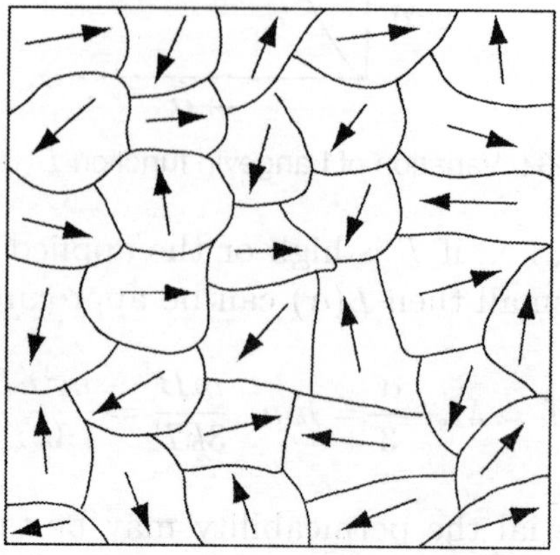

Fig. 9.7-4 : Ferromagnetic domain structure

The theory of ferromagnetism was first developed by P. Weiss using Langevin's theory of paramagnetism and introducing the idea of domains and molecular field. He assumed that the effective magnetic field experienced by a dipole is the sum of external magnetizing field $\vec{H}$ and the molecular field arising from the interaction of neighbouring dipoles. This molecular field was assumed to be proportional to the magnetization $\vec{M}$. Thus effective magnetizing field is may be written as,

$$\vec{H}_e = \vec{H} + \gamma\vec{M}, \tag{9.7-18}$$

where γ is a constant independent of temperature called the *Weiss constant* or *molecular field constant.*

Assuming that the domains obey the Langevin's theory of paramagnetism we can extend the paramagnetic theory to ferromagnetic material just by replacing the external field $\vec{H}$ by the effective field $\vec{H}_e$. Thus,

$$\frac{M}{M_s} = \coth\alpha - \frac{1}{\alpha} = L(\alpha), \tag{9.7-19}$$

where

$$\alpha = \frac{\mu_0 m H_e}{kT} \approx \frac{\mu_0 m}{kT}(H + \gamma M)$$

and $M_s = nm$ represents saturation magnetization.

Since the effective magnetizing field depends on $\vec{M}$, any increase in $\vec{M}$ will cause increase in magnetizing field and thus a cumulative effect may be expected. In fact for a ferromagnetic material it is possible to have magnetization even in the absence of external magnetizing field. This magnetization is known as *spontaneous magnetization.*

For spontaneous magnetization $H = 0$ so that we can write

$$\alpha = \frac{\mu_0 m\gamma}{kT} M \quad \text{or} \quad M = \frac{kT}{\mu_0 m\gamma}\alpha.$$

Using $M_s = nm$ we can write

$$\frac{M}{M_s} = \frac{\alpha kT}{\mu_0 \gamma nm^2} = \frac{\alpha T}{3\theta}, \tag{9.7-20}$$

where $\theta = \mu_0 \gamma nm^2/3k$.

The condition for spontaneous magnetization is given by the solutions of the simultaneous Eqs. (9.7-19) and (9.7-20). The value of M at any particular temperature may be obtained from the point of intersection of the two M versus α curves representing (9.7-19) and (9.7-20). An examination of Fig. 9.7-5 shows that for $T < \theta$ a non-vanishing value of M is obtained, i.e., spontaneous magnetization becomes possible. For $T > \theta$ there is no point of intersection other than at the origin, which means that spontaneous magnetization is not possible. The critical temperature θ beyond which spontaneous magnetization vanishes is known as *Curie temperature*. Above θ a ferromagnetic material behaves like a paramagnetic material. In this paramagnetic region ($T > \theta$) for weak magnetic fields $L(\alpha)$ may be approximated by $\alpha/3$. Then from Eq. (9.7-19)

$$\frac{M}{M_s} = \frac{\alpha}{3} = \frac{\mu_0 m}{3kT}(H + \gamma M)$$

$$\text{or} \quad M = \frac{\mu_0 nm^2}{3kT}(H + \gamma M) \quad [\because M_s = nm]$$

$$\text{or} \quad M = \frac{\mu_0 nm^2 H}{3k\left(T - \frac{\mu_0 \gamma nm^2}{3k}\right)}. \tag{9.7-21}$$

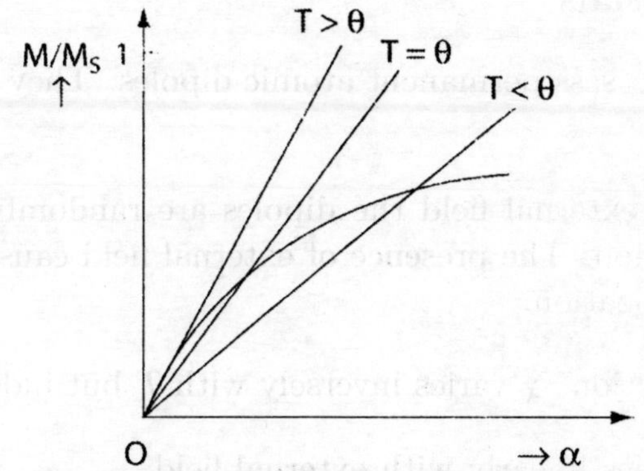

Fig. 9.7-5: Spontaneous magnetization

Therefore, susceptibility

$$\chi = \frac{M}{H} = \frac{C}{T-\theta}, \tag{9.7-22}$$

where $C = \dfrac{\mu_0 n m^2}{3k}$ and $\theta = \dfrac{\mu_0 \gamma n m^2}{3k}$.

This result is known as *Curie-Weiss law.* It describes fairly satisfactorily the observed susceptibility variation in the paramagnetic region well above the Curie temperature but not in the region close to the Curie temperature.

9.8 Comparison of Dia-, Para- and Ferromagnetic Materials

Diamagnetic materials

1. There is no permanent intrinsic atomic magnetic dipole moment. These materials are feebly repelled by magnetic field.
2. Induced dipole moments always oppose the applied field and susceptibility $\chi < 0$.
3. χ is practically independent temperature (T) and magnetic field.
4. Magnetization (M) varies linearly with external magnetic field.
5. Diamagnetism does exit in all substances. However, in materials with permanent atomic dipoles diamagnetic effects are masked by the effects of permanent dipoles.
6. Example: Bi, Au, Ag, Cu, H_2

Paramagnetic materials

1. These materials possess permanent atomic dipoles. They are feebly attracted by magnetic field.
2. In the absence of external field the dipoles are randomly oriented and there is no net magnetization. The presence of external field causes alignment of dipoles and hence magnetization.
3. χ is a positive fraction. χ varies inversely with T but independent of H.
4. Magnetization varies linearly with external field.
5. Examples: Pt, Al, Na, O_2.

Ferromagnetic materials

1. It contains permanent atomic dipoles which interact strongly with each other and may cause spontaneous magnetization within a single domain even in the absence of external field. They are strongly attracted by magnetic fields.

2. A small external field can cause large magnetization.

3. $\chi \gg 1$ and depends on temperature according to the Curie-Weiss law: $\chi \propto \frac{1}{T-\theta}$. Here χ is a function of H.

4. For each ferromagnetic material there is a particular temperature called *Curie temperature* θ. There is spontaneous magnetization only for $T < \theta$. For $T > \theta$ ferromagnetic material becomes paramagnetic.

5. Magnetization M depends on the past history of the specimen and it shows the phenomenon of hysteresis.

6. Example: Fe, Ni, Co, Gd.

9.9 Hysteresis Loss

When a specimen of ferromagnetic material is subjected to cycles of magnetization and demagnetization some amount of energy is dissipated in the specimen. This is known as *hysteresis loss.* As the magnetic domains try to line up with the magnetizing field first in one way and then the other it gives rise to mechanical stresses and produce heating. In fact the energy expended during magnetization is not totally recovered because of irreversible changes in domain structure of the material.

To calculate the hysteresis loss let a typical domain of magnetic moment $\vec{m}$ makes an angle θ with the direction of the magnetizing field $\vec{H}$ at an instant of time when the magnetization in the specimen is $\vec{M}$. The sum of the components of magnetic moments perpendicular to $\vec{H}$ over unit volume must be zero. Therefore, we can write

$$\sum m \sin\theta = 0, \quad M = \sum m\cos\theta \quad \text{and} \quad dM = -\sum m \sin\theta d\theta, \tag{9.9-1}$$

where the sum refers to unit volume. The negative sign indicates that if θ decreases to $\theta - d\theta$, M increases to $M + dM$. Now the torque acting on the domain of dipole moment $\vec{m}$, when it is making an angle θ with $\vec{H}$, is of magnitude

$$\tau = \left|\vec{m} \times \vec{B}_{\text{ext}}\right| = m\mu_0 H \sin\theta,$$

where $\vec{B}_{\text{ext}} = \mu_0\vec{H}$ is the magnetic induction due to external magnetizing field $\vec{H}$. So the work done by the external source in turning all the domains in unit volume through a small angle $d\theta$ towards the field direction would be

$$dW = -\sum m\mu_0 H \sin\theta d\theta = \mu_0 H dM. \tag{9.9-2}$$

Therefore, work done per unit volume in taking the specimen through a complete cycle of magnetization would be

$$W = \mu_0 \oint HdM = \mu_0 \times \text{ area enclosed by } M\text{-}H \text{ loop.} \tag{9.9-3}$$

Since

$$\vec{B} = \mu_0 \left(\vec{H} + \vec{M}\right)$$

$$\text{or} \quad d\vec{B} = \mu_0 \left(d\vec{H} + d\vec{M}\right)$$

$$\text{or} \quad \vec{H} \cdot d\vec{B} = \mu_0 \vec{H} \cdot d\vec{H} + \mu_0 \vec{H} \cdot d\vec{M}$$

$$\text{or} \quad \oint HdB = \mu_0 \oint HdH + \mu_0 \oint HdM.$$

$\oint HdH = 0$ because the plot of H against H is a straight line and area enclosed by it is zero. Therefore, from Eq. (9.9-3)

$$W = \mu_0 \oint HdM = \oint HdB = \text{area enclosed by } B\text{-}H \text{ loop.}$$

Thus the work done per unit volume in taking the specimen once through a complete cycle of magnetization is equal to the area enclosed by the B-H loop or μ_0 times the area enclosed by the M-H loop.

The Steinmetz formula

Steinmetz proposed an empirical formula relating the hysteresis loss (W) and the maximum value $B_{\max}$ of B attained in the hysteresis loop. According to this formula the work done per unit volume in taking a ferromagnetic specimen through one cycle of magnetization can be expressed by

$$W = \eta B_{\max}^n,$$

where the value of n can be taken as 1.6 for a large number of materials; η is a characteristic constant of the given material and it is called *Steinmetz coefficient.* This formula can be used for order of magnitude calculation for the hysteresis loss.

9.10 Magnetic Circuits

We know that the magnetic flux lines form closed loops. If all the magnetic flux associated with a particular distribution of currents is confined to a rather well-defined path then we speak of a magnetic circuit in close analogy with electric circuits. This idea of magnetic circuit helps in solving some magnetic field problems by using circuit approach. Magnetic devices such as electromagnets, transformers, motors, generators etc. may be considered as magnetic circuits. Calculation of magnetic field in these cases becomes very simple if an analogy between electric circuit and magnetic circuit is exploited.

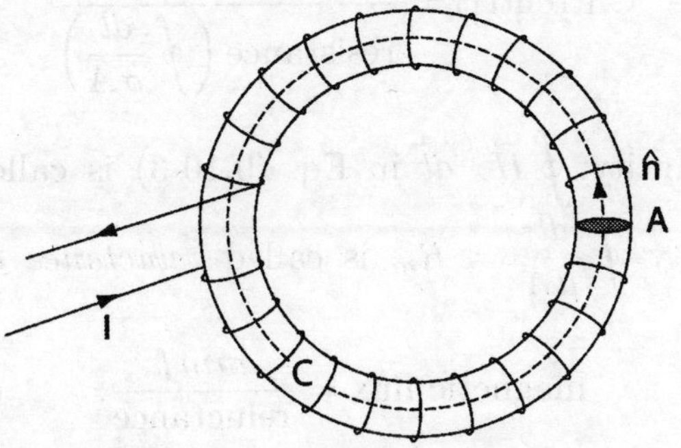

Fig 9.10-1

The idea of a magnetic circuit can be well understood by considering the example of a ring of ferromagnetic material which is wound over closely with a coil of insulated wire carrying current (Fig. 9.10-1). In this case magnetic flux lines are confined to the region inside the windings and hence form a magnetic circuit. Let N be the total number of turns and I the current through it. Now applying Ampere's circuital law to the path C as shown by dotted lines in Fig. 9.10-1 we get

$$\oint \vec{H} \cdot d\vec{l} = NI. \tag{9.10-1}$$

The product NI is known as the *number of ampere-turns.* Suppose we are interested in calculating magnetic flux Φ over any cross-sectional area A of the circuit. Using

$$\vec{B} = \mu\vec{H} \quad \text{and} \quad \Phi = \vec{B} \cdot \hat{n}A = BA.$$

We can write $H = \Phi/\mu A$. Therefore, from (9.10-1),

$$\oint \frac{\Phi dl}{\mu A} = NI. \tag{9.10-2}$$

As we are dealing with a magnetic circuit we expect Φ to be essentially the same over any cross-section of the circuit. Therefore, we may write

$$\Phi = \frac{NI}{\oint \frac{dl}{\mu A}} = \frac{\oint \vec{H} \cdot d\vec{l}}{\oint \frac{dl}{\mu A}}. \tag{9.10-3}$$

In drawing analogy with electric circuit we note that magnetic flux is the magnetic counterpart of electric current, which has the same value around a loop. Equation

(9.10-3) is the basic equation for a magnetic circuit and enables us to solve for magnetic flux in terms of circuit parameters. The equation is similar to Ohm's law for an electric circuit:

$$\text{Current(I)} = \frac{emf\left(\oint \vec{E}\cdot d\vec{l}\right)}{\text{resistance}\left(\oint \frac{dl}{\sigma A}\right)}. \tag{9.10-4}$$

By analogy the quantity $\oint \vec{H}\cdot d\vec{l}$ in Eq. (9.10-3) is called *magnetomotive force* (mmf) and the quantity $\oint \frac{dl}{\mu A} = R_m$ is called *reluctance of the magnetic circuit.* Thus,

$$\text{magnetic flux} = \frac{mmf}{\text{reluctance}}. \tag{9.10-5}$$

This relation is often referred to as the *Ohm's law for magnetic circuit.* It is used by engineers for calculation of the number of ampere-turns required to produce a desired magnetic flux in motors, dynamos, electromagnets, transformers etc.

The magnetomotive force in any magnetic circuit tends to drive magnetic flux through the circuit. It is defined as the line integral of magnetic intensity $\vec{H}$ once around the magnetic circuit. It is measured in units of ampere-turns.

The reluctance is a property of magnetic circuit by which it opposes the creation of magnetic flux in it. Its unit is ampere-turns/weber. Because of the similarity between the expressions for reluctance and resistance we expect that the series and parallel reluctances may be combined in the same manner as resistances. Thus, if a magnetic circuit is made from several homogeneous pieces of various materials all connected in series then total reluctance will be

$$R_m = \sum_i \frac{l_i}{\mu_i A_i} = \sum_i R_{mi}, \tag{9.10-6}$$

where R_{mi} is the reluctance of the ith section of the circuit.

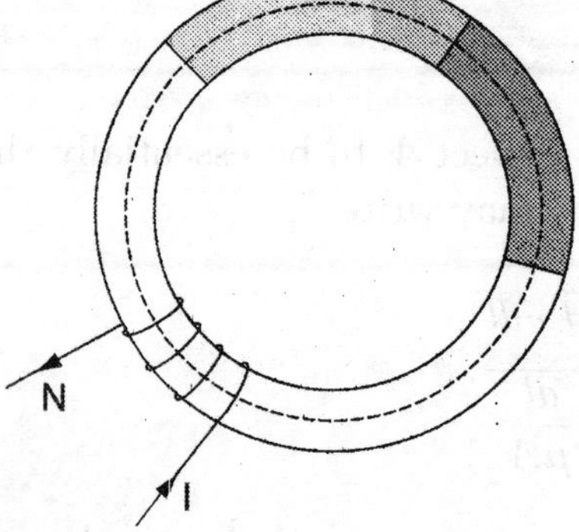

Fig. 9.10-2 : Reluctances in series

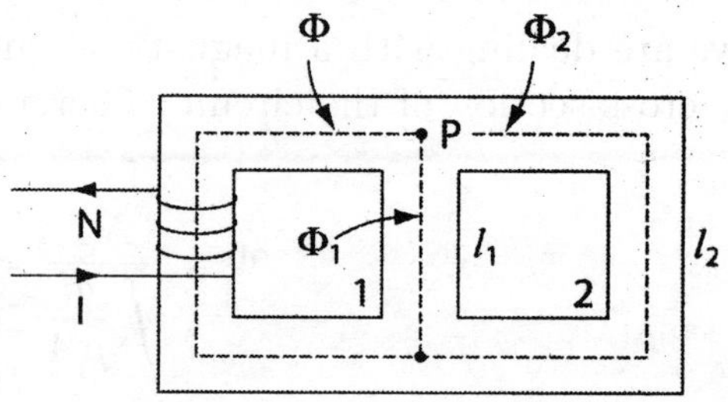

Fig. 9.10-3 : Reluctances in parallel

In Fig. 9.10-3 we show a magnetic circuit with reluctance connected in parallel. The magnetomotive force NI sets up a total magnetic flux Φ in the circuit which encounters two paths at the point P. Suppose Φ_1 is flux along path 1 and Φ_2 along path 2. Then

$$\Phi = \Phi_1 + \Phi_2 \quad \text{or} \quad \frac{NI}{R_m} = \frac{NI}{R_{m1}} + \frac{NI}{R_{m2}} \quad \text{or} \quad \frac{1}{R_m} = \frac{1}{R_{m1}} + \frac{1}{R_{m2}}, \tag{9.10-7}$$

where R_m is the equivalent reluctance, R_{m1} and R_{m2} are the reluctance of the paths 1 and 2, respectively. Note that $\Phi_2/\Phi_1 = R_{m1}/R_{m2}$, which means that the flux prefers passing through the path of lower reluctance. Ferromagnetic materials having high μ offers very low reluctance. So most of the magnetic flux can be channeled through ferromagnetic materials. This fact indicates that magnetic shielding can be obtained with such materials. For example if a ferromagnetic cylindrical shell is placed in a magnetic field with its axis normal to the field, most of the flux lines will pass through the material of the cylinder and the hollow region will be almost free of magnetic field.

Analogy between electric and magnetic circuits

Following are the points of similarities between electric and magnetic circuits:

1. In electric circuit current (I)$= \dfrac{\text{emf } (\mathcal{E})}{\text{resistance(R)}}$. In magnetic circuit there is a similar relation, magnetic flux $(\Phi) = \dfrac{\text{mmf } (\mathcal{E}_m)}{\text{reluctance}(\text{R}_m)}$.

2. Electrical resistance is expressed by $R = \oint \dfrac{dl}{\sigma A}$. The reluctance of magnetic circuit is expressed by a similar expression $R_m = \oint \dfrac{dl}{\mu A}$.

3. The series and parallel reluctances combine in the same manner as resistances.

The above analogy may be further extended but it is important to note the points of differences also.

1. In electric circuit current is produced due to flow of charge carriers but in magnetic circuit magnetic flux does not indicate the flow of any kind of particles.

2. An electric circuit may be closed or opened. But a magnetic circuit is always closed.

3. In an electric circuit energy is required for setting up and maintenance of current but in magnetic circuit energy is needed only to set up the flux initially and not to maintain it.

4. Resistance of a conductor is independent of current at a given temperature but reluctance varies considerably with magnetic flux because μ depends on magnetization.

5. The analogy between electric and magnetic circuits demands that μ corresponds to conductivity σ. But actually μ corresponds to permittivity ε.

Magnetic circuit with a small air gap

In order to access the magnetic field it is a common practice to have small air gaps in a magnetic circuit. Consider an iron ring of mean radius a, area of cross-section A and with a small air gap of length Δl as shown in Fig. 9.10-5. The reluctance of iron path of the circuit is $(2\pi a - \Delta l)/\mu A$, where μ is the permeability of iron. The reluctance of the small air gap is $\Delta l/\mu_0 A$, where we assume that $\Delta l \ll a$ so that the spreading of magnetic flux in the air gap is negligible and the effective cross-section of the air gap may be taken as that of the iron core.

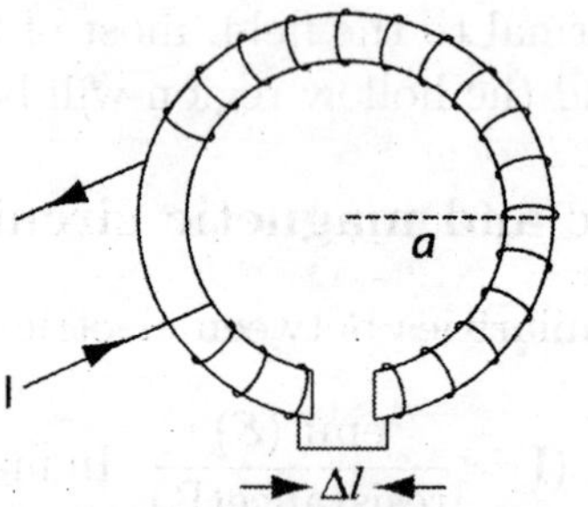

Fig 9.10-5

So the total reluctance of the circuit is

$$R_m = \frac{2\pi a - \Delta l}{\mu A} + \frac{\Delta l}{\mu_0 A}.$$

Now the flux Φ through the circuit is given by

$$\Phi = \frac{mmf}{R_m} = \frac{NI}{\frac{2\pi a - \Delta l}{\mu A} + \frac{\Delta l}{\mu_0 A}},$$

where I is the current passing through the windings having N turns.

As $\mu_0 \ll \mu$, the presence of air gap increases the reluctance of the circuit and causes enormous drop in magnetic flux. In this case larger number of ampere-turns will be required to produce the same flux as without air gap.

Note that for small air gap the value of $\vec{B}$, i.e.,

$$\frac{\Phi}{A} = B = \frac{NI}{\frac{2\pi a - \Delta l}{\mu} + \frac{\Delta l}{\mu_0}}$$

will be same both in the material of the ring and within the air gap. *The value of $\vec{H}$ within the material will be given by B/μ and that within the air gap will be given by B/μ_0.*

9.11 Boundary Conditions on $\vec{B}$ and $\vec{H}$

When the magnetic field exists in more than one medium then it becomes important to know how the field vectors $\vec{B}$ and $\vec{H}$ change in passing the interface of two media.

Let us consider two different magnetic media 1 and 2, in contact, as shown in Fig. 9.11-1. We construct a small shallow pill box which is partly in medium 1 and partly in medium 2. Integrating the relation $\vec{\nabla} \cdot \vec{B} = 0$ over the volume V of the pill box we can write

$$\int_V \vec{\nabla} \cdot \vec{B}\, dV = 0.$$

Applying divergence theorem we get

$$\int_S \vec{B} \cdot d\vec{S} = 0, \tag{9.11-1}$$

where S is the surface of the pill box.

Since the contribution from the vertical sides is negligible when the height of the pill box tends to zero, we can write

$$\vec{B}_2 \cdot \hat{n}_2 \Delta S + \vec{B}_1 \cdot \hat{n}_1 \Delta S = 0, \tag{9.11-2}$$

where ΔS is the area of the top or bottom flat surfaces; $\hat{n}_1$ and $\hat{n}_2$ are the unit vectors normal to the lower and upper flat surfaces (Fig. 9.11-1).

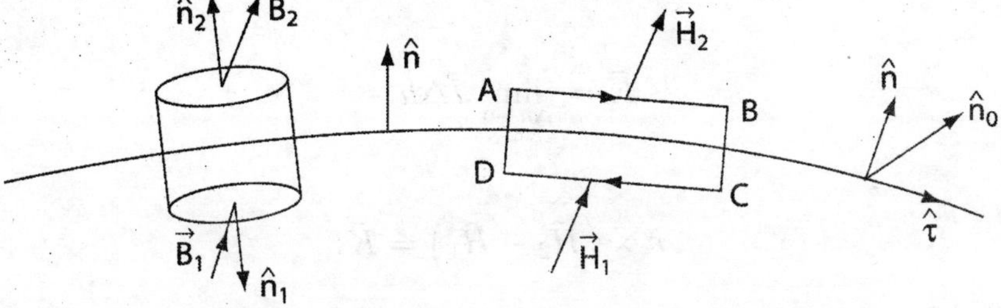

Fig. 9.11-1 : Obtaining boundary conditions on $\vec{B}$ and $\vec{H}$

Since $\hat{n}_2 = -\hat{n}_1 = \hat{n}$ as shown in Fig. 9.11-1; we can write

$$\left(\vec{B}_2 - \vec{B}_1\right) \cdot \hat{n} = 0 \quad \text{or} \quad B_{2n} = B_{1n}. \tag{9.11-3}$$

Thus, *the normal components of $\vec{B}$ are continuous across the interface.*

The boundary condition on $\vec{H}$ can be obtained by integrating the relation $\vec{\nabla} \times \vec{H} = \vec{J}$ over a small rectangular circuit $ABCD$ whose sides AD and BC are very small compared with AB or CD :

$$\int_{\Delta S} \vec{\nabla} \times \vec{H} \cdot d\vec{S} = \int_{\Delta S} \vec{J} \cdot d\vec{S}, \tag{9.11-4}$$

where ΔS is the area bounded by the rectangle $ABCD$.

Applying Stokes' theorem we get,

$$\oint_C \vec{H}\cdot d\vec{l} = \int_{\Delta S} \vec{J}\cdot d\vec{S}.$$

Now taking $AB = CD = \Delta l$ and assuming that $BC = DA = \Delta h \to 0$ we can write

$$\vec{H}_2\cdot\Delta\vec{l} - \vec{H}_1\cdot\Delta\vec{l} = \lim_{\Delta h\to 0} \vec{J}\cdot\Delta\vec{S}$$

$$\text{or } \left(\vec{H}_2 - \vec{H}_1\right)\cdot\Delta l\hat{\tau} = \lim_{\Delta h\to 0} \vec{J}\cdot\hat{n}_0\Delta l\cdot\Delta h$$

From Fig. 9.11-1, $\hat{n}_0\times\hat{n} = \hat{\tau}$, therefore,

$$\left(\vec{H}_2 - \vec{H}_1\right)\cdot\hat{n}_0\times\hat{n} = \lim_{\Delta h\to 0} \vec{J}\cdot\hat{n}_0\Delta h$$

$$\text{or } \hat{n}_0\cdot\hat{n}\times\left(\vec{H}_2 - \vec{H}_1\right) = \lim_{\Delta h\to 0} \hat{n}_0\cdot\vec{J}\Delta h$$

$$\text{or } \hat{n}\times\left(\vec{H}_2 - \vec{H}_1\right) = \lim_{\Delta h\to 0} \vec{J}\Delta h.$$

If the current density $\vec{J}$ is finite then the right-hand side is zero. However, if the current through the rectangle, $I = \vec{J}\cdot\hat{n}_0\Delta l\Delta h$, is concentrated in a very thin layer on the surface, then as $\Delta h\to 0$ it is convenient to introduce the concept of surface current density $\vec{K}$ (i.e., current per unit length in the surface layer). In this case we can write

$$I = \vec{J}\cdot\vec{n}_0\Delta h\Delta l = \vec{K}\cdot\vec{n}_0\Delta l,$$

where

$$\vec{K} = \lim_{\substack{\Delta h\to 0\\ \vec{J}\to\infty}} \vec{J}\Delta h.$$

Thus,

$$\hat{n}\times\left(\vec{H}_2 - \vec{H}_1\right) = \vec{K}. \tag{9.11-5}$$

If $\vec{K} = 0$ then

$$H_{2t} = H_{1t}. \tag{9.11-6}$$

Thus, *in the absence of any surface current the tangential components of magnetic intensity are continuous across the interface.*

Boundary conditions on scalar and vector potentials can be deduced from the boundary conditions on $\vec{B}$ and $\vec{H}$. Since $\vec{H} = -\vec{\nabla}\phi$, the potential difference between two close points can be written as

$$\Delta\phi = \vec{\nabla}\phi\cdot\Delta\vec{r} = -\vec{H}\cdot\Delta\vec{r}, \tag{9.11-7}$$

where $\Delta\vec{r}$ is the separation between the points. Because $H_{2t} = H_{1t}$ across an interface having no surface current then Eq. (9.11-7) implies that as $\Delta\vec{r}\to 0$, $\Delta\phi\to 0$ or $\phi_2 = \phi_1$. Thus, *magnetic scalar potential is continuous across the interface where $\vec{H}$ remains finite.*

To find the boundary condition of $\vec{A}$ we note that the magnetic flux through the small rectangle $ABCD$ of Fig. 9.11-1 is

$$\int_{\Delta S} \vec{B}\cdot\hat{n}\,dS = \int_{\Delta S}\left(\vec{\nabla}\times\vec{A}\right)\cdot\hat{n}dS = \oint_C \vec{A}\cdot d\vec{l}.$$

Now if the rectangle is allowed to shrink to the surface the flux through the surface vanishes and we can write

$$\oint_C \vec{A}\cdot d\vec{l} = \vec{A}_2\cdot\Delta\vec{l} - \vec{A}_1\cdot\Delta\vec{l} = 0 \quad \text{or} \quad A_{2t} = A_{1t}.$$

Thus, *the tangential components of $\vec{A}$ are continuous across the boundary.*

9.12 Magnetostatic Boundary Value Problems

Basic equations of magnetostatics in presence of magnetizable materials are

$$\vec{\nabla}\cdot\vec{B} = 0 \tag{9.12-1}$$

$$\text{and} \quad \vec{\nabla}\times\vec{H} = \vec{J}_{\text{free}}. \tag{9.12-2}$$

There is a special class of problems which concern with the calculation of magnetic field inside magnetic materials in which there is no free current $\left(\vec{J}_{\text{free}} = 0\right)$. In this case $\vec{\nabla}\times\vec{H} = 0$ and we can introduce a magnetic scalar potential ϕ by

$$\vec{H} = -\vec{\nabla}\phi. \tag{9.12-3}$$

For a linear magnetic material $\vec{B} = \mu\vec{H}$ and for a uniformly magnetized specimen $\vec{\nabla}\cdot\vec{M} = 0$. In both the cases Eq. (9.12-1) reduced to $\vec{\nabla}\cdot\vec{H} = 0$. Combining this result with (9.12-3) we get

$$\nabla^2\phi = 0, \tag{9.12-4}$$

which is Laplace's equation. Thus in such cases the magnetic problem reduces to finding a solution of Laplace's equation with proper boundary conditions. If ϕ is known $\vec{H}$ is obtained from $\vec{H} = -\vec{\nabla}\phi$ and $\vec{B}$ is obtained from $\vec{B} = \mu\vec{H}$ or $\vec{B} = \mu_0\left(\vec{H}+\vec{M}\right)$. To illustrate the usefulness of the method just described above we shall consider a few problems in the following sections.

9.13 A Magnetic Sphere in a Uniform Magnetic Field

Let us consider a sphere of linear magnetic material of radius a and permeability μ_1 placed in a medium of permeability μ_2, where there is an initially uniform magnetic field of strength $\vec{H}_0 = H_0\hat{z}$ along z-axis. We choose the centre of the sphere as the origin of a spherical polar coordinates (r, θ, φ). The magnetic field magnetizes the sphere and this distorts the field near the sphere. Let the magnetic scalar potential at any external point be

$$\phi_e = \phi_0 + \phi_2, \tag{9.13-1}$$

where ϕ_0 is the potential associated with $\vec{H}_0$ and ϕ_2 is due to induced magnetization on the sphere. Therefore,

$$\vec{H}_0 = -\vec{\nabla}\phi_0$$

$$\text{or} \quad H_0\hat{z} = -\hat{z}\frac{d\phi_0}{dz}$$

$$\text{or} \quad \phi_0 = -\int H_0 dz = -H_0 z + C = -H_0 r\cos\theta + C, \tag{9.13-2}$$

where C is a constant.

Symmetry of the problem about z-axis suggests that ϕ_2 will be a function of r and θ only. Hence it is given by the solution of Laplace's equation as

$$\phi_2(r,\theta) = \sum_{n=0}^{\infty}\left(a_n r^n + \frac{b_n}{r^{n+1}}\right)P_n(\cos\theta). \tag{9.13-3}$$

Since the induced magnetization will have no effect at infinity, $\phi_2 \to 0$ as $r \to \infty$. So we must choose $a_n = 0$. Thus, total potential at any external point may be written as

$$\phi_e(r,\theta) = C - H_0 r\cos\theta + \sum_{n=0}^{\infty}\frac{b_n}{r^{n+1}}P_n(\cos\theta). \tag{9.13-4}$$

For any internal point the potential ϕ_i satisfies Laplace's equation and requires the condition that ϕ_i is finite at $r = 0$. So coefficients like b_n must be zero and we can write for ϕ_i,

$$\phi_i(r,\theta) = \sum_{n=0}^{\infty} c_n r^n P_n(\cos\theta). \tag{9.13-5}$$

Now, to find the constants b_n and c_n we use the following boundary conditions:

(i) The continuity of the tangential component of magnetic intensity across the boundary, which is equivalent to the continuity of potential on the surface, i.e.,

$$\phi_i = \phi_e \text{ at } r = a.$$

(ii) The normal component of magnetic induction is continuous across the surface. This implies that

$$\mu_1\frac{\partial\phi_i}{\partial r} = \mu_2\frac{\partial\phi_e}{\partial r} \text{ at } r = a.$$

Applying these conditions to the Eqs. (9.13-4) and (9.13-5) and then equating coefficients of $P_n(\cos\theta)$ from both sides and solving the resulting equations it can be shown that $c_0 = c$, $b_0 = 0$, $c_n = b_n = 0$ for $n \geq 2$ and

$$c_1 = -\frac{3\mu_2}{\mu_1 + 2\mu_2} H_0, \quad b_1 = \frac{\mu_1 - \mu_2}{\mu_1 + 2\mu_2} H_0 a^3. \tag{9.13-6}$$

Thus, we get from Eqs. (9.13.4) and (9.13-5)

$$\phi_i(r,\theta) = c - \frac{3\mu_2}{\mu_1 + 2\mu_2} H_0 r \cos\theta \quad \text{for } r < a \tag{9.13-7}$$

$$\phi_e(r,\theta) = c - H_0 r\cos\theta + \frac{\mu_1 - \mu_2}{\mu_1 + 2\mu_2} H_0 a^3 \frac{\cos\theta}{r^2} \quad \text{for } r > a. \tag{9.13-8}$$

The corresponding magnetic intensities are given by

$$\vec{H}_i = -\vec{\nabla}\phi_i = -\hat{z}\frac{d\phi_i}{dz} \quad (\because r\cos\theta = z)$$

$$= \frac{3\mu_2}{\mu_1 + 2\mu_2} H_0 \hat{z} \tag{9.13-9}$$

$$\vec{H}_e = -\vec{\nabla}\phi_e = H_0\hat{z} - \frac{\mu_1 - \mu_2}{\mu_1 + 2\mu_2} H_0 a^3 \vec{\nabla}\left(\frac{\cos\theta}{r^2}\right)$$

$$= H_0\hat{z} + \frac{\mu_1 - \mu_2}{\mu_1 + 2\mu_2}\frac{H_0 a^3}{r^3}\left(2\cos\theta\,\hat{r} + \sin\theta\,\hat{\theta}\right) \tag{9.13-10}$$

Corresponding magnetic inductions will be given by

$$\vec{B}_i = \mu_1 \vec{H}_i \quad \text{and} \quad \vec{B}_e = \mu_2 \vec{H}_e. \tag{9.13-11}$$

From Eq. (9.13-8) it is clear that the potential due to induced magnetization is that of a magnetic dipole situated at the centre of the sphere and is of moment [See Eq. (8.12-4)],

$$\vec{m} = 4\pi a^3 \frac{\mu_1 - \mu_2}{\mu_1 + 2\mu_2} H_0 \hat{z}. \tag{9.13-12}$$

Magnetization, i.e., induced dipole moment per unit volume is

$$\vec{M} = \frac{\vec{m}}{\frac{4}{3}\pi a^3} = \frac{3(\mu_1 - \mu_2)}{\mu_1 + 2\mu_2} H_0 \hat{z} \tag{9.13-13}$$

The field strength $\vec{H}_i$ inside the sphere can be regarded as the sum of external field $\vec{H}_0$ and a field strength $\vec{H}_d$ associated with the magnetization of the sphere. Let

$$\vec{H}_i = \vec{H}_0 + \vec{H}_d$$

$$\text{or} \quad \vec{H}_d = \vec{H}_i - \vec{H}_0 = -\frac{\mu_1 - \mu_2}{\mu_1 + 2\mu_2} H_0 \hat{z} = -\frac{1}{3}\vec{M} = -\frac{1}{4\pi}\left(\frac{4\pi}{3}\right)\vec{M} \tag{9.13-14}$$

When $\mu_1 > \mu_2$, $\vec{H}_d$ is in opposite direction to $\vec{H}_0$ and hence, it is a *demagnetization field*. Obviously $\vec{H}_d$ is proportional to $\vec{M}$. The factor $\frac{4\pi}{3}$ in Eq. (9.13-14) depends explicitly on the spherical geometry. The quantity $4\pi/3$ is known as the *demagnetizing factor* of a sphere. This factor will have different values for different geometries. For example in the case of cylindrical geometry it will have the value of 2π.

9.14 Magnetic Field due to a Uniformly Magnetized Sphere

Consider a sphere of radius a that has a uniform magnetization $\vec{M}$ along z-axis. We choose the centre of the sphere as the origin of an (r, θ, φ) spherical polar coordinates. The magnetic scalar potential ϕ, in this case, satisfies Laplace's equation $\nabla^2\phi = 0$. Because of symmetry about z-axis ϕ will be function of r and θ only. Further keeping in mind that ϕ is finite at $r = 0$ and $\phi \to 0$ as $r \to \infty$ we can write the solutions of Laplace's equation as

$$\phi_1(r,\theta) = \sum_{n=0}^{\infty} a_n r^n P_n(\cos\theta) \quad \text{for } r < a \tag{9.14-1}$$

$$\phi_2(r,\theta) = \sum_{n=0}^{\infty} \frac{b_n}{r^{n+1}} P_n(\cos\theta) \quad \text{for } r > a. \tag{9.14-2}$$

To find the constants a_n and b_n we are to apply the following boundary conditions:

(i) The continuity of the tangential component of $\vec{H}$ across the surface, which is equivalent to $\phi_1 = \phi_2$ at $r = a$

(ii) The normal component of $\vec{B}$ is continuous across the surface. This implies that

$$B_{1n} = B_{2n}$$

$$\text{or} \quad \mu_0\left(\vec{H}_1 + \vec{M}\right)\cdot\hat{n} = \mu_0\vec{H}_2\cdot\hat{n}$$

$$\text{or} \quad -\frac{\partial\phi_1}{\partial r} + M\cos\theta = -\frac{\partial\phi_2}{\partial r}$$

Simultaneous satisfaction of these two conditions requires that $a_n = b_n = 0$ for $n \neq 1$ and for $n = 1$,

$$a_1 = \frac{M}{3} \quad \text{and} \quad b_1 = \frac{1}{3}Ma^3.$$

Therefore, the solutions (9.14-1) and (9.14-2) become

$$\phi_1(r,\theta) = \frac{M}{3} r\cos\theta \tag{9.14-3}$$

$$\phi_2(r,\theta) = \frac{Ma^3}{3r^2}\cos\theta. \tag{9.14-4}$$

The corresponding magnetic intensities are

$$\vec{H}_1 = -\vec{\nabla}\phi_1 = -\frac{M}{3}\hat{z} = -\frac{1}{4\pi}\left(\frac{4\pi}{3}\right)\vec{M} \tag{9.14-5}$$

$$\vec{H}_2 = -\vec{\nabla}\phi_2 = -\frac{Ma^3}{3}\vec{\nabla}\left(\frac{\cos\theta}{r^2}\right) = \frac{Ma^3}{3r^3}\left(2\cos\theta\,\hat{r} + \sin\theta\,\hat{\theta}\right). \tag{9.14-6}$$

Corresponding magnetic inductions may be calculated as

$$\vec{B}_1 = \mu_0\left(\vec{H}_1 + M\hat{z}\right) = \frac{2}{3}\mu_0 M\hat{z} \tag{9.14-7}$$

$$\text{and}\quad \vec{B}_2 = \mu_0\vec{H}_2 = \frac{\mu_0 Ma^3}{3r^3}\left(2\cos\theta\,\hat{r} + \sin\theta\,\hat{\theta}\right). \tag{9.14-8}$$

Note that the potential ϕ_2 as given by the Eq. (9.14-4), at any external point is that of a dipole of moment $\vec{m} = (4\pi/3)a^3M\hat{z}$ placed at the centre of the sphere. The magnetic intensity $\vec{H}_1$ inside the sphere is proportional to magnetization and is in a direction opposite to the magnetization. Hence, it results in a demagnetization effect and is called a *demagnetizing field.* The factor $4\pi/3$ in Eq. (9.14-5) depends explicitly on the spherical geometry and is called the *demagnetizing factor* of a sphere. Fig. 9.14 shows the lines of $\vec{H}$ and the lines of $\vec{B}$. The lines of $\vec{B}$ are continuous closed curves whereas the lines of $\vec{H}$ are discontinuous; they originate from the effective magnetic pole surface charges situated on the surface of the sphere.

At this point we may point out that in the interior of a permanently magnetized body (e.g., a permanent magnet) $\vec{B}$ and $\vec{H}$ have opposite directions. Such a magnetic state can be found in any ferromagnetic substance in the second and fourth quadrant of the hysteresis loop.

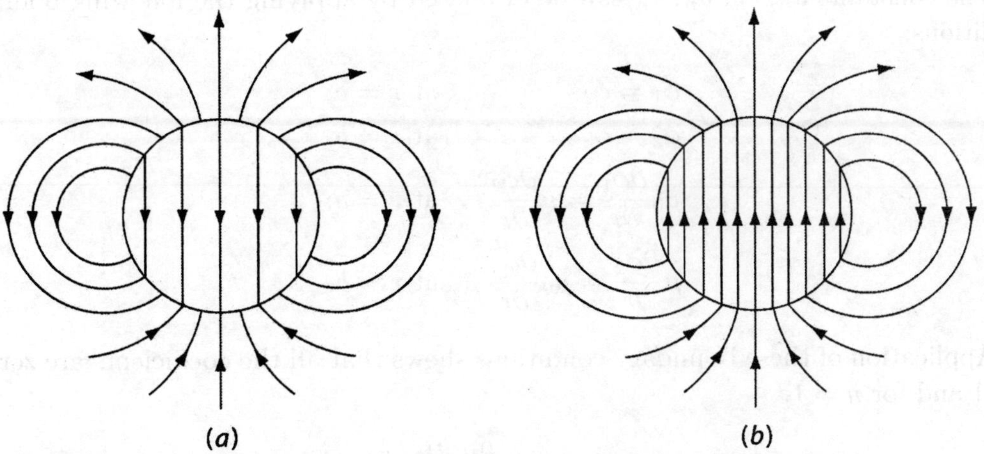

Fig. 9.14: (a) Lines of magnetic intensity $\vec{H}$ and (b) Lines of magnetic induction $\vec{B}$

9.15 Spherical Shell of Magnetic Material Placed in a Uniform Magnetic Field: Magnetic Shielding

Suppose a spherical shell of inner radius a and outer radius b, made of magnetic material of permeability μ is placed in an initially uniform magnetic field $\vec{H}_0 = H_0\hat{z}$ along z-axis. The magnetic field gets modified in the vicinity of the shell. The field lines tend to be normal to the surface and more lines pass through the material of the shell. This causes the field inside the cavity much smaller than the external field for large values of μ. Thus, if the shell is made of ferromagnetic material the space inside the cavity will be almost free of magnetic field. This is called *magnetic shielding* by the material of the shell. It is of considerable practical importance because in many experiments and for faithful working of some electronic devices magnetic field free regions are desirable.

To find the fields everywhere particularly within the cavity we note that the magnetic scalar potentials ϕ in the regions $r < a$, $a < r < b$ and $r > b$ satisfy Laplace's equation. Because of symmetry about z-axis potentials are functions of r and θ only.

For the region $r < a$, ϕ must be finite at $r = 0$. So we can expect a solution as

$$\phi_1 = \sum_n a_n r^n P_n(\cos\theta) \tag{9.15-1}$$

For $a < r < b$,

$$\phi_2 = \sum_n \left(c_n r^n + \frac{d_n}{r^{n+1}} \right) P_n(\cos\theta). \tag{9.15-2}$$

For $r > b$,

$$\phi_3 = -H_0 r\cos\theta + \sum_n \frac{f_n}{r^{n+1}} P_n(\cos\theta), \tag{9.15-3}$$

where $-H_0 r\cos\theta$ is the potential due to external field.

The constants a_n, c_n, d_n, f_n can be evaluated by applying the following boundary conditions:

$$\begin{aligned}
\phi_1 &= \phi_2 && \text{at } r = a;\\
\phi_2 &= \phi_3 && \text{at } r = b;\\
\mu_0 \frac{\partial \phi_1}{\partial r} &= \mu \frac{\partial \phi_2}{\partial r} && \text{at } r = a;\\
\mu \frac{\partial \phi_2}{\partial r} &= \mu_0 \frac{\partial \phi_3}{\partial r} && \text{at } r = b.
\end{aligned}$$

Application of these boundary conditions shows that all the coefficients are zero for $n \neq 1$ and for $n = 1$,

$$a_1 = -\frac{9\mu_r H_0}{(2\mu_r + 1)(\mu_r + 2) - 2\frac{a^3}{b^3}(\mu_r - 1)^2}.$$

For large μ_r,

$$a_1 \to -\frac{9H_0}{2\mu_r\left(1-\frac{a^3}{b^3}\right)}.$$

Now the magnetic field inside the cavity is given by

$$\vec{B}_1 = \mu_0 \vec{H}_1 = -\mu_0 \vec{\nabla}\phi_1 = -\mu_0 \vec{\nabla}(a_1 r \cos\theta) = -\mu_0 \vec{\nabla}(a_1 z) = -\mu_0 a_1 \hat{z}. \qquad (9.15\text{-}4)$$

As $\mu_r \to \infty$, $a_1 \to 0$, $\vec{B}_1 \to 0$. Thus, we may conclude that a high-permeability material with $\mu/\mu_0 \sim 10^3$ to 10^6 causes a great reduction in the field within the cavity. So there is magnetic shielding effect. This is found to be true even with a relatively thin shell. Fig. 9.15-1 shows the magnetic shielding.

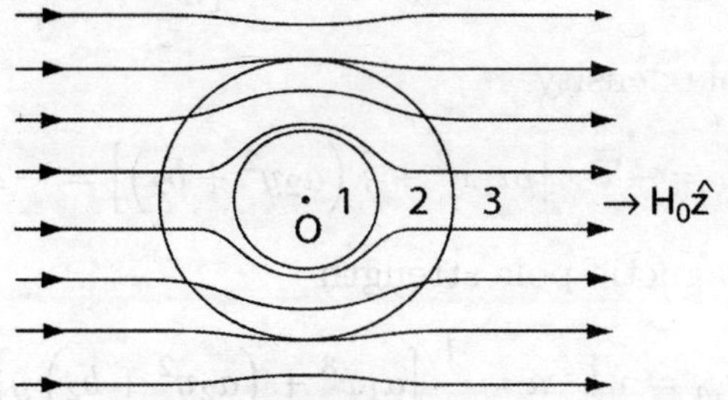

Fig 9.15-1

SOLVED PROBLEMS

▶ Problem 1. *A sphere of radius a is so magnetized that its magnetization at any inside point (x, y, z) with respect to its centre as origin is given by*

$$\vec{M} = a_1 x^2 \hat{\imath} + \left(a_2 y^2 + b_2\right)\hat{\jmath}.$$

Find the magnetization current densities and also all magnetic pole densities.

Solution. The volume density of magnetization current is

$$\vec{J}_m = \vec{\nabla} \times \vec{M} = \begin{vmatrix} \hat{\imath} & \hat{\jmath} & \hat{k} \\ \frac{\partial}{\partial x} & \frac{\partial}{\partial y} & \frac{\partial}{\partial z} \\ a_1 x^2 & a_2 y^2 + b_2 & 0 \end{vmatrix} = 0.$$

The surface density of magnetization current is

$$\vec{K}_m = \vec{M} \times \hat{n},$$

where $\hat{n}$ is the outward unit normal to the surface of the sphere. The equation of the surface of the sphere is

$$\psi(x, y, z) = x^2 + y^2 + z^2 - a^2 = 0.$$

$$\therefore \quad \hat{n} = \frac{\vec{\nabla}\psi}{\left|\vec{\nabla}\psi\right|} = \frac{1}{a}\left(\hat{i}x + \hat{j}y + \hat{k}z\right)$$

$$\therefore \quad \vec{K}_m = \left[a_1x^2\hat{i} + \left(a_2y^2 + b_2\right)\hat{j}\right] \times \frac{1}{a}\left(\hat{i}x + \hat{j}y + \hat{k}z\right)$$
$$= \hat{i}\cdot\frac{z}{a}\left(a_2y^2 + b_2\right) - \hat{j}\frac{z}{a}a_1x^2 + \hat{k}\left[\frac{y}{a}\cdot a_1x^2 - \frac{x}{a}\left(a_2y^2 + b_2\right)\right].$$

Now the magnetic pole density

$$\rho_m = -\vec{\nabla}\cdot\vec{M} = -\vec{\nabla}\cdot\left[\hat{i}a_1x^2 + \hat{j}\left(a_2y^2 + b_2\right)\right] = -2(a_1x + a_2y).$$

and surface density of magnetic pole strength

$$\sigma_m = \vec{M}\cdot\hat{n} = \frac{1}{a}\left[a_1x^3 + \left(a_2y^2 + b_2\right)y\right]$$

► **Problem 2.** *Show that the flux of the field vector $\vec{B}$ is continuous everywhere. Is it so for the vector $\vec{H}$?*

Solution. Let us imagine a volume whose sides are bounded by the magnetic field lines and whose end faces are S_1 and S_2 as shown in Fig. 9.P-2.

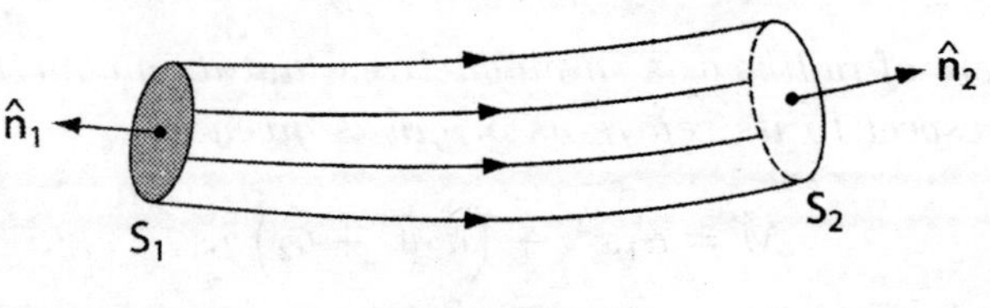

Fig 9.P-2

Integrating $\vec{\nabla}\cdot\vec{B} = 0$ over this volume we get

$$\int_V \vec{\nabla}\cdot\vec{B}\,dV = \int_{S_2}\vec{B}_2\cdot\hat{n}_2dS - \int_{S_1}\vec{B}_1\cdot\hat{n}_1dS = 0,$$

where the side surface does not contribute because $\vec{B}$ is tangential to it. So

$$\Phi(S_2) - \Phi(S_1) = 0 \quad \text{or} \quad \Phi(S_2) = \Phi(S_1),$$

which establishes the continuity of magnetic flux. Thus, the same flux enters the volume through S_1 as leaves through S_2. The flux lines can never terminate and must form closed loops. The above statement is not true for the flux of $\vec{H}$. It is easy to show that there may be discontinuity in the flux of $\vec{H}$. Since

$$\vec{\nabla} \cdot \vec{B} = 0 \quad \text{or} \quad \vec{\nabla} \cdot \left(\vec{H} + \vec{M}\right) = 0 \quad \text{or} \quad \vec{\nabla} \cdot \vec{H} = -\vec{\nabla} \cdot \vec{M},$$

which is not zero everywhere. Now the above procedure leads to

$$\int_{S_2} \vec{H}_2 \cdot \hat{n}_2 dS - \int_{S_1} \vec{H}_1 \cdot \hat{n}_1 dS = -\int \vec{\nabla} \cdot \vec{M}\, dV = \int \rho_m\, dV.$$

▶ **Problem 3.** *A sample of gold having magnetic susceptibility* -3.6×10^{-5} *is placed in a magnetizing field of strength* 60×10^3 *A. turn/m. Find the magnetic induction within the sample.*

Solution. Magnetization $M = \chi H$ and $B = \mu_0(H + M)$

$$\therefore \quad B = \mu_0(1+\chi)H = 4\pi \times 10^{-7}\left(1 - 3.6 \times 10^{-5}\right) \times 60 \times 10^3\,\mathrm{T} = 0.075\,\mathrm{T}$$

▶ **Problem 4.** *Suppose a spherical shell of inner radius a and outer radius b is magnetized in such a way that the magnetization is given by* $\vec{M} = K\vec{r}$, *where K is a constant and* $\vec{r}$ *is the position vector of the point in question with respect to the centre of the shell. Calculate all magnetic pole densities and show that the total magnetic pole strength is zero.*

Solution. Magnetic pole density,

$$\rho_m = -\vec{\nabla} \cdot \vec{M} = -\vec{\nabla} \cdot (K\vec{r}) = -3K.$$

Surface density of pole strength,

$$\sigma_m = \vec{M} \cdot \hat{n} = K\vec{r} \cdot \hat{r} = Kr \quad \text{(on the outer surface)}$$

$$\sigma_m = K\vec{r} \cdot (-\hat{r}) = -Kr \quad \text{(on the inner surface)}$$

$$\text{Total pole strength} = Kb \times 4\pi b^2 - Ka \times 4\pi a^2 + \int_a^b 4\pi r^2 dr \cdot \rho_m$$

$$= 4\pi K\left(b^3 - a^3\right) - 4\pi \cdot 3K \int_a^b r^2 dr = 0.$$

▶ **Problem 5.** *A specimen of iron of density* 7700 kg/m^3 *and specific heat* 462 J.kg^{-1}. K^{-1} *is magnetized by an ac field of frequency* 50 Hz. *Assuming no loss of heat calculate rise in temperature of the specimen per minute. Given that the area enclosed by the B-H loop of the specimen is equivalent to* 5000 J.m^{-3}.cycle^{-1}.

Is the area of the toop temperature dependent? **[cf. C.U. 2010, 2018, 2019]**

Solution. Energy loss per unit volume per cycle = 5000 J.

Therefore, energy loss per minute $= 5000 \times 50 \times 60 \times \frac{\text{m}}{7700}$ J, where m is mass of the specimen in kg.

If θ be the rise in temperature in °C per min. then

$$m \times 462 \times \theta = 5000 \times 50 \times 60 \times \frac{\text{m}}{7700} \quad \text{or} \quad \theta = 4.22° \text{ C/min.}$$

Yes, because the permeability of a ferromagnetic material changes with temperature.

▶ **Problem 6.**

(a) *A closed ring of ferromagnetic material having mean diameter* 20 cm *and cross sectional area* 3 cm^2 *is wound over by* 100 *turns of wire. What current must be passed through these turns to produce a magnetic flux of* 6×10^{-4} *Wb in the ring? Take the relative permeability of the material* = 1500.

(b) *If the above ring is cut into two pieces and kept separated from each other by two equal gaps of width* 0.5mm. *Find the current required to produce the same flux as before.*

Solution. (a) We know that

$$\text{magnetic flux}(\Phi) = \frac{mmf}{\text{reluctance}} = \frac{NI}{\frac{2\pi r}{\mu A}}.$$

Here $N = 1000$, $r = 10$ cm $= 0.1$m, $\mu = \mu_0\mu_r = 1500\,\mu_0$, $A = 3$ cm^2 $= 3 \times 10^{-4}$ m^2, $\Phi = 6 \times 10^{-4}$ Wb. Therefore, the required current,

$$I = \frac{\Phi \cdot 2\pi r}{N \cdot \mu A} = \frac{6 \times 10^{-4} \times 2\pi \times 0.1}{1000 \times 1500 \times 4\pi \times 10^{-7} \times 3 \times 10^{-4}} \text{ A} = 0.66 \text{ A}$$

(b) In this case,

$$\text{reluctance} = \frac{2\pi r}{\mu A} + \frac{2 \times \Delta l}{\mu_0 A},$$

where $\Delta l = 0.5$ mm $= 0.5 \times 10^{-3}$ m is the air gap.

Therefore, in this case required current would be

$$I = \frac{\Phi}{N} \cdot \left[\frac{2\pi r}{\mu A} + \frac{2 \times \Delta l}{\mu_0 A}\right]$$

$$= \frac{6 \times 10^{-4}}{1000}\left[\frac{2\pi \times 0.1}{1500 \times 4\pi \times 10^{-7} \times 3 \times 10^{-4}} + \frac{2 \times 0.5 \times 10^{-3}}{4\pi \times 10^{-7} \times 3 \times 10^{-4}}\right] \text{A}$$

$$= 2.26 \text{ A}.$$

▶ **Problem 7.** *Assuming the earth to be a uniformly magnetized sphere calculate the angle of dip (α) at a place having a magnetic latitude λ°N.*

Solution. If the earth is considered to be a sphere of radius a having a uniform magnetization $\vec{M}$ parallel to its axis of rotation then the field at any external point (r, θ) will be the same as that due to a dipole of moment $\vec{m} = \frac{4}{3}\pi a^3 \vec{M}$ placed at its centre. Thus, we can write [See, Eq. (9.14-8)]

$$\vec{B}(r,\theta) = \frac{\mu_0 M a^3}{3r^3}\left(2\cos\theta\hat{r} + \sin\theta\hat{\theta}\right) = \frac{\mu_0}{4\pi}\left(\frac{2m\cos\theta}{r^3}\hat{r} + \frac{m\sin\theta}{r^3}\hat{\theta}\right) = B_r\hat{r} + B_\theta\hat{\theta}.$$

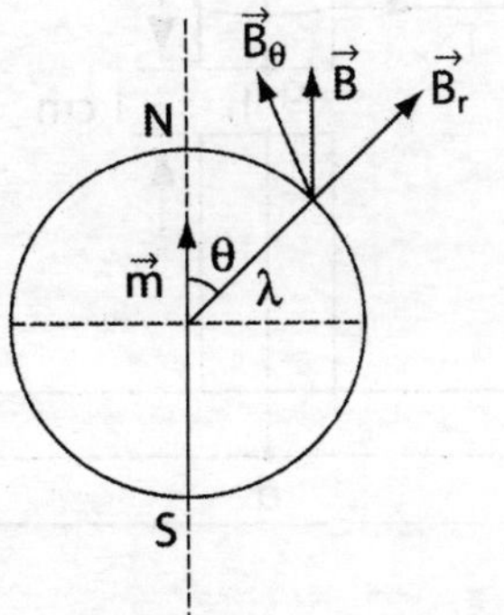

Fig. 9.P-7: The earth as a magnetized sphere

So the angle of dip, i.e., the angle made by the resultant $\vec{B}$ with $B_\theta\hat{\theta}$ will be given by

$$\tan\alpha = \frac{B_r}{B_\theta} = 2\cot\theta = 2\cot(90^\circ - \lambda) = 2\tan\lambda$$

$$\text{or}\quad \alpha = \tan^{-1}(2\tan\lambda).$$

▶ **Problem 8.** *The magnetic intensity in a certain material having permeability $\mu = 5\mu_0$ is given by $\vec{H} = 2\hat{i} + 5\hat{j} - 8\hat{k}$. Calculate the susceptibility of the material, magnetization $\vec{M}$ and the magnetic induction $\vec{B}$.* **[N.B.U. 2017]**

Solution. We know that $\mu = \mu_0(1+\chi)$

$$\therefore\quad \chi = \frac{\mu}{\mu_0} - 1 = 5 - 1 = 4.$$

Magnetization $\vec{M} = \chi\vec{H} = 8\hat{i} + 20\hat{j} - 32\hat{k}$.

Magnetic induction

$$\vec{B} = \mu\vec{H} = 5 \times 4\pi \times 10^{-7}\left(2\hat{i} + 5\hat{j} - 8\hat{k}\right)\text{ T} = 2\pi\left(2\hat{i} + 5\hat{j} - 8\hat{k}\right)\ \mu\text{T}.$$

▶ **Problem 9.** *Consider the magnetic circuit of* Fig. 9.P-9 *and find the current required to produce a magnetic field of* 0.1 *T in the air gap. Assume that the permeability of the core is* $\mu = 1000\mu_0$ *and all branches have the same cross-sectional area of* 10 cm^2.

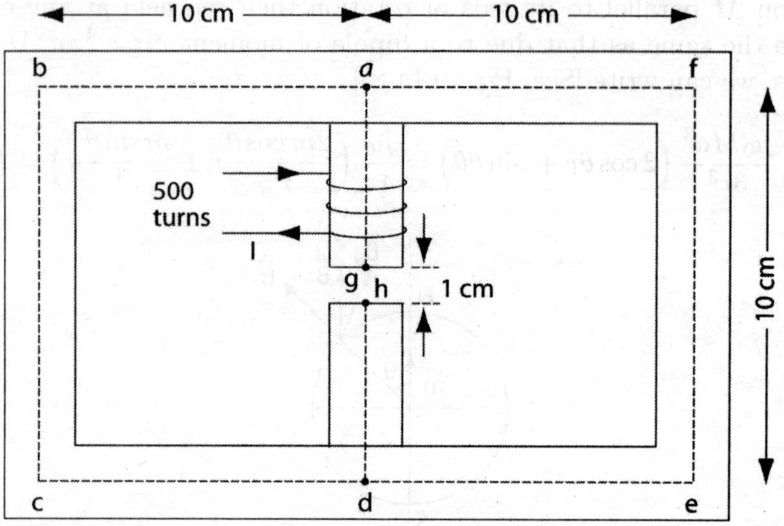

Fig 9.P-9

Solution.

$$\text{Reluctance of the path } abcd(R_1) = \text{Reluctance of path } afed\ (R_2)$$

$$= \frac{0.30}{1000\mu_0 \cdot 10 \times 10^{-4}} \text{ A-turn/Wb}$$

$$= \frac{0.3}{\mu_0} \text{ A-turn/Wb.}$$

Combining these two as resistors in parallel we get the equivalent reluctance as

$$R = \frac{R_1 R_2}{R_1 + R_2} = \frac{R_1}{2} = \frac{0.15}{\mu_0} \text{ A-turn/Wb}$$

Reluctance of the path *aghd*

= sum of reluctance of 9 cm iron path and 1 cm air path

$$= \frac{0.09}{1000\mu_0 \times 10 \times 10^{-4}} + \frac{0.01}{\mu_0 \times 10 \times 10^{-4}} = \frac{0.09}{\mu_0} + \frac{10}{\mu_0}$$

$$= \frac{10.09}{\mu_0} \text{ A-turn/Wb.}$$

Therefore, total reluctance of the magnetic circuit

$$R_m = \frac{10.09}{\mu_0} + \frac{0.15}{\mu_0} = \frac{10.24}{\mu_0} \text{ A-turn/Wb.}$$

Now, magnetic flux $\quad (\text{BA}) = \dfrac{mmf(\text{NI})}{\text{reluctance}(R_m)}$

$$\text{or} \quad 0.1 \times 10 \times 10^{-4} = \frac{500 \times I}{\frac{10.24}{4\pi \times 10^{-7}}}$$

Therefore,

$$I = \frac{0.1 \times 10^{-3} \times 10.24}{500 \times 4\pi \times 10^{-7}} \text{ A} = 1.63 \text{ A}.$$

▶ **Problem 10.** *Consider a long cylinder of radius a made of a magnetic material having permeability μ. It is placed in free space and a line current I of infinite extent is flowing along its axis. Find $\vec{B}$, $\vec{H}$ and $\vec{M}$ inside the cylinder.*

Solution. Considering a circular loop of radius r $(r < a)$ around the current we can write from Ampere's law,

$$\oint \vec{H} \cdot d\vec{l} = I \quad \text{or} \quad H \cdot 2\pi r = I \quad \text{or} \quad H = \frac{I}{2\pi r}.$$

By symmetry $\vec{H}$ is tangential to the loop at every point. In vector form we can write

$$\vec{H} = \frac{I}{2\pi r}\hat{\theta}.$$

The magnetic induction

$$\vec{B} = \mu\vec{H} = \frac{\mu I}{2\pi r}\hat{\theta} \quad \text{for } 0 < r < a.$$

Magnetization $\vec{M}$ is given by the relation,

$$\vec{B} = \mu_0\left(\vec{H} + \vec{M}\right) \quad \text{or} \quad \vec{M} = \frac{\vec{B}}{\mu_0} - \vec{H} = \left(\frac{\mu}{\mu_0} - 1\right) \cdot \frac{I}{2\pi r}\hat{\theta} \quad \text{for } 0 < r < a.$$

▶ **Problem 11.** *Consider a plane interface of a magnetic medium of permeability μ and vacuum. If the magnetic field $\vec{B}$ in the medium in the vicinity of a point on the interface makes an angle θ with the normal to the interface, find the magnetic field $\vec{B}_0$ in vacuum in the vicinity of the above point.* **[N.B.U. 2017]**

Solution. Continuity of the normal component of $\vec{B}$ indicates that $B_{0n} = B\cos\theta$.
From the continuity of the tangential component of $\vec{H}$ we get,

$$H_{0t} = H_t \quad \text{or} \quad \frac{B_{0t}}{\mu_0} = \frac{B_t}{\mu} \quad \text{or} \quad B_{0t} = \frac{\mu_0 B_t}{\mu} = \mu_0 B\sin\theta/\mu.$$

Therefore, the magnitude of $\vec{B}$-field in vacuum would be

$$B_0 = \sqrt{B_{0n}^2 + B_{0t}^2} = B^2\sqrt{\cos^2\theta + \mu_0^2 \sin^2\theta/\mu^2}.$$

The angle θ_0 made by $\vec{B}_0$ with the normal to the interface is given by

$$\tan\theta_0 = \frac{B_{ot}}{B_{on}} = \frac{\mu_0}{\mu}\tan\theta.$$

▶ **Problem 12.** *Consider a thin permanently magnetized toroid of mean radius a having a small air gap Δl between its pole faces. If it has a magnetization $\vec{M}$ of constant magnitude in circumferential direction find $\vec{B}$ and $\vec{H}$ both in the air gap and within the material.*

Solution. Applying Ampere's circuital law to the dotted path in Fig. 9.P-12 we get

$$\oint \vec{H}\cdot d\vec{l} = 0 \quad \text{or} \quad H_i(2\pi a - \Delta l) + H_0\cdot\Delta l = 0.$$

Now we note that the magnetic flux (BA) or flux density (B) may be assumed to be the same throughout. Magnetization inside the air gap is zero. Therefore, we can write,

$$H_0 = \frac{B}{\mu_0} \quad \text{and} \quad H_i = \frac{B}{\mu_0} - M.$$

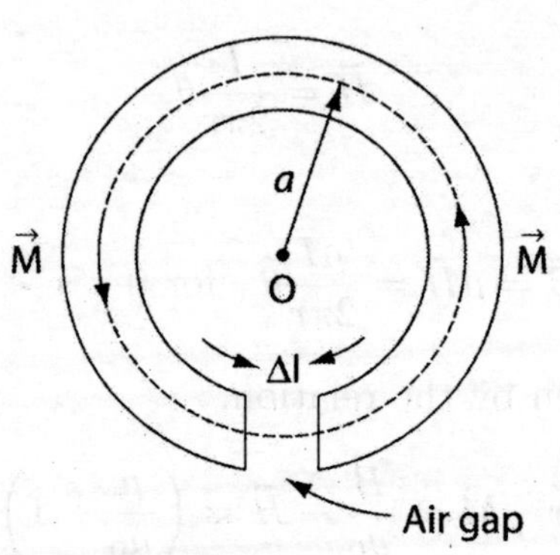

Fig 9.P-12

Therefore,

$$\left(\frac{B}{\mu_0} - M\right)(2\pi a - \Delta l) + \frac{B}{\mu_0}\cdot\Delta l = 0 \quad \text{or} \quad B = \frac{\mu_0 M(2\pi a - \Delta l)}{2\pi a}$$

both in the material and in the air gap. Now the magnetic intensity in air gap is $H_0 = B/\mu_0$ and in the material,

$$H_i = \frac{B}{\mu_0} - M = -\frac{M\cdot\Delta l}{2\pi a}.$$

▶ **Problem 13.** *Suppose a long circular cylinder of radius a is uniformly magnetized in a direction perpendicular to its axis. Find the magnetic scalar potentials at any inside and outside points in cylindrical polar coordinates.*

Solution. We choose axis of the cylinder as z-axis. Magnetization $\vec{M}$ is along x-axis, which represents $\theta = 0$ direction.

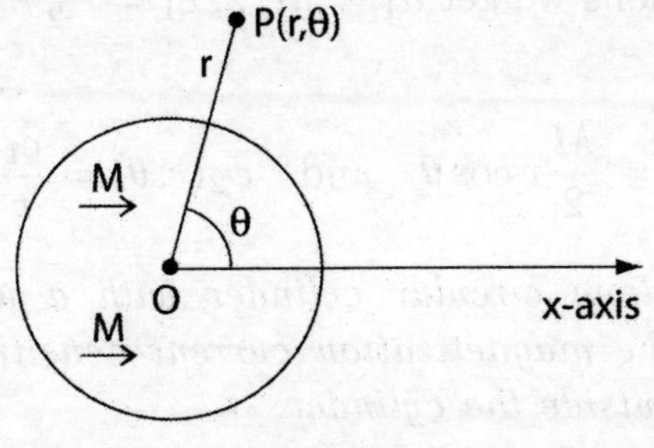

Fig 9.P-13

As the cylinder is long the potential may be assumed to be independent of z-coordinates and can be taken as a function of r and θ of a polar coordinate system. Therefore, the magnetic scalar potential ϕ will be given by the solution of Laplace's equation as

$$\phi(r,\theta) = a_0 \ln r + \sum_{n=1}^{\infty} (a_n \cos n\theta + b_n \sin n\theta)\, r^n + \sum_{n=1}^{\infty} (c_n \cos n\theta + d_n \sin n\theta) \cdot \frac{1}{r^n} + c_0.$$

Now for any external point $(r > a)$ $\phi \to 0$ as $r \to \infty$, so we must take $a_n = b_n = a_0 = c_0 = 0$. Further, the symmetry about x-axis requires that ϕ is an even function of θ. This requires that $d_n = 0$. Thus,

$$\phi = \phi_2(r,\theta) = \sum_{n=1}^{\infty} \frac{c_n}{r^n} \cos n\theta \quad \text{for } r > a.$$

For $r < a$, ϕ must be finite at $r = 0$. Thus, we may write

$$\phi = \phi_1(r,\theta) = \sum_{n=1}^{\infty} a_n \cdot r^n \cdot \cos n\theta \quad \text{for } r < a.$$

Now we require the boundary conditions (i) $\phi_1 = \phi_2$ at $r = a$

and (ii) $B_{1n} = B_{2n}$

$$\text{or } \mu_0 \left(\vec{H}_1 + \vec{M}\right) \cdot \hat{n} = \mu_0 \vec{H}_2 \cdot \hat{n}$$

$$\text{or } -\frac{\partial \phi_1}{\partial r} + M\cos\theta = -\frac{\partial \phi_2}{\partial r} \quad \text{at } r = a.$$

Applying these conditions we get

$$\sum_n a_n \cdot a^n \cos n\theta = \sum_n \frac{c_n}{a^n} \cos n\theta$$

$$-\sum n a_n \cdot a^{n-1} \cos n\theta + M\cos\theta = \sum \frac{n c_n}{a^{n+1}} \cos n\theta.$$

Solving these two equations we get $a_1 = M/2$, $c_1 = \frac{Ma^2}{2}$ and $a_n = c_n = 0$ for $n \geq 2$. Thus,

$$\phi_1(r,\theta) = a_1 \cdot r\cos\theta = \frac{M}{2} r\cos\theta \quad \text{and} \quad \phi_2(r,\theta) = \frac{c_1}{r}\cos\theta = \frac{Ma^2}{2r}\cos\theta.$$

▶ **Problem 14.** *Consider a long circular cylinder with a uniform magnetization $\vec{M}$ parallel to its axis. Find the magnetization current densities and the magnetic field due to $\vec{M}$ both inside and outside the cylinder.* **[C.U. 2016]**

Solution. The volume density of magnetization current is $\vec{J}_m = \vec{\nabla} \times \vec{M} = 0$ and surface density of magnetization current is $\vec{K}_m = \vec{M} \times \hat{n} = M\hat{\theta}$.

So the system is equivalent to a solenoid with $nI = K_m = M$, where n is the number of terms per unit length and I, the current. Therefore, magnetic field inside is $\vec{B} = \mu_0 \vec{M}$ and that outside is simply zero.

▶ **Problem 15.** *Suppose a long cylinder of radius a carries a magnetization $\vec{M} = Kr^2\hat{\theta}$, where K is a constant, r is the distance from axis and $\hat{\theta}$ is the usual unit vector in (r,θ,z) cylindrical coordinate system. Find the magnetic field due to $\vec{M}$ both inside and outside the cylinder.*

Solution. Magnetization volume current density

$$\vec{J}_m = \vec{\nabla} \times \vec{M} = \frac{1}{r}\begin{vmatrix} \hat{r} & r\hat{\theta} & \hat{z} \\ \frac{\partial}{\partial r} & \frac{\partial}{\partial \theta} & \frac{\partial}{\partial z} \\ 0 & r \cdot Kr^2 & 0 \end{vmatrix} = \hat{z}\frac{1}{r}\frac{\partial}{\partial r}\left(r \cdot Kr^2\right) = \hat{z}3Kr.$$

Magnetization surface current density

$$\vec{K}_m = \vec{M} \times \hat{n} = Ka^2\hat{\theta} \times \hat{r} = -Ka^2\hat{z}.$$

Total volume current is

$$I_V = \int_0^a J_m 2\pi r dr = 6\pi K \int_0^a r^2 dr = 2\pi K a^3.$$

Total surface current is

$$I_S = -Ka^2 \cdot 2\pi a = -2\pi K a^3.$$

Obviously total magnetization current is zero.

Now considering an Amperian loop in the form of a circle of radius $r < a$, we can write from Ampere's circuital law

$$\oint \vec{B} \cdot d\vec{l} = \mu_0 I_{\text{enclosed}} = \mu_0 \int_0^r J_m \cdot 2\pi r dr = \mu_0 \cdot 2\pi K r^3$$

$$\text{or} \quad B \cdot 2\pi r = \mu_0 2\pi K r^3$$

$$\text{or} \quad B = \mu_0 K r^2.$$

Vectorially, $\vec{B} = \mu_0 K r^2 \hat{\theta}$ for $r < a$ (i.e., inside the cylinder)

Considering a circular Amperian loop of radius $r > a$, we get

$$\oint \vec{B} \cdot d\vec{l} = \mu_0 I_{\text{enclosed}} = 0.$$

Therefore, $B = 0$ for $r > a$ (i.e., outside the cylinder).

▶ **Problem 16.** *When a sample of nickel is subjected to a large magnetizing field it gets saturated. The corresponding values of $\vec{B}$ and $\vec{H}$ are 1.7 T and 5×10^5 A/m, respectively. If the relative density of nickel is 9 and relative atomic mass 58.71, calculate the magnetic moment per atom of nickel in units of Bohr magneton.*

Solution. We know that $\vec{B} = \mu_0 \left(\vec{H} + \vec{M}\right)$

$$\therefore \quad M = \frac{B}{\mu_0} - H = \frac{1.7}{4\pi \times 10^{-7}} - 5 \times 10^5 \text{ A-m}^2/\text{m}^3$$

$$= 8.53 \times 10^5 \text{ A-m}^2/\text{m}^3.$$

Now 58.71 g or $\frac{58.71}{9}$ cm^3 of Ni contains 6.023×10^{23} atoms.

Therefore, 1 m^3 of Ni contains $\dfrac{6.023 \times 10^{23} \times 9 \times 10^6}{58.71}$ atoms

Thus, the magnetic moment per atom is

$$\frac{8.53 \times 10^5}{\frac{6.023 \times 10^{23} \times 9 \times 10^6}{58.71}} \text{ A-m}^2 = 9.24 \times 10^{-24} \text{ A-m}^2 \approx \text{one Bohr magneton.}$$

▶ **Problem 17.** *A long wire of radius a made of linear magnetic material (say, copper) of susceptibility χ_m is carrying a current I, which is uniformly distributed over its cross-section. Find $\vec{B}$ at a distance r from the axis. Show that the magnetization volume current density $\vec{J}_m = \chi_m \vec{J}_f$, where $\vec{J}_f$ is the free current density. Show the directions of $\vec{J}_m$ and $\vec{K}_m$ (the surface current density). Also, show that the net bound current flowing down the wire is zero.*

Solution. Considering an Amperial loop of radius r we get

$$\oint \vec{H} \cdot d\vec{l} = I_{\text{enclosed}} \quad \text{or} \quad H \cdot 2\pi r = \left(\frac{I}{\pi a^2}\right) \cdot \pi r^2$$

$$\therefore \quad H = \frac{Ir}{2\pi a^2} \quad \text{for } r < a.$$

For $r > a$, $\oint \vec{H} \cdot d\vec{l} = I$. This gives

$$H \cdot 2\pi r = I \quad \text{or} \quad H = \frac{I}{2\pi r} \quad \text{for } r > a.$$

Therefore,

$$\vec{B} = \mu \vec{H} = \mu_0 (1 + \chi_m) \frac{Ir}{2\pi a^2} \hat{\theta} \quad \text{for } r < a.$$

$$\vec{B} = \mu_0 \vec{H} = \frac{\mu_0 I}{2\pi r} \hat{\theta} \quad \text{for } r > a.$$

$$\vec{J}_m = \vec{\nabla} \times \vec{M} = \vec{\nabla} \times \left(\chi_m \vec{H} \right) = \chi_m \vec{\nabla} \times \vec{H} = \chi_m \vec{J}_f$$

$\vec{J}_m = \chi_m \cdot \dfrac{I}{\pi a^2} \hat{z}$, where $\hat{z}$ is in the direction of current flow.

$\vec{K}_m = \vec{M} \times \hat{n} = \chi_m \vec{H} \times \hat{n} = \chi_m \cdot \dfrac{I}{2\pi a} \hat{\theta} \times \hat{n}$, which is opposite to the direction of I.

Net bound/magnetization current $= J_m \cdot \pi a^2 - K_m \cdot 2\pi a = 0$.

QUESTIONS

1. (a) What do you mean by free current and bound current? What is meant by magnetization $\vec{M}$ of a magnetized matter?

(b) Show that the vector potential due to a magnetized body is the same as would be produced by a bound volume current current density $\vec{J}_m = \vec{\nabla} \times \vec{M}$ and a bound surface current density $\vec{K}_m = \vec{M} \times \hat{n}$, where $\hat{n}$ is a unit vector normal to the surface and $\vec{M}$ is the magnetization vector.

OR, Determine the magnetization current density due to nonuniform magnetization of matter. **[C.U. 2014, 2016]**

2. (a) Define the magnetic field vectors $\vec{B}$ and $\vec{H}$. Which one is the fundamental magnetic vector?

(b) Show that for a magnetized material Ampere's law takes the form $\vec{\nabla} \times \vec{H} = \vec{J}_{\text{free}}$, where $\vec{J}_{\text{free}}$ is the free current density.

3. What is magnetic scalar potential? Find an expression for the magnetic scalar potential due to a magnetized body and hence, introduce the idea of *magnetic pole density* and *surface density of magnetic pole strength.* Show that the total pole strength of any magnetized body is always zero.

4. (a) Define susceptibility and permeability for a linear magnetic material. Establish a relation between them. **[J.U. 2011, cf. B.U. 2015]**

   (b) Plot and explain the nature of the variation of relative permeability of a magnetic material with magnetic field intensity during magnetization. **[C.U. 2015]**

5. What do you mean by hysteresis in a ferromagnetic material? What is a *B*-*H* loop? Show that the hysteresis loss per unit volume per cycle of magnetization is equal to the area enclosed by the *B*-*H* loop (or μ_0 times the area enclosed by *M*-*H* loop).

6. (a) Draw the hysteresis loops for soft iron and steel on the same graph. Comment on the effect of temperature on area of the loops. **[C.U. 2014]**

   (b) With reference to the above curves, compare their properties in respect of residual magnetization, coercive force and hysteresis loss. **[C.U. 2019]**

   (c) Discuss why soft iron is suitable for use as the core of a transformer (or an electromagnet) while steel is preferred for making permanent magnets.

   (d) Explain with the help of hysteresis loop the process of demagnetizing a magnetic material. **[C.U 2015]**

7. (a) Distinguish between dia-, para, and ferro-magnetic substances.

   (b) On the basis of Langevin's theory show that the diamagnetic susceptibility is independent of temperature.

8. Show that the magnetic moment associated with the orbital motion of an electron in an atom must be an integral multiple of some basic unit. What is the name of this unit? Find its numerical value.

9. An electron is rotating in a fixed orbit of radius *r* in an atom. A magnetic field *B* is applied in a direction perpendicular to the plane of the orbit.

   (a) Calculate the change in dipole moment of the orbiting electron. **[J.U. 2011]**

   (b) Hence answer the question: Is diamagnetism a property possessed by all materials?

10. Give an account of Langevin's theory of paramagnetism for a paramagnetic gas. What is Curie law? **[cf. J.U. 2012]**

11. (a) What is ferromagnetism? How can it be explained on the basis of domain theory?

    (b) On the basis of Weiss molecular field theory and Langevin's theory of paramagnetism explain the phenomenon of spontaneous magnetization in a ferromagnetic sample. What is Curie temperature? What is Curie-Weiss law?

12. (a) What do you mean by a magnetic circuit? What is its practical significance? Establish the fundamental equation for a magnetic circuit. What are the units of the three quantities involved in this equation? **[cf. C.U. 2015]**

(b) Make a comparative study between a magnetic circuit and an electric circuit. **[cf. C.U. 2017]**

(c) Find an expression for magnetic flux in an iron ring of cross-section A and permeability μ when a current I is passed through a coil of N turns wound over the ring. What would be the change in the flux if a short air gap is cut in the ring? **[cf. C.U. 2013]**

13. (a) Establish the boundary conditions satisfied by $\vec{B}$ and $\vec{H}$ at the interface of two media of different permeabilities. Assume no free surface current. **[cf. C.U. 2015]**

(b) What do you mean by magnetic scalar potential? Show that it satisfies the Laplace's equation in absence of free current in linear material and for uniform magnetization.

OR, Show that in a magnetized material where there is no free current $\vec{H}$ can be written as the gradient of a scalar which satisfies Laplace's equation. **[C.U. 2013, 2015]**

14. A sphere of radius a made of linear magnetic material of permeability μ is placed in an external uniform field of strength $\vec{H}_0$. Solving the Laplace's equation satisfied by the magnetic scalar potential with proper boundary conditions find the magnetic field at any inside and outside points. Find the demagnetization factor of the sphere.

15. Consider a uniformly magnetized sphere of radius a. Solving the Laplace's equation satisfied the magnetic scalar potential with proper boundary conditions show that the potential at any external point is the same as that due to a magnetic dipole placed at the centre of the sphere. Also, show that the magnetic intensity inside is a demagnetizing field. Find the demagnetizing factor.

16. Consider a spherical shell made of a material having high permeability. It is placed in a uniform magnetic field. Treating it as a boundary value problem show that the magnetic field in the space inside the cavity is much smaller than the external field. Comment on the practical importance of the result.

17. Show that the magnetization volume density of current $\left(\vec{J}_m\right)$ is related to the free current density $\left(\vec{J}_f\right)$ in a linear homogeneous medium by $\vec{J}_m = \chi_m \vec{J}_f$, where χ_m is the magnetic susceptibility.

[**Hints.** $\vec{J}_m = \vec{\nabla} \times \vec{M} = \vec{\nabla} \times \chi_m \vec{H} = \chi_m \left(\vec{\nabla} \times \vec{H}\right) = \chi_m \vec{J}_f$.]

18. An electron (charge $-e$, mass m) revolves around nucleus (charge e) in a circular orbit of radius r with speed v. The electrostatic force supplies the necessary centripetal force.

(a) Calculate the current and magnetic dipole moment due to the orbital motion of the electron.

(b) Write down the force equation.

(c) When the electron is placed in a uniform magnetic field B perpendicular to the plane of the orbit, how is the force equation modified? **[C.U. 2010]**

[**Hints.** See, section 9.7A]

EXERCISES

1. Consider a thin disk of magnetic material having radius a and thickness $d \ll a$. Its magnetization $\vec{M} = M\hat{i}$ is parallel to x-axis. Assuming the top flat surface of the disk as xy-plane and centre of the disk as origin, find the magnetization current densities.
[*Ans.* $\vec{J}_m = 0$, $\vec{K}_m = -M\hat{j}$ on the top face, $+M\hat{j}$ on the bottom face and $\hat{k}M\sin\theta$ on the curved face, where θ is the polar angle with respect to x-axis]

2. (a) Suppose a right circular cylinder of length L and radius a has a uniform magnetization $\vec{M} = M\hat{k}$ directed along its axis. Find the magnetization current densities. [*Ans.* $\vec{J}_m = 0$, $\vec{K}_m = 0$ on flat faces and $\vec{K}_m = \frac{M}{a}\left(x\hat{j} - y\hat{i}\right)$ on curved faces]

(b) An infinitely long cylinder of radius R carries a 'frozen-in' magnetization parallel to the axis, $\vec{M} = Kr\hat{z}$ where K is a constant and r is the distance from the axis. There is no free current anywhere. Find the bound currents.
[*Ans.* $\vec{J}_m = 0$, $\vec{K}_m = KR\hat{\theta}$] **[C.U. 2011]**

3. (a) Consider a plane interface of two media of permeabilities μ_1 and μ_2. If the $\vec{B}$-fields on either side make angles θ_1 and θ_2 with the normal to the interface show that $\mu_1 \cot\theta_1 = \mu_2 \cot\theta_2$. **[cf. C.U. 2010]**

(b) Two magnetic media are separated by a plane interface. Show that the angles between the normal to the boundary and the $\vec{B}$-field on either side satisfy $\mu_2 \tan\theta_1 = \mu_1 \tan\theta_2$ where the symbols have their usual meaning.
[**Hints.** See Worked out Problem 11] **[C.U. 2013]**

4. A current carrying solenoid produces a magnetizing field of 150 A/m. If an iron core of susceptibility 2000 is placed inside the solenoid what would be the magnetic induction B within the core? [*Ans.* 0.38 T]

5. The core of a transformer is made of soft iron of total mass 10 kg and density 7500 kg/m^3. If the hysteresis loss of soft iron is 250 $J.m^{-3}\cdot cycle^{-1}$, find the hourly loss of energy when the transformer is used for operation in ac of frequency 50 Hz.

[cf. C.U. 2006] [*Ans.* 6×10^4 J]

6. An electromagnet consists of an iron circuit of total length 80 cm and average area of cross-section 3 cm^2 with an air gap of 3 cm. Find the number of ampere-turns required to produce a magnetic field of 0.1 T in the air gap. Assume $\mu_r = 2500$ and there is no fringing effect. [*Ans.* 2414 A-turns]

7. The electron in a hydrogen atom is rotating in an orbit of radius 0.51Å. Calculate the change in magnetic moment associated with the orbital motion of the electron when a magnetic field of 1 T is applied at right angle to the plane of the orbit.

[*Ans.* 18.5×10^{-28} A-m^2]

8. An element has relative density 9, relative atomic mass 58.71 and each atom has a magnetic moment of one Bohr magneton (i.e., 9.27×10^{-24}A-m^2). A sample of this element is subjected to a strong magnetizing field such that it gets saturated. Find the corresponding magnetization. [*Ans.* 8.56×10^5 A-m^{-1}]

9. An iron ring of mean diameter d is wound over by N turns of wire carrying current I. A narrow gap of width Δl is cut from the core. If B is the magnetic field at the gap show that the permeability μ of iron under these conditions is given by

$$\mu = \frac{\mu_0 \pi d B}{\mu_0 N I - dB}$$

10. (a) In a magnetic medium having relative permeability $\mu_r = 4$ the magnetic field is given by $\vec{B} = 0.01e^{-y}\hat{z}\,T$. Calculate the susceptibility and magnetization.

[*Ans.* $\chi = 3$, $\vec{M} = 5971\hat{z}$ A/m]

(b) A sample of gold having magnetic susceptibility -3.6×10^{-5} is placed in a magnetizing field of strength 60×10^3 A-turn/m. Find the magnetic induction within the sample. [*Ans.* 0.075 T] **[C.U. 2019]**

11. Consider a toroidal core of mean radius 10 cm and a circular cross-section of π cm^2. A coil of 1000 turns is wound on it. What current is required to establish a magnetic flux of 0.5 mWb? Assume that the core is made of a ferromagnetic material having permeability $\mu = 500\,\mu_0$. What would be the value of corresponding current if the core is made of nonmagnetic material?

[*Ans.* 1.59 A; 796.18 A]

12. The magnetization curve for a ferromagnetic sample is approximately given by $B = \left(300\text{H} + \text{H}^2\right)\,\mu$T, where H is in units of A/m. Find the relative permeability of the sample when $H = 200$A/m. [*Ans.* $\mu_r = 398$]

13. (a) A ring of ferromagnetic material has a mean diameter of 0.6 m. A coil wound on the ring carries a current of 6 A. Find the number of turns required to establish an average magnetic flux density of 1.5 T in the ring. Take $\mu_r = 600$. [*Ans.* 625]

(b) An air gap of length 0.4 mm is cut across the section of an iron ring of diameter 25 cm and cross-sectional area 3 cm^2. A current of $2A$ is sent through a coil of 200 turns wound on the ring. If the magnetic flux generated is 2×10^{-4} Wb, find the relative permeability of iron.

[C.U. 2014] [*Ans.* 2019]

14. An electromagnet is made by using a C-shaped piece of iron of total length 80 cm with an air gap of 2 cm between its end faces. A wire carrying 1 A current is wrapped around it. If the relative permeability of iron is 3000, find the number of turns required to produce a field of 0.01 T in the air gap. [*Ans.* 161 turns]

15. A toroidal iron core of mean radius a is wound with N closely spaced turns of wire carrying a current I. Show that the magnetization inside the iron core is given by

$$M = \frac{\mu - 1}{\mu_0} \cdot \frac{NI}{2\pi a},$$

where μ is the permeability of iron.

16. Consider a cube of magnetic material defined by $0 \leq x,\ y,\ z \leq a$. It has magnetization $\vec{M} = Ax\hat{k}$, where A is a constant. Find the magnetization current densities.

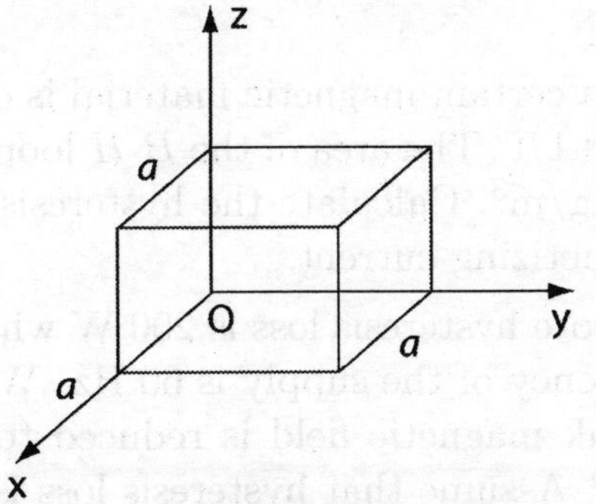

Fig 9.E-16

[*Ans.* $\vec{J}_m = -A\hat{j}$, $\vec{K}_m = 0$ at $x = 0$ face; $Aa\hat{j}$ at $x = a$; $Ax\hat{i}$ at $y = 0$; $-Ax\hat{i}$ at $y = a$; 0 at $z = 0$ and $z = a$]

17. Consider a toroidal magnet of total length l with a small air gap Δl between its end faces. If H_0 is the magnetic intensity in the air gap show that the magnetic intensity H_i inside the magnet is given by $H_i = -H_0 \cdot \Delta l / l$.

[**Hints.** Use Ampere's law: $\oint \vec{H} \cdot d\vec{l} = H_i \cdot l + H_\circ \Delta l = NI = 0$.]

18. Consider a plane interface of a medium of relative permeability 5000 and vacuum. If the $\vec{B}$-field in the medium makes an angle (a) 0°, (b) 85° with the normal to the interface, what would be the corresponding angles made by the $\vec{B}$-field in vacuum with the normal. Comment on the result.

[*Ans.* (a) 0° (b) 0.13°; $\vec{B}$-field in vacuum or low permeable medium in contact with a high permeable medium may always be assumed to be nearly normal to the interface]

19. Consider a toroidal ring of uniform cross-sectional area A and of total length $l_1 + l_2$. l_1 length of it is made of a material of permeability μ_1 and l_2 length made of a material of permeability μ_2. Find the total reluctance of the circuit.

[*Ans.* $\frac{l_1}{\mu_1 A} + \frac{l_2}{\mu_2 A}$]

20. In order to demagnetize a ferromagnetic sample in the form of a bar, it is placed inside a long solenoid having 10 turns per cm. If the coercive force of the sample is 10^3 A.m^{-1}, what current must be passed through the solenoid? [*Ans.* 1 A]

21. The core of a generator armature is taken in the form of an iron cylinder of length 40 cm and radius 10 cm. If it rotates with a speed 1200 rpm find the rate of heat generated in it. Assume that the area of its B-H loop is equal to 10^4 J.m$^{-3}\cdot$ cycle^{-1}. [*Ans.* 800π J.m$^{-3}\cdot$ cycle^{-1}]

22. Suppose an iron rod having permeability 1000 μ_0 is placed along the axis of a solenoid of length 50 cm with 1000 turns carrying a current of 1 A. Find the values of $\vec{B}$, $\vec{H}$ and $\vec{M}$ inside the iron rod.[*Ans.* 2.5 T, 2000 A $\cdot$ m^{-1}, 1.99×10^6 A $\cdot$ m^{-1}]

23. Show that the unit of BH in SI system is J.m^{-3}. Find the dimensions of magnetic flux. [*Ans.* $ML^2T^{-2}I^{-1}$]

24. The B-H loop of a certain magnetic material is drawn with the scale 1 cm = 100 A/m and 1 cm = 0.1 T. The area of the B-H loop is 50 cm^2 and the density of the material is 7800 kg/m^3. Calculate the hysteresis loss in watt/kg for a frequency 50 Hz of the magnetizing current. [*Ans.* 3.2 W/kg]

25. In a transformer core hysteresis loss is 200 W when the peak magnetic field B_{max} is 1.1 T and frequency of the supply is 60 Hz. What would be the corresponding loss when the peak magnetic field is reduced to 0.9 T and supply frequency is reduced to 50 Hz? Assume that hysteresis loss is proportional to $B_{max}^{1.6}$.

[*Ans.* 120 W]

26. A long cylinder of radius a carries a *frozen-in* magnetization $\vec{M} = Kr\hat{z}$ parallel its axis, where K is a constant and r is the distance from axis. Applying Ampere's circuital law find $\vec{H}$ and then find magnetic field $\vec{B}$ both inside and outside the cylinder.

[**Hints.** As there is no free current $\oint \vec{H}\cdot d\vec{l} = I_{free}$ indicates that $\vec{H} = 0$. Therefore, inside the cylinder $\vec{B} = \mu_0\left(\vec{H} + \vec{M}\right) = \mu_0\vec{M}$. Outside cylinder $\vec{H}=0$, $\vec{M}=0$ and hence $\vec{B}=0$.]

27. An infinitely long solenoid having n turns per unit length and carrying current I, is filled with a linear magnetic material of susceptibility χ_m. Find the magnetic field $\vec{B}$ inside the cylinder. Also, find the magnetization surface current density $\vec{K}_m$.

[**Hints.** Ampere's circuital law gives $\vec{H} = nI\hat{z}$ (inside).

$\therefore\ \vec{B}_{\text{inside}} = \mu_0(1+\chi_m)nI\hat{z}$; $\vec{K}_m = \vec{M}\times\hat{n} = \chi_m\vec{H}\times\hat{n} = \chi_m\cdot nI\hat{z}\times\hat{r} = \chi_m nI\hat{\theta}$.]

28. A cylindrical wire of radius a and permeability μ carries a uniformly distributed steady free current I. Find $\vec{B}$ and $\vec{H}$ both inside and outside the wire.

[*Ans.* $\vec{H} = \frac{I}{2\pi a^2}r\hat{\theta}$, $\vec{B} = \frac{\mu I r}{2\pi a^2}\hat{\theta}$ for $r < a$; $\vec{H} = \frac{I}{2\pi r}\hat{\theta}$, $\vec{B} = \frac{\mu_0 I}{2\pi r}\hat{\theta}$ for $r > a$.]

29. A toroid of mean radius 12 cm has 800 turns of wire wound on it. The wire can safely carry a maximum current of 4 A. Find the maximum value of H that can be generated in the volume of the toroid. What is the corresponding value of B for an air core toroid of the above description?

[*Ans.* $H = 4246$ A.turns/m; $B = 5.3\times10^{-3}$ T]

30. A toroid having an iron core of square cross-section and permeability μ is wound with N closely spaced turns of wire carrying a current I. Find the magnitude of magnetization everywhere inside the iron.

[**Hints.** $\oint\vec{H}\cdot d\vec{l} = NI$ gives $H = NI/2\pi r$. So, at a radial distance r, magnetization.

$M = \left|\frac{\vec{B}}{\mu_0} - \vec{H}\right| = \left(\frac{\mu}{\mu_0} - 1\right)\cdot\frac{NI}{2\pi r}$.]

31. Consider an electromagnet made by wrapping a current carrying coil N times around a C-shaped piece of iron having permeability μ as shown in Fig. 9.E-31. If the width of the gap is Δl, length of each side of the C is l, find the B-field in the gap for a current I. [*Ans.* $B = \frac{NI}{\frac{4l-\Delta l}{\mu}+\frac{\Delta l}{\mu_0}}$]

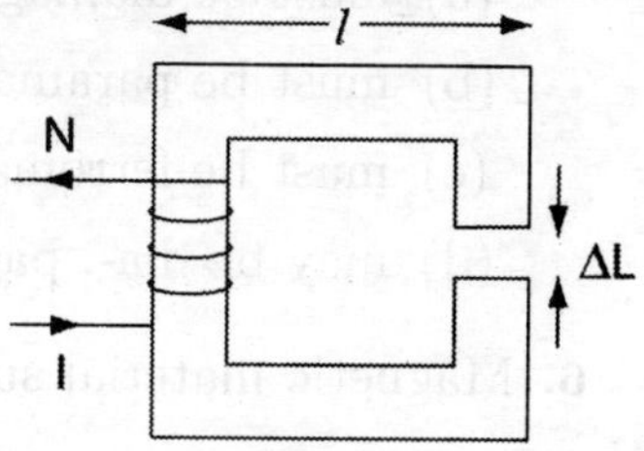

Fig 9.E-31

32. A coil of N turns is wrapped around an iron ring of radius a and cross-section $A(a \gg \sqrt{A})$. Assuming a constant relative permeability $\mu_r \gg 1$ for the iron, find the flux as a function of current I. If a gap of width $l(l^2 \ll A)$ is cut in the ring, what would be the flux for the same current I? Hence find the ratio of the magnitudes of the magnetic induction in the two cases. **[C.U. 2006]**

[*Ans.* $2\pi a/(2\pi a + \mu_r l)$]

OBJECTIVE TEST

1. The intrinsic magnetic moment of the atoms of a material is not zero. The material

(a) must be paramagnetic,

(b) must be diamagnetic,

(c) must be ferromagnetic,

(d) may be paramagnetic or ferromagnetic.

2. The relative permeability of a material is 0.98. The material must be

(a) paramagnetic, (c) ferromagnetic,

(b) diamagnetic, (d) none of these.

3. Susceptibility χ of a paramagnetic gas varies with temperature as

(a) $\chi \propto T$, (c) $\chi \propto T^{-1}$,

(b) $\chi \propto T^{-2}$, (d) $\chi \propto T^{0}$.

4. The unit of reluctance in a magnetic circuit is:

(a) ampere-turns, (c) ampere/metre,

(b) ampere-turns/weber, (d) none of these.

5. The permanent magnetic moment of the atoms of a material is zero. The material

(a) must be diamagnetic,

(b) must be paramagnetic,

(c) must be ferromagnetic,

(d) may be dia-, para- or ferromagnetic.

6. Magnetic material suitable for making permanent magnetics must have

(a) very small B-H loop,

(b) high retentivity and low coercivity,

(c) low retentivity and high coercivity,

(d) high retentivity and high coercivity.

7. Magnetization current can

(a) transport charge over macroscopic distances,

(b) produce Joule heating,

(c) produce magnetic field,

(d) none of these.

8. Which of the following materials requires least magnetizing field to magnetize it?

(a) Gold, (c) Tungsten,

(b) Silver, (d) Cobalt.

9. Which of the following statements is not applicable to ferromagnetic materials?

(a) Susceptibility $\chi_m \gg 1$,

(b) It shows the phenomenon of hysteresis,

(c) Its permeability has a fixed value,

(d) It can show spontaneous magnetization below certain temperature.

10. When a ferromagnetic material is taken through a hysteresis loop its susceptibility (χ) defined through $\vec{M} = \chi \vec{H}$ may not take up a value which is

(a) zero,

(b) infinity,

(c) negative,

(d) constant.

ANSWERS

1. (d) **2.** (b) **3.** (c) **4.** (b) **5.** (a) **6.** (d) **7.** (c)

8. (d) **9.** (c) **10.** (d)

Chapter **10**

Electromagnetic Induction

10.1 Introduction

So far we have limited our discussions to the static electric and magnetic fields. In this chapter we shall consider the effects of time varying magnetic field. We have learnt about magnetic induction, a phenomenon in which magnetic field is induced by electric current. Naturally a question arises: can a magnetic field produce an electric current? The answer came from Faraday in 1831. He showed experimentally that whenever the magnetic flux linked with a closed circuit changes with time an electric current is induced in the circuit. This phenomenon is known as *electromagnetic induction.* It has played an important role in the technological development as we find today.

10.2 Faraday's Laws of Electromagnetic Induction

Faraday performed a series of experiments with a closed circuit and varying magnetic field and observed the following points: (i) When the magnetic flux linked with the circuit changes with time a current is induced in the circuit. (ii) The direction of induced current is reversed if the magnetic field begins to decay instead of growing with time (iii) The magnitude of induced emf is proportional to the rate of change of the magnetic field. These observations led to the development of the following two laws of electromagnetic induction:

(i) *Induced emf in a circuit is proportional to the rate of change of magnetic flux linked with the circuit.*

(ii) *The direction of induced emf is such that it tries to oppose the cause of its generation, i.e., the variation of magnetic flux producing it.*

The second law is known as *Lenz's law.* It helps to understand the direction of induced emf or current. It follows from the law of conservation of energy and ensures that perpetual motion machine is impossible. Suppose the induced current disobeys Lenz's law and it flows in a direction to aid the change in flux. It would produce a current, which would further increase the flux, which would in turn induce still more current. It thus, gives rise to a regenerative process creating more and more energy in the system. It violates the principle of conservation of energy. So we conclude that the induced current must obey Lenz's law, which is consistent with the principle of conservation of energy. Lenz's law is a particular case of *Le Chatelier's principle* which states that if a stress is applied to a system in equilibrium, the equilibrium is so displaced as to reduce the effect of the stress.

If Φ be the magnetic flux linked with a circuit at any time t then the laws of electromagnetic induction can be expressed mathematically as

$$\mathcal{E} = -\frac{d\Phi}{dt}, \tag{10.2-1}$$

where $\mathcal{E}$ is the induced emf. Here it is important to point out that Eq. (10.2-1) does not depend on how the flux Φ is changed.

If the electric field in space is denoted by $\vec{E}$ then, by definition, the emf around a closed path C is

$$\mathcal{E} = \oint_C \vec{E} \cdot d\vec{l}. \tag{10.2-2}$$

If S is an open surface bounded by the curve C placed in a magnetic field $\vec{B}$ then the magnetic flux through the surface is

$$\Phi = \int_S \vec{B} \cdot d\vec{S}. \tag{10.2-3}$$

So Eq. (10.2-1) can be written as

$$\oint_C \vec{E} \cdot d\vec{l} = -\frac{d}{dt} \int_S \vec{B} \cdot d\vec{S}. \tag{10.2-4}$$

This is the *integral form of Faraday's law.*

If the circuit is rigid and stationary in our reference system then neither S nor C depends on t and as a result the time derivative can be taken inside the integral, i.e.,

$$\oint_C \vec{E} \cdot d\vec{l} = -\int_S \frac{\partial \vec{B}}{\partial t} \cdot d\vec{S}, \tag{10.2-5}$$

where partial derivative has been used because $\vec{B}$ is a function of position and time. Using Stokes' theorem we can write,

$$\int_S \vec{\nabla} \times \vec{E} \cdot d\vec{S} = -\int_S \frac{\partial \vec{B}}{\partial t} \cdot d\vec{S}. \tag{10.2-6}$$

Since this must be true for any arbitrary fixed surface S, it follows that

$$\vec{\nabla} \times \vec{E} = -\frac{\partial \vec{B}}{\partial t}, \tag{10.2-7}$$

which is the *differential form of Faraday's law.* It relates the space derivatives of $\vec{E}$ at a particular point to the time rate of change of $\vec{B}$ at the same point.

10.3 Vector and Scalar Potentials in Time Varying Electromagnetic Field

For static electric field $\vec{\nabla} \times \vec{E} = 0$ and we can introduce scalar potential ϕ by the relation $\vec{E} = -\vec{\nabla}\phi$. But for time varying field $\vec{\nabla} \times \vec{E} \neq 0$ and we cannot do so. The magnetic vector potential $\vec{A}$ was defined from the fact that $\vec{\nabla} \cdot \vec{B} = 0$, which still holds for time varying fields. Hence the relation

$$\vec{B} = \vec{\nabla} \times \vec{A} \tag{10.3-1}$$

holds also for time varying situations. Using it in Faraday's law in Eq. (10.2-7) we get

$$\vec{\nabla} \times \vec{E} = -\frac{\partial}{\partial t}\left(\vec{\nabla} \times \vec{A}\right) \quad \text{or} \quad \vec{\nabla} \times \left(\vec{E} + \frac{\partial \vec{A}}{\partial t}\right) = 0.$$

Since the curl of the gradient of any scalar is always zero, we can write

$$\vec{E} + \frac{\partial \vec{A}}{\partial t} = -\vec{\nabla}\phi \quad \text{or} \quad \vec{E} = -\vec{\nabla}\phi - \frac{\partial \vec{A}}{\partial t}. \tag{10.3-2}$$

From Eqs. (10.3-1) and (10.3-2) we can find the vectors $\vec{E}$ and $\vec{B}$ provided that the potentials $\vec{A}$ and ϕ are known. Equation (10.3-2) suggests that for varying fields $\vec{E}$ can be considered as the sum of two terms $\vec{E}_q$ and $\vec{E}_i$, where

$$\vec{E}_q = -\vec{\nabla}\phi \quad \text{and} \quad \vec{E}_i = -\frac{\partial \vec{A}}{\partial t}.$$

$\vec{E}_q$ is the field generated by charges; it is conservative and nonsolenoidal. $\vec{E}_i$ is the field generated by varying magnetic field. It is nonconservative and solenoidal.

10.4 Proof of Faraday's Law for any Loop Moving in a Static Magnetic Field

Let us consider a close loop C of wire of any arbitrary shape moving in a static magnetic field $\vec{B}$ with a nonrelativistic velocity $\vec{v}$. Fig. 10.4-1 shows two positions of the loop at times t and $t + dt$. If the velocity of a conducting electron relative to the wire is $\vec{u}$ then

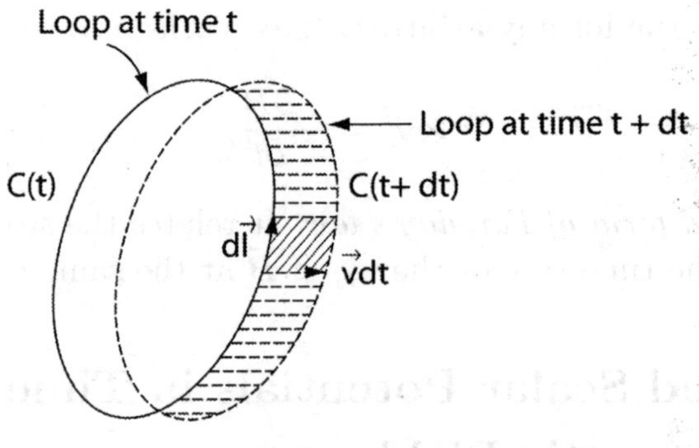

Fig 10.4-1

its velocity relative to the field is $\vec{u}+\vec{v}$. The Lorentz force on the electron of charge q is $q\left(\vec{u}+\vec{v}\right)\times\vec{B}$. We define force per unit charge as the motional induced electric field, i.e., $\vec{E}=\left(\vec{u}+\vec{v}\right)\times\vec{B}$. Therefore, the induced emf

$$\mathcal{E}=\oint_C \vec{E}\cdot d\vec{l}=\oint\left[\left(\vec{u}+\vec{v}\right)\times\vec{B}\right]\cdot d\vec{l}=\oint\left(\vec{v}\times\vec{B}\right)\cdot d\vec{l}\quad\left[\because\ \vec{u}\parallel^{l} d\vec{l}\right] \tag{10.4-1}$$

Now, the element $d\vec{l}$ sweeps in time dt an infinitesimal area $\vec{v}dt\times d\vec{l}$. The magnetic flux through this elementary area is

$$\left(\vec{v}dt\times d\vec{l}\right)\cdot\vec{B}.$$

So the flux change, i.e., the flux over the area swept out by the loop in time dt would be

$$d\Phi=\int\vec{B}\cdot\left(\vec{v}dt\times d\vec{l}\right)$$

$$\text{or}\quad\frac{d\Phi}{dt}=\oint\vec{B}\cdot\left(\vec{v}\times d\vec{l}\right)=-\oint\left(\vec{v}\times\vec{B}\right)\cdot d\vec{l}=-\mathcal{E}.$$

Thus, induced emf $\mathcal{E}=-d\Phi/dt$. This emf, which arises due to motion of conductors through a magnetic field is known as *motional emf.* Generators typically exploit this motional emf. Note that Lorentz force plays the key role in inducing motional emf.

10.5 Emf Induced in Rotating Loop in a Fixed Magnetic Field

Suppose we consider a vertical loop of wire rotating with angular velocity ω in a uniform steady magnetic field $\vec{B}$ acting in the horizontal direction. If at any time instant t the normal $(\hat{n})$ to the plane of the coil makes an angle $\theta=\omega t$ with $\vec{B}$ then flux linked with the coil of area S at the instant t would be

$$\Phi=\vec{B}\cdot\hat{n}S=BS\cos\omega t.$$

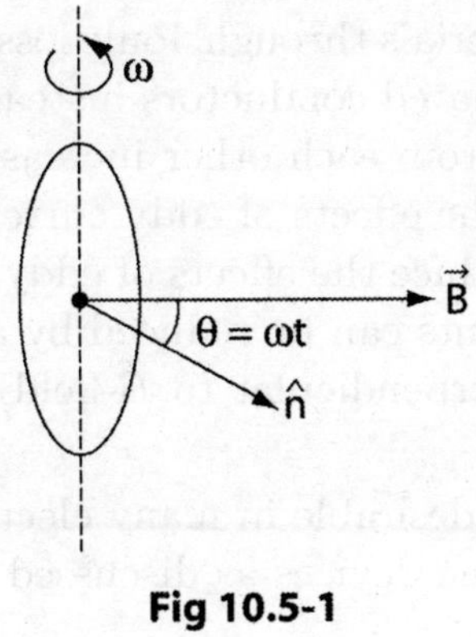

Fig 10.5-1

Therefore, according to Faraday's law the induced emf $\mathcal{E} = -d\Phi/dt = BS\omega \sin \omega t$.

This principle is used in earth inductor for the measurement of the earth's horizontal magnetic field and the angle of dip.

10.6 Emf Induced in a Stationary Loop in a Time-Varying Magnetic Field

Suppose we consider a loop of wire of area A placed normal to a varying magnetic field $\vec{B} = \vec{B}_0 \sin \omega t$. If $\vec{E}$ be the electric field induced in the coil then induced emf,

$$\mathcal{E} = \oint_C \vec{E} \cdot d\vec{l} = \int_S \left(\vec{\nabla} \times \vec{E}\right) \cdot d\vec{S}.$$

Using Faraday's law (10.2-7) we can write

$$\mathcal{E} = -\int_S \frac{\partial \vec{B}}{\partial t} \cdot d\vec{S} \qquad (10.6\text{-}1)$$

$$= -B_0 \omega S \cos \omega t \qquad \left[\because\ \vec{B}_0 \parallel^l d\vec{S}\right]$$

This emf induced by time-varying magnetic field in a stationary loop is often called *transformer emf.*

10.7 Eddy Currents

Whenever a conducting body moves in a magnetic field or is placed in a varying magnetic field currents are induced in the volume of the conductor. Closed circuital paths for such currents are formed in the interior of the conductor. The direction of circulation of these currents is governed by Lenz's law. These currents are called *eddy currents.* These currents are often undesirable in electromagnetic machinery because they heat up the conducting materials through Joule losses. The effects of eddy currents can be minimized by using laminated conductors instead of a single large body. Use of thin sheets of lamina insulated from each other increases the resistance to the flow of eddy current and consequently the effects of eddy current are minimized. Transformer cores are usually laminated to reduce the effects of eddy currents. It can be shown that the power loss due to eddy currents can be reduced by a factor proportional to $1/n$ by subdividing the cross-section (perpendicular to $\vec{B}$-field) of the conductor into n equal parts by lamination.

Though eddy currents are undesirable in many electromagnetic devices, it can also be used to our advantages in some devices as discussed below:

(i) Induction furnace

Heating effect of eddy current is used in induction furnace, where a substance is heated by placing it in a rapidly changing magnetic field. High frequency magnetic field can be used in melting metals or extracting metals from ore. Food can be uniformly cooked by placing it in a region of intense alternating magnetic fields.

(ii) Induction motor

If a conducting cylinder is placed in a rotating magnetic field it rotates due to eddy current induced in it. This principle is used in induction motor.

(iii) Electric brakes

If a strong magnetic field is suddenly applied to a rotating conductor the induced eddy current tries to oppose the relative motion between the conductor and the magnetic field. This principle used in electric brakes.

(iv) Galvanometer

In a suspended coil dead-beat galvanometer the coil wound on a light metallic frame moves in a magnetic field and eddy current is induced in the frame. This current opposes the motion of the frame. The motion of the frame is thus, damped. Such dampings are deliberately used in some electrical instruments to control the motion of the moving parts of the instruments.

10.8 Self-Inductance

When a current flows in a circuit it produces a magnetic field in the surrounding space and hence there will be a magnetic flux linked with the circuit. If the current through the circuit varies with time the flux linked with the circuit also varies with time and gives rise to an induced emf in the circuit. This phenomenon is known as *self-inductance.* According to Lenz's law the induced emf opposes the change in current in the circuit. For this reason it is called a *back emf.* Self-inductance plays somewhat the same role in electrical circuits that mass plays in mechanical system. Thus, self-inductance of a circuit is a measure of its inertia to changes of current. If the circuit is in a nonferromagnetic medium the flux Φ linked with it will be proportional to the current I in the circuit. Thus,

$$\Phi = LI, \tag{10.8-1}$$

where L is a constant, which depends on the geometry of the circuit, its dimensions and the permeability of the surrounding medium. The quantity L is called the *coefficient of self-inductance* or simply the *self-inductance* of the circuit. *Numerically self-inductance may be defined as the total flux linked with the circuit when unit current flows through it.*

The self induced emf in the circuit is given by

$$\mathcal{E} = -\frac{d\Phi}{dt} = -\frac{d}{dt}(LI)$$

If the medium is not ferromagnetic and the circuit is rigid then L is a constant and we can write

$$\mathcal{E} = -L\frac{dI}{dt}. \tag{10.8-2}$$

Thus, *self-inductance may also be defined as numerically equal to the emf induced in the circuit for a unit rate of change of current in it.*

In case of ferromagnetic material Φ is not linearly related to I. Then for a rigid stationary circuit where flux change is caused by change in current only, we can write

$$\mathcal{E} = -\frac{d\Phi}{dt} = -\frac{d\Phi}{dI}\cdot\frac{dI}{dt} = -L\frac{dI}{dt}, \tag{10.8-3}$$

where $L = \dfrac{d\Phi}{dI}$ is called *incremental self-inductance.*

The SI unit of self-inductance is called *henry* (H). *A circuit is said to have a self-inductance of* 1 henry *if* 1 weber *of flux is linked with the circuit when* 1 ampere *of current flows through it. Alternatively, a circuit is said to have a self-inductance of* 1 henry *if an emf of* 1 V *is induced in the circuit when current through it changes at the rate of* 1 ampere per sec.

10.9 Calculation of Self-inductance in Simple Cases

1. A solenoid

Suppose we consider an air-cored long solenoid of length l containing N number of turns. If a current I flows through it the axial magnetic field at any inside point is given by

$$B = \mu_0 nI = \frac{\mu_0 NI}{l}.$$

If A be the area of cross-section of the solenoid and the magnetic field B is assumed to have the same value over the entire cross-section then total magnetic flux linking the N-turns of the solenoid would be

$$\Phi = N \cdot B \cdot A = \frac{\mu_0 N^2 A}{l} I.$$

Now the self-inductance L is defined as the flux linkage per unit current. Therefore, the self-inductance of the solenoid is

$$L = \frac{\Phi}{I} = \frac{\mu_0 N^2 A}{l}. \qquad (10.9\text{-}1)$$

Note that from this equation *the units of μ_0 can be considered as henry/metre.*

If the solenoid is wound on a material having constant permeability μ then

$$L = \frac{\mu N^2 A}{l}. \qquad (10.9\text{-}2)$$

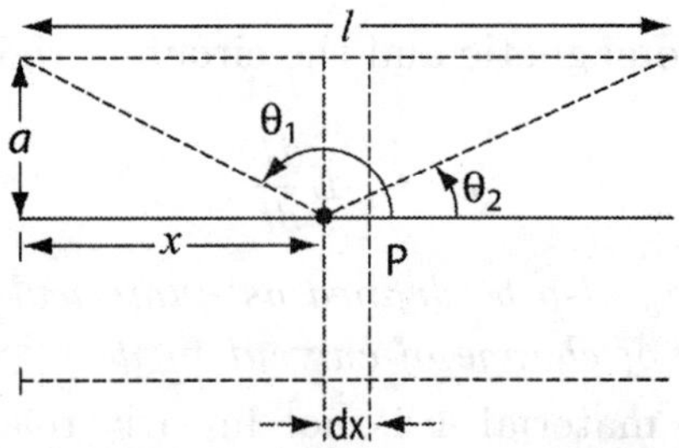

Fig 10.9-1

If the solenoid is not very long then the axial magnetic field at any axial point P (Fig. 10.9-1) is given by the Eq. (8.4-8) as

$$B = \frac{\mu_0 NI}{2}(\cos\theta_2 - \cos\theta_1) = \frac{\mu_0 NI}{2l}\left[\frac{l-x}{\sqrt{(l-x)^2+a^2}} + \frac{x}{\sqrt{x^2+a^2}}\right].$$

In a length dx about P there are $(N/l)\,dx$ number of turns and hence the flux linking these turns is

$$d\Phi = \left(\frac{N}{l}dx\right) BA,$$

where we assume that B is uniform over the entire cross-section.

Therefore, total magnetic flux linking the solenoid is

$$\Phi = \int d\Phi = \int_0^l \frac{BAN}{l} dx = \frac{\mu_0 N^2 AI}{2l^2}\int_0^l \left[\frac{l-x}{\sqrt{(l-x)^2+a^2}} + \frac{x}{\sqrt{x^2+a^2}}\right]$$

$$\text{or} \quad \Phi = \frac{\mu_0 N^2 AI}{2l^2}\left[-\sqrt{(l-x)^2+a^2} + \sqrt{x^2+a^2}\right]_0^l$$

$$= \frac{\mu_0 N^2 A}{l^2}\left[\sqrt{l^2+a^2} - a\right] \cdot I.$$

Therefore,

$$L = \frac{\mu_0 N^2 A}{l^2}\left[\sqrt{l^2+a^2}-a\right] \tag{10.9-3}$$

$$= \frac{\mu_0 N^2 A}{l}\left[\sqrt{1+(a/l)^2}-a/l\right]. \tag{10.9-4}$$

Note that for $l \gg a$, Eq. (10.9-3) reduces to Eq. (10.9-1). If, instead of air, the solenoid is wound on a core having permeability μ then μ_0 in Eq. (10.9-3) is to be replaced by μ.

2. A toroidal coil

The magnetic field inside a toroidal coil having N number of turns, mean length l, cross-sectional area A and carrying current I is given by the Eq. (8.7-4) as

$$B = \mu_0 n I = \mu_0 \cdot \frac{N}{l} \cdot I.$$

Neglecting any variation of B over the cross-sectional area we can write for the flux linking the N turns of the toroid as

$$\Phi = N \cdot B \cdot A = \frac{\mu_0 N^2 A}{l} I.$$

Therefore, the self-inductance is

$$L = \frac{\Phi}{I} = \frac{\mu_0 N^2 A}{l}. \tag{10.9-5}$$

3. Long coaxial cable

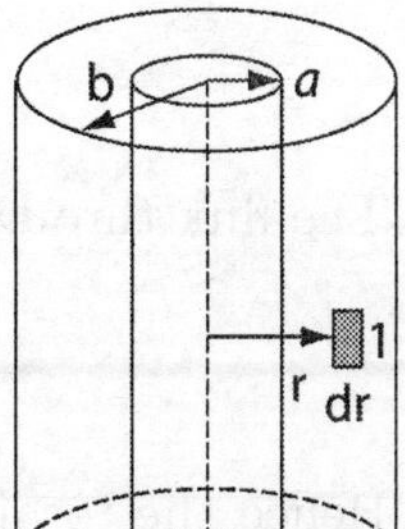

Fig 10.9-2

Consider a long coaxial cable consisting of two concentric cylinders. The inner one has radius a and carries a current I while the outer one has radius b and carries a return current I. Applying Ampere's circuital law it is easy to show that the magnetic field outside the cable is zero and at an internal point at a distance r from the axis ($a < r < b$) the field is given by

$$B = \frac{\mu_0 I}{2\pi r}.$$

The lines of $\vec{B}$ are circles about the axis of the cable. If we consider an elemental area of radial width dr and axial length unity then flux through it would be

$$d\Phi = B \cdot dr \times 1 = \frac{\mu_0 I}{2\pi r} dr.$$

Therefore, total flux per unit axial length is

$$\Phi = \int d\Phi = \frac{\mu_0 I}{2\pi}\int_a^b \frac{dr}{r} = \frac{\mu_0 I}{2\pi}\ln\frac{b}{a}.$$

So the self-inductance per unit length is

$$L = \frac{\Phi}{I} = \frac{\mu_0}{2\pi}\ln\frac{b}{a}. \tag{10.9-6}$$

In the above discussion we *have neglected the flux within the materials of the two cylinders. This is justified when* $b \gg a$.

4. Two-wire transmission line

Let us consider two long parallel wires separated by a distance d in air (Fig. 10.9-3). The wires carry the same current I in opposite directions. We shall assume that the *radius a of each wire is much less than d, so that the flux inside the material of the wires*

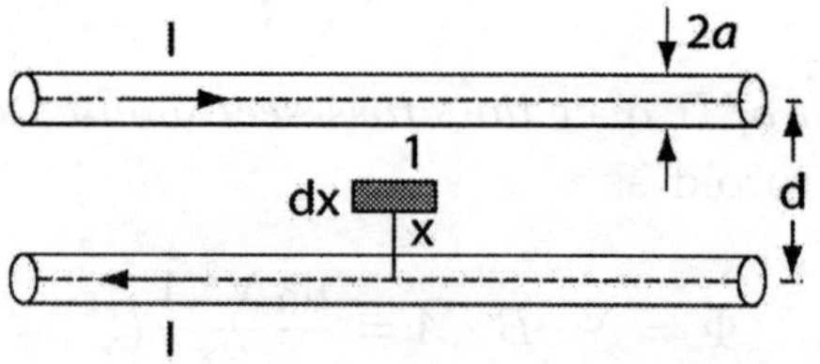

Fig 10.9-3

may be neglected. The flux lines will be mainly concentrated in the region between the wires. Total magnetic field at any point at a distance x from one wire is

$$B = \frac{\mu_0 I}{2\pi}\left[\frac{1}{x} + \frac{1}{d-x}\right].$$

The flux through an elemental area of width dx and length unity is

$$d\Phi = B \cdot dx \times 1 = \frac{\mu_0 I}{2\pi}\left[\frac{1}{x} + \frac{1}{d-x}\right] dx.$$

Hence the total flux through the entire area between the two wires of unit length is

$$\begin{aligned}\Phi = \int d\Phi &= \frac{\mu_0 I}{2\pi}\int_a^{d-a}\left[\frac{1}{x} + \frac{1}{d-x}\right] dx \\ &= \frac{\mu_0 I}{2\pi}\left[\ln\frac{d-a}{a} - \ln\frac{a}{d-a}\right] \\ &= \frac{\mu_0 I}{\pi}\ln\frac{d-a}{a}.\end{aligned}$$

Therefore, self-inductance per unit length is

$$L = \frac{\Phi}{I} = \frac{\mu_0}{\pi} \ln \frac{d-a}{a}.$$

If the wires are kept close together then $d \approx 2a$ and $L \approx 0$. Thus, if an insulated wire is doubled on itself and then wound on an insulated bobbin and a current is passed through it there will be no self-inductive effect. Such windings are called *non-inductive windings,* which are used in the construction of resistance coils.

10.10 Mutual Inductance

Suppose we have two fixed loops C_1 and C_2 placed sufficiently close to each other. If a current I_1 flows in loop C_1 it will produce a magnetic $\vec{B}_1$ at the position of loop C_2. Thus, some flux of $\vec{B}_1$ pass through C_2. If now I_1 is varied the flux will also vary and

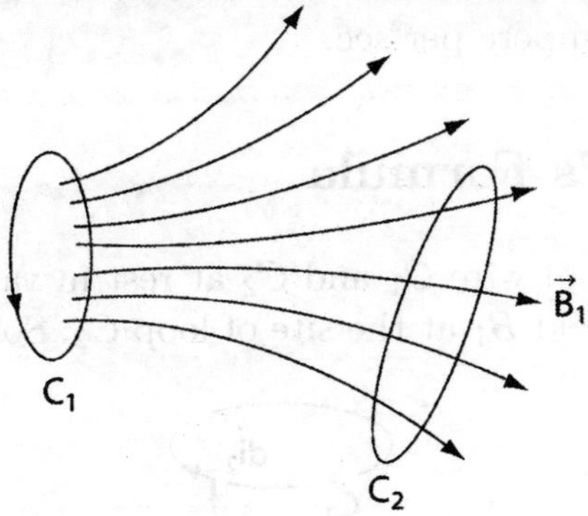

Fig 10.10-1

according to Faraday's law there will be an emf induced in loop C_2. This phenomenon is known as *mutual inductance.* Thus *mutual inductance refers to the induction of an emf in a circuit due to current changes in other circuits placed in its vicinity.*

Let Φ_2 be the flux through loop C_2 due to a current I_1 in loop C_1. In the absence of ferromagnetic material Φ_2 is proportional to I_1. Let

$$\Phi_2 = M_{21} I_1. \tag{10.10-1}$$

Similarly, the flux Φ_1 through loop C_1 due to a current I_2 in loop C_2 will be given by

$$\Phi_1 = M_{12} I_2. \tag{10.10-2}$$

These proportionality constants M_{12} and M_{21} are called the *mutual inductance* of the two loops. These coefficients depend on the geometry of the circuits, their dimensions and permeability of the surrounding medium. It will be shown that

$M_{12} = M_{21}$. From Eq. (10.10-1) or (10.10-2) we find that *numerically mutual inductance of two loops is equal to the flux linked with one loop due to unit current in the other loop.*

The emf induced in the loop C_2 due to current changes in C_1 is given by

$$\mathcal{E}_2 = -\frac{d\Phi_2}{dt} = -\frac{d}{dt}(M_{21}I_1) = -M_{21}\frac{dI_1}{dt}. \tag{10.10-3}$$

Similarly, emf induced in loop C_1 due to current changes in C_2 is given by

$$\mathcal{E}_1 = -\frac{d\Phi_1}{dt} = -\frac{d}{dt}(M_{12}I_2) = -M_{12}\frac{dI_2}{dt}. \tag{10.10-4}$$

Thus, *mutual inductance of two loops is numerically equal to the emf induced in one loop due to unit rate of change of current in the other.*

The unit of mutual inductance is also henry (H). *The mutual inductance of two loops is said to be* 1 H *if* 1 V *emf is induced in one loop when current through the second loop changes at the rate of* 1 ampere per sec.

10.11 Neumann's Formula

Let us consider two loops of wire C_1 and C_2 at rest in vacuum. A current I_1 through C_1 produces a magnetic field $\vec{B}_1$ at the site of loop C_2. So the flux linked with loop C_2

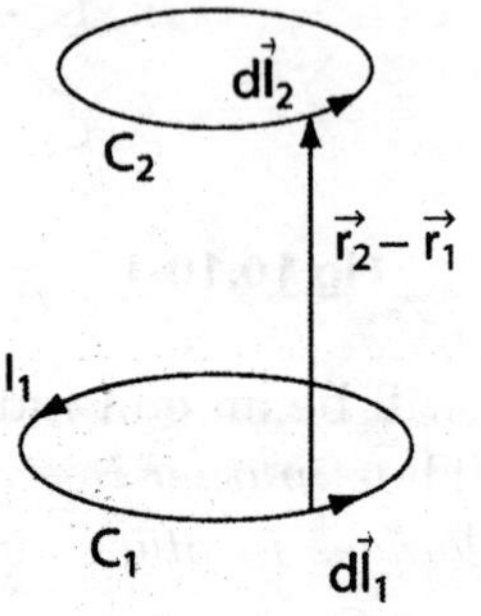

Fig 10.11-1

due current in C_1 is

$$\Phi_2 = \int_{S_2} \vec{B}_1 \cdot d\vec{S}_2 = \int_{S_2} \left(\vec{\nabla} \times \vec{A}_1\right) \cdot d\vec{S}_2,$$

where $\vec{A}_1$ is the magnetic vector potential corresponding to $\vec{B}_1$. Using Stokes' theorem we get

$$\phi_2 = \oint_{C_2} \vec{A}_1 \cdot d\vec{l}_2. \tag{10.11-1}$$

Now the vector potential $\vec{A}_1$ is given by

$$\vec{A}_1 = \frac{\mu_0}{4\pi} \oint_{C_1} \frac{I_1 d\vec{l}_1}{|\vec{r}_2 - \vec{r}_1|}, \tag{10.11-2}$$

where $\vec{r}_2$ and $\vec{r}_1$ are the position vectors of the line elements $d\vec{l}_1$ and $d\vec{l}_2$ with respect to some arbitrary origin.

Therefore,

$$\Phi_2 = \frac{\mu_0 I_1}{4\pi} \oint_{C_2} \oint_{C_1} \frac{d\vec{l}_1 \cdot d\vec{l}_2}{|\vec{r}_2 - \vec{r}_1|} = M_{21} I_1,$$

where

$$M_{21} = \frac{\mu_0}{4\pi} \oint_{C_2} \oint_{C_1} \frac{d\vec{l}_1 \cdot d\vec{l}_2}{|\vec{r}_2 - \vec{r}_1|}. \tag{10.11-3}$$

Since $|\vec{r}_2 - \vec{r}_1| = |\vec{r}_1 - \vec{r}_2|$ and the order of integration may be interchanged we can write that

$$M_{21} = M_{12} = \frac{\mu_0}{4\pi} \oint_{C_2} \oint_{C_1} \frac{d\vec{l}_1 \cdot d\vec{l}_2}{|\vec{r}_2 - \vec{r}_1|}. \tag{10.11-4}$$

This is known as *Neumann's formula* for the mutual inductance of two arbitrary loops. The double integral (10.11-4) is not easy to work with except for circuits with simple geometry but it does illuminate two important points:

(i) M_{12} or M_{21} is a purely geometrical quantity. It depends on the shapes, sizes and relative positions of the two loops.

(ii) $M_{12} = M_{21}$. So we may drop the subscripts and call them both M as the mutual inductance of the two loops. The meaning of this equality is that *in any case the flux Φ_1 through loop C_1 when a current I flows around C_2 is exactly equal to the flux Φ_2 through loop C_2 when the same current I flows around C_1.* This remarkable property is usually called the *reciprocity theorem.*

10.12 Calculation of Mutual Inductances in Simple Cases

1. Two solenoids

Let us consider a very short solenoid S wound over a long narrow air-cored solenoid P at its centre (Fig. 10.12-1). It is known as a *standard solenoid.* P is called a *primary solenoid* and S is called a *secondary solenoid.* If a current I flows through the primary the magnetic induction produced at the centre would be

$$B = \mu_0 n I = \mu_0 \frac{N_1}{l} I,$$

where N_1 is the total number of turns on the primary solenoid of length l. If A be the cross-sectional area of P then the total flux linked with the secondary solenoid of total number of turns N_2 would be

$$\Phi = B \cdot A \cdot N_2 = \frac{\mu_0 N_1 N_2 A}{l} I.$$

Therefore, the mutual inductance between the primary and secondary solenoids is

$$M = \frac{\Phi_2}{I} = \frac{\mu_0 N_1 N_2 A}{l}. \qquad (10.12\text{-}1)$$

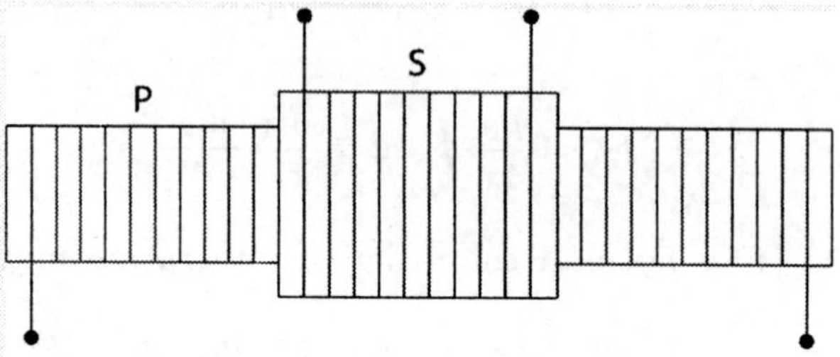

Fig 10.12-1: Standard solenoid.

2. Two parallel circular coaxial coils

Let us consider two parallel circular coaxial coils C_1 and C_2 of radii a and b, respectively and the distance between their centres O_1 and O_2 be x (Fig. 10.12-2). We assume that

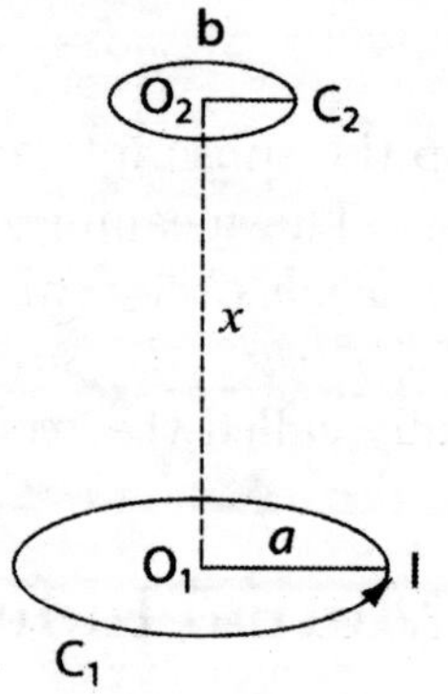

Fig 10.12-2

the coil C_2 is so small ($b \ll a$) that the magnetic field all over it may be assumed to be uniform. If I be the current through coil C_1, the magnetic field at the site of the coil C_2 due to coil C_1 would be

$$B = \frac{\mu_0}{4\pi} \frac{2\pi N_1 I a^2}{(a^2 + x^2)^{3/2}},$$

where N_1 is the number of turns in coil C_1.

Now the total flux linked with the coil C_2 is

$$\Phi_2 = B \cdot \pi b^2 \cdot N_2 = \frac{\mu_0 \pi N_1 N_2 a^2 b^2}{2\,(a^2 + x^2)^{3/2}} I, \tag{10.12-2}$$

where N_2 is the number of turns in coil C_2.

Therefore, the mutual inductance between the two coils is

$$M = \frac{\Phi_2}{I} = \frac{\mu_0 \pi N_1 N_2 a^2 b^2}{2\,(a^2 + x^2)^{3/2}}. \tag{10.12-3}$$

If the coils are coplanar then $x = 0$ and

$$M = \frac{\mu_0 \pi N_1 N_2 b^2}{2a}.$$

If the coil C_2 is not small then the value of M can be determined by using Neumann's formula.

10.13 Coefficient of Coupling

Coils used to obtain inductance for use in electrical circuits are called *inductors.* When two such coils are used together they get coupled to each other and in addition to their self-inductances mutual inductive effects also come into play. The mutual inductance between two coils is related to their self-inductances through a factor called *coefficient of coupling. It may be defined as the fraction of the magnetic flux generated by one coil that gets linked up with the other.*

Let us consider two coils C_1 and C_2 having self-inductances L_1 and L_2 and coupled together. Let a current I_1 through C_1 produces a flux Φ_1. Therefore, $\Phi_1 = L_1 I_1$. Let a fraction k_1 of this flux gets linked up with coil C_2. So flux linked with coil C_2 will be

$$\Phi_2 = k_1 \Phi_1 = k_1 L_1 I_1.$$

If the mutual inductance between the coils is M then we can write

$$\Phi_2 = M I_1.$$

Therefore,

$$M I_1 = k_1 L_1 I_1 \quad \text{or} \quad M = k_1 L_1. \tag{10.13-1}$$

Similarly, if a current I_2 flowing through C_2 produces a flux Φ_2 and a fraction k_2 of it gets linked up with C_1 we can write for the flux through C_1 as,

$$\Phi_1 = k_2 \cdot \Phi_2 = k_2 L_2 I_2 \quad \text{and} \quad \Phi_1 = M I_2.$$

Thus,

$$M = k_2 L_2. \tag{10.13-2}$$

From Eqs. (10.13-1) and (10.13-2) we find that

$$M^2 = k_1 k_2 L_1 L_2$$

$$\text{or } M = k\sqrt{L_1 L_2}, \tag{10.13-3}$$

where $k = \sqrt{k_1 k_2}$ is called the *coefficient of coupling* between the two coils. It can have values ranging from 0 to 1. It purely depends on the geometry of the coils and their relative positions. k can be varied by varying the geometry and then a variable inductance can be constructed. If whole of the flux generated by a coil gets linked up with the second coil then $k = 1$ and the coupling is said to be perfect. If $k \lesssim 1$ the coils are said to be *close-coupled.* On the other hand, if k is much loss than unity the coils are said to be *loose-coupled.*

From Eq. (10.13-3) we can define the coefficient of coupling as the ratio of mutual inductance actually present between two coils to its maximum possible value.

10.14 Series and Parallel Combinations of Inductances

Series connection

Fig. 10.14-1 shows the series combination of two self-inductances L_1 and L_2. Let M be the mutual inductance of the two coils and I be the current flowing through the coils at time t. In Fig. 10.14-1(a) the mutual flux is aiding the self-flux. So the flux linked with L_1 at time t is $L_1 I + MI$ and that linked with L_2 is $L_2 I + MI$. Thus, total flux for both the coil is $(L_1 + L_2 + 2M)I$. Therefore, the effective self-inductance, i.e., flux per unit current is

$$L_{\text{eff}} = L_1 + L_2 + 2M. \tag{10.14-1}$$

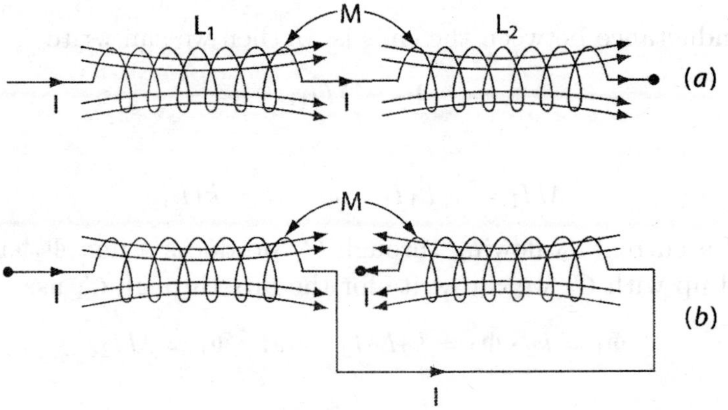

Fig. 10.14-1 : Inductances in series: (a) Mutual flux aiding self-flux, (b) Mutual flux opposing self-flux

In Fig. 10.14-1(*b*) mutual flux oppose the self-flux and hence total flux linked with the coils at time t would be

$$(L_1 I - MI) + (L_2 I - MI) = (L_1 + L_2 - 2M)I.$$

Therefore, in this case, effective self-inductance is

$$L_{\text{eff}} = \text{flux per unit current} = L_1 + L_2 - 2M. \tag{10.14-2}$$

Parallel connection

Fig. 10.14-2 shows the parallel combination of two self-inductances L_1 and L_2.

Let M be the mutual inductance between the two coils. Let at time t the main current I is divided into branch currents I_1 and I_2. So

$$I = I_1 + I_2 \quad \text{or} \quad \frac{dI}{dt} = \frac{dI_1}{dt} + \frac{dI_2}{dt}. \tag{10.14-3}$$

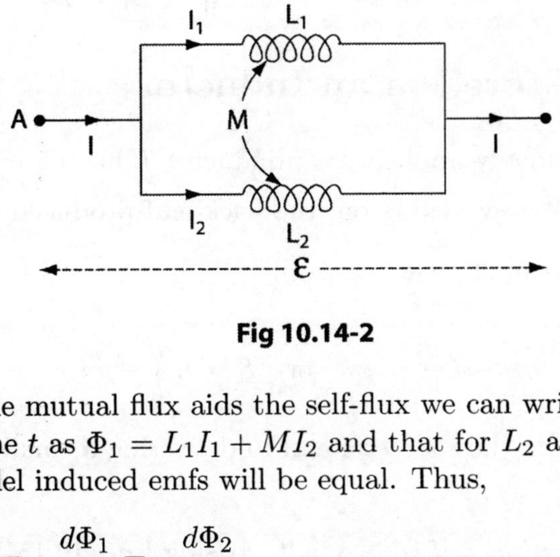

Fig 10.14-2

Assuming that the mutual flux aids the self-flux we can write for the effective flux linked with L_1 at time t as $\Phi_1 = L_1 I_1 + MI_2$ and that for L_2 as $\Phi_2 = L_2 I_2 + MI_1$. As the coils are in parallel induced emfs will be equal. Thus,

$$\mathcal{E} = -\frac{d\Phi_1}{dt} = -\frac{d\Phi_2}{dt}$$

$$\text{or } L_1 \frac{dI_1}{dt} + M\frac{dI_2}{dt} = -\mathcal{E} \quad \text{and} \quad L_2 \frac{dI_2}{dt} + M\frac{dI_2}{dt} = -\mathcal{E}.$$

Solving these two equations we get

$$\frac{dI_1}{dt} = \frac{-\mathcal{E}(L_2 - M)}{L_1 L_2 - M^2} \quad \text{and} \quad \frac{dI_2}{dt} = \frac{-\mathcal{E}(L_1 - M)}{L_1 L_2 - M^2}.$$

Substituting these values in Eq. (10.14-3) we get

$$\frac{dI}{dt} = \frac{-\mathcal{E}(L_1 + L_2 - 2M)}{L_1 L_2 - M^2}. \tag{10.14-4}$$

If L_{eff} be the effective self-inductance then we can write

$$\mathcal{E} = -L_{\text{eff}}\frac{dI}{dt} \tag{10.14-5}$$

Comparing Eqs. (10.14-4) and (10.14-5) we get

$$L_{\text{eff}} = \frac{L_1L_2 - M^2}{L_1 + L_2 - 2M}. \tag{10.14-6}$$

If the inductances are so connected that mutual flux opposes the self-flux then

$$L_{\text{eff}} = \frac{L_1L_2 - M^2}{L_1 + L_2 + 2M}. \tag{10.14-7}$$

If there is no magnetic coupling between the coils then $M = 0$ and we have

$$L_{\text{eff}} = \frac{L_1L_2}{L_1 + L_2} \quad \text{or} \quad \frac{1}{L_{\text{eff}}} = \frac{1}{L_1} + \frac{1}{L_2}. \tag{10.14-8}$$

10.15 Energy Stored in an Inductor

Suppose an emf $\mathcal{E}$ is suddenly applied to an inductor L having resistance r. If i be the current at any time t after switching on, the back emf produced is $-L\frac{di}{dt}$ and net emf is $\mathcal{E} - L\frac{di}{dt}$. So from Kirchhoff's law,

$$\mathcal{E} - L\frac{di}{dt} = ri \quad \text{or} \quad \mathcal{E} = L\frac{di}{dt} + ri. \tag{10.15-1}$$

Now the work done by the source of emf in delivering a small charge $dq = Idt$ in a further time dt is

$$dU = \mathcal{E}idt = Li\,di + i^2rdt \quad \text{[Using Eq. (10.15-1)]}$$

Therefore, total work done in time T when current increases from zero to its final value I is

$$U = L\int_0^I idi + r\int_0^T i^2dt = \frac{1}{2}LI^2 + r\int_0^T i^2dt.$$

The second term on the right-hand side represents the Joule heat loss in the resistance. The first term represents the work done against back emf; it can be regarded as energy stored in the magnetic field of the inductor. This amount of energy is completely recovered when the current is switched off. Thus, magnetic energy stored in an inductor is $U_m = \frac{1}{2}LI^2$. This expression can be used to define the self-inductance of a circuit or it can be used to calculate L.

10.16 To Prove $M^2 \leq L_1 L_2$ from Energy Consideration

Consider two fixed closed circuits 1 and 2 as shown in Fig. 10.16-1, where $\mathcal{E}_1$ and $\mathcal{E}_2$ are the instantaneous emfs and i_1 and i_2 are the instantaneous currents.

The instantaneous emf equations are

$$\mathcal{E}_1 = R_1 i_1 + L_1 \frac{di_1}{dt} + M \frac{di_2}{dt} \tag{10.16-1}$$

$$\mathcal{E}_2 = R_2 i_2 + L_2 \frac{di_2}{dt} + M \frac{di_1}{dt}, \tag{10.16-2}$$

where we assume positive coupling, i.e., mutual fluxes aid self-fluxes. Now the rate of energy supplied from the source must be equal to the rate of Joule heat dissipation plus the rate of energy stored in the magnetic field. Therefore,

$$\mathcal{E}_1 i_1 + \mathcal{E}_2 i_2 = i_1^2 R_1 + i_2^2 R_2 + \frac{dU}{dt}$$

$$\text{or} \quad \frac{dU}{dt} = \mathcal{E}_1 i_1 + \mathcal{E}_2 i_2 - \left(i_1^2 R_1 + i_2^2 R_2\right)$$

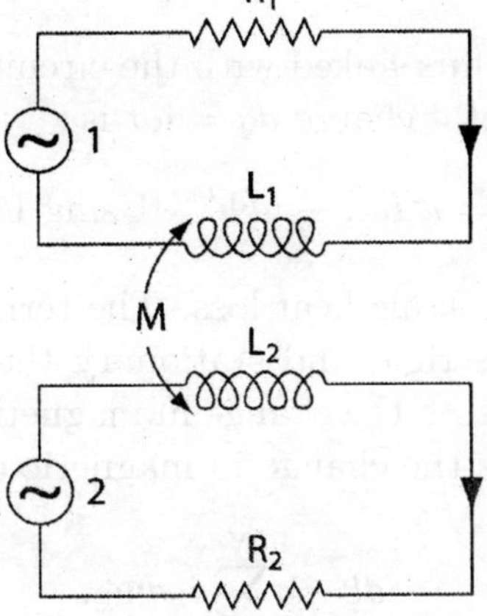

Fig 10.16-1

Using Eqs. (10.16-1) and (10.16-2) we can write

$$\frac{dU}{dt} = L_1 i_1 \frac{di_1}{dt} + L_2 i_2 \frac{di_2}{dt} + M i_1 \frac{di_2}{dt} + M i_2 \frac{di_1}{dt}$$

$$= \frac{1}{2} L_1 \frac{di_1^2}{dt} + \frac{1}{2} L_2 \frac{di_2^2}{dt} + M \frac{d}{dt}(i_1 i_2).$$

Assuming $i_1 = i_2 = 0$ at $t=0$ and integrating the equation from $t=0$ to $t=t$ we get

$$U = \tfrac{1}{2} L_1 i_1^2 + \tfrac{1}{2} L_2 i_2^2 + M i_1 i_2 \tag{10.16-3}$$

$$\text{or} \quad \frac{2U}{L_1} = \left(i_1 + \frac{M}{L_1} i_2\right)^2 + \left(\frac{L_2}{L_1} - \frac{M^2}{L_1^2}\right) i_2^2. \tag{10.16-4}$$

This is valid for all i_1 and i_2. Let $i_1 = -\dfrac{M}{L_1}i_2$. Then, since U is positive or zero for all values of i_1 and i_2 we must have $M^2 \leq L_1L_2$.

Note that Eq. (10.16-3) gives the magnetic energy of two coupled circuits. Depending on the sign of coupling the term Mi_1i_2 may be positive or negative but total energy U is always positive or zero.

10.17 Energy Stored in a Magnetic Field

Establishing a magnetic field involves expenditure of energy which is stored in the associated magnetic field. The creation of a steady state configuration of currents and associated magnetic fields passes through initial transient period. For such varying fields there are induced emfs against which the source of current has to do work. Suppose an instantaneous emf $\mathcal{E}$ is applied to a circuit having resistance R. Then instantaneous current i will be given by the equation

$$\mathcal{E} - \frac{d\Phi'}{dt} = iR, \tag{10.17-1}$$

where Φ' is the instantaneous flux linked with the circuit. The work done by the source in delivering a small incremental charge $dq = idt$ is

$$dU = \mathcal{E}idt = i^2Rdt + id\Phi' \quad \text{[Using Eq. (10.17-1)]}$$

The term i^2Rdt represents Joule heat loss. The term $id\Phi'$ is the work done against the back emf. If the circuit is rigid and stationary there will be no mechanical work and the term $id\Phi'$ will be equal to the change in magnetic energy of the circuit. If there are N stationary rigid circuits the change in magnetic energy of the system would be

$$dU = \sum_{k=1}^{N} i_k d\Phi'_k. \tag{10.17-2}$$

In presence of linear magnetic media Φ'_k will be linearly related to the current in the circuit and magnetic energy will be independent of the way in which all currents are brought to their final values. Let us assume that all currents and hence all fluxes are brought to their final values in such a way that at any instant of time t all currents and all fluxes are at the same fraction α of their final values. Then at time t,

$$i_k = \alpha I_k \quad \text{and} \quad \Phi'_k = \alpha\Phi_k,$$

where I_k and Φ_k are the final currents and final fluxes.

Therefore,

$$dU = \sum_{k=1}^{N} I_k\Phi_k \cdot \alpha d\alpha$$

and total magnetic energy

$$U = \sum_{k=1}^{N} I_k \Phi_k \int_0^1 \alpha d\alpha = \frac{1}{2}\sum_{k=1}^{N} I_k \Phi_k. \tag{10.17-3}$$

Assuming each circuit to consist of a single loop we can write

$$\Phi_k = \int_{S_k} \vec{B}\cdot d\vec{S} = \int_{S_k} \vec{\nabla}\times\vec{A}\cdot d\vec{S} = \oint_{C_k} \vec{A}\cdot d\vec{l_k} \quad \text{[Using Stokes' law]}$$

where $\vec{A}$ is the local magnetic vector potential.

$$\therefore\ U = \frac{1}{2}\sum_k \oint_{C_k} I_k \vec{A}\cdot d\vec{l_k}.$$

For a general type of volume distribution of currents (instead of current circuits defined by wires) we can replace $I_k d\vec{l_k}$ by $\vec{J}\,dV$ and $\sum_k \oint_{C_k}$ by $\int_V\cdot$

Therefore,

$$U = \frac{1}{2}\int_V \vec{A}\cdot\vec{J}\,dV, \tag{10.17-4}$$

where V is the volume occupied by the current distribution.

Now using the relation $\vec{\nabla}\times\vec{H} = \vec{J}$ and the vector identity

$$\vec{\nabla}\cdot\left(\vec{A}\times\vec{H}\right) = \vec{H}\cdot\left(\vec{\nabla}\times\vec{A}\right) - \vec{A}\cdot\left(\vec{\nabla}\times\vec{H}\right).$$

We get

$$\begin{aligned} U &= \frac{1}{2}\int_V \vec{H}\cdot\left(\vec{\nabla}\times\vec{A}\right) dV - \frac{1}{2}\int_V \vec{\nabla}\cdot\left(\vec{A}\times\vec{H}\right) dV \\ &= \frac{1}{2}\int_V \vec{H}\cdot\vec{B}\,dV - \frac{1}{2}\int_S \left(\vec{A}\times\vec{H}\right)\cdot d\vec{S} \ \text{[Using divergence theorem]}, \end{aligned} \tag{10.17-5}$$

where S is the surface bounding the volume V.

Note that integration is to be carried out over the entire volume occupied by the current distribution. For convenience we can move the surface S to infinity. This will not affect the integration (10.17-4) because $\vec{J} = 0$ outside the region occupied by the current distribution. Now at large distances $\vec{A}$ and $\vec{H}$ will fall at least as rapidly as those of dipoles; hence $A \sim \frac{1}{r^2}$, $H \sim \frac{1}{r^3}$. So the integrand $\vec{A}\times\vec{H}$ falls off at least as $1/r^3$ or faster. As the surface element $d\vec{S}$ goes as r^2, the surface integral vanishes as $1/r$ or faster as r goes to infinity. Therefore, Eq. (10.17-5) becomes

$$U = \frac{1}{2}\int_{\text{all space}} \vec{H}\cdot\vec{B}\,dV. \tag{10.17-6}$$

This equation suggests that we may consider the magnetic energy as stored in the magnetic field with an energy density (i.e., energy per unit volume)

$$u = \tfrac{1}{2}\vec{H} \cdot \vec{B}. \tag{10.17-7}$$

In case of linear isotropic magnetic media of permeability μ, $\vec{B} = \mu\vec{H}$ and

$$u = \frac{1}{2}\mu H^2 = \frac{1}{2} \cdot \frac{B^2}{\mu}. \tag{10.17-8}$$

For vacuum, $u = \frac{1}{2}\mu_0 H^2$.

10.18 Forces and Torques from Magnetostatic Energy

From a knowledge of the magnetostatic energy it is possible to calculate the forces and torques acting on a system of current circuits. Suppose we have a set of rigid, fixed current circuits. These current circuits produce magnetic fields and hence exert forces on each other. Let us allow one of the parts of the system to make a virtual rigid displacement $d\vec{r}$ while all the currents are maintained constant. The mechanical work done by the magnetic force $\vec{F}$ acting on the part in question is

$$dW = \vec{F} \cdot d\vec{r}.$$

Let the electrical work done by external energy sources against the induced emfs to keep the currents constants be dW_B and the increase in magnetic energy of the system be dU. So for the conservation of energy,

$$dW_B = dU + dW \tag{10.18-1}$$

If the change in flux passing through the k-th circuit is $d\Phi_k$ then dU is given by Eq. (10.17-3) as

$$dU = \frac{1}{2}\sum I_k d\Phi_k. \tag{10.18-2}$$

Now the work done by the batteries to maintain all currents constant is given as

$$dW_B = \sum_k \frac{d\Phi_k}{dt} \cdot I_k dt = \sum_k I_k d\Phi_k. \tag{10.18-3}$$

Therefore,

$$dW_B = 2dU \tag{10.18-4}$$

and hence from Eq. (10.18-1),

$$dW = dU \quad \text{(constant currents)} \tag{10.18-5}$$

Thus, $dW = \vec{F} \cdot d\vec{r} = dU = \vec{\nabla}U \cdot d\vec{r}$.

Therefore, $\vec{F} = \vec{\nabla}U|_I$. In component forms,

$$F_x = \left.\frac{\partial U}{\partial x}\right|_I, \quad F_y = \left.\frac{\partial U}{\partial y}\right|_I \quad \text{and} \quad F_z = \left.\frac{\partial U}{\partial z}\right|_I. \tag{10.18-6}$$

If the part of the system under question is constrained to rotate about an axis and it undergoes a small rotation $d\vec{\theta}$ under the action of a torque $\vec{\tau}$ then

$$dW = \vec{\tau} \cdot d\vec{\theta} = dU = \vec{\nabla}_\theta U \cdot d\vec{\theta}.$$

Therefore, $\vec{\tau} = \vec{\nabla}_\theta U|_I$. In component forms,

$$\tau_x = \left.\frac{\partial U}{\partial \theta_x}\right|_I, \quad \tau_y = \left.\frac{\partial U}{\partial \theta_y}\right|_I \quad \text{and} \quad \tau_z = \left.\frac{\partial U}{\partial \theta_z}\right|_I. \tag{10.18-7}$$

Another important physical situation arises when the circuits are isolated and the fluxes through these circuits remain constant. In this case, $dW_B = 0$ and we have

$$\vec{F} \cdot d\vec{r} = dW = -dU = -\vec{\nabla}U \cdot d\vec{r}.$$

Thus, $\vec{F} = -\vec{\nabla}U|_\Phi$. In component forms

$$F_x = -\left.\frac{\partial U}{\partial x}\right|_\Phi, \quad F_y = -\left.\frac{\partial U}{\partial y}\right|_\Phi \quad \text{and} \quad F_z = -\left.\frac{\partial U}{\partial z}\right|_\Phi. \tag{10.18-8}$$

In a similar way, $\vec{\tau} = -\vec{\nabla}_\theta U|_\Phi$ and in component forms

$$\tau_x = -\left.\frac{\partial U}{\partial \theta_x}\right|_\Phi, \quad \tau_y = -\left.\frac{\partial U}{\partial \theta_y}\right|_\Phi \quad \text{and} \quad \tau_z = -\left.\frac{\partial U}{\partial \theta_z}\right|_\Phi. \tag{10.18-9}$$

An example of the above energy method

Suppose we have a long solenoid of length l, number of turns N and cross-sectional area A. A constant current I is being passed through the solenoid. An iron rod of constant permeability μ and cross-sectional area A is partially inserted in the solenoid along its axis as shown in Fig. 10.18-1. We are to calculate the force acting on the rod. To solve the problem we assume that the field structure remains the same when the rod is moved by a small distance Δx from its position. The only difference is that a

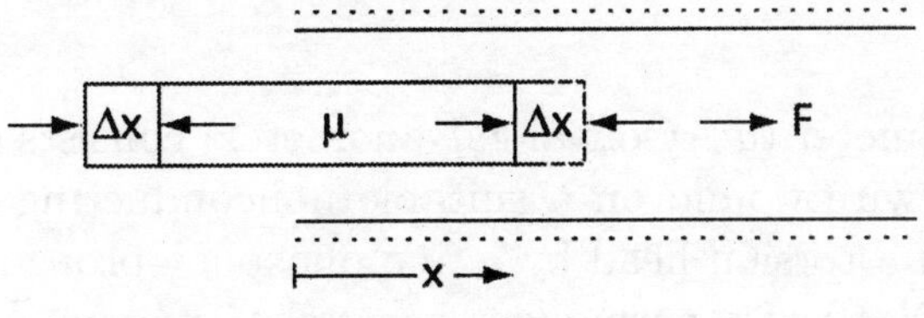

Fig 10.18-1

length Δx of the rod is effectively transferred from a region outside the field to a region having uniform field inside the solenoid. So the energy difference between these two configurations as a function of x can be calculated as

$$\Delta U = U(x+\Delta x) - U(x) \approx \frac{1}{2}(\mu - \mu_0)\int_{A\Delta x} H^2\, dV,$$

where $A\Delta x$ is the change in volume caused by the displacement. Now inside the solenoid $H = NI/l$, which is constant. Therefore,

$$\Delta U = \frac{1}{2}(\mu - \mu_0)\left(\frac{NI}{l}\right)^2 \cdot A\Delta x.$$

The force on the rod is then given by

$$F_x = \left.\frac{\partial U}{\partial x}\right|_I = \frac{1}{2}(\mu - \mu_0)\left(\frac{NI}{l}\right)^2 A$$

which is in the direction of increasing x.

10.19 Ballistic Galvanometer

Galvanometer is an electrical instrument used for the detection and measurement of currents and charges. Ammeters and voltmeters are the modified forms of the galvanometers. Most commonly used galvanometers are of moving coil suspended type. Here a movable coil is suspended in the magnetic field of a permanent magnet. When current passes through the coil it experiences a torque and rotates until deflecting torque is balanced by the restoring torque of the suspension fibre. The angle of deflection gives a measure of the current. For steady current measurement the galvanometer is designed that a steady deflection is obtained without any oscillation. Such a galvanometer is called a *dead-beat galvanometer*. For the measurement of a small amount of charge the galvanometer is used ballistically. The quantity of charge to be measured is passed through the galvanometer before the coil moves appreciably. The coil receives an impulse and the consequent throw gives a measure of the charge. Here damping in the system is kept as small as possible such that an oscillatory motion is obtained.

Construction

A moving coil galvanometer (*d* Arsonval galvanometer) consists of a rectangular coil of fine insulated copper wire wound on a suitable nonconducting heavy frame (F). The coil is suspended from a torsion head by a fine phosphor-bronze fibre between the two concave pole pieces (N, S) of a permanent horse-shoe magnet. A cylindrical soft iron core is used within the coil to increase the magnetic flux in the gap. A mirror (M) is

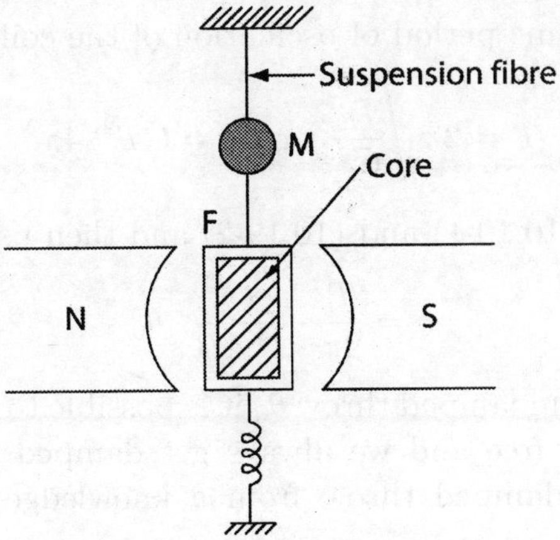

Fig. 10.19-1 : Moving-coil galvanometer

attached to the suspension fibre, which helps to note the deflection of the coil by lamp and scale arrangement.

Theory

Let N be the number of turns and A the area of the galvanometer coil. The coil is assumed to be suspended by a torsion fibre in a uniform radial magnetic field B. Let $i = dq/dt$ be the instantaneous current through the coil. As a result of this current flow the coil experiences an instantaneous deflecting torque $\vec{\tau} = Ni\vec{A} \times \vec{B}$ or, $\tau = NABi$ because the plane of the coil is parallel to $\vec{B}$. The quantity of charge to be measured is passed through the coil before it moves appreciably. So the current through the coil lasts for a very short time Δt and we get an impulsive motion. The angular impulse given to the coil is equal to the change in angular momentum and is given by

$$NAB \int_0^{\Delta t} idt = NABq = I\omega_1, \tag{10.19-1}$$

where I is the moment of inertia of the coil about the axis of suspension and ω_1 is the sudden angular velocity acquired by the coil, which is assumed to start from rest in the position $\theta = 0$. Now the initial kinetic energy $\frac{1}{2}I\omega_1^2$ of the coil is used up in twisting the suspension fibre. If θ_0 be the maximum deflection when the coil is momentarily stopped then work done in twisting the fibre is

$$\int_0^{\theta_0} C\theta d\theta = \frac{1}{2}C\theta_0^2,$$

where C is the couple per unit angle of twist. Thus,

$$\frac{1}{2}I\omega_1^2 = \frac{1}{2}C\theta_0^2. \tag{10.19-2}$$

Assuming no damping, the equation of motion of the coil after the charge has passed off is

$$I\frac{d^2\theta}{dt^2} + C\theta = 0.$$

This shows that the time period of oscillation of the coil is

$$T = 2\pi\sqrt{\frac{I}{C}} \quad \text{or} \quad I = CT^2/4\pi^2. \tag{10.19-3}$$

Eliminating ω_1 from (10.19-1) and (10.19-2) and then using this value of I we get

$$q = \frac{T}{2\pi} \cdot \frac{C}{NAB} \cdot \theta_0. \tag{10.19-4}$$

Thus, noting the first undamped throw θ_0 it is possible to measure q. But in practice the motion is not totally free and we always get damped throw. Thus, it becomes necessary to find the undamped throw from a knowledge of the observed damped throw.

The motion of the coil is always slightly damped due to air friction and it can be taken as proportional to the angular velocity. Let it be $a \cdot d\theta/dt$. As the coil rotates in a magnetic field there is induced emf in it, which tries to oppose the motion of the coil. This is known as *electromagnetic damping*, which can be calculated as follows:

If the coil rotates through an angle $d\theta$ in time dt then the flux cut by the vertical sides of the coil will be given by the fluxes through the cylindrical surfaces $PP'S'S$ and $QQ'R'R$ as shown in Fig. 10.19-2.

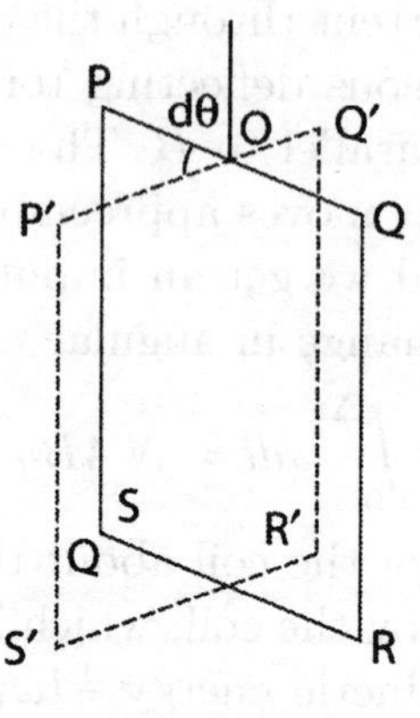

Fig 10.19-2

So the induced emf

$$\begin{aligned} \mathcal{E} &= -\frac{d}{dt}\left[2 \cdot PS \cdot (OPd\theta) \cdot B\right] N \\ &= -\frac{d}{dt}[NABd\theta] \qquad [\because\ PS \times 2.OP = PS \times PQ = A, \text{ area of the coil}] \\ &= -NAB\frac{d\theta}{dt} = -G\frac{d\theta}{dt} \text{ where } G = NAB. \end{aligned}$$

If R be the total resistance in the galvanometer circuit it will produce a current $i = \mathcal{E}/R$, which will try to oppose the motion of the coil.

The opposing torque due to this current is

$$-NiA \cdot B = \frac{G^2}{R}\frac{d\theta}{dt}.$$

Now including the air damping and electromagnetic damping the equation of motion of the coil after the charge has passed off would be

$$I\frac{d^2\theta}{dt^2} + \left(a + \frac{G^2}{R}\right)\frac{d\theta}{dt} + C\theta = 0 \quad \text{or} \quad \frac{d^2\theta}{dt^2} + 2b\frac{d\theta}{dt} + \omega_0^2\theta = 0, \tag{10.19-5}$$

where $2b = \frac{1}{I}\left(a + \frac{G^2}{R}\right)$ and $\omega_0^2 = \frac{C}{I}$.

Eq. (10.19-5) has a solution of the form

$$\theta = e^{-bt}\left[A_1 e^{\sqrt{b^2-\omega_0^2}t} + A_2 e^{-\sqrt{b^2-\omega_0^2}t}\right]. \tag{10.19-6}$$

In ballistic galvanometer the damping is kept very low, i.e., $b^2 < \omega_0^2$ and we can put the solution in the form

$$\theta = \theta_0 e^{-bt}\sin(\omega t + \alpha), \tag{10.19-7}$$

where $\omega = \sqrt{\omega_0^2 - b^2}$ and α is a constant. At $t = 0$, $\theta = 0$, therefore, $\alpha = 0$. Thus,

$$\theta = \theta_0 e^{-bt}\sin\omega t. \tag{10.19-8}$$

Thus, the coil undergoes damped oscillatory motion (Fig. 10.19-3) with a time period

$$T = \frac{2\pi}{\omega} = \frac{2\pi}{\sqrt{\omega_0^2 - b^2}}.$$

The deflection is maximum on either side of central position at times given by

$$\sin\omega t = \pm 1 = \sin(2m+1)\frac{\pi}{2}, \text{i.e.,} \quad t = \frac{T}{4}, \frac{3T}{4}, \frac{5T}{4} \ldots \text{etc.}$$

Corresponding throws are

$$\theta_1 = \theta_0 e^{-bT/4}, \quad \theta_2 = \theta_0 e^{-3bT/4}, \quad \theta_3 = \theta_0 e^{-5bT/4} \ldots$$

So the ratios of successive amplitudes are

$$\frac{\theta_1}{\theta_2} = \frac{\theta_2}{\theta_3} = \cdots = \frac{\theta_n}{\theta_{n+1}} = e^{\frac{bT}{2}} = d\,(\text{say})$$

Here d is called *decrement* and $\lambda = \log_e d = bT/2$ is called *log decrement.* Now

$$\theta_0 = \theta_1 e^{\frac{bT}{4}} = \theta_1 e^{\lambda/2} \approx \theta_1 \left(1 + \frac{\lambda}{2}\right) \quad [\because \lambda \text{ is small}]$$

Thus, undamped throw can be obtained from a knowledge of the observed first throw θ_1 and log decrement λ, which can be measured by noting the first throw θ_1 and the $(n+1)th$ throw as indicated below:

$$\frac{\theta_1}{\theta_{n+1}} = \frac{\theta_1}{\theta_2} \cdot \frac{\theta_2}{\theta_3} \cdots \frac{\theta_n}{\theta_{n+1}} = (d)^n \quad \text{or} \quad \lambda = \log_e d = \frac{1}{n} \log_e \frac{\theta_1}{\theta_{n+1}}. \tag{10.19-9}$$

Thus, the correct relation for the charge q is

$$q = \frac{T}{2\pi} \cdot \frac{C}{nAB} \theta_1 \left(1 + \frac{\lambda}{2}\right). \tag{10.19-10}$$

Now q can be determined by noting θ_1 provided we know the galvanometer constants. These constants can, however, be eliminated by noting steady deflection θ_d for a steady current I_d. For such a steady current the deflecting torque $NABI_d$ = Restoring torque $C\theta_d$

$$\therefore \quad \frac{C}{NAB} = \frac{I_d}{\theta_d}. \tag{10.19-11}$$

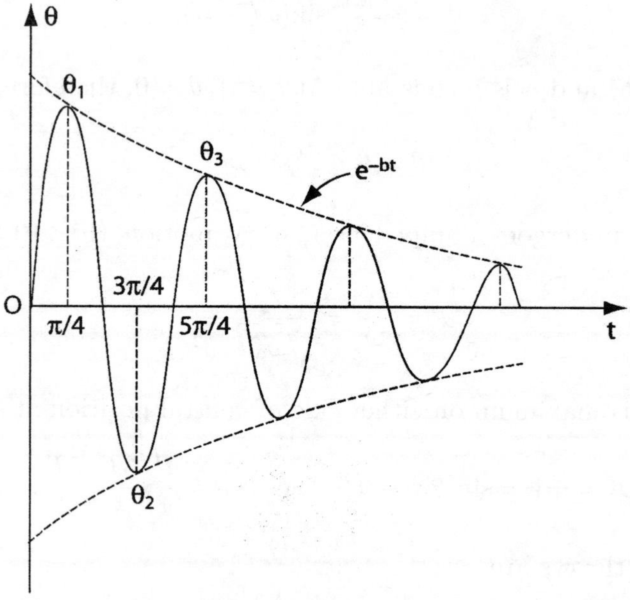

Fig. 10.19-3: Variation of θ with t

The open circuit time period of oscillation T can be determined by noting the time for several oscillations with the help of a stop watch.

Now all the quantities on the right-hand side of Eq. (10.19-10) being known q can be determined.

Critical damping resistance (CDR)

The damping constant

$$2b = \frac{1}{I}\left(a + \frac{G^2}{R}\right),$$

where $R = R_g + R_{\text{ext}}$ is the total resistance in the galvanometer circuit. Obviously the damping in the system can be controlled by varying the resistance R_{ext} in the external circuit. As R_{ext} is increased the induced current in the coil and hence the electromagnetic damping decreases. In order to make the motion of the coil oscillatory we must make $b < \omega_0$. This indicates that the value of R_{ext} should be chosen greater than certain critical value (R_c) in order to have oscillatory motion. Otherwise the motion would be dead-beat type. This critical resistance is known as *external critical damping resistance*(CDR) and is given by the condition

$$b = \omega_0 \quad \text{or} \quad \frac{1}{2I}\left(a + \frac{G^2}{R}\right) = \sqrt{\frac{C}{I}}.$$

If air damping is small compared to the term G^2/R then CDR is given by

$$R_c = \frac{G^2}{2\sqrt{IC}} - R_g. \tag{10.19-12}$$

Sensitivity of the galvanometer

A galvanometer is said to be sensitive if it is able to measure small current, voltage or charge:

(i) The *current sensitivity* (S_i) of a galvanometer is defined as the deflection of the light spot in mm on a scale placed at a distance of 1 m from the galvanometer mirror when a current of 1 μA is passed through the galvanometer. Steady deflection θ produced by a current i is given by

$$i = \frac{C}{NAB} \cdot \theta.$$

Therefore,

$$S_i = \frac{d\theta}{di} = \frac{NAB}{C}. \tag{10.19-13}$$

So for given values of A, B and C, $S_i \propto N$. If the number of turns N is increased by keeping area A constant then the length of the wire in the winding will increase N times and the cross-sectional area of the wire must be necessarily reduced by a factor of $1/N$. So the resistance R_g of the wire in the winding will increase by N^2 ($\because R_g = \rho \cdot l_g/A_g$).

Thus, $R_g \propto N^2$, $S_1 \propto N$ and hence $Si \propto \sqrt{R_g}$. So a current sensitive galvanometer should have high resistance. Such a galvanometer is used for current measurement.

(ii) The *voltage sensitivity* (S_V) of a galvanometer is defined as the deflection of the light spot in mm on a scale placed at a distance 1 m from the galvanometer mirror when a pd of 1 μV is applied across the galvanometer. The pd across the galvanometer for a steady current i is $V = iR_g$. Therefore,

$$S_V = \frac{d\theta}{dV} = \frac{1}{R_g}\frac{d\theta}{di} = \frac{S_i}{R_g}. \tag{10.19-14}$$

As $S_i \propto \sqrt{R_g}$, therefore, $S_V \propto 1/\sqrt{R_g}$.

So a voltage sensitive galvanometer should have low resistance. Such a galvanometer is used for null point work.

(iii) The *charge sensitivity* (S_q) *of a galvanometer* is defined as the deflection of the light spot in mm on a scale placed at a distance of 1 m from the galvanometer mirror when a charge of 1 μC is passed through the galvanometer. As

$$q = \frac{T}{2\pi}\cdot\frac{C}{NAB}\theta_0.$$

$$\therefore \quad S_q = \frac{d\theta_0}{dq} = \frac{2\pi}{T}\cdot\frac{NAB}{C} = \frac{2\pi}{T}\times S_i. \tag{10.19-15}$$

Difference between a deat-beat and a ballistic galvanometer

1. In dead-beat galvanometer damping is large and the motion is non-oscillatory. In ballistic galvanometer damping is small and the motion is oscillatory.

2. Dead-beat galvanometer measures steady current in terms of steady deflection, whereas ballistic galvanometer usually measures small amount of charges in terms of first throw.

3. In dead-beat galvanometer the coil moves under a constant torque. But in ballistic galvanometer we have impulsive angular motion. Here the quantity of charge to be measured is passed through the coil before it moves appreciably.

4. To enhance electromagnetic damping in a dead-beat galvanometer the coil is wound on a metallic frame and external resistance is kept below CDR. On the other hand, in a ballistic galvanometer to reduce damping the coil is wound on a nonmetallic frame and external resistance is kept above CDR.

5. In dead-beat galvanometer the time period of the moving system is kept small but in a ballistic galvanometer the time period is made large (10-20 seconds) by choosing large moment inertia of the moving system and keeping the torsional couple per unit twist of the suspension fibre small.

10.20 Measurement of Magnetic Field by Using a Search Coil and a Ballistic Galvanometer

A search coil is a closely wound small coil of fine insulated copper wire having about 100 turns and a face of about 1 cm^2 in area. This coil together with a properly calibrated ballistic galvanometer can be used to measure magnetic field.

Suppose we are to find the magnetic field between the pole pieces N and S of a magnet. The search coil is placed within the gap with its plane perpendicular to the magnetic field. The search coil is connected in series with a ballistic galvanometer (BG) and very often with the secondary coil (S_1) of a standard solenoid which helps to calibrate the ballistic galvanometer.

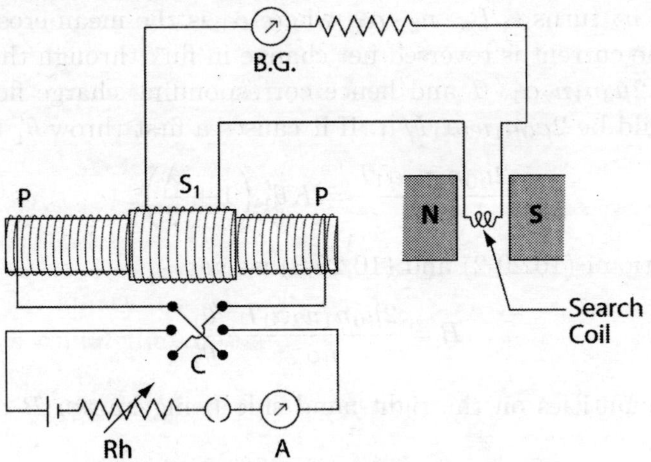

Fig. 10.20-1 : Measurement of magnetic field

If the search coil is quickly withdrawn from the field there will be change in flux, some amount of charge will flow through the galvanometer and a throw will be obtained in the galvanometer. If Φ is the flux linked with the search coil at any instant t then induced emf in it is

$$\mathcal{E} = -\frac{d\Phi}{dt}.$$

If R is the total resistance of the galvanometer circuit then induced current is

$$i = \frac{\mathcal{E}}{R} = -\frac{1}{R}\frac{d\Phi}{dt}.$$

If Δt is the time in which the search coil is withdrawn then total charge-flow through the galvanometer is given by

$$q = \int_0^{\Delta t} idt = -\frac{1}{R}\int_{\Phi_1}^{\Phi_2} d\Phi = \frac{1}{R}(\Phi_1 - \Phi_2), \qquad (10.20\text{-}1)$$

where Φ_1 and Φ_2 are the initial and final flux linked with the search coil. Now $\Phi_2 = 0$ and $\Phi_1 = n\alpha B$, where n is the number of turns and α is the face area of the search coil. Thus,

$$q = \frac{N\alpha B}{R}.$$

If θ_1 be the first throw of the galvanometer due to this charge then

$$q = \frac{N\alpha B}{R} = K\theta_1\left(1 + \frac{\lambda}{2}\right), \tag{10.20-2}$$

where K is the galvanometer constant and λ is the log decrement. To find the constant K a known current I passing through the primary P of a standard solenoid is suddenly reversed by means of a Phol's commutator C and corresponding ballistic throw θ_1' is noted. If n_1 be the number of turns per unit length of the primary then magnetic field at its middle region is $B_s = \mu_0 n_1 I$. The flux linked with the secondary having n_2 turns is $B_s \cdot n_2 \cdot \alpha_1$, where α_1 is the mean cross-sectional area of the primary. If the current is reversed net change in flux through the secondary would be $2B_s \cdot n_2\alpha_1 = 2\mu_0 n_1 n_2 \alpha_1 \cdot I$ and hence corresponding charge flowing through the galvanometer would be $2\mu_0 n_1 n_2 \alpha_1 I/R$. If it causes a first throw θ_1' then

$$\frac{2\mu_0 n_1 n_2 \alpha_1 I}{R} = K\theta_1'\left(1 + \frac{\lambda}{2}\right). \tag{10.20-3}$$

Taking the ratio of (10.20-2) and (10.20-3) we get

$$B = \frac{2\mu_0 n_1 n_2 \alpha_1 I}{n\alpha} \cdot \frac{\theta_1}{\theta_1'}. \tag{10.20-4}$$

Now all the quantities on the right-hand side being known B can be determined from this relation.

Note : Here it is important to mention that nowadays a simple and convenient method of measuring magnetic field has been developed by using *Hall probe*, which is based on a phenomenon known as *Hall effect*. If a current carrying conductor is placed in a magnetic field at right angles to the direction of current flow then a voltage appears in a direction normal to both the directions of current flow and the magnetic field. This phenomenon is known as *Hall effect* and the voltage developed is called *Hall voltage*. As the Hall voltage is found to be proportional to the magnetic field, it can be used to measure unknown magnetic field.

SOLVED PROBLEMS

► Problem 1.

(a) *A coil of N turns and cross-sectional area A is rotated at a constant angular velocity ω about a diameter in a magnetic field $B = B_0 \sin\omega t$ directed normal to the axis of rotation. Find the induced emf in the coil. Assume at $t = 0$ the coil axis coincides with the direction of the magnetic field.*

(b) *A uniform magnetic field $\vec{B}(t)$ pointing along the z-axis, fills up a circular region centred at the origin and lying on the xy-plane. Show that the induced electric field at a distance r from the origin is given by*

$$\vec{E} = -\frac{r}{2}\frac{dB}{dt}\hat{\theta},$$

where $\hat{\theta}$ is the unit cross-radial vector.

Solution. (a) At any instant t the coil axis makes an angle ωt with the direction of magnetic field. Hence, flux linked with the coil at any instant t is

$$\Phi = NAB\cos\omega t = NAB_0\sin\omega t\cos\omega t = \tfrac{1}{2}NAB_0\sin 2\omega t.$$

Now from Faraday's law, the induced emf is

$$\mathcal{E} = -\frac{d\Phi}{dt} = -NAB_0\omega\cos 2\omega t.$$

(b) Considering an Amperian loop of radius r we can write from Faraday's law, induced emf

$$\mathcal{E} = -\frac{d\Phi}{dt} \quad \text{or} \quad \oint \vec{E}\cdot d\vec{l} = -\frac{d}{dt}\left[\pi r^2 B(t)\right]$$

$$\text{or} \quad E\cdot 2\pi r = -\pi r^2\frac{dB}{dt} \qquad [\because \vec{E} \text{ points in the circumferential direction}]$$

Therefore, $\vec{E} = -\dfrac{r}{2}\cdot\dfrac{dB}{dt}\hat{\theta}.$

► **Problem 2.** *A small magnet of moment M is approaching toward a circular loop of radius a along its axis with a constant speed v. Show that when the magnet is at a distance x from the coil the current induced in the loop is given by*

$$i = \frac{3\mu_0 Ma^2vx}{(a^2+x^2)^{5/2}},$$

where R is the resistance of the loop.

Solution. The magnetic field at a point at a distance z from the centre of the loop as shown in Fig. 10.P-2 is given by

$$\vec{B} = \frac{\mu_0}{4\pi}\left[-\frac{\vec{M}}{r^3} + \frac{3\left(\vec{M}\cdot\vec{r}\right)\vec{r}}{r^5}\right].$$

Its radial and transverse components are

$$B_r = \frac{\mu_0}{4\pi}\cdot\frac{2M\cos\theta}{r^3} \quad \text{and} \quad B_\theta = \frac{\mu_0}{4\pi}\cdot\frac{M\sin\theta}{r^3}.$$

So the component of $\vec{B}$ normal to the plane of the loop is

$$B_n = B_r \cos\theta + B_\theta \cos(90^\circ + \theta) = \frac{\mu_0}{4\pi} \cdot \frac{M}{r^3} \left(3\cos^2\theta - 1\right).$$

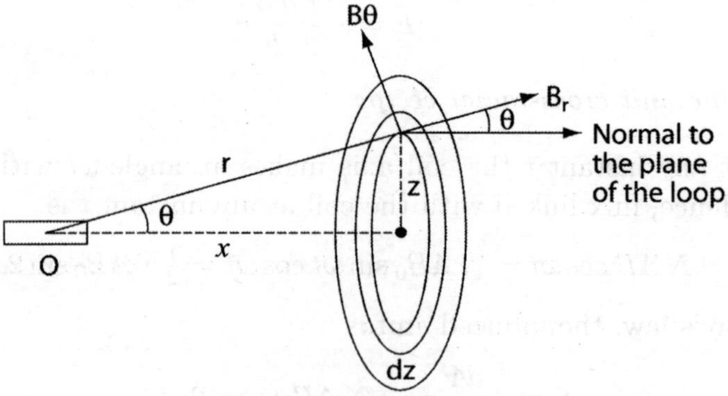

Fig 10-P-2

Now considering an elementary ring of radii z and $z + dz$ in the plane of the loop we find that the magnetic flux through this ring is $B_n \cdot 2\pi z\, dz$. Therefore, total flux through the loop is

$$\Phi = \frac{\mu_0}{4\pi} \cdot \int_0^a \frac{M}{r^3} \left(3\cos^2\theta - 1\right) \cdot 2\pi z\, dz = \frac{\mu_0 M}{2} \int_0^a \frac{(2x^2 - z^2)}{(x^2 + z^2)^{5/2}} dz,$$

where we use $\cos\theta = x/\sqrt{x^2 + z^2}$.

Substituting $x^2 + z^2 = y^2$ and integrating we get

$$\Phi = \frac{\mu_0 M}{2} \cdot \frac{a^2}{(a^2 + x^2)^{3/2}}.$$

Now the induced emf $\mathcal{E} = -\dfrac{d\Phi}{dt}$ and induced current

$$i = \frac{\mathcal{E}}{R} = -\frac{1}{R}\frac{d\Phi}{dt} = -\frac{1}{R}\frac{d\Phi}{dx} \cdot \frac{dx}{dt} = -\frac{v}{R}\frac{d\Phi}{dx} = \frac{3\mu_0 M v a^2 x}{2R\,(a^2 + x^2)^{5/2}}.$$

► **Problem 3.** *A conducting disc of radius a is rotating about its own axis with a constant angular velocity ω in a uniform magnetic field $\vec{B}$ acting perpendicular to the plane of the disc. Find the value of emf induced between the rim and the centre of the disc.*

[B.U. 2016]

Solution. Suppose at any time instant t any radius OP of the disc makes an angle θ with y-axis as shown in Fig. 10.P-3. Let in time dt it rotates through an angle $d\theta$. The area swept out by OP is $\frac{1}{2}a \cdot a d\theta$ and the flux cut by OP in this time is

$$d\Phi = B \cdot \tfrac{1}{2} a^2 d\theta.$$

Hence the value of emf induced between the centre and the rim would be

$$\mathcal{E} = \left|\frac{d\Phi}{dt}\right| = \frac{1}{2}Ba^2\frac{d\theta}{dt} = \frac{1}{2}Ba^2\omega.$$

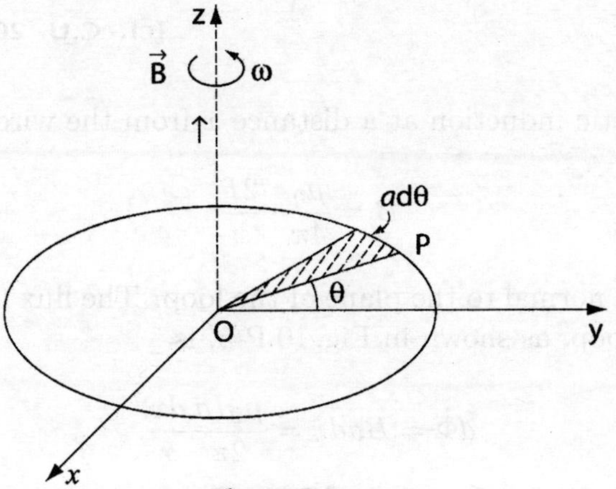

Fig 10.P-3

▶ **Problem 4.** *A thin straight conducting rod of length l is placed perpendicular to a long straight wire carrying current I such that the nearest end of the rod is at a distance a from the wire. If the rod now moves with a constant velocity $\vec{v}$ at right angles to its own length and parallel to the wire show that the emf induced in the rod is given by*

$$\mathcal{E} = \frac{\mu_0 I v}{2\pi}\ln\left(1+\frac{l}{a}\right).$$

Solution. Let us consider an element dx of the rod at a distance x from the wire. The magnetic field at the element is

$$B = \frac{\mu_0}{4\pi}\cdot\frac{2I}{x},$$

which is in a direction normal to the element.

Now as the element moves with a velocity v the rate of flux cut by the element would be $Bvdx$. So from Faraday's law the emf induced in this element is $d\mathcal{E} = Bvdx$.

Therefore, total emf across the rod is

$$\mathcal{E} = \int Bvdx = \frac{\mu_0 I v}{2\pi}\int_a^{a+l}\frac{dx}{x} = \frac{\mu_0 I v}{2\pi}\ln\left(1+\frac{l}{a}\right).$$

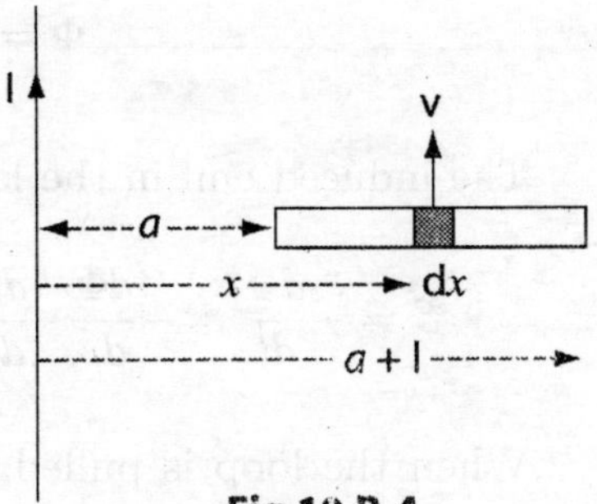

Fig 10.P-4

► **Problem 5.** *Suppose a square loop of side a is placed in the plane of a long straight wire carrying current I. The nearest side of the loop is at a distance r from the wire. Find the magnetic flux through the loop. If someone pulls the loop directly away from the wire at a constant speed, what would be the emf generated in the loop? What is the value of emf generated when the loop is pulled parallel to the wire?*

[cf. C.U. 2012, 2016, B.U. 2017]

Solution. The magnetic induction at a distance x from the wire is

$$B = \frac{\mu_0}{4\pi} \cdot \frac{2I}{x},$$

which is in a direction normal to the plane of the loop. The flux through an elementary area adx within the loop, as shown in Fig. 10.P-5, is

$$d\Phi = Badx = \frac{\mu_0 I a}{2\pi}\frac{dx}{x}.$$

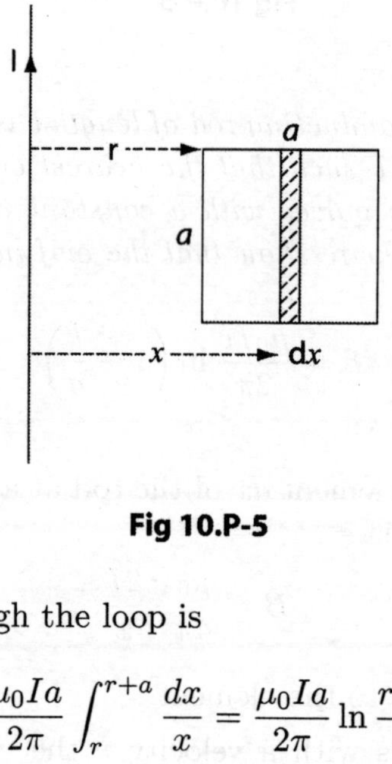

Fig 10.P-5

Therefore, total flux through the loop is

$$\Phi = \frac{\mu_0 I a}{2\pi}\int_r^{r+a}\frac{dx}{x} = \frac{\mu_0 I a}{2\pi}\ln\frac{r+a}{r}.$$

The induced emf in the loop is

$$\mathcal{E} = -\frac{d\Phi}{dt} = -\frac{d\Phi}{dr}\cdot\frac{dr}{dt} = -v\frac{d\Phi}{dr} = -\frac{\mu_0 I a v}{2\pi}\left[\frac{1}{r+a} - \frac{1}{r}\right] = \frac{\mu_0 I v a^2}{2\pi r(r+a)}.$$

When the loop is pulled parallel to the wire the flux does not change and hence $\mathcal{E} = 0$.

▶ **Problem 6.** *Consider two coplanar concentric flat circular coils of average radii r_1 and r_2 and consisting of N_1 and N_2 turns respectively ($r_1 \ll r_2$). If the smaller coil is moved along the common axis with a constant speed v keeping its plane parallel to the plane of the larger coil show that the induced emf in it will be maximum when it is at a distance $r_2/2$ from the larger coil.*

Solution. The magnetic field at the centre of the smaller coil due to a current I_2 in larger coil is

$$B = \frac{\mu_0}{4\pi} \cdot \frac{2\pi N_2 I_2 r_2^2}{(r_2^2 + x^2)^{3/2}},$$

where x is the distance between the coils. Assuming B to be uniform over the area of the smaller coil we can write for the flux linked with the smaller coil as

$$\Phi = B \cdot \pi r_1^2 \cdot N_1 = \frac{\mu_0 \pi N_1 N_2 r_1^2 r_2^2 I_2}{2\,(r_2^2 + x^2)^{3/2}}.$$

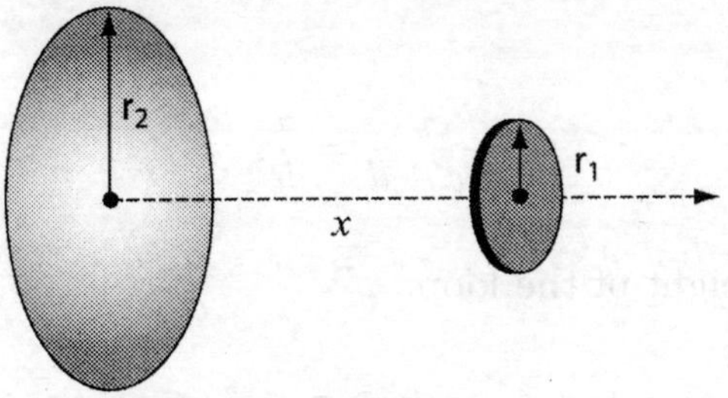

Fig 10.P-6

Therefore, induced emf in the smaller coil is

$$\mathcal{E} = -\frac{d\Phi}{dt} = -\frac{d\Phi}{dx}\frac{dx}{dt} = -v\frac{d\Phi}{dx} = \frac{3\mu_0 \pi N_1 N_2 r_1^2 r_2^2 I_2 v \cdot x}{2\,(r_2^2 + x^2)^{5/2}}.$$

Now for $\mathcal{E}$ to be maximum,

$$\frac{d\mathcal{E}}{dx} = 0 \quad \text{or} \quad \frac{d}{dx}\left[\frac{x}{(r_2^2 + x^2)^{5/2}}\right] = 0.$$

This requires that $x = r_2/2$.

▶ **Problem 7.** *Consider a wire loop in the form an equilateral triangle of side a. A long straight wire carrying current I is placed in the plane of the loop at a distance d from the nearest vertex of the triangle and parallel to the farthest side of the triangle. Find the total magnetic flux through the loop.*

Solution. Consider an element of thickness dx and length y in the loop. The magnetic field at this element due to the straight wire is

$$B = \frac{\mu_0}{4\pi} \cdot \frac{2I}{x}.$$

Therefore, the flux through the element is

$$d\Phi = B \cdot y dx.$$

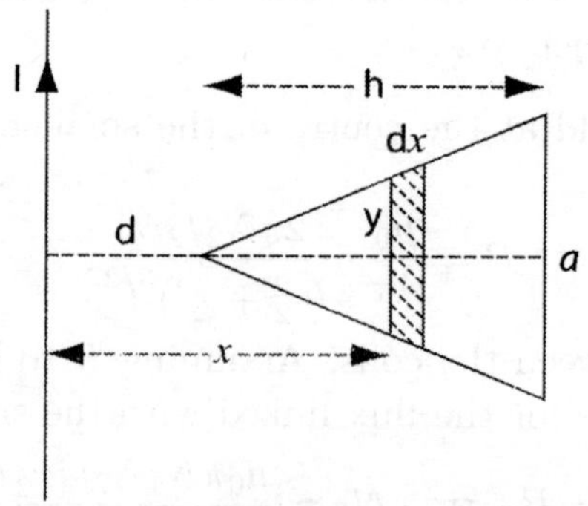

Fig 10.P-7

From Fig. 10.P-7,

$$\frac{y}{x-d} = \frac{a}{h},$$

where $h = \dfrac{\sqrt{3}}{2}a$ is the height of the loop.

Therefore,

$$d\Phi = \frac{\mu_0 I}{2\pi x} \cdot \frac{a(x-d)}{h} \cdot dx = \frac{\mu_0 I a}{2\pi h} \cdot \frac{x-d}{x} dx.$$

So the total flux through the loop is

$$\begin{aligned}\Phi &= \frac{\mu_0 I a}{2\pi h} \int_d^{d+h} \frac{x-d}{x} dx = \frac{\mu_0 I a}{2\pi h} \cdot [x - d \ln x]_d^{d+h} \\ &= \frac{\mu_0 I a}{2\pi h}\left[h - d \ln \frac{d+h}{d}\right] = \frac{\mu_0 I a}{2\pi}\left[1 - \frac{d}{h} \ln\left(1 + \frac{h}{d}\right)\right].\end{aligned}$$

► **Problem 8.** *Consider two long parallel rails of negligible resistance separated by a distance d and connected to a cell of emf $\mathcal{E}$ at one end. A jumper of mass m and resistance R slides without friction along the rails. An external uniform magnetic field B is applied normal to the plane of the rails. Find the velocity of the jumper as a function of time. Find its terminal velocity and also current through it when it attains the terminal velocity.* **[cf. C.U. 2006]**

Solution. Let at any instant of time t the jumper is moving with velocity v. The emf induced in it is Bvd. So from Ohm's law current in the circuit is

$$i = \frac{\mathcal{E} - Bvd}{R}. \tag{1}$$

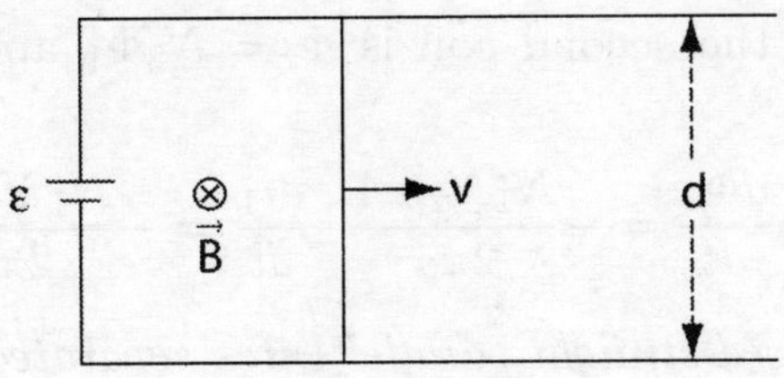

Fig 10.P-8

As the force on the conductor is Bid, its equation of motion is

$$m\frac{dv}{dt} = Bid = \frac{Bd}{R}(\mathcal{E} - Bvd) \quad \text{or} \quad \frac{dv}{v - \frac{\mathcal{E}}{Bd}} = -\frac{B^2d^2}{mR}dt.$$

Integrating,

$$\int_0^v \frac{dv}{v - \frac{\mathcal{E}}{Bd}} = -\frac{B^2d^2}{mR}\int_0^t dt \quad \text{or} \quad \ln\frac{v - \mathcal{E}/Bd}{-\mathcal{E}/Bd} = -\frac{B^2d^2}{mR}t$$

$$\text{or} \quad v = \frac{\mathcal{E}}{Bd}\left(1 - e^{-B^2d^2t/mR}\right).$$

It gives the velocity v as a function of time t. As $t \to \infty$, v tends to a constant value $v_T = \mathcal{E}/Bd$, which is called the *terminal velocity.* Equation (1) shows that at $v = v_T = \mathcal{E}/Bd$ the current $i = 0$. Note that the current due to $\mathcal{E}$ tries to move the jumper away with an acceleration whereas the current due to induced emf tries to oppose this motion. When the induced emf equals the cell emf net current in the jumper becomes zero and the jumper moves with a constant terminal velocity.

▶ **Problem 9.** *Consider an iron ring of mean radius a and cross-sectional area A. There is a coil of N_1 turns would over it, through which a current $i_1 = I_0 \sin\omega t$ is following. There is another coil of N_2 turns wound over the ring. Find the emf induced in this second coil.*

Solution. The flux in the magnetic circuit is

$$\Phi_1 = \frac{N_1 i_1}{\frac{2\pi a}{\mu A}} = \frac{N_1 i_1 \mu A}{2\pi a}.$$

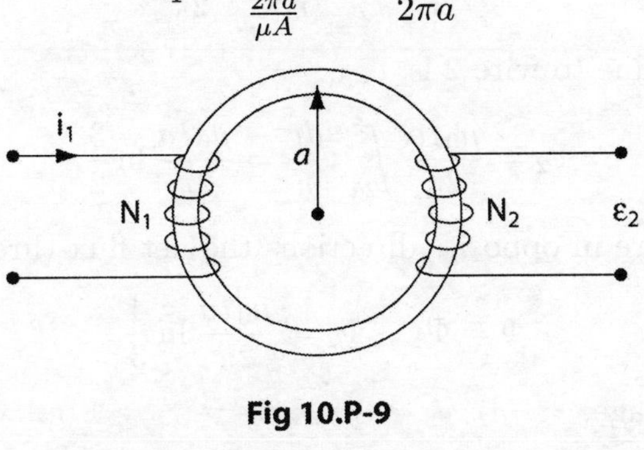

Fig 10.P-9

Total flux linked with the second coil is $\Phi = N_2\Phi_1$ and the emf induced in the second coil is

$$\mathcal{E} = -\frac{d\Phi}{dt} = -N_2\frac{d\Phi_1}{dt} = -\frac{N_1N_2\mu A}{2\pi a}\cdot\frac{di_1}{dt} = -\frac{N_1N_2\mu A}{2\pi a}\cdot\omega I_0\cos\omega t.$$

► **Problem 10.** *Two long fixed straight parallel wires separated by a distance a carry the same current I in opposite directions with I increasing at the rate $\frac{dI}{dt}$. A square loop of side a is placed in the plane of the wires with its near side at a distance a from one of the wires. Find the induced emf in the square loop. What is the direction of the induced current?*

Solution. Magnetic field produced by the straight wire 1 (Fig. 10.P-10) at a distance r from it is

$$B_1 = \frac{\mu_0}{4\pi}\cdot\frac{2I}{r}.$$

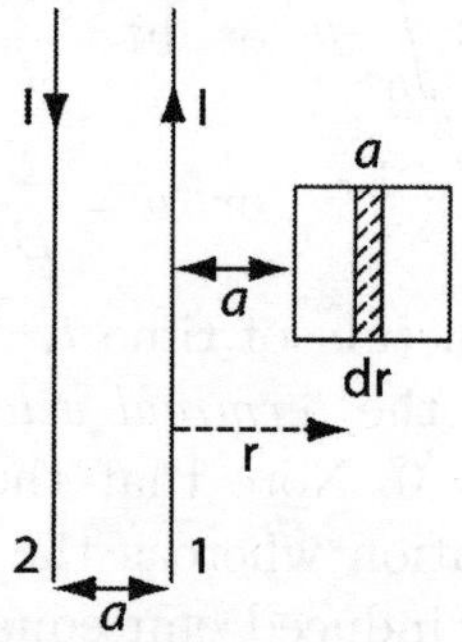

Fig 10.P-10

So the flux through an elementary area adr is

$$d\Phi_1 = B_1\cdot adr = \frac{\mu_0 Ia}{2\pi r}dr.$$

Total magnetic flux through the loop due to wire 1 is

$$\Phi_1 = \frac{\mu_0 Ia}{2\pi}\int_a^{2a}\frac{dr}{r} = \frac{\mu_0 Ia}{2\pi}\ln 2.$$

Similarly, the flux due to wire 2 is

$$\Phi_2 = \frac{\mu_0 Ia}{2\pi}\int_{2a}^{3a}\frac{dr}{r} = \frac{\mu_0 Ia}{2\pi}\ln\frac{3}{2}.$$

As the two fluxes are in opposite directions the net flux through the loop is

$$\Phi = \Phi_1 - \Phi_2 = \frac{\mu_0 Ia}{2\pi}\ln\frac{4}{3}$$

pointing towards the page.

Therefore, the induced emf in the loop is

$$\mathcal{E} = -\frac{d\Phi}{dt} = -\frac{\mu_0 a}{2\pi}\ln\frac{4}{3}\cdot\frac{dI}{dt}.$$

According to Lenz's law the direction of induced current will be such as to oppose the change in flux, i.e., it must produce a flux pointing out from the page. So, in Fig. 10.P-10, the induced current will be anticlockwise as seen from the above.

▶ **Problem 11.** *Consider a long cylindrical conductor of radius a carrying a current I with uniform current distribution over its cross-section. Find the self-inductance per unit length by considering the flux linked or energy stored only within the material of the conductor. Assume the permeability of the material to be equal to* μ_0.

Solution. The magnetic field at a distance r from the axis is given by

$$B(r) = \frac{\mu_0 I}{2\pi a^2}r \quad \text{for } r < a$$

Considering a concentric cylindrical shell of radii r and $r + dr$ we can calculate the energy stored per unit length as

$$U = \frac{1}{2}\int \vec{H}\cdot\vec{B}\,dV = \frac{1}{2\mu_0}\int B^2\cdot 1.2\pi r dr = \frac{1}{2\mu_0}\cdot\frac{\mu_0^2 I^2}{4\pi^2 a^4}\int r^2\cdot 2\pi r dr = \frac{\mu_0 I^2}{16\pi}.$$

Since

$$U = \frac{1}{2}LI^2,\ L = \frac{2U}{I^2} = \frac{\mu_0}{8\pi}.$$

▶ **Problem 12.** *Consider a toroidal coil of N turns wound uniformly on a form of nonmagnetic material with square cross-section of side a. If the mean radius of toroid is R show by taking into account the variation of B over the cross-sectional area that its self-inductance is given by*

$$L = \frac{\mu_0 a N^2}{2\pi}\ln\frac{R + a/2}{R - a/2}.$$

Solution. Considering a circular loop of radius r the magnetic field at a distance r from the centre can be calculated as

$$\oint \vec{B}\cdot d\vec{l} = \mu_0 NI \quad \text{or} \quad B\cdot 2\pi r = \mu_0 NI \quad \text{or} \quad B = \frac{\mu_0 NI}{2\pi r},$$

where I is the current through the coil. Considering an elementary section of area $a\cdot dr$ we find that the total flux through the coil is

$$\Phi = \int NBadr = \frac{\mu_0 N^2 Ia}{2\pi}\int_{R-a/2}^{R+a/2}\frac{dr}{r} = \frac{\mu_0 N^2 aI}{2\pi}\ln\frac{R + a/2}{R - a/2}$$

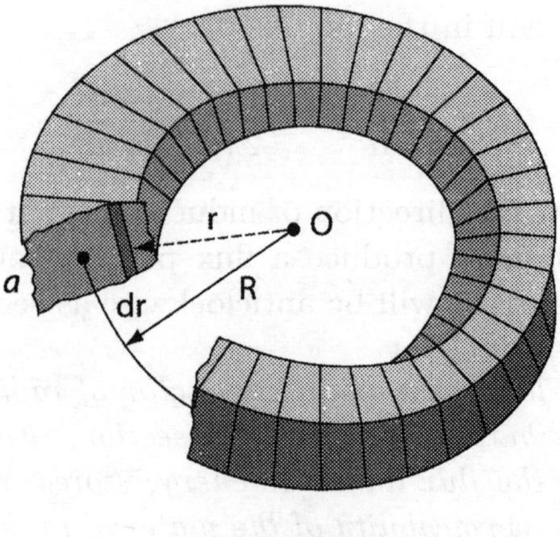

Fig 10.P-12

Therefore, self-inductance

$$L = \frac{\Phi}{I} = \frac{\mu_0 a N^2}{2\pi} \ln \frac{R + a/2}{R - a/2}.$$

► **Problem 13.** *In a certain region of space the electric field is given by*

$$\vec{E} = \hat{\jmath} E_0 \cos(\omega t - kx).$$

Using the differential form of Faraday's law, find the corresponding magnetic field $\vec{B}$.

Solution. We have

$$\vec{\nabla} \times \vec{E} + \frac{\partial \vec{B}}{\partial t} = 0$$

$$\text{or} \quad \frac{\partial \vec{B}}{\partial t} = -\vec{\nabla} \times \vec{E} = -\begin{vmatrix} \hat{\imath} & \hat{\jmath} & \hat{k} \\ \frac{\partial}{\partial x} & \frac{\partial}{\partial y} & \frac{\partial}{\partial z} \\ 0 & E_0 \cos(\omega t - kx) & 0 \end{vmatrix} = -\ \hat{k} E_0 k \sin(\omega t - kx).$$

Integrating,

$$\vec{B} = \hat{k} \frac{E_0 k}{\omega} \cos(\omega t - kx)$$

► **Problem 14.** *The primary of a standard solenoid consists of* 1000 *turns wound uniformly over a length of* 1 m *with mean cross-sectional radius of* 2 cm. *The secondary consists of* 400 *turns and is connected in series with a ballistic galvanometer. Total resistance of this secondary circuit is* 100 Ω. *On reversing a current of* 2 *A through the primary a deflection of* 10 cm *is obtained in the galvanometer. Find the charge sensitivity of the galvanometer.* **[C.U. 1997]**

Solution. From Eq. (10.20-3), charge flowing through the galvanometer is given by

$$q = \frac{2\mu_0 n_1 n_2 \alpha_1 I}{R}.$$

Here $n_1 = \frac{1000}{1}$ turns/m, $n_2 = 400$, $\alpha_1 = \pi(.02)^2\ \text{m}^2$, $I = 2\text{A}$ and $R = 100\,\Omega$

$$\therefore \quad q = \frac{2 \times 4\pi \times 10^{-7} \times 1000 \times 400 \times \pi(.02)^2 \times 2}{100}\ \text{C} = 25.6\,\mu\text{C}$$

This charges causes a deflection of 10 cm = 100 mm.

Therefore, charge sensitivity $S_q = \frac{100}{25.6}\ \text{mm}/\mu\text{C} = 3.9\,\text{mm}/\mu\text{C}$

▶ **Problem 15.** *A certain moving coil galvanometer with its scale at a distance of* 1 m *from the mirror has a current sensitivity of* 50 mm/μA. *Time period of its free oscillation is* 3.14 second. *The galvanometer becomes dead-beat when the total circuit resistance falls below* 2 kΩ. *Calculate the moment inertia of the moving system about the axis of suspension. Assume that the damping is entirely electromagnetic.*

[C.U. 1992]

Solution. If a steady current $i\,\mu$A produces a steady deflection θ radian then

$$i \times 10^{-6} = \frac{C}{NAB}\theta = \frac{C}{G} \cdot \frac{d \times 10^{-3}}{2D},$$

where d is the deflection in mm on a scale placed at a distance of D metre. Therefore, current sensitivity in mm/μA,

$$50 = \frac{d}{i} = \frac{2G}{C} \times 10^{-3} \qquad [\because\ D = 1\ \text{m}]$$

$$\text{or} \quad G = 25 \times 10^3\, C.$$

Now time period

$$T = 2\pi\sqrt{\frac{I}{C}} \quad \text{or} \quad C = 4\pi^2 I/T^2 = 4I.$$

Therefore, $G = 10^5 I$. From Eq. (10.19-12) we have for critical damping

$$R_c + R_g = 2000 = \frac{G^2}{2\sqrt{IC}} = \frac{G^2}{4I} \qquad (\because\ C = 4I)$$

$$\text{or} \quad I = \frac{G^2}{8000} = \frac{10^{10} I^2}{8000}$$

$$\text{or} \quad I = 8 \times 10^{-7}\ \text{kg-m}^2.$$

► **Problem 16.** *A search coil having 60 turns and face area of* 3 mm^2 *is placed between the poles of an electromagnet with its plane being perpendicular to the direction of magnetic field. When the coil is turned through* 180° *about a diameter, a ballistic galvanometer connected to the coil indicates a charge of* 4.5μC *flowing through it. Find the magnetic field at the location of the coil. Assume that the total resistance in the galvanometer circuit is* 40 Ω.

Solution. Change in flux through the coil is $2 \times nAB$, where n =number of turns and A =area.

$\therefore$ Charge flowing through the galvanometer is

$$q = \frac{2nAB}{R} \quad \text{or} \quad B = \frac{qR}{2nA} = \frac{4.5 \times 10^{-6} \times 40}{2 \times 60 \times 3 \times 10^{-6}}\,\text{T} = 0.5\,\text{T}.$$

QUESTIONS

1. (a) State Faraday's laws of electromagnetic induction. Express it in integral and differential forms. What is Lenz's law? **[cf. C.U. 2016, 2019]**

   (b) Show that in an electromagnetic field we can write $\vec{E} = -\vec{\nabla}\phi - \frac{\partial \vec{A}}{\partial t}$, where the symbols have their usual significance. OR, **[cf. B.U. 2012; J.U. 2014]**
   Obtain the expression of electric field of an electromagnetic field in terms of scalar and vector potentials. **[B.U. 2011]**

2. (a) What do you mean by motional emf? Show that the motional emf generated in a loop of arbitrary shape moving in a static magnetic field is given by Faraday's law of electromagnetic induction.

   (b) Find an expression for emf induced in a vertical loop of wire placed in a horizental magnetic field and rotating with constant angular velocity about a vertical axis.

3. What do you mean by eddy current? How can you reduce the losses due to eddy current? Mention a few cases where eddy current can be used to our advantages.

4. (a) What do you mean by self and mutual inductances? Self-inductances of two coils are L_1 and L_2, respectively and their mutual inductance is M. Show from energy consideration that in general $M^2 \leq L_1 L_2$. Define the coefficient of coupling of two circuits. **[B.U. 2017]**

   (b) Two coils of self-inductances L_1 and L_2 are connected in parallel. If their mutual inductance is M show that the effective/equivalent self-inductance is given by

$$L_{\text{ef}} = \frac{L_1 L_2 - M^2}{L_1 + L_2 \mp 2M}$$ **[C.U. 2013, 2016]**

5. State and prove the reciprocity theorem in mutual inductance. Derive Neumann's formula for the mutual inductance between two arbitrary loops. Discuss the result.

6. (a) What is a solenoid? Find an expression for the coefficient of self-inductance of a long solenoid. **[cf. C.U. 2012]**

(b) Two long parallel wires carrying the same current I in the opposite direction are separated by a distance d in air. The lengths of the wires are much larger than d. Find the self-inductance per unit length. **[C.U. 2011, 2017]**

(c) Determine the self-inductance of a coaxial cable. **[C.U. 2016]**

7. Find the mutual inductance:

(a) between the primary and secondary of a standard solenoid,

(b) between two parallel coaxial loops, one loop being small compared with the second loop.

8. Two coils having self-inductances L_1 and L_2 and mutual inductance M between them are joined in (a) series (b) parallel. Find the effective self-inductance.

9. (a) Show that the magnetic energy stored in an inductor is $\frac{1}{2}LI^2$, where I is the current through it.

(b) Show that the energy density in a magnetic field is given by $u = \frac{1}{2}\vec{H}\cdot\vec{B}$.

10. Two coils with self inductances L_1 and L_2, respectively, have mutual inductance M. Find an expression for their coefficient of coupling (k). **[C.U. 2011]**

11. (a) What is a ballistic galvanometer? How does it differ from a dead-beat galvanometer?

(b) Work out the theory of working of a suspended coil type ballistic galvanometer. Explain the meaning of critical damping and log decrement. **[N.B.U. 2017]**

12. Briefly discuss how a magnetic field can be measured by using a search coil and a ballistic galavanometer. You may use a standard solenoid to calibrate the galvanometer.

13. Show that the self-inductance of a long solenoid of length l, radius a and with n turns per unit length is approximately given by

$$L = \mu_0 n^2 \cdot \pi a^2 \left[\sqrt{l^2 + a^2} - a\right]$$

EXERCISES

1. A uniform magnetic field $\vec{B}$ is suddenly established normal to the plane of a coil of N turns and area of cross-section A. Show that the total charge that would flow through the coil is NAB/R, where R is the resistance of the coil.

2. A coil of N turns, cross-sectional area A and resistance R is rotated with uniform angular velocity ω about a diameter in a uniform magnetic field B directed normal to the axis of rotation. Show that the mean power required to maintain the coil in motion is $N^2A^2B^2\omega^2/2R$.

3. A straight conductor of length l is placed at right angles to a uniform magnetic field $\vec{B}$. The conductor now moves with a constant velocity $\vec{v}$ at right angles to its own length and also to the magnetic field. From a detailed consideration of the Lorentz force on the electrons show that the emf induced between the ends of the conductor is Bvl. **[cf. C.U. 2011]**

4. A conducting rod of length l rotates with a constant angular velocity ω about an axis passing through one end and perpendicular to its length in a uniform magnetic B which acts at right angles to the rod. Calculate the emf developed between the ends of the rod. [*Ans.* $\frac{1}{2}\omega Bl^2$]

5. The rails of a railway track are 1.5 m apart. Calculate the emf that would be generated between the rails when a train passes at a constant speed 100 km/hr. Assume that the horizontal component of the earth's magnetic field is 36 μT and angle of dip $\theta = 45°$. [*Ans.* 1.5 mV]

6. (a) The magnetic field in a certain region is given by $\vec{B} = a_0 xy^2 \hat{k}$ Wb/m^2, where a_0 is a constant. Find the magnetic flux through the portion of xy-plane lying between the lines $x = 0$, $x = 1$, $y = 0$ and $y = 1$. [*Ans.* $a_0/6$ Wb]

   (b) In cylindrical coordinates $\vec{B} = (2/r)\hat{\phi}$ T. Determine the magnetic flux crossing the plane surface defined by $0.5 \le r \le 2.5$ m and $0 \le z < 2.0$ m. **[C.U. 2016]**

   [**Hints.** $\phi = \int \vec{B} \cdot d\vec{S} = \int_{z=0}^{2} \int_{r=0.5}^{2.5} \left(\frac{2}{r}\right) \cdot dr\, dz = 4 \ln 5 = 6.437\,\text{Wb}$]

7. A circular conducting disc of radius 25 cm is rotating 50 times per second about its own axis. If the plane of the disc is perpendicular to a uniform magnetic field of 7 mT, calculate the current through a wire of resistance 22 Ω when its two ends touch the centre and the rim of the disc. [*Ans.* 3.125 mA]

8. A large electromagnet has an inductance of 5 H. Suppose the current flowing through the electromagnet is interrupted and it drops from 10 A to 1 A in 1/20 second. Estimate the mean back emf generated. [*Ans.* 900 V]

9. A square loop of side a is so placed that its top side is in a uniform horizontal magnetic field B normal to the plane of the loop and the lower side is outside the field and horizontal. The resistance of the loop is R and its mass is m. If the loop is allowed to fall under the action of gravity what would be the terminal velocity of the loop? [*Ans.* mgR/a^2B^2]

10. Consider a wire in shape of a parabola $y = Kx^2$ placed in a uniform magnetic field $\vec{B}$ perpendicular to the xy-plane. A straight conducting wire is placed on the parabolic wire with its length parallel to x-axis. If now this straight wire moves with constant acceleration a starting from rest at the apex of the parabola, find the emf induced in the parabolic wire as a function of y.

[**Hints.** $d\Phi = B \cdot 2xdy$. $\quad \therefore \mathcal{E} = -\dfrac{d\Phi}{dt} = -2B\sqrt{\dfrac{y}{K}}\dfrac{dy}{dt}$.

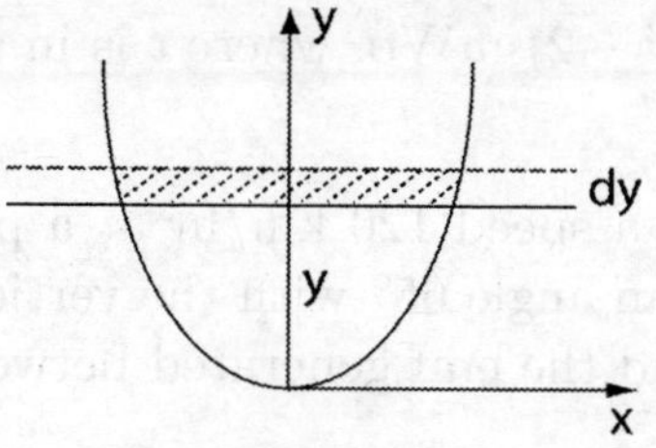

Fig 10.E-10

For constant acceleration a,

$$\left(\frac{dy}{dt}\right)^2 = 0^2 + 2 \cdot a \cdot y \quad \text{or} \quad \frac{dy}{dt} = \sqrt{2ay}.$$

Thus, $\mathcal{E} = -By\sqrt{8a/K}$.]

11. Suppose a square loop of side a is placed in the plane of a long straight wire carrying current I. The nearest side of the loop is at a distance r from the wire (Fig. 10.P-5). If the square loop is rotated through 180° about its side farthest from the straight wire show that the quantity of charge that passes through the loop is given by

$$\frac{\mu_0 Ia}{2\pi R} \ln \frac{r+2a}{r}.$$

12. Two long parallel rails of negligible resistance separated by a distance d are shunted at one end through a resistor R. A conducting bar of mass m placed across the rails can slide without friction along the rails. An external uniform magnetic field $\vec{B}$ is applied perpendicular to the plane of the circuit. The conducting bar is given an initial velocity v_0 along the rails away from the closed end. Neglecting resistance of the conducting bar find its velocity as a function of time. Also, find the distance travelled by the bar before coming to rest. Where does the initial kinetic energy $\frac{1}{2}mv_0^2$ of the bar go?

[*Ans.* $v = v_0 e^{-\frac{B^2d^2}{mR}\cdot t}$, $\frac{v_0 mR}{B^2d^2}$, lost as Joule heat in the circuit]

13. A conducting rod of length l rotates with a constant angular velocity ω about an axis perpendicular to its length and passing through a point on the rod, which is at a distance $l/3$ from one end. A uniform magnetic field $\vec{B}$ is applied parallel to the axis of rotation. Show that the emf induced between the ends of the rod is $B\omega l^2/6$.

14. The current through a long solenoid having n turns per unit length changes with time as $i(t) = Kt$, where K is a constant. Find the magnetic and electric field inside the solenoid. [*Ans.* $\vec{B} = \mu_0 nKt\hat{z}, \vec{E} = -\frac{\mu_0 nKr}{2}\hat{\theta}$]

15. The flux through each turn of a 100 turn coil varies with time in accordance with the equation $\Phi = t(t^2 - 2)$ mWb, where t is in seconds. Find the induced emf at $t = 2$ sec. [*Ans.* -1 V]

16. A train is moving at a speed 120 km/hr at a place where earth's magnetic field is 43 μT and makes an angle 65° with the vertical. If the separation between the two rails is 1.6 m, find the emf generated between the two wheels of the train.

[*Ans.* 0.97 mV]

17. Consider a V-shaped rail lying in the xy-plane with one arm along x-axis and the other making an angle 45° with it. A jumper starts from rest at $t = 0$ from the origin and slides along the rail with a velocity $v\hat{i}$. If there is a uniform magnetic field $B\hat{k}$ find the emf induced in the V-shaped rail as a function of time t.

[*Ans.* $-Bv^2t$]

18. Consider a circular coil of N_1 turns and radius a carrying a current I placed on a horizontal table. A small circular coil of N_2 turns and radius $b(b \ll a)$ is placed coaxial at a vertical distance x from the centre of the coil. Find an expression for the induced emf when the loop is allowed to fall freely. **[C.U. 2003, 2006]**

[**Hints.** See Eq. (10.12-2); $\mathcal{E} = -\frac{d\Phi_2}{dt} = -\frac{3\mu_0\pi N_1 N_2 a^2 b^2 I}{2\left(a^2 + x^2\right)^{5/2}} \cdot \frac{dx}{dt}$.]

19. An iron ring of mean radius a and cross-sectional area A is uniformly wound with N turns of insulated wire. There is a small air gap of length Δl in the ring. Find the magnetic field energy stored in the iron and the air gap. Hence show that the self-inductance of the system is given by

$$L = \frac{N^2 A}{\frac{2\pi a - \Delta l}{\mu} + \frac{\Delta l}{\mu_0}},$$

where μ is the permeability of iron.

[**Hints.** Using the idea of magnetic circuit show that

$$B = \frac{NA}{\frac{2\pi a - \Delta l}{\mu} + \frac{\Delta l}{\mu_0}}.$$

Energy within the iron $= \frac{1}{2} \cdot \vec{B} \cdot \vec{H} \times \text{volume} = \frac{B^2}{2\mu} \cdot (2\pi a - \Delta l)A$.

$$\text{Energy within the air gap} = \frac{B^2}{2\mu_0} \cdot \Delta l A.$$

$$\therefore \text{Total energy} = \frac{B^2 A}{2}\left(\frac{2\pi a - \Delta l}{\mu} + \frac{\Delta l}{\mu_0}\right) = \tfrac{1}{2}LI^2.]$$

20. A circular loop of radius a, made of insulated wire, is placed between two long straight parallel wires separated by a distance $2a$. Assuming that each wire carries the same current but in opposite directions, calculate the mutual inductance of the system. [*Ans.* $2\mu_0 a$]

21. The mutual inductance between two coils is 10 mH. If the current through one coil changes from 2.5 A to 0.5 A in 0.1 sec, what would be the emf induced in the second coil? **[J.U. 2014]** [*Ans.* 0.2 V]

22. An air-cored solenoid with length 1 m, radius 2 m has 1000 turns. Calculate its self-inductance. [*Ans.* 1.6 mH]

23. Suppose a square loop of side a is placed in the plane of a long straight wire carrying current I. The nearest side of the loop is at a distance r from the wire. Find the mutual inductance. [*Ans.* $\frac{\mu_0 a}{2\pi} \ln \frac{a+r}{r}$]

24. A toroidal coil of N turns is wound on a ferromagnetic core having mean radius 0.5 m and cross-sectional area 10 cm^2. The permeability of the core is 200 μ_0. Find the value of N required to obtain an inductance of 2.5 H. [*Ans.* 5590 turns]

25. Two coaxial circular loops 1 and 2 lying in the same plane have radii r_1 and r_2 ($r_1 \ll r_2$). Find the magnetic flux linked with loop 1 when a current I flows in loop 2 (neglect field variation over the small loop). Will this be equal to the flux in the loop 2 due to a current I in loop 1? [*Ans.* $\frac{\mu_0 \pi I r_1^2}{2r_2}$]

[**Hints.** Use reciprocity theorem.]

26. Consider a flat coil with large number of turns wound in the form of a spiral of external radius a. It is placed in a uniform magnetic field $B = B_0 \sin \omega t$ normal to the plane of the coil. If the number of turns per unit radial distance is n show that the maximum value of emf induced in the coil is $\frac{1}{3}\pi n B_0 \omega a^3$.

27. Consider a rail in the form of a right-angle with one arm along x-axis and the other along y-axis. A rod placed on the rail slides at a constant velocity $\vec{v} = v_0\left(\hat{i} + \hat{j}\right)$. There is a uniform magnetic field $B\hat{k}$. Assuming that the rod starts from rest at the origin at time $t = 0$, find the induced emf as a function of time.

[*Ans.* $-2Bv_0^2 t$]

28. (a) A long coaxial cable carries a current I. The current flows down the surface of the inner cylinder of radius a and back along the outer cylinder of radius b.

Find the magnetic energy stored in a section of length l. Hence find the self-inductance.

[**Hints.** $\displaystyle\int_a^b \frac{1}{2\mu_0}\left(\frac{\mu_0 I}{2\pi r}\right)^2 \cdot 2\pi r l dr = \frac{\mu_0 I^2 l}{4\pi}\ln(b/a) = \frac{1}{2}LI^2. \quad \therefore\ L = \frac{\mu_0 l}{2\pi}\ln(b/a)$]

(b) Consider a long coaxial cable consisting of an inner solid cylinder of radius a and a thin outer cylinder of radius b. The same current passes through the cylinders but in opposite directions. The current through the inner cylinder is distributed uniformly over its cross-section. Calculate the total magnetic energy stored per unit length and hence show that self-inductance per unit length is $L = \frac{\mu_0}{2\pi}\left(\frac{1}{4} + \ln\frac{b}{a}\right)$ [*Ans.* $\frac{\mu_0 I^2}{4\pi}\left(\frac{1}{4} + \ln\frac{b}{a}\right)$]

29. Suppose the radius r of a circular loop increases linearly with time as $r = vt$, where v is the constant radial velocity of a point on the loop. There is a uniform magnetic field B perpendicular to the plane of the loop. If B varies with time t as $B = B_0(1 + kt)$, show that the emf induced in the coil is given by

$$\mathcal{E} = -\pi B_0 v^2(2 + 3kt).$$

30. A circular loop described by the equation $x^2 + y^2 = 1$ is located in the xy-plane with it centre at the origin. If a magnetic field $\vec{B} = \hat{z}\sqrt{x^2 + y^2}\cos\omega t$ is applied what would be the emf induced in the loop?

[**Hints.** $\displaystyle\mathcal{E} = -\frac{d}{dt}\int_S \vec{B}\cdot d\vec{S} = -\int_S \frac{\partial\vec{B}}{\partial t}\cdot d\vec{S} = \iint \omega\sin\omega t\sqrt{x^2 + y^2}\,dxdy$

$\displaystyle = \omega\sin\omega t\int_{r=0}^{1}\int_{\theta=0}^{2\pi} r\cdot r d\theta dr$ (Using cylindrical coordinates)

$\displaystyle = \frac{2\pi}{3}\omega\sin\omega t.$]

31. In a certain region of space the magnetic field varies with time in accordance with the law $\vec{B} = \hat{z}B_0 e^{\alpha t}$, where B_0 and α are constants. Find the corresponding electric field $\vec{E}$. [*Ans.* $\vec{E} = -\hat{\theta}\frac{\alpha}{2}B_0 e^{\alpha t}\cdot r$ in cylindrical coordinates]

32. A coil of self-inductance 100 mH is connected in series with another coil of self-inductance 170 mH. The effective self-inductance of the combination is found to be 70 mH. Find the coefficient of coupling between the two coils. [*Ans.* 0.76]

33. A coil of self-inductance 1 H and resistance 10 Ω is connected to a dc source of 10 V. Find the energy stored in it after the current reaches its maximum value. What is the magnetic flux linked with the coil? [*Ans.* 0.5 J; 1 Wb]

34. Two coils have self-inductances 1 H and 4 H and a coefficient of coupling 0.9 between them. If current through one coil changes at the rate 10 A/s what would the emf induced in the second coil? [*Ans.* 18 V]

35. Find the mutual inductance between two coplanar concentric circular loops of radii 1 m and 0.1 m. [*Ans.* 0.02μ H]

36. A ballistic galvanomater gives a deflection of 10 cm when a steady current of 20 μA is passed through it. The open circuit time period of its moving part is 8 s. Find the value of transient charge which causes a ballistic throw of 10 cm.

[*Ans.* $25.4\mu C$]

37. The coil of a ballistic galvanometer has 30 turns and is suspended by phosphor-bronze strip whose torsion constant is 2×10^{-7}N-m/rad, in a radial magnetic field of 0.1 T. The coil is 2 cm wide, 2.5 cm long and has a moment of inertial 1.5×10^{-7}kg-m^2 about the axis of rotation. If the galvanometer resistance is $R_g = 100\,\Omega$, find the external CDR. [*Ans.* $549.5\,\Omega$]

38. A toroidal coil of N turns wound uniformly on a form of nonmagnetic material with circular cross-section of radius a. If the mean radius of the toroid is R show by taking into account the variation of $\vec{B}$ over the cross-sectional area, that its self-inductance is given by $L = \mu_0 N^2(R - \sqrt{R^2 - a^2})$.

Given

$$\int_{-a}^{+a} \frac{\sqrt{a^2 - x^2}}{R + x} dx = \pi(R - \sqrt{R^2 - a^2}).$$

39. An infinite number of surfaces can be constructed, which are bounded by the same closed boundary line. Show that the magnetic flux through this closed line is independent of the choice of the surface.

[**Hints.** $\vec{\nabla} \cdot \vec{B} = 0.\ \therefore\ \int_V \left(\vec{\nabla} \cdot \vec{B}\right) dV = \oint_S \vec{B} \cdot d\vec{S} = 0$ or $\int_{S_1} \vec{B} \cdot d\vec{S} = -\int_{S_2} \vec{B} \cdot d\vec{S},$

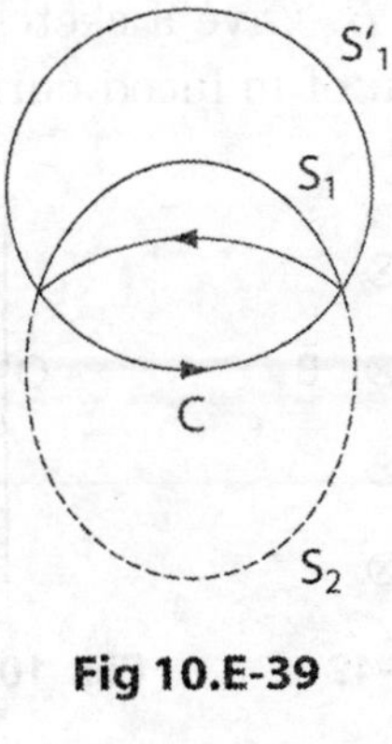

Fig 10.E-39

where S_1 and S_2 together forms the closed surface S with a common boundary C (Fig. 10.E-39). So, keeping S_2 fixed if S_1 is changed to a new surface S_1' having the same contour, the integral on the left-hand side cannot change because right-hand integral is fixed.]

40. A long solenoid is driven by an ac so that the axial field inside varies with time according to $B(t) = B_0 \cos \omega t$. A small circular loop of radius a and resistance R is placed coaxial inside the solenoid. Find the current induced in the loop as a function of time. [*Ans.* $\frac{\pi a^2 \omega B_0}{R} \cdot \sin \omega t$]

41. (a) A square loop of wire of side a lies in the first quadrant of the xy-plane with one corner at the origin. In this region there is a nonuniform time varying magnetic field given by $\vec{B} = Ky^3t^2\hat{k}$, where K is a constant. Find the emf induced in the loop as a function of time.

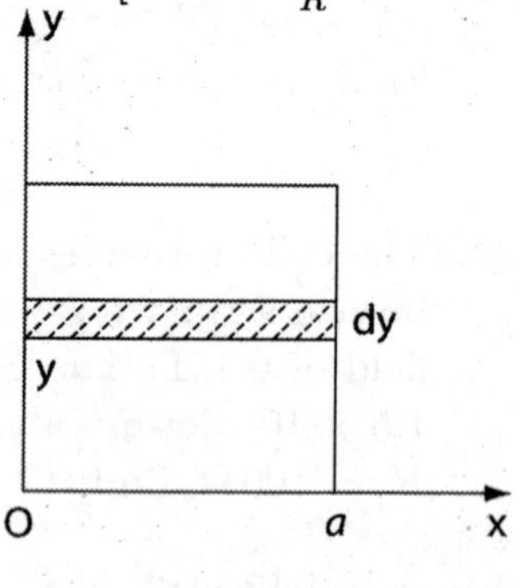

Fig 10.E-41

[**Hints.** Total flux linked with the loop at time t is

$$\Phi(t) = \int_0^a B \cdot a dy = Kat^2 \int_0^a y^3 dy = \frac{Ka^5}{4}t^2.$$

$$\therefore\ \mathcal{E} = -\frac{d\Phi}{dt} = -\frac{1}{2}Ka^5t.]$$

(b) An infinitely long wire carrying current $i = I_0 \cos \omega t$ is placed at a distance 'a' from a square loop of side 'a'. If the resistance of the loop is R, then determine the rms value of the induced current in the loop. **[C.U. 2018]**

[**Hints.** See Solved Problem 5.

$$\Phi = \frac{\mu_0 I a}{2\pi} \ln 2$$

$$\text{Induced current} = \frac{\mathcal{E}}{R} = -\frac{1}{R}\frac{d\Phi}{dt} = \frac{\mu I_0 a \omega \ln 2}{2\pi R} \sin \omega t]$$

42. A semicircular wire loop of radius a is located at the boundary of a uniform magnetic field $\vec{B}$ as shown in Fig. 10.E-42. At time $t = 0$ the loop is set rotating in the clockwise direction at a constant angular velocity ω about an axis passing through O and parallel to $\vec{B}$. Give a sketch of the induced emf as a function of time. Mention the direction of induced current.

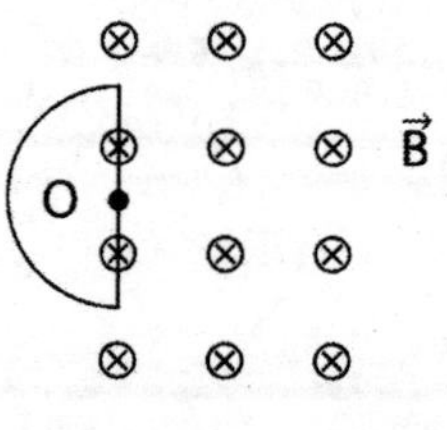

Fig. 10.E-42

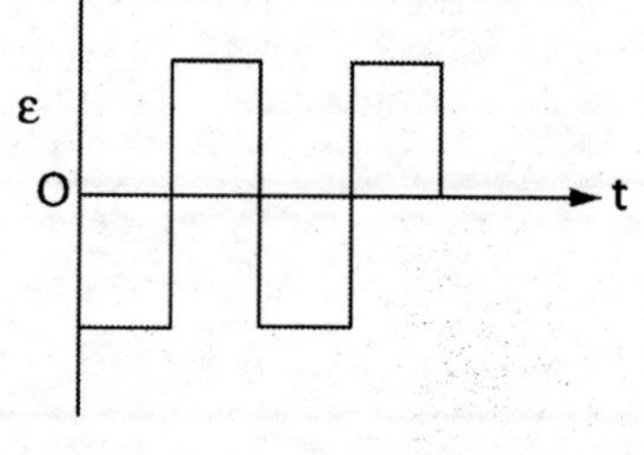

Fig. 10.E-42 (a): Nature of induced emf.

[**Hints.** If the loop rotates through an angle $\theta (< \pi)$ in time t then

$$\Phi(t) = B \cdot \frac{1}{2}a^2\theta = \frac{1}{2}Ba^2\omega t$$

$$\therefore\ \mathcal{E} = -\frac{1}{2}B\omega a^2.$$

As the flux is increasing the induced emf will produce a current in the anticlockwise sense so that the magnetic field due to it opposes the original field.

After half a rotation the flux linked with the loop begins to decrease and hence the induced emf will have the same magnitude but opposite sense.]

43. The vector potential $\vec{A}$ and the scalar potential ϕ in a certain region of space are given by

$$\vec{A} = \tfrac{1}{2}\alpha t(-y\hat{i} + x\hat{j}); \quad \phi = \tfrac{1}{4}\alpha(x^2 + y^2)$$

where α is a constant. Find the electric and magnetic fields corresponding to these potentials. **[B.U. 2011]**

[**Hints.** $\vec{B} = \vec{\nabla} \times \vec{A}$ and $\vec{E} = -\vec{\nabla}\phi - \frac{\partial \vec{A}}{\partial t}$]

44. A coil of insulated wire of 10^4 turns is wound on wooden core 1 m long and 2 cm in diameter. Another coil of 500 turns is wound closely round the middle of the first coil. Calculate the mutual inductance between the coils. If the flux through the first coil changes from $+10^{-3}$ Wb perturn to -10^{-3} Wb per turn in 0.1 s, determine the emf induced in the second coil.

[C.U. 2014] [*Ans.* 1.97 mH, 10 V]

OBJECTIVE TEST

1. The electric field produced by a uniformly moving charge is

(a) conservative,
(b) non-conservative,
(c) irrotational,
(d) none of these.

2. A conducting rod of length l rotates with a constant angular velocity ω about its perpendicular bisector. An external uniform magnetic field $\vec{B}$ is applied parallel to the axis of rotation. The pd between the two ends of the rod is

(a) $\frac{1}{2}B\omega l^2$,
(b) $B\omega l^2$,
(c) $\frac{1}{4}B\omega l^2$,
(d) zero.

3. A small bar magnet is allowed to fall freely from rest along the axis of a very long vertical conducting tube. After sometime the magnet will be found to

(a) come to rest,
(b) move with an acceleration g,
(c) oscillate,
(d) move with a constant speed.

[JEST 2012]

4. A small wire loop is placed inside a solenoid with its axis perpendicular to the axis of the solenoid. Now, if the current through the solenoid increases with time at a constant rate the emf induced in the loop would

(a) increase with time,
(b) decrease with time,
(c) remain at a constant value,
(d) be zero.

5. The differential form of Faraday's law of electromagnetic induction is

(a) $\vec{\nabla} \times \vec{E} = -\dfrac{\partial \vec{B}}{\partial t}$,
(b) $\vec{\nabla} \times \vec{B} = -\dfrac{\partial \vec{E}}{\partial t}$,
(c) $\vec{\nabla} \times \vec{B} = \mu_0 \vec{J}$,
(d) $\vec{\nabla} \times \vec{H} = \vec{J} + \dfrac{\partial \vec{D}}{\partial t}$.

6. A circular coil of a single turn of thin conducting wire has self-inductance L. If the number of turns is increased to 8 the self-inductance would be

(a) $8L$,
(b) $L/8$,
(c) $64L$,
(d) $2\sqrt{2}L$.

7. A vertical loop placed in a horizontal varying magnetic field is rotating about a vertical diameter. The emf induced in the coil may be identified as

(a) a motional emf only,
(b) a transformer emf only,
(c) a combination of motional and transformer emfs,
(d) none of these.

8. Lenz's law is a consequence of the law of conservation of

(a) charge,
(b) momentum,
(c) energy,
(d) mass.

9. What are the dimensions of $\Delta\Phi/R$, where $\Delta\Phi$ is the change in magnetic flux and R is the resistance?

(a) $[IT]$,
(b) $[IT^{-1}]$,
(c) $[MLT^{-3}I^{-1}]$,
(d) $[ML^2T^{-2}I^{-1}]$.

10. For null-point work a galvanometer should have

(a) high resistance,
(b) low resistance,
(c) any resistance,
(d) high current sensitivity.

11. A circular loop of radius r is lying in the xy-plane with its centre at the origin. If there exists a time varying magnetic field $\vec{B}(t) = B_0 e^{-\alpha t}\hat{z}$ $(B_0 > 0, \alpha > 0)$ then the induced emf in the loop is

(a) $-\pi r^2 \alpha B_0 e^{-\alpha t}$, (c) $-\pi r^2 B_0 e^{-\alpha t}$,

(b) $\pi r^2 B_0 e^{-\alpha t}$, (d) $\pi r^2 \alpha B_0 e^{-\alpha t}$.

12. The current passing through a choke coil of 1 H is decreasing at the rate of 2A/s. Then, the emf developed in the coil is

(a) 2 V, (b) −2 V, (c) 0.5 V, (d) −0.5 V.

13. A conductor of length 0.4 m is moving with a speed 5 m/s perpendicular to a magnetic field of 1 T. The emf induced across the conductor is

(a) 2 V, (b) 1 V, (c) 0.2 V, (d) 0.1 V.

14. A copper ring is held horizontally and a bar magnet is dropped through the ring with its length along the axis of the ring. Then, the acceleration of the falling magnet is

(a) equal to g, (c) greater than g,

(b) less than g, (d) zero.

15. The equation $\vec{\nabla} \times \hat{H} = \vec{J} + \frac{\delta \vec{D}}{\delta t}$ is the differential form of

(a) Faraday's law, (c) Gauss's law,

(b) Ampere's law, (d) Coulomb's law.

16. A circular conducting ring of radius 'a' rotates with constant angular velocity ω about its diameter placed along the x-axis. A uniform magnetic field B is applied along the y-axis. If at time $t = 0$ the ring is entirely in the xy-plane, the emf induced in the ring at time $t > 0$ would be

(a) $B\omega^2 \pi a^2 t$, (c) $B\omega\pi a^2 \tan \omega t$

(b) $B\omega\pi a^2 \sin \omega t$, (d) $B\omega\pi a^2 \cos \omega t$ **[JEST 2012]**

17. A conducting loop of area $1\,\text{cm}^2$ and resistance 1Ω is placed in a perpendicular time varying magnetic field $B(t) = 2e^{-0.5t}\cos(2\pi t)$ tesla. Then net charge that flows through any section of the conductor from $t = 0$ to $t = \infty$ is

(a) 0, (b) $100\,\mu$C, (c) $200\,\mu$C, (d) $300\,\mu$C.

[**Hints.** $Q = \dfrac{\Phi_1 - \Phi_2}{R}$]

ANSWERS

1. (b)	**2.** (d)	**3.** (d)	**4.** (d)	**5.** (a)	**6.** (c)	**7.** (c)
8. (c)	**9.** (a)	**10.** (b)	**11.** (d)	**12.** (a)	**13.** (a)	**14.** (b)
15. (b)	**16.** (b)	**17.** (c)				

Chapter **11**

Transient Phenomena in Electric Circuits

11.1 Introduction

If an electrical circuit containing inductors or capacitors is suddenly connected to a source of steady voltage the current in the inductors or charge on the capacitors does not attain the final steady value instantaneously. The attainment of steady state is somehow delayed. Similarly, if the source is suddenly disconnected the current or the charge does not fall to zero immediately but the process takes some time. During the time interval between switching operation and the final steady state of current or charge, the current and voltage in the circuit change with time in a non-periodic manner. Such variations usually exist for a short time and this is called the *transient phenomenon* or simply *transients in electric circuits.* Obviously no transients can be produced in a purely resistive circuit. The purpose of this chapter is to investigate the transient response of electric circuits excited by constant voltages.

11.2 Growth and Decay of Currents in Series LR Circuit

(i) Growth of current

Fig. 11.2-1 shows a circuit containing an inductor L and a resistor R connected in series to a source of constant voltage V via a switch K. Suppose the switch K is suddenly thrown to the position 1 at time $t = 0$. As the current in the circuit begins to grow the magnetic flux linked with the inductor also increases and consequently according to Faraday's law of induction a back emf $-L\frac{di}{dt}$ is generated in the circuit which opposes the driving emf V and thus, delays the process of the growth of current.

So the net emf in the circuit is $V-L\frac{di}{dt}$ and hence, from Ohm's law, the instantaneous current

$$i = \frac{V - L\frac{di}{dt}}{R}, \quad \text{or} \quad L\frac{di}{dt} + Ri = V, \quad \text{or} \quad \frac{di}{i - V/R} = -\frac{R}{L}dt \tag{11.2-1}$$

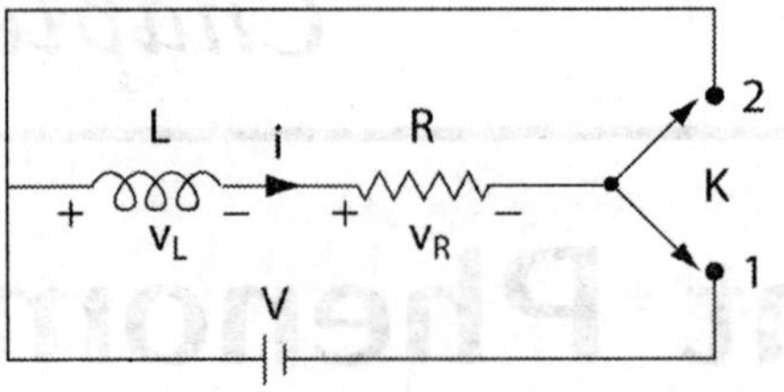

Fig 11.2-1

Assuming that $i = 0$ at $t = 0$ we get on integration,

$$\int_0^i \frac{di}{i - V/R} = -\frac{R}{L}\int_0^t dt$$

$$\text{or} \quad \ln\frac{i - V/R}{-V/R} = -\frac{R}{L}t$$

$$\text{or} \quad i = \frac{V}{R}\left(1 - e^{-\frac{R}{L}t}\right) \tag{11.2-2}$$

$$= I_0\left(1 - e^{-\frac{R}{L}t}\right), \tag{11.2-3}$$

where $I_0 = V/R$ is the maximum steady current which is reached when $t \to \infty$. Equation (11.2-3) describes the growth of current in a series LR circuit. Though theoretically the current reaches the steady value only after a long time, but in fact, for all practical purposes the steady value is attained in a very short time. The rate of growth of current is

$$\frac{di}{dt} = I_0\frac{R}{L}e^{-\frac{R}{L}t} = \frac{R}{L}(I_0 - i). \tag{11.2-4}$$

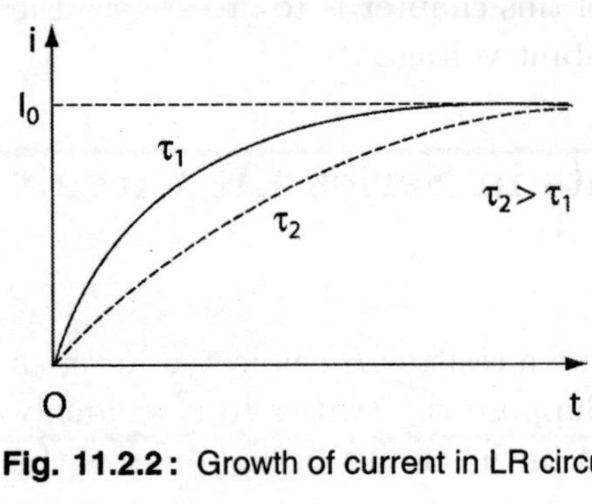

Fig. 11.2.2: Growth of current in LR circuit

It depends on the value of R/L. The quantity $L/R = \tau$ has the dimension of time and is called the *time constant of the circuit.* At $t = \tau$, the current is

$$i = I_0\left(1 - \frac{1}{e}\right) \approx 0.63I_0.$$

So *the time constant* τ *may be defined as the time in which the current in the circuit grows from zero to 63% of its final steady value.* Greater is the value of τ (i.e., the greater the L and the smaller the R) longer is the time taken to attain the steady state.

When the current reaches its maximum steady value the variation of current with time stops and so the back emf becomes zero. At this stage the circuit behaves as if there were no inductor in the circuit. The instantaneous voltage drops across L and R are given by

$$v_R = Ri = RI_0\left(1 - e^{-\frac{R}{L}t}\right) \tag{11.2-5}$$

$$\text{and } v_L = L\frac{di}{dt} = RI_0 e^{-\frac{R}{L}t}. \tag{11.2-6}$$

The variations of v_R and v_L with time are shown in Fig. 11.2-3.

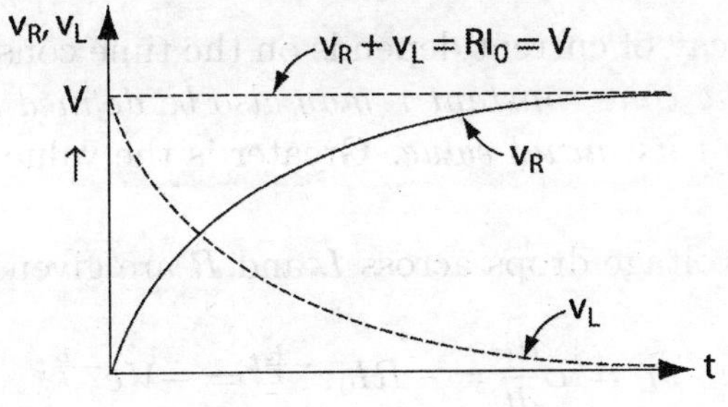

Fig 11.2-3

(i) Decay of current

Suppose a steady current has been set up in the LR circuit of Fig. 11.2-1 by connecting the switch K to the position 1 for a sufficient time. Now the key is suddenly thrown from position 1 to position 2. As a result, the source is cut off and current in the circuit begins to fall. The flux linked with the inductor decreases with time and consequently an emf $-L\frac{di}{dt}$ is generated which tries to oppose the decay of current. In this case, $V = 0$ and the instantaneous current

$$i = \frac{0 - L\frac{di}{dt}}{R}, \quad \text{or} \quad L\frac{di}{dt} + Ri = 0, \quad \text{or} \quad \frac{di}{i} = -\frac{R}{L}dt. \tag{11.2-7}$$

Assuming that $i = I_0$ at $t = 0$, we get on integration,

$$\int_{I_0}^{i} \frac{di}{i} = -\frac{R}{L}\int_0^t dt, \quad \text{or} \quad \ln\frac{i}{I_0} = -\frac{R}{L}t, \quad \text{or} \quad i = I_0 e^{-\frac{R}{L}t}. \tag{11.2-8}$$

This equation describes the decay of current in a series LR circuit after the sudden removal of the driving source. Obviously the current decays exponentially with time

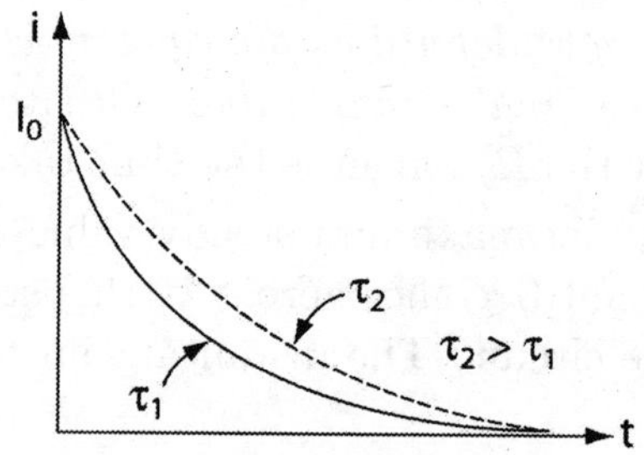

Fig. 11.2-4: Decay of current in LR circuit

(Fig. 11.2-4). The rate of decay of current is

$$\frac{di}{dt} = -\frac{R}{L}I_0e^{-\frac{R}{L}t} = -\frac{I_0}{\tau}e^{-t/\tau}, \quad \text{or} \quad \left.\frac{di}{dt}\right|_{t=0} = -\frac{I_0}{\tau}. \tag{11.2-9}$$

Thus, the rate of decay of current depends on the time constant $\tau = L/R$. At $t = \tau$, $i = I_0/e \approx 0.37I_0$. *So the time constant τ may also be defined as the time in which the current decays to 37% of its initial value.* Greater is the value of τ slower is the decay of current.

The instantaneous voltage drops across L and R are given by

$$v_L = L\frac{di}{dt} = -RI_0e^{-\frac{R}{L}t} = -Ve^{-\frac{R}{L}t} \tag{11.2-10}$$

$$v_R = Ri = RI_0e^{-\frac{R}{L}t} = Ve^{-\frac{R}{L}t}. \tag{11.2-11}$$

Thus, v_L and v_R are always equal and opposite as shown in Fig. 11.2-5. The negative sign in Eq. (11.2-10) indicates that the polarity of v_L is opposite to that assumed in Fig. 11.2-1. This means that it tries to send current in the same direction in which it was set up by the source V.

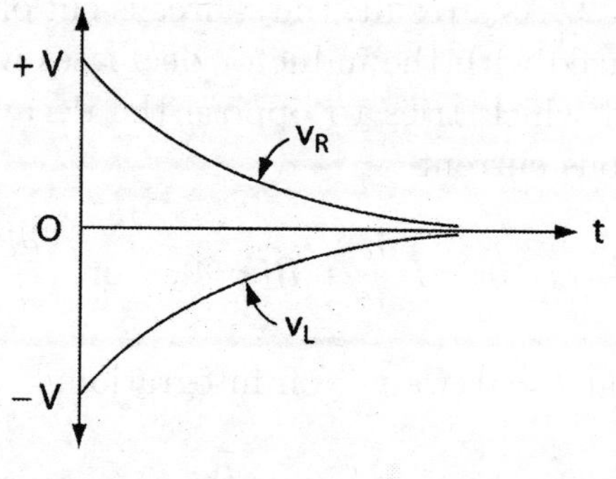

Fig 11.2-5

When an inductive circuit is suddenly broken the air gap at the point of breaking makes the effective resistance of the circuit very large and time constant τ very small.

The current decreases very rapidly which in turn produces a large back emf. Since the air gap at the point of breaking is usually small its insulation may breakdown and electric sparks may appear. This is our common experience when we quickly switch off and on an electric circuit. This sparking can be reduced by connecting a capacitor across the points of breaking. The capacitor takes up the charge and allows it to pass at a slow rate between the points of breaking.

In terms of energy the growth and decay of current may be understood as follows: During growth of current the source V supplies energy a part of which is lost as Joule heat in R and another part goes to build up magnetic field around L. During decay of current the stored magnetic field energy is used up in driving current in the circuit.

11.3 Charging and Discharging of a Capacitor in Series CR Circuit

(i) Charging of the capacitor

Fig. 11.3-1 shows a circuit containing a capacitor C and a resistor R connected in series to a source of constant voltage V via a switch K. Suppose the switch is suddenly thrown to the position 1 at time $t = 0$. The capacitor is now charged by the source V. No steady current can flow through the dielectric of a capacitor. But, if an emf is suddenly applied across it some electrons from the negative terminal of the source flow to one plate of C and equal number of electrons flow out of the other plate into the positive terminal of the source. This is known as the *charging of the capacitor*. The source tries to introduce more charge on the capacitor but the charges already acquired by the plates of C opposes the introduction of further charges on the plates. So the source has to do some work in charging the capacitor. A part of the energy supplied by the source is lost as Joule heat in R and rest goes to build up electrostatic field around C. When the pd across C becomes equal to the supply emf V then charging becomes complete and current in the circuit becomes zero.

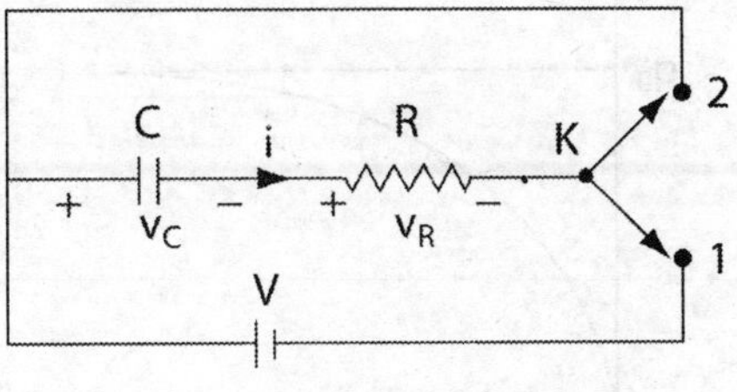

Fig 11.3-1

Let q be the charge on the positive plate of the capacitor and i be the current at any time instant t.

So the instantaneous emf equation of the circuit is

$$v_C + v_R = V, \quad \text{or} \quad \frac{q}{C} + Ri = V,$$

$$\text{or} \quad \frac{q}{C} + R\frac{dq}{dt} = V,$$

$$\text{or} \quad \frac{dq}{q - CV} = -\frac{1}{CR}dt. \tag{11.3-1}$$

Assuming that $q = 0$ at $t = 0$ we get on integration,

$$\int_0^q \frac{dq}{q - CV} = -\frac{1}{CR}\int_0^t dt,$$

$$\text{or} \quad \ln\frac{q - CV}{-CV} = -\frac{t}{CR},$$

$$\text{or} \quad q = CV\left(1 - e^{-t/CR}\right) = q_0\left(1 - e^{-t/CR}\right), \tag{11.3-2}$$

where $q_0 = CV$ is the final charge on the capacitor when $t \to \infty$. Equation (11.3-2) describes the growth of charge on the capacitor. It is shown graphically in Fig. 11.3-2. The rate of charging or instantaneous current in the circuit is given by

$$i = \frac{dq}{dt} = \frac{q_0}{CR}e^{-t/CR} = I_0 e^{-t/CR}, \tag{11.3-3}$$

where $I_0 = q_0/CR = \frac{V}{R}$ is the initial charging current. Note that initially there is no pd across C and the total applied emf is across R.

The charging rate depends on the value of CR. The quantity $CR = \tau$ has the dimension of time and is called the *time constant* of the circuit.

At $t = \tau$, $q = q_0\ (1 - 1/e) \approx 0.63q_0$. So the time constant τ may be defined as the time in which the charge on the capacitor grows from zero to 63% of its steady value. Greater is the time constant (i.e., the greater C and R) larger is the time taken to attain the steady state.

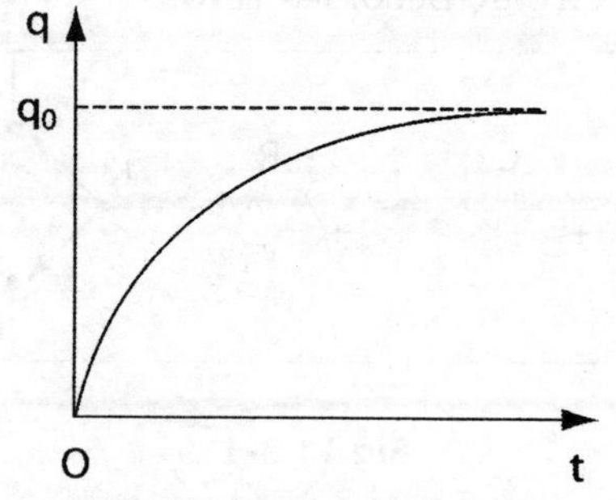

Fig. 11.3-2: Variation of charge on the capacitor with time during charging

Eq. (11.3-3) shows that the charging current i is maximum initially and then it decreases exponentially with time (Fig. 11.3-3). When the capacitor attains maximum

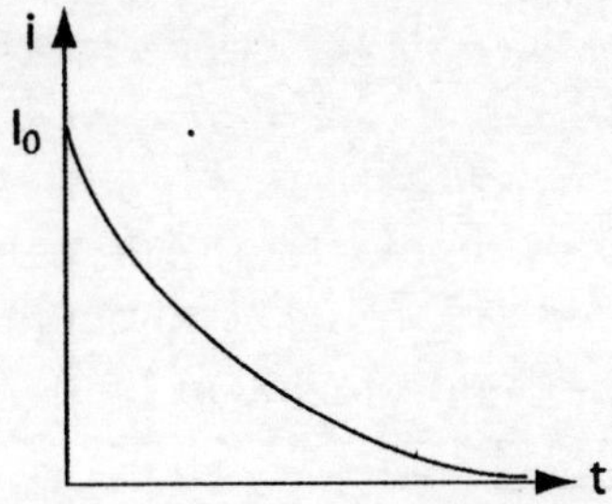

Fig. 11.3-3 : Variation of charging current with time

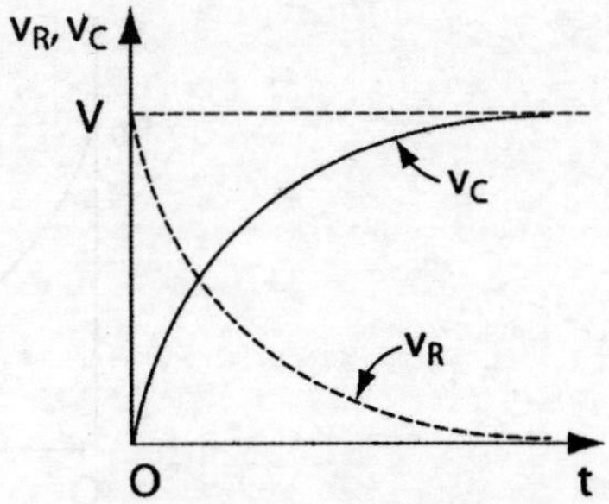

Fig. 11.3-4 : Variation of V_C and V_R with time

charge the charging current becomes zero. Initially there is no pd across C and the total applied voltage V is across R and finally when the current is zero the pd across C becomes equal to the applied voltage V. The expressions for v_R and v_C are

$$v_R = Ri = Ve^{-t/CR} \tag{11.3-4}$$

$$\text{and } v_C = V - v_R = V(1 - e^{-t/CR}). \tag{11.3-5}$$

The variations of v_R and v_C with time are shown in Fig. 11.3-4.

(ii) Discharging of the capacitor

Suppose the capacitor in Fig. 11.3-1 has been fully charged by connecting the switch K to the position 1 for a sufficient time. Now the key is suddenly thrown from position 1 to position 2. As a result, the source is cut off and the charge stored in the capacitor begins to discharge through R. Physically, the excess electrons from the negative plate of the capacitor flow through R to its positive plate where there is deficiency of electrons. The process continues till the pd across C becomes zero. This phenomenon is known as the *discharging of the capacitor*. In this process the electrostatic energy stored in the capacitor is dissipated as a Joule heat in the resistor R.

Let $q = q_0$ be the full charge on C at time $t = 0$; q be the charge on the capacitor at any time t and $i = dq/dt$ be the instantaneous current. Then the emf equation of the circuit is

$$\frac{q}{C} + Ri = 0, \quad \text{or} \quad \frac{q}{C} + R\frac{dq}{dt} = 0, \quad \text{or} \quad \frac{dq}{q} = -\frac{1}{CR}dt. \tag{11.3-6}$$

Integrating we get

$$\int_{q_0}^{q} \frac{dq}{q} = -\frac{1}{CR}\int_0^t dt, \quad \text{or} \quad \ln\frac{q}{q_0} = -\frac{1}{CR}t, \quad \text{or} \quad q = q_0 e^{-t/CR}. \tag{11.3-7}$$

This equation shows that the charge on the capacitor decays exponentially with time (Fig. 11.3-5).

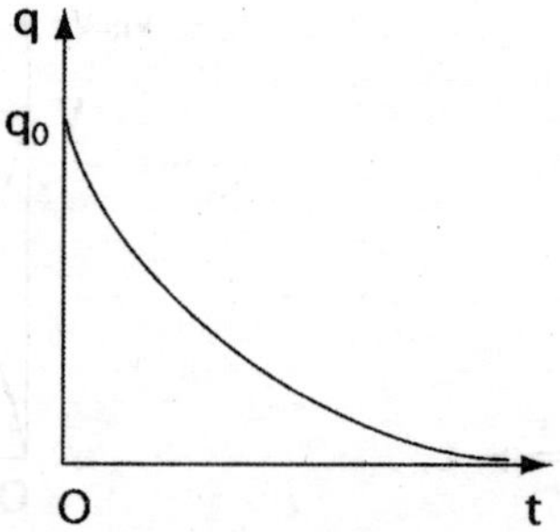

Fig. 11.3-5: Variation of charge on the capacitor with time during discharging

The rate of discharge or discharging current is

$$\frac{dq}{dt} = i = -\frac{q_0}{CR}e^{-t/CR} = -I_0 e^{-t/CR}. \tag{11.3-8}$$

This negative sign indicates that the discharging current flows in a direction opposite to that shown in Fig. 11.3-1. Obviously the rate of discharge depends on the time constant $\tau = CR$. At $t = \tau$, $q = q_0/e \approx 0.37q_0$. So *the time constant* τ *may also be defined as the time in which the charge on the capacitor decays to 37% of its initial value.*

Equations (11.3-3) and (11.3-8) show that the nature of variation of current is the same both during charging and discharging of the capacitor. However, their directions are opposite to each other. The expressions for v_C and v_R are

$$v_C = \frac{q}{C} = Ve^{-t/CR} \tag{11.3-9}$$

$$\text{and } v_R = Ri = -Ve^{-t/CR} \tag{11.3-10}$$

Thus, v_C and v_R are always equal and opposite as shown in Fig. 11.3-6. The negative sign in Eq. (11.3-10) indicates that the polarity of v_R is opposite to that assumed in Fig. 11.3-1.

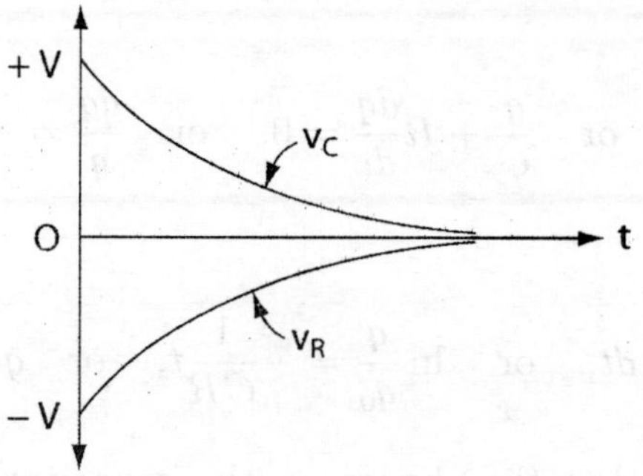

Fig. 11.3-6: Variations of v_C and v_R with time

11.4 Series LCR Circuit

Charging of the capacitor

Let us consider a circuit consisting of an inductor L, capacitor C and a resistor R in series as shown in Fig. 11.4-1. Suppose the switch K is thrown to the position 1 at time $t = 0$. As a result, the source of voltage V gets connected in the circuit and the capacitor begins to charge. Let q be the instantaneous charge on the capacitor and i be the instantaneous current in the circuit. So the instantaneous emf equation of the circuit is

$$L\frac{di}{dt} + \frac{q}{C} + Ri = V. \tag{11.4-1}$$

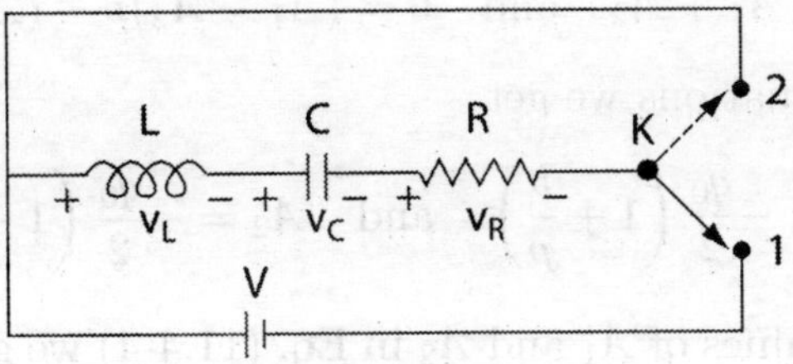

Fig 11.4-1

Since $i = dq/dt$ we can rewrite the equation as

$$L\frac{d^2q}{dt^2} + R\frac{dq}{dt} + \frac{1}{C}q = V,$$

$$\text{or}\quad \frac{d^2q}{dt^2} + 2b\frac{dq}{dt} + \omega_0^2\left(q - \frac{V}{L\omega_0^2}\right) = 0, \tag{11.4-2}$$

where $2b = R/L$ and $\omega_0^2 = 1/LC$. Introducing a new variable $Q = q - V/L\omega_0^2$ we get

$$\frac{d^2Q}{dt^2} + 2b\frac{dQ}{dt} + \omega_0^2 Q = 0. \tag{11.4-3}$$

To solve the equation let $Q = Ae^{\alpha t}$ be a trial solution, where A and α are arbitrary constants. Substituting it in Eq. (11.4-2) we get

$$\left(\alpha^2 + 2b\alpha + \omega_0^2\right)Ae^{\alpha t} = 0.$$

For this to be valid for all t we must have

$$\alpha^2 + 2b\alpha + \omega_0^2 = 0.$$

It has two roots

$$\alpha_1 = -b + \sqrt{b^2 - \omega_0^2} \quad \text{and} \quad \alpha_2 = -b - \sqrt{b^2 - \omega_0^2}.$$

So the general solution of Eq. (11.4-3) is

$$Q = A_1 e^{\alpha_1 t} + A_2 e^{\alpha_2 t}$$
$$= e^{-bt}\left[A_1 e^{\sqrt{b^2-\omega_0^2}\cdot t} + A_2 e^{-\sqrt{b^2-\omega_0^2}\cdot t}\right]$$

$$\text{or} \quad q = Q + \frac{V}{L\omega_0^2} = q_0 + e^{-bt}\left[A_1 e^{pt} + A_2 e^{-pt}\right], \tag{11.4-4}$$

where $q_0 = V/L\omega_0^2 = V \cdot C$ and $p = \sqrt{b^2 - \omega_0^2}$. The arbitrary constants A_1 and A_2 can be determined by applying the initial conditions: at $t = 0$, $q = 0$ and $i = \frac{dq}{dt} = 0$. Therefore,

$$0 = q_0 + A_1 + A_2 \quad \text{and} \quad 0 = (A_1 - A_2)p - (A_1 + A_2)b.$$

Solving these two equations we get

$$A_1 = -\frac{q_0}{2}\left(1 + \frac{b}{p}\right) \quad \text{and} \quad A_2 = -\frac{q_0}{2}\left(1 - \frac{b}{p}\right).$$

Substituting these values of A_1 and A_2 in Eq. (11.4-4) we get the complete solution:

$$q = q_0 - \frac{q_0}{2}\cdot e^{-bt}\left[\left(1 + \frac{b}{p}\right)e^{pt} + \left(1 - \frac{b}{p}\right)e^{-pt}\right]. \tag{11.4-5}$$

The physical nature of this solution depends on the relative values of b and ω_0. Let us consider three possible cases of interest separately.

Case I. $b^2 > \omega_0^2$

In this case, $p = \sqrt{b^2 - \omega_0^2}$ is a positive quantity and the solution (11.4-5) shows that the charge on the capacitor grows gradually and ultimately acquires a steady state value q_0 as shown by the curve I in Fig. 11.4-2. The circuit, in this case, is said to be *overdamped.*

Case II. $b^2 = \omega_0^2$

In this case, $p = 0$ and b/p tends to infinity. To study the behaviour of the circuit in this case, we assume $p \ll 1$ and make the following approximations:

$$e^{pt} \simeq 1 + pt \quad \text{and} \quad e^{-pt} \simeq 1 - pt.$$

So the solution (11.4-5) takes the form

$$q = q_0 - \frac{q_0}{2}\cdot e^{-bt}\left[\left(1 + \frac{b}{p}\right)(1 + pt) + \left(1 - \frac{b}{p}\right)(1 - pt)\right]$$
$$= q_0 - q_0(1 + bt)e^{-bt}. \tag{11.4-6}$$

Thus, in this case, also, the charge on the capacitor grows gradually and ultimately acquires a steady state value q_0 as shown by the curve II in Fig. 11.4-2. But compared to the overdamped case here smaller time is required to reach the steady state. The circuit, in this case, is said to be *critically damped.* The resistance, in this case, is given by $b^2 = \omega_0^2$ or $R = 2\sqrt{L/C}$. This is known as *critical damping resistance.*

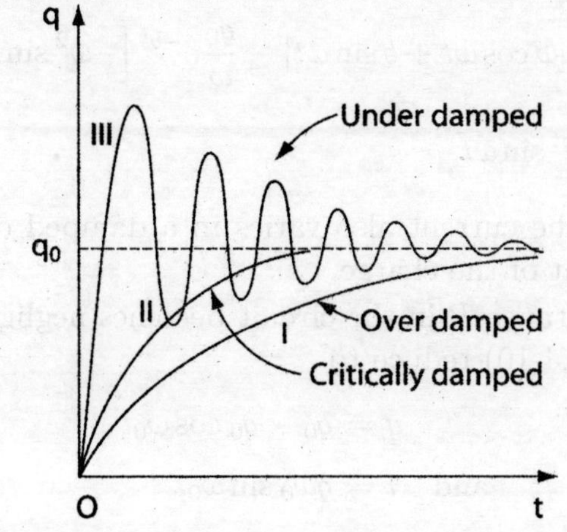

Fig. 11.4-2 : Variation of charge on the capacitor with time in a series LCR circuit

Case III. $b^2 < \omega_0^2$

In this case, $p = \sqrt{b^2 - \omega_0^2}$ is imaginary and we can write $p = j\sqrt{\omega_0^2 - b^2} = j\omega$, where $\omega = \sqrt{\omega_0^2 - b^2}$. The solution (11.4-5) now takes the form

$$\begin{aligned} q &= q_0 - \frac{q_0}{2}e^{-bt}\left[\left(e^{j\omega t} + e^{-j\omega t}\right) + \frac{b}{j\omega}\left(e^{j\omega t} - e^{-j\omega t}\right)\right] \\ &= q_0 - q_0 e^{-bt}\left[\cos\omega t + \frac{b}{\omega}\sin\omega t\right] \\ &= q_0 - \frac{q_0}{\omega}e^{-bt}\left[\omega\cos\omega t + b\sin\omega t\right]. \end{aligned} \tag{11.4-7}$$

Putting $\omega = A\cos\theta$ and $b = A\sin\theta$, where $A = \sqrt{\omega^2 + b^2} = \omega_0$ and $\theta = \tan^{-1} b/\omega$, we can write the solution (11.4-7) as

$$q = q_0 - q_0\frac{\omega_0}{\omega}e^{-bt}\cos(\omega t - \theta). \tag{11.4-8}$$

Thus, in this case, the charge on the capacitor grows in a damped oscillatory manner. The capacitor charge oscillates about the final value q_0 with decaying amplitude as shown by the curve III in Fig. 11.4-2. The circuit, in this case, is said to be *underdamped* or *oscillatory.* Note that in the steady state the maximum charge on C is q_0 but during transient moments it may go above q_0.

The frequency of oscillation of the capacitor charge is

$$f = \frac{\omega}{2\pi} = \frac{\sqrt{\omega_0^2 - b^2}}{2\pi} = \frac{1}{2\pi}\sqrt{\frac{1}{LC} - \frac{R^2}{4L^2}}. \tag{11.4-9}$$

The current in the circuit is obtained by differentiating Eq. (11.4-7) as

$$\begin{aligned} i = \frac{dq}{dt} &= \frac{q_0}{\omega} b e^{-bt}\left[\omega \cos \omega t + b \sin \omega t\right] - \frac{q_0}{\omega} e^{-bt}\left[-\omega^2 \sin \omega t + b\omega \cos \omega t\right] \\ &= \frac{q_0 \omega_0^2}{\omega} e^{-bt} \sin \omega t. \end{aligned} \tag{11.4-10}$$

Thus, the current in the current also varies in a damped oscillatory manner with the same frequency as that of the charge.

If in any case the resistance R in the circuit becomes negligible then $b = 0$, $\omega = \omega_0$ and Eqs. (11.4-7) and (11.4-10) reduce to

$$q = q_0 - q_0 \cos \omega_0 t \tag{11.4-11}$$

$$\text{and} \quad i = q\omega_0 \sin \omega_0 t \tag{11.4-12}$$

So, in this case, the charge and current oscillate in an undamped manner with the angular frequency $\omega_0 = 1/\sqrt{LC}$. For this ω_0 is called the *undamped natural frequency* of the circuit.

(ii) Discharging of the capacitor

Suppose the capacitor C in Fig. 11.4-1 is charged by connecting the switch K to the position 1 for a sufficient time. Now the key is suddenly thrown from position 1 to the position 2. As a result, the source is cut off and the stored charge on the capacitor begins to discharge through L and R. Let q be the charge on C at any time t after the key is thrown to the position 2. The instantaneous current is $i = \frac{dq}{dt}$. So the emf equation of the circuit is

$$L\frac{di}{dt} + \frac{q}{C} + Ri = 0 \tag{11.4-13}$$

$$\text{or} \quad L\frac{d^2q}{dt^2} + R\frac{dq}{dt} + \frac{1}{C}q = 0$$

$$\text{or} \quad \frac{d^2q}{dt^2} + 2b\frac{dq}{dt} + \omega_0^2 q = 0, \tag{11.4-14}$$

where $2b = R/L$ and $\omega_0^2 = 1/LC$.

As before, the general solution of Eq. (11.4-14) can be written in the form

$$q = e^{-bt}\left[A_1 e^{pt} + A_2 e^{-pt}\right], \tag{11.4-15}$$

where $p = \sqrt{b^2 - \omega_0^2}$.

To find the arbitrary constants we apply the initial conditions that at $t = 0$, $q = q_0 = VC$ and $\frac{dq}{dt} = 0$. Thus, we get

$$q_0 = A_1 + A_2 \quad \text{and} \quad 0 = (A_1 - A_2)p - (A_1 + A_2)b$$

Solving these two equations we get

$$A_1 = \frac{q_0}{2}\left(1 + \frac{b}{p}\right) \quad \text{and} \quad A_2 = \frac{q_0}{2}\left(1 - \frac{b}{p}\right).$$

Substituting these values of A_1 and A_2 in Eq. (11.4-15) we get the complete solution as

$$q = \frac{q_0}{2}e^{-bt}\left[\left(1 + \frac{b}{p}\right)e^{pt} + \left(1 - \frac{b}{p}\right)e^{-pt}\right]. \tag{11.4-16}$$

The physical nature of this solution depends on the relative values of b and ω_0. Let us consider three different cases of interest separately.

Case I: $b^2 > \omega_0^2$

In this case, $p = \sqrt{b^2 - \omega_0^2}$ is a positive quantity and the solution (11.4-16) indicates a nonoscillatory discharge of the capacitor; the charge on C decreases gradually till at $t = \infty$, it becomes zero. This is shown by the curve I in Fig. 11.4-3. The circuit in this case, is said to be *overdamped.*

Case II: $b^2 = \omega_0^2$

In the case $p = 0$ and b/p tends to infinity. To study the behaviour of the circuit in this case, we assume $p \ll 1$ and make the following approximations: $e^{\pm pt} \approx 1 \pm pt$. Then the solution (11.4-16) reduces to

$$q = q_0(1 + bt)e^{-bt}. \tag{11.4-17}$$

Thus, in this case, also, the charge on the capacitor decreases gradually in a nonosicllatory manner. This is shown by the curve II in Fig. 11.4-3. Here the charge reduces to zero in a time which is less than that taken in the overdamped case. The circuit, in this case, is said to be *critically damped.*

Case III: $b^2 < \omega_0^2$

In this case, $p = \sqrt{b^2 - \omega_0^2}$ is imaginary and we can write $p = j\sqrt{\omega_0^2 - b^2} = j\omega$, where $\omega = \sqrt{\omega_0^2 - b^2}$. The solution (11.4-16) now takes the form

$$q = \frac{q_0\omega_0}{\omega}e^{-bt}\cos(\omega t - \theta), \tag{11.4-18}$$

where $\theta = \tan^{-1} b/\omega$.

Thus, in this case, the charge on the capacitor decreases in a damped oscillatory manner. This is shown by the curve III in Fig. 11.4-3. The circuit, in this case, is said to be *oscillatory.*

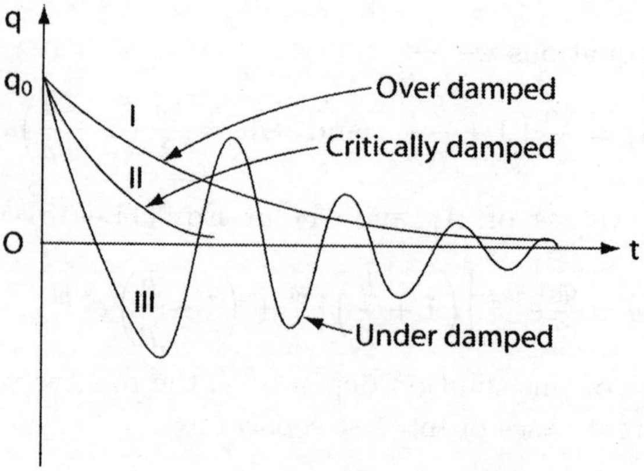

Fig. 11.4-3: Discharge of a capacitor in a series LCR circuit

The frequency of oscillation of the capacitor charge is

$$f = \frac{\omega}{2\pi} = \frac{\sqrt{\omega_0^2 - b^2}}{2\pi} = \frac{1}{2\pi}\sqrt{\frac{1}{LC} - \frac{R^2}{4L^2}}. \qquad (11.4\text{-}19)$$

The current in the circuit is given by

$$i = \frac{dq}{dt} = -\frac{q_0\omega_0^2}{\omega} e^{-bt} \sin \omega t. \qquad (11.4\text{-}20)$$

Obviously, the current also varies in a damped oscillatory manner with the same frequency as that of the charge.

If, in any case, the resistance R in the circuit becomes negligible then $b = 0$, $\omega = \omega_0$ and Eqs. (11.4-18) and (11.4-20) reduce to

$$q = q_0 \cos \omega_0 t \qquad (11.4\text{-}21)$$

$$\text{and} \quad i = -q_0\omega_0 \sin \omega_0 t. \qquad (11.4\text{-}22)$$

In this case, the charge and current both oscillate in an undamped manner with the angular frequency ω_0.

11.5 Discharge of a Capacitor Through a Pure Inductor and Electrical Oscillations

Suppose a capacitor C is charged by connecting it to a source of voltage V as shown in Fig. 11.5-1. Now the switch K is thrown suddenly from the position 1 to the position 2. As a result, the source is cut off and the charge stored on C begins to discharge through a

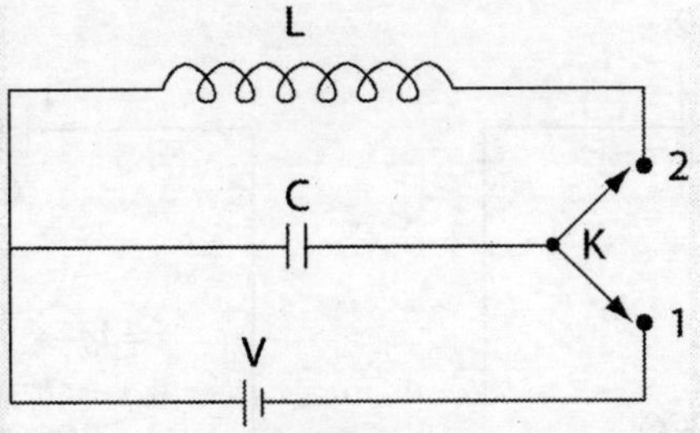

Fig 11.5-1

pure inductor L. Let q be instantaneous charge on C and $i = dq/dt$ be the instantaneous current in the circuit. Then the instantaneous emf equation of the circuit is

$$L\frac{di}{dt} + \frac{q}{C} = 0, \quad \text{or} \quad L\frac{d^2q}{dt^2} + \frac{q}{C} = 0, \quad \text{or} \quad \frac{d^2q}{dt^2} + \omega_0^2 = 0, \tag{11.5-1}$$

where $\omega_0^2 = \frac{1}{LC}$.

The solution of the Eq. (11.5-1) is

$$q = A_1 \cos \omega_0 t + A_2 \sin \omega_0 t,$$

where A_1 and A_2 are two arbitrary constant. To find these constants we apply the initial conditions that at $t = 0$, $q = q_0 = VC$ and $\frac{dq}{dt} = 0$. Thus, we get

$$q_0 = A_1 \quad \text{and} \quad A_2 = 0.$$

Therefore,

$$q = q_0 \cos \omega_0 t. \tag{11.5-2}$$

Thus, the charge on the capacitor varies in an oscillatory manner with constant amplitude q_0 and frequency $f = \frac{\omega_0}{2\pi} = \frac{1}{2\pi\sqrt{LC}}$. The current in the circuit is

$$i = \frac{dq}{dt} = -q_0\omega_0 \sin \omega_0 t \tag{11.5-3}$$

$$\text{or} \quad i = q_0\omega_0 \cos(\omega_0 t + \pi/2). \tag{11.5-4}$$

It also oscillates with the same angular frequency ω_0.

Physics of electric oscillation

To start with ($t = 0$), let us suppose that the capacitor is fully charged with the plate 1 being charged positively and the plate 2 negatively [Fig. 11.5-2(a)]. At this moment current in the circuit is zero the pd across the capacitor is maximum, an electric field is set up between the plates and the entire energy of the oscillator circuit is stored as the electrostatic energy ($q_0^2/2C$) in the capacitor. Now as the circuit is closed the capacitor begins to discharge, a current grows through the inductor L and it builds up a magnetic field around.

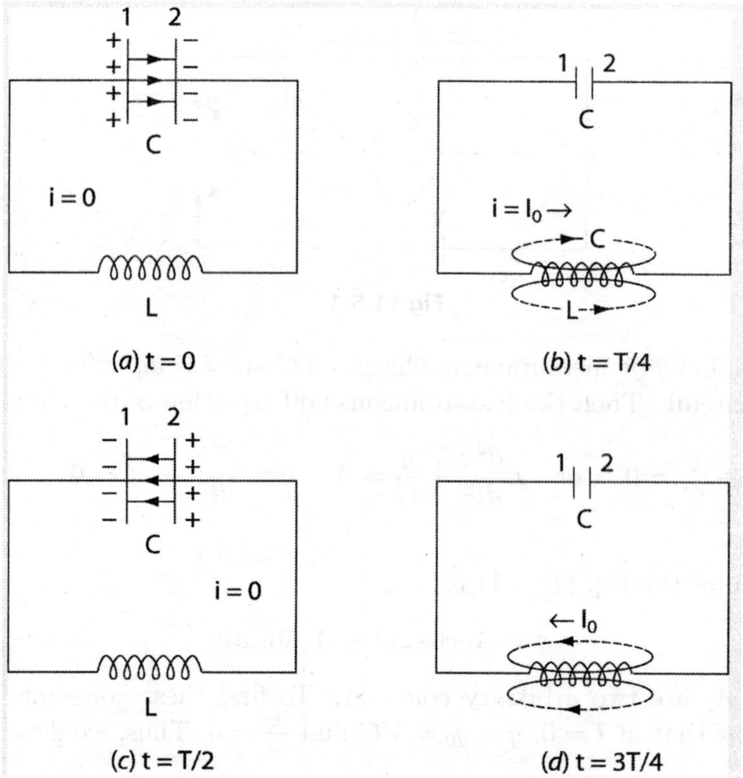

Fig. 11.5-2: Electric oscillation in LC circuit

In this process the electrical energy of the capacitor decreases and the magnetic energy ($\frac{1}{2}Li^2$) of the inductor increases. At time $t = \frac{T}{4}$, where T is the time period, the charge on the capacitor reduces to zero and the current in the circuit becomes maximum [See Eqs. (11.5-2) and (11.5-4)]. At this stage the electrical energy of the capacitor is completely converted into the magnetic energy of the inductor [Fig. 11.5-2(b)]. Starting from this stage the current begins to decrease retaining its direction. It causes an induced emf in the inductor which tries to maintain the current in the same direction as before. In this attempt the capacitor is charged in the opposite direction, i.e., the plate 2 is charged positively and the plate 1 negatively [Fig. 11.5-2(c)]. The appearance of charge on the capacitor tends to reduce the current. Finally at $t = T/2$ the current ceases while the capacitor attain maximum charge. At this stage magnetic energy of inductor is completely converted into the electrical energy of the capacitor. From this moment, the capacitor starts to discharge again, current flows in the opposite direction and the process is repeated.

The state of the system at $t = 3T/4$ will be as shown in Fig. 11.5-2(d) and at $t = T$ the system returns to the initial state as shown in Fig. 11.5-2(a). If there is no resistance in the circuit this charging and discharging of C will go on indefinitely. However, in practice, the circuit always contains some finite resistance. As a result, in each cycle a part of the energy is dissipated as Joule heat in the resistance and the oscillation cannot continue indefinitely. If we continuously supply energy equal to that dissipated in the circuit then an electric oscillation of constant amplitude can be sustained even in practical circuits.

11.6 Measurement of High Resistance by Leakage

The discharge of a capacitor through a high resistance offers a method of measuring the value of the resistance. When a capacitor C with an initial charge q_0 is allowed to discharge through a resistance R for a time t the charge remaining on C is given by

$$q = q_0 e^{-t/CR}, \quad \text{or} \quad R = \frac{t}{C \ln \frac{q_0}{q}}. \tag{11.6-1}$$

So R can be determined if q_0/q is measured experimentally. Fig. 11.6-1 shows a circuit diagram for measuring q_0/q with a ballistic galvanometer. The capacitor C of known value is first charged by joining the key K to a and keeping the key K_1 open. The full charge on C is passed through the ballistic galvanometer (B.G.) immediately by connecting K to b and the first ballistic throw d_0 is noted. The capacitor C is again charged and disconnected from a and b and then discharged through R for a known time t by pressing the key K_1. The charge left on C is then discharged through the ballistic galvanometer and the first ballistic throw d is noted. Since the ballistic throw is proportional to the charge we can write $q_0/q = d_0/d$ and hence, from (11.6-1)

$$R = \frac{t}{C \ln \frac{d_0}{d}}. \tag{11.6-2}$$

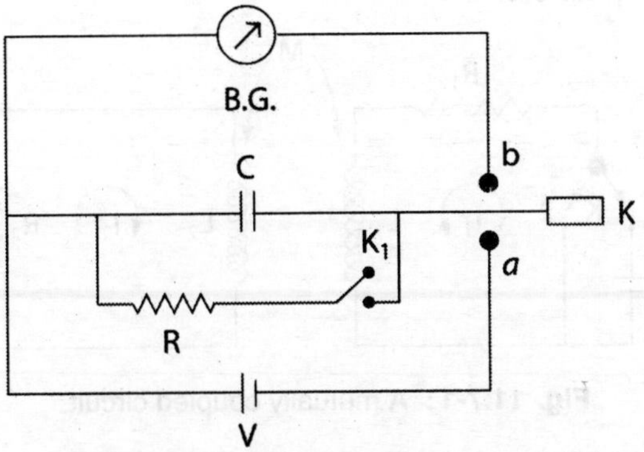

Fig. 11.6-1 : High resistance by leakage

From this equation R can be determined provided we use a perfect capacitor. It often happens that a charged practical capacitor loses charge by leakage through its dielectric even when it is kept insulated. It can be taken into account by considering a practical capacitor as a perfect capacitor with a high resistance R_0 connected in parallel with it. R_0 is called the *natural leakage resistance*. To find R_0 the capacitor is charged by a charge q_0 and is then allowed for natural leakage for t_1 sec, then the remaining charge is passed through the galvanometer. If d_1 be the first ballistic throw due to this then as before

$$R_0 = \frac{t_1}{C \ln \frac{d_0}{d_1}}. \tag{11.6-3}$$

In this case, the value of R determined from Eq. (11.6-2) is the equivalent resistance of the parallel combination of the actual resistance R_a to be measured and the natural leakage resistance R_0. Therefore,

$$\frac{1}{R} = \frac{1}{R_a} + \frac{1}{R_0} \quad \text{or} \quad R_a = \frac{R \cdot R_0}{R_0 - R}. \tag{11.6-4}$$

Thus, knowing R and R_0 from Eqs. (11.6-2) and (11.6-3), actual resistance R_a can be calculated from (11.6.4).

Note that the time constant CR should have appreciable value otherwise the method cannot be applied. If CR is small then leakage time t will be too low to measure. For this the method is unsuitable for the measurement of low resistances.

11.7 Transients in Mutually Coupled Circuits

Let us consider the mutually coupled circuit of Fig. 11.7-1, where the primary has self-inductance L_1 and resistance R_1 while the secondary has self-inductance L_2 and resistance R_2. M is the mutual inductance between the two circuits. Our purpose is to study the variation of current in the secondary due to growth or decay of current in the primary. Suppose a source of voltage V in connected to the primary at $t = 0$ by throwing the key to the position 1.

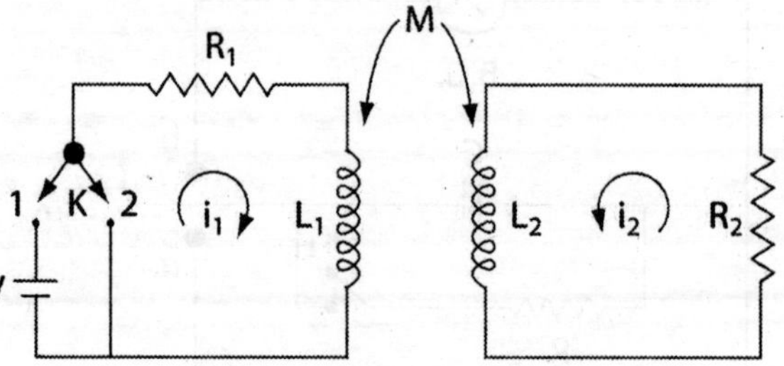

Fig. 11.7-1 : A mutually coupled circuit

As the current in the primary grows there will be an induced emf and current i_2 in the secondary due to mutual induction. To study how these currents grow with time let us consider the instantaneous emf equations

$$L_1 \frac{di_1}{dt} + M \frac{di_2}{dt} + R_1 i_1 = V \tag{11.7-1}$$

$$\text{and } L_2 \frac{di_2}{dt} + M \frac{di_1}{dt} + R_2 i_2 = 0 \tag{11.7-2}$$

Putting $i_1 - V/R_1 = i_1'$ in these equations we get

$$L_1\frac{di_1'}{dt} + M\frac{di_2}{dt} + R_1 i_1' = 0 \tag{11.7-3}$$

$$\text{and } L_2\frac{di_2}{dt} + M\frac{di_1'}{dt} + R_2 i_2 = 0. \tag{11.7-4}$$

To solve these coupled equation let us assume that both i_1' and i_2 vary with time as $e^{\alpha t}$, where α is a constant to be determined. Then,

$$(L_1\alpha + R_1)i_1' + M\alpha i_2 = 0 \quad \text{and} \quad M\alpha i_1' + (L_2\alpha + R_2)i_2 = 0.$$

For nontrivial solutions we must have

$$\begin{vmatrix} L_1\alpha + R_1 & M\alpha \\ M\alpha & L_2\alpha + R_2 \end{vmatrix} = 0$$

$$\text{or } (L_1L_2 - M^2)\alpha^2 + (L_1R_2 + L_2R_1)\alpha + R_1R_2 = 0$$

Therefore, the roots are

$$\alpha = \frac{-(L_1R_2 + L_2R_1) \pm \sqrt{(L_1R_2 + L_2R_1)^2 - 4(L_1L_2 - M^2)R_1R_2}}{2(L_1L_2 - M^2)}. \tag{11.7-5}$$

Since $L_1L_2 > M^2$ both the roots are negative. Let us call these roots to be $-\alpha_1$ and $-\alpha_2$. Corresponding to the upper + and lower − sign, respectively. So we can write

$$i_1' = A_1e^{-\alpha_1 t} + B_1e^{-\alpha_2 t} \tag{11.7-6}$$

$$\text{and } \ i_2' = A_2e^{-\alpha_1 t} + B_2e^{-\alpha_2 t} \tag{11.7-7}$$

To determine the constants A_1, B_1, A_2 and B_2 we use the following initial conditions:

(i) At $t = 0$, $i_1 = 0$ or $i_1' = -V/R_1$;

(ii) At $t = 0$, $i_2 = 0$.

Applying these conditions we get from Eqs. (11.7-6) and (11.7-7),

$$A_1 + B_1 = -V/R_1 \tag{11.7-8}$$

$$\text{and } \quad A_2 + B_2 = 0 \quad \text{or} \quad B_2 = -A_2. \tag{11.7-9}$$

Therefore,

$$i_2 = A_2\left(e^{-\alpha_1 t} - e^{-\alpha_2 t}\right) \tag{11.7-10}$$

Now substituting i_1' and i_2 from Eqs. (11.7-6) and (11.7-10) in Eq. (11.7-4) we get

$$L_2A_2\left[-\alpha_1e^{-\alpha_1 t} + \alpha_2e^{-\alpha_2 t}\right] + M\left[-\alpha_1A_1e^{-\alpha_1 t} - \alpha_2B_1e^{-\alpha_2 t}\right] + R_2A_2\left(e^{-\alpha_1 t} - e^{-\alpha_2 t}\right) = 0.$$

Since this is valid for all t we can equate to zero the coefficients of $e^{-\alpha_1 t}$ and $e^{-\alpha_2 t}$ separately. Thus, we get

$$(R_2 - L_2\alpha_1)A_2 - M\alpha_1 A_1 = 0 \tag{11.7-11}$$

$$\text{and } -(R_2 - L_2\alpha_2)A_2 - M\alpha_2 B_1 = 0. \tag{11.7-12}$$

Solving Eqs. (11.7-8), (11.7-11) and (11.7-12) we get

$$A_1 = \frac{V\alpha_2(R_2 - L_2\alpha_1)}{R_1R_2(\alpha_1 - \alpha_2)},$$

$$B_1 = -\frac{V\alpha_1(R_2 - L_2\alpha_2)}{R_1R_2(\alpha_1 - \alpha_2)},$$

$$A_2 = \frac{VM\alpha_1\alpha_2}{R_1R_2(\alpha_1 - \alpha_2)} \tag{11.7-13}$$

Substituting these values of A_1, B_1 and A_2 we get from Eqs. (11.7-6) and (11.7-10),

$$i_1 = \frac{V}{R_1} + i_1' = \frac{V}{R_1} \cdot \left[1 + \frac{\alpha_2(R_2 - L_2\alpha_1)}{R_2(\alpha_1 - \alpha_2)}e^{-\alpha_1 t} - \frac{\alpha_1(R_2 - L_2\alpha_2)}{R_2(\alpha_1 - \alpha_2)}e^{-\alpha_2 t}\right] \tag{11.7-14}$$

and

$$i_2 = \frac{VM\alpha_1\alpha_2}{R_1R_2(\alpha_1 - \alpha_2)}\left(e^{-\alpha_1 t} - e^{-\alpha_2 t}\right). \tag{11.7-15}$$

It is now clear that $i_1 \to V/R_1$ when $t \to \infty$. The secondary current, on the other hand, begins from zero at $t = 0$ and ends at zero when $t \to \infty$. This is what is expected from physical consideration. When the primary current becomes constant the emf and current induced in the secondary must be zero. The secondary current attains a maximum value at some time t_0 given by $\left.\frac{di_2}{dt}\right|_{t_0} = 0$. The nature of variation of i_1 and i_2 with time is shown in Fig. 11.7-2. Note that i_2 is negative. The primary current grows exponentially to a limiting value of V/R_1. The secondary current grows in the opposite direction, attains a maximum value and then decreases to zero again.

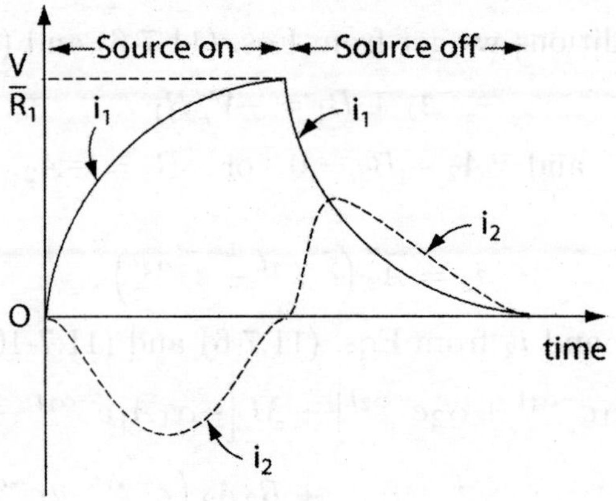

Fig. 11.7-2: Growth and decay of currents in mutually coupled circuits

After attainment of steady state, let the switch be thrown from position 1 to position 2. As a result, the source is cut off and the primary current decays with time. This induces an emf and current in the secondary. Let us investigate how the primary and secondary currents decay with time. The differential equations satisfied by i_1 and i_2 will be the Eqs. (11.7-1) and (11.7-2) with $V = 0$. As before, solutions will be given by (11.7-6) and (11.7-7) with $i_1' = i_1$. However, to find the constants A_1, B_1, A_2 and B_2 we are to apply the following initial conditions:

(i) At $t = 0$, $i_1 = i_1' = V/R_1$;

(ii) At $t = 0$, $i_2 = 0$.

As a result, Eq. (11.7-9) remains same and Eq. (11.7-8) modifies to

$$A_1 + B_1 = V/R_1.$$

Therefore, in this case, A_1, B_1 and B_2 will be given by the Eqs. (11.7-13) with V replaced by $-V$ and hence, solutions will be

$$i_1 = -\frac{V}{R_1}\left[\frac{\alpha_2(R_2 - L_2\alpha_1)}{R_2(\alpha_1 - \alpha_2)}e^{-\alpha_1 t} - \frac{\alpha_1(R_2 - L_2\alpha_2)}{R_2(\alpha_1 - \alpha_2)}e^{-\alpha_2 t}\right] \tag{11.7-16}$$

$$\text{and } i_2 = -\frac{VM\alpha_1\alpha_2}{R_1R_2(\alpha_1 - \alpha_2)}\left(e^{-\alpha_1 t} - e^{-\alpha_2 t}\right). \tag{11.7-17}$$

Thus, the primary current decays exponentially from the value V/R_1 to zero, the secondary current increases from zero to a maximum and then dies away to zero (Fig. 11.7-2). Note that in this case, the primary and the secondary currents are in the same direction.

11.8 Ruhmkorff's Induction Coil or Induction Coil

The Ruhmkorff's induction coil or the induction coil is an electrical device based on the principle of mutual inductance, which can increase a small dc voltage to a very high value, which may be used to study the electric discharge through gases. Essential parts of such an induction coil are shown in Fig. 11.8-1. It has a primary coil which consists of a few turns of thick copper wire wound over a tube of insulating material. A bundle of soft iron wires insulated from each other is placed inside the tube. It acts as a core. The primary coil is connected to a high capacity dc source and an automatic *make and break* arrangement. A soft iron hammer H is attached to a metallic plate P whose lower part is fixed. There is a platinum contact point A_1 at its back. There is an adjustable screw S provided with a platinum tip A_2. When this tip A_2 touches the tip A_1 current from the battery B flows through the primary coil. It magnetizes the soft iron core, which then attracts the hammer H toward itself. As a result, the contact between A_1 and A_2 breaks and the primary current stops. The core loses its magnetization and hence the attractive power. So the hammer H jumps back to the original position due to the spring action of the plate P. The primary circuit is thus, again made on and the whole process is repeated over and over again.

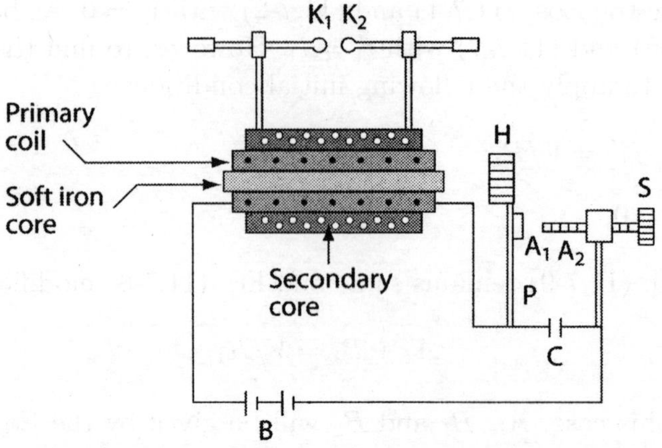

Fig. 11.8-1 : Induction coil

The primary coil is surrounded by a hollow tube of nonconducting material over which thin insulated copper wire of large number of turns is wound. It is called *secondary coil.* Its terminals are connected to two adjustable conductors carrying the discharge knobs K_1 and K_2. A large capacitor C is connected parallel to the contact breaker.

When the tips A_1 and A_2 are made to touch each other with the help of the screw S, current grows through the primary coil at a rate determined by the time constant of the circuit and an emf is induced in the secondary coil. The soft iron core acts as an electromagnet. It attracts the hammer H towards it and the primary circuit is broken. As a result, the current in the primary falls from a maximum value to zero in a short time, because the time constant of the primary circuit during *break* is much smaller than that during *make.* Consequently a large emf is induced in the secondary coil. The automatic make and break arrangement makes the primary circuit alternatively on

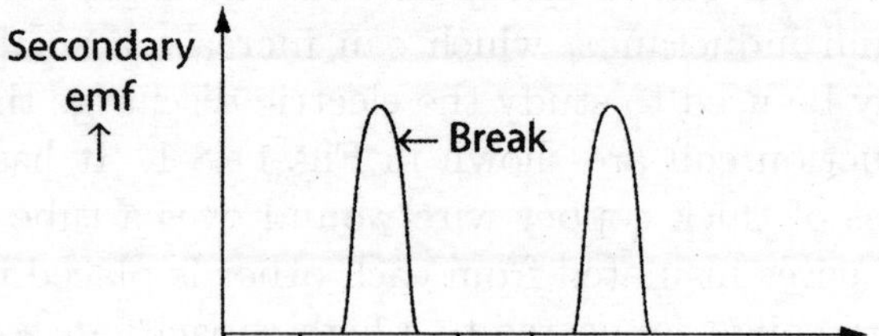

Fig 11.8-2

and off and the whole process is repeated over and over again. As the primary current goes through one make and one break, the secondary emf also undergoes a reversal of direction (Fig. 11.8-2).

The faster rate of fall of primary current during *break* makes the emf induced in the secondary much larger during *break* than that during *make* of the primary. So over a complete cycle of make-and-break an average unidirectional voltage appears across the secondary.

When the primary circuit is broken there is a tendency of sparking at the small air gap between A_1 and A_2. The capacitor C helps to reduce the possibility of such sparking. The capacitor takes up most of the charges on its plates and allows the charge to pass between the points A_1 and A_2 at a much slower rate. Thus, it reduces the possibility of sparking. The use of the capacitor also increases the secondary emf during the break of primary. The capacitor gets fully charged during *make* of the primary. During *break* the capacitor discharges itself and causes a current flow in a direction opposite to the normal primary current. As a result, the primary current falls much more rapidly and the secondary voltage increases considerably.

SOLVED PROBLEMS

▶ **Problem 1.** *Set up the instantaneous emf equation of series LCR circuit driven by a source of voltage V.*

Solution. Suppose at any time t the source is sending a current i through the circuit. Work done or energy supplied by the source in time dt is $Vidt$. This energy is partly dissipated as a Joule heat, i^2Rdt, in the resistor, partly stored as magnetic energy, $L\frac{di}{dt}\cdot idt$, in the inductor and the rest is stored as electrostatic energy $\frac{q}{C}\cdot idt$ in the capacitor. So from the law of conservation of energy,

$$Vidt = i^2Rdt + L\frac{di}{dt}\cdot idt + \frac{q}{C}\cdot idt, \quad \text{or} \quad V = iR + L\frac{di}{dt} + \frac{q}{C},$$

which is the required emf equation.

▶ **Problem 2.** *A relay has an inductance of 10 H and a resistance of 100 Ω and operates with a current of 2 mA. How long will the relay take to operate when a pd of 0.5V is suddenly applied across it?*

Solution. The instantaneous current $i = I_0\left(1 - e^{-\frac{R}{L}t}\right)$, where $I_0 = \frac{0.5}{100}$ A$=5\times10^{-3}$ A, $R = 100\,\Omega$ $L = 10$H, $i = 2\times10^{-3}$A.

We are to find the value of t.

$$\therefore\ 2\times10^{-3} = 5\times10^{-3}\left(1 - e^{-\frac{100}{10}t}\right), \text{ or } \frac{2}{5} = 1 - e^{-10t} \text{ or } t = \frac{1}{10}\ln\frac{5}{3}\,\text{s} = 51\,\text{ms}.$$

► **Problem 3.** *Suppose, a series LCR oscillatory circuit of low resistance contains an air capacitor enclosed in a container. When the container is evacuated the frequency of current oscillation is* 10 kHz *and when the container is filled with* CO_2 *the frequency of oscillation changes by* 50 Hz. *Find the dielectric constant of* CO_2. **[C.U. 1992]**

Solution. As R is low, the frequency of oscillation is $f = \dfrac{1}{2\pi\sqrt{LC}}$.

If the value of capacitance is C_0 with vacuum and C_m with CO_2 then $C_m = KC_0$ where K is the dielectric constant of CO_2. Now

$$100\,\text{kHz} = \frac{1}{2\pi\sqrt{LC_0}} \quad \text{and} \quad 100\,\text{kHz} - 50\,Hz = \frac{1}{2\pi\sqrt{LC_m}}.$$

Dividing we get

$$\frac{10^5}{10^5 - 50} = \sqrt{\frac{C_m}{C_0}} = \sqrt{K}.$$

Therefore, $K = \dfrac{10^{10}}{(10^5 - 50)^2} \approx 1.001.$

► **Problem 4.** *A capacitor* C *is charged through a resistance* R *by a battery of emf* $\mathcal{E}$. *The energy supplied by the battery is* $q_0\mathcal{E}$ *(where* $q_0 = C\mathcal{E}$*) whereas the energy gained by the capacitor is* $\frac{1}{2}q_0\mathcal{E}$. *Account for the difference in energy.* **[C.U. 1990]**

Solution. The energy supplied by the battery is partly stored in the capacitor as electrostatic energy and rest is dissipated in R as Joule heat. The charge on C at any time t is given by $q = q_0\left(1 - e^{-t/CR}\right)$.

So the instantaneous current

$$i = \frac{dq}{dt} = -\frac{q_0}{CR} \cdot e^{-t/CR}.$$

So the energy dissipated in R

$$= \int_0^\infty i^2 R dt = \frac{q_0^2 \cdot R}{C^2R^2}\int_0^\infty e^{-2t/CR}dt = \frac{1}{2}\frac{q_0^2}{C} = \frac{1}{2}q_0\mathcal{E}.$$

Note that energy dissipated in R is the same as the energy stored in C.

► **Problem 5.** *Examine whether the discharge of a charged capacitor 0.1μF through an inductive coil of inductance 100 mH and resistance* 200 Ω *is oscillatory or not. If it is oscillatory, find the frequency of oscillation.*

Solution. The critical damping resistance is given by

$$2\sqrt{\frac{L}{C}} = 2\sqrt{\frac{100 \times 10^{-3}}{0.1 \times 10^{-6}}} = 2000\,\Omega.$$

As the coil resistance is smaller than this value, the discharge will be oscillatory.

The frequency of oscillation is

$$f = \frac{1}{2\pi}\sqrt{\frac{1}{LC} - \frac{R^2}{4L^2}} = \frac{1}{2\pi}\sqrt{\frac{1}{0.1 \times 0.1 \times 10^6} - \frac{200^2}{4 \times (0.1)^2}} = 1584\,\text{Hz}.$$

► **Problem 6.** *Suppose a coil of inductance L and resistance R is connected in parallel to a series combination of capacitance C and a resistance R. The whole combination is suddenly connected to a voltage source of voltage V. If $L/R = CR$ show that the current drawn from the source is independent of time.*

Solution. From Fig. 11.P-6 the current drawn from the source at any time t is

$$i = i_1 + i_2$$

when i_1 is the current through CR branch and i_2 is the current through LR branch.

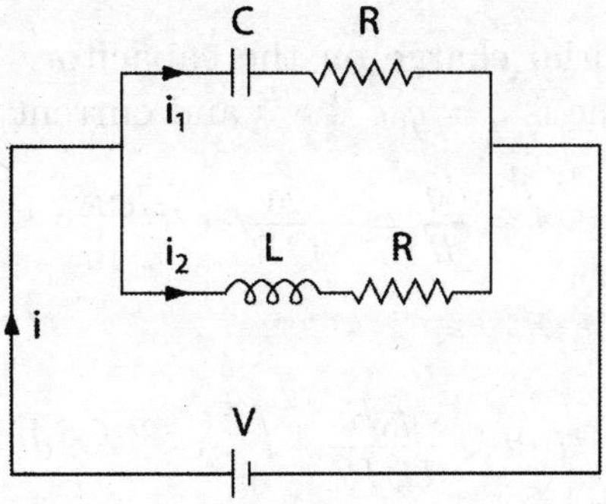

Fig 11.P-6

Now, if q be the instantaneous charge on C then

$$q = q_0\left(1 - e^{-t/CR}\right) \quad \text{where } q_0 = CV$$

$$\text{or} \quad i_1 = \frac{dq}{dt} = \frac{q_0}{CR}e^{-t/CR} = \frac{V}{R}e^{-t/CR}.$$

Also

$$i_2 = \frac{V}{R}\left(1 - e^{-\frac{R}{L}t}\right) = \frac{V}{R}\left(1 - e^{-t/CR}\right).$$

Therefore, $i = i_1 + i_2 = \frac{V}{R}$, which is independent of t.

► **Problem 7.** *A charged capacitor of capacitance $2\mu F$ when left alone loses 2% of its charge in 1 min. But when connected across a high resistance R it loses 25% its charge in 1 min. Find the value of R.*

Solution. The natural leakage resistance R_0 is given by

$$R_0 = \frac{t}{C \ln \frac{q_0}{q}} = \frac{60}{2 \times 10^{-6} \ln \frac{100}{98}}\,\Omega = 1485\,\text{M}\Omega.$$

If $R_{\parallel}$ is the equivalent resistance of R_0 and R in parallel then

$$\frac{1}{R_{\parallel}} = \frac{1}{R_0} + \frac{1}{R}$$

and
$$R_{\parallel} = \frac{60}{2 \times 10^{-6} \ln \frac{100}{75}} \,\Omega = 104\,\text{M}\Omega$$

$$\therefore \quad R = \frac{R_{\parallel} R_0}{R_0 - R_{\parallel}} = \frac{1485 \times 10^4}{1485 - 10^4} \text{M}\Omega = 111\,\text{M}\Omega.$$

► **Problem 8.** *A charged capacitor of capacitance C is suddenly connected across a resistor R. Show that the energy which was stored in the capacitor is now entirely dissipated in R.* **[cf. C.U. 2017]**

Solution. Let q_0 be the initial charge on the capacitor. So energy stored is $q_0^2/2C$. Now charge on C at any time is $q = q_0 e^{-t/CR}$ and current through the circuit is

$$i = \frac{dq}{dt} = -\frac{q_0}{CR} \cdot e^{-t/CR}.$$

Total energy dissipated in R

$$= \int_0^\infty i^2 R dt = \frac{q_0^2 R}{C^2 R^2} \cdot \int_0^\infty e^{-2t/CR} dt = \frac{q_0^2}{2C},$$

which is equal to the initial energy of the capacitor.

► **Problem 9.** *Suppose a series combination of a capacitor C and a resistor R_1 is connected in parallel to a resistor R_2. Then a cell of emf $\mathcal{E}$ and internal resistance R_3 is suddenly connected across R_2 at time $t = 0$. Find the charge on C as a function of time t.*

Solution. Let q be the charge on C at any time t after the cell is connected. From Fig. 11.P-9(a)

$$i = i_1 + i_2 \qquad (1)$$

$$\mathcal{E} = iR_3 + i_2 R_2 \qquad (2)$$

$$\mathcal{E} = iR_3 + i_1 R_1 + \frac{q}{C} \qquad (3)$$

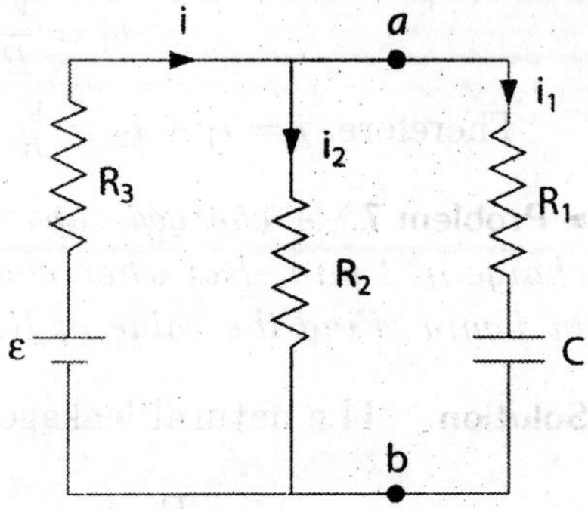

Fig 11.P-9(a)

From (1) and (2) we get

$$\mathcal{E} = (i_1 + i_2)R_3 + i_2 R_2 = i_1 R_3 + i_2(R_2 + R_3)$$

$$\text{or} \quad i_2 = \frac{\mathcal{E} - i_1 R_3}{R_2 + R_3} \qquad (4)$$

From (2) and (3),

$$i_1 R_1 + \frac{q}{C} = i_2 R_2 = \frac{\mathcal{E} - i_1 R_3}{R_2 + R_3} \cdot R_2$$

$$\text{or} \quad i_1 \left(R_1 + \frac{R_2 R_3}{R_2 + R_3} \right) + \frac{q}{C} = \frac{\mathcal{E} R_2}{R_2 + R_3}.$$

Now using $i_1 = \dfrac{dq}{dt}$ we can write

$$\frac{dq}{dt} = -Aq + B, \tag{5}$$

where $A = \dfrac{R_2 + R_3}{C(R_1 R_2 + R_2 R_3 + R_3 R_1)}$ and $B = \dfrac{\mathcal{E} R_2}{R_1 R_2 + R_2 R_3 + R_3 R_1}$.

From (5),

$$\frac{dq}{q - B/A} = -A dt.$$

Integrating,

$$\ln(q - B/A) = -At + \ln K, \quad \text{or} \quad q = \frac{B}{A} + K e^{-At}, \tag{6}$$

where K is a constant. To find K we apply the initial condition:

$$q = 0 \quad \text{at}\, t = 0.$$

Thus, $K = -B/A = -\dfrac{\mathcal{E} R_2}{R_2 + R_3}$.

$$\therefore \quad q = \frac{B}{A}\left(1 - e^{-At}\right) = \frac{\mathcal{E} R_2 C}{R_2 + R_3} \left[1 - \exp\left\{ -\frac{(R_2 + R_3)t}{R_1 R_2 + R_2 R_3 + R_3 R_1} \right\} \right].$$

Note: The solution of the problem can be greatly simplified by applying Thevenin's theorem to the circuit lying left of the points a and b. In the simplified circuit of Fig. 11.P-9(b)

$$V_{Th} = \frac{\mathcal{E} R_2}{R_2 + R_3} \quad \text{and} \quad R_{Th} = \frac{R_2 R_3}{R_2 + R_3}.$$

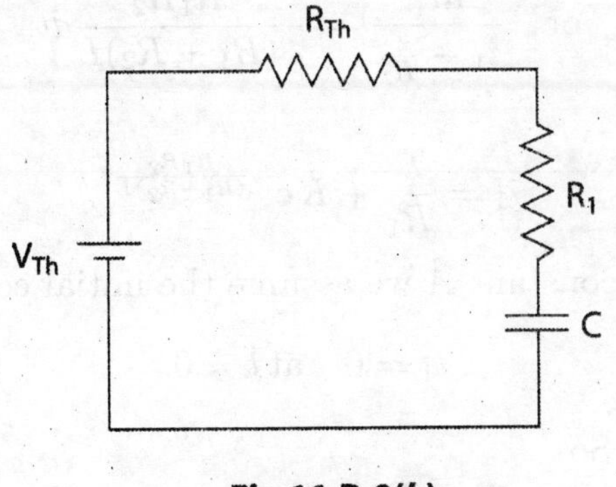

Fig 11.P-9(*b*)

The circuit can now be analysed as in Section 11.3.

▶ **Problem 10.** *A battery of emf $\mathcal{E}$ and internal resistance R_1 is connected across a parallel combination of a pure inductance L and a resistance R_2. If the battery is switched on at time $t = 0$, find the current through L as a function of time.*

Solution. From Fig. 11.P-10(a)

$$i = i_1 + i_2 \tag{1}$$

$$\mathcal{E} = R_1 i + R_2 i_2 \tag{2}$$

$$\mathcal{E} = R_1 i + L\frac{di_1}{dt}. \tag{3}$$

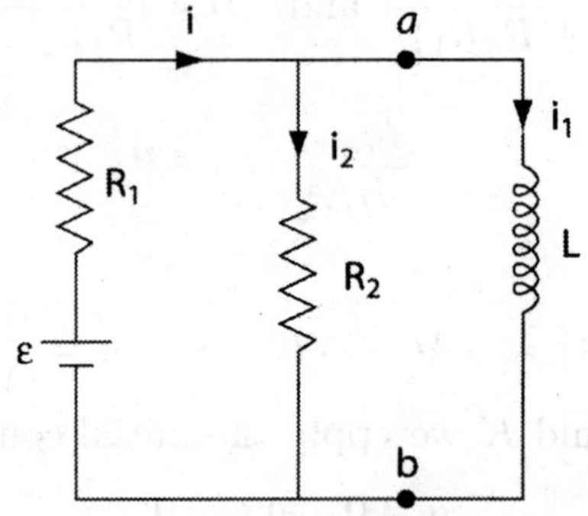

Fig 11.P-10(*a*)

From (2) and (3)

$$i_2 = \frac{L}{R_2}\frac{di_1}{dt}. \tag{4}$$

Now from (1) and (2),

$$\mathcal{E} = R_1(i_1 + i_2) + R_2 i_2 = (R_1 + R_2)i_2 + R_1 i_1 = \frac{(R_1 + R_2)L}{R_2}\frac{di_1}{dt} + R_1 i_1$$

$$\text{or} \quad \frac{di_1}{i_1 - \frac{\mathcal{E}}{R_1}} = -\frac{R_1 R_2}{(R_1 + R_2)L}dt.$$

Integrating,

$$i_1 = \frac{\mathcal{E}}{R_1} + Ke^{-\frac{R_1 R_2}{(R_1+R_2)L}t}.$$

To find the arbitrary constant A we assume the initial condition:

$$i_1 = 0 \quad \text{at}\, t = 0.$$

Therefore, $K = -\frac{\mathcal{E}}{R_1}$. So

$$i_1 = \frac{\mathcal{E}}{R_1}\left(1 - e^{-\frac{R_1 R_2}{(R_1+R_2)L}t}\right).$$

Note: The solution of the problem can be greatly simplified by applying Thevenin's theorem to the circuit lying left of the points a and b. In the simplified circuit shown in Fig. 11.P-10(b),

$$V_{Th} = \frac{\mathcal{E} \cdot R_2}{R_1 + R_2}$$

and $$R_{Th} = \frac{R_1 R_2}{R_1 + R_2}.$$

Now the circuit can be analysed as in Section 11.2.

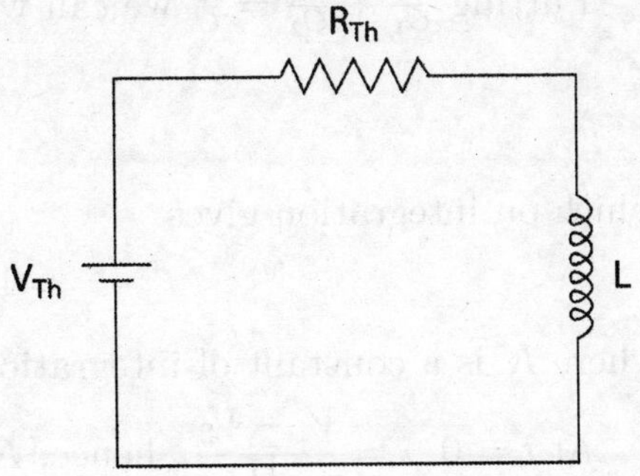

Fig 11.P-10(b)

▶ **Problem 11.** *Two capacitors of capacitances C_1 and C_2 are charged to potentials V_1 and V_2, respectively. They are then disconnected from the source and connected in parallel through a resistor R. Find an expression for current in R as a function of time. Show that the energy dissipated in R is independent of the value of R. Explain the result. Assume that $V_1 > V_2$.* **[C.U. 2004]**

Solution. Let i be the instantaneous current through R; q_1 and q_2 be the instantaneous charges, v_1 and v_2 be the instantaneous voltages across C_1 and C_2, respectively. Then,

$$v_1 - v_2 = Ri, \quad \text{or} \quad \frac{q_1}{C_1} - \frac{q_2}{C_2} = Ri, \quad \text{or} \quad \frac{1}{C_1}\frac{dq_1}{dt} - \frac{1}{C_2}\frac{dq_2}{dt} = R\frac{di}{dt}.$$

As $V_1 > V_2$, C_1 discharges and C_2 charges. Hence,

$$i = -\frac{dq_1}{dt} = \frac{dq_2}{dt}.$$

Thus,

$$-\left(\frac{1}{C_1} + \frac{1}{C_2}\right) i = R\frac{di}{dt}.$$

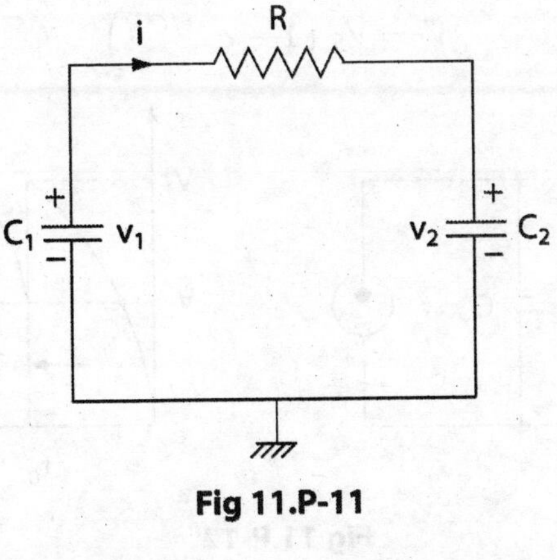

Fig 11.P-11

Putting $\frac{1}{C_1} + \frac{1}{C_2} = \frac{1}{C}$ we can write

$$\frac{di}{i} = -\frac{dt}{CR},$$

which on integration gives

$$\ln i = -\frac{t}{CR} + K,$$

where K is a constant of integration.

At $t = 0$, $i = \frac{V_1 - V_2}{R}$; hence, $K = \ln \frac{V_1 - V_2}{R}$.

Thus,

$$i = \frac{V_1 - V_2}{R} \cdot e^{-t/CR}.$$

Total energy dissipated in R is

$$\int_0^\infty i^2 R dt = \left(\frac{V_1 - V_2}{R}\right)^2 \cdot R \int_0^\infty e^{-\frac{2t}{CR}} dt$$

$$= \frac{1}{2} C (V_1 - V_2)^2 = \frac{C_1 C_2}{2(C_1 + C_2)} (V_1 - V_2)^2.$$

The energy dissipated in R equals the difference between the initial and final energies stored in the capacitors. As these energies are independent of R so the above result is an expected one.

► **Problem 12.** *A neon lamp, shunted by a capacitor C, is connected to a dc source of voltage V_S through a resistance R. The neon lamp glows when the pd across it grows to $V_1 (V_1 < V_S)$ and is extinguished when this pd falls to V_2. Find the frequency of the flicker produced by the above arrangement. Assume that the resistance of the lamp when glowing is zero.*

Solution. The voltage across C increases with time according to the law

$$V = V_S \left(1 - e^{-t/CR}\right) \tag{1}$$

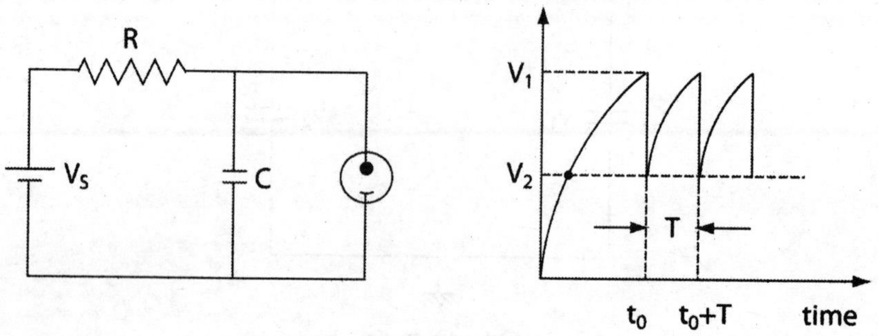

Fig 11.P-12

As this voltage reaches the value V_1 (called the *striking voltage*) the lamp glows and acts as a short circuit. Then the charge on C discharges through the neon lamp almost instaneously causing the voltage V to fall quickly. The lamp is extinguished when the voltage drops to V_2. The lamp now acts as an open circuit and consequently the capacitor is charged again through R according to the law (1) and the cycle repeats. As the discharging time is very small compared to the charging time the time period T is given by the time required by V to increase from V_2 to V_1. Let V_2 be the voltage at time t_0 and V_1 be the voltage across C at time $t_0 + T$ then,

$$V_2 = V_S\left(1 - e^{-t_0/CR}\right) \quad \text{and} \quad V_1 = V_S\left(1 - e^{-(t_0+T)/CR}\right)$$

Eliminating t_0 we get

$$\frac{V_S - V_2}{V_S - V_1} = e^{T/CR} \quad \text{or} \quad T = CR\ln\frac{V_S - V_2}{V_S - V_1}.$$

Therefore, the frequency of the flicker is

$$f = \frac{1}{T} = \frac{1}{CR\ln\frac{V_S - V_2}{V_S - V_1}}$$

QUESTIONS

1. (a) A dc source of voltage V is suddenly applied to a circuit consisting of a resistor R and an inductor L in series. Write down the instantaneous emf equation and hence, find the instantaneous current. Calculate the maximum energy stored in the inductor. **[cf. C.U. 2018]**

(b) A dc source of voltage V is connected for a sufficient time to a circuit consisting of a resistor R and an inductor L in series. The source is now suddenly removed keeping the circuit closed. Find how the current in the circuit decays with time. What do you mean by the time constant of the circuit? Show graphically how the voltages across L and R vary with time.

(c) Consider a circuit consisting of an inductor R, a battery and a key all in series. Explain why the self-induced emf in the circuit is greater during break of the circuit than that during make of the circuit.

2. (a) A dc source of voltage V is suddenly applied to a circuit consisting of a resistor R and a capacitor C in series. Write down the instantaneous emf equation and hence find how the charge on the capacitor grows with time. How does the charging current vary with time?

(b) A capacitor C in series with a resistor R is fully charged by connecting it to a dc source of voltage V. The source is now suddenly removed keeping the circuit closed. Investigate how the charge on C decays with time. What is the time constant of the circuit?

3. A dc source of emf is suddenly applied to a circuit consisting of a resistor R, a capacitor C and an inductor L in series. Investigate the growth of charge on the capacitor pointing out clearly the conditions for (a) a steady growth and (b) an oscillatory growth. Calculate the frequency of oscillating growth of the charge. What type of growth of charge is expected when the circuit is dipped in a medium of temperature near absolute zero?

4. A capacitor C in series with a resistor R and an inductor L is fully charged by connecting it to a dc source of voltage V. The source is now suddenly removed keeping the circuit closed. Investigate how the charge on C decays with time pointing out clearly the conditions for (a) a steady decay and (b) an oscillatory decay. Calculate the frequency of oscillation. What do you mean by critical damping resistance? Discuss the case with $R = 0$. OR

   Investigate the discharge of a capacitor in series LCR circuit driven by a dc source. When is the discharge critically damped? **[C.U. 2011]**

5. Describe the method for the measurement of high resistance by the method of leakage. Deduce the theory required in this connection. Can you measure low resistance by this method?

6. Suppose a fully charged capacitor is suddenly connected to a pure inductor L. Show that the charge on the capacitor varies with time in an oscillatory manner. Find the frequency of oscillation. Try to explain physically the mechanism of oscillation. **[cf. N.B.U. 2017]**

7. Consider a mutually coupled circuit consisting of a primary of inductance L_1, resistance R_1 and a secondary of inductance L_2, resistance R_2

   (a) A source of voltage V is suddenly connected to the primary. Investigate how the currents in the primary and the secondary grow with time.

   (b) Suppose after attainment of steady state, the source is suddenly removed from the primary keeping the circuit closed. Investigate how the currents in the primary and the secondary decay with time.

8. What is the principle of operation of an induction coil? Discuss the working of an induction coil. Cite one practical example of its use.

EXERCISES

1. A coil of inductance 10 H and resistance 10 Ω is connected to a steady voltage of 100 V at time $t = 0$. Find the value of current at $t = 0.1$ s. What is the time taken for the current to reach one half of its steady value?

   [cf. B.U. 2011] [*Ans.* 0.952 A, 0.693 s]

2. A capacitor of 0.1μF is charged by a certain voltage. It is then disconnected from the source and connected across a high resistance of 10 MΩ. Find the time the pd across the capacitor takes to fall to half of its initial value. Also, find the time in which energy stored in the capacitor decreases to one half of its initial value.
[*Ans.* 0.693 s, 0.346 s]

3. A capacitor of $10\mu F$ is charged and then allowed to discharge through a resistor of 1 MΩ. Find the time in which the charge on the capacitor reduces to 36.78% of its initial value. Also find the time in which energy stored in the capacitor decreases to half of its initial value. [*Ans.* 10 s, 12.3 s]

4. Show that the quantities CR and L/R have the dimensions of time. What is its physical significance?

5. A capacitor of $1\mu F$ is charged and then allowed to discharge through a resistor of R for 10s. The charge remaining on the capacitor is found to be half of its initial value. Find the value of R. [*Ans.* 14.43 MΩ]

6. Consider a relay of inductance 0.5 H and resistance 20 Ω. It becomes on when current through it grows to a value 200 mA and becomes off when the current falls to 100 mA. Suppose a voltage pulse defined by

$$V(t) = \begin{cases} 10\,V, & \text{for } 0 \le t \le 0.1\,s \\ 0, & \text{for } t < 0 \text{ and } t > 0.1\,s \end{cases}$$

is suddenly applied to the relay. Find the time duration for which the relay remains operative. [*Ans.* 0.1276 s]

7. A charged capacitor of $0.1\mu F$ is being discharged through a coil of 10 mH. What must be the resistance of the coil in order that the discharge shall be critically damped? **[cf. C.U. 2011]** [*Ans.* 632 Ω]

8. A capacitor of $20\mu F$ is charged to a pd of 30 V and is suddenly short circuited by a coil of inductance 2 mH and negligible resistance. Find the frequency and amplitude of the resultant current oscillation. [*Ans.* 796 HZ, 3 A]

9. A capacitor of capacitance $0.5\mu F$ and natural leakage resistance 10 MΩ is charged and then kept insulated. Find the time taken by the charge on the capacitor to fall to half of its initial value. [*Ans.* 3.46 s]

10. A capacitor $C = 2\mu F$ is first charged and then allowed to discharge through a coil of inductance 150 mH and resistance 150 Ω. Examine whether the discharge is oscillatory or not. [*Ans.* Oscillatory]

11. (a) An inductive coil of inductance L and resistance R is suddenly connected to a battery of emf $\mathcal{E}$. Show that the current after a time t is given by $\mathcal{E}(1-\alpha)/R$, where $L \ln \alpha + Rt = 0$. **[N.B.U. 2017]**

(b) Also show that if the contact with the battery is now broken keeping the circuit closed then the current, after a further time t seconds after the break, will be $\mathcal{E}\cdot\alpha(1-\alpha)/R$.

[**Hints.** For the second part use the initial condition: At $t = 0$, $i = \mathcal{E}\alpha(1-\alpha)/R$]

12. Show that the number of oscillations after which the amplitude of current decreases to $\frac{1}{e}$ of its initial value in a series LCR circuit is given by $(1/2\pi)\sqrt{4L/CR^2 - 1}$.

13. A charged capacitor of capacitance 1μF loses 10% of its charge in 10 min. When a high resistance is connected between its terminals it loses 50% of its charge in 2 min. Calculate the value of the high resistance. [*Ans.* 178 MΩ]

14. The time constant of a coil is 2 millisecond. If a resistance of 40 Ω is connected in series with the coil the time constant changes to 1 millisecond. Find the self-inductance and resistance of the coil. [*Ans.* 80 mH, 40 Ω]

15. A dc source of 200 V is suddenly applied at time $t = 0$ to a series circuit consisting of a capacitor 40μF and a resistor 2 kΩ. Find the energy stored in the capacitor at time $t = 0.04$ sec. [*Ans.* 0.124 J]

16. A source of constant voltage is suddenly applied to a series LR circuit at $t = 0$. The voltage across L is 25 V at $t = 0$ and drops to 5 V at $t = 25$ millisec. If $R = 120\,\Omega$, find the value of L. [*Ans.* 1.86 H]

17. A source of constant voltage 40 V is suddenly applied to a series LR circuit. The current in the circuit, 0.3s after the switching on the source, is found to be 1 A. After the current in the circuit reaches a steady value the source is switched off keeping the circuit closed. The current is again found to be 1 A, 0.3s after the switching off the source. Find the values of L and R. [*Ans.* 8.656 H, 20 Ω]

18. A capacitor charged to a pd of 200 V is suddenly connected to a voltmeter having an internal resistance of 20 MΩ at time $t = 0$. The voltmeter reading is found to drop to 25 V in a time of 2 mins. Find the capacitance of the capacitor.

[*Ans.* 2.88 μF]

19. Suppose a capacitor 10μF and a resistor 2 MΩ are connected in series, the capacitor being shunted by a neon lamp, which just begins to glow when the pd across it is 120 V. If a dc source of 200 V is suddenly applied across the whole combination find the time after the connection when the neon lamp will begin to glow. [*Ans.* 18.3 s]

20. A capacitor of capacitance C is charged by a cell of emf V_1 through a resistance R for a sufficient time. Then the cell is suddenly replaced by another cell of lower emf V_2. Show that the voltage across C, t sec after the replacement, is given by

$$v(t) = V_1 e^{-t/CR} + V_2\left(1 - e^{-t/CR}\right).$$

21. A capacitor of capacitance $10\,\mu$F is connected through a $1\,\text{M}\Omega$ resistor to a dc source. After being charged for 10 sec the capacitor is disconnected. Calculate the percentage of the energy input from the source, that is stored in the capacitor. [*Ans.* 31.6%]

22. A capacitor of capacitance C having a charge q_0 on it is suddenly connected to a cell of emf V through a resistance R. Show that the charge on C, t sec after connection, is given by

$$q(t) = q_0 e^{-t/CR} + CV\left(1 - e^{-t/CR}\right).$$

23. The electrical energy stored in a charged capacitor of 0.4μF is the same as the magnetic energy in a 1 mH inductor carrying a current of 100 mA. How much charge is stored in the capacitor? [*Ans.* 2 μC]

24. A coil of inductance $L = 100$ mH having a resistance $R = 1\Omega$ is connected across a cell of e.m.f. $E = 2V$ with an internal resistance $r = 9\Omega$. If the source is removed, find the current through the inductor as a function of time.
[**Hints.** Use Eq. (11.4-20); $i = -0.2e^{-5t}\sin 1000t$ A] **[C.U. 2007]**

25. (*a*) A series RC circuit is excited by a source of constant voltage V switched at time $t = 0$. What is the maximum growth rate of charge on the capacitor? **[C.U. 2009]** [*Ans.* V/R]

(*b*) In an oscillator circuit $L = 0.1$ H, $C = 0.047\,\mu$F, find the maximum value of resistance so that the circuit may oscillate. **[C.U. 2009]**
[**Hints.** $b < \omega_0$ or $R_{\text{max}} = 2\sqrt{L/C} = 2917\Omega$]

26. Check whether the LCR series combination with $L = 0.2$ H, $C = 2\mu$ F, $R = 100\Omega$ connected to a DC source will be oscillatory in nature? Plot the current through the circuit as a function of time when the source is suddenly connected.
[**Hints.** Show that $b < \omega_0$ or $< 2\sqrt{L/C}$. So oscillatory] **[C.U. 2014, 2017]**

27. A battery of emf 10 V and internal resistance 200Ω is suddenly connected across the parallel combination of a resistor of 200Ω and an inductor of 0.01 mH. Find the current through the inductor as a function of time
[*Ans.* $i_L(t) = 0.05(1 - e^{-10^7 t})A$]

28. A capacitance C shunted by a resistance R is connected to a dc voltage V_0 through a resistance $2R$ and a switch all in series. Show that the voltage across C, t time after the switch is made on, is given by

$$V(t) = \frac{V_0}{3}\left(1 - e^{-3t/2CR}\right)$$

[**Hints.** $i_R \cdot R = q/C$

or $\left(i - \frac{dq}{dt}\right) R = q/C$ or $i = \frac{q}{CR} + \frac{dq}{dt}$

Now $V_0 = 2iR + q/C$

$$= 2R\left(\frac{q}{CR} + \frac{dq}{dt}\right) + q/C$$

or $\frac{dq}{dt} + \frac{3q}{2CR} = \frac{V_0}{2R}$

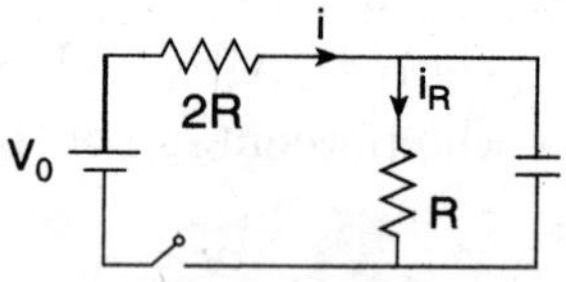

Fig. 11.E-28

Integrating with the initial condition that at $t = 0$, $q = 0$ we get the result.]

29. A charged capacitor $C = 1\,\mu$F when connected in parallel with an inductor $L = 10\,$mH the charge on C oscillates with amplitude $10\,\mu$C. Find the maximum energy stored in L during oscillation. [*Ans.* $50\,\mu$J]

30. A series LR circuit is supplied with a voltage pulse defined by $v(t) = V_0$ for $0 < t < T$ and $v(t) = 0$ for $t > T$. Find the current in the circuit for $t > T$.

[**Hints.** $i(t) = \frac{V_0}{R}\left(1 - e^{-\frac{R}{L}t}\right)$ for $0 < t < T$

For $t > T$, $L\frac{di}{dt} + Ri = 0$

Integrating we get $\ln i = -\frac{R}{L}t + Constant.$

At $t = T$, $i(T) = \frac{V_0}{R}\left(1 - e^{-\frac{R}{L}T}\right)$

$\therefore$ $Constant = \ln i(T) + \frac{R}{L}T$

Thus $i = i(T)e^{-\frac{R}{L}(t-T)}$]

OBJECTIVE TEST

1. Of various combination of resistance R, inductance L and capacitance C which one of the following has the dimension of time?

(a) RC, (b) R/L, (c) LC/R, (d) $1/\sqrt{LC}$.

2. A capacitor is charged to a pd of 100 V and is then connected across a resistor. It is found that the pd decays to 80 V after 1 s. The pd across the capacitor after $2s$ will be

(a) 40 V, (b) 60 V, (c) 64 V, (d) almost zero.

3. No transient is produced in pure resistive circuit, because it

(a) is linear (c) is passive,

(b) is non-linear, (d) has no energy storing mechanism.

4. Time constant of a CR circuit may be defined as the time in which the initial charge on the capacitor decays to

(a) 63%, (b) 37%, (c) 0.63%, (d) 0.37%.

5. If the discharge of a fully charged capacitor C through a coil of self-inductance L and resistance R is oscillatory then the frequency of oscillation is given by

(a) $\frac{1}{2\pi}\sqrt{\frac{1}{LC}}$, (b) $\frac{1}{2\pi}\sqrt{\frac{1}{LC}-\frac{R^2}{4L^2}}$, (c) $\frac{1}{2\pi}\sqrt{\frac{1}{LC}-\frac{R}{2L}}$,

(d) none of these.

6. By leakage method it is possible to measure

(a) high resistances only, (c) low resistances only,

(b) high as well as low resistances, (d) none of these.

7. Time constant of a CR circuit is

(a) $1/CR$, (b) CR, (c) R/C, (d) C/R.

8. Time constant of a LR circuit is

(a) LR, (b) $1/LR$, (c) L/R, (d) R/L.

9. In the following circuit C_1 and C_2 are ideal capacitors. c_1 has been charged to 12 V before the ideal switch 5 is closed at $t = 0$. The current $i(t)$ for all t is

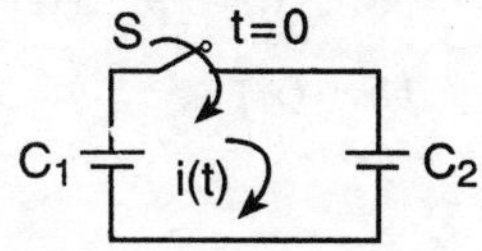

(a) zero, (b) a step function,

(c) an exponentially decaying function,

(d) an impulse function. **[GATE 2013]**

10. In the following circuit the switch s is opened at time $t = 0$. Current through the 1μF capacitor at time $t = 0+$ is

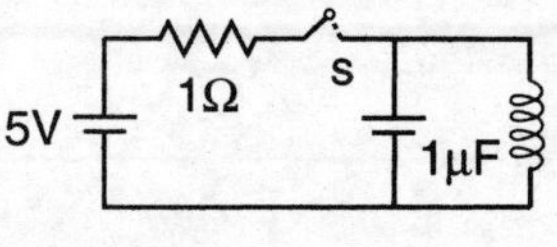

(a) 0 A, (b) 1 A, (c) 1.25 A, (d)] 5 A **[GATE 2010]**

11. In the following circuit when switch S is closed a steady current $I = V/R$ is established and an energy $\frac{1}{2}LI^2$ is stored in the magnetic field of inductor L. If the switch is suddenly opened what happens to this energy?

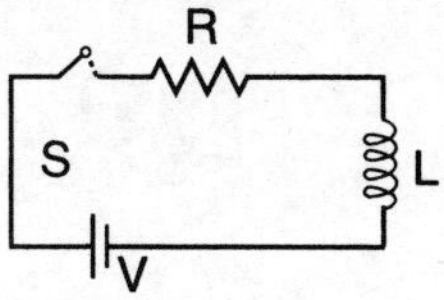

(a) dissipates as Joule heat in R (c) radiates as EM waves

(b) goes back to the battery, (d) remains as it is.

12. A rectangular voltage pulse of magnitude V and duration T is applied to a series CR circuit. The maximum voltage developed across C is

(a) $\frac{VT}{CR}$, (b) V, (c) $Ve^{-T/CR}$, (d) $V\left[1 - e^{-T/CR}\right]$

13. A time varying voltage $v(t) = kT^2$ is suddenly applied across a pure inductor L at time $t = 0$. Then current through L at time t is

(a) $\frac{kt}{L}$, (b) $\frac{2kt}{L}$, (c) $\frac{kt^3}{2L}$, (d) $\frac{kt^3}{3L}$

14. The time constant of the following circuit is

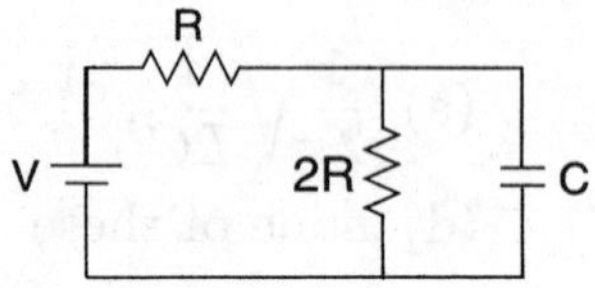

(a) $2CR$, (c) $\frac{1}{2}CR$,

(b) $3CR$, (d) $\frac{2}{3}CR$

15. A capacitor when connected to an ideal dc voltage source will be charged

(a) instantaneously, (c) exponentially

(b) in infinite time, (d) linearly

ANSWERS

1. (a) **2.** (c) **3.** (d) **4.** (b) **5.** (b) **6.** (a) **7.** (b)
8. (c) **9.** (d) **10.** (b) **11.** (c) **12.** (d) **13.** (d) **14.** (d)
15. (a)

Chapter 12

Alternating Currents

12.1 Alternating Currents and Some Terms Associated with it

An alternating current or voltage (*ac*) is one which passes through a complete cycle of changes both in magnitude and direction at regular intervals of time. For a symmetric ac the positive half-cycle is exactly similar to the negative half-cycle and its average over a complete cycle is zero. Calculations on ac are usually done by considering sinusoidal variations with time. This is very useful because even if the current or voltage varies in a nonsinusoidal arbitrary periodic manner it can always be considered as the superposition of a number of sinusoidal variations. Fig. 12.1-1 shows the waveform of a sinusoidal ac voltage. Such a voltage may be generated by rotating a coil in a uniform magnetic field. A pure sinusoidal voltage can be represented by

$$v = V_0 \sin \omega t, \quad \text{or} \quad v = V_0 \cos \omega t. \tag{12.1-1}$$

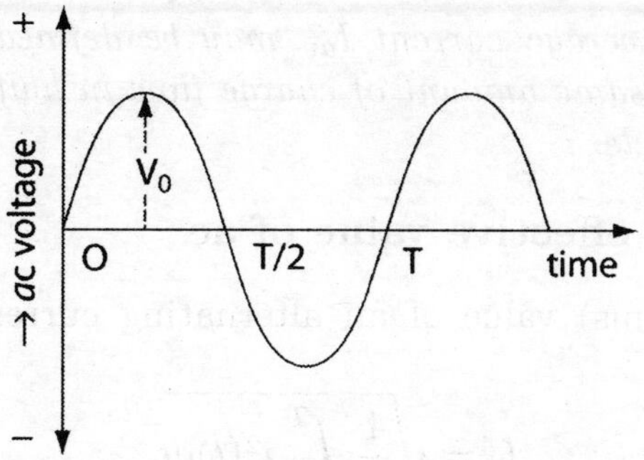

Fig. 12.1-1 : Sinusoidal ac voltage

with a suitable choice of the origin of time. Here v is the instantaneous value, V_0 is the maximum value of the voltage called *peak value* and ω is the *angular frequency.* $\omega/2\pi = f$ is the frequency and $2\pi/\omega = T$ is the *time period* of variation. In India the commercial electric power supply has the sinusoidal waveform with a frequency of 50 Hz.

When an alternating voltage is applied to a circuit the current through the circuit also varies with time at the frequency of the applied ac voltage. This alternating current can be represented similarly as

$$i = I_0 \sin(\omega t + \phi), \tag{12.1-2}$$

where I_0 is the peak current and ϕ is the phase difference between the applied voltage v and the current i. Note that in steady dc circuit current and voltage remain in phase but in ac circuits this is, in general, not so.

Average or mean value of ac

The average of any symmetric ac wave with period T over a complete cycle is zero. When we speak of the average of ac we mean average over a half-cycle. Thus, average current

$$I_{av} = \frac{1}{T/2}\int_{t_1}^{t_1+T/2} i(t)dt, \tag{12.1-3}$$

where t_1 is a time instant when $i(t) = 0$.

For a simple sinusoidal current $i = I_0 \sin \omega t$ we get

$$I_{av} = \frac{1}{T/2}\int_0^{T/2} I_0 \sin \omega t \cdot dt = \frac{2I_0}{T\omega}\left[-\cos \omega t\right]_0^{T/2} = \frac{2I_0}{\pi} \quad \left[\because\ \omega = \frac{2\pi}{T}\right]$$

In a similar way the average of a sinusoidal voltage is given by

$$I_{av} = \frac{2V_0}{\pi}.$$

Equation (12.1-3) may be rewritten in the form,

$$I_{av} \cdot \frac{T}{2} = \int_{t_1}^{t_1+T/2} i(t)dt$$

Thus, *physically the average current I_{av} may be defined as the dc equivalent of current, which causes the same amount of charge flow in half time period as is caused by the ac during a half-cycle.*

Root mean square or effective value of ac

The root mean square (rms) value of an alternating current $i(t)$ with period T is defined by

$$I_r = \sqrt{\frac{1}{T}\int_0^T i^2(t)dt}. \tag{12.1-4}$$

This rms value I_r is also known as *effective* or *virtual value.* It is an important parameter of ac. It can be related to the heating effect of ac. Heat produced by ac in a resistance R over a time period T is

$$\int_0^T i^2(t) \cdot R \cdot dt = I_r^2 \cdot R \cdot T \quad \text{[Using Eq. (12.1-4)]}$$

Thus, *physically the rms value of ac may be defined as the dc equivalent of current which produces the same heating effect as is caused by the ac during one time period.*

For a simple sinusoidal current $i = I_0 \sin \omega t$, the rms value is given by

$$I_r^2 = \frac{1}{T}\int_0^T I_0^2 \sin^2 \omega t dt = \frac{I_0^2}{2T}\int_0^T (1 - \cos 2\omega t) dt$$

$$= \frac{I_0^2}{2T}\left[t - \frac{\sin 2\omega t}{2\omega}\right]_0^T = \frac{I_0^2}{2}.$$

Therefore,

$$I_r = \frac{I_0}{\sqrt{2}}.$$

Similarly for a sinusoidal voltage $V_r = V_0/\sqrt{2}$. Measuring instruments based on the heating effect of ac are generally calibrated in terms of rms values. When we speak of 220 V ac supply we mean its rms value. Thus, peak value of a 220 V ac supply is $V_0 = \sqrt{2} \times V_r = \sqrt{2} \times 220\,\text{V} = 311\,\text{V}$. A 220 V dc supply always remains at 220 V whereas a 220 V ac supply passes through a peak value of 311 V. For this 220 V ac *is more dangerous than* 220 V dc.

Form factor

The ratio of the rms value to the average value of ac is called its *form factor* (K_f). For a pure sinusoidal ac,

$$K_f = \frac{I_r}{I_{av}} = \frac{I_0/\sqrt{2}}{2I_0/\pi} = 1.11 \tag{12.1-5}$$

Form factor gives an idea about the waveshape. Any deviation in the value of K_f from 1.11 indicates deviation from sinusoidal nature. A knowledge of form factor helps to find the rms value from the average value and vice versa.

Peak factor

The ratio of the peak value to the rms value of any ac waveform is called its *peak factor.* For a pure sinusoidal ac,

$$\text{peak factor } (K_p) = \frac{I_0}{I_0/\sqrt{2}} = \sqrt{2}. \tag{12.1-6}$$

Since waves of same rms value may have different peak values a knowledge of peak factor is necessary to get an idea of peak value. Again a knowledge of peak value is important while testing dielectric insulation or hysteresis loss.

rms value of nonsinusoidal ac

If the alternating current is not sinusoidal but an arbitrary periodic function of time then it may be represented by a Fourier series such as

$$i = I_0 + \sum_{n=1}^{\infty} I_n \sin(n\omega t + \phi_n). \tag{12.1-7}$$

Now its rms value I_r is given by

$$\begin{aligned}
I_r^2 &= \frac{1}{T}\int_0^T i^2 dt \\
&= \frac{1}{T}\int_0^T \left[I_0^2 + \left\{ \sum_{n=1}^{\infty} I_n \sin(n\omega t + \phi_n) \right\}^2 + 2I_0 \sum_{n=1}^{\infty} I_n \sin(n\omega t + \phi_n) \right] dt \\
&= I_0^2 + \frac{1}{T}\int_0^T \sum_{n=1}^{\infty}\sum_{m=1}^{\infty} I_n I_m \sin(n\omega t + \phi_n)\sin(m\omega t + \phi_m) dt \\
&\quad + \frac{2I_0}{T}\int_0^T \sum_{n=1}^{\infty} I_n \sin(n\omega t + \phi_n) dt.
\end{aligned}$$

Using the orthogonal properties of sine and cosine functions we get

$$\begin{aligned}
I_r^2 &= I_0^2 + \sum_{n=1}^{\infty}\sum_{m=1}^{\infty} \frac{1}{2} I_n I_m \delta_{mn} + 0 \quad \text{[where } \delta_{mn} \text{ is the Kronecker delta]} \\
&= I_0^2 + \frac{1}{2}\sum_{n=1}^{\infty} I_n^2 = I_0^2 + \frac{I_1^2}{2} + \frac{I_2^2}{2} + \frac{I_3^2}{2} + \cdots .
\end{aligned}$$

$$\therefore \quad I_r = \sqrt{I_0^2 + I_{1r}^2 + I_{2r}^2 + \cdots}, \tag{12.1-8}$$

where I_{1r}, I_{2r}, etc. represent the rms values of harmonic components.

Power in ac circuits

The instantaneous power $P(t)$ in ac circuits is given by the product of instantaneous voltage $v(t)$ and instantaneous current $i(t)$, i.e.,

$$p(t) = v(t) \cdot i(t).$$

As the voltage and current reverse theirs signs periodically and there is a phase difference between them, $P(t)$ may be positive, zero or negative. A positive value of $p(t)$ indicates that the circuit is receiving energy from the source and negative value of $p(t)$ indicates that the source is receiving energy from the circuit. The instantaneous power is not of any physical sense. The real power is given by the average over a time period (T). Thus,

$$\text{average power } P = \frac{1}{T}\int_0^T v(t)\cdot i(t)dt.$$

Usually in ac circuits voltage and current differ by some phase angle (ϕ). Let

$$v(t) = V_0 \sin\omega t \quad \text{and} \quad i(t) = I_0 \sin(\omega t + \phi)$$

$$\therefore \quad P = \frac{V_0 I_0}{T}\int_0^T \sin\omega t \cdot \sin(\omega t + \phi)dt = \frac{V_0 I_0}{2T}\int_0^T [\cos\phi - \cos(2\omega t + \phi)]\, dt$$

$$= \frac{V_0 I_0}{2}\cos\phi = \frac{V_0}{\sqrt{2}}\cdot\frac{I_0}{\sqrt{2}}\cos\phi = V_r \cdot I_r \cdot \cos\phi, \qquad (12.1\text{-}9)$$

where V_r is the rms voltage and I_r, the rms current. The term $\cos\phi$ is known as *power factor.* The product $V_r I_r$ does not, except for the case $\phi = 0$, give the true power dissipated in the circuit as does the product in dc circuits. The product $V_r I_r$ is called *apparent power.* Power factor is thus, the factor by which the apparent power is to be multiplied to get the *real power* in ac circuits. Therefore,

$$\textit{real power} = \textit{apparent power} \times \textit{power factor}.$$

If in any case voltage and current differ in phase by 90° then $\cos\phi = 0$ and hence no power is dissipated in the circuit. The current flowing in such circuits is called *wattless current.* It will be shown latter that current flowing through a pure inductor or a pure capacitor is a wattless current.

Power and wattless components of current

The instantaneous power

$$p(t) = v(t)i(t) = V_0 \sin\omega t \cdot I_0 \sin(\omega t + \phi)$$
$$= V_0 \sin\omega t \cdot (I_0 \cos\phi) \sin\omega t + V_0 \sin\omega t \cdot (I_0 \sin\phi) \cos\omega t.$$

The first term on the right-hand side represents instantaneous real power. Its average value is $\frac{1}{2}V_0 I_0 \cos\phi$. It may be considered as the product of a voltage $V_0 \sin\omega t$ and a current $(I_0 \cos\phi)\sin\omega t$, which are in phase. For this the term $I_0 \cos\phi$ is called *power component* of current.

The second term on the right-hand side has the average value zero. It may be considered as the product of the voltage $V_0 \sin\omega t$ and the current $(I_0 \sin\phi)\sin(\omega t + \pi/2)$, which are 90° out of phase. For this the term $I_0 \sin\phi$ is called *wattless component* of current.

12.2 Current-Voltage Relationship in Pure Resistance, Inductor and Capacitor

(i) Pure resistance

Fig. 12.2-1 shows a circuit in which a sinusoidal voltage source $v = V_0 \sin \omega t$ has been applied to a resistance R. Considering the polarities and assumed current direction at any time instant t, as shown in Fig. 12.2-1 we can write from Ohm's law,

$$v = Ri \quad \text{or} \quad i = \frac{V_0}{R} \sin \omega t = I_0 \sin \omega t,$$

where $I_0 = V_0/R$ is the peak current.

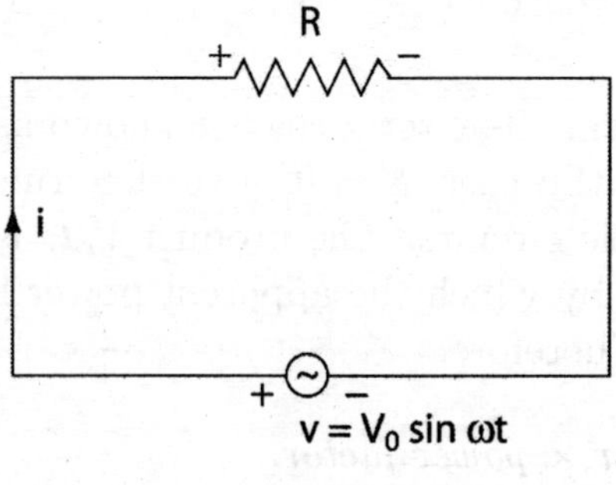

Fig. 12.2-1

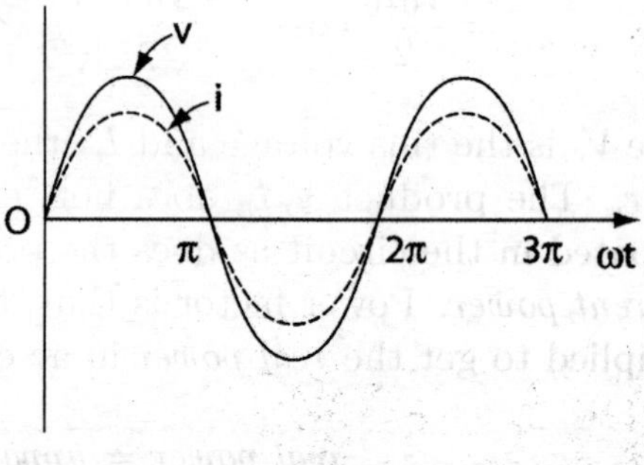

Fig. 12.2-2: Current and voltage in pure R

Obviously, in *pure resistance current* is in *phase with the applied voltage.* Power dissipation in R is given by

$$P = V_r \cdot I_r \cdot \cos \phi = \frac{V_0 I_0}{2} \cos 0° = I_r^2 R.$$

(ii) Pure inductance

Suppose an ac voltage $v = V_0 \sin \omega t$ is applied across an inductor of inductance L (Fig. 12.2-3). If i be the instantaneous current then

$$v = V_0 \sin \omega t = L\frac{di}{dt} \quad \text{or} \quad di = \frac{V_0}{L} \sin \omega t dt.$$

Integrating,

$$i = \frac{V_0}{L} \int \sin \omega t dt = -\frac{V_0}{\omega L} \cos \omega t$$

$$= -I_0 \cos \omega t = I_0 \sin(\omega t - \pi/2),$$

where $I_0 = \frac{V_0}{\omega L}$ is the peak current. The quantity ωL has the dimensions of resistance and is called *inductive reactance.* It is a measure of the opposition offered by L to the flow of ac current through it.

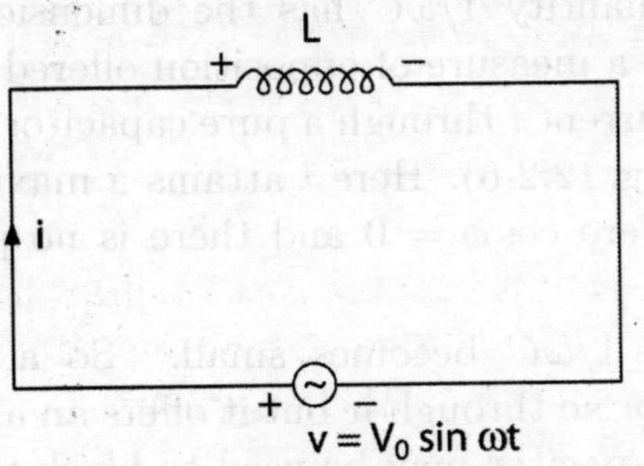

Fig. 12.2-3

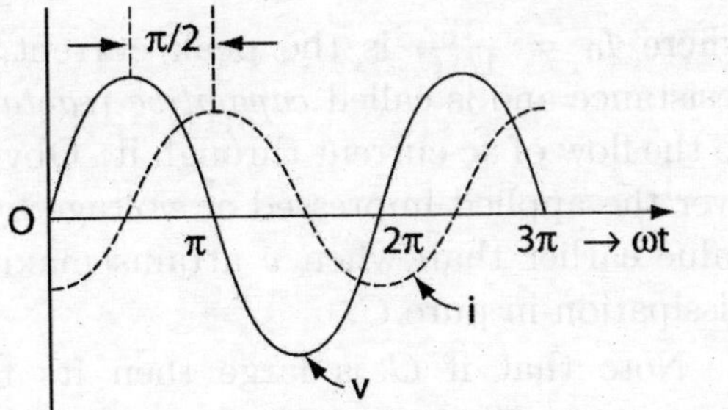

Fig. 12.2-4 : Current and voltage in pure L

Obviously, *current i through a pure inductance lags behind the impressed ac voltage by* 90° (Fig. 12.2-4). i attains a maximum value later than when v attains maximum. Here $\cos\phi = 0$ and no power is dissipated in L. A coil of high inductance L and negligible resistance offers a high reactance (ωL) to the flow of ac and hence, it can be used to restrict the current in ac circuits. An inductor used in this way is known as *choke coil.* Thus, a choke coil can be used to reduce current in ac circuits without introducing considerable wastage of energy. On the other hand, if a resistance is used to reduce current there is a considerable wastage of energy as a Joule heat. Thus, use of a choke coil is more economic than a resistance in controlling current in an ac circuit.

Note that the inductive reactance ωL increases with the frequency of ac. Physically we expect it to be so. As the frequency increases the rate of change of current with time, i.e., di/dt increases. This induces more back emf in L and hence gives rise to more opposition or reactance to the flow of ac.

(iii) Pure capacitance

Suppose an ac voltage $v = V_0 \sin\omega t$ is applied across a capacitor C (Fig. 12.2-5). The instantaneous charge on the capacitor is $q = Cv = CV_0 \sin\omega t$.

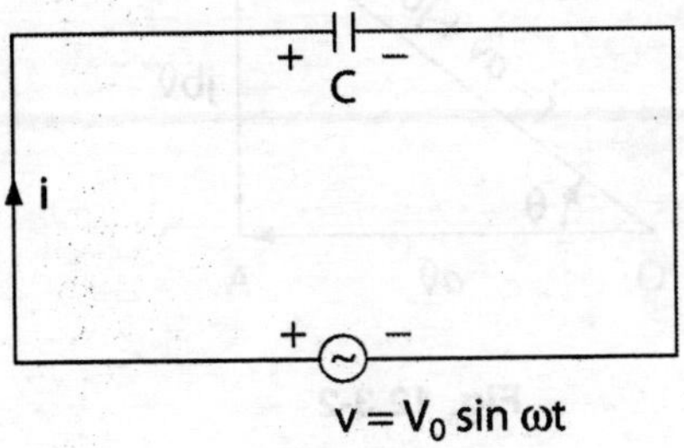

Fig. 12.2-5

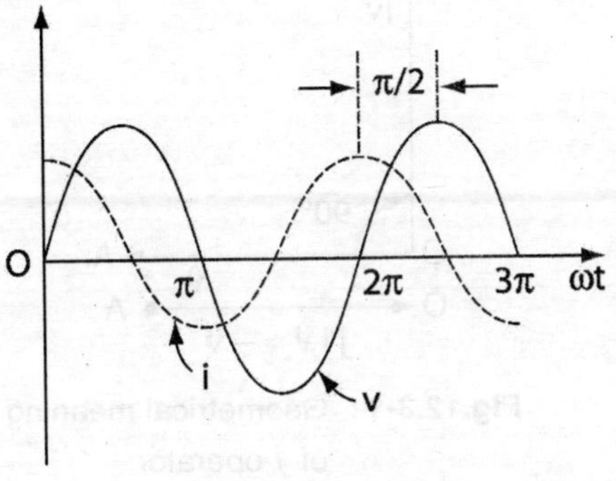

Fig. 12.2-6 : Current and voltage in pure C

and the instantaneous current is given by

$$i = \frac{dq}{dt} = \omega CV_0 \cos\omega t = I_0 \sin(\omega t + \pi/2),$$

where $I_0 = \frac{V_0}{1/\omega C}$ is the peak current. The quantity $1/\omega C$ has the dimensions of resistance and is called *capacitive reactance.* It is a measure of opposition offered by C to the flow of ac current through it. Obviously, current i through a pure capacitor leads over the applied impressed ac voltage by 90° (Fig. 12.2-6). Here i attains a maximum value earlier than when v attains maximum. Here $\cos\phi = 0$ and there is no power dissipation in pure C.

Note that if C is large then its reactance $1/\omega C$ becomes small. So a large capacitance offers almost a short-circuited path for ac through it but it offers an almost infinite resistance to the flow of dc. So a large capacitor may be used to block the dc component and allow easy flow to ac components. A capacitor used in this way is called a *blocking capacitor.*

Note that the capacitive reactance $1/\omega C$ decreases with increases in frequency of ac. As the frequency increases the rate of change of charge on C with time increases. This increases current $i = dq/dt$ through C and hence in effect decreases the capacitive reactance.

12.3 Representation of Sinusoidal ac by Using j-Operator and Rotating Vectors

Mathematical analysis of ac circuits becomes very much simplified if complex numbers and the concept of rotating vectors are used. Let us first find the geometrical meaning of $j = \sqrt{-1}$. Suppose $\overrightarrow{OA} = \hat{\nu}$ be a unit vector in some definite direction in space. Then $-\hat{\nu} = jj\hat{\nu}$ can be represented by a vector $\overrightarrow{AO}$. So the effect of multiplying $\hat{\nu}$ by jj is to rotate the vector through 180°. Thus, j may be considered as on operator whose effect is to rotate the operant through an angle 90° without altering its magnitude. Conventionally we take the rotation in anticlockwise direction.

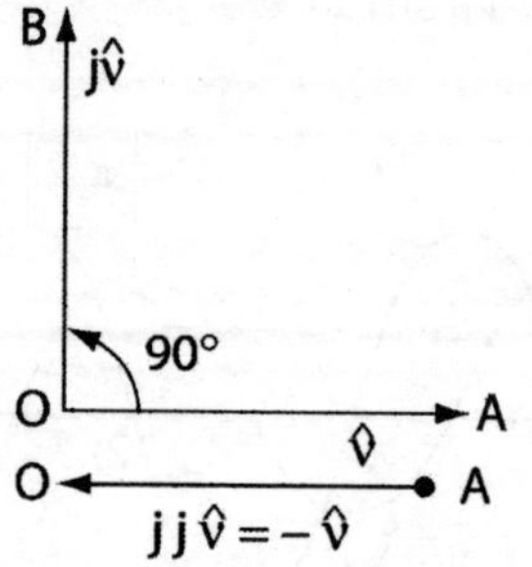

Fig.12.3-1 : Geometrical meaning of j operator

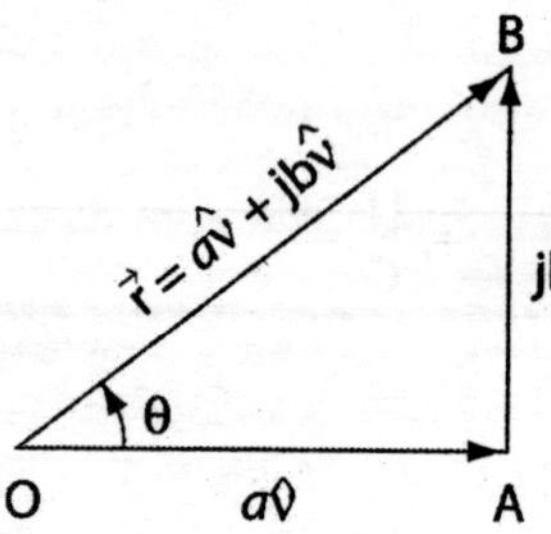

Fig. 12.3-2

In Fig. 12.3-2 let $\overrightarrow{OA} = a\hat{\nu}$ and $\overrightarrow{AB} = jb\hat{\nu}$ then

$$\vec{r} = \overrightarrow{OB} = \overrightarrow{OA} + \overrightarrow{AB} = (a + jb)\hat{\nu}$$

$$\text{and} \quad |\vec{r}| = \sqrt{a^2 + b^2} \quad \text{and} \quad \theta = \tan^{-1}\tfrac{b}{a}.$$

Thus, $a + jb$ is an operator whose effect on $\hat{\nu}$ is to increase its magnitude by a factor $r = \sqrt{a^2 + b^2}$ and to rotate it anticlockwise though an angle $\theta = \tan^{-1}(b/a)$. From Fig. 12.3-2 we can write

$$a + jb = r(\cos\theta + j\sin\theta) = re^{j\theta}.$$

Thus, we can say that the vector $\vec{r} = (a + jb)\hat{\nu} = re^{j\theta}\hat{\nu}$ makes an angle θ with $\hat{\nu}$. Now if $r =$ constant and $\theta = \omega t$ where ω is constant then

$$\vec{r} = re^{j\omega t}\hat{\nu}. \tag{12.3-1}$$

It represents a vector of constant magnitude r which rotates with constant angular velocity ω about its own tail. At time $t = 0$, $\vec{r}$ is in the direction of $\hat{\nu}$ and at time t, $\vec{r}$ makes an angle ωt with $\hat{\nu}$.

A vector like $\vec{r} = re^{j(\omega t+\phi)}\hat{\nu}$ also represents a rotating vector. At time $t = 0$ it makes an initial angle ϕ with $\hat{\nu}$. The vector $\vec{r} = re^{-j\omega t}\hat{\nu}$ represents a clockwise rotating vector. Such rotating vectors when used to represent ac voltages or currents are called *phasors*. Its magnitude represents the peak or rms value of ac and the angle θ which the vector makes with a specified direction represents the phase difference with respect to the impressed ac voltage.

We know that sinusoidal alternating voltages can be represented by $v = V_0 \sin\omega t$ or by $v = V_0 \cos\omega t$ with proper choice of the origin of time. Since

$$V_0 \cos\omega t = \text{Real part of } V_0 e^{j\omega t}$$

$$\text{and} \quad V_0 \sin\omega t = \text{Imaginary part of } V_0 e^{j\omega t}$$

we can express the impressed voltage as a complex quantity $v = V_0 e^{j\omega t}$, provided we remember that the actual voltage is given by the real or the imaginary part of this complex voltage. As the mathematical operations of complex numbers do not affect the identities of real and complex parts we can perform the circuit analysis by using the complex voltage and finally the actual currents or potential drops in different branches are determined either by taking the real or the complex part of the corresponding complex quantities. This is done for mathematical convenience.

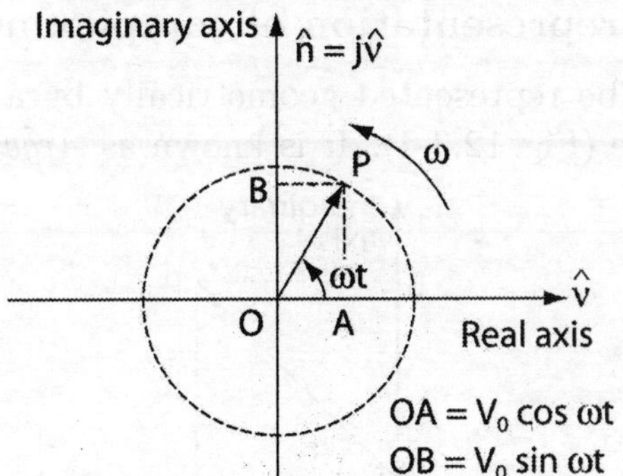

Fig. 12.3-3: Representation of complex voltage by a rotating vector

Now the complex voltage $v = V_0 e^{j\omega t}$ can be represented by a vector

$$\overrightarrow{OP} = \vec{V} = V_0 e^{j\omega t}\hat{\nu}, \tag{12.3-2}$$

which has a constant magnitude V_0 and rotates with constant angular velocity ω; at time $t = 0$, $\vec{V}$ is directed along the unit vector $\hat{\nu}$ and at any time t it makes an angle ωt with $\hat{\nu}$ (Fig. 12.3-3).

$$\vec{V}\cdot\hat{\nu} = V_0\cos\omega t \quad \text{and} \quad \vec{V}\cdot\hat{n} = V_0\sin\omega t. \tag{12.3-3}$$

Thus, the actual instantaneous voltage is given by the projection of $\overrightarrow{OP}$ either on the direction $\hat{\nu}$ (real axis) or on the direction $\hat{n}$ normal to $\hat{\nu}$ (imaginary axis). Similarly, the complex current $i = I_0 e^{j(\omega t+\phi)}$ can also be represented by a vector

$$\vec{I} = I_0 e^{j(\omega t+\phi)}\hat{\nu}, \tag{12.3-4}$$

which has a constant magnitude I_0 and rotates with constant angular velocity ω. At time $t = 0$, the vector $\vec{I}$ makes an angle ϕ with the unit vector $\hat{\nu}$.

$$\vec{I}\cdot\hat{\nu} = I_0\cos(\omega t+\phi) \quad \text{and} \quad \vec{I}\cdot\hat{n} = I_0\sin(\omega t+\phi). \tag{12.3-5}$$

Thus, the actual instantaneous current is given by the projection of $\vec{I}$ on the direction of $\hat{\nu}$ or on its perpendicular direction $\hat{n}$.

Differentiation and integration of a rotating vector

For vectors of constant magnitude rotating with constant angular velocity ω we can write, for example, from (12.3-1)

$$\frac{d\vec{r}}{dt} = j\omega\vec{r}. \tag{12.3-6}$$

Thus, here the operator $\frac{d}{dt}$ is equivalent to the operator $j\omega$. Also, we find that

$$\vec{r} = \int d\vec{r} = j\omega\int\vec{r}dt \quad \text{or} \quad \int\vec{r}dt = \frac{1}{j\omega}\vec{r}. \tag{12.3-7}$$

Thus, here the integration operator $\int\cdots dt$ is equivalent to the operator $1/j\omega$.

Vectorial and polar representation of complex numbers

A complex number can be represented geometrically by a directed line segment like vectors on complex plane (Fig. 12.3-4). It is known as *Argand's diagram.*

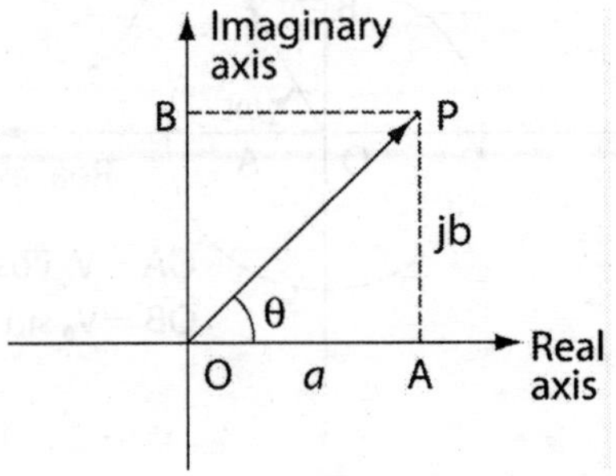

Fig. 12.3-4: Representation of a complex number by a vector

The complex number $Z = a + jb$ is represented by the line $\overrightarrow{OP}$ whose projection on the real axis gives the real part a and the projection on the imaginary axis gives the imaginary part b. The modulus or magnitude of Z is $|Z| = \left|\overrightarrow{OP}\right| = \sqrt{a^2 + b^2}$ and the angle $\theta = \tan^{-1}(b/a)$ is the argument or angle of Z. Writing

$$a = |Z| \cos\theta \quad \text{and} \quad b = |Z| \sin\theta$$

we can express Z in the following polar form:

$$Z = |Z| \cos\theta + j|Z| \sin\theta = |Z| \cdot e^{j\theta}.$$

12.4 Alternating Voltage Applied to a Series LR Circuit

Suppose an alternating emf $v = V_0 \cos\omega t$ is applied to a circuit containing an inductance L and a resistance R in series (Fig. 12.4-1). We shall analyse the circuit by using the concept of rotating vectors. The instantaneous emf $v = V_0 \cos\omega t$ may be represented by a rotating vector

$$\vec{V} = V_0 e^{j\omega t} \hat{\nu}, \tag{12.4-1}$$

which has a constant magnitude V_0 and rotates with constant angular velocity ω; at $t = 0$, $\vec{V}$ is directed along the unit vector $\hat{\nu}$. At time t, $\vec{V}$ makes an angle ωt with $\hat{\nu}$ and hence, the instantaneous value v is given by $v = \vec{V} \cdot \hat{\nu}$.

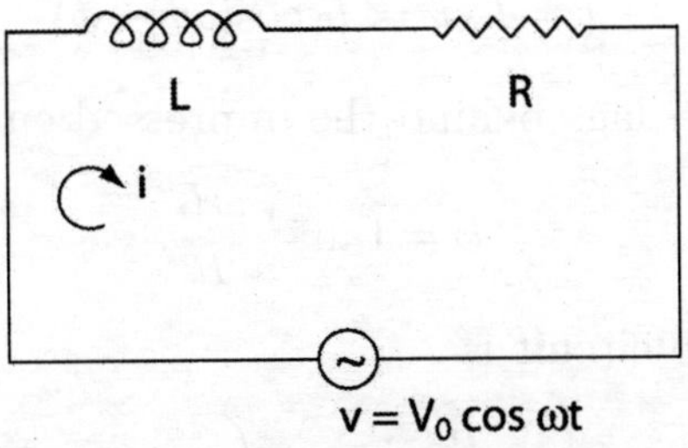

Fig. 12.4-1 : A series LR circuit

In the steady state the instantaneous current i is expected to vary with the same frequency ω. So i can also be represented by some vector $\vec{I}$ rotating with the same angular frequency ω. Instantaneous current will be given by $i = \vec{I} \cdot \hat{\nu}$. The phase relationship between $\vec{V}$ and $\vec{I}$ can be determined from the following instantaneous emf equation

$$Ri + L\frac{di}{dt} = v \tag{12.4-2}$$

Note that Ohm's law and Kirchhoff's laws are valid for ac circuits also.

In terms of rotating vectors Eq. (12.4-2) can be written as

$$R\left(\vec{I}\cdot\hat{\nu}\right)+L\frac{d}{dt}\left(\vec{I}\cdot\hat{\nu}\right)=\vec{V}\cdot\hat{\nu}$$

$$\text{or}\quad R\vec{I}\cdot\hat{\nu}+j\omega L\vec{I}\cdot\hat{\nu}=\vec{V}\cdot\hat{\nu}\qquad\left[\because\ \frac{d\vec{I}}{dt}=j\omega\vec{I}\right]$$

$$R\vec{I}+j\omega L\vec{I}=\vec{V}\tag{12.4-3}$$

$$\text{or}\quad \vec{I}=\frac{\vec{V}}{R+j\omega L}=\frac{\vec{V}}{Z},\tag{12.4-4}$$

where $Z=R+j\omega L$ is known as the complex *impedance* of the circuit. The reciprocal of impedance Z is called the *admittance* Y. Thus, $Y=1/Z$. The quantity ωL is known as the *inductive reactance of the inductor.* Now

$$Z=\sqrt{R^2+\omega^2L^2}\cdot e^{j\phi}\quad\text{where }\phi=\tan^{-1}\frac{\omega L}{R}\tag{12.4-5}$$

Therefore, using (12.4-1) we can write from (12.4-4),

$$\vec{I}=\frac{V_0e^{j\omega t}\hat{\nu}}{\sqrt{R^2+\omega^2L^2}\cdot e^{j\phi}}=I_0e^{j(\omega t-\phi)}\hat{\nu},\tag{12.4-6}$$

where $I_0=V_0/\sqrt{R^2+\omega^2L^2}$ is the *peak value of current.*

Now the instantaneous current is given by

$$i=\vec{I}\cdot\hat{\nu}=I_0\cos(\omega t-\phi).\tag{12.4-7}$$

Obviously, the current i lags behind the impressed emf by an angle

$$\phi=\tan^{-1}\frac{\omega L}{R}.$$

The power factor of the circuit is

$$\cos\phi=\frac{R}{\sqrt{R^2+\omega^2L^2}}\qquad\left(\because\ \tan\phi=\frac{\omega L}{R}\right)$$

and power in the circuit is given by

$$\begin{aligned}P&=\text{rms voltage}\times\text{rms current}\times\text{power factor}\\&=V_r\cdot I_r\cdot\cos\theta\\&=I_r^2\cdot\sqrt{R^2+\omega^2L^2}\cdot\frac{R}{\sqrt{R^2+\omega^2L^2}}\qquad\left[\because\ I_r=\frac{V_r}{\sqrt{R^2+\omega^2L^2}}\right]\\&=I_r^2R.\end{aligned}$$

This indicates that the power is dissipated only in R.

Phasor diagram

The rotating vectors $\vec{V}$, $\vec{I}$ that are employed to represent sinusoidally varying voltages and currents in a coplanar system are called *phasors.*

From Eq. (12.4-3)

$$\vec{V} = \vec{V}_R + \vec{V}_L, \tag{12.4-8}$$

where $\vec{V}_R = R\vec{I}$ and $\vec{V}_L = j\omega L\vec{I}$ are the phasors for voltages across R and L, respectively. $\vec{V}_R$ is in phase with the current phasor $\vec{I}$ and $\vec{V}_L$ leads over $\vec{I}$ by 90°. The sum of $\vec{V}_R$ and $\vec{V}_L$ gives the phasor for the applied voltage. Let us represent the phasors $\vec{V}$, $\vec{V}_R$ and $\vec{V}_L$ in a vector diagram, which is known as *phasor diagram* (Fig. 12.4-2). In Fig. 12.4-2

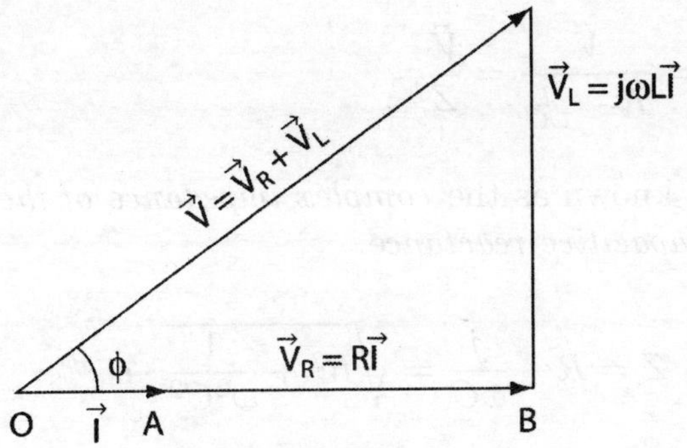

Fig 12.4-2

$$\overrightarrow{OA} = \vec{I}, \quad \overrightarrow{OB} = R\vec{I} = \vec{V}_R, \quad \overrightarrow{BC} = j\omega L\vec{I} = \vec{V}_L,$$

which is perpendicular to $\overrightarrow{OB}$ and rotated 90° in anticlockwise sense with respect to $\vec{I}$. The vector sum of $\overrightarrow{OB}$ and $\overrightarrow{BC}$ gives the phasor $\vec{V}$ for the applied emf. The diagram shows that $\vec{I}$ lags behind $\vec{V}$ by an angle

$$\phi = \tan^{-1}\frac{BC}{OB} = \tan^{-1}\frac{\omega L}{R}.$$

12.5 Alternating Voltage Applied to a Series CR Circuit

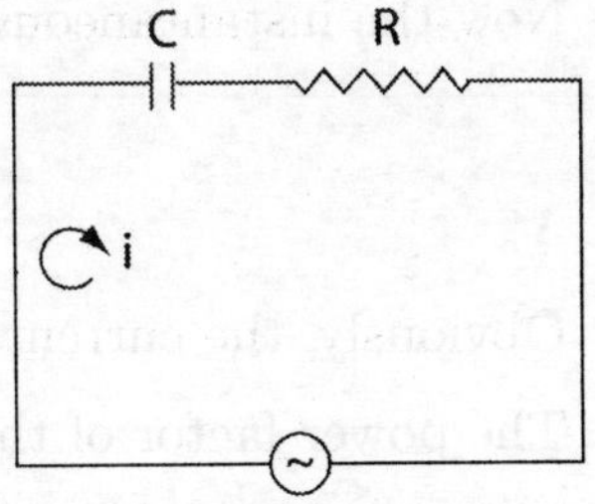

Fig 12.5-1

Suppose an alternating emf $v = V_0 \cos\omega t$ is applied to a circuit containing a capacitance C and a resistance R in series (Fig. 12.5.1) we shall analyse the circuit by using the concept of rotating vectors, as *done in case of series LR circuit* (Section 12.4). *Thus, introducing the idea of rotating vectors* $\vec{V}$ *and* $\vec{I}$ *as in* Section 12.4

we can determine their phase relationship from the following instantaneous emf equation,

$$Ri + \frac{q}{C} = v \quad \text{or} \quad Ri + \frac{1}{C}\int idt = v. \tag{12.5-1}$$

In terms of rotating vectors the equation can be written as

$$R\vec{I}\cdot\hat{\nu} + \frac{1}{C}\int\left(\vec{I}\cdot\hat{\nu}\right)dt = \vec{V}\cdot\hat{\nu}$$

$$\text{or } R\vec{I}\cdot\hat{\nu} + \frac{1}{j\omega C}\vec{I}\cdot\hat{\nu} = \vec{V}\cdot\hat{\nu} \qquad \left[\because \int\vec{I}dt = \frac{1}{j\omega}\vec{I}\right]$$

$$\text{or } R\vec{I} - \frac{j}{\omega C}\vec{I} = \vec{V} \tag{12.5-2}$$

$$\text{or } \vec{I} = \frac{\vec{V}}{R - \frac{j}{\omega C}} = \frac{\vec{V}}{Z}, \tag{12.5-3}$$

where $Z = R - j/\omega C$ is known as the *complex impedance of the circuit*. The quantity $1/\omega C$ is known as the *capacitive reactance*.

Now

$$Z = R - \frac{j}{\omega C} = \sqrt{R^2 + \frac{1}{\omega^2C^2}}\cdot e^{-j\phi},$$

where

$$\phi = \tan^{-1}\frac{1}{\omega CR}. \tag{12.5-4}$$

Therefore, putting $\vec{V} = V_0e^{j\omega t}\hat{\nu}$ in Eq. (12.5-3) we get

$$\vec{I} = \frac{V_0e^{j\omega t}\hat{\nu}}{\sqrt{R^2 + \frac{1}{\omega^2C^2}}\cdot e^{-j\phi}} = I_0e^{j(\omega t+\phi)}\hat{\nu}, \tag{12.5-5}$$

where $I_0 = \dfrac{V_0}{\sqrt{R^2 + \frac{1}{\omega^2C^2}}}$ is the *peak value* of current.

Now the instantaneous current,

$$i = \vec{I}\cdot\hat{\nu} = I_0\cos(\omega t - \phi). \tag{12.5-6}$$

Obviously, the current i leads over the applied emf v by an angle $\phi = \tan^{-1}\dfrac{1}{\omega CR}$.

The power factor of the circuit is

$$\cos\phi = \frac{R}{\sqrt{R^2 + \frac{1}{\omega^2C^2}}} \qquad \left[\because \tan\phi = \frac{1}{\omega CR}\right]$$

and power in the circuit is given by

$$\begin{aligned} P &= \text{rms voltage} \times \text{rms current} \times \cos\phi \\ &= V_r \cdot I_r \cdot \cos\theta \\ &= I_r^2 \cdot \sqrt{R^2 + \frac{1}{\omega^2 C^2}} \cdot \frac{R}{\sqrt{R^2 + \frac{1}{\omega^2 C^2}}} \\ &= I_r^2 R. \end{aligned}$$

This indicates that the power is dissipated only in R.

Phasor diagram

From Eq. (12.5-2)

$$\vec{V} = \vec{V}_R + \vec{V}_C, \tag{12.5-7}$$

where $\vec{V}_R = R\vec{I}$ and $\vec{V}_C = -\frac{j}{\omega C}\vec{I}$ are the phasors for the voltages across R and C, respectively. $\vec{V}_R$ is in phase with the current phasor $\vec{I}$ and $\vec{V}_c$ lags behind $\vec{I}$ by 90°. The sum of $\vec{V}_R$ and $\vec{V}_C$ gives the phasor for the applied emf. Let us represent it by a vector diagram which is known as *phasor diagram* (Fig. 12.5-2).

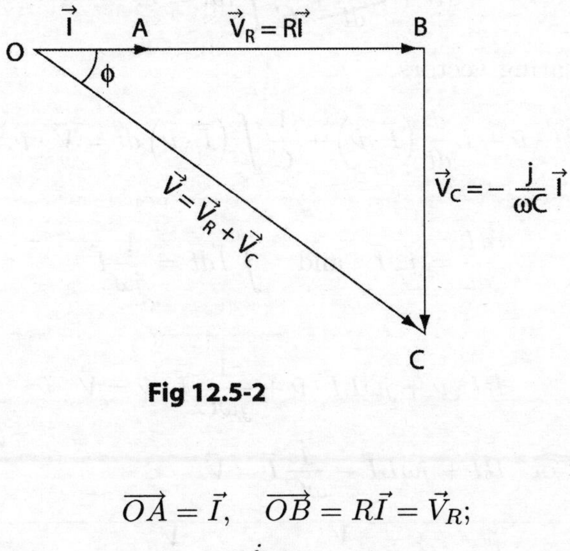

Fig 12.5-2

In figure

$$\overrightarrow{OA} = \vec{I}, \quad \overrightarrow{OB} = R\vec{I} = \vec{V}_R;$$

$$\overrightarrow{BC} = -\frac{j}{\omega C}\vec{I} = \vec{V}_C,$$

which is perpendicular to $\overrightarrow{OB}$ and rotated 90° in clockwise sense with respect to $\vec{I}$. The sum of $\overrightarrow{OB}$ and $\overrightarrow{BC}$ gives the phasor $\vec{V}$ for the applied emf. The diagram shows that $\vec{I}$ leads over $\vec{V}$ by an angle

$$\phi = \tan^{-1}\frac{BC}{OB} = \tan^{-1}\frac{1}{\omega CR}.$$

12.6 Alternating Voltage Applied to a Series LCR Circuit

Suppose an alternating emf $v = V_0 \cos \omega t$ is applied to a circuit containing an inductance L, a capacitance C and a resistance R in series (Fig. 12.6-1). To analyse the circuit we shall use the concept of rotating vectors $\vec{V}$ and $\vec{I}$ for the voltage and current as done

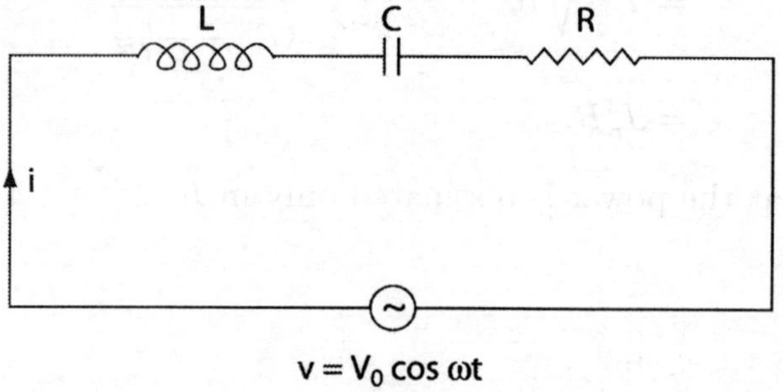

Fig 12.6-1

in Section 12.4. The phase relationship between $\vec{V}$ and $\vec{I}$ can be determined from the following instantaneous emf equation:

$$Ri + L\frac{di}{dt} + \frac{1}{C}\int idt = v.$$

In terms of the rotating vectors,

$$R\vec{I}\cdot\hat{\nu} + L\frac{d}{dt}\left(\vec{I}\cdot\hat{\nu}\right) + \frac{1}{C}\int\left(\vec{I}\cdot\hat{\nu}\right)dt = \vec{V}\cdot\hat{\nu}.$$

Since

$$\frac{d\vec{I}}{dt} = j\omega\vec{I} \quad \text{and} \quad \int\vec{I}dt = \frac{1}{j\omega}\vec{I}$$

we can write

$$R\vec{I}\cdot\hat{\nu} + j\omega L\vec{I}\cdot\hat{\nu} + \frac{1}{j\omega C}\vec{I}\cdot\hat{\nu} = \vec{V}\cdot\hat{\nu}$$

$$\text{or} \quad R\vec{I} + j\omega L\vec{I} - \frac{j}{\omega C}\vec{I} = \vec{V} \tag{12.6-1}$$

$$\text{or} \quad \vec{I} = \frac{\vec{V}}{R + j\left(\omega L - \frac{1}{\omega C}\right)} = \frac{\vec{V}}{Z}, \tag{12.6-2}$$

where $Z = R + j\left(\omega L - \frac{1}{\omega C}\right)$ is known as the *complex impedance of the circuit.* The quantity ωL is called *inductive reactance* and $\frac{1}{\omega C}$ is called *capacitive reactance.* Now

$$Z = R + j\left(\omega L - \frac{1}{\omega C}\right) = \sqrt{R^2 + \left(\omega L - \frac{1}{\omega C}\right)^2}\cdot e^{j\phi}, \tag{12.6-3}$$

where

$$\phi = \tan^{-1} \frac{\omega L - \frac{1}{\omega C}}{R}. \tag{12.6-4}$$

Now putting $\vec{V} = V_0 e^{j\omega t} \hat{\nu}$ in Eq. (12.6-2) and using Eq. (12.6-3) we get

$$\vec{I} = \frac{V_0 e^{j\omega t} \hat{\nu}}{\sqrt{R^2 + \left(\omega L - \frac{1}{\omega C}\right)^2}\, e^{j\phi}} = I_0 e^{j(\omega t - \phi)} \hat{\nu}, \tag{12.6-5}$$

where

$$I_0 = V_0 \Bigg/ \sqrt{R^2 + \left(\omega L - \frac{1}{\omega C}\right)^2} \tag{12.6-6}$$

is the peak value of current.

The instantaneous current i is given by

$$i = \vec{I} \cdot \hat{\nu} = I_0 \cos(\omega t - \phi). \tag{12.6-7}$$

From Eq. (12.6-4) we find that three cases may arise:

Case I. *If $\omega L > 1/\omega C$ then ϕ is positive and current i lags behind the applied emf by the angle ϕ.*

Case II. *If $\omega L < 1/\omega C$ then ϕ is negative and current i leads over the applied emf by the angle $|\phi|$.*

Case III. *If $\omega L = 1/\omega C$ then $\phi = 0$ and current i is in phase with the applied emf. The circuit is then purely resistive in effect and is said to be in resonance.*

Power factor and power

The power factor of the circuit is give by

$$\cos\phi = \frac{R}{\sqrt{R^2 + \left(\omega L - \frac{1}{\omega C}\right)^2}} \qquad \text{[Using Eq. (12.6.4)]} \tag{12.6-8}$$

and power in the circuit is

$$P = \text{rms voltage} \times \text{rms current} \times \cos\phi$$

$$= V_r \cdot I_r \cdot \cos\theta.$$

From Eq. (12.6-6) we have

$$V_r = I_r \sqrt{R^2 + \left(\omega L - \frac{1}{\omega C}\right)^2}.$$

Now using (12.6-8) we can get

$$P = I_r^2 R.$$

This indicates that the power is dissipated only in R.

Phasor diagram

From Eq. (12.6-1) we can write

$$\vec{V} = \vec{V}_R + \vec{V}_L + \vec{V}_C,$$

where $\vec{V}_R = \vec{I}R$, $\vec{V}_L = j\omega L\vec{I}$ and $\vec{V}_C = -(j/\omega C)\vec{I}$ are the phasors for the voltages across R, L and C, respectively. $\vec{V}_R$ is in phase with the current phasor $\vec{I}$, $\vec{V}_L$ leads over $\vec{I}$ by 90° and $\vec{V}_C$ lags behind $\vec{I}$ by 90°. The sum of $\vec{V}_R$, $\vec{V}_L$ and $\vec{V}_C$ gives the phasor $\vec{V}$ for the applied emf. Let us represent it by a vector diagram (Fig. 12.6-2), which is known as *phasor diagram.* Depending on the relative values of ωL and $1/\omega C$, we have three distinct cases as shown in Figs. 12.6-2.

In figures $\overrightarrow{OA} = \vec{I}$, $\overrightarrow{OB} = R\vec{I} = \vec{V}_R$, $\overrightarrow{BC} = j\omega L\vec{I} = \vec{V}_L$, $\overrightarrow{BD} = -\frac{j}{\omega C}\vec{I} = \vec{V}_C$.

Note that $\overrightarrow{BC}$ and $\overrightarrow{BD}$ are drawn perpendicular to $\overrightarrow{OB}$ and rotated with respect to $\vec{I}$ through 90° in anticlockwise and clockwise sense, respectively. $\overrightarrow{BE}$ represents the sum $\vec{V}_L + \vec{V}_C$. In Fig. 12.6-2(c) $\vec{V}_L$ and $\vec{V}_C$ are equal and opposite and cancel each other. For this B and E denote the same point in Fig. 12.6-2(c). The sum $\overrightarrow{OB} + \overrightarrow{BE}$ gives the phasor $\vec{V}$ for the applied emf. In the case of $\omega L > \frac{1}{\omega C}$ we find from Fig. 12.6-2(a)

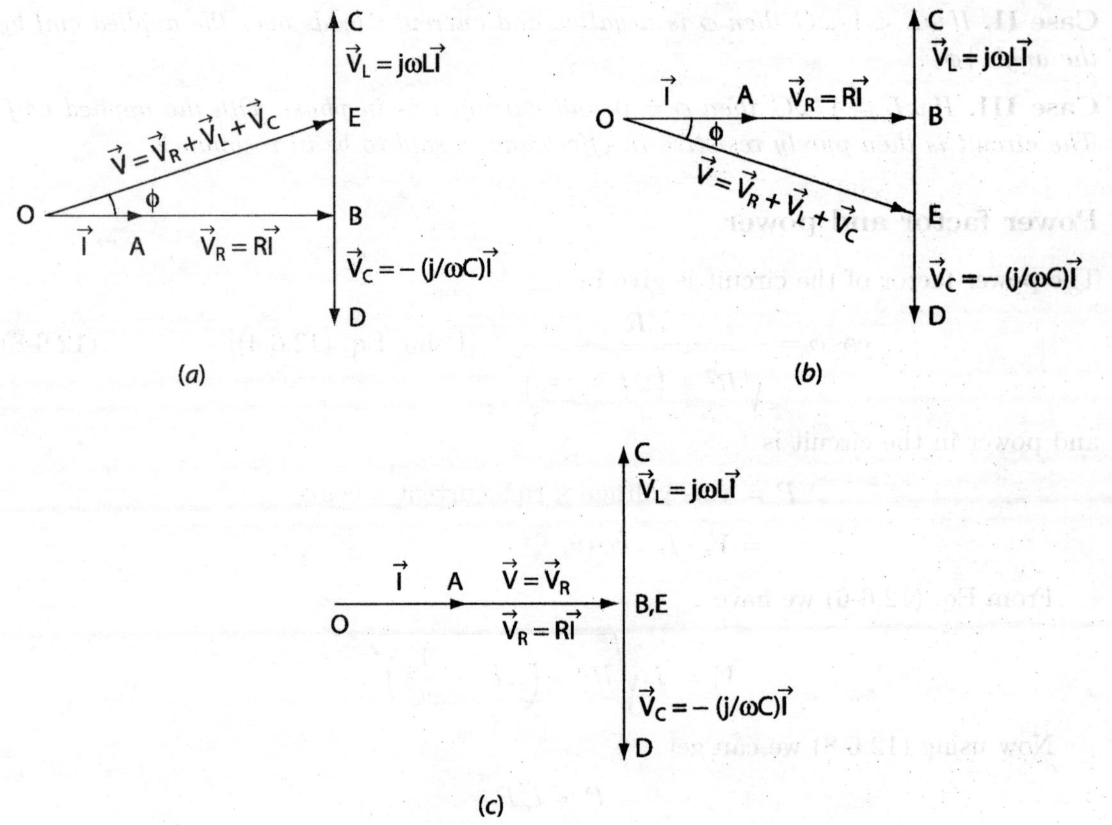

Fig. 12.6-2: Phasor diagram for (a) $\omega L > 1/\omega C$, (b) $\omega L < 1/\omega C$, (c) $\omega L = 1/\omega C$

that $\vec{I}$ lags behind $\vec{V}$ by an angle ϕ given by

$$\phi = \tan^{-1}\frac{BE}{OB} = \frac{\omega L - \frac{1}{\omega C}}{R}.$$

Fig. 12.6-2(b) shows that for the case $\omega L < \frac{1}{\omega C}$, $\vec{I}$ leads over $\vec{V}$ by the angle ϕ given by

$$\phi = \tan^{-1}\frac{BE}{OB} = \tan^{-1}\frac{\frac{1}{\omega C} - \omega L}{R}.$$

For the case $\omega L = \frac{1}{\omega C}$, $\vec{I}$ is in phase with $\vec{V}$.

12.7 Resonance in Series LCR Circuit

In Section 12.6 we have seen that the current in a series LCR circuit impressed by an ac emf $v = V_0 \cos\omega t$ is given by

$$i = I_0 \cos(\omega t - \phi), \tag{12.7-1}$$

where

$$I_0 = \frac{V_0}{\sqrt{R^2 + \left(\omega L - \frac{1}{\omega C}\right)^2}} \quad \text{and} \quad \phi = \tan^{-1}\frac{\omega L - \frac{1}{\omega C}}{R}. \tag{12.7-2}$$

The rms current is

$$I_r = \frac{V_r}{\sqrt{R^2 + \left(\omega L - \frac{1}{\omega C}\right)^2}}, \tag{12.7-3}$$

where V_r is the rms voltage. Thus, I_r depends on the frequency ω of the impressed ac emf. For certain value of ω, I_r becomes maximum and we say there is *resonance.* If I_r is plotted as a function of ω we get a curve like that shown in Fig. 12.7-1. Obviously the current I_r attain a maximum value when

$$\omega L = \frac{1}{\omega C} \quad \text{or} \quad \omega = \frac{1}{\sqrt{LC}} = \omega_0 \text{ (say)}. \tag{12.7-4}$$

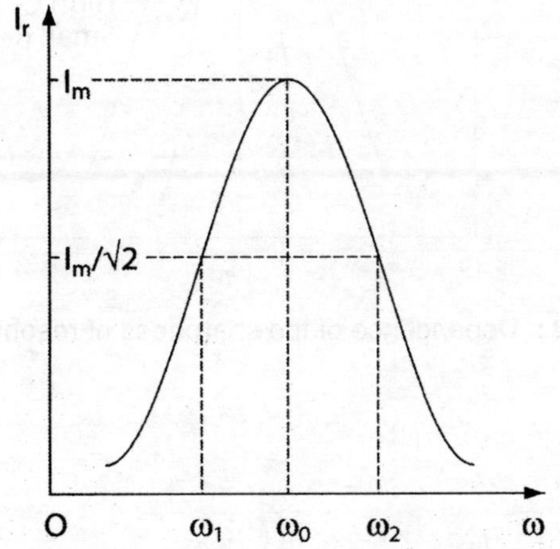

Fig. 12.7-1 : Current response curve

This is known as *current resonance* and ω_0 is called the *resonant angular frequency.* At resonance $\phi = 0$, i.e., current and voltage are in phase. As the current response of the circuit is maximum at resonance the series LCR circuit is sometimes called an *acceptor circuit.* If the input consists of a number of frequency components the circuit will have good response only for frequencies near the resonant frequency. The circuit has thus, *selective properties.* It can be used as a filter to select a narrow band of frequencies about resonant frequency.

Sharpness of resonance

As the frequency ω of the applied emf differs from the resonant frequency ω_0, the current response of the circuit diminishes on either side of ω_0. The rate of fall of response with departure from the equality between the two frequencies gives a measure of the *sharpness of resonance.* If R is smaller then a slight deviation of ω from ω_0 causes the response to decrease appreciably (Fig. 12.7-2).

To get a quantitative idea about the sharpness of resonance let us calculate the bandwidth, which is defined as the frequency separation between two half-power points. The average power in the circuit is given by

$$\langle P \rangle = V_r \cdot I_r \cdot \cos\phi = I_r^2 R = \frac{V_r^2 R}{R^2 + \left(\omega L - \frac{1}{\omega C}\right)^2}. \tag{12.7-5}$$

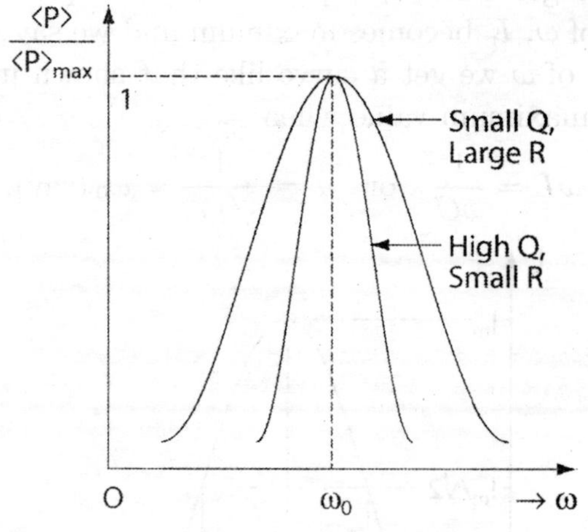

Fig. 12.7-2: Dependence of the sharpness of resonance on R

$$\therefore\ \langle P \rangle_{\max} = \frac{V_r^2}{R} \text{ and we can write}$$

$$\langle P \rangle = \frac{\langle P \rangle_{\max}}{1 + \frac{1}{R^2}\left(\omega L - \frac{1}{\omega C}\right)^2}. \tag{12.7-6}$$

At half-power points, $\langle P\rangle = \langle P\rangle_{\max}/2$ and, therefore,

$$\left(\omega L - \frac{1}{\omega C}\right)^2 = R^2, \text{ or } \omega L - \frac{1}{\omega C} = \pm R,$$

$$\text{or } \omega^2 \mp \frac{R}{L}\omega - \omega_0^2 = 0,$$

$$\text{or } \omega^2 \mp \frac{\omega_0}{Q}\omega - \omega_0^2 = 0, \qquad (12.7\text{-}7)$$

where $Q = \omega_0 L/R$ is called the *quality factor* of the resonant circuit.

The acceptable positive roots of Eq. (12.7-7) are

$$\omega_1 = \omega_0\left[-\frac{1}{2Q} + \sqrt{1 + \frac{1}{4Q^2}}\right] \text{ and } \omega_2 = \omega_0\left[\frac{1}{2Q} + \sqrt{1 + \frac{1}{4Q^2}}\right]. \qquad (12.7\text{-}8)$$

Therefore, the bandwidth is

$$\Delta\omega = \omega_2 - \omega_1 = \frac{\omega_0}{Q}. \qquad (12.7\text{-}9)$$

Thus, the Q-factor gives a measure of the sharpness of resonance. Higher is the value of Q smaller will be the value of $\Delta\omega$ and hence, sharper will be the resonance curve. This can be clearly understood by plotting $\langle P\rangle/\langle P\rangle_{\max}$ as a function of ω from Eq. (12.7-6). This plot is shown in Fig. 12.7-2.

Voltage magnification

At resonance rms current $I_r = V_r/R$ and the rms voltages across L, C and R are given by

$$V_L = \omega_0 L I_r, \quad V_C = \frac{I_r}{\omega_0 C} \quad \text{and} \quad V_R = RI_r.$$

At resonance V_L and V_C are equal and 180° out of phase and hence, cancel out. The entire applied voltage appears across R. Therefore,

$$\frac{\text{voltage across } L \text{ or } C \text{ at resonance}}{\text{applied input voltage}} = \frac{\omega_0 L}{R} \quad \text{or} \quad \frac{1}{\omega_0 CR} = Q.$$

If the resistance R is small compared to $\omega_0 L$ or $1/\omega_0 C$ then Q may be large compared to unity. Thus, at resonance the voltage across L or C may be many times larger than the input voltage, i.e., in series circuit there is *voltage magnification.*

Quality factor

The sharpness of resonance and the degree of selectivity of a series LCR circuit is usually expressed in terms of a quantity Q, called the *quality factor*. Different forms of definition of Q appear in the literature. However, they all intend to convey the same meaning. A few of the definitions are mentioned :

1. The quality factor Q may be defined as the ratio of inductive or capacitive reactance at resonance to the resistance of the circuit, i.e.,

$$Q = \frac{\omega_0 L}{R} = \frac{1}{\omega_0 CR}.$$

2. Q may be defined as the ratio of the voltage across L or C at resonance to the applied voltage. Thus, Q may be called the *voltage magnification factor.*

3. Q may be defined as the ratio of the resonant frequency and the bandwidth, i.e., $Q = \frac{\omega_0}{\Delta\omega}$. So the sharpness of resonance and selectivity increase with Q.

4. The circuit can store energy in the magnetic field around L or in the electric field around C. The efficiency with which the circuit performs it is given by the Q-factor. By definition

$$Q = 2\pi \times \frac{\text{maximum energy stored}}{\text{energy dissipated per cycle}} = 2\pi \times \frac{\frac{1}{2}LI_0^2}{I_r^2 \cdot R \cdot T},$$

where T is the time period. Since $2\pi/T = \omega_0$ and rms current $I_r = I_0/\sqrt{2}$ we get

$$Q = \frac{\omega_0 L}{R} = \frac{1}{\omega_0 CR},$$

which agrees with the definition 1.

Variation of impedance and phase with frequency

The impedance of the series circuit is $|Z| = \sqrt{R^2 + \left(\omega L - \frac{1}{\omega C}\right)^2}$. It attains a minimum value R at the resonant frequency $\omega = \omega_0 = 1/\sqrt{LC}$ and increases on either side of ω_0 (Fig. 12.7-3).

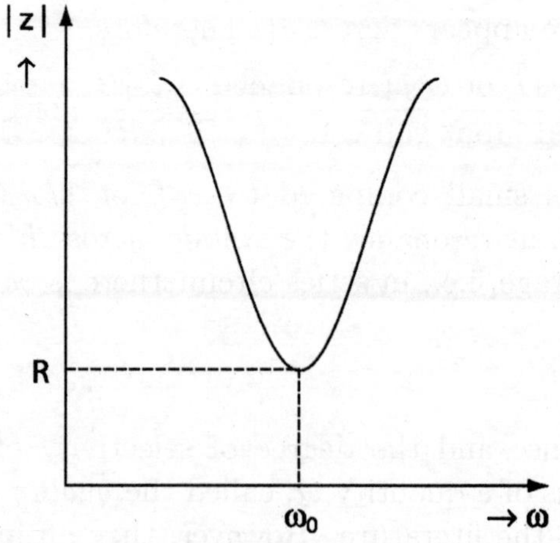

Fig. 12.7-3 : Variation of impedance with frequency

Eq. (12.7-1) shows that the current in a series LCR circuit lags behind the impressed emf by an angle ϕ given by

$$\tan\phi = \frac{\omega L - \frac{1}{\omega C}}{R} = \frac{\omega_0 L}{R}\left(\frac{\omega}{\omega_0} - \frac{\omega_0}{\omega}\right) = Q\left(\frac{\omega}{\omega_0} - \frac{\omega_0}{\omega}\right). \qquad (12.7\text{-}10)$$

Obviously $\phi = 0$ for $\omega = \omega_0$, ϕ is negative for $\omega < \omega_0$ and ϕ is positive for $\omega > \omega_0$. As $\omega \to 0$, $\phi \to -\pi/2$ and as $\omega \to \infty$, $\phi \to +\pi/2$. Thus, ϕ changes from $-90°$ to $+90°$ as ω is changed from 0 to ∞ (Fig. 12.7-4).

Note that for $\omega > \omega_0$ the circuit reactance is inductive in nature and for $\omega < \omega_0$ it is capacitive.

Differentiating (12.7-10) with respect to ω it is easy to show that

$$\frac{d\phi}{d\omega} = \frac{2Q}{\omega_0} \quad \text{at } \omega = \omega_0.$$

So the rate of change of phase angle with frequency near resonance is greater for higher values of Q. This fact becomes very important while considering the frequency stability of an electronic oscillator.

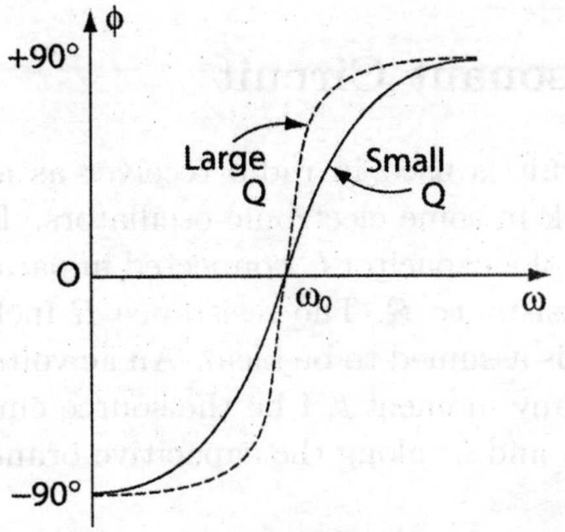

Fig. 12.7-4 : Variation of phase with frequency

Voltage resonance

The phenomenon of resonance in the series LCR circuit can also be studied by examining how the pd across C or L varies with the frequency of the input voltage. The rms value of the pd across the capacitor is given by

$$V_C = I_r \times \frac{1}{\omega C},$$

where rms current I_r is given by Eq. (12.7-3). Therefore,

$$V_C = \frac{V_r}{\sqrt{\omega^2 C^2 R^2 + (\omega^2 LC - 1)^2}}. \qquad (12.7\text{-}11)$$

For V_C to be maximum we must have

$$\frac{dV_C}{d\omega} = 0, \quad \text{or} \quad \frac{d}{d\omega}\left\{\omega^2 C^2 R^2 + \left(\omega^2 LC - 1\right)^2\right\} = 0.$$

This gives

$$\omega = \sqrt{\frac{1}{LC} - \frac{R^2}{2L^2}} = \omega_c\,(\text{say}). \tag{12.7-12}$$

Therefore, the voltage across C becomes maximum at a frequency slightly smaller than the frequency ω_0 at which current resonance occurs. Thus, we require to speak of *current resonance* and *voltage resonance.* The frequency ω_c given by Eq. (12.7-12) may be called *voltage resonant frequency* and the phenomenon of capacitor voltage being maximum at ω_c is called the *voltage resonance.* Note that if $R \to 0$ then voltage resonant frequency and current resonant frequency become equal. The voltage resonance in electrical system corresponds to the amplitude resonance in mechanical system whereas the current resonance corresponds to the velocity resonance in mechanical system.

12.8 Parallel Resonant Circuit

The parallel resonant circuit is used in radio receiver as a tuned circuit and as the frequency selective network in some electronic oscillators. Fig. 12.8-1 shows a parallel resonant circuit consisting of a capacitor C connected in parallel to a series combination of an inductor L and a resistance R. The resistance R includes the resistance of the inductor L. The capacitor is assumed to be ideal. An ac voltage $v = V_0 \cos\omega t$ is applied across the circuit. Let at any moment t, i be the source current which divides into i_L along the inductive branch and i_C along the capacitive branch. Thus, from Kirchhoff's current law,

$$i = i_L + i_C. \tag{12.8-1}$$

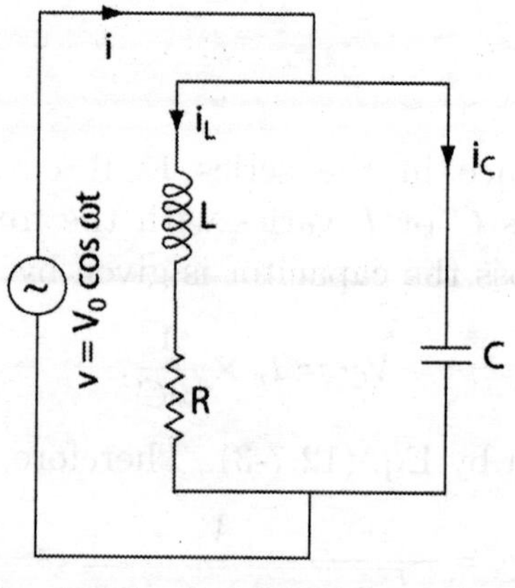

Fig. 12.8-1 : Parallel resonant circuit

To analyse the circuit let us use the concept of rotating vectors as done in Section 12.4. In terms of the rotating vectors for the currents we can write Eq. (12.8-1) as

$$\vec{I}\cdot\hat{\nu} = \vec{I}_L\cdot\hat{\nu} + \vec{I}_C\cdot\hat{\nu}, \quad \text{or} \quad \vec{I} = \vec{I}_L + \vec{I}_C. \tag{12.8-2}$$

For the inductive branch,

$$v = Ri_L + L\frac{di_L}{dt}$$

$$\text{or} \quad \vec{V}\cdot\hat{\nu} = R\vec{I}_L\cdot\hat{\nu} + L\frac{d}{dt}\left(\vec{I}_L\cdot\hat{\nu}\right)$$

$$\text{or} \quad \vec{V} = R\vec{I}_L + j\omega L\vec{I}_L \qquad \left(\because \frac{d\vec{I}_L}{dt} = j\omega\vec{I}_L\right)$$

$$\text{or} \quad \vec{I}_L = \frac{\vec{V}}{R + j\omega L}. \tag{12.8-3}$$

For the capacitive branch,

$$v = \frac{1}{C}\int i_C dt$$

$$\text{or} \quad \vec{V}\cdot\hat{\nu} = \frac{1}{C}\int\left(\vec{I}_C\cdot\hat{\nu}\right)dt = \frac{1}{j\omega C}\vec{I}_C\cdot\hat{\nu} \qquad \left(\because \int\vec{I}_C dt = \frac{1}{j\omega}\vec{I}_C\right)$$

$$\text{or} \quad \vec{I}_C = j\omega C\vec{V}. \tag{12.8-4}$$

From Eqs. (12.8-2)–(12.8-4) we get

$$\vec{I} = \frac{\vec{V}}{R + j\omega L} + j\omega C\vec{V} = \frac{\vec{V}}{Z}, \tag{12.8-5}$$

where Z is the complex impedance of the circuit. Therefore,

$$\frac{1}{Z} = \frac{1}{R + j\omega L} + j\omega C = \frac{R - j\omega L}{R^2 + \omega^2 L^2} + j\omega C$$

$$= \frac{R}{R^2 + \omega^2 L^2} + j\left(\omega C - \frac{\omega L}{R^2 + \omega^2 L^2}\right) \tag{12.8-6}$$

The source current $\vec{I}$ and the source voltage $\vec{V}$ will be in phase when the imaginary part of $1/Z$ vanishes. That is, when

$$\omega C = \frac{\omega L}{R^2 + \omega^2 L^2} \quad \text{or} \quad \omega = \sqrt{\frac{1}{LC} - \frac{R^2}{L^2}} = \omega_0 \text{ (say)}, \tag{12.8-7}$$

the source current and voltage are in phase, power factor is unity and the circuit is said to be in *(phase) resonance.* ω_0 is known as *resonant (angular) frequency.* For the occurrence of parallel resonance in a practical circuit ω_0 must be real, i.e., we must have

$$\frac{R^2}{L^2} < \frac{1}{LC}, \quad \text{or} \quad R < \sqrt{\frac{L}{C}}.$$

The impedance the circuit at resonance is purely resistive and has a value

$$\frac{R^2 + \omega_0^2 L^2}{R} = R + \frac{L^2}{R}\left(\frac{1}{LC} - \frac{R^2}{L^2}\right) = \frac{L}{CR} = R_d \text{ (say)}, \tag{12.8-8}$$

where R_d is called *dynamic resistance.* At resonance the peak current from the supply is called *make up* current and is given by

$$I_0 = \frac{V_0}{R_d}. \tag{12.8-9}$$

Note that when $R \to 0$ when $R_d \to \infty$ meaning that no current is drawn from the source at resonance. For this the parallel resonant circuit is known as a *rejector circuit.* If the input voltage contains a number of frequency components the circuit rejects the current at the resonant frequency of the circuit. At zero frequency (*dc*) the effective impedance of the circuit is equal to the resistance R, which is usually small and hence, the current has a large value. At high frequencies the reactance of the capacitive branch is very low and gives rise to a large value of current. The minimum value of source current occurs at some intermediate frequency ω_m (Fig. 12.8-2).

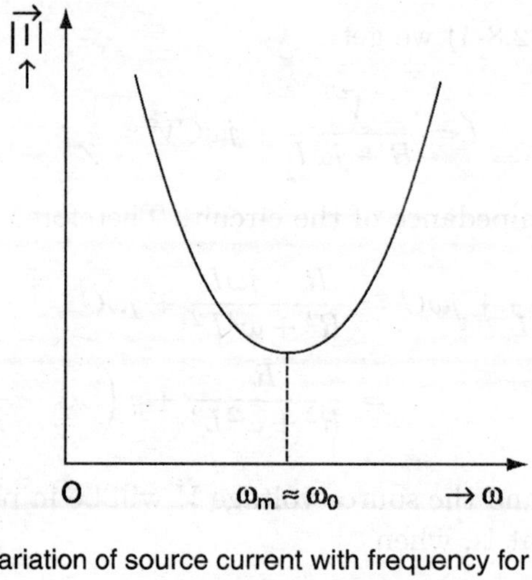

Fig. 12.8-2: Variation of source current with frequency for a high Q circuit

Current magnification

The current flowing through the capacitive or inductive branch at resonance may be many times greater than the source current. Thus, in parallel resonant circuit *current magnification* is possible. The rms value of source current at resonance is

$$I_r = \frac{V_r}{R_d},$$

where V_r is the rms source voltage. The rms value of capacitor branch current at resonance is given by

$$I_{cr} = \frac{V_r}{1/\omega_0 C}.$$

Therefore,

$$\frac{I_{cr}}{I_r} = \omega_0 C R_d = \omega_0 C \cdot \frac{L}{CR} = \frac{\omega_0 L}{R} = Q \text{ (say)}, \tag{12.8-10}$$

where Q is known as the *quality factor of the circuit.* Usually Q is greater than unity and we have current through the capacitive branch Q-times the current drawn from the source, i.e., we get current magnification.

Variation of impedance with frequency

The impedance Z or admittance Y is given by (12.8-6) as

$$|Y|^2 = \left|\frac{1}{Z}\right|^2 = \frac{R^2}{(R^2+\omega^2 L^2)^2} + \left(\omega C - \frac{\omega L}{R^2+\omega^2 L^2}\right)^2.$$

If the frequency alone is variable then the condition for maximum impedance or minimum admittance is given by

$$\frac{d}{d\omega}|Y|^2 = 0$$

This gives

$$\omega = \sqrt{\frac{\sqrt{1+2R^2C/L}}{LC} - \frac{R^2}{L^2}} = \omega_m \text{ (say)}. \tag{12.8-11}$$

Comparing (12.8-7) and (12.8-11) we note that the impedance of the circuit becomes maximum and hence current becomes minimum at a frequency ω_m, which is slightly greater than the phase resonant frequency ω_0. However, if Q is high (i.e., R is low) then we can assume $\omega_m \approx \omega_0$. The phenomenon of current being minimum at $\omega = \omega_m$ is known as *antiresonance*.

Selectivity of a parallel resonant circuit

The nature of variation of impedance of a parallel circuit with frequency is shown in Fig. 12.8-3, where we assume $\omega_m \approx \omega_0$ for a high-Q circuit. As the impedance is maximum at resonance all the frequency components of the input having frequencies near about ω_0 will be *blocked* by the parallel circuit. The range of frequency components

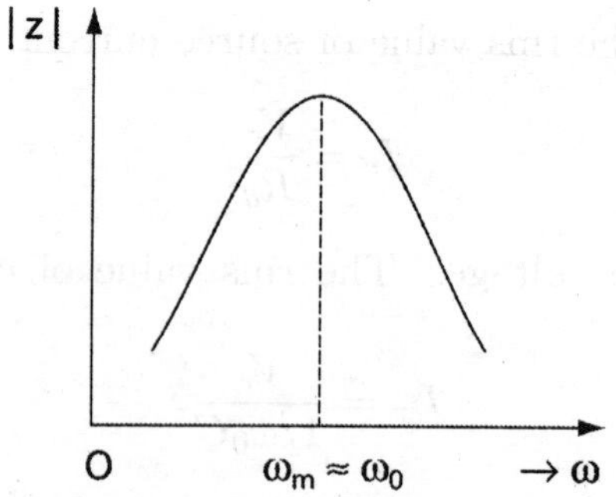

12.8-3: Variation of impedance with frequency for a high Q circuit

blocked by the circuit is determined by the rate of fall of $|Z|$ with departure from the equality between the input frequency (ω) and resonant frequency (ω_0) of the circuit. If a slight departure of ω from ω_0 causes large drop in $|Z|$ then only a narrow selected band of frequencies will be blocked by the circuit and we say that the circuit has high selectivity.

To get a quantitative idea about the selectivity let us calculate the bandwidth $\Delta\omega$ of the impedance curve. $\Delta\omega$ corresponds to the frequency separation $\omega_2 - \omega_1$ between the points on the curve, where $|Z|$ falls to $1/\sqrt{2}$ times its maximum value $Z_0 = R_d = L/CR$ at resonance. In practical cases R is small, Q is high and Z is given by

$$\frac{1}{Z} = \frac{1}{R + j\omega L} + j\omega C = \frac{(1 - \omega^2 LC) + j\omega CR}{R + j\omega L}$$

$$\text{or} \quad Z \simeq \frac{j\omega L}{(1 - \omega^2 LC) + j\omega CR} \quad \text{(In the limit } R \to 0\text{)}$$

Now at $\omega = \omega_0$, $Z = Z_0 = \frac{L}{CR}$.
Therefore, we can write

$$Z = \frac{Z_0}{1 - j\frac{1-\omega^2 LC}{\omega CR}} = \frac{Z_0}{1 + \frac{j}{R}\left(\omega L - \frac{1}{\omega C}\right)} \quad \text{or} \quad |Z| = \frac{Z_0}{\sqrt{1 + \frac{1}{R^2}\left(\omega L - \frac{1}{\omega C}\right)^2}}. \tag{12.8-12}$$

Now for $|Z| = \dfrac{Z_0}{\sqrt{2}}$,

$$\frac{1}{R^2}\left(\omega L - \frac{1}{\omega C}\right)^2 = 1 \quad \text{or} \quad \omega L - \frac{1}{\omega C} = \pm R, \quad \text{or} \quad \omega^2 \mp \frac{R}{L}\omega - \frac{1}{LC} = 0.$$

For very low R we may write $\omega_0^2 \approx \frac{1}{LC}$ and then

$$\omega^2 \mp \frac{\omega_0}{Q}\omega - \omega_0^2 = 0.$$

The acceptable positive roots of this equation are

$$\omega_1 = \omega_0 \left[-\frac{1}{2Q} + \sqrt{1 + \frac{1}{4Q^2}} \right] \quad \text{and} \quad \omega_2 = \omega_0 \left[\frac{1}{2Q} + \sqrt{1 + \frac{1}{4Q^2}} \right]. \tag{12.8-13}$$

Therefore, the bandwidth is

$$\Delta\omega = \omega_2 - \omega_1 = \frac{\omega_0}{Q}. \tag{12.8-14}$$

Thus, higher is the value of Q, smaller is the bandwidth $\Delta\omega$ and hence, better is the selectivity. Thus, the quality factor Q may be considered as a measure of the selectivity of the circuit.

Phasor diagram

From Eq. (12.8-2), $\vec{I} = \vec{I}_L + \vec{I}_C$, where $\vec{I}_C = j\omega CV$ is the phasor for current in the capacitive branch. It leads over $\vec{V}$ by 90° The phasor for current in the inductive branch is

$$\vec{I}_L = \frac{\vec{V}}{R + j\omega L}$$

It lags behind $\vec{V}$ by an angle $\phi = \tan^{-1}\frac{\omega L}{R}$. The sum of $\vec{I}_C$ and $\vec{I}_L$ gives the phasor $\overrightarrow{OD} = \vec{I}$ for source current. It can be represented by a vector diagram as shown in Fig. 12.8-4. At resonance $\vec{I}_C$ and $\vec{I}_L$ combine to give a resultant source current $\vec{I}$, which is in phase with the voltage phasor $\vec{V}$.

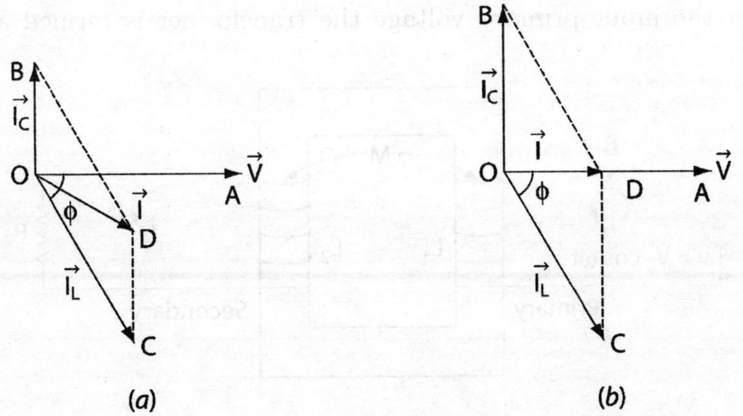

Fig. 12.8-4: Phasor diagram: (a) off-resonance, (b) at resonance

Parallel circuit driven by a current source

Suppose a parallel LCR circuit is driven by a source having a constant current amplitude instead of a source having constant voltage amplitude. In this case, the voltage across the circuit is given by $\vec{V} = Z\vec{I}$, where $\vec{V}$ and $\vec{I}$ are the phasor for voltage and current respectively. For high-Q circuits we know that $|Z|$ is maximum at the resonant frequency ω_0 and falls off on either side of resonance. As $\vec{I}$ is constant *here the voltage across the circuit is maximum at resonance and falls off on either side of resonance* (Fig. 12.8-5).

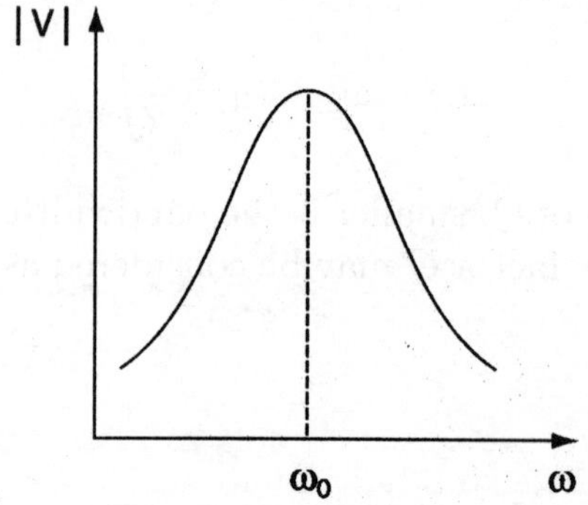

Fig. 12.8-5 : Variation of voltage with frequency in a parallel circuit excited by a constant current source

12.9 Ideal Transformer

A transformer is essentially a device consisting of two magnetically coupled coils, one is called *primary* and the other *secondary.* If an ac voltage is applied to the primary the variation of magnetic flux linked with the secondary induces an emf in the secondary. If the secondary output voltage is greater than the primary input voltage the transformer is called a *step-up transformer.* On the other hand, if the output secondary voltage is smaller than the input primary voltage the transformer is termed as a *step-down transformer.*

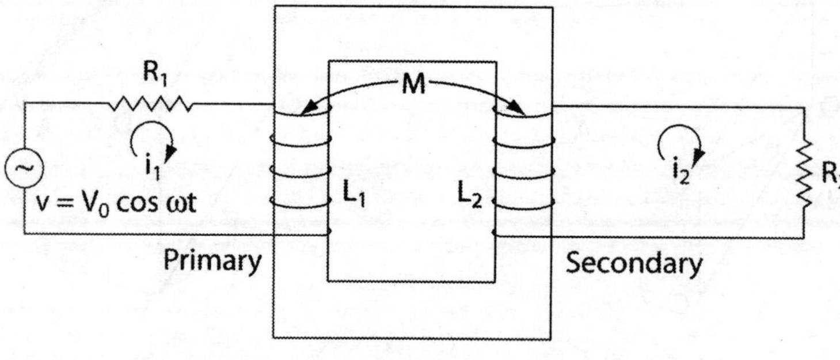

Fig 12.9-1

To develop the theory of a transformer let us consider an ideal transformer in which there is no leakage of magnetic flux or loss of energy. Fig. 12.9-1 shows the circuit diagram of a transformer. Let L_1 and L_2 be the self-inductances of the primary and secondary coils respectively; R_1 be the resistance of the primary coil, R_2 be the total resistance of the secondary circuit and M be the mutual inductance between the primary and the secondary. Let us represent the input voltage $v = V_0 \cos \omega t$ by a rotating vector $\vec{V} = V_0 e^{j\omega t}\hat{\nu}$, which has a constant magnitude V_0 and rotates with constant angular velocity ω; at time $t = 0$, $\vec{V}$ is along the unit vector $\hat{\nu}$. Instantaneous value of i is given by $i = \vec{V} \cdot \hat{\nu}$. Similarly, primary and secondary currents i_1 and i_2 will also be represented by rotating vectors, $\vec{I_1}$ and $\vec{I_2}$ (say) respectively. Now the instantaneous emf equations are,

$$R_1 i_1 + L_1 \frac{di_1}{dt} + M \frac{di_2}{dt} = v \tag{12.9-1}$$

$$R_2 i_2 + L_2 \frac{di_2}{dt} + M \frac{di_1}{dt} = 0. \tag{12.9-2}$$

In terms of rotating vectors these equations become

$$R_1 \vec{I_1} + j\omega L_1 \vec{I_1} + j\omega M \vec{I_2} = \vec{V} \text{ or, } Z_1 \vec{I_1} + j\omega M \vec{I_2} = \vec{V} \tag{12.9-3}$$

$$R_2 \vec{I_2} + j\omega L_2 \vec{I_2} + j\omega M \vec{I_1} = 0 \text{ or, } Z_2 \vec{I_2} + j\omega M \vec{I_1} = 0 \tag{12.9-4}$$

Solving these equations we get

$$\vec{I_1} = \frac{\vec{V}}{Z_1'} \tag{12.9-5}$$

$$\text{and } \vec{I_2} = \frac{-j\omega M \vec{V}/Z_1}{Z_2'}, \tag{12.9-6}$$

where

$$Z_1' = Z_1 + \frac{\omega^2 M^2}{Z_2} \quad \text{and} \quad Z_2' = Z_2 + \frac{\omega^2 M^2}{Z_1}$$

are the effective primary and secondary impedances; the quantity $\omega^2 M^2 / Z_2$ is the impedance reflected from the secondary to the primary and $\omega^2 M^2 / Z_1$ is the impedance reflected from the primary to the secondary.

Now

$$Z_1' = R_1 + j\omega L_1 + \frac{\omega^2 M^2}{R_2 + j\omega L_2} = R_1 + j\omega L_1 + \frac{\omega^2 M^2 (R_2 - j\omega L_2)}{R_2 + \omega^2 L_2^2} = R_1' + j\omega L_1',$$

where

$$R_1' = R_1 + \frac{\omega^2 M^2 R_2}{R_2^2 + \omega^2 L_2^2} \quad \text{and} \quad L_1' = L_1 - \frac{\omega^2 M^2 L_2}{R_2^2 + \omega^2 L_2^2} \tag{12.9-7}$$

are the effective primary resistance and effective primary inductance respectively. The quantity $\omega^2 M^2 R_2 / \left(R_2^2 + \omega^2 L_2^2\right)$ is called the *reflected resistance* from the secondary to the primary and $\omega^2 M^2 L_2 / \left(R_2^2 + \omega^2 L_2^2\right)$ is the inductance reflected from the secondary to the primary.

Usually the primary and secondary coil reactances are much larger than the connected impedances. Assuming ωL_2 large compared with R_2 we can write from (12.9-7)

$$R_1' \simeq R_1 + \frac{M^2}{L_2^2} R_2 \quad \text{and} \quad L_1' \simeq L_1 - \frac{M^2}{L_2}.$$

If there is no magnetic leakage, we can put $M = \sqrt{L_1 L_2}$, then

$$R_1' = R_1 + \frac{L_1}{L_2} R_2 \quad \text{and} \quad L_1' = 0.$$

Thus, when a load of low resistance is connected across the secondary and there is no magnetic leakage the primary inductance is effectively zero. That is the reflected reactance neutralises the inductive reactance of the primary. The effective impedance across the primary source of supply is then purely resistive and is given by

$$Z_1' = R_1' = R_1 + \frac{L_1}{L_2} R_2 = R_1 + \left(\frac{N_1}{N_2}\right)^2 R_2, \tag{12.9-8}$$

where N_1 and N_2 are the number of primary and secondary turns respectively. Here we use the fact that $L_1 \propto N_1^2$ and $L_2 \propto N_2^2$.

In practice the primary and secondary coil resistances are often small and the presence of a resistive load R_L across the secondary is considered. Then we may put $R_1 \approx 0$, $R_2 \approx R_L$ and for approximate calculations we can consider a resistance $R_1' \approx \left(\frac{N_1}{N_2}\right)^2 R_L$ across the source.

From Eq. (12.9-5),

$$\vec{I}_1 = \frac{\vec{V}}{R_1' + j\omega L_1'} = \frac{V_0 e^{j(\omega t - \phi)} \hat{\nu}}{\sqrt{R_1'^2 + \omega^2 L_1'^2}} = I_0 e^{j(\omega t - \phi)} \hat{\nu},$$

where

$$\phi = \tan^{-1} \frac{\omega L_1'}{R_1'} \quad \text{and} \quad I_0 = \frac{V_0}{\sqrt{R_1'^2 + \omega^2 L_1'^2}}.$$

Therefore, instantaneous primary current is

$$i_i = \vec{I}_1 \cdot \hat{\nu} = I_0 \cos(\omega t - \phi) \tag{12.9-9}$$

If the secondary load is very small then $L_1' \approx 0$ and hence, $\phi = 0$. This means that power factor $\cos\phi = 1$ and power drawn from the source becomes maximum.

From Eq. (12.9-4) secondary current is given by

$$\vec{I}_2 = \frac{-j\omega M}{Z_2} \vec{I}_1 = \frac{-j\omega M}{R_2 + j\omega L_2} \vec{I}_1 = \frac{\omega M e^{-j\pi/2}}{\sqrt{R_2^2 + \omega^2 L_2^2} e^{j\theta}} \vec{I}_1$$

$$= \frac{\omega M I_0}{\sqrt{R_2^2 + \omega^2 L_2^2}} e^{j(\omega t - \phi - \frac{\pi}{2} + \theta)} \hat{\nu},$$

where $\theta = \tan^{-1} \dfrac{\omega L_2}{R_2}$.

The instantaneous secondary current is

$$i_2 = \vec{I}_2 \cdot \hat{\nu} = \frac{\omega M I_0}{\sqrt{R_2^2 + \omega^2 L_2^2}} \cos\left(\omega t - \phi - \frac{\pi}{2} + \theta\right) \tag{12.9-10}$$

From Eqs. (12.9-9) and (12.9-10) it is clear that the secondary current lags behind the primary current by the angle $\pi/2+\theta$.

Voltage and current transformation ratios

To find these ratios let us consider an ideal transformer, which is defined as a transformer having no losses like Joule heating loss, hysteresis loss, eddy current loss, leakage of magnetic flux etc. Here, $M=\sqrt{L_1L_2}$ and coil resistances are negligible compared with their reactances.

Now from Eq. (12.9-4) we have,

$$\vec{I}_2=-j\omega M\vec{I}_1/Z_2.$$

In terms of rms values we have for an ideal transformer

$$\frac{I_2}{I_1}=\frac{\omega M}{\sqrt{R_2^2+\omega^2L_2^2}}\approx\frac{M}{L_2}=\sqrt{\frac{L_1}{L_2}}=\frac{N_1}{N_2}. \qquad (12.9\text{-}11)$$

Equation (12.9-6) indicates that the emf induced in the secondary is

$$\vec{V}_2=-j\omega M\vec{V}_1/Z_1,$$

where $\vec{V}_1=\vec{V}$ is the primary applied voltage.

In terms of rms values the voltage transformation ratio for an ideal transformer becomes

$$\frac{V_2}{V_1}=\frac{\omega M}{\sqrt{R_1^2+\omega^2L_1^2}}\approx\frac{M}{L_1}=\sqrt{\frac{L_2}{L_1}}=\frac{N_2}{N_1}. \qquad (12.9\text{-}12)$$

Thus, for $N_2>N_1$, $V_2>V_1$ and we have a step-up transformer. On the other hand, for $N_2<N_1$, $V_2<V_1$ and we have a step-down transformer.

Equations (12.9-11) and (12.9-12) show that a step-up transformer increases voltage but decreases current and a step-down transformer decreases voltage but increases current.

Combining Eqs. (12.9-11) and (12.9-12) we get

$$V_2I_2=V_1I_1. \qquad (12.9\text{-}13)$$

This shows that *for an ideal transformer output power equals input power.*

Transformer losses

In the above discussion we have considered ideal transformer which is far from practical realisation. A practical transformer is always associated with some energy losses and the output power is always to some extent smaller than the input power. These losses are (i) the copper losses arising from the Joule heating in the primary and secondary coil resistance, (ii) the core or iron losses due to hysteresis and eddy current heating of the core (iii) the losses associated with the leakage of magnetic flux. Copper losses will be small if the primary and secondary coil resistances are made small. Eddy current losses are reduced by using laminated core material. Hysteresis losses can be reduced by choosing a core material having small hysteresis loop area. Use of iron core increases coupling coefficient. But in transformers used in the radio frequency range such a core cannot be used because of large core losses. Radio frequency transformers use *ferrite* cores which have high permeability and low electrical conductivity. Transformers without core are known as *air-cored* transformers. Because of various kinds of losses in practical transformers output power is smaller than the input power and we define the efficiency η of a transformer as

$$\eta = \frac{\text{output power}}{\text{input power}} \times 100\%.$$

The rating of a transformer is usually expressed in volt-ampere and not in watt because copper loss depends on current, iron loss depends on voltage and the total transformer loss depends on current × voltage and not on the phase angle as is the usual case in ac circuits.

Types of transformers

Depending on the frequency of operation transformers may be classified as mains frequency transformers, audio frequency transformers and radio-frequency transformers. Depending on the structure transformers may be classified as *core type* with single magnetic circuit, *shell type* with double magnetic circuit and *berry type* with distributed magnetic circuit.

Uses of transformers

Transformers are used to step-up or step-down ac voltage in electrical power distribution system. They are also used for the construction of power supplies, impedance matching between two independent circuits, welding, measuring large alternating current, high pressure etc.

Auto-transformers

An autotransformer consists of a single winding only, part of which is common to both primary and secondary. It can also step-up or step-down ac voltages (Fig. 12-9-2).

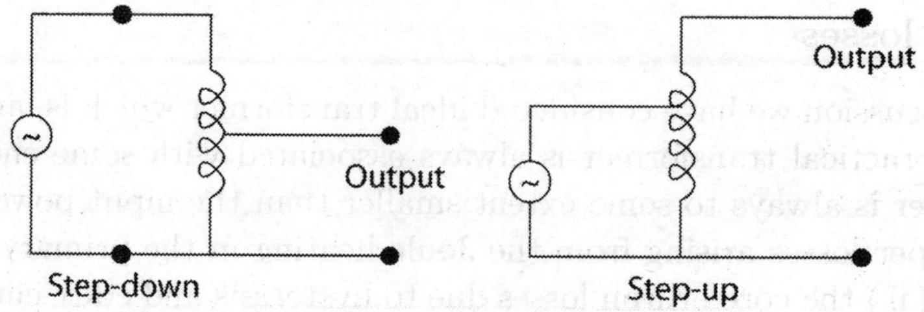

Fig. 12.9-2 : Auto-transformer

12.10 Coupled LCR Circuits

Suppose an ac voltage $v = V_0 \cos \omega t$ is applied to a circuit of resistance R_1, inductance L_1 and capacitance C_1 which is inductively coupled to a second circuit of resistance R_2, inductance L_2 and capacitance C_2. Let M be the mutual inductance of the two coils (Fig. 12.10-1). As before, we shall represent the voltage v by a rotating vector $\vec{V} = V_0 e^{j\omega t}\hat{\nu}$, the primary and secondary currents will also be represented by rotating

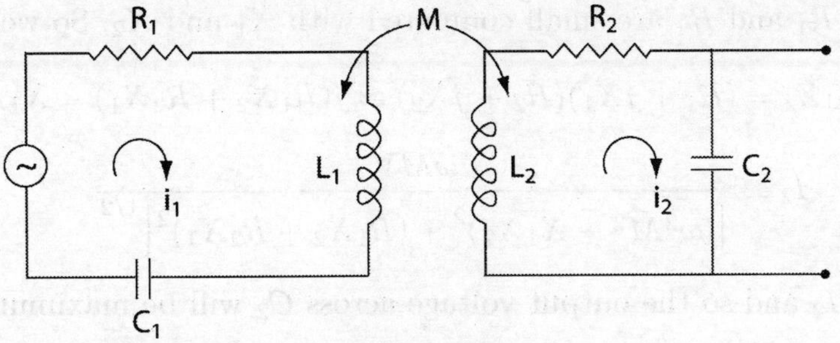

Fig 12.10-1

vectors $\vec{I}_1$ and $\vec{I}_2$, respectively. Now the instantaneous emf equations are

$$R_1 i_1 + L_1 \frac{di_1}{dt} + \frac{1}{C_1}\int i_1 dt + M\frac{di_2}{dt} = v \qquad (12.10\text{-}1)$$

$$\text{and} \quad R_2 i_2 + L_2 \frac{di_2}{dt} + \frac{1}{C_2}\int i_2 dt + M\frac{di_1}{dt} = 0. \qquad (12.10\text{-}2)$$

In terms of rotating vectors these equations become

$$\left[R_1 + j\left(\omega L_1 - \frac{1}{\omega C_1}\right)\right]\vec{I}_1 + j\omega M\vec{I}_2 = \vec{V} \quad \text{or} \quad Z_1\vec{I}_1 + j\omega M\vec{I}_2 = \vec{V} \qquad (12.10\text{-}3)$$

$$\left[R_2 + j\left(\omega L_2 - \frac{1}{\omega C_2}\right)\right]\vec{I}_2 + j\omega M\vec{I}_1 = 0 \quad \text{or} \quad Z_2\vec{I}_2 + j\omega M\vec{I}_1 = 0, \qquad (12.10\text{-}4)$$

where

$$Z_1 = R_1 + j\left(\omega L_1 - \frac{1}{\omega C_1}\right) = R_1 + jX_1$$

$$\text{and} \quad Z_2 = R_2 + j\left(\omega L_2 - \frac{1}{\omega C_2}\right) = R_2 + jX_2.$$

Solving Eqs. (12.10-3) and (12.10-4) we get

$$\vec{I}_1 = \frac{\vec{V}}{Z_1'} \quad \text{and} \quad \vec{I}_2 = \frac{-j\omega M\vec{V}/Z_1}{Z_2'}, \qquad (12.10\text{-}5)$$

where

$$Z_1' = Z_1 + \frac{\omega^2 M^2}{Z_2} \quad \text{and} \quad Z_2' = Z_2 + \frac{\omega^2 M^2}{Z_1}$$

are the effective primary and secondary impedances. The quantity ω^2M^2/Z_2 is the impedance reflected from the secondary to the primary and ω^2M^2/Z_1 is the impedance reflected to the secondary from the primary. Now, in terms of rms values we can write for the secondary current as

$$I_2 = \frac{\omega MV}{|Z_1Z_2'|} = \frac{\omega MV}{|Z_1Z_2 + \omega^2M^2|}. \tag{12.10-6}$$

In practice R_1 and R_2 are small compared with X_1 and X_2. So we can write

$$Z_1Z_2 = (R_1 + jX_1)(R_2 + jX_2) \approx j(R_1X_2 + R_2X_1) - X_1X_2$$

$$\therefore \quad I_2 \simeq \frac{\omega MV}{\left[(\omega^2M^2 - X_1X_2)^2 + (R_1X_2 + R_2X_1)^2\right]^{1/2}}. \tag{12.10-7}$$

Obviously, I_2 and so the output voltage across C_2 will be maximum when

$$\omega^2M^2 - X_1X_2 = 0. \tag{12.10-8}$$

Let us now consider the most commonly encountered case in radio communication, where $L_1 = L_2 = L$ and $C_1 = C_2 = C$. That is, each circuit has the same natural resonant frequency

$$\omega_0 = \frac{1}{\sqrt{L_1C_1}} = \frac{1}{\sqrt{L_2C_2}}.$$

Therefore, from Eq. (12.10-8)

$$\left(\omega L - \frac{1}{\omega C}\right)^2 = \omega^2M^2$$

$$\text{or} \quad \left(\frac{\omega^2}{\omega_0^2} - 1\right)^2 = \omega^4M^2C^2 \quad \left[\because\ \omega_0^2 = \frac{1}{LC}\right]$$

$$\text{or} \quad \frac{\omega^2}{\omega_0^2} - 1 = \pm\omega^2MC$$

$$= \pm\omega^2kLC \quad \left[\because\ M = k\sqrt{L_1L_2} = kL\right]$$

$$= \pm\frac{\omega^2}{\omega_0^2}k.$$

$$\therefore \quad \omega = \frac{\omega_0}{\sqrt{1+k}} \quad \text{or} \quad \frac{\omega_0}{\sqrt{1-k}}. \tag{12.10-9}$$

Thus, there are two resonant frequencies at which secondary current becomes maximum (Fig. 12.10-2)—one frequency is slightly smaller and the other slightly greater than the natural resonant frequency of the either circuit considered separately.

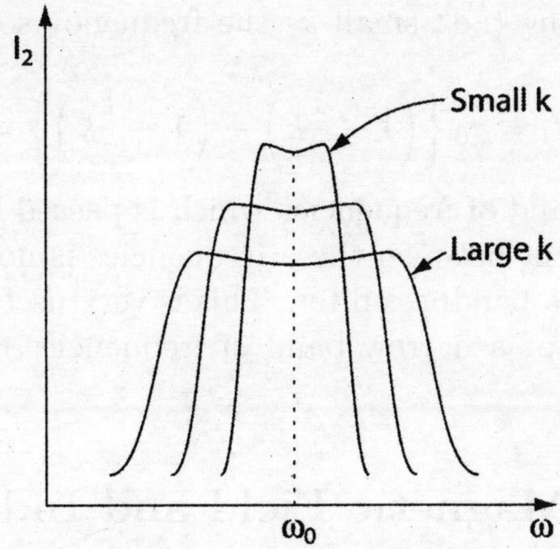

Fig. 12.10-2: Current response in the secondary of a coupled circuit

This phenomenon of showing two peaks by the secondary current against supply frequency is termed as *double-humped tuning* by radio engineers.

The separation between the resonant frequencies depends on the coupling coefficient k. If k is small the coupling is said to be *loose* and if k is large the coupling is said to be *tight*. There is a value of k for which secondary current becomes maximum. The system is then said to have an optimum coupling. To find the condition for optimum coupling let us consider the case when both the primary and secondary circuits are resonant to the supply with frequency ω_0. Then $X_1 = X_2 = 0$ and from Eq. (12.10-6) we can write

$$I_2 = \frac{\omega M V}{R_1 R_2 + \omega_0^2 M^2}. \tag{12.10-10}$$

For I_2 to be maximum, $dI_2/dM = 0$.

This gives the value of M for optimum coupling between the two tuned circuits as,

$$\omega_0 M = \sqrt{R_1 R_2}. \tag{12.10-11}$$

If k_0 be the corresponding coupling coefficient, then

$$k_0 = \frac{M}{\sqrt{L_1 L_2}} = \frac{\sqrt{R_1 R_2}}{\omega_0 \sqrt{L_1 L_2}} = \frac{1}{\sqrt{Q_1 Q_2}}, \tag{12.10-12}$$

where $Q_1 = \omega_0 L_1/R_1$ and $Q_2 = \omega_0 L_2/R_2$ are the quality factors of the primary and secondary circuits, respectively.

Bandwidth

We find that the secondary current becomes maximum corresponding to two frequencies f_1 and f_2 on either side of the resonant frequence $f_0 (= \omega_0/2\pi)$, where f_1 and f_2 are given by (12.10-9) as

$$f_1 = \frac{f_0}{\sqrt{1+k}} \quad \text{and} \quad f_2 = \frac{f_0}{\sqrt{1-k}}.$$

In case of loose coupling (i.e., small k) the frequency separation between the two peaks is

$$f_2 - f_1 = f_0\left\{\left(1+\frac{1}{2}k\right) - \left(1-\frac{1}{2}k\right)\right\} = f_0 k. \qquad (12.10\text{-}13)$$

This determines the band of frequencies which is passed by the system because the fall of the secondary current between these frequencies is not much for small k. Thus, the circuit can be used as a bandpass filter. This is very useful in radio communication when it is desired to accept a narrow band of frequencies and reject any frequencies outside it.

12.11 Rotating Magnetic Field and Induction Motor

A rotating magnetic field is one in which the magnitude of the magnetic field remains constant while its direction continuously rotates at a constant rate about a fixed axis. It can be produced by using a system of stationary coils supplied with single, two, three or polyphase alternating currents. Its importance mainly lies in the fact that the design of electric motors using it becomes very simple and the motors themselves become more robust. By rotating a single coil with constant angular velocity in a uniform magnetic field we can generate the simplest *single-phase* ac, which may be described by a sine or cosine function. If two identical coils are kept at right angles to each other and the combination is rotated at a constant rate in a uniform magnetic field then the emfs generated in the coils differ in phase by $\pi/2$. Such an arrangement is known as *two-phase ac generator*. When a system of three identical coils with their planes fixed at $2\pi/3$ angles apart is rotated at a constant rate in a uniform magnetic field, the emfs generated in the coils differ in phase by $2\pi/3$ with respect to each other. Such emfs may be represented as $v_1 = V_0 \cos\omega t$, $v_2 = V_0\cos(\omega t - 2\pi/3)$ and $v_3 = V_0\cos(\omega t - 4\pi/3)$. Such arrangement is known as *three-phase ac generator*. This three-phase system is found advantageous and has been universally adopted for the generation and transmission of electrical energy.

Rotating magnetic field with two-phase supply

Suppose we have two identical coils C_1C_2 and C_3C_4 arranged at right angles to each other and fed by a two-phase ac supply (Fig. 12.11-1). The currents in these coils are of equal frequency and amplitude but differ in phase by $\pi/2$. The magnetic fields produced at the centre o of the coils will be in phase with the currents and acts in a direction normal to the plane of the respective coils. So we may represent these magnetic fields by

$$B_1 = B_0\cos\omega t \quad \text{and} \quad B_2 = B_0\cos(\omega t - \pi/2) = B_0\sin\omega t, \qquad (12.11\text{-}1)$$

where B_1 is due to the coil C_1C_2 acting along oC_4 and B_2 is due to the coil C_3C_4 acting along oC_1.

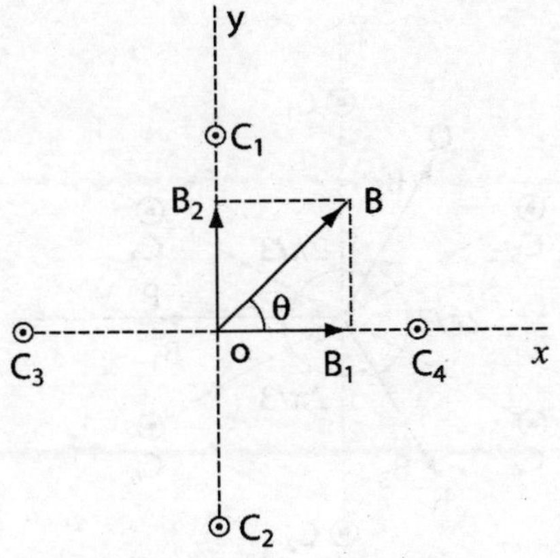

Fig. 12.11-1 : Rotating magnetic field produced from two-phase ac supply

The resultant magnetic field at o is of magnitude

$$B = \sqrt{B_1^2 + B_2^2} = B_0$$

and makes an angle θ with ox given by

$$\tan\theta = \frac{B_2}{B_1} = \tan\omega t, \quad \text{or} \quad \theta = \omega t.$$

Hence, the resultant magnetic field $\vec{B}$ has a constant magnitude and rotates with a constant angular velocity ω in anticlockwise direction (as θ increases with time t). If on the other hand, the fields are given by

$$B_1 = B_0 \cos(\omega t - \pi/2) \quad \text{and} \quad B_2 = B_0 \cos\omega t,$$

then again $B = B_0$ but

$$\tan\theta = \frac{B_2}{B_1} = \cot\omega t = \tan\left(\frac{\pi}{2} - \omega t\right) \quad \text{or} \quad \theta = \frac{\pi}{2} - \omega t.$$

Under such conditions θ decreases with time, i.e., the resultant field $\vec{B}$ rotates in clockwise direction.

Rotating magnetic field with three-phase supply

Suppose we have three identical coils C_1C_2, C_3C_4 and C_5C_6 arranged with a common centre o and their planes fixed at $2\pi/3$ angles apart (Fig. 12.11-2). Each coil is supplied with one phase of a three-phase alternating current supply. So the currents in these coils are of equal frequency and amplitude but differ in phase by $2\pi/3$ with respect to each other. The magnetic fields produced at the centre o of the coils will be in phase with the currents and act in a direction normal to the plane of the respective coils. So

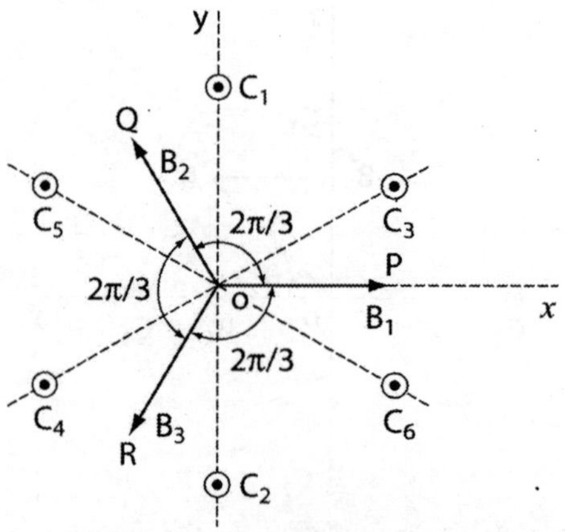

Fig. 12.11-2: Rotating magnetic field-produced from three-phase ac supply

we may represent these magnetic fields by

$$B_1 = B_0 \cos \omega t, \quad B_2 = B_0 \cos(\omega t - 2\pi/3) \quad \text{and} \quad B_3 = B_0 \cos(\omega t - 4\pi/3), \tag{12.11-2}$$

where B_1 is due to coil C_1C_2 and acts along oP or ox, B_2 is due to C_3C_4 and acts along OQ, B_3 is due to coil C_5C_6 and acts along oR. In Fig. 12.11-2 the mutual angles between B_1, B_2 and B_3 is $2\pi/3$. To find the resultant magnetic field let us resolve the magnetic fields along ox and its perpendicular direction oy. The resultant x-component is

$$\begin{aligned} B_x &= B_1 - B_2 \cos\frac{\pi}{3} - B_3 \cos\frac{\pi}{3} \\ &= B_0 \cos \omega t - \frac{B_0}{2} \cos(\omega t - 2\pi/3) - \frac{B_0}{2} \cos(\omega t - 4\pi/3) \\ &= \frac{3}{2} B_0 \cos \omega t. \end{aligned}$$

Similarly, the resultant y-component is

$$\begin{aligned} B_y &= 0 + B_2 \cos\frac{\pi}{6} - B_3 \cos\frac{\pi}{6} \\ &= B_0 \cdot \frac{\sqrt{3}}{2} \cos(\omega t - 2\pi/3) - B_3 \frac{\sqrt{3}}{2} \cos(\omega t - 4\pi/3) \\ &= \frac{3}{2} B_0 \sin \omega t. \end{aligned}$$

The resultant magnetic field, therefore, is of magnitude

$$B = \sqrt{B_x^2 + B_y^2} = \frac{3}{2} B_0$$

and makes an angle θ with ox given by

$$\tan\theta = \frac{B_y}{B_x} = \tan\omega t \quad \text{or} \quad \theta = \omega t.$$

Thus, in this case, the resultant magnetic field has a constant magnitude and rotates with a constant angular velocity ω.

Couple acting on a small coil placed in a rotating magnetic field

Suppose a small coil C_1C_2 of N turns and effective area A is placed at o, where there is a rotating magnetic field of constant magnitude B, which rotates with constant angular velocity ω_1. According to Faraday's law of electromagnetic induction there will be an induced current in the coil. According to Lenz's law the direction of induced current will be such as to oppose the relative motion between the coil and the magnetic field. So a couple acts on the coil tending it to rotate in the direction of the field. Let us calculate the moment of this couple. If the coil rotates with constant angular velocity ω_2 the relative angular velocity of the magnetic field with respect to the coil is $\omega_1 - \omega_2 = \omega$ (say). Now the magnetic flux linked with the coil at any instant t is given by

$$\Phi = NAB\sin\omega t. \tag{12.11-3}$$

Therefore, instantaneous induced emf in the coil is

$$\mathcal{E} = -\frac{d\Phi}{dt} = -NAB\omega\cos\omega t = NAB\omega\cos(\omega t - \pi). \tag{12.11-4}$$

If L and R are respectively the self-inductance and resistance of the coil, the current in the coil at that instant will be

$$i = \frac{NAB\omega}{\sqrt{R^2 + \omega^2L^2}}\cos(\omega t - \pi - \phi),$$

where

$$\tan\phi = \frac{\omega L}{R}. \tag{12.11-5}$$

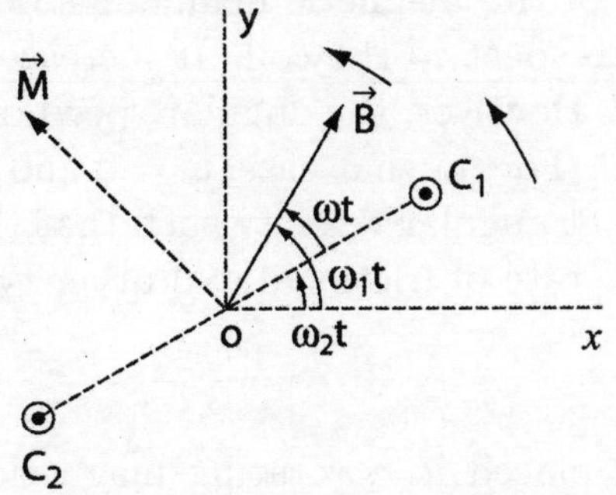

Fig. 12.11-3 : Coil in a rotating magnetic field

Now the coil with a current i flowing though it behaves like a magnetic dipole of moment $\vec{M} = \hat{n}NiA$, where $\hat{n}$ is a unit normal to the plane of the coil in the usual right-hand sense. The torque acting on the coil tending it to turn into the direction $\vec{B}$ is given by

$$\begin{aligned}\vec{\tau} = \vec{M} \times \vec{B} &= \hat{k}MB \sin\left(\frac{\pi}{2} - \omega t\right) = \hat{k}NiAB\cos\omega t \\ &= \hat{k}\frac{N^2A^2B^2\omega}{\sqrt{R^2+\omega^2L^2}}\cos\omega t\cos(\omega t - \pi - \phi) \\ &= -\hat{k}\frac{N^2A^2B^2\omega}{\sqrt{R^2+\omega^2L^2}}\cos\omega t\cos(\omega t - \phi) \\ &= -\hat{k}\frac{N^2A^2B^2\omega}{\sqrt{R^2+\omega^2L^2}}\left[\cos^2\omega t\cos\phi + \cos\omega t\sin\omega t\sin\phi\right],\end{aligned}$$

where $\hat{k}$ is a unit vector normal to the plane containing $\vec{M}$ and $\vec{B}$. Since the average value of $\cos^2\omega t$ over a cycle is $\frac{1}{2}$ and that of $\cos\omega t\sin\omega t$ is zero, the average torque acting on the coil is given by

$$\vec{\tau}_{av} = -\hat{k}\frac{N^2A^2B^2\omega}{2\sqrt{R^2+\omega^2L^2}}\cos\phi = -\hat{k}\frac{N^2A^2B^2\omega R}{2(R^2+\omega^2L^2)}. \qquad (12.11\text{-}6)$$

From diagram it is clear that the torque tends to rotate the coil in the direction of rotation of the magnetic field. From Eq. (12.11-6) we find that $\vec{\tau}_{av} = 0$ when $\omega = 0$ or $\omega = \infty$. The average torque will have a maximum value for some value of ω given by

$$\frac{d\vec{\tau}_{av}}{d\omega} = 0 \quad \text{or} \quad \frac{d}{d\omega}\left[\frac{\omega}{R^2+\omega^2L^2}\right] = 0$$

This requires that $\omega L = R$ for $\vec{\tau}_{av}$ to be maximum. The value of maximum average couple is

$$\vec{\tau}_{av} = \frac{N^2A^2B^2}{4L}. \qquad (12.11\text{-}7)$$

When the coil is placed in the rotating magnetic field, the coil begins to rotate in the direction of rotation of the magnetic field. This decreases the relative angular velocity ω of the field with respect to the coil. It increases the torque and hence, the speed of rotation of the coil. However, the angular speed of the coil can never be equal to that of the magnetic field. For, in that case, $\omega = 0$ and $\vec{\tau}_{av} = 0$. In the steady state the coil moves with a constant angular velocity such that the rate of doing work by the torque becomes equal to the rate of frictional and other type of losses.

Induction motor

The fact that a closed coil placed in a rotating magnetic field experiences a torque and rotates in the direction of rotation of the field is used in the construction of ac motor known as *induction motor.* This motor is found to be comparatively simple in construction, low in cost, robust and efficient. Depending on the type of supply available an induction motor may be a single, two or three phase motor. The motor has a stationary part called the *stator* and a rotating part called the *rotor.* In a three phase motor the stator consists of three coils 120° apart in space, each coil being supplied with one phase of a three-phase ac power supply (Fig. 12.11-4). It generates a rotating magnetic field. In this field is placed the rotor. The rotor of an induction motor usually consists of a core of laminated soft iron with heavy copper bars placed in slots. The bars are welded at both ends to two massive copper rings. The whole arrangement resembles the wheel of a squirrel cage and the rotor is, therefore, known as the *squirrel-cage rotor.* As the flux of the rotating magnetic field linked with the rotor changes with time a current is induced in the rotor. According to Lenz's law this induced current tries to oppose the relative motion between the rotor and the magnetic field. So a couple acts on the rotor tending it to rotate in the direction of the magnetic field. The angular speed of the rotor can never be equal to that of the magnetic field. For in that case there will be no relative motion and no induced current in it or torque on it. For this induction motor is sometimes referred to as *asynchronous motor.* The

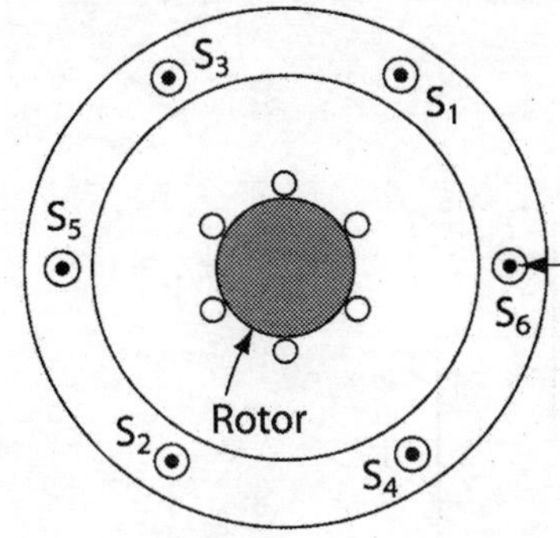

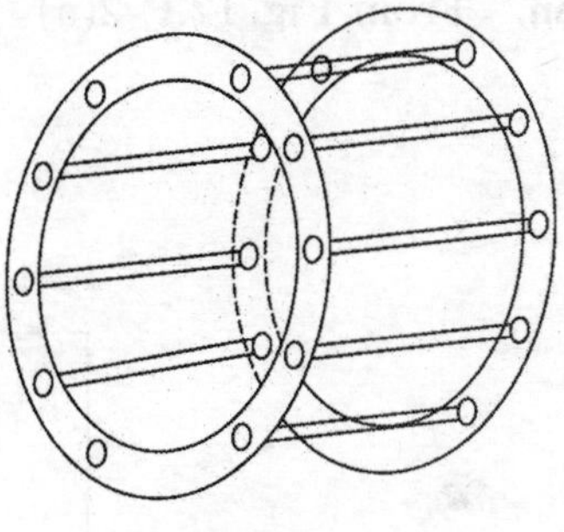

Fig. 12.11-4: A simple three-phase induction motion: (a) schematic diagram, (b) Squirrel-cage rotor structure

difference between the speed of the rotating field and the speed of the rotor is called the *slip.* The frequency of the current induced in the rotor windings depends on this slip and is known as *slip frequency.* Induction motor is substantially a constant speed machine. Its speed changes slightly with the increase of load.

Note that in most other types of motors current is supplied to both the stator and rotor windings but in induction motor current is supplied only to the stator and current is induced in the rotor by the rotating magnetic field produced by the stator winding current. For this reason the motor is called an *induction motor.*

SOLVED PROBLEMS

► **Problem 1.** *A steady current of 2 A flows through a coil of self-inductance* 30 mH *when connected to a* 20 V *dc supply. Calculate the current, power factor and power dissipation in the coil when it is reconnected to a* 200 V, 50 Hz *ac supply.*

Solution. The resistance of the coil is $R = 20V/2A = 10\,\Omega$.

When connected to ac, the rms current is given by

$$I_r = \frac{V_r}{\sqrt{R^2 + \omega^2 L^2}} = \frac{200}{\sqrt{10^2 + (2\pi \times 50 \times 30 \times 10^{-3})^2}}\,A = \frac{200}{13.74}\,A = 14.55\,A$$

Power factor $\cos\phi = \dfrac{R}{\sqrt{R^2 + \omega^2 L^2}} = \dfrac{10}{13.74} = 0.73$

Power dissipation $P = I_r \cdot V_r \cdot \cos\phi = 14.55 \times 200 \times 0.73\,W = 2118\,W$.

► **Problem 2.** *Consider a resistor of* 3 Ω, *an inductance of reactance* 5 Ω *and a capacitor of reactance* 8 Ω *all connected in parallel and the combination is supplied with an ac source of 120 V. Draw the phasor diagram for the currents through each branch and hence, find the magnitude and phase of the current drawn from the source.*

Solution. From Fig. 12.P-2(a)

$$i = i_R + i_L + i_C.$$

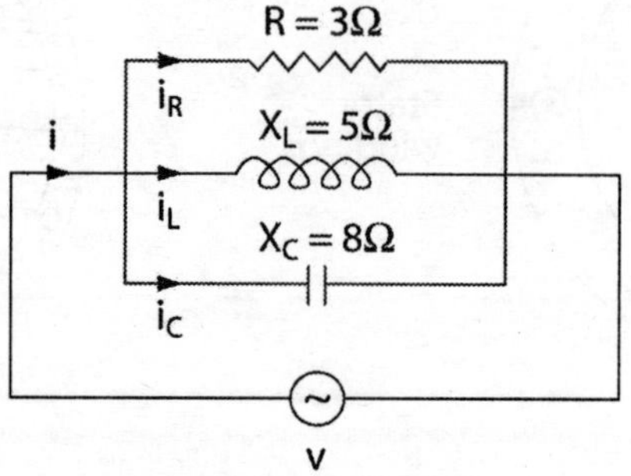

Fig 12.P-2(*a*)

In terms of rotating vectors or phasor currents,

$$\vec{I} = \vec{I}_R + \vec{I}_L + \vec{I}_C.$$

Here

$$I_R = \frac{V}{R} = \frac{120\,V}{3\,\Omega} = 40A \text{ which is in phase with the voltage phasor } \vec{V}$$

$$I_L = \frac{V}{X_L} = \frac{120\,V}{5\,\Omega} = 24\text{A which lags behind } \vec{V} \text{ by } 90^\circ$$

$$I_C = \frac{V}{X_C} = \frac{120\,V}{8\,\Omega} = 15\text{A which leads over } \vec{V} \text{ by } 90^\circ.$$

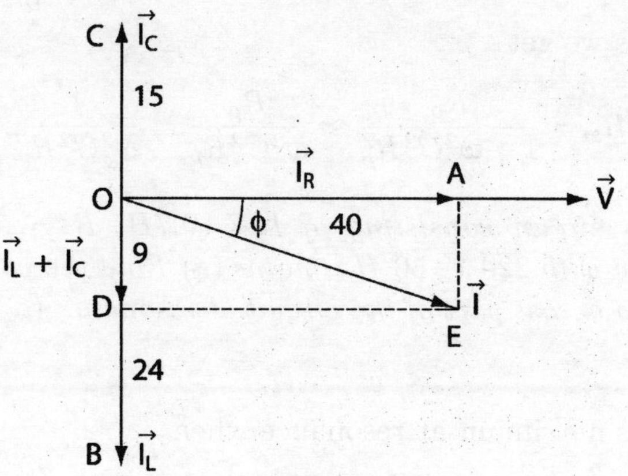

Fig 12.P-2(b)

In Fig. 12.P-2(b), $\overrightarrow{OA} = \vec{I}_R$, $\overrightarrow{OB} = \vec{I}_L$, $\overrightarrow{OC} = \vec{I}_C$, $\overrightarrow{OD} = \vec{I}_L + \vec{I}_C$ and $\overrightarrow{OE} = \vec{I}$.

Obviously, $I = \left|\overrightarrow{OE}\right| = \sqrt{9^2 + 40^2}A = 41A$.

This resultant current lags behind the applied emf phasor $\vec{V}$ by the angle

$$\phi = \tan^{-1}\frac{AE}{AO} = \tan^{-1}\frac{9}{40} = 12.68°.$$

► **Problem 3.** *Show that an imperfect capacitor can be represented either by a perfect capacitor C with a high leakage resistance R_p in parallel or by a perfect capacitor C with a low resistance R_s in series, where $R_s \approx 1/\omega^2 C^2 R_p$. Here ω is the frequency of operation, which is assumed to be high enough such that $1/\omega C \ll R_p$.*

Solution. From Fig. 12.P-3, equating the impedance for the two representations we get

$$R_s + \frac{1}{j\omega C} = \frac{R_p \times \frac{1}{j\omega C}}{R_p + \frac{1}{j\omega C}} = \frac{R_p(1 - j\omega C R_p)}{1 + \omega^2 C^2 R_p^2}.$$

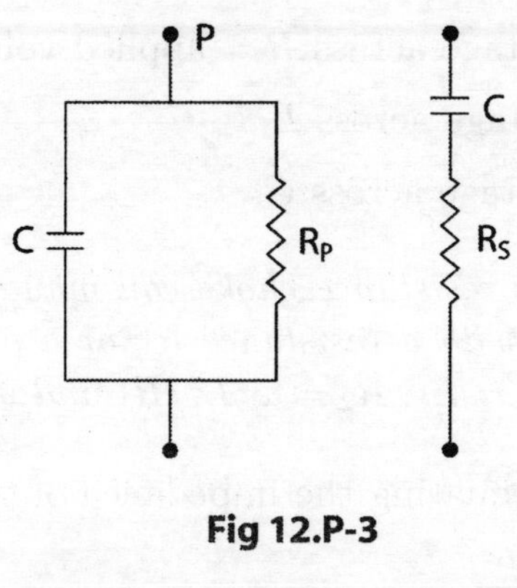

Fig 12.P-3

Equating real parts we get

$$R_s = \frac{R_p}{1+\omega^2 C^2 R_p^2} \approx \frac{R_p}{\omega^2 C^2 R_p} = \frac{1}{\omega^2 C^2 R_p}.$$

► **Problem 4.** *A series circuit consisting of $L = 0.7H$, $R = 100\ \Omega$ and a variable capacitor C, is supplied with 220 V, 50 Hz mains* (a) *Find the value of C for maximum current* (b) *Find the pd across part of the circuit at resonance* (c) *Also, draw the phasor diagram at resonance.*

Solution. Current I is maximum at resonance when

$$\omega L = \frac{1}{\omega C}, \quad \text{or} \quad C = \frac{1}{\omega^2 L} = \frac{1}{(2\pi \times 50)^2 \times 0.7} F = 14.5\,\mu F.$$

At resonance, current $I = \frac{200\ V}{100\ \Omega} = 2.2\ A$

pd across R, $V_R = IR = 2.2 \times 100\,\text{V} = 220\,\text{V}$

pd across L, $V_L = I \cdot \omega L = 2.2 \times 2\pi \times 50 \times 0.7V = 483.5V$

pd across C, $V_C = I \cdot \frac{1}{\omega C} = 483.5\,\text{V}$

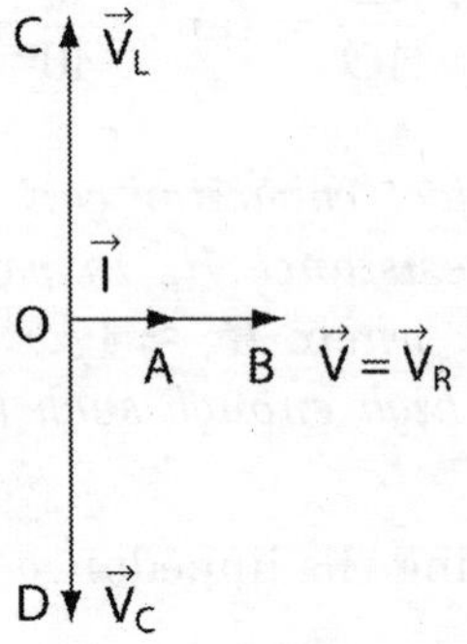

Fig 12.P-4

Fig. 12.P-4 shows the phasor diagram, where $\overrightarrow{OA} = \vec{I}$, current phasor.

$\overrightarrow{OB} = \vec{V}_R$, phasor for voltage across R = applied voltage phasor $\vec{V}$.

$\overrightarrow{OC} = \vec{V}_L$, phasor for voltage across L

$\overrightarrow{OD} = \vec{V}_C$, phasor for voltage across C.

► **Problem 5.** *Show that a low resistance choke coil may be regarded as equivalent to a pure inductance L_1 in series with a resistance R_1 or a pure inductance L_2 in parallel with a resistance R_2, where $L_2 = L_1$, $R_2 = \omega^2 L_1^2/R_1$ and ω is the frequency of operation.*

Solution. From Fig. 12.P-5, equating the impedance of the two equivalent circuits we get

$$R_1 + j\omega L_1 = \frac{R_2 \cdot j\omega L_2}{R_2 + j\omega L_2}$$

$$\text{or } R_1R_2 - \omega^2 L_1 L_2 + j\omega(L_1R_2 + L_2R_1) = j\omega L_2 R_2.$$

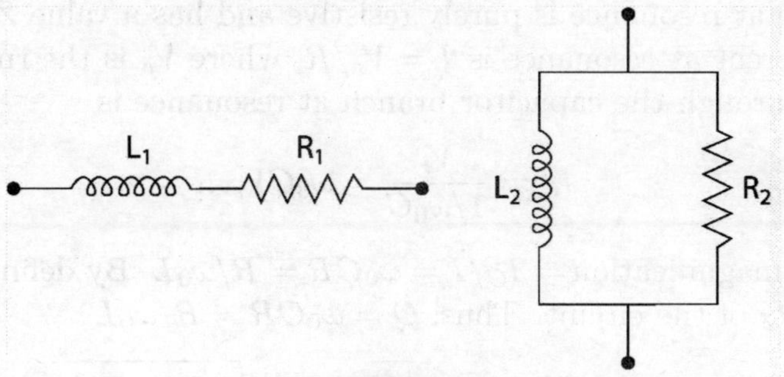

Fig 12.P-5

Equating real and imaginary parts separately we get

$$R_1R_2 - \omega^2 L_1 L_2 = 0 \tag{1}$$

$$\text{and} \quad L_1R_2 + L_2R_1 = L_2R_2 \quad \text{or} \quad L_1R_2 = L_2R_2\left(1 - \frac{R_1}{R_2}\right) \tag{2}$$

As R_1 is small we can write. $L_1 \approx L_2$. Now from (1).

$$R_2 \approx \omega^2 L_1^2 / R_1.$$

▶ **Problem 6.** *An alternating emf* $v = V_0 \cos\omega t$ *is applied across an inductor* L*, capacitor* C *and a resistor* R *all connected in parallel. Find the resonant frequency and* Q *value of the circuit. If* $R = 1\ k\Omega$*,* $C = 0.4\mu F$ *and* $L = 4mH$*, what would be the value of* Q*?*

Solution. The impedance of the circuit is given by

$$\frac{1}{Z} = \frac{1}{R} + \frac{1}{j\omega L} + j\omega C = \frac{1}{R} + j\left(\omega C - \frac{1}{\omega L}\right).$$

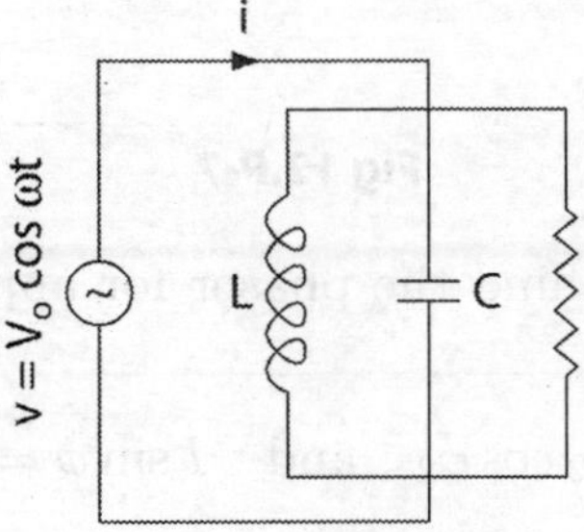

Fig 12.P-6

For the current i to be in phase with the voltage v, the imaginary part of $\frac{1}{Z}$ must be zero. That is

$$\omega C = \frac{1}{\omega L} \quad \text{or} \quad \omega = \frac{1}{\sqrt{LC}} = \omega_0 \text{ (say)}.$$

This is the resonant angular frequency of the circuit. So the resonant frequency is

$$f_0 = \frac{\omega_0}{2\pi} = \frac{1}{2\pi\sqrt{LC}}.$$

The impedance at resonance is purely resistive and has a value $Z = R$. So the rms value of source current at resonance is $I = V_r/R$, where V_r is the rms value of source voltage. Current through the capacitor branch at resonance is

$$I_C = \frac{V_r}{1/\omega_0 C} = \omega_0 C V_r.$$

So the current magnification $= I_C/I = \omega_0 CR = R/\omega_0 L$. By definition it represents the quality factor Q of the circuit. Thus, $Q = \omega_0 CR = R/\omega_0 L$

$$Q = \omega_0 CR = \frac{CR}{\sqrt{LC}} = R\sqrt{\frac{C}{L}} = 10^3\sqrt{\frac{0.4\times10^{-6}}{4\times10^{-3}}} = 10.$$

▶ **Problem 7.** *Two refrigerators are supplied from the same ac mains. The first one draws a current of* 3A *at a power factor* 0.6 *and the second one draws a current of* 2A *at a power factor of* 0.7. *Assuming that the current lags behind the emf in both the cases calculate the total current drawn by the two refrigerators and the corresponding power factor.*

Solution. From Fig. 12.P-7, we can write in terms of current phasors as

$$\vec{I} = \vec{I_1} + \vec{I_2}.$$

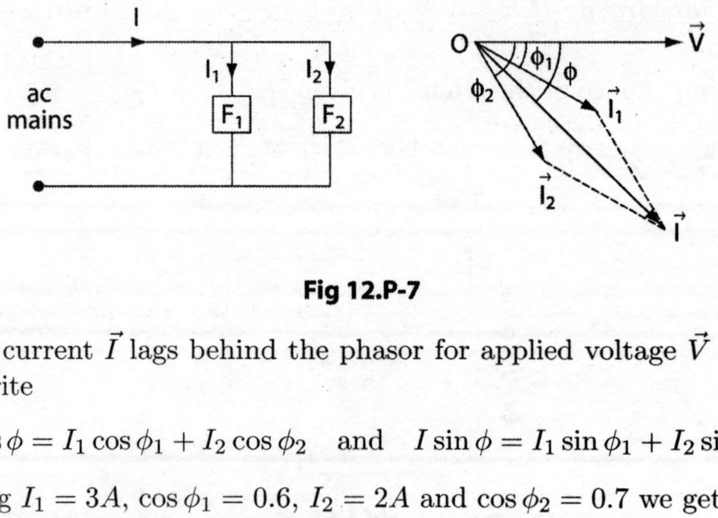

Fig 12.P-7

If the total current $\vec{I}$ lags behind the phasor for applied voltage $\vec{V}$ by the angle ϕ then we can write

$$I\cos\phi = I_1\cos\phi_1 + I_2\cos\phi_2 \quad \text{and} \quad I\sin\phi = I_1\sin\phi_1 + I_2\sin\phi_2$$

Now putting $I_1 = 3A$, $\cos\phi_1 = 0.6$, $I_2 = 2A$ and $\cos\phi_2 = 0.7$ we get

$$I\cos\phi = 3\times0.6 + 2\times0.7 = 3.2$$

$$I\sin\phi = 3\times0.8 + 2\times0.714 = 3.8 \quad [\because\ \sin\phi_1 = 0.8 \text{ and } \sin\phi_2 = 0.714]$$

$$\therefore\quad I = \sqrt{(3.2)^2 + (3.8)^2} = 4.96A \text{ and hence, } \cos\phi = \frac{3.2}{I} = \frac{3.2}{4.96} = 0.64$$

▶ **Problem 8.** *A* 100 V, 60 W *lamp is to be operated on* 220 V, 50 Hz *ac supply. What* (a) *pure resistance* (b) *pure inductance placed in series with the lamp will enable it to be correctly used? Which method will be more economical and why?* **[C.U. 2004, 2006]**

Solution. Resistance of the lamp is $r = \dfrac{V^2}{W} = \dfrac{100 \times 100}{60}\,\Omega = 166.66\,\Omega$.

(a) Let a resistance R is to be connected in series. Then pd across the lamp must be 100 V for its normal operation. Therefore,

$$\frac{200 \times r}{r + R} = 100 \quad \text{or} \quad R \simeq 200\,\Omega.$$

(b) With an inductance L connected in series we must have

$$I_r^2 \cdot r = 60 \quad \text{or} \quad \left(\frac{220}{\sqrt{r^2 + \omega^2 L^2}}\right)^2 \cdot r = 60.$$

Putting $r = 166.66\,\Omega$ and $\omega = 2\pi \times 50$ we get $L = 1.03$ H.

If we use a resistance there will be additional energy dissipation in the resistance. But if we use a pure inductance there will be no additional energy dissipation. So use of pure inductance will be economical.

▶ **Problem 9.** *A coil of inductance L_1 and resistance R_1 is connected in parallel with another coil of inductance L_2 and resistance R_2. The combination is connected to a source of sinusoidal emf. Prove that the currents in the two coils will be in phase provided that the mutual inductance M between the coils is given by*

$$M = \frac{R_1 L_2 - R_2 L_1}{R_1 - R_2}$$

[cf. C.U. 2015, 2018]

Solution. If $\vec{V}$ be the phasor for the applied voltage $\vec{I}_1$ and $\vec{I}_2$ be the phasors for the currents through coil 1 and 2, respectively then

$$\vec{V} = R_1\vec{I}_1 + j\omega L_1\vec{I}_1 + j\omega M\vec{I}_2 \quad \text{and} \quad \vec{V} = R_2\vec{I}_2 + j\omega L_2\vec{I}_2 + j\omega M\vec{I}_1$$

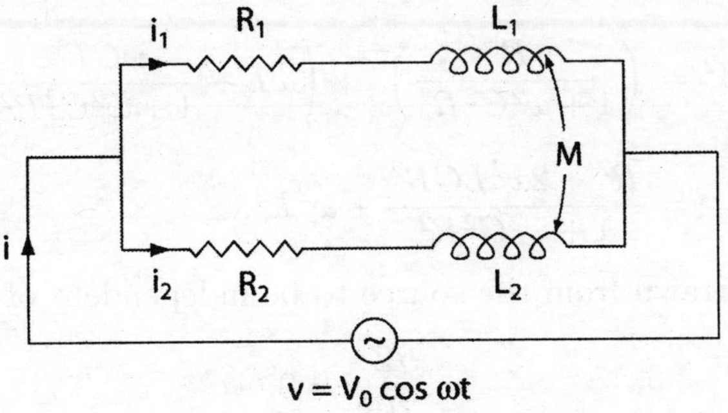

Fig 12.P-9

Equating these two equations we get

$$[R_1 + j\omega(L_1 - M)]\,\vec{I_1} = [R_2 + j\omega(L_2 - M)]\,\vec{I_2}$$

$$\text{or}\quad \sqrt{R_1^2 + \omega^2(L_1 - M)^2}\,e^{j\theta_1}\vec{I_1} = \sqrt{R_2^2 + \omega^2(L_2 - M)^2}\,e^{j\theta_2}\vec{I_2},$$

where $\tan\theta_1 = \omega(L_1 - M)/R_1$ and $\tan\theta_2 = \omega(L_2 - M)/R_2$.

Obviously, for $\vec{I_1}$ and $\vec{I_2}$ to be phase we must have

$$\theta_1 = \theta_2 \quad\text{or}\quad \frac{\omega(L_1 - M)}{R_1} = \frac{\omega(L_2 - M)}{R_2} \quad\text{or}\quad M = \frac{R_1L_2 - R_2L_1}{R_1 - R_2}$$

► **Problem 10.** *Suppose an ac emf $v = V_0\cos\omega t$ is applied to a circuit consisting of a pure inductor of inductance L and a capacitor of variable capacitance C in series. The capacitor is shunted by a resistance R. Find the value of C which makes the current through the coil independent of R.*

Solution. The complex impedance of the circuit is

$$Z = j\omega L + \frac{R \times \frac{1}{j\omega C}}{R + \frac{1}{j\omega C}} = j\omega L + \frac{R}{1 + j\omega CR}$$

$$= \frac{R}{1 + \omega^2C^2R^2} + j\left(\omega L - \frac{\omega CR^2}{1 + \omega^2C^2R^2}\right).$$

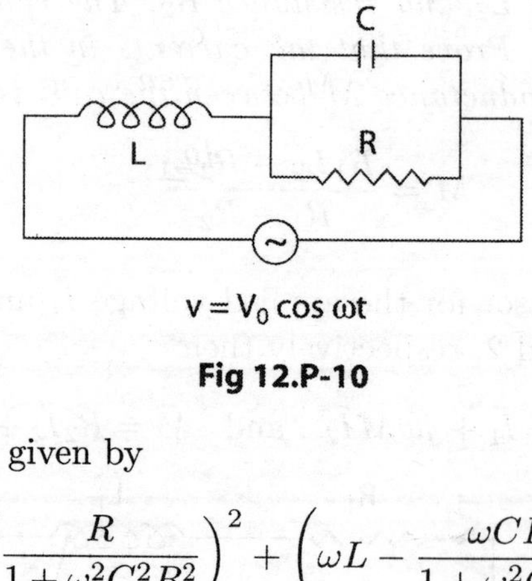

Fig 12.P-10

The magnitude of Z is given by

$$Z^2 = \left(\frac{R}{1 + \omega^2C^2R^2}\right)^2 + \left(\omega L - \frac{\omega CR^2}{1 + \omega^2C^2R^2}\right)^2$$

$$= \frac{R^2 - 2\omega^2LCR^2}{1 + \omega^2C^2R^2} + \omega^2L^2.$$

For the current drawn from the source to be independent of R we must have

$$\frac{\partial Z^2}{\partial R} = 0.$$

This requires that $C = 1/2\omega^2L$.

▶ **Problem 11.** *Find an expression for the impedance Z of a parallel resonant circuit as a function of source frequency ω, resistance R, inductance L and capacitance C. Keeping other parameters constant if C is varied show that the impedance of the circuit will be maximum when $C = L/(R^2 + \omega^2 L^2)$. What is the power factor of the circuit in this condition? Show graphically how current in the circuit varies with C.*

Solution. From Section 12.8 we find that impedance Z is given by

$$\frac{1}{Z} = \frac{1}{R + j\omega C} + j\omega C = \frac{R^2}{R^2 + \omega^2 L^2} + j\left(\omega C - \frac{\omega L}{R^2 + \omega^2 L^2}\right)$$

$$\therefore \quad \left|\frac{1}{Z}\right| = \sqrt{\left(\frac{R^2}{R^2 + \omega^2 L^2}\right)^2 + \left(\omega C - \frac{\omega L}{R^2 + \omega^2 L^2}\right)^2}.$$

If C is varied then $|Z|$ will be maximum or $\dfrac{1}{|Z|}$ will be minimum when

$$\omega C = \frac{\omega L}{R^2 + \omega^2 L^2} \quad \text{or} \quad C = \frac{L}{R^2 + \omega^2 L^2} = C_0 \text{ (say)}.$$

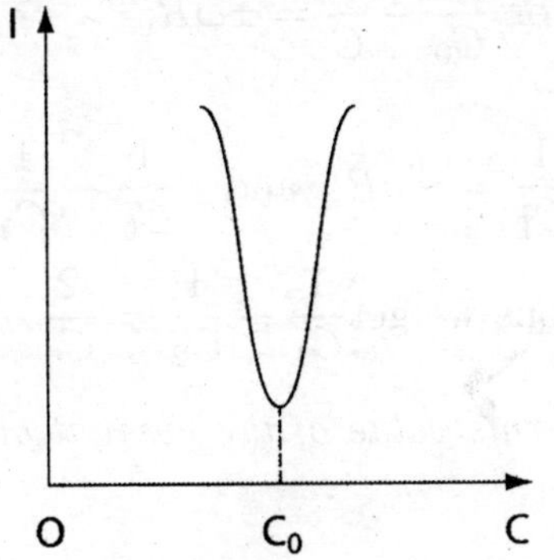

Fig 12.P-11

At this value of C the power factor of the circuit will be unity. Fig. 12.P-11 shows the variation of source current with C. The current is minimum for $C = C_0$ and it increases on either side of C_0.

▶ **Problem 12.** *In a series LCR circuit supplied with an ac source of constant angular frequency ω the value of C is varied continuously. Find the value C_0 of C for which current in the circuit is maximum. Also find the values C_1 and C_2 of C for which the current falls to $\frac{1}{\sqrt{2}}$ of its maximum value. Hence, show that*

$$\frac{1}{C_1} + \frac{1}{C_2} = \frac{2}{C_0}.$$

Solution. If V_r be the rms value of supplied voltage then rms value of current is given by

$$I_r = \frac{V_r}{\sqrt{R^2 + \left(\omega L - \frac{1}{\omega C}\right)^2}} \tag{1}$$

Now I_r will be maximum when

$$\omega L = \frac{1}{\omega C} \quad \text{or} \quad C = \frac{1}{\omega^2 L} = C_0 \text{ (say)}$$

For $C = C_0$, I_r attains a maximum value $I_m = V_r/R$.
So we can write from Eq. (1),

$$I_r^2 = \frac{I_m^2}{1 + \frac{1}{R^2}\left(\omega L - \frac{1}{\omega C}\right)^2} = \frac{I_m^2}{1 + \frac{1}{\omega^2 R^2}\left(\frac{1}{C_0} - \frac{1}{C}\right)^2}$$

For $I_r = I_m/\sqrt{2}$, we have

$$2 = 1 + \frac{1}{\omega^2 R^2}\left(\frac{1}{C_0} - \frac{1}{C}\right)^2$$

$$\text{or} \quad \frac{1}{C_0} - \frac{1}{C} = \pm\omega R.$$

Let

$$\frac{1}{C_0} - \frac{1}{C_1} = -\omega R \quad \text{and} \quad \frac{1}{C_0} - \frac{1}{C_2} = +\omega R.$$

Adding these two equations we get $\dfrac{1}{C_1} + \dfrac{1}{C_2} = \dfrac{2}{C_0}$.

▶ **Problem 13.** *Calculate the rms value of the current given by $i = I_0 + I_1 \cos(\omega t + \theta)$*

[cf. C.U. 2004]

Solution. By definition

$$I\text{rms} = \sqrt{\frac{1}{T}\int_0^T i^2 dt.}$$

$$\therefore \quad I_{\text{rms}}^2 = \frac{1}{T}\int_0^T \left[I_0^2 + I_1^2\cos^2(\omega t + \theta) + 2I_0 I_1 \cos(\omega t + \theta)\right] dt$$

$$= I_0^2 + I_1^2 \cdot \frac{1}{2} + 2I_0 I_1 \times 0$$

$$\therefore \quad I_{\text{rms}}^2 = \sqrt{I_0^2 + I_1^2/2}.$$

▶ **Problem 14.** *Find the form factor of the waveform given by*

$$v(t) = V_0 \cdot t, \quad 0 < t < T.$$

Solution. Average value $V_a = \frac{1}{T}\int_0^T v dt = \frac{1}{T}\int_0^T V_0 t dt = \frac{V_0}{T}\cdot\frac{T^2}{2} = \frac{V_0 T}{2}$

rms value V_r is given by

$$V_r^2 = \frac{1}{T}\int_0^T v^2 dt = \frac{V_0^2}{T}\int_0^T t^2 dt = \frac{V_0^2}{T}\cdot\frac{T^3}{3} = \frac{V_0^2 T^2}{3}$$

$$\therefore \quad V_r = \frac{V_0 T}{\sqrt{3}}.$$

Now by definition form factor $K_f = \frac{V_r}{V_a} = \frac{2}{\sqrt{3}} \approx 1.15$

▶ **Problem 15.** *A capacitor of capacitance C is connected by leads of resistance r to a coil of inductance L and resistance R. An ac emf $v = V_0 \cos \omega t$ is now applied across the coil* (a) *Find the resonant frequency of the circuit.* (b) *Show that the impedance of the circuit will be independent of the frequency if $R = r = \sqrt{L/C}$.*

Solution. (a) Referring to Fig. 12.P-15, we find that the complex impedance Z of the circuit is given by

$$\frac{1}{Z} = \frac{1}{R + j\omega L} + \frac{1}{r + \frac{1}{j\omega C}}$$

$$= \frac{R - j\omega L}{R^2 + \omega^2 L^2} + \frac{j\omega C(1 - j\omega C r)}{1 + \omega^2 C^2 r^2}$$

$$= \frac{R}{R^2 + \omega^2 L^2} + \frac{\omega^2 C^2 r}{1 + \omega^2 C^2 r^2} + j\left(\frac{\omega C}{1 + \omega^2 C^2 r^2} - \frac{\omega L}{R^2 + \omega^2 L^2}\right).$$

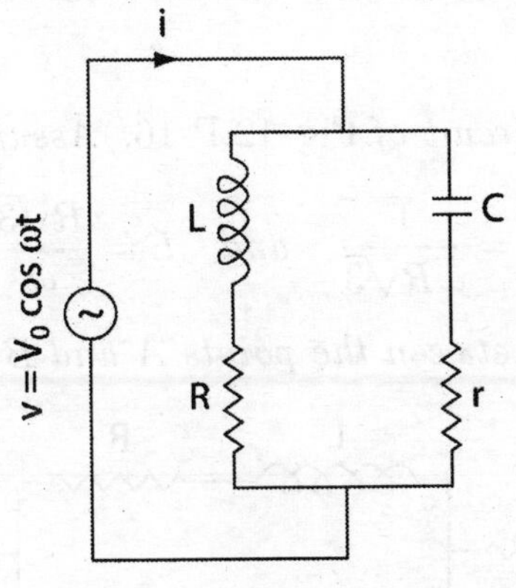

Fig 12.P-15

At resonance the source current i will be in phase with the source voltage when the imaginary part of $1/Z$ vanishes. So for resonance,

$$\frac{\omega C}{1 + \omega^2 C^2 r^2} = \frac{\omega L}{R^2 + \omega^2 L^2}.$$

This gives the resonant frequency as

$$f = \frac{\omega}{2\pi} = \frac{1}{2\pi}\sqrt{\frac{\frac{1}{LC} - \frac{R^2}{L^2}}{1 - \frac{r^2C}{L}}}$$

(b) $\dfrac{1}{Z} = \dfrac{1}{R + j\omega L} + \dfrac{1}{r + \frac{1}{j\omega C}}$

For Z to be independent of the frequency ω we must have

$$\frac{d}{d\omega}\left(\frac{1}{Z}\right) = 0, \text{ or } \frac{-jL}{(R + j\omega L)^2} + \frac{\frac{1}{j\omega^2 C}}{\left(r + \frac{1}{j\omega C}\right)^2} = 0,$$

$$\text{or } \left(\frac{R + j\omega L}{r + \frac{1}{j\omega C}}\right)^2 = -\omega^2 LC,$$

$$\text{or } \frac{R + j\omega L}{r + \frac{1}{j\omega C}} = j\omega\sqrt{LC},$$

$$\text{or } R + j\omega L = \sqrt{\frac{L}{C}} + j\omega r\sqrt{LC}.$$

Equating real and imaginary parts separately from both sides we get

$$R = r \quad \text{and} \quad \frac{L}{R} = Cr.$$

So the resistances of the two branches must be equal as well as their time constants must be equal.

► **Problem 16.** *Consider the circuit of* Fig. 12.P-16. *Assuming*

$$C = \frac{1}{\omega R\sqrt{3}} \quad \textit{and} \quad L = \frac{R\sqrt{3}}{\omega},$$

calculate the total impedance between the points A and B. **[C.U. 2003]**

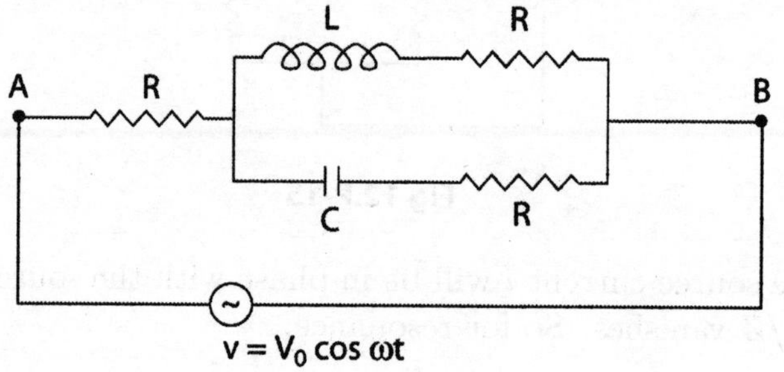

Fig 12.P-16

Solution. Total complex impedance

$$Z = R + \frac{(R + j\omega L)\left(R + \frac{1}{j\omega C}\right)}{R + j\omega L + R + \frac{1}{j\omega C}} = R + \frac{R^2 + \frac{L}{C} + jR\left(\omega L - \frac{1}{\omega C}\right)}{2R + j\left(\omega L - \frac{1}{\omega C}\right)}.$$

Now as

$$C = \frac{1}{\omega R\sqrt{3}} \quad \text{and} \quad L = \frac{R\sqrt{3}}{\omega}$$

we have

$$\omega L = \frac{1}{\omega C} \quad \text{and} \quad \frac{L}{C} = 3R^2.$$

$$\therefore \quad Z = R + \frac{R^2 + 3R^2}{2R} = R + 2R = 3R.$$

▶ **Problem 17.** *A parallel combination of an ideal inductor L and an ideal capacitor C is connected in series with a resistance R and the circuit is then supplied with a sinusoidal voltage $v = V_0 \sin \omega t$.*

(a) *Find the voltage across the LC combination.* **[C.U. 2013]**

(b) *Show that the voltage attains a maximum value when $\omega = 1/\sqrt{LC}$. Calculate the frequencies at which this voltage reduces to $1/\sqrt{2}$ of its maximum value. Find the average power dissipated in the circuit.*

Solution. Taking the impressed voltage as the imaginary part (ImP) of $V_0 e^{j\omega t}$ we can write for the complex impedance of the circuit as

$$Z = R + \frac{j\omega L \times \frac{1}{j\omega C}}{j\omega L + \frac{1}{j\omega C}} = R + \frac{\omega L}{j(\omega^2 LC - 1)}.$$

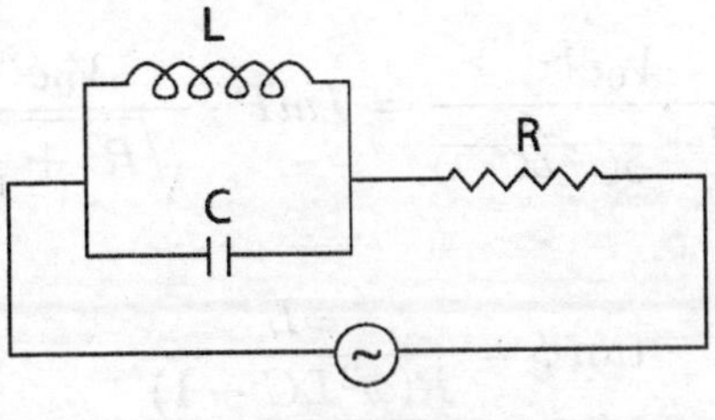

Fig 12.P-17

Voltage across LC combination is

$$V_{LC} = ImP \frac{V_0 e^{j\omega t}}{R + \frac{j\omega L \times \frac{1}{j\omega C}}{j\omega L + \frac{1}{j\omega C}}} \cdot \frac{j\omega L \times \frac{1}{j\omega C}}{j\omega L + \frac{1}{j\omega C}} = ImP \frac{\omega L V_0 e^{j\omega t}}{\omega L + jR(\omega^2 LC - 1)}.$$

$$\therefore \quad |V_{LC}| = \frac{\omega L V_0}{\sqrt{\omega^2 L^2 + R^2(\omega^2 LC - 1)^2}} = \frac{V_0}{\sqrt{1 + \frac{R^2}{L^2}\left(\frac{\omega^2 LC-1}{\omega}\right)^2}}.$$

$\therefore$ For $|V_{LC}|$ to be maximum we must have

$$\omega^2 LC - 1 = 0 \quad \text{or} \quad \omega = \frac{1}{\sqrt{LC}} = \omega_0 \text{ (say)}.$$

Note that at $\omega = \omega_0$, $|Z| = \infty$, $|V_{LC}| = V_0$. Thus, at this frequency no current flows through R and the circuit acts as a rejector circuit. Frequencies at which $|V_{LC}|$ reduces to $|V_{LC}|_{\max}/\sqrt{2} = V_0/\sqrt{2}$ is given by,

$$2 = 1 + \frac{R^2}{L^2}\left(\frac{\omega^2 LC - 1}{\omega}\right)^2, \text{ or } \omega^2 \pm \frac{1}{CR}\omega - \omega_0^2 = 0$$

$$\text{or } \omega = \mp\frac{1}{2CR} \pm \sqrt{\frac{1}{4C^2R^2} + \omega_0^2}.$$

Admissible values of half-power frequencies are,

$$\omega_1 = -\frac{1}{2CR} + \sqrt{\frac{1}{4C^2R^2} + \omega_0^2} \quad \text{and} \quad \omega_2 = +\frac{1}{2CR} + \sqrt{\frac{1}{4C^2R^2} + \omega_0^2}$$

We may take $\omega_2 - \omega_1 = 1/CR$ as the band of frequencies that will be rejected by the circuit and define the Q-factor of the circuit by

$$Q = \frac{\omega_0}{(\omega_2 - \omega_1)} = \omega_0 CR.$$

The current in the circuit is

$$i = ImP \cdot \frac{V_0 e^{j\omega t}}{R + \frac{\omega L}{j(\omega^2 LC - 1)}} = ImP \cdot \frac{V_0 e^{j(\omega t + \phi)}}{\sqrt{R^2 + \frac{\omega^2 L^2}{(\omega^2 LC-1)^2}}},$$

where

$$\tan\phi = \frac{\omega L}{R(\omega^2 LC - 1)}.$$

The average power

$$P = V_{\text{rms}} \cdot I_{\text{rms}} \cdot \cos\phi = \frac{V_0^2 R}{2\left[R^2 + \frac{\omega^2 L^2}{(\omega^2 LC-1)^2}\right]}.$$

► **Problem 18.** *A complex voltage* $(10+j0)$ *V is applied to a series LR circuit of complex impedance* $(\sqrt{3} + j1)\,\Omega$. *Calculate the power factor and the power consumed by the circuit.*

Solution. Complex current

$$I = \frac{10 + j0}{\sqrt{3} + j1} = \frac{10}{\sqrt{(\sqrt{3})^2 + 1^2}} \cdot e^{-j\phi} = 5e^{-j\phi},$$

where $\tan\phi = 1/\sqrt{3}$.

So the current lags behind the emf by the angle $\phi = \tan^{-1}\frac{1}{\sqrt{3}} = 30°$. Therefore, power factor $\cos\phi = \cos 30° = \frac{\sqrt{3}}{2} = 0.866$.

Power consumed $P = V \cdot I \cdot \cos\phi = 10 \times 5 \times 0.866\,W = 43.3\,W$.

▶ **Problem 19.** *Consider a single-phase ideal step-down transformer with turn ratio 5:1. If a resistance of* $15\,\Omega$ *is connected to the secondary, what would be the effective primary resistance?*

Solution. Let V_1, I_1 and N_1 be respectively the primary voltage, primary current and number of primary turns. Let V_2, I_2 and N_2 be the corresponding quantities for the secondary. Now for an ideal transformer,

$$\frac{V_2}{V_1} = \frac{N_2}{N_1} \quad \text{and} \quad \frac{I_2}{I_1} = \frac{N_1}{N_2}.$$

If R_2 be the load connected to the secondary then

$$I_2 = \frac{V_2}{R_L}.$$

Now effective primary resistance $R_1' = \frac{V_1}{I_1} = \frac{V_2 N_1/N_2}{I_2 N_2/N_1} = R_L\left(\frac{N_1}{N_2}\right)^2$

Here $N_1 : N_2 = 5 : 1$, $R_L = 15\,\Omega$. Therefore $R_1' = 375\,\Omega$.

▶ **Problem 20.** *A ideal step-up transformer with a turns ratio of 1:20 is connected to the circuit, shown in Fig. 12.P-20. Calculate the primary and secondary currents as well as the power delivered by the source.*

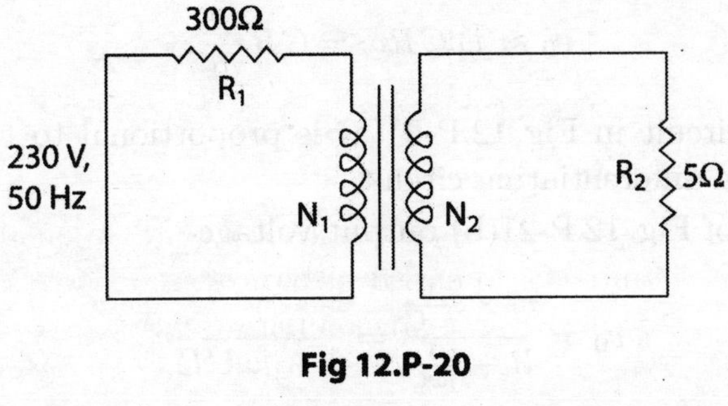

Fig 12.P-20

Solution. Effective primary impedance

$$Z_1' = R_1' = R_1 + \left(\frac{N_1}{N_2}\right)^2 R_2 = 300 + \left(\frac{1}{20}\right)^2 \cdot 5\,\Omega \approx 300.01\,\Omega$$

$\therefore$ Primary current $I_1 = \frac{230}{300.01}$A=0.77A.
The secondary voltage V_2 is given by

$$\frac{V_2}{V_1} = \frac{N_2}{N_1} \quad \text{or} \quad V_2 = \frac{20}{1} \times 230\,V = 4600\,V$$

The secondary current I_2 is given by

$$V_2 I_2 = V_1 I_1 \quad \text{or} \quad I_2 = \frac{V_1 I_1}{V_2} = \frac{230 \times 0.77}{4600} A = 38.5\,mA.$$

► **Problem 21.** *Consider the circuits shown in* Figs. 12.P-21. (a) *Assuming RC time constant small compared with the time period of the input ac voltage v_i find the output voltage v_0 in* Fig. 12.P-21(a) (b) *Assuming RC time constant large compared with the time period of the input ac voltage v_i find the output voltage v_0 in* Fig. 12.P-21(b). *Comment on the results.*

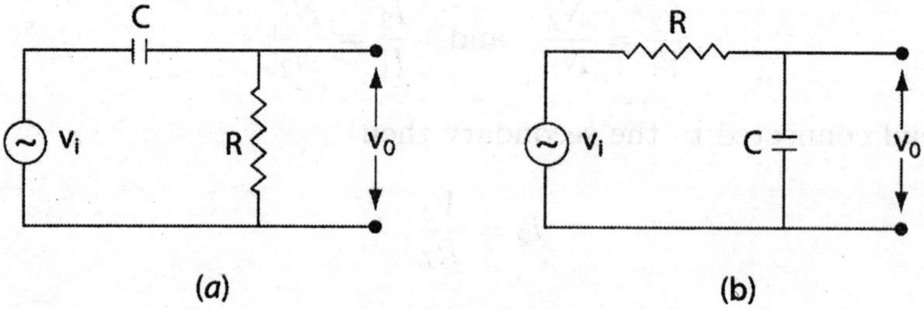

Fig 12.P-21

Solution. (a) If the input voltage is of the form $v_i = V_0 e^{j\omega t}$ then the output voltage

$$v_0 = \frac{v_i R}{R + \frac{1}{j\omega C}} = \frac{j\omega C R v_i}{1 + j\omega CR}.$$

Now, if $\omega CR \ll 1$ or $CR \ll T$ (time period) then

$$v_0 \approx j\omega CR v_i = CR\frac{dv_i}{dt}.$$

Thus, output of circuit in Fig. 12.P-21(a) is proportional to the derivative of the input signal. So it is a differentiating circuit.

(b) In the circuit of Fig. 12.P-21(b) output voltage

$$v_0 = \frac{v_i \times \frac{1}{j\omega C}}{R + \frac{1}{j\omega C}} = \frac{1}{1 + j\omega CR} v_i.$$

Now, if $\omega CR \gg 1$ or $CR \gg T$ then

$$v_0 \approx \frac{1}{j\omega CR} v_i = \frac{1}{CR} \int v_i dt.$$

Thus, output of the circuit in Fig. 12.P-21(b) is proportional to the integral of the input voltage. So, it is an integrating circuit.

These circuits may thus be used as *differentiator* or *integrator*, which provides a straight forward means of converting one type of waveshape to another type. For example, the differentiator can produce a square wave from a triangular wave; the integrator can be used to get a triangular wave from a square wave.

QUESTIONS

1. (a) Distinguish between the mean value and root-mean-square (rms) value of a sinusoidal emf and obtain expressions for them. Which value do you get from an ac voltmeter? Give the physical meaning of the rms value of current.

(b) Explain the terms *power factor* and *wattless current* in connection with ac circuits.

(c) What do you mean by the term *form factor* of an ac waveform. Find an expression for it for a pure sinusoidal waveform.

(d) Define and explain the terms *reactance* and *impedance* of an ac circuit.

2. (a) Show that the current through a pure inductor lags by 90° behind the impressed ac emf and that through a pure capacitor leads by 90° over it.

(b) What is a choke coil? Explain that the use of a choke coil is more economic than a resistor in controlling current in ac circuits.

(c) Explain physically why the reactance of an inductor increases and that of a capacitor decreases with the increase in frequency. **[N.B.U. 2017]**

3. A sinusoidal emf is applied to a circuit containing

(a) an inductor and a resistor in series

(b) a capacitor and a resistor in series

(c) an inductor, a capacitor and a resistor in series.

Derive an expression for the instantaneous current and power factor. Draw the phasor diagram. Show that the power is dissipated only in the resistance.

4. A sinusoidal voltage $v = V_0 \cos \omega t$ is applied to a series LCR circuit.

(a) Find an expression for the instantaneous current in the circuit. **[cf. C.U. 2016]**

(b) How do the rms value of current and its phase vary with ω? Under what condition does the voltage lead the current?

(c) Find the value of ω at which power consumed by the circuit becomes maximum.

(d) Find the two half-power frequencies and hence, the Q of the circuit.

(e) Show that voltage across L at resonance is Q times V_0. Can you use this result for amplification of practical signal. **[B.U. 2016]**

[*Ans.* No, because here voltage magnification is a function of ω]

5. (a) What do you mean by resonance in a series LCR circuit? What are current and voltage resonances? Find the corresponding resonant frequencies?

(b) A series resonant circuit is called an *acceptor circuit* —Explain. **[B.U. 2017]**

(c) Write an expression for the half power bandwidth of a series LCR circuit in terms of circuit parameters. **[B.U. 2001]**

(d) Explain the term *sharpness of resonance.* From current response curve for a series LCR circuit show how Q-factor of the circuit quantifies the sharpness of resonance. **[C.U. 2019]**

6. (a) What is a parallel resonant circuit? Why is it called a *rejector circuit*?

(b) Find an expression for the impedance of the circuit. How does it vary with the frequency of the input source? **[N.B.U. 2017]**

(c) What do you mean by the Q of the circuit? Find the Q value as a current magnification factor.

(d) Distinguish between the functions of a series resonant circuit and a parallel resonant circuit.

7. (a) What is a transformer? In an ideal transformer show that the ratio of output voltage to the input voltage is equal to the ratio of the number of secondary turns to the number of primary turns.

(b) In a step-up transformer output voltage is greater than the input voltage. Is the output power greater than the input power? Discuss it by considering an ideal transformer.

(c) Show how the primary of a transformer is affected by the increase of load in the secondary.

(d) Enumerate different types of losses in a practical transformer. How are these losses reduced? **[cf. B.U. 2017]**

(e) What is autotransformer?

8. Two tuned circuits each containing inductor, capacitor and a small resistor, are coupled inductively. An ac emf $v = V_0 \cos \omega t$ is set up in one of them by an oscillator of variable frequency. Obtain an expression for the current in the second circuit. Discuss how it varies with the frequency ω. What is optimum coupling? Cite one important application of the above circuit.

9. (a) What do you mean by a rotating magnetic field? How can such a field be generated with a two-phase/three-phase electric supply?

(b) Derive an expression for the torque acting on a small coil placed in a rotating magnetic field. Find the maximum value of average torque.

10. What is an induction motor? Why is it so-called? Explain how the principle of rotating magnetic field is utilized in the construction of an induction motor. What do you mean by the slip of the motor?

11. Show how two alternating magnetic fields at right angles to each other may produce a rotating magnetic field. How is this principle utilized in the construction of an induction motor?

12. Write short notes on:

(a) Rotating magnetic field;

(b) Tuned coupled circuits;

(c) Induction motor.

EXERCISES

1. (a) A coil of inductance 0.7 H and resistance $R = 50\,\Omega$ is connected to 220 V, 50 Hz ac supply. Find the wattless and power components of current.

[*Ans.* 0.95 A, 0.22 A]

(b) A 100 W, 200 V lamp is to be operated on 400 V, 50 Hz ac supply. What capacitor placed in series with the lamp will enable it to be correctly used?

[C.U. 2014] [*Ans.* 4.59 μF]

(c) A current wave has form factor 1.2 and peak factor 1.7. Find the average and rms values of current if peak current is 100 A.

[C.U. 2019] [*Ans.* 49.0 A, 58.8 A]

2. An ac emf 230 V, 50 Hz is applied to a circuit consisting of a resistance $100\,\Omega$ and a capacitance 5μF in series. Find the rate at which energy is consumed in the circuit.

[*Ans.* 12.9 W]

3. An ac source of 10 V, 1 kHz is applied to a parallel combination of a resistance of $10\,\Omega$ and a capacitance of $10\,\mu$F. Find the magnitude and phase of the current drawn from the source. [*Ans.* 1.18 A, 32.1° lead over the voltage]

4. A series circuit consisting of $L = 0.3$ H, $C = 4\,\mu$F and $R = 60\,\Omega$ is connected to 220 V, 50 Hz mains. Find the rms value of current, power factor and power dissipation in the circuit. **[cf. C.U. 2017]** [*Ans.* 0.31 A, 0.0852, 5.8 W]

5. A series LCR circuit is supplied with a 250 V, 50 Hz ac supply. Now C is varied until the current attains a maximum value of 0.75 A. Then the voltage across C

is 600 V. Calculate the values of L, C and R. What is the power factor in this condition? [*Ans.* $R = 333.3\,\Omega, C = 4\,\mu\text{F}$, $L = 2.5\,\text{H}$, power factor $= 1$]

6. Suppose we have two series combinations of inductors and capacitors, one with L_1, C_1 and another with L_2, C_2. The resonant frequencies of these combinations are equal. Show that if these combinations are connected in series the resonant frequency remains the same.

7. A capacitor of capacitance C is connected by leads of resistance r to a coil of inductance L and resistance R. Show that for operation with sinusoidal ac of angular frequency ω the above arrangement can be replaced by a wire of resistance $(Rr + L/C)/(R + r)$ and no inductance if $R^2 - L/C = \omega^2 LC(r^2 - L/C)$.

8. Show that the magnitude of the impedance of a series LCR circuit at an angular frequency ω may be expressed as

$$|Z| = R\sqrt{1 + Q^2\left(\frac{\omega}{\omega_0} - \frac{\omega_0}{\omega}\right)^2},$$

where Q is the quality factor of the circuit at the resonant frequency ω_0.

9. (a) A parallel combination of an inductance L and a resistance R is connected to an ac source of emf $v = V_0 \cos \omega t$. Find the phase difference between the main current drawn from the source and the source emf.

[*Ans.* Current lags by $\tan^{-1}(R/\omega L)$]

(b) A coil of inductance L and resistance R is connected in parallel with another coil of inductance $2L$ and resistance R. The combination is connected to an ac source of emf $v = V_0 \cos \omega t$. Find the phase difference between the currents in the two coils. [*Ans.* $\tan^{-1} \frac{\omega(2L-M)}{R} - \tan^{-1} \frac{\omega(L-M)}{R}$]

[**Hints.** See *Solved Problem* 9.]

10. (a) In a series LCR circuit supplied with an ac emf, the pds across L, C and R are measured to be 90 V, 50 V and 30 V, respectively. Find the applied emf. [*Ans.* 50 V]

(b) A 230 V, 50 Hz ac voltage is applied to a coil of $L = 5\,\text{H}$ and $R = 2\,\Omega$ in series with a capacior C. Find C for which voltage across the coil will be 400 V. **[C.U. 2019]** [*Ans.* $4.77\,\mu\text{F}$]

11. An ac source of frequency 50 Hz is applied to series LR circuit with $L = 10\text{mH}$ and $R = 2\,\Omega$. Calculate the power factor. What capacitance placed in the circuit will make the power factor unity? [*Ans.* 0.53, $1000\,\mu\text{F}$]

12. A capacitor of capacitance $0.001\,\mu\text{F}$ is connected in parallel with a coil of inductance 100 mH and resistance $10\,\Omega$. Calculate the resonant frequency and dynamic resistance of the circuit. [*Ans.* 50.35 kHz, 1 MΩ]

13. (a) A series LCR circuit has $L = 1\text{mH}$, $C = 0.1\,\mu\text{F}$ and $R = 1\,\Omega$. Find the resonant frequency and bandwidth of the circuit. [*Ans.* 15.9 kHz, 1 kHz]

(b) A series LCR circuit has $L = 200\,\text{mH}$, $C = 0.5\,\mu\text{F}$ and $R = 10\,\Omega$. Find voltage magnification at resonant frequency. **[C.U. 2017]** [*Ans.* 63.2]

14. A step-up ideal transformer has primary to secondary turns ratio as $2:25$. If the primary voltage is 220 V and the transformers supplies 1 A current to an external load, find the output power. [*Ans.* 2750 W]

15. (a) Consider an ideal transformer with negligible primary and secondary coil resistance. It is used to match a source of internal resistance $1\,\text{k}\Omega$ to a load of $10\,\Omega$ connected across the secondary. Find the turn ratio required.
[**Hints.** Use $R_1' = 1\,\text{k}\Omega = \left(\frac{N_1}{N_2}\right)^2 R_L = \left(\frac{N_1}{N_2}\right)^2 \cdot 10\,\Omega$. This gives $N_1 : N_2 = 10 : 1$.

(b) Suppose a transformer is designed to have a core of uniform cross-section of 12 cm^2 and total volume of 840 cm^3. The maximum value of B-field in the core is required to be 0.8 T, which corresponds to $H = 210$ A-turns/m. If the peak value of the primary current is 200 mA, calculate the number of turns required in the primary.
[**Hints.** Length of the core $= \frac{840}{12}$cm=70 cm. Now $H = nI = \frac{N}{L}I$.
$\therefore\quad 210 = \frac{N\times 0.2}{0.70}$ or $N = \frac{210\times 0.70}{0.2} = 735$ turns.

16. Two electrical machines are supplied from the same ac mains. The first one draws a current of 5A at a power factor 0.7. The two machines together draws a current of 8A at a power factor 0.6. Assuming that the current lags in each case, calculate the current drawn and power factor for the second machine.

[*Ans.* 3.1 A, 0.42 (lag)]

17. In an ac circuit the complex impedance is $Z = 1 + 2j\ \Omega$ and complex voltage is $-4 + 7j$V. Find the complex current in the circuit. [*Ans.* 2+3 j A]

18. In a series LCR circuit the measured voltage across L, C and R are respectively 220 V, 100 V and 185 V. Calculate the net voltage across the circuit.

[*Ans.* 220 V]

19. A coil of inductance 6 mH and resistance $100\,\Omega$ is supplied with an ac emf $v = 310\cos 100\pi t$ volt. Find the power factor of the circuit. Also, find the value of capacitance, which when connected parallel to the coil increases the power factor to unity. [*Ans.* 0.9998, $0.6\,\mu$F]

20. A series LCR circuit consists of an inductance 0.3 H, a resistance $20\,\Omega$ and an unknown capacitor C. When the circuit is supplied with an ac source 230 V, 50 Hz resonance is found to occur. Find the value of C and Q-factor of the circuit.

[*Ans.* $33.8\,\mu$F, 4.713]

21. Suppose a parallel combination of a resistance $R = 10\,\Omega$ and an inductance $L =$ 50 mH is supplied with and ac source of 220 V, 50 Hz. Find the magnitude and phase of the current drawn from the supply.

[*Ans.* 26.08 A, 32.47° lag behind the supply voltage]

22. Suppose a parallel combination of a capacitor $C = 10\mu$F and a resistance $R = 100\,\Omega$ is supplied with a 220 V, 50 Hz ac supply. Find the magnitude and phase of the current drawn from the supply.

[*Ans.* 2.3 A, 17.41° lead over the supply voltage]

23. A coil of inductance 120 mH and resistance $12\,\Omega$ is connected in parallel with a capacitor of capacitance 60 μF. The combination is supplied with an ac source of 10 V. Find the resonant frequency, dynamic resistance, current from supply at resonance and Q-factor of the circuit. [*Ans.* 57.1 Hz, $166.7\,\Omega$, 60 mA, 3.59]

24. Consider the circuit of Fig. 12.E-24. Find Z for which the source voltage v and current i will be in phase. Given rms values of v and i are respectively 100 V and 10 A.

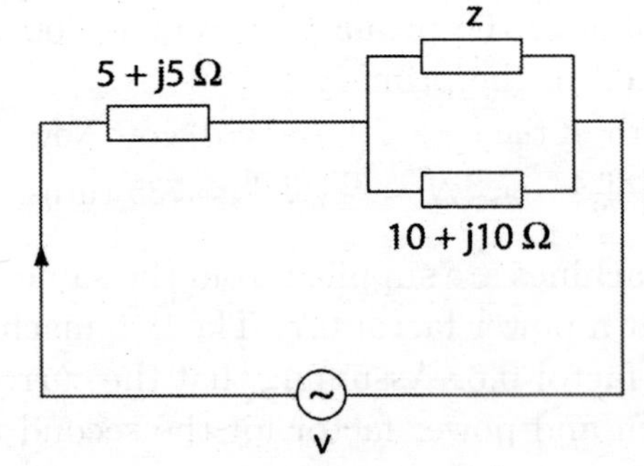

Fig 12.E-24

[**Hints.** $\frac{100}{10} = 5 + j5 + \frac{Z(10+j10)}{Z+10+J10}$. Equate real and imaginary parts and show that $Z = 2 - j6\,\Omega$.

25. The emfs induced in two series connected coils are given by $v_1 = 3\sin\omega t$ and $v_2 = 2\sin(\omega t + \pi/6)$. Determine the resultant emf.

[**Hints.** The resultant is

$$v = v_1 + v_2 = 3\sin\omega t + 2\left(\sin\omega t \cdot \frac{\sqrt{3}}{2} + \cos\omega t \cdot \frac{1}{2}\right)$$

$$= (3 + \sqrt{3})\sin\omega t + \cos\omega t = A\sin(\omega t + \theta),$$

where $A\cos\theta = 3 + \sqrt{3}$ and $A\sin\theta = 1$. Thus, $v = 4.83\sin(\omega t + 11.93°)$)

26. Find the average and rms values of the ac voltage represented by

$$v(t) = \begin{cases} V_0 \sin\omega t, & 0 \le \omega t \le \pi \\ 0, & \pi \le \omega t \le 2\pi \end{cases}$$

[*Ans.* $\frac{V_0}{\pi}$, $\frac{V_0}{2}$]

27. A series LCR circuit is supplied with an ac source whose frequency can be altered without changing its amplitude. If the current amplitude becomes maximum at a frequency f_0 and it falls to $1/\sqrt{2}$ times the maximum value at frequencies f_1 and f_2 then show that $f_0 = \sqrt{f_1 f_2}$.

28. Consider the network of Fig. 12.E-28. Assuming $L < C_1 r^2$ show that the network acts as a resistance for a current of angular frequency $\omega = (C_1 - L/r^2)/LC_1(C_1 + C_2)$.

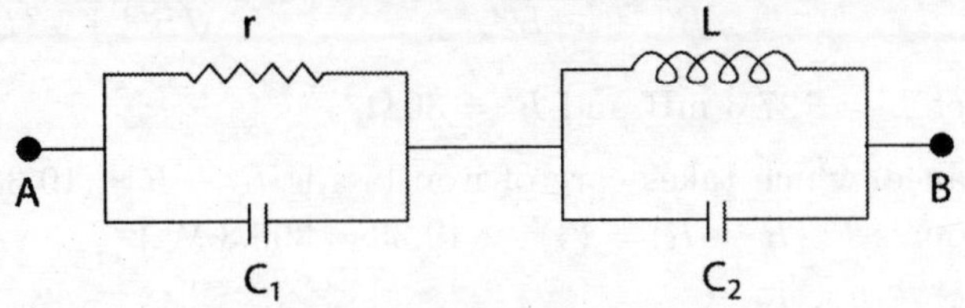

Fig 12.E-28

29. Suppose a voltage pulse defined by

$$v(t) = \begin{cases} V_0 & \text{for } 0 < t < T \\ 0 & \text{for } t > T \end{cases}$$

is applied to a series LR circuit. Determine the current in the circuit for $0 < t < T$. Also find the voltage across the inductance for $0 < t < T$ and for $t > T$.

[*Ans.* $i(t) = \frac{V_0}{R}\left(1 - e^{-Rt/L}\right)$, $v(t) = V_0 e^{-Rt/L}$, $v(t) = V_0\left(e^{-Rt/L} - e^{-R(t-T)/L}\right)$]

30. Consider two neighbouring circuits. Instantaneous current in one circuit is $i = I_0 \cos \omega t$. The other circuit consists of an inductor L in series with a capacitor C. If M be the mutual inductance between the two circuits show that the instantaneous charge on the capacitor C is given by

$$q = \frac{\omega M C I_0}{(\omega^2 LC - 1)} \sin \omega t.$$

31. An emf $v = V_0 + V_1 \cos \omega t$ is applied across a coil of inductance L and resistance R. Show that the average power consumed in the coil is

$$\frac{V_0^2}{R} + \frac{V_1^2}{2\sqrt{R^2 + \omega^2 L^2}} \cos \phi \quad \text{where } \tan \phi = \omega L/R.$$

32. A coil with an iron core takes 4 A current at a power factor of 0.6, when connected to a 200 V, 50 Hz supply. If the iron core is removed and the voltage is reduced to 80 V the current changes to 6A at a power factor of 0.8. Calculate the iron loss in the core and the inductance in each case. **[C.U. 2003]**

[**Hints.** If the coil resistance is R and inductance without core is L then

$$I = 6 = \frac{80}{\sqrt{R^2 + \omega^2 L^2}} \quad \text{and} \quad \cos\phi = 0.8 = \frac{R}{\sqrt{R^2 + \omega^2 L^2}},$$

where $\omega = 2\pi f = 100\pi$.

Solving these two we get $L = 25.47$mH and $R = 10.67\,\Omega$.

Similarly, if with the core, effective resistance is R' and inductance is L' then

$$4 = \frac{200}{\sqrt{R'^2 + \omega^2 L'^2}} \quad \text{and} \quad 0.6 = \frac{R'}{\sqrt{R'^2 + \omega^2 L'^2}}.$$

Solving we get $L' = 127.3$ mH and $R' = 30\,\Omega$.

So the resistance, which takes care of iron loss is $R' - R = 19.33\,\Omega$ and hence iron loss in the core equals $I'^2(R' - R) = (4)^2 \times 19.33 = 309.3$ W.]

33. An alternating current $i = I_2 \cos\omega t$ is superposed on a direct current I_1. Show that its rms value is $\sqrt{I_1^2 + I_2^2/2}$.

34. (a) A single-phase half-wave rectifier allows current to flow from time $t_1 = \pi/6\omega$ to $t_2 = 5\pi/6\omega$, the applied ac voltage being given by $v = 100 \sin\omega t$. Estimate the average and rms value of the output voltage.

[**Hints.** $V_{av} = \frac{1}{T}\int_{t_1}^{t_2} v \cdot dt = 27.56$ V, $V_{\text{rms}} = \sqrt{\frac{1}{T}\int_{t_1}^{t_2} v^2 dt} = 48.5$ V.]

(b) The current in a circuit is given by $i(t) = I_0 e^{-t/\tau}$ for $0 \le t \le \tau$. Find the rms value of current for the period $t = 0$ to $t = \tau$.

[NBU 2004] [*Ans.* $\frac{I_0}{\sqrt{2}}\sqrt{1 - \frac{1}{e^2}}$]

35. An ac supply of 50 V, 1000 Hz is applied to a series LR circuit with $L = 100$ mH and $R = 2000\,\Omega$. Find the rms current in the circuit, pd across L and the pd across R. Does the sum of these rms pds give the supply emf? How do you explain the difference, if there is any? **[C.U. 2006]**

[**Hints.** $I_{\text{rms}} = 23.8\,$mA, $V_L = 15\,V$, $V_R = 47.6\,V$, $V_L + V_R > 50\,V$. Actually the phasor sum of V_L and V_R will be equal to 50 V also, the sum of instantaneous pds across L and R will be equal to the instantaneous value of supply voltage.]

36. A 100 W lamp of rating 110 V is to be operated with an ac source of 220 V, 50 Hz. Calculate the value of pure inductance to be connected with it so that the lamp can efficiently glow. [**Hints.** See solved problem 8; $L = 624$ mH] **[C.U. 2007]**

37. A coil of inductive reactance 3Ω and resistance 4Ω is connected in parallel with a capacitor branch having a capacitive reactance 8Ω and series resistance 6Ω. The applied line voltage is 100 V at a frequency of 50 Hz. Calculate the power factor. **[C.U. 2009]**

[**Hints.** $z = (4+3j)\|^{l}(6-8j) = \frac{2}{5}(11+2j)$, $\therefore$ $\tan\phi = \frac{2}{11}$ and power factor $\cos\phi = 0.984$]

38. A capacitor and inductor have equal reactances at 750 Hz. What is the ratio of their respective reactances at 50 Hz? [*Ans.* 225 : 1] **[C.U. 2010]**

39. A parallel LC combination is connected in series with a resistance R. If a sinusoidal voltage $V_0 \sin \omega t$ is applied to this circuit, calculate the output voltage, taken across the LC combination. Show that this voltage reaches a maximum value when $\omega = 1/\sqrt{LC}$. **[C.U. 2008]**

[**Hints.** See, solved problem 17]

40. Consider the circuit shown in Fig. 12-E-40 where v is an *ac* voltage source and A_1, A_2, A_3 are three rms current meters. Is the reading of A_3 equal to the sum of the readings of A_1 and A_2? Explain your answer. **[B.U. 2011]**

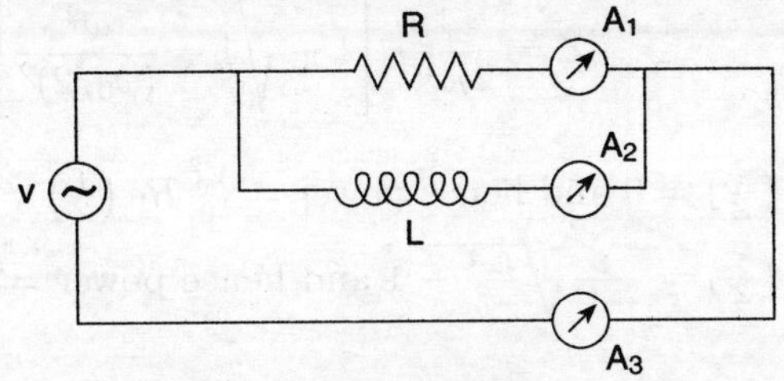

Fig. 12.E-40

41. In the circuit shown in Fig. 12.E-41 the three voltmeter readings are $V_1 = 220\,\text{V}$, $V_2 = 122\,\text{V}$ and $V_3 = 136\,\text{V}$. Find the power factor and the power consumption of the load $R_L = 5\Omega$. **[GATE 2012]**

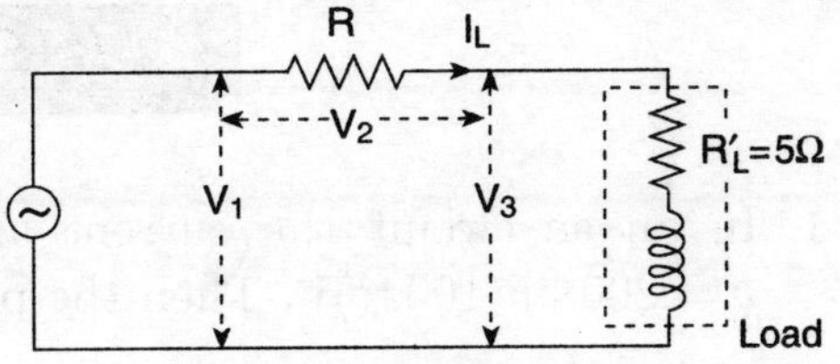

Fig. 12.E-41

[**Hints.** $\vec{V}_1 = \vec{V}_2 + \vec{V}_3$

$$\cos\theta = \frac{V_1^2 - V_2^2 - V_3^2}{2V_1V_2} = 0.45$$

Again, $\cos\theta = \dfrac{R_L}{Z_L}$

$$\therefore \quad Z_L = \frac{5}{0.45} = 11.11\Omega$$

$$\therefore \quad P_L = I_L^2 R_L = \left|\frac{V_3}{Z_L}\right|^2 \cdot R_L = \left(\frac{13.6}{11.11}\right)^2 \times 5\,\text{W} = 749\,\text{W}]$$

42. Consider the semi-infinite network shown in Fig. 12.E-42. Compute the average power that is fed into the circuit by the ac source in the regimes $\omega > \omega_0/2$ and $\omega < \omega_0/2$ where $\omega_0 = 1/\sqrt{LC}$.

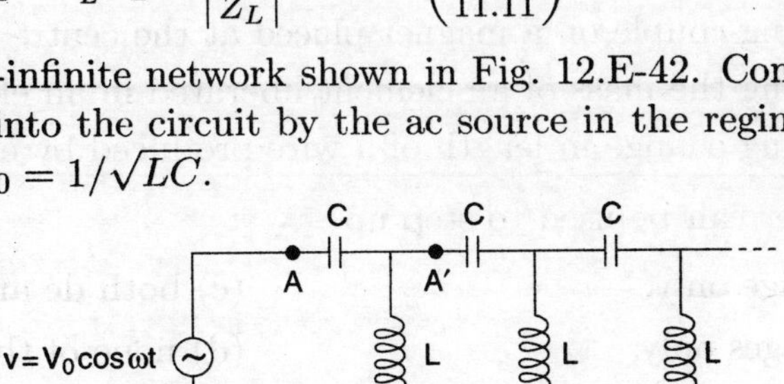

Fig. 12.E-42

[**Hints.** It the impedance between the points A, B is Z, then that between the points $A;, B'$ will also be Z. Therefore,

$$Z = \frac{1}{j\omega C} + j\omega L \|^{l} Z$$
$$= \frac{1}{j\omega C} + \frac{j\omega L \cdot Z}{Z + j\omega L}$$

Simplifying we get for $|Z| > 0$,

$$Z = \frac{1}{2j\omega C}\left[1 + \sqrt{1 - \frac{\omega^2}{(\omega_0/2)^2}}\right]$$

For $\omega < \omega_0/2$, $Re\left(\frac{1}{Z}\right) = 0$ and hence power $= \frac{V_0^2}{2} Re\left(\frac{1}{Z}\right) = 0$

For $\omega > \omega_0/2$, $Re\left(\frac{1}{Z}\right) = \frac{1}{2\omega C}\sqrt{\frac{4\omega^2}{\omega_0^2} - 1}$ and hence power $= \frac{V_0^2}{4\omega C}\sqrt{\frac{4\omega^2}{\omega_0^2} - 1}$]

OBJECTIVE TEST

1. In an ac circuit the current and the emf are $i = 5\sin(100\pi t - \pi/2)A$ and $v = 200\sin(100\pi t)V$. Then the power dissipation in the circuit is

(a) 1000 W, (b) 40 W, (c) 500 W, (d) zero.

2. The rms value of ac of any waveform having a peak value I_0

(a) is always $I_0/\sqrt{2}$,
(b) is always $I_0/2$,
(c) is always I_0/π,
(d) depends on the waveform.

3. Which of the following method is suitable for the measurement of alternating current?

(a) measuring couple on a coil placed in a magnetic field,
(b) measuring couple on a magnet placed at the centre of a coil,
(c) measuring the mass of an element liberated in an electrolyte,
(d) measuring change in length of a wire produced by electrical heating.

4. A transformer can be used to step up

(a) dc voltage only,
(b) ac voltages only,
(c) both dc and ac voltages,
(d) none of these.

5. In a pure inductive circuit impressed with an ac emf $\mathcal{E}$ current

(a) lags behind $\mathcal{E}$ by 90°,
(b) leads over $\mathcal{E}$ by 90°,
(c) is in phase with $\mathcal{E}$,
(d) is 180° out of phase with $\mathcal{E}$.

6. In a series LCR circuit the current lags the applied ac voltage. The phase lag can be reduced to zero by

(a) increasing C,
(b) increasing L,
(c) increasing the frequency of the ac voltage,
(d) decreasing the frequency of the ac voltage.

7. Which of the following equations represent correct current-voltage relationship for a capacitor with respect to a square wave excitation.

(a) $i = C\dfrac{dV}{dt}$,
(b) $i = j\omega CV$,
(c) $i = \dfrac{V}{j\omega C}$,
(d) $i = \dfrac{1}{C}\dfrac{dV}{dt}$.

8. A capacitor acts as an open circuit for

(a) ac,
(b) dc,
(c) both ac and dc,
(d) none of these.

9. A series ac circuit has a resistance of $3\,\Omega$ and a reactance of $4\,\Omega$. The impedance of the circuit in units of Ω is

(a) 7, (b) 12/7, (c) 5, (d) 1.

10. An ac source of frequency 50 Hz is supplied to a series LR circuit. The time taken by the magnetic field energy of the inductor to change from maximum to minimum value is

(a) 5 ms, (b) 10 ms, (c) 20 ms, (d) none of these.

11. An ac source of 220 V (rms) supplies a current of 1 A (rms) to circuit. The average power P delivered by the source is such that

(a) $P = 220\text{W}$, (b) $P \geq 220\text{W}$, (c) $P \leq 220\text{W}$, (d) none of these.

12. A series LCR circuit becomes oscillatory if

(a) $R > 2\sqrt{\dfrac{L}{C}}$, (b) $R < 2\sqrt{\dfrac{L}{C}}$, (c) $R = 2\sqrt{\dfrac{L}{C}}$, (d) none of these.

13. A series LCR circuit with the quality factor Q is impressed with an ac emf V (rms). At resonance the rms voltage across C is

(a) V, (b) 0, (c) QV, (d) V/Q.

14. The average value of the ac voltage $v = V_0 \sin \omega t$ over a complete period is

(a) $\dfrac{2V_0}{\pi}$, (b) $\dfrac{V_0}{\pi}$, (c) $\dfrac{V_0}{2}$, (d) zero.

15. An ac voltage is applied to the parallel combination of a pure C and a pure L. Then at resonance main current from the source is

(a) zero,
(b) infinite,
(c) 180° out of phase with the applied voltage,
(d) amplified.

16. A complex current consists of a dc of $3A$ and an ac of $3\sin(100\pi t + \pi/6)A$. Its average value in ampere is

(a) $\sqrt{18}$, (b) 6, (c) 4.5, (d) 3.

17. A pure LC parallel circuit is impressed with an ac source of frequency ω. If $\omega > \frac{1}{\sqrt{LC}}$, then the circuit reactance is

(a) inductive, (b) capacitive, (c) resistive, (d) zero.

18. The rms value of current given by $i = I_0 + I_1 \cos\omega t$ is

(a) $\frac{1}{\sqrt{2}}(I_0 + I_1)$, (b) $I_0 + \frac{I_1}{\sqrt{2}}$, (c) $\sqrt{I_0^2 + \frac{I_1^2}{2}}$, (d) I_0

19. Suppose you have two series resonant circuits $L_1C_1R_1$ and $L_2C_2R_2$, having equal resonant frequencies f_0. If these two circuits are again connected in series to make a single series circuit its resonant frequency would be

(a) f_0, (b) $2f_0$, (c) $f_0/2$, (d) $f_0/\sqrt{2}$.

20. An electric bulb of resistance $R\,\Omega$ connected to an ac voltage source, is being operated at P W. The peak value of voltage is

(a) $\sqrt{PR}$, (b) $\sqrt{2PR}$, (c) $2\sqrt{PR}$, (d) $\frac{1}{2}\sqrt{PR}$.

21. In an ac circuit the voltage is given by $v = 220\sin(\omega t + \pi/6)$ V and the current by $i = 10\sin(\omega t - \pi/6)$ A. The impedance of the circuit is

(a) $22\,\Omega$, (b) $11(1+j\sqrt{3})\,\Omega$, (c) $11(1-j\sqrt{3})\,\Omega$, (d) $22(1+j\sqrt{3})\,\Omega$.

ANSWERS

1. (d)	**2.** (d)	**3.** (d)	**4.** (b)	**5.** (a)	**6.** (d)	**7.** (a)
8. (b)	**9.** (c)	**10.** (a)	**11.** (c)	**12.** (b)	**13.** (c)	**14.** (d)
15. (a)	**16.** (d)	**17.** (b)	**18.** (c)	**19.** (a)	**20.** (b)	**21.** (b)

Chapter **13**

AC Bridges and AC Network Analysis

13.1 Introduction

In Chapter 7 we have discussed Wheatstone bridge for the comparison of resistances using dc sources, Kirchhoff's laws and some other network theorems for the analysis of dc networks. The purpose of the present chapter is to consider a few ac bridges with the help of which we can measure self-inductance, capacitance, mutual inductance, frequency of ac source, etc. Also, we shall discuss how Kirchhoff's laws and various other network theorems as discussed in chapter 7 with dc circuits can be used for the analysis of ac networks. Kirchhoff's laws are also applicable to ac networks with slight change in nomenclature. Here, instead of resistance we have impedance and instead of taking algebraic sum of currents and voltages we are to take phasor sum or vector sum.

Hence, Kirchhoff's laws for ac networks can be stated as follows:

1. ***Kirchhoff's current law***—According to this law, the phasor or vector sum of the currents meeting at a junction point in a network is zero. Mathematically,

$$\sum_k \vec{I}_k = 0.$$

2. ***Kirchhoff's voltage law***—According to this law, the phasor or vector sum of the potential drops in different parts of any loop in a network is equal to the phasor or vector sum of the emfs acting in the same loop. Mathematically,

$$\sum_k \vec{I}_k Z_k = \sum_j \vec{V}_j.$$

13.2 Four-Branch Impedance Bridge or a Generalised Wheatstone Bridge

Fig. 13.2-1 shows an ac bridge consisting of four impedances arranged as in the Wheatstone bridge. The dc source and the galvanometer of the familiar dc bridge have been replaced by an ac source and a suitable null detector d. The detector is usually a pair of headphones giving minimum sound at balance with an ac source of about 1 kHz in frequency or an ac amplifier with an indicating meter (called a *null detector*).

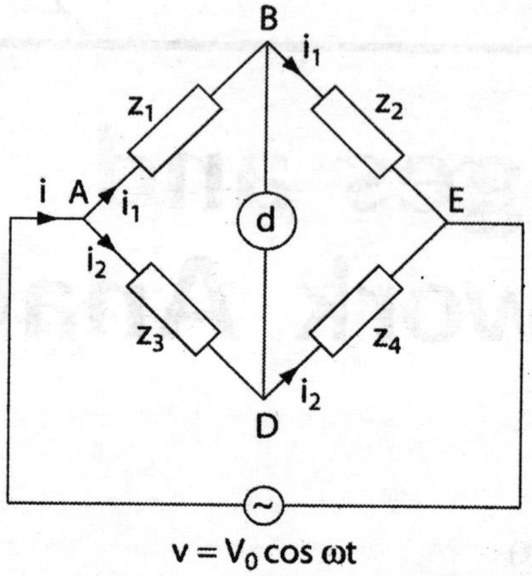

Fig. 13.2-1 : Impedance bridge

When the bridge is balance no current passes through the detector d. Then the current distribution can be assumed to be as shown in Fig. 13.2-1. In terms of phasor currents and voltages we can write at balance,

$$\vec{I}_1 = \frac{\vec{V}}{Z_1 + Z_2} \quad \text{and} \quad \vec{I}_2 = \frac{\vec{V}}{Z_3 + Z_4}, \tag{13.2-1}$$

where $\vec{V}$ is the phasor for the applied voltage. At balance potentials of the points B and D are equal. Therefore,

$$\vec{V}_{AB} = \vec{V}_{AD} \quad \text{or} \quad \vec{I}_1 Z_1 = \vec{I}_2 Z_3 \quad \text{or} \quad \vec{I}_1 Z_1 - \vec{I}_2 Z_3 = 0. \tag{13.2-2}$$

From Eq. (13.2-1) we get

$$\vec{I}_1(Z_1 + Z_2) = \vec{I}_2(Z_3 + Z_4).$$

Using (13.2-2) it reduces to

$$\vec{I}_1 Z_2 - \vec{I}_2 Z_4 = 0. \tag{13.2-3}$$

For nontrivial solution of Eqs. (13.2-2) and (13.2-3) we must have

$$\begin{vmatrix} Z_1 & -Z_3 \\ Z_2 & -Z_4 \end{vmatrix} = 0 \quad \text{or} \quad Z_1 Z_4 = Z_2 Z_3 \quad \text{or} \quad \frac{Z_1}{Z_2} = \frac{Z_3}{Z_4}. \tag{13.2-4}$$

All the impedances are, in general, complex and may have both resistive and reactive parts. Let

$$Z_k = |\,Z_k\,|\, e^{j\phi_k} \quad (k = 1, 2, 3, 4),$$

where $|\,Z_k\,|$ is the magnitude and ϕ_k is the phase angle of Z_k.

Therefore, from Eq. (13.2-4) we get

$$\frac{|\,Z_1\,|}{|\,Z_2\,|} e^{j(\phi_1 - \phi_2)} = \frac{|\,Z_3\,|}{|\,Z_4\,|} e^{j(\phi_3 - \phi_4)}.$$

It shows that for balance of the bridge, i.e., for zero current in the detector at all times, following two conditions must be satisfied simultaneously:

$$\frac{|\,Z_1\,|}{|\,Z_2\,|} = \frac{|\,Z_3\,|}{|\,Z_4\,|} \tag{13.2-5}$$

and

$$\phi_1 - \phi_2 = \phi_3 - \phi_4. \tag{13.2-6}$$

Unlike dc bridges all ac bridges, except in special circumstances, require a double balance. This is because the current through the detector, in general, has two components—one is in phase and the other 90° out of phase with the ac supply. These two components must be balanced out separately. The first one is usually a balance of resistive components of the bridge and could be made with a dc source and a galvanometer if the bridge arms do not contain capacitors. The second one is obtained by using an ac source and a suitable null detector.

If the balance conditions are independent of frequency of the supply then a nonsinusoidal source may be used without much error because different frequency components will balance out separately. However, an ac source of single frequency is desirable because the values of the circuit components may depend on the frequency. For example, the resistance of a conductor increases with frequency due to an effect known as *skin effect* (see, Section 15.7). In all cases the resistance boxes used in ac bridges must be non-inductive.

13.3 AC Bridges for Measuring Self-inductance

1. Maxwell's L/C bridge

The circuit arrangement of Maxwell's L/C bridge is shown in Fig. 13.3-1. With its help self-inductance L can be measured in terms of a variable capacitance C and two standard resistances R_2 and R_3. Resistance R_4 is a variable resistance, which includes the resistance of the inductance L.

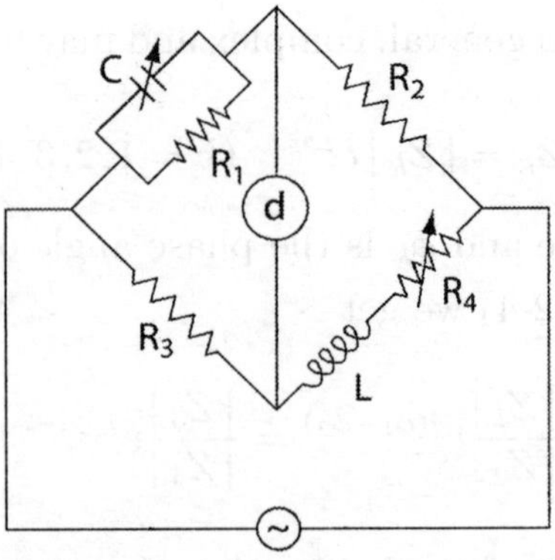

Fig. 13.3-1 : Maxwell's L/C bridge for measuring self-inductance

For the balance of the bridge the general Wheatstone bridge balance conditions apply :

$$\frac{Z_1}{Z_2} = \frac{Z_3}{Z_4}, \tag{13.3-1}$$

where Z_1, Z_2, Z_3 and Z_4 are the impedances of the four arms of the bridge. Using the notation given in Fig. 13.3-1, we can write

$$Z_1 = \frac{R_1 \times \frac{1}{j\omega C}}{R_1 + \frac{1}{j\omega C}} = \frac{R_1}{1 + j\omega C R_1},$$

$Z_2 = R_2$, $Z_3 = R_3$ and $Z_4 = R_4 + j\omega L$.

$$\therefore \quad \frac{R_1}{R_2(1 + j\omega C R_1)} = \frac{R_3}{R_4 + j\omega L}$$

$$\text{or} \quad R_1 R_4 + j\omega L R_1 = R_2 R_3 + j\omega C R_1 R_2 R_3.$$

Equating real and imaginary parts we get conditions for balance as

$$\frac{R_1}{R_2} = \frac{R_3}{R_4} \tag{13.3-2}$$

$$\text{and} \quad L = C R_2 R_3. \tag{13.3-3}$$

Note that the balance conditions are independent of the frequency of the ac supply and hence, of the waveform of the supply. If a variable C is used then the balance conditions become mutually independent. In practice R_1/R_2 is kept as a simple ratio and at first R_4 is varied to satisfy the in phase minimum condition (13.3-2) and then, C is varied to satisfy the quadrature minimum condition (13.3-3). Two or three alternate variations of R_4 and C give a good balance. Then, unknown L is obtained from the relation (13.3-3).

Here, it is important to point out that the inductor coil will always have some self-capacitance and so it will not be possible to obtain a perfect balance. Moreover, the perfect balance may be disturbed by the mutual coupling between various branches of the circuit. So this mutual coupling should be kept as small as possible. The capacitor used must be of good quality having very small power factor, otherwise the expression (13.3-3) for L would require modification. For example, if C is an imperfect capacitor and a small series resistance r accounts for its loss then the expression (13.3-3) modifies to

$$L = \frac{CR_2R_3}{1+\omega^2C^2r^2}. \tag{13.3-4}$$

This condition involves ω, so in this case the source must be free from harmonics.

2. Owen's bridge

Owen's bridge as shown in Fig. 13.3-2, is used to measure self-inductance in terms of a standard fixed capacitor. It is perhaps the most suitable ac bridge for the accurate measurement of inductance. For balance of the bridge we must have

$$\frac{Z_1}{Z_2} = \frac{Z_3}{Z_4},$$

where $Z_1 = R_1 + 1/j\omega C_1$, $Z_2 = 1/j\omega C_2$, $Z_3 = R_3 + j\omega L$ and $Z_4 = R_4$ are the impedances of the four arms of the bridge. Here R_3 includes the resistance of the inductor coil L.

$$\therefore \quad \left(R_1 + \frac{1}{j\omega C_1}\right) j\omega C_2 = (R_3 + j\omega L)\cdot\frac{1}{R_4}.$$

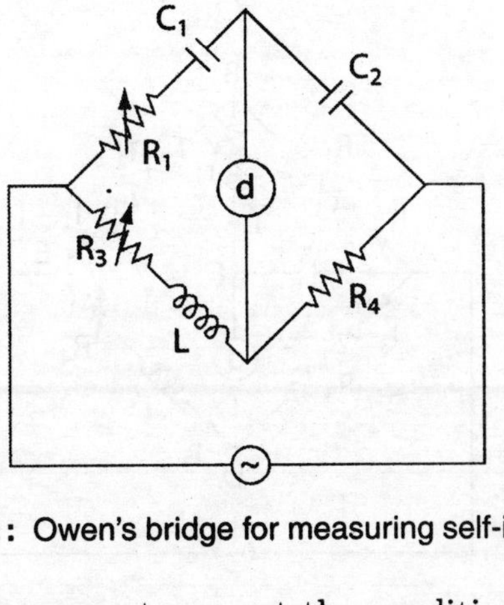

Fig. 13.3-2: Owen's bridge for measuring self-inductance

Equating real and imaginary parts we get the conditions for balance as

$$\frac{C_2}{C_1} = \frac{R_3}{R_4} \tag{13.3-5}$$

$$\text{and} \quad R_1C_2 = L/R_4 \quad \text{or} \quad L = C_2R_1R_4 \tag{13.3-6}$$

Note that the balance conditions are independent of the frequency of the supply and hence of the waveform of the supply. The conditions are mutually independent. The condition (13.3-5) is satisfied by varying R_3 and condition (13.3-6) by varying R_1. The bridge possesses great practical advantage that the balance is obtained by varying resistances only. The sensitivity of the bridge is very high over the frequency range 100–1000 Hz. The bridge can be used to measure L over a wide range.

3. Anderson bridge

Anderson bridge is a modified form of Maxwell's L/C bridge. It is most commonly used in the laboratory for the measurement of self-inductance in terms of a standard fixed capacitance. One of its advantages is that the double balance can be obtained by adjusting resistances only. The circuit arrangement of Anderson bridge is shown in Fig. 13.3-3. Let in the balanced condition the phasor current flowing in different parts of the bridge be as shown in Fig. 13.3-3. Note that at balance there is no current through the detector d. Now, applying Kirchhoff's laws to the closed meshes $ABFA$, $AFDA$ and $FBEDF$ we get

$$R_1\vec{I}_1 - r\vec{I} - \frac{1}{j\omega C}\vec{I} = 0 \qquad (13.3\text{-}7)$$

$$\frac{1}{j\omega C}\vec{I} - R_3\vec{I}_3 = 0 \qquad (13.3\text{-}8)$$

$$\text{and} \quad r\vec{I} + R_2\left(\vec{I}_1 + \vec{I}\right) - (R_4 + j\omega L)\vec{I}_3 = 0. \qquad (13.3\text{-}9)$$

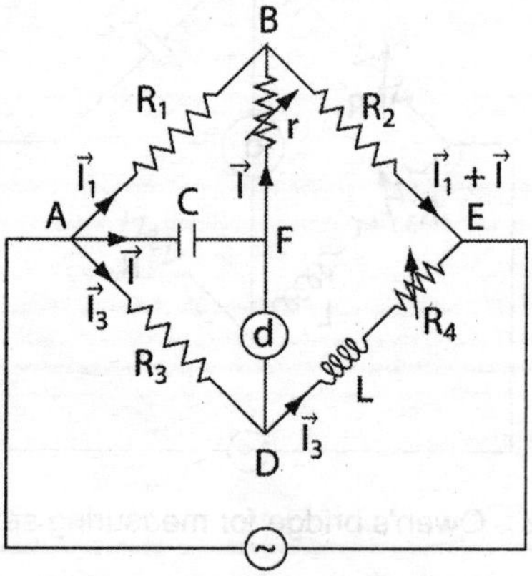

Fig. 13.3-3 : Anderson bridge for measuring self-inductance

Substituting $\vec{I}_1$ and $\vec{I}_3$ from Eqs. (13.3-7) and (13.3-8) in Eq. (13.3-9) we get for nonzero $\vec{I}$,

$$r + R_2 + R_2 \cdot \frac{r + \frac{1}{j\omega C}}{R_1} - (R_4 + j\omega L) \cdot \frac{1}{j\omega C R_3} = 0.$$

Equating real and imaginary parts we get

$$r + R_2 + \frac{rR_2}{R_1} - \frac{L}{CR_3} = 0 \quad \text{or} \quad L = CR_3\left[r\left(1 + \frac{R_2}{R_1}\right) + R_2\right] \qquad (13.3\text{-}10)$$

$$\text{and} \quad \frac{R_2}{\omega CR_1} - \frac{R_4}{\omega CR_3} = 0 \quad \text{or} \quad \frac{R_1}{R_2} = \frac{R_3}{R_4}. \qquad (13.3\text{-}11)$$

Note that the balance conditions are independent of the frequency of the supply and hence, of the waveform of the supply. The conditions are mutually independent and can be satisfied by adjusting r and R_4. Usually the balance condition (13.3-11) is obtained by using a dc source, a dc detector and adjusting the resistance R_4. It is called *dc balance.* The balance condition (13.3-10) is obtained by using an ac source, an ac detector and adjusting the resistance r. It is called *ac balance.* Eq. (13.3-10) shows that ac balance is possible for any positive value of r only if $L > CR_2R_3$. It should also be noted that as there is no other inductance all mutual coupling effects between neighbouring inductances are avoided here. Unlike Owen's bridge it is found that in Anderson bridge the imperfection of the capacitor has no effect on the measured value of L. It has only a small effect on the measured value of R_4. This shows that Anderson bridge is superior to Owen's bridge. In Anderson bridge a fixed capacitor may be used instead of a costly variable capacitor as in the case of Maxwell's L/C bridge.

Butterworth investigated the sensitivity of the Anderson bridge with respect to change in r in a nearly balanced condition and showed that the bridge becomes most sensitive when

$$R_1 = R_2 = \frac{R_3}{2} = \frac{R_4}{2} \quad \text{and} \quad \frac{L}{C} = 2R_4^2$$

13.4 AC Bridges for Measuring Capacitance

1. De Sauty's bridge

One of the simplest methods of comparing capacitances is to use De Sauty's bridge as shown in Fig. 13.4-1. If the power factors of the capacitor's are neglected then for

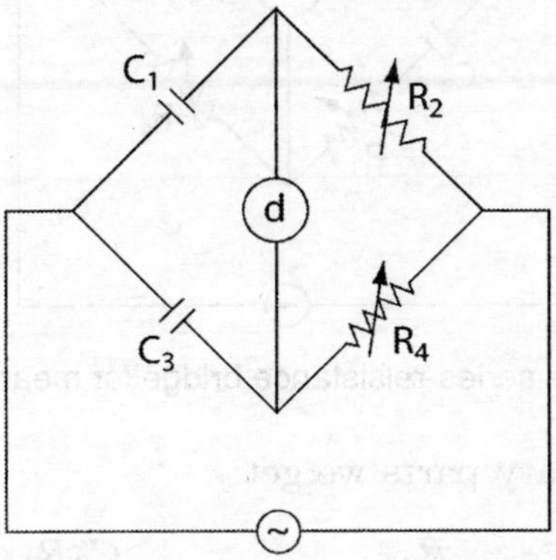

Fig.13.4-1 : De Sauty's bridge for comparing capacitances

balance we have

$$\frac{Z_1}{Z_2} = \frac{Z_3}{Z_4},$$

where $Z_1 = 1/j\omega C_1$, $Z_2 = R_2$, $Z_3 = 1/j\omega C_3$ and $Z_4 = R_4$. Therefore,

$$\frac{C_3}{C_1} = \frac{R_2}{R_4}. \tag{13.4-1}$$

Thus, here we have only one condition of balance, which is independent of frequency and can be obtained by adjusting the resistances R_2 and R_4. With practical imperfect capacitors perfect balance is difficult to obtain with this bridge. In this case we require to balance out the associated resistances accounting for the associated loss.

2. Wien's series-resistance bridge

It is a simple ac bridge for the measurement of capacitance and power factor of a capacitor. The bridge circuit is shown in Fig. 13.4-2. C_1 is the unknown capacitor. r_1 represents a small series resistance, which accounts for loss in C_1. C_3 is a good quality capacitor having negligible loss, which is arranged in series with a known adjustable resistor R_3. For balance of the bridge we must have

$$\frac{Z_1}{Z_2} = \frac{Z_3}{Z_4},$$

where $Z_1 = r_1 + 1/j\omega C_1$, $Z_2 = R_2$, $Z_3 = R_3 + 1/j\omega C_3$ and $Z_4 = R_4$.

$$\therefore \quad \frac{1}{R_2}\left(r_1 + \frac{1}{j\omega C_1}\right) = \frac{1}{R_4}\left(R_3 + \frac{1}{j\omega C_3}\right).$$

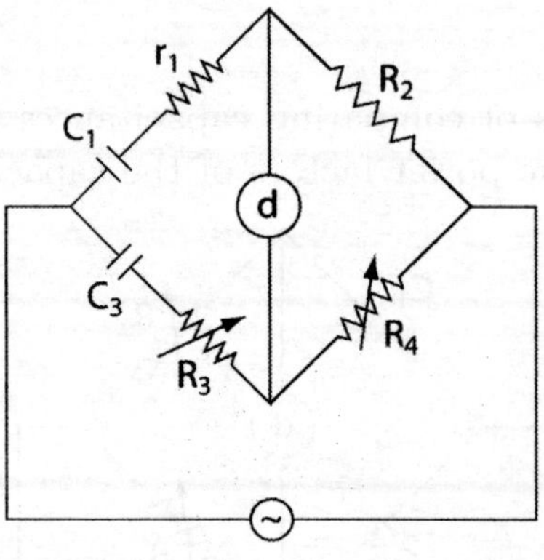

Fig. 13.4-2 : Wien's series-reisistance bridge for measuring capacitance

Equating real and imaginary parts we get

$$\frac{r_1}{R_2} = \frac{R_3}{R_4} \quad \text{and} \quad C_1 = \frac{C_3 R_4}{R_2}. \tag{13.4-2}$$

These conditions may be satisfied, though not quite independently, by alternate adjustment of R_3 and R_4.

The power factor of the capacitor C_1 is now given by

$$\cos\phi = \frac{r_1}{\sqrt{r_1^2 + \frac{1}{\omega^2 C_1^2}}} \simeq \omega_1 C_1 r_1. \tag{13.4-3}$$

3. Schering bridge

It provides one of the best methods for the measurement of capacitance and power factor of a capacitor in terms of standard capacitors. The bridge circuit is shown in Fig. 13.4-3. C_1 is the unknown capacitor and r_1 represents a small series resistance, which accounts for loss in C_1. C_3 is a standard capacitor and C_4 is a variable air capacitor having negligible power factor. For balance of the bridge we must have

$$\frac{Z_1}{Z_2} = \frac{Z_3}{Z_4},$$

where

$$Z_1 = r_1 + \frac{1}{j\omega C_1}, \quad Z_2 = R_2, \quad Z_3 = \frac{1}{j\omega C_3} \quad \text{and} \quad Z_4 = \frac{R_4 \cdot \frac{1}{j\omega C_4}}{R_4 + \frac{1}{j\omega C_4}} = \frac{R_4}{1 + j\omega C_4 R_4}.$$

Therefore,

$$\frac{1}{R_2}\left(r_1 + \frac{1}{j\omega C_1}\right) = \frac{1}{j\omega C_3 R_4}(1 + j\omega C_4 R_4).$$

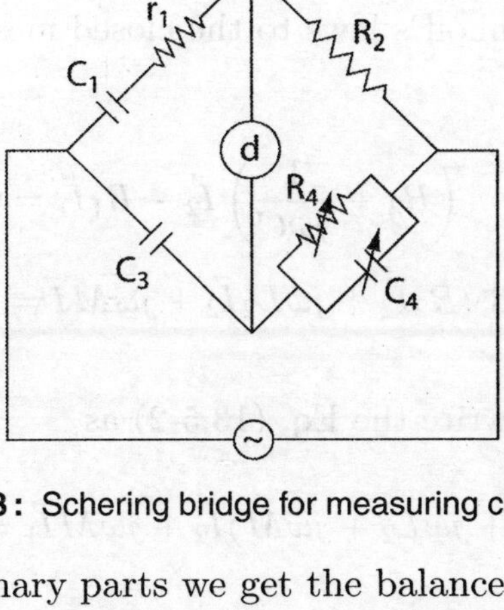

Fig. 13.4-3: Schering bridge for measuring capacitance

Equating real and imaginary parts we get the balance conditions as

$$\frac{r_1}{R_2} = \frac{C_4}{C_3} \quad \text{or} \quad r_1 = R_2 \cdot \frac{C_4}{C_3} \tag{13.4-4}$$

$$\text{and} \quad C_1 R_2 = C_3 R_4 \quad \text{or} \quad C_1 = C_3 \cdot \frac{R_4}{R_2}. \tag{13.4-5}$$

These conditions are quite independent of one another and of the frequency of the supply. These conditions can be satisfied by varying R_4 and C_4. The power factor of the capacitor C_1 is given by

$$\cos\phi = \frac{r_1}{\sqrt{r_1^2 + \frac{1}{\omega^2 C_1^2}}} \approx \omega C_1 r_1 = \omega C_3 \cdot \frac{R_4}{R_2} \cdot R_2 \frac{C_4}{C_3} = \omega C_4 R_4. \tag{13.4-6}$$

Note that if the power factor is not required, then for the measurement C_1 only the value of the variable capacitor C_4 need not be known. The Schering bridge is also used for the measurement of dielectric constant, especially of liquids, for the determination of very low capacitances, the measurement of cable capacitances at high voltage and the study of insulation at high voltages.

13.5 AC Bridges for Measuring Mutual Inductance

1. Modified Carey Foster bridge(or Campbell bridge)

The Carey Foster bridge modified by Heydweiller and Campbell for the measurement of mutual inductance using ac source and detector is shown in Fig. 13.5-1. In the modified form of the original Carey Foster bridge a variable resistance R_3 is included in series with the standard capacitor C. The resistance R_2 includes the resistance of the coil L_2.

Let in the balanced condition the phasor current flowing in different parts of the bridge be as shown in Fig. 13.5-1. Note that at balance current through the detector d is zero. Now applying Kirchhoff's laws to the closed meshes $ABDA$ and $BEDB$ we get

$$\left(R_3 + \frac{1}{j\omega C}\right)\vec{I}_2 - R_1\vec{I}_1 = 0 \tag{13.5-1}$$

$$\text{and } R_2\vec{I}_2 + j\omega L_2\vec{I}_2 + j\omega M\vec{I} = 0 \tag{13.5-2}$$

Since $\vec{I} = \vec{I}_1 + \vec{I}_2$ we can write the Eq. (13.5-2) as

$$(R_2 + j\omega L_2 + j\omega M)\vec{I}_2 + j\omega M\vec{I}_1 = 0 \tag{13.5-3}$$

Substituting $\vec{I}_1$ from Eq. (13.5-1) in Eq. (13.5-3) we get for nonzero $\vec{I}_2$,

$$R_2 + j\omega(L_2 + M) + \frac{j\omega M}{R_1}\left(R_3 + \frac{1}{j\omega C}\right) = 0.$$

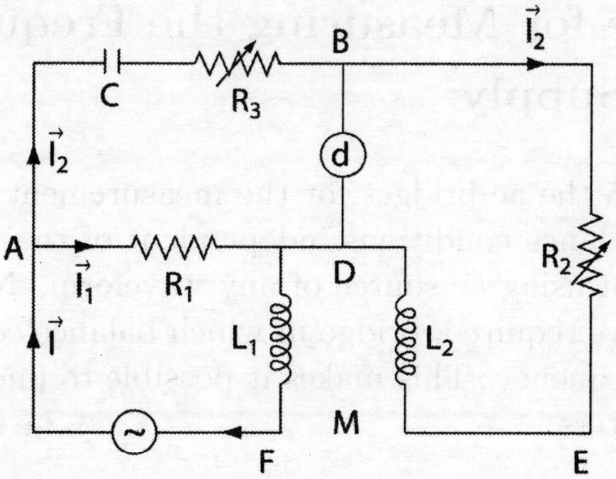

Fig. 13.5-1 : Modified Carey Foster or Campbell bridge

Equating real and imaginary parts we get

$$R_1R_2 = -\frac{M}{C} \quad \text{and} \quad R_1(L_2 + M) = -MR_3.$$

These conditions require that M must be negative, i.e., $M = -|M|$. So we must have

$$|M| = CR_1R_2 \quad \text{and} \quad L_2 = |M|\left(1 + \frac{R_3}{R_1}\right). \qquad (13.5\text{-}4)$$

The second condition requires that $L_2 > |M|$. Note that the balance conditions are independent of one another and the frequency of the source. The balance is obtained by varying R_2 and R_3 alternatively. Note that the mutual inductance must be inserted with correct sign otherwise balance cannot be obtained. Also a necessary condition for balance is that L_2 must be greater than $|M|$.

2. Use of self-inductance bridges for the determination of mutual inductance

The mutual inductance M between two coils with self-inductances L_1 and L_2 can be found by connecting the coils in series and measuring their equivalent self-inductance by any self-inductance measuring bridges like Anderson bridge, Owen's bridge, etc. The equivalent self-inductance of the two coils is given by

$$L_{\text{eff}} = L_1 + L_2 \pm 2M,$$

where the sign before M depends on the mode in which the coils are connected. Let

$$L'_{\text{eff}} = L_1 + L_2 + 2M \quad \text{and} \quad L''_{\text{eff}} = L_1 + L_2 - 2M.$$

Therefore,

$$M = 14\left(L'_{\text{eff}} - L''_{\text{eff}}\right).$$

Thus, measuring L'_{eff} and L''_{eff} with the coils connected in series and antiseries mode by any self-inductance measuring bridge M can be determined.

13.6 AC Bridge for Measuring the Frequency of an AC Supply

So far in connection with the ac bridges for the measurement of L, C and M our aim has been to make the balance conditions independent of the frequency of the supply. This gives the freedom of using ac source of any waveform. Now in the measurement of the source frequency we require a bridge in which balance conditions can be fulfilled only at one particular frequency. This makes it possible to find the frequency in terms of known circuit parameters.

Robinson's frequency bridge/Wien's parallel bridge

Robinson used a bridge (Fig. 13.6-1) for the measurement of ac frequency in terms of known capacitances and resistances. The bridge was originally used by Wien for the comparison of two capacitances but later on Robinson used it to measure frequency.

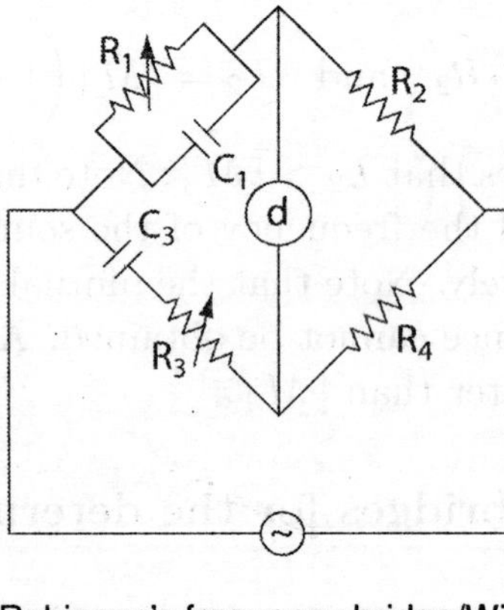

Fig. 13.6-1 : Robinson's frequency bridge/Wien parallel bridge

In the bridge shown in Fig. 13.6-1, C_1, C_3 are two good quality capacitors and R_1, R_2, R_3, R_4 are non-inductive resistors. For balance, i.e., zero current through the detector d the general Wheatstone bridge balance conditions apply:

$$\frac{Z_1}{Z_2} = \frac{Z_3}{Z_4},$$

where

$$Z_1 = \frac{R_1 \cdot \frac{1}{j\omega C_1}}{R_1 + \frac{1}{j\omega C_1}} = \frac{R_1}{1 + j\omega C_1 R_1},$$

$$Z_2 = R_2, \quad Z_3 = R_3 + \frac{1}{j\omega C_3} = \frac{1 + j\omega C_3 R_3}{j\omega C_3} \quad \text{and} \quad Z_4 = R_4$$

are the impedances of the four arms of the bridge. Therefore,

$$\frac{R_1}{R_2(1+j\omega C_1R_1)} = \frac{1+j\omega C_3R_3}{j\omega C_3R_4}$$

$$\text{or}\quad j\omega C_3R_1R_4 = R_2(1+j\omega C_1R_1)(1+j\omega C_3R_3)$$

$$= R_2(1-\omega^2C_1R_1C_3R_3 + j\omega C_1R_1 + j\omega C_3R_3).$$

Equating real and imaginary parts we get

$$\omega^2 = \frac{1}{C_1C_3R_1R_3} \quad\text{and}\quad \frac{R_4}{R_2} = \frac{R_3}{R_1} + \frac{C_1}{C_3} \tag{13.6-1}$$

Robinson simplified these conditions for balance by making $C_1 = C_3 = C$ (say) and $R_1 = R_3 = R$ (say). Then the above conditions become

$$\omega = \frac{1}{CR} \quad\text{or}\quad f = \frac{1}{2\pi CR} \quad\text{and}\quad R_4 = 2R_2, \tag{13.6-2}$$

where f is the required frequency of the supply. Thus, the bridge can be used for the measurement of frequency. The bridge is balanced by keeping $R_4 = 2R_2$ and varying R_1 and R_3 together but maintaining their equality with the help of a ganged potentiometer.

A major difficulty with the bridge arises when the source voltage is not purely sinusoidal but contains higher harmonics. These harmonics will appear in the detector since the balance conditions involve ω. So special care is needed to detect the balancing out of the fundamental in the presence of harmonics.

Many other frequency measuring bridges are available but Robinson's bridge has the great merit that its arms are made of resistors and capacitors only. So there is no inductive coupling between the components of the bridge or even between the bridge and other neighbouring sources.

13.7 Some useful Network Theorems Applied to AC Circuits

In Section 7.8 we discussed some network theorems in connection with the analysis of dc circuits. These theorems can also the applied to ac circuits. Here, instead of resistances we have impedances and instead of taking algebraic sum of voltages and currents we have to consider the phasor sum. The statements of some useful network theorems as applicable to ac networks are given in next page:

1. Superposition theorem

It states that in any linear bilateral network containing several energy sources, the current or voltage for any element is the phasor sum of the effects produced by each source acting separately; while considering the effect of a particular source other sources are replaced by their internal impedances. The theorem is the result of linear relationship between voltage and current in a linear circuit.

2. Thevenin's theorem

According to this theorem any two-terminal linear network containing energy sources and impedances can be replaced by a simple equivalent circuit containing a voltage source V_T together with a series impedance Z_T, where V_T is the open circuit voltage between the terminals of the network and Z_T is the impedance measured between the terminals with all the energy sources replaced by their internal impedances.

3. Norton's theorem

It states that any two terminal linear network containing energy sources and impedances can be replaced by a simple equivalent circuit containing a current source I_N in parallel with an impedance Z_N, where I_N is the short-circuit current between the terminals of the network and Z_N is the impedance measured between the terminals with all the energy sources replaced by their internal impedances.

All these theorems can be illustrated by using the circuits of Figs. 7.8-1, 7.8-2 and 7.8-3, with all the resistances R_1, R_2, R_3, R_L, R_T and R_N replaced by the impedances Z_1, Z_2, Z_3, Z_L, Z_T and Z_N, respectively; the dc sources replaced by ac sources; various currents and voltages replaced by their respective phasors.

4. Maximum power transfer theorem

According to this theorem maximum power is delivered from a source to a load impedance Z_L when Z_L is the complex conjugate of the internal impedance of the source.

To prove the theorem let us consider the Thevenin's equivalent of any linear network or an energy source and a load impedance Z_L connected to it (Fig. 13.7-1). Let $Z_T = R_T + jX_T$ be fixed and $Z_L = R_L + jX_L$, where R_L and X_L can be varied separately. The load current phasor is

$$\vec{I}_L = \frac{\vec{V}_T}{Z_T + Z_L} = \frac{\vec{V}_T}{(R_T + R_L) + j(X_T + X_T)}.$$

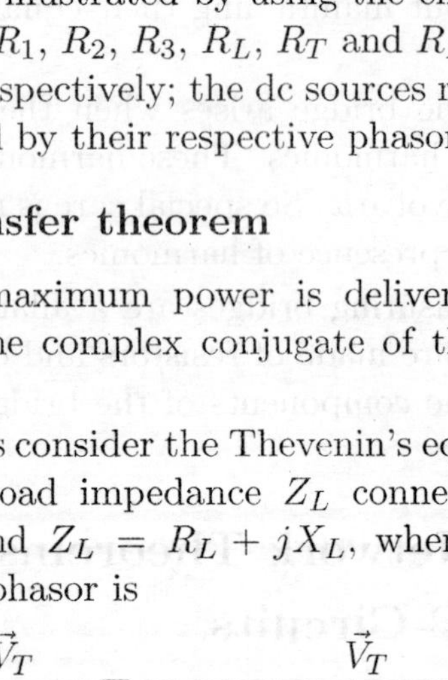

Fig. 13.7-1 : Thevenin's equivalent source connected to a load impedance

The power delivered to the load is

$$P_L = \left| \vec{I}_L \right|^2 R_L = \frac{V_T^2 R_L}{(R_T + R_L)^2 + (X_T + X_L)^2}. \tag{13.7-1}$$

(i) Let us now consider the case where R_L is fixed and X_L is varied. For maximum power $\dfrac{\partial P_L}{\partial X_L} = 0$. This requires that

$$X_L = -X_T. \tag{13.7-2}$$

Substituting this value in Eq. (13.7-1) we get

$$P_L = \frac{V_T^2 R_L}{(R_T + R_L)^2}.$$

Now if R_L is varied then for maximum power, $\dfrac{\partial P_L}{\partial R_L} = 0$. This requires that $R_L = R_T$. This for maximum power we must have

$$Z_L = R_L + jX_L = R_T - jX_T = Z_T^*, \tag{13.7-3}$$

where Z_T^* is the complex conjugate of the source impedance. The process of adjusting the load impedance to match the internal impedance of a source or network is known as *impedance matching.* It is an important consideration for many electronic circuits.

The maximum power delivered to the load is obtained from Eq. (13.7-1) by putting $R_L = R_T$ and $X_L = -X_T$. Thus,

$$P_{L,\max} = \frac{V_T^2}{4R_L} = \frac{V_T^2}{4R_T}. \tag{13.7-4}$$

(ii) If X_L is fixed and R_L is variable only. Then for maximum power we have from Eq. (13.7-1),

$$\frac{\partial P_L}{\partial R_L} = 0.$$

It requires that for maximum power

$$R_L = \sqrt{R_T^2 + (X_T + X_L)^2}. \tag{13.7-5}$$

(iii) If the load is a pure variable resistance, then $X_L = 0$ and we have from (13.7-1)

$$P_L = \frac{V_T^2 R_L}{(R_T + R_L)^2 + X_T^2}.$$

For maximum power $\dfrac{\partial P_L}{\partial R_L} = 0$. This requires that

$$R_L = \sqrt{R_T^2 + X_T^2} = |\, Z_T \,| \tag{13.7-6}$$

13.8 T- and π-Networks

In Section 7.10 we have seen how a two-port or four-terminal linear resistive network can be replaced by an equivalent T- or π-network of resistances. For the general case of a two-port network containing linear resistive as well as reactive components we can also find similar T- and π- equivalent networks containing impedances instead of resistances. So all the discussions made in Section 7.10 for resistive networks are equally applicable to the general type of network containing impedances, the only difference is that in this case we have to use impedances instead of resistances.

SOLVED PROBLEMS

▶ **Problem 1.** *Consider an ac Wheatstone bridge containing resistances R_1, R_2, R_3 and R_4 in its four arms. A coil of inductance L and resistance r_L is connected in series with R_4 and a high-quality variable capacitor C is connected in parallel with R_1. $R_1 = 500\,\Omega$, $R_2 = 600\,\Omega$, $R_3 = 400\,\Omega$ and R_4 is a variable resistor. If the balance is obtained with $C = 0.5\,\mu$F and $R_4 = 400\,\Omega$, find the values of L and r_L.*

Solution. The circuit appropriate to the problem is the same as shown in Fig. 13.3-1 with R_4 replaced by $R_4 + r$. The conditions of balance are given by the Eqs. (13.3-2) and (13.3-3) as

$$\frac{R_1}{R_2} = \frac{R_3}{R_4 + r_L} \quad \text{and} \quad L = CR_2R_3.$$

Substituting the values of the given circuit parameters we get

$$r_L = \frac{R_2R_3}{R_1} - R_4 = \frac{600 \times 400}{500} - 400\,\Omega = 80\,\Omega$$

$$\text{and} \quad L = 0.5 \times 10^{-6} \times 600 \times 400\,\text{H} = 0.12\,\text{H}.$$

▶ **Problem 2.** *Consider the Anderson bridge of* Fig. 13.3-3 *with $R_1 = R_2 = 1\,k\Omega$, $R_3 = 500\,\Omega$, $C = 2\,\mu$F. If balance is obtained with $r = 200\,\Omega$ and $R_4 = 480\,\Omega$, determine the values of the inductance L and resistance (r_L) of the coil L.* **[N.B.U. 2017]**

Solution. The balance conditions are given by the Eqs. (13.3-10) and (13.3-11), where R_4 is to be replaced by $R_4 + r_L$.

$$\therefore \quad \frac{R_1}{R_2} = \frac{R_3}{R_4 + r_L} \quad \text{or} \quad r_L = \frac{R_2R_3}{R_1} - R_4 = 500\,\Omega - 480\,\Omega = 20\,\Omega$$

and

$$L = CR_3\left[r\left(1 + \frac{R_2}{R_1}\right) + R_2\right] = 2 \times 10^{-6} \times 500[200(1+1) + 1000]\text{H} = 1.4\,\text{H}.$$

► **Problem 3.** *Suppose a capacitor is being tested with a Schering bridge. The bridge is constructed with the capacitor under test in the first arm, a pure resistor $R_2 = 160\,\Omega$ in the second arm, a high quality capacitor $C_3 = 0.1\,\mu$F in the third arm and a parallel combination of a variable resistance R_4 and a variable capacitor C_4 in the fourth arm. The bridge is supplied with an ac source of frequency 50 Hz. The balance is obtained with $R_4 = 500\,\Omega$ and $C_4 = 3300$ pF. Find the equivalent series capacitance and resistance of the lossy capacitor. Also, find the power factor of the capacitor at the operating frequency.*

Solution. Referring to Fig. 13.4-3 and the balance conditions (13.4-4) and (13.4-5) we can write for the equivalent series resistance r_1 and capacitance C_1 as

$$r_1 = R_2 \cdot \frac{C_4}{C_3} = 160 \times \frac{3300 \times 10^{-12}}{0.1 \times 10^{-6}}\,\Omega = 5.28\,\Omega$$

$$\text{and } C_1 = C_3 \cdot \frac{R_4}{R_2} = 0.1\frac{500}{160}\,\mu\text{F} = 0.3125\,\mu\text{F}.$$

$$\text{Power factor} = \frac{r_1}{\sqrt{r_1^2 + \frac{1}{\omega^2 C_1^2}}} \approx \omega C_1 r_1 = 2\pi \times 50 \times 0.3125 \times 10^{-6} \times 5.28 = 0.0005$$

► **Problem 4.** *Suppose an ac Wheatstone bridge contains a coil of inductance L and resistance R_1 in its first arm, a resistance R_2 in the second arm, a parallel combination of a perfect capacitor C and a resistance R_3 in the third arm and a similar perfect capacitor C in the fourth arm. Find the frequency at which balance is obtained. Also, find the relationship between L, C, R_1, R_2 and R_3 at balance.*

Solution. The circuit arrangement of the bridge is shown in Fig. 13.P-4. For balance of the bridge

$$\frac{Z_1}{Z_2} = \frac{Z_3}{Z_4},$$

where $Z_1 = R_1 + j\omega L$, $Z_2 = R_2$, $Z_3 = \dfrac{R_3 \times \frac{1}{j\omega C}}{R_3 \times \frac{1}{j\omega C}} = \dfrac{R_3}{1 + j\omega C R_3}$ and $Z_4 = \dfrac{1}{j\omega C}$.

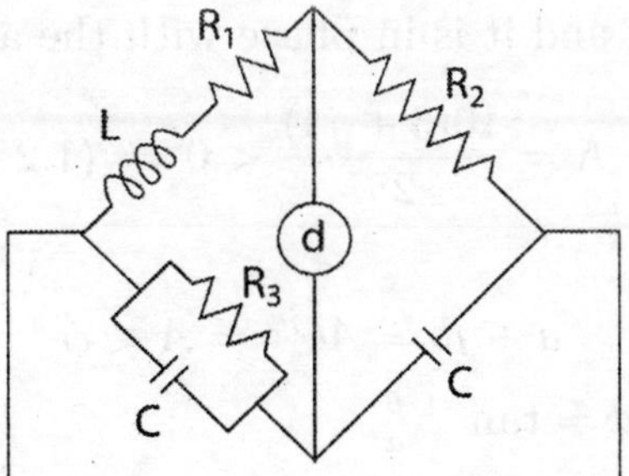

Fig 13.P-4

Therefore,

$$\frac{R_1 + j\omega L}{R_2} = \frac{j\omega C R_3}{1 + j\omega C R_3} \quad \text{or} \quad (R_1 + j\omega L)(1 + j\omega C R_3) = j\omega C R_3 R_2.$$

Equating real and imaginary parts we get

$$R_1 - \omega^2 L C R_3 = 0 \quad \text{and} \quad L + C R_3 R_1 = C R_3 R_2$$

$$\text{or } \omega^2 = \frac{R_1}{L C R_3} \quad \text{and} \quad L = C R_3 (R_2 - R_1).$$

The frequency at which balance is obtained is

$$f = \frac{\omega}{2\pi} = \frac{1}{2\pi}\sqrt{\frac{R_1}{L C R_3}}.$$

► **Problem 5.** *Consider the circuit of* Fig. 13.P-5 *and determine the current through each branch of it.*

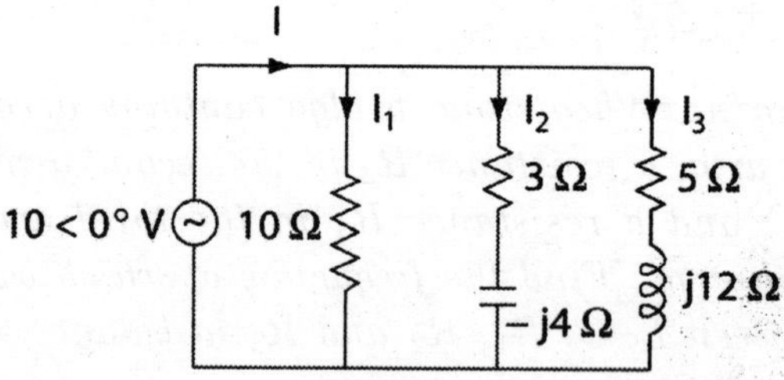

Fig 13.P-5

Solution. Taking the applied emf $10 < 0°$V as reference we can calculate the complex branch currents as follows:

$$I_1 = \frac{10 < 0°\text{V}}{10\,\Omega} = 1 < 0°.$$

So rms value of I_1 is 1 A and it is in phase with the applied emf

$$I_2 = \frac{10 < 0°}{3 - j4}\text{A} = \frac{10(3 + j4)}{25} < 0° = (1.2 + j1.6) < 0°,$$

where we use the notation

$$a + jb = Ae^{j\phi} = A < \phi$$

in which $A = \sqrt{a^2 + b^2}$ and $\phi = \tan^{-1}\frac{b}{a}$.

Thus, the rms value of I_2 is $2A$ and it leads over the applied emf by 53.1°.

$$I_3 = \frac{10 < 0°}{5 + j12} = \frac{10(5 - j12)}{169} < 0° = (0.305 - j0.71) < 0° = 0.77 < -67.1°.$$

Thus, the rms value of I_3 is 0.77 A and it lags behind the applied emf by 67.1°. The complex source current is

$$I = I_1 + I_2 + I_3 = 1 + 1.2 + j1.6 + 0.30 - j0.71 = 2.5 + j0.89 = 2.65 < 19.6°.$$

The rms value of source current is 2.65 A and it leads over the applied emf by 19.6°.

▶ **Problem 6.** *Calculate the Thevenin voltage, Thevenin impedance and load current for the circuit shown in* Fig. 13.P-6.

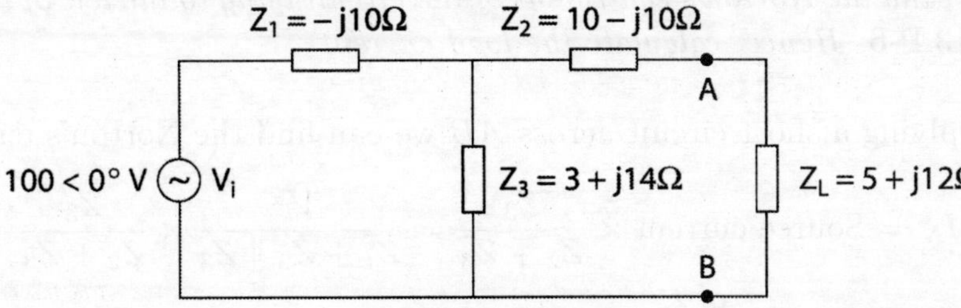

Fig 13.P-6

Solution. Disconnecting Z_L we can find the open circuit voltage between A and B, i.e., the Thevenin voltage V_T as

$$V_T = \frac{V_i Z_3}{Z_1 + Z_3} = \frac{100(3+j14)}{-j10+3+j14} = \frac{100(3+j14)}{3+j4}$$

$$= \frac{100 \times 14.32 < 77.90°}{5 < 53.13°} = 286.4 < 24.77° \text{V}.$$

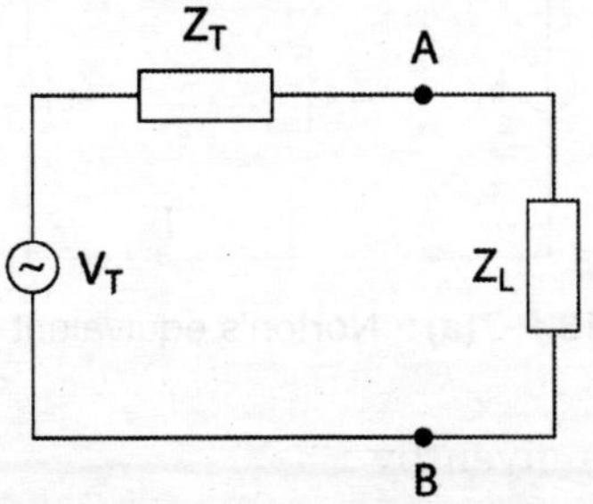

Fig. 13.P-6(a): Thevenin's equivalent circuit

The impedance measured between A and B with Z_L open and V_1 short circuited, i.e., the Thevenin impedance is

$$Z_T = Z_2 + \frac{Z_1 Z_3}{Z_1 + Z_3} = 10 - j10 - \frac{j10 \cdot (3+j14)}{-j10+3+j14}$$

$$= 10 - j10 - \frac{j10 \cdot (3+j14)}{3+j4} = 10 - j10 - \frac{j10(3+j14)(3-j4)}{25}$$

$$= 22 - j36\Omega.$$

∴ The complex load current

$$I_L = \frac{V_T}{Z_T + R_L} = \frac{286.4 < 24.77^\circ}{27 - j24} = \frac{286.4 < 24.77^\circ}{36.12 < -41.63^\circ} = 7.93 < 66.40^\circ\ A.$$

Thus, the rms value of load current is 7.93A and it leads over the applied emf by 66.40°.

▶ **Problem 7.** *Find the Norton's equivalent of the circuit lying to the left of the terminals A, B in* Fig. 13.P-6. *Hence, calculate the load current.*

Solution. Applying a short circuit across AB we can find the Norton's current:

$$I_N = \text{Source current} \times \frac{Z_3}{Z_2 + Z_3} = \frac{V_i}{Z_1 + Z_2 \parallel Z_3} \times \frac{Z_3}{Z_2 + Z_3}$$

$$= \frac{V_i Z_3}{Z_1 Z_2 + Z_2 Z_3 + Z_3 Z_1}$$

$$= \frac{100(3 + j14)}{-j10(10 - j10) + (10 - j10)(3 + j14) - j10(3 + j14)}$$

$$= \frac{100(3 + j14)}{210 - j20} = \frac{10}{89}(7 + 60j) = 6.79 < 83.34^\circ\ A$$

$$Z_N = 22 - j36\,\Omega \text{ (See, \textit{Solved Problem 6})}$$

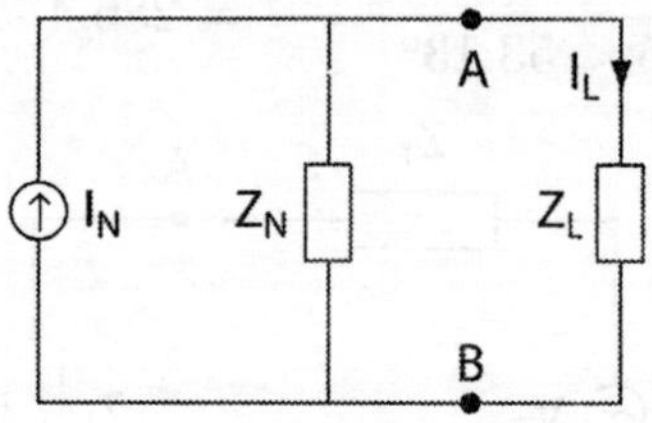

Fig. 13.P-7(a): Norton's equivalent circuit

Now current through Z_L is given by

$$I_L = I_N \times \frac{Z_N}{Z_N + Z_L} = \frac{10}{89}(7 + 60j) \times \frac{22 - j36}{22 - j36 + 5 + j12}$$

$$= \frac{10}{89} \cdot \frac{60.40 < 83.34^\circ \times 42.19 < -58.57^\circ}{36.12 < -41.63} = 7.93 < 66.40^\circ\ A.$$

▶ **Problem 8.** *An ac source of* 100 V *having internal impedance* $Z_i = 3 - j10\,\Omega$ *is connected to a load impedance* $Z_L = R_L + j14\,\Omega$, *where* R_L *is a variable resistive component of the load. Find the value of* R_L *for which the power transferred to the load becomes maximum. Also, find the value of this maximum power.*

Solution. From maximum power transfer, we have from the condition (13.7-5)

$$R_L = \sqrt{R_T^2 + (X_T + X_L)^2}.$$

Here $R_T = 3\,\Omega$, $X_T = -10\,\Omega$, $X_L = 14\,\Omega$.

$$\therefore \quad R_L = \sqrt{3^2 + (-10+14)^2}\ \Omega = 5\,\Omega.$$

Now power delivered to R_L is obtained from Eq. (13.7-1)

$$P_L = \frac{V_T^2 R_L}{(R_T + R_L)^2 + (X_T + X_L)^2} = \frac{100^2 \times 5}{(3+5)^2 + (-10+14)^2}\ \text{W} = 625\ \text{W}.$$

▶ **Problem 9.** *Consider the Wheatstone bridge of* Fig. 13.P-9 *and find its Thevenin's equivalent between the terminals A and B.*

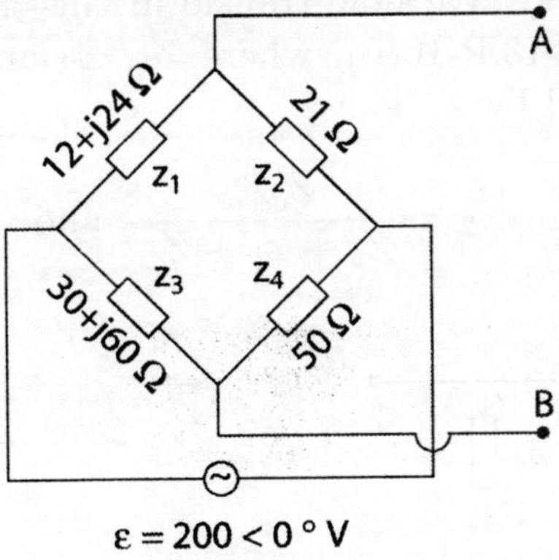

Fig 13.P-9

Solution. Thevenin's voltage

$$V_T = \frac{\mathcal{E}\cdot Z_2}{Z_1 + Z_2} - \frac{\mathcal{E}\cdot Z_4}{Z_3 + Z_4} = \frac{200\times 21}{33 + j24} - \frac{200 \times 50}{80 + j60}$$

$$= 3.2 - 0.5j \text{ (on simplification)}$$

$$= 3.24 < -8.8^\circ\ \text{V (with reference to source voltage)}.$$

Thevenin's impedance is obtained by short circuiting the source. Thus,

$$Z_T = Z_1 \parallel Z_2 + Z_3 \parallel Z_4 = \frac{Z_1 Z_2}{Z_1 + Z_2} + \frac{Z_3 Z_4}{Z_3 + Z_4}$$

$$= \frac{(12 + j24)\cdot 21}{33 + j24} + \frac{(30 + j60)\cdot 50}{80 + j60}$$

$$= 42.15 + j21.3\Omega \quad \text{(on simplification)}$$

► **Problem 10.** *Consider the bridged T-network of* Fig. 13.P-10 *and transform it into an equivalent T-network.*

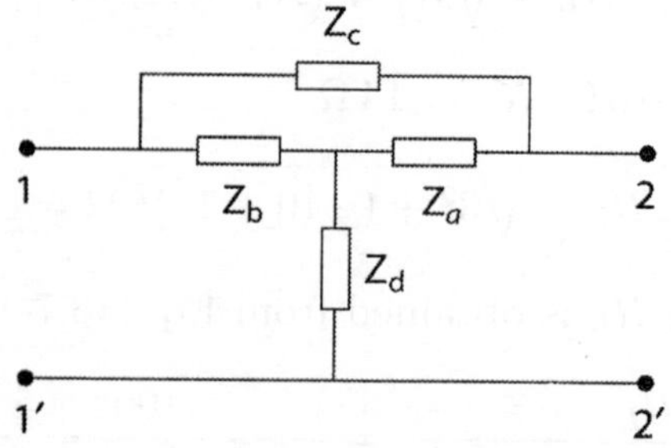

Fig 13.P-10

Solution. At first we redraw the circuit as shown in Fig. 13.P-10(a). Obviously Z_a, Z_b and Z_c form a π-network. We now transform this π-network into an equivalent T-network as shown in Fig. 13.P-10(b), where Z_1, Z_2 and Z_3 are given by the Eqs. (7.10-10) with all R replaced by Z. Thus,

$$Z_1 = \frac{Z_b Z_c}{Z_a + Z_b + Z_c}, \quad Z_2 = \frac{Z_c Z_a}{Z_a + Z_b + Z_c} \quad \text{and} \quad Z_3 = \frac{Z_a Z_b}{Z_a + Z_b + Z_c}$$

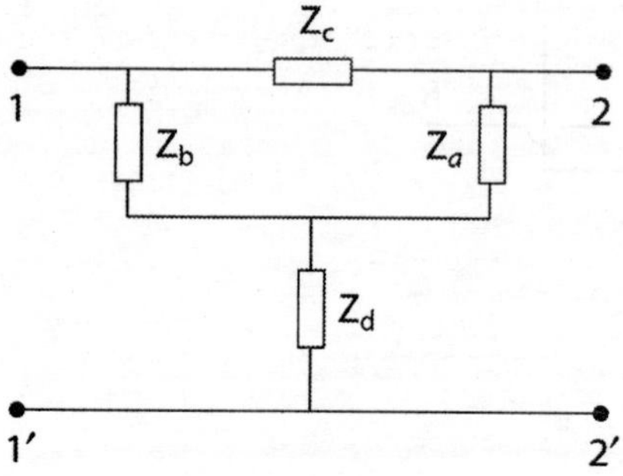

Fig 13.P-10(*a*)

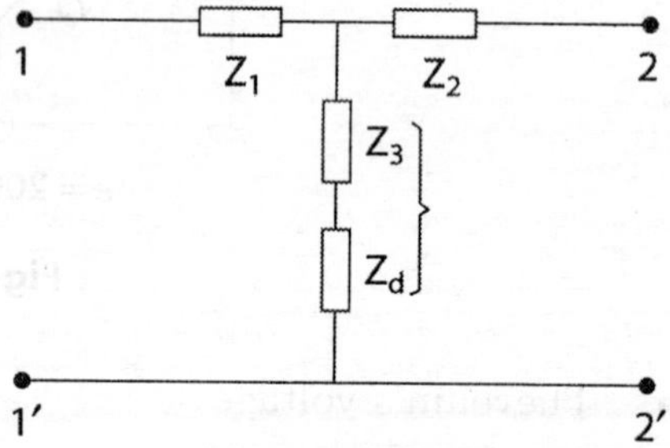

Fig 13.P-10(*b*)

QUESTIONS

1. (a) Find the conditions of balance in a generalised Wheatstone bridge.
   (b) What is the advantage of a bridge having balance conditions independent of source frequency?

2. (a) Give the principle of measurement of self-inductance by using
   (i) Maxwell's L/C bridge,

(ii) Owen's bridge,
(iii) Anderson bridge.

(b) How can you measure mutual inductance with such a bridge?

3. Consider a Wheatstone bridge containing resistances R_1, R_2, R_3 and R_4 in its four arms. An inductance of negligible resistance is connected in series with R_3 and a high quality capacitor C is connected in series with R_1. Find the conditions of balance of the bridge when operated with ac.

4. (a) Draw the circuit arrangement of Anderson bridge. Find the conditions of balance for the bridge.

(b) Why do you get two conditions for balance in an ac bridge?

(c) Write down the conditions for the Anderson bridge to be most sensitive.

(d) How would you measure the mutual inductance between two coils using Anderson's bridge? **[C.U. 2009]**

5. Describe with the help of a circuit diagram the working of an Anderson bridge mentioning how dc and ac balances are obtained. **[C.U. 2011]**

6. (a) Give the principle of measurement of capacitance by using

(i) De Sauty's bridge,
(ii) Wien's series-resistance bridge,
(iii) Schering bridge.

(b) What do you mean by the power factor of a capacitor? How can it be measured with a Schering bridge?

7. Give the principle of measurement of mutual inductances by using the modified Carey Foster bridge (or Campbell bridge) with ac source.

8. (a) Discuss the principle of measurement of the frequency of an ac source by using Robinson's bridge/Wien's parallel bridge.

(b) What difficulty would arise if the source is not purely sinusoidal?

9. (a) State maximum power transfer theorem. Prove it for a source driving a complex load.

(b) What do you mean by impedance matching? What is its importance?

10. State superposition theorem as applied to network analysis. Considering a simple T-network of impedances and two ac sources verify the validity of the theorem.

11. (a) State Thevenin's theorem and Norton's theorem as applied to ac network analysis.

(b) Consider a four-terminal network in the form of T-network of impedances Z_1, Z_2 and Z_3. An ac source of voltage V and negligible internal impedance is connected between two terminals and a load impedance Z_L between other two terminals. Find the current through Z_L by applying (i) Kirchhoff's laws, (ii) Thevenin's theorem and (iii) Norton's theorem.

12. A source of voltage V_s and internal impedance $R_s + jX_s$ is supplying power to a load impedance $Z_L = R_L + jX_L$. Find the condition for maximum power transfer to the load when (a) X_L is fixed and R_L is variable (b) R_L is fixed and X_L is variable (c) both R_L and X_L are variable.

EXERCISES

1. Consider an ac Wheatstone bridge containing resistances R_1, R_2, R_3 and R_4 in its four arms. A coil of inductance L and negligible resistance is connected in series with R_4 and a high quality capacitor C is connected in parallel with R_1. Show that the bridge is balanced if $R_1R_4 = R_2R_3 = L/C$.

2. Consider an ac Wheatstone bridge containing resistance R_1, R_2, R_3 and R_4 in its four arms. A coil of inductance L and negligible resistance is connected in series with R_4 and a high quality capacitor C is connected in series R_1. Show that the bridge is balanced if $R_2R_3 - R_1R_4 = L/C$ and the frequency of the ac source is

$$f = \frac{1}{2\pi}\sqrt{\frac{R_4}{LCR_1}}.$$ **[cf. C.U. 2000]**

3. Consider an ac Wheatstone bridge containing resistances $R_1 = 100\,\Omega$, $R_2 = 400\,\Omega$, $R_3 = 100\,\Omega$ and $R_4 = 800\,\Omega$ in its four arms. A capacitor of $1\mu F$ is connected in parallel with R_1 and another similar capacitor is connected in series with R_3. Find the frequency at which the bridge is balanced.

[*Ans.* 1.59 kHz]

4. Consider the Hay bridge as shown in Fig. 13.E-4. Assuming the balance condition for a generalised ac bridge show that under the balance condition of the bridge,

$$L_x = \frac{C_1R_2R_3}{1+\omega^2C_1^2R_1^2} \quad \text{and} \quad R_x = \omega^2 R_1C_1L_x.$$

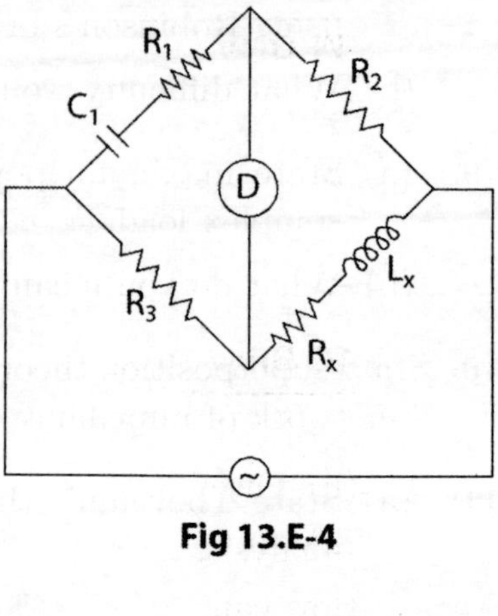

Fig 13.E-4

5. Show that the complex output voltage v_0 in the circuit of Fig. 13.E-5 is given by

$$\frac{v_0}{v_i} = \frac{1}{3 + j\left(\frac{\omega}{\omega_0} - \frac{\omega_0}{\omega}\right)} - \frac{1}{1 + \frac{R_1}{R_2}},$$

where $\omega_0 = 1/CR$ and ω is the angular frequency of the input ac.

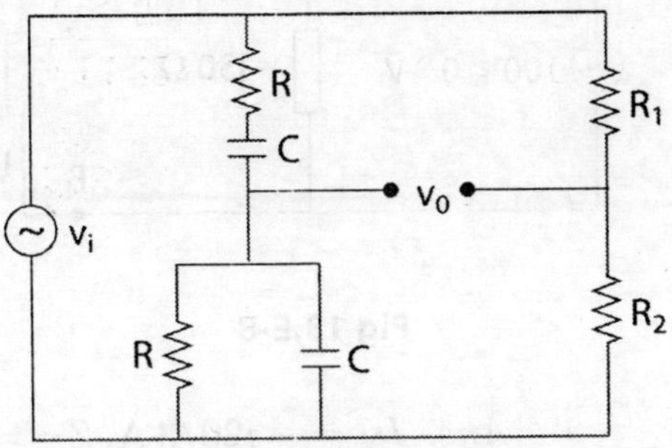

Fig 13.E-5

6. Calculate the Thevenin voltage, Thevenin impedance and the load current for the circuit of Fig. 13.E-6. **[C.U. 2002]**

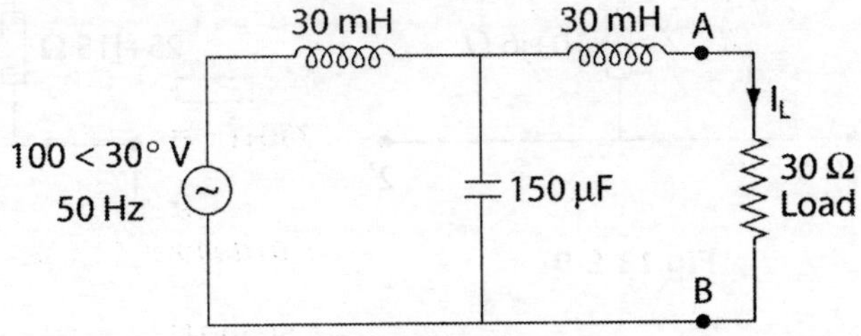

Fig 13.E-6

[*Ans.* $V_T = 179.89 < 30°\text{V}, Z_T = j26.37\,\Omega, I_L = 4.5 < -11.3°$]

7. Find the Thevenin's equivalent of the circuit lying to the left of the terminals A and B in Fig. 13.E-7. Hence, find the load current.

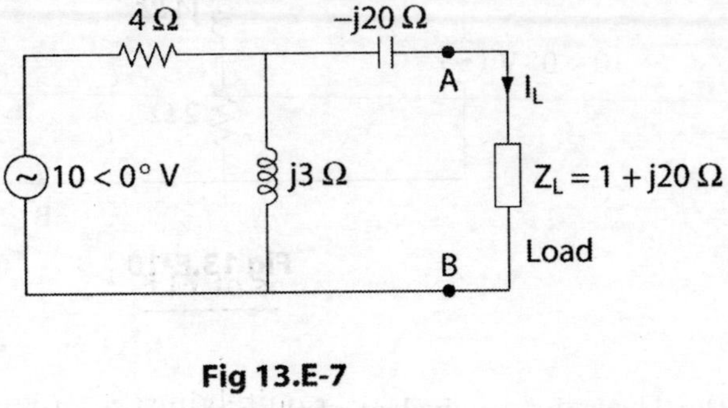

Fig 13.E-7

[*Ans.* $V_T = 3.6 + j4.8\,\text{V}, Z_T = 1.44 - j18.08\,\Omega, I_L = 1.94 < 14.93°$ A]

8. Find the Norton's equivalent of the circuit lying to the left of the terminals A and B in Fig. 13.E-8. Hence, find the load current.

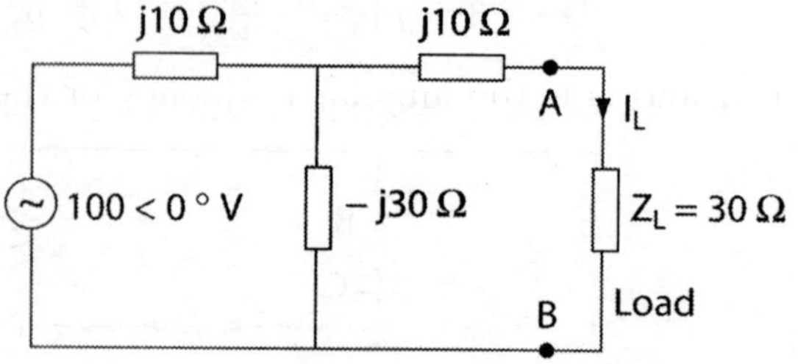

Fig 13.E-8

[*Ans.* $I_N = -j20/3\,\text{A}, Z_N = j30\,\Omega, I_L = 4.71 < -45°\ \text{A}$]

9. Consider the T-network of Fig. 13.E-9 and find its equivalent π-network.

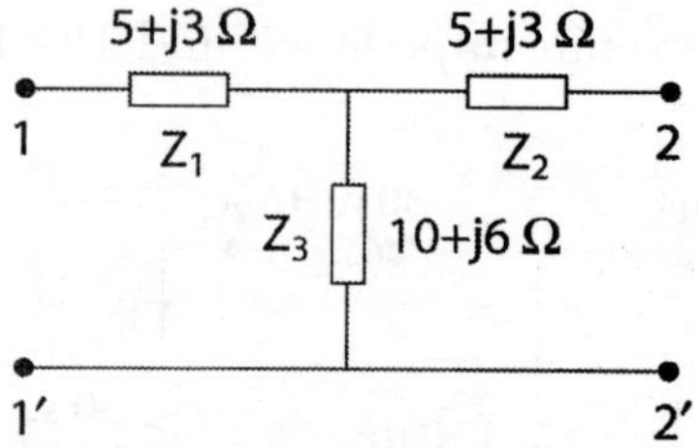

Fig 13.E-9

Ans.

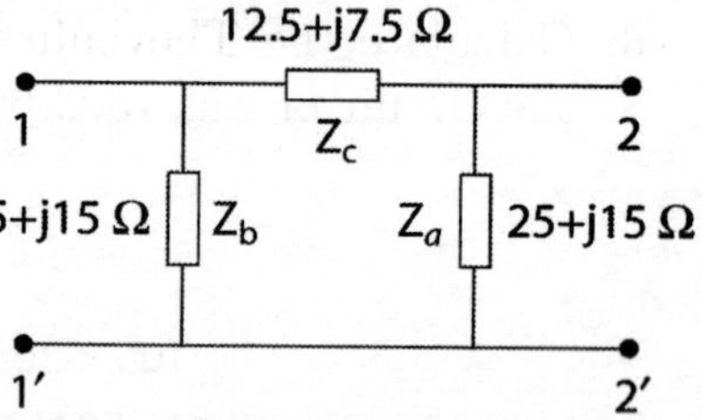

Fig 13.E-9(a)

[**Hints.** Use Eq. (7.10-11) with all R replaced by Z.]

10. Consider the circuit of Fig. 13.E-10 and find the value of R_L for maximum power transfer. Also, calculate the value of this maximum power delivered to R_L.

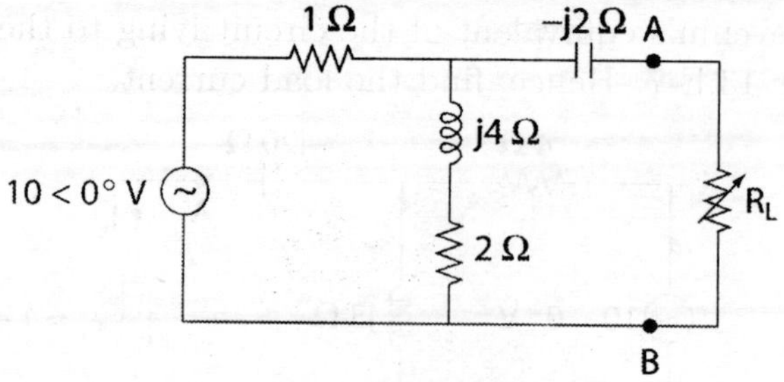

Fig 13.E-10

[*Ans.* $2.04\,\Omega, 17.12\,\text{W}$]

[**Hints.** Find the Thevenin's equivalent of the circuit lying left of the terminals of A and B and then proceed.]

11. Consider the circuit of Fig. 13.E-11 and find its Thevenin's equivalent between the terminals A and B.

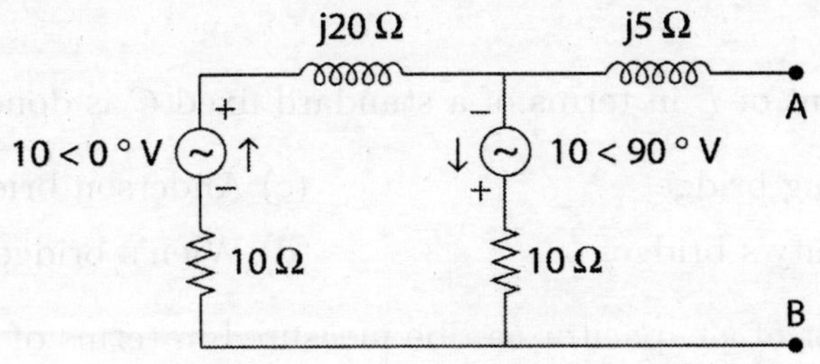

Fig 13.E-11

[*Ans.* $V_T = -5 + j10\,\text{V} = 11.18 < 63.43°\text{V}, Z_T = 7.5 + 7.5\,j$]

12. Using superposition theorem find the current I_3 through the resistor R_3 in the circuit of Fig. 13.E-12.

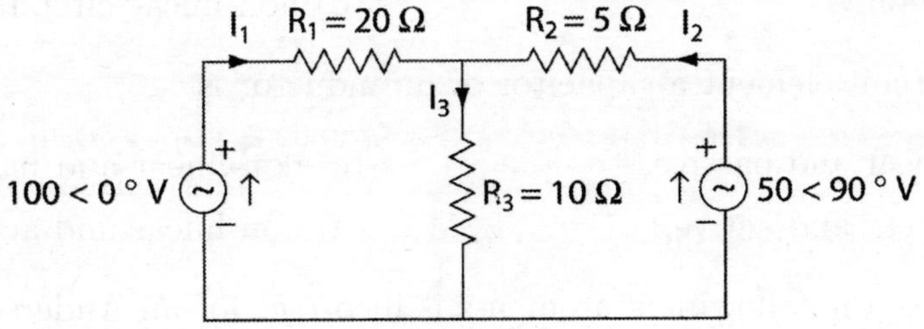

Fig 13.E-12

[*Ans.* $I_3 = 3.19 < 63.43°$ A]

13. Using Thevenin's theorem, find the current through the $2\,\text{k}\Omega$ resistor in the circuit of Fig. 13.E-13.

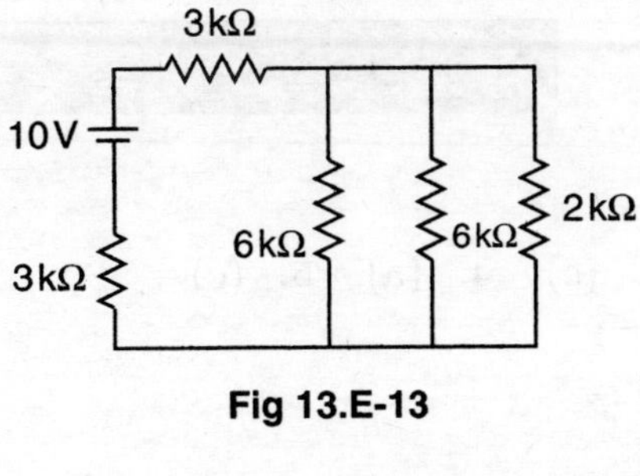

Fig 13.E-13

[*Ans.* 0.83 mA] **[C.U. 2007]**

OBJECTIVE TEST

1. Measurement of L in terms of a standard fixed C is done by

(a) Schering bridge,
(b) De Sauty's bridge,
(c) Anderson bridge,
(d) Wien's bridge.

2. Power factor of a capacitor can be measured in terms of a standard capacitor by using

(a) Schering bridge,
(b) Anderson bridge,
(c) Owen's bridge,
(d) Maxwell's bridge.

3. Kirchhoff's laws are applicable to

(a) dc only,
(b) ac only,
(c) both dc and ac,
(d) non-linear circuits only.

4. As a circuit element a capacitor or an inductor is

(a) linear and passive,
(b) linear and active,
(c) non-linear and passive,
(d) non-linear and active.

5. Which of the following statements is incorrect for an Anderson bridge?

(a) Balance conditions are independent of the frequency of the supply.
(b) Balance can be obtained by using ac source of arbitrary waveform.
(c) Balance can be obtained only with sinusoidal source.
(d) The dc and ac balance conditions are mutually independent.

ANSWERS

1. (c) **2.** (a) **3.** (c) **4.** (a) **5.** (c)

Chapter 14

Thermoelectricity

14.1 Seebeck Effect

Seebeck, in 1821, discovered that if two wires of dissimilar metals are joined end to end and the two junctions are maintained at different temperatures then a current flows through the circuit. This is known as *Seebeck effect.* Such a current is called *thermoelectric current* and the emf producing it is called *thermo-emf.* The direction of flow and the magnitude of thermoelectric current depend on the nature of the metals and also on the temperature difference between the two junctions. The circuit formed by the two wires of dissimilar metals is called a *thermocouple.* For example, in an Fe-Cu thermocouple current flows from Fe to Cu through the cold junction (Fig. 14.1-1).

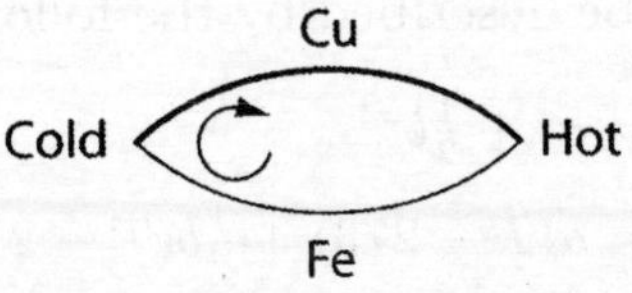

Fig. 14.1-1 : Fe-Cu thermocouple

Variation of thermo-emf with temperature

Suppose in a thermocouple one junction is kept at low fixed temperature and the temperature of the other junction is gradually increased. Experiments show that the thermo-emf generated at first increases with the increase in the difference of temperatures until it reaches a maximum value. On further increasing the difference in temperature the emf decreases till it is zero and then it is reversed in direction.

Typical variation of thermo-emf with the temperature of the hot junction, keeping the cold junction at 0°C, is shown in Fig. 14.1-2. The temperature of the hot junction at which thermo-emf becomes maximum is called *neutral temperature* (t_n) of the given thermocouple. It is constant for a particular thermocouple. The temperature of the hot junction at which thermo-emf is zero and about to change direction is called *inversion temperature* (t_i). Its value depends on the couple as well as the temperature of the cold junction. It is as much above the neutral temperature as the cold junction is below the neutral temperature, i.e.,

$$t_i - t_n = t_n - t_{\text{cold}} \quad \text{or} \quad t_n = \frac{t_i + t_{\text{cold}}}{2}.$$

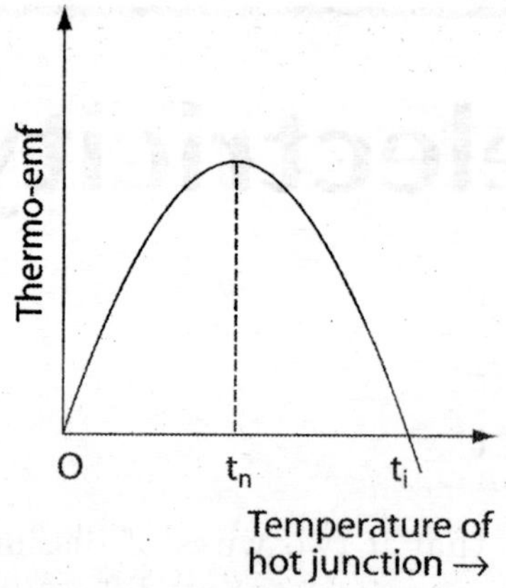

Fig. 14.1-2: Variation of thermo-emf with temperature

For example, in case of Fe-Cu thermocouple with the cold junction at 0°C, $t_n = 275$°C and $t_i = 550$°C. If the cold junction is at 10°C then t_n remains at 275°C but t_i reduces to 550°C − 10°C = 540°C.

Suppose in a thermocouple the cold junction is kept at 0°C and the temperature t°C of the hot junction is varied over a limited range. Then experiments show that the thermo-emf $\mathcal{E}$ generated may be described by the following almost parabolic function:

$$\mathcal{E} = at + \tfrac{1}{2}bt^2 \tag{14.1-1}$$

$$= a(T - 273) + \tfrac{1}{2}b(T - 273)^2, \tag{14.1-2}$$

where a and b are called *thermoelectric constants* and T is the temperature of the hot junction on Kelvin scale.

At neutral temperature t_n the emf is maximum. Therefore,

$$\left.\frac{d\mathcal{E}}{dt}\right|_{t_n} = 0 \quad \text{or} \quad a + bt_n = 0 \quad \text{or} \quad t_n = -a/b. \tag{14.1-3}$$

At the inversion temperature t_i the emf $\mathcal{E} = 0$. Therefore, from (14.1-1),

$$at_i + \tfrac{1}{2}bt_i^2 = 0.$$

Since $t_i = 0$ implies that both the junctions are at the same temperature, the inversion temperature is given by

$$t_i = -2a/b. \tag{14.1-4}$$

Thermoelectric power

The rate of change of thermo-emf of a thermocouple with the change in temperature of one junction is called the *thermoelectric power.* It is determined by the slope of the thermoelectric curve (i.e., $\mathcal{E}$-T or $\mathcal{E}$-t curve) at any specified temperature. Thus, thermoelectric power P at temperature T is defined by

$$P = \left.\frac{d\mathcal{E}}{dT}\right|_T = \left.\frac{d\mathcal{E}}{dt}\right|_t.$$

If $\mathcal{E}$ is given by the Eq. (14.1-1) then

$$P = \frac{d\mathcal{E}}{dT} = \frac{d\mathcal{E}}{dt} = a + bt.$$

Thermoelectric series

The direction of thermoelectric current depends on the pair of metals forming the thermocouple. Seebeck performed a series of experiments with a large number of metals and arranged them in a particular order giving rise to a series called *thermoelectric series.* The series is such that when two metals of the series are used to form a thermocouple then thermoelectric current flows from one earlier to the one occurring latter in the series through the cold junction. The series reads as,

Antimony, Fe · ·Zn · ·Ag · ·Au · ·Sn · ·Pb · ·Cu · ·Pt · ·Ni · ·Bismuth

Thus, in an Fe-Cu couple current flows from Fe to Cu through cold junction (Fig. 14.1-1). It may be helpful to remember that current flows from A (Antimony) to B (Bismuth) through C (cold) junction. It is found that the more the metals are separated in the series the higher is the thermo-emf produced for a given temperature difference between the junctions.

14.2 Peltier Effect

In 1834 Peltier discovered that if a current is made to flow through a thermocouple by using an external source then one junction gets heated whereas the other gets cooled. This phenomenon is known as *Peltier effect.* The effect is just the reverse of the Seebeck effect. The heating or cooling at the two junctions are reversed on reversal of the direction of the current. The rate of heating or cooling at a junction is found to be proportional to the strength of the current.

In an Fe-Cu couple we have seen that thermo-current flows from Fe to Cu through cold junction due to Seebeck effect (Fig. 14.1-1). Now suppose that to begin with both the junction are kept at the same temperature and by using a cell current is driven in the same direction as the Seebeck current. It is found that the junction which was hot during Seebeck current flow now gets cooled and the other junction which was cold during Seebeck current flow now becomes hot.

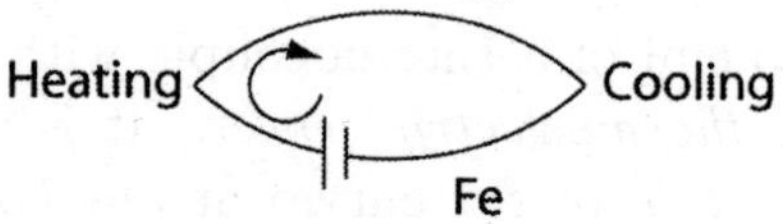

Fig 14.2-1

Difference between Peltier effect and Joule effect

1. In Joule effect there is always heating but Peltier effect produces both heating and cooling.
2. Joule heating is proportional to the square of the current whereas heat generated or absorbed in Peltier effect is proportional to the current.
3. Joule heating is an irreversible phenomenon whereas Peltier effect is reversible.
4. Joule heating occurs at all parts of the circuit, whereas the Peltier effect is restricted to the junctions only.
5. Joule heating depends on the resistance of the conductor but Peltier effect has nothing to do with the resistance.

Peltier coefficient (π)

The amount of energy absorbed or produced at a junction of a thermocouple when a unit charge passes across the junction is known as *Peltier coefficient.* Its unit is joule/coulomb, i.e., volt. Its value depends on the nature of the two metals forming the couple and also on the temperature of the junction.

14.3 Thomson Effect

If two points of a conductor of same material are maintained at different temperatures then there exists an emf between these points. So there will be evolution or absorption of heat in excess of Joule heat when a direct current passes through a non-uniformly heated conductor. This effect was first predicted by Thomson and is known as *Thomson effect.* It is also a reversible phenomenon.

If in a conductor we consider two points with temperatures T and $T+dT$ then the pd between these two points is given by σdT, where σ is called *Thomson coefficient.* Thus, Thomson coefficient of a metal may be defined as the pd between two points having unit difference of temperature. Kelvin called it as the *specific heat of electricity.* Its unit is $V/°C$. It is not constant for a particular metal but depends on its temperature. The coefficient σ can be positive or negative.

If σ is positive then Thomson effect is said to be positive. It happens when a point at a higher temperature is at a higher potential than a point at a lower temperature.

On the other hand, if σ is negative Thomson effect is said to be negative. This happens when a point at a higher temperature is at a lower potential than a point at a lower temperature. If a current flows in an unequally heated conductor from the point of lower temperature to the point of higher temperature heat is absorbed from the lattice for positive σ. If current flows in the direction of decreasing temperature then heat is evolved for positive σ. The opposite is the situation when σ is negative (Fig. 14.3-1).

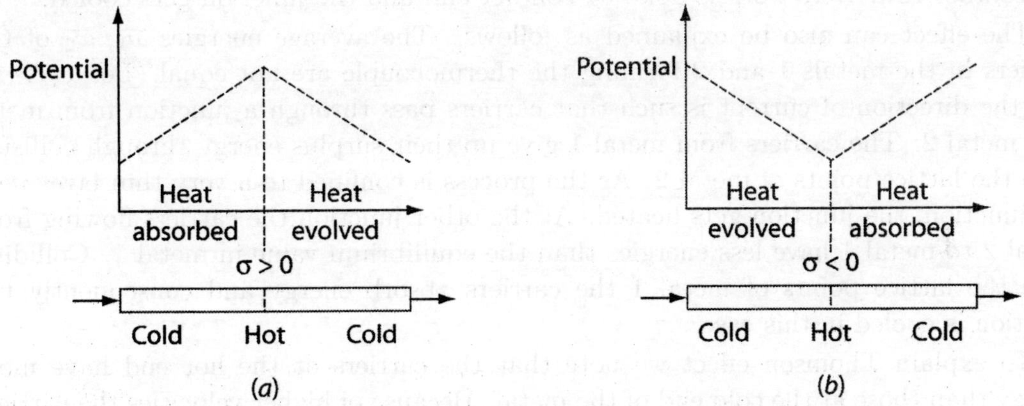

Fig. 14.3-1 : (a) Positive Thomson effect, (b) Negative Thomson effect

Experiments show that σ is positive for metals like Cu, Zn, Ag, Cd, etc. and negative for metals like Fe, Pt, Ni, Bi, etc. For lead σ is almost zero.

14.4 Origin of Thermoelectric Effects

The origin of the thermoelectric effects can be understood in the following way. The concentration of free electrons is different in different metals. So when two dissimilar metals are placed in contact with each other free electrons from the metal having higher electron concentration diffuse into the other. In this process one metal gets positively charged and the other negatively. As a result, an emf is built up across the junction, which tries to prevent further electron diffusion.

In the equilibrium condition there is no net electron flow across the junction and an emf known as *contact emf* exists across the junction. This equals the difference in work functions of the two metals. If the two junctions of a thermocouple are at the same temperature the contact emfs at the two junctions are equal and act in the oppositive sense and hence net emf becomes zero. However, the contact emf depends slightly on temperature and a net emf is produced when the junctions are at different temperatures. There are also other causes of thermo-emf. The current carriers at the hot end of the metal have more energy than those at the cold end. Because of their higher velocities they diffuse down the metal from hot end to cold end more rapidly than those from the cold end. This gives rise to the volume component of thermo-emf. At low temperatures the phonons (quantized lattice vibrations) move predoninantly from hot end to cold end. They interact with the free electron exert drag on them and give rise to another component of thermo-emf.

To explain Peïtier effect we suppose that an external battery is being used to send current through a thermocouple whose junctions are initially at the same temperature. Current flows through one junction against the contact emf, work is done against the contact emf and the junction is heated. At the other junction the current flows down the contact emf. Here work is done by contact emf and the junction gets cooled.

The effect can also be explained as follows. The average energies $\overline{\omega_1}$, $\overline{\omega_2}$ of the carriers in the metals 1 and 2 forming the thermocouple are not equal. Let $\overline{\omega_1} > \overline{\omega_2}$ and the direction of current is such that carriers pass through a junction from metal 1 to metal 2. The carriers from metal 1 give up their surplus energy through collision with the lattice points of metal 2. As the process is confined to a very thin layer near the junction, the junction gets heated. At the other junction the carriers flowing from metal 2 to metal 1 have less energies than the equilibrium value in metal 1. Colliding with the lattice points of metal 1 the carriers absorb energy and consequently the junction is cooled in this case.

To explain Thomson effect we note that the carriers at the hot end have more energy than those at the cold end of the metal. Because of higher velocities the carriers from the hot end diffuse down the metal more rapidly than those from the cold end. These carriers give up their excess energy on making collisions with the lattice in the colder part and causes evolution of Thomson heat. However, this simple theory cannot explain why Thomson effect is negative for some metals and positive for some others.

Using *band theory of solids* we can get a plausible explanation of it. Electrons from the hot end of the metal have higher energies (E) and also higher wave number (k) of the associated wave. But their velocities depend on the shape of $E - k$ curve near the top of a band. If d^2E/dk^2 is negative then the velocity. $v = (2\pi/h)(dE/dk)$, of the electrons from the hot end will be lower than that of the electrons from the cold end, instead of being greater. The net effect in this case, is that the electrons diffuse more rapidly from cold end to the hot end setting up an emf of opposite sign to that we should expect from free electrons.

14.5 Application of Thermodynamics to Thermoelectric Effects

Let us consider a thermocouple made of two metals A and B with one junction 1 at temperature T_1 and the other junction 2 at temperature T_2 ($T_2 > T_1$). Let σ_a and σ_b be the Thomson coefficients of the metals A and B; π_1 and π_2 be the Peltier coefficients at temperatures T_1 and T_2, respectively. Due to Seebeck effect a current flows in the circuit. Due to Peltier effect heat is absorbed at the hot junction and evolved at the cold junction and due to Thomson effect heat is absorbed and evolved all along the conductors. So the source of energy in a thermocouple may be regarded as due to heat absorbed being greater than the heat evolved. If we consider an infinitesimally small current flowing through the circuit then irreversible effects such as Joule heating may be neglected. Since the thermoelectric effects are reversible in such a situation a thermoelectric circuit may be regarded as a reversible heat engine.

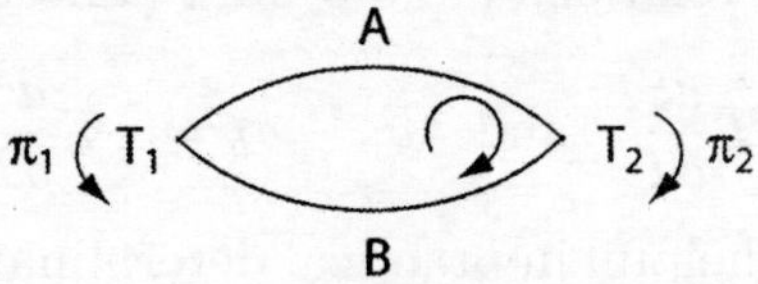

Fig 14.5-1

Let the current passes from A to B through the hot junction. The energy gained by unit charge in passing round the circuit at an infinitesimally slow rate equals the Seebeck emf $\mathcal{E}$ in the circuit. Thus,

$$\mathcal{E} = \int_{T_1}^{T_2} \sigma_a dT + \pi_2 + \int_{T_2}^{T_1} \sigma_b dT - \pi_1 = \pi_2 - \pi_1 + \int_{T_1}^{T_2} (\sigma_a - \sigma_b) dT. \qquad (14.5\text{-}1)$$

From second law of thermodynamics total change in entropy in traversing one cycle of a perfectly reversible engine is zero. Therefore,

$$\frac{\pi_2}{T_2} - \frac{\pi_1}{T_1} + \int_{T_1}^{T_2} \frac{\sigma_a - \sigma_b}{T} dT = 0. \qquad (14.5\text{-}2)$$

Considering the temperature T_1 of the cold junction as constant and the temperature T_2 of the hot junction as variable (i.e., π_1 constant and π_2 variable) we get from Eq. (14.5-1) after differentiation with respect to T_2,

$$\frac{d\mathcal{E}}{dT_2} = \frac{d\pi_2}{dT_2} + (\sigma_a - \sigma_b). \qquad (14.5\text{-}3)$$

Also differentiating Eq. (14.5-2) with respect to T_2 we get

$$-\frac{\pi_2}{T_2^2} + \frac{1}{T_2}\frac{d\pi_2}{dT_2} + \frac{\sigma_a - \sigma_b}{T_2} = 0 \quad \text{or} \quad \frac{d\pi_2}{dT_2} + (\sigma_a - \sigma_b) = \frac{\pi_2}{T_2}. \qquad (14.5\text{-}4)$$

Eliminating $\sigma_a - \sigma_b$ from Eqs. (14.5-3) and (14.5-4) we get

$$\frac{d\mathcal{E}}{dT_2} = \frac{\pi_2}{T_2} \quad \text{or} \quad \pi_2 = T_2 \frac{d\mathcal{E}}{dT_2}. \qquad (14.5\text{-}5)$$

Differentiating Eq. (14.5-5) with respect to T_2 we can write

$$\frac{d\pi_2}{dT_2} = T_2\frac{d^2\mathcal{E}}{dT_2^2} + \frac{d\mathcal{E}}{dT_2}.$$

Substituting $\frac{d\mathcal{E}}{dT_2}$ from Eq. (14.5-3) we get

$$\frac{d\pi_2}{dT_2} = T_2\frac{d^2\mathcal{E}}{dT_2^2} + \frac{d\pi_2}{dT_2} + (\sigma_a - \sigma_b) \quad \text{or} \quad \sigma_a - \sigma_b = -T_2\frac{d^2\mathcal{E}}{dT_2^2}. \tag{14.5-6}$$

For any value T of T_2 the relations (14.5-5) and (14.5-6) take the form

$$\pi = T\frac{d\mathcal{E}}{dT} \quad \text{and} \quad \sigma_a - \sigma_b = -T\frac{d^2\mathcal{E}}{dT^2}. \tag{14.5-7}$$

The relations (14.5-7) are helpful in practical determination of the Peltier coefficient π and the difference between the two Thomson coefficients, $\sigma_a - \sigma_b$. Experimentally the thermo-emf $\mathcal{E}$ is plotted as a function of the temperature difference. From the graph we can find the slope $d\mathcal{E}/dT$ and the second derivative $d^2\mathcal{E}/dT^2$ and hence, π and $\sigma_a - \sigma_b$ can be determined.

Discussions

The above analysis based on Kelvin's original theory, neglects the irreversible Joule heating and also the irreversible heat conduction due to temperature gradient. One can justify neglecting heat conduction by assuming sufficiently thin conductors. But this attempt is automatically offset by the consequent increase in resistance and hence Joule heating. This drawback has been removed in theory involving statistical methods, which employ Onsager's principle of microscopic reversibility. A rigorous treatment confirms that Eqs. (14.5-7) are valid. They have been tested experimentally over a large temperature range for many substances.

14.6 Experimental Laws of Thermoelectric Circuits

There are two simple experimentally established laws of thermoelectric effects:

1. Law of intermediate temperatures or Law of successive temperatures

It states that for a given thermocouple the thermo-emf for any specified temperature difference is equal to the sum of the emfs for any number of smaller intervals into which the given range of temperature may be subdivided. Symbolically it may be expressed as

$$\mathcal{E}_{T_1}^{T_n} = \mathcal{E}_{T_1}^{T_2} + \mathcal{E}_{T_2}^{T_3} + \mathcal{E}_{T_3}^{T_4} + \cdots + \mathcal{E}_{T_{n-1}}^{T_n}, \tag{14.6-1}$$

where $\mathcal{E}_{T_i}^{T_j}$ is the thermo-emf for a temperature difference $T_j - T_i$ and $T_1 < T_2 < T_3 < \ldots < T_n$.

This law can be easily established by using Eq. (14.5-1).

$$\mathcal{E}_{T_1}^{T_2} = \pi_2 - \pi_1 + \int_{T_1}^{T_2} (\sigma_a - \sigma_b) dT$$

$$\mathcal{E}_{T_2}^{T_3} = \pi_3 - \pi_2 + \int_{T_2}^{T_3} (\sigma_a - \sigma_b) dT$$

....

$$\mathcal{E}_{T_{n-1}}^{T_n} = \pi_n - \pi_{n-1} + \int_{T_{n-1}}^{T_n} (\sigma_a - \sigma_b) dT.$$

Adding all these equations we get

$$\mathcal{E}_{T_1}^{T_2} + \mathcal{E}_{T_2}^{T_3} + \cdots + \mathcal{E}_{T_{n-1}}^{T_n} = \pi_n - \pi_1 + \int_{T_1}^{T_n} (\sigma_a - \sigma_b) dT = \mathcal{E}_{T_1}^{T_n}.$$

2. Law of intermediate metals

It states that the insertion of an additional metal into any thermoelectric circuit does not alter the effective thermo-emf of the circuit provided that the additional metal is at the same temperature as that of the point of insertion.

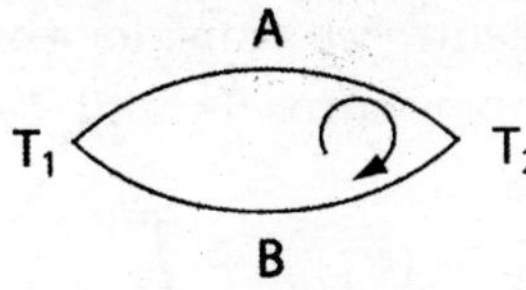

Fig 14.6-1

To prove the law let us consider a thermocouple (Fig. 14.6-1) made of two metals A and B with their junctions at temperatures T_1 and $T_2 (T_2 > T_1)$. The Seebeck emf in the circuit is

$$\mathcal{E}_{ab} = (\pi_2)_{ab} - (\pi_1)_{ab} + \int_{T_1}^{T_2} (\sigma_a - \sigma_b) dT. \qquad (14.6\text{-}2)$$

Now let a third metal C is inserted between A and B at the junction of temperature T_1 (Fig. 14.6-2). As there is no temperature difference between the ends of C total Seebeck emf in the circuit is now given by

$$\mathcal{E}_{abc} = (\pi_2)_{ab} + \int_{T_2}^{T_1} \sigma_b dT + (\pi_1)_{bc} + 0 + (\pi_1)_{ca} + \int_{T_1}^{T_2} \sigma_a dT$$

$$= (\pi_2)_{ab} + + [(\pi_1)_{bc} + +(\pi_1)_{ca}] + \int_{T_1}^{T_2} (\sigma_a - \sigma_b) dT. \qquad (14.6\text{-}3)$$

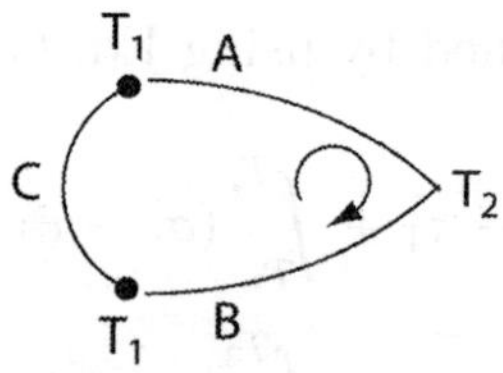

Fig 14.6-2

To evaluate the second term on the right-hand side of the above equation we consider the circuit of Fig. 14.6-2 with all the points of the circuit being at the same temperature $T_1 = T_2$. In this situation net emf in the circuit must be zero, for otherwise a current would flow and energy be dissipated without any source of energy being present. Obviously, this is not permitted by principle of conservation of energy. Thus,

$$(\pi_1)_{ab} + (\pi_1)_{bc} + (\pi_1)_{ca} = 0 \tag{14.6-4}$$

$$\text{or} \quad (\pi_1)_{bc} + (\pi_1)_{ca} = -(\pi_1)_{ab}. \tag{14.6-5}$$

So from Eq. (14.6-3) we get

$$\mathcal{E}_{abc} = (\pi_2)_{ab} - (\pi_1)_{ab} + \int_{T_1}^{T_2} (\sigma_a - \sigma_b)dT = \mathcal{E}_{ab} \text{ [Using Eq. (14.6-2)]}$$

This establishes the law of intermediate metals.

To derive another important result, we consider three thermocouples AB, BC and CA junctions of each being at temperatures T_1 and T_2, where $T_2 > T_1$. Then

$$\mathcal{E}_{ac} = (\pi_2)_{ac} - (\pi_1)_{ac} + \int_{T_1}^{T_2} (\sigma_a - \sigma_c)dT$$

$$\mathcal{E}_{bc} = (\pi_2)_{bc} - (\pi_1)_{bc} + \int_{T_1}^{T_2} (\sigma_b - \sigma_c)dT.$$

$$\therefore \quad \mathcal{E}_{ac} - \mathcal{E}_{bc} = (\pi_2)_{ac} - (\pi_2)_{bc} + (\pi_1)_{bc} - (\pi_1)_{ac} + \int_{T_1}^{T_2} (\sigma_a - \sigma_b)dT.$$

From Eq. (14.6-4) and a similar result for π_2 we have

$$(\pi_1)_{bc} - (\pi_1)_{ac} = -(\pi_1)_{ab} \quad \text{and} \quad (\pi_2)_{bc} - (\pi_2)_{ac} = -(\pi_2)_{ab}. \tag{14.6-6}$$

Therefore,

$$\mathcal{E}_{ac} - \mathcal{E}_{bc} = (\pi_2)_{ab} - (\pi_1)_{ab} + \int_{T_1}^{T_2} (\sigma_a - \sigma_b)dT = \mathcal{E}_{ab}. \tag{14.6-7}$$

Thus, if the thermo-emfs for each of two metals in combination with a third one are known then the thermo-emf of the thermocouple formed by combining the two metals is obtained by subtraction as given by the Eq. (14.6-7). Usually the thermo-emf for various metals are measured and tabulated with respect to lead (Pb) as standard. Lead is taken as standard because Thomson effect in lead is either zero or negligibly small. From the table we can find the thermo-emf for any two metals by using the relation (14.6-7). Thus, $\mathcal{E}_{ab} = \mathcal{E}_{a-Pb} - \mathcal{E}_{b-Pb}$. The *sign* convention is such that $\mathcal{E}_{a-Pb}$ is positive if conventional current flows from metal A to Pb through the cold junction.

The law of intermediate metals has very important consequences. Introduction of measuring instruments and connecting wires in a thermoelectric circuit does not influence the emf obtained provided that the temperatures are the same at each end of the instruments or wires. Also, the wires to form a junction may be soldered without changing the thermo-emf provided that the whole junction together with the solder as the intermediate metal is at the same temperature.

14.7 Thermoelectric Diagram

Various thermoelectric effects in a thermocouple can be represented on a diagram obtained by plotting thermoelectric power as a function of the temperature T of the hot junction of the couple, keeping the cold junction at a fixed temperature. Such diagrams were first constructed by Tait and are known as *Tait diagrams* or *thermoelectric diagram.*

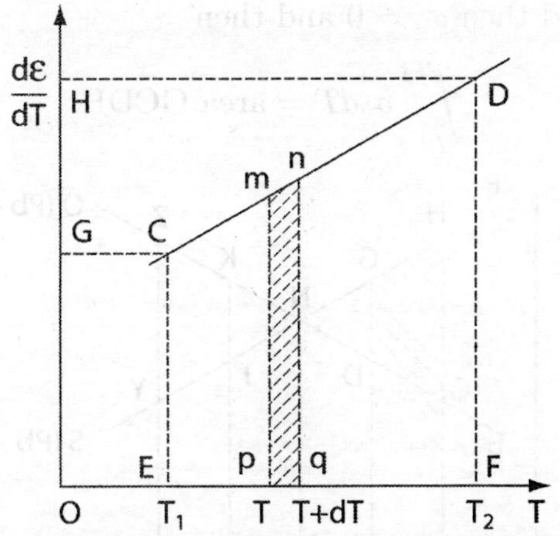

Fig.14.7-1 : Thermoelectric diagram

In most cases the Seebeck emf is almost a parabolic function of temperature and hence, thermoelectric diagram will be almost straight lines. Suppose Fig. 14.7-1 represents the thermoelectric diagram of a metal A with respect to metal B. Tait has shown that it is possible to calculate the thermo-emf produced in a thermocouple in terms of area under the thermoelectric diagram. Let us consider a small strip *pqnm* of thickness dT at a temperature T. Then the area of the strip is

$$mp \times pq = \left.\frac{d\mathcal{E}}{dT}\right|_T \cdot dT = d\mathcal{E},$$

which equals the thermo-emf available in the couple with the junctions at temperatures T and $T + dT$. Therefore, the area $EFDC$ represents the total Seebeck emf available with the junctions at temperatures T_1 and T_2. Peltier coefficient π_1 at temperature T_1 is given by

$$\pi_1 = T_1 \left.\frac{d\mathcal{E}}{dT}\right|_{T_1} = OE \times EC = \text{area of the rectangle}\, OECG$$

Similarly, Peltier coefficient π_2 at temperature T_2 is

$$\pi_2 = T_2 \left.\frac{d\mathcal{E}}{dT}\right|_{T_2} = OF \times FD = \text{area of the rectangle}\, OFDH.$$

We have

$$\mathcal{E} = \pi_2 - \pi_1 + \int_{T_1}^{T_2} (\sigma_a - \sigma_b) dT$$

or $$\int_{T_1}^{T_2} (\sigma_b - \sigma_a) dT = \pi_2 - \pi_1 - \mathcal{E}$$

$$= \text{area OFDH} - \text{area OECG} - \text{area EFDC}$$

$$= \text{area GCDH}.$$

If the metal A be lead then $\sigma_a = 0$ and then

$$\int_{T_1}^{T_2} \sigma_b dT = \text{area GCDH}$$

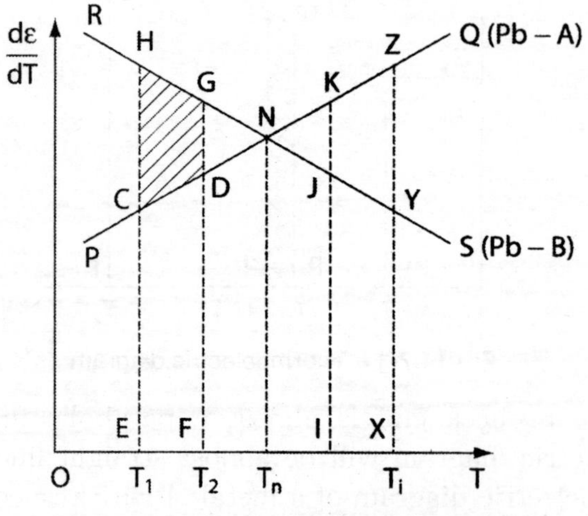

Fig.14.7-2 : Two thermoelectric power lines

Let us now consider thermoelectric power lines (PQ and RS) of two metals A and B both relative to lead (Pb) as reference metal (Fig. 14.7-2). The Seebeck emf generated in the couple A-B with junction temperature T_1 and T_2 is given by Eq. (14.6-7) as

$$\mathcal{E}_{ab} = \mathcal{E}_{Pb-b} - \mathcal{E}_{Pb-a} = \text{area EFGH} - \text{area EFDC} = \text{area CDGH}.$$

*This emf is such that the current flows from **a** to **b** through the hot junction*

Keeping T_1 fixed if T_2 is increased the area CDGH increases and so the Seebeck emf of the couple A-B increases till the point of intersection N of the two power lines is reached. At temperature T_n corresponding to the point N thermo-emf in the couple A-B becomes maximum. Hence, T_n is the neutral temperature of the thermocouple A-B. Beyond the point N the Seebeck emf $\mathcal{E}_{ab}$ decreases in magnitude. For example, at any temperature T beyond T_n,

$$\mathcal{E}_{ab} = \text{area EIJH} - \text{area EIKC} = \Delta CNH - \Delta JNK,$$

which is obviously less than ΔCNH. If for any temperature T_i,

$$\Delta CNH = \Delta YNZ$$

then $\mathcal{E}_{ab} = 0$. By definition this T_i is the inversion temperature. Thus, from the thermoelectric diagram we can find total thermo-emf, Peltier and Thomson coefficients, neutral and inversion temperatures, etc. In fact from this diagram we can get all the necessary informations about the thermocouple.

14.8 Applications of Thermoelectric Effects

Thermoelectric effects have many important practical applications. Some of the important applications are mentioned below:

1. ***Thermoelectric thermometers*:** The most common application of thermoelectric effects is the measurement of unknown temperatures, both low and high. One junction of a previously calibrated thermocouple is kept at a reference temperature and the other junction is placed in contact of the body whose temperature is to be measured. Measuring the thermo-emf generated it is possible to determine the unknown temperature.

2. ***Thermopile*:** A thermopile consists of a number of thermocouples usually of antimony-bismuth, connected in series. It can be used to detect small amount of radiant heat.

3. ***Boy's radio-micrometer*:** It is a combination of a thermocouple and a sensitive galvanometer. It is used as a sensitive radiant heat detector.

4. ***Duddell thermo-galvanometer*:** It is a modification of Boy's radiomicrometer and can measure small alternating or varying currents.

5. ***Thermo-milliammeter*:** Sir J. A. Fleming devised a sensitive form of ammeter by using thermoelectric effects. It is suitable for measuring small currents both dc and ac.

6. ***Thermoelectric refrigerators*:** Much higher thermo-emf is developed in semiconductor thermocouples. The cooling caused by Peltier effect in semiconductor thermocouples has been successfully used in the construction of refrigerators.

SOLVED PROBLEMS

▶ **Problem 1.** *The emf in a thermocouple with cold junction at $0°C$ and hot junction at $t°C$ is given by $\mathcal{E} = at + \frac{1}{2}bt^2$, where a and b are thermoelectric constants. Find the neutral and inversion temperatures, the Peltier and Thomson coefficients.*

Solution. In terms of absolute temperature we have thermo-emf

$$\mathcal{E} = a(T - 273) + \tfrac{1}{2}b(T - 273)^2.$$

At the neutral temperature T_n, $\mathcal{E}$ is maximum and hence,

$$\left.\frac{d\mathcal{E}}{dT}\right|_{T_n} = 0, \quad \text{or} \quad a + b\,(T_n - 273) = 0, \quad \text{or} \quad T_n - 273 = -\frac{a}{b}.$$

In °C, neutral temperature $t_n = -\frac{a}{b}$.

At the inversion temperature T_i we have $\mathcal{E} = 0$.

$$\therefore \quad a(T_i - 273) + \tfrac{1}{2}b(T_i - 273)^2 = 0.$$

As $T_i = 273$ implies that both the junctions are at the same temperature, the inversion temperature is given by

$$a + \tfrac{1}{2}b(T_i - 273) = 0 \quad \text{or} \quad T_i - 273 = -\frac{2a}{b}.$$

In °C, inversion temperature $t_i = -\frac{2a}{b}$.

The Peltier coefficient

$$\pi = T\frac{d\mathcal{E}}{dT} = T[a + b(T - 273)].$$

The difference of Thomsom coefficient for the metals forming the couple is

$$\sigma_a - \sigma_b = -T\frac{d^2\mathcal{E}}{dT^2} = -bT.$$

▶ **Problem 2.** *Thermo emf in a* Cu-Fe *couple working between* 0°C *and* t°C *is given by* $\mathcal{E} = at + \frac{1}{2}bt^2\ \mu\text{V}$. *The values of the thermoelectric constants a and b with respect to lead are given as*

$$a_{\text{Cu-Pb}} = 2.76\,\mu\text{V}/°\text{C} \quad b_{\text{Cu-Pb}} = 0.012\,\mu\text{V}/(°\text{C})^2$$

$$a_{\text{Fe-Pb}} = 16.65\,\mu\text{V}/°\text{C} \quad b_{\text{Fe-Pb}} = -0.030\,\mu\text{V}/(°\text{C})^2$$

Find the total thermo-emf in the couple for $t = 100°C$. Also, calculate the neutral and inversion temperatures, Peltier coefficients and the total Thomsom emf generated in the couple. **[cf. C.U. 2005]**

Solution. From the law of intermediate metals

$$\begin{aligned}\mathcal{E}_{\text{Cu-Fe}} &= \mathcal{E}_{\text{Cu-Pb}} + \mathcal{E}_{\text{Pb-Fe}} \\ &= \mathcal{E}_{\text{u-Pb}} - \mathcal{E}_{\text{Fe-Pb}} \\ &= a_{\text{Cu-Pb}}t + \tfrac{1}{2}b_{\text{Cu-Pb}}t^2 - a_{\text{Fe-Pb}}t - \tfrac{1}{2}b_{\text{Fe-Pb}}t^2 \\ &= (a_{\text{Cu-Pb}} - a_{\text{Fe-Pb}})\,t + \tfrac{1}{2}(b_{\text{Cu-Pb}} - b_{\text{Fe-Pb}})\,t^2 \\ &= at + \tfrac{1}{2}bt^2.\end{aligned}$$

∴ For the Cu-Fe couple,

$$a = a_{\text{Cu-Fe}} = a_{\text{Cu-Pb}} - a_{\text{Fe-Pb}} = 2.76 - 16.65 = -13.89\,\mu\text{V}/°\text{C}$$
$$b = b_{\text{Cu-Fe}} = b_{\text{Cu-Pb}} - b_{\text{Fe-Pb}} = 0.012 + 0.030 = 0.042\,\mu\text{V}/(°\text{C})^2$$

∴ The thermo-emf is

$$\begin{aligned}\mathcal{E}_{\text{Cu-Fe}} &= -13.89 \times 100 + \tfrac{1}{2} \times 0.042 \times 100^2\,\mu\text{V}, \\ &= -1.179\,\text{mV},\end{aligned}$$

where the negative sign indicates that the conventional current flows from Fe to Cu through the cold junction. The neutral temperature

$$t_n = -\frac{a}{b} = \frac{13.89}{0.042} = 330.7°\text{C}$$

The inversion temperature $t_i = -2a/b = 661.4°\text{C}$.

Peltier coefficient at temperature T is

$$\pi = T\frac{d\mathcal{E}}{dT} = T[a + b(T - 273)]$$

∴ Peltier coefficient at 100°C or 373 K,

$$\pi_{373} = 373[-13.89 + 0.042 \times 100] = -3614.37\,\mu\text{V}$$

and at 0°C or 273 K,

$$\pi_{273} = 273[-13.89 + 0.042(273 - 273)] = -3791.97\,\mu\text{V}.$$

The difference in Thomson coefficient is

$$\sigma_{\text{Cu}} - \sigma_{\text{Fe}} = -T\frac{d^2\mathcal{E}}{dT^2} = -bT.$$

Now the Thomson emf in the couple is

$$\int_{T_1}^{T_2} (\sigma_{\text{Cu}} - \sigma_{\text{Fe}})dT = -\int_{T_1}^{T_2} bT\,dT = -\frac{b}{2}\left(T_2^2 - T_1^2\right)$$
$$= -\frac{0.042}{2}\left(373^2 - 273^2\right)\ \mu\text{V}$$
$$= -1356.6\,\mu\text{V}.$$

Total thermo-emf in the couple can also be calculated as follows.

$$\mathcal{E} = \pi_{373} - \pi_{273} + \int_{273}^{373} (\sigma_{\text{Cu}} - \sigma_{\text{Fe}})dT$$
$$= -3614.37 + 3791.97 - 1356.6\,\mu\text{V}$$
$$= -1179\,\mu\text{V}$$
$$= -1.179\,\text{mV}$$

which is the same as calculated directly.

► **Problem 3.** *Thermoelectric power lines of Cu and Fe with respect to Pb are given by*

$$P_{\text{Cu}} = \frac{d\mathcal{E}}{dT} = \frac{d\mathcal{E}}{dt} = 2.76 + 0.012\,t\,\mu\text{V}/°\text{C}$$
$$\text{and}\quad P_{\text{Fe}} = 16.65 - 0.030t\,\mu\text{V}/°\text{C}.$$

Find the neutral temperature of the Cu-Fe couple. Also, calculate the emf in a Cu-Fe couple with the junctions at 0°C *and* 100°C.

Solution. We know that the temperature corresponding to the point of intersection of the power lines is the neutral temperature. Therefore, if t_n be the neutral temperature then

$$2.76 + 0.012t_n = 16.65 - 0.030t_n, \quad \text{or} \quad t_n = 330.7°\text{C}.$$

The emf,

$$\mathcal{E}_{\text{Cu-Fe}} = \mathcal{E}_{\text{Cu-Pb}} - \mathcal{E}_{\text{Fe-Pb}}$$

$$\therefore \quad \frac{d\mathcal{E}_{\text{Cu-Fe}}}{dt} = \frac{d\mathcal{E}_{\text{Cu-Fe}}}{dT} = P_{\text{Cu}} - P_{\text{Fe}} = -13.89 + 0.042t\,\mu\text{V}/°\text{C}$$

$$\therefore \quad \mathcal{E}_{\text{Cu-Fe}} = \int_0^{100} (-13.89 + 0.042t)dt = -13.89 \times 100 + \frac{0.042 \times 100^2}{2}\,\mu\text{V}$$
$$= -1179\,\mu\text{V}$$

▶ **Problem 4.** *A certain thermocouple with one junction at 0°C and the other junction at t°C has an emf given by* $\mathcal{E} = \alpha t - \beta\left(1 - e^{-\gamma t}\right)$, *where* α, β *and* γ *are constants. Determine the Peltier coefficient at 0°C and show that the difference between the Thomson coefficients is practically stationary in the neighbourhood of the temperature* $(1/\gamma - 273)$°C.

Solution. $\dfrac{d\mathcal{E}}{dT} = \dfrac{d\mathcal{E}}{dt} = \alpha - \beta\gamma e^{-\gamma t}$ and $\dfrac{d^2\mathcal{E}}{dT^2} = \dfrac{d^2\mathcal{E}}{dt^2} = \beta\gamma^2 e^{-\gamma t}$.

Now Peltier coefficient at 0°C is

$$\pi = \left(T\frac{d\mathcal{E}}{dT}\right)_{0^\circ\mathrm{C}} = 273(\alpha - \beta\gamma).$$

The difference between the Thomson coefficients is

$$\sigma_a - \sigma_b = -T\frac{d^2\mathcal{E}}{dT^2} = -(t + 273)\beta\gamma^2 e^{-\gamma t}.$$

For $\sigma_a - \sigma_b$ to be stationary we must have

$$\frac{d}{dt}(\sigma_a - \sigma_b) = 0$$

$$\text{or} \quad -\beta\gamma^2 e^{-\gamma t} - (t + 273)\beta\gamma^2 \cdot (-\gamma)e^{-\gamma t} = 0$$

$$\text{or} \quad t = \left(\frac{1}{\gamma} - 273\right)^\circ\mathrm{C}$$

▶ **Problem 5.** *The emf in a thermocouple with one junction at 0°C and the other at t°C is given by* $\mathcal{E} = at + \frac{1}{2}bt^2$. *Show that the Peltier coefficient can be expressed as* $\pi = \alpha + \beta t + \gamma t^2$, *where* α, β *and* γ *are constants. Determine the values of these constants in terms of a and b.* **[cf. C.U. 1990]**

Solution. Peltier coefficient

$$\pi = T\frac{d\mathcal{E}}{dT} = T\frac{d\mathcal{E}}{dt} = T(a + bt) = (t + 273)(a + bt)$$

$$= 273a + (a + 273b)t + bt^2 = \alpha + \beta t + \gamma t^2,$$

where $\alpha = 273a$, $\beta = (a + 273b)$ and $\gamma = b$.

▶ **Problem 6.** *The emf in a thermocouple with cold junction at 0°C and hot junction at t°C is given by* $\mathcal{E} = at + bt^2$, *where* $a = 10\,\mu\mathrm{V}/^\circ\mathrm{C}$ *and* $b = -\frac{1}{40}\,\mu\mathrm{V}/^\circ\mathrm{C}^2$. *Calculate the temperature at which Peltier coefficient is maximum. Also, calculate the temperature at which thermo-emf.* $\mathcal{E}$ *is solely due to Thomson effect.* **[cf. C.U. 2002]**

Solution. Peltier coefficient at temperature t°C is

$$\pi = T\frac{d\mathcal{E}}{dT} = T\frac{d\mathcal{E}}{dt} = T(a+2bt).$$

For π to be maximum,

$$\frac{d\pi}{dT} = \frac{d\pi}{dt} = 0, \text{ or, } T\cdot 2b + (a+2bt) = 0$$

$$\text{or, } (t+273)\cdot 2b + (a+2bt) = 0$$

$$\text{or, } t = -\frac{a}{4b} - \frac{273}{2} = 100 - 136.5 = -36.5°\text{C}.$$

At 0°C, $\pi = \pi_1 = 273 \times a$

At t°C, $\pi = \pi_2 = (t+273)(a+2bt)$

For $\mathcal{E}$ to be solely due to Thomson effect we must have

$$\pi_2 = \pi_1 \quad \text{or} \quad t = -\frac{a}{2b} - 273 = -73°\text{C}.$$

▶ **Problem 7.** *The thermoelectric power of iron with respect to lead is* 17.5 μV/°C *at* 0°C *and zero at* 360°C *that of copper is* 5 μV/°C *at* 450°C *and zero at* −50°C. *Draw the thermoelectric power lines of these metals. Hence, or otherwise calculate the emf of a* Cu-Fe *couple with the junctions at* 0°C *and at the neutral temperature.*

[cf. C.U. 2004]

Solution. Let the thermoelectric power of iron be given by

$$P_{\text{Fe}} = a_{\text{Fe-Pb}} + b_{\text{Fe-Pb}}t.$$

Using the conditions that $P_{\text{Fe}} = 17.5\,\mu\text{V/°C}$ at $t = 0$°C we get $a_{\text{Fe-Pb}} = 17.5\,\mu\text{V/°C}$. Now for $t = 360$°C, $P_{\text{Fe}} = 0$. Therefore,

$$b_{\text{Fe-Pb}} = -\frac{a_{\text{Fe-Pb}}}{360} = -0.0486\,\mu\text{V/°C}^2.$$

Similarly for copper, $a_{\text{Cu-Pb}} = 0.5\,\mu\text{V/°C}$ and $b_{\text{Cu-Pb}} = 0.01\,\mu\text{V/°C}^2$.

$$\therefore\ a_{\text{Cu-Fe}} = a_{\text{Cu-Pb}} - a_{\text{Fe-Pb}} = -17\,\mu\text{V/°C}$$

$$\text{and } b_{\text{Cu-Fe}} = b_{\text{Cu-Pb}} - b_{\text{Fe-Pb}} = 0.0586\,\mu\text{V/°C}^2.$$

The power lines are, therefore, given by

$$P_{\text{Fe}} = 17.5 - 0.0486t\,\mu\text{V/°C}$$

$$P_{\text{Cu}} = 0.5 + 0.01t\,\mu\text{V/°C}.$$

At neutral temperature t_n, $P_{\text{Fe}} = P_{\text{Cu}}$. Therefore,

$$17.5 - 0.0486t_n = 0.5 + 0.01t_n, \quad \text{or} \quad t_n = 290.04°\text{C}.$$

Now

$$a = a_{\text{Cu-Fe}} = a_{\text{Cu-Pb}} - a_{\text{Fe-Pb}} = -17\,\mu\text{V}/°\text{C}$$

$$\text{and} \quad b = b_{\text{Cu-Fe}} = b_{\text{Cu-Pb}} - b_{\text{Fe-Pb}} = 0.0586\,\mu\text{V}/°\text{C}^2$$

∴ The required emf

$$\mathcal{E}_{\text{Cu-Fe}} = at_n + \tfrac{1}{2}bt_n^2 = -17 \times 290.04 + \tfrac{1}{2} \times 0.0586 \times (290.04)^2$$

$$= -2465\,\mu\text{V}.$$

emf by using thermoelectric diagram

Fig. 14.P-7 shows the thermoelectric power lines of Cu and Fe. They intersect at the neutral point N. The magnitude of thermo-emf in a Cu-Fe couple with the junctions at 0°C and t_n°C is given by the area ANC. So the magnitude of the required emf is

$$\tfrac{1}{2}AC \times t_n = \tfrac{1}{2}(P_{\text{Fe}} - P_{\text{Cu}})_0 \times t_n$$

$$= \tfrac{1}{2}(17.5 - 0.5) \times 290.04\,\mu\text{V}$$

$$= 2465\,\mu\text{V}.$$

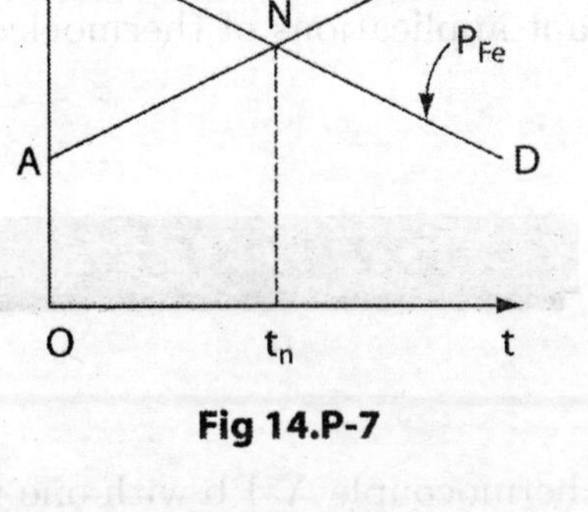

Fig 14.P-7

QUESTIONS

1. (a) What is thermoelectricity? Show graphically how thermo-emf varies with temperature. What do you mean by *neutral and inversion temperatures* of a thermocouple? What is thermoelectric power? What is its unit?
(b) What is thermoelectric series?

2. (a) What is Peltier effect? Differentiate between Peltier effect and Joule heating effect.

   (b) What do you mean by Peltier coefficient? Use thermodynamics to find a relationship between the Peltier coefficient of a junction of a thermocouple, the absolute temperature of the junction and the rate of change of thermo-emf with the change in temperature of the junction.

3. (a) State and explain the two laws of thermoelectricity. Mention the important consequence of the law of intermediate metals.

   (b) What are Peltier and Thomson coefficients?

   (c) Applying thermodynamics to a thermocouple establish the following relations:

$$\pi = T\frac{d\mathcal{E}}{dT} \quad \text{and} \quad \sigma_a - \sigma_b = -T\frac{d^2\mathcal{E}}{dT^2},$$

   where the symbols have their usual significance. **[N.B.U. 2017]**

4. What do you mean by Seebeck, Peltier and Thomson effects? What do you mean by positive and negative Thomson effects? Briefly explain the origin of these effects. Explain the significance of negative Thomson coefficient of a metal.

5. What is a thermoelectric diagram? Show how total Seebeck emf, Peltier and Thomson coefficients, neutral and inversion temperatures can be represented on this diagram.

6. Mention some important applications of thermoelectric effects.

EXERCISES

1. The thermo-emf in a thermocouple X-Pb with one junction at 0°C and the other at t°C is given by $\mathcal{E} = at + bt^2$, where a and b are constants. Find the Peltier coefficient at t°C and the Thomson coefficient of the metal X.

   [*Ans.* $(t + 273)(a + 2bt)$ and $2b(t + 273)$]

2. The thermoelectric power of copper with respect to lead is 2.76 μV/°C at 0°C and 3.96 μV/°C at 100°C. Corresponding quantities for iron are 16.65 μV/°C and 13.65 μV/°C. Calculate the emf of an Fe-Cu couple with junctions at 0°C and 100°C. [*Ans.* 1179 μV]

3. Thermoelectric power of silver and iron with respect lead are given by $3.34 + 0.008t\,\mu V/°C$ and $16.65 - 0.030t\,\mu V/°C$, respectively, where t is the temperature in °C. Calculate the emf of a Ag-Fe thermocouple with the junctions at 20°C and 80°C. [*Ans.* $-685\,\mu V$]

[**Hints.** $a_{Ag-Pb} = 3.34\,\mu V/°C$, $b_{Ag-Pb} = 0.008\,\mu V/°C^2$, $a_{Fe-Pb} = 16.65\,\mu V/°C$, $b_{Fe-Pb} = -0.030\,\mu V/°C^2$. Find a_{Ag-Fe}, b_{Ag-Fe} and then use $\mathcal{E}_{Ag-Pb} = a_{Ag-Fe}(t_2 - t_1) + \frac{1}{2}b_{Ag-Fe}(t_2^2 - t_1^2)$.]

4. A certain thermocouple with one junction at 0°C and the other junction at t°C has an emf given by $\mathcal{E} = [100t - 15000\{1 - \exp(-0.0025t)\}]\,\mu V$.

Determine the Peltier coefficient at 0°C and show that the difference between the Thomson coefficients is practically stationary in the neighbouring of 127°C. [*Ans.* 17.06 mV]

5. Thermo emf in a thermocouple with one junction at 0°C and the other at t°C is given by $\mathcal{E} = (17.35t - 0.025t^2)\,\mu V$. Find Seebeck emf in the couple for $t = 100°C$, Peltier coefficients and the difference of Thomson coefficients.

[*Ans.* $1485\,\mu V$, $\pi_0 = 4.74\,mV$, $\pi_{100} = 4.61 mV$, $\sigma_a - \sigma_b = 0.05T$.]

6. The thermo-emf in a thermocouple with cold junction at 0°C and hot junction at t°C is given by $\mathcal{E} = at + \frac{1}{2}bt^2$, where a and b are constants. Using the law of intermediate temperatures show that the emf of the couple with cold junction at t_1°C and hot junction at t_2°C is given by $\mathcal{E} = a(t_2 - t_1) + \frac{1}{2}b(t_2^2 - t_1^2)$.

7. Thermoelectric powers of two metals with respect to lead are given by $P_1 = 16 - 0.04t\,\mu V/°C$ and $P_2 = 2 + 0.01t\,\mu V/°C$, where t is the temperature in °C. Find the neutral temperature of the couple formed by the two metals. [*Ans.* 280°C]

8. Thermo-emf in a thermocouple with cold junction at 0°C and hot junction at t°C is given by $\mathcal{E} = 16t - 0.04t^2\,\mu V$. Find the neutral temperature, inversion temperature and the thermoelectric power at inversion temperature.

[*Ans.* 200°C, 400°C, $-16\,\mu V/°C$]

9. Two similar semi-circular rings, one of copper and the other of constantan, are joined together to form a ring of mean diameter 14 cm. Each metal has a cross-sectional area 1 cm^2. If the junctions are kept at temperatures 0°C and 100°C estimate the thermoelectric current in the ring. Take for the copper-constantan couple $a = 40.80\,\mu V/°C$, $b = 0.09\,\mu V/°C^2$; resistivity of copper $= 2 \times 10^{-8}\Omega\,m$ and resistivity of constantan $= 50 \times 10^{-8}\Omega\,m$. [*Ans.* 3.96 A]

10. For a certain thermocouple neutral temperature is 282°C and thermoelectric power at 100°C is $6.68\,\mu V/°C$. Find the thermoelectric power of the couple at 0°C. [*Ans.* $10.35\,\mu V/°C$]

OBJECTIVE TEST

1. Thermo emf developed in a thermocouple is measured by using a potentiometer and not by a moving coil millivoltmeter because: **[N.B.U. 2017]**

   (a) the potentiometer is more sensitive than the voltmeter,

   (b) the potentiometer makes measurement without drawing any current from the thermocouple,

   (c) measurement with a potentiometer is simpler than with a voltmeter,

   (d) none of these.

2. Heat developed in a circuit is found to be proportional to the current through it. Then which of the following statements is incorrect?

   (a) It may be due to Seebeck effect,

   (b) It may be due to Peltier effect,

   (c) It may be due to Thomson effect,

   (d) It may be due to Joule effect.

3. The Peltier coefficient π at temperature T is related to the thermo-emf by the relation

   (a) $\pi = -T\dfrac{d^2\mathcal{E}}{dT^2}$,

   (b) $\pi = T\dfrac{d\mathcal{E}}{dT}$,

   (c) $\pi = T^2\dfrac{d^2\mathcal{E}}{dT^2}$,

   (d) none of these.

4. Thomson coefficient σ for lead (Pb) is

   (a) almost zero,

   (b) negative,

   (c) positive,

   (d) none of these.

5. Thermoelectric diagram is a plot between

   (a) Thermo emf and temperature,

   (b) Thermoelectric power and temperature,

   (c) Peltier coefficient and temperature,

   (d) Thomson coefficient and temperature.

ANSWERS

1. (b) **2.** (d) **3.** (b) **4.** (a) **5.** (b)

Chapter **15**

Maxwell's Equations and Electromagnetic Wave

15.1 Introduction

So far in our study of electricity and magnetism we have encountered the following basic equations governing the electric and magnetic fields:

$$\vec{\nabla} \cdot \vec{E} = \frac{1}{\varepsilon_0}\rho \quad \text{(Gauss's law in electrostatics)} \tag{15.1-1}$$

$$\vec{\nabla} \cdot \vec{B} = 0 \quad \text{(Gauss's law in magnetostatics)} \tag{15.1-2}$$

$$\vec{\nabla} \times \vec{E} = -\frac{\partial \vec{B}}{\partial t} \quad \text{(Faraday's law)} \tag{15.1-3}$$

$$\vec{\nabla} \times \vec{H} = \vec{J} \quad \text{(Ampere's law)} \tag{15.1-4}$$

Faraday's law shows that a time varying magnetic field gives rise to an electric field. From symmetry consideration we may expect that a time varying electric field can create a magnetic field. In fact Maxwell proved it to be true. He modified the above field equations and removed the incompleteness of the interdependence of the field equations. The equations thus obtained are called *Maxwell's equations,* which govern the behaviour of the classical electromagnetic field as we believe it today.

In this chapter we shall consider the formulation of the Maxwell's equations and the electromagnetic wave phenomena predicted by these equations.

15.2 Generalisation of Ampere's Law and the Concept of Displacement Current

Maxwell pointed out that the Eqs. (15.1-1)–(15.1-4), representing the state of electromagnetism before Maxwell, are not consistent. As the divergence of curl of any vector is always zero, by taking divergence of Eq. (15.1-4) we find that $\vec{\nabla} \cdot \vec{J} = 0$. For steady currents this is true. But for time varying cases the continuity equation, namely,

$$\frac{\partial \rho}{\partial t} + \vec{\nabla} \cdot \vec{J} = 0 \quad \text{or} \quad \vec{\nabla} \cdot \vec{J} = -\frac{\partial \rho}{\partial t}$$

indicates that $\vec{\nabla} \cdot \vec{J}$ is not zero in general.

There is another way to prove the inconsistency of Ampere's law as expressed by the Eq. (15.1-4). Suppose we are in the process of charging a capacitor by a constant current I as shown in Fig. 15.2-1. The integral form of Ampere's law is

$$\oint_C \vec{H} \cdot d\vec{l} = \int_S \vec{J} \cdot d\vec{S} = I, \tag{15.2-1}$$

where I is the total current through the surface S whose periphery is the closed curve C. Now S may or may not be chosen to intersect the wire carrying current. In Fig. 15.2-1, the surface S_1 chosen in the plane of the curve C intersects the wire but the balloon-shaped surface S_2 does not intersect the wire. Thus,

$$\oint_C \vec{H} \cdot d\vec{l} = \int_{S_1} \vec{J} \cdot d\vec{S} = I \tag{15.2-2}$$

$$\text{and} \oint_C \vec{H} \cdot d\vec{l} = \int_{S_2} \vec{J} \cdot d\vec{S} = 0. \tag{15.2-3}$$

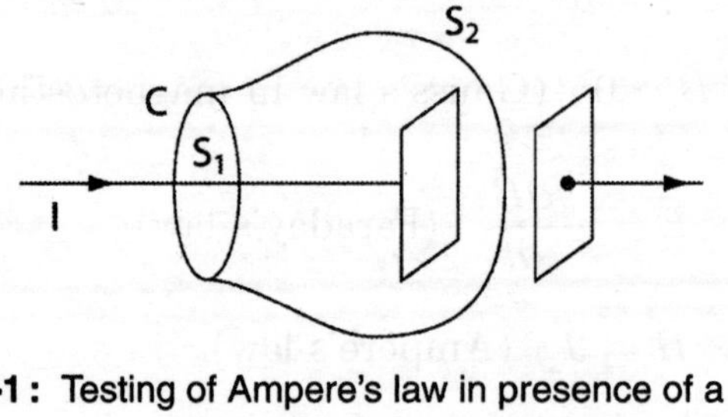

Fig. 15.2-1 : Testing of Ampere's law in presence of a capacitor

Obviously these equations are contradictory. Equation (15.2-2) is expected to be correct as it does not involve new feature namely the capacitor, while Eq. (15.2-3) is incorrect and requires consideration of the capacitor.

When the capacitor is being charged, charges are piling on the plates of the capacitor. This causes the charge density and associated electric field to vary with time. For such time varying fields Ampere's circuital law requires modification so as to make it consistent with the equation of continuity. Maxwell investigated the situation mathematically and assumed that the definition of current density $\vec{J}$ is incomplete and we must replace it by $\vec{J} + \vec{J}_d$, where $\vec{J}_d$ is to be so chosen as to make the total current divergenceless, i.e.,

$$\vec{\nabla} \cdot \left(\vec{J} + \vec{J}_d\right) = 0 \quad \text{or} \quad \vec{\nabla} \cdot \vec{J}_d = -\vec{\nabla} \cdot \vec{J} = \frac{\partial \rho}{\partial t} = \frac{\partial}{\partial t}\left(\vec{\nabla} \cdot \vec{D}\right) = \vec{\nabla} \cdot \frac{\partial \vec{D}}{\partial t}.$$

This gives

$$\vec{J}_d = \frac{\partial \vec{D}}{\partial t}. \tag{15.2-4}$$

Therefore, Ampere's law (15.1-4) modifies to

$$\vec{\nabla} \times \vec{H} = \vec{J} + \frac{\partial \vec{D}}{\partial t}. \tag{15.2-5}$$

Note that this generalised form of Ampere's law makes $\vec{\nabla} \cdot \left(\vec{\nabla} \times \vec{H}\right) = 0$ always.

Also, it does not change anything in so far as fields do not vary with time. For if $\vec{E}$ is steady, we still have $\vec{\nabla} \times \vec{H} = \vec{J}$ as Ampere said. The quantity $\partial\vec{D}/\partial t$ introduced by Maxwell in the Ampere's law is known as *displacement current density.* It is a current in the sense that it can produce a magnetic field. As it is not linked with the motion of free charges, it has none of the other properties of current. It can have a finite value even in vacuum, where there are no charges at all. It cannot be detected directly but its existence is confirmed indirectly by the electromagnetic theory of light.

15.3 Maxwell's Equations

There are four fundamental equations of electromagnetism known as *Maxwell's equations.* These equations represent the generalisation of experimental observations. In *SI* units the equations are

$$\vec{\nabla} \cdot \vec{D} = \rho \tag{15.3-1}$$

$$\vec{\nabla} \cdot \vec{B} = 0 \tag{15.3-2}$$

$$\vec{\nabla} \times \vec{E} = -\frac{\partial \vec{B}}{\partial t} \tag{15.3-3}$$

$$\vec{\nabla} \times \vec{H} = \vec{J} + \frac{\partial \vec{D}}{\partial t}, \tag{15.3-4}$$

where $\vec{E}$ and $\vec{D}$ are electric field vectors, $\vec{B}$ and $\vec{H}$ are magnetic field vectors, ρ is the free charge density and $\vec{J}$ is the free current density. Equation (15.3-1) is the differential form of Gauss's law in electrostatics and Eq. (15.3-2) is the corresponding law in magnetostatics. Equation (15.3-3) is the differential form of Faraday's law of electromagnetic induction. Equation (15.3-4) is the Maxwell's modification of Ampere's law.

Maxwell's equations are supplemented by the following two *constitutive relations,*

$$\vec{D} = \varepsilon \vec{E} \quad \text{and} \quad \vec{B} = \mu \vec{H}, \tag{15.3-5}$$

where ε is the permittivity and μ the permeability of surrounding medium.

Properties of Maxwell's equations

1. *Maxwell's equations are linear.* It is directly related to the principle of superposition. Thus, if any two fields satisfy Maxwell's equations their sum will also satisfy these equations.

2. *Maxwell's equations include the equation of continuity.* Taking the divergence of the Eq. (15.3-4) we get

$$\vec{\nabla} \cdot \vec{J} + \frac{\partial}{\partial t}\left(\vec{\nabla} \cdot \vec{D}\right) = 0$$

   Now using Eq. (15.3-1) we get the equation of continuity,

$$\frac{\partial \rho}{\partial t} + \vec{\nabla} \cdot \vec{J} = 0$$

   Thus, we conclude that Maxwell's equations are consistent with the local conservation of charge.

3. *Maxwell's equations are relativistic invariants.* That is, the form of the equations remains unchanged under Lorentz transformations. Thus, Maxwell's equations are correct relativistic equations.

4. *Maxwell's equations are not symmetric with respect to electric and magnetic fields.* This is due to the fact that electric charges exist in nature and magnetic charges as far as it is known at present, do not exist. Maxwell's equations take up symmetric form in free space with $\rho = 0$ and $\vec{J} = 0$.

5. *Maxwell's equations predict the existence of electromagnetic waves.* Any time variation of a magnetic field induces an electric field. Again a variation of electric field, in its turn, induces a magnetic field. This continuous interconversion or interaction of the fields preserves them and causes an electromagnetic wave propagating in space. It was confirmed by Hertz's experiment on electromagnetic radiation.

15.4 Conservation of Electromagnetic Energy —Poynting's Theorem

Poynting's theorem is a statement of conservation of energy applied to electromagnetic fields. It helps to interpret the flow of energy with the motion of electromagnetic waves in space. To establish the theorem let us start from the Maxwell's Eqs. (15.3-3) and (15.3-4). Now

$$\vec{\nabla}\cdot\left(\vec{E}\times\vec{H}\right)=\vec{H}\cdot\left(\vec{\nabla}\times\vec{E}\right)-\vec{E}\cdot\left(\vec{\nabla}\times\vec{H}\right)$$

$$=-\vec{H}\cdot\frac{\partial\vec{B}}{\partial t}-\vec{E}\cdot\vec{J}-\vec{E}\cdot\frac{\partial\vec{D}}{\partial t} \tag{15.4-1}$$

For any linear medium $\vec{D}=\varepsilon\vec{E}$ and $\vec{B}=\vec{\mu}H$. So we can rewrite Eq. (15.4-1) as

$$\vec{\nabla}\cdot\left(\vec{E}\times\vec{H}\right)=-\frac{\partial}{\partial t}\left(\frac{1}{2}\vec{E}\cdot\vec{D}+\frac{1}{2}\vec{B}\cdot\vec{H}\right)-\vec{E}\cdot\vec{J}. \tag{15.4-2}$$

Integrating this equation over a fixed volume V bounded by a closed surface S and applying Gauss's divergence theorem we get

$$\oint_S\left(\vec{E}\times\vec{H}\right)\cdot d\vec{S}=-\frac{d}{dt}\int_V\frac{1}{2}\left(\vec{E}\cdot\vec{D}+\vec{B}\cdot\vec{H}\right)dV-\int_V\vec{E}\cdot\vec{J}\,dV$$

$$\text{or}\quad -\frac{d}{dt}\int_V\frac{1}{2}\left(\vec{E}\cdot\vec{D}+\vec{B}\cdot\vec{H}\right)dV=\int_V\vec{E}\cdot\vec{J}\,dV+\oint_S\left(\vec{E}\times\vec{H}\right)\cdot d\vec{S} \tag{15.4-3}$$

To understand the physical significance of Eq. (15.4-3) let us interpret each term in it. Suppose we have some charge and current configuration which at time t produces the $\vec{E}$ and $\vec{B}$ fields. The rate of work done by the electromagnetic forces on an element of charge $dq=\rho\,dV$ is given by

$$\frac{dW}{dt}=dq\left(\vec{E}+\vec{v}\times\vec{B}\right)\cdot\vec{v}$$

$$=dq\vec{E}\cdot\vec{v}=\vec{E}\cdot\vec{v}(\rho\,dV)=\vec{E}\cdot\vec{J}\,dV,$$

where $\vec{v}$ is the velocity of the charge element and $\vec{J}=\rho\vec{v}$.

Thus, the term

$$\int_V\vec{E}\cdot\vec{J}\,dV$$

in Eq. (15.4-3) represents the rate of doing work on the charges in volume V by the electromagnetic field.

We know that $\frac{1}{2}\vec{E}\cdot\vec{D}$ is the electrostatic energy density and $\frac{1}{2}\vec{B}\cdot\vec{H}$ is the magnetostatic energy density. Hence,

$$\tfrac{1}{2}\left(\vec{E}\cdot\vec{D}+\vec{B}\cdot\vec{H}\right)$$

may be interpreted as the electromagnetic energy density.

Thus, the term

$$-\frac{d}{dt}\int_V\frac{1}{2}\left(\vec{E}\cdot\vec{D}+\vec{B}\cdot\vec{H}\right)dV$$

in Eq. (15.4-3) represents the rate at which the total electromagnetic energy in volume V is decreasing. Physical meaning of the term

$$\oint_S \left(\vec{E} \times \vec{H}\right) \cdot d\vec{S}$$

now follows from the principle of conservation of energy. *The rate of decrease of electromagnetic field energy within a certain volume is equal to the rate of work done by the field on the charges inside the given volume plus the rate of outflow of electromagnetic energy through the surface bounding the volume. This statement is known as Poynting' theorem.* So the term

$$\oint_S \left(\vec{E} \times \vec{H}\right) \cdot d\vec{S}$$

represents the rate of flow of electromagnetic energy outward through the surface S. The vector $\vec{s} = \vec{E} \times \vec{H}$ represents the amount of electromagnetic energy flowing out normally through unit area per unit time. This vector $\vec{s}$ is known as *Poynting's vector.*

Differential form of Poynting's theorem

The work done by the electromagnetic field on the charges increases their mechanical energy. If we denote the mechanical energy density by u_M then we can write $\frac{\partial u_M}{\partial t}$ = rate of work done on charges per unit volume $= \vec{E} \cdot \vec{J}$.

Now denoting the electromagnetic energy density $\frac{1}{2}\left(\vec{E} \cdot \vec{D} + \vec{B} \cdot \vec{H}\right)$ by u_{em} we can write Eq. (15.4-2) in the following form:

$$\frac{\partial}{\partial t}(u_M + u_{em}) + \vec{\nabla} \cdot \vec{s} = 0. \qquad (15.4\text{-}4)$$

This is the *differential form of Poynting's theorem.* This has the same form as the equation of continuity expressing the conservation of charge, with the total energy density taking the place of charge density ρ and $\vec{s}$ taking the place of current density $\vec{J}$. Therefore, from analogy with $\vec{J}$ the *Poynting's vector can be interpreted as the energy flowing through unit area per unit time.*

15.5 Wave Equation in Free Space

In free space, where there is no charge ($\rho = 0$) or current $\left(\vec{J} = 0\right)$ Maxwell's equations take the following form:

$$\vec{\nabla} \cdot \vec{E} = 0 \qquad (15.5\text{-}1)$$

$$\vec{\nabla} \cdot \vec{H} = 0 \qquad (15.5\text{-}2)$$

$$\vec{\nabla} \times \vec{E} = -\mu_0 \frac{\partial \vec{H}}{\partial t} \qquad (15.5\text{-}3)$$

$$\vec{\nabla} \times \vec{H} = \varepsilon_0 \frac{\partial \vec{E}}{\partial t} \qquad (15.5\text{-}4)$$

Taking curl of Eq. (15.5-3) we get

$$\vec{\nabla}\left(\vec{\nabla}\cdot\vec{E}\right)-\nabla^2\vec{E}=-\mu_0\frac{\partial}{\partial t}\left(\vec{\nabla}\times\vec{H}\right)$$

$$\text{or}\quad -\nabla^2\vec{E}=-\varepsilon_0\mu_0\frac{\partial^2\vec{E}}{\partial t^2}\quad \text{[Using Eqs. (15.5-1) and (15.5-4)]}$$

$$\therefore\quad \nabla^2\vec{E}-\varepsilon_0\mu_0\frac{\partial^2\vec{E}}{\partial t^2}=0. \tag{15.5-5}$$

Similarly taking curl of Eq. (15.5-4) and using Eqs. (15.5-2) and (15.5-3) we can get

$$\nabla^2\vec{H}-\varepsilon_0\mu_0\frac{\partial^2\vec{H}}{\partial t^2}=0. \tag{15.5-6}$$

Thus, both $\vec{E}$ and $\vec{H}$ satisfy the well-known wave equation. Both $\vec{E}$ and $\vec{H}$ propagate in free space in the form of a wave. Comparing with the standard wave equation

$$\nabla^2\psi-\frac{1}{c^2}\frac{\partial^2\psi}{\partial t^2}=0 \tag{15.5-7}$$

we can find that the velocity of propagation of electromagnetic wave is

$$c=\frac{1}{\sqrt{\varepsilon_0\mu_0}}\approx 3\times 10^8\ \text{m/s},$$

which is precisely the speed of light in free space. This indicates that the light is an electromagnetic wave.

Plane wave solutions

The simplest solution of Eqs. (15.5-5) and (15.5-6) are plane wave solutions:

$$\vec{E}\left(\vec{r},t\right)=\vec{E}_0e^{j\left(\vec{k}\cdot\vec{r}-\omega t\right)}\quad\text{and}\quad \vec{H}\left(\vec{r},t\right)=\vec{H}_0e^{j\left(\vec{k}\cdot\vec{r}-\omega t\right)}, \tag{15.5-8}$$

where $\vec{E}_0$ and $\vec{H}_0$ are complex amplitudes, which are constants in space and time; $\vec{k}$ is the wavevector determining the direction of propagation of the wave. $\vec{k}$ is defined as

$$\vec{k}=\frac{2\pi}{\lambda}\hat{n}=\frac{2\pi\nu}{c}\hat{n}=\frac{\omega}{c}\hat{n}, \tag{15.5-9}$$

where $\hat{n}$ is the unit vector along the direction of propagation.

Relative directions of $\vec{E}$, $\vec{H}$ and $\vec{k}$

Substituting the solutions (15.5-8) in Eqs. (15.5-1) and (15.5-2) we get[1]

$$\vec{k}\cdot\vec{E}=0\quad\text{and}\quad\vec{k}\cdot\vec{H}=0. \tag{15.5-10}$$

---

[1] Here operator $\vec{\nabla}$ is equivalent to $j\vec{k}$ while $\dfrac{\partial}{\partial t}$ is equivalent to $-j\omega$.

Thus, $\vec{E}$ and $\vec{H}$ are both perpendicular to the direction of propagation vector $\vec{k}$. This implies that *electromagnetic waves are transverse in nature.*

Again substituting the solutions (15.5-8) in Eqs. (15.5-3) and (15.5-4) we get

$$j\vec{k} \times \vec{E} = -\mu_0 \left(-j\omega\vec{H}\right) \text{ or } \vec{k} \times \vec{E} = \mu_0\omega\vec{H} \tag{15.5-11}$$

$$\text{and} \quad j\vec{k} \times \vec{H} = \varepsilon_0 \left(-j\omega\vec{E}\right) \text{ or } \vec{k} \times \vec{H} = -\varepsilon_0\omega\vec{E}. \tag{15.5-12}$$

Equation (15.5-11) implies that $\vec{H}$ is perpendicular to both $\vec{k}$ and $\vec{E}$. Equation (15.5-12) implies that $\vec{E}$ is perpendicular to both $\vec{k}$ and $\vec{H}$. Thus, the field vectors $\vec{E}$ and $\vec{H}$ are *mutually perpendicular and also they are perpendicular to the direction of propagation* $\vec{k}$. The vectors $\vec{E}$, $\vec{H}$ and $\vec{k}$ form a set of orthogonal vectors, which constitutes a right-handed system in that order.

Relative phase of $\vec{E}$ and $\vec{H}$

Substituting the solutions (15.5-8) in Eq. (15.5-5) or (15.5-6) we get

$$k^2 = \varepsilon_0\mu_0\omega^2 \tag{15.5-13}$$

Thus, in free space k is a real quantity and hence, Eq. (15.5-11) implies that $\vec{E}$ *and* $\vec{H}$ *are in phase.*

Wave impedance

The ratio of the magnitudes of $\vec{E}$ and $\vec{H}$ is

$$Z_0 = \frac{\left|\vec{E}\right|}{\left|\vec{H}\right|} = \frac{\mu_0\omega}{k} = \sqrt{\frac{\mu_0}{\varepsilon_0}} \quad \text{[Using equation (15.5-13)]} \tag{15.5-14}$$

This quantity Z_0 has the dimensions of impedance and is known as the *wave impedance* in free space. Its value is $376.6\,\Omega$.

Poynting's vector

The Poynting's vector for the plane electromagnetic wave in free space is

$$\begin{aligned} \vec{s} = \vec{E} \times \vec{H} &= \frac{1}{\mu_0\omega}\vec{E} \times \left(\vec{k} \times \vec{E}\right) \quad \text{[Using Eq. (15.5-11)]} \\ &= \frac{1}{\mu_0\omega}\left[\vec{k}\left(\vec{E}\cdot\vec{E}\right) - \vec{E}\left(\vec{E}\cdot\vec{k}\right)\right] \\ &= \frac{E^2}{\mu_0\omega}\vec{k} = \frac{E^2}{Z_0}\hat{n}. \end{aligned} \tag{15.5-15}$$

Thus, the energy flow is in the direction of wave propagation.

Since $\vec{E}$ is normal to $\vec{k}$, from Eq. (15.5-11) we can write in terms of magnitude,

$$kE = \mu_0 \omega H \quad \text{or} \quad \sqrt{\varepsilon_0}E = \sqrt{\mu_0}H \quad \text{or} \quad \tfrac{1}{2}\varepsilon_0 E^2 = \tfrac{1}{2}\mu_0 H^2. \tag{15.5-16}$$

This shows that *in case of electromagnetic waves in free space electromagnetic energy is equally shared between electric and magnetic fields.*

Total electromagnetic energy density is

$$u = \tfrac{1}{2}\varepsilon_0 E^2 + \tfrac{1}{2}\mu_0 H^2 = \varepsilon_0 E^2. \tag{15.5-17}$$

So Eq. (15.5-15) can also be written as

$$\vec{s} = uc\hat{n}, \tag{15.5-18}$$

where $c = 1/\sqrt{\mu_0\varepsilon_0} = \omega/k$ is the speed of the wave.

Thus, Poynting's vector equals the energy density (u) times the velocity of the wave. This means that the energy associated with the wave propagates with the same velocity with which the field vectors $\vec{E}$ and $\vec{H}$ propagate.

Electromagnetic momentum

An electromagnetic wave carries both energy and momentum. From quantum theory electromagnetic radiations consist of photons, each of energy $E = h\nu$ where ν is the frequency and h is Planck's constant. Photons move with speed c and hence if m is the mass of a moving photon then from mass energy equivalence $E = h\nu = mc^2$. So the momentum of a photon is $mc = \frac{h\nu}{c} = E/c$. So we can say that electromagnetic energy density u is associated with a momentum density $P_{\text{em}} = \frac{u}{c} = \frac{1}{c^2}|\vec{s}|$. In vector form

$$\vec{P}_{\text{em}} = \frac{1}{c^2}\vec{s} = \mu_0\varepsilon_0\vec{s} = \mu_0\varepsilon_0(\vec{E}\times\vec{H})$$

Time average Poynting's vector

So far we have freely used the complex solutions for the field vectors $\vec{E}$ and $\vec{H}$ with the understanding that the actual quantities are given by the real parts of the complex solutions. As $\vec{s}$ is non-linear in the fields it is essential to take real parts of the fields before multiplying them. Thus, the real Poynting's vector $\vec{s} = Re\vec{E} \times Re\vec{H}$.

Now

$$Re\vec{E} = \tfrac{1}{2}\left(\vec{E} + \vec{E}^*\right) \quad \text{and} \quad Re\vec{H} = \tfrac{1}{2}\left(\vec{H} + \vec{H}^*\right),$$

where * indicates complex conjugate.

Therefore, real Poynting's vector

$$\vec{s} = \frac{1}{4}\left[\vec{E}\times\vec{H} + \vec{E}^*\times\vec{H}^* + \vec{E}^*\times\vec{H} + \vec{E}\times\vec{H}^*\right]$$

As the origin is arbitrary we may calculate the representative values at $\vec{r} = 0$. Now writing $\vec{E} = \vec{E}_0 e^{-j\omega t}$ and $\vec{H} = \vec{H}_0 e^{-j\omega t}$ and averaging over a period $T = 2\pi/\omega$ we find that the first two terms on the right-hand side becomes zero, since

$$\int_0^T e^{\pm 2j\omega t} dt = 0.$$

Thus, the time average Poynting's vector

$$\langle \vec{s} \rangle = \tfrac{1}{4}\left[\vec{E} \times \vec{H}^* + \vec{E}^* \times \vec{H}\right] = \tfrac{1}{2} Re\left(\vec{E} \times \vec{H}^*\right) = \tfrac{1}{2} Re\left[\vec{E}^* \times \vec{H}\right]. \qquad (15.5\text{-}19)$$

The magnitude of the time average of the Poynting's vector is called the *intensity of radiation* (I). Thus, the intensity

$$I = |\langle \vec{s} \rangle| = \tfrac{1}{2} E_0 H_0 = \frac{E_0}{\sqrt{2}} \cdot \frac{H_0}{\sqrt{2}} = E_{\text{rms}} \times H_{\text{rms}} \qquad (15.5\text{-}20)$$

Similarly, we can show that the time average energy density

$$\begin{aligned} \langle u \rangle = \langle \varepsilon_0 E^2 \rangle &= \langle \varepsilon_0 \vec{E} \cdot \vec{E} \rangle \quad \text{[from Eq. (15.5-17)]} \\ &= \tfrac{1}{2}\varepsilon_0 Re \vec{E} \cdot \vec{E}^* = \tfrac{1}{2}\varepsilon_0 E_0^2 = \varepsilon_0 E_{\text{rms}}^2. \end{aligned} \qquad (15.5\text{-}21)$$

From Eq. (15.5-16) we note that $\sqrt{\varepsilon_0} E_{\text{rms}} = \sqrt{\mu_0} H_{\text{rms}}$. Hence, from (15.5-20) and (15.5-21) we can write

$$\frac{I}{\langle u \rangle} = \frac{1}{\varepsilon_0} \cdot \frac{H_{\text{rms}}}{E_{\text{rms}}} = \frac{1}{\sqrt{\varepsilon_0 \mu_0}} = c \quad \text{or} \quad I = \langle u \rangle c. \qquad (15.5\text{-}22)$$

Spherical wave solutions

Because of the vector nature of $\vec{E}$ and $\vec{H}$ it is very difficult to find the spherical wave solutions of the wave Eqs. (15.5-5) and (15.5-6). However, as each component of $\vec{E}$ and $\vec{H}$ vectors satisfies the scalar wave Eq. (15.5-7), some important informations about spherical waves can be obtained by solving the scalar wave Eq. (15.5-7) with the assumption that the field parameter ψ depends only on the radial coordinate r of spherical coordinate system. Then Eq. (15.5-7) reduces to

$$\frac{1}{r^2}\frac{\partial}{\partial r}\left(r^2 \frac{\partial \psi}{\partial r}\right) - \frac{1}{c^2}\frac{\partial^2 \psi}{\partial t^2} = 0.$$

Putting $\psi = \psi'/r$ we get

$$\frac{\partial^2 \psi'}{\partial r^2} - \frac{1}{c^2}\frac{\partial^2 \psi'}{\partial t^2} = 0.$$

This equation is of the form of a one dimensional wave equation. Hence, its general solution may be written as

$$\psi' = f_1(ct - r) + f_2(ct + r)$$

$$\text{or} \quad \psi = \frac{1}{r} f_1(ct - r) + \frac{1}{r} f_2(ct + r). \qquad (15.5\text{-}23)$$

First term on the right-hand side of Eq. (15.5-23) represents a spherical wave diverging from the origin of the coordinate system with a constant speed c and having an arbitrary functional form. The second term represents a similar wave converging to the origin. A diverging spherical harmonic wave can be represented by

$$\psi(r,t) = A \cdot \frac{1}{r} \cos(kr - \omega t) + B \frac{1}{r} \sin(kr - \omega t). \qquad (15.5\text{-}24)$$

Note that unlike plane waves, the amplitude of a spherical wave falls as $1/r$ and thereby changing its profile as it moves on. The wavefronts of such waves are concentric spheres. For enough away from the source a small portion of the wavefront may be considered as a portion of a plane wave.

15.6 Plane Electromagnetic Waves in an Isotropic Dielectric Medium

Let us consider a linear, homogeneous and isotropic dielectric (nonconducting) medium with no free charge. Maxwell's equations, in this case, take the following form:

$$\vec{\nabla} \cdot \vec{E} = 0 \qquad (15.6\text{-}1)$$

$$\vec{\nabla} \cdot \vec{H} = 0 \qquad (15.6\text{-}2)$$

$$\vec{\nabla} \times \vec{E} = -\mu \frac{\partial \vec{H}}{\partial t} \qquad (15.6\text{-}3)$$

$$\vec{\nabla} \times \vec{H} = \varepsilon \frac{\partial \vec{E}}{\partial t} \qquad (15.6\text{-}4)$$

Taking curl of Eq. (15.6-3) and using Eq. (15.6-4) we get

$$\vec{\nabla}\left(\vec{\nabla} \cdot \vec{E}\right) - \nabla^2 \vec{E} = -\mu \frac{\partial}{\partial t}\left(\vec{\nabla} \times \vec{H}\right) = -\varepsilon\mu \frac{\partial^2 \vec{E}}{\partial t^2}$$

Using Eq. (15.6-1) we get

$$\nabla^2 \vec{E} - \varepsilon\mu \frac{\partial^2 \vec{E}}{\partial t^2} = 0. \qquad (15.6\text{-}5)$$

Similarly staring from Eq. (15.6-4) we can show that

$$\nabla^2 \vec{H} - \varepsilon\mu \frac{\partial^2 \vec{H}}{\partial t^2} = 0. \qquad (15.6\text{-}6)$$

Thus, both $\vec{E}$ and $\vec{H}$ satisfy the well-known wave equation. Comparing with the standard wave equation we can say that both $\vec{E}$ and $\vec{H}$ propagate in the form of a wave with a speed

$$v = \frac{1}{\sqrt{\varepsilon\mu}} = \frac{1}{\sqrt{\varepsilon_0\varepsilon_r \cdot \mu_0\mu_r}} = \frac{c}{\sqrt{\varepsilon_r\mu_r}}, \tag{15.6-7}$$

where $c = 1/\sqrt{\varepsilon_0\mu_0}$ is the speed of electromagnetic waves in free space

ε_r = relative permittivity or dielectric constant (K)

μ_r = relative permeability of the medium.

For a nonmagnetic dielectric medium $\mu_r = 1$. Therefore,

$$v = \frac{c}{\sqrt{K}}.$$

Refractive index (n) of the medium is defined by

$$n = \frac{c}{v} = \sqrt{K}. \tag{15.6-8}$$

Plane wave solutions

The plane wave solutions of Eqs. (15.6-5) and (15.6-6) are given by

$$\vec{E}(\vec{r}, t) = \vec{E}_0 e^{j(\vec{k}\cdot\vec{r}-\omega t)} \quad \text{and} \quad \vec{H}(\vec{r}, t) = \vec{H}_0 e^{j(\vec{k}\cdot\vec{r}-\omega t)}, \tag{15.6-9}$$

where $\vec{E}_0$ and $\vec{H}_0$ are complex amplitudes, which are constants both in space and time, and $\vec{k}$ is the wavevector given by

$$\vec{k} = k\hat{n} = \frac{2\pi}{\lambda}\hat{n} = \frac{2\pi\nu}{c}\hat{n} = \frac{\omega}{c}\hat{n}, \tag{15.6-10}$$

where $\hat{n}$ is the unit vector in the direction of propagation.

Relative directions of $\vec{E}$, $\vec{H}$ and $\vec{k}$

Substituting the solutions (15.6-9) in Eqs. (15.6-1) and (15.6-2) we get

$$\vec{k} \cdot \vec{E} = 0 \quad \text{and} \quad \vec{k} \cdot \vec{H} = 0. \tag{15.6-11}$$

Thus, $\vec{E}$ and $\vec{H}$ are both perpendicular to the direction of propagation. *This implies that electromagnetic waves are transverse in nature.*

Again substituting the solutions (15.6-9) in Eqs. (15.6-3) and (15.6-4) we get

$$j\vec{k} \times \vec{E} = -\mu\left(-j\omega\vec{H}\right) \quad \text{or} \quad \vec{k} \times \vec{E} = \mu\omega\vec{H} \tag{15.6-12}$$

$$\text{and} \quad j\vec{k} \times \vec{H} = \varepsilon\left(-j\omega\vec{E}\right) \quad \text{or} \quad \vec{k} \times \vec{H} = -\varepsilon\omega\vec{E} \tag{15.6-13}$$

Equation (15.6-12) indicates that $\vec{H}$ is perpendicular to both $\vec{k}$ and $\vec{E}$. Equation (15.6-13) indicates that $\vec{E}$ is perpendicular to both $\vec{k}$ and $\vec{H}$.

Thus, the field vectors $\vec{E}$ *and* $\vec{H}$ *are mutually perpendicular and also they are perpendicular to the direction of propagation vector* $\vec{k}$. The vectors $\vec{E}$, $\vec{H}$ and $\vec{k}$ form a set of orthogonal vectors which constitute a right-handed system.

Relative Phase of $\vec{E}$ and $\vec{H}$

Substituting the solutions (15.6-9) in Eq. (15.6-5) or (15.6-6) we get

$$k^2 = \varepsilon\mu\omega^2. \tag{15.6-14}$$

It indicates that for a linear homogeneous isotropic dielectric k is a real quantity. Therefore, from Eq. (15.6-12) we can say that $\vec{E}$ *and* $\vec{H}$ *are in phase.*

Wave impedance

The ratio of the magnitudes of $\vec{E}$ and $\vec{H}$ is

$$Z = \frac{|\vec{E}|}{|\vec{H}|} = \frac{\mu\omega}{k} = \sqrt{\frac{\mu}{\varepsilon}} \quad \text{[Using Eqs. (15.6-12) and (15.6-14)]} \tag{15.6-15}$$

The quantity Z has the dimensions of impedance and is called the *wave impendence* of isotropic dielectric medium. In terms of the wave impendence of free space, i.e., $Z_0 = \sqrt{\mu_0/\varepsilon_0}$, we can write

$$Z = Z_0\sqrt{\frac{\mu_r}{\varepsilon_r}} = \frac{Z_0}{\sqrt{K}}. \tag{15.6-16}$$

Poynting's vector

The Poynting's vector

$$\vec{s} = \vec{E}\times\vec{H} = \frac{1}{\mu\omega}\vec{E}\times\left(\vec{k}\times\vec{E}\right) \quad \text{[Using Eq. (15.6-12)]}$$

$$= \frac{1}{\mu\omega}\left[\vec{k}\left(\vec{E}\cdot\vec{E}\right) - \vec{E}\left(\vec{E}\cdot\vec{k}\right)\right] = \frac{E^2}{\mu\omega}\vec{k} = \frac{E^2}{Z}\hat{n}. \tag{15.6-17}$$

This indicates that the energy flow is in the direction of wave propagation. Since $\vec{E}$ is perpendicular to $\vec{k}$, from Eq. (15.6-12) we can write in terms of magnitude,

$$kE = \mu\omega H \text{or} \quad \sqrt{\varepsilon}E = \sqrt{\mu}H \quad \text{[Using Eq. (15.6-14)]}$$

$$\text{or} \quad \tfrac{1}{2}\varepsilon E^2 = \tfrac{1}{2}\mu H^2. \tag{15.6-18}$$

This shows that *in case of electromagnetic waves in isotropic dielectric electromagnetic energy is equally shared between electric and magnetic fields.*

Total electromagnetic energy density is

$$u = \tfrac{1}{2}\varepsilon E^2 + \tfrac{1}{2}\mu H^2 = \varepsilon H^2. \tag{15.6-19}$$

So Eq. (15.6-17) can also be written as

$$\vec{s} = \frac{u}{\varepsilon\mu\omega} \cdot k\hat{n} = uv\hat{n}, \tag{15.6-20}$$

where $v = 1/\sqrt{\varepsilon\mu} = \omega/k$ is the speed of the wave.

Thus, Poynting's vector equals the energy density (u) times the velocity of the wave. This means that the energy associated with wave propagates with the same velocity with which the field vectors $\vec{E}$ and $\vec{H}$ propagate.

15.7 Plane Electromagnetic Waves in a Conducting Medium

Let us consider a linear, homogeneous and isotropic conducting medium characterised by constant permittivity ε, permeability μ and conductivity σ. To simplify the discussion we assume that the medium is charge free ($\rho = 0$) and external-current free such that the currents existing in the medium are induced only by the electromagnetic wave itself. Thus, we take $\vec{J} = \sigma\vec{E}$. Any initial charge distribution within the conductor dies out quickly. The charges move to the surface and make $\rho = 0$ inside.

For such a medium Maxwell's equations take the following form:

$$\vec{\nabla} \cdot \vec{E} = 0 \tag{15.7-1}$$

$$\vec{\nabla} \cdot \vec{H} = 0 \tag{15.7-2}$$

$$\vec{\nabla} \times \vec{E} = -\mu\frac{\partial \vec{H}}{\partial t} \tag{15.7-3}$$

$$\vec{\nabla} \times \vec{H} = \sigma\vec{E} + \varepsilon\frac{\partial \vec{E}}{\partial t} \tag{15.7-4}$$

Taking curl of Eq. (15.7-3) and using Eq. (15.7-4) we get

$$\vec{\nabla}\left(\vec{\nabla} \cdot \vec{E}\right) - \nabla^2\vec{E} = -\mu\frac{\partial}{\partial t}\left(\vec{\nabla} \times \vec{H}\right) = -\sigma\mu\frac{\partial \vec{E}}{\partial t} - \varepsilon\mu\frac{\partial^2 \vec{E}}{\partial t^2}.$$

Now using Eq. (15.7-1) we get

$$\nabla^2\vec{E} - \sigma\mu\frac{\partial \vec{E}}{\partial t} - \varepsilon\mu\frac{\partial^2 \vec{E}}{\partial t^2} = 0. \tag{15.7-5}$$

Similarly, taking curl of Eq. (15.7-4) and using Eqs. (15.7-3) and (15.7-1) we can get

$$\nabla^2\vec{H} - \sigma\mu\frac{\partial \vec{H}}{\partial t} - \varepsilon\mu\frac{\partial^2 \vec{H}}{\partial t^2} = 0. \tag{15.7-6}$$

Suppose we are interested in the plane wave solutions and assume that

$$\vec{E}(\vec{r},t) = \vec{E}_0 e^{j(\vec{k}\cdot\vec{r}-\omega t)} \quad \text{and} \quad \vec{H}(\vec{r},t) = \vec{H}_0 e^{j(\vec{k}\cdot\vec{r}-\omega t)}, \tag{15.7-7}$$

where $\vec{E}_0$ and $\vec{H}_0$ are complex amplitudes which are constants in space and time, $\vec{k} = k\hat{n}$ is the propagation vector.

Substituting the solution (15.7-7) in Eq. (15.7-5) or (15.7-6) wet get

$$\left[-k^2 + j\omega\sigma\mu + \omega^2\varepsilon\mu\right]\vec{E} = 0.$$

For nonzero solution,

$$k^2 = \omega^2\varepsilon\mu + j\omega\sigma\mu. \tag{15.7-8}$$

k is thus, complex in character here. Let $k = \alpha + j\beta$. Then

$$k^2 = \alpha^2 - \beta^2 + j2\alpha\beta. \tag{15.7-9}$$

Comparing Eqs. (15.7-8) and (15.7-9) we get,

$$\alpha^2 - \beta^2 = \omega^2\varepsilon\mu \quad \text{and} \quad 2\alpha\beta = \omega\sigma\mu.$$

On solving these two equations we get

$$\alpha = \omega\sqrt{\frac{\varepsilon\mu}{2}}\left[\sqrt{1+\left(\frac{\sigma}{\omega\varepsilon}\right)^2}+1\right]^{1/2},$$

$$\beta = \omega\sqrt{\frac{\varepsilon\mu}{2}}\left[\sqrt{1+\left(\frac{\sigma}{\omega\varepsilon}\right)^2}-1\right]^{1/2}. \tag{15.7-10}$$

In terms of α and β the field vectors become

$$\vec{E}(\vec{r},t) = \vec{E}_0 \cdot e^{-\beta\hat{n}\cdot\vec{r}} \cdot e^{j(\alpha\hat{n}\cdot\vec{r}-\omega t)}$$

$$\text{and} \quad \vec{H}(\vec{r},t) = \vec{H}_0 \cdot e^{-\beta\hat{n}\cdot\vec{r}} \cdot e^{j(\alpha\hat{n}\cdot\hat{r}-\omega t)}. \tag{15.7-11}$$

These solutions show that the field amplitudes are spatially attenuated. The physical reason of the attenuation is that the wave sets up electric current in the medium, which causes dissipation of energy in the form of Joule heating. The quantity β, the imaginary part of wave number k, is a measure of attenuation and is called ***attenuation constant.*** It depends on the frequency ω and conductivity σ.

For nonconductors $\sigma = 0$ and then we have $\beta = 0$ meaning that there is no attenuation of the field vectors.

For goods conductors $\sigma/\omega\varepsilon \gg 1$ and then we have

$$\alpha \simeq \beta \approx \sqrt{\frac{\omega\sigma\mu}{2}}. \tag{15.7-12}$$

The quantity $1/\beta$ measures the depth at which electromagnetic wave entering a conductor is attenuated to $1/e$ of its initial amplitude at the surface. It is known as *skin depth* or *penetration depth* into a conducting medium. Thus, skin depth δ for a good conductor is

$$\delta = \frac{1}{\beta} = \sqrt{\frac{2}{\omega\sigma\mu}}. \tag{15.7-13}$$

Obviously, skin depth δ decreases with increase in frequency and conductivity. For good conductors at high frequencies δ is very small. That is why in high frequency circuits current flows only through the surface of good conductors. This phenomenon is called *skin effect.* Due to this effect the ac resistance of a conductor is greater than its dc resistance. For this in high frequency circuits it is better to use a number of fine stranded wires instead of a thick wire. It increases surface area for a given area of cross-section and reduces resistance. At microwave frequencies δ for Ag is very small ($\sim 10^{-3}$ mm). As a result, in the microwave region the performance of a waveguide made of pure Ag and another waveguide made of Ag-coated brass would appear to be identical. This technique is used to reduce the material cost of good conductors.

The attenuation of electromagnetic waves in the conducting sea water creates problem in radiocommunication with submerged submarine.

Wavelength, propagation speed and the index of refraction

The real part of k, i.e., α determines the wavelength, the propagation speed of the wave and the index of refraction of the conductor, in the usual way. Thus,

$$\lambda_c = \frac{2\pi}{\alpha}, \quad v = \frac{\omega}{\alpha} \quad \text{and} \quad n = \frac{c\alpha}{\omega}$$

For a good conductor $\alpha = \sqrt{\dfrac{\omega\sigma\mu}{2}}$ and hence,

$$\lambda_c = 2\pi\sqrt{\frac{2}{\omega\sigma\mu}}, \quad v = \omega\sqrt{\frac{2}{\omega\sigma\mu}} = \sqrt{\frac{2\omega}{\sigma\mu}} \quad \text{and} \quad n = c\sqrt{\frac{\sigma\mu}{2\omega}}.$$

Note that the skin depth δ as given by (15.7-13) for a good conductor can be expressed in terms of the wavelength λ_c in the conductor as

$$\delta = \frac{\lambda_c}{2\pi}.$$

Relative directions of $\vec{E}$, $\vec{H}$ and $\vec{k}$

Substituting the solutions (15.7-7) in Eqs. (15.7-1) and (15.7-2) we get

$$\vec{k}\cdot\vec{E} = 0 \quad \text{and} \quad \vec{k}\cdot\vec{H} = 0. \tag{15.7-14}$$

These equations indicates that $\vec{E}$ and $\vec{H}$ are both perpendicular to the direction of propagation. So *electromagnetic waves in a conducting medium are transverse in nature.*

Again, the substitution of the solutions (15.7-7) in Eqs. (15.7-3) and (15.7-4) gives

$$j\vec{k} \times \vec{E} = -\mu\left(-j\omega\vec{H}\right) \quad \text{or} \quad \vec{k} \times \vec{E} = \mu\omega\vec{H} \tag{15.7-15}$$

$$\text{and} \quad j\vec{k} \times \vec{H} = \sigma\vec{E} - j\omega\varepsilon\vec{E} \quad \text{or} \quad \vec{k} \times \vec{H} = -\left(\omega\varepsilon + j\sigma\right)\vec{E}. \tag{15.7-16}$$

These two equations imply that $\vec{E}$ and $\vec{H}$ are mutually perpendicular and also they are perpendicular to the direction of propagation vector $\vec{k}$.

Relative phase of $\vec{E}$ and $\vec{H}$

From Eq. (15.7-15) we have

$$\vec{H} = \frac{1}{\mu\omega}\left(\vec{k} \times \vec{E}\right) = \frac{k}{\mu\omega}\left(\hat{n} \times \vec{E}\right) = \frac{\alpha + j\beta}{\mu\omega}\left(\hat{n} \times \vec{E}\right). \tag{15.7-17}$$

This equation shows that $\vec{E}$ *and* $\vec{H}$ *are not in phase in a conductor.*

Writing $\alpha + j\beta = \sqrt{\alpha^2 + \beta^2}e^{j\phi}$; $\phi = \tan^{-1}(\beta/\alpha)$ and using (15.7-7), Eq. (15.7-17) may be rewritten as

$$\vec{H} = \frac{\sqrt{\alpha^2 + \beta^2}}{\mu\omega}\left(\hat{n} \times \vec{E}_0\right) \cdot e^{j\left(\vec{k}\cdot\vec{r} - \overline{\omega t - \phi}\right)}, \tag{15.7-18}$$

where

$$\sqrt{\alpha^2 + \beta^2} = \omega\sqrt{\varepsilon\mu}\left[1 + \left(\frac{\sigma}{\omega\varepsilon}\right)^2\right]^{1/4}.$$

Thus, $\vec{H}$ *lags behind* $\vec{E}$ *in time by the phase angle*

$$\phi = \tan^{-1}\frac{\beta}{\alpha} = \frac{1}{2}\tan^{-1}\left(\frac{\sigma}{\omega\varepsilon}\right).$$

For good conductors $\alpha \approx \beta$ and $\phi = 45°$. Therefore, *the phase difference between the* $\vec{E}$ *and* $\vec{H}$ *fields in a perfect conductor is* 45°.

Relative magnitudes of $\vec{E}$ and $\vec{H}$ is

$$\frac{\left|\vec{H}\right|}{\left|\vec{E}\right|} = \frac{H_0}{E_0} = \frac{\sqrt{\alpha^2 + \beta^2}}{\mu\omega} = \sqrt{\frac{\varepsilon}{\mu}}\left[1 + \left(\frac{\sigma}{\omega\varepsilon}\right)^2\right]^{1/4}. \tag{15.7-19}$$

For good conductors

$$\frac{\left|\vec{H}\right|}{\left|\vec{E}\right|} = \sqrt{\frac{\sigma}{\omega\mu}}.$$

Thus, in this case,

$$\left|\vec{H}\right| \gg \left|\vec{E}\right|,$$

which indicates that in *a good conducting medium the field energy is not equally shared between* $\vec{E}$*- and* $\vec{H}$*-fields but it is almost entirely magnetic in nature.*

Poynting's vector

The time average Poynting's vector is

$$\langle \vec{s} \rangle = \tfrac{1}{2} Re \left(\vec{E} \times \vec{H}^* \right).$$

Using Eq. (15.7-17),

$$\begin{aligned}\langle \vec{s} \rangle &= \tfrac{1}{2} \frac{\sqrt{\alpha^2 + \beta^2}}{\mu\omega} \cdot Re \left[\vec{E} \times \left(\hat{n} \times \vec{E}^* \right) \right] e^{-j\phi} \\ &= \frac{\sqrt{\alpha^2 + \beta^2}}{2\mu\omega} Re \left[\hat{n} \left(\vec{E} \cdot \vec{E}^* \right) - \vec{E}^* \left(\vec{E} \cdot \hat{n} \right) \right] e^{-j\phi}\end{aligned}$$

Now putting $\vec{E} \cdot \hat{n} = 0$ and using Eq. (15.7-11) we get

$$\langle \vec{s} \rangle = \frac{\sqrt{\alpha^2 + \beta^2}}{2\mu\omega} \cdot E_0^2 e^{-2\beta \hat{n} \cdot \vec{r}} \cdot \cos\phi \cdot \hat{n}. \tag{15.7-20}$$

For good conductors $\alpha \simeq \beta \approx \sqrt{\dfrac{\omega\sigma\mu}{2}}$ and $\phi = 45°$. Then

$$\langle \vec{s} \rangle = \frac{1}{2} \cdot \sqrt{\frac{\sigma}{2\mu\omega}} E_0^2 e^{-2\beta \hat{n} \cdot \vec{r}} \hat{n}. \tag{15.7-21}$$

Thus, *energy flow is along the direction of propagation of the wave and is damped off exponentially.*

Average electric energy density is

$$\langle u_e \rangle = \tfrac{1}{2} Re \tfrac{1}{2} \left(\vec{E} \cdot \vec{D}^* \right) = \tfrac{1}{4} \varepsilon Re \left(\vec{E} \cdot \vec{E}^* \right) = \tfrac{1}{4} \varepsilon E_0^2 \cdot e^{-2\beta \hat{n} \cdot \vec{r}}.$$

Average magnetic energy density is

$$\langle u_m \rangle = \tfrac{1}{2} Re \tfrac{1}{2} \left(\vec{B} \cdot \vec{H}^* \right) = \tfrac{1}{4} \mu Re \left(\vec{H} \cdot \vec{H}^* \right) = \tfrac{1}{4} \mu H_0^2 e^{-2\beta \hat{n} \cdot \vec{r}}.$$

Thus,

$$\frac{\langle u_m \rangle}{\langle u_e \rangle} = \frac{\mu H_0^2}{\varepsilon E_0^2} = \left[1 + \left(\frac{\sigma}{\omega\varepsilon} \right)^2 \right]^{1/2} \tag{15.7-22}$$

Obviously, for good conductors $\langle u_m \rangle \gg \langle u_e \rangle$, i.e., *the field energy inside is almost entirely magnetic in nature.*

Total time averaged energy density is

$$\langle u \rangle = \langle u_e \rangle + \langle u_m \rangle = \tfrac{1}{4} e^{-2\beta \hat{n} \cdot \vec{r}} \left[\varepsilon E_0^2 + \mu H_0^2 \right]$$

Using Eq. (15.7-19) we can write

$$\langle u \rangle = \tfrac{1}{4} e^{-2\beta \hat{n}\cdot\vec{r}} \cdot \varepsilon E_0^2 \left[1 + \sqrt{1 + \left(\frac{\sigma}{\varepsilon\omega}\right)^2} \right]. \tag{15.7-23}$$

In term of α, the real part of wave number as given by (15.7-10), it can be expressed as

$$\langle u \rangle = \frac{\alpha^2}{2\mu\omega^2} E_0^2 e^{-2\beta \hat{n}\cdot\vec{r}}. \tag{15.7-24}$$

15.8 Electromagnetic Waves in Ionised Gases

Let us consider a dilute medium of ionised gases such as the ionosphere. The motion of positive ions may be ignored as they are massive as compared to the electrons. Also, the damping of the motion of free electrons due to collisions is negligible here. So the nonrelativistic ($v \ll c$) equation of motion of an electron of mass m and charge e under the action of the incident electromagnetic field will be

$$m\frac{d\vec{v}}{dt} = e\vec{E}, \tag{15.8-1}$$

where $\vec{v}$ is the instantaneous velocity of the electron. Here we have neglected the force due to magnetic field which is only v/c times the force due to electric field.

Assuming $\vec{E} = \vec{E}(\vec{r})\, e^{-j\omega t}$, we get on integration,

$$\vec{v} = \frac{e\vec{E}(\vec{r})}{-jm\omega} = \frac{je\vec{E}(\vec{r})}{m\omega}.$$

If n_0 is the number of free electrons per unit volume then the complex current density,

$$\vec{J} = n_0 e\vec{v} = \frac{jn_0 e^2 \vec{E}(\vec{r})}{m\omega}. \tag{15.8-2}$$

Comparing this equation with $\vec{J} = \sigma\vec{E}$, we find that the complex conductivity of the medium is given by

$$\sigma = \frac{jn_0 e^2}{m\omega}. \tag{15.8-3}$$

In case of dilute ionised gaseous medium we can take $\rho = 0$, $\varepsilon \simeq \varepsilon_0$ and $\mu \simeq \mu_0$. So here Maxwell's equations take the following form.

$$\vec{\nabla}\cdot\vec{E} = 0, \quad \vec{\nabla}\cdot\vec{H} = 0, \quad \vec{\nabla}\times\vec{E} = -\mu_0\frac{\partial\vec{H}}{\partial t} \quad \text{and} \quad \vec{\nabla}\times\vec{H} = \sigma\vec{E} + \varepsilon_0\frac{\partial\vec{E}}{\partial t}. \tag{15.8-4}$$

Following the procedure as adopted in Section 15.7, we can arrive at the wave equation,

$$\nabla^2 \vec{E} - \sigma\mu_0 \frac{\partial \vec{E}}{\partial t} - \varepsilon_0\mu_0 \frac{\partial^2 \vec{E}}{\partial t^2} = 0. \tag{15.8-5}$$

A similar equation for $\vec{H}$ can also be obtained.

Assuming a plane wave solution, $\vec{E}(\vec{r}, t) = \vec{E}_0 e^{j(\vec{k}\cdot\vec{r} - \omega t)}$ and substituting it in Eq. (15.8-5) we get

$$-k^2 + j\omega\sigma\mu_0 + \omega^2\varepsilon_0\mu_0 = 0 \quad \text{or} \quad k^2 = \varepsilon_0\mu_0\omega^2 \left[1 + \frac{j\sigma}{\varepsilon_0\omega}\right]. \tag{15.8-6}$$

Using the relation (15.8-3) and putting $1/\sqrt{\varepsilon_0\mu_0} = c$ we get

$$k^2 = \frac{\omega^2}{c^2}\left[1 - \frac{\omega_p^2}{\omega^2}\right], \tag{15.8-7}$$

where $\omega_p = \sqrt{n_0 e^2/\varepsilon_0 m}$ is known as *electron plasma frequency.*

As ω/k is the speed of the electromagnetic wave through the medium, the refractive index of the medium is given by

$$n = \frac{c}{(\omega/k)} = \sqrt{1 - \frac{\omega_p^2}{\omega^2}} \tag{15.8-8}$$

For $\omega > \omega_p$, k is real and waves can propagate freely. For example, the typical electron number density of metals is $n_0 \approx 10^{28}\,\text{m}^{-3}$ and the corresponding $\omega_p \sim 10^{16}\text{s}^{-1}$. For ultraviolet light $\omega > 10^{16}\,\text{s}^{-1}$. For this ultraviolet light can generally propagate in metals. For $\omega < \omega_p$, k is purely imaginary, say jk_1. In this case,

$$\vec{E}(\vec{r}, t) = \vec{E}_0 e^{-k_1(\hat{n}\cdot\vec{r})} \cdot e^{-j\omega t}$$

$$\text{and} \quad \vec{H}(\vec{r}, t) = \vec{H}_0 e^{-k_1(\hat{n}\cdot\vec{r})} \cdot e^{-j\omega t}. \tag{15.8-9}$$

Thus, *for $\omega < \omega_p$, the field vectors fall off exponentially as they penetrate into the medium and cannot pass through it.*

In case of ionosphere electron concentration increases and hence, refractive index decreases as we go deeper into it. This may cause total internal reflection of the radio waves from the ionosphere, depending on its frequency and angle of incidence.

Phase velocity of electromagnetic waves through the ionised gas is, from (15.8-8),

$$\frac{\omega}{k} = \frac{c}{\sqrt{1 - \frac{\omega_p^2}{\omega^2}}}. \tag{15.8-10}$$

Obviously, $\omega/k > c$ for $\omega > \omega_p$. This does not violate the principle of relativity because it is the group velocity not the phase velocity with which signals move from one place to other and this group velocity is always less than c, the speed of light in free space.

15.9 Polarization of Plane Electromagnetic Waves

We have seen that in a plane electromagnetic wave the $\vec{E}$-field vibrates in a plane perpendicular to the direction of propagation vector $\vec{k}$. In order to completely specify the wave we are to specify moment-by-moment direction of $\vec{E}$. Such a description corresponds to the polarization of the wave. The state of polarization of an electromagnetic wave is customarily given by the orientation of its $\vec{E}$-vector. If the $\vec{E}$-field of a plane wave is aligned along a fixed direction in space although its magnitude and sign may vary in time. It is said to be *linearly polarized.*

The plane containing the $\vec{E}$-vector and the direction of propagation $\left(\vec{k}\right)$ is called the *plane of vibration.* Suppose a linearly polarized electromagnetic wave is propagating along z-direction. The $\vec{E}$-vector of the wave may be in any direction in the xy-plane. So it can be considered as the superposition of two independent orthogonal plane polarized components having same amplitude and phase. Symbolically, a linearly polarized wave may be represented by

$$\begin{aligned}\vec{E}(z,t) &= \vec{E}_x(z,t) + \vec{E}_y(z,t) \\ &= \hat{\imath}E_{0x}e^{j(kz-\omega t)} + \hat{\jmath}E_{0y}e^{j(kz-\omega t)} \\ &= \left(\hat{\imath}E_{0x} + \hat{\jmath}E_{0y}\right)e^{j(kz-\omega t)}. \end{aligned} \tag{15.9-1}$$

The $\vec{E}$-vector has the amplitude $\sqrt{E_{0x}^2 + E_{0y}^2}$ and its direction makes an angle $\theta = \tan^{-1}(E_{0y}/E_{0x})$ with x-direction. θ is called *polarization angle.*

Other types of polarizations can be constructed by the superposition of two plane polarized waves, which are not in phase. Let us consider two plane polarized waves both moving along z-direction, one wave polarized along x-direction and the other along y-direction. Let us represent two such waves by

$$\vec{E}_x = \hat{\imath}E_{0x}e^{j(kz-\omega t+\phi_x)} \quad \text{and} \quad \vec{E}_y = \hat{\jmath}E_{0y}e^{j(kz-\omega t+\phi_y)}, \tag{15.9-2}$$

where ϕ_x and ϕ_y are some phase angles. Corresponding real fields are obtained by taking real parts. Thus,

$$\vec{E}_x = \hat{\imath}E_{0x}\cos(kz - \omega t + \phi_x) \quad \text{and} \quad \vec{E}_y = \hat{\imath}E_{0y}\cos(kz - \omega t + \phi_y)$$

Considering the special case when $\phi_x = \phi$ and $\phi_y = \phi \pm \pi/2$ we can write for the resultant wave by superposition as

$$\begin{aligned}\vec{E} = \vec{E}_x + \vec{E}_y &= \hat{\imath}E_{0x}\cos(kz - \omega t + \phi) \mp \hat{\jmath}E_{0y}\sin(kz - \omega t + \phi) \\ &= \hat{\imath}E_x \mp \hat{\jmath}E_y, \end{aligned} \tag{15.9-3}$$

where

$$\frac{E_x^2}{E_{0x}^2} + \frac{E_y^2}{E_{0y}^2} = 1.$$

This is equation of an ellipse for components of $\vec{E}$, with major and minor axes E_{0x} and E_{0y}, respectively. Thus, at any fixed plane z = constant the tip of $\vec{E}$-vector

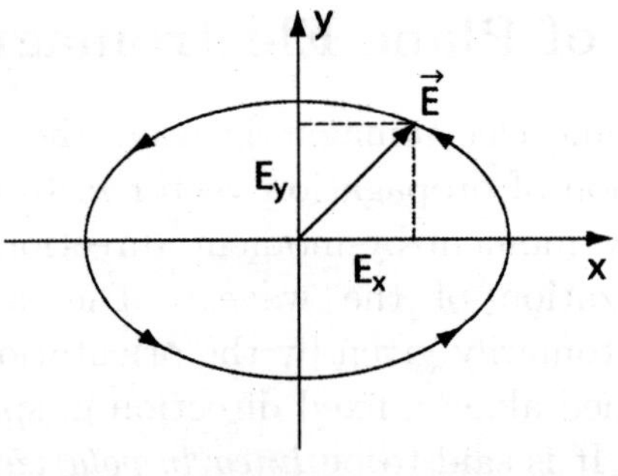

Fig. 15.9-1 : Left elliptically polarized wave

describes an ellipse. The sense of rotation—clockwise or anticlockwise—depends on the phase difference and can be seen by taking successive increasing values of ωt. Thus, the resultant of the components

$$E_x = E_{0x}\cos(kz - \omega t + \phi)$$

$$\text{and} \quad E_y = E_{0y}\sin(kz - \omega t + \phi) \tag{15.9-4}$$

represents a vector $\vec{E}$, which rotates clockwise at an angular frequency ω, as seen by an observer towards whom the wave is moving. Such a wave is *right-elliptically polarized.* On the other hand, the resultant of the components

$$E_x = E_{0x}\cos(kz - \omega t + \phi) \quad \text{and} \quad E_y = -E_{0y}\sin(kz - \omega t + \phi) \tag{15.9-5}$$

is *left-elliptically polarized* (Fig. 15.9-1).

If we let $E_{0x} = E_{0y} = E_0$ and keep the same phase conditions as above then $\vec{E} = \hat{\imath}E_x \mp \hat{\jmath}E_y$ with $E_x^2 + E_y^2 = E_0^2$.

This shows that the resultant wave is *circularly polarized.* The magnitude of $\vec{E}$ remains constant, although its direction rotates. The components

$$E_x = E_0\cos(kz - \omega t + \phi) \quad \text{and} \quad E_y = E_0\sin(kz - \omega t + \phi) \tag{15.9-6}$$

represent a wave, which is *right-circularly polarized.* On the other hand, the components

$$E_x = E_0\cos(kz - \omega t + \phi) \quad \text{and} \quad E_y = -E_0\sin(kz - \omega t + \phi) \tag{15.9-7}$$

represent a *left-circularly polarized wave.*

It is interesting to note that a circularly polarized wave may be regarded as the superposition of two orthogonal linearly polarized waves having a phase difference of $\pi/2$. Also, a linearly polarized wave can be regarded as the superposition of two oppositely polarized circular waves of equal amplitude. For example, if we add the right-circular wave of Eq. (15.9-6) to the left-circular wave of Eq. (15.9-7), we get for the resultant wave

$$E_x = 2E_0\cos(kz - \omega t + \phi) \quad \text{and} \quad E_y = 0$$

and the resultant $\vec{E}$-vector is

$$\vec{E} = \hat{\imath}E_x + \hat{\jmath}E_y = \hat{\imath}2E_0\cos(kz - \omega t + \phi).$$

Obviously, it is a linearly polarized wave.

Angular momentum density

An *EM*-wave impinging on an object can impart energy, linear momentum as well as angular momentum. If the incident plane wave is circularly polarized the rotating $\vec{E}$-field drives the electrons in the material into circular motion with angular velocity ω equal to the frequency of the *EM*-wave. Thus angular momentum is also imparted by the *EM*-wave to the material. If $\frac{dW}{dt}$ is the rate of *EM*-energy absorbed by the body then

$$\frac{dW}{dt} = \Gamma \cdot \omega$$

where Γ is the torque generated. Since Γ is equal to the rate of change of angular momentum (L), it follows that on the coverage

$$\frac{dW}{dt} = \omega \cdot \frac{dL}{dt}$$

Thus if a body absorbs W amount of energy from the incident left circularly polarized *EM*-wave then it also simultaneously absorbs an angular momentum $L = W/\omega$ directed along the direction of propagation. If the incident *EM*-wave is right circularly polarized then angular momentum vector is taken to point in the direction opposite to the direction of propagation.

Thus we can say that *EM*-energy density $\langle u \rangle$ is associated with an *angular momentum density* $\langle u \rangle/\omega$.

According to quantum-mechanical description an *EM*-wave consists of a stream of photons, each photon carrying an energy $\hbar\omega$ $(\hbar = h/2\pi)$, linear momentum $\hbar\omega/c$ and intrinsic angular momentum $+\hbar$ or $-\hbar$ where the sign indicates either the direction of propagation or opposite to it. If n be the number of photons per unit volume then *EM*-energy density is $\langle u \rangle = n\hbar\omega$ and angular momentum density is $\langle u \rangle/\omega = n\hbar$.

15.10 Wave equation of Potentials with Sources—Gauge Transformations

So far we have considered EM waves without inquiring how these waves are produced. We did not consider any external charge and current sources. But it is found that *accelerated charges can produce EM radiations.* There are a number of procedures for the calculation of radiation from accelerated charges. However, the most fruitful one is the potential formulation. For a prescribed time dependent charge and current distributions one finds the scalar and vector potentials from which the fields are then obtained.

Since $\vec{\nabla} \cdot \vec{B} = 0$ we can always write $\vec{B}$ as the curl of some vector function $\vec{A}$, i.e.,

$$\vec{B} = \vec{\nabla} \times \vec{A}, \tag{15.10-1}$$

where $\vec{A}$ is called the *vector potential.* Now from Maxwell's Eq. (15.3-3),

$$\vec{\nabla} \times \left(\vec{E} + \frac{\partial \vec{A}}{\partial t} \right) = 0.$$

So $\vec{E} + \frac{\partial \vec{A}}{\partial t}$ can be written as the gradient of some scalar function ϕ, i.e.,

$$\vec{E} + \frac{\partial \vec{A}}{\partial t} = -\vec{\nabla}\phi \quad \text{or} \quad \vec{E} = -\vec{\nabla}\phi - \frac{\partial \vec{A}}{\partial t}, \tag{15.10-2}$$

where ϕ is known as *scalar potential.*

Now, substituting $\vec{H} = \vec{B}/\mu$, $\vec{E} = \vec{D}/\varepsilon$ and Eqs. (15.10-1) and (15.10-2) into Maxwell's Eq. (15.3-4) we get

$$\frac{1}{\mu}\vec{\nabla} \times \left(\vec{\nabla} \times \vec{A}\right) = \vec{J} - \varepsilon\frac{\partial}{\partial t}\left(\vec{\nabla}\phi + \frac{\partial \vec{A}}{\partial t}\right)$$

$$\text{or} \quad \nabla^2 \vec{A} - \mu\varepsilon\frac{\partial^2 \vec{A}}{\partial t^2} = -\mu\vec{J} + \vec{\nabla}\left[\vec{\nabla}\cdot\vec{A} + \mu\varepsilon\frac{\partial\phi}{\partial t}\right]. \tag{15.10-3}$$

Using Eq. (15.10-2) in the Maxwell's equation, $\vec{\nabla}\cdot\vec{D} = \rho$, we get

$$-\varepsilon\left[\vec{\nabla}\cdot\vec{\nabla}\phi + \vec{\nabla}\cdot\frac{\partial \vec{A}}{\partial t}\right] = \rho \quad \text{or} \quad \nabla^2\phi + \frac{\partial}{\partial t}\left(\vec{\nabla}\cdot\vec{A}\right) = -\frac{\rho}{\varepsilon}. \tag{15.10-4}$$

We now note that $\vec{A}$ defined by Eq. (15.10-1) is not unique because $\vec{B}$ remains unchanged if we replace $\vec{A}$ by

$$\vec{A}' = \vec{A} + \vec{\nabla}\psi, \tag{15.10-5a}$$

where ψ is a scalar function. To keep $\vec{E}$ unchanged as well, ϕ is to be simultaneously changed to

$$\phi' = \phi - \frac{\partial\psi}{\partial t}. \tag{15.10-5b}$$

The transformation (15.10-5) is called a *gauge transformation* and the invariance of the fields under such transformation is called *gauge invariance.* The freedom implied by (15.10-5) means that we can always choose a set of potentials $\vec{A}$, ϕ such that

$$\vec{\nabla}\cdot\vec{A} + \mu\varepsilon\frac{\partial\phi}{\partial t} = 0. \tag{15.10-6}$$

This choice is referred to as the *Lorentz gauge.* Substituting this Lorentz gauge in Eqs. (15.10-3) and (15.10-4) we get uncoupled wave equations for each of $\vec{A}$ and ϕ.

$$\nabla^2\vec{A} - \mu\varepsilon\frac{\partial^2\vec{A}}{\partial t^2} = -\mu\vec{J} \tag{15.10-7}$$

$$\nabla^2\phi - \mu\varepsilon\frac{\partial^2\phi}{\partial t^2} = -\frac{\rho}{\varepsilon}. \tag{15.10-8}$$

It may be pointed out here that even the potentials, satisfying Lorentz condition, are not unique. If $\vec{A}$ and ϕ satisfy the Lorentz condition (15.10-6), $\vec{A}'$ and ϕ' will also satisfy the same condition namely

$$\vec{\nabla}\cdot\vec{A}' + \mu\varepsilon\frac{\partial\phi'}{\partial t} = 0,$$

if the gauge function ψ satisfies the homogeneous scalar wave equation

$$\nabla^2\psi - \mu\varepsilon\frac{\partial^2\psi}{\partial t} = 0.$$

All potentials in this restricted class are said to belong to the Lorentz gauge.

Many other gauge transformations are found in the literature. Common purpose of all these transformations is to simplify Eqs. (15.10-3) and (15.10-4). Suitability of a gauge depends on the problem at hand. Let us consider another useful gauge-the so-called *Coulomb gauge.* In this gauge one takes

$$\vec{\nabla}\cdot\vec{A} = 0. \tag{15.10-9}$$

With this choice Eqs. (15.10-3) and (15.10-4) become

$$\nabla^2\vec{A} - \mu\varepsilon\frac{\partial^2\vec{A}}{\partial t^2} = -\mu\vec{J} + \mu\varepsilon\vec{\nabla}\frac{\partial\phi}{\partial t} \tag{15.10-10}$$

$$\nabla^2\phi = -\frac{\rho}{\varepsilon}. \tag{15.10-11}$$

Eq. (15.10-11) can be easily solved to find ϕ as in electrostatics

$$\phi(\vec{r},t) = \frac{1}{4\pi\varepsilon}\int\frac{\rho(\vec{r}\,',t)}{|\vec{r}-\vec{r}\,'|}\,dV.$$

But here it is difficult to find $\vec{A}$ by solving (15.10-10). The Coulomb gauge is often utilized in cases where there are no charge or current distributions, i.e., $\rho = 0$, $\vec{J} = 0$. In this case, one takes $\phi = 0$ and $\vec{A}$ satisfies the equation

$$\nabla^2\vec{A} - \mu\varepsilon\frac{\partial^2\vec{A}}{\partial t^2} = 0.$$

In this case the fields are derivable from a single potential

$$\vec{B} = \vec{\nabla}\times\vec{A} \quad \text{and} \quad \vec{E} = -\frac{\partial\vec{A}}{\partial t}.$$

15.11 Retarted Potential

An electric charge at rest or in uniform motion cannot generate EM radiations. But an accelerating charge can do it. To know how it radiates we must calculate electric and magnetic fields of moving charges. As usual it is easier to begin with the potentials $\vec{A}$ and ϕ and then calculate fields from them. In the Lorentz gauge defined by the condition (15.10-6), the potentials $\vec{A}$ and ϕ satisfy the inhomogeneous wave Eqs. (15.10-7) and (15.10-8). In the static case and in vacuum (15.10-7) and (15.10-8) reduce to Poisson's equation,

$$\nabla^2\vec{A} = -\mu_0\vec{J} \tag{15.11-1}$$

$$\nabla^2\phi = -\frac{\rho}{\varepsilon_0} \tag{15.11-2}$$

with the familiar solutions

$$\vec{A}(\vec{r}) = \frac{\mu_0}{4\pi}\int \frac{\vec{J}(\vec{r}\,')\,dV}{|\vec{r}-\vec{r}\,'|} \quad \text{and} \quad \phi(\vec{r}) = \frac{1}{4\pi\varepsilon_0}\int \frac{\rho(\vec{r}\,')\,dV}{|\vec{r}-\vec{r}\,'|}, \tag{15.11-3}$$

where $R = |\vec{r}-\vec{r}\,'|$ is the distance from the source to the point of observation. Since the electromagnetic disturbances travel in vacuum with the finite speed of light c, in the nonstatic case it takes a time equal R/c to reach the observer. Therefore, what the observer measures at time t is correlated to the conditions of the source at some earlier time $t' = t - R/c$ when the disturbance *left.* This time t' is called *retarded time.*

Hence, for nonstatic case Eqs. (15.11-3) generalise to

$$\vec{A}(\vec{r},t) = \frac{\mu_0}{4\pi}\int \frac{\vec{J}(\vec{r}\,',t')\,dV}{|\vec{r}-\vec{r}\,'|} \quad \text{and} \quad \phi(\vec{r},t) = \frac{1}{4\pi\varepsilon_0}\int \frac{\rho(\vec{r}\,',t')\,dV}{|\vec{r}-\vec{r}\,'|} \tag{15.11-4}$$

Here the quantities $\vec{J}(\vec{r}\,',t')$ and $\rho(\vec{r}\,',t')$ are respectively, the current density and the charge density at the source point $\vec{r}\,'$ at the retarded time t'. Corresponding potentials are called *retarded potentials.*

Having determined the retarded vector and scalar potentials for a given charge and current distributions, one can determine $\vec{E}$ and $\vec{B}$-fields by using the relations (15.10-1) and (15.10-2). As an example, let us calculate the $\vec{E}$-field due to a point charge q, located at the origin ($\vec{r}\,' = 0$) at time t, whose magnitude is a function of time. The retarded potential in this case is given by

$$\phi(\vec{r},t) = \frac{1}{4\pi\varepsilon_0}\cdot\frac{q(t-r/c)}{r}.$$

As the charge is not moving $\vec{J}(\vec{r}\,',t') = 0$.

Therefore, the electric field $\vec{E}$ is given by

$$\vec{E} = -\vec{\nabla}\phi = -\hat{r}\frac{1}{4\pi\varepsilon_0}\frac{\partial}{\partial r}\frac{q(t-r/c)}{r} = \frac{\hat{r}}{4\pi\varepsilon_0}\left[\frac{q(t-r/c)}{r^2} + \frac{\dot{q}}{cr}\right],$$

where $\dot{q}$ is the derivative of q with respect to its argument $(t - r/c)$.

15.12 Radiation from an Oscillating Electric Dipole

Suppose we consider two tiny metal spheres separated by a small distance l and connected by a fine wire (Fig. 15.12-1). The system is assumed to be as a whole electrically neutral. So if at any time t the charge on the upper sphere is $q(t)$, then the charge on the lower sphere is $-q(t)$. Suppose further that some external agent causes the charge to move back and forth through the wire from one end to the other, at an angular frequency ω. Thus, we may write,

$$q(t) = q_0\cos\omega t. \tag{15.12-1}$$

Obviously, the system is a simple model for an oscillating electric dipole of dipole moment

$$\vec{p}(t) = q(t) \cdot \vec{l} = \hat{k} p_0 \cos \omega t, \qquad (15.12\text{-}2)$$

where $p_0 = q_0 l$ is the maximum value of the dipole moment.

The retarded scalar potential at P at an instant t is given by

$$\phi(\vec{r}, t) = \frac{1}{4\pi\varepsilon_0} \left[\frac{q_0 \cos \omega (t - R_+/c)}{R_+} - \frac{q_0 \cos \omega (t - R_-/c)}{R_-} \right]; \qquad (15.12\text{-}3)$$

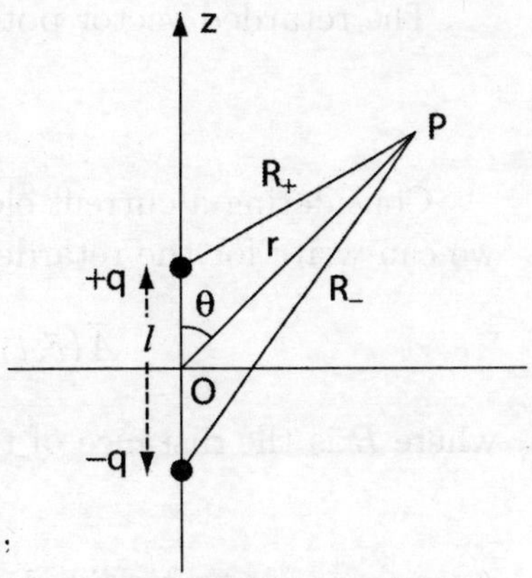

Fig. 15.12-1 : Oscillating electric dipole

where from the geometry of Fig. 15.12-1,

$$R_{\pm} = \sqrt{r^2 + (l/2)^2 \mp rl \cos \theta}.$$

Now for a perfect dipole $l \ll r$, so we can make the following approximations:

$$R_{\pm} \simeq r \left(1 \mp \frac{l}{2r} \cos \theta \right) \quad \text{or} \quad \frac{1}{R_{\pm}} \simeq \frac{1}{r} \left(1 \pm \frac{l}{2r} \cos \theta \right). \qquad (15.12\text{-}4)$$

Further we assume that $l \ll \lambda$ or $l \ll c/\omega$, i.e., l is much smaller than the wavelength of emitted radiation.

Therefore,

$$\begin{aligned} \cos \omega \left(t - \frac{R_{\pm}}{c} \right) &= \cos \left[\omega \left(t - \frac{r}{c} \right) \pm \frac{\omega l}{2c} \cos \theta \right] \\ &= \cos \omega \left(t - \frac{r}{c} \right) \cos \left(\frac{\omega l}{2c} \cos \theta \right) \mp \sin \omega \left(t - \frac{r}{c} \right) \sin \left(\frac{\omega l}{2c} \cos \theta \right) \\ &\approx \cos \omega \left(t - \frac{r}{c} \right) \mp \frac{\omega l}{2c} \cos \theta \sin \omega \left(t - \frac{r}{c} \right). \end{aligned} \qquad (15.12\text{-}5)$$

Putting (15.12-4) and(15.12-5) in Eq. (15.12-3) we get

$$\phi(\vec{r}, t) = \frac{p_0 \cos \theta}{4\pi\varepsilon_0 r} \left[-\frac{\omega}{c} \sin \omega (t - r/c) + \frac{1}{r} \cos \omega (t - r/c) \right]. \qquad (15.12\text{-}6)$$

Note that in the static limit $\omega \to 0$, $\phi \to \frac{p_0 \cos \theta}{4\pi\varepsilon_0 r^2}$, which is the potential for a stationary dipole. We are interested in the fields at large distances from the source. In this region, known as *radiation zone*, $r \gg \lambda$ or $r \gg c/\omega$ and the potential (15.12-6) reduces to

$$\phi(\vec{r}, t) = -\frac{p_0 \omega}{4\pi\varepsilon_0 c} \left(\frac{\cos \theta}{r} \right) \sin \omega (t - r/c). \qquad (15.12\text{-}7)$$

The retarded vector potential is determined by the current in the wire,

$$I(t) = \frac{dq}{dt} = -q_0\omega \sin \omega t. \tag{15.12-8}$$

Considering a current element $\hat{k}I(t)dz$ at a distance z from the centre of the dipole we can write for the retarded vector potential as

$$\vec{A}(\vec{r}, t) = \hat{k}\frac{\mu_0}{4\pi}\int_{-l/2}^{+l/2} \frac{-q_0\omega \sin \omega(t - R/c)dz}{R}, \tag{15.12-9}$$

where R is the distance of the point P from the current element, i.e.,

$$R = \sqrt{r^2 + z^2 - 2rz\cos\theta}.$$

As we are interested in determining $\vec{A}$ at points for which $r \gg l$, to a first order approximation we can neglect the variation over the length of the wire and can replace the integrand in (15.12-9) by its value at the the centre. Thus,

$$\vec{A}(\vec{r}, t) = -\hat{k}\frac{\mu_0 q_0 \omega}{4\pi r}\sin \omega(t - r/c)\int_{-l/2}^{+l/2} dz = -\hat{k}\frac{\mu_0 p_0 \omega}{4\pi r}\sin \omega(t - r/c). \tag{15.12-10}$$

In spherical polar coordinates (r, θ, φ), $\hat{k} = \cos\theta\,\hat{r} - \sin\theta\,\hat{t}\theta$.
Therefore,

$$\vec{A}(\vec{r}, t) = -\frac{\mu_0 p_0 \omega}{4\pi r}\sin \omega(t - r/c)\left(\cos\theta\,\hat{r} - \sin\theta\,\hat{t}\theta\right) \tag{15.12-11}$$

Having determined $\vec{A}$ and ϕ we can now calculate the $\vec{E}$ and $\vec{B}$-fields. From (15.12-7),

$$\begin{aligned}\vec{\nabla}\phi &= \left(\hat{r}\frac{\partial \phi}{\partial r} + \hat{\theta}\frac{1}{r}\frac{\partial \phi}{\partial \theta}\right)\\ &= -\frac{p_0\omega}{4\pi\varepsilon_0 c}\left[\hat{r}\cos\theta\left\{-\frac{1}{r^2}\sin\omega(t - r/c) - \frac{\omega}{rc}\cos\omega(t - r/c)\right\}\right.\\ &\qquad\left. - \frac{\sin\theta}{r^2}\sin\omega(t - r/c)\hat{\theta}\right]\end{aligned} \tag{15.12-12}$$

For fields at large distances $r \gg c/\omega$. So we can drop the first and the last terms. Thus, we get

$$\vec{\nabla}\phi = \hat{r}\frac{p_0\omega^2}{4\pi\varepsilon_0 c^2}\left(\frac{\cos\theta}{r}\right)\cos\omega(t - r/c) \tag{15.12-13}$$

Similarly, from (15.12-11),

$$\frac{\partial \vec{A}}{\partial t} = -\frac{\mu_0 p_0 \omega^2}{4\pi r}\cos\omega(t - r/c)\left(\cos\theta\,\hat{r} - \sin\theta\,\hat{\theta}\right) \tag{15.12-14}$$

Therefore, electric field

$$\vec{E} = -\vec{\nabla}\phi - \frac{\partial \vec{A}}{\partial t} = -\hat{\theta}\frac{\mu_0 p_0 \omega^2}{4\pi}\left(\frac{\sin\theta}{r}\right)\cos\omega(t - r/c) \tag{15.12-15}$$

Now $$\vec{\nabla}\times\vec{A} = \hat{\varphi}\frac{1}{r}\left[\frac{\partial}{\partial r}(rA_\theta) - \frac{\partial A_r}{\partial\theta}\right]$$
$$= -\hat{\varphi}\frac{\mu_0 p_0 \omega}{4\pi r}\left[\frac{\omega}{c}\sin\theta\cos\omega(t-r/c) + \frac{\sin\theta}{r}\sin\omega(t-r/c)\right].$$

For $r \gg c/\omega$ we can neglect the second term again and get

$$\vec{B} = \vec{\nabla}\times\vec{A} \simeq -\hat{\varphi}\frac{\mu_0 p_0 \omega^2}{4\pi c}\left(\frac{\sin\theta}{r}\right)\cos\omega(t-r/c). \qquad (15.12\text{-}16)$$

Equations (15.12-15) and (15.12-16) represent monochromatic waves of frequency ω travelling in the radial direction from the dipole with the speed of light. Obviously, $\vec{E}$ and $\vec{B}$ are in phase, they are transverse to the direction of propagation and also mutually perpendicular to each other. The ratio of their amplitude is c. In fact $\vec{E}$ and $\vec{B}$ represent spherical waves.

The power radiated by the oscillating dipole is given by the Poynting's vector as

$$\vec{s} = \vec{E}\times\vec{H} = \frac{1}{\mu_0}\left(\vec{E}\times\vec{B}\right) = \hat{r}\frac{\mu_0}{c}\left[\frac{p_0\omega^2}{4\pi}\left(\frac{\sin\theta}{r}\right)\cos\omega(t-r/c)\right]^2. \qquad (15.12\text{-}17)$$

The intensity of radiation is obtained by taking time average over a complete cycle. Therefore,

$$\langle\vec{s}\rangle = \hat{r}\cdot\frac{\mu_0 p_0^2\omega^4}{32\pi^2 c}\frac{\sin^2\theta}{r^2}. \qquad (15.12\text{-}18)$$

Note that there is no radiation along the axis ($\theta = 0$) of the dipole and the intensity is maximum in the equatorial plane ($\theta = \pi/2$). The intensity varies as $\sin^2\theta$ and hence, its profile looks like a figure of eight (Fig. 15.12-2).

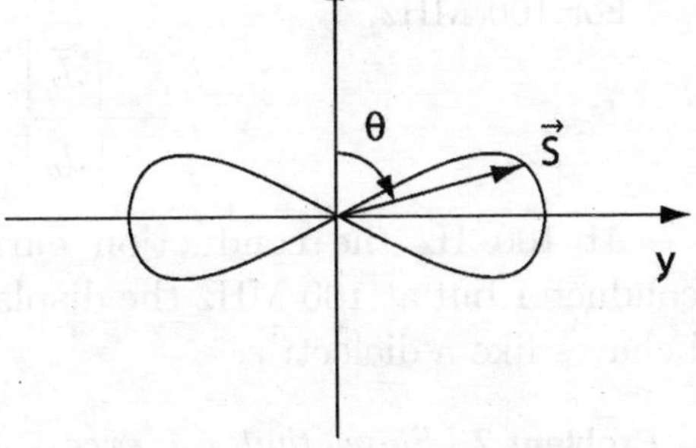

Fig. 15.12.2: Intensity profile of dipole radiation

Unit area on the surface of a sphere of radius r subtends a solid angle $1/r^2$ at the dipole. So the time average power radiated per unit solid angle is given by

$$\left\langle\frac{dP}{d\Omega}\right\rangle = |\langle\vec{s}\rangle| \Big/ \frac{1}{r^2} = \frac{\mu_0 p_0^2\omega^4}{32\pi^2 c}\sin^2\theta = \frac{p_0^2\omega^4\sin^2\theta}{32\pi^2\varepsilon_0 c^3}. \qquad (15.12\text{-}19)$$

The total power radiated can be obtained by integrating $\langle\vec{S}\rangle$ over a sphere of radius r or integrating (15.12-19) over all solid angles. Thus,

$$P = \int\langle\vec{s}\rangle\cdot d\vec{S} = \frac{\mu_0 p_0^2\omega^4}{32\pi^2 c}\int_{\theta=0}^{\pi}\int_{\varphi=0}^{2\pi}\frac{\sin^2\theta}{r^2}r^2\sin\theta\, d\theta\, d\varphi = \frac{1}{4\pi\varepsilon_0}\cdot\frac{p_0^2\omega^4}{3c^3}. \qquad (15.12\text{-}20)$$

It is independent of the radius of the sphere as expected from conservation of energy. Note that the total power radiated by an oscillating dipole is proportional to the fourth power of the frequency.

SOLVED PROBLEMS

► **Problem 1.** *Consider a medium of dielectric constant* $K = 80$ *and conductivity* $\sigma = 10^{-3}$ S/m. *Compare the value of conduction and displacement current densities at frequencies* 100 Hz *and* 100 MHz. *Comment on the result.*

Solution. Conduction current density $\vec{J}_c = \sigma\vec{E}$ and displacement current density

$$\vec{J}_d = \frac{\partial\vec{D}}{\partial t} = \varepsilon\frac{\partial\vec{E}}{\partial t}.$$

Assuming harmonic variation $\vec{E} \sim \vec{E}_0 e^{-j\omega t}$ we can write

$$\left|\vec{J}_d\right|_{\max} = \varepsilon\omega\left|\vec{E}\right|.$$

Therefore,

$$\frac{\left|\vec{J}_c\right|}{\left|\vec{J}_d\right|_{\max}} = \frac{\sigma}{\varepsilon\omega} = \frac{10^{-3}}{80 \times 8.854 \times 10^{-12} \times 2\pi \times 100} \quad \text{(for 100 Hz)}$$

$$= 2248.$$

For 100 MHz,

$$\frac{\left|\vec{J}_c\right|}{\left|\vec{J}_d\right|_{\max}} = \frac{2248}{10^6} = 0.002248.$$

At 100 Hz the conduction current dominates and the medium behaves like a conductor but at 100 MHz the displacement current dominates and hence, the medium behaves like a dielectric.

► **Problem 2.** *Show that an excess charge placed at any point in a medium of conductivity* σ *and permittivity* ε *decays exponentially with a time constant* ε/σ. *Find this characteristic time for a conductor having conductivity* $\sigma = 3 \times 10^7(\Omega\text{-m})^{-1}$ *and dielectric constant* $\varepsilon_r = K = 1$. **[cf. C.U. 2017]**

Solution. We have the Maxwell's equation

$$\vec{\nabla} \times \vec{H} = \vec{J} + \frac{\partial\vec{D}}{\partial t} = \sigma\vec{E} + \frac{\partial\vec{D}}{\partial t}.$$

Taking divergence we get

$$\sigma\vec{\nabla}\cdot\vec{E} + \frac{\partial}{\partial t}\left(\vec{\nabla}\cdot\vec{D}\right) = 0 \quad \text{or} \quad \sigma\cdot\frac{\rho}{\varepsilon} + \frac{\partial\rho}{\partial t} = 0 \quad \left[\because\ \vec{\nabla}\cdot\vec{D} = \rho \text{ and } \vec{D} = \varepsilon\vec{E}\right].$$

Integrating we get $\rho = \rho_0 e^{-\frac{\sigma t}{\varepsilon}}$, where $\rho = \rho_0$ at $t = 0$.
Thus, the charge density decays exponentially with time constant $\tau = \varepsilon/\sigma$.

For the given conductor,

$$\tau = \frac{\varepsilon}{\sigma} = \frac{1}{3 \times 10^7}\,\text{s} = 33.3\,\text{ns}.$$

▶ **Problem 3.** *Suppose we have two plane waves represented by*

$$f(\vec{r}, t) = A\cos\left(\vec{k}\cdot\vec{r} - \omega t + \delta_1\right) \quad \textit{and} \quad g(\vec{r}, t) = B\cos\left(\vec{k}\cdot\vec{r} - \omega t + \delta_2\right).$$

Show that the time average value of fg can be represented as $\langle fg\rangle = \frac{1}{2}Re\left[f_c g_c^*\right]$, where $f_c = A_c e^{j(\vec{k}\cdot\vec{r}-\omega t)}$; $A_c = Ae^{j\delta_1}$
$g_c = B_c e^{j(\vec{k}\cdot\vec{r}-\omega t)}$; $B_c = Be^{j\delta_2}$

Solution.

$$\begin{aligned}\langle fg\rangle &= \frac{1}{T}\int_0^T A\cos\left(\vec{k}\cdot\vec{r} - \omega t + \delta_1\right)\cdot B\cos\left(\vec{k}\cdot\vec{r} - \omega t + \delta_2\right) dt \\ &= \frac{AB}{2T}\int_0^T \left[\cos\left(2\vec{k}\cdot\vec{r} - 2\omega t + \delta_1 + \delta_2\right) + \cos\left(\delta_1 - \delta_2\right)\right] dt \\ &= \frac{AB}{2}\cos\left(\delta_1 - \delta_2\right).\end{aligned}$$

Again,

$$\tfrac{1}{2} f_c g_c^* = \tfrac{1}{2} A_c B_c^* = \tfrac{1}{2} AB e^{j(\delta_1 - \delta_2)}.$$

$$\therefore \quad \tfrac{1}{2} Re\left[f_c g_c^*\right] = \tfrac{1}{2} AB\cos\left(\delta_1 - \delta_2\right) = \langle fg\rangle.$$

▶ **Problem 4.** *Two sinusoidal waves are described by the following wave functions:*

$$\psi_1(z,t) = A_1\cos(kz - \omega t + \delta_1) \quad \textit{and} \quad \psi_2(z,t) = A_2\cos(kz - \omega t + \delta_2).$$

Show that by superposing these waves we again get a sinusoidal wave.

Solution. Method 1:

$$\begin{aligned}\psi(z,t) &= \psi_1(z,t) + \psi_2(z,t) \\ &= (A_1\cos\delta_1 + A_2\cos\delta_2)\cos(kz - \omega t) - (A_1\sin\delta_1 + A_2\sin\delta_2)\sin(kz - \omega t).\end{aligned}$$

Putting $A_1\cos\delta_1 + A_2\cos\delta_2 = A\cos\delta$ and $A_1\sin\delta_1 + A_2\sin\delta_2 = A\sin\delta$ we get

$$\psi(z,t) = A\cos(kz - \omega t + \delta).$$

Obviously, the superposed wave is a sinusoidal one. Here

$$A^2 = A_1^2 + A_2^2 + 2A_1A_2\cos(\delta_1 - \delta_2) \quad \text{and} \quad \tan\delta = \frac{A_1\sin\delta_1 + A_2\sin\delta_2}{A_1\cos\delta_1 + A_2\cos\delta_2}$$

Method 2: Let us represent the waves as the real part of complex waves as

$$\psi_1(z,t) = ReA_{1c}e^{j(kz-\omega t)} \quad \text{and} \quad \psi_2(z,t) = ReA_{2c}e^{j(kz-\omega t)},$$

where $A_{1c} = A_1e^{j\delta}$ and $A_{2c} = A_2e^{j\delta_2}$.

Now the superposed wave is

$$\begin{aligned}\psi(z,t) &= \psi_1(z,t) + \psi_2(z,t) = Re\left[A_{1c}e^{j(kz-\omega t)} + A_{2c}e^{j(kz-\omega t)}\right]\\ &= Re\,(A_{1c} + A_{2c})\,e^{j(kz-\omega t)} = ReAe^{j(kz-\omega t+\delta)}\end{aligned}$$

Obviously, it is a complex representation of a sinusoidal wave. Here we have put $A_{1c} + A_{2c} = Ae^{j\delta}$.

$\therefore\ A_1e^{j\delta_1} + A_2e^{j\delta_2} = Ae^{j\delta}$. Equating real and imaginary parts we get the same values of A and δ as in method 1.

▶ **Problem 5.** *Describe the state of polarization of the EM wave represented by*

$$\vec{E}(z,t) = \hat{\imath}E_0\cos(kz-\omega t) - \hat{\jmath}E_0\cos(kz-\omega t)$$ **[C.U. 2011]**

Solution.

$$\vec{E}(z,t) = \left(\hat{\imath} - \hat{\jmath}\right)E_0\cos(kz-\omega t) = \hat{n}\cdot\sqrt{2}E_0\cos(kz-\omega t),$$

where $\hat{n} = (\hat{\imath} - \hat{\jmath})/\sqrt{2} = \cos 45°\hat{\imath} - \sin 45°\hat{\jmath}$ is a constant unit vector in the xy-plane. So we conclude that the given equation represents a linearly polarized wave with its electric vector making an angle of 45° with both the x- and y-axes and lying in the second and fourth quadrant.

This can also be easily proved by writing

$$\vec{E}(z,t) = E_x\hat{\imath} + E_y\hat{\jmath},$$

where $E_x = E_0\cos(kz-\omega t)$ and $E_y = -E_0\cos(kz-\omega t)$.

Thus, $E_y = -E_x$, which is an equation for a straight line in the E_x, E_y-plane making an angle of 45° with the axes and the line lies in the second and fourth quadrants.

▶ **Problem 6.** *Find an expression for a linearly polarized EM wave of angular frequency ω, propagating along the positive z-direction with its plane of vibration at 30° to the xz-plane.*

Solution. Let the electric vector associated with the wave be represented by

$$\vec{E}(z,t) = \vec{E}_0\cos(kz-\omega t),$$

where $\vec{E}_0$ lies in the xy-plane and makes an angle 30° with x-axis. So we can write

$$\vec{E}_0 = E_0\cos 30°\hat{\imath} + E_0\sin 30°\hat{\jmath} = \frac{E_0}{2}\left(\sqrt{3}\,\hat{\imath} + \hat{\jmath}\right).$$

Thus, the expression

$$\vec{E}(z,t) = \frac{E_0}{2}\left(\sqrt{3}\,\hat{\imath} + \hat{\jmath}\right)\cos(kz - \omega t)$$

may be taken to represent the wave in question.

▶ **Problem 7.** *A light bulb of 20 W radiates energy isotropically, mostly in infrared region. Assuming it to be a point source find the irradiance (average energy per unit area per unit time or intensity) at a distance 1 m away. Also, calculate the strength of $\vec{E}$-field associated with the radiation at this distance.*

Solution. Irradiance

$$I = \frac{20}{4\pi \cdot 1^2}\,\mathrm{W/m^2} = 1.59\,\mathrm{W/m^2}.$$

Now $I = \dfrac{E_{\mathrm{rms}}^2}{c\mu_0}$. Therefore,

$$E_{\mathrm{rms}} = \sqrt{Ic\mu_0} = \sqrt{1.59 \times 3 \times 10^8 \times 4\pi \times 10^{-7}} = 24.5\ \mathrm{Vm^{-1}}.$$

▶ **Problem 8.** *A parallel plate capacitor with circular plates of radius $a = 5.5\,\mathrm{cm}$ is being charged at a uniform rate so that the electric field between the plates changes at a constant rate $dE/dt = 1.5 \times 10^{12}\mathrm{V \cdot m^{-1} \cdot s^{-1}}$. Find the displacement current for the capacitor. Find expressions for the induced magnetic field at various radial distances r from the centre of the capacitor, $r < a$, $r = a$, $r > a$. Find the numerical value of B at $r = a$. Comment on the magnitude of the field and compare it with the horizontal component of the earth's magnetic field near equator.* **[cf. C.U. 2006]**

Solution. The displacement current density is

$$J_d = \varepsilon_0 \frac{\partial E}{\partial t}$$

and total displacement current

$$I_d = \pi a^2 \cdot J_d = \pi a^2 \cdot \varepsilon_0 \frac{\partial E}{\partial t} = \pi(0.055)^2 \times 8.854 \times 10^{-12} \times 1.5 \times 10^{12} = 0.13\ \mathrm{A}.$$

Considering a circular path of radius $r \le a$, we can write from Ampere's law,

$$\oint \vec{B} \cdot d\vec{l} = \mu_0 \int \vec{J}_d \cdot d\vec{S} \quad \text{or} \quad B \cdot 2\pi r = \mu_0 \varepsilon_0 \frac{\partial E}{\partial t} \cdot \pi r^2$$

$$\text{or}\ B = \frac{1}{2}\mu_0\varepsilon_0 \cdot r \frac{\partial E}{\partial t}$$

For $r \ge a$, we have

$$B \cdot 2\pi r = \mu_0 \cdot J_d \pi a^2 = \mu_0 \varepsilon_0 \frac{\partial E}{\partial t}\pi a^2 \quad \text{or} \quad B = \frac{1}{2}\mu_0\varepsilon_0 \cdot \frac{a^2}{r} \cdot \frac{\partial E}{\partial t}.$$

The magnitude of the magnetic field at $r = a$ is

$$B = \tfrac{1}{2}\mu_0\varepsilon_0 \cdot a\frac{\partial E}{\partial t}$$
$$= \tfrac{1}{2} \times 4\pi \times 10^{-7} \times 8.854 \times 10^{-12} \times 0.055 \times 1.5 \times 10^{12}$$
$$= 0.46 \times 10^{-6}\,\text{T} = 0.46\,\mu\text{T}.$$

Note that the displacement current though significant it produces a magnetic field which is much smaller than even the horizontal component of the earth's magnetic field ($\sim 36\,\mu$T at Kolkata, India).

▶ **Problem 9.** *The intensity of sunlight reaching the earth's surface is about* 1300 W.m^{-2}. *Calculate the strength of electric and magnetic fields of the incoming sunlight.* **[C.U. 2008]**

Solution. The time average Poynting's vector is

$$\langle \vec{s} \rangle = \tfrac{1}{2} Re\left(\vec{E} \times \vec{H}^*\right).$$

Assuming harmonic variations

$$\vec{E} = \vec{E}_0 e^{-j\omega t} \quad \text{and} \quad \vec{H} = \vec{H}_0 e^{-j\omega t}$$

we get

$$s_{av} = |\langle \vec{s} \rangle| = \frac{1}{2} E_0 \cdot H_0 = \frac{E_0}{\sqrt{2}} \cdot \frac{H_0}{\sqrt{2}} = E_{\text{rms}} \cdot H_{\text{rms}}.$$

Now

$$B = \mu_0 H = \frac{E}{c} \quad \text{or} \quad H_{\text{rms}} = \frac{E_{\text{rms}}}{c\mu_0}.$$

$$\therefore \quad s_{av} = \frac{E_{\text{rms}}^2}{c\mu_0}$$

$$\text{or} \quad E_{\text{rms}} = \sqrt{c\mu_0 \cdot S_{av}} = \sqrt{3 \times 10^8 \times 4\pi \times 10^{-7} \times 1300}\ \text{V} \cdot \text{m}^{-1} = 700\,\text{Vm}^{-1}.$$

$$\therefore \quad B_{\text{rms}} = E_{\text{rms}}/c = 700/(3 \times 10^8)\,\text{T} = 2.33 \times 10^{-6}\,\text{T}.$$

▶ **Problem 10.** *Suppose a wire of radius a, length l and resistance R is carrying a steady current I with a pd V across its length. Find* $\vec{E}$ *and* $\vec{H}$ *on the surface of the wire. Calculate the Poynting's vector* $\vec{s}$ *and show that it represents a flow of energy into the wire from the surrounding space. Interpret the result.* **[cf. C.U. 2017]**

Solution. Let the axis of the wire is along z-axis. The electric field is parallel to the wire and is given as

$$\vec{E} = \hat{z}\frac{V}{l} = \hat{z}\frac{IR}{l}.$$

The magnetic field on the surface is tangential to the curved surface and is given by Ampere's law as

$$H \cdot 2\pi a = I \quad \text{or} \quad H = \frac{I}{2\pi a}.$$

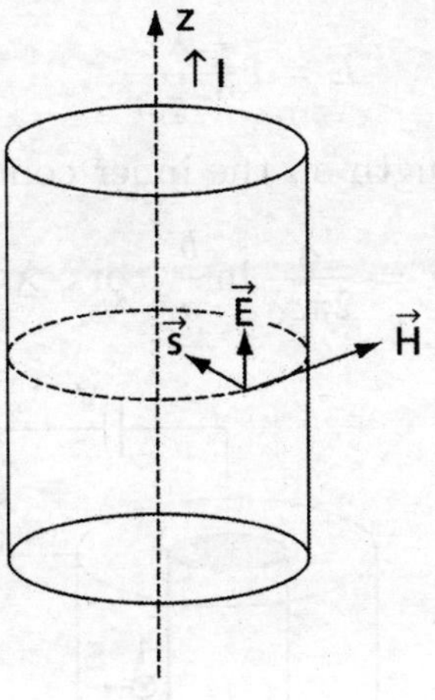

Fig 15.P-10

In vector form

$$\vec{H} = \hat{\theta}\frac{I}{2\pi a},$$

where $\hat{\theta}$ is a unit vector tangential to the surface at the point under consideration.

Now the Poynting's vector $\vec{s} = \vec{E} \times \vec{H}$ is obviously directed radially into the surface of the wire and is of magnitude

$$s = EH = \frac{I^2 R}{2\pi a l}.$$

So we can write

$$\vec{s} = -\hat{n}\frac{I^2 R}{2\pi a l},$$

where $\hat{n}$ is a unit outward normal to the surface of the wire. The rate of passage of electromagnetic energy through the curved surface of the wire is

$$s \cdot 2\pi a l = I^2 R = \text{rate of Joule heating.}$$

The result may be interpreted as follows: The energy from the voltage source is transmitted through the space surrounding the wire in the form of electromagnetic energy. It then flows from outside into the wire and is completely converted into Joule's heat. Thus *Poynting's theorem can explain Joule heating.*

► **Problem 11.** *Suppose energy is being transmitted from a constant voltage source V to a load R via a long coaxial cable of negligible resistance. As a result, a steady current I flows down the cable. Show by explicit calculation of Poynting's vector that electromagnetic energy flows down the cable at the rate VI.* **[C.U. 2016]**

Solution. Let the radii of the inner and outer conductors be a and b, respectively and the space between them is a free space. The axis of the cable is chosen as the z-axis of an $(\hat{r}, \theta, z)$ cylindrical coordinate system. The electric field at a distance r from the axis of the cable can be easily found by using Gauss's theorem as

$$\vec{E} = \hat{r}\frac{\lambda}{2\pi\varepsilon_0 r},$$

where λ is the charge per unit length on the inner conductor. The pd

$$V = \int_a^b E dr = \frac{\lambda}{2\pi\varepsilon_0}\ln\frac{b}{a} \quad \text{or} \quad \lambda = \frac{2\pi\varepsilon_0 V}{\ln\frac{b}{a}}.$$

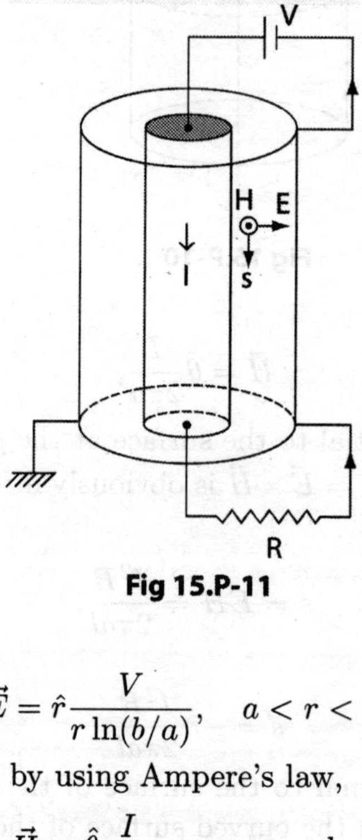

Fig 15.P-11

Therefore,

$$\vec{E} = \hat{r}\frac{V}{r\ln(b/a)}, \quad a < r < b$$

The magnetic field is obtained by using Ampere's law,

$$\vec{H} = \hat{\theta}\frac{I}{2\pi r}, \quad a < r < b.$$

Therefore, Poynting's vector

$$\vec{s} = \vec{E} \times \vec{H} = \hat{z}\frac{VI}{2\pi r^2 \ln(b/a)}.$$

As $\vec{s}$ is directed along the axis of the cable, the electromagnetic energy flows down the cable. As $\vec{E}$ is zero everywhere except for $a < r < b$, we can calculate total rate of flow of electromagnetic energy by integrating $\vec{s}$ over the cross-section of the cable. Thus, the rate of flow of energy is

$$\int \vec{s} \cdot \hat{z}\, 2\pi r dr = \frac{VI}{\ln(b/a)}\int_a^b \frac{dr}{r} = VI.$$

▶ **Problem 12.** *Consider a spherical capacitor with an inner sphere of radius a and outer grounded sphere of radius b. The space between them is filled with a dielectric of dielectric constant 6. A voltage $v = 200 \sin 100\pi t$ volt is applied across the capacitor. Calculate the displacement current through the dielectric.*

Solution. The capacitance of the capacitor is

$$C = 4\pi \times 6\varepsilon_0 \times \frac{ab}{b-a}.$$

The instantaneous charge on the capacitor is

$$q = Cv = C \times 200 \sin 100\pi t.$$

Therefore, conduction current

$$I = \frac{dq}{dt} = 200\,\mathrm{C} \times 100\,\pi \cos 100\pi t$$

$$= 200 \times 4\pi \times 6\varepsilon_0 \times \frac{ab}{b-a} \times 100\pi \cos 100\pi t.$$

Putting $\varepsilon_0 = 8.854 \times 10^{-12}\,\mathrm{F/m}$ we get

$$I = 4.19 \times 10^{-5} \times \frac{ab}{b-a} \times \cos 100\pi t \text{ ampere.}$$

Continuity equation demands that the displacement current within the dielectric will be equal to this conduction current.

▶ **Problem 13.** *The current flowing through a long solenoid of radius a is varied in such a manner that the axial magnetic field inside increases with time according to the law $\vec{B} = \hat{r} B_0 t^2$, where B_0 is a constant. Find the displacement current density as a function of the distance r from the axis of the solenoid ($r < a$).*

Solution. From Maxwell's equation

$$\vec{\nabla} \times \vec{E} = -\frac{\partial \vec{B}}{\partial t} = -\hat{r} 2B_0 t.$$

Integrating over a cross-section of radius r we get

$$\int \vec{\nabla} \times \vec{E} \cdot d\vec{S} = -\int 2B_0 t \hat{r} \cdot \hat{r}\, dS \quad \text{or} \quad \oint \vec{E} \cdot d\vec{l} = -2B_0 t \cdot \pi r^2$$

$$\text{or,} \quad E \cdot 2\pi r = -2B_0 t \pi r^2$$

$$\text{or,} \quad E = -B_0 t r.$$

$\therefore$ The magnitude of displacement current density is

$$\left| \vec{J}_d \right| = \varepsilon_0 \left| \frac{\partial \vec{E}}{\partial t} \right| = \varepsilon_0 B_0 r.$$

▶ **Problem 14.**

(i) *Show that for a good conductor skin depth* $\delta = \lambda_c/2\pi$, *where* λ_c *is the wavelength of electromagnetic waves in the conductor.* **[C.U. 2003]**

(ii)) *Show that for an electromagnetic wave incident on a good conductor the electric vector reduces to about* 1% *at a depth of* $0.73\ \lambda_c$.

(iii) *Find the wavelength and the propagation speed in copper for radio waves at* 1 MHz. *For copper assume* $\mu = \mu_0$, $\varepsilon = \varepsilon_0$ *and conductivity* $\sigma = 5.8 \times 10^7 (\Omega\text{-}m)^{-1}$.

Solution. (i) Suppose the wave is incident normally on the surface of the conductor along z-axis. Then the electric field inside the conductor is given by Eq. (15.7-11) as

$$\vec{E}(z,t) = \vec{E}_0 e^{-\beta z} \cdot e^{j(\alpha z - \omega t)}.$$

For a good conductor

$$\alpha = \beta = \sqrt{\frac{\omega\mu\sigma}{2}}$$

and skin depth

$$\delta = \frac{1}{\beta} = \sqrt{\frac{2}{\omega\mu\sigma}}.$$

The wavelength of the wave in the conductor is

$$\lambda_c = \frac{2\pi}{\alpha} = \frac{2\pi}{\beta} = 2\pi\delta.$$

$\therefore$ Skin depth $\delta = \lambda_c/2\pi$.

(ii) Let us write

$$E(z) = E_0 e^{-\beta z} = E_0 e^{-z/\delta} = E_0 e^{-2\pi z/\lambda_c}.$$

Now putting

$$\frac{E(z)}{E_0} = \frac{1}{100}$$

we get

$$z = \frac{\lambda_c}{2\pi} \ln \frac{E_0}{E(z)} = \frac{\lambda_c}{2\pi} \ln 100 = 0.73\,\lambda_c.$$

(iii) The required wavelength

$$\lambda_c = \frac{2\pi}{\alpha} = 2\pi\sqrt{\frac{2}{\omega\mu\sigma}}$$

$$= 2\pi\sqrt{\frac{2}{2\pi \times 10^6 \times 4\pi \times 10^{-7} \times 5.8 \times 10^7}}\ \text{m}$$

$$= 4.15 \times 10^{-4}\ \text{m}$$

and the propagation speed

$$v = \frac{\omega}{k_c} = \frac{\omega}{2\pi} \times \lambda_c = \frac{2\pi \times 10^6}{2\pi} \times 4.15 \times 10^{-4}\,\text{ms}^{-1} = 415\,\text{ms}^{-1}.$$

▶ **Problem 15.** *The electric field associated with an electromagnetic wave is*

$$\vec{E} = \hat{x}E_0 \cos(kz - \omega t) + \hat{y}E_0 \sin(kz - \omega t),$$

where E_0 is a constant. Find the corresponding magnetic field $\vec{H}$ and the Poynting's vector $\vec{s}$.

Solution. We know that

$$\vec{H} = \frac{\vec{k} \times \vec{E}}{\mu\omega}.$$

Here $\vec{k} = \hat{z}k$.

$$\begin{aligned}\therefore \quad \vec{H} &= \frac{k}{\mu\omega}\hat{z} \times [\hat{x}E_0 \cos(kz - \omega t) + \hat{y}E_0 \sin(kz - \omega t)] \\ &= \hat{y}\frac{kE_0}{\mu\omega}\cos(kz - \omega t) - \hat{x}\frac{kE_0}{\mu\omega}\sin(kz - \omega t).\end{aligned}$$

Poynting's vector

$$\vec{s} = \vec{E} \times \vec{H} = \hat{z}\frac{kE_0^2}{\mu\omega}\left[\cos^2(kz - \omega t) + \sin^2(kz - \omega t)\right] = \hat{z}\frac{kE_0^2}{\mu\omega}.$$

▶ **Problem 16.** *An electromagnetic wave is propagating through a nonconducting medium characterised by permittivity $5\varepsilon_0$ and permeability $2\mu_0$. The magnetic field associated with the wave is*

$$\vec{H} = \hat{y}2\cos(3z - \omega t)\ A/m,$$

where z is in metre. Find the value of ω. Also find the electric field associated with the wave. What is the direction of propagation of the wave?

Solution. Speed of the wave is given by

$$v = \frac{\omega}{k} = \frac{1}{\sqrt{\mu\varepsilon}}$$

$$\begin{aligned}\text{or} \quad \omega &= \frac{k}{\sqrt{\mu\varepsilon}} = \frac{3}{\sqrt{2\mu_0 \cdot 5\varepsilon_0}} = \frac{3}{\sqrt{10}}c \\ &= \frac{3}{\sqrt{10}} \times 3 \times 10^8\,\text{rad/s} = 2.846 \times 10^8\,\text{rad/s}.\end{aligned}$$

From the equation

$$\vec{\nabla} \times \vec{H} = \frac{\partial \vec{D}}{\partial t},$$

we get

$$j\vec{k} \times \vec{H} = -j\omega\varepsilon\vec{E}.$$

$$\therefore \quad \vec{E} = -\frac{\vec{k} \times \vec{H}}{\varepsilon\omega} = \hat{x}\frac{k \cdot 2}{\varepsilon\omega}\cos(3z - \omega t)$$

$$= \hat{x}\frac{3 \cdot 2 \cdot \cos(3z - \omega t)}{5 \times 8.854 \times 10^{-12} \times 2.846 \times 10^{8}}$$

$$= \hat{x}476.2\cos(3z - 2.846 \times 10^{8}t)\ \text{V/m}$$

The constant phase planes are

$$\phi = 3z - \omega t = \text{constant}.$$

The direction of propagation is perpendicular to these planes, hence,

$$\frac{\vec{k}}{k} = \frac{\vec{\nabla}\phi}{\left|\vec{\nabla}\phi\right|} = \hat{z},$$

i.e., the wave propagates along $+z$-direction.

► **Problem 17.** *In a certain region of space through which an EM wave is propagating the Poynting's vector is given by* $\vec{s} = \hat{z}\ 0.16\cos^2(kz - \omega t)\ \text{W/m}^2$. *Find the total time average power carried by the wave through* 100 cm^2 *of area on the plane* $y + 2z = 5$.

Solution. The time average Poynting's vector is

$$\vec{s}_{av} = \frac{1}{T}\int_0^T \vec{s}dt = \hat{z}\,0.16\frac{1}{T}\int_0^T \cos^2(kz - \omega t)dt\ \text{W/m}^2 = \hat{z}\,0.08\ \text{W/m}^2.$$

The unit normal to the plane $\phi = y + 2z - 5 = 0$ is

$$\hat{n} = \frac{\vec{\nabla}\phi}{\left|\vec{\nabla}\phi\right|} = \frac{\hat{y} + 2\hat{z}}{\sqrt{5}}.$$

Hence, the required power

$$= \int \vec{s}_{av} \cdot \hat{n}\,dS \quad (\hat{n}\,dS = \text{surface element})$$

$$= 0.08\hat{z} \cdot \left(100 \times 10^{-4}\right)\hat{n}$$

$$= 8 \times 10^{-4} \times \frac{2}{\sqrt{5}}\ \text{W}$$

$$= 0.72\,\text{mW}.$$

► **Problem 18.** *Suppose the radiation fields of a point like source are given in (r, θ, φ) spherical coordinates by*

$$E_\theta = 377 H_\varphi, \quad \vec{H} = \hat{\varphi}\frac{1}{r}\sin\theta \cdot \cos(kr - \omega t)$$

$$and \ \vec{E} = \hat{\theta}\frac{377}{r}\sin\theta\cos(kr - \omega t).$$

Find the Poynting's vector and hence, calculate the time average power radiated by the source (assume to be at the origin).

Solution. Poynting's vector $\vec{s} = \vec{E} \times \vec{H} = \hat{r}\frac{377}{r^2}\sin^2\theta\cos^2(kr - \omega t)$

$\therefore$ Time average Poynting's vector

$$\vec{s}_{av} = \hat{r}\frac{377}{r^2}\sin^2\theta \frac{1}{T}\int_0^T \cos^2(kr - \omega t)dt = \hat{r}\frac{377}{2r^2}\sin^2\theta.$$

Considering a sphere of radius r with its centre at the source we can calculate the total power radiated by the source as $P = \int \vec{s}_{av} \cdot d\vec{S}$, where $d\vec{S} = \hat{r}r^2 \sin\theta d\theta d\varphi$ is a surface element.

Therefore,

$$P = \frac{377}{2}\int_0^\pi \sin^3\theta d\theta \int_0^{2\pi} d\varphi \ W = 1578 \ W.$$

► **Problem 19.** *Consider the propagation of EM waves through dilute ionised gases such as the ionosphere. Hence, show that the critical frequency below which wave propagation through it is not possible is given by $f_c \simeq 9\sqrt{n_0}$, where n_0 is the number of electrons per metre^3.*

Solution. The effective refractive index of the medium is given by (15.8-8),

$$n = \sqrt{1 - \left(\frac{\omega_p}{\omega}\right)^2},$$

where $\omega_p = \sqrt{\frac{n_0 e^2}{\varepsilon_0 m}}$ is the electron plasma frequency.

For free wave propagation n should be real, i.e., $\omega \geq \omega_p$.

$\therefore$ The critical frequency

$$f_c = \frac{\omega_p}{2\pi} = \frac{1}{2\pi}\sqrt{\frac{n_0 e^2}{\varepsilon_0 m}} = \frac{1}{2\pi} \cdot \frac{1.6 \times 10^{-19}}{\sqrt{8.854 \times 10^{-12} \times 9.1 \times 10^{-31}}}\sqrt{n_0} \approx 9\sqrt{n_0} \ \text{Hz}.$$

► **Problem 20.** *Calculate the retarded potentials of an infinite, straight wire in which a constant current I_0 is turned on abruptly at $t = 0$. Also, find the resulting electric and magnetic fields.*

Solution. Consider an infinite straight wire in which a constant current I_0 is started at $t = 0$. Thus, current $I(t) = 0$ for $t < 0$ and $I(t) = I_0$ for $t \geq 0$. Since the wire is electrically neutral charge density $\rho = 0$ everywhere in the wire and hence, scalar potential produced by the wire at the point of observation P is zero. Assuming the wire to be along z-axis (Fig. 15.P-20), we can write for the vector potential at P as

$$\vec{A}(\vec{r}, t) = \hat{k}\frac{\mu_0}{4\pi}\int_{-\infty}^{+\infty}\frac{I(t - R/c)dz}{R}.$$

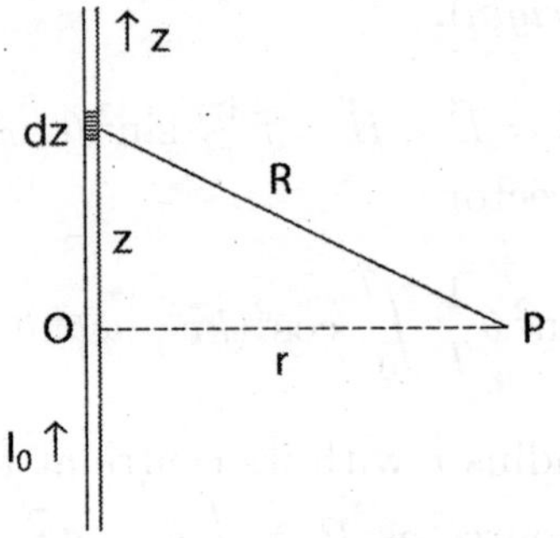

Fig 15.P-20

Because of the finite speed of disturbance, the retarded current is restricted. A given element dz will contribute only if its distance from P is less than ct. Thus, $I(t - R/c) = 0$ when $R^2 \geq c^2t^2$, i.e., when $|z|^2 \geq c^2t^2 - r^2$ and $I(t - R/c) = I_0$ when $|z|^2 < c^2t^2 - r^2$.

Thus, $\vec{A}(\vec{r}, t) = 0$ for $r \geq ct$ and

$$\vec{A}(\vec{r}, t) = \hat{k}\frac{\mu_0 I_0}{4\pi}\int_{-\sqrt{c^2t^2-r^2}}^{+\sqrt{c^2t^2-r^2}}\frac{dz}{\sqrt{z^2+r^2}} \quad \text{for } r < ct$$

$$= \hat{k}\frac{2\mu_0 I_0}{4\pi}\ln\left[z + \sqrt{z^2+r^2}\right]\Big|_0^{\sqrt{c^2t^2-r^2}} = \hat{k}\frac{\mu_0 I_0}{2\pi}\ln\left(\frac{\sqrt{c^2t^2-r^2}+ct}{r}\right).$$

Note that in the limit $t \to \infty$ this potential reduces to the corresponding result of the static case; thus,

$$\vec{A}(\vec{r}, t) = -\hat{k}\frac{\mu_0 I_0}{2\pi}\ln r + \text{a very large quantity.}$$

The electric field is

$$\vec{E} = -\frac{\partial \vec{A}}{\partial t} = -\hat{k}\frac{\mu_0 I_0 c}{2\pi\sqrt{c^2t^2-r^2}} \qquad \left[\text{Use: } \frac{d}{dz}\ln\left(z + \sqrt{z^2+a^2}\right) = \frac{1}{\sqrt{z^2+a^2}}\right]$$

The magnetic field is

$$\vec{B} = \vec{\nabla}\times\vec{A} = \frac{1}{r}\begin{vmatrix} \hat{r} & r\hat{\theta} & \hat{z} \\ \frac{\partial}{\partial r} & \frac{\partial}{\partial \theta} & \frac{\partial}{\partial z} \\ 0 & 0 & A_z \end{vmatrix} = -\hat{\theta}\frac{\partial A_z}{\partial r} = \hat{\theta}\frac{\mu_0 I_0}{2\pi r}\frac{ct}{\sqrt{c^2t^2-r^2}}.$$

Note that in the limit $t \to \infty$, $\vec{E} = 0$ and $\vec{B} = \hat{\theta}\dfrac{\mu_0 I_0}{2\pi r}$, which corresponds to the static case.

▶ **Problem 21.** *Suppose the $\vec{E}$-field associated with a simple spherical wave is given by*

$$\vec{E}(r, \theta, \varphi, t) = A\frac{\sin\theta}{r}\left[\cos(kr - \omega t) - \frac{1}{kr}\sin(kr - \omega t)\right]\hat{\varphi}$$

with $\omega/k = c$.

(i) *Show that it obeys all the four Maxwell's equations in vacuum. Find the associated magnetic field.*

(ii) *Calculate the Poynting's vector and its time average value.*

(iii) *Integrating the time average Poynting's vector over a spherical surface calculate the total power radiated.*

Solution.

(i) $$\vec{\nabla} \cdot \vec{E} = \frac{1}{r\sin\theta}\frac{\partial E_\varphi}{\partial \varphi} = 0$$

$$\vec{\nabla} \times \vec{E} = \frac{1}{r^2 \sin\theta} \cdot \begin{vmatrix} \hat{r} & r\hat{\theta} & r\sin\theta\hat{\varphi} \\ \frac{\partial}{\partial r} & \frac{\partial}{\partial \theta} & \frac{\partial}{\partial \varphi} \\ E_r & rE_\theta & r\sin\theta E_\varphi \end{vmatrix}$$

$$= \frac{1}{r^2\sin\theta}\frac{\partial}{\partial\theta}(r\sin\theta E_\varphi)\,\hat{r} - \frac{1}{r\sin\theta}\frac{\partial}{\partial r}(r\sin\theta E_\varphi)\,\hat{\theta}$$

$$= \frac{1}{r\sin\theta} \cdot \frac{\partial}{\partial\theta}\left[\frac{A\sin^2\theta}{r}\left(\cos\psi - \frac{1}{kr}\sin\psi\right)\right]\hat{r}$$

$$-\frac{1}{r} \cdot \frac{\partial}{\partial r}\left[A\sin\theta\left(\cos\psi - \frac{1}{kr}\sin\psi\right)\right]\hat{\theta},$$

where we have put $kr - \omega t = \psi$. So

$$\frac{\partial}{\partial r}\cos\psi = -k\sin\psi \quad \text{and} \quad \frac{\partial}{\partial r}\sin\psi = k\cos\psi.$$

Thus,

$$\vec{\nabla} \times \vec{E} = \frac{2A\cos\theta}{r^2}\left(\cos\psi - \frac{1}{kr}\sin\psi\right)\hat{r}$$

$$-\frac{A\sin\theta}{r}\left(-k\cos\psi + \frac{1}{kr^2}\sin\psi - \frac{1}{r}\cos\psi\right)\hat{\theta}$$

$$= -\frac{\partial\vec{B}}{\partial t} \quad \text{(By Faraday's law)}$$

Integrating with respect to t

$$\vec{B} = \frac{2A\cos\theta}{\omega r^2}\left(\sin\psi + \frac{1}{kr}\cos\psi\right)\hat{r} + \frac{A\sin\theta}{\omega r}\left(-k\cos\psi + \frac{1}{kr^2}\cos\psi + \frac{1}{r}\sin\psi\right)\hat{\theta}$$

$$\vec{\nabla}\cdot\vec{B} = \frac{1}{r^2}\frac{\partial}{\partial r}(r^2 B_r) + \frac{1}{r\sin\theta}\frac{\partial}{\partial\theta}(\sin\theta B_\theta) + \frac{1}{r\sin\theta}\frac{\partial B_\varphi}{\partial\varphi}$$

$$= \frac{1}{r^2}\frac{\partial}{\partial r}\left[\frac{2A\cos\theta}{\omega}\left(\sin\psi + \frac{1}{kr}\cos\psi\right)\right]$$

$$+\frac{1}{r\sin\theta}\frac{\partial}{\partial\theta}\left[\frac{A\sin^2\theta}{\omega r}\left(-k\cos\psi + \frac{1}{kr^2}\cos\psi + \frac{1}{r}\sin\psi\right)\right]$$

$$= \frac{1}{r^2}\cdot\frac{2A\cos\theta}{\omega}\left(k\cos\psi - \frac{1}{kr^2}\cos\psi - \frac{1}{r}\sin\psi\right)$$

$$+\frac{1}{r\sin\theta}\cdot\frac{2A\sin\theta\cos\theta}{\omega r}\left(-k\cos\psi + \frac{1}{kr^2}\cos\psi + \frac{1}{r}\sin\psi\right)$$

$$= \frac{2A\cos\theta}{\omega r^2}\left(k\cos\psi - \frac{1}{kr^2}\cos\psi - \frac{1}{r}\sin\psi - k\cos\psi + \frac{1}{kr^2}\cos\psi + \frac{1}{r}\sin\psi\right) = 0.$$

$$\vec{\nabla}\times\vec{B} = \frac{1}{r^2\sin\theta}\begin{vmatrix} \hat{r} & r\hat{\theta} & r\sin\theta\hat{\varphi} \\ \frac{\partial}{\partial r} & \frac{\partial}{\partial\theta} & \frac{\partial}{\partial\varphi} \\ B_r & rB_\theta & 0 \end{vmatrix}$$

$$= \frac{1}{r^2\sin\theta} r\sin\theta\hat{\phi}\left[\frac{\partial}{\partial r}(rB_\theta) - \frac{\partial B_r}{\partial\theta}\right] \quad \left[\because\ \frac{\partial}{\partial\varphi}(rB_\theta) = \frac{\partial}{\partial\varphi}(B_r)\right]$$

$$= \hat{\varphi}\left[\frac{1}{r}\frac{\partial}{\partial r}\cdot\frac{A\sin\theta}{\omega}\left(-k\cos\psi + \frac{1}{kr^2}\cos\psi + \frac{1}{r}\sin\psi\right)\right.$$

$$\left.-\frac{2A}{\omega r^3}\frac{\partial}{\partial\theta}\cdot\cos\theta\left(\sin\psi + \frac{1}{kr}\cos\psi\right)\right]$$

$$= \hat{\varphi}\left[\frac{A\sin\theta}{\omega r}\left(k^2\sin\psi - \frac{2}{kr^3}\cos\psi - \frac{1}{r^2}\sin\psi - \frac{1}{r^2}\sin\psi + \frac{k}{r}\cos\psi\right)\right.$$

$$\left.+\frac{2A\sin\theta}{\omega r^3}\left(\sin\psi + \frac{1}{kr}\cos\psi\right)\right]$$

$$= \hat{\varphi}\cdot\frac{A\sin\theta}{r}\cdot\frac{k}{\omega}\left(k\sin\psi + \frac{1}{r}\cos\psi\right)$$

or $$\vec{\nabla}\times\vec{H} = \hat{\varphi}\frac{A\sin\theta}{r}\cdot\frac{k^2}{\mu_0\omega}\left(\sin\psi + \frac{1}{kr}\cos\psi\right).$$

Now

$$\frac{\partial \vec{D}}{\partial t} = \varepsilon_0 \frac{\partial \vec{E}}{\partial t} = \hat{\varphi} \frac{A \sin\theta}{r} \cdot \varepsilon_0 \omega \left(\sin\psi + \frac{1}{kr} \cos\psi \right).$$

Since

$$\frac{k^2}{\mu_0 \omega} = \frac{\omega}{\mu_0} \cdot \frac{1}{c^2} = \frac{\omega}{\mu_0} \mu_0 \varepsilon_0 = \varepsilon_0 \omega,$$

it is proved that

$$\vec{\nabla} \times \vec{H} = \frac{\partial \vec{D}}{\partial t}.$$

(ii) Poynting's vector

$$\begin{aligned}
\vec{s} &= \vec{E} \times \vec{H} \\
&= \frac{A \sin\theta}{r} \left(\cos\psi - \frac{1}{kr} \sin\psi \right) \left[\frac{2A\cos\theta}{\mu_0 \omega r^2} \left(\sin\psi + \frac{1}{kr} \cos\psi \right) \hat{\theta} \right. \\
&\quad \left. + \frac{A \sin\theta}{\mu_0 \omega r} \left(-k \cos\varphi + \frac{1}{kr^2} \cos\psi + \frac{1}{r} \sin\psi \right) (-\hat{r}) \right] \\
&\quad \left[\because \hat{\varphi} \times \hat{r} = \hat{\theta} \quad \text{and} \quad \hat{\varphi} \times \hat{\theta} = -\hat{r} \right]
\end{aligned}$$

$$\begin{aligned}
&= \frac{A^2 \sin\theta}{\mu_0 \omega r^2} \left[\frac{2\cos\theta}{r} \left(\sin\psi \cos\psi + \frac{1}{kr} \left(\cos^2\psi - \sin^2\psi \right) - \frac{1}{k^2 r^2} \sin\psi \cos\psi \right) \hat{\theta} \right. \\
&\quad - \sin\theta \left(-k \cos^2\psi + \frac{1}{kr^2} \cos^2\psi + \frac{1}{r} \sin\psi \cos\psi + \frac{1}{r} \sin\psi \cos\psi \right. \\
&\quad \left. \left. - \frac{1}{k^2 r^3} \sin\psi \cos\psi - \frac{1}{kr^2} \sin^2\psi \right) \hat{r} \right] \\
&= \frac{A^2 \sin\theta}{\mu_0 \omega r^2} \left[\frac{2\cos\theta}{r} \left\{ \left(1 - \frac{1}{k^2 r^2} \right) \sin\psi \cos\psi + \frac{1}{kr} \left(\cos^2\psi - \sin^2\psi \right) \right\} \hat{\theta} \right. \\
&\quad \left. - \sin\theta \left\{ -k \cos^2\psi + \left(\frac{2}{r} - \frac{1}{k^2 r^3} \right) \sin\psi \cos\psi + \frac{1}{kr^2} \left(\cos^2\psi - \sin^2\psi \right) \right\} \hat{r} \right]
\end{aligned}$$

Note the time average values are

$$\langle \sin\psi \cos\psi \rangle = 0, \quad \langle \sin^2\psi \rangle = \langle \cos^2\psi \rangle = \frac{1}{2}.$$

$$\therefore \quad \langle \vec{s} \rangle = \frac{A^2 \sin\theta}{\mu_0 \omega r^2} \cdot k \sin\theta \cdot \frac{1}{2} \hat{r} = \frac{A^2 \sin^2\theta}{2\mu_0 c r^2} \hat{r}.$$

(iii) $\displaystyle \int \langle \vec{s} \rangle \cdot \hat{r} r^2 \sin\theta d\theta d\varphi = \frac{A^2}{2\mu_0 c} \int_0^{\pi} \sin^3\theta d\theta \cdot \int_0^{2\pi} d\varphi = \frac{4\pi A^2}{3\mu_0 c}.$

▶ **Problem 22.** *Given that* $\phi = 0$,

$$\vec{A} = \begin{cases} \frac{\mu_0 \alpha}{4c}(ct - |x|)^2 \hat{k}, & \textit{for } |x| < ct \\ 0, & \textit{for } |x| > ct \end{cases}$$

Find $\vec{E}$ *and* $\vec{B}$. *Also find the volume charge and current distributions that can, for* $|x| < ct$, *give rise to such potentials.*

Solution. $\vec{E} = -\dfrac{\partial \vec{A}}{\partial t} = -\dfrac{\mu_0 \alpha}{2}(ct - |x|)\hat{k}$.

$$\vec{B} = \vec{\nabla} \times \vec{A} = \begin{vmatrix} \hat{i} & \hat{j} & \hat{k} \\ \frac{\partial}{\partial x} & \frac{\partial}{\partial y} & \frac{\partial}{\partial z} \\ 0 & 0 & A_z \end{vmatrix} = -\hat{j}\frac{\partial A_z}{\partial x} = -\hat{j}\frac{\mu_0 \alpha}{4c}\frac{\partial}{\partial x}(ct - |x|)^2$$

$$= \pm \hat{j}\frac{\mu_0 \alpha}{2c}(ct - |x|) \quad \begin{bmatrix} + \text{ sign for } x > 0 \\ - \text{ sign for } x < 0 \end{bmatrix}$$

For $|x| > ct$, $\phi = \vec{A} = 0$.
Therefore, $\vec{E} = \vec{B} = 0$.
Now $\vec{\nabla} \cdot \vec{E} = 0$, which indicates that charge density $\rho = 0$.

$$\vec{\nabla} \times \vec{B} = \begin{vmatrix} \hat{i} & \hat{j} & \hat{k} \\ \frac{\partial}{\partial x} & \frac{\partial}{\partial y} & \frac{\partial}{\partial z} \\ 0 & B_y & 0 \end{vmatrix} = -\frac{\mu_0 \alpha}{2c}\hat{k} \quad \text{or} \quad \vec{\nabla} \times \vec{H} = -\frac{\alpha}{2c}\hat{k}.$$

$$\frac{\partial \vec{E}}{\partial t} = -\frac{\mu_0 \alpha c}{2}\hat{k} \quad \text{or} \quad \varepsilon_0 \frac{\partial \vec{E}}{\partial t} = -\frac{\mu_0 \varepsilon_0 \alpha c}{2}\hat{k} = -\frac{\alpha}{2c}\hat{k}.$$

Thus,

$$\vec{\nabla} \times \vec{H} = \varepsilon_0 \frac{\partial \vec{E}}{\partial t} = \frac{\partial \vec{D}}{\partial t}.$$

Comparing it with the Maxwell's equation

$$\vec{\nabla} \times \vec{H} = \vec{J} + \frac{\partial \vec{D}}{\partial t}$$

we find that the volume current density is $\vec{J} = 0$.

QUESTIONS

1. What is displacement current? Explain why and how Ampere's circuital law for steady current was modified to include displacement current.

OR

Explain how Maxwell generalised Ampere's circuital law. Hence, discuss the concept of displacement current. Distinguish between conduction and displacement currents.

2. (a) Write down Maxwell's electromagnetic field equations. Explain the physical significance of each.

(b) Show that the equation of continuity is contained in Maxwell's equations. OR, Derive the law of conservation of charge from Maxwell's field equations. **[C.U. 2009]**

OR, Show that Maxwell's equations are consistent with the equation of contnuity. **[C.U. 2014]**

(c) Using Maxwell's equations obtain Coulomb's law in electrostatic. **[B.U. 2013]**

[**Hints.** Integrate Eq. (15.3-1) over a closed volume and get Gauss's law. Then follow Section 1.9]

3. (a) Starting from Maxwell's equation derive the wave equation for electric and magnetic fields in free space.

(b) Hence, show that electric, magnetic and propagation vectors are all mutually perpendicular to each other. **[C.U. 2011]**

(c) Show that for electromagnetic waves in free space, energy is equally shared between electric and magnetic fields.

(d) Show that $\vec{B} = \frac{1}{c}(\hat{n} \times \vec{E})$ where $\hat{n}$ is the unit vector in the direction of propagation and c is the speed of light in free space. **[C.U. 2006]**

4. (a) State and establish Poynting's theorem in electromagnetism. **[C.U. 2013]**

(*b*) Express the theorem in the following differential form

$$\frac{\partial u}{\partial t} + \vec{\nabla} \cdot \vec{s} = 0,$$

where $\vec{s}$ is the Poynting's vector and u is the total electromagnetic energy density. Compare it with the equation of continuity and give an interpretation of $\vec{s}$. **[cf. C.U. 2008]**

(c) Show that in a harmonically varying electromagnetic field represented by complex notations the time average Poynting's vector can be represented by *Real part of* $\frac{1}{2}\left(\vec{E} \times \vec{H}^*\right)$, where $\vec{H}^*$ is the complex conjugate of $\vec{H}$.

(d) Show that the momentum density stored in EM field is given by $\vec{P}_{\text{em}} = \vec{s}/c^2$ in vacuum. **[C.U. 2014]**

5. Starting from Maxwell's equations show that in a homogeneous isotropic dielectric medium of permittivity ε and permeability μ

(a) the velocity of an electromagnetic wave is given by $v = 1/\sqrt{\mu\varepsilon}$; **[B.U. 2013]**

(b) electric and magnetic fields are in phase;

(c) energy is equally shared between electric and magnetic fields.

6. Consider the propagation of a plane electromagnetic wave through a linear homogeneous isotropic conducting medium having no accumulated charge or externally induced current. Write down the appropriate Maxwell's equations and derive the wave equations for electric and magnetic fields. Hence, show that

(a) the field amplitudes are spatially attenuated;

(b) electric and magnetic fields are not in phase; **[C.U. 2009, 2011, 2014]**

(c) field energy is not equally shared between electric and magnetic fields.

7. (a) Starting from Maxwell's equations derive the wave equation in a conducting medium. Hence, show that a plane electromagnetic wave is attenuated as it propagates through the medium. What is skin depth?

(b) In the microwave region the surface of a pure silver waveguide and that of a silver coated brass waveguide appear to be identical.–Explain.

(c) Show that for a good conductor skin depth can be expressed as $\lambda_c/2\pi$ where λ_c is the wavelength inside the conductor.

8. Consider the propagation of electromagnetic waves through a dilute ionised gas such as the ionosphere. Show that it behaves like a medium of refractive index $n = \sqrt{1 - \omega_p^2/\omega^2}$, where ω_p is the so-called electron plasma frequency and ω is the angular frequency of the wave. **[cf. B.U. 2017]**

Discuss the possibility of total internal reflection of radiowaves from the ionosphere.

OR, How is electromagnetic wave reflected from ionosphere? Obtain the expression for the refractive index of the ionosphere medium. **[B.U. 2011]**

9. (a) What do you mean by the polarization of a plane electromagnetic wave?

(b) Show how by the superposition of two plane polarized waves an elliptically or a circularly polarized light can be produced.

(c) Show that a plane polarized wave can be considered as the superposition of two oppositely polarized circular waves of equal amplitude.

10. Show that for any scalar and vector potentials representing a particular EM field we can always find another set of potentials representing the same field. What is Lorentz gauge condition? Is the set of potentials satisfying Lorentz gauge condition unique?

11. Starting from Maxwell's equations find the condition on the magnetic vector potential $\vec{A}$ that leads to the Poisson's equation for the scalar potential ϕ.

12. Starting from Maxwell's equation find the differential equation satisfied by the magnetic vector potential $\vec{A}$. Derive the condition for which the equation simplifies to a wave equation.

13. (a) What do you mean by retarded potentials?

(b) Find the retarded potentials of an oscillating electric dipole.

EXERCISES

1. (a) In a medium of dielectric constant 5, the maximum displacement current is equal to the maximum conduction current at a frequency of 1 MHz. Find the conductivity of the medium. [*Ans.* 2.78×10^{-4} S/m]

(b) At frequency $\nu = 4 \times 10^8$ Hz, sea water has permittivity $\varepsilon = 81\varepsilon_0$, permeability $\mu = \mu_0$ and resistivity $\rho = 0.23\,\Omega m$. What is the ratio of conduction current to displacement current? **[B.U. 2011]** [*Ans.* 2.4]

(c) A 50 pF capacitor is being charged at such a rate that its voltage is increasing at the rate 300 V/s. Find the displacement current generated.
[**Hints.** $I_d = AJ_d = A\frac{\partial D}{\partial t} = A\cdot\frac{1}{A}\frac{\partial q}{\partial t} = C\frac{\partial V}{\partial t}$] [*Ans.* 0.015 μA]

(d) The current density in a region is given by $\vec{J} = \hat{i}J_0\exp(-x^2)$. Find the rate of change of charge density at the point $(1, 0, 0)$. [*Ans.* $2J_0\exp(-1)$]

2. (a) The intensity of sunlight reaching the earth's surface is about 1.86 calorie $\cdot$ cm^{-2} $\cdot$ min^{-1}. Calculate the strength of electric field of incoming sunlight.
[*Ans.* $E_{\text{rms}} = 700$ V.m^{-1}]

(b) A laser beam has a diameter of 2 mm. What is the amplitude of the electric field and the magnetic field if the power of the laser is 1.5 mW? **[C.U. 2007]**
[**Hints.** See solved Problem 7; $I = 1.5 \times 10^{-3}/\pi(10^{-3})^2$ W/m^2; $E_0 = E_{\text{rms}}\sqrt{2} =$ 600 V/m; $B_0 = E_0/c = 2\mu$T]

3. A plane EM wave traveling in vacuum has an electric field amplitude $E_0 =$ 50 μV/m. Calculate the average value of Poynting's vector and average energy density of the wave. **[cf. C.U. 2011]**

[*Ans.* 3.31×10^{-12} W.m^{-2}, 11.06×10^{-21} J.m^{-3}]

4. Calculate the strength of electric field at a distance of 10 m from a 60 W lamp. [*Ans.* $4.24\,\text{V.m}^{-1}$]

5. A space probe is radiating EM waves isotropically with a power of 20 W. Find the amplitude of the $\vec{E}$-field of the signal received by a detector at a distance 2×10^9m from the probe. Assume that the waves incident on the detector are plane waves. [*Ans.* $\sqrt{3} \times 10^{-8}\,\text{V.m}^{-1}$]

6. The maximum value of $\vec{E}$-field in an EM wave in free space is $1000\ \text{V.m}^{-1}$. Find the maximum value of $\vec{H}$-field in the wave. [*Ans.* $H_0 = 2.65\,\text{A.m}^{-1}$]

7. Show that in a conducting medium the displacement current leads over the conduction current by 90°.

8. Suppose the sun radiates energy isotropically at the rate $3.8 \times 10^{26}\ \text{J.s}^{-1}$. The average distance between the sun and the earth is 1.5×10^8 km. Calculate the magnitude of Poynting's vector on the surface of the earth. Also, calculate the rms amplitude of the electric field of the radiant energy reaching the earth's surface. [*Ans.* $1345\,\text{W.m}^{-2}$; $712\,\text{V.m}^{-1}$]

9. Find the unit, dimensions and magnitude of the quantity $\sqrt{\mu_0/\varepsilon_0}$.
[C.U. 2013, cf. C.U. 2017] [*Ans.* $376.6\,\Omega$]

10. The electric field associated with an electromagnetic wave in free space is given by $\vec{E} = \hat{x}E_0 \cos(kz - \omega t)$. Find the corresponding $\vec{H}$-field. [*Ans.* $\hat{y}\frac{E_0}{\mu_0 c}\cos(kz - \omega t)$]

11. Consider a parallel plate capacitor with its plates in the form of large circular discs and the space between them being filled up with a homogeneous weakly conducting medium of conductivity σ and permittivity ε. An alternating electric field $E = E_0 \cos \omega t$ is applied across the plates. Show that the magnitude of the magnetic field $\vec{H}$ between the plates at a distance r from the axis of system is given by

$$H = \frac{rE_0}{2}\sqrt{\sigma^2 + (\varepsilon\omega)^2}\cos(\omega t + \phi),$$

where $\tan\phi = \varepsilon\omega/\sigma$.

[**Hints.** $\oint \vec{H} \cdot d\vec{l} = \int_S \vec{\nabla} \times \vec{H} \cdot d\vec{S} = \int \left(\vec{J} + \frac{\partial \vec{D}}{\partial t}\right) \cdot d\vec{S} = \left(J_n + \varepsilon_0 \frac{\partial E_n}{\partial t}\right)\pi r^2$

$= (\sigma E_0 \cos\omega t - \varepsilon\omega E_0 \sin\omega t)\,\pi r^2$ or $H \cdot 2\pi r = \pi r^2 E_0\,(\sigma\cos\omega t - \varepsilon\omega\sin\omega t)$

or $H = \dfrac{rE_0}{2}\,(\sigma\cos\omega t - \varepsilon\omega\sin\omega t) =$ Required result]

12. A linearly polarized plane EM wave of angular frequency ω is propagating along $+z$-direction with its plane of vibration making an angle 45° to the yz-plane. Write an expression for $\vec{E}$ and $\vec{B}$ fields

[*Ans.* $\vec{E} = \frac{E_0}{\sqrt{2}}\left(\hat{\imath} + \hat{\jmath}\right)\cos(kz - \omega t)$; $\vec{B} = \frac{E_0}{c\sqrt{2}}\left(\hat{\jmath} - \hat{\imath}\right)\cos(kz - \omega t)$]

13. A laser emits pulses of EM radiation in vacuum lasting for 2 ns each. What is the length of each wavetrain in space? Assuming the beam diameter to be 2.5 mm and energy of each pulse as 6 J, calculate the average energy per unit volume for such a pulse

[**Hints.** $\Delta l = c \cdot \Delta t = 3 \times 10^8 \times 2 \times 10^{-9}\,\text{m} = 0.6\,\text{m}$

$$u = \frac{\text{Energy of each pulse}}{\text{Volume of each pulse}} = \frac{6}{0.6 \times \pi \left(\frac{2.5\times10^{-3}}{2}\right)^2} = 2 \times 10^6\,\text{J/m}^3]$$

14. Consider a parallel plate capacitor immersed in sea water. The charge on the capacitor is varied according to $q = q_0 \sin 2\pi\nu t$, where $\nu = 4 \times 10^8$ Hz. At this frequency sea water has permittivity $\varepsilon = 81\varepsilon_0$ permeability $\mu = \mu_0$ and resistivity $\rho = 0.23\,\Omega$-m. Compare the amplitudes of the conduction and the displacement currents between the plates.

[**Hints.** $E = \dfrac{V}{d} = \dfrac{q}{Cd} = \dfrac{q_0}{Cd}\sin 2\pi\nu t;\ J_c = \sigma E = \dfrac{\sigma q_0}{Cd}\sin 2\pi\nu t;\ J_d = \varepsilon\dfrac{\partial E}{\partial t} = \dfrac{\varepsilon q_0}{Cd}\cdot 2\pi\nu\cos 2\pi\nu t.$

∴ The ratio of amplitudes is

$$\frac{J_c}{J_d} = \frac{\sigma}{2\pi\nu\varepsilon} = \frac{1}{2\pi\nu\varepsilon\rho} = \frac{1}{2\pi \times 4 \times 10^8 \times 81 \times 8.85 \times 10^{-12} \times 0.23} = 2.41]$$

15. Suppose the $\vec{E}$-field associated with a linearly polarized plane EM wave in free space is given by $\vec{E} = \hat{x}E_0\cos(kz - \omega t)$. Calculate electromagnetic energy density (u_{em}) and Poynting's vector $(\vec{s})$. Show that the energy moves along the wave with a speed $c = \omega/k$.

$$[\textit{Ans.}\ u_{\text{em}} = \varepsilon_0 E_0^2\cos^2(kz - \omega t)\cdot \vec{s} = \hat{z}c\varepsilon_0 E_0^2\cos^2(kz - \omega t),\ \vec{s} = u_{\text{em}}c\hat{z}]$$

16. (a) Calculate the frequency at which skin depth in sea water is 1 m. Given for sea water, conductivity $\sigma = 4.3(\Omega\text{-m})^{-1}$ and relative permeability equal to 1. [59 kHz]

(b) Find the skin depth in sea water for low frequency radio waves of wavelength 3×10^3 m in air. Assume that the sea water has a conductivity $\sigma = 4(\Omega\text{-m})^{-1}$ and permeability $\mu = \mu_0$. [*Ans.* 0.796 m]

17. (a) Calculate the skin depth for silver at a frequency of 10 GHz. Given the conductivity $\sigma = 3 \times 10^7(\Omega\text{-m})^{-1}$. **[cf. (C.U. 2010]** [*Ans.* 9.2×10^{-7} m]

(b) Show that the skin depth in a poor conductor ($\sigma \ll \omega\varepsilon$) is independent of frequency and is given by

$$\delta = \frac{2}{\sigma}\sqrt{\frac{\varepsilon}{\mu}}.$$

[**Hints.** Make Binomial expansions in Eq. (15.7-10) and then find that $\delta = 1/\beta$.]

18. Find out the attenuation of an electromagnetic wave of frequency 1 kHz by a metal plate of thickness 0.1 mm. Given the conductivity of the metal $\sigma = 5.8 \times 10^7 (\Omega\text{-m})^{-1}$. [*Ans.* $E_{\text{transmitted}}/E_{\text{incident}} = 0.953$]

19. For the propagation of electromagnetic waves a medium is considered to be a good conductor if $\sigma/\omega\varepsilon \geq 100$ and a good dielectric if $\sigma/\omega\varepsilon \leq 0.01$, where σ is the conductivity, ε the permittivity and ω the angular frequency. For sea water $\sigma = 4.3(\Omega\text{-m})^{-1}$ and dielectric constant 80. Find the frequency range for which sea water can be considered as a good conductor. [Note that $\sigma/\omega t$ is the ratio of conduction and displacement current densities and it depends on frequency. Thus, the behaviour of a medium is frequency dependent.] [*Ans.* $f \leq 9.7\,\text{MHz}$]

20. Show that the phase difference between electric and magnetic fields in a good conductor is 45°.

21. The components of the electric field associated with a plane EM wave in free space are given by: $E_x = E_0 \sin(kz - \omega t)$, $E_y = E_0 \cos(kz - \omega t)$, $E_z = 0$.

(a) Find the state of polarization of the wave.

(b) Using Maxwell's equations find the wave's magnetic field $\vec{H}$.

(c) Show explicitly that $\vec{E} \cdot \vec{H} = 0$.

[*Ans.* (a) Circularly polarized, (b) $E_0\sqrt{\frac{\varepsilon_0}{\mu_0}}\,[\hat{y}\sin(kz - \omega t) - \hat{x}\cos(kz - \omega t)]$]

22. The electric field associated with an EM wave is given by

$$\vec{E}(\vec{r}, t) = \left(-\hat{\imath} + \sqrt{2}\hat{\jmath}\right) \cdot 10^4 \cdot \exp\left[j\left(\frac{\sqrt{2}x + y}{3}\pi \times 10^7 - \pi\sqrt{3} \times 10^{15} t\right)\right] \text{ V.m}^{-1}$$

(a) Show that it is consistent with a wave travelling in charge-free space.

(b) What is the direction of propagation of the wave?

(c) Find the associated magnetic field.

(d) Calculate the time average Poynting's vector for the field.

[**Hints.** (a) Show that $\rho = \vec{\nabla} \cdot \vec{E} = 0$.

(b) The direction of propagation $\hat{n}$ is $\perp^r$ to the planes of constant phases;

$$\frac{\sqrt{2}x + y}{3}\pi \times 10^7 - \pi\sqrt{3} \times 10^{15} t = \phi = \text{constant.}$$

$$\therefore \quad \hat{n} = \frac{\vec{\nabla}\phi}{\left|\vec{\nabla}\phi\right|} = \frac{\sqrt{2}\hat{\imath} + \hat{\jmath}}{\sqrt{3}}.$$

(c) $\dfrac{\sqrt{2}x+y}{3}\pi\times 10^7 = \vec{k}\cdot\vec{r} = k\hat{n}\cdot\vec{r} = k\dfrac{\sqrt{2}x+y}{\sqrt{3}}$ or $k = \dfrac{\pi}{\sqrt{3}}\times 10^7$.

Thus,

$$\vec{k} = \left(\sqrt{2}\,\hat{\imath}+\hat{\jmath}\right)\frac{\pi}{3}\times 10^7;\quad \omega = \pi\sqrt{3}\times 10^{15}.$$

Now

$$\vec{H} = \frac{\vec{k}\times\vec{E}}{\mu_0\omega}\quad\text{or}\quad \vec{B} = \frac{\vec{k}\times\vec{E}}{\omega} = \hat{k}\frac{10^{-4}}{\sqrt{3}}.\exp\left[j\left(\frac{\sqrt{2}x+y}{3}\pi\times 10^7 - \pi\sqrt{3}\times 10^{15}t\right)\right].$$

(d) Time average Poynting's vector

$$\langle\vec{s}\rangle = \tfrac{1}{2}Re\left[\vec{E}^*\times\vec{H}\right] = \tfrac{1}{2\sqrt{3}\mu_0}\left(\sqrt{2}\,\hat{\imath}+\hat{\jmath}\right)\ \text{W.m}^{-2}.]$$

23. (a) The magnetic field associated with an EM wave in free space is

$$\vec{H} = \hat{y}\ 0.2\cos(kz-\omega t)\ \text{A/m}.$$

Find the total electromagnetic power passing through a circular area of radius 1 cm on a plane $z = 1$. [*Ans.* 2.37 mW]

(b) A plane EM wave travels in free space in the negative Z-direction with a propagation constant 20 rad/m, the amplitude of the magnetic field being (40/377) A/m. At $t = 0$, $z = 0$ the magnetic field has the direction $-\vec{j}$. Give expression for electric field. [*Ans.* $\vec{E} = \hat{\imath}40\cos(\omega t + 20z)$ V/m] **[C.U. 2009]**

24. The magnetic field associated with an EM wave propagating through a nonmagnetic medium ($\mu = \mu_0$) is

$$\vec{H} = \hat{y}\ 0.025\sin(6z + 2\times 10^8 t)\ \text{A/m},$$

where z is in metre. Find the direction of propagation of the wave, the dielectric constant of the medium and the associated $\vec{E}$-field.

[*Ans.* Negative z-direction, 81, $\vec{E} = \hat{x}\ 1.04\cos(6z + 2\times 10^8 t)$ V/m]

25. Consider a simple magnetic field, which increases exponentially with time,

$$\vec{B} = \hat{z}B_0 e^{\alpha t},$$

where B_0 and α are constants. Using Maxwell's equation find the electric field produced by this time varying magnetic field.

[*Ans.* $\vec{E} = -\hat{\theta}\frac{\alpha B_0\cdot r}{2}e^{\alpha t}$ in (r,θ,z) cylindrical coordinates]

26. An EM wave is passing through a nonconducting medium characterised by permittivity $10\varepsilon_0$ and permeability $5\mu_0$. Find the wave impedance.

[*Ans.* 266.58 Ω]

27. Find the phase velocity of an EM wave passing through a medium characterised by a relative permittivity of 4 and a relative permeability of unity. Also, calculate the refractive index of the medium. [*Ans.* $1.5 \times 10^8\,\mathrm{ms}^{-1}, 2$]

28. A plane EM wave whose $\vec{E}$-field is represented by $\vec{E} = \hat{x}\ 10\cos(kz - 6\pi \times 10^9 t)\ \mathrm{Vm}^{-1}$, is passing through a medium characterised by conductivity $\sigma = 0.1$ S/m, permittivity ε_0 and permeability μ_0. Find the total current density induced in the medium. [*Ans.* $\vec{J} = \hat{x}\sqrt{\frac{34}{9}}\cos\left(kz - 6\pi \times 10^9 t - \tan^{-1} 5/3\right)$]

29. (a) A plane EM wave with $\vec{E}$ field linearly polarized along $\hat{k}$ (z-axis) is travelling along the direction $\left(\hat{\imath} + \hat{\jmath}\right)/\sqrt{2}$ in free space. Assuming the wavelength to be λ and angular frequency ω, write down expressions for $\vec{E}$ and $\vec{B}$.

(b) Consider the following electromagnetic fields

$$E_x = 0, \quad E_y = E_0 \sin(kx + \omega t), \quad E_z = 0$$

$$B_x = 0, \quad B_y = 0, \quad B_z = -B_0 \sin(kx + \omega t).$$

Find the relation between ω and k for which these fields will satisfy Maxwell's equations.

[*Ans.* $\vec{E} = \hat{k}E_0 e^{j\left(\frac{2\pi}{\lambda}\cdot\frac{x+y}{\sqrt{2}} - \omega t\right)}, \vec{B} = \frac{\hat{i}-\hat{j}}{\sqrt{2}}\sqrt{\mu_0\varepsilon_0}E_0 \cdot e^{j\left(\frac{2\pi}{\lambda}\cdot\frac{x+y}{\sqrt{2}} - \omega t\right)}, \omega^2 = c^2k^2$]

30. The maximum electron density of an ionospheric layer is $10^{12}\mathrm{m}^{-3}$. Find the critical frequency for reflecting radio waves. [*Ans.* 9 MHz]

31. Brass waveguides are often silver coated to reduce the material cost. If the thickness of silver required for this purpose is at least five times the skin depth, find the minimum thickness of silver coating required for a waveguide operating at 10 GHz. Assume for silver $\mu = \mu_0$ and $\sigma = 6 \times 10^7$ S/m. [*Ans.* 0.0032 mm]

32. Suppose an EM wave in vacuum polarized with its electric vector along y-axis is incident on an interface at $z = 0$ at an angle of incidence 30°. Taking the peak magnitude of the $\vec{E}$-field as 10 V.m^{-1} and its frequency as 6×10^{14} Hz, write down a complete expression for the incident wave.

[*Ans.* $\vec{E} = \hat{y}\ 10e^{j\left\{\left(\frac{1}{2}x + \frac{\sqrt{3}}{2}z\right)4\pi\times10^6 - 12\pi\times10^{14}t\right\}}$ V.m^{-1}]

33. Suppose an electron is under the action of an EM wave. Show that the magnetic force on the electron is negligible compared with that due to the electric field of the wave.

[**Hints.** $\dfrac{\left|e\left(\vec{v} \times \vec{B}\right)\right|}{\left|e\vec{E}\right|} = \dfrac{v}{c} \ll 1.$]

34. Show that Kirchhoff's laws can be derived from Maxwell's equations.

[C.U. 2015, 2017]

[**Hints.** Integrating Eq. (15.3-4) over a closed surface having no contour we get

$$0 = \oint \vec{J} \cdot d\vec{S} + \oint_S \vec{J}_d \cdot d\vec{S}$$

Assuming that the surface S consists of several small sections (S_k) we can write

$$\sum_k \int_{S_k} \vec{J} \cdot d\vec{S} + \sum_K \int_{S_k} \vec{J}_d \cdot d\vec{S} = 0$$

$$\text{or} \quad \sum_k (I_k + I_{d_k}) = 0. \quad \text{This is KCL}$$

Integrating Eq. (15.3-3) over an open surface S bounded by a contour C we get

$$\oint_C \vec{E} \cdot d\vec{l} = -\frac{d}{dt} \int_S \vec{B} \cdot d\vec{S}$$

Assuming that the contour consists of several conducting sections (C_k) and forms a closed circuit we can write

$$\sum_k \int_{C_k} \vec{E} \cdot d\vec{l} = \text{Induced emf acting in the circuit}$$

Note that LHS represents the sum of pds across different conducting sections. Thus it represents KVL.]

35. In a region of space the Poynting vector is given by $\vec{s} = \hat{k} \cdot 0.8 \cos^2(kz - \omega t)\ \mathrm{W/m^2}$. Find the time average EM power passing through an area $0.01\ \mathrm{m^2}$ on a plane $3y + 4z = 7$.

[**Hints.** $\langle \vec{s} \rangle = \frac{1}{T} \int_0^T \vec{s}\, dl = 0.4\hat{k}\ \mathrm{W/m^2}$. Unit normal to the plane $\phi = 3y + 4z - 7$ is

$$\hat{n} = \frac{\vec{\nabla}\phi}{|\vec{\nabla}\phi|} = \frac{3\hat{j} + 4\hat{k}}{5}$$

$$\begin{aligned} \therefore \text{ Required power } &= \int \langle \vec{s} \rangle \cdot \hat{n}\, dS \\ &= 0.4\hat{k} \cdot \frac{3\hat{j} + 4\hat{k}}{5} \times 0.01 \\ &= 3.2 \times 10^{-3}\ \mathrm{W} \end{aligned}$$]

36. In a region of space the magnetic field intensity is given by $\vec{H} = \hat{j} H_0 \cos(kz - \omega t)$. Assuming that there is no free current find the displacement current density.

[**Hints.** Use $\vec{J}_d = \vec{\nabla} \times \vec{H} = \hat{i} H_0 k \sin(kz - \omega t)$]

37. The electric field of an EM wave is $\vec{E}(\vec{r},t) = \hat{k}\sqrt{2}E_0\cos(x-y-\omega t)$. Find the associated magnetic field $\vec{B}(\vec{r},t)$.

[**Hints.** $x-y-\omega t = \vec{k}\cdot\vec{r}-\omega t$ where $\vec{k} = \hat{\imath}-\hat{\jmath}$

$$\therefore\ \vec{B}|\vec{r},t| = \frac{\vec{k}\times\vec{E}}{\omega} = -\frac{\sqrt{2}\cdot E_0}{\omega}(\hat{\imath}+\hat{\jmath})\cos(x-y-\omega t)]$$

38. The $\vec{E}$-field of an EM wave propagating in certain medium is given by $\vec{E} = \hat{k}E_0\cos(10^7x+10^7y-10^{15}t)$ V/m. Find the refractive index of the medium.

[*Ans.* 4.2]

[**Hints.** Here $\vec{k} = 10^7(\hat{\imath}+\hat{\jmath})$. Phase speed $V = \frac{\omega}{|\vec{k}|}$ and refractive index $n = \frac{c}{V}$]

39. A steady current I is flowing down a wire of length l, radius a and conductivity σ. Calculate $\int_V \vec{J}\cdot\vec{E}\,dV$ and hence find an expression for the resistance R of the wire. Also calculate $\oint \vec{s}\cdot d\vec{S}$ and show that Poynting's theorem is obeyed by the system.

[cf. C.U. 2015]

[**Hints.** See Solved Problem 10.

$$\int_V \vec{J}\cdot\vec{E}\,dV = \sigma\int E^2\,dV = \frac{\sigma\varepsilon^2}{l^2}\cdot\pi a^2 l$$

$$= \text{Joule heat loss}$$

$$= \frac{\varepsilon^2}{R} \quad (\varepsilon = \text{emf} \quad \text{across the length of the wire})$$

$$\therefore\ R = \frac{l}{\sigma\cdot\pi a^2}$$

$$\oint \vec{s}\cdot d\vec{S} = -\oint \frac{I^2R}{2\pi al}\hat{n}\cdot d\vec{S}$$

$$= -\frac{I^2R}{2\pi al}\cdot 2\pi al$$

$$= -I^2R = -\frac{\varepsilon^2}{R} \quad (\because\ \varepsilon = IR)$$

Since the current and hence fields are steady the rate of change of EM energy i.e., LHS of Eq. (15.4-3) is zero. Thus Eq. (15.4-3) i.e., Poynting's theorem is obeyed by the system,]

40. Given the scalar and vector potentials as

$$\phi(\vec{r},t) = 0$$

$$\vec{A}(\vec{r},t) = \begin{cases} \hat{k}\frac{\mu_0 I_0}{2\pi}\ln\left(\frac{\sqrt{c^2t^2-r^2}+ct}{r}\right) & \text{for } r < ct \\ 0 & \text{for } r \geq ct \end{cases}$$

Find out $\vec{E}$ and $\vec{B}$.

$$\left[Ans.\ \vec{E}(\vec{r},t) = -\hat{k}\frac{\mu_0 I_0 c}{2\pi\sqrt{c^2t^2 - r^2}},\quad \vec{B}(\vec{r},t) = \hat{\theta}\frac{\mu_0 I_0}{2\pi r}\frac{ct}{\sqrt{c^2t^2 - r^2}} \text{ in } (r,\theta,z) \text{ cylindrical coordinates}\right]$$

41. Show that in free space with $\rho = 0$, $\vec{J} = 0$, EM fields may be correctly described by a single vector function $\vec{A}$ satisfying

$$\vec{\nabla}\cdot\vec{A} = 0 \quad \text{and} \quad \nabla^2\vec{A} - \frac{1}{c^2}\frac{\partial^2\vec{A}}{\partial t^2} = 0.$$

42. Find a suitable gauge condition for which each of ϕ and $\vec{A}$ satisfies the damped wave equation

$$\nabla^2\psi - \mu\varepsilon\frac{\partial^2\psi}{\partial t^2} - \sigma\mu\frac{\partial\psi}{\partial t} = 0 \quad \left(\psi = \vec{A} \text{ or } \phi\right)$$

in a linear conducting medium of conductivity σ.

$$\left[Ans.\ \vec{\nabla}\cdot\vec{A} + \mu\varepsilon\frac{\partial\phi}{\partial t} + \mu\sigma\phi = 0\right]$$

43. Given the scalar and vector potentials as

$$\phi(\vec{r},t) = 0$$

$$\vec{A}(\vec{r},t) = \hat{j}A_0\sin(kx - \omega t),$$

where A_0, k and ω are constants. Find out $\vec{E}$ and $\vec{B}$. Show that these fields satisfy Maxwell's equations in free space, provided we choose $\omega = ck$, $c = 1/\sqrt{\mu_0\varepsilon_0}$ being the speed of light in free space.

$$\left[Ans.\ \vec{E} = \hat{j}\omega A_0\cos(kx - \omega t), \vec{B} = \hat{k}kA_0\cos(kx - \omega t)\right]$$

44. What physical system is represented by the scalar and vector potentials

$$\phi(\vec{r},t) = 0 \quad \text{and} \quad \vec{A}(\vec{r},t) = -\frac{qt}{4\pi\varepsilon_0 r^2}\hat{r}$$

Use the gauge transformation function $\psi = -\dfrac{qt}{4\pi\varepsilon_0 r}$ to transform these potentials. Comment on the result.

[**Hints.** $\vec{E} = -\vec{\nabla}\phi - \dfrac{\partial\vec{A}}{\partial t} = \dfrac{q}{4\pi\varepsilon_0}\dfrac{\hat{r}}{r^2}$, $\vec{B} = \vec{\nabla}\times\vec{A} = 0$

These are the fields produced by a stationary point charge q at the origin.

$$\phi' = \phi - \frac{\partial\psi}{\partial t} = \frac{1}{4\pi\varepsilon_0}\frac{q}{r}; \quad \vec{A}' = \vec{A} + \vec{\nabla}\psi = 0.$$

These potentials also yield the same physical fields $\vec{E}$ and $\vec{B}$. In fact, potentials in these forms are commonly chosen.]

OBJECTIVE TEST

1. The quantity $\mu_0\varepsilon_0$ has the dimensions of

(a) $[L^{-2}T^2]$, (b) $[L^2T^{-2}]$, (c) $[L^{-1}T]$, (d) $[LT^{-1}]$.

2. Electromagnetic waves can be produced by

(a) a uniformly moving charge, (c) a static charge,
(b) an accelerating charge, (d) none of these.

3. The energy of an EM wave of frequency ν oscillates with a frequency

(a) ν, (b) $\nu/2$, (c) 2ν, (d) zero.

4. The quantity $\varepsilon_0 \dfrac{\partial \vec{E}}{\partial t}$ has the dimensions of

(a) $[IL^{-2}]$, (b) $[I]$, (c) $[IL^2]$, (d) none of these.

5. For a plane EM wave propagating in a linear homogeneous dielectric medium $\vec{E}$ and $\vec{H}$ fields differ in phase by

(a) 90°, (b) 180°, (c) 0°, (d) none of these.

6. Refractive index n of a dielectric medium is related to its dielectric constant K by

(a) $n = \sqrt{K}$, (b) $n = 1/\sqrt{K}$, (c) $1/K$, (d) none of these.

7. If an EM wave is represented by $\vec{E}(\vec{r}, t) = \vec{E}_0 e^{j(\vec{k}\cdot\vec{r}-\omega t)}$ and $\vec{B}(\vec{r}, t) = \vec{B}_0 e^{j(\vec{k}\cdot\vec{r}-\omega t)}$ then

(a) $\vec{k} \cdot \left(\vec{E}_0 \times \vec{B}_0\right) = 0$, (c) $\dfrac{d\omega}{dk} = 0$,
(b) $\vec{k} \times \left(\vec{E}_0 \times \vec{B}_0\right) = 0$, (d) none of these.

8. The dispersion relation for EM waves in a certain medium is given by $\omega^2 = ak$ where ω = angular frequency, a = constant and k = magnitude of wave vector. The velocity of energy propagation in this medium is

(a) $\dfrac{a}{2\omega}$, (b) $\dfrac{a}{\omega}$, (c) $\dfrac{a}{4\omega}$, (d) $\dfrac{2a}{\omega}$.

9. In good conductors $\vec{E}$ and $\vec{H}$ differ in phase by

(a) 0°, (b) 45°, (c) 90°, (d) 180°.

10. Which of the following statements is true of displacement current?

(a) It can produce Joule heating,
(b) It involves motion of free charges,

(c) The concept of displacement was introduced by Ampere,
(d) It can produce magnetic field.

11. The electric field associated with an EM wave in a certain medium is given by $\vec{E} = \hat{\imath}10\cos(3z - 10^8 t)\,\text{V/m}$. The medium must be a

(a) free space,
(b) perfect conductor,
(c) dielectric,
(d) a lossy dielectric.

12. The intensity of radiation from an oscillating electric dipole is

(a) maximum along its axis,
(b) maximum in its equatorial plane,
(c) minimum at an angle 45° with its axis,
(d) the same in all directions.

13. A sinusoidal EM wave in free space is propagating with a speed c along the direction defined by the unit vector $\hat{n}$. Then the associated $\vec{E}$- and $\vec{B}$- vectors are related by

(a) $\vec{B} = c\left(\hat{n} \times \vec{E}\right)$,
(b) $\vec{B} = \frac{1}{c}\left(\hat{n} \times \vec{E}\right)$,
(c) $\vec{E} = \frac{1}{c}\left(\hat{n} \times \vec{B}\right)$,
(d) $\vec{E} = c\left(\hat{n} \times \vec{B}\right)$.

14. The ratio of the magnitudes of $\vec{E}$ and $\vec{H}$ has the dimensions of

(a) impedance,
(b) capacitance,
(c) inductance,
(d) velocity.

15. The r.m.s value of displacement current in a parallel plate air capacitor with the plate area A, plate separation d and an applied ac voltage $v = V_0 \sin\omega t$

(a) zero, (b) $\frac{\varepsilon_0 A V_0 \omega}{d}$, (c) $\frac{\varepsilon_0 A V_0 \omega}{\sqrt{2}d}$, (d) $\frac{\varepsilon_0 A V_0}{\omega d}$.

16. The magnetic field associated with electric field $\vec{E} = \hat{\jmath}E_0\sin(kz - \omega t)$ of an EM wave as

(a) $\vec{B} = -\hat{\imath}\frac{E_0}{c}\sin(kz - \omega t)$,
(b) $\vec{B} = +\hat{\imath}\frac{E_0}{c}\sin(kz - \omega t)$,
(c) $\hat{k}\frac{E_0}{c}\cos(kz - \omega t)$
(d) $\vec{B} = \hat{\jmath}\frac{E_0}{c}\sin(kz - \omega t)$

17. For an EM wave travelling in vacuum $\vec{E}$-field amplitude is $30\pi\,\text{V/m}$. Then amplitude of its $\vec{H}$-field in A/m unit must be

(a) $\pi \times 10^{-7}$, (b) 2π, (c) 0.25, (d) 367.

[**Hints.** $E_0/H_0 = Z_0 = 376.6\Omega$]

18. An EM wave propagating in a dielectric medium is given by $\vec{E}(\vec{r},t) = (\hat{i} + A\hat{k})E_0 \exp j\, k_0\{(x+\sqrt{3}z) - ct\}$ where A, E_0, k_0 are constants and c is the speed of light in free space. The relative dielectric constant ε_r of the medium and the constant A are

(a) $2, -1/\sqrt{3}$, (b) $2, +1/\sqrt{3}$, (c) $4, -1/\sqrt{3}$, (d) $4, +1/\sqrt{3}$.

[NET 2011]

[**Hints.** $\vec{E}\perp^r$ to the direction of propagation. Here $\vec{k} = k_0(\hat{i}+\sqrt{3}\hat{k})$

$$\therefore \quad (\hat{i}+A\hat{k})\cdot(\hat{i}+\sqrt{3}\hat{k}) = 0, \quad \therefore \quad A = -1/\sqrt{3}$$

Now $c_m = \dfrac{\omega}{k} = \dfrac{k_0 c}{k_0 \cdot 2}$, $\quad \therefore \quad \dfrac{c}{e_m} = 2 = \sqrt{\dfrac{\mu\varepsilon}{\mu_0\varepsilon_0}} \approx \sqrt{\varepsilon_r}$]

19. Suppose a parallel plate capacitor with circular plates parallel to xy-plane and its centre at origin is being discharged. Then the direction of energy flow in terms of Poynting's vector in space between the plates is

(a) along a direction normal to the plates,
(b) radially outward,
(c) radially inward,
(d) circumferential.

[GATE 2008]

20. The $\vec{E}$-field of a circuarly polarized EM wave propagating along $+z$-direction may be expressd as

(a) $E_0(\hat{i}+\hat{j})\exp j(kz+\omega t)$, (c) $E_0(\hat{i}+\hat{j})\exp j(kz-\omega t)$
(b) $E_0(\hat{i}+j\hat{j})\exp j(kz-\omega t)$ (d) $E_0(\hat{i}+j\hat{j})\exp j(kz+\omega t)$

21. If $(\phi, \vec{A})$ and $(\phi', \vec{A'})$ denote two sets of scalar and vector potentials and ψ is a scalar. Then which of the following transformation leave $\vec{E}$- and $\vec{B}$- fields unchanged?

(a) $\vec{A'} = \vec{A} + \vec{\nabla}\psi,\ \phi' = \phi - \dfrac{\partial\psi}{\partial t}$, (c) $\vec{A'} = \vec{A} - \vec{\nabla}\psi,\ \phi' = \phi - \dfrac{\partial\psi}{\partial t}$,
(b) $\vec{A'} = \vec{A} - \vec{\nabla}\psi,\ \phi' = \phi + \dfrac{\partial\psi}{\partial t}$, (d) $\vec{A'} = \vec{A} + \vec{\nabla}\psi,\ \phi' = \phi - 2\dfrac{\partial\psi}{\partial t}$

[NET 2013]

ANSWERS

1. (a)	**2.** (b)	**3.** (c)	**4.** (a)	**5.** (c)	**6.** (a)	**7.** (b)
8. (a)	**9.** (b)	**10.** (d)	**11.** (c)	**12.** (b)	**13.** (b)	**14.** (a)
15. (c)	**16.** (a)	**17.** (c)	**18.** (c)	**19.** (b)	**20.** (b)	**21.** (a)

Chapter 16

Electromagnetic Boundary Value Problems

16.1 Introduction

In the previous chapter we have discussed the plane wave solutions of Maxwell's equations and showed how electromagnetic waves propagate through different homogeneous media. The purpose of the present chapter is to consider the behaviour of electromagnetic waves when it is incident on the interface of two different media. The details depend on the boundary conditions satisfied by the electromagnetic fields. So we shall first find these boundary conditions. The laws of reflection and refraction of electromagnetic waves will then be established by solving the homogeneous wave equation with appropriate boundary conditions. The problem of wave propagation through waveguides will also be considered.

16.2 Boundary Conditions

When electromagnetic waves exist in more than one medium the field vectors must satisfy certain conditions at the interface of two media. These conditions are known as *boundary conditions.* These conditions can be deduced from Maxwell's equations.

To find the boundary conditions for $\vec{D}$ and $\vec{B}$ we note that Maxwell's divergence equations, i.e.,

$$\vec{\nabla} \cdot \vec{D}(t) = \rho(t) \quad \text{and} \quad \vec{\nabla} \cdot \vec{B}(t) = 0$$

are exactly of the same form as the corresponding equations in electrostatics and magnetostatics except that here $\vec{D}$, $\vec{B}$ and ρ are time dependent. Hence following the same procedure as done in Section 2.13 and 9.11 we can deduce the boundary conditions on $\vec{D}$ and $\vec{B}$.

To find the boundary condition on $\vec{D}$ we consider Fig. 2.13-1. Integrating the equation $\vec{\nabla} \cdot \vec{D} = \rho$ over the volume of the pill-box and using divergence theorem we get

$$\int_S \vec{D} \cdot \hat{n}\, dS = \sigma \Delta S.$$

Now following the procedure of Section 2.13 we get

$$\vec{D}_2 \cdot \hat{n} - \vec{D}_1 \cdot \hat{n} = \sigma \tag{16.2-1}$$

So the normal components of $\vec{D}$ are discontinuous across the interface by an amount equal to the free surface charge density.

Following the procedure of Section 9.11 we get the boundary condition on $\vec{B}$:

$$\vec{B}_2 \cdot \hat{n} = \vec{B}_1 \cdot \hat{n}. \tag{16.2-2}$$

Thus, *the normal components of $\vec{B}$ are continuous across the interface.*

Maxwell's curl equations,

$$\vec{\nabla} \times \vec{H} = \vec{J} + \frac{\partial \vec{D}}{\partial t} \quad \text{and} \quad \vec{\nabla} \times \vec{E} = -\frac{\partial \vec{B}}{\partial t}$$

are different from the corresponding static equations. However, it turns out that the boundary conditions remains exactly the same as in static case with the time dependence retained in the fields and the current. To find the conditions on the tangential components of $\vec{H}$ we integrate

$$\vec{\nabla} \times \vec{H} = \vec{J} + \frac{\partial \vec{D}}{\partial t}$$

over the surface of the small rectangle shown in Fig. 9.11-1. Thus,

$$\int_{\Delta S} \vec{\nabla} \times \vec{H} \cdot d\vec{S} = \int_{\Delta S} \vec{J} \cdot d\vec{S} + \int_{\Delta S} \frac{\partial \vec{D}}{\partial t} \cdot d\vec{S}.$$

If $\partial \vec{D}/\partial t$ is bounded, then its integral over ΔS vanishes in the limit $\Delta h \to 0$. So following the procedure of Section 9.11 we get

$$\hat{n} \times \vec{H}_2 - \hat{n} \times \vec{H}_1 = \vec{K}. \tag{16.2-3}$$

So the tangential components of $\vec{H}$ are discontinuous across the interface by an amount equal to the free surface current density.

A similar integration of the equation

$$\vec{\nabla} \times \vec{E} = -\frac{\partial \vec{B}}{\partial t}$$

over the surface of the small rectangle of Fig. 2.13-2 gives

$$\int_{\Delta S} \vec{\nabla} \times \vec{E} \cdot d\vec{S} = -\int_{\Delta S} \frac{\partial \vec{B}}{\partial t} \cdot d\vec{S}$$

$$\text{or} \oint \vec{E} \cdot d\vec{l} = -\int_{\Delta S} \frac{\partial \vec{B}}{\partial t} \cdot d\vec{S} \quad \text{(Using Stokes' theorem)}$$

Now, if $\partial \vec{B}/\partial t$ is bounded, then its integral over ΔS vanishes in the limit $\Delta h \to 0$. So following the procedure of Section 2.13 we get

$$\vec{E}_2 \cdot \hat{\tau} = \vec{E}_1 \cdot \hat{\tau}. \tag{16.2-4}$$

Thus, *the tangential components of $\vec{E}$ are continuous across the boundary.*

Equations (16.2-1)–(16.2-4) constitute the most general boundary conditions for electromagnetic fields. In case of linear medium the boundary conditions can be expressed in terms of $\vec{E}$ and $\vec{B}$ only:

$$\varepsilon_2 E_{2n} - \varepsilon_1 E_{1n} = \sigma, \quad B_{2n} = B_{1n}$$
$$\frac{1}{\mu_2}\hat{n} \times \vec{B}_2 - \frac{1}{\mu_1}\hat{n} \times \vec{B}_1 = \vec{K}, \quad E_{2t} = E_{1t} \tag{16.2-5}$$

Special cases

1. ***Dielectric-dielectric interface:*** In this case, there is no free charge or free current at the interface and we have

$$\varepsilon_2 E_{2n} = \varepsilon_1 E_{1n}, \quad B_{2n} = B_{1n}$$
$$\frac{1}{\mu_2} B_{2t} = \frac{1}{\mu_1} B_{1t}, \quad E_{2t} = E_{1t} \tag{16.2-6}$$

2. ***Conductor-dielectric interface:*** Suppose the medium 1 is highly conducting. The fields in this region are related by

$$\vec{\nabla} \times \vec{H}_1 = \vec{J}_1 + \frac{\partial \vec{D}_1}{\partial t}.$$

Taking $\vec{D}_1 = \varepsilon_1 \vec{E}_1 = \varepsilon_1 \vec{E}_{10} e^{-j\omega t}$ and $\vec{J}_1 = \sigma_1 \vec{E}_1$ we can write

$$\vec{E}_1 = \frac{\vec{\nabla} \times \vec{H}_1}{\sigma_1 - j\omega\varepsilon_1}.$$

If $\vec{H}_1$ is differentiable and bounded then in the limit $\sigma_1 \to \infty$, $\vec{E}_1$ is zero.

Now

$$\vec{\nabla} \times \vec{E}_1 = -\frac{\partial \vec{B}_1}{\partial t} = -\mu_1 \frac{\partial \vec{H}_1}{\partial t}.$$

So taking $\vec{H}_1 = \vec{H}_{10}e^{-j\omega t}$ we can write

$$\vec{H}_1 = \frac{\vec{\nabla} \times \vec{E}_1}{j\omega\mu}.$$

This indicates that inside a perfect conductor, where $\vec{E}_1 = 0$, $\vec{H}_1$ is also zero. So both the tangential and normal components of $\vec{E}_1$ and $\vec{H}_1$ are zero, i.e.,

$$E_{1n} = E_{1t} = 0 \quad \text{and} \quad H_{1n} = H_{1t} = 0.$$

So in this case, the boundary conditions become

$$\varepsilon_2 E_{2n} = \sigma, \quad B_{2n} = B_{1n} = 0$$
$$\frac{1}{\mu_2}\hat{n} \times \vec{B}_2 = \vec{K}, \quad E_{2t} = E_{1t} = 0. \tag{16.2-7}$$

16.3 Reflection and Refraction of Electromagnetic Waves at the Interface of Two Dielectric Media

Reflection and refraction of electromagnetic waves (light) at a plane interface of two different media are familiar phenomena. Various aspects of these phenomena including ordinary laws of reflection and refraction, amplitude relations, phase changes, polarization etc. can be well-described by electromagnetic theory.

A. Ordinary laws of reflection and refraction

Suppose that the xy-plane forms the interface ($z = 0$) between two linear homogeneous dielectric (nonconducting) media characterised by permittivities ε_1 and ε_2, and permeabilities μ_1 and μ_2, respectively as shown in Fig. 16.3-1. Let a plane electromagnetic wave be incident on the interface at the point O. The wave is partly reflected and partly refracted. Let us represent the electric fields associated with these waves by

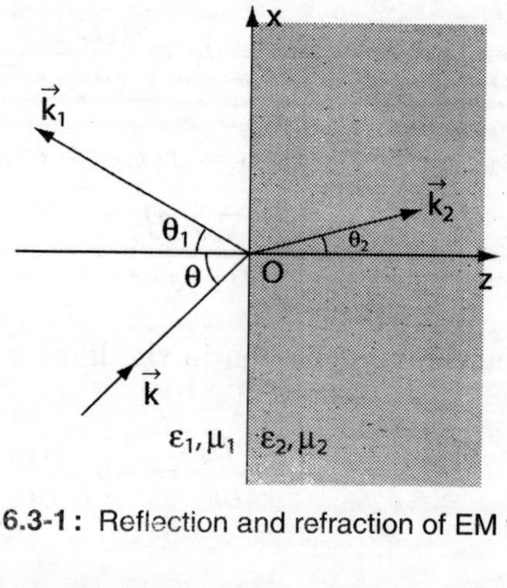

Fig. 16.3-1 : Reflection and refraction of EM waves

$$\vec{E} = \vec{E}_0 e^{j(\vec{k}\cdot\vec{r}-\omega t)},$$

$$\vec{E}_1 = \vec{E}_{10} e^{j(\vec{k}_1\cdot\vec{r}-\omega_1 t)},$$

$$\text{and} \quad \vec{E}_2 = \vec{E}_{20} e^{j(\vec{k}_2\cdot\vec{r}-\omega_2 t)}, \tag{16.3-1}$$

where $\vec{E}_0$, $\vec{E}_{10}$ and $\vec{E}_{20}$ are the amplitudes, $\vec{k}$, $\vec{k}_1$ and $\vec{k}_2$ are the propagation vectors, ω, ω_1 and ω_2 are the frequencies of the incident, reflected and refracted waves respectively.

Let us now apply the boundary condition that the tangential components of electric field are continuous across the interface $z = 0$. Thus,

$$\left(\vec{E}_0\right)_t \cdot e^{j(\vec{k}\cdot\vec{r}-\omega t)} + \left(\vec{E}_{10}\right)_t \cdot e^{j(\vec{k}_1\cdot\vec{r}-\omega_1 t)} = \left(\vec{E}_{20}\right)_t \cdot e^{j(\vec{k}_2\cdot\vec{r}-\omega_2 t)}, \tag{16.3-2}$$

where the subscript t is used to denote tangential components. The condition (16.3-2) must be valid for all values of time and it immediately follows that

$$\omega = \omega_1 = \omega_2. \tag{16.3-3}$$

Thus, we find that *the frequency of electromagnetic wave does not change on reflection and refraction.*

The condition (16.3-2) must also hold at all points on the interface ($z = 0$). So we must have

$$\vec{k}\cdot\vec{r} = \vec{k}_1\cdot\vec{r} = \vec{k}_2\cdot\vec{r} \quad \text{at } z = 0$$

Therefore,

$$k_x = k_{1x} = k_{2x} \quad \text{and} \quad k_y = k_{1y} = k_{2y}. \tag{16.3-4}$$

If the incident beam is assumed to be in the xz-plane then $k_y = 0$. Consequently $k_{1y} = k_{2y} = 0$ and both $\vec{k}_1$ and $\vec{k}_2$ lie in the xz-plane.

Since z-axis is normal to the interface we may conclude that *the incident, reflected, refracted waves and normal to the interface at the point of incidence all lie in the same plane.*

From Fig. 16.3-1,

$$k_x = k\sin\theta, \quad k_{1x} = k_1\sin\theta_1 \quad \text{and} \quad k_{2x} = k_2\sin\theta_2. \tag{16.3-5}$$

Therefore, from the condition $k_x = k_{1x}$ we get

$$k\sin\theta = k_1\sin\theta_1. \tag{16.3-6}$$

Since the vectors $\vec{k}$ and $\vec{k}_1$ lie in the same medium we can write for the phase velocity in medium 1 as

$$\frac{\omega}{k} = \frac{\omega_1}{k_1} = \frac{1}{\sqrt{\mu_1\varepsilon_1}}. \tag{16.3-7}$$

As $\omega = \omega_1$, therefore, $k = k_1$ and from Eq. (16.3-6) we get

$$\theta = \theta_1. \tag{16.3-8}$$

That is, *the angle of incidence is equal to the angle of reflection.*

Again from Eqs. (16.3-4) and (16.3-5),

$$k_{1x} = k_{2x}$$

$$\text{or} \quad k_1 \sin\theta_1 = k_2 \sin\theta_2$$

$$\text{or} \quad \frac{\sin\theta_1}{\sin\theta_2} = \frac{k_2}{k_1} = \frac{\omega/k_1}{\omega/k_2} = \frac{v_1}{v_2} = \frac{c/v_2}{c/v_1} = \frac{n_2}{n_1}, \tag{16.3-9}$$

where $v_1 = \omega/k_1$ and $v_2 = \omega/k_2$ represent the phase velocities of the waves in media 1 and 2, respectively; n_1 and n_2 are the refractive indices of the media 1 and 2, respectively. Obviously,

$$n_1 \sin\theta_1 = n_2 \sin\theta_2, \tag{16.3-10}$$

which is the well-known *Snell's law of refraction.*

B. Fresnel's equations

The ordinary laws of reflection and refraction as discussed above tells nothing about the fractions of incident wave-amplitude reflected or refracted. The equations which relate the amplitudes of the reflected and refracted waves with that of the incident wave are known as *Fresnel's equations.*

We know that $\vec{E}$-vector in a plane electromagnetic wave is perpendicular to the direction of propagation of the wave. However, it may lie in arbitrary directions in the plane normal to the direction of propagation. It is convenient to consider two extreme cases separately, one in which the $\vec{E}$-vector of the incident wave is parallel to the plane of incidence (the plane defined by $\vec{k}$ and $\hat{n}$); this is called *p polarization* (*p* for parallel). The other in which the $\vec{E}$-vector is perpendicular to the plane of incidence is called *s polarization* (*s* for a German word *senkrecht* meaning perpendicular). The general case of arbitrary polarization can then be obtained by a suitable linear combination of the two extreme results.

Case IV. *$\vec{E}$ is perpendicular to the plane of incidence (s-polarized wave).*

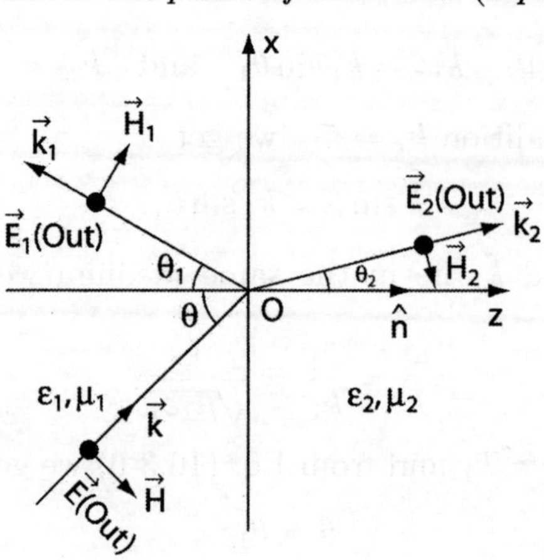

Fig. 16.3-2: Reflection and refraction with $\vec{E}$-vector normal to the plane of incidence

Suppose that the $\vec{E}$-vector of the incident wave lies perpendicular to the plane of incidence (xz-plane) as shown in Fig. 16.3-2. As the media on two sides of the interface $z = 0$ are isotropic the $\vec{E}$-vectors of reflected and refracted waves will also be perpendicular to the plane of incidence. In Fig. 16.3-2 we assume that all the $\vec{E}$-vectors are directed normally out of the plane of the paper. Magnetic vectors $\vec{H}$, $\vec{H}_1$ and $\vec{H}_2$ are in the plane of incidence and their directions are chosen so as to give a positive flow of energy in the direction of the respective propagation vectors. Let us represent the incident, reflected and refracted waves by

$$\vec{E} = \vec{E}_0 e^{j(\vec{k}\cdot\vec{r}-\omega t)}, \quad \vec{H} = \frac{\vec{k}\times\vec{E}}{\omega\mu_1} \tag{16.3-11}$$

$$\vec{E}_1 = \vec{E}_{10} e^{j(\vec{k}_1\cdot\vec{r}-\omega t)}, \quad \vec{H}_1 = \frac{\vec{k}_1\times\vec{E}_1}{\omega\mu_1} \tag{16.3-12}$$

$$\vec{E}_2 = \vec{E}_{20} e^{j(\vec{k}_2\cdot\vec{r}-\omega t)}, \quad \vec{H}_1 = \frac{\vec{k}_2\times\vec{E}_2}{\omega\mu_2} \tag{16.3-13}$$

Since the electric vectors are all parallel to the interface $z = 0$, the continuity of the tangential components of $\vec{E}$-field gives

$$E_0 e^{j(\vec{k}\cdot\vec{r}-\omega t)} + E_{10} e^{j(\vec{k}_1\cdot\vec{r}-\omega t)} = E_{20} e^{j(\vec{k}_2\cdot\vec{r}-\omega t)}.$$

This condition must be valid for all values of t and for all points on the interface. This leads to the cancellations of the exponential factors and gives rise to ordinary laws of reflection and refraction. Thus,

$$E_0 + E_{10} = E_{20} \tag{16.3-14}$$

The continuity of the tangential components of $\vec{H}$-vectors across the interface $z = 0$ requires that

$$-H_0\cos\theta + H_{10}\cos\theta_1 = -H_{20}\cos\theta_2. \tag{16.3-15}$$

Now

$$\vec{H} = \frac{\vec{k}\times\vec{E}}{\omega\mu_1} = \sqrt{\frac{\varepsilon_1}{\mu_1}}\cdot\frac{\vec{k}\times\vec{E}}{k} \quad \text{[using Eq. (16.3-7)]}$$

$$\therefore H_0 = \sqrt{\frac{\varepsilon_1}{\mu_1}} E_0$$

Similarly,

$$H_{10} = \sqrt{\frac{\varepsilon_1}{\mu_1}} E_{10} \quad \text{and} \quad H_{20} = \sqrt{\frac{\varepsilon_2}{\mu_2}} E_{20}.$$

Therefore, Eq. (16.3-15) may be rewritten as

$$(-E_0\cos\theta + E_{10}\cos\theta)\sqrt{\frac{\varepsilon_1}{\mu_1}} = -E_{20}\cos\theta_2\sqrt{\frac{\varepsilon_2}{\mu_2}}, \tag{16.3-16}$$

where we use $\theta_1 = \theta$.

Solving Eqs. (16.3-14) and (16.3-16) we get

$$r_\perp = \frac{E_{10}}{E_0} = \frac{\sqrt{\dfrac{\varepsilon_1}{\mu_1}}\cos\theta - \sqrt{\dfrac{\varepsilon_2}{\mu_2}}\cos\theta_2}{\sqrt{\dfrac{\varepsilon_1}{\mu_1}}\cos\theta + \sqrt{\dfrac{\varepsilon_2}{\mu_2}}\cos\theta_2} \tag{16.3-17}$$

and

$$t_\perp = \frac{E_{20}}{E_0} = \frac{2\sqrt{\dfrac{\varepsilon_1}{\mu_1}}\cos\theta}{\sqrt{\dfrac{\varepsilon_1}{\mu_1}}\cdot\cos\theta + \sqrt{\dfrac{\varepsilon_2}{\mu_2}}\cos\theta_2}, \tag{16.3-18}$$

where $r_\perp$ is called *amplitude reflection coefficient* and $t_\perp$, the *amplitude transmission coefficient.* The subscript $\perp$ is used to denote that we are dealing with the case in which $\vec{E}$ is perpendicular to the plane of incidence. Equations (16.3-17) and (16.3-18) are known as *Fresnel's equations.*

Note that these equations give the amplitudes of the reflected and transmitted waves relative to the amplitude of the incident wave. Most dielectrics are essentially nonmagnetic and we can use $\mu_1 = \mu_2 \approx \mu_0$ and $n_1 = \sqrt{\varepsilon_1/\varepsilon_0}$ as the refractive index of medium 1 and $n_2 = \sqrt{\varepsilon_2/\varepsilon_0}$ as the refractive index of medium 2. In this case, Fresnel's Eqs. (16.3-17) and (16.3-18) take the following common from:

$$r_\perp = \frac{E_{10}}{E_0} = \frac{n_1\cos\theta - n_2\cos\theta_2}{n_1\cos\theta + n_2\cos\theta_2} \tag{16.3-19}$$

and

$$t_\perp = \frac{E_{20}}{E_0} = \frac{2n_1\cos\theta}{n_1\cos\theta + n_2\cos\theta_2}. \tag{16.3-20}$$

Using Snell's law

$$\frac{\sin\theta}{\sin\theta_2} = \frac{n_2}{n_1},$$

the Fresnel's equations can be rewritten as

$$r_\perp = \frac{E_{10}}{E_0} = -\frac{\sin(\theta - \theta_2)}{\sin(\theta + \theta_2)} \tag{16.3-21}$$

and

$$t_\perp = \frac{E_{20}}{E_0} = \frac{2\cos\theta\sin\theta_2}{\sin(\theta + \theta_2)} \tag{16.3-22}$$

Case V. *$\vec{E}$ is parallel to the plane of incidence (p-polarized wave).*

Now we consider the case when $\vec{E}$-vector of the incident light is parallel to the plane of incidence (xz-plane) as shown in Fig. 16.3-3. Since the media on the two sides of the interface $z = 0$ are isotropic the $\vec{E}$-vectors of reflected and refracted waves will

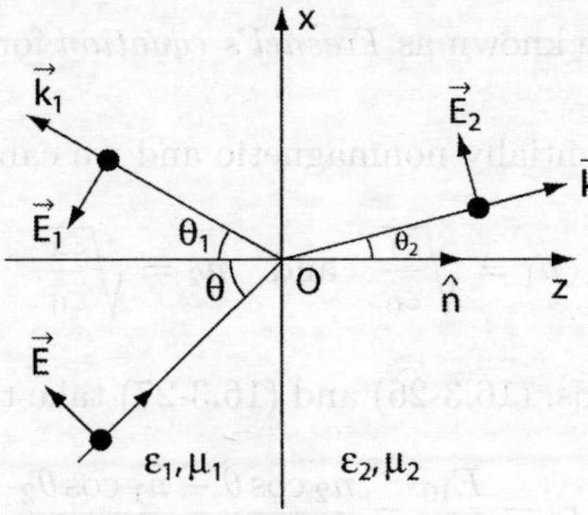

Fig. 16.3.3 : Reflection and refraction with $\vec{E}$-vector parallel to the plane of incidence

also be in the plane of incidence. Corresponding magnetic vectors $\vec{H}$, $\vec{H}_1$ and $\vec{H}_2$ are perpendicular to the plane of incidence. Their directions are chosen so as to give a positive flow of energy in the direction of respective propagation vectors. Thus, in Fig. 16.3-3 the vectors $\vec{H}$, $\vec{H}_1$ and $\vec{H}_2$ are all directed normally out of the plane of the paper and parallel to the interface $z = 0$.

The incident, reflected and refracted waves may be represented by the Eqs.(16.3-11)–(16.3-13) as before. Now the continuity of the tangential components of $\vec{E}$-fields gives

$$E_0 \cos\theta - E_{10}\cos\theta = E_{20}\cos\theta_2, \tag{16.3-23}$$

where we use $\theta_1 = \theta$.

The continuity of the tangential components of $\vec{H}$-fields gives

$$H_0 + H_{10} = H_{20}. \tag{16.3-24}$$

As before,

$$H_0 = \sqrt{\frac{\varepsilon_1}{\mu_1}}E_0, \quad H_{10} = \sqrt{\frac{\varepsilon_1}{\mu_1}}E_{10} \quad \text{and} \quad H_{20} = \sqrt{\frac{\varepsilon_2}{\mu_2}}E_{20}.$$

Therefore, Eq. (16.3-24) becomes

$$(E_0 + E_{10})\sqrt{\frac{\varepsilon_1}{\mu_1}} = E_{20}\sqrt{\frac{\varepsilon_2}{\mu_2}} \tag{16.3-25}$$

Solving Eqs. (16.3-23) and (16.3-25) we get

$$r_\parallel = \frac{E_{10}}{E_0} = \frac{\sqrt{\dfrac{\varepsilon_2}{\mu_2}}\cos\theta - \sqrt{\dfrac{\varepsilon_1}{\mu_1}}\cos\theta_2}{\sqrt{\dfrac{\varepsilon_2}{\mu_2}}\cos\theta + \sqrt{\dfrac{\varepsilon_1}{\mu_1}}\cos\theta_2} \tag{16.3-26}$$

and

$$t_\parallel = \frac{E_{20}}{E_0} = \frac{2\sqrt{\dfrac{\varepsilon_1}{\mu_1}}\cos\theta}{\sqrt{\dfrac{\varepsilon_2}{\mu_2}}\cos\theta + \sqrt{\dfrac{\varepsilon_1}{\mu_1}}\cos\theta_2}. \tag{16.3-27}$$

These two equations are known as *Fresnel's equation* for the case of polarization in the plane of incidence.

Most dielectrics are essentially nonmagnetic and we can use $\mu_1 = \mu_2 \approx \mu_0$,

$$n_1 = \sqrt{\frac{\varepsilon_1}{\varepsilon_0}} \quad \text{and} \quad n_2 = \sqrt{\frac{\varepsilon_2}{\varepsilon_0}}.$$

In this case, Fresnel's Eqs. (16.3-26) and (16.3-27) take the following common form:

$$r_{\|} = \frac{E_{10}}{E_0} = \frac{n_2 \cos\theta - n_1 \cos\theta_2}{n_2 \cos\theta + n_1 \cos\theta_2} \tag{16.3-28}$$

and

$$t_{\|} = \frac{E_{20}}{E_0} = \frac{2n_1 \cos\theta}{n_2 \cos\theta + n_1 \cos\theta_2} \tag{16.3-29}$$

Using Snell's law

$$\frac{\sin\theta}{\sin\theta_2} = \frac{n_2}{n_1},$$

Fresnel's equation can be rewritten as

$$\begin{aligned} r_{\|} = \frac{E_{10}}{E_0} &= \frac{\sin 2\theta - \sin 2\theta_2}{\sin 2\theta + \sin 2\theta_2} = \frac{2\cos(\theta+\theta_2)\cdot\sin(\theta-\theta_2)}{2\sin(\theta+\theta_2)\cdot\cos(\theta-\theta_2)} \\ &= \frac{\tan(\theta-\theta_2)}{\tan(\theta+\theta_2)} \end{aligned} \tag{16.3-30}$$

$$\text{and} \quad t_{\|} = \frac{E_{20}}{E_0} = \frac{4\sin\theta_2 \cdot \cos\theta}{\sin 2\theta + \sin 2\theta_2} = \frac{2\sin\theta_2 \cdot \cos\theta}{\sin(\theta+\theta_2)\cos(\theta-\theta_2)}. \tag{16.3-31}$$

16.4 Physical Implications of Fresnel's Equations

Various important aspects of electromagnetic waves can be described by using Fresnel's equations. In particular, we are interested in polarizing effects, phase changes, total internal reflection, reflectance and transmittance.

1. Polarization by reflection

From Eq. (16.3-30) we find that $r_{\|} = 0$ for $\theta + \theta_2 = \pi/2$. This means that if $\theta + \theta_2 = \pi/2$ then electric field polarized parallel to the plane of incidence is not reflected at all. Equation (16.3-21) shows that under this condition $r_{\perp} \neq 0$, i.e., the electric field polarized normal to the plane of incidence is partly reflected. Thus, an unpolarized light consisting of both types of $\vec{E}$-fields incident at an angle $\theta = \theta_p$ satisfying the condition $\theta_p + \theta_2 = \pi/2$, will be plane polarized normal to the plane of incidence.

This angle of incidence θ_p for which $r_{\parallel} = 0$ is known as *Brewster's angle or angle of polarization.* Snell's law in this case, gives

$$\frac{n_2}{n_1} = \frac{\sin\theta}{\sin\theta_2} = \frac{\sin\theta_p}{\sin\left(\frac{\pi}{2} - \theta_p\right)} = \tan\theta_p. \tag{16.4-1}$$

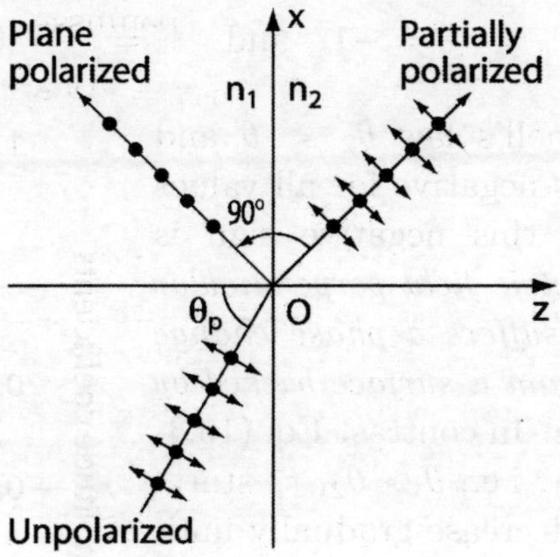

Fig. 16.4.1 : Brewster's angle

Thus, *the tangent of the angle of polarization is equal to the refractive index of the reflecting medium relative to the incident medium.* This is known as *Brewster's law.* The law is illustrated in Fig. 16.4-1. At the angle of polarization the reflected and refracted rays are 90° apart.

2. Amplitude coefficients and phase change

Fresnel's equations show that the amplitude coefficients depend on the value of the angle of incidence θ. Let us examine this dependence over the entire range of values of θ. For nearly normal incidence θ and θ_2 tend to zero. Then Eqs. (16.3-19), (16.3-20), (16.3-28) and (16.3-29) give

$$[r\perp]_{\theta=0} = -\left[r_{\parallel}\right]_{\theta=0} = \frac{n_1 - n_2}{n_1 + n_2} \tag{16.4-2}$$

and

$$\left[t_{\parallel}\right]_{\theta=0} = [t\perp]_{\theta=0} = \frac{2n_1}{n_1 + n_2}. \tag{16.4-3}$$

As an example, for air-glass interface $n_1 = 1$ and $n_2 = 1.5$, then

$$[r\perp]_{\theta=0} = -\left[r_{\parallel}\right]_{\theta=0} = -0.2$$

and

$$\left[t_{\parallel}\right]_{\theta=0} = [t\perp]_{\theta=0} = 0.8.$$

For normal incidence the plane of incidence becomes undefined and any distinction between the parallel and perpendicular components vanishes. So the physical result must be independent of polarization. The difference in signs of $r_{\|}$ and $r_{\perp}$ arises only because $\vec{E}$ and $\vec{E}_1$ are antiparallel in Fig. 16.3-3 when θ goes to zero, where as $\vec{E}$ and $\vec{E}_1$ in Fig. 16.3-2 point in the same direction. For grazing incidence θ tends to 90° then

$$r_{\|} = r_{\perp} = -1 \quad \text{and} \quad t_{\|} = t_{\perp} = 0. \tag{16.4-4}$$

For $n_2 > n_1$, from Snell's law $\theta_2 < \theta$ and hence from (16.3-21), $r_{\perp}$ is negative for all values of θ. The significance of this negative sign is that *the component of electric field perpendicular to the plane of incidence suffers a phase change* $\Delta\phi_{\perp} = \pi$ *when reflected from a surface backed by an optically denser medium.* In contrast Eq. (16.3-30) shows that for $n_2 > n_1$, i.e., $\theta > \theta_2$, $r_{\|}$ starts out positive at $\theta = 0$ and decrease gradually until it becomes zero at the polarizing angle $\theta = \theta_p$. As θ increases beyond θ_p, $r_{\|}$ becomes progressively more negative and reaches the value -1 at $\theta = 90°$ (Fig. 16.4-2). Thus, for $n_2 > n_1$ and the electric field polarized parallel to the plane of incidence $r_{\|}$ is positive (i.e., no phase change, $\Delta\phi_{\|} = 0$) for $\theta < \theta_p$ and $r_{\|}$ is negative (i.e., a phase change of π) for $\theta > \theta_p$ (Fig. 16.4-2).

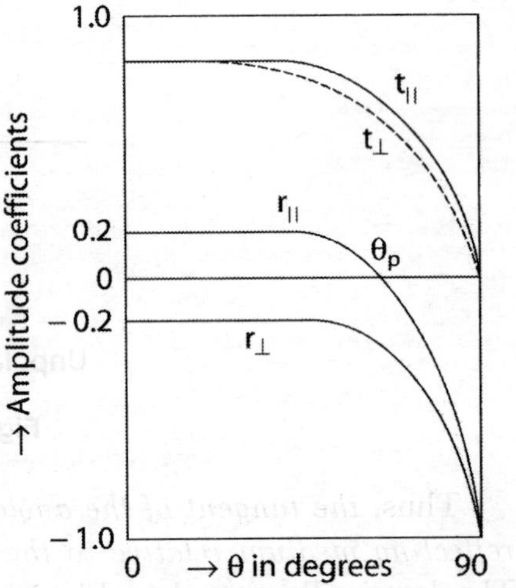

Fig. 16.4-2: Amplitude coefficients as a function of the angle of incidence for $n_2 > n_1$

$t_{\|}$ and $t_{\perp}$ are always positive meaning that the *refracted waves do not suffer any phase change.*

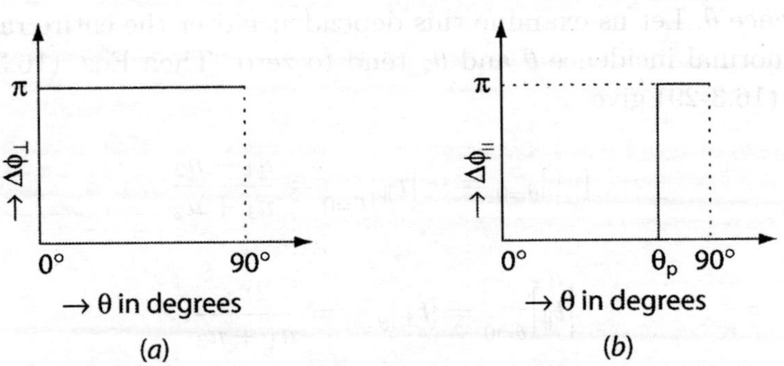

Fig. 16.4-3: Phase changes on reflection for; (a) normal component (b) parallel component ($n_2 > n_1$).

Here it is important to mention that we have chosen $\vec{E}$, $\vec{H}$ and $\vec{k}$-vectors, (whether the ray was incident, reflected or refracted), to have the same relative orientation when looking down the propagation vector. The $\vec{E}$-fields normal to the plane of incidence are said to be in phase, if parallel; π-out of phase, if antiparallel. The $\vec{E}$-fields parallel to the plane of incidence are said to be in phase if their z-components are parallel and out-of-phase if the components are antiparallel. Note that the phase difference between two $\vec{E}$-fields are the same as that between two associated $\vec{H}$-fields. So we need only to look at the vectors normal to the plane of incidence, whether they be $\vec{E}$ or $\vec{H}$ to determine the relative phase of accompanying fields in the plane of incidence.

Total internal reflection

Suppose we consider the case of internal reflection in which the incident medium is more dense, i.e., $n_1 > n_2$. In this case, $\theta_2 > \theta$ and Eq. (16.3-21) shows that $r_\perp$ is always positive. Fig. 16.4-4 shows that $r_\perp$ starts from the value $(n_1 - n_2)/(n_1 + n_2)$ at $\theta = 0$ and then increases with θ and attains the value $+1$ for $\theta = \theta_c$, where θ_c is called *critical angle.* θ_c is the value of the angle of incidence for which the angle of refraction θ_2 becomes 90°. Similarly, from Eqs. (16.3-28) and (16.3-30) it can be shown that in the case of $n_1 > n_2$, i.e., $\theta_2 > \theta$, $r_\parallel$ starts from a negative value $-(n_1 - n_2)/(n_1 + n_2)$ at $\theta = 0$, passes through zero at the polarizing angle $\theta = \theta_p$ and thereafter increases to $+1$ at $\theta = \theta_c$.

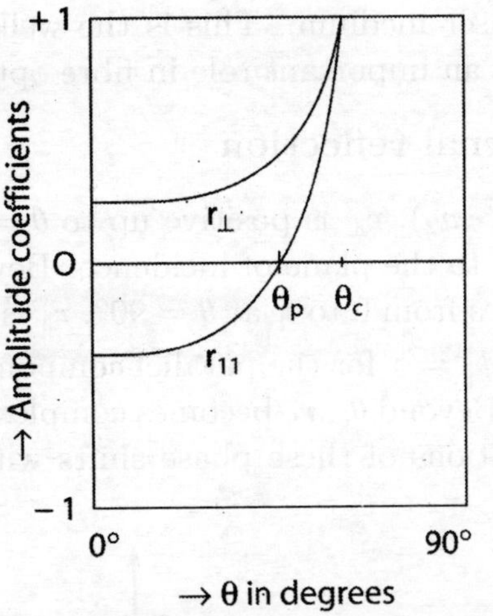

Fig. 16.4-4 : Amplitude coefficients as a function of the angle of incidence for internal reflection ($n_2 > n_1$)

The critical angle is given by Snell's law as

$$\frac{\sin\theta_c}{\sin\pi/2} = \frac{n_2}{n_1} \quad \text{or} \quad n\sin\theta_c = 1, \tag{16.4-5}$$

where $n = n_1/n_2$ is the refractive index of the incident denser medium relative to the lighter second medium. For any angle of incidence θ,

$$\frac{\sin\theta}{\sin\theta_2} = \frac{n_2}{n_1} \quad \text{or} \quad \sin\theta_2 = n\sin\theta. \tag{16.4-6}$$

Hence

$$\cos\theta_2 = \sqrt{1-\sin^2\theta_2} = \sqrt{1-n^2\sin^2\theta} \tag{16.4-7}$$

When $\theta > \theta_c$ Eq. (16.4-5) indicates that $n\sin\theta > 1$ and $\cos\theta_2$ as given by Eq. (16.4-7) becomes imaginary. Let

$$\cos\theta_2 = j\sqrt{n^2\sin^2\theta - 1}. \tag{16.4-8}$$

We now examine Fresnel's equations for $\theta > \theta_c$. Using the values of $\sin\theta_2$ and $\cos\theta_2$ as given by Eqs. (16.4-6) and (16.4-8) we can write from Eq. (16.3-19),

$$r_\perp = \frac{E_{10}}{E_0} = \frac{n\cos\theta - \cos\theta_2}{n\cos\theta + \cos\theta_2} = \frac{n\cos\theta - j\sqrt{n^2\sin^2\theta - 1}}{n\cos\theta + j\sqrt{n^2\sin^2\theta - 1}}. \tag{16.4-9}$$

Obviously the magnitude of the amplitude reflection coefficient is $|r_\perp| = 1$. $r_\perp$ is complex, which implies a change of phase $(\Delta\phi_\perp)$ on reflection. This change of phase depends on the angle of incidence greater than θ_c but the magnitude of $r_\perp$ remains 1.

Similarly, we can show that for $\theta > \theta_c$ the coefficient $r_\parallel$ is complex and $\left|r_\parallel\right| = 1$. Again the change of phase $\left(\Delta\phi_\parallel\right)$ depends on the angle of incidence greater than θ_c but the magnitude of $r_\parallel$ remains 1. Thus, both the components of electric field is totally reflected back into the denser medium. This is the well-known phenomenon of *total internal reflection.* It plays an important role in fibre optics.

Phase changes in internal reflection

For internal reflection $(n_1 > n_2)$, $r_\perp$ is positive up to $\theta = \theta_c$ indicating no phase shift for the electric field normal to the plane of incidence. Beyond θ_c, $r_\perp$ becomes complex and $\Delta\phi_\perp$ gradually increases from 0 to π at $\theta = 90°$. $r_\parallel$ is negative up to $\theta = \theta_p$, which indicates a phase change $\Delta\phi_\parallel = \pi$ for the parallel component of $\vec{E}$-field. From θ_p to θ_c, $r_\parallel$ is positive and $\Delta\phi_\parallel = 0$. Beyond θ_c, $r_\parallel$ becomes complex and $\Delta\phi_\parallel$ gradually increases to π at $\theta = 90°$. The variations of these phase shifts with the angle of incidence are shown in Fig. 16.4-5.

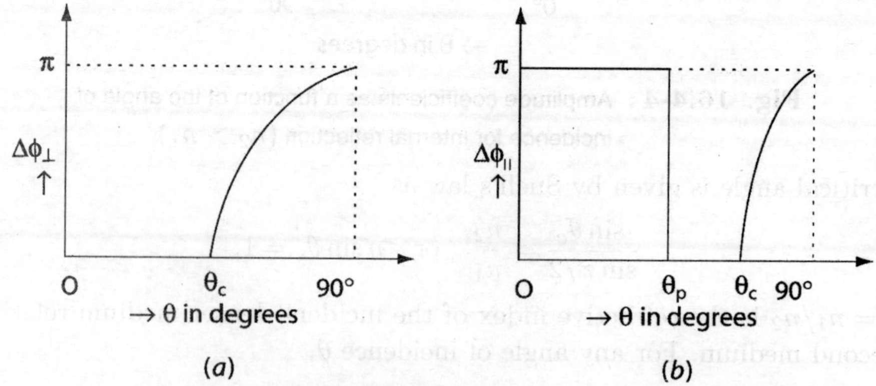

Fig. 16.4-5 : Phase change on internal reflection for; (a) normal component, (b) parallel component $(n_1 . n_2)$

The components of electric field of the reflected wave parallel and perpendicular to the plane of incidence have a relative phase difference $\left(\Delta\phi_{\parallel} - \Delta\phi_{\perp}\right)$, which depends upon the angle of incidence greater than the critical angle. Thus, by proper choice of the angle of incidence any desired type of elliptically polarized wave can be conveniently produced by the method of total internal reflection.

Evanescent wave

In the total internal reflection the phase difference between the incident and reflected wave is not π. As a result, some resultant $\vec{E}$-field exists at the denser side of the boundary. In order to satisfy boundary conditions, there must exist a disturbance in the rarer medium. The wave function of the transmitted $\vec{E}$-field is

$$\vec{E}_2 = \vec{E}_{20} e^{j(\vec{k}_2 \cdot \vec{r} - \omega t)},$$

where $\vec{k}_2 \cdot \vec{r} = k_{2x} \cdot x + k_{2z} \cdot z$, there being no y-component of $\vec{k}_2$. Now $k_{2x} = k_2 \sin\theta_2$ and $k_{2z} = k_2 \cos\theta_2$. Using values of $\sin\theta_2$ and $\cos\theta_2$ as given by (16.4-6) and (16.4-8) we can write

$$\begin{aligned} \vec{E}_2 &= \vec{E}_{20} e^{j(k_2 \sin\theta_2 \cdot x + k_2 \cos\theta_2 \cdot z - \omega t)} \\ &= \vec{E}_{20} e^{-\beta z} e^{j(k_2 n \sin\theta \cdot x - \omega t)}, \end{aligned} \tag{16.4-10}$$

where $\beta = k_2\sqrt{n^2 \sin^2\theta - 1}$.

This equation shows that for $\theta > \theta_c$ the refracted wave advances in the x-direction as a surface or *evanescent wave*. But its amplitude decays exponentially with the depth of penetration into the rarer medium, becoming negligible at a small distance (of only a few wavelengths) into the second medium. Although some fields exist on the rarer side of the interface there is no energy flow through the interface. This can be verified by calculating the time averaged normal component of the Poynting's vector just inside the surface:

$$\vec{s}_2 \cdot \hat{n} = \tfrac{1}{2} Re\left(\vec{E}_2 \times \vec{H}_2^*\right) \cdot \hat{n}.$$

Using

$$\vec{H}_2 = \frac{\vec{k}_2 \times \vec{E}_2}{\mu_2 \omega},$$

we find that

$$\begin{aligned} \vec{s}_2 \cdot \hat{n} &= \tfrac{1}{2} \cdot Re\left[\vec{k}_2\left(\vec{E}_2 \cdot \vec{E}_2^*\right) - \vec{E}_2^*\left(\vec{E}_2 \cdot \vec{k}_2\right)\right] \cdot \hat{n} \\ &= \tfrac{1}{2} \cdot Re\left[\left(\hat{n} \cdot \vec{k}_2\right)\left|\vec{E}_2\right|^2\right] \end{aligned}$$

Now $\hat{n} \cdot \vec{k}_2 = k_2 \cos\theta_2$ is purely imaginary so $\vec{s}_2 \cdot \hat{n} = 0$. A more extensive treatment shows that the energy actually circulates back and forth across the boundary and on the average, results in a zero net flow through the boundary into the rarer medium. Thus, for $\theta > \theta_c$ the incoming energy is totally reflected back into the incident medium.

16.4.1 Reflectance and Transmittance

The usual measurable quantities are not the amplitudes of the reflected and transmitted $\vec{E}$-fields, but the intensities of the reflected and transmitted waves. The intensity of a wave defined as the average energy per unit time crossing a unit area imagined in the plane normal to the direction of flow. This is related to the time average Poynting's vector,

$$I = \langle s \rangle = \tfrac{1}{2}\varepsilon_1 E_0^2 \cdot v_1, \tag{16.4-11}$$

where $v_1 = 1/\sqrt{\varepsilon_1 \mu_1}$ is the velocity of the wave. Let the incident beam falls on an area A of the interface. Then the cross-sectional area of the incident, reflected and transmitted beams are respectively, $A\cos\theta$, $A\cos\theta_1$ and $A\cos\theta_2$. So the energy per unit time flowing in the incident beam is $IA\cos\theta$ and it is the power incident on the area A of the interface. Similarly, the power being reflected from area A is $I_1 A\cos\theta_1$ and the power being transmitted through A is $I_2 A\cos\theta_2$. We now define the reflectance (or the reflection coefficient) R of the surface as the fraction of incident power that is reflected and the *transmittance* (or the transmission coefficient) T as the fraction of incident power that is transmitted. Thus,

$$R = \frac{I_1 A\cos\theta_1}{IA\cos\theta} = \frac{I_1}{I} = \left(\frac{E_{10}}{E_0}\right)^2 \tag{16.4-12}$$

[$\because$ Incident and reflected waves are in the same medium]

and

$$T = \frac{I_2 A\cos\theta_2}{IA\cos\theta} = \frac{I_2\cos\theta_2}{I\cos\theta} = \left(\frac{E_{20}}{E_0}\right)^2 \cdot \frac{\varepsilon_2 v_2}{\varepsilon_1 v_1} \cdot \frac{\cos\theta_2}{\cos\theta}. \tag{16.4-13}$$

For nonmagnetic dielectric medium,

$$T = \left(\frac{E_{02}}{E_0}\right)^2 \cdot \frac{n_2\cos\theta_2}{n_1\cos\theta}. \tag{16.4-14}$$

The principle of conservation of energy requires that the rate of energy flowing into area A must be equal to the rate of energy being reflected from it plus the rate of energy being transmitted through A, provided there is no loss at the interface. This requires that

$$R + T = 1. \tag{16.4-15}$$

For the case of waves with $\vec{E}$-vector parallel to the plane of incidence, we can write for R and T, using Eqs. (16.3-28) and (16.3-29) as

$$R_{\|} = \left(\frac{n_2\cos\theta - n_1\cos\theta_2}{n_2\cos\theta + n_1\cos\theta_2}\right)^2 \tag{16.4-16}$$

and

$$T_{\|} = \left(\frac{2n_1\cos\theta}{n_2\cos\theta + n_1\cos\theta_2}\right)^2 \cdot \frac{n_2\cos\theta_2}{n_1\cos\theta}. \tag{16.4-17}$$

They are plotted as a function of the angle of incidence in Fig. 16.4-6. For waves polarized normal to the plane of incidence one can write, using Eqs. (16.3-19) and (16.3-20),

$$R_\perp = \left(\frac{E_{10}}{E_0}\right)^2 = \left(\frac{n_1 \cos\theta - n_2 \cos\theta_2}{n_1 \cos\theta + n_2 \cos\theta_2}\right)^2 \tag{16.4-18}$$

and

$$T_\perp = \left(\frac{E_{02}}{E_0}\right)^2 \frac{n_2 \cos\theta_2}{n_1 \cos\theta} = \left(\frac{2n_1 \cos\theta}{n_1 \cos\theta + n_2 \cos\theta_2}\right)^2 \cdot \frac{n_2 \cos\theta_2}{n_1 \cos\theta} \tag{16.4-19}$$

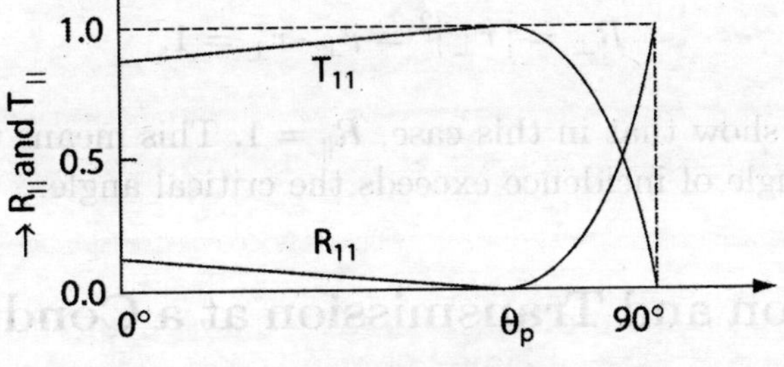

Fig. 16.4-6 : Reflectance and tramittance as a function of the angle of incidence for $\vec{E}$-field polarized parallel to the plane of incidence

A plot of $R_\perp$ and $T_\perp$ as a function of the angle of incidence is shown in Fig. 16.4-7.

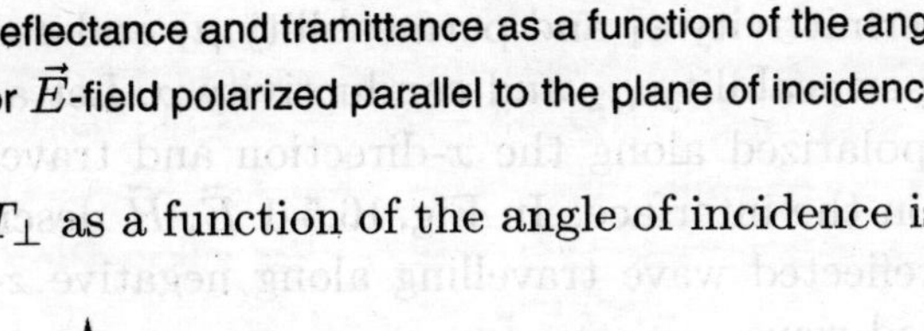

Fig. 16.4-7 : Reflectance and tramittance as a function of the angle of incidence for $\vec{E}$-field polarized normal to the plane of incidence

Note that the value of reflectance is high near grazing incidence. This explains why a glass, polished metal and smooth wet surfaces appear shining when looked at them tangentially.

For normal incidence which is a situation of great practical importance

$$\theta = \theta_2 = 0$$

$$\text{and } R = R_\perp = R_\parallel = \left(\frac{n_1 - n_2}{n_1 + n_2}\right)^2$$

$$\text{and } T = T_\perp = T_\parallel = \frac{4n_1 n_2}{(n_1 + n_2)^2}.$$

Obviously, $R + T = 1$, which is in accordance with the principle of conservation of energy. For an air-glass surface $n_2/n_1 = 1.5$ and $R = 1/25$. This means that 4% of light incident normally on an air-glass interface is reflected back. For $n_1 > n_2$ and angle of incidence greater than θ_c we note from Eq. (16.4-9) that reflectance

$$R_\perp = |r_\perp|^2 = r_\perp \cdot r_\perp^* = 1.$$

Similarly, we can show that in this case, $R_\parallel = 1$. This means that all the energy is reflected when the angle of incidence exceeds the critical angle.

16.5 Reflection and Transmission at a Conducting Surface

Suppose xy-plane forms an interface ($z = 0$) between a dielectric medium (1) characterised by the permittivity ε_1 and permeability μ_1 and a conducting medium (2) with permittivity ε_2, permeability μ_2 and conductivity σ. Let a plane electromagnetic wave of frequency ω polarized along the x-direction and travelling along z-direction be incident normally on the interface. In Fig. 16.5-1 $\vec{E}$, $\vec{H}$ describe the incident wave; $\vec{E}_1$, $\vec{H}_1$ describe the reflected wave travelling along negative z-direction and $\vec{E}_2$, $\vec{H}_2$ describe the transmitted wave.

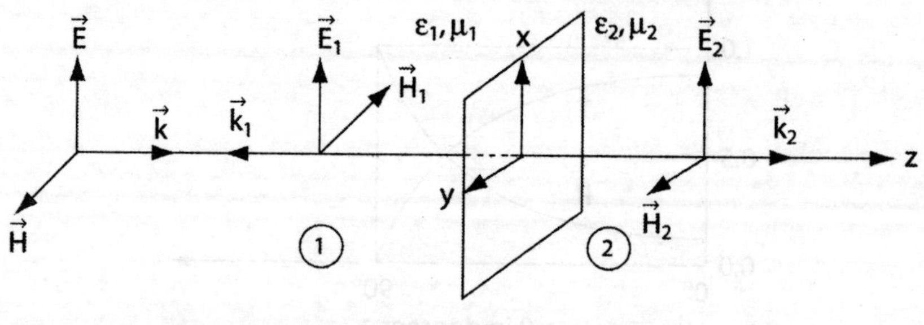

Fig. 16.5-1 : Reflection and transmission at normal incidence from a conducting surface

Let us describe the wave fields by:

$$\begin{aligned} &\text{Incident wave:} \quad \vec{E} = \hat{\imath} E_0 e^{j(kz-\omega t)}, \quad \vec{H} = \hat{\jmath} H_0 e^{j(kz-\omega t)} \\ &\text{Reflected wave:} \quad \vec{E}_1 = \hat{\imath} E_{10} e^{-j(k_1 z+\omega t)}, \quad \vec{H}_1 = -\hat{\jmath} H_{10} e^{-j(kz+\omega t)} \qquad (16.5\text{-}1) \\ &\text{Transmitted wave:} \quad \vec{E}_2 = \hat{\imath} E_{20} e^{j(k_2 z-\omega t)}, \quad \vec{H}_2 = \hat{\jmath} H_{20} e^{j(k_2 z-\omega t)}, \end{aligned}$$

where the amplitudes of $\vec{H}$-fields are related to the amplitudes of $\vec{E}$-field by

$$H_0 = \sqrt{\frac{\varepsilon_1}{\mu_1}} E_0, \quad H_{10} = \sqrt{\frac{\varepsilon_1}{\mu_1}} E_{10} \quad \text{and} \quad H_{20} = \sqrt{\frac{\varepsilon_2}{\mu_2}} E_{20}. \tag{16.5-2}$$

The reflected and transmitted waves are assumed to have the same frequency as that of the incident wave otherwise the boundary conditions at $z = 0$ cannot be satisfied for all t. The minus sign in $\vec{H}_1$ is required by the fact that the Poynting's vector must aim in the direction of propagation. Now we apply the boundary conditions. The continuity of the tangential components of $\vec{E}$-fields across the boundary requires that

$$E_0 + E_{10} = E_{20}. \tag{16.5-3}$$

Similarly, the continuity of $\vec{H}$-fields across the boundary gives

$$\sqrt{\frac{\varepsilon_1}{\mu_1}} (E_0 - E_{10}) = \sqrt{\frac{\varepsilon_2}{\mu_2}} E_{20}. \tag{16.5-4}$$

Taking the media to be nonmagnetic we can use $\mu_1 = \mu_2 \approx \mu_0$. Then in terms of refractive index $n = c/v = \sqrt{\varepsilon/\varepsilon_0}$, Eq. (16.5-4) can be rewritten as

$$n_1(E_0 - E_{10}) = n_2 E_{20}. \tag{16.5-5}$$

Note that for the conducting medium refractive index n_2 is a complex quantity. Solving Eqs. (16.5-3) and (16.5-5) we get

$$r_{12} = \frac{E_{10}}{E_0} = \frac{n_1 - n_2}{n_1 + n_2} \quad \text{and} \quad t_{12} = \frac{E_{20}}{E_0} = \frac{2n_1}{n_1 + n_2}, \tag{16.5-6}$$

where r_{12} is the *amplitude reflection coefficient* and t_{12} is the *amplitude transmission coefficient*.

Note that these results are formally identical to those, Eqs. (16.4-2) and (16.4-3), obtained at the interface between two dielectric media. But here inside the metallic medium the refractive index n_2 is a complex quantity and is given by

$$n_2 = \frac{c}{(\omega/k_2)} = \frac{c}{\omega}(\alpha + j\beta), \tag{16.5-7}$$

where α and β are given by Eqs. (15.7-10) with $\varepsilon = \varepsilon_2$ and $\mu = \mu_2 = \mu_0$. Obviously, n_2 is complex, let

$$n_2 = n_r + jn_i, \tag{16.5-8}$$

where n_r and n_i are respectively the real and imaginary parts of n_2.

For good conductors $\sigma \gg \omega\varepsilon_2$ and then

$$\alpha = \beta \approx \sqrt{\frac{\omega\sigma\mu_0}{2}}$$

and

$$n_2 = \frac{c}{\omega}\sqrt{\frac{\omega\sigma\mu_0}{2}}(1+j) = \sqrt{\frac{\sigma}{2\omega\varepsilon_0}}(1+j) \quad \left[\because\ c = \frac{1}{\sqrt{\mu_0\varepsilon_0}}\right]$$

In this case,

$$n_r = n_i = \sqrt{\frac{\sigma}{2\omega\varepsilon_0}}. \tag{16.5-9}$$

Obviously, $n_r = n_i$. Using (16.5-8) the relations (16.5-6) can be written in the following form:

$$r_{12} = \frac{E_{10}}{E_0} = -\frac{(n_r - n_1) + jn_i}{(n_r + n_1) + jn_i} \quad \text{and} \quad t_{12} = \frac{E_{20}}{E_0} = \frac{2n_1}{(n_r + n_1) + jn_i}. \tag{16.5-10}$$

The complex nature of r_{12} and t_{12} indicates that the reflected and transmitted waves may be phase shifted relative to the incident wave. We can write the complex quantities r_{12} and t_{12} as, $r_{12} = |\, r_{12}\,|\, e^{j\phi_1}$ and $t_{12} = |\, t_{12}\,|\, e^{j\phi_2}$

$$\text{where } |\, r_{12}\,| = \left[\frac{(n_r - n_1)^2 + n_i^2}{(n_r + n_1)^2 + n_i^2}\right]^{1/2},$$

$$|\, t_{12}\,| = \frac{2n_1}{\left[(n_r + n_1)^2 + n_i^2\right]^{1/2}},$$

$$\phi_1 = \tan^{-1}\frac{n_i}{n_r - n_1} - \tan^{-1}\frac{n_i}{n_r + n_1} = \tan^{-1}\frac{2n_1 n_i}{n_r^2 - n_1^2 + n_i^2}$$

$$\text{and} \quad \phi_2 = -\tan^{-1}\frac{n_i}{n_r + n_1}. \tag{16.5-11}$$

These results show that the reflected and the transmitted $\vec{E}$-fields are phase shifted relative to the incident $\vec{E}$-field.

For a perfect conductor $\sigma = \infty$, $n_r = n_i = \infty$ and hence $r_{12} = -1$. It indicates *a phase shift of* 180° *due to reflection from a perfect conducting surface.* So the components of $\vec{E}$ and $\vec{E}_1$ cancel at the surface of a perfect conductor. This is expected because fields in a perfect conductor must vanish and there can be no transmitted wave $\left(\vec{E}_2 = 0\right)$. The totally reflected wave combines with the incident wave and forms a standing or stationary wave with a node on the surface. This has been detected by using photographic plates. The standing wave is given by

$$\vec{E}_s = \vec{E} + \vec{E}_1 = \hat{\imath}E_0\left[e^{j(kz-\omega t)} - e^{-j(kz+\omega t)}\right]$$

$$= \hat{\imath}2E_0\,[j\sin kz]\,e^{-j\omega t} \quad [\because\ E_{01} = -E_0 \text{ for a perfect conductor}].$$

Taking real parts we can write

$$\vec{E}_s = \hat{\imath}2E_0 \sin kz \sin\omega t. \tag{16.5-12}$$

Now the reflectance of the metal surface is

$$R = r_{12} \cdot r_{12}^* = |r_{12}|^2 = \frac{(n_r - n_1)^2 + n_i^2}{(n_r + n_1)^2 + n_i^2} \tag{16.5-13}$$

and the transmittance is

$$T = 1 - R = \frac{4n_r n_1}{(n_r + n_1)^2 + n_i^2}. \tag{16.5-14}$$

The transmitted wave is eventually absorbed in the conducting medium. For this it is customary to call T the *absorptance* A. Therefore,

$$T = A = 1 - R.$$

For good conductors $n_r = n_i = \sqrt{\sigma/2\omega\varepsilon_0} \gg 1$. If the incident medium is air, then $n_1 = 1$. Thus, for incidence from air on a good conducting surface we can write from (16.5-14)

$$T = A = \frac{2}{n_i} \simeq 2\sqrt{\frac{2\omega\varepsilon_0}{\sigma}}$$

and

$$R = 1 - T = 1 - \frac{2}{n_i} = 1 - 2\sqrt{\frac{2\omega\varepsilon_0}{\sigma}}. \tag{16.5-15}$$

This approximate relation is known as *Hagen-Rubens formula.* This relation is found to be valid for metals at frequencies as high as 10^{15} Hz. For moderately good conductors the relation is valid for frequencies below the microwave frequencies.

For a perfect conductor $\sigma = \infty$ and hence $R = 1$ and $T = 0$. Thus, in this case, the incident wave is totally reflected. This explains why metals are opaque to light and why excellent conductors, such as silver, make good mirrors. In practice a thin coating of silver on to the back of a piece of glass is used as a reflector. Glass has nothing to do with the reflection. It just supports the silver and keep it from tarnishing.

Let us use the relation (16.5-15) to determine the reflectance of air-to-silver surface. Taking $\sigma = 6 \times 10^7/\Omega\text{–m}$ we find that for silver at an optical frequency $f = 6{\cdot}4 \times 10^{14}$ Hz.

$$R = 1 - \sqrt{\frac{8 \times 2\pi \times 6.4 \times 10^{14} \times 8.854 \times 10^{-12}}{6 \times 10^7}} = 0.9311.$$

This means that about 93% of the incident energy is reflected and about 7% is absorbed by silver at the optical frequency 6.4×10^{14} Hz. For lower frequencies the reflection is even more complete. The high reflectance of sea-water level for radio waves causes problems in radio communication from land with submerged submarines.

16.6 Propagation of Electromagnetic Waves Between Parallel Conducting Plates

A structure consisting of highly conducting walls in the form of a long open pipe is used to guide electromagnetic waves in the microwave region (3–300 GHz) from one place to other. Such a structure is called a *waveguide.* It may be studied as a boundary value problem with multiple boundaries. As a preliminary to the study of such waveguides we consider here a simpler boundary value problem-namely, the propagation of electromagnetic waves in the region between two parallel, perfectly conducting plates.

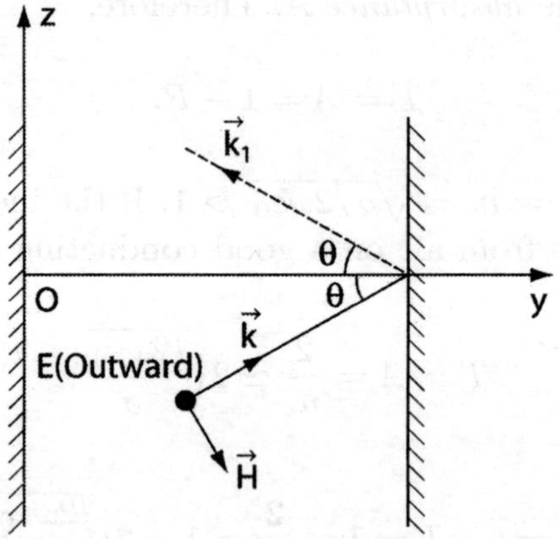

Fig. 16.6-1 : Wave propagateion between two parallel conducting plates

Suppose we consider two perfectly conducting plates located at $y = 0$ and $y = a$ as shown in Fig. 16.6-1. The space between the plates is assumed to be free of charges, currents and have permittivity and permeability equal to those of vacuum. Let an electromagnetic wave of frequency ω with wave vector $\vec{k}$ in the yz-plane be incident on the $y = a$ plate with an angle of incidence θ. Note that in Fig. 16.6-1 the x- and z-directions are physically indistinguishable and hence no generality is lost by considering only waves with wave vectors in the yz-plane. The wave incident on $y = a$ plate will be reflected between the plates by the process of multiple reflections and in effect it will propagate in a direction parallel to the plates. For such waves there are two possible polarizations. The electric field $\vec{E}$ may be parallel to x-axis, i.e., at right angles to the direction of propagation (s-polarization). This is called *transverse electric (TE) wave.* If the $\vec{H}$-field is along x-axis, i.e., perpendicular to the direction of propagation (p-polarization) then corresponding wave is called *transverse magnetic (TM) wave.* Here we shall consider only the TE-waves. The electric field in the region between the plates is given by the sum of incident and reflected waves. Thus, the electric field in the region between the plates for TE waves may be described by

$$\vec{E} = \hat{i}E'_0 e^{j(\vec{k}\cdot\vec{r}-\omega t)} + \hat{i}E'_{10} e^{j(\vec{k}_1\cdot\vec{r}-\omega t)}. \tag{16.6-1}$$

Since the plates are assumed to be perfectly conducting the tangential components of $\vec{E}$-field must vanish at the boundary $y = 0$. For this condition to be true for all values of z and t we must have

$$E_0' = -E_{10}'. \tag{16.6-2}$$

This indicates that the reflected wave will have the same amplitude as the incident wave but with a phase change of π. Therefore,

$$\vec{E} = \hat{i}E_0'\left[e^{j(\vec{k}\cdot\vec{r}-\omega t)} - e^{j(\vec{k}_1\cdot\vec{r}-\omega t)}\right]. \tag{16.6-3}$$

Now we have $\vec{k}\cdot\vec{r} = k\cos\theta\cdot y + k\sin\theta\cdot z$ and $\vec{k}_1\cdot\vec{r} = -k\cos\theta\cdot y + k\sin\theta\cdot z$. Thus,

$$\begin{aligned}\vec{E} &= \hat{i}E_0'\left[e^{jky\cos\theta} - e^{-jky\cos\theta}\right]\cdot e^{j(kz\sin\theta-\omega t)}\\ &= \hat{i}E_0\sin(ky\cos\theta)e^{j(kz\sin\theta-\omega t)},\end{aligned} \tag{16.6-4}$$

where $E_0 = 2jE_0'$. We now apply the boundary condition that the $\vec{E}$-field must vanish at $y = a$ for all values z and t. This requires that

$$\sin(ka\cos\theta) = 0 \quad \text{or} \quad ka\cos\theta = n\pi, \tag{16.6-5}$$

where n is a positive integer. Thus, for a given k and a, there are a number of discrete directions or modes possible for the propagation of waves between the plates.

For convenience let us rewrite Eq. (16.6-4) as

$$\vec{E} = \hat{i}E_0\sin(k_c y)e^{j(k_g z-\omega t)}, \tag{16.6-6}$$

where

$$k_c = k\cos\theta \quad \text{and} \quad k_g = k\sin\theta. \tag{16.6-7}$$

Equation (16.6-6) may how be considered to represent a disturbance propagating along z-axis with an effective wavelength

$$\lambda_g = \frac{2\pi}{k_g} = \frac{2\pi}{k\sin\theta} = \frac{\lambda_0}{\sin\theta}, \tag{16.6-8}$$

where $\lambda_0 = 2\pi/k$ is the free space wavelength. The disturbance also varies sinusoidally along y-axis with an effective wavelength

$$\lambda_c = \frac{2\pi}{k_c} = \frac{2\pi}{k\cos\theta} = \frac{\lambda_0}{\cos\theta}. \tag{16.6-9}$$

From Eqs. (16.6-8) and (16.6-9) it follows immediately that

$$\frac{1}{\lambda_g^2} + \frac{1}{\lambda_c^2} = \frac{1}{\lambda_0^2}. \tag{16.6-10}$$

From Eq. (16.6-5) we can write

$$k_c \cdot a = n\pi \quad \text{or} \quad \lambda_c = \frac{2a}{n}. \tag{16.6-11}$$

Physical importance of λ_g and λ_c can be derived as follows. Suppose for a given n and a, λ_0 is increased gradually. Equation (16.6-10) indicates that if λ_0 becomes greater than λ_c then λ_g becomes imaginary. In this case the exponent $jk_g z$ in Eq. (16.6-6) becomes negative for $z > 0$, and the wave becomes exponentially damped in z, instead of propagating. Thus, for the waves to propagate between the plates we must have

$$\lambda_0 < \lambda_c = \frac{2a}{n}.$$

For this λ_c is called *cut off wavelength.* It is the longest wavelength that can propagate for a given mode (n value).

λ_g is known as *guide wavelength.* From Eq. (16.6-10),

$$\lambda_g = \frac{\lambda_0}{\sqrt{1 - \left(\frac{\lambda_0}{\lambda_c}\right)^2}}. \tag{16.6-12}$$

Obviously λ_g is greater than corresponding free space wavelength.

Equation (16.6-6) indicates that the wave propagates along z-axis with an apparent phase velocity

$$v_p = \frac{\omega}{k_g} = \frac{\omega}{k \sin\theta} = \frac{c}{\sin\theta}. \tag{16.6-13}$$

As $\sin\theta \leq 1$, $v_p \geq c$. For $\theta = 0$, i.e., $\lambda_0 = \lambda_c$, v_p becomes infinite. This result appears to contradict the principle of relativity. In reality it does not violate the principle of relativity because signals are always transmitted with group velocity and not with phase velocity. Phase velocity is not the velocity of propagation of any physical quantity. It simply represents the velocity of a point of constant phase on the wave. The energy propagates down the guide with group velocity, which is always smaller than c.

The group velocity v_g is given by

$$v_g = \frac{d\omega}{dk_g}.$$

From Eqs. (16.6-7) and (16.6-11) we can write

$$k_g^2 = k^2 \sin^2\theta = k^2(1 - \cos^2\theta) = k^2 - k_c^2 = k^2 - \frac{n^2\pi^2}{a^2}$$

Therefore,

$$k_g^2 = \frac{\omega^2}{c^2} - \frac{n^2\pi^2}{a^2} \qquad \left[\because c = \frac{\omega}{k}\right] \tag{16.6-14}$$

Differentiating with respect to k_g we get

$$2k_g = \frac{2\omega}{c^2}\frac{d\omega}{dk_g} \quad \text{or} \quad \frac{d\omega}{dk_g} = \frac{c^2 k_g}{\omega} = \frac{c^2 \cdot k \sin\theta}{ck} = c\sin\theta. \tag{16.6-15}$$

Therefore, group velocity $v_g = c\sin\theta$, which is smaller than the speed of light in vacuum. Moreover, it is interesting to note that

$$v_p \cdot v_g = \frac{c}{\sin\theta} \cdot c\sin\theta = c^2, \tag{16.6-16}$$

which is generally true for propagation in a waveguide.

16.7 Rectangular Waveguide

Waveguides may be in the form of hollow pipes with highly conducting walls (usually of copper) having arbitrary shape but of uniform cross-section. For simplicity we shall consider the special case of a rectangular waveguide with air as dielectric inside. Basically it is a hollow pipe having uniform rectangular cross-section. A wire antenna is often used to launch electromagnetic waves inside the waveguide. For effective transmission through the waveguide the transverse dimensions of the guide must be comparable to the wavelength of the electromagnetic waves to be propagated. The waves launched into the guide propagate by a process of multiple internal reflections. The resulting field configurations can be obtained by solving Maxwell's equations with proper boundary conditions. We shall assume the walls to be made of perfect conductors. So the tangential components of $\vec{E}$-fields and normal components of $\vec{B}$-field must vanish at the boundary.

Assuming the space inside the waveguide to be air or vacuum it is easy to show by using Maxwell's equations that $\vec{E}$ and $\vec{H}$ satisfy the homogenous wave equation [see Section 15.5]

$$\nabla^2 \vec{X} - \frac{1}{c^2}\frac{\partial^2 \vec{X}}{\partial t^2} = 0; \quad \left(\vec{X} \equiv \vec{E} \text{ or } \vec{H}\right). \tag{16.7-1}$$

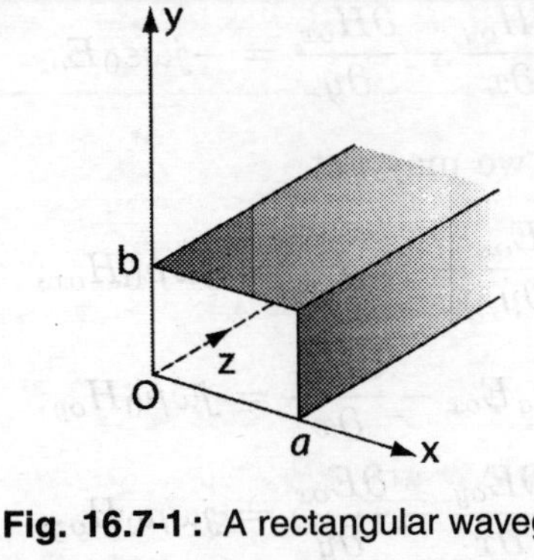

Fig. 16.7-1 : A rectangular waveguide

We consider a rectangular waveguide of sides a and b as shown in Fig. 16.7-1. Let the wave is propagating along z-axis. If the waves to be propagated are monochromatic with time dependence $e^{-j\omega t}$ then Eq. (16.7-1) takes the form

$$\left(\nabla^2 + \frac{\omega^2}{c^2}\right)\vec{X} = 0. \tag{16.7-2}$$

and Maxwell's equations

$$\vec{\nabla} \times \vec{H} = \varepsilon_0 \frac{\partial \vec{E}}{\partial t} \quad \text{and} \quad \vec{\nabla} \times \vec{E} = -\mu_0 \frac{\partial \vec{H}}{\partial t}$$

become

$$\vec{\nabla} \times \vec{H} = -j\omega\varepsilon_0 \vec{E} \tag{16.7-3}$$

$$\text{and } \vec{\nabla} \times \vec{E} = j\omega\mu_0 \vec{H}. \tag{16.7-4}$$

We seek solutions to Eq. (16.7-2) in the form

$$\vec{X}(x, y, z, t) = \vec{X}_0(x, y) e^{j(k_g z - \omega t)}. \tag{16.7-5}$$

Equation (16.7-3) may be written in component forms as

$$\frac{\partial H_z}{\partial y} - \frac{\partial H_y}{\partial z} = -j\omega\varepsilon_0 E_x;$$

$$\frac{\partial H_x}{\partial z} - \frac{\partial H_z}{\partial x} = -j\omega\varepsilon_0 E_y;$$

$$\frac{\partial H_y}{\partial x} - \frac{\partial H_x}{\partial y} = j\omega\varepsilon_0 E_z.$$

Using solutions (16.7-5) we may write

$$\frac{\partial H_{oz}}{\partial y} - jk_g H_{oy} = -j\omega\varepsilon_0 E_{ox}; \tag{16.7-6}$$

$$jk_g H_{ox} - \frac{\partial H_{oz}}{\partial x} = -j\omega\varepsilon_0 E_{oy}; \tag{16.7-7}$$

$$\frac{\partial H_{oy}}{\partial x} - \frac{\partial H_{ox}}{\partial y} = -j\omega\varepsilon_0 E_{oz}. \tag{16.7-8}$$

Similarly from Eq. (16.7-4) we may get

$$\frac{\partial E_{oz}}{\partial y} - jk_g E_{oy} = j\omega\mu_0 H_{ox}; \tag{16.7-9}$$

$$jk_g E_{ox} - \frac{\partial E_{oz}}{\partial x} = j\omega\mu_0 H_{oy}; \tag{16.7-10}$$

$$\frac{\partial E_{oy}}{\partial x} - \frac{\partial E_{ox}}{\partial y} = j\omega\mu_0 H_{oz}. \tag{16.7-11}$$

Eliminating H_{oy} from Eqs. (16.7-6) and (16.7-10) we get

$$E_{ox} = \frac{j}{h^2}\left[k_g \frac{\partial E_{oz}}{\partial x} + \omega\mu_0 \frac{\partial H_{oz}}{\partial y}\right], \tag{16.7-12}$$

where

$$h^2 = \frac{\omega^2}{c^2} - k_g^2. \tag{16.7-13}$$

Similarly considering other equations appropriately we can express E_{oy}, H_{ox} and H_{oy} in terms of E_{oz} and H_{oz} only.

$$E_{oy} = \frac{j}{h^2}\left[k_g \frac{\partial E_{oz}}{\partial y} - \omega\mu_0 \frac{\partial H_{oz}}{\partial x}\right] \tag{16.7-14}$$

$$H_{ox} = \frac{j}{h^2}\left[k_g \frac{\partial H_{oz}}{\partial x} - \omega\varepsilon_0 \frac{\partial E_{oz}}{\partial y}\right] \tag{16.7-15}$$

$$H_{oy} = \frac{j}{h^2}\left[k_g \frac{\partial H_{oz}}{\partial y} + \omega\varepsilon_0 \frac{\partial E_{oz}}{\partial x}\right] \tag{16.7-16}$$

Thus, all the transverse components are being completely determined by the longitudinal components and hence the problem of determining fields inside the waveguide reduces to that of determining longitudinal field components only. In general both E_{oz} and H_{oz} are present. However, we shall consider two separate situations:

(i) If $E_{oz} = 0$ then entire field is determined by H_{oz}. For this corresponding waves are called *H-waves.* Since here only transverse components of electric field exist, it is also called *transverse electric or TE waves.*

(ii) If $H_{oz} = 0$ then entire field is determined by E_{oz}. For this corresponding waves are called *E-waves.* Since here only transverse components of magnetic field exist, it is also called *transverse magnetic or TM waves.*

As Maxwell's equations are linear, by studying TE and TM waves separately we can then superpose them to get the actual picture. There may be a third choice, where both E_{oz} and H_{oz} are zero. Corresponding waves are called *transverse electromagnetic waves or TEM waves.* But Eqs. (16.7-6) to (16.7-11) shows that in this case, the entire field inside the waveguide vanishes identically. It indicates that *pure TEM waves cannot exist in a rectangular waveguide.*

TE modes

In this case, $E_{oz} = 0$ and all other field components may be determined in terms of H_{oz}. From (16.7-2) we find that the equation satisfied by the z-component of $\vec{H}$ is

$$\left(\nabla^2 + \frac{\omega^2}{c^2}\right) H_z = 0.$$

Taking the z-dependence of H_z of the form $e^{jk_g z}$ this equation can be transformed to

$$\left(\frac{\partial^2}{\partial x^2} + \frac{\partial^2}{\partial y^2} + h^2\right) H_{oz} = 0. \tag{16.7-17}$$

To solve this equation by the method of separation of variables we assume

$$H_{oz}(x, y) = F_1(x) \cdot F_2(y).$$

Substituting it in Eq. (16.7-17) and dividing throughout by $F_1 F_2$ we get

$$\frac{1}{F_1}\frac{d^2 F_1}{dx^2} + \frac{1}{F_2}\frac{d^2 F_2}{dy^2} = -h^2.$$

Let

$$\frac{1}{F_1}\frac{d^2 F_1}{dx^2} = -k_x^2 \quad \text{and} \quad \frac{1}{F_2}\frac{d^2 F_2}{dy^2} = -k_y^2,$$

where $k_x^2 + k_y^2 = h^2$.

So the most general solution for $H_{oz}(x, y)$ is

$$H_{oz}(x, y) = (A_1 \cos k_x x + A_2 \sin k_x x)(A_3 \cos k_y y + A_4 \sin k_y y), \tag{16.7-18}$$

where A_i are arbitrary constants to be evaluated from the given boundary conditions. The boundary conditions are

$$E_y = 0 \text{ at } x = 0 \quad \text{and} \quad x = a,$$

$$E_x = 0 \text{ at } y = 0 \quad \text{and} \quad y = b$$

For TE mode $E_{oz} = 0$ and, therefore, from Eq. (16.7-14),

$$E_{oy} \propto \frac{\partial H_{oz}}{\partial x} = 0 \text{ at } x = 0.$$

This condition requires that $A_2 = 0$ in Eq. (16.7-18). From Eq. (16.7-12),

$$E_{ox} \propto \frac{\partial H_{oz}}{\partial y} = 0 \text{ at } y = 0.$$

This condition requires that $A_4 = 0$ in Eq. (16.7-18). Thus,

$$H_{oz}(x, y) = H_0 \cos k_x x \cdot \cos k_y y, \tag{16.7-19}$$

where $H_0 = A_1 A_3$ and is determined by the introduced field.

Now the condition

$$E_{oy} \propto \frac{\partial H_{oz}}{\partial x} = 0 \quad \text{at } x = a$$

requires that

$$\sin k_x a = 0 \quad \text{or} \quad k_x = \frac{m\pi}{a}; \quad m = 0, 1, 2, 3 \cdots$$

Similarly, the condition

$$E_{ox} \propto \frac{\partial H_{oz}}{\partial y} = 0 \quad \text{at } y = b$$

requires that

$$\sin k_y b = 0 \quad \text{or} \quad k_y = \frac{n\pi}{b}; \quad n = 0, 1, 2, 3, \ldots$$

Thus, finally we get

$$H_{oz}(x, y) = H_0 \cos\frac{m\pi x}{a} \cdot \cos\frac{n\pi y}{b} \tag{16.7-20a}$$

Since H_{oz} is now known other field components can be easily calculated by using the relations (16.7-12)–(16.7-16).

$$E_{ox} = -\frac{j\omega\mu_0}{h^2} H_0 \left(\frac{n\pi}{b}\right) \cos\frac{m\pi x}{a} \sin\frac{n\pi y}{b}$$

$$E_{oy} = \frac{j\omega\mu_0}{h^2} H_0 \left(\frac{m\pi}{a}\right) \sin\frac{m\pi x}{a} \cos\frac{n\pi y}{b}$$

$$H_{ox} = -\frac{jk_g}{h^2} H_0 \left(\frac{m\pi}{a}\right) \sin\frac{m\pi x}{a} \cos\frac{n\pi y}{b}$$

$$H_{oy} = -\frac{jk_g}{h^2} H_0 \left(\frac{n\pi}{b}\right) \cos\frac{m\pi x}{a} \sin\frac{n\pi y}{b}. \tag{16.7-20b}$$

Now

$$h^2 = k_x^2 + k_y^2 \quad \text{or} \quad \frac{\omega^2}{c^2} - k_g^2 = \left(\frac{m\pi}{a}\right)^2 + \left(\frac{n\pi}{b}\right)^2$$

$$\text{or} \quad k_g^2 = \frac{\omega^2}{c^2} - \left[\left(\frac{m\pi}{a}\right)^2 + \left(\frac{n\pi}{b}\right)^2\right]. \tag{16.7-21}$$

If

$$\omega^2 < c^2 \left[\left(\frac{m\pi}{a}\right)^2 + \left(\frac{n\pi}{b}\right)^2\right]$$

then k_g is imaginary. As the z-dependence of the field quantities is assumed to be of the form $e^{jk_g z}$, the fields will be exponentially damped in z, instead of propagating. Thus, for the waves to propagate along the guide the frequency must be greater than some critical value given by

$$\omega_c = c\sqrt{\left(\frac{m\pi}{a}\right)^2 + \left(\frac{n\pi}{b}\right)^2}. \tag{16.7-22}$$

This is called *cut-off frequency.* All frequency components greater than ω_c can propagate through the guide and below it propagation is not possible. The *cut-off wavelength* is given by

$$\lambda_c = \frac{c}{\omega_c/2\pi} = \frac{2}{\sqrt{\left(\frac{m}{a}\right)^2 + \left(\frac{n}{b}\right)^2}}. \tag{16.7-23}$$

Propagation is possible for wavelength $\lambda < \lambda_c$.

Now that the values of ω_c or λ_c depend on the transverse dimensions of the waveguide and also on the mode (m, n) of propagation. TE mode for arbitrary values of m and n is denoted by $(TE)_{mn}$. A mode for which λ_c is highest is called a *dominant mode.* Note that $(TE)_{00}$ mode is not possible because for $m = n = 0$, $H_{oz} = H_0$ and all other field components vanish; further, the condition $\vec{\nabla} \cdot \vec{H} = 0$ requires that $ik_g H_z = 0$ or $H_z = 0$.

From (16.7-21) and (16.7-22) we can write

$$k_g^2 = \frac{\omega^2}{c^2} - \frac{\omega_c^2}{c^2}. \tag{16.7-24}$$

The quantity $\frac{2\pi}{k_g} = \lambda_g$ is called *guide wavelength.* $\lambda_0 = \frac{c}{\omega/2\pi}$ is the free space wavelength. So we can write from (16.7-24)

$$\frac{1}{\lambda_g^2} = \frac{1}{\lambda_c^2} - \frac{1}{\lambda_c^2} \quad \text{or} \quad \lambda_g = \frac{\lambda_0}{\sqrt{1 - \left(\frac{\lambda_0}{\lambda_c}\right)^2}}. \tag{16.7-25}$$

Thus, guide wavelength is greater than corresponding free space wavelength. The phase velocity of the wave along z-direction is

$$v_p = \frac{\omega}{k_g} = \frac{\omega}{\sqrt{\frac{\omega^2}{c^2} - \frac{\omega_c^2}{c^2}}} = \frac{c}{\sqrt{1 - \left(\frac{\lambda_0}{\lambda_c}\right)^2}}. \tag{16.7-26}$$

Obviously phase velocity of the wave is greater than the velocity of light in free space. This does not violate the principle of relativity because the group velocity with which energy propagates along the guide is always smaller than c. Using (16.7-24) we can show that the group velocity

$$v_g = \frac{d\omega}{dk_g} = c\sqrt{1 - (\lambda_0/\lambda_c)^2}. \tag{16.7-27}$$

It is interesting to note that

$$v_p v_g = c^2. \tag{16.7-28}$$

Power flow associated with TE waves

The time average power flowing through the wavelength along z-axis is given by

$$
\begin{aligned}
P &= \int_0^a \int_0^b \frac{1}{2} Re\left(\vec{E} \times \vec{H}^*\right) \cdot \hat{k} dx dy \\
&= \frac{1}{2} Re \int_0^a \int_0^b \left(E_x H_y^* - E_y H_x^*\right) dx dy \\
&= \frac{1}{2} Re \int_0^a \int_0^b \left(E_{ox} H_{oy}^* - E_{oy} H_{ox}^*\right) dx dy \\
&= \frac{1}{2} \frac{\omega \mu_0 k_g}{h^4} H_0^2 \int_0^a \int_0^b \left[\left(\frac{n\pi}{b}\right)^2 \cdot \cos^2 \frac{m\pi x}{a} \sin^2 \frac{n\pi y}{b}\right. \\
&\qquad \left. + \left(\frac{m\pi}{a}\right)^2 \sin^2 \frac{m\pi x}{a} \cos^2 \frac{n\pi y}{b}\right] dx dy \\
&= \frac{H_0^2 \omega \mu_0 k_g}{2h^4} \cdot \frac{a}{2} \cdot \frac{b}{2} \cdot \left[\left(\frac{m\pi}{a}\right)^2 + \left(\frac{n\pi}{b}\right)^2\right] \\
&= \frac{ab}{8} \cdot \frac{\omega \mu_0 k_g}{h^2} \cdot H_0^2 \qquad \left[\because\ h^2 = \left(\frac{m\pi}{a}\right)^2 + \left(\frac{n\pi}{b}\right)^2\right]
\end{aligned}
\tag{16.7-29}
$$

For either $m = 0$ or $n = 0$, we have

$$
P = \frac{ab}{4} \cdot \frac{\omega \mu_0 k_g}{h^2} \cdot H_0^2.
$$

Thus, for a particular waveguide power transmission will be greater for larger H_0. A waveguide, however, cannot transmit any arbitrarily large power. The maximum value is limited by the puncture of the dielectric filling the waveguide.

TM modes

In this case, the waveguide contains no z-component of magnetic field, i.e., $H_{oz} = 0$ but $E_{oz} \neq 0$. Here we start from the equation satisfied by E_z and proceeding as before we can write for the most general solution of E_{oz} as

$$
E_{oz}(x, y) = (A_1 \cos k_x x + A_2 \sin k_x x)(A_3 \cos k_y y + A_4 \sin k_y y).
$$

To find the constants A_i we apply the boundary conditions $E_{oz} = 0$ at $x = 0$, $x = a$, $y = 0$, $y = b$.

$E_{oz} = 0$ at $x = 0$ requires that $A_1 = 0$.

$E_{oz} = 0$ at $y = 0$ requires that $A_3 = 0$.

Thus,

$$
E_{oz} = E_0 \sin k_x x \sin k_y y,
$$

where $E_0 = A_2 A_4$ is determined by introduced field.

Now the condition $E_{oz} = 0$ at $x = a$ requires that

$$\sin k_x a = 0 \quad \text{or} \quad k_x = \frac{m\pi}{a}; \quad m = 0, 1, 2, \ldots$$

Similarly, the condition $E_{oz} = 0$ at $y = b$ implies that

$$\sin k_y b = 0 \quad \text{or} \quad k_y = \frac{n\pi}{b}; \quad n = 0, 1, 2, \ldots$$

Thus, finally we get

$$E_{oz}(x, y) = E_0 \sin \frac{m\pi x}{a} \sin \frac{n\pi y}{b}. \tag{16.7-30a}$$

Now as E_{oz} is known other field components can be easily calculated in terms of it.

$$E_{ox} = \frac{jk_g}{h^2} \frac{\partial E_{oz}}{\partial x} = \frac{jk_g}{h^2} \cdot E_0 \left(\frac{m\pi}{a}\right) \cos \frac{m\pi x}{a} \sin \frac{n\pi y}{b}$$

$$E_{oy} = \frac{jk_g}{h^2} \frac{\partial E_{oz}}{\partial y} = \frac{jk_g}{h^2} E_0 \left(\frac{m\pi}{b}\right) \sin \frac{m\pi x}{a} \cos \frac{n\pi y}{b}$$

$$H_{ox} = -\frac{j\omega\varepsilon_0}{h^2} \frac{\partial E_{oz}}{\partial y} = -\frac{j\omega\varepsilon_0}{h^2} E_0 \left(\frac{n\pi}{b}\right) \sin \frac{m\pi x}{a} \cos \frac{n\pi y}{b}$$

$$H_{oy} = \frac{j\omega\varepsilon_0}{h^2} \cdot \frac{\partial E_{oz}}{\partial x} = \frac{j\omega\varepsilon_0}{h^2} E_0 \left(\frac{m\pi}{a}\right) \cos \frac{m\pi x}{a} \sin \frac{n\pi y}{b} \tag{16.7-30b}$$

As before we can calculate the cut off wavelength, which is the same as given by (16.7-23). In TM mode all the field components vanish for both m and n zero or any one of them zero. Thus, $(TM)_{00}$, $(TM)_{10}$ or $(TM)_{01}$ modes are not possible. For TM mode the longest wavelength available for transmission though the guide is

$$(\lambda_c)_{11} = \frac{2}{\sqrt{\left(\frac{1}{a}\right)^2 + \left(\frac{1}{b}\right)^2}},$$

which is definitely smaller than that for TE mode, where the longest wavelength is

$$(\lambda_c)_{10} = \frac{2}{\sqrt{\left(\frac{1}{a}\right)^2}} \quad (\text{assuming}\, a > b).$$

This result shows that the frequency band available for transmission through the waveguide is smaller in TM mode than that in TE mode. In this respect, TE mode is superior to TM mode of propagation.

Power flow associated with TM modes

Time average power transmitted through the waveguide in TM modes is given by

$$P = \int_0^a \int_0^b \tfrac{1}{2} Re\left(\vec{E} \times \vec{H}^*\right) \cdot \hat{k} dx\, dy.$$

As before,

$$P = \tfrac{1}{2} Re \int_0^a \int_0^b \left(E_{ox} H_{oy}^* - E_{oy} H_{ox}^*\right) dx\, dy.$$

Using the field components given by (16.7-30) we can get

$$P = \frac{ab}{8} \cdot \frac{\omega \varepsilon_0 k_g}{h^2} \cdot E_0^2.$$

Impossibility of TEM mode in a hollow waveguide

Suppose a TEM mode, for which there is no axial component of either $\vec{E}$ or $\vec{H}$, is assumed to exist in a hollow waveguide. In presence of nonmagnetic material $\vec{\nabla} \cdot \vec{H} = 0$. So if a TEM mode exists lines of $\vec{H}$ will be closed loops lying entirely in transverse plane. Now, from Maxwell's equation,

$$\vec{\nabla} \times \vec{H} = \vec{J} + \frac{\partial \vec{D}}{\partial t},$$

we can write

$$\oint_C \vec{H} \cdot d\vec{l} = \int_S \left(\vec{J} + \frac{\partial \vec{D}}{\partial t}\right) \cdot d\vec{S},$$

i.e., the magnetomotive force around these closed loops must be equal to axial current through the loops. For a hollow waveguide having no inner conductor conduction current is zero. Further, any axial displacement current requires an axial component of $\vec{E}$, something not present in TEM mode. Thus, we conclude that TEM mode cannot exist in a hollow waveguide.

But such a mode can exist in coaxial cable and in parallel line transmission system.

16.8 Cavity Resonator

A cavity resonator is an enclosure which completely surrounds electromagnetic fields by means of highly conducting walls. Such a cavity can sustain stationary waves of only certain discrete frequencies, called its *eigen frequencies*. It has properties similar to ordinary LC resonant circuits. In microwave region, where ordinary LC circuits cannot work cavity resonators are invariably used as resonant circuits.

The simplest cavity resonator is a rectangular cavity with perfectly conducting walls. It may be regarded as a rectangular waveguide of dimensions a along x-axis,

b along y-axis and short circuited by means of perfectly conducting plates at $z = 0$ and $z = l$ along its axis. Thus, we have a cavity of size $a \times b \times l$. An electromagnetic wave propagating along z-axis is reflected by the planes at $z = 0$, l and superposing on the incident wave produces a stationary wave. Similar to the case of a waveguide we may classify these waves as TE and TM modes. In case of TE mode the initial field components will be similar to that of a waveguide. Let us represent the z-component of the magnetic field going to the positive z-direction as (see, Section 16.7)

$$H'_z = H'_0 \cos\frac{m\pi x}{a}\cos\frac{n\pi y}{b}\cdot e^{j(k_g z-\omega t)};\quad m, n = 0, 1, 2, 3, \ldots,$$

where m and n are not zero simultaneously.

This wave is reflected at $z = l$ and goes to the negative z-direction. It may be represented by

$$H''_z = H''_0 \cos\frac{m\pi x}{a}\cos\frac{n\pi y}{b}\cdot e^{-j(k_g z+\omega t)}.$$

Since the walls are assumed to be perfectly conducting we must have

$$H'_z + H''_z = 0 \quad \text{at } z = 0 \quad \text{and} \quad z = l.$$

The condition at $z = 0$ requires that $H''_0 = -H'_0$.
Now the condition at $z = l$ requires that

$$H''_0 \cos\frac{m\pi x}{a}\cos\frac{n\pi y}{b}e^{-j\omega t}\cdot\left[e^{jk_g\cdot l} - e^{-jk_g l}\right] = 0.$$

This is true for all x, y and t. Hence

$$e^{jk_g l} - e^{-jk_g l} = 0 \text{ or } \sin k_g l = 0$$

$$\text{or } k_g = \frac{p\pi}{l};\quad p = 0, 1, 2, 3, \ldots .$$

Thus, total z-component of $\vec{H}$-field is

$$H_z = H'_z + H''_z = 2jH'_0 \cos\frac{m\pi x}{a}\cos\frac{n\pi y}{b}\sin\frac{p\pi z}{l}e^{-j\omega t}. \tag{16.8-1}$$

From this solution it is clear that $p = 0$ is not allowed, since it makes H_z and other field components identically zero.

Similarly other field components can be calculated.

$$H'_x = \frac{jk_g}{h^2}\frac{\partial H'_z}{\partial x} = -\frac{jk_g}{h^2}H'_0\left(\frac{m\pi}{a}\right)\sin\frac{m\pi x}{a}\cdot\cos\frac{n\pi y}{b}\cdot e^{j(k_g z-\omega t)}$$

$$H''_x = -\frac{jk_g}{h^2}\frac{\partial H''_z}{\partial x} = +\frac{jk_g}{h^2}H''_0\left(\frac{m\pi}{a}\right)\sin\frac{m\pi x}{a}\cdot\cos\frac{n\pi y}{b}e^{-j(k_g z+\omega t)}$$

$$\therefore \text{ Total } H_x = H'_x + H''_x = -\frac{2jk_g}{h^2}H'_0\left(\frac{m\pi}{a}\right)\sin\frac{m\pi x}{a}\cos\frac{n\pi y}{b}\cos\frac{p\pi z}{l}e^{-j\omega t}.$$

Similarly,

$$H_y = H'_y + H''_y = -\frac{2jk_g}{h^2} H'_0 \left(\frac{n\pi}{b}\right) \cos \frac{m\pi x}{a} \sin \frac{n\pi y}{b} \cos \frac{p\pi z}{l} \cdot e^{-j\omega t}$$

$$E_x = E'_x + E''_x = \frac{2\mu_0\omega}{h^2} H'_0 \left(\frac{n\pi}{b}\right) \cos \frac{m\pi x}{a} \sin \frac{n\pi y}{b} \cdot \sin \frac{p\pi z}{l} e^{-j\omega t}$$

$$E_y = E'_y + E''_y = -\frac{2\mu_0\omega}{h^2} H'_0 \left(\frac{m\pi}{a}\right) \sin \frac{m\pi x}{a} \cos \frac{n\pi y}{b} \cdot \sin \frac{p\pi z}{l} e^{-j\omega t} \quad (16.8\text{-}2)$$

To find the resonant or eigen frequencies we note that H_z satisfies the wave equation

$$\left(\nabla^2 - \frac{1}{c^2}\frac{\partial^2}{\partial t^2}\right) H_z = 0 \quad \text{or} \quad \left(\frac{\partial^2}{\partial x^2} + \frac{\partial^2}{\partial y^2} + \frac{\partial^2}{\partial z^2} - \frac{1}{c^2}\frac{\partial^2}{\partial t^2}\right) H_z = 0.$$

Putting the expression for H_z in this equation we get

$$\left[-\left\{\left(\frac{m\pi}{a}\right)^2 + \left(\frac{n\pi}{b}\right)^2 + \left(\frac{p\pi}{l}\right)^2\right\} + \frac{\omega^2}{c^2}\right] H_z = 0.$$

As this is valid for all x, y, z and t, we must have

$$\frac{\omega^2}{c^2} = \left(\frac{m\pi}{a}\right)^2 + \left(\frac{n\pi}{b}\right)^2 + \left(\frac{p\pi}{l}\right)^2$$

$$\text{or} \quad \omega = \omega_{mnp} = c\sqrt{\left(\frac{m\pi}{a}\right)^2 + \left(\frac{n\pi}{b}\right)^2 + \left(\frac{p\pi}{l}\right)^2}. \quad (16.8\text{-}3)$$

This relation gives the frequencies that can be sustained in the cavity resonator. If $p = 0$ or $m = n = 0$ or any two of m, n, l are zero then all the components of $\vec{E}$ and $\vec{H}$ vanish or they do not satisfy Maxwell's equation. This means that such modes are non-existent. Due to finite conductivity of the walls a practical cavity resonator dissipates a fraction of the stored-energy in each cycle. So the concept of Q-factor as used to described the quality of resonance in conventional LC circuit can also be applied to cavity resonators. Cavity resonators are characterised by higher Q-values. Cavities of practical size have resonant frequencies in the range where it is almost impossible to construct ordinary LC resonant circuits.

SOLVED PROBLEMS

▶ **Problem 1.** *The regions of space $z < 0$ and $z > 0$ are filled with materials having permeabilities $2\mu_0$ and $5\mu_0$, respectively. The magnetic field in the region $z > 0$ is $\vec{B}_2 = \mu_0 (75\hat{x} + 40\hat{z})$ T and there is a surface current distribution $\vec{K} = -10\hat{y}$ A/m at $z = 0$. Find the possible magnetic field in the region $z < 0$.* **[cf. C.U. 2015]**

Solution. From the boundary condition we have

$$\vec{B}_2 \cdot \hat{n} = \vec{B}_1 \cdot \hat{n} \quad \text{or} \quad B_{1z} = B_{2z} = 40\mu_0.$$

$$\therefore \; \left(\vec{B}_1\right)_{\text{normal}} = 40\mu_0 \hat{z}.$$

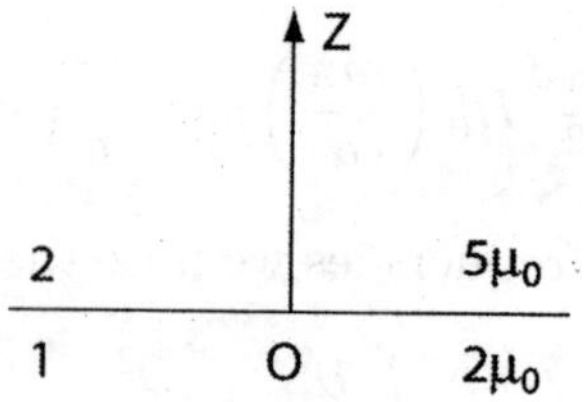

Fig 16.P-1

Also,

$$\hat{n} \times \vec{H}_2 - \hat{n} \times \vec{H}_1 = \vec{K}$$

$$\text{or} \quad \hat{z} \times \frac{\vec{B}_2}{5\mu_0} - \hat{z} \times \frac{\vec{B}_1}{2\mu_0} = -10\hat{y}$$

$$\text{or} \quad \hat{z} \times (15\hat{x} + 8\hat{z}) - \hat{z} \times \frac{\vec{B}_1}{2\mu_0} = -10\hat{y}$$

$$\text{or} \quad \hat{z} \times \frac{\vec{B}_1}{2\mu_0} = 25\hat{y} = 25\hat{z} \times \hat{x}$$

$$\text{or} \quad \left(\vec{B}_1\right)_{\text{tangential}} = 50\mu_0 \hat{x}$$

$$\therefore \; \vec{B}_1 = \left(\vec{B}_1\right)_{\text{normal}} + \left(\vec{B}_1\right)_{\text{tangential}} = \mu_0 \left(50\hat{x} + 40\hat{z}\right).$$

► **Problem 2.** *Find the angular altitude (angle above the horizontal) of the sun for which sunlight reflected from a still take becomes plane polarized. Given refractive index of water* = 4/3. *What is the direction of* $\vec{E}$*-field in the reflected plane polarized light? How can you cut off the reflected light by using a polaroid?* **[cf. C.U. 2007]**

Solution. It happens so at Brewster's angle θ_p, which is given by

$$\theta_p = \tan^{-1}(4/3) = 53.13^\circ.$$

θ_p is the angle of incidence, i.e., the angle with the normal. So the angular altitude above the horizontal $= 90^\circ - 53.13^\circ = 36.87^\circ$. The $\vec{E}$-field is perpendicular to the vertical plane of incidence. So to cut off the glare of this light by a polaroid, its pass-axis may be kept vertical.

▶ **Problem 3.** *An EM wave is incident normally from air on air-glass interface. Taking refractive index of the glass as* 1.5, *find the amplitude reflection coefficient and the percentage of total incident energy that is transmitted into glass.*

Solution. The amplitude reflection coefficient, i.e., the ratio of the amplitude of the reflected wave to that of the incident wave has the magnitude

$$|r| = \frac{n_2 - n_1}{n_2 + n_1} = \frac{1.5 - 1}{1.5 + 1} = \frac{1}{5}.$$

Now reflectance $R = |r|^2 = 1/25$ and transmittance $T = 1 - R = 24/25$. So the percentage of incident energy transmitted into glass is $(24/25) \times 100$ or 96%.

▶ **Problem 4.** *Suppose a plane EM-wave with electric field*

$$\vec{E} = \hat{x}10\cos(kz - \omega t)\,\text{V.m}^{-1}$$

is incident from air on a dielectric occupying the region $z \geq 0$. Assuming that the permittivity of the medium is $4\varepsilon_0$ and permeability μ_0 determine the electric field of the reflected and transmitted waves.

Solution. Retractive index of the medium is

$$n_2 = \sqrt{\frac{\mu}{\mu_0} \cdot \frac{\varepsilon}{\varepsilon_n}} = \sqrt{4} = 2.$$

Therefore, amplitude reflection coefficient will have the magnitude

$$|r| = \frac{n_2 - n_1}{n_2 + n_1} = \frac{2 - 1}{2 + 1} = \frac{1}{3}.$$

Now considering the phase reversal of the $\vec{E}$-field in the reflected wave we can write for the reflected wave (propagating in the negative z-direction),

$$\vec{E_1} = -\hat{x}\frac{10}{3}\cos(kz + \omega t)\,\text{V.m}^{-1}.$$

The amplitude transmission coefficient

$$t_{12} = \frac{2n_1}{n_2 + n_1} = \frac{2}{3}.$$

Therefore, the transmitted wave may be represented by

$$\vec{E_2} = \hat{x}\frac{20}{3}\cos(kz - \omega t)\,\text{V.m}^{-1}.$$

▶ **Problem 5.** *An EM wave polarized perpendicular to the plane of incidence impinges at* 30° *on a glass slab having refractive index* 1.5. *Find the amplitude reflection and transmission coefficients.*

Solution. From Snell's law

$$n_1 \sin\theta = n_2 \sin\theta_2 \quad \text{or} \quad \sin 30° = 1.5 \times \sin\theta_2 \quad \text{or} \quad \sin\theta_2 = \tfrac{1}{3}$$

$$\text{and} \quad \cos\theta_2 = \sqrt{1-\sin^2\theta_2} = \sqrt{\tfrac{8}{9}} = 0.943.$$

Now

$$r_\perp = \frac{n_1\cos\theta - n_2\cos\theta_2}{n_1\cos\theta + n_2\cos\theta_2} = \frac{1\times\frac{\sqrt{3}}{2} - 1.5\times 0.943}{1\times\frac{\sqrt{3}}{2} + 1.5\times 0.943} = -0.2403$$

$$t_\perp = \frac{2n_1\cos\theta}{n_1\cos\theta + n_2\cos\theta_2} = \frac{2\times\frac{\sqrt{3}}{2}}{1\times\frac{\sqrt{3}}{2} + 1.5\times 0.943} = 0.7596.$$

▶ **Problem 6.** *A plane EM wave falls obliquely on air* ($n_1 = 1.0$)*-glass* ($n_2 = 1.5$) *interface. Find the angle of incidence for which the reflection and transmission coefficients are each equal to* 0.5, *i.e.,* $R = T = 0.5$.

Solution. For waves polarized parallel to the plane of incidence we can write from Eq. (16.4-16)

$$R_\parallel = \left(\frac{n_2\cos\theta - n_1\cos\theta_2}{n_2\cos\theta + n_1\cos\theta_2}\right)^2 = 0.5 = \frac{1}{2}$$

$$\text{or} \quad \frac{1.5\cos\theta - \cos\theta_2}{1.5\cos\theta + \cos\theta_2} = \pm\frac{1}{\sqrt{2}}$$

$$\text{or} \quad \frac{\cos\theta_2}{1.5\cos\theta} = \frac{\sqrt{2}\mp 1}{\sqrt{2}\pm 1}$$

$$\text{or} \quad \frac{\cos\theta_2}{\cos\theta} = 1.5\times\frac{\sqrt{2}\mp 1}{\sqrt{2}\pm 1} = 8.7428 \quad \text{or} \quad 0.2573.$$

As $n_2 > n_1$, $\theta_2 < \theta$ and hence $\cos\theta_2 > \cos\theta$. So we take

$$\frac{\cos\theta_2}{\cos\theta} = 8.7428$$

$$\text{or} \quad (8.7428)^2\cos^2\theta = \cos^2\theta_2 = 1 - \sin^2\theta_2 = 1 - \frac{\sin^2\theta}{1.5^2} \quad \left[\because \frac{\sin\theta}{\sin\theta_2} = 1.5\right]$$

$$\text{or} \quad (8.7428)^2(1-\sin^2\theta) = 1 - 0.4444\sin^2\theta$$

$$\text{or} \quad 8.7428^2 - 1 = (8.7428^2 - 0.4444)\sin^2\theta$$

$$\text{or} \quad 75.4365 = 75.9921\sin^2\theta$$

$$\text{or} \quad \sin\theta = 0.9926 \quad \text{or} \quad \theta = 83.06°.$$

Since $R_{\|} + T_{\|} = 1$, for $R_{\|} = 0.5$ we have also $T_{\|} = 1 - 0.5 = 0.5$. Thus, with this angle of incidence 83.06° we get $R_{\|} = T_{\|} = 0.5$. Similarly using Eqs. (16.4-18) and (16.4-19) we can show that for $R_{\perp} = T_{\perp} = 0.5$ the angle of incidence is $\theta = 78.77°$.

▶ **Problem 7.** *A plane polarized EM wave falls obliquely at the boundary of two dielectrics. The plane of polarization of the wave is perpendicular to the plane of incidence. Show that there is always a phase change of π in the reflected wave for reflection at the interface backed by denser medium, but there is no such phase change if the interface is backed by rarer medium. Also, show that there is no Brewster's angle here as long as the media have different refractive indices. Assume that the amplitude reflection coefficient is given by* (16.3-19) *or* (16.3-21).

Solution. If $n_2 > n_1$ then the angle of refraction $\theta_2 <$ the angle of incidence θ and hence in this case,

$$r_{\perp} = \frac{\sin(\theta - \theta_2)}{\sin(\theta + \theta_2)}$$

is a negative quantity, which indicates a phase change of π. But if $n_2 < n_1$ then $\theta_2 > \theta$ and $r_{\perp}$ becomes positive, which indicates no phase change. For any Brewster's angle to exist we must have

$$r_{\perp} = 0 \quad \text{or} \quad \sin(\theta - \theta_2) = 0 \quad \text{or} \quad \sin\theta\cos\theta_2 - \cos\theta\sin\theta_2 = 0.$$

Since

$$\frac{\sin\theta}{\sin\theta_2} = \frac{n_2}{n_1}$$

we can write

$$n_2 \cos\theta_2 = n_1 \cos\theta$$

$$\text{or} \quad \frac{n_2^2}{n_1^2}\left(1 - \sin^2\theta_2\right) = \cos^2\theta$$

$$\text{or} \quad \frac{n_2^2}{n_1^2}\left(1 - \frac{n_1^2}{n_2^2}\sin^2\theta\right) = \cos^2\theta$$

$$\text{or} \quad \sin^2\theta + \cos^2\theta = \frac{n_2^2}{n_1^2}.$$

So, if n_1 and n_2 are different this relation cannot be satisfied. This means that Brewster's angle cannot exist in this case.

▶ **Problem 8.** *An EM wave of wavelength* 653 nm *in air is incident normally on a metal surface having complex reflective index* $n = 0.17 + j3.14$. *Find the reflectivity of the surface.*

Solution. From Eq. (16.5-13) the reflectivity

$$R = \frac{(n_r - n_1)^2 + n_i^2}{(n_r + n_1)^2 + n_i^2}.$$

Here $n_1 = 1$, $n_r = 0.17$ and $n_i = 3.14$

Therefore,

$$R = \frac{(0.17 - 1)^2 + (3.14)^2}{(0.17 + 1)^2 + (3.14)^2} \approx 0.94.$$

So about 94% of the incident energy will be reflected back.

► **Problem 9.** *A plane polarized EM wave with its electric vector making an angle of 45° with the plane of incidence, is incident at angle θ on an interface of two dielectrics of refractive indices n_1 and n_2. Show that the electric vector of the wave transmitted into medium 2 makes an angle ϕ with the plane of incidence, which is given by*

$$\tan\phi = \cos\left[\theta - \sin^{-1}\left(\frac{n_1}{n_2}\sin\theta\right)\right].$$

Solution. $\vec{E}$-vector of the incident wave can be resolved into two components having amplitudes $(E_0)_\parallel = E_0 \cos 45°$ parallel to the plane of incidence and $(E_0)_\perp = E_0 \sin 45°$ perpendicular to the plane of incidence. Now the amplitude transmission coefficients are given by the Eqs. (16.3-22) and (16.3-31) as

$$t_\parallel = \left(\frac{E_{20}}{E_0}\right)_\parallel = \frac{2\sin\theta_2\cos\theta}{\sin(\theta+\theta_2)\cos(\theta-\theta_2)} \tag{1}$$

$$t_\perp = \left(\frac{E_{20}}{E_0}\right)_\perp = \frac{2\cos\theta\sin\theta_2}{\sin(\theta+\theta_2)}, \tag{2}$$

where θ_2 is the angle of refraction, given by Snell's law as

$$n_1 \sin\theta = n_2 \sin\theta_2 \quad \text{or} \quad \theta_2 = \sin^{-1}\left(\frac{n_1}{n_2}\sin\theta\right)$$

From (1) and (2) we can get for the amplitudes of the transmitted waves as

$$(E_{20})_\parallel = \frac{2\sin\theta_2\cos\theta}{\sin(\theta+\theta_2)\cos(\theta-\theta_2)} \cdot \frac{E_0}{\sqrt{2}} \quad \text{and} \quad (E_{20})_\perp = \frac{2\cos\theta\sin\theta_2}{\sin(\theta+\theta_2)} \cdot \frac{E_0}{\sqrt{2}}.$$

Therefore, the angle ϕ that the $\vec{E}$-vector of the transmitted wave makes with the plane of incidence is given by

$$\tan\phi = \frac{(E_{20})_\perp}{(E_{20})_\parallel} = \cos(\theta - \theta_2) = \cos\left[\theta - \sin^{-1}\left(\frac{n_1}{n_2}\sin\theta\right)\right]$$

▶ **Problem 10.** *An EM wave is incident normally at the boundary of two dielectric media. Show that the electric field of the reflected and transmitted waves must have the same polarization as the incident wave.*

Solution. Suppose xy-plane forms the interface ($z = 0$) of the two media and an EM wave travelling along z-axis is incident on the interface (Fig. 16.P-10). Let the incident wave the represented by

$$\vec{E} = \hat{\imath} E_0 e^{j(kz-\omega t)},$$

which is assumed to be plane polarized along x-axis. If possible, let the reflected and transmitted waves have arbitrary polarization in xy-plane and be represented by

$$\vec{E}_1 = \left(\hat{\imath} E_{1x} + \hat{\jmath} E_{1y}\right) e^{j(-k_1 z-\omega t)} \quad \text{(Reflected wave)}$$

$$\vec{E}_2 = \left(\hat{\imath} E_{2x} + \hat{\jmath} E_{2y}\right) e^{j(k_2 z-\omega t)} \quad \text{(Transmitted wave)}$$

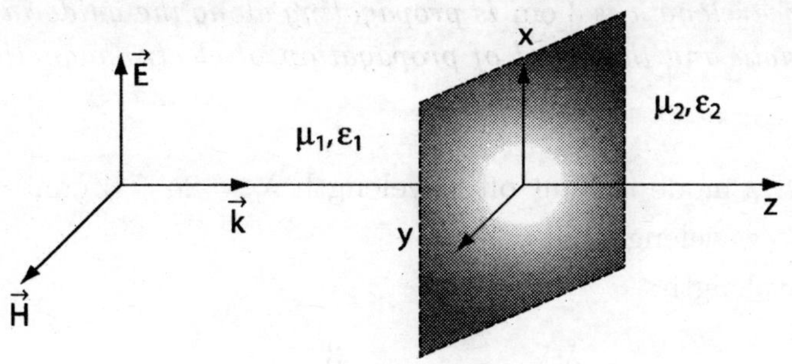

Fig 16.P-10

Corresponding magnetic fields are

$$\vec{H} = \frac{\vec{k} \times \vec{E}}{\mu_1 \omega} = \frac{\hat{k} \times \vec{E}}{\mu_1 \left(\frac{\omega}{k}\right)} = \frac{1}{\mu_1 v_1} \hat{\jmath} E_0 e^{j(kz-\omega t)} \quad [v_1 = \text{phase velocity}]$$

$$\vec{H}_1 = \frac{\vec{k}_1 \times \vec{E}_1}{\mu_1 \omega} = \frac{-\hat{k} \times \vec{E}_1}{\mu_1 \left(\frac{\omega}{k_1}\right)} = \frac{1}{\mu_1 v_1} \cdot \left(-\hat{\jmath} E_{1x} + \hat{\imath} E_{1y}\right) e^{j(-k_1 z-\omega t)}$$

$$\vec{H}_2 = \frac{\vec{k}_2 \times \vec{E}_2}{\mu_2 \omega} = \frac{\hat{k} \times \vec{E}_2}{\mu_2 \left(\frac{\omega}{k_2}\right)} = \frac{1}{\mu_2 v_2} \left(\hat{\jmath} E_{2x} - \hat{\imath} E_{2y}\right) e^{j(k_2 z-\omega t)}$$

Now the continuity of the tangential components of $\vec{E}$ and $\vec{H}$-fields across the interface gives

$$\hat{\imath} E_0 + \hat{\imath} E_{1x} + \hat{\jmath} E_{1y} = \hat{\imath} E_{2x} + \hat{\jmath} E_{2y} \tag{1}$$

$$\text{and} \quad \frac{1}{\mu_1 v_1} \cdot \left[\hat{\jmath} E_0 - \hat{\jmath} E_{1x} + \hat{\imath} E_{1y}\right] = \frac{1}{\mu_2 v_2} \left(\hat{\jmath} E_{2x} - \hat{\imath} E_{2y}\right). \tag{2}$$

Equating y-components of Eq. (1) we get

$$E_{1y} = E_{2y} \tag{3}$$

and equating x-components of Eq. (2) we get

$$\frac{1}{\mu_1 v_1} E_{1y} = -\frac{1}{\mu_2 v_2} E_{2y} \tag{4}$$

Solving (3) and (4) we get $E_{1y} = E_{2y} = 0$. This means that both the reflected and transmitted waves will have the same polarization of the incident wave, i.e., along the x-direction.

▶ **Problem 11.** *Consider a perfectly conducting rectangular waveguide infinitely long along z-axis with a width* 2 cm *along x-axis and* 1 cm *along y-axis. An EM wave whose free space wavelength is* 3 cm *is propagating along the guide in TE_{10} mode. Find the guide wavelength and the speed of propagation of electromagnetic energy along the guide.*

Solution. For TE_{10} mode the cut off wavelength $\lambda_c = 2a = 4$ cm.

The free space wavelength $\lambda_0 = 3$ cm

The guide wavelength

$$\lambda_g = \frac{\lambda_0}{\sqrt{1-\left(\frac{\lambda_0}{\lambda_c}\right)^2}} = \frac{3}{\sqrt{1-\left(\frac{3}{4}\right)^2}} \text{ cm.} = 4.5 \text{ cm}$$

The speed of propagation of electromagnetic energy

$$= \text{group velocity of the wave}$$

$$= c\sqrt{1-\left(\frac{\lambda_0}{\lambda_c}\right)^2} = c\sqrt{1-\left(\frac{3}{4}\right)^2} = 0.66\,c,$$

where c is the speed of light in free space.

▶ **Problem 12.** *Consider three dielectric media A, B and C forming two interfaces A-B and B-C. The refractive indices of the media are n_1, n_2 and n_3, respectively. An EM wave from medium A is incident on the medium B normally. Show that for the amplitude reflection coefficients at A-B and B-C interfaces will be equal if $n_2 = \sqrt{n_1 n_3}$.*

Solution. For equality of amplitude reflection coefficients we must have

$$\frac{n_2 - n_1}{n_2 + n_1} = \frac{n_3 - n_2}{n_3 + n_2} \quad \text{or} \quad n_2 = \sqrt{n_1 n_3}.$$

▶ **Problem 13.** (i) *Consider a waveguide of square cross-section having side a. It is to transmit an EM wave of wavelength* λ *in* TE_{10} *mode but not in the* TE_{11} *or* TM_{11} *modes. Find the limiting range of values of a.*

(ii) *Consider a waveguide of rectangular cross-section having dimensions* 2.3 cm × 1.0 cm. *Find the TE modes that can propagate in this waveguide, if the driving frequency is* 1.7×10^{10} Hz. **[C.U. 2014, 2018]**

If you are to excite only the TE_{10} *mode what frequency range you can use? What is the corresponding free space wavelength range?*

Solution. (i) The cut off wavelength

$$\lambda_{c,mn} = \frac{2}{\sqrt{\left(\frac{m}{a}\right)^2 + \left(\frac{n}{a}\right)^2}}.$$

$$\therefore\ \lambda_{c,10} = 2a \quad \text{and} \quad \lambda_{c,11} = \sqrt{2}a.$$

So for the propagation of EM wave in TE_{10} mode its wavelength

$$\lambda < \lambda_{c,10} \quad \text{or} \quad \lambda < 2a \quad \text{or} \quad \lambda/2 < a$$

and for the non-propagation of TE_{11} or TM_{11} modes

$$\lambda > \lambda_{c,11} \quad \text{or} \quad \lambda > \sqrt{2}a \quad \text{or} \quad a < \lambda/\sqrt{2}.$$

Therefore, the required range of values of a is

$$\frac{\lambda}{2} < a < \frac{\lambda}{\sqrt{2}}$$

$$\text{(ii)}\ \nu_{c,mn} = \frac{\omega_{c,mn}}{2\pi} = \frac{c}{2\pi}\sqrt{\left(\frac{m\pi}{a}\right)^2 + \left(\frac{n\pi}{b}\right)^2}.$$

$$\therefore\ \nu_{c,10} = \frac{c}{2a} = \frac{3 \times 10^8}{2 \times 2.3 \times 10^{-2}}\ \text{Hz} = 0.65 \times 10^{10}\ \text{Hz}.$$

$$\nu_{c,20} = 2 \times \frac{c}{2a} = 1.30 \times 10^{10}\ \text{Hz},\ \nu_{c,30} = 3 \times \frac{c}{2a} = 1.95 \times 10^{10}\ \text{Hz}.$$

$$\nu_{c,01} = \frac{c}{2b} = 1.5 \times 10^{10}\ \text{Hz},\ \nu_{c,02} = 2 \times \frac{c}{2b} = 3.0 \times 10^{10}\ \text{Hz}.$$

$$\nu_{c,11} = \frac{c}{2}\sqrt{\frac{1}{a^2} + \frac{1}{b^2}} = 1.63 \times 10^{10}\ \text{Hz}.$$

As the driving frequency is greater than $\nu_{c,10}$, $\nu_{c,20}$, $\nu_{c,01}$ and $\nu_{c,11}$ these modes can propagate in the waveguide.

In order to excite only the TE_{10} mode exciting frequency ν must be greater than $\nu_{c,10}$ but smaller than the critical frequency of the adjacent mode. So we must have $\nu_{c,10} < \nu < \nu_{c,20}$. As $\lambda = c/\nu$, so

$$\nu_{c,10} = \frac{c}{\nu_{c,10}} = 2a \quad \text{and} \quad \lambda_{c,20} = \frac{c}{\nu_{c,20}} = a.$$

Therefore, we must have

$$\frac{c}{\lambda_{c,10}} < \frac{c}{\lambda} < \frac{c}{\lambda_{c,20}} \quad \text{or} \quad \lambda_{c,20} < \lambda < \lambda_{c,10}$$

$$\text{or} \quad a < \lambda < 2a$$

$$\text{or} \quad 2.3 \text{ cm} < \lambda < 4.6 \text{ cm}$$

► **Problem 14.** *Consider a plane polarized EM wave of amplitude E_0 in air is incident normally on a good conducting metal foil of thickness d. The complex refractive index of the metal is $n = n_0(1+j)$ and skin depth δ. Neglecting multiple reflections show that the amplitude of the transmitted wave is given by*

$$\frac{4e^{-d/\delta} \cdot E_0}{\left(1 + n_0\sqrt{2}e^{j\pi/4}\right)\left(1 + \frac{1}{n_0\sqrt{2}e^{j\pi/4}}\right)}.$$

Solution. Let the amplitude of the wave just after entering the foil be E_{20}, just before leaving the foil be E'_{20} and just after leaving the foil be E_{30}. Now from Eq. (16.5-6)

$$\frac{E_{20}}{E_0} = \frac{2n_1}{n_1 + n_2} = \frac{2}{1 + n_0(1+j)} = \frac{2}{n_0\sqrt{2}e^{j\pi/4}}.$$

As the wave travels through the foil its amplitude decays exponentially at a rate determined by skin depth δ,

$$E'_{20} = E_{20}e^{-d/\delta} \quad \text{or} \quad \frac{E'_{20}}{E_{20}} = e^{-d/\delta}.$$

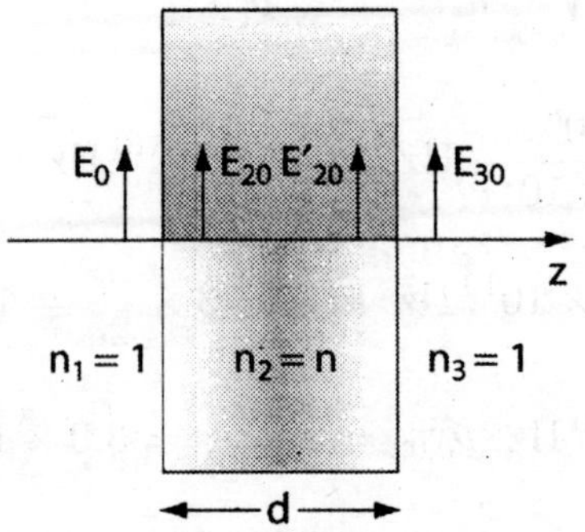

Fig 16.P-14

Again, using Eq. (16.5-6),

$$\frac{E_{30}}{E'_{20}} = \frac{2n_2}{n_2 + n_3} = \frac{2}{1 + \frac{1}{n_0(1+j)}} = \frac{2}{1 + \frac{1}{n_0\sqrt{2}e^{j\pi/4}}}$$

Now

$$\frac{E_{30}}{E_0} = \frac{E_{30}}{E'_{20}} \times \frac{E'_{20}}{E_{20}} \times \frac{E_{20}}{E_0} = \frac{4e^{-d/\delta}}{\left(1 + n_0\sqrt{2}e^{j\pi/4}\right)\left(1 + \frac{1}{n_0\sqrt{2}e^{j\pi/4}}\right)}.$$

▶ **Problem 15.** *In a rectangular waveguide having dimensions $a \times b$, show that the cut off wavelength for the lowest TE mode is higher than that of the lowest TM mode by a factor $\sqrt{1 + a^2/b^2}$.*

Solution. We know that

$$\lambda_{c,mn} = \frac{2}{\sqrt{\left(\frac{m}{a}\right)^2 + \left(\frac{n}{b}\right)^2}}.$$

For the lowest TE mode $m = 1$, $n = 0$. Therefore, $\lambda_{c,10}^{TE} = 2a$.

For the lowest TM mode $m = 1$, $n = 1$. Therefore,

$$\lambda_{c,11}^{TM} = \frac{2}{\sqrt{\frac{1}{a^2} + \frac{1}{b^2}}} = \frac{2a}{\sqrt{1 + a^2/b^2}} \quad \text{or} \quad \lambda_{c,10}^{TE} = \sqrt{1 + \frac{a^2}{b^2}}\lambda_{c,11}^{TM}.$$

QUESTIONS

1. (a) Establish the boundary conditions satisfied by the electromagnetic field vectors $\vec{E}$, $\vec{D}$, $\vec{B}$ and $\vec{H}$ at the plane interface between two different media. **[cf. C.U. 2014]**

   (b) An electromagnetic plane wave undergoes reflection and refraction at the interface of two isotropic dielectric media. Using Maxwell's equations and proper boundary conditions obtain (i) the Snell's law (ii) the laws of reflection. **[B.U. 2011]**

2. An electromagnetic wave is incident on the plane interface between two different media.

   (a) Show that the frequency of the wave remains unchanged upon reflection or refraction.

   (b) Find the relation between the angles of incidence, reflection and refraction.

   (c) Show that the wave vectors of the incident, reflected and refracted waves all lie on the same plane.

3. A plane electromagnetic wave falls obliquely on the interface between two simple dielectrics. Assuming the electric vector to be a parallel to the plane of incidence, (b) perpendicular to the plane incidence obtain expressions for the amplitude reflection and transmission coefficients.

   Assuming the incident medium less dense discuss how these coefficients depend on the angle of incidence. Also, show graphically the variation of phase change with the angle of incidence.

4. A plane polarized electromagnetic wave with its electric vector parallel to the plane of incidence is incident obliquely on the interface between two simple dielectrics. Obtain expressions for the amplitude reflection and transmission coefficients. Hence prove Brewster's law.

5. Fresnel's formula for amplitude reflection and transmission coefficients for a plane electromagnetic wave incident obliquely on the interface between two simple dielectrics having refractive indices n_1 and n_2, are given by

$$r_\perp = \frac{n_1 \cos\theta - n_2 \cos\theta_2}{n_1 \cos\theta + n_2 \cos\theta_2}, \quad t_\perp = \frac{2n_1 \cos\theta}{n_1 \cos\theta + n_2 \cos\theta_2}$$

$$r_\parallel = \frac{n_2 \cos\theta - n_1 \cos\theta_2}{n_2 \cos\theta + n_1 \cos\theta_2}, \quad t_\parallel = \frac{2n_1 \cos\theta}{n_2 \cos\theta + n_1 \cos\theta_2},$$

   where θ and θ_2 are respectively the angles of incidence and refraction; the subscripts $\perp$ and $\parallel$ refer to the plane of polarization of $\vec{E}$-field.

   (a) Assuming the incident medium (n_1) denser than the other (n_2), explain the phenomenon of total internal reflection.

   (b) Does any $\vec{E}$-field exist in the rarer medium during total internal reflection?

   (c) Show that on the average there is no flow of energy through the interface during total internal reflection.

   (d) What is the phase change of the reflected wave when there is no transmission?

6. Explain the terms reflectance (R) and transmittance (T) in connection with the reflection and refraction of an electromagnetic wave at the interface of two different media. Considering the case of normal incidence show that $R + T = 1$. What does it imply?

7. Suppose a plane electromagnetic wave is incident normally on an air-metal interface. Introducing the idea of complex refractive index obtain an approximate expression for reflecting power of the metal. Hence show that a perfect conductor is a perfect reflector of electromagnetic waves.

8. Consider the propagation of electromagnetic waves between two parallel perfectly conducting plates. Considering TE mode show that the waves can propagate only if the wavelength is smaller than a certain critical value λ_c for a particular mode. What is a dominant mode? If λ_g is the guide wavelength and λ_0, the free space wavelength then show that

$$\frac{1}{\lambda_0^2} = \frac{1}{\lambda_g^2} + \frac{1}{\lambda_c^2}.$$

Show that the phase velocity of the wave is greater than the speed of light in vacuum. Does it violate the principle of relativity?

9. (a) What is a waveguide? Considering TE or TM waves propagating along a rectangular waveguide with perfectly conducting walls, find the cut off wavelength (λ_c) and guide wavelength (λ_g).

(b) If λ_0 be the corresponding free space wavelength, show that

$$\frac{1}{\lambda_0^2} = \frac{1}{\lambda_g^2} + \frac{1}{\lambda_c^2}.$$

(c) What is the velocity of energy propagation through the waveguide?

(d) Show that TE_{00} and TEM modes are not possible in a rectangular waveguide.

10. (a) Starting from Maxwell's equations determine the field components associated with (a) TE_{10} mode (b) TM_{11} mode in a rectangular waveguide with perfectly conducting walls.

(b) Show that the frequency band available for transmission through the waveguide is greater in TE mode than TM mode.

11. (a) What do you mean by TE, TM nd TEM waves?

(b) Show that TEM waves cannot occur in the hollow waveguide. **[C.U. 2010]**

12. (a) What are cavity resonators? How do they resemble LC resonant circuits? Discuss its superiority over LC resonant circuits.

(b) Consider a cavity resonator as a rectangular waveguide of size a along x-axis, b along y-axis and short circuited by conducting plates at $z = 0$ and $z = l$. Considering the superposition of incident and reflected waves along z-axis find the resultant z-component of magnetic field associated with TE mode. Hence find the eigen frequencies that the resonator can sustain. Assume that the z-component of magnetic field going to the $+z$-direction in a rectangular waveguide is given by

$$H_z = H_0 \cos\frac{m\pi x}{a} \cos\frac{n\pi y}{b} \cdot e^{j(k_g z - \omega t)},$$

where the symbols have their usual meaning.

EXERCISES

1. The refractive index of water is 4/3 for visible light and 9 for radiowaves. Compare the reflectance of air-water interface in the two cases. [*Ans.* 1 : 3.14]

2. Refractive index of glass is 1.5. Show that about 4% of the light incident normally on glass from air is reflected back. What would be the result if the light comes from glass and passes into air? [*Ans.* The same % of light will be reflected back]

3. A light wave that is linearly polarized in the plane of incidence impinges at 30° on a glass ($n = 1.52$) plate placed in air. Calculate the amplitude reflection and transmission coefficients. [*Ans.* $r_{\parallel} = 0.165$, $t_{\parallel} = 0.766$]

4. An EM wave is incident normally at the boundary of two dielectrics having refractive indices n_1 and $n_2 (n_1 < n_2)$. If the transmission coefficient is required to have a value 0.80, what should be the value of n_2/n_1?

   [C.U. 2011] [*Ans.* 2.618]

5. (a) Show that the polarizing angles for internal and external reflections at a given interface are complementary to each other.

   (b) Show that the angle between reflected and transmitted beams is 90° when beam is incident at Brewster's angle. **[C.U. 2010]**

6. Using Hagen-Rubens formula for reflection of EM waves from metallic surface show that radar waves of frequency 1 MHz may get almost totally reflected from the surface of an aeroplane made of a metal for which conductivity σ is 5.8×10^7 S/m and permeability μ_0.

7. A plane EM wave with the $\vec{E}$-field given by $\vec{E} = 30\cos(kz - \omega t)\hat{x}$ is incident normally from air on a medium characterised by permittivity $4\epsilon_0$ and permeability μ_0. Find the electric field in the reflected wave. [*Ans.* $-10\cos(kz + \omega t)\hat{x}$]

8. Consider an EM wave incident normally on a dielectric-dielectric interface for which $\mu_1 = \mu_2 = \mu_0$ and refractive indices n_1 and n_2, respectively. Show that the reflectance R and transmittance T are given by

$$R = \left(\frac{n_2 - n_1}{n_2 + n_1}\right)^2 \quad \text{and} \quad T = \frac{4n_1 n_2}{(n_2 + n_1)^2}$$

9. Suppose Fresnel's formula for amplitude reflection and transmission coefficients for $\vec{E}$-field polarized parallel to the plane of incidence are given by

$$r_{\parallel} = \frac{n_2\cos\theta - n_1\cos\theta_2}{n_2\cos\theta + n_1\cos\theta_2} \quad \text{and} \quad t_{\parallel} = \frac{2n_1\cos\theta}{n_2\cos\theta + n_1\cos\theta_2},$$

   where θ and θ_2 are respectively the angles of incidence and refraction; n_1 and n_2 are respectively the refractive index of the incident (1) and refracted (2) media.

Show that

$$1 + r_{\parallel} r'_{\parallel} = t_{\parallel} t'_{\parallel},$$

where $r'_{\parallel}$ and $t'_{\parallel}$ are respectively the amplitude reflection and transmission coefficients when an EM wave from medium (2) is incident on medium (1) at angle θ_2.

[**Hints.** To find $r'_{\parallel}$ and $t'_{\parallel}$ interchange n_1 and n_2; θ and θ_2 in the given formula for $r_{\parallel}$ and $t_{\parallel}$.]

10. The wave impedance of any medium is defined by $Z = \sqrt{\mu/\epsilon}$. Suppose an EM wave is incident normally from air on a medium whose wave impedance is equal to that of air. Show that in this case, reflection coefficient is identically zero.

11. (a) What will be the reflection and transmission coefficients for normal incidence from vacuum to water which has a refractive index 1.33?

[C.U. 2006] [*Ans.* 0.1416, 0.8584]

(b) Light waves fall normally on water-glass interface. If the refractive indices of water and glass are respectively 1.33 and 1.52, find the percentage of incident light energy reflected from the interface. [*Ans.* 0.44%]

12. Find the Brewster's angle for a glass-oil boundary for external and internal reflections. Given refractive index for glass 1.52 and that of oil 1.81.[*Ans.* $50°, 40°$]

13. A unpolarized light is incident on an air-glass surface at an angle of incidence 45°. Compare the reflectivities of the surface for the components polarized parallel and perpendicular to the plane of incidence. Take the refractive index of the glass as 1.5 Comment on the result.

[*Ans.* $R_{\parallel}/R_{\perp} = 0.60$; Reflected wave becomes predominantly polarized perpendicular to the plane of incidence]

14. The $\vec{E}$-field associated with an EM wave in free space is described by

$$\vec{E} = 3\cos(kz - \omega t)\hat{x} + 4\sin(kz - \omega t)\hat{y}\,\text{V.m}^{-1}.$$

The wave is incident normally on the surface ($z = 0$) of a perfect conductor and gets reflected. Find the $\vec{E}$-field in the reflected wave for $z \leq 0$.

[*Ans.* $\vec{E}_1 = -3\cos(kz + \omega t)\hat{x} - 4\sin(kz + \omega t)\hat{y}\,\text{V.m}^{-1}$]

15. An EM is incident normally on an interface between air and a dielectric characterised by permeability μ_0 and dielectric constant K. Show that the reflectivity of the interface is

$$R = \left(\frac{1 - \sqrt{K}}{1 + \sqrt{K}}\right)^2.$$

16. A plane polarized EM wave in free space is incident on the surface of a conductor having complex refractive index $n = 1 + jn_i$. Show that the phase shift of the electric vector of the reflected wave with respect to that of the incident wave is $\tan^{-1}(2/n_i)$.

17. Find the cut off wavelength for the TE_{10} mode in a waveguide of dimensions 5 cm × 2.5 cm. Also, find the phase velocity and group velocity of the wave along the waveguide at 6 GHz. [*Ans.* $\lambda_c = 10$ cm, $v_p = 1.15$ c, $v_g = 0.87$ c]

18. Consider a dielectric-perfect conductor interface ($z = 0$). The electric field in the dielectric $z < 0$ characterised by permeability μ and permittivity ϵ is given by

$$\vec{E} = \hat{x}E_1 \cos(kz - \omega t) + \hat{x}E_2 \cos(kz + \omega t),$$

where $k = \omega\sqrt{\mu\epsilon}$. Applying boundary conditions show that $E_2 = -E_1$.

19. Consider a dielectric-conductor interface at $z = 0$. The magnetic field in the dielectric $z \leq 0$ is given by $\vec{H} = \hat{y}H_1 \cos kz \cos \omega t$.

Applying the boundary conditions show that the surface current density is

$$\vec{K} = \hat{x}H_1 \cos \omega t.$$

[**Hints.** $\hat{n} \times \left(\vec{H}_2 - \vec{H}_1\right)\Big|_{z=0} = \vec{K}$ or $\hat{z} \times \left(0 - \vec{H}\right)\Big|_{z=0} = \vec{K}$.]

20. An EM wave polarized normal to the plane of incidence is incident from free space onto a dielectric slab at the Brewster's angle. Show that the reflectance is given by

$$R = \left(\frac{n^2 - 1}{n^2 + 1}\right)^2,$$

where n is the refractive index of the dielectric.

21. Consider a waveguide of square cross-section having side a and filled with a dielectric medium of permeability μ and permittivity ε. Find the range of frequencies of EM waves which can propagate through the guide only as a TE_{10} or TE_{01} mode. [*Ans.* $\frac{1}{2a\sqrt{\mu\varepsilon}} < \nu < \frac{1}{\sqrt{2}a\sqrt{\mu\varepsilon}}$]

22. (a) Suppose $z = 0$ represents the interface between two different media. The magnetic fields in the regions $z < 0$ and $z > 0$ are given by $\vec{H}_1 = 3\hat{y} + 10\hat{z}$ A/m and $\vec{H}_2 = -4\hat{x} + 5\hat{y} + 2\hat{z}$ A/m, respectively. Find the surface current distribution at $z = 0$. [*Ans.* $-2\hat{x} - 4\hat{y}$ A/m]

(b) A current sheet of surface density $\vec{K} = 10\hat{\jmath}$ A/m is located in the $z = 0$ plane. The permeabilities of the material for $z > 0$ and z,0 are $5\mu_0$ and $2\mu_0$ respectively. The magnetic field intensity in the region $z > 0$ is $\vec{H}_> = 15\hat{\imath} - 8\hat{k}$ A/m. Find $\vec{B}_>$, $\vec{H}_<$ and $\vec{B}_<$. **[C.U. 2015]**

[*Ans.* $(75\hat{\imath} - 40\hat{k})\mu_0$ T, $(5\hat{\imath} - 20\hat{k})$ A/m, $(10\hat{\imath} - 40\hat{k})\mu_0$ T]

[**Hints.** See Solved Problem 1]

23. Consider the propagation of TM waves between two perfectly conducting parallel plates at $y = 0$ and $y = a$. Determine the $\vec{H}$-field of the wave propagating in the yz-plane between the plates.

[*Ans.* $\vec{H} = \hat{\imath} H_0 \cos(ky\cos\theta) \cdot e^{j(kz\sin\theta - \omega t)}$, where $\cos\theta = \frac{n\pi}{ka}$ and n is an integer]

24. Considering the TE mode show that energy travels down the waveguide with group velocity.

[**Hints.** Show that average power flow P =time average energy density $u \times ab \times v_g$. Average energy per unit length of the waveguide is

$$u \cdot ab = \frac{1}{4}\int_0^a\int_0^b\int_0^1 dxdydz\left(\vec{E}\cdot\vec{D}^* + \vec{H}\cdot\vec{B}^*\right)$$

$$= \frac{1}{4}\int_0^a\int_0^b dxdy\left[\varepsilon_0\left(E_{ox}^2 + E_{oy}^2\right) + \mu_0\left(H_{ox}^2 + H_{oy}^2 + H_{oz}^2\right)\right]$$

$$= \frac{ab}{8}\cdot\frac{H_0^2}{h^2}\cdot\frac{\mu_0\omega^2}{c^2}.$$

$\therefore\ v_g = \frac{P}{u\cdot ab} = \frac{c^2 k_g}{\omega}$ = group velocity.]

25. A TEM wave of frequency 300 GHz propagating along the positive x-direction, has an $\vec{E}$-field of amplitude $20\sqrt{2}$ V/m. The wave is linearly polarized with the plane of vibration of the electric field at an angle of 45° to the xz-plane. Write down the expressions for $\vec{E}$ and $\vec{B}$. **[B.U. 2014]**

[**Hints.** Here $E_{0x} = B_{0x} = 0$, $E_{0y} = E_{0z} = E_0\cos 45° = 20$ V/m

$$\omega = 2\pi\times 300\times 10^9 = 6\pi\times 10^{11}\ \text{rad/s},\quad k = \frac{\omega}{c} = \frac{6\pi\times 10^{11}}{3\times 10^8} = 2\pi\times 10^3\ \text{m}^{-1},$$

$B_0 = E_0/c = 20\sqrt{2}/cT$. As $\vec{B}\perp^r\vec{E}$ and $\vec{E}\times\vec{B}$ is along +ve x-direction. $B_{0y} = B_{0z} = B_0\cos 45° = \frac{20}{c}T$.

$$\therefore\ \vec{E}_x = 0,$$

$$\vec{E}_y = \hat{\jmath}E_{oy}\cos(kx - \omega t),$$
$$\vec{E}_z = \hat{k}E_{oz}\cos(kx - \omega t)$$
$$\vec{B}_x = 0$$
$$\vec{B}_y = -\hat{\jmath}B_{0y}\cos(kx - \omega t)$$
$$\vec{B}_z = \hat{k}B_{0z}\cos(kx - \omega t)\ ]$$

26. Consider a hollow copper box of dimensions 2 cm × 3 cm × 1 cm. For this cavity resonator find the wavelength of the stationary wave mode $(1, 3, 0)$.

[*Ans.* $4/\sqrt{5}$ cm]

OBJECTIVE TEST

1. The dominant mode in a rectangular wave guide is

(a) TE_{00}, (b) TE_{10}, (c) TE_{11}, (d) TM_{11}.

2. Frequency band available for transmission in a rectangular waveguide is

(a) smaller in TE mode, (c) the same for TE and TM modes,
(b) smaller in TM mode, (d) infinite.

3. Which of the following modes cannot exist in a rectangular resonant cavity?

(a) TE_{011}, (b) TE_{101}, (c) TE_{110}, (d) TE_{111}.

4. If θ_p be the angle of polarization then the refractive index of the reflecting medium relative to the incident medium is given by

(a) $\cot\theta_p$, (c) $\tan\theta_p$,
(b) $\cos\theta_p$, (d) none of these.

5. Which of the following modes is possible with longest cut off wavelength?

(a) TM_{00}, (b) TM_{01}, (c) TM_{10}, (d) TM_{11}.

6. For an EM wave the component of $\vec{E}$-field perpendicular to the plane of incidence, when reflected from a surface backed by an optically denser medium, suffers a phase change of

(a) 0°, (b) 45°, (c) 90°, (d) 180°.

7. The $\vec{E}$-field of an EM wave incident normally on the surface of a good conductor suffers due to reflection a phase shift of

(a) 0°, (b) 45°, (c) 90°, (d) 180°.

8. For an EM wave incident on a medium at the polarizing angle, the angle between the reflected and refracted rays is

(a) 45°, (b) 90°, (c) 180°, (d) of any value.

9. Dispersion relation for an EM wave propagating in a waveguide is $\omega^2 = c^2(k_c^2 + k_g^2)$ where k_c = cut-off wavenumber and k_g = guide wavenumber. The product of phase and group velocities in such a system is

(a) c^2, (c) $c^2/(1 + k_g^2/k_e^2)$,

(b) $c^2(1 + k_g^2/k_e^2)$, (d) $c^2/(1 + k_e^2/k_g^2)$.

10. A beam of light propagating in air reflects normally from an interface with a medium of refractive index n. The fraction of light energy reflected is

(a) $\dfrac{4n}{(n+1)^2}$, (b) $\left(\dfrac{n-1}{n+1}\right)^2$, (c) $\dfrac{2n}{n+1}$, (d) $\dfrac{n}{n+1}$.

ANSWERS

1. (b) **2.** (b) **3.** (c) **4.** (c) **5.** (d) **6.** (d) **7.** (d)
8. (b) **9.** (a) **10.** (b)

Chapter **17**

Dispersion and Scattering

17.1 Introduction

So far we have considered the propagation of EM waves through a medium characterised by the constant values of permittivity ε, permeability μ and conductivity σ. Actually, these parameters depend to some extent on the frequency of the EM waves under consideration. If ε and μ were truly constant then the speed of EM waves in a dielectric, i.e., $v = 1/\sqrt{\mu\varepsilon}$ and the refractive index $n = c/v = \sqrt{\varepsilon_r \cdot \mu_r}$ would likewise be constant. But it is well-known from optics that n depends on frequency. For example, under dc or static condition $\varepsilon \approx 80\varepsilon_0$ and $\mu \approx \mu_0$ for water. So we expect the refractive index of water to be $n = \sqrt{80} = 8.94$. But we know from optics that n for water is only about 1.33 in the visible region. The reason for this is that the value of dielectric constant reflects the tendency of molecular electric dipoles to be aligned along the oscillating EM field. As the frequency increases it becomes increasingly more difficult for the dipoles to *follow the oscillating field* and the value of dielectric constant and hence of refractive index decreases.

This variation of refractive index with the frequency or wavelength of light is known as *dispersion.* When the speed of a wave varies with its frequency, the supporting medium is said to be *dispersive.* This property arises due to dependence of the dielectric constant on frequency.

Whenever an EM wave falls on matter the electrons in the matter are set into oscillation under the action of the oscillating fields in the EM wave. These electrons in turn emit EM waves into space. This phenomenon is known as *scattering.* Thus to explain various optical and electrical phenomena associated with the propagation of EM waves through matter we need to consider the interaction between matter and EM waves. Such interactions are well explained by quantum theory. However, many such interesting effects can be explained even classically. In this chapter we shall study the phenomenon of dispersion and scattering of EM radiation by considering the interaction of EM radiation with discrete charges classically.

17.2 Electromagnetic Theory of Dispersion

Normal and anomalous dispersion

The variation of refractive index (n) of a medium with the wavelength (λ) is known as *dispersion.* Usually the refractive index decreases with increases in wavelength. This is known as *normal dispersion.* Mathematically it may be represented by $dn/d\lambda$. But over a small wavelength range (absorption band) there is often increase in refractive index with increase of wavelength due to an increased absorption of radiation by the medium. This is known as *anomalous dispersion.*

Cauchy's formula

The normal dispersion can be represented with considerable accuracy by Cauchy's empirical formula,

$$n = A + \frac{B}{\lambda^2} + \frac{C}{\lambda^4}, \tag{17.2-1}$$

where A, B, C are constants that depend on the medium and decreases rapidly in magnitude as we proceed to the higher order terms. For some purposes the normal dispersion can be represented with sufficient accuracy by considering the first two terms only. Thus,

$$n = A + \frac{B}{\lambda^2}. \tag{17.2-2}$$

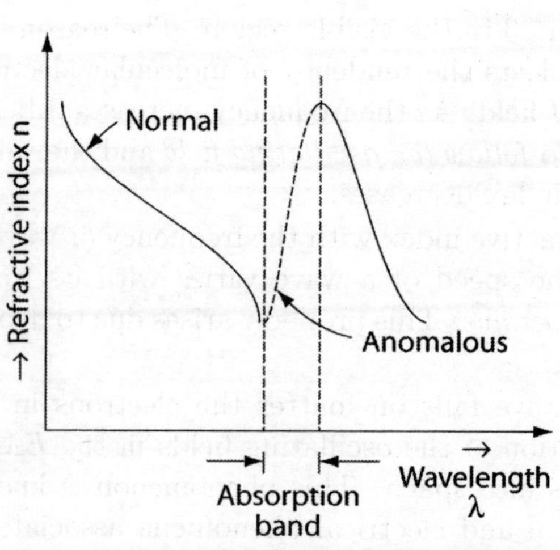

Fig. 17.2-1 : Normal and anomalous dipersion

As A and B are positive constants it is obvious that refractive index decreases with increase in wavelength. Differentiating Eq. (17.2-2) we get for the dispersion,

$$\frac{dn}{d\lambda} = -\frac{2B}{\lambda^3}, \tag{17.2-3}$$

which shows that the dispersion varies inversely as the cube of the wavelength.

Sellmeier's formula

A dispersion formula of general application was first derived by Sellmeier on the basis of elastic solid theory of light. He assumed that the medium contains elastically bound particles capable of vibrating with some natural frequencies. Under the action of the incident EM radiation the particles undergo forced oscillatory motion. If the frequency of the incident EM radiation approaches any one of these natural frequencies the particles resonate and the energy is absorbed. It causes a change in the velocity of the EM wave. On the basis of above mechanism Sellmeier arrived at the following formula,

$$n^2 = 1 + \sum_k \frac{A_k \lambda^2}{\lambda^2 - \lambda_k^2}, \tag{17.2-4}$$

where A_k is a constant proportional to the number of particles per unit volume whose natural frequency corresponds to the wavelength λ_k of EM radiation in vacuum.

Although Sellemeier's formula represents a great improvement over Cauchy's formula, it fails in the region of absorption band. It shows a discontinuity at $\lambda = \lambda_k$. However, in regions away from the absorption band it gives better results than Cauchy's formula. In fact Cauchy's formula is merely an approximation to Sellemeier's formula. If there is only one absorption band about λ_0 then for wavelengths $\lambda \gg \lambda_0$ formula (17.2-4) reduces to

$$n^2 = 1 + \frac{A_0}{1 - \frac{\lambda_0^2}{\lambda^2}} \simeq 1 + A_0 + \frac{A_0 \lambda_0^2}{\lambda^2} + \frac{A_0 \lambda_0^4}{\lambda^4}, \tag{17.2-5}$$

which has the same form as Cauchy's formula (17.2-1).

Electromagnetic theory of dispersion

An adequate theory of dispersion based on the electromagnetic theory was given by H. A. Lorentz. In the case of gases he assumed the following:

1. When an EM wave passes through a medium atomic (or molecular) dipoles are created due to relative displacement of electrons and nuclei of neutral atoms due to electric field of the wave. These dipoles undergo forced oscillation under the action of the electric field $\left(\vec{E}\right)$ of the passing wave.

2. The molecules of a gas are apart and, therefore, the force on an electron due to local electric intensity produced by the dipoles is negligible than that due to the electric field of the EM wave. Also, the magnetic force on the electron on account of its motion is negligible.

3. The electrons are bound to the nucleus by an elastic restoring force proportional to its displacement $\vec{r}$ from its equilibrium position.

4. A damping force proportional to velocity acts on the electrons.

5. The nucleus is assumed to be heavy compared to the electrons and hence it can be assumed to be at rest.

6. The atoms are assumed to be small in size compared to the wavelength of the wave. So the electric field may be assumed to be constant in space over an atom or molecule.

In the light of the above assumptions the equation of motion of an electron of mass m and charge e may be written as

$$m\frac{d^2\vec{r}}{dt^2} = e\vec{E} - s\vec{r} - R\frac{d\vec{r}}{dt}, \tag{17.2-6}$$

where s is the restoring force constant and R is the damping constant.

Now substituting $s/m = \omega_0^2$, $R/m = \gamma$ and $\vec{E} = \vec{E}_0 e^{-j\omega t}$ we can rewrite Eq. (17.2-6) as

$$\frac{d^2\vec{r}}{dt^2} + \gamma\frac{d\vec{r}}{dt} + \omega_0^2\vec{r} = \frac{e}{m}\vec{E} = \frac{e}{m}\vec{E}_0 e^{-j\omega t}. \tag{17.2-7}$$

It represents forced vibration of the electrons having natural frequency ω_0. To solve the equation we note that the complementary function represents a damped motion which dies out with time. The steady state solution is given by the particular integral,

$$\vec{r} = \frac{e\vec{E}}{m\left(\omega_0^2 - \omega^2 - j\gamma\omega\right)}. \tag{17.2-8}$$

The instantaneous dipole moment resulting due to displacement of the electron is $\vec{p} = e\vec{r}$. If there are N dispersion electrons per unit volume, then polarization

$$\vec{P} = Ne\vec{r} = \frac{Ne^2\vec{E}}{m\left(\omega_0^2 - \omega^2 - j\gamma\omega\right)}. \tag{17.2-9}$$

The imaginary term in the denominator indicates that $\vec{P}$ is not in phase with $\vec{E}$. Here we have assumed that all the oscillators in the medium have the same natural frequency. In general, differently situated electrons within a given molecule exhibit different natural frequencies and damping constants. Let there are N molecules per unit volume and in each molecule there are f_1 oscillators characterised by constants ω_1 and γ_1; f_2 oscillators of constants ω_2 and γ_2 and so on. Then Eq. (17.2-9) takes the form

$$\vec{P} = Ne\sum_k f_k\vec{r}_k = \frac{Ne^2}{m}\sum_k \frac{f_k\vec{E}}{\omega_k^2-\omega^2-j\gamma_k\omega}. \tag{17.2-10}$$

The complex electric susceptibility $\mathcal{X}_e$ is given by $\vec{P} = \varepsilon_0\mathcal{X}_e\vec{E}$. Therefore,

$$\mathcal{X}_e = \frac{Ne^2}{m\varepsilon_0}\sum_k \frac{f_k}{\omega_k^2-\omega^2-j\gamma_k\omega}$$

and dielectric constant

$$K = \frac{\varepsilon}{\varepsilon_0} = 1+\mathcal{X}_e = 1+\frac{Ne^2}{m\varepsilon_0}\sum_k \frac{f_k}{\omega_k^2-\omega^2-j\gamma_k\omega}. \tag{17.2-11}$$

For a non-magnetic medium permeability $\mu \approx \mu_0$ and, therefore, refractive index $n = \sqrt{K}$. Thus,

$$n^2 = 1+\frac{Ne^2}{m\varepsilon_0}\sum_k \frac{f_k}{\omega_k^2-\omega^2-j\gamma_k\omega} \tag{17.2-12}$$

This equation shows that the refraction index is frequency dependent. It is known as *Lorentz dispersion equation.*

Normal dispersion

In the region well away from the natural frequencies of the oscillators damping is very small for transparent substances and the term containing $\gamma_k\omega$ may be neglected in comparison with $\omega_k^2-\omega^2$. Then Eq. (17.2-12) reduces to

$$n^2 = 1+\frac{Ne^2}{m\varepsilon_0}\sum_k \frac{f_k}{\omega_k^2-\omega^2}. \tag{17.2-13}$$

This equation shows that refractive index n increases with increase in frequency ω. This is *normal dispersion.* For most substances the natural frequencies ω_k lie scattered all over the spectrum in a rather chaotic fashion. An an example, we note that for visible light in glass the nearest significant resonant frequencies lie in the ultraviolet region, i.e., $\omega < \omega_k$ and n is real.

In terms of free space wavelengths

$$\lambda = \frac{2\pi c}{\omega} \quad \text{and} \quad \lambda_k = \frac{2\pi c}{\omega_k},$$

Eq. (17.2-13) may be expressed as

$$n^2 = 1+\sum_k \frac{A_k\lambda^2}{\lambda^2-\lambda_k^2}, \tag{17.2-14}$$

where

$$A_k = \frac{Ne^2}{m\varepsilon_0} \cdot \frac{f_k \lambda_k^2}{4\pi^2 c^2}.$$

This is *Sellmeier's dispersion formula.*

If we assume $\lambda \gg \lambda_k$ then, from Eq. (17.2-14),

$$n^2 = 1 + \sum_k A_k \left(1 - \frac{\lambda_k^2}{\lambda^2}\right)^{-1} = A + \frac{B}{\lambda^2} + \frac{C}{\lambda^4}, \tag{17.2-15}$$

where

$$A = 1 + \sum_k A_k, \quad B = \sum_k A_k \lambda_k^2 \quad \text{and} \quad C = \sum_k A_k \lambda_k^4.$$

This is of the same form as the Cauchy's dispersion formula.

Anomalous dispersion

If the frequency of the EM wave lies close to one of the natural frequencies ω_k then we obtain the phenomenon of anomalous dispersion. In this case dielectric constant K and refractive index n as given by Eqs. (17.2-11) and (17.2-12) are complex. The wave number k of the plane EM wave represented by

$$\vec{E}(\vec{r}, t) = \vec{E}_0 e^{j(\vec{k}\cdot\vec{r} - \omega t)},$$

is also complex. Let us write $\vec{k} = \vec{k}_r + j\vec{k}_i$. Then

$$\vec{E}(\vec{r}, t) = \vec{E}_0 e^{-\vec{k}_i\cdot\vec{r}} e^{j(\vec{k}_r\cdot\vec{r} - \omega t)}.$$

Obviously k_i measures the attenuation of the wave. As the intensity is proportional to E^2, the quantity $\alpha = 2k_i$ is called the *absorption coefficient.* The wave velocity is ω/k_r and, therefore, the refractive index $n = c/(\omega/k_r)$. For gases the second term in Eq. (17.2-11) is usually small and we can write approximately (by using binomial expansion),

$$\begin{aligned} k = \frac{\omega}{c}\sqrt{K} &\approx \frac{\omega}{c}\left[1 + \frac{Ne^2}{2m\varepsilon_0} \sum_k \frac{f_k}{\omega_k^2 - \omega^2 - j\gamma_k \omega}\right] \\ &= \frac{\omega}{c}\left[1 + \frac{Ne^2}{2m\varepsilon_0} \sum_k \frac{f_k\left(\omega_k^2 - \omega^2 + j\gamma_k\omega\right)}{(\omega_k - \omega^2)^2 + \gamma_k^2\omega^2}\right] \\ &= k_r + jk_i. \end{aligned} \tag{17.2-16}$$

Therefore, the real refractive index

$$n = \frac{c}{\omega} k_r = 1 + \frac{Ne^2}{2m\varepsilon_0} \sum_k \frac{f_k\left(\omega_k^2 - \omega^2\right)}{\left(\omega_k^2 - \omega^2\right)^2 + \gamma_k^2\omega^2} \tag{17.2-17}$$

and

$$\alpha = 2k_i = \frac{Ne^2\omega^2}{m\varepsilon_0 c}\sum_k \frac{f_k\gamma_k}{\left(\omega_k^2-\omega^2\right)^2+\gamma_k^2\omega^2}. \tag{17.2-18}$$

The variations of the refractive index n and the absorption coefficient α as a function of frequency near one of the resonant frequencies are shown in Fig. 17.2-2. Obviously in the neighbourhood of the resonant frequency ω_0 the real refractive index n decreases with the increase in frequency. As this behaviour is contrary to the normal behaviour it is called *anomalous dispersion.*

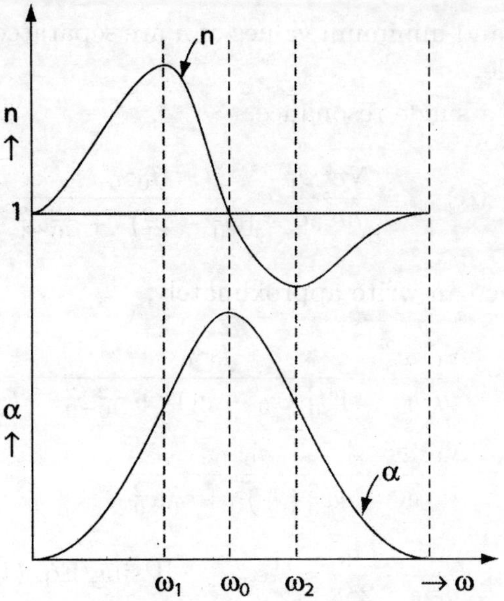

Fig. 17.2.2: Variation refractive index and absorption coefficient with frequency

The absorption coefficient α shows a typical resonance shape and attains a maximum value at resonance frequency $\omega = \omega_0$. Thus for the case of a single resonant frequency at $\omega = \omega_0$,

$$\alpha_{\max} = \frac{Ne^2}{m\varepsilon_0 c}\cdot\frac{f_0}{\gamma_0}. \tag{17.2-19}$$

Now, n is maximum or minimum when $\partial n/\partial\omega = 0$. From Eq. (17.2-17), for the case of single resonance,

$$\frac{\partial}{\partial\omega}\left[\frac{\omega_0^2-\omega^2}{\left(\omega_0^2-\omega^2\right)^2+\gamma_0^2\omega^2}\right] = 0.$$

Near the absorption band $\omega \to \omega_0$ and we can write $\omega_0^2-\omega^2 \approx 2\omega_0(\omega_0-\omega)$ and $\omega = \omega_0$ in the term with $\gamma_0\omega$. Thus,

$$\frac{\partial}{\partial\omega}\left[\frac{2\omega_0(\omega_0-\omega)}{4\omega_0^2\left(\omega_0-\omega\right)^2+\gamma_0^2\omega_0^2}\right] = 0.$$

This gives

$$\omega_0 - \omega = \pm\frac{\gamma_0}{2}.$$

If ω_1 and ω_2 correspond to maximum and minimum values of n then

$$\omega_0 - \omega_1 = +\frac{\gamma_0}{2} \quad \text{and} \quad \omega_0 - \omega_2 = -\frac{\gamma_0}{2}$$

$$\text{or} \quad \omega_1 = \omega_0 - \frac{\gamma_0}{2} \quad \text{and} \quad \omega_2 = \omega_0 + \frac{\gamma_0}{2}. \tag{17.2-20}$$

Thus the maximum and minimum values of n are separated by a distance $\omega_2 - \omega_1 = \gamma_0$ on the frequency scale.

Now, for the case of a single resonance,

$$(\alpha)_{\omega_1} = \frac{Ne^2\omega_1^2}{m\varepsilon_0 c} \cdot \frac{f_0\gamma_0}{\left(\omega_0^2 - \omega_1^2\right)^2 + \gamma_0^2\omega_1^2}.$$

As ω_1 lies near ω_0 we can write approximately,

$$\begin{aligned}(\alpha)_{\omega_1} &\approx \frac{Ne^2\omega_0^2}{m\varepsilon_0 c} \cdot \frac{f_0\gamma_0}{4\omega_0^2\left(\omega_0 - \omega_1\right)^2 + \gamma_0^2\omega_0^2} \\ &= \frac{Ne^2\omega_0^2}{m\varepsilon_0 c} \cdot \frac{f_0\gamma_0}{4\omega_0^2\left(\frac{\gamma_0}{2}\right)^2 + \gamma_0^2\omega_0^2} \\ &= \frac{Ne^2}{m\varepsilon_0 c} \cdot \frac{f_0}{2\gamma_0} = \tfrac{1}{2}\alpha_{\max} \quad \text{[Using Eq. (17.2-19)]}\end{aligned}$$

Similarly, we can show that

$$(\alpha)_{\omega_2} = \tfrac{1}{2}\alpha_{\max}.$$

Thus for gases the refractive index assumes its maximum and minimum values at points where the absorption coefficient is at half-maximum. From Fig. 17.2-2 it is obvious that the region of anomalous dispersion coincides with the region of maximum absorption. In this region the electrons are forced to oscillate at their natural frequency, the amplitude of their oscillation becomes relatively large and consequently a large amount of energy is dissipated by the damping force. As a result the material may appear practically opaque in this frequency range. This indicates that an optically transparent material cannot have a resonant frequency in the optical band.

From Fig. 17.2-2 it may be noted that $n > 1$ for $\omega < \omega_0$ and $n < 1$ for $\omega > \omega_0$. These can be explained physically as follows. For frequencies $\omega < \omega_0$ the electrons can *follow* the oscillatory force $e\vec{E}(t)$ due to the incident wave. Then charges move in phase with the driving force and build up a field that tends to cancel the original field. This reduced field gives a reduced return force and a reduced phase velocity or greater refractive index. On the other hand, when $\omega > \omega_0$, the charges cannot follow the field and move in out of phase with the field. In this case the field due to relative displacement of charges tends to increase the original field. That gives rise to an enhanced return force and hence a phase velocity greater than that in vacuum or a refractive index less than 1. For example, extreme UV radiation in glass shows $n < 1$ or phase velocity greater c. This does not violate the principle of relativity because it is the group velocity, not the phase velocity, with which energy propagates and in fact, we can show that group velocity is always smaller than c.

Assuming just one resonant frequency (ω_0) and neglecting the damping we can write from Eq. (17.2-16)

$$k \simeq \frac{\omega}{c}\left[1 + \frac{Ne^2}{2m\varepsilon_0} \cdot \frac{f_0}{\omega_0^2 - \omega^2}\right]. \tag{17.2-21}$$

Near the absorption band we can write $\omega_0^2 - \omega^2 \simeq 2\omega_0(\omega_0 - \omega)$. Thus,

$$k = \frac{\omega}{c}\left[1 + \frac{Ne^2 f_0}{4m\varepsilon_0\omega_0(\omega_0 - \omega)}\right]. \tag{17.2-22}$$

Therefore, phase velocity

$$v_p = \frac{\omega}{k} = \frac{c}{1 + \frac{Ne^2 f_0}{4m\varepsilon_0\omega_0(\omega_0 - \omega)}}. \tag{17.2-23}$$

Thus v_p can be greater than c or less than c depending on ω.

Differentiating (17.2-22) with respect to k we get

$$1 = \frac{v_g}{c}\left[1 + \frac{Ne^2 f_0}{4m\varepsilon_0\omega_0(\omega_0 - \omega)}\right] + \frac{\omega}{c} \cdot \frac{Ne^2 f_0}{4m\varepsilon_0\omega_0(\omega_0 - \omega)^2} \cdot v_g,$$

where $v_g = d\omega/dk$ is the group velocity. Thus

$$v_g = \frac{c}{1 + \frac{Ne^2 f_0}{4m\varepsilon_0(\omega_0 - \omega)^2}}. \tag{17.2-24}$$

This shows that group velocity v_g is smaller than c.

Absorptive and dispersive modes

Assuming the electric field of the passing EM wave as

$$\vec{E} = \hat{x}E_0\cos(kz - \omega t) = Re\ \hat{x}E_0 e^{j(kz - \omega t)},$$

the steady state solution (17.2-8) can be expressed as

$$x = Re \frac{eE_0 e^{j(kz-\omega t+\phi)}}{m\sqrt{\left(\omega_0^2-\omega^2\right)^2+\gamma^2\omega^2}}$$

$$= A' \cos(kz - \omega t + \phi)$$

$$= A_1 \cos(kz - \omega t) - A_2 \sin(kz - \omega t),$$

where

$$\phi = \tan^{-1}\frac{\gamma\omega}{\omega_0^2-\omega^2}, \quad A' = \frac{eE_0}{m\sqrt{\left(\omega_0^2-\omega^2\right)^2+\gamma^2\omega^2}},$$

$$A_1 = A'\cos\phi \quad \text{and} \quad A_2 = A'\sin\phi.$$

The instantaneous velocity

$$\frac{dx}{dt} = A_1\omega \sin(kz-\omega t) + A_2\omega\cos(kz-\omega t).$$

Now the time average power absorbed by the electron from the $\vec{E}$-field is

$$\langle P\rangle = \left\langle eE_0\cos(kz-\omega t)\cdot\frac{dx}{dt}\right\rangle = \tfrac{1}{2}eE_0\omega A_2.$$

Obviously it is proportional to the amplitude (A_2) of that part of the steady displacement which is 90° out of phase with the $\vec{E}$-field. For this the constant A_2 is called *absorptive amplitude.* Near resonance it plays the dominant role. The in phase term of the steady displacement does not contribute to the time average power. At resonance it is zero but at driving frequencies far from resonance this term is the dominant one. The amplitude A_1 is called *elastic amplitude* or sometimes *dispersive amplitude.*

Thus it is possible to describe the whole process in terms of two independent modes of vibrations namely, the absorptive mode and the dispersive or elastic mode.

17.3 Dispersion in Solids and Liquids

In case of solids and liquids the molecules are close enough and hence their mutual interactions cannot be neglected. In this case total electric field experienced by an electron is the sum of the $\vec{E}$-field of the passing EM wave and the local field $\vec{P}/3\varepsilon_0$ produced due to the polarization of the medium. So Eq. (17.2-10) in this case takes the form

$$\vec{P} = \frac{Ne^2}{m}\sum_k \frac{f_k\left(\vec{E}+\vec{P}/3\varepsilon_0\right)}{\omega_k^2-\omega^2-j\gamma_k\omega}. \tag{17.3-1}$$

Using the relation $\vec{D} = \varepsilon_0\vec{E} + \vec{P}$ or $\vec{P} = \varepsilon_0(K-1)\vec{E}$ we get

$$\frac{K-1}{K+2} = \frac{Ne^2}{3\varepsilon_0 m}\sum_k \frac{f_k}{\omega_k^2 - \omega^2 - j\gamma_k\omega}. \tag{17.3-2}$$

This is a generalisation of *Clausius-Mossotti equation* and from this equation the frequency dependence of the dielectric constant K or the refractive index can be studied as done in Section 17.2.

For transparent substances the damping factor may be neglected. In this case K and refractive index $n = \sqrt{K}$ both are real. For a particular frequency the right-hand side of Eq. (17.3-2) is constant. Therefore, in this case we get

$$\frac{n^2-1}{n^2+2} = \frac{Ne^2}{3\varepsilon_0 m}\sum_k \frac{f_k}{\omega_k^2 - \omega^2}. \tag{17.3-3}$$

Since the number of molecules per unit volume, N, is related to the molecule weight M, Avogadro's number N_A and density ρ by $N{=}\rho N_A/M$ we can write from Eq. (17.3-3), for a particular frequency,

$$\frac{n^2-1}{n^2+2}\cdot\frac{M}{\rho} = \text{constant}. \tag{17.3-4}$$

This equation is known as *Lorentz-Lorentz formula.* It shows the dependence of refractive index on the density of the medium and is found to hold for many substances. The quantity on the left-hand side of Eq. (17.3-4) refers to a gram molecule and hence is called *molar refraction.*

17.4 Scattering of *EM* waves by a Harmonically Bound Electron

Scattering of EM radiation is a process in which EM radiation interacts with matter and in course of the interaction, matter absorbs incident radiation and subsequently re-emits it as EM radiation into space. The scattering process can be easily explained on the basis of classical electromagnetic theory of light. When an EM wave is incident on matter dipole moment is induced in the atoms or molecules due to relative displacement of bound electrons and nuclei of neutral atoms under the action of the electric field of the incident wave. These dipoles execute forced oscillatory motion under the action of the electric field of the incident wave. These oscillatory dipoles radiate EM waves which are known as *scattered radiations.* The processes of transmission, reflection and refraction of EM waves are actually the macroscopic manifestations of the scattering process occurring on atomic level.

Suppose a plane monochromatic EM wave represented by $\vec{E} = \vec{E}_0 e^{j(kz-\omega t)}$ is incident on an electron which is bound to the nucleus by an elastic restoring force $-sx$. The nucleus is assumed to be heavy and at rest. The motion of the electron is assumed to be confined to the $z = 0$ plane and the centre of oscillation of the electron is chosen as origin as shown in Fig. 17.4-1.

The incident wave is plane polarized with its electric vector along x-axis. A damping force proportional to velocity is assumed to act on the electrons. Furthermore, we assume non-relativistic motion, so that we can neglect the force on the electron due to the magnetic field compared with that due to the electric field.

In the light of the above assumptions the differential equation of motion of an electron of mass m and charge e may be written as

$$m\frac{d^2x}{dt^2} = eE_0 e^{-j\omega t} - sx - R\frac{dx}{dt},$$

where R is the damping force constant. It can be rewritten as

$$\frac{d^2x}{dt^2} + \gamma\frac{dx}{dt} + \omega_0^2 x = \frac{eE_0}{m}e^{-j\omega t}, \tag{17.4-1}$$

where $\gamma = R/m$ and $\omega_0 = \sqrt{s/m}$ is the natural frequency of oscillation of the electron. The solution of Eq. (17.4-1) consists of two parts-complementary function and the particular integral. The complementary function represents damped natural oscillatory motion which dies out after some time. In the steady state the electron continues to vibrate with the frequency of the incident wave. This steady state motion is given by the particular integral. Consider a solution of the from $x = Ae^{-j\omega t}$. Therefore,

$$\frac{dx}{dt} = -j\omega x \quad \text{and} \quad \frac{d^2x}{dt^2} = -\omega^2 x.$$

Substituting all these in (17.4-1) we get the steady state solution,

$$x = \frac{eE_0}{m} \cdot \frac{e^{-j\omega t}}{\omega_0^2 - \omega^2 - j\gamma\omega}. \tag{17.4-2}$$

The instantaneous induced dipolemoment is

$$p = ex = \frac{e^2 E_0}{m} \cdot \frac{e^{-j\omega t}}{\omega_0^2 - \omega^2 - j\gamma\omega} = p_0 e^{-j(\omega t - \delta)}, \tag{17.4-3}$$

where

$$p_0 = \frac{e^2 E_0}{m} \cdot \frac{1}{\sqrt{(\omega_0^2 - \omega^2)^2 + \gamma^2\omega^2}} \quad \text{and} \quad \tan\delta = \frac{\gamma\omega}{\omega_0^2 - \omega^2}. \tag{17.4-4}$$

Now the time average power radiated per unit solid angle by such an oscillating dipole is given by Eq. (15.12-19)

$$\left\langle \frac{dP}{d\Omega} \right\rangle = \frac{p_0^2 \omega^4 \sin^2\psi}{32\pi^2 \varepsilon_0 c^3}, \tag{17.4-5}$$

where ψ is the angle between the direction of $\vec{E}$-field and the direction of observation.

Using Eq. (17.4-4) we can write Eq. (17.4-5) as

$$\left\langle \frac{dP}{d\Omega} \right\rangle = \frac{1}{2} c\varepsilon_0 E_0^2 \cdot \frac{e^4 \sin^2 \psi}{16\pi^2 \varepsilon_0^2 m^2 c^4} \cdot \frac{\omega^4}{\left(\omega_0^2 - \omega^2\right)^2 + \gamma^2 \omega^2}. \tag{17.4-6}$$

Now the scattering process is described in terms of a quantity known as *scattering cross-section.* It may be visualized as the effective area that a target atom or molecule presents to the incident wave for undergoing scattering. The differential scattering cross-section $d\sigma/d\Omega$ is defined as the ratio of the amount of energy scattered per unit solid angle per second to the average energy incident normally per unit area per second. Thus

$$\frac{d\sigma}{d\Omega} = \frac{\left\langle \frac{dP}{d\Omega} \right\rangle}{\frac{1}{2} c\varepsilon_0 E_0^2} = r_0^2 \cdot \frac{\omega^4}{\left(\omega_0^2 - \omega^2\right)^2 + \gamma^2 \omega^2} \sin^2 \psi, \tag{17.4-7}$$

where $r_0 = e^2/(4\pi\varepsilon_0 mc^2)$ is known as the *classical electron radius.*

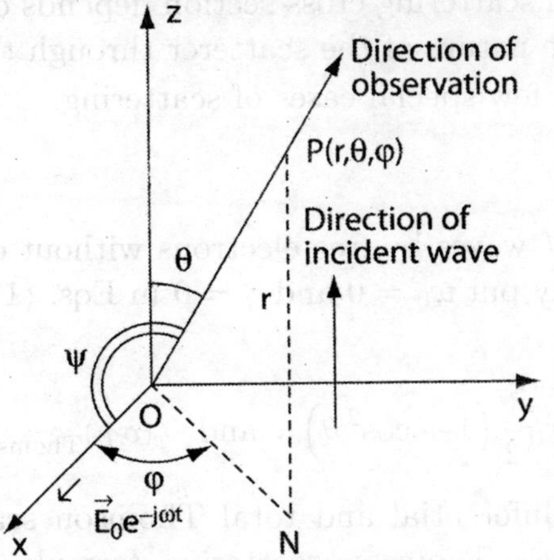

Fig. 17.4-1 : Scattering of EM wave

Sometimes it becomes desirable to express $d\sigma/d\Omega$ in terms of the angle of scattering θ and the polarizing angle φ as shown in Fig. 17.4-1. Drawing perpendicular (PN) from the point of observation P on the plane containing $\vec{E}$, we can write from the figure

$$r\cos\psi = x = r\sin\theta\cos\varphi \quad \text{or} \quad \cos\psi = \sin\theta\cos\varphi.$$

Therefore, $\sin^2\psi = 1 - \cos^2\psi = 1 - \sin^2\theta\cos^2\varphi$

Thus

$$\left(\frac{d\sigma}{d\Omega}\right)_{\text{Polarized}} = r_0^2 \frac{\omega^4 \left(1 - \sin^2\theta\cos^2\varphi\right)}{\left(\omega_0^2 - \omega^2\right)^2 + \gamma^2\omega^2}. \tag{17.4-8}$$

If the incident wave is unpolarized we have to take average of $\cos^2\varphi$ which is simply 1/2. Therefore,

$$\langle \sin^2\psi \rangle = 1 - \tfrac{1}{2}\sin^2\theta = \tfrac{1}{2}\left(1 + \cos^2\theta\right).$$

Thus,

$$\left(\frac{d\sigma}{d\Omega}\right)_{\text{Unpolarized}} = r_0^2 \frac{\omega^4}{\left(\omega_0^2 - \omega^2\right)^2 + \gamma^2\omega^2} \cdot \frac{1}{2}\left(1 + \cos^2\theta\right). \tag{17.4-9}$$

It shows that the scattering depends on the frequency of the incident wave, nature of the scatterer through the factors ω_0, γ and also on the angle of scattering.

The total scattering cross-section (σ_T) for unpolarized light is then obtained by integrating the differential scattering cross-section.

$$\begin{aligned}\sigma_T &= r_0^2 \frac{\omega^4}{\left(\omega_0^2 - \omega^2\right)^2 + \gamma^2\omega^2} \int_0^{\pi}\int_0^{2\pi} \frac{1}{2}\left(1 + \cos^2\theta\right) \sin\theta d\theta d\varphi \\ &= \frac{8\pi}{3} r_0^2 \cdot \frac{\omega^4}{\left(\omega_0^2 - \omega^2\right)^2 + \gamma^2\omega^2}\end{aligned} \tag{17.4-10}$$

It shows that the total scattering cross-section depends on the frequency (ω) of the incident radiation and the nature of the scatterer through the factors ω_0 and γ.

Let us now consider a few special cases of scattering.

Thomson scattering

It is the scattering of *EM* waves by free electrons without change in frequency. Since the electron is free we may put $\omega_0 = 0$ and $\gamma = 0$ in Eqs. (17.4-9) and (17.4-10).

Thus,

$$\left(\frac{d\sigma}{d\Omega}\right)^{\text{Thomson}}_{\text{Unpolarized}} = r_0^2 \frac{1}{2}\left(1 + \cos^2\theta\right) \quad \text{and} \quad (\sigma_T)_{\text{Thomson}} = \frac{8\pi}{3} r_0^2. \tag{17.4-11}$$

Obviously, both the differential and total Thomson scattering cross-sections are independent of frequency. Thomson scattering formula (17.4-11) is found to be appropriate for the scattering of X-rays by electrons or γ-rays by protons. The formula is found to be valid only at low frequencies where the energy of the incident photon ($h\nu$) is very much smaller than the rest mass energy (mc^2) of the electron.

Resonant scattering

If $\omega = \omega_0$ then corresponding scattering is called *resonant scattering.* In this case scattering cross-section can become very large and the scatterer may appear luminous.

$$(\sigma_T)_{\text{Resonance}} = \frac{8\pi}{3} r_0^2 \left(\frac{\omega}{\gamma}\right)^2 = (\sigma_T)_{\text{Thomson}} \cdot \left(\frac{\omega}{\gamma}\right)^2. \tag{17.4-12}$$

Rayleigh scattering

It is the scattering of *EM* waves by bound charges. Here the frequency of incident wave is very much less than the natural frequency of the oscillating charge, i.e., $\omega \ll \omega_0$.

Furthermore $\gamma \ll \omega_0$. Thus, from Eqs. (17.4-9) and (17.4-10) we get

$$\left(\frac{d\sigma}{d\Omega}\right)^{\text{Rayleigh}}_{\text{Unpolarized}} = r_0^2 \left(\frac{\omega}{\omega_0}\right)^4 \cdot (1 + \cos^2\theta),$$

$$(\sigma_T)_{\text{Rayleigh}} = \frac{8\pi}{3} r_0^2 \left(\frac{\omega}{\omega_0}\right)^4 . \tag{17.4-13}$$

Thus, we find that the Rayleigh scattering cross-sections are proportional to the fourth power of frequency or inversely proportional to the fourth power of wavelength of the incident radiation.

On the basis of Rayleigh scattering we can explain a few important phenomena like blue colour of sky, red colour during sunrise and sunset, use of red light for danger signals etc.

1. ***Blue Colour of sky***: According to Rayleigh the total scattering cross-section $\sigma_T \propto 1/\lambda^4$ where λ is the wavelength of incident light. This implies that scattering will be stronger in the smaller wavelength side of the spectrum. Thus the white light coming from the sun will suffer more scattering by the air molecules in the blue end (i.e., smaller wavelength side) of the spectrum than the long wavelength red end. This makes the sky to appear blue.

2. ***Red colour of sunrise and sunset***: As $\sigma_T \propto 1/\lambda^4$, red colour having longest wavelength in the visible range scatters least. On the other hand, blue colour suffers strong scattering. During sunrise or sunset the sun rays have to traverse a longer path through the atmosphere in the way to reach our eyes. As a result blue light is almost scattered away and in the remaining light the red colour becomes prominent. This makes the rising and setting sun coloured crimson red.

3. ***Red light for danger signal***: Red light has longest wavelength in the visible region and hence according to Rayleigh scattering theory it scatters least. So red light can travel longer distances with least disturbance in comparison to other colours and can be seen from longer distances. For this red light is used for danger signals even though our eye is most sensitive to yellow-green colour.

Note that the colouring effect of the sky is due to the scattering of sunlight by the atmospheric particles. If the earth had no atmosphere, the sky would appear as black as the void of space even in the day time.

Polarization by scattering

Scattering of light involves interaction of radiation with matter. The atomic dipoles produced due to the $\vec{E}$-field of the incident wave have their axes parallel to the $\vec{E}$-field and power radiated has a simple dipole radiation pattern. Thus there is no scattered radiation along the direction of dipole axis; the intensity is maximum in the direction normal to the dipole axis; the scattered waves propagating normal the axis of the dipole will be plane polarized with its electric vector along the dipole axis. For example, if a plane polarized EM wave with its $\vec{E}$-field along x-axis propagates along z-axis, then there will be no scattered light along x-axis, the scattered wave propagating along y-axis will have maximum intensity and will be plane polarized with its electric vector still along x-axis.

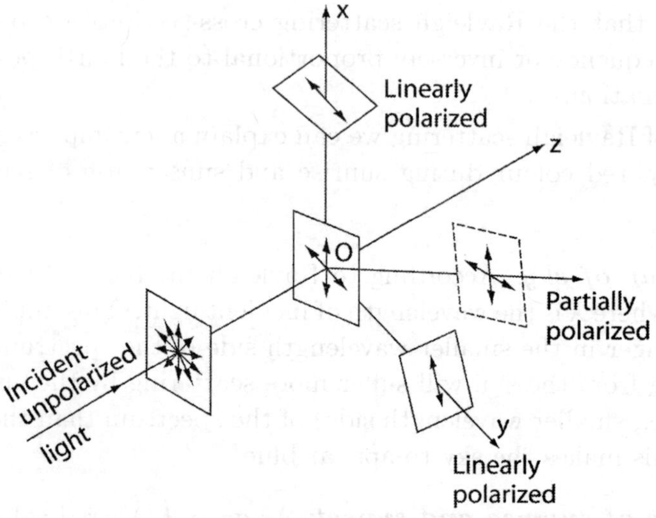

Fig. 17.4-2: Polarization by scattering

If the incident wave travelling along z-axis is unpolarized its $\vec{E}$-vector lying in xy-plane can be resolved into two components—one along x-axis and the other along y-axis. So the molecule at the origin will behave like two oscillating dipoles with their axes along x- and y-axes. An observer looking along y-direction (or x-direction) will find the scattered light plane polarized with $\vec{E}$-field vibrating along x-axis (or y-axis) due to dipole components along x-axis (or y-axis). Since choice of x-, y-axes is arbitrary, *the light scattered in any direction normal to the direction of propagation of the incident light is completely plane polarized.* In all other directions the scattered light will be only partially polarized.

Mie scattering

Theoretical analysis of scattering from spherical particles of any size (containing large number of molecules) was first done by G. Mie in 1908. When the size of the particle is only a fraction of wavelength and the number of molecules in a particle is small, they act in unison and scattering becomes strong. But as the size of the particle approaches the wavelength the molecules do not act in unison and scattering decreases. This happens first at short wavelength range. So as the particle size increases it scatters proportionately more of long wavelength light. When the particle size exceeds the wavelength scattering becomes almost independent of wavelength. Thus, cloud droplets with diameter of 20μm or so are large enough to scatter all visible wavelengths more or less equally. As a result clouds appear white. When the cloud becomes very deep less and less of the incoming solar radiation can penetrate through the cloud. This gives a darker appearance of the cloud.

Note that Rayleigh scattering may be considered as the small-size limiting case of Mie scattering. Air molecules like oxygen and nitrogen, for example, are small in size and causes more Rayleigh scattering at shorter wavelength. This is responsible for producing blue sky on a clear sunny day.

SOLVED PROBLEMS

▶ **Example 1.** *Calculate the value of total Thomson scattering cross-section.*

Solution. We know that

$$(\sigma_T)_{\text{Thomson}} = \frac{8\pi}{3} r_0^2 = \frac{8\pi}{3}\left(\frac{e^2}{4\pi\varepsilon_0 mc^2}\right)^2$$

$$= \frac{8\pi}{3}\left[\frac{(1.6 \times 10^{-19})^2}{4\pi \times 8.854 \times 10^{-12} \times 9.1 \times 10^{-31} \times (3 \times 10^8)^2}\right] m^2$$

$$= 0.67 \times 10^{-28}\,\text{m}^2$$

$= 0.67$ barn. [1 barn $= 10^{-28}\,\text{m}^2$, a unit used to express nuclear reaction cross-section]

▶ **Example 2.** *Using dimensional analysis establish the λ^{-4} dependence of intensity in Rayleigh scattering. Assume that the scattered amplitude at a distance r is inversely proportional to r, directly proportional to the incident amplitude and the volume of the scatterer.* **[cf. C.U. 2015]**

Solution. Let E_{io} be the incident amplitude, E_{so} the scattered amplitude at a distance r and V is the volume of the scatterer. Since E_{so} is expected to be directly proportional to E_{io} and V; inversely proportional to r and in some way on the wavelength λ of incident light we may write

$$E_{so} = \beta \cdot \frac{E_{io} \cdot V}{r} \cdot \lambda^x,$$

where β is a dimensionless constant and x is to be determined. As E_{so} and E_{io} have the same dimensions, we can write dimensionally

$$[r] = [V]\,[\lambda^x] \quad \text{or} \quad [L] = \left[L^3\right][\lambda^x].$$

From dimensional homogeneity $1 = 3 + x$ or $x = -2$.

Therefore, $E_{xo} \propto \lambda^{-2}$ and hence

$$\begin{aligned}\text{scattered intensity} &\propto (E_{so})^2\\ &\propto \lambda^{-4}.\end{aligned}$$

► **Example 3.** *About* 10.05% *of incident sunlight is lost at* 694.3 nm *wavelength due to Rayleigh scattering in the atmosphere. Find the percentage of sunlight lost due to scattering in the atmosphere at the wavelength* 1060 nm.

Solution. From Rayleigh scattering theory, intensity of scattered light is proportional to $1/\lambda^4$ where λ is the wavelength. Therefore, if x be the required percentage of loss of sunlight at 1060 nm then

$$\frac{x}{10.05} = \left(\frac{694.3}{1060}\right)^4 \quad \text{or} \quad x = 1.85\%.$$

► **Example 4.** *The static dielectric constant of diamond is* 5.50 *and its refractive index at the wavelength* 589.3 nm *is* 2.417. *Fit these data into Sellmeier's dispersion formula with a single natural frequency. Find the free space wavelength corresponding to this natural frequency.*

Solution. Sellmeier's dispersion formula with a single natural frequency is

$$n^2 = 1 + \frac{A_0\lambda^2}{\lambda^2 - \lambda_0^2} = 1 + \frac{A_0}{1 - \frac{\lambda_0^2}{\lambda^2}}$$

In the static case $\omega \to 0$ or $\lambda \to \infty$ and we have $n^2 = 1 + A_0$.

Therefore, $n^2 = \dfrac{\varepsilon}{\varepsilon_0} = 5.50 = 1 + A_0$ or $A_0 = 4.50$

For $\lambda = 589.3$ nm,

$$(2.417)^2 = 1 + \frac{A_0}{1 - \left(\frac{\lambda_0}{589.3}\right)^2}.$$

Putting the value of A_0 we get $\lambda_0 = 156$ nm.

QUESTIONS

1. (a) What do you mean by normal and anomalous dispersion?

OR

Draw the variation of refractive index of a medium with wavelength. Hence explain normal and anomalous dispersion. **[C.U. 2010]**

(b) A monochromatic plane *EM* wave of angular frequency ω is incident on a gaseous medium containing N molecules per unit volume and each molecule having f_k electrons with natural frequency ω_k and damping constant γ_k. Set up the equation of motion for the bound electrons. Mention the underlying assumptions. Show that the dielectric constant of the medium is given by

$$K = \frac{\varepsilon}{\varepsilon_0} = 1 + \frac{Ne^2}{m\varepsilon_0}\sum_k \frac{f_k}{\omega_k^2 - \omega^2 - j\gamma_k\omega},$$ **[cf. C.U. 2014, 2015]**

where the symbols have their usual meaning.

(b) The refractive index n for a gaseous system is given by Eq. (17.2-12). Show that away from resonance it reduces to the Cauchy's equation (17.2-15). **[cf. C.U. 2015]**

2. Suppose the complex dielectric constant of a dilute gaseous medium is given as found in question 1 above.

(a) From this expression find the refraction index and absorption coefficient of the medium.

(b) Show graphically the variations of these quantities with frequency. Hence explain the phenomenon of anomalous dispersion. *An optically transparent material cannot have a resonant frequency in the optical band.*—Justify.

(c) Assuming the case of a single resonant frequency ω_0 and $\gamma_0 \ll \omega_0$ show that the refractive index assumes its maximum and minimum values at points where the absorption coefficient is at half-maximum.

(d) Show that the maximum and minimum values of refractive index are separated by a distance equal to γ_0 on the frequency scale. **[cf. C.U. 2015]**

3. In a dielectric medium the refractive index at a frequency greater than the natural resonant frequency of the oscillators of the medium becomes less than 1. Consequently phase velocity becomes greater than the velocity of light in free space (c). How do you explain this? Assuming no damping show explicitly that group velocity is smaller than c. **[cf. C.U. 2008]**

4. (a) Suppose the complex dielectric constant of a dielectric medium is given as found in question 1 above. From this deduce Sellmeier's dispersion formula:

$$n^2 = 1 + \sum_k \frac{A_k\lambda^2}{\lambda^2 - \lambda_k^2},$$

where A_k are constants and λ_k is the vacuum wavelength associated with the natural frequency ω_k.

(b) Show that under suitable conditions Sellmeier's formula reduces to Cauchy's dispersion formula. **[cf. B.U. 2015]**

5. The dielectric constant of water under static condition is about 80. So we expect its refractive index (n) to be about $\sqrt{80} \approx 8.94$. But from optics we know that $n \approx 1.33$. Explain the discrepancy.

6. On the basis of Lorentz theory of dispersion show that for a transparent dielectric solid at a particular frequency,

$$\frac{n^2 - 1}{n^2 + 2} \cdot \frac{M}{\rho} = \text{constant},$$

where n is the refractive index, ρ is the density and M is the molecular weight of the dielectric.

7. (a) What do you mean by the scattering of light and scattering cross-section?

   (b) Derive the formula for the differential and total scattering cross-sections for the scattering of EM waves by a harmonically bound electron. You may assume the formula for the time average power radiated per unit solid angle by an oscillating dipole. Hence obtain expressions for the scattering cross-sections for Rayleigh scattering.

   OR, Briefly outline the theory of scattering of electromagnetic radiation by a bound electron and use it to derive the condition for Rayleigh scattering. **[B.U. 2014]**

8. Distinguish between:

   (a) Thomson scattering and Rayleigh scattering;

   (b) Rayleigh scattering and Mie scattering.

9. What is differential scattering cross-section? Assuming the time average radiated power per unit solid angle by an oscillating dipole, obtain the expression for the total scattering cross-section in case of Thomson scattering. **[B.U. 2012, 2016]**

10. On the basis of Rayleigh scattering explain why

    (a) the sky appears blue on a clear sunny day;

    (b) the rising and setting sun is coloured crimson red; **[C.U. 2009, 2013]**

    (c) red light is used for danger signals even though our eye is most sensitive to yellow-green colour.

11. (a) In which direction is the light totally linearly polarized in Rayleigh scattering? **[C.U. 2011]**

    (b) How is it possible to get plane polarized light by using scattering phenomenon?

12. (a) *Clear sky appears blue but a thin cloud appears white.* Explain these on the basis of scattering.

(b) *Without an atmosphere the day time sky would appear black.*—Explain.

[C.U. 2011]

EXERCISES

1. Intensity of light scattered by a polarizable substance is found to be 10 arbitrary units at a wavelength 546 nm. What would be the scattered intensity at a wavelength 436 nm? [*Ans.* 24.59 units]

2. 10% of the incident sunlight is lost at wavelength 690 nm due to Rayleigh scattering in the atomsphere. What per cent of sunlight is so lost at the wavelength $\lambda = 450$ nm? **[B.U. 2014]**

3. A plane EM wave travelling through a transparent medium is given by

$$\vec{E} = \vec{E}_0 \cos\left(0.4\pi \times 10^7 x - 6\pi \times 10^{14} t\right)$$

in SI units. Determine the refractive index of the medium. [*Ans.* 2]

4. The constants A and B in the Cauchy's dispersion formula $n = A + B/\lambda^2$, for a certain glass are 1.5020 and 5.2×10^3 nm^2. Find the refractive index and dispersion of the glass at a wavelength 500 nm. [*Ans.* $1.5228, 8.32 \times 10^{-5}\text{nm}^{-1}$]

5. Taking first two terms of the Cauchy's equation of refractive index show that dispersion varies inversely as the cube of wavelength. **[B.U. 2013]**

OBJECTIVE TEST

1. A pulse of EM radiation gets broadened as it passes through a normally dispersive medium. This is because

(a) the medium amplifies the pulse,
(b) HF components travel faster than LF components,
(c) LF components travel faster than HF components,
(d) all frequency components travel at the same speed.

2. Normally the refractive index of a transparent medium

(a) is independent of frequency,
(b) decreases with increases in wavelength,

(c) increases with increases in wavelength,

(d) at first decreases and then increases with frequency.

3. For anomalous dispersion refractive index

(a) increases with increase in frequency,

(b) decreases with increase in frequency,

(c) does not depend on frequency,

(d) at first increases and then decreases with frequency.

4. Rayleigh scattering is the scattering of EM waves by

(a) free electrons,

(b) spherical particles of any size,

(c) bound electrons,

(d) both free and bound electrons.

5. Thomson scattering cross-section

(a) is independent of frequency,

(b) depends on the fourth power of frequency,

(c) depends on the square of frequency,

(d) none of these.

6. Regarding phase velocity (v_p) of EM waves in dielectric medium, which of the following statements is true?

(a) v_p is always less than c,

(b) v_p may be greater or smaller than c,

(c) v_p is always equal to c,

(d) Energy propagates with the velocity v_p.

7. An EM wave passing through vacuum is described by $\vec{E} = \vec{E}_0 e^{j(\vec{k}\cdot\vec{r}-\omega t)}$, which of the following is independent of the wavelength?

(a) ω, (b) k, (c) ωk, (d) ω/k.

ANSWERS

1. (c) **2.** (b) **3.** (b) **4.** (c) **5.** (a) **6.** (b) **7.** (d)

Appendices

Appendix A

Units and Dimensions for Electromagnetic Quantities

In this book we have used throughout the well-known *SI* system of units. But if one examines the history of development of electricity and magnetism one would find the use of various other system of units at various stages of development of the subject. The electrostatic system of units (*esu*) was introduced to suit the formulation of electrostatics and the electromagnetic system of units (*emu*) was introduced to suit the formulation of magnetism, a field that was initially studied separately from electricity. Later, when the connection between electricity and magnetism was established, the subject of electromagnetism emerged. The need for a new system of units was felt. In this way Gaussian and *SI* systems appeared. Gaussian system is a hybrid one, which uses *esu* for electrical quantities and *emu* for magnetic quantities. *SI* system uses the practical units of different quantities and hence, is found to be convenient for use in large-scale practical phenomena and in engineering applications. This *SI* system is being extensively used in electricity and magnetism particularly by electrical engineers. On the other hand, the Gaussian system is widely used in other branches of physics. The electromagnetic interactions play some important roles in the other branches of physics and hence, a physicist must be familiar with the formulation of electricity and magnetism in both the *SI* and Gaussian systems and should be able to convert from *SI* to Gaussian system and vice versa.

All the electromagnetic quantities cannot be expressed in terms of the three fundamental units, namely mass (M), length (L) and time (T) alone and we require a fourth fundamental unit. Consequently, different system of units were developed in electromagnetism. In order to see the origin of different system of units in electromagnetism let us examine the Coulomb's law of force in electrostatics and the magnetic force between two current carrying wires per unit length:

$$F_e = k_e \cdot \frac{q_1 q_2}{r^2} \quad \text{and} \quad \frac{dF_m}{dl} = k_m \cdot \frac{2I_1 I_2}{r}, \qquad \text{(A-1)}$$

where q and I stand for charge and current, F for force, r for distance, l for length; k_e and k_m are proportionality constants whose numerical values and dimensions can be chosen properly to develop a particular system of units. However, there is a restriction that $k_e/k_m = c^2$, the square of the speed of light in free space. This restriction is necessary, otherwise the same physical quantity will have different dimensions in different systems. Also, it is a direct consequence of the field equations that the quantity $\sqrt{k_e/k_m} = 1/\sqrt{\varepsilon_0\mu_0}$ shall have the dimensions of a velocity. In fact, the velocity of light in free space (c) appears as a fundamental constant in electromagnetic theory.

Note that to define charge q or current I from Eq. (A-1) we require to know or fix the numerical values and dimensions of k_e and k_m. This freedom of choice led to different system of units in electromagnetism.

In the electrostatic system of units (*esu*) the units of mass, length and time are chosen as in CGS system and here one chooses k_e as dimensionless and equal to unity (i.e., $k_e = 1$) and $k_m = 1/c^2$. The two force laws then become

$$F_e = \frac{q_1 q_2}{r^2} \quad \text{and} \quad \frac{dF_m}{dl} = \frac{2I_1 I_2}{c^2 r} \quad \text{(in esu)} \qquad \text{(A-2)}$$

The first equation is used to define the unit of charge in this system. It is known as *statcoulomb* and defined as a charge such that two such charges at 1 cm apart in air exert a force of 1 dyne upon each other. The equation also gives the dimensions of charge q as $\left[M^{1/2}L^{3/2}T^{-1}\right]$. The dimensions of all other electromagnetic quantities can then be found by considering the appropriate fundamental equations.

In the electromagnetic system of units (*emu*) the units of mass length and time are chosen as in CGS system and here one chooses k_m as dimensionless and equal to unity (i.e., $k_m = 1$) and $k_e = c^2$. The two force laws then become

$$F_e = c^2 \cdot \frac{q_1 q_2}{r^2} \quad \text{and} \quad \frac{dF_m}{dl} = \frac{2I_1 I_2}{r} \quad \text{(in emu)} \qquad \text{(A-3)}$$

The second equation is used to define the unit of current in this system. It is known as *abampere*. The equation also gives the dimensions of current I as $\left[M^{1/2}L^{1/2}T^{-1}\right]$. The dimensions of all other electromagnetic quantities can then be found by considering the appropriate fundamental equations. Comparing the dimensions of any electromagnetic quantity in *esu* and *emu*, it is seen that they are different. But it is unphysical to have different dimensions for the same quantity. Thus, we are forced to consider dimensions of k_e and k_m (or, equivalently of ε_0 and μ_0) and we take $k_e/k_m = c^2$ or equivalently $1/\varepsilon_0\mu_0 = c^2$.

Historically *esu* was used primarily for electrostatic problems and *emu* for electromagnetic problems. Subsequently, a hybrid system, known as Gaussian system, was developed in which electrical quantities ($\rho, \vec{J}, \vec{E}$ etc.) are expressed in *esu* and magnetic quantities ($\vec{B}, \vec{H}$ etc.) in *emu*. Like *esu* and *emu* systems the Gaussian system uses the CGS units of mass, length and time. Thus it is also a CGS system. The other principal point of contact between *esu* and *emu* systems is the current density $\vec{J}$, where $\vec{J}_{emu} = \vec{J}_{esu}/c$. In Gaussian system one uses $\vec{J}_{esu}$ and naturally the velocity of light c appears explicitly in electromagnetic equations. Thus, the two force laws in Gaussian system appear as

$$F_e = \frac{q_1 q_2}{r^2} \quad \text{and} \quad \frac{dF_m}{dl} = \frac{2I_1 I_2}{c^2 r} \tag{A-4}$$

In Gaussian units, Maxwell's equations are

$$\begin{gathered} \vec{\nabla}\cdot\vec{D} = 4\pi\rho, \quad \vec{\nabla}\cdot\vec{B} = 0 \\ \vec{\nabla}\times\vec{E} + \frac{1}{c}\frac{\partial\vec{B}}{\partial t} = 0, \quad \vec{\nabla}\times\vec{H} = \frac{4\pi}{c}\vec{J} + \frac{1}{c}\frac{\partial\vec{D}}{\partial t} \end{gathered} \tag{A-5}$$

The field vectors are related to the potentials in this system as

$$\vec{B} = \vec{\nabla}\times\vec{A} \quad \text{and} \quad \vec{E} = -\vec{\nabla}\phi - \frac{1}{c}\frac{\partial\vec{A}}{\partial t} \tag{A-6}$$

The field vectors $\vec{D}$ and $\vec{B}$ are related to $\vec{E}$ and $\vec{H}$ by

$$\vec{D} = \vec{E} + 4\pi\vec{P} \quad \text{and} \quad \vec{B} = \vec{H} + 4\pi\vec{M}. \tag{A-7}$$

The constitutive relations in simple materials are

$$\vec{D} = \varepsilon\vec{E}, \quad \vec{B} = \mu\vec{H}, \quad \vec{J} = \sigma\vec{E}, \quad \vec{P} = \mathcal{X}\vec{E} \quad \text{and} \quad \vec{M} = \mathcal{X}_m\vec{H}. \tag{A-8}$$

Thus, we have the relations

$$\varepsilon = 1 + 4\pi\mathcal{X} \quad \text{and} \quad \mu = 1 + 4\pi\mathcal{X}_m \tag{A-9}$$

in Gaussian units.

The Lorentz force law is

$$\vec{F} = q\left(\vec{E} + \frac{\vec{v}}{c}\times\vec{B}\right). \tag{A-10}$$

The electromagnetic energy density associated with the fields and the Poynting's vector in this system are

$$u = \frac{1}{8\pi}\left(\vec{E}\cdot\vec{D} + \vec{B}\cdot\vec{H}\right) \quad \text{and} \quad \vec{s} = \frac{c}{4\pi}\left(\vec{E}\times\vec{H}\right). \tag{A-11}$$

The continued use of Gaussian system in many branches of physics may be probably due to the following advantages of the system:

(i) The factor $1/4\pi\varepsilon_0$ does not occur in Coulomb's law and in many equations of atomic physics.

(ii) The velocity v always enters in the dimensionless form v/c as prescribed by Lorentz transformation.

(iii) In relativity it is advantageous that in Gaussian system $\vec{E}$ and $\vec{B}$ have the same dimensions.

(iv) It is convenient that the permittivity and susceptibility of a material medium are dimensionless.

In *SI* system of units (also, known as rationalized MKSA system) the units of mass, length and time are chosen as in MKS system. Here we do not choose a dimensionless constant in the two force laws, instead choose one of the electrical quantities, current (I) arbitrarily and assign to it a dimension. Thus, in SI system one chooses current (I) as the fourth fundamental dimension in addition to M, L, T. In this system the two force laws become

$$F_e = \frac{1}{4\pi\varepsilon_0} \cdot \frac{q_1 q_2}{r^2} \tag{A-12}$$

$$\text{and} \quad \frac{dF_m}{dl} = \frac{\mu_0}{4\pi} \cdot \frac{2I_1 I_2}{r}. \tag{A-13}$$

The explicit use of the factor 4π in Eq. (A-13) in SI system is a convenience that eliminates the factor 4π in the Maxwell's field equations, which involve sources and makes the equations simple looking. Another object of using this factor is to avoid the occurrence of 4π in systems without spherical symmetry and of 2π in systems without cylindrical symmetry.

Here $1/4\pi\varepsilon_0$ is so chosen that $F_e = 1$ newton (N) when $r = 1\text{m}$ and $q_1 = q_2 = 1$ coulomb (C). This is found to be true if $\varepsilon_0 = 1/(36\pi \times 10^9)\, \text{C}^2/\text{N-m}^2$. As $I = dq/dt$, the dimensions of charge are $[IT]$ and that of ε_0 are given by

$$[\varepsilon_0] = \frac{[q^2]}{[L^2]\,[MLT^{-2}]} = \left[M^{-1}L^{-3}T^4I^2\right]$$

In SI system the value of μ_0 is chosen as $4\pi \times 10^{-7}$ so that $1/\sqrt{\mu_0\varepsilon_0} = 3 \times 10^8 \text{m/s}$. Then Eq. (A-13) leads to a new definition of the ampere (the practical unit of current):

One ampere is that constant current, which when flows through two long parallel straight wires separated by a distance of 1 m *from each other in free space, results in a force of* $2 \times 10^{-7}\text{N/m}$.

The dimensions of μ_0 are obtained from (A-13) as

$$[\mu_0] = \frac{[MLT^{-2}]}{[L]} \cdot \frac{[L]}{[I^2]} = \left[MLT^{-2}I^{-2}\right]$$

and the unit of μ_0 as N/A^2.

The dimensions of all other electromagnetic quantities in SI system can now be found by considering the appropriate fundamental equation:

Potential (ϕ), pd (V) or emf ($\mathcal{E}$): Work = pd × charge

$$\therefore \quad [\phi] \quad \text{or} \quad [V] \quad \text{or} \quad [\mathcal{E}] = \frac{[\text{Work}]}{[\text{Charge}]} = \frac{[MLT^{-2} \cdot L]}{[IT]} = \left[ML^2T^{-3}I^{-1}\right]$$

Electric Field ($\vec{E}$) : $\vec{E} = -\vec{\nabla}\phi$.

$$\therefore [E] = \frac{[\phi]}{[L]} = \left[MLT^{-3}I^{-1}\right]$$

Electric displacement ($\vec{D}$) : $\vec{D} = \varepsilon\vec{E}$.

$$\therefore \quad [D] = [\varepsilon]\,[E] = \left[M^{-1}L^{-3}T^4I^2\right]\left[MLT^{-3}I^{-1}\right] = \left[L^{-2}TI\right].$$

Resistance (R) : $R = \dfrac{\text{pd}}{I}$.

$$\therefore \quad [R] = \frac{[\phi]}{[I]} = \left[ML^2T^{-3}I^{-2}\right].$$

Capacitance (C) : $C = \dfrac{Q}{V}$

$$\therefore [C] = \frac{[Q]}{[V]} = \frac{[IT]}{[ML^2T^{-3}I^{-1}]} = \left[M^{-1}L^{-2}T^4I^2\right].$$

Inductance (L): emf $(\mathcal{E}) = -L\dfrac{dI}{dt}$.

$$\therefore \quad [L] = \frac{[\mathcal{E}]\cdot[T]}{[I]} = \left[ML^2T^{-2}I^{-2}\right].$$

Magnetic flux (Φ) : $\mathcal{E} = -\dfrac{d\Phi}{dt}$.

$$\therefore \quad [\Phi] = [\mathcal{E}]\,[T] = \left[ML^2T^{-2}I^{-1}\right].$$

Magnetic inductance ($\vec{B}$): Flux $\Phi = \oint \vec{B} \cdot d\vec{S}$.

$$\therefore \quad [B] = \frac{[\Phi]}{[\text{area}]} = \frac{[ML^2T^{-2}I^{-1}]}{[L^2]} = \left[MT^{-2}I^{-1}\right].$$

Magnetic field strength ($\vec{H}$) : $\vec{\nabla} \times \vec{H} = \vec{J}$.

$$\therefore \quad \frac{[H]}{[L]} = \frac{[I]}{[L^2]} \quad \text{or} \quad [H] = \left[L^{-1}I\right].$$

$\dfrac{\mathbf{1}}{\sqrt{\varepsilon_0 \mu_0}}$: Since $[\varepsilon_0] = \left[M^{-1}L^{-3}T^4I^2\right]$ and $[\mu_0] = \left[MLT^{-2}I^{-2}\right]$,

$\left[\dfrac{1}{\sqrt{\varepsilon_0 \mu_0}}\right] = \left[LT^{-1}\right]$, which has the dimensions of velocity.

$\sqrt{\dfrac{\mu_0}{\varepsilon_0}}$: $\left[\sqrt{\dfrac{\mu_0}{\varepsilon_0}}\right] = \left[\dfrac{MLT^{-2}I^{-2}}{M^{-1}L^{-3}T^4I^2}\right]^{1/2} = \left[ML^2T^{-3}I^{-2}\right]$, which is the dimensions of resistance R.

$\vec{B} \cdot \vec{H}$: Since $[B] = \left[MT^{-2}I^{-1}\right]$ and $[H] = \left[L^{-1}I\right]$,

$\left[\vec{B} \cdot \vec{H}\right] = \left[ML^{-1}T^{-2}\right] = \dfrac{[ML^2T^{-2}]}{[L^3]}$, which is the dimensions of energy density.

$\dfrac{\varepsilon}{\sigma}$: Conductivity σ is related to resistance by the relation $R = \dfrac{l}{\sigma A}$, where l is the length and A the area.

$$\therefore [\sigma] = \frac{1}{[R]\,[L]} = \frac{1}{[ML^3T^{-3}I^{-2}]} = \left[M^{-1}L^{-3}T^3I^2\right].$$

$$\therefore \left[\frac{\varepsilon}{\sigma}\right] = \frac{[\varepsilon]}{[\sigma]} = \frac{[M^{-1}L^{-3}T^4I^2]}{[M^{-1}L^{-3}T^3I^2]} = [T].$$

Thus, ε/σ has the dimensions of time.

CR : $CR = \dfrac{Q}{V} \cdot \dfrac{V}{I} = \dfrac{Q}{I}$. $\therefore$ $[CR] = \dfrac{[IT]}{[I]} = [T]$, dimensions of time.

$\dfrac{L}{R}$: $\left[\dfrac{L}{R}\right] = \dfrac{[L]}{[R]} = \dfrac{[ML^2T^{-2}I^{-2}]}{[ML^2T^{-3}I^{-2}]} = [T]$, dimensions of time.

Sometimes we require numerical conversion between Gaussian and SI units. For convenience the SI and Gaussian units of some important physical quantities and their conversion formula are given in *Table* A1.

Table A1 : Numerical relationship between SI and Gaussian Units

Quantity	SI Unit	Gaussian Unit	Conversion Formula
Length	metre (m)	Centimetre (cm)	$1\text{ m} = 10^2\text{ cm}$
Mass	kilogram (kg)	gram (g)	$1\text{ kg} = 10^3\text{ g}$
Time	second (s)	second (s)	
Force	newton (N)	dyne (dyn)	$1\text{ N} = 10^5\text{ dyn}$
Work	joule (J)	erg	$1\text{ J} = 10^7\text{ erg}$
Power	watt (W)	erg/s	$1\text{ W} = 10^7\text{ erg/s}$
Charge (Q)	coulomb (C)	statcoulomb	$1\text{ C} = 3 \times 10^9$ statcoulomb
Current (I)	ampere (A)	statampere/ab ampere (emu)	$1\text{ A} = 3 \times 10^9$ statampere $= 10^{-1}$ abampere
Electric field (E)	volt per metre (V/m)	statvolt per cm	$1\text{ V/m} = \frac{1}{3} \times 10^{-4}$ statvolt/cm
Potential (ϕ)	volt (V)	statvolt	$1\text{ V} = \frac{1}{300}$ statvolt
Displacement (D)	coulomb per metre^2	statcoulomb per cm^2	$1\text{ C/m}^2 = 12\pi \times 10^5$ stat C/cm^2
Resistance (R)	ohm (Ω)	statohm	$1\ \Omega = \frac{1}{9} \times 10^{-11}$ statohm
Capacitance (C)	farad (F)	statfarad	$1\text{ F} = 9 \times 10^{11}$ statfarad
Inductance (L)	henry (H)	stathenry	$1\text{ H} = \frac{1}{9} \times 10^{-11}$ stathenry
Magnetic flux (Φ)	weber (Wb)	maxwell (Mx)	$1\text{ Wb} = 10^8\text{ Mx}$
Magnetic Induction (B)	tesla (T)	gauss (G)	$1\text{ T} = 10^4\text{ G}$
Magnetic intensity (H)	ampere/m	oersted (Oe)	$1\text{ A/m} = 4\pi \times 10^{-3}\text{ Oe}$
Magnetization (M)	ampere/m	oersted (Oe)	

Table A2 gives the conversion table for symbols and formulas. It is helpful for quick conversion from SI to Gaussian or Gaussian to SI. Symbols for mass, length, time and other non-electromagnetic quantities are unchanged. To convert any equation in SI variables to the corresponding equation in Gaussian system, replace the relevant symbols listed under the heading "SI" by the corresponding "Gaussian" symbols listed under the heading "Gaussian". The reverse transformation is also allowed.

Table A2 : Conversion table for symbols and formulas

Quantity	SI	Gaussian
Electric field, potential, voltage	$\vec{E}$, ϕ, V	$\frac{1}{\sqrt{4\pi\varepsilon_0}}\left(\vec{E}, \phi, V\right)$
Electric displacement	$\vec{D}$	$\sqrt{\frac{\varepsilon_0}{4\pi}}\vec{D},$
Charge density, current density, current, polarization	ρ, $\vec{J}$, I, $\vec{P}$	$\sqrt{4\pi\varepsilon_0}\left(\rho, \vec{J}, I, \vec{P}\right)$
Magnetic induction, vector potential	$\vec{B}$, $\vec{A}$	$\sqrt{\frac{\mu_0}{4\pi}}\left(\vec{B}, \vec{A}\right)$
Magnetic intensity	$\vec{H}$	$\frac{1}{\sqrt{4\pi\mu_0}}\vec{H}$
Magnetization	$\vec{M}$	$\sqrt{\frac{4\pi}{\mu_0}}\vec{M}$
Resistance, Inductance	R, L	$\frac{1}{4\pi\varepsilon_0}(R, L)$
Capacitance	C	$4\pi\varepsilon_0 C$
Permittivity	ε	$K\varepsilon_0$
Permeability	μ	$K_m\mu_0$
Speed of light	$\frac{1}{\sqrt{\mu_0\varepsilon_0}}$	c
Susceptibility	$\mathcal{X}, \mathcal{X}_m$	$4\pi(\mathcal{X}, \mathcal{X}_m)$

For convenience we show some important equations of electromagnetism with SI and Gaussian units side-by-side in *Table* A3.

Table A3: Some important equations of electromagnetism with SI and Gaussian units

Equation	**SI**	**Gaussian**
Coulomb's law	$\vec{F}_e = \frac{1}{4\pi\varepsilon_0}\frac{q_1 q_2}{r^2}\hat{r}$	$\vec{F}_e = \frac{q_1 q_2}{r^2}\hat{r}$
Magnetic force per unit length on parallel current	$\frac{d\vec{F}_m}{dl} = \frac{\mu_0}{4\pi}\cdot\frac{2I_1 I_2}{r}\hat{r}$	$\frac{d\vec{F}_m}{dl} = \frac{2I_1 I_2}{c^2 r}\hat{r}$
Magnetic force on a current element	$d\vec{F} = I\left(d\vec{l}\times\vec{B}\right)$	$d\vec{F} = \frac{I}{c}\left(d\vec{l}\times\vec{B}\right)$
Magnetic field due to a current element	$d\vec{B} = \frac{\mu_0}{4\pi}\frac{Id\vec{l}\times\vec{r}}{r^3}$	$d\vec{B} = \frac{Id\vec{l}\times\vec{r}}{cr^3}$
Maxwell's equation	$\vec{\nabla}\cdot\vec{D} = \rho$	$\vec{\nabla}\cdot\vec{D} = 4\pi\rho$
Maxwell's equation	$\vec{\nabla}\cdot\vec{B} = 0$	$\vec{\nabla}\cdot\vec{B} = 0$
Maxwell's equation	$\vec{\nabla}\times\vec{E} + \frac{\partial\vec{B}}{\partial t} = 0$	$\vec{\nabla}\times\vec{E} + \frac{1}{c}\frac{\partial\vec{B}}{\partial t} = 0$
Maxwell's equation	$\vec{\nabla}\times\vec{H} = \vec{J} + \frac{\partial\vec{D}}{\partial t}$	$\vec{\nabla}\times\vec{H} = \frac{4\pi\vec{J}}{c} + \frac{1}{c}\frac{\partial\vec{D}}{\partial t}$
Lorentz force	$\vec{F} = q\left(\vec{E} + \vec{v}\times\vec{B}\right)$	$\vec{F} = q\left(\vec{E} + \frac{\vec{v}}{c}\times\vec{B}\right)$
Field potential relations	$\vec{B} = \vec{\nabla}\times\vec{A}$	$\vec{B} = \vec{\nabla}\times\vec{A}$
Field potential relations	$\vec{E} = -\vec{\nabla}\phi - \frac{\partial\vec{A}}{\partial t}$	$\vec{E} = -\vec{\nabla}\phi - \frac{1}{c}\frac{\partial\vec{A}}{\partial t}$
Relation between $\vec{D}$, $\vec{E}$ and $\vec{P}$	$\vec{D} = \varepsilon_0\vec{E} + \vec{P}$	$\vec{D} = \vec{E} + 4\pi\vec{P}$
Relation between $\vec{H}$, $\vec{B}$ and $\vec{M}$	$\vec{B} = \mu_0\left(\vec{H} + \vec{M}\right)$	$\vec{B} = \vec{H} + 4\pi\vec{M}$
Poynting's vector	$\vec{s} = \vec{E}\times\vec{H}$	$\vec{s} = \frac{c}{4\pi}\left(\vec{E}\times\vec{H}\right)$
Electromagnetic energy density	$u = \frac{1}{2}\left(\vec{E}\cdot\vec{D} + \vec{B}\cdot\vec{H}\right)$	$u = \frac{1}{8\pi}\left(\vec{E}\cdot\vec{D} + \vec{B}\cdot\vec{H}\right)$

PROBLEMS

1. Find the dimensions of $\vec{B}$ and μ_0. Show that the equation $\vec{\nabla} \times \vec{B} = \mu_0 \vec{J}$ is dimensionally correct.

2. Find the dimensions of resistance and $\sqrt{\mu_0/\varepsilon_0}$ and hence, find the value of $\sqrt{\mu_0/\varepsilon_0}$ in proper units.

3. Find the dimension of the quantity ε/σ, where ε is the permittivity and σ is the conductivity of a medium.

4. Suppose we develop a particular system of units with a choice $\varepsilon_0 = 3 \times 10^{-11}$ units, keeping the mechanical units same as in MKS system. What should be the value of μ_0 in this particular system?

   [**Hints.** Use $\frac{1}{\sqrt{\mu_0 \varepsilon_0}} = c$.]

Appendix **B**

Some Useful Formulas from Vector Analysis

It is expected that the students using this book on electromagnetism are acquainted with vector analysis. However, in this appendix we quote some results from vector analysis as a ready reference.

B.1 Dot or scalar product

$\vec{A} \cdot \vec{B} = AB\cos\theta$, where θ is the angle between the vectors $\vec{A}$ and $\vec{B}$.

If $\hat{i}$, $\hat{j}$ and $\hat{k}$ be unit vectors along the Cartesian x-, y- and z-axes, respectively then the vectors $\vec{A}$ and $\vec{B}$ can be expanded in terms of their components as

$$\vec{A} = \hat{i}A_x + \hat{j}A_y + \hat{k}A_z \quad \text{and} \quad \vec{B} = \hat{i}B_x + \hat{j}B_y + \hat{k}B_z.$$

Since $\hat{i}$, $\hat{j}$ and $\hat{k}$ are mutually perpendicular unit vectors,

$$\hat{i}\cdot\hat{i} = \hat{j}\cdot\hat{j} = \hat{k}\cdot\hat{k} = 1 \quad \text{and} \quad \hat{i}\cdot\hat{j} = \hat{j}\cdot\hat{k} = \hat{k}\cdot\hat{i} = 0.$$

Accordingly,

$$\vec{A}\cdot\vec{B} = A_xB_x + A_yB_y + A_zB_z = \vec{B}\cdot\vec{A}.$$

B.2 Cross or vector product

$$\vec{A} \times \vec{B} = AB \sin\theta\, \hat{n},$$

where $\hat{n}$ is a unit vector normal to the plane containing $\vec{A}$ and $\vec{B}$; it is in the direction of motion of a right-handed screw when rotated from the direction of $\vec{A}$ to that of $\vec{B}$.

$$\hat{\imath} \times \hat{\imath} = \hat{\jmath} \times \hat{\jmath} = \hat{k} \times \hat{k} = 0$$

$$\hat{\imath} \times \hat{\jmath} = \hat{k}, \quad \hat{\jmath} \times \hat{k} = \hat{\imath}, \quad \hat{k} \times \hat{\imath} = \hat{\jmath}$$

Accordingly,

$$\vec{A} \times \vec{B} = \begin{vmatrix} \hat{\imath} & \hat{\jmath} & \hat{k} \\ A_x & A_y & A_z \\ B_x & B_y & B_z \end{vmatrix} \quad \text{and} \quad \vec{A} \times \vec{B} = -\vec{B} \times \vec{A}.$$

Unit vector normal to the plane containing $\vec{A}$ and $\vec{B}$ is given by

$$\hat{n} = \pm \frac{\vec{A} \times \vec{B}}{\left|\vec{A} \times \vec{B}\right|}.$$

B.3 Triple product

$$\vec{A} \cdot \left(\vec{B} \times \vec{C}\right) = \vec{B} \cdot \left(\vec{C} \times \vec{A}\right) = \vec{C} \cdot \left(\vec{A} \times \vec{B}\right)$$

$$\vec{A} \times \left(\vec{B} \times \vec{C}\right) = \vec{B}\left(\vec{A} \cdot \vec{C}\right) - \vec{C}\left(\vec{A} \cdot \vec{B}\right)$$

B.4 Gradient of a scalar function

The gradient of a scalar function $\phi(\vec{r})$ in Cartesian coordinates (x, y, z) is defined by

$$\vec{\nabla}\phi = \left(\hat{\imath}\frac{\partial}{\partial x} + \hat{\jmath}\frac{\partial}{\partial y} + \hat{k}\frac{\partial}{\partial z}\right)\phi = \hat{\imath}\frac{\partial \phi}{\partial x} + \hat{\jmath}\frac{\partial \phi}{\partial y} + \hat{k}\frac{\partial \phi}{\partial z}$$

The differential change in the function ϕ can be expressed as

$$d\phi = \frac{\partial \phi}{\partial x}dx + \frac{\partial \phi}{\partial y}dy + \frac{\partial \phi}{\partial z}dz = \vec{\nabla}\phi \cdot d\vec{r} \quad \left[\because \quad d\vec{r} = \hat{\imath}dx + \hat{\jmath}dy + \hat{k}dz\right].$$

$\vec{\nabla}\phi$ is a vector perpendicular to the surface represented by ϕ =constant. Thus,

$$\vec{\nabla}\phi \Big/ \left|\vec{\nabla}\phi\right| = \hat{n}$$

is a unit vector normal to the surface $\phi =$ constant. $\vec{\nabla}\phi$ represents the maximum rate of change of ϕ with respect to space variables and has the direction of that change.

B.5 Divergence of a vector field

The divergence of a vector field $\vec{A}(\vec{r})$, written as $\vec{\nabla}\cdot\vec{A}$ or div $\vec{A}$, is defined by

$$\vec{\nabla}\cdot\vec{A} = \left(\hat{i}\frac{\partial}{\partial x} + \hat{j}\frac{\partial}{\partial y} + \hat{k}\frac{\partial}{\partial z}\right)\cdot\left(\hat{i}A_x + \hat{j}A_y + \hat{k}A_z\right)$$

$$= \frac{\partial A_x}{\partial x} + \frac{\partial A_y}{\partial y} + \frac{\partial A_z}{\partial z} \text{ (in Cartesian coordinates)}$$

It is shown that

$$\vec{\nabla}\cdot\vec{A} = \lim_{\Delta V\to 0}\frac{\oint_{\Delta S}\vec{A}\cdot d\vec{S}}{\Delta V},$$

where ΔV is a small volume enclosed by the surface ΔS. Thus, physically $\vec{\nabla}\cdot\vec{A}$ at a given point represents the limit of net flux or out flow of $\vec{A}$ per unit volume. It is a measure of how much the vector $\vec{A}$ spreads out (diverges) at the point in question. A point of large positive divergence is called a source [Fig. B-1(a)] and a point of large negative divergence is a sink [Fig. B-1(b)]

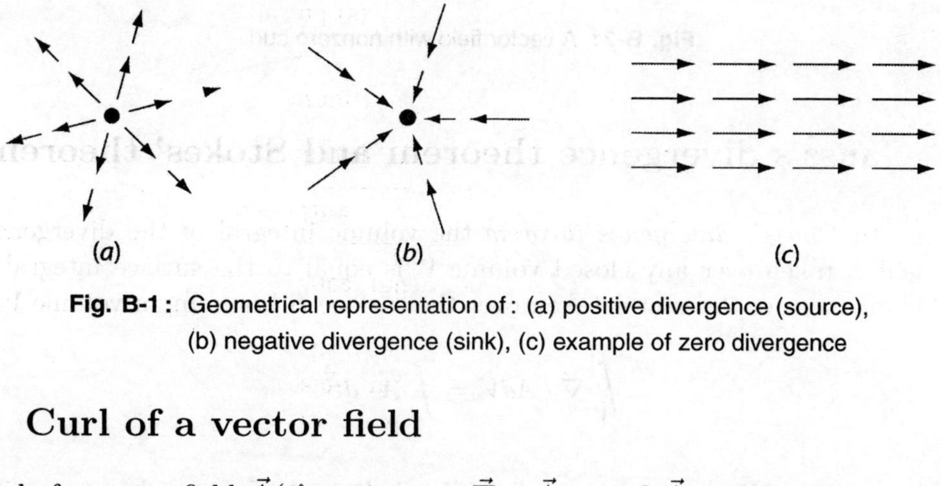

Fig. B-1 : Geometrical representation of : (a) positive divergence (source), (b) negative divergence (sink), (c) example of zero divergence

B.6 Curl of a vector field

The curl of a vector field $\vec{A}(\vec{r})$, written as $\vec{\nabla}\times\vec{A}$ or curl $\vec{A}$, is defined by

$$\vec{\nabla}\times\vec{A} = \left(\hat{i}\frac{\partial}{\partial x} + \hat{j}\frac{\partial}{\partial y} + \hat{k}\frac{\partial}{\partial z}\right)\times\left(\hat{i}A_x + \hat{j}A_y + \hat{k}A_z\right)$$

$$= \begin{vmatrix} \hat{i} & \hat{j} & \hat{k} \\ \dfrac{\partial}{\partial x} & \dfrac{\partial}{\partial y} & \dfrac{\partial}{\partial z} \\ A_x & A_y & A_z \end{vmatrix} \text{ (in Cartesian coordinates)}$$

It is shown that

$$\left(\vec{\nabla}\times\vec{A}\right)\cdot\hat{n} = \lim_{\Delta S\to 0}\frac{\oint_C\vec{A}\cdot d\vec{l}}{\Delta S},$$

where C is the contour bounding the small surface ΔS. Thus, the normal component of $\vec{\nabla} \times \vec{A}$ physically represents the limit of circulation per unit area. The circulation is greatest when the normal to the surface element is along the direction of $\vec{\nabla} \times \vec{A}$. Thus, physically $\vec{\nabla} \times \vec{A}$ is the greatest circulation per unit area and is directed along the normal to the exploring area. The name 'curl' is justified in the sense that geometrically $\vec{\nabla} \times \vec{A}$ is a measure of how much the vector $\vec{A}$ curls around the point in question. As a physical example, suppose there is turbulent water flow along a river. If a paddle-wheel is floated on the water it is found to rotate as it flows down stream. Then we say the velocity vector field has a nonzero curl.

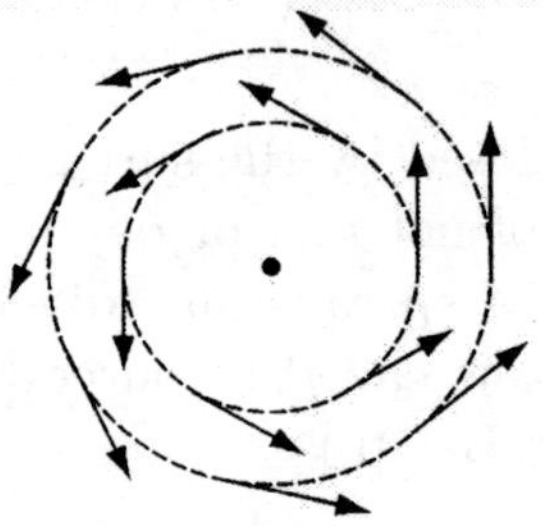

Fig. B-2: A vector field with nonzero curl

B.7 Gauss's divergence theorem and Stokes' theorem

According to *Gauss's divergence theorem* the volume integral of the divergence of a vector field $\vec{A}$ taken over any closed volume V is equal to the surface integral of the normal component of the vector taken over the surface S bounding a volume V, i.e.,

$$\int_V \vec{\nabla} \cdot \vec{A} dV = \oint_S \vec{A} \cdot d\vec{S}.$$

Note that here $d\vec{S}$ represents an infinitesimal element of area, whose magnitude is equal to the area of the element and whose direction is normal to the surface pointing outward.

According to *Stokes' theorem* the surface integral of the normal component of the curl of any vector $\vec{A}$ over any (open) surface S bonded by the curve C is equal to the line integral of $\vec{A}$ around the periphery of the surface, i.e.,

$$\int_S \left(\vec{\nabla} \times \vec{A}\right) \cdot d\vec{S} = \oint_C \vec{A} \cdot d\vec{l},$$

where $d\vec{l}$ is a line element taken in the sense of rotation of a right-handed screw, the tip of which advances in the direction of positive normal to the surface.

B.8 Some important vector identities

1. $\vec{\nabla} \times \vec{\nabla}\phi = 0.$

2. $\vec{\nabla} \cdot \left(\vec{\nabla} \times \vec{A}\right) = 0.$

3. $\vec{\nabla} \times \left(\vec{\nabla} \times \vec{A}\right) = \vec{\nabla}\left(\vec{\nabla} \cdot \vec{A}\right) - \nabla^2 \vec{A}.$

4. $\vec{\nabla} \cdot \left(\phi\vec{A}\right) = \phi\vec{\nabla} \cdot \vec{A} + \vec{\nabla}\phi \cdot \vec{A}.$

5. $\vec{\nabla} \times \left(\phi\vec{A}\right) = \phi\vec{\nabla} \times \vec{A} + \vec{\nabla}\phi \times \vec{A}.$

6. $\vec{\nabla} \cdot \left(\vec{A} \times \vec{B}\right) = \vec{B} \cdot \left(\vec{\nabla} \times \vec{A}\right) - \vec{A} \cdot \left(\vec{\nabla} \times \vec{B}\right).$

7. $\vec{\nabla} \cdot \left(\vec{A} \times \vec{B}\right) = \vec{A}\left(\vec{\nabla} \cdot \vec{B}\right) - \vec{B}\left(\vec{\nabla} \cdot \vec{A}\right) + \left(\vec{B} \cdot \vec{\nabla}\right)\vec{A} - \left(\vec{A} \cdot \vec{\nabla}\right)\vec{B}.$

8. $\vec{\nabla}\left(\vec{A} \cdot \vec{B}\right) = A \times \left(\vec{\nabla} \times \vec{B}\right) + \vec{B} \times \left(\vec{\nabla} \times \vec{A}\right) + \left(\vec{B} \cdot \vec{\nabla}\right)\vec{A} + \left(\vec{A} \cdot \vec{\nabla}\right)\vec{B}.$

9. $\vec{\nabla} \cdot \vec{\nabla}\phi = \nabla^2\phi.$

10. $\vec{\nabla}(\phi\psi) = \phi\vec{\nabla}\psi + \psi\vec{\nabla}\phi.$

B.9 Some important relations

If $\vec{r}$ be the radius vector from origin to the point (x, y, z) then

1. $\vec{\nabla} \cdot \vec{r} = 3.$

2. $\vec{\nabla} \times \vec{r} = 0.$

3. $\vec{\nabla}\frac{1}{r} = -\frac{\vec{r}}{r^3}, \ \vec{\nabla}\frac{1}{r^3} = -\frac{3\vec{r}}{r^5}.$

4. $\nabla^2\frac{1}{r} = 0$ for $r \neq 0.$

If V is a volume bounded by a closed surface S then

5. $\oint_S \phi d\vec{S} = \int_V \left(\vec{\nabla}\phi\right) dV.$

6. $\oint_S \hat{n} dS = 0.$

7. $\oint_S \vec{r} \times d\vec{S} = 0.$

8. $\int_V \left(\vec{\nabla}\times\vec{A}\right) dV = \oint_S d\vec{S}\times\vec{A}.$

If S is an open surface bounded by the contour C then

9. $\oint_C \phi d\vec{r} = \int_S d\vec{S}\times\vec{\nabla}\phi.$

10. $\oint_C \vec{r}\times d\vec{r} = 2\int_S d\vec{S}.$

B.10 Basic unit vectors, position vector, velocity and acceleration in different coordinate systems

The cylindrical coordinates (r, θ, z) are related to rectangular Cartesian coordinates by [Fig. B-3(a)]

$$x = r\cos\theta, \quad y = r\sin\theta, \quad z = z$$

where r is the distance from z-axis to the point of interest P. Note that here $r \neq |\vec{r}|$, $\vec{r} = \overrightarrow{OP}$ being the position vector.

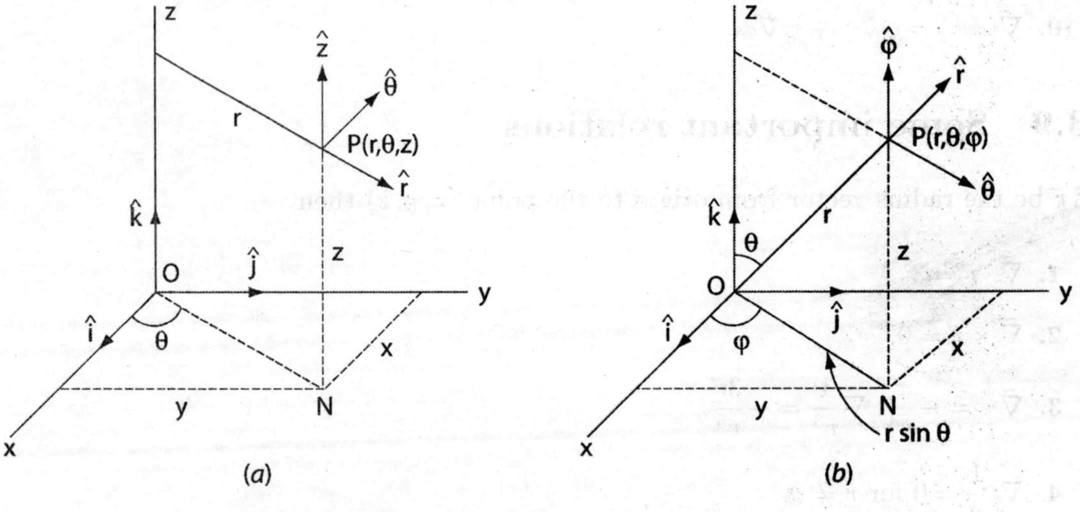

Fig. B-3: (a) Cylindrical and (b) Spherical coordinates systems

From geometry of Fig. B-3(a) we can write

$$\hat{r} = \cos\theta\,\hat{\imath} + \sin\theta\,\hat{\jmath}$$
$$\hat{\theta} = -\sin\theta\,\hat{\imath} + \cos\theta\,\hat{\jmath}$$
$$\hat{z} = \hat{k}$$

These equations can be solved simultaneously. Thus, one gets

$$\hat{i} = \cos\theta\,\hat{r} - \sin\theta\,\hat{\theta}$$
$$\hat{j} = \sin\theta\,\hat{r} + \cos\theta\,\hat{\theta}$$
$$\hat{k} = \hat{z}$$

Position vector

$$\begin{aligned}\vec{r} = \overrightarrow{OP} &= x\hat{i} + y\hat{j} + z\hat{k}\\ &= r\cos\theta\hat{i} + r\sin\theta\hat{j} + z\hat{k}\\ &= r\hat{r} + z\hat{z}.\end{aligned}$$

Line element

$$d\vec{r} = dr\hat{r} + rd\theta\hat{\theta} + dz\hat{z}.$$

Velocity

$$\vec{v} = \frac{d\vec{r}}{dt} = \dot{r}\hat{r} + r\dot{\theta}\hat{\theta} + \dot{z}\hat{z}.$$

Acceleration

$$\vec{a} = \frac{d\vec{v}}{dt} = \left(\ddot{r} - r\dot{\theta}^2\right)\hat{r} + \left(r\ddot{\theta} + 2\dot{r}\dot{\theta}\right)\hat{\theta} + \ddot{z}\hat{z}.$$

Note that putting $z = 0$, one can get the corresponding results in plane polar coordinates (r, θ). Also note that $\hat{r}$, $\hat{\theta}$ and $\hat{z}$ form a rectangular system.

$$\hat{r}\cdot\hat{\theta} = \hat{\theta}\cdot\hat{z} = \hat{z}\cdot\hat{r} = 0 \quad \text{and} \quad \hat{r}\times\hat{\theta} = \hat{z},\ \hat{\theta}\times\hat{z} = \hat{r},\ \hat{z}\times\hat{r} = \hat{\theta}.$$

The spherical coordinates (r, θ, φ) are related to the rectangular Cartesian coordinates by [Fig. B-3(b)]

$$x = r\sin\theta\cdot\cos\varphi, \quad y = r\sin\theta\sin\varphi, \quad z = r\cos\theta,$$

where $r = |\vec{r}|$, $\vec{r}$ being the position vector $\overrightarrow{OP}$.

From geometry of Fig. B-3(b) we can write

$$\hat{r} = \sin\theta\cos\varphi\hat{i} + \sin\theta\sin\varphi\hat{j} + \cos\theta\hat{k}$$
$$\hat{\theta} = \cos\theta\cdot\cos\varphi\hat{i} + \cos\theta\sin\varphi\hat{j} - \sin\theta\hat{k}$$
$$\hat{\varphi} = -\sin\varphi\hat{i} + \cos\varphi\hat{j}.$$

On solving these equations we get

$$\hat{i} = \sin\theta\cos\varphi\hat{r} + \cos\theta\cos\varphi\hat{\theta} - \sin\varphi\hat{\varphi}$$
$$\hat{j} = \sin\theta\sin\varphi\hat{r} + \cos\theta\sin\varphi\hat{\theta} + \cos\varphi\hat{\varphi}$$
$$\hat{k} = \cos\theta\hat{r} - \sin\theta\hat{\theta}.$$

Position vector $\vec{r} = \overrightarrow{OP} = x\hat{i} + y\hat{j} + z\hat{k}$

$$= \sin\theta\cos\varphi\hat{i} + \sin\theta\sin\varphi\hat{j} + r\cos\theta\hat{k}$$

$$= r\hat{r}$$

Line element $\vec{r} = dr\hat{r} + rd\theta\hat{\theta} + r\sin\theta \cdot d\varphi\hat{\varphi}$

Velocity

$$\vec{v} = \frac{d\vec{r}}{dt} = \dot{r}\hat{r} + r\dot{\theta}\hat{\theta} + r\dot{\varphi}\sin\theta\hat{\varphi}.$$

Acceleration $\vec{a} = \dfrac{d\vec{v}}{dt} = \left(\ddot{r} - r\dot{\theta}^2 - r\sin^2\theta\dot{\varphi}^2\right)\hat{r} + \left(2\dot{r}\dot{\theta} + r\ddot{\theta} - r\dot{\varphi}^2\sin\theta\cos\theta\right)\hat{\theta}$

$$+ \left(2r\dot{\theta}\dot{\varphi} + 2\dot{r}\dot{\varphi}\sin\theta + r\ddot{\varphi}\sin\theta\right)\hat{\varphi}$$

B.11 Vector differential operators in different coordinate systems

Rectangular Cartesian coordinates (x, y, z) :

$$\vec{\nabla}\phi = \hat{i}\frac{\partial\phi}{\partial x} + \hat{j}\frac{\partial\phi}{\partial y} + \hat{k}\frac{\partial\phi}{\partial z}$$

$$\vec{\nabla}\cdot\vec{A} = \hat{i}\frac{\partial A_x}{\partial x} + \hat{j}\frac{\partial A_y}{\partial y} + \hat{k}\frac{\partial A_z}{\partial z}$$

$$\vec{\nabla}\times\vec{A} = \begin{vmatrix} \hat{i} & \hat{j} & \hat{k} \\ \dfrac{\partial}{\partial x} & \dfrac{\partial}{\partial y} & \dfrac{\partial}{\partial z} \\ A_x & A_y & A_z \end{vmatrix}$$

$$\nabla^2\phi = \frac{\partial^2\phi}{\partial x^2} + \frac{\partial^2\phi}{\partial y^2} + \frac{\partial^2\phi}{\partial z^2}$$

Cylindrical coordinates (r, θ, z) :

$$\vec{\nabla}\phi = \hat{r}\frac{\partial\phi}{\partial r} + \hat{\theta}\frac{1}{r}\frac{\partial\phi}{\partial\theta} + \hat{z}\frac{\partial\phi}{\partial z}$$

$$\vec{\nabla}\cdot\vec{A} = \frac{1}{r}\frac{\partial}{\partial r}(rA_r) + \frac{1}{r}\frac{\partial A_\theta}{\partial\theta} + \frac{\partial A_z}{\partial z}$$

$$\vec{\nabla}\times\vec{A} = \frac{1}{r}\begin{vmatrix} \hat{r} & r\hat{\theta} & \hat{z} \\ \dfrac{\partial}{\partial r} & \dfrac{\partial}{\partial\theta} & \dfrac{\partial}{\partial z} \\ A_r & rA_\theta & A_z \end{vmatrix}$$

$$\nabla^2\phi = \frac{1}{r}\frac{\partial}{\partial r}\left(r\frac{\partial\phi}{\partial r}\right) + \frac{1}{r^2}\frac{\partial^2\phi}{\partial\theta^2} + \frac{\partial^2\phi}{\partial z^2}$$

Spherical coordinates (r, θ, φ) :

$$\vec{\nabla}\phi = \hat{r}\frac{\partial \phi}{\partial r} + \hat{\theta}\frac{1}{r}\frac{\partial \phi}{\partial \theta} + \hat{\varphi}\frac{1}{r\sin\theta}\frac{\partial \phi}{\partial \varphi}$$

$$\vec{\nabla} \cdot \vec{A} = \frac{1}{r^2}\frac{\partial}{\partial r}\left(r^2 A_r\right) + \frac{1}{r\sin\theta}\frac{\partial}{\partial \theta}\left(A_\theta \sin\theta\right) + \frac{1}{r\sin\theta}\frac{\partial A_\varphi}{\partial \varphi}$$

$$\vec{\nabla} \times \vec{A} = \frac{1}{r^2\sin\theta}\begin{vmatrix} \hat{r} & r\hat{\theta} & r\sin\theta\hat{\varphi} \\ \dfrac{\partial}{\partial r} & \dfrac{\partial}{\partial \theta} & \dfrac{\partial}{\partial \varphi} \\ A_r & rA_\theta & r\sin\theta A_\varphi \end{vmatrix}$$

$$\nabla^2\phi = \frac{1}{r^2}\frac{\partial}{\partial r}\left(r^2\frac{\partial \phi}{\partial r}\right) + \frac{1}{r^2\sin\theta}\frac{\partial}{\partial \theta}\left(\sin\theta\frac{\partial \phi}{\partial \theta}\right) + \frac{1}{r^2\sin^2\theta}\frac{\partial^2 \phi}{\partial \varphi^2}.$$

Appendix **C**

Dirac Delta Function

Discrete sources such as point charges, filamentary currents, current shells, etc. constitute singularities of electromagnetic fields. One convenient method to treat these discrete sources is the use of Dirac delta function.

In one dimension Dirac delta function at a point $x = x_0$ is designated by $\delta(x - x_0)$ and at $x = 0$ by $\delta(x)$. It has the following properties:

1. $\delta(x - x_0) = \begin{cases} 0, & \text{for } x \neq x_0 \\ \infty, & \text{for } x = x_0 \end{cases}$

2. $\int_a^b \delta(x - x_0)dx = \begin{cases} 1, & \text{if } x_0 \text{ is in } (a, b) \\ 0, & \text{if } x_0 \text{ is not in } (a, b) \end{cases}$

3. $\int_a^b f(x)\delta(x - x_0)dx = \begin{cases} f(x_0), & \text{if } x_0 \text{ is in } (a, b) \\ 0, & \text{if } x_0 \text{ is not in } (a, b) \end{cases}$

4. $\int_a^b f(x)\delta'(x - x_0)dx = \begin{cases} -f'(x_0), & \text{if } x_0 \text{ is in } (a, b) \\ 0, & \text{if } x_0 \text{ is not in } (a, b) \end{cases}$

where a prime denotes differentiation with respect to the argument. In three dimensions the delta function can be expressed in terms of one dimensional deltas as

5. $\delta(\vec{r} - \vec{r}_0) = \delta(x - x_0)\delta(y - y_0)\delta(z - z_0)$

which vanishes at all points except at $\vec{r} = \vec{r}_0$ and is such that

6. $\int_V \delta(\vec{r}-\vec{r}_0)\,dV = \begin{cases} 1, & \text{if } \vec{r}_0 \text{ is in } V \\ 0, & \text{if } \vec{r}_0 \text{ is not in } V \end{cases}$

Also, we have the property,

7. $\int_V f(\vec{r})\,\delta(\vec{r}-\vec{r}_0)\,dV = \begin{cases} f(\vec{r}_0), & \text{if } \vec{r}_0 \text{ is in } V \\ 0, & \text{if } \vec{r}_0 \text{ is not in } V \end{cases}$

Note that the delta function is not continuous one, but it can be considered as the limit of an ordinary function. For example, the delta function may be represented as a vanishingly thin sharply peaked Gaussian function of unit area (Fig. C-1). Thus, we could define

$$\delta(x) = \lim_{\varepsilon\to 0} \frac{1}{\sqrt{\pi}\varepsilon} e^{-x^2/\varepsilon^2}.$$

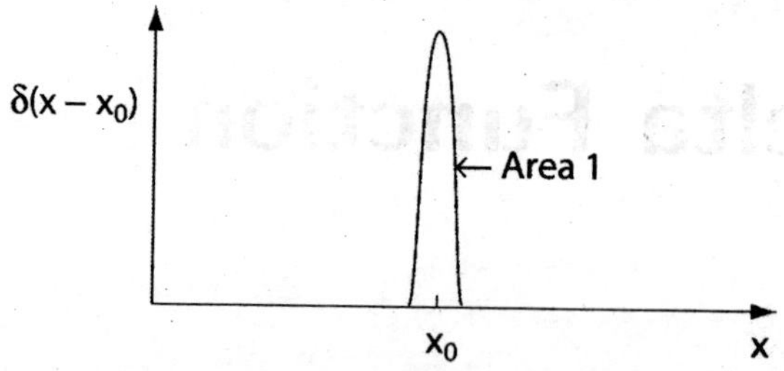

Fig. C-1 : Approximate form of a delta function

From this definition we immediately find the following properties of delta function:

$$\delta(x) = \delta(-x)$$

$$\delta'(x) = -\delta'(-x)$$

$$\delta(ax) = \frac{\delta(x)}{a}$$

Another important property of the Dirac delta function comes from its relation to the divergence or Laplacian operator.

$$\vec{\nabla}\cdot\frac{\vec{r}}{r^3} = \vec{\nabla}\frac{1}{r^3}\cdot\vec{r} + \frac{1}{r^3}\vec{\nabla}\cdot\vec{r} = -\frac{3\vec{r}\cdot\vec{r}}{r^5} + \frac{3}{r^3} = 0,$$

for $r \neq 0$ and becomes indeterminate as $r \to 0$. Applying divergence theorem to a small sphere of radius R around the origin we get

$$\int_V \vec{\nabla}\cdot\frac{\vec{r}}{r^3}dV = \oint_S \frac{\vec{r}\cdot\hat{n}dS}{r^3} = \frac{1}{R^2}\oint_S dS = \frac{4\pi R^2}{R^2} = 4\pi$$

Since this result is true regardless of how small the radius R, one can replace $\vec{\nabla}\cdot\frac{\vec{r}}{r^3}$ by $4\pi\delta(\vec{r})$. Thus,

$$\vec{\nabla}\cdot\frac{\vec{r}}{r^3} = \vec{\nabla}\cdot\vec{\nabla}\left(-\frac{1}{r}\right) = -\nabla^2\frac{1}{r} = 4\pi\delta(\vec{r}) \quad \text{or} \quad \delta(\vec{r}) = \nabla^2\left(-\frac{1}{4\pi r}\right).$$

Appendix D

Concept of Tensors

Most physical quantities may be classified as scalars and vectors. A scalar F is a quantity completely defined by one component, which does not depend on the coordinate system, i.e., it transforms according to the law $F' = F$. A vector $\vec{A}$ may be defined as a quantity whose components transform under an orthogonal transformation as the components of the position vector. Thus,

$$A'_i = \sum_{k=1}^{3} \alpha_{ik} A_k,$$

where A'_i and A_i are components of the vector in S' and S coordinate systems, respectively; α_{ik} are the direction consines relating to the axes of S and S' systems. There are certain entities, which cannot be properly represented either by scalars or by vectors. For example, in an anisotropic dielectric medium the displacement vector $\vec{D}$ is not, in general, in the direction of $\vec{E}$. An electric field $\vec{E}$ applied in one direction may result in a vector $\vec{D}$ having components in all three directions. So in the relation $\vec{D} = \varepsilon\vec{E}$, ε cannot be a scalar or a vector. Thus, the scalar and vector concepts become inadequate in this case. In a three dimensional continuum each component of $\vec{D}$ can be expressed as a linear function of all the components of $\vec{E}$. Thus,

$$D_1 = \varepsilon_{11}E_1 + \varepsilon_{12}E_2 + \varepsilon_{13}E_3$$

$$D_2 = \varepsilon_{21}E_1 + \varepsilon_{22}E_2 + \varepsilon_{23}E_3$$

$$D_3 = \varepsilon_{31}E_1 + \varepsilon_{32}E_2 + \varepsilon_{33}E_3$$

In compact notation,

$$D_i = \sum_{k=1}^{3} \varepsilon_{ik} E_k \quad (i = 1, 2, 3).$$

The quantity represented by the nine components ε_{ik}, which carry the vector $\vec{E}$ into $\vec{D}$ is known as a *tensor of rank two.* In order that the association of the components of $\vec{D}$ with the components of $\vec{E}$ is preserved in different coordinate systems, the nine components must transform in a specific manner. Formally, by a tensor of rank 2 in three dimensions is meant a quantity uniquely specified by nine components which transform under change of coordinate systems according to the law

$$T'_{ij} = \sum_{l=1}^{3} \sum_{m=1}^{3} \alpha_{il} \alpha_{jm} T_{lm},$$

where T'_{ij} and T_{ij} are components of the tensor in S' and S coordinate systems respectively, α_{il} are the direction cosines relating to the axes of S and S' systems.

The number of different indices attached to the symbol representing a tensor is known as the *rank of the tensor.* Thus, T_{ij} is a tensor of rank 2. A tensor of rank n in three dimensions has 3^n components. In this hierarchy a scalar with 3° component is a tensor of rank zero and a vector with 3^1 components is a tensor of rank 1. A tensor T_{ij} is said to be symmetric if $T_{ij} = T_{ji}$ and skew symmetric if $T_{ij} = -T_{ji}$. Kronecker delta δ_{ij} is a tensor of rank 2. A familiar example of tensor from mechanics is the moment of inertia tensor. Tensors/vectors may be classified as covariant and contravariant. The vector whose components transform in a change of coordinates as the partial derivatives of the coordinates $(\partial/\partial x_i)$ is called a *covariant vector* and the vector whose components change as the coordinate differentials (dx_i) is called a *contravariant vector.*

The laws of transformation of the coordinate differentials are

$$dx'_i = \sum_{k=1}^{3} \frac{\partial x'_i}{\partial x_k} dx_k$$

and the laws of transformation of the derivatives are

$$\frac{\partial}{\partial x'_i} = \sum_{k=1}^{3} \frac{\partial x_k}{\partial x'_i} \frac{\partial}{\partial x_k},$$

where (x_1, x_2, x_3) and (x'_1, x'_2, x'_3) are the coordinates in S and S' coordinate systems, respectively.

Appendix **E**

Some Important Physical Constants

Constant	Symbol	Approximate value
Speed of light in free space	c	3×10^8 m/s
Electronic charge	e	1.6×10^{-19} C
Electronic rest mass	m	9.1×10^{-31} kg
Proton rest mass	m_p	1.67×10^{-27} kg
Avogadro number	N	6.02×10^{23}/mol
Molar volume of a gas at STP	V_m	22.4 litre/mol
Universal gas constant	R	8.31 J/K-mol
Boltzmann constant	k	1.38×10^{-23} J/K
Gravitational constant	G	6.67×10^{-11} N.m^2/kg^2
Planck's constant	h	6.62×10^{-34} J-s
Permittivity of free space	ε_0	8.85×10^{-12} F/m
Permeability of free space	μ_0	$4\pi \times 10^{-7}$ H/m
Faraday constant	F	96500 C/mol
Mechanical equivalent of heat	J	4.2 J/cal = 1(in SI)
Bohr magneton	μ_B	9.27×10^{-24} J/T

Appendix F

List of Some Symbols and their principal Meanings

Symbol	Meaning
$\vec{a}$	Acceleration
$\vec{A}$	Vector potential
$\vec{B}$	Magnetic induction
c	Speed of light in free space
C	Capacitance
$\vec{D}$	Electric displacement
$d\vec{l}$, $d\vec{r}$	Length element
$d\vec{S}$	Surface element
dV	Volume element
e	Electronic charge,
	Naperian base
$\vec{E}$	Electric field
f	Frequency
$\vec{F}$	Force
g	Acceleration due to gravity
$\vec{H}$	Magnetic field intensity
i	Instantaneous current
I	Steady current
	x, y and z-axes
j	$\sqrt{-1}$
$\vec{J}$	Volume current density
k	Boltzmann constant
	Wavenumber

Symbol	Meaning
$\vec{k}$	Wave vector
K	Dielectric constant
$\vec{K}$	Surface current density
L	Self-inductance
m	Mass
$\vec{m}$	Magnetic dipole moment
$\vec{M}$	Magnetization
M	Mutual inductance,
	Molecular weight
$[M, L, T]$	[Mass, Lenght, Time]
N	Avogadro number
n	Refractive index
$\hat{n}$	Unit normal to a
	surface element
$\vec{p}$	Electric dipole moment
$\vec{P}$	Polarization
P	Power
q	Charge
	Quality factor
Q_{ij}	Quadrupole moment
$\vec{r}$	Position vector
$\hat{r}$	Unit vector along $\vec{r}$
(r, θ, z)	Cylindrical coordinates

Symbol	Meaning
$(\hat{r}, \hat{\theta}, \hat{z})$	Unit vectors in cylindrical system
(r, θ, φ)	Spherical coordinates
θ	An angle
$(\hat{r}, \hat{\theta}, \hat{\varphi})$	Unit vectors in spherical system
R	Resistance, Reflectance
R_m	Reluctance
$\vec{R}$	Vector from source point to field point
$\vec{s}$	Poynting's vector
S	Surface area
t	Time
T	Time period, Absolute temperature, Transmittance
u	Energy density
U	Potential energy
v	Instantaneous voltage
$\vec{v}$	Velocity
V	Potential difference
W	Work
X	Reactance
(x, y, z)	Field point
(x', y', z')	Source point
$(\hat{x}, \hat{y}, \hat{z})$	Unit vectors along x, y and z-axes
Z	Impedance
α	Polarizability
β	Attenuation constant
δ	Skin depth
δ_{ij}	Kronecker delta
ε	Permittivity

Symbol	Meaning
ε_r	Relative permittivity
ε_{ij}	Dielectric tensor components
ε	Electromotive force
λ	Wavelength Linear charge density
μ	Permeability
μ_r	Relative permeability
μ_B	Bohr magneton
ν	Frequency
π	Peltier coefficient
ρ	Volume charge density Resistivity
ρ_m	Volume pole density
σ	Surface charge density Electrical conductivity Thomson coefficient Scattering cross-section
σ_m	Surface pole density
τ	Time constant Torque Strength of magnetic shell Mean free time
ϕ	Electric scalar potential Magnetic scalar potential
Φ	Electric flux Magnetic flux
φ	Azimuthal angle
χ	Electric susceptibility
χ_m	Magnetic susceptibility
ω	Angular frequency Angular velocity
Ω	Solid angle Unit ohm

Index

D

E

N

O

P

U

V

W

Z